MW00843682

Developments in Mathematics

Volume 56

More information about this series at http://www.springer.com/series/5834

Yoshihiro Sawano

Theory of Besov Spaces

Yoshihiro Sawano
Department of Mathematics and Information Science
Tokyo Metropolitan University
Tokyo, Japan

ISSN 1389-2177 ISSN 2197-795X (electronic)
Developments in Mathematics
ISBN 978-981-13-0835-2 ISBN 978-981-13-0836-9 (eBook)
https://doi.org/10.1007/978-981-13-0836-9

Library of Congress Control Number: 2018949641

Printed on acid-free paper

This Springer imprint is published by the registered company Springer Nature Singapore Pte Ltd.
The registered company address is: 152 Beach Road, #21-01/04 Gateway East, Singapore 189721, Singapore

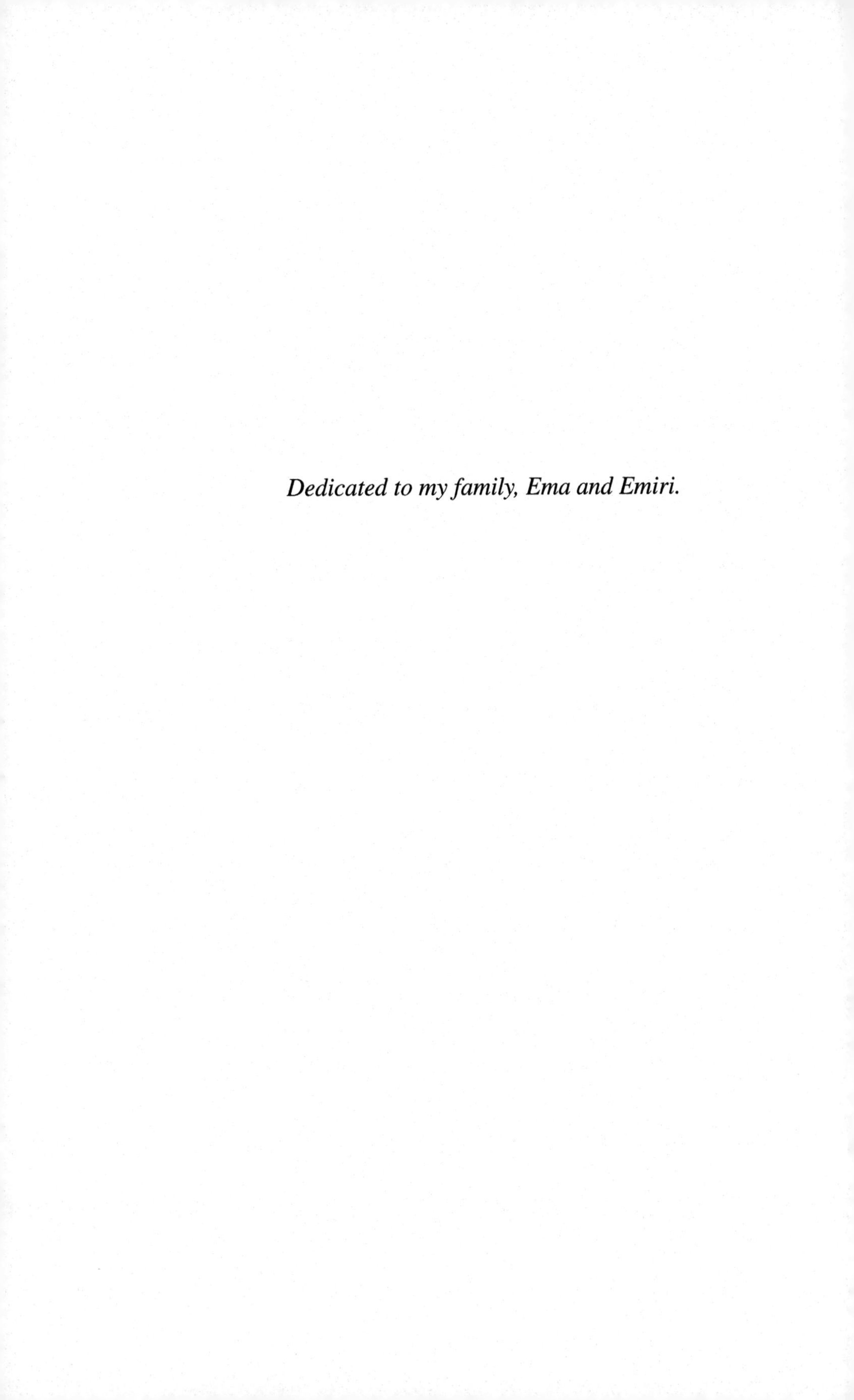

Dedicated to my family, Ema and Emiri.

Preface

This book is intended to be as an exhaustive and self-contained treatment of Besov spaces and Triebel–Lizorkin spaces. In this book I aim to explain Besov spaces and Triebel–Lizorkin spaces from the start and to apply them to partial differential equations (PDEs).

I mean here by "a function space" a linear subspace of the space of all functions in a set X. We will often work in the setting of $\mathbb{R}^n$ but in the last part of the book we consider open sets in $\mathbb{R}^n$. For example, the set of all functions on $X = \mathbb{R}^n$ is an example of function spaces. However, it is too vague and too hard to grasp. As more typical examples, let us consider the subspace of all Borel measurable functions, in particular the space of all continuous functions. Or, readers may envisage the space $\mathrm{BC}(\mathbb{R}^n)$ the set of all bounded continuous functions which readers will have encountered in lectures on topological spaces. If readers are familiar with the theory of integrals, we can think that we are going to study a new framework containing the set of all integrable or measurable functions.

There are too many continuous or measurable functions and, therefore, their various linear subspaces. In this book, we are going to propose new frameworks called Besov spaces and Triebel–Lizorkin spaces and then discuss their properties in detail.

When we were high school students or undergraduate students, our main concern lay in the C^∞-functions. Therefore, readers cannot understand why we are going to investigate functions in a very subtle manner. However, there do exist many examples where the non-differentiable functions play a fundamental role in the rule of the nature.

As an example, I take up Brownian motion. This is familiar because we learnt in chemistry lessons that this describes the motion of particles. Here by Brownian motion, I mean the "mathematical" one which grew out of the chemical one. This mathematical Brownian motion is a fundamental concept of stochastic integrals initiated by Kiyosi Itô and hence we are convinced that the mathematical Brownian motion plays a fundamental role in economics. As well as the chemical one, the mathematical one moves in a very complicated way. Namely, each path is continuous but it is too complicated and non-differentiable. When we want to describe properties of continuous functions which are not differentiable, Besov spaces and Triebel–Lizorkin spaces are useful. As this example shows, Besov spaces and Triebel–Lizorkin spaces play an important role.

Another aim of this book is to apply Besov spaces and Triebel–Lizorkin spaces to PDEs.

In PDEs, we are led to consider the equations beyond the framework of $C^k(\mathbb{R}^n)$ functions. As the example of the functions

$$f(x, y) = (x^2 - y^2) \log \log \frac{1}{x^2 + y^2} \in C^2(B(1) \setminus \{(0, 0)\}),$$

defined in $B(1)$, together with computation

$$\frac{\partial^2 f}{\partial x^2}(x, y) + \frac{\partial^2 f}{\partial y^2}(x, y) = \frac{4(x^2 - y^2)(2 \log(x^2 + y^2) - 1)}{(x^2 + y^2)(\log(x^2 + y^2))^2} \in C^2(B(1))$$

shows, it does not suffice to consider functions which are k-times differentiable and whose partial derivatives up to order k are all continuous when we consider the Poisson equation $-\Delta u = f$. For example, when we consider the elliptic differential equations, we use not C^2 but $\mathscr{C}^{2+\varepsilon}(\mathbb{R}^n)$ with $\varepsilon \in (0, 1)$, where $\mathscr{C}^{2+\varepsilon}(\mathbb{R}^n)$ denotes the Hölder–Zygmund space of order $2 + \varepsilon$. As we establish later in this book, the space $\mathscr{C}^{2+\varepsilon}(\mathbb{R}^n)$ is realized as a special case of Besov spaces and Triebel–Lizorkin spaces. Hence we see that Besov spaces and Triebel–Lizorkin spaces are useful in PDEs. The branch of PDE being too wide, we cannot take up all of them, but we seek to investigate the wave equations, the Schrödinger equations, the heat equations, and the elliptic differential equations in the context of applications of Besov spaces and Triebel–Lizorkin spaces.

Let us consider the complexity and what we obtain from that. We learnt in high school that $f : \mathbb{R} \to \mathbb{R}$ is convex if f is twice differentiable, and $f'' \geq 0$. However, in undergraduate course, this is equivalent to $f((1-t)x+ty) \leq (1-t)f(x)+tf(y)$ for all $x, y \in \mathbb{R}$ and $t \in [0, 1]$. With this in mind, let us reconsider the definition of convex functions and discuss the properties of this notion.

First of all, compare "$f'' \geq 0$" with "$f((1 - t)x + ty) \leq (1 - t)f(x) + tf(y)$ for all $x, y \in \mathbb{R}$ and $t \in [0, 1]$"; occasionally the latter is not easy to prove. We can clearly say that the former is simple. However, the latter enjoys properties that the former does not enjoy. For example, due to the fact that the differentiation comes into play and that the latter is readily extended to the functions on linear spaces, the former is harder to generalize than the latter. Thus, we can say that the latter is general. In addition, when we define the convexity by way of the differentiation, we cannot say that the convexity is stable under taking the modulus. Also, in entrance examinations, we frequently encounter inequalities of the form $f((1 - t)x + ty) \leq (1 - t)f(x) + tf(y)$ because this inequality contains three parameters x, y, and t. Thus, once $f((1 - t)x + ty) \leq (1 - t)f(x) + tf(y)$ is proved, it should be useful.

For PDEs, we must be familiar with functional analysis. However, we need not be so serious: we provide some facts on functional analysis in this book and this is sufficient.

Many mathematicians hate Besov spaces and Triebel–Lizorkin spaces. Why? I now consider the reason here. As I mentioned earlier, there are many function spaces. Although I content myself with listing them briefly, there are $L^p(\mathbb{R}^n)$-spaces, Sobolev spaces, Morrey spaces, Orlicz spaces, Besov spaces, Triebel–Lizorkin spaces, and so on. Among them I am led to the following conclusion: The easier to define the function spaces are, the fewer good properties they have. Ideally, from a good notion of function spaces, we want that: it is simple enough to memorize the definition and it has a rich structure. The spaces $L^p(\mathbb{R}^n)$-spaces, Sobolev spaces, Morrey spaces, Orlicz spaces, which I listed earlier, are easy to describe but they cannot cover differentiability and integrability. Meanwhile, Besov spaces and Triebel–Lizorkin spaces describe very well differentiability and integrability but their definitions are very complicated. For example, we can consider Besov-Morrey spaces which are defined by mixing Besov spaces and Morrey spaces. You can easily guess that the definition of Besov-Morrey spaces is very complicated. Many people hate Besov spaces and Triebel–Lizorkin spaces because of the complexity of the definition which arises as a price of the good properties, I think.

However, let me stress that these function spaces have many big advantages. As I mentioned before, it is important to be able to grasp many other function spaces. For example, $L^p(\mathbb{R}^n)$-spaces, the BMO space, Sobolev spaces, and Hardy spaces fall under the unified framework of Besov spaces and Triebel–Lizorkin spaces. Here I content myself with mentioning that the atomic decomposition is extremely important. For we can learn much more once we establish the theory of the atomic decomposition. The details are left to Chap. 5.

Let me describe the structure of this book while comparing the content of this book with the existing literature. We have to prepare a lot in order to define Besov spaces and Triebel–Lizorkin spaces and in order to establish the theory of these spaces because the definitions of Besov spaces and Triebel–Lizorkin spaces are very complicated. So in Chap. 1, I explain the Fourier transform, the maximal operator, and the singular integral operators, which are of fundamental importance in harmonic analysis. The theory of singular integral operators being too wide with a large amount of literature, I content myself with its brief introduction. I also kept the description of the singular integral operators to the minimum. See [22, 31, 32, 86] for more about the theory of singular integral operators. In Chap. 2, for the purpose of a survey of the book, I investigate $B^s_{pq}(\mathbb{R}^n)$ with $1 \leq p, q \leq \infty$ and $s \in \mathbb{R}$. If we are limited to $B^s_{pq}(\mathbb{R}^n)$ with $1 \leq p, q \leq \infty$ and $s \in \mathbb{R}$, we definitely want to make it clear that we can define the space and investigate the property without using any heavy tool except $\mathscr{S}(\mathbb{R}^n)$ and $\mathscr{S}'(\mathbb{R}^n)$. Chapter 2 presents one of my primary aims in this book. We want to define Besov spaces and Triebel–Lizorkin spaces and then we investigate fundamental properties: density of the spaces, completeness of the spaces, and so on. As a by-product of the Plancherel–Polya–Nikolskii inequality, I described the modulation spaces. See [35] for the exhaustive description of the time frequency analysis. There are recent textbooks dealing with Besov spaces. For example Grafakos gave a description in [33]. If I compare at this book with my book,

I believe that some equivalent expressions are described in more detail. Chapter 3 is devoted to showing that these spaces cover many other spaces as special cases. This chapter somewhat overlaps [22, 31, 32, 86] and supplements [97, 99]. We paid attention to dyadic analysis which has developed rapidly in this decade. Atomic decomposition is taken up in Chap. 4 together with applications to the boundedness of operators. Bourdaud has written a book on the paraproduct [12]. I did manage to formulate the model cases. For more about the paraproduct see [12]. Ma'zya dealt with Sobolev spaces in the context of multipliers [57, 58]. Applications to PDEs are contained in Chap. 5. Besov together with his collaborators has written books [8, 9]. Although Besov and I dealt commonly with the function spaces defined on domains, I can say that Besov concentrated more on the function spaces on domains, while I placed myself mainly in the Euclidean space. Recently theory of function spaces underwent a major diversification. So, in Chap. 6 we describe how what we have been doing can be generalized and applied. Each section of the last chapter will play a role of an introduction of various function spaces and it will also play the role of the brief introduction of the subject. See [1] for Morrey spaces, [16] for variable Lebesgue spaces, and [51] Weighted Sobolev spaces. Triebel has written many recent books [106–110]. The books [23, 101, 106] deal with fractals. I will content myself with introducing the function spaces. My book does not overlap [107, 108]. In [109, 110] Triebel dealt with Morrey-type function spaces. In this book, I mentioned briefly this type of function spaces in a different context. As an application, I chose the solution to the Kato problem. However, I could manage to allude to its solution and some related facts. See the book [5] for more exhaustive details.

The key theorems in this book are Theorem 1.49 and its corollary Theorem 1.53, as well as Theorem 2.34 and its corollary Theorem 4.1. Needless to say, the definition is the most important in mathematics. However, these four theorems appear repeatedly and we cannot follow the proof without them. By this I do not mean that we have to memorize these theorems but we will use them quite often.

Finally, we describe how different this book is from the existing literature. As is mentioned above, we heavily depend on Theorems 1.49, 1.53, 2.34, and 4.1. Among the textbooks I listed in the references the first two theorems are exhaustively investigated in [99] as well as [97]. So we can say that our book will cover these two books largely. Let me also mention that our book will cover [2]. Our book will also cover or overlap [4, 7] in some sense. However, the approach will be quite different. For example, I did my best to explain how we must be careful when we use the complex interpolation functor defined in [7]. Theorem 2.34 appeared in [100]. The atomic decomposition appeared in [23, 101]. Although there is strong overlap with these three books in our book, I did not describe the compactness of the embedding in depth, unlike [23, 101]. Its variant, quarkonial decomposition, appeared in [103, 104]. In the context of the atomic decomposition, I have included quarkonial decomposition. However, I present a different application of the quarkonial decomposition from [103, 104].

Tokyo, Japan
May 2018

Yoshihiro Sawano

Acknowledgments

I would like to offer my deepest gratitude to the following individuals: Kôzo Yabuta, Denny Ivanal Hakim, Takahiro Ono, Alexander Meskhi, Akihiro Shimomura, Baoxiang Wang, Kentaro Hirata, Kwok-Pun Ho, Hendra Gunawan, Akihiko Miyachi, Akira Kaneko, Neal Bez, Yiyu Liang, Sanghyuk Lee, Komori-Yasuo Furuya, Shohei Nakamura, Eiichi Nakai, Mieczysław Mastyło, Koichi Saka, Yoshifumi Ito, Naoko Ogata, Naohito Tomita, Kei Morii, Kunio Yoshino, Mitsuo Izuki, Tino Ullrich, Yohei Tsutsui, Tsukamoto Masaki, Shohei Nakamura, Takahiro Noi, Hidemitsu Wadade, Enji Sato, Hiroki Saito, Gaku Sadasue, Toru Nogayama, Dachun Yang, Wen Yuan, Ciqiang Zhuo, Maria Alessandra Ragusa, Giuseppe di Fazio, Hitoshi Tanaka, Tetsu Shimomura, Xuan Thinh Duong, Alan McIntosh, Takuya Sobukawa, Akaho Manabu, Tomorou Asai, Yuzuru Inahama, Yuusuke Ochiai, Tsukasa Iwabuchi, Tokio Matsuyama, Kouichi Taniguchi, Okihiro Sawada, Yukihiro Seki, Liu Liguang, Shin-ichiro Matsuo, Tsuyoshi Yoneda, Hiroshi Watanabe, Hidetoshi Saitoh, Yoshio Tsutsumi, Masanobu Oshima, and Keisuke Suzuki.

I am thankful to Professors Dachun Yang, Wen Yuan, Kunio Yoshino, Yiyu Liang, Xing Fu, and Akira Kaneko for fruitful discussion of the topology of $\mathscr{S}_\infty(\mathbb{R}^n)$.

I am thankful to Mr. Stephan Horochoryn for checking the first chapter of the manuscript carefully.

My earlier manuscript of 2005 was not so polished, and because of this my presentation of the seminar was poor. I apologize here for confusing the participants of my seminar and thank them for their attendance.

Notation in This Book

Sets and Set Functions

1. The metric open ball defined by ℓ^2 is usually called a *ball*. We denote by $B(x,r)$ the *ball centered at x of radius r*. Namely, we write

$$B(x,r) \equiv \{y \in \mathbb{R}^n : \|x-y\| < r\}$$

when $x \in \mathbb{R}^n$ and $r > 0$. Given a ball B, we denote by $c(B)$ its *center* and by $r(B)$ its *radius*. We write $B(r)$ instead of $B(o,r)$, where $o \equiv (0,0,\ldots,0)$.
2. By a "cube" we mean a compact cube whose edges are parallel to the coordinate *axes*. The metric closed ball defined by ℓ^∞ is called a *cube*. If a cube has center x and radius r, we denote it by $Q(x,r)$. Namely, we write

$$Q(x,r) \equiv \left\{ y = (y_1, y_2, \ldots, y_n) \in \mathbb{R}^n : \max_{j=1,2,\ldots,n} |x_j - y_j| \le r \right\}$$

when $x = (x_1, x_2, \ldots, x_n) \in \mathbb{R}^n$ and $r > 0$. From the definition of $Q(x,r)$, its volume is $(2r)^n$. We write $Q(r)$ instead of $Q(o,r)$. Given a cube Q, we denote by $c(Q)$ the *center of Q* and by $\ell(Q)$ the *sidelength of Q*: $\ell(Q) = |Q|^{1/n}$, where $|Q|$ denotes the volume of the cube Q.
3. Given a cube Q and $k > 0$, $k\,Q$ means the *cube concentric to Q with sidelength $k\,\ell(Q)$*. Given a ball B and $k > 0$, we denote by $k\,B$ the *ball concentric to B with radius $k\,r(B)$*.
4. For $\nu \in \mathbb{Z}$ and $m = (m_1, m_2, \ldots, m_n) \in \mathbb{Z}^n$, we define $Q_{\nu m} \equiv \prod_{j=1}^{n} \left[\frac{m_j}{2^\nu}, \frac{m_j+1}{2^\nu}\right)$. Denote by $\mathscr{D} = \mathscr{D}(\mathbb{R}^n)$ the set of such cubes. The elements in $\mathscr{D}$ are called dyadic cubes.
5. Let E be a measurable set. Then we denote its *indicator function* by χ_E. If E has positive measure and E is integrable over f, then denote by $m_E(f)$ the *average of f* over E. $|E|$ denotes the *volume of E*.

6. If we are working on $\mathbb{R}^n$, then $\mathscr{B}$ denotes the set of all balls in $\mathbb{R}^n$, while $\mathscr{Q}$ denotes the set of all cubes in $\mathbb{R}^n$. Be careful because $\mathscr{B}$ can be used for a different purpose: When we are working on a measure space $(X, \mathscr{B}, \mu)$, then $\mathscr{B}$ stands for the set of all Borel sets.
7. The symbol $\sharp A$ means the *cardinality of the set* A.
8. The symbol 2^X denotes the set of all subsets in X.
9. Let X be a topological space. Then $\mathscr{K}_X$ is the set of all compact subsets of X, and $\mathscr{O}_X$ is the set of all open subsets of X.
10. The set $\mathscr{I}(\mathbb{R})$ denotes the set of all closed intervals in $\mathbb{R}$.
11. A tacit understanding is that by a "cube" we mean a closed cube whose edges are parallel to the coordinate axes. However, we say that dyadic cubes are also cubes.
12. We define the *upper half space* $\mathbb{R}^n_+$ and the *lower half space* $\mathbb{R}^n_-$ by

$$\mathbb{R}^n_\pm \equiv \{(x', x_n) \in \mathbb{R}^n \,:\, \pm x_n > 0\}. \tag{1}$$

Numbers

1. Let $a \in \mathbb{R}$. Then write $a_+ = a \vee 0 \equiv \max(a, 0)$ and $a_- = a \wedge 0 \equiv \min(a, 0)$. Correspondingly, given an $\mathbb{R}$-valued function f, f_+ and f_- are functions given by $f_+(x) \equiv \max(f(x), 0)$ and $f_-(x) \equiv \min(f(x), 0)$, respectively.
2. Let $a, b \in \mathbb{R}$. Then write $a \vee b \equiv \max(a, b)$ and $a \wedge b \equiv \min(a, b)$. Correspondingly, given $\mathbb{R}$-valued functions f, g, $f \vee g$ and $f \wedge g$ are functions given by $f \vee g(x) \equiv \max(f(x), g(x))$ and $f \wedge g(x) \equiv \min(f(x), g(x))$, respectively.
3. The constants C and c denote positive constants that may change from one occurrence to another. Because the two constants c can be different, the inequality $0 < 2c < c$ is by no means a contradiction. When we add a subscript, for example, this means that the constant c depends upon the parameter. It can happen that the constants with subscript differ according to the above rule. In particular, we prefer to use c_n for various constants that depend on n, when we do not want to specify its precise value.
4. Let $A, B \geq 0$. Then $A \lesssim B$ and $B \gtrsim A$ mean that there exists a constant $C > 0$ such that $A \leq CB$, where C depends only on the parameters of importance. The symbol $A \sim B$ means that $A \lesssim B$ and $B \lesssim A$ happen simultaneously, while $A \simeq B$ means that there exists a constant $C > 0$ such that $A = CB$.
5. When we need to emphasize or keep in mind that the constant C depends on the parameters α, β, γ, etc:

   (a) Instead of $A \lesssim B$, we write $A \lesssim_{\alpha,\beta,\gamma,\dots} B$.
   (b) Instead of $A \gtrsim B$, we write $A \gtrsim_{\alpha,\beta,\gamma,\dots} B$.
   (c) Instead of $A \sim B$, we write $A \sim_{\alpha,\beta,\gamma,\dots} B$.
   (d) Instead of $A \simeq B$, we write $A \simeq_{\alpha,\beta,\gamma,\dots} B$.

6. We define

$$\mathbb{N} \equiv \{1, 2, \ldots\}, \quad \mathbb{Z} \equiv \{0, \pm 1, \pm 2, \ldots\}, \quad \mathbb{N}_0 \equiv \{0, 1, \ldots\}.$$

7. We denote by $\mathbb{K}$ either $\mathbb{R}$ or $\mathbb{C}$, the coefficient field under consideration.
8. For $a \in \mathbb{R}^n$, we write $\langle a \rangle \equiv \sqrt{1 + |a|^2}$.
9. When $0 < p, q \leq \infty$, we define $\sigma_p \equiv n \left(\frac{1}{p} - 1 \right)_+$, $\sigma_{p,q} \equiv \max(\sigma_p, \sigma_q)$.
10. We sometimes identify $\mathbb{R}^{m+n}$ with $\mathbb{R}^m \times \mathbb{R}^n$.

Function Spaces

1. We use $\star$ for functions; $f = f(\star)$.
2. The function spaces are tacitly on $\mathbb{R}^n$. But sometimes, we will work on an open set Ω with C^∞-boundary.
3. Let X be a Banach space. We denote its norm by $\|\star\|_X$. However, we sometimes denote the $L^p(\mathbb{R}^n)$-norm by $\| \star \|_p$.
4. Let Ω be an open set in $\mathbb{R}^n$. Then $C^\infty_{\mathrm{c}}(\Omega)$ denotes the set of smooth functions with compact support in Ω.
5. Let $1 \leq j \leq n$. The symbol x_j denotes not only the j-th coordinate but also the function $x = (x_1, \ldots, x_n) \mapsto x_j$.
6. Suppose that $\{f_j\}_{j=1}^\infty$ is a sequence of measurable functions. Then we write

$$\|f_j\|_{L^p(\ell^q)} \equiv \left(\int_{\mathbb{R}^n} \left(\sum_{j=1}^\infty |f_j(x)|^q \right)^{\frac{p}{q}} \mathrm{d}x \right)^{\frac{1}{p}} \quad (0 < p, q \leq \infty)$$

and

$$\|f_j\|_{\ell^q(L^p)} \equiv \left(\sum_{j=1}^\infty \left(\int_{\mathbb{R}^n} |f_j(x)|^p \mathrm{d}x \right)^{\frac{q}{p}} \right)^{\frac{1}{q}} \quad (0 < p, q \leq \infty).$$

7. The space $L^2(\mathbb{R}^n)$ is the Hilbert space of square integrable functions on $\mathbb{R}^n$ whose inner product is given by

$$\langle f, g \rangle = \int_{\mathbb{R}^n} f(x)\overline{g}(x)\mathrm{d}x. \tag{2}$$

8. In view of (2), the inner product of $L^2(\mathbb{R}^n)$, it seems appropriate that we define the embedding $L^1_{\mathrm{loc}}(\mathbb{R}^n) \cap \mathscr{S}'(\mathbb{R}^n) \hookrightarrow \mathscr{S}'(\mathbb{R}^n)$ by

$$f \in L^1_{\rm loc}(\mathbb{R}^n) \cap \mathcal{S}'(\mathbb{R}^n) \mapsto F_f \equiv \left[g \in \mathcal{S}(\mathbb{R}^n) \mapsto \int_{\mathbb{R}^n} g(x)\overline{f(x)}{\rm d}x \right].$$

However, in order that $f \mapsto F_f$ be linear, we will define it later by

$$f \in L^1_{\rm loc}(\mathbb{R}^n) \cap \mathcal{S}'(\mathbb{R}^n) \mapsto F_f \equiv \left[g \in \mathcal{S}(\mathbb{R}^n) \mapsto \int_{\mathbb{R}^n} g(x)f(x){\rm d}x \right].$$

9. Let E be a measurable set and f be a measurable function with respect to the Lebesgue measure. Then write $m_E(f) \equiv \dfrac{1}{|E|}\displaystyle\int_E f$.
10. Let $0 < \eta < \infty$, E be a measurable set, and f be a positive measurable function with respect to the Lebesgue measure. Then write $m_E^{(\eta)}(f) \equiv m_E(f^\eta)^{\frac{1}{\eta}}$.
11. Let $0 < \eta < \infty$. We define the *powered Hardy–Littlewood maximal operator* $M^{(\eta)}$ by

$$M^{(\eta)}f(x) \equiv \sup_{R>0} \left(\frac{1}{|B(x,R)|} \int_{B(x,R)} |f(y)|^\eta {\rm d}y \right)^{\frac{1}{\eta}}.$$

12. For $x \in \mathbb{R}^n$, we define $\langle x \rangle \equiv \sqrt{1+|x|^2}$.
13. The space C denotes the set of all continuous functions on $\mathbb{R}^n$.
14. The space ${\rm BC}(\mathbb{R}^n)$ denotes the set of all bounded continuous functions on $\mathbb{R}^n$.
15. The space ${\rm BUC}(\mathbb{R}^n)$ denotes the set of all bounded uniformly continuous functions on $\mathbb{R}^n$.
16. Occasionally we identify the value of functions with functions. For example, $\sin x$ denotes the function on $\mathbb{R}$ defined by $x \mapsto \sin x$.
17. Given a Banach space X, we denote by X^* its dual space. The set X_1 is the closed unit ball in X.
18. Let μ be a measure on a measure space $(X, \mathcal{B}, \mu)$. Given a μ-measurable set A with positive μ-measure and a function f, we write

$$m_Q(f) \equiv \frac{1}{\mu(A)} \int_A f(x){\rm d}\mu(x).$$

Let $0 < \eta < \infty$. Then define $m_Q^{(\eta)}(f) \equiv m_Q(f^\eta)^{\frac{1}{\eta}}$ whenever f is positive.
19. For $x \in \mathbb{R}^n$, we define $\mathcal{Q}_x$ to be the set of all cubes containing x. Given a measurable function, Mf denotes the uncentered Hardy–Littlewood maximal operator and $M'f$ denotes the centered Hardy–Littlewood maximal operator.

$$Mf(x) \equiv \sup_{Q \in \mathcal{Q}_x} m_Q(|f|),$$

$$M'f(x) \equiv \sup_{r>0} m_{Q(x,r)}(|f|).$$

We sometimes use balls instead of cubes.

20. If notational confusion seems likely, Then we use [ ] to denote $Mf(x) = M[f](x)$, $\mathscr{F}\varphi(\xi) = \mathscr{F}[\varphi](\xi)$, etc.
21. If notational confusion seems likely, for the Fourier transform $\mathscr{F}$ and the Hardy–Littlewood maximal operator M, we use [ ] to denote $\mathscr{F}[f+g+h]$, $M[f+g+h]$.
22. For $j = 1, 2, \ldots, n$, $\partial_{x_j} = \frac{\partial}{\partial x_j}$ stands for the partial derivative. The symbol Δ stands for the Laplacian $\sum_{j=1}^{n} \frac{\partial^2}{\partial {x_j}^2}$.
23. We denote the $L^p(\mathbb{R}^n)$-norm by $\| \star \|_p$. For other function spaces such as the Hölder continuous function space $\mathscr{C}^s(\mathbb{R}^n)$, we use $\| \star \|_{\mathscr{C}^s}$ to stress the function spaces.
24. When we consider function spaces on a domain Ω, we denote by $\mathscr{C}^s(\Omega)$ the *Hölder continuous function space of order s*.
25. A quasi-norm over a linear space X enjoys positivity, homogeneity, and quasi-triangle inequality: for some $\alpha \geq 1$, $\|f+g\|_X \leq \alpha(\|f\|_X + \|g\|_X) \quad (f, g \in X)$. However, to simplify, we frequently omit the word "quasi". Likewise we abbreviate the word "quasi-Banach space" to Banach space.
26. Let $j \in \mathbb{Z}$ and $\varphi \in \mathscr{S}(\mathbb{R}^n)$. Then define $\varphi_j \equiv \varphi(2^{-j}\star)$ and $\varphi^j \equiv 2^{jn}\varphi(2^j\star)$.
27. The Kronecker delta function is given by $\delta_{jk} \equiv \begin{cases} 1 & (j = k), \\ 0 & (j \neq k). \end{cases}$ for $j, k \in \mathbb{Z}$.
28. When two normed space X, Y are isomorphic, we write $X \approx Y$.
29. For subsets A, B of a linear space V and $v \in V$, define the Minkovski sum by

$$v + A \equiv \{v + a \,:\, a \in A\}, \quad A + B \equiv \{a + b \,:\, a \in A,\, b \in B\}.$$

30. When two topological spaces X, Y are homeomorphic, we write $X \approx Y$.
31. When A and B are sets, $A \subset B$ stands for the inclusion of sets. If, in addition, both A and B are topological spaces, and if the natural embedding mapping $A \to B$ is continuous, we write $A \hookrightarrow B$ in the sense of continuous embedding.

Contents

1 Elementary Facts on Harmonic Analysis 1
1.1 Measure Theory 1
1.1.1 $L^p(\mu)$-Spaces for $0 < p \leq \infty$ 2
1.1.2 Covering Lemma and Carleson Tent 12
1.1.3 Hausdorff Capacity 19
1.1.4 Choquet Integral 30
1.1.5 Fundamental Facts on Functional Analysis 35
1.2 Schwartz Function Space $\mathscr{S}(\mathbb{R}^n)$ and Function Spaces $\mathscr{D}(\Omega)$ on Domains 41
1.2.1 Definition of the Schwartz Function Space $\mathscr{S}(\mathbb{R}^n)$ and Its Topology 41
1.2.2 Definition of the Schwartz Distribution Space $\mathscr{S}'(\mathbb{R}^n)$ and Its Topology 48
1.2.3 Definition of the Fourier Transform and Its Elementary Properties 65
1.2.4 The Space $\mathscr{D}$–$\mathscr{D}(\mathbb{T}^n)$ and $\mathscr{D}(\Omega)$ 75
1.2.5 Some Functional Equations in $\mathscr{S}(\mathbb{R}^n)$ 83
1.2.6 Schwartz's Kernel Theorem 92
1.3 Difference/Oscillation Operators 96
1.3.1 Elementary Formulas 96
1.3.2 Oscillation 100
1.4 Boundedness of the Hardy–Littlewood Maximal Operator 106
1.4.1 Hardy–Littlewood Maximal Inequality 106
1.4.2 Fefferman–Stein Vector-Valued Maximal Inequality 119
1.4.3 Properties of Band-Limited Distributions 124
1.4.4 Some Integral Inequalities 134
1.4.5 Carleson Measure 141
1.5 Singular Integral Operators 146
1.5.1 Dyadic Maximal Operator and the Calderón–Zygmund Decomposition 146
1.5.2 Singular Integral Operators 156

1.6 Harmonic Functions 171
1.6.1 Harmonic Polynomials 172
1.6.2 Harmonic Functions on the Unit Ball and the Half-Plane 182
1.6.3 Subharmonic Functions 189
1.7 Notes for Chap. 1 195

2 Besov Spaces, Triebel–Lizorkin Spaces and Modulation Spaces 205
2.1 Definition of the Nikolskii–Besov Space $B^s_{pq}(\mathbb{R}^n)$ with $1 \le p, q \le \infty$ and $s \in \mathbb{R}$ 206
2.1.1 Definition of Nikolskii–Besov Spaces 206
2.1.2 Elementary Properties of the Besov Space $B^s_{pq}(\mathbb{R}^n)$ with $1 \le p, q \le \infty$ and $s \in \mathbb{R}$ 211
2.2 Besov Spaces in Analysis 221
2.2.1 Sobolev Spaces and Besov Spaces 222
2.2.2 Hölder–Zygmund Spaces and Besov Spaces 224
2.2.3 Applications to Fractals and the Fourier Transform 231
2.3 Besov Spaces, Triebel–Lizorkin Spaces and Modulation Spaces with $0 < p, q \le \infty$ and $s \in \mathbb{R}$ 237
2.3.1 Definition of $A^s_{pq}(\mathbb{R}^n)$ with $0 < p, q \le \infty$ and $s > 0$ 238
2.3.2 Fundamental Properties of Function Spaces 244
2.3.3 Modulation Spaces 262
2.4 Homogeneous Besov Spaces and Homogeneous Triebel-Lizorkin Spaces 267
2.4.1 $\mathscr{S}_\infty(\mathbb{R}^n)$ and its Dual $\mathscr{S}'_\infty(\mathbb{R}^n)$ 269
2.4.2 Function Spaces of Homogeneous Type 279
2.4.3 Realization of $\dot{A}^s_{pq}(\mathbb{R}^n)$ 283
2.5 Local Means 289
2.5.1 Maximal Inequality Adapted to the Local Means 290
2.5.2 Local Means 294
2.5.3 Characterizations by Means of the Difference and the Oscillation 300
2.6 Notes for Chap. 2 307

3 Relation with Other Function Spaces 321
3.1 $L^p(\mathbb{R}^n)$ Spaces and Sobolev Spaces 321
3.1.1 Rademacher Sequence 322
3.1.2 The $L^p(\mathbb{R}^n)$ Space and the Triebel–Lizorkin Spaces $F^0_{p2}(\mathbb{R}^n)$ and $\dot{F}^0_{p2}(\mathbb{R}^n)$ with $1 < p < \infty$ 325
3.2 Hardy Spaces 331
3.2.1 Definition of Hardy Spaces 332
3.2.2 Singular Integral Operators on Hardy Spaces 341
3.2.3 Atoms for Hardy Spaces 349
3.2.4 Hilbert Space $\mathscr{H}_j$ 355
3.2.5 Calderón–Zygmund Decomposition for Distributions 363
3.2.6 Atomic Decomposition of Hardy Spaces 372

3.2.7 Characterization of Hardy Spaces via Riesz Transforms 382
3.2.8 Local Hardy Spaces 387
3.2.9 The Hardy Space $H^p(\mathbb{R}^n)$ and the Triebel–Lizorkin Space $\dot{F}^0_{p2}(\mathbb{R}^n)$ 392
3.3 BMO($\mathbb{R}^n$) 398
3.3.1 BMO($\mathbb{R}^n$): Definition and Fundamental Properties 398
3.3.2 Local bmo($\mathbb{R}^n$) Space 407
3.3.3 Function Spaces $F^s_{\infty q}(\mathbb{R}^n)$ and $\dot{F}^s_{\infty q}(\mathbb{R}^n)$ 410
3.4 Notes for Chap. 3 423

4 Decomposition of Function Spaces and Its Applications 429
4.1 Decomposition of Function Spaces 430
4.1.1 Atomic Decomposition and Molecular Decomposition 430
4.1.2 Wavelet Decomposition 447
4.1.3 Quarkonial Decomposition 454
4.1.4 Applications of the Atomic Decomposition to the Embedding Theorems 464
4.2 Interpolation Theory 472
4.2.1 Topological Vector Spaces and Compatible Couple 473
4.2.2 Real Interpolation 476
4.2.3 Complex Interpolation 483
4.3 Paraproduct and Pointwise Multipliers 507
4.3.1 Paraproduct 509
4.3.2 Hölder's Inequality for Besov Spaces and Triebel–Lizorkin Spaces 513
4.3.3 Characteristic Function of the Upper Half Plane as a Pointwise Multiplier 523
4.3.4 Applications: Div-Curl Lemma, Kato–Ponce Inequality and Riemann–Stieltjes Integral 530
4.4 Fundamental Theorems on Function Spaces 537
4.4.1 Diffeomorphism 538
4.4.2 Trace Operator 541
4.4.3 Fubini's Property 549
4.5 Notes for Chap. 4 554

5 Applications: PDEs, the $T1$ Theorem and Related Function Spaces 565
5.1 Function Spaces on Domains 565
5.1.1 Function Spaces on the Half Space 565
5.1.2 Function Spaces on Bounded C^∞-Domains 576
5.1.3 Function Spaces on Uniformly C^m-Open Sets 580
5.1.4 Function Spaces on Lipschitz Domains 583
5.2 Pseudo-differential Operators on Besov Spaces and Triebel–Lizorkin Spaces 589
5.2.1 Pseudo-differential Operators 589

5.2.2 Boundedness of Pseudo-differential Operators on Besov Spaces and Triebel–Lizorkin Spaces ... 607
5.2.3 Applications to Partial Differential Equations ... 613
5.2.4 Examples and Classical Results ... 617
5.3 Semi-groups: Applications to Heat Equations, Schrödigner Equations and Wave Equations ... 622
5.3.1 Bounded Holomorphic Calculus ... 623
5.3.2 The Square Root of the Sectorial Operators ... 634
5.3.3 Applications of Function Spaces to the Heat Semi-group ... 642
5.3.4 Applications of Function Spaces to the Wave Equations ... 646
5.3.5 Applications of Modulation Spaces to the Schrödinger Propagator ... 652
5.4 Elliptic Differential Equations of the Second Order ... 655
5.4.1 A Priori Estimate on the Whole Space ... 655
5.4.2 A Priori Estimate in the Half Space ... 669
5.4.3 A Priori Estimate on Domains with Smooth Boundary ... 677
5.5 $T1$ Theorem and Its Applications ... 679
5.5.1 $T1$-Theorem ... 679
5.5.2 Applications of $T1$-Theorem ... 690
5.6 Notes for Chap. 5 ... 695

6 Various Function Spaces ... 709
6.1 Various Function Spaces ... 709
6.1.1 Function Norms ... 710
6.1.2 Weighted Lebesgue Spaces ... 713
6.1.3 Mixed Lebesgue Spaces ... 726
6.1.4 Variable Lebesgue Spaces ... 728
6.1.5 Morrey Spaces ... 745
6.1.6 Orlicz Spaces ... 751
6.1.7 Herz Spaces ... 766
6.2 Hardy Spaces Based on Ball Quasi-Banach Function Spaces ... 770
6.2.1 General Definition of Hardy-Type Spaces ... 770
6.2.2 Hardy–Orlicz Spaces and Their Applications to Pointwise Multipliers ... 772
6.3 Besov Spaces and Triebel–Lizorkin Spaces Based on Ball Quasi-Banach Function Spaces ... 774
6.3.1 Besov Spaces and Triebel–Lizorkin Spaces Based on Ball Quasi-Banach Function Spaces ... 775
6.3.2 Besov Spaces and Triebel–Lizorkin Spaces Based on Morrey Spaces and Herz Spaces ... 782
6.4 Besov-Type Spaces and Triebel–Lizorkin-Type Spaces ... 785
6.4.1 The Spaces $F\dot{W}^{s,\tau}_{pq}(\mathbb{R}^{n+1}_+)$ and $F\dot{T}^{s,\tau}_{pq}(\mathbb{R}^{n+1}_+)$ ... 785
6.4.2 Besov-Type Spaces and Triebel–Lizorkin-Type Spaces ... 791
6.4.3 Besov–Hausdorff-Type Spaces and Triebel–Lizorkin–Hausdorff Spaces ... 802

6.5 Weighted Besov Spaces and Triebel–Lizorkin Spaces 805
6.5.1 Besov Spaces and Triebel–Lizorkin Spaces with A_p-Weights 806
6.5.2 Microlocal Besov Spaces and Triebel–Lizorkin Spaces 808
6.5.3 Function Spaces with Variable Exponents 811
6.5.4 Function Spaces with Mixed Smoothness 812
6.5.5 Anisotropic Function Spaces 814
6.6 Function Spaces on Various Sets 817
6.6.1 Function Spaces on the Torus 817
6.6.2 Function Spaces on Fractals 821
6.7 Applications of Function Spaces to the Kato Theorem 832
6.7.1 Kato Conjecture 832
6.7.2 Kato Conjecture (Kato Theorem): Some Reductions 837
6.7.3 Kato Conjecture for Other Lebesgue Spaces 855
6.8 Notes for Chap. 6 864

References 891

Index 939

Chapter 1
Elementary Facts on Harmonic Analysis

Now we elaborate a fundamental theory of harmonic analysis. We aim here to collect fundamental facts on analysis and the tools which we use throughout this book.

First we review some fundamental facts on integrals including complex analysis in Sect. 1.1.

Before we go into the detailed study of function spaces, let us consider the properties of the space $\mathscr{S}'(\mathbb{R}^n)$. Since we rely upon Fourier analysis on $\mathbb{R}^n$ to consider function spaces, we collect elementary facts on Fourier analysis in Sect. 1.2. One of the ways to study the property of functions is to consider the difference and oscillation, which we do in Sect. 1.3. We consider the boundedness of the Hardy–Littlewood maximal operator in Sect. 1.4. The Hardy–Littlewood maximal operator is used to control the average of the functions. We consider the boundedness of the singular integral operators in Sect. 1.5. The singular integral operators are important because they arise naturally in harmonic analysis. Finally, we study the boundedness of the harmonic functions and the subharmonic functions in Sect. 1.6.

1.1 Measure Theory

The space $L^p(\mu)$ with $1 \le p \le \infty$ is familiar. But to familiarize ourselves with $L^p(\mathbb{R}^n)$ with $0 < p < 1$, we discuss it here. In Sect. 1.1.1, we deal with L^p-spaces for $0 < p \le \infty$. We give a brief review of measure theory. In Sect. 1.1.2 we present a set theoretical consideration. We discuss the covering lemmas here, as well as the Carleson tents. We include here the fundamental property of Hausdorff capacity in Sect. 1.1.3 and the theory of Choquet integral in Sect. 1.1.4. We do not use Sects. 1.1.3 and 1.1.4 except in Chap. 6. More precisely, we do not use Sects. 1.1.3 and 1.1.4 up to Sect. 6.4.3. Finally, unlike other sections in this chapter, we review some fundamental facts on functional analysis needed in this book in Sect. 1.1.5.

Y. Sawano, *Theory of Besov Spaces*, Developments in Mathematics 56,
https://doi.org/10.1007/978-981-13-0836-9_1

The couple $(X, \mathscr{B}, \mu)$ is a measure space if X is a set, $\mathscr{B}$ is a σ-algebra and $\mu : \mathscr{B} \to [0, \infty]$ is a measure.

1.1.1 $L^p(\mu)$-Spaces for $0 < p \le \infty$

1.1.1.1 Fundamental Theorems in Integration Theory

We recall the following theorems which are used everywhere in this book. The proof is omitted; see [92], for example.

Theorem 1.1 (Convergence theorems on integration) *Let $(X, \mathscr{B}, \mu)$ be a measure space and let $\{f_j\}_{j=1}^\infty$ be a sequence of μ-measurable functions taking the value in $\overline{\mathbb{R}} \equiv \mathbb{R} \cup \{\pm\infty\}$ or $\mathbb{C}$.*

1. *(Monotone convergence theorem) Let $\{f_j\}_{j=1}^\infty$ be a nonnegative increasing sequence; that is, $0 \le f_j \le f_{j+1}$ for all $j \in \mathbb{N}$, μ-almost everywhere. Then*

$$\lim_{j\to\infty} \int_X f_j(x) \mathrm{d}\mu(x) = \int_X \lim_{j\to\infty} f_j(x) \mathrm{d}\mu(x). \tag{1.1}$$

2. *(Fatou's lemma) Suppose that we have a sequence $\{f_j\}_{j=1}^\infty$ of nonnegative functions. Then $\displaystyle\int_X \liminf_{j\to\infty} f_j(x)\mathrm{d}\mu(x) \le \liminf_{j\to\infty} \int_X f_j(x)\mathrm{d}\mu(x)$.*
3. *(Dominated convergence theorem, Lebesgue's convergence theorem) Suppose that a sequence $\{f_j\}_{j=1}^\infty$ of μ-measurable functions converges μ-almost everywhere to f. Also assume that there exists $g \in L^1(\mu)$ such that $|f_j| \le g$ μ-almost everywhere for all $j \in \mathbb{N}$. Then (1.1) holds.*

Theorem 1.2 (Change of integration and differentiation) *Let $(X, \mathscr{B}, \mu)$ be a measure space. Suppose that a function $f : X \times (a, b) \to \mathbb{C}$ satisfies the following conditions:*

1. *Fix $t \in (a, b)$. Then $f(\star, t)$ is a μ-integrable function.*
2. *For μ-almost all $x \in X$, the mapping $f(x, \star)$ is differentiable on (a, b).*
3. *There exists a μ-integrable function g such that the differential inequality*

$$\left|\frac{\partial f}{\partial t}(x, t)\right| \le g(x), \quad \text{for all } t \in (a, b)$$

holds for μ-almost all $x \in X$.

Then we can interchange differentiation and integral:

$$\frac{\mathrm{d}}{\mathrm{d}t} \int_X f(x, t)\mathrm{d}\mu(x) = \int_X \frac{\partial f}{\partial t}(x, t)\mathrm{d}\mu(x).$$

Recall that a measure space $(X, \mathscr{B}, \mu)$ is σ-finite, if X is expressed as the countable sum of measurable sets of finite μ-measure. A *measurable rectangular* is a cross product of two measurable products of the sets in $\mathscr{M}$ and $\mathscr{N}$.

Theorem 1.3 (Fubini's theorem) [88, Theorem 3.1] *Let $(X, \mathscr{M}, \mu)$ and $(Y, \mathscr{N}, \nu)$ be σ-finite measure spaces. We let $(X \times Y, \mathscr{M} \otimes \mathscr{N}, \mu \otimes \nu)$ be a measure space over $X \times Y$, which is induced by a product outer measure*

$$\mu^*(E) = \inf \left\{ \sum_{j=1}^{\infty} \lambda_0(C_j) \; : \; \text{each } C_j \text{ is a measurable rectangular such that } E \subset \bigcup_{j=1}^{\infty} C_j \right\}.$$

Assume that f is a nonnegative measurable function or an integrable function.
Then;

1. $f_x \equiv f(x, \star)$ *is integrable for μ-almost every $x \in X$.*
2. $\int_Y f(\star, y) \mathrm{d}\nu(y)$ *is an integrable function.*
3. *Equalities*

$$\int_{X \times Y} f(x, y) \mathrm{d}\mu \otimes \nu(x, y) = \int_X \left(\int_Y f(x, y) \mathrm{d}\nu(y) \right) \mathrm{d}\mu(x) = \int_Y \left(\int_X f(x, y) \mathrm{d}\mu(x) \right) \mathrm{d}\nu(y)$$

hold.

Next, we investigate the norm inequalities which are fundamental tools to measure the size of the functions. We consider $L^p(\mu)$-spaces for $0 < p \le \infty$ not only for $1 \le p \le \infty$. As an undergraduate student, the author learnt the theory of Banach spaces, where the topology induced by the triangle inequality played an elementary role. What was new at that time was the completeness of the normed spaces. However, after learning the completeness, we feel that the triangle inequality is also difficult to show. As is shown, $L^p(\mu)$ with $0 < p < 1$ does not satisfy the triangle inequality. So, it will turn out that $L^p(\mu)$ with $0 < p < 1$ is a nasty object. However, the space $L^p(\mu)$ with $0 < p < 1$ arises naturally. When we consider the product $f \cdot g$ for $f, g \in L^1(\mu)$, the product will be no better than an $L^{1/2}(\mu)$-function.

1.1.1.2 Norm Inequalities

These prerequisites are elementary results in integration theory. We also use some integral inequalities, which we now recall.

When $p < 1$, it should be more appropriate to use the term "quasi-norm". But we may still use the term "norm" for short. The space $L^\infty(\mathbb{R}^n)$ is the set of all essentially bounded measurable functions.

Theorem 1.4 (Integral inequalities) *Let $(X, \mathscr{B}, \mu)$ be a measure space. Then the following inequalities hold;*

1. (Triangle inequality) *Let $1 \le p \le \infty$. Then, $\|f+g\|_{L^p(\mu)} \le \|f\|_{L^p(\mu)} + \|g\|_{L^p(\mu)}$ for all $f, g \in L^p(\mu)$.*
2. (Hölder's inequality) *Let $0 < p, q, r \le \infty$ satisfy $\frac{1}{r} = \frac{1}{p} + \frac{1}{q}$. Then for $f \in L^p(\mu)$ and $g \in L^q(\mu)$, $\|fg\|_{L^r(\mu)} \le \|f\|_{L^p(\mu)}\|g\|_{L^q(\mu)}$. See* [4, Theorem 2.4] *for example.*
3. (Minkovski's inequality) *Let $1 \le p \le \infty$. Then for $f, g \in L^p(\mu)$,*
$$\|f+g\|_{L^p(\mu)} \le \|f\|_{L^p(\mu)} + \|g\|_{L^p(\mu)}.$$
See [4, Theorem 2.8] *for example.*
4. (Chebyshev's inequality) *Let $f \in L^1(\mu), \lambda > 0$. Then; $\lambda\mu\{|f| > \lambda\} \le \|f\|_{L^1(\mu)}$.*

The space $L^p(\mu)$ with $p < 1$ fails the triangle inequality:

$$\|f+g\|_{L^p(\mu)} \le \|f\|_{L^p(\mu)} + \|g\|_{L^p(\mu)}; \tag{1.2}$$

hence it is not a normed space (see Exercise 1.2). Let $0 < p \le 1$. But it satisfies the inequality similar to the triangle inequality:

$$\|f+g\|_{L^p(\mu)}{}^p \le \|f\|_{L^p(\mu)}{}^p + \|g\|_{L^p(\mu)}{}^p, \tag{1.3}$$

$$\|f+g\|_{L^p(\mu)} \le 2^{1/p-1}(\|f\|_{L^p(\mu)} + \|g\|_{L^p(\mu)}). \tag{1.4}$$

See Exercise 1.2. We make $L^p(\mu)$ into a distance space using the triangle inequality (1.2) with $1 \le p \le \infty$ and the p-triangle inequality (1.3) with $0 < p < 1$.

We characterize the bounded linear functionals from $L^1(\mu)$ to $L^\infty(\nu)$, where $(X, \mathscr{B}, \mu)$ and $(Y, \mathscr{C}, \nu)$ are σ-finite measure spaces. This is used in Chap. 6.

Proposition 1.1 *Let $T \in B(L^1(\mu), L^\infty(\nu))$. Then there exists a function $K \in L^\infty(\mu \otimes \nu)$ such that $Tf(x) = \int_X K(x,y)f(y)\mathrm{d}\nu(y)$ for almost every $x \in X$.*

Proof We consider the bilinear operator $T^\star : L^1(\mu) \times L^1(\nu) \mapsto \mathbb{C}$ defined by

$$T^\star(f,g) \equiv \int_X g(x)Tf(x)\mathrm{d}\mu(x).$$

Then since $\mathscr{X} \equiv \{f \otimes g : f \in L^1(\mu), g \in L^1(\nu)\}$ densely spans $L^1(\mu \otimes \nu)$, we see that there exists a bounded linear operator $Z : L^1(\mu \otimes \nu) \to \mathbb{C}$ such that $Z(f \otimes g) = T^\star(f, g)$ for all $f \in L^1(\mu)$ and $g \in L^1(\nu)$. Thus, by the duality $L^1(\mu \otimes \nu)$-$L^\infty(\mu \otimes \nu)$ (see [89, Theorem 4.1]), there exists a function $K \in L^\infty(\mu \otimes \nu)$ such that

$$Z(F) = \int_{X \times Y} F(y, x) K(x, y) \mathrm{d}\mu(x) \mathrm{d}\nu(y)$$

for all $F \in L^1(\mu \otimes \nu)$. If we let $F \equiv f \otimes g$ with $f \in L^1(\mu)$, $g \in L^1(\nu)$, we obtain

$$\int_{X \times Y} f(y) g(x) K(x, y) \mathrm{d}\mu(x) \mathrm{d}\nu(y) = \int_X g(x) T f(x) \mathrm{d}\mu(x).$$

Thus, we have the desired expression.

1.1.1.3 Layer Cake Representation

Let (X, μ) be a σ-finite measure space. One of the important understandings in the notion of L^p spaces is that we measure the size of functions. Usually, we do not want to find the precise value of integrals. For example, when we are faced with general functions we hope that we are able to calculate their L^1 or L^2 norms. However, when $p = \frac{7}{4}$, it is almost impossible to handle the L^p-norm because $|f(x)|^p$ is difficult to expand. The next theorem in some sense linearizes the integral in terms of f and will pave a way to estimate the various norms, not to say, find their precise value.

Theorem 1.5 (Layer cake representation) *Let $0 < p < \infty$ and f be a μ-measurable function.*

1. *We have*

$$\|f\|_{L^p(\mu)} = \left(p \int_0^\infty \lambda^{p-1} \mu\{x \in X : |f(x)| > \lambda\} \mathrm{d}\lambda \right)^{\frac{1}{p}}. \tag{1.5}$$

2. *More generally, if $\Phi : [0, \infty) \to [0, \infty)$ is a C^1-increasing function such that $\Phi(0) = 0$, then $\int_X \Phi(|f(x)|) \mathrm{d}\mu(x) = \int_0^\infty \Phi'(\lambda) \mu\{x \in X : |f(x)| > \lambda\} \mathrm{d}\lambda$.*

Proof We concentrate on 1. the proof of 2. is merely a slight adaptation of 1. We observe

$$|f(x)|^p = \int_0^{|f(x)|} p \lambda^{p-1} \mathrm{d}\lambda = \int_0^\infty p \lambda^{p-1} \chi_{(\lambda, \infty)}(|f(x)|) \mathrm{d}\lambda$$

for $x \in X$. By Fubini's theorem (Theorem 1.3) for nonnegative functions, we calculate that

$$\begin{aligned}(\|f\|_{L^p(\mu)})^p &= \int_X |f(x)|^p \mathrm{d}\mu(x) \\ &= \int_0^\infty \left(\int_X p\,\lambda^{p-1} \chi_{(\lambda,\infty)}(|f(x)|)\mathrm{d}\mu(x) \right) \mathrm{d}\lambda \\ &= \int_0^\infty p\,\lambda^{p-1} \left(\int_X \chi_{(\lambda,\infty)}(|f(x)|)\mathrm{d}\mu(x) \right) \mathrm{d}\lambda \\ &= \int_0^\infty p\lambda^{p-1}\mu\{x \in X : |f(x)| > \lambda\}\mathrm{d}\lambda.\end{aligned}$$

By taking the p-th root, we obtain Theorem 1.51.

1.1.1.4 Inequality for Sequences

The next inequality is useful for the study of Besov spaces and Triebel–Lizorkin spaces.

Proposition 1.2 (**l^p-boundedness for special convolutions, Hardy-type inequality**) *Let $0 < p \le \infty$ and $\delta > 0$, and let $\{a_j\}_{j=1}^\infty$ be a nonnegative sequence.*

1. *We have* $\left(\sum_{j=1}^\infty \left(\sum_{k=1}^\infty 2^{-|k-j|\delta} a_k \right)^p \right)^{\frac{1}{p}} \lesssim_{\delta,p} \left(\sum_{j=1}^\infty a_j{}^p \right)^{\frac{1}{p}}$.
2. *When $p = \infty$, we have* $\sup_{j \in \mathbb{N}} \left(\sum_{k=1}^\infty 2^{-|k-j|\delta} a_k \right) \lesssim_\delta \sup_{j \in \mathbb{N}} a_j$.

We explain why Proposition 1.2 deserves its name; note that an $\ell^p(\mathbb{Z})$-analogue to Proposition 1.2 is

$$\left(\sum_{j=-\infty}^\infty \left(\sum_{k=-\infty}^\infty 2^{-|k-j|\delta} a_k \right)^p \right)^{\frac{1}{p}} \lesssim_\delta \left(\sum_{j=-\infty}^\infty a_j{}^p \right)^{\frac{1}{p}} \tag{1.6}$$

for nonnegative sequences $\{a_j\}_{j=-\infty}^\infty$. If we equip $\mathbb{Z}$ with the counter measure, then we learn that (1.6) is a convolution estimate.

We can easily prove 2. and we leave the proof to interested readers; see Exercise 1.5. When $1 \le p < \infty$, Proposition 1.21 can be proved by Hölder's inequality (Theorem 1.4) or the Minkovski inequality (Theorem 1.4).

It counts that Proposition 1.2 still holds even for $0 < p \le 1$, although the proof is not that difficult.

Proof Let $0 < p \le 1$. Then

$$\left(\sum_{j=1}^{\infty}\left(\sum_{k=1}^{\infty}2^{-|k-j|\delta}a_k\right)^p\right)^{\frac{1}{p}} \le \left(\sum_{j=1}^{\infty}\sum_{k=1}^{\infty}2^{-|k-j|\delta p}{a_k}^p\right)^{\frac{1}{p}} \lesssim_\delta \left(\sum_{j=1}^{\infty}{a_j}^p\right)^{\frac{1}{p}}.$$

For the sake of completeness we supply the proof for the case where $1 \le p < \infty$. By the Hölder inequality, we have

$$\begin{aligned}\sum_{j=1}^{\infty}\left(\sum_{k=1}^{\infty}2^{-|k-j|\delta}a_k\right)^p &\le \sum_{j=1}^{\infty}\left(\left(\sum_{k=1}^{\infty}2^{-|k-j|\delta}\right)^{\frac{p}{p'}}\cdot\sum_{k=1}^{\infty}2^{-|k-j|\delta}{a_k}^p\right)\\ &\lesssim_\delta \sum_{j=1}^{\infty}\sum_{k=1}^{\infty}2^{-|k-j|\delta}{a_k}^p \lesssim_\delta \sum_{j=1}^{\infty}{a_j}^p.\end{aligned}$$

Thus, the conclusion follows.

We need the following inequality in Sect. 4.2.3:

Lemma 1.1 *Let $\{a_j\}_{j=1}^{\infty}$ be a nonnegative sequence. Then for $\kappa > 0$,*

$$\sum_{j=1}^{\infty}a_j\left(\sum_{k=1}^{j}a_k\right)^{\kappa-1} \le \frac{1}{\kappa\wedge 1}\left(\sum_{j=1}^{\infty}a_j\right)^{\kappa}.$$

It is understood that $0^{\kappa-1} = 0$ for $0 < \kappa < 1$.

Proof If $0 < \kappa < 1$, we have $a_j\left(\sum_{k=1}^{j}a_k\right)^{\kappa-1} \le \int_{\sum_{k=1}^{j-1}a_k}^{\sum_{k=1}^{j}a_k}t^{\kappa-1}dt$ for all $j \ge 2$. If $\kappa \ge 1$, then Lemma 1.1 is trivial. Thus, the proof is complete.

When $0 < \kappa < 1$, the proof above shows that the inequality is reversed:

$$\left(\sum_{j=1}^{\infty}a_j\right)^{\kappa} \le \kappa\sum_{j=1}^{\infty}a_j\left(\sum_{k=1}^{j-1}a_k\right)^{\kappa-1}. \tag{1.7}$$

See Exercise 1.6.

Lemma 1.2 (Power trick for sequences growing mildly) *Let* $0 < r < 1$ *and let* $\{C_N\}_{N=1}^{\infty}$, $\{\beta_j\}_{j=-\infty}^{\infty}$ *and* $\{\gamma_j\}_{j=-\infty}^{\infty}$ *be nonnegative sequences.*

1. *Assume that* $\{\gamma_j\}_{j=-\infty}^{\infty}$ *grows mildly in the sense that*

$$\sup_{j \in \mathbb{Z}} \frac{\log \gamma_j}{|j|+1} < \infty. \tag{1.8}$$

Furthermore, assume that

$$\gamma_j \le C_N \sum_{k=-\infty}^{\infty} 2^{-|k|N} \beta_{j+k} {\gamma_{j+k}}^{1-r} \tag{1.9}$$

for any $j \in \mathbb{Z}$ *and* $N \in \mathbb{N}$. *Then*

$$\gamma_j{}^r \le C_N \sum_{l=-\infty}^{\infty} 2^{-|j-l|Nr} \beta_l. \quad (j \in \mathbb{N}). \tag{1.10}$$

2. *Assume that* $\{\gamma_j\}_{j=1}^{\infty}$ *grows mildly in the sense that* $\sup\limits_{j \in \mathbb{N}} \dfrac{\log \gamma_j}{j} < \infty$ *and that* $\gamma_j \le C_N \sum\limits_{k=0}^{\infty} 2^{-kN} \beta_{j+k} {\gamma_{j+k}}^{1-r}$ *for any* $j \in \mathbb{N}_0$ *and* $N \in \mathbb{N}$. *Then*

$$\gamma_j{}^r \le C_N \sum_{l=j}^{\infty} 2^{-(j-l)Nr} \beta_l. \quad (j \in \mathbb{N}). \tag{1.11}$$

Proof We concentrate on (1.10); the proof of (1.11) is similar. Let us set $\Gamma_{j,N} \equiv \sup\limits_{k \in \mathbb{Z}} 2^{-|j-k|N} \gamma_k$ for $j \in \mathbb{N}_0$, $N \in \mathbb{N}$. Then from (1.9)

$$\begin{aligned}
\Gamma_{j,N} &\le C_N \sup_{k \in \mathbb{Z}} \sum_{l=-\infty}^{\infty} 2^{-|j-k|N-|l|N} \beta_{k+l} {\gamma_{k+l}}^{1-r} \\
&\le C_N {\Gamma_{j,N}}^{1-r} \sup_{k \in \mathbb{Z}} \sum_{l=-\infty}^{\infty} 2^{-|j-k-l|Nr} \beta_{k+l} \\
&= C_N {\Gamma_{j,N}}^{1-r} \sum_{l=-\infty}^{\infty} 2^{-|j-l|Nr} \beta_l.
\end{aligned}$$

Thus, if $\Gamma_{j,N} < \infty$, $\gamma_j{}^r \le \Gamma_{j,N}{}^r \le C_N \sum\limits_{l=-\infty}^{\infty} 2^{-|j-l|Nr} \beta_l$. This is possible when N is large enough due to (1.8), say N_0. Suppose that N is small; $N < N_0$. To prove (1.10), we can assume that $\sum_{l=-\infty}^{\infty} 2^{-|j-l|Nr}\beta_l < \infty$. Then

$$
\begin{aligned}
\Gamma_{j,N}{}^r = \sup_{l\in\mathbb{Z}} 2^{-|j-l|Nr}\gamma_l{}^r &\le C_{N_0} \sup_{l\in\mathbb{Z}} \sum_{k=-\infty}^{\infty} 2^{-|j-l|Nr-|k|N_0 r}\beta_{l+k} \\
&\le C_{N_0} \sup_{l\in\mathbb{Z}} \sum_{k=-\infty}^{\infty} 2^{-|j-l|Nr-|k|Nr}\beta_{l+k} \\
&= C_{N_0} \sum_{k=-\infty}^{\infty} 2^{-|j-k|Nr}\beta_k < \infty \quad (j \in \mathbb{N}_0,\ N \in \mathbb{N})
\end{aligned}
$$

as was to be shown.

1.1.1.5 Integration Inequalities Over $\mathbb{R}^n$

We recall the Young inequality. See [4, Theorem 2.24], for example, for its proof.

Theorem 1.6 (Young's inequality) *Let* $1 \le p, q, r \le \infty$ *satisfy* $\frac{1}{r} = \frac{1}{p} + \frac{1}{q} - 1$. *Define a convolution* $f * g$ *by* $f * g(x) \equiv \int_{\mathbb{R}^n} f(x-y)g(y)\mathrm{d}y$ *as long as the integral makes sense. Then* $\|f * g\|_r \le \|f\|_p \|g\|_q$ *for* $f \in L^p(\mathbb{R}^n)$ *and* $g \in L^q(\mathbb{R}^n)$.

Recall the *inner regularity* of the Lebesgue measure. For any measurable set A, we have the following property called the inner regularity of the Lebesgue measure:

$$
|A| = \sup\{|K| \,:\, K \text{ is a compact set contained in } A\}. \tag{1.12}
$$

Recall that the support of a continuous function $u : \mathbb{R}^n \to \mathbb{C}$ is the closure of $\{x \in \mathbb{R}^n : u(x) \ne 0\}$. Denote by $C_{\mathrm{c}}^{\infty}(\mathbb{R}^n)$ the set of all smooth functions having compact support. A typical example of the function in this class is

$$
\varphi(t) = \chi_{(-1,1)}(t) \exp\left(-\frac{1}{1-t^2}\right)
$$

when $n = 1$. For higher dimensions, consider its tensor product. As an immediate consequence of (1.12), we have the following density property:

Proposition 1.3 *Let* $0 < p < \infty$. *Then* $C_{\mathrm{c}}^{\infty}(\mathbb{R}^n)$ *is dense in the* $L^p(\mathbb{R}^n)$*-topology.*

Proof When $1 \le p < \infty$, this is well known. By Lebesgue's convergence theorem we can prove that we can approximate any function in $L^p(\mathbb{R}^n)$ with the space $L_{\mathrm{c}}^{\infty}(\mathbb{R}^n)$ of all the measurable functions that are bounded and assume 0 for almost every point outside a compact set. The same argument works for $0 < p < 1$. See Exercise 1.1.

Exercises

Exercise 1.1 Show that $L^\infty_{\rm c}(\mathbb{R}^n)$ and $C^\infty_{\rm c}(\mathbb{R}^n)$ are dense in the $L^p(\mathbb{R}^n)$-topology for $0 < p < 1$.

Exercise 1.2 Let $0 < p < 1$.

1. For $a, b \ge 0$, prove that
$$(a+b)^p \le a^p + b^p \tag{1.13}$$
using
$$\left(\frac{a}{a+b}\right)^p \ge \frac{a}{a+b}, \quad \left(\frac{b}{a+b}\right)^p \ge \frac{b}{a+b} \tag{1.14}$$
or the Taylor expansion of order 2.
2. [30, Theorem 5.1.1.(i)] Prove (1.3).
3. [137, Example 1], [4, Lemma 2.2] Prove (1.4) using $a^p + b^p \le 2^{1-p}(a+b)^p$.
4. In (1.4), present an example showing that $2^{\frac{1}{p}-1}$ cannot be replaced by 1.

Exercise 1.3 Let $0 < p < 1$ and set $q \equiv \dfrac{p}{p-1} < 0$. Let $f, g : X \to [0, \infty]$ be μ-measurable functions defined on a measure space $(X, \mathscr{B}, \mu)$.

1. Show that $\|fg\|_1 \ge \|f\|_p \left(\int_X g(x)^q {\rm d}\mu(x)\right)^{\frac{1}{q}}$ using the Hölder inequality carefully. See [4, Theorem 2.12].
2. When $f(x) > 0$ for all $x \in X$, look for a function g such that $g(x) > 0$ for all $x \in X$ and that $\|fg\|_1 = \|f\|_p \left(\int_X g(x)^q {\rm d}\mu(x)\right)^{\frac{1}{q}}$.
3. Show that $\|f+g\|_p \ge \|f\|_p + \|g\|_p$ mimicking a proof of the Hölder inequality.

Exercise 1.4 [287, Lemma 3.7], [648, Lemma 2], [715] Let $s_0 \ne s_1$ be real numbers and $0 < q < \infty$. Define $s \equiv (1-\theta)s_0 + \theta s_1$ with $\theta \in (0, 1)$. Then show that
$$\|\{2^{js}a_j\}_{j=-\infty}^\infty\|_{\ell^q} \lesssim_{\theta,s_0,s_1,q} (\|\{2^{js_0}a_j\}_{j=-\infty}^\infty\|_{\ell^\infty})^{1-\theta}(\|\{2^{js_1}a_j\}_{j=-\infty}^\infty\|_{\ell^\infty})^\theta$$
for any complex sequence $\{a_j\}_{j=-\infty}^\infty$.

Exercise 1.5

1. Prove Proposition 1.21 using Hölder's inequality.

2. Prove Proposition 1.22 using $a_j \le \sup_{j' \in \mathbb{N}} a_{j'}$ for all $j \in \mathbb{N}$.

Exercise 1.6 Reexamine the proof of Lemma 1.1 to prove (1.7) when $\kappa \in (0, 1)$.

Exercise 1.7 [30] Let $(X, \mathscr{B}, \mu)$ be a σ-finite measure space. Let F and G be nonnegative μ-measurable functions. Assume that F and G satisfy

$$\alpha \mu\{F > \alpha\} \le \int_{\{F>\alpha\}} G(x) \mathrm{d}\mu(x) \tag{1.15}$$

for all $\alpha > 0$.

1. Let $\alpha, \beta > 0$. Show that

$$\alpha \mu\{\min(F, \beta) > \alpha\} \le \int_{\{\min(F,\beta)>\alpha\}} G(x) \mathrm{d}\mu(x).$$

(Hint: Maybe two cases must be considered; $\alpha \le \beta$ and otherwise.)

2. Let $1 < p \le \infty$. Using the layer cake representation, Theorem 1.5, prove that

$$\|F\|_{L^p(\mu)} \le p' \|G\|_{L^p(\mu)}. \tag{1.16}$$

Note that (1.16) requires no further assumption but (1.15).

Exercise 1.8 Let $(X, \mathscr{B}, \mu)$ be a measure space. Let $F : X \to [0, \infty)$ be a positive measurable function. Then show that

$$\sum_{j=-\infty}^{\infty} 2^{jp} \mu\{F > 2^j\} \sim_p \int_X F(x)^p \mathrm{d}\mu(x)$$

using

$$\sum_{j=-\infty}^{\infty} 2^j \chi_{\{F>2^j\}} \sim F.$$

Exercise 1.9 Find $I(a, b; A, B) \equiv \displaystyle\int_0^\infty \min\left(A t^a, \frac{B}{t^b}\right) \frac{\mathrm{d}t}{t}$ for $A, B, a, b > 0$. Hint: Let $t_0 > 0$ be the unique solution to $At^a = Bt^{-b}$.

Exercise 1.10 [526] Let $1 < p < \infty$, and let ρ be a positive measurable function such that

$$\int_0^\infty \min\left(1, \frac{x^{p'}}{t^{p'}}\right) \rho(t)^{p'} \frac{\mathrm{d}t}{t} \lesssim \rho(x)^{p'}.$$

Then show that for any positive sequence $\{u_j\}_{j=-\infty}^{\infty}$

$$\sum_{j=-\infty}^{\infty} u_j \lesssim R\left(\sum_{j=-\infty}^{\infty}\left(\frac{u_j}{\rho(2^j)}\right)^p, \sum_{j=-\infty}^{\infty}\left(\frac{2^j u_j}{\rho(2^j)}\right)^p\right),$$

where $R(U, V) \equiv U\rho(U/V)$. Hint: Decompose the sum according to $(-\infty, \log_2 V/U)$ and $[\log_2 V/U, \infty)$, where $U \equiv \sum_{j=-\infty}^{\infty}\left(\frac{u_j}{\rho(2^j)}\right)^p$ and $V \equiv \sum_{j=-\infty}^{\infty}\left(\frac{2^j u_j}{\rho(2^j)}\right)^p$.

Exercise 1.11 Let $f \in \bigcup_{q\in[1,\infty)} L^q(\mathbb{R}^n)$. Show that there exists a sequence $\{f_k\}_{k=1}^{\infty}$ in $C_c^{\infty}(\mathbb{R}^n)$ such that $f_k \to f$ in $L^r(\mathbb{R}^n)$ for any $r \in [1, \infty)$ such that $f \in L^r(\mathbb{R}^n)$. Hint: Use the truncation and the mollification simultaneously.

Exercise 1.12 (Strauss's lemma [732, 1011]) Let $f \in C_c^{\infty}(\mathbb{R}^n)$ be radial: $f = \varphi(|\star|)$ for some $\varphi \in C(\mathbb{R})$. Write $S^{n-1} \equiv \{\omega = (\omega_1, \omega_2, \dots, \omega_n) \in \mathbb{R}^n : |\omega| = 1\}$.

1. Using $\nabla f(x) = \varphi'(|x|)\frac{x}{|x|}$ for $x \neq 0$, show that

$$\int_{\mathbb{R}^n} |\nabla f(x)|^2 dx = |S^{n-1}| \int_0^{\infty} t^{n-1}|\varphi'(t)|^2 dt.$$

2. Using $(t^{n-1}\varphi(t)^2)' \geq 2t^{n-1}\varphi'(t)\varphi(t)$ for $t > 0$ and $\varphi \in C^1(0, \infty)$, show that

$$|x|^{n-1}|f(x)|^2|S^{n-1}| \leq 2\|f\|_2\|\nabla f\|_{(L^2)^n}.$$

Exercise 1.13 Let $x = \{x_j\}_{j=1}^{\infty} \in \ell^1(\mathbb{N})$ be a positive sequence. Denote by $x^{\downarrow} = \{x_j^{\downarrow}\}_{j=1}^{\infty}$ be a sequence defined by sorting x in decreasing order. Then show that $\|x^{\downarrow}\|_{\ell^1} = \|x\|_{\ell^1}$.

1.1.2 Covering Lemma and Carleson Tent

1.1.2.1 Dyadic Cubes

Now we define dyadic cube, which plays a central role in harmonic analysis.

A dyadic cube is a cube in $\mathbb{R}^n$ taking a special form. The dyadic cubes, which we are going to define, have nothing to do with the functions. However, they have the nested structure. So, using the nested structure, we will be able to decide whether we stop or not when we investigate the property of the functions.

Definition 1.1 (Dyadic cube)

1. For $j \in \mathbb{Z}$ and $k = (k_1, k_2, \dots, k_n) \in \mathbb{Z}^n$ define the set Q_{jk} by

$$Q_{jk} \equiv \left[\frac{k_1}{2^j}, \frac{k_1+1}{2^j}\right) \times \cdots \times \left[\frac{k_n}{2^j}, \frac{k_n+1}{2^j}\right) = \prod_{l=1}^{n} \left[\frac{k_l}{2^j}, \frac{k_l+1}{2^j}\right).$$

2. A dyadic cube is a set of the form Q_{jk} for some $j \in \mathbb{Z}, k = (k_1, k_2, \dots, k_n) \in \mathbb{Z}^n$.
3. The set of all dyadic cubes is denoted by $\mathscr{D}$; $\mathscr{D} = \mathscr{D} \equiv \left\{Q_{jk} : j \in \mathbb{Z}, k \in \mathbb{Z}^n\right\}$.
4. For $j \in \mathbb{Z}$ the set of dyadic cubes of the j-th generation is given by

$$\mathscr{D}_j = \mathscr{D}_j(\mathbb{R}^n) \equiv \{Q_{jk} : k \in \mathbb{Z}^n\} = \{Q \in \mathscr{D} : \ell(Q) = 2^{-j}\}.$$

The next lemma is almost trivial and we omit its proof.

Lemma 1.3 (The nested structure of dyadic cubes) *When two dyadic cubes intersect, one is contained in the other.*

Based on Lemma 1.3, we present the following terminology.

Definition 1.2 (Dyadic parent, dyadic child)

1. A *dyadic parent* of a dyadic cube Q is the smallest dyadic cubes which properly contains Q.
2. A *dyadic child* of a dyadic cube Q is the largest dyadic cubes which is properly contained in Q.

Remark 1.1

1. Various other terminology is used for dyadic cubes. Some authors prefer to name the dyadic parent a dyadic double, a dyadic mother or a dyadic father.
2. Given a dyadic cube Q, there are 2^n dyadic children.

We can define the dyadic cubes relative to a fixed cube Q. Fix a cube Q. Denote by $\mathscr{D}(Q)$ the set of all cubes obtained by bisecting Q finitely many times. Note that any two cubes in $\mathscr{D}$ do not have an interior point in common unless one is included in the other, that is, they are not overlapping.

1.1.2.2 Covering Lemma

The following theorem concerns the triple of balls but originally it was related with the five times expansion. This is why the next theorem deserves its name, the $5r$-covering lemma.

Theorem 1.7 (5r **- covering lemma (1))** *Let* $\{B_\lambda\}_{\lambda\in\Lambda}$ *be a finite collection of* $J \equiv \sharp\Lambda$ *balls in a metric space. If we relabel* $\{B_\lambda\}_{\lambda\in\Lambda}$ $B_1, B_2, \dots, B_J$, *then for the newly labeled family* $B_1, B_2, \dots, B_J$, *there exists* $1 \le K \le J$ *such that the following condition is fulfilled:*

1. $B_1, B_2, \dots, B_K$ *are disjoint.*
2. *There exists a mapping* $\iota : \{1, 2, \dots, J\} \to \{1, 2, \dots, K\}$ *such that* $B_j \subset 3\, B_{\iota(j)}$.

Proof We induct on J. When $J = 1$, $K = 1$ trivially does the job and Theorem 1.7 is true in this case.

Let $J_0 \in \mathbb{N}$. For a finite family $\{B^*_\lambda\}_{\lambda\in\Lambda}$ of $J \le J_0$ balls, we suppose that we can relabel $1, 2, \dots, J$ to have $B^*_1, B^*_2, \dots, B^*_J$ so that:

1. $B^*_1, B^*_2, \dots, B^*_K$ are disjoint.
2. There exists a mapping $\iota : \{1, 2, \dots, J\} \to \{1, 2, \dots, K\}$ such that $B^*_j \subset 3\, B^*_{\iota(j)}$.

Now suppose we have $J_0 + 1$ balls. We relabel $B_1, B_2, \dots, B_{J_0+1}$ so that B_1 has the largest radius. Again, if necessary, we change the label of $B_2, B_3, \dots, B_{J_0+1}$. If there are more than one ball having the maximum radius, then choose one of them among $B_2, B_3, \dots, B_{J_0+1}$. If B_1 intersects B_j for some $j = 2, 3, \dots, J_0 + 1$, Then $B_j \subset 3\, B_1$ since $r(B_j) \le r(B_1)$.

Among $\{B_j\}_{j=1}^{J_0+1}$, we can find at most J_0 balls such that $B_j \subset 3\, B_1$, since B_1 is automatically excluded. Then such a ball B_j does not intersect B_1. Thus, among $\{B_j\}_{j=1}^{J_0+1}$, let $\{B_j\}_{j=2}^{J+1}$ be a family of balls such that $B_j \cap B_1 = \emptyset$ for all $j \in [2, J+1] \cap \mathbb{N}$. Note that $B_j \subset 3B_1$ for $j > J$ in view of the definition of J.

If we apply the induction assumption to $\{B_j\}_{j=2}^{J+1}$, a relabeling yields $B_2, \dots, B_K$ and $J(j) = 2, 3, \dots, K$ such that $B_j \subset 3\, B_{J(j)}$ unless $2 \le j \le J$. Then define a function $\iota : \mathbb{N} \to \{1, 2, \dots, n\}$ by

$$\iota(j) \equiv \begin{cases} 1, & j \in \{J+1, J+2, \dots\}, \\ J(j), & 2 \le j \le J+1. \end{cases}$$

Thus, the balls $B_1, B_2, \dots, B_K$ satisfy the condition of the theorem and the proof is complete.

We mean by "the 5r-covering lemma" the following theorem (due to Wiener):

Theorem 1.8 (5r **- covering lemma (2))** *Suppose we have a family* $\{B_\lambda\}_{\lambda\in\Lambda}$ *of balls in a metric measure space of general cardinality. If the radii of the balls in the family is bounded from above by a positive constant, then we can find* $\Lambda_0 \subset \Lambda$ *such that:*

1. $\{B_\lambda\}_{\lambda\in\Lambda_0}$ *is disjoint.*
2. *There exists a mapping* $\iota : \Lambda_0 \to \Lambda$ *such that* $B_\lambda \subset 5\, B_{\iota(\lambda)}$.

We leave the proof of Theorem 1.8; see Exercise 1.16.

1.1.2.3 Whitney Covering Lemma

The following theorem is known as the Whitney covering lemma, which plays the role of the coordinate of a given proper open set of $\mathbb{R}^n$. By definition a packing is a family of nonoverlapping cubes.

Theorem 1.9 (Whitney decomposition/Whitney covering) *Let Ω be a proper open subset of $\mathbb{R}^n$. Then there exists a packing $\{Q_j\}_{j=1}^{\infty}$ which satisfies the following conditions:*

1. *The cubes do not overlap so much; more precisely,*

$$\chi_\Omega \sim_n \sum_{j=1}^{\infty} \chi_{10Q_j} \sim_n \sum_{j=1}^{\infty} \chi_{1000Q_j}. \tag{1.17}$$

In particular, the covering $\{Q_j\}_{j=1}^{\infty}$ is locally finite in the following sense:

$$\sum_{j=1}^{\infty} \chi_{1000Q_j} \leq N\chi_\Omega. \tag{1.18}$$

We can consider a variant of using balls instead of cubes.

2. *The distance* $\mathrm{dist}(\partial\Omega, Q_j)$ *between each cube Q_j and the boundary of Ω satisfies* $\ell(Q_j) \leq \mathrm{dist}(\partial\Omega, Q_j) \leq 10000\ell(Q_j)$*; more precisely,*

$$10Q_j \subset \Omega, \quad 10000Q_j \cap (\mathbb{R}^n \setminus \Omega) \neq \emptyset. \tag{1.19}$$

Proof We will make use not of the Euclidean distance but of the ℓ^∞-distance $\|x\|_\infty \equiv \max(|x_1|, |x_2|, \ldots, |x_n|)$ in the proof. Let $Q(x)$ be a cube centered at x having the sidelength $\dfrac{1}{8000}$ of the ℓ^∞-distance from x to $\partial\Omega$. Note that the sidelength of $Q(x)$ does not exceed twice that of $Q(y)$ if $Q(x) \cap Q(y) \neq \emptyset$. In fact, since

$$\ell(Q(x)) = \frac{\mathrm{dist}(x, \partial\Omega)}{8000}, \quad \ell(Q(y)) = \frac{\mathrm{dist}(y, \partial\Omega)}{8000}, \quad \|x-y\|_\infty \leq \ell(Q(x)) + \ell(Q(y)),$$

we have $8000\ell(Q(x)) \leq 8000\ell(Q(y)) + \ell(Q(x)) + \ell(Q(y))$ by the triangle inequality $\mathrm{dist}(x, \partial\Omega) \leq \mathrm{dist}(y, \partial\Omega) + \|x-y\|_\infty$. Thus, $2^{-1}\ell(Q(x)) \leq \ell(Q(y)) \leq 2\ell(Q(x))$.

Choose a disjoint maximal family from $\{Q(x)\}_{x\in\Omega}$. Note that such a family is made up of infinite cubes, which we label $\{Q_j\}_{j=1}^{\infty}$ to have a packing, So, (1.19) follows from the above observation.

It remains to prove (1.17). The proof of (1.18) is left as an exercise, see Exercise 1.17. Recall that $Q(x)$ is centered at x and the sidelength is $\dfrac{1}{8000}$ times the ℓ^∞-distance from x to $\partial\Omega$. We have $10000Q_j \cap (\mathbb{R}^n \setminus \Omega) \neq \emptyset$. Let $x \in \Omega$. By the maximality, we can find a cube Q_j such that $Q(x) \cap Q_j \neq \emptyset$. Then from the above observation, we have $Q(x) \subset 10Q_j$. Hence we have $\chi_\Omega = \chi_{\bigcup_{x\in\Omega} Q(x)} \leq \sum_{j=1}^{\infty} \chi_{10Q_j}$. Furthermore, we have

$$J(x) \equiv \{j \in \mathbb{N} : x \in 1000Q_j\}, \quad x \in \Omega. \tag{1.20}$$

The ratio of the distance from the center of Q_j to $\partial\Omega$ and the distance from x to $\partial\Omega$ does not exceed five times. Since $1000Q_j$, $j \in J(x)$ is contained in $5000Q(x)$, we have a bound N of $\sharp J(x)$ with a constant that depends on n. As a result (1.17) follows. Therefore, the proof is complete.

Remark 1.2 If one uses the Euclidean distance, one will have a collection of balls with similar properties.

We cannot grasp the Whitney decomposition on general domains easily. But in the model case $(0, \infty)$, we have a concrete construction; see Exercise 1.18. Intuitively, we refer to Exercise 1.18 for the Whitney decomposition. So we can say that the Whitney covering is a version of $\{I_j\}_{j=1}^{\infty}$ generalized to Ω. The space $C_c(\Omega)$ stands for the space of continuous functions having compact support. We write $C_c^{\infty}(\Omega) \equiv C^{\infty}(\Omega) \cap C_c(\Omega)$. For $\alpha = (\alpha_1, \alpha_2, \dots, \alpha_n) \in {\mathbb{N}_0}^n$, we define

$$|\alpha| \equiv \sum_{j=1}^{n} \alpha_j, \quad \partial^\alpha \equiv {\partial_{x_1}}^{\alpha_1} {\partial_{x_2}}^{\alpha_2} \cdots {\partial_{x_n}}^{\alpha_n}.$$

Recall also that the *support* of a continuous function f is defined to be the closure of the set $\{f \neq 0\}$.

Proposition 1.4 (A partition of unity subordinate to Whitney decomposition) *Maintain the same notation as Theorem* 1.9. *Choose* $\psi \in C^{\infty}(\mathbb{R}^n)$ *so that*

$$\chi_{Q(10)} \leq \psi \leq \chi_{Q(50)}. \tag{1.21}$$

Define a collection $\{\psi^{(j)}\}_{j=1}^{\infty} \cup \{\Psi\}$ *and* $\{\varphi^{(j)}\}_{j=1}^{\infty}$ *of* $C_c^{\infty}(\Omega)$*-functions by*

$$\psi^{(j)} \equiv \psi\left(\frac{\star - c(Q_j)}{\ell(Q_j)}\right), \quad \Psi \equiv \sum_{l=1}^{\infty} \psi^{(l)}, \quad \varphi^{(j)} \equiv \frac{\psi^{(j)}}{\Psi}.$$

Then the following holds:

1. $\sum_{l=1}^{\infty} \varphi^{(l)} \equiv \chi_\Omega$.
2. $\sup_{j\in\mathbb{N}} \sharp\{l \in \mathbb{N} : \mathrm{supp}(\varphi^{(j)}) \cap 1000Q_l\} < \infty$.
3. $|\partial^\alpha \varphi^{(l)}(x)| \lesssim_\alpha \ell(Q_l)^{-|\alpha|}$ *for all* $l \in \mathbb{N}$.

Note that $\{\psi^{(j)}\}_{j=1}^{\infty}$ satisfies the second condition and the third condition. We remark that $\varphi^{(j)}$ and $\varphi^{j} = 2^{jn}\varphi(2^{j}\star)$.

Proof The first two properties follow from the conditions in the Whitney decomposition (see Theorem 1.9); for the last property, we use

$$\Psi(x) \geq 1, \quad |\partial^{\alpha}\varphi^{(j)}(x)| \lesssim_{\alpha} \ell(Q)^{-|\alpha|} \quad x \in \mathbb{R}^{n}.$$

Therefore, the proof is complete.

In the Whitney decomposition $\{Q_j\}_{j=1}^{\infty}$ of the domain Ω, we fix $z_j \in 10000Q_j \setminus \Omega$ for each j. The point z_j is called the reference point of Q_j.

1.1.2.4 Carleson Tent

Although we work on $\mathbb{R}^n = \mathbb{R}^n_x$, matters can become simpler if we introduce another variable $t > 0$. For given a ball B, the Carleson tent $\hat{B}$ will correspond to a ball in $\mathbb{R}^{n+1} = \mathbb{R}^{n+1}_{x,t}$.

Here and below we denote by $\mathrm{Int}(E)$ the set of all interior points of E and $\mathrm{conv}(E)$ denotes the convex hull of $E \subset \mathbb{R}^{n+1}$.

Definition 1.3 (Carleson box, Carleson tent) Let O be an open set in $\mathbb{R}^n$. Define $\hat{O} \equiv \bigcup_{B \subset O} \mathrm{Int}(\mathrm{conv}((c(B), r(B)) \cup B \times \{0\}))$, where B runs over all balls contained in O. The set $\mathrm{Int}(\mathrm{conv}((c(B), r(B)) \cup B \times \{0\}))$ appearing in the definition of $\hat{O}$ is called the *Carleson tent*.

By definition, we have $\hat{B} = \mathrm{Int}(\mathrm{conv}((c(B), r(B)) \cup B \times \{0\}))$ for any ball B, where $\mathrm{conv}(E)$ denotes the convex hull of the set E.

For a cube Q, it is sometimes convenient to consider $Q \times (0, \ell(Q)]$ which plays the role of $\hat{B}$ above. In fact, we can prove the following proposition:

Lemma 1.4 *Let Q be a dyadic cube, and let $\mathcal{G} \subset \mathcal{D}(Q)$ be a disjoint collection of dyadic cubes. If we set*

$$\mathcal{G}^* \equiv \{R \in \mathcal{D}(Q) : \text{there does not exist } S \in \mathcal{G} \text{ such that } R \subset S\},$$

then we have a partition: $Q \times (0, \ell(Q)] = \bigcup_{R \in \mathcal{G}} R \times (0, \ell(R)] \cup \bigcup_{R \in \mathcal{G}^*} R \times (\ell(R)/2, \ell(R)]$.

Proof It is trivial that $Q \times (0, \ell(Q)] \supset \bigcup_{R \in \mathcal{G}} R \times (0, \ell(R)] \cup \bigcup_{R \in \mathcal{G}^*} R \times (\ell(R)/2, \ell(R)]$. Let us prove the reverse inclusion. Let $(x, t) \in Q \times (0, \ell(Q)]$. Three different cases must be considered.

1. If $(x, t) \in \bigcup_{R \in \mathcal{G}} R \times (0, \ell(R)]$, then there is nothing to prove.

2. If $x \in \bigcup_{R \in \mathscr{G}} R$ but $(x,t) \in \bigcup_{R \in \mathscr{G}} R \times (0, \ell(R)]$, we have a unique cube R such that $x \in R \in \mathscr{G}$ and that $\ell(R) < t \le \ell(Q)$. Thus there exists $S \in \mathscr{D}(Q)$ such that $x \in S$ and that $\ell(S)/2 < t \le \ell(S)$. Since $S \in \mathscr{G}^*$, we have the desired result.
3. If $x \notin \bigcup_{R \in \mathscr{G}} R$, we go through the same argument.

Exercises

Exercise 1.14 Draw $\mathscr{D}_0(\mathbb{R})$ in the line $\mathbb{R}$.

Exercise 1.15 Show that each $\mathscr{D}_l$ is partioned into 3^n subfamilies $\{\mathscr{D}_l^k\}_{k=1}^{3^n}$ so that the following properties hold, if we set $\tilde{\mathscr{D}}^k \equiv \bigcup_{l=-\infty}^{\infty} \mathscr{D}_l^k$:

- For all $Q_1, Q_2 \in \tilde{\mathscr{D}}^k$, $3Q_1$ and $3Q_2$ are either disjoint or one is contained in the other.
- For all $Q \in \mathscr{D} = \bigcup_{k' \in \mathbb{Z}} \mathscr{D}^{k'}$ and $k = 1, 2, \ldots, 3^n$, there exists $Q' \in \mathscr{D}^k$ such that $Q \subset 3Q' \subset 5Q$.

Hint: Let $n = 1$ and consider

$$\mathscr{D}_0^1 = \{\ldots, [-3,-2), [0,1), [3,4), [6,7), \ldots\}$$

$$\mathscr{D}_{-1}^1 = \{\ldots, [-2,0), [4,6), [10,12), [16,18), \ldots\}$$

$$\mathscr{D}_{-2}^1 = \{\ldots, [0,4), [12,16), [24,28), [36,40), \ldots\}.$$

Exercise 1.16 Prove Theorem 1.8 by mimicking the proof of Theorem 1.7.

Exercise 1.17 By the volume argument, show that N in (1.18) is less than or equal to $10{,}000^n$.

Exercise 1.18 (Whitney decomposition of $(0,\infty)$**)** Let $I_j \equiv (2^{j-1}, 2^{j+1}) \subset \mathbb{R}$ for each $j \in \mathbb{Z}$. For an open covering $\{I_j\}_{j=1}^{\infty}$ in $(0,\infty)$ show that $\chi_{(0,\infty)} \le \sum_{j=1}^{\infty} \chi_{I_j} \le 2\chi_{(0,\infty)}$ and that $\operatorname{dist}(0, I_j) = 2^{j-1}$ for each j. Hint: Draw a picture.

Exercise 1.19 (Dyadic Whitney decomposition) Denote by $\operatorname{diam}(E)$ the diameter of a set E. Use Lemma 1.3 to prove the following:

1. Let $\Omega \subsetneq \mathbb{R}^n$ be a nonempty open set. For each $x \in \Omega$, we choose a dyadic cube Q_x containing x uniquely so that $\frac{1}{4} d(x, \Omega^c) \le \operatorname{diam}(Q_x) < \frac{1}{2} d(x, \Omega^c)$.
   (a) Show that $d(x, \Omega^c) \ge d(Q_x, \Omega^c)$.
   (b) Show that $d(x, \Omega^c) \le d(Q_x, \Omega^c) + \operatorname{diam}(Q_x)$.

(c) Show that $\mathrm{diam}(Q_x) \le d(Q_x, \Omega^{\mathrm{c}}) \le 4\mathrm{diam}(Q_x)$.

2. If $Q_x = Q_y$ for some distinct $x, y \in \Omega$, then discard one of them. Then we will obtain a packing $\mathscr{D}_\Omega = \{Q_j\}_{j \in J}$.

   (a) If $\Omega \subset \Omega'$, then for any $Q \in \mathscr{D}_\Omega$, there exists $R \in \mathscr{D}_{\Omega'}$ such that $Q \subset R$.
   (b) Show that such an R is unique.

Exercise 1.20 In the proof of Theorem 1.9, show that $2^{-1}\ell(Q(x)) \le \ell(Q(y)) \le 2\ell(Q(x))$ if $1000Q(x)$ and $1000Q(y)$ intersects. Hint: Use the maximality.

Exercise 1.21 Let K be a compact set and U be an open set containing K.

1. Find a finite collection $\{B_j\}_{j=1}^J$ of balls such that $K \subset \bigcup_{j=1}^{J} B_j \subset U$.
2. For $x \in K$, define $\varphi(x) \equiv \sup_{j=1,\dots,J} \mathrm{dist}(x, \mathbb{R}^n \setminus B_j)$. Show that φ is continuous and $\varphi(x) > 0$ for all $x \in K$.
3. Show that $K^\varepsilon = \{x \in \mathbb{R}^n : \mathrm{dist}(K, x) < \varepsilon\} \subset U$ for $\varepsilon \equiv \inf_K \varphi$.

Exercise 1.22 (Regularized distance) [8, 834] Here we use the following notation: For $x = (x_1, x_2, \dots, x_n) \in \mathbb{R}^n$, we set $x' = (x_1, x_2, \dots, x_{n-1}) \in \mathbb{R}^{n-1}$. Let $\Omega \subsetneq \mathbb{R}^n$ be an open set. Then there exists a function $r \in C^\infty(\Omega)$ such that $r(x) \sim \mathrm{dist}(x, \partial\Omega)$ and that $|\partial^\alpha r(x)| \lesssim \mathrm{dist}(x, \partial\Omega)^{1-|\alpha|}$ for all multi-indexes α. If Ω is given by a Lipschitz graph: $\Omega = \{x \in \mathbb{R}^n : x_n > \gamma(x')\}$, then show that one can arrange that $\partial_{x_n} r(x) \gtrsim 1$ for all $x \in \Omega$.

Exercise 1.23 Let O be an open set in $\mathbb{R}^n$. Denote by Λ_0 the set of all balls $B \subset O$ such that $r(B) \in \mathbb{Q}$ and that $c(B) \in \mathbb{Q}^n$.

1. For any ball B contained in O, show that we can find $B' \in \Lambda_0$ such that $B \subset B'$.
2. Show that $\bigcup_{B \in \Lambda_0} B = O$ and $\bigcup_{B \in \Lambda_0} \hat{B} = \hat{O}$.

1.1.3 Hausdorff Capacity

This section is used in Sect. 6.4.3, where we will consider Besov–Hausdorff-type spaces and Triebel–Lizorkin–Hausdorff spaces. It turns out that these spaces are the duals of Besov-type spaces and Triebel–Lizorkin-type spaces, which we also define in Chap. 6.

1.1.3.1 Hausdorff Capacity

Since we want to collect some more preliminary facts on measure theory, we recall the theory of capacity briefly.

Definition 1.4 (Hausdorff capacity) Let $0 < d \le n$. The *d-dimensional Hausdorff capacity of* the set $E \subset \mathbb{R}^n$ is defined by setting

$$H^d(E) \equiv \inf \left\{ \sum_{j=1}^{\infty} |B(x_j, r_j)|^{\frac{d}{n}} : E \subset \bigcup_{j=1}^{\infty} B(x_j, r_j) \right\},$$

where the infimum is taken over all covers $\{B(x_j, r_j)\}_{j=1}^{\infty}$ of E by countable families of open balls.

Here E need not be measurable.

We can show that the set function H^d is monotone, countably subadditive and vanishes on empty sets. Moreover, the notion of H^d can be extended to $d = 0$; for any subset $E \subset \mathbb{R}^n$, define

$$H^0(E) \equiv \inf \left\{ N \in \mathbb{N} : E \subset \bigcup_{j=1}^{N} B(x_j, r_j) \right\},$$

where the infimum is taken over all covers $\{B(x_j, r_j)\}_{j=1}^{N}$ of E by at most countable families of open balls.

As is the case with the Lebesgue integral, it is difficult to calculate $H^d(E)$ for general sets. Here is a special example of E for which we can find $H^d(E)$ exactly.

Example 1.1 Let us show that $H^d(B(1)) = |B(1)|^{\frac{d}{n}}$ for $0 < d \le n$. It is easy to see that $H^d(B(1)) \le |B(1)|^{\frac{d}{n}}$. Thus, we need to show $H^d(B(1)) \ge |B(1)|^{\frac{d}{n}}$. Let $\{B_j\}_{j=1}^{\infty}$ be a ball covering of $B(1)$. Then using (1.13) we have

$$\sum_{j=1}^{\infty} |B_j|^{\frac{d}{n}} \ge \left(\sum_{j=1}^{\infty} |B_j| \right)^{\frac{d}{n}} \ge |B(1)|^{\frac{d}{n}}.$$

The capacity and the tent go well as the following proposition shows:

Proposition 1.5 (Canonical ball cover with respect to Hausdorff capacity) *Let $d \in (0, n]$ and O be an open set. Then there exists a ball cover $\{B_\lambda\}_{\lambda \in \Lambda}$ of O such that*

$$\sum_{\lambda \in \Lambda} |B_\lambda|^{\frac{d}{n}} \lesssim H^d(O) \tag{1.22}$$

and that

$$\hat{O} \subset \bigcup_{\lambda \in \Lambda} \widehat{10^{10} B_\lambda}. \tag{1.23}$$

Furthermore, when $O \ne \mathbb{R}^n$, we can arrange that for any ball B contained in O there exist $\lambda_B \in \Lambda$ such that $B \subset 10^{10} B_{\lambda_B}$.

Proof Let O be an open set such that $H^d(O) < \infty$; otherwise the proof is easy including the case of $H^d(O) = \infty$ and $O \neq \mathbb{R}^n$. The proof is made up of two steps. First we find an intermediate collection $\mathscr{V} \equiv \{10^{-4}B_x\}_{x \in E}$ of balls. After that we prove (1.22) and (1.23).

1. For each $x \in O$, denote by $d(x)$ the distance between O^c and x. Let $B_x \equiv B(x, d(x))$. By definition of B_x, $x \in O$ an $\hat{O}$,

$$\hat{O} = \bigcup_{x \in O} \widehat{B_x}.$$

Form the Whitney covering $\mathscr{V} \equiv \{10^{-4}B_x\}_{x \in E}$ of O. Then if $10^{-4}B_x$ with $x \in E$ and $10^{-4}B_y$ with $y \in O$ intersect, we have $10B_x \supset 10^{-4}B_y$; hence $10^5 B_x \supset B_y$. Thus,

$$\hat{O} \subset \bigcup_{x \in E} \widehat{10^5 B_x}.$$

2. Let $\{B_j\}_{j=1}^{\infty}$ be a sequence of balls such that

$$O \subset \bigcup_{j=1}^{\infty} B_j, \quad \sum_{j=1}^{\infty} |B_j|^{\frac{d}{n}} \leq 2H^d(O).$$

We let $J_1 \equiv \bigcup_{x \in E} \{j \in \mathbb{N} : 10^{6n}|B_j| \geq |B_x| \text{ if } B_x \cap B_j \neq \emptyset\}$ and $J_2 \equiv \mathbb{N} \setminus J$.

Then

$$\begin{aligned}
\sum_{j=1}^{\infty} |B_j|^{\frac{d}{n}} &= \sum_{j \in J_1} |B_j|^{\frac{d}{n}} + \sum_{j \in J_2} |B_j|^{\frac{d}{n}} = \sum_{j \in J_1} |B_j|^{\frac{d}{n}} + \sum_{x \in E} \left(\sum_{j \in J_2,\, B_j \cap B_x \neq \emptyset} |B_j|^{\frac{d}{n}} \right) \\
&\geq \sum_{j \in J_1} |B_j|^{\frac{d}{n}} + \sum_{x \in E} \left(\sum_{j \in J_2,\, B_j \cap B_x \neq \emptyset} |B_j| \right)^{\frac{d}{n}} \\
&\geq \sum_{j \in J_1} |B_j|^{\frac{d}{n}} + \sum_{x \in E} |B_x|^{\frac{d}{n}}
\end{aligned}$$

and

$$\bigcup_{j=1}^{\infty} B_j \cap O = \bigcup_{j \in J_1} (B_j \cap O) \cup \bigcup_{j \in J_2} (B_j \cap O) \subset \bigcup_{j \in J_1} B_j \cup \bigcup_{x \in E} 100B_x.$$

Thus, the family $\{B_\lambda\}_{\lambda \in \Lambda} = \{B_j\}_{j \in J_1} \cup \{100B_x\}_{x \in E}$ enjoys the following properties:

$$O \subset \bigcup_{j=1}^{\infty} B_j \cap O_j \subset \bigcup_{j \in J_1} B_j \cup \bigcup_{x \in E} 100B_x,$$

$$\sum_{j \in J_1} |B_j|^{\frac{d}{n}} + \sum_{x \in E} |100B_x|^{\frac{d}{n}} \lesssim_{n,d} H^d(O),$$

Finally, if $B \subset O$ is a ball, then $B \subset B(c(B), d(c(B))) = B_{c(B)}$. Thus, using the definition of J_1 and J_2, we learn that $B(c(B), d(c(B))) \subset 10^{10} B_{\lambda_B}$ for some $\lambda_B \in \Lambda$ and

$$\hat{O} = \bigcup_{x \in O} \widehat{B_x} \subset \bigcup_{\lambda \in \Lambda} \widehat{10^{10} B_\lambda}.$$

It remains to relabel and redefine the balls suitably.

1.1.3.2 Dyadic Hausdorff Capacity

We want to define a functional by the Hausdorff capacity. This functional is known to be nonsublinear as Example 1.2 shows, so sometimes we need to use an equivalent integral with respect to the d-dimensional dyadic Hausdorff capacity $\widetilde{H}^d$, which is sublinear according to [1151].

Definition 1.5 (**d-dimensional modified dyadic Hausdorff capacity**) Let $E \subset \mathbb{R}^n$ and $0 < d \le n$. One defines *d-dimensional modified dyadic Hausdorff capacity* by

$$\tilde{H}_0^d(E) \equiv \inf \left\{ \sum_{j=1}^{\infty} \ell(Q_j)^d \right\},$$

where $\{Q_j\}_{j=1}^{\infty}$ moves over all countable collections of dyadic cubes satisfying

$$E \subset \mathrm{Int}\left(\bigcup_{j=1}^{\infty} Q_j \right). \tag{1.24}$$

Example 1.2 Let $n = 2$ and $a \in \mathbb{R}$.

We prove $\tilde{H}_0^1([0,1) \times \{a\}) = 1$. In fact, $\tilde{H}_0^1([0,1) \times \{a\}) \le 1$ since $[0,1) \times \{a\} \subset Q_{0m}$ for some $m \in \mathbb{Z}$. Let us prove the reverse conclusion. Let $\{Q_\lambda\}_{\lambda \in \Lambda}$ be a covering of $[0, 1) \times \{a\}$. By Lemma 1.3 we may assume that $\{Q_\lambda\}_{\lambda \in \Lambda}$ is disjoint. Then denoting by $\ell([x_0, x_1) \times \{a\}) = x_1 - x_0$ for $-\infty < x_0 < x_1 < \infty$, we obtain

$$\ell([0, 1) \times \{a\}) = \sum_{\lambda \in \Lambda} \ell(([0, 1) \times \{a\}) \cap Q_\lambda) \le \sum_{\lambda \in \Lambda} \ell(Q_\lambda),$$

showing $\tilde{H}^1_0([0, 1) \times \{a\}) \ge 1$. Likewise we can show $\tilde{H}^1_0([0, 1) \times \{3, \pi\}) = 1(< 2)$.

Based on a fundamental geometric observation, we see that H^d and $\tilde{H}^d_0$ are not so different.

Proposition 1.6 *Let $0 < d \le n$. Then $H^d \lesssim_n \tilde{H}^d_0 \sim H^d$.*

Proof Let $\{Q_j\}_{j=1}^\infty$ be a countable collection of dyadic cubes satisfying (1.24). We write $B_j \equiv B(c(Q_j), 2n\ell(Q_j))$. Then $E \subset \text{Int}\left(\bigcup_{j=1}^\infty Q_j\right) \subset \bigcup_{j=1}^\infty Q_j \subset \bigcup_{j=1}^\infty B_j$. As a result, we see that $H^d \lesssim \tilde{H}^d_0$. To end the proof, we may thus assume $H^d(E) < \infty$.

When we have a ball $B = B(x, r)$, we can find 6^n dyadic cubes $Q_1, Q_2, \dots, Q_{6^n}$ such that

$$\frac{1}{2} \le \frac{\ell(Q_j)}{r} \le 2, \quad B(x, r) \subset \bigcup_{j=1}^{6^n} Q_j, \tag{1.25}$$

we see that $B(x, r) = \text{Int}(B(x, r)) \subset \text{Int}\left(\bigcup_{j=1}^{6^n} Q_j\right)$. With this in mind, we suppose that $M \in (H^d(E), \infty)$. Then we have a covering $\{B(x_j, r_j)\}_{j=1}^\infty$ such that $M > \sum_{j=1}^\infty |B(x_j, r_j)|^{\frac{d}{n}}$ and that $E \subset \bigcup_{j=1}^\infty B(x_j, r_j)$. Then we can find dyadic cubes $\{Q_{j,k}\}_{j \in \mathbb{N}, k \in \mathbb{N} \cap [1, 6^n]}$ such that

$$B(x_j, r_j) \subset \text{Int}\left(\bigcup_{k=1}^{6^n} Q_{j,k}\right), \quad \frac{1}{2} \le \frac{\ell(Q_{j,k})}{r_j} \le 2, \qquad (j \in \mathbb{N}).$$

As a result, we have $M \gtrsim \tilde{H}^d_0(E)$, as was to be shown.

From the definition of the covering, it is easy to see that

$$\tilde{H}^d_0(E_1 \cup E_2) \le \tilde{H}^d_0(E_1) + \tilde{H}^d_0(E_2). \tag{1.26}$$

According to Example 1.2, we do not have $\tilde{H}^d_0(E_1 \cup E_2) + \tilde{H}^d_0(E_1 \cap E_2) = \tilde{H}^d_0(E_1) + \tilde{H}^d_0(E_2)$. But we have the following weaker version.

Theorem 1.10 (Strong subadditivity) *We have $\tilde{H}^d_0(E_1 \cup E_2) + \tilde{H}^d_0(E_1 \cap E_2) \le \tilde{H}^d_0(E_1) + \tilde{H}^d_0(E_2)$ for any sets $E_1, E_2 \subset \mathbb{R}^n$ and $0 < d \le n$.*

Proof From (1.26), we may assume that $\tilde{H}^d_0(E_1) + \tilde{H}^d_0(E_2) < \infty$ and that $E_1 \cap E_2 \neq \emptyset$ without any loss of generality. From the definition of the $\tilde{H}^d_0(E_1)$ and $\tilde{H}^d_0(E_2)$, we have only to establish that

$$\tilde{H}^d_0(E_1 \cup E_2) + \tilde{H}^d_0(E_1 \cap E_2) \le \sum_{j=1}^{\infty} \ell(Q_{1,j})^d + \sum_{k=1}^{\infty} \ell(Q_{2,k})^d, \tag{1.27}$$

whenever we have two collections $\{Q_{1,j}\}_{j=1}^{\infty}$ and $\{Q_{2,k}\}_{k=1}^{\infty}$ of dyadic cubes satisfying

$$E_1 \subset \mathrm{Int}\left(\bigcup_{j=1}^{\infty} Q_{1,j}\right), \quad E_2 \subset \mathrm{Int}\left(\bigcup_{k=1}^{\infty} Q_{2,k}\right).$$

Observe first that

$$E_1 \cap E_2 \subset \mathrm{Int}\left(\bigcup_{j=1}^{\infty} Q_{1,j}\right) \cap \mathrm{Int}\left(\bigcup_{k=1}^{\infty} Q_{2,k}\right) = \mathrm{Int}\left(\bigcup_{j,k=1}^{\infty} Q_{1,j} \cap Q_{2,k}\right) \tag{1.28}$$

and that

$$E_1 \cup E_2 \subset \mathrm{Int}\left(\bigcup_{j=1}^{\infty} Q_{1,j}\right) \cup \mathrm{Int}\left(\bigcup_{k=1}^{\infty} Q_{2,k}\right) = \mathrm{Int}\left(\bigcup_{j,k=1}^{\infty} Q_{1,j} \cup Q_{2,k}\right). \tag{1.29}$$

Let

$$\mathfrak{A}_1 \equiv \{(j,k) \,:\, Q_{1,j} \subsetneq Q_{2,k}\}, \quad \mathfrak{A}_2 \equiv \{(j,k) \,:\, Q_{1,j} \supsetneq Q_{2,k}\},$$

$$\mathfrak{A}_3 \equiv \{(j,k) \,:\, Q_{1,j} = Q_{2,k}\}, \quad \mathfrak{A}_4 \equiv \mathbb{N} \times \mathbb{N} \setminus \bigcup_{l=1}^{3} \mathfrak{A}_l.$$

Then

$$\bigcup_{j,k=1}^{\infty} Q_{1,j} \cup Q_{2,k} = \bigcup_{(j,k) \in \mathfrak{A}_2 \cup \mathfrak{A}_3 \cup \mathfrak{A}_4} Q_{1,j} \cup \bigcup_{(j,k) \in \mathfrak{A}_1 \cup \mathfrak{A}_4} Q_{2,k} \tag{1.30}$$

and

$$\bigcup_{j,k=1}^{\infty} Q_{1,j} \cap Q_{2,k} = \bigcup_{(j,k)\in\mathfrak{A}_1} Q_{1,j} \cup \bigcup_{(j,k)\in\mathfrak{A}_2\cup\mathfrak{A}_3} Q_{2,k}. \tag{1.31}$$

Consequently, we deduce from (1.29) and (1.30)

$$E_1 \cup E_2 \subset \mathrm{Int}\left(\bigcup_{(j,k)\in\mathfrak{A}_2\cup\mathfrak{A}_3\cup\mathfrak{A}_4} Q_{1,j} \cup \bigcup_{(j,k)\in\mathfrak{A}_1\cup\mathfrak{A}_4} Q_{2,k} \right)$$

and from (1.28) and (1.31)

$$E_1 \cap E_2 \subset \mathrm{Int}\left(\bigcup_{(j,k)\in\mathfrak{A}_1} Q_{1,j} \cup \bigcup_{(j,k)\in\mathfrak{A}_2\cup\mathfrak{A}_3} Q_{2,k} \right).$$

Thus,

$$\tilde{H}_0^d(E_1 \cup E_2) \le \sum_{(j,k)\in\mathfrak{A}_2\cup\mathfrak{A}_3\cup\mathfrak{A}_4} |Q_{1,j}|^{\frac{d}{n}} + \sum_{(j,k)\in\mathfrak{A}_1\cup\mathfrak{A}_4} |Q_{2,k}|^{\frac{d}{n}}$$

and

$$\tilde{H}_0^d(E_1 \cup E_2) \le \sum_{(j,k)\in\mathfrak{A}_1} |Q_{1,j}|^{\frac{d}{n}} + \sum_{(j,k)\in\mathfrak{A}_2\cup\mathfrak{A}_3} |Q_{2,k}|^{\frac{d}{n}}.$$

Therefore, (1.27) follows from these two inequalities.

We have the following property for decreasing sequences of compact sets.

Theorem 1.11 *Let $0 < d \le n$. If $\{K_j\}_{j=1}^{\infty}$ is a sequence of compact sets decreasing to K, then* $\lim_{j\to\infty} \tilde{H}_0^d(K_j) = \tilde{H}_0^d\left(\bigcap_{j=1}^{\infty} K_j\right)$.

Proof We have only to show that

$$\lim_{j\to\infty} \tilde{H}_0^d(K_j) \le \tilde{H}_0^d(K), \quad K \equiv \bigcap_{j=1}^{\infty} K_j. \tag{1.32}$$

To this end, we choose a sequence of dyadic cubes $\{Q_j\}_{j=1}^{\infty}$ arbitrarily so that $K \subset \mathrm{Int}\left(\bigcup_{j=1}^{\infty} Q_j\right)$. Since K is compact, we can choose $\varepsilon > 0$ so that

$$K^{\varepsilon} \equiv \{x \in \mathbb{R}^n \, : \, \mathrm{dist}(K, x) < \varepsilon\} \subset \mathrm{Int}\left(\bigcup_{j=1}^{\infty} Q_j\right). \tag{1.33}$$

Thus, there exists $j_0 \in \mathbb{N}$ such that $K_l \subset \mathrm{Int}\left(\bigcup_{j=1}^{\infty} Q_j\right)$ for $l \geq j_0$. As a result, $\lim\limits_{j\to\infty} \tilde{H}_0^d(K_j) \leq \sum\limits_{j=1}^{\infty} \ell(Q_j)^d$. Since the sequence $\{Q_j\}_{j=1}^{\infty}$ is arbitrary, we obtain (1.32).

Unlike the Lebesgue measure, essentially the case where $d = n$, the compactness of the sets is absolutely necessary as an example similar to Example 1.5 shows.

Lemma 1.5 (Increasing property for open sets) *Let* $0 < d \leq n$, *and let* $\{O_j\}_{j=1}^{\infty}$ *be a sequence of open sets in* $\mathbb{R}^n$ *expanding to* O. *Then* $\lim\limits_{j\to\infty} \tilde{H}_0^d(O_j) = \tilde{H}_0^d\left(\bigcup_{j=1}^{\infty} O_j\right)$.

Proof We have only to show that

$$\lim_{j\to\infty} \tilde{H}_0^d(O_j) \geq \tilde{H}_0^d(O). \tag{1.34}$$

To this end, we may assume that $\sup\limits_{j\in\mathbb{N}} \tilde{H}_0^d(O_j) < \infty$; otherwise (1.34) is trivial. Let $\varepsilon > 0$ be fixed. We choose $\{Q_{j,k}\}_{k=1}^{\infty} \subset \mathscr{D}$ so that

$$O_j \subset \mathrm{Int}\left(\bigcup_{k=1}^{\infty} Q_{j,k}\right) \tag{1.35}$$

and

$$\sum_{j=1}^{\infty} \ell(Q_{j,k})^d \leq \tilde{H}_0^d(O_j) + 2^{-j}\varepsilon. \tag{1.36}$$

Note that $\sup\limits_{j,k\in\mathbb{N}} \ell(Q_{j,k})^d \leq \varepsilon + \sup\limits_{j\in\mathbb{N}} \tilde{H}_0^d(O_j)$. This means that the size of $Q_{j,k}$ is bounded by a constant independent of j and k. Denote by $\{Q_i\}_{i\in I}$ the disjoint maximal cubes in $\{Q_{j,k}\}_{j,k\in\mathbb{N}}$.

Then $O = \bigcup\limits_{j=1}^{\infty} O_j \subset \mathrm{Int}\left(\bigcup\limits_{j=1}^{\infty}\bigcup\limits_{k=1}^{\infty} Q_{j,k}\right) = \mathrm{Int}\left(\bigcup\limits_{i\in I} Q_i\right)$ from (1.35). Let $m \in \mathbb{N}$ be fixed. We write

$$\{Q_i\}_{i\in I} \cap \{Q_{1,j}\}_{j=1}^{\infty} = \{Q_i^{(1)}\}_{i\in I_1}. \tag{1.37}$$

Define

$$O_{1,m} \equiv O_m \cap \mathrm{Int}\left(\bigcup_{i\in I_1} Q_i^{(1)}\right) \quad (m \in \mathbb{N}). \tag{1.38}$$

We set $\{Q_{m,k}^{(1)}\}_{k\in K_{1,m}} \equiv \{Q_{m,k} : k \in \mathbb{N}, O_{1,m} \cap Q_{m,k} \neq \emptyset\}$. Then

$$O_{1,m} \subset \bigcup_{k\in K_{1,m}} Q_{m,k}^{(1)}. \tag{1.39}$$

For each $k \in K_{1,m}$, there exists $i \in I_1$ such that $Q_i^{(1)} \cap Q_{m,k}^{(1)} \neq \emptyset$. Since $Q_i^{(1)}$ is maximal in inclusion among $\{Q_{j,k}\}_{j,k\in\mathbb{N}}$ and $Q_{m,k}^{(1)} = Q_{J,K}$ for some $J, K \in \mathbb{N}$, we have $Q_{m,k}^{(1)} \subset Q_i^{(1)}$. As a result, we have

$$\bigcup_{k\in K_{1,m}} Q_{m,k}^{(1)} \subset \bigcup_{i\in I_1} Q_i^{(1)}. \tag{1.40}$$

We let

$$\{Q_{1,k}\}_{k=1}^{\infty} \setminus \{Q_i^{(1)}\}_{i\in I_1} = \{Q_{1,k,m}^{*}\}_{k\in J_1}. \tag{1.41}$$

From (1.35) with $j = 1$, and (1.41), we learn

$$O_1 = \left(O_1 \cap O_m \cap \mathrm{Int}\left(\bigcup_{k\in I_1} Q_i^{(1)}\right)\right) \cup \left(O_1 \cap \bigcup_{k\in J_1} Q_{1,k,m}^{*}\right), \tag{1.42}$$

since we are assuming that O_1 is open. From (1.39) and (1.42), we have

$$O_1 \subset \bigcup_{k\in K_{1,m}} Q_{m,k}^{(1)} \cup \bigcup_{k\in J_1} Q_{1,k,m}^{*}. \tag{1.43}$$

As a result, we have

$$\tilde{H}_0^d(O_1) \leq \sum_{k\in K_{1,m}} \ell(Q_{m,k}^{(1)})^d + \sum_{k\in J_1} \ell(Q_{1,k,m}^{*})^d. \tag{1.44}$$

Note that (1.36) with $j = 1$ reads as:

$$\sum_{i\in I_1}\ell(Q_i^{(1)})^d+\sum_{i\in J_1}\ell(Q_{1,k,m}^*)^d\le \tilde{H}_0^d(O_1)+\frac{\varepsilon}{2}, \tag{1.45}$$

by virtue of (1.37) and (1.41). By combining (1.44) and (1.45), we obtain

$$\sum_{i\in I_1}\ell(Q_i^{(1)})^d\le \tilde{H}_1^d(O_1)+\frac{\varepsilon}{2}-\sum_{k\in J_1}\ell(Q_{1,k,m}^*)^d\le \sum_{k\in K_{1,m}}\ell(Q_{m,k}^{(1)})^d+\frac{\varepsilon}{2}. \tag{1.46}$$

For $j=2$, we mimic the argument above to have a sequence $\{Q_{m,k}^{(2)}\}_{k\in K_{2,m}}$ having properties similar to (1.43) and (1.46) above in addition to the property

$$\{Q_{m,k}^{(1)}\}_{k\in K_{1,m}}\cap\{Q_{m,k}^{(2)}\}_{k\in K_{2,m}}=\emptyset. \tag{1.47}$$

More precisely, we set

$$(\{Q_i\}_{i\in I}\setminus\{Q_{m,k}^{(1)}\}_{k\in K_{1,m}})\cap\{Q_{2,k}\}_{k=1}^\infty=\{Q_{i,m}^{(2)}\}_{i\in I_{2,m}}. \tag{1.48}$$

Define

$$O_{2,m}\equiv O_m\cap \operatorname{Int}\left(\bigcup_{i\in I_{2,m}}Q_{i,m}^{(2)}\right).$$

We write $\{Q_{m,k}^{(2)}\}_{k\in K_{2,m}}\equiv\{Q_{m,k}\,:\,O_{2,m}\cap Q_{m,k}\neq\emptyset\}$. As before, for any $k\in K_{2,m}$, $Q_{m,k}^{(2)}$ intersects $Q_{i,m}^{(2)}$ for some $i\in I_2$. In view of (1.48), we have (1.47). Rearrange the cubes to have

$$\{Q_{2,k}\}_{k=1}^\infty\setminus\{Q_{i,m}^{(2)}\}_{i\in I_{2,m}}=\{Q_{2,k}^*\}_{k\in J_1}.$$

Going through an argument above, we have $O_2\subset\bigcup_{k\in K_{2,m}}Q_{m,k}^{(2)}\cup\bigcup_{k\in J_2}Q_{2,k}^*$ corresponding to (1.43), and $\sum_{i\in I_{2,m}}\ell(Q_i^{(2)})^d\le\sum_{k\in K_{2,m}}\ell(Q_{m,k}^{(2)})^d+\frac{1}{4}\varepsilon$, corresponding to (1.46).

Continuing this procedure, we can find a collection $\{Q_{m,k}^{(j)}\}_{k\in I_{j,m}}$ for $j=1,2,\ldots,m$ and a sequence of subsets $\{J_j\}_{j=1}^m$ of $\mathbb{N}$ such that $\{Q_{m,k}^{(j_1)}\}_{k\in I_{j_1,m}}\cap\{Q_{m,k}^{(j_2)}\}_{k\in I_{j_2,m}}=\emptyset$ for any $j_1<j_2\le m$, that $O_2\subset\bigcup_{k\in K_{j,m}}Q_{m,k}^{(j)}\cup\bigcup_{k\in J_j}Q_{j,k}$, and that

$$\sum_{i\in I_{j,m}}\ell(Q_i^{(j)})^d\le\sum_{k\in K_{j,m}}\ell(Q_{m,k}^{(j)})^d+2^{-j}\varepsilon.$$

Notice that $\sum_{j=1}^{m}\sum_{i\in I_{j,m}}\ell(Q_i^{(j)})^d \le 2\varepsilon+\tilde{H}_0^d(O_m)$. Observe that, for any finite set I_0, there exists $M=M(I_0)$ such that $\{Q_i\}_{i\in I_0}\subset\{Q_{m,k} : m=1,2,\ldots,M, k=1,2,\ldots\}$. Thus, if we let $m\to\infty$, we have $\sum_{i\in I}\ell(Q_i)^d \le \lim_{m\to\infty}\tilde{H}_0^d(O_m)+2\varepsilon$, as was to be shown.

Theorem 1.12 (Monotone property of $\tilde{H}_0^d$) *Let $0<d\le n$. Whenever we have an increasing sequence $\{E_j\}_{j=1}^\infty$ of sets in $\mathbb{R}^n$, we have*

$$\lim_{j\to\infty}\tilde{H}_0^d(E_j)=\tilde{H}_0^d\left(\bigcup_{j=1}^{\infty}E_j\right).$$

Proof Again, we need to prove

$$\lim_{j\to\infty}\tilde{H}_0^d(E_j)\ge\tilde{H}_0^d\left(\bigcup_{j=1}^{\infty}E_j\right). \tag{1.49}$$

To this end, we may assume that $\sup_{j\in\mathbb{N}}\tilde{H}_0^d(E_j)<\infty$. Let $\varepsilon>0$ be fixed. Then we can choose an open set O_j such that $E_j\subset O_j$ and that $\tilde{H}^d(O_j)\le\tilde{H}_0^d(E_j)+2^{-j}\varepsilon$. Since $E_1\subset O_1\cap O_2$, and

$$\tilde{H}_0^d(O_1\cup O_2)+\tilde{H}_0^d(O_1\cap O_2)\le\tilde{H}_0^d(O_1)+\tilde{H}_0^d(O_2)\le\tilde{H}_0^d(E_1)+\tilde{H}_0^d(E_2)+\frac{3}{4}\varepsilon,$$

we have $\tilde{H}^d(O_1\cup O_2)\le\tilde{H}_0^d(E_2)+\frac{3}{4}\varepsilon$. Going through an inductive argument, we obtain $\tilde{H}^d(O_1\cup O_2\cup\cdots\cup O_j)\le\tilde{H}_0^d(E_j)+(1-2^{-j})\varepsilon$. Letting $j\to\infty$, we obtain

$$\tilde{H}_0^d\left(\bigcup_{j=1}^{\infty}E_j\right)\le\tilde{H}_0^d\left(\bigcup_{j=1}^{\infty}O_j\right)=\lim_{J\to\infty}\tilde{H}_0^d\left(\bigcup_{j=1}^{J}O_j\right)\le\lim_{J\to\infty}\tilde{H}_0^d(E_J)+\varepsilon$$

thanks to Lemma 1.5, proving (1.49).

1.1.3.3 Others

We followed the textbook [100, Section 3.4] in Sect. 1.3.1.

1.1.4 Choquet Integral

1.1.4.1 Choquet Integral

Motivated by the layer cake representation, Theorem 1.5, we define the Choquet integral against H^d [363]. We adopt the idea in Theorem 1.5 because H^d is not additive. As is the case with the Lebesgue integral, we need to consider $f^{-1}(\lambda, \infty) = \{x \in \mathbb{R}^n : f(x) > \lambda\}$; see Theorem 1.5.

Definition 1.6 (Choquet integral against H^d) For any function $f : \mathbb{R}^n \mapsto [0, \infty]$, its *Choquet integral* $\int_{\mathbb{R}^n} f(x)\mathrm{d}H^d(x)$ *against* H^d is defined by

$$\int_{\mathbb{R}^n} f(x)\mathrm{d}H^d(x) \equiv \int_0^\infty H^d(f^{-1}(\lambda, \infty))\mathrm{d}\lambda,$$

where the right-hand side is understood as the infinite Riemannian integral.

We do not have to assume that f is measurable in some sense.

Example 1.3 Let $d \in (0, n]$, $N \geq 0$ and $a > 0$. Define $f(x) \equiv (a + |x|)^{-N}$ for $x \in \mathbb{R}^n$. Then $f^{-1}(\lambda, \infty) = B((\lambda^{-1/N} - a)_+)$. Here we used $B(0) = \emptyset$. Hence

$$|\{f > \lambda\}| \sim ((\lambda^{-1/N} - a)_+)^d.$$

If we integrate this over $(0, \infty)$, we see that

$$\int_{\mathbb{R}^n} (a + |x|)^{-N}\mathrm{d}H^d(x) \sim \int_0^\infty ((\lambda^{-1/N} - a)_+)^d \mathrm{d}\lambda = C_{N,d}a^{1-\frac{d}{N}}.$$

Here $C_{N,d}$ is a constant such that $C_{N,d}$ is finite if and only if $d > N$.

1.1.4.2 Choquet Integral Against $\tilde{H}_0^d$

One considers the Choquet integral against $\tilde{H}_0^d$ analogously to the one against H^d keeping in mind that H^d and $\tilde{H}_0^d$ are equivalent according to Proposition 1.6.

Definition 1.7 (Choquet integral against $\tilde{H}_0^d$) For any function $f : \mathbb{R}^n \mapsto [0, \infty]$, its *Choquet integral* $\int_{\mathbb{R}^n} f(x)\mathrm{d}\tilde{H}_0^d(x)$ *against* $\tilde{H}_0^d$ is defined by

$$\int_{\mathbb{R}^n} f(x)\mathrm{d}\tilde{H}_0^d(x) \equiv \int_0^\infty \tilde{H}_0^d(f^{-1}(\lambda, \infty))\mathrm{d}\lambda,$$

where the right-hand side is understood as the infinite Riemann integral.

As is the case with the Lebesgue integral, we are not interested in the value of the integral itself. Let $f : \mathbb{R}^n \mapsto [0, \infty]$ be any function and let $a > 0$, We mention that we readily obtain

$$\int_{\mathbb{R}^n} a \cdot f(x) \mathrm{d}\tilde{H}_0^d(x) = a \int_{\mathbb{R}^n} f(x) \mathrm{d}\tilde{H}_0^d(x) \tag{1.50}$$

by a change of variables.

The following formulas are easy to verify:

Proposition 1.7 *For any increasing sequences* $\{f_k\}_{k=1}^{\infty}$ *satisfying* $0 \le f_k \le f_{k+1}$,

$$\lim_{k\to\infty} \int_{\mathbb{R}^n} f_k(x) \mathrm{d}\tilde{H}_0^d(x) = \int_{\mathbb{R}^n} \lim_{k\to\infty} f_k(x) \mathrm{d}\tilde{H}_0^d(x).$$

Proof By the monotone convergence theorem and Theorem 1.12, we have

$$\begin{aligned}
\lim_{k\to\infty} \int_{\mathbb{R}^n} f_k(x) \mathrm{d}\tilde{H}_0^d(x) &= \lim_{k\to\infty} \int_0^{\infty} \tilde{H}_0^d \left(f_k{}^{-1}(\lambda, \infty) \right) \mathrm{d}\lambda \\
&= \int_0^{\infty} \lim_{k\to\infty} \tilde{H}_0^d \left(f_k{}^{-1}(\lambda, \infty) \right) \mathrm{d}\lambda \\
&= \int_0^{\infty} \tilde{H}_0^d \left\{ x \in \mathbb{R}^n : \lim_{k\to\infty} f_k(x) > \lambda \right\} \mathrm{d}\lambda \\
&= \int_{\mathbb{R}^n} \lim_{k\to\infty} f_k(x) \mathrm{d}\tilde{H}_0^d(x),
\end{aligned}$$

as was to be shown.

One of the important properties of the Choquet integral is the subadditivity given in the next theorem:

Theorem 1.13 (Subadditivity) *Let* $0 < d \le n$. *Then for any functions* $f, g : \mathbb{R}^n \mapsto [0, \infty]$,

$$\int_{\mathbb{R}^n} (f(x) + g(x)) \mathrm{d}\tilde{H}_0^d(x) \le \int_{\mathbb{R}^n} f(x) \mathrm{d}\tilde{H}_0^d(x) + \int_{\mathbb{R}^n} g(x) \mathrm{d}\tilde{H}_0^d(x). \tag{1.51}$$

We prove Theorem 1.13 step by step. The next lemma, showing that additivity is available for some special cases, is a key to our observation:

Lemma 1.6 *Let* $E_1 \supset E_2 \supset \cdots \supset E_N$ *be a finite decreasing sequence of sets in* $\mathbb{R}^n$. *Let us set* $f \equiv \sum_{k=1}^{N} \chi_{E_k}$ *and* $h \equiv \sum_{k=1}^{N-1} \chi_{E_k} = f - \chi_{E_N}$. *Then*

$$\int_{\mathbb{R}^n} f(x) \mathrm{d}\tilde{H}_0^d(x) = \int_0^{\infty} \tilde{H}_0^d(h^{-1}(\lambda, \infty)) \mathrm{d}\lambda + \tilde{H}_0^d(E_N).$$

Proof We decompose the integral in the definition of $\int_{\mathbb{R}^n} f(x)\mathrm{d}\tilde{H}_0^d(x)$ into three parts to have:

$$\begin{aligned}
&\int_{\mathbb{R}^n} f(x)\mathrm{d}\tilde{H}_0^d(x)\\
&= \int_0^\infty \tilde{H}_0^d(f^{-1}(\lambda,\infty))\mathrm{d}\lambda\\
&= \int_0^{N-1} \tilde{H}_0^d(f^{-1}(\lambda,\infty))\mathrm{d}\lambda + \int_{N-1}^{N} \tilde{H}_0^d(f^{-1}(\lambda,\infty))\mathrm{d}\lambda + \int_N^\infty \tilde{H}_0^d(f^{-1}(\lambda,\infty))\mathrm{d}\lambda\\
&= \int_0^{N-1} \tilde{H}_0^d(h^{-1}(\lambda,\infty))\mathrm{d}\lambda + \int_{N-1}^{N} \tilde{H}_0^d(E_N)\mathrm{d}\lambda + \int_N^\infty \tilde{H}_0^d(\emptyset)\mathrm{d}\lambda\\
&= \int_0^\infty \tilde{H}_0^d(h^{-1}(\lambda,\infty))\mathrm{d}\lambda + \tilde{H}_0^d(E_N),
\end{aligned}$$

as was to be shown.

Lemma 1.7 *Inequality* (1.51) *remains true when f and g assume their value in $\mathbb{N}_0$ and* $\sup g \le 1$.

Proof In view of Proposition 1.7, we may assume that f is bounded. So, we can assume that $f = \sum_{k=1}^{N} \chi_{E_k}$ and $g = \chi_{F_1}$, where $E_1 \supset E_2 \supset \cdots \supset E_N$. If $N = 1$, then we readily have (1.51) thanks to Theorem 1.10:

$$\begin{aligned}
\int_{\mathbb{R}^n} (f(x)+g(x))\mathrm{d}\tilde{H}_0^d(x) &= \tilde{H}_0^d(E_1 \cup F_1) + \tilde{H}_0^d(E_1 \cap F_1)\\
&\le \tilde{H}_0^d(E_1) + \tilde{H}_0^d(F_1)\\
&= \int_{\mathbb{R}^n} f(x)\mathrm{d}\tilde{H}_0^d(x) + \int_{\mathbb{R}^n} g(x)\mathrm{d}\tilde{H}_0^d(x).
\end{aligned}$$

Suppose that (1.51) is true for $m = N-1 \ge 1$. Set $h \equiv \sum_{k=1}^{N-1} \chi_{E_k}$. Then according to Lemma 1.6, we have

$$\int_{\mathbb{R}^n} f(x)\mathrm{d}\tilde{H}_0^d(x) + \int_{\mathbb{R}^n} g(x)\mathrm{d}\tilde{H}_0^d(x) = \int_{\mathbb{R}^n} h(x)\mathrm{d}\tilde{H}_0^d(x) + \tilde{H}^d(F_1) + \tilde{H}_0^d(E_N).$$

We have

$$\int_{\mathbb{R}^n} f(x)\mathrm{d}\tilde{H}_0^d(x)+\int_{\mathbb{R}^n} g(x)\mathrm{d}\tilde{H}_0^d(x)$$
$$\geq \int_{\mathbb{R}^n} h(x)\mathrm{d}\tilde{H}_0^d(x)+\tilde{H}^d(F_1\cup E_N)+\tilde{H}_0^d(F_1\cap E_N)$$

by Theorem 1.10. Thus, we have

$$\int_{\mathbb{R}^n} f(x)\mathrm{d}\tilde{H}_0^d(x)+\int_{\mathbb{R}^n} g(x)\mathrm{d}\tilde{H}_0^d(x)$$
$$\geq \int_{\mathbb{R}^n} (h(x)+\chi_{F_1\cup E_N}(x))\mathrm{d}\tilde{H}_0^d(x)+\tilde{H}_0^d(F_1\cap E_N)$$

from the induction assumption to the function h, which satisfies $\sup h \leq N-1$. Finally, using Lemma 1.6, we obtain

$$\begin{aligned}\int_{\mathbb{R}^n} f(x)\mathrm{d}\tilde{H}_0^d(x)+\int_{\mathbb{R}^n} g(x)\mathrm{d}\tilde{H}_0^d(x) &\geq \int_{\mathbb{R}^n} (h(x)+\chi_{F_1\cup E_N}(x)+\chi_{F_1\cap E_N}(x))\mathrm{d}\tilde{H}_0^d(x)\\ &\geq \int_{\mathbb{R}^n} (h(x)+\chi_{F_1}(x)+\chi_{E_N}(x))\mathrm{d}\tilde{H}_0^d(x)\\ &\geq \int_{\mathbb{R}^n} (f(x)+g(x))\mathrm{d}\tilde{H}_0^d(x),\end{aligned}$$

which proves (1.51).

Lemma 1.8 *Inequality* (1.51) *remains true when f and g assume its value in* $\mathbb{N}_0$.

Proof We first assume that f and g are bounded. So, we can assume that

$$f=\sum_{k=1}^N \chi_{E_k}, \quad g=\sum_{k=1}^N \chi_{F_k}, \tag{1.52}$$

where $E_1 \supset E_2 \supset \cdots \supset E_N$ and $F_1 \supset F_2 \supset \cdots \supset F_N$.

We prove (1.51) by the induction on N. If $N=1$, then (1.51) readily follows from Lemma 1.7. Suppose that (1.51) is true for $N=m-1\geq 1$. Consider a function given by (1.52). Let us set $h \equiv \sum_{k=1}^{N-1} \chi_{E_k}$, and $k \equiv \sum_{k=1}^{N-1} \chi_{F_k}$. Then according to Lemma 1.6 and Theorem 1.10, we have

$$\begin{aligned}
&\int_{\mathbb{R}^n} f(x)\mathrm{d}\tilde{H}_0^d(x) + \int_{\mathbb{R}^n} g(x)\mathrm{d}\tilde{H}_0^d(x)\\
&= \int_{\mathbb{R}^n} h(x)\mathrm{d}\tilde{H}_0^d(x) + \int_{\mathbb{R}^n} k(x)\mathrm{d}\tilde{H}_0^d(x) + \tilde{H}_0^d(E_N) + \tilde{H}_0^d(F_N)\\
&\geq \int_{\mathbb{R}^n} h(x)\mathrm{d}\tilde{H}_0^d(x) + \int_{\mathbb{R}^n} k(x)\mathrm{d}\tilde{H}_0^d(x) + \tilde{H}_0^d(E_N \cup F_N) + \tilde{H}_0^d(E_N \cap F_N).
\end{aligned}$$

Now we invoke the induction assumption to have

$$\begin{aligned}
&\int_{\mathbb{R}^n} f(x)\mathrm{d}\tilde{H}_0^d(x) + \int_{\mathbb{R}^n} g(x)\mathrm{d}\tilde{H}_0^d(x)\\
&\geq \int_{\mathbb{R}^n} (h(x)+k(x))\mathrm{d}\tilde{H}_0^d(x) + \tilde{H}_0^d(E_N \cup F_N) + \tilde{H}_0^d(E_N \cap F_N)\\
&\geq \int_{\mathbb{R}^n} (h(x)+k(x))\mathrm{d}\tilde{H}_0^d(x) + \int_{\mathbb{R}^n} \chi_{E_N \cup F_N}(x)\mathrm{d}\tilde{H}_0^d(x) + \tilde{H}_0^d(E_N \cap F_N).
\end{aligned}$$

Next, by using Lemma 1.7 twice, we obtain

$$\begin{aligned}
&\int_{\mathbb{R}^n} f(x)\mathrm{d}\tilde{H}_0^d(x) + \int_{\mathbb{R}^n} g(x)\mathrm{d}\tilde{H}_0^d(x)\\
&\geq \int_{\mathbb{R}^n} (h(x)+k(x)+\chi_{E_N \cup F_N}(x))\mathrm{d}\tilde{H}_0^d(x) + \tilde{H}_0^d(E_N \cap F_N)\\
&\geq \int_{\mathbb{R}^n} (h(x)+k(x)+\chi_{E_N \cup F_N}(x)+\chi_{E_N \cap F_N}(x))\mathrm{d}\tilde{H}_0^d(x)\\
&\geq \int_{\mathbb{R}^n} (f(x)+g(x))\mathrm{d}\tilde{H}_0^d(x),
\end{aligned}$$

which proves (1.51).

Finally, we prove (1.51) for the general case. Set

$$f_N \equiv 2^N \min([2^{-N} f], 2^{-N} N), \quad g_N \equiv 2^N \min([2^{-N} g], 2^{-N} N).$$

Then according to (1.50) and Proposition 1.7, we have

$$\int_{\mathbb{R}^n} (f(x)+g(x))\mathrm{d}\tilde{H}_0^d(x) = \lim_{N\to\infty} \int_{\mathbb{R}^n} (f_N(x)+g_N(x))\mathrm{d}\tilde{H}_0^d(x)$$

and

$$\begin{aligned}
&\int_{\mathbb{R}^n} f(x)\mathrm{d}\tilde{H}_0^d(x) = \lim_{N\to\infty} \int_{\mathbb{R}^n} f_N(x)\mathrm{d}\tilde{H}_0^d(x), \int_{\mathbb{R}^n} g(x)\mathrm{d}\tilde{H}_0^d(x)\\
&\quad = \lim_{N\to\infty} \int_{\mathbb{R}^n} g_N(x)\mathrm{d}\tilde{H}_0^d(x).
\end{aligned}$$

This proves (1.51).

Exercises

Exercise 1.24 For all nonempty sets $E \subset \mathbb{R}^n$, show that $H^0(E) \geq 1$, and that $H^0(E) = 1$ if and only if E is bounded. Hint: When $H^0(E) = 1$, then we can find a ball that covers E.

Exercise 1.25 Let $0 < d \leq n$, and let $E \subset \mathbb{R}^n$. Then show that $\tilde{H}_0^d(E) = \inf\{\tilde{H}_0^d(O) : E \subset O\}$.

Exercise 1.26 By a change of variables, prove (1.50).

1.1.5 Fundamental Facts on Functional Analysis

1.1.5.1 Helly's Theorem

We invoke a result on linear algebra.

Theorem 1.14 (Helly's theorem) *Let V be a complex linear space, and let $\{f_j\}_{j=1}^N$ be a finite collection of linear mappings from V to $\mathbb{C}$. Then a linear mapping $f : V \to \mathbb{C}$ is expressed as a linear combination of $\{f_j\}_{j=1}^N$ if and only if*

$$\bigcap_{j=1}^{N} \ker(f_j) \subset \ker(f). \tag{1.53}$$

Proof The necessity is clear; let us prove the sufficiency. Let f be a linear mapping satisfying (1.53), and let $W \equiv \mathrm{Im}(f_1 \times f_2 \times \cdots \times f_N)$, where for a linear mapping $f : V_1 \to V_2$ from a linear space V_1 to another linear space V_2, $\mathrm{Im}(f)$ stands for the image. Then the mapping

$$\ell : (f_1(v), f_2(v), \ldots, f_N(v)) \in W \mapsto f(v) \in \mathbb{C}$$

is a well-defined linear mapping despite the ambiguity of the choice of v. Extend the above mapping from $\mathbb{C}^N$ to $\mathbb{C}$ so that $\ell(z_1, z_2, \ldots, z_N) = \sum_{j=1}^{N} a_j z_j$ for all $(z_1, z_2, \ldots, z_N)$, where $a_1, a_2, \ldots, a_N$ are complex constants. Then we have $f = \sum_{j=1}^{N} a_j f_j$, as required.

1.1.5.2 Density Argument

A *quasi-normed space* X is a complex linear space which is equipped with a mapping $\| \star \|_X : X \to [0, \infty)$ satisfying:

1. Let $x \in X$. Then $\|x\|_X = 0$ if and only if $x = 0$.
2. $\|\alpha x\|_X = |\alpha| \cdot \|x\|_X$ for all $\alpha \in \mathbb{C}$ and $x \in X$.
3. There exists $K > 0$ such that $\|x + y\|_X \le K(\|x\|_X + \|y\|_X)$ for all $x, y \in X$.

A *quasi-Banach space* is a quasi-normed space for which any Cauchy sequence converges. Generally speaking, the next theorem is of use when we construct bounded linear mappings. Since the result is well known and the proof is elementary, we omit the proof.

Theorem 1.15 (Density argument) *Let X be a quasi-normed space, and let Y be a quasi-Banach space. If $A : D \to Y$ is a bounded mapping defined on a dense subset D of X such that $\|A x\|_Y \le M \|x\|_X$ for all $x \in D$, then A extends uniquely to a bounded linear mapping from X to Y.*

1.1.5.3 Baire's Category Theorem

Next, we recall Baire's category theorem, which was necessary to prove the open mapping theorem, the closed graph theorem and the uniformly bounded principle. See [89, p. 159] for the proof.

Theorem 1.16 (Baire's category theorem) *For any closed covering $\{F_j\}_{j=1}^{\infty}$ of a complete metric space X, at least one of the F_j's contains an interior point.*

We recall that a quasi-Banach space $(X, \| \cdot \|_X)$ is a Banach space with the axiom of the triangle inequality replaced by the *quasi-triangle inequality*: $\|x + y\|_X \le \alpha(\|x\|_X + \|y\|_X)$ for all $x, y \in X$. According to [119, p. 79, Theorem 1], we have the closed graph theorem, for example, for quasi-Banach spaces. In this connection we also remark that the quasi-norm $\| \star \|_X$ satisfies the η-triangle inequality for $\eta > 0$ if $\|x + y\|_X{}^{\eta} \le \|x\|_X{}^{\eta} + \|y\|_X{}^{\eta}$ for all $x, y \in X$. For later consideration, we content ourselves with stating the following lemma, which will be necessary for the proof of the closed graph theorem.

Lemma 1.9 *Let X and Y be Banach spaces, and let $T : X \to Y$ be a bounded linear mapping. If for all $y \in Y$, there exists $x \in X$ such that* $\min(2\|Tx - y\|_Y, M\|x\|_X) \le \|y\|_Y$ *for some $M > 0$ which depends only on T, then T is surjective.*

We do not recall the proof; see Exercise 1.27.

1.1.5.4 Banach–Alaoglu Theorem

We state the Banach–Alaoglu theorem. We refer to [60] for the proof.

Theorem 1.17 (Banach–Alaoglu theorem) *Let X be a separable Banach space, that is, X is realized as the closure of a countable subset of X, and let $\{x_j^*\}_{j=1}^\infty$ be a sequence in the unit ball of X^*; $B_{X^*} \equiv \{x^* \in X^* : \|x^*\|_{X^*} \le 1\}$. Then we can find a subsequence $\{x_{j(k)}^*\}_{k=1}^\infty$ and $x \in B_{X^*}$ such that $x^*(x) = \lim_{k\to\infty} x_{j(k)}^*(x)$ for all $x \in X$.*

1.1.5.5 Hahn–Banach Theorem

There are many forms for the Hahn–Banach theorem. Here we recall what we need about this in this book. For the proof, we refer to [30, 60] and [89, Chapter 6], for example.

Theorem 1.18 (Hahn–Banach theorem) *Let V be a complex normed space, let U be a subspace which is not always a closed set, and let $u^* : U \to \mathbb{C}$ be a continuous $\mathbb{C}$-linear mapping. Then there exists a continuous $\mathbb{C}$-linear mapping $v^* : V \to \mathbb{C}$ such that $v^*|U = u^*$.*

1.1.5.6 A Duality Result

Let X be a Banach space. We let $\mathrm{Lip}(\mathbb{R}, X)$ be the set of all Lipschitz continuous functions $f : \mathbb{R} \to X$ for which the quantity $\|f\|_{\mathrm{Lip}(\mathbb{R},X)} \equiv \sup_{-\infty<s<t<\infty} |t - s|^{-1}\|f(t) - f(s)\|_X$ is finite. If $X = \mathbb{C}$, then abbreviate $\mathrm{Lip}(\mathbb{R}, X)$ to $\mathrm{Lip}(\mathbb{R})$.

Lemma 1.10 *Let X be a Banach space.*

1. *Let $f \in L^1(\mathbb{R}; X) \cap C(\mathbb{R}; X)$ and $g \in \mathrm{Lip}(\mathbb{R}; X^*)$. Then the limit*

$$L_g(f) \equiv \lim_{N\to\infty} \sum_{j=-\infty}^{\infty} \langle g(N^{-1}(j+1)) - g(N^{-1}j), f(N^{-1}j)\rangle$$

   exists and satisfies $|L_g(f)| \le \|f\|_{L^1(\mathbb{R};X)}\|g\|_{\mathrm{Lip}(\mathbb{R};X^*)}$. Thus, $g \in \mathrm{Lip}(\mathbb{X}^*; X)$ induces a bounded linear functional on $L^1(\mathbb{R}; X)$.
2. *Conversely, any bounded linear functional L on $L^1(\mathbb{R}; X)$ is realized as $L = L_g$ for some $g \in \mathrm{Lip}(\mathbb{R}; X^*)$.*

Proof The first assertion is straightforward and we omit the proof. For the second assertion, we define

$$\langle g(t), a\rangle \equiv \begin{cases} L(\chi_{[0,t]}a) & t \ge 0 \\ -L(\chi_{[t,0]}a) & t < 0. \end{cases}$$

It is not so hard to check $L = L_g$, since functions of the form $\chi_{[a,b]}x$ with $a, b \in \mathbb{R}$ and $x \in X$ span a dense subspace in $L^1(\mathbb{R}; X)$.

1.1.5.7 Operators on Banach Spaces

Although we do not use unbounded operators much in this book, we give a brief review of the terminologies of (unbounded) linear operators. Let X and Y be Banach spaces. An operator $L : X \to X$ is said to be a *densely defined closed operator* if L is a linear mapping defined on a dense linear subspace $Z = \mathrm{Dom}(L)$ on X, whose graph $\{(x, Lx) : x \in Z\}$ is a closed set in $X \times Y$. The set $\mathrm{Dom}(L)$ is called the *domain* of L. The *range* of L is defined by $\mathrm{Ran}(L) \equiv \{Lx : x \in \mathrm{Dom}(L)\} \subset Y$. The set $\{x \in \mathrm{Dom}(L) : Lx = 0 \in Y\}$ is called the *kernel* of L and is denoted by $\ker(L)$. Let $L_1 : X \to Y$ and $L_2 : X \to Y$ be unbounded operators. The *sum* $L_1 + L_2$ is a linear operator defined on $\mathrm{Dom}(L_1 + L_2) = \mathrm{Dom}(L_1) \cap \mathrm{Dom}(L_2)$ satisfying $(L_1 + L_2)x = L_1x + L_2x$. If $\mathrm{Dom}(L_1) \subset \mathrm{Dom}(L_2)$ and $L_2x = L_1x$ for all $x \in \mathrm{Dom}(L_1)$, then we write $L_1 \subset L_2$. Let X, Y, Z be Banach spaces, and let $A : X \to Y$ and $B : Y \to Z$ be linear mappings. Then BA is defined by $BAx \equiv B(Ax)$ for $x \in \mathrm{Dom}(BA) \equiv \{x \in \mathrm{Dom}(A) : Ax \in \mathrm{Dom}(B)\}$.

Let $X = Y$ be a complex Banach space for the time being, so that the operator $\mathrm{id}_X : X \to X = Y$ makes sense. For such an operator L, we denote by $\rho(L)$ the set of all complex numbers $\lambda \in \mathbb{C}$ such that $\lambda \mathrm{id}_X - L : Z \to X$ is bijective and that the mapping $(\lambda \mathrm{id}_X - L)^{-1} : X \to X(\supset Z)$ is bounded. We call $\rho(L)$ the *resolvent* set. The complement of $\rho(L)$ in $\mathbb{C}$ is called the *spectrum set* and we denote it by $\sigma(L)$. One can readily check that the *resolvent equation* holds: For $z, z' \in \rho(L)$,

$$(z' - z)(L - z\mathrm{id}_X)^{-1}(L - z'\mathrm{id}_X)^{-1} = (L - z'\mathrm{id}_X)^{-1} - (L - z\mathrm{id}_X)^{-1}. \tag{1.54}$$

1.1.5.8 Functional Calculus

Like the Riemannian integral and the Lebesgue integral, we can define integrals of the functions whose values are in Banach spaces. We can make use of the Lebesgue integration theorem, the Fubini theorem and so on. For quasi-Banach spaces X and Y, we denote by $B(X, Y)$ the set of all linear operators $L : X \to X$ defined on X such that the operator norm $\|L\|_{B(X,Y)} \equiv \sup\{\|Lx\|_Y : \|x\|_X = 1\}$ is finite. If $X = Y$, we abbreviate $B(X, Y)$ to $B(X)$. If X, Y are Banach spaces, then $B(X, Y)$ is also a Banach space.

For a compact set K, define $\mathscr{O}(K)$ to be the set of all holomorphic functions defined on an open set U containing K. We use the following theorem in Sect. 5.3.1. We admit that we have properties similar to the Lebesgue integral and the Riemannian integral in the wide sense; see [119].

Theorem 1.19 *Let $L : X \to X$ be a bounded linear operator on a Banach space X. Then*

$$f \in \mathscr{O}(\sigma(L)) \mapsto f(L) \equiv \frac{1}{2\pi i}\int_C f(z)(z\mathrm{id}_X - L)^{-1}\mathrm{d}z \in B(X)$$

does not depend on a smooth curve C which surrounds $\sigma(L)$ and is contained in the open set U on which f is defined. Furthermore, $f \in \mathscr{O}(\sigma(L)) \mapsto f(L) \in B(X)$ is a ring homomorphism; that is, for $f, g \in \mathscr{O}(\sigma(L))$,

$$(f+g)(L) = f(L) + g(L), \quad (f \cdot g)(L) = f(L)g(L).$$

Proof By the Cauchy integration theorem, we see that $f(L)$ does not depend on a smooth curve C which surrounds $\sigma(L)$ and is contained in the open set U on which f is defined. The additivity $(f+g)(L) = f(L) + g(L)$ can be proved by taking a smooth curve C which surrounds $\sigma(L)$ and is contained in the open set U on which f and g are defined. Let us prove the multiplicativity $(f \cdot g)(L) = f(L)g(L)$. Let C be a curve contained in the open set U which surrounds $\sigma(L)$ and is on which f and g are defined. Denote by D the domain which C surrounds. Choose a smooth curve C' that surrounds $\overline{D}$. Then

$$f(L) = \frac{1}{2\pi i}\int_C f(z)(z\mathrm{id}_X - L)^{-1}\mathrm{d}z, \quad g(L) = \frac{1}{2\pi i}\int_{C'} g(w)(w\mathrm{id}_X - L)^{-1}\mathrm{d}w.$$

Thus,

$$f(L)g(L) = \frac{1}{(2\pi i)^2}\int_C\int_{C'} f(z)g(w)(z\mathrm{id}_X - L)^{-1}(w\mathrm{id}_X - L)^{-1}\mathrm{d}z\mathrm{d}w.$$

By the resolvent equation (1.54), we have

$$\begin{aligned} f(L)g(L) = &\frac{1}{(2\pi i)^2}\int_C\int_{C'} \frac{f(z)g(w)}{z-w}(w\mathrm{id}_X - L)^{-1}\mathrm{d}z\mathrm{d}w \\ &- \frac{1}{(2\pi i)^2}\int_C\int_{C'} \frac{f(z)g(w)}{z-w}(z\mathrm{id}_X - L)^{-1}\mathrm{d}z\mathrm{d}w. \end{aligned}$$

We note that $\displaystyle\int_C \frac{f(\mu)}{\mu - w}\mathrm{d}\mu = 0, \int_{C'} \frac{g(\mu)}{z-\mu}\mathrm{d}\mu = -2\pi i g(z)$ for all $z \in C$ and $w \in C'$. Thus,

$$f(L)g(L) = \frac{1}{2\pi i}\int_C f(z)g(z)(z\mathrm{id}_X - L)^{-1}\mathrm{d}z\mathrm{d}w = (f \cdot g)(L),$$

proving multiplicativity.

If we combine Theorem 1.19 with Lemma 1.11, we can say that $f \mapsto f(L)$ is an algebraic homomorphism.

Lemma 1.11 *Let $L : X \to X$ be a bounded linear operator. Then for any smooth curve C which surrounds $\sigma(L)$,* $\dfrac{1}{2\pi i}\displaystyle\int_C (z\mathrm{id}_X - L)^{-1}\mathrm{d}z = \mathrm{id}_X$.

Proof Since $\sigma(L) \neq \emptyset$, by considering $L - \alpha\mathrm{id}_X$ with $\alpha \in \sigma(L)$, we may assume that $0 \in \sigma(L)$. Then change variables: $z \mapsto w = z^{-1}$. See Exercise 1.28.

Exercises

Exercise 1.27 [653, Lemma 2.4] Reconstruct the proof of the closed graph theorem and then prove Lemma 1.9.

Exercise 1.28 Let $L : X \to X$ be a bounded linear operator on a Banach space X. Using $w = z^{-1}$ and letting C be a big circle containing $\sigma(L)$, prove Lemma 1.11.

Exercise 1.29 Show that the definition of $(a\mathrm{id}_X + T)^{-1}$ is consistent for $a \notin \sigma(T)$. More precisely show that $(a\mathrm{id}_X + T)^{-1}$ obtained in Theorem 1.19 is the inverse of $a\mathrm{id}_X + T : X \to X$ using Lemma 1.11.

Textbooks in Sect. 1.1

L^p Spaces with $0 < p < \infty$

We refer to [30, Chapter 5] and [89, Chapter 1], for example, for fundamental facts on Lebesgue spaces used in this book. For example, Garling dealt with $L^p(\mathbb{R}^n)$ for $0 < p \leq 1$ in his textbook [30, Chapter 5].

Layer Cake Representation: Theorem 1.5

See [22, Proposition 2.3], [32, Proposition 1.1.4] and [117, Proposition 6.1].

Whitney Covering: Theorem 1.9

See [33, Appendix J] and [86, p. 15, Lemma 2].

Functional Analysis

See [60] for example.

1.2 Schwartz Function Space $\mathcal{S}(\mathbb{R}^n)$ and Function Spaces $\mathcal{D}(\Omega)$ on Domains

The space $C_c^\infty(\mathbb{R}^n)$, the function space of all compactly supported $C^\infty(\mathbb{R}^n)$-functions, is supposed to be the function space that enjoys the "nicest" properties. However, this space has a disadvantage; it is not stable under the Fourier transform. In fact, the Fourier transform of any nonzero function in $C_c^\infty(\mathbb{R}^n)$ is not compactly supported. Although from the definition of the Fourier transform it seems natural to consider $L^1(\mathbb{R}^n)$, the image of the functions in this space is much better than $L^\infty(\mathbb{R}^n)$.

One of the fundamental ideas in this book is that we pay attention to the size of the Fourier transform of the functions. Therefore, it is quite natural that we look for linear spaces that are invariant under the Fourier transform. What is the suitable space invariant under the Fourier transform?

In Sect. 1.2.1, we consider the Schwartz function space $\mathcal{S}(\mathbb{R}^n)$. Section 1.2.2 is concerned with the investigation of $\mathcal{S}'(\mathbb{R}^n)$. In Sect. 1.2.3, we discuss the Fourier transform. The space $\mathcal{D}(\Omega)$ on a domain Ω is covered in Sect. 1.2.4. In Sect. 1.2.5, we present some nontrivial formulas on $\mathcal{S}(\mathbb{R}^n)$, $\mathcal{D}(\mathbb{R}^n)$. In Sect. 1.3, we consider some formulas on difference operators.

1.2.1 Definition of the Schwartz Function Space $\mathcal{S}(\mathbb{R}^n)$ and Its Topology

To answer the above question, we define and investigate the Schwartz function space $\mathcal{S}(\mathbb{R}^n)$.

1.2.1.1 Definition of $\mathcal{S}(\mathbb{R}^n)$ and Some Examples

Let us recall the notation of multi-indexes to define the Schwartz space $\mathcal{S}(\mathbb{R}^n)$. By "a multi-index", we mean an element in ${\mathbb{N}_0}^n \equiv \{0, 1, 2, \ldots\}^n$. A tacit understanding is that all functions assume their value in $\mathbb{C}$. For multi-indexes $\beta = (\beta_1, \beta_2, \ldots, \beta_n) \in {\mathbb{N}_0}^n$ and $f \in C^\infty(\mathbb{R}^n)$, we set

$$\partial^\beta f \equiv \left(\frac{\partial}{\partial x_1}\right)^{\beta_1} \left(\frac{\partial}{\partial x_2}\right)^{\beta_2} \cdots \left(\frac{\partial}{\partial x_n}\right)^{\beta_n} f.$$

For $x = (x_1, x_2, \ldots, x_n) \in \mathbb{R}^n$, we define $x^\beta \equiv {x_1}^{\beta_1} {x_2}^{\beta_2} \cdots {x_n}^{\beta_n}$.

Definition 1.8 (Schwartz function space $\mathcal{S}(\mathbb{R}^n)$) For multi-indexes $\alpha, \beta \in {\mathbb{N}_0}^n$ and a function φ, write $\varphi_{(\alpha,\beta)}(x) \equiv x^\alpha \partial^\beta \varphi(x)$, $x \in \mathbb{R}^n$ temporarily. The *Schwartz function space* $\mathcal{S}(\mathbb{R}^n)$ is the set of all the functions satisfying

$$\mathscr{S}(\mathbb{R}^n) \equiv \bigcap_{\alpha,\beta \in \mathbb{N}_0{}^n} \left\{\varphi \in C^\infty(\mathbb{R}^n) \,:\, \varphi_{(\alpha,\beta)} \in L^\infty(\mathbb{R}^n)\right\}.$$

The elements in $\mathscr{S}(\mathbb{R}^n)$ are called the *test functions.*

To simplify the notation, for $N \in \mathbb{N}_0$, we define

$$p_N(\varphi) \equiv \sum_{\substack{\alpha \in \mathbb{N}_0{}^n \\ |\alpha| \le N}} \left(\sup_{x \in \mathbb{R}^n} \langle x \rangle^N |\partial^\alpha \varphi(x)|\right), \quad \varphi \in \mathscr{S}(\mathbb{R}^n). \tag{1.55}$$

As one of other functionals than $p_N(\varphi)$, we define

$$p'_N(\varphi) \equiv \sum_{\substack{\alpha \in \mathbb{N}_0{}^n \\ |\alpha| \le N}} \left(\sup_{x \in \mathbb{R}^n} (1+|x|)^N |\partial^\alpha \varphi(x)|\right), \quad \varphi \in \mathscr{S}(\mathbb{R}^n).$$

Example 1.4

1. The function φ, given by $\varphi(x) \equiv \exp(-|x|^2)$, $x \in \mathbb{R}^n$, is called *Gaussian* and it belongs to $\mathscr{S}(\mathbb{R}^n)$.
2. If we define $\varphi(x) \equiv \exp(-\sqrt[3]{|x|^2+1})$ for $x \in \mathbb{R}^n$, then we have $\varphi \in \mathscr{S}(\mathbb{R}^n)$. However, $\psi(x) = \exp(a|x|)\varphi(x)$ is not bounded for all $a > 0$.
3. The space $\mathscr{S}(\mathbb{R}^n)$ is made up of functions enjoying good properties. For example the Hölder inequality yields $L^p(\mathbb{R}^n) \hookrightarrow \mathscr{S}'(\mathbb{R}^n)$ for $1 \le p \le \infty$. In fact, letting $N = [p^{-1}(n+1)] + 1$, we have $\|\varphi\|_p \lesssim p_N(\varphi)$ by the Hölder inequality.

We conclude our examples of Schwartz functions by presenting the following function g:

Lemma 1.12 *There exists $g \in \mathscr{S}(\mathbb{R})$ supported in $[1,\infty)$ such that* $\displaystyle\int_{-\infty}^{\infty} g(t)\mathrm{d}t = \delta_{k0}$ *for all $k \in \mathbb{N}$.*

Proof Choose a branch of $\log : \mathbb{C} \setminus [0,\infty) \to \mathbb{C}$ so that $\log i = \dfrac{\pi}{2} i$. Define $\sqrt[8]{z} \equiv \exp\left(\dfrac{1}{8}\log z\right)$ for $z \in \mathbb{C} \setminus [0,\infty) \to \mathbb{C}$ and

$$G(z) \equiv \exp\left(-\sqrt[8]{z-1} - \frac{1}{(\sqrt[8]{z-1})^7}\right), \quad z \in \mathbb{C} \setminus [0,\infty). \tag{1.56}$$

Let $0 < \varepsilon \ll 1$. Define a curve Γ_ε by the continuous mapping $\gamma_\varepsilon : \mathbb{R} \to \mathbb{C}$ given by:

$$\gamma_\varepsilon(t) \equiv \begin{cases} 1 - t - \varepsilon i, & t \le -\varepsilon, \\ 1 + \varepsilon \exp\left(\pi i - \dfrac{t\pi}{2\varepsilon} i\right), & |t| \le \varepsilon, \\ 1 + t + \varepsilon i, & t \ge \varepsilon. \end{cases}$$

The curve γ_ε partitions $\mathbb{C}$ into two domains. Denote by Ω_ε the component of $-1 \in \mathbb{C}$.

We let $\tilde{\gamma}_\varepsilon$ be a curve with counterclockwise orientation given by the boundary of $\Omega_\varepsilon \cap \{|z| > \varepsilon^{-1}\}$. Since $\sqrt{2}\Re(\sqrt[8]{z}) \ge \sqrt[8]{|z|}$ for any $z \in \mathbb{C} \setminus [0, \infty)$, we have

$$\lim_{\varepsilon \downarrow 0} \int_{\Gamma_\varepsilon} z^{j-1} G(z) dz = \lim_{\varepsilon \downarrow 0} \int_{\tilde{\Gamma}_\varepsilon} z^{j-1} G(z) dz = 2\pi i G(0) \delta_{0j}.$$

Let $C \neq 0$ be a constant. We set $g(t) \equiv \dfrac{C(G(t + i0) - G(t - i0))}{t}$ for $t \in \mathbb{R}$. Then we have $g \in \mathscr{S}(\mathbb{R}^n)$ and $\displaystyle\int_1^\infty t^j g(t) dt = \lim_{\varepsilon \downarrow 0} \int_{\Gamma_\varepsilon} z^j G(z) dz$. Hence choosing C suitably, we obtain the desired function.

1.2.1.2 Topology of $\mathscr{S}(\mathbb{R}^n)$

Here we topologize $\mathscr{S}(\mathbb{R}^n)$. See [60], for example, for general topology needed in this book. We equip $\mathscr{S}(\mathbb{R}^n)$ with a natural topology defined from the semi-norms. A linear topology on a linear space X is a topology of X in which the multiplication and the addition are continuous. Such a space X will be called a topological vector space.

Definition 1.9 (Topology of $\mathscr{S}(\mathbb{R}^n)$) For multi-indexes $\alpha, \beta \in {\mathbb{N}_0}^n$, define the norm $p_{\alpha,\beta}$ on $\mathscr{S}(\mathbb{R}^n)$ by $p_{\alpha,\beta}(\varphi) \equiv \sup_{x \in \mathbb{R}^n} |x^\alpha \partial^\beta \varphi(x)| = \|\varphi_{(\alpha,\beta)}\|_\infty$ for $\varphi \in \mathscr{S}(\mathbb{R}^n)$. Equip $\mathscr{S}(\mathbb{R}^n)$ with the weakest linear topology under which each $p_{\alpha,\beta}$ is continuous for all $\varphi \in \mathscr{S}(\mathbb{R}^n)$.

For $a \in \mathbb{R}^n$, we write $\langle a \rangle \equiv \sqrt{1 + |a|^2}$, which is called the *Japanese bracket* of a.

We use an elementary inequality for $\langle x \rangle \equiv \sqrt{1 + |x|^2}$, $x \in \mathbb{R}^n$ due to Peetre.

Lemma 1.13 (Peetre's inequality) *For $x, y \in \mathbb{R}^n$, we have $\langle x + y \rangle \le \sqrt{2} \langle x \rangle \langle y \rangle$.*

The proof is left as Exercise 1.30. In this book, we mainly use $p_N(\varphi)$ because this is the largest one among the related functionals above and the plausible definitions.

From Definition 1.9 and (1.55), we have $p_{\alpha,\beta}(\varphi) \le p_N(\varphi)$ for all $N \in \mathbb{N}$ and multi-indexes α, β satisfying $|\alpha|, |\beta| \le N$.

By way of p_N, we can redefine the topology of $\mathscr{S}(\mathbb{R}^n)$, as follows. A nonempty subset $U \subset \mathscr{S}(\mathbb{R}^n)$ is an open set if and only if, for any $\varphi \in U$, there exist finite sequences $N_1, N_2, \ldots, N_L \in \mathbb{N}$ and $\varepsilon_1, \varepsilon_2, \ldots, \varepsilon_L > 0$ such that

$$\bigcap_{l=1}^{L} \{\psi \in \mathscr{S}(\mathbb{R}^n) \, : \, p_{N_l}(\psi - \varphi) < \varepsilon_l\} \subset U. \tag{1.57}$$

The next useful example characterizes open sets: We may assume that $\varepsilon_1 = (N_1)^{-1}$ and $\ell = 1$ in the above.

Proposition 1.8 *Let $U \subset \mathscr{S}(\mathbb{R}^n)$ be an open set, and let $\varphi \in U$ be a point in U; that is, a "function" in U. Then there exists $N \gg 1$ such that*

$$\{\psi \in \mathscr{S}(\mathbb{R}^n) \, : \, N \, p_N(\varphi - \psi) < 1\} \subset U.$$

Proof Since $\varphi \in U$, there exist $\alpha_1, \alpha_2, \dots, \alpha_L \in \mathbb{N}_0{}^n$, $\beta_1, \beta_2, \dots, \beta_L \in \mathbb{N}_0{}^n$ and $\varepsilon_1, \varepsilon_2, \dots, \varepsilon_L > 0$ satisfying $\bigcap_{l=1}^{L}\{\psi \in \mathscr{S}(\mathbb{R}^n) \, : \, p_{\alpha_l, \beta_l}(\psi - \varphi) < \varepsilon_l\} \subset U$. By replacing ε_j with $\min(\varepsilon_j, 1)$, we may assume that $\varepsilon_j \le 1$ for each $j = 1, 2, \dots, L$. Define

$$N \equiv \max\left\{\left[\frac{|\alpha_j| + |\beta_j| + 1}{\varepsilon_j}\right] \, : \, j = 1, 2, \dots, L\right\} + 1.$$

Let $j = 1, 2, \dots, L$. Since $\varepsilon_j \le 1$, $|\alpha_j| \le N$ implies $|x^{\alpha_j}| \le \langle x \rangle^N$. If $\psi \in \mathscr{S}(\mathbb{R}^n)$ satisfies $N \, p_N(\varphi - \psi) < 1$, then $p_{\alpha_j, \beta_j}(\varphi - \psi) < \varepsilon_j$. Hence N is the desired number.

It is easy to see that there is the continuous embedding $\mathscr{S}(\mathbb{R}^n) \hookrightarrow L^p(\mathbb{R}^n)$ for all $1 \le p \le \infty$. This embedding is dense for $p < \infty$. But when $p = \infty$, this is not the case as Exercise 1.32 shows.

To familiarize ourselves with the definitions, we prove the translation invariance; for $\varphi \in \mathscr{S}(\mathbb{R}^n)$ and $y \in \mathbb{R}^n$ we define $\varphi_y \equiv \varphi(\star - y)$, which we call the translation of φ by y.

Example 1.5 Fix $\varphi \in \mathscr{S}(\mathbb{R}^n)$. Then the mapping $y \in \mathbb{R}^n \mapsto \varphi_y \in \mathscr{S}(\mathbb{R}^n)$ is continuous. That is, for all $\varphi \in \mathscr{S}(\mathbb{R}^n)$, $\lim_{y \to y_0} \varphi_y = \varphi_{y_0}$.

Proof By the translation invariance of $\mathscr{S}(\mathbb{R}^n)$, we have only to prove $\lim_{y \to 0} \varphi_y = \varphi$. This amounts to the proof of

$$\lim_{y \to 0} p_{\alpha, \beta}(\varphi - \varphi_y) = 0 \tag{1.58}$$

for all multi-indexes $\alpha, \beta \in \mathbb{N}_0{}^n$. In view of the definition of $p_{\alpha,\beta}$ (see Definition 1.9), by replacing $\partial^\beta \varphi$ with φ, we can assume that $\beta = 0$. For the purpose of taking the limit as $y \to 0$, let $|y| \le 1$. Recall that $p_N(\varphi)$ is given by (1.55). By the mean-value theorem and the binomial expansion, we have

$$\sup_{x \in \mathbb{R}^n} |x^\alpha(\varphi(x) - \varphi(x-y))| \le |y| \sup_{x \in \mathbb{R}^n} \int_0^1 |x^\alpha \nabla\varphi(x - u\, y)| du \lesssim |y|\, p_{|\alpha|+1}(\varphi),$$

proving (1.58).

Let us investigate more about the topology of $\mathscr{S}(\mathbb{R}^n)$.

Definition 1.10 (Metrizable topological space) A topological space $(X, \mathscr{O}_X)$ is said to be *metrizable* if there exists a distance function $d : X \times X \to [0, \infty)$ such that the d-topology agrees with $\mathscr{O}_X$.

Proposition 1.9 *The space $\mathscr{S}(\mathbb{R}^n)$ is metrizable.*

The proof being long, we first define the distance function d and then prove that the d-topology agrees with the original topology.

Proof **The definition of the distance function** d. Define a distance function d : $\mathscr{S}(\mathbb{R}^n) \times \mathscr{S}(\mathbb{R}^n) \to [0, \infty)$ so that

$$d(\varphi, \psi) \equiv \sum_{j=1}^{\infty} \frac{1}{2^j} \min(p_j(\varphi - \psi), 1) \quad (\varphi, \psi \in \mathscr{S}(\mathbb{R}^n)). \tag{1.59}$$

In the proof, by an "open set", we mean an open set with respect to the original topology of $\mathscr{S}(\mathbb{R}^n)$ and by a "d-open set" we mean the topology induced by d. We content ourselves with defining d as above; we leave to readers the proof that d actually defines a distance function; see Exercise 1.34.

Any d-open set is an $\mathscr{S}(\mathbb{R}^n)$ open set. Choose ψ from $B_d(\varphi, r)$, the d-ball of radius $r > 0$ and centered at a point (function) $\varphi \in \mathscr{S}(\mathbb{R}^n)$. Choose $N \gg 1$ so that

$$\sum_{j=N+1}^{\infty} \frac{1}{2^j} < r_0 \equiv 2^{-1}(r - d(\varphi, \psi)).$$

Suppose that a point (function) $\eta \in \mathscr{S}(\mathbb{R}^n)$ satisfies $p_N(\eta - \psi) < r_0$. For all $a, b, c > 0$, note that the inequality $\min(a + b, c) \le \min(a, c) + \min(b, c)$ holds. Hence

$$\begin{aligned}
d(\varphi, \eta) &= \sum_{j=1}^{N} \frac{1}{2^j} \min(p_j(\varphi - \eta), 1) + \sum_{j=N+1}^{\infty} \frac{1}{2^j} \min(p_j(\varphi - \eta), 1) \\
&\le \sum_{j=1}^{N} \frac{1}{2^j} \min(p_j(\varphi - \psi) + p_j(\eta - \psi), 1) + \sum_{j=N+1}^{\infty} \frac{1}{2^j} \\
&\le \sum_{j=1}^{N} \frac{1}{2^j} \min(p_j(\varphi - \psi), 1) + \sum_{j=1}^{N} \frac{1}{2^j} \min(p_j(\psi - \eta), 1) + \sum_{j=N+1}^{\infty} \frac{1}{2^j} \\
&< d(\varphi, \psi) + r_0 + r_0 = d(\varphi, \psi) + 2r_0 = r.
\end{aligned}$$

This implies that $d(\varphi,\eta) < r_0$ whenever $2p_N(\varphi-\eta) < d(\varphi,\psi)$. Thus, for all $\psi \in B_d(\varphi,r)$, there exist $N \gg 1$ and $r_0 = 2^{-1}(r - d(\varphi,\psi))$ such that

$$\{\eta \in \mathscr{S}(\mathbb{R}^n) \,:\, p_N(\psi-\varphi) < r_0\} \subset B_d(\varphi,r).$$

Hence any d-ball is an open set with respect to the original topology.

Any open set is a d-open set. From Proposition 1.8, we see that

$$A \equiv \{\psi \in \mathscr{S}(\mathbb{R}^n) \,:\, N\,p_N(\varphi-\psi) < 1\} \quad (\varphi \in \mathscr{S}(\mathbb{R}^n), N \in \mathbb{N}) \tag{1.60}$$

is an open base of the original $\mathscr{S}(\mathbb{R}^n)$-topology. It suffices to show that the set given by (1.60) is a d-open set.

In order to show that A is made up of interior points, we choose a function $\psi \in A$ arbitrarily, so that $p_N(\varphi-\psi) \le N^{-1}$. Let $r \equiv 4^{-N-1}(1 - N\,p_N(\varphi-\psi)) > 0$. Then we claim $B(\psi,r) \subset A$. Indeed, we let $\eta \in B(\psi,r)$. Then (1.59) yields

$$2^{-N}p_N(\psi-\eta) < d(\psi,\eta) < r = 4^{-N-1}(1 - N\,p_N(\varphi-\psi)).$$

Thus, $p_N(\psi-\eta) \le N \cdot 2^{-N-2}(1 - N\,p_N(\varphi-\psi)) \le 2^{-1}(1 - N\,p_N(\varphi-\psi))$. If we combine this with the triangle inequality for p_N, we obtain

$$N\,p_N(\varphi-\eta) < \frac{1}{2}(1 - N\,p_N(\varphi-\psi)) + N\,p_N(\varphi-\psi) = \frac{1 + N\,p_N(\varphi-\psi)}{2} < 1.$$

Consequently, we have $\eta \in A$. Hence being made up of interior points, A is a d-open set.

In functional analysis we learned that the closed unit ball is not compact in the norm topology. However, in our setting the family $p_N, N \in \mathbb{N}$ is nested. So, the "closed ball" can become compact as the following theorem shows:

Theorem 1.20 *Let A be a bounded set in $\mathscr{S}(\mathbb{R}^n)$; that is, $a_N \equiv \sup_{f\in A} p_N(f) < \infty$ for all $N \in \mathbb{N}$. Then A is a relatively compact set.*

Proof Since $\mathscr{S}(\mathbb{R}^n)$ is metrizable, we have only to show that any sequence $\{f_j\}_{j=1}^\infty$ in A has a convergent subsequence. Let $N \in \mathbb{N}$ be fixed. Since $a_{N+1} < \infty$, we can use the Ascoli–Arzelá theorem to have a subsequence convergent from $\{f_j\}_{j=1}^\infty$ in the closure of $\mathscr{S}(\mathbb{R}^n)$ with respect to p_N. Cantor's diagonal argument yields a subsequence convergent in $\mathscr{S}(\mathbb{R}^n)$. Thus, A is relatively compact.

Exercises

Exercise 1.30 With the triangle inequality $|x+y| \le |x| + |y|, x, y \in \mathbb{R}^n$ in mind, prove Peetre's inequality.

Exercise 1.31 Show that $C_c^\infty(\mathbb{R}^n) \subsetneq \mathscr{S}(\mathbb{R}^n)$ from Example 1.4.

Exercise 1.32

1. Show that $\mathrm{BUC}(\mathbb{R}^n)$ is a closed set in $\mathrm{BC}(\mathbb{R}^n)$. Hint: Write down the uniform continuity first.
2. If f is in $\mathrm{BUC}(\mathbb{R}^n)$, then show that there exists a sequence $\{f_k\}_{k=1}^\infty$ in $C_c^\infty(\mathbb{R}^n)$ convergent to f in $L^\infty(\mathbb{R}^n)$. Hint: Use the mollifier.
3. Show that $\overline{\mathscr{S}(\mathbb{R}^n)}^{L^\infty(\mathbb{R}^n)}$, the $L^\infty(\mathbb{R}^n)$-closure of $\mathscr{S}(\mathbb{R}^n)$, is given by

$$X = \left\{\varphi \in \mathrm{BUC}(\mathbb{R}^n) \, : \, \lim_{|x|\to\infty} \varphi(x) = 0\right\}.$$

Hint: Show first that X is closed.

Exercise 1.33 Let $\varphi \in \mathscr{S}(\mathbb{R}^n)$. Denote by $\mathbf{e}_j \equiv (\delta_{jk})_{k=1}^n$ the j-th elementary vector.

1. For all $N \in \mathbb{N}$, prove that $\sup\limits_{y\in[0,1]^n} p_N(\varphi_y) < \infty$ using the compactness of $[0,1]$.
2. Prove that $\lim\limits_{t\to 0} \frac{1}{t}(\varphi - \varphi_{t\,\mathbf{e}_j}) = \partial_{x_j}\varphi$ in the topology of $\mathscr{S}(\mathbb{R}^n)$. Hint: First assume $t \in [-1, 1]$. Then use

$$\begin{aligned}\varphi(x) - \varphi(x - t\mathbf{e}_j) - t\partial_{x_j}\varphi(x) &= \int_0^t \left(\partial_{x_j}\varphi(x - s\mathbf{e}_j) - \partial_{x_j}\varphi(x)\right) \mathrm{d}s \\ &= -\int_0^t \left(\int_0^s \partial_{x_j}^2\varphi(x - v\mathbf{e}_j)\mathrm{d}v\right) \mathrm{d}s\end{aligned}$$

and use the Peetre inequality, keeping in mind $|t| \le 1$.

Exercise 1.34 Show that the function d given in (1.59) is a distance function by showing the following:

1. (Positivity) For all $\varphi, \psi \in \mathscr{S}(\mathbb{R}^n)$, $0 \le d(\varphi, \psi) < \infty$ and the left equality holds if and only if $\varphi \equiv \psi$.
2. (Symmetry) For all $\varphi, \psi \in \mathscr{S}(\mathbb{R}^n)$, $d(\varphi, \psi) = d(\psi, \varphi)$.
3. (Triangle inequality) For all $\varphi, \psi, \eta \in \mathscr{S}(\mathbb{R}^n)$, $d(\varphi, \eta) \le d(\varphi, \psi) + d(\psi, \eta)$. Hint: For $a, b \ge 0$, can we prove $\min(a+b, 1) \le \min(a, 1) + \min(b, 1)$?

Exercise 1.35 An advantage of the metric is that we can characterize continuity in terms of the convergence of sequence. Let (X, d) be a metric space. Show that a mapping $f : X \to \mathbb{C}$ is continuous if and only if $\lim\limits_{j\to\infty} f(x_j) = f\left(\lim\limits_{j\to\infty} x_j\right)$ for any convergent sequence $\{x_j\}_{j=1}^\infty$ in X.

Exercise 1.36 Let $m, n \in \mathbb{N}$. For $f \in \mathscr{S}(\mathbb{R}^{n+m})$, show that $g \in \mathscr{S}(\mathbb{R}^n)$, where $g : \mathbb{R}^n \to \mathbb{C}$ is defined by setting:

$$g(x) \equiv \int_{\mathbb{R}^m} f(x, y)\mathrm{d}y \quad (x \in \mathbb{R}^n).$$

Hint: Note that $\|(x, y)\| = \sqrt{|x|^2 + |y|^2}$ is the norm on $\mathbb{R}^{n+m}$, which can be used to control the information on $|x|$ and $|y|$. It may be helpful to use $\|(x, y)\| \geq \max(|x|, |y|)$.

Exercise 1.37 **[85, Section 5.1.3]** Let $\gamma(t) \equiv \dfrac{e\chi_{[1,\infty)}(t)}{\pi t}\Im[\exp(-e^{\pi i/4}(t-1)^{\frac{1}{4}})]$ for $t \in \mathbb{R}$. Then show that $\gamma \in \mathscr{S}(\mathbb{R})$ and that $\displaystyle\int_{\mathbb{R}} t^k \gamma(t)\mathrm{d}t = \delta_{k0}$. This function is called the *Stein function*.

1.2.2 *Definition of the Schwartz Distribution Space $\mathscr{S}'(\mathbb{R}^n)$ and Its Topology*

We feel that the elements in $\mathscr{S}(\mathbb{R}^n)$ are good, or rather, too good because the functions are infinitely differentiable and the partial derivatives decay rapidly. However, these functions are too good to accommodate functions we usually consider. In this book, by considering its topological dual $\mathscr{S}'(\mathbb{R}^n)$, we study the property of the functions.

1.2.2.1 Schwartz Distribution Space $\mathscr{S}'(\mathbb{R}^n)$

The distributions are closely related to physics. In fact, the most closely connected distributions are the Dirac delta. In the late 1919s, Dirac deduced $\dfrac{\mathrm{d}}{\mathrm{d}x}\log x = \dfrac{1}{x} - i\pi\delta(x)$. We note that δ is a function defined on $\mathbb{R}$ satisfying $\delta(0) = \infty$, $\displaystyle\int_{-\infty}^{\infty} \delta(x)\,\mathrm{d}x = 1$, or more generally for all continuous functions f defined on the real line, $\displaystyle\int_{-\infty}^{\infty} f(x)\delta(x)\,\mathrm{d}x = f(0)$. The useful distribution δ is called the Dirac distribution. But this is not a usual function. Thus, we need a wider framework to include this Dirac delta. The spaces $\mathscr{S}'(\mathbb{R}^n)$ and $\mathscr{D}'(\mathbb{R}^n)$ are such attempts. As we will define just below, the space $\mathscr{S}'(\mathbb{R}^n)$ denotes the topological dual of $\mathscr{S}(\mathbb{R}^n)$; that is, the set of all continuous linear functionals. In this book we propose a way of "measuring" the functions in $\mathscr{S}'(\mathbb{R}^n)$ and discuss their properties. In the sequel we denote by $\mathrm{Hom}_{\mathbb{C}}(V, W)$ the set of all $\mathbb{C}$-linear mappings from a complex linear space V to a complex linear space W.

Definition 1.11 (Schwartz distribution (space)) Denote by $\mathscr{S}'(\mathbb{R}^n)$ the set of all continuous linear mappings from $\mathscr{S}(\mathbb{R}^n)$ to $\mathbb{C}$. That is,

$$\mathscr{S}'(\mathbb{R}^n) \equiv \{f \in \mathrm{Hom}_{\mathbb{C}}(\mathscr{S}(\mathbb{R}^n), \mathbb{C}) \,:\, f \text{ is continuous}\}.$$

Denote by $\langle f, \varphi \rangle$ the value of φ evaluated at f; $\langle f, \varphi \rangle \equiv f(\varphi)$.

We do not have to stick to complex vector spaces; analogously we sometimes define $\mathscr{S}(\mathbb{R}^n)$ as a linear space and its dual.

In general, the dual of "small sets" is large. As the following example shows, $\mathscr{S}'(\mathbb{R}^n)$ contains elements other than "functions" and it is indeed very large:

Example 1.6 (Dirac distribution, Dirac delta) For $a \in \mathbb{R}^n$, the evaluation mapping $\delta_a : \varphi \in \mathscr{S}(\mathbb{R}^n) \mapsto \varphi(a) \in \mathbb{C}$ belongs to $\mathscr{S}'(\mathbb{R}^n)$. In fact, since $| \langle \delta_a, \varphi \rangle | \le p_0(\varphi)$, the continuity from $\mathscr{S}(\mathbb{R}^n)$ to $\mathbb{C}$ is ensured. The distribution δ_a is important in physics and, in particular, when $a = 0$, we write $\delta_a = \delta$.The distribution δ_a is called the Dirac delta (massed) at a.

The following characterization of $\mathscr{S}'(\mathbb{R}^n)$ is concrete:

Theorem 1.21 (Characterization of $\mathscr{S}'(\mathbb{R}^n)$) *Let F be a $\mathbb{C}$-linear mapping from $\mathscr{S}(\mathbb{R}^n)$ to $\mathbb{C}$. Then $F \in \mathscr{S}'(\mathbb{R}^n)$ if and only if there exists $N \in \mathbb{N}$ such that*

$$|F(\varphi)| \le N\, p_N(\varphi) \tag{1.61}$$

for all $\varphi \in \mathscr{S}(\mathbb{R}^n)$.

Since we need to guarantee that F is continuous, we prefer to use $F(\varphi)$ instead of $\langle F, \varphi \rangle$.

Proof Condition (1.61) being trivially sufficient for $F \in \mathscr{S}'(\mathbb{R}^n)$, we will prove that there exists $N \in \mathbb{N}$ satisfying (1.61) for $F \in \mathscr{S}'(\mathbb{R}^n)$. Suppose that a linear mapping $F : \mathscr{S}(\mathbb{R}^n) \to \mathbb{C}$ is continuous; our task is to find $N \in \mathbb{N}$ such that (1.61) holds. Let $\Delta(1) \equiv \{z \in \mathbb{C} \,:\, |z| < 1\}$. By the continuity of F, the set

$$F^{-1}(\Delta(1)) = \{\varphi \in \mathscr{S}(\mathbb{R}^n) \,:\, |F(\varphi)| < 1\}$$

is an open set of $\mathscr{S}(\mathbb{R}^n)$ that contains 0.

Therefore, if $L \in \mathbb{N}$ is sufficiently large, then

$$\{\varphi \in \mathscr{S}(\mathbb{R}^n) \,:\, L\, p_L(\varphi) < 1\} \subset \{\varphi \in \mathscr{S}(\mathbb{R}^n) \,:\, |F(\varphi)| < 1\} \tag{1.62}$$

from Proposition 1.8. Thus, $|F(\varphi)| \le 1$ if $\varphi \in \mathscr{S}(\mathbb{R}^n)$ satisfies $2L\, p_{2L}(\varphi) = 1$.

Now we suppose that we have $\varphi \in \mathscr{S}(\mathbb{R}^n) \setminus \{0\}$. Then $\psi \equiv \dfrac{\varphi}{2L\, p_{2L}(\varphi)}$ satisfies $2L\, p_{2L}(\psi) = 1$. Thus, $|F(\psi)| \le 1$. In view of the definition of ψ, we have $|F(\varphi)| \le 2L\, p_{2L}(\varphi), \varphi \in \mathscr{S}(\mathbb{R}^n) \setminus \{0\}$. The case where $\varphi = 0$ can be readily incorporated. Thus, by letting $N = 2L$, we can choose N satisfying (1.61). So, we conclude (1.61) is also a necessary condition of the continuity of F.

1.2.2.2 Regular Elements in $\mathscr{S}'(\mathbb{R}^n)$

The space $L^1_{\rm loc}(\mathbb{R}^n)$ is the set of all the measurable functions such that $\chi_K \cdot f \in L^1(\mathbb{R}^n)$ for all compact sets K. By Hölder's inequality (Theorem 1.4), for all $1 \le p < \infty$, we have $L^p(\mathbb{R}^n) \subset L^1_{\rm loc}(\mathbb{R}^n)$. Since $L^1_{\rm loc}(\mathbb{R}^n)$ is made up of functions, the elements are sometimes easier to handle than those in $\mathscr{S}'(\mathbb{R}^n)$.

Definition 1.12 (Regular elements in $\mathscr{S}'(\mathbb{R}^n)$) A locally integrable function $f \in L^1_{\rm loc}(\mathbb{R}^n)$ is said to belong to $\mathscr{S}'(\mathbb{R}^n)$ or is *regular*, if the mapping

$$\varphi \in C^\infty_{\rm c}(\mathbb{R}^n) \mapsto \int_{\mathbb{R}^n} f(x)\varphi(x){\rm d}x \in \mathbb{C}$$

extends to a bounded linear functional on $\mathscr{S}(\mathbb{R}^n)$. In this case, we identify f with the following bounded linear functional $F_f \in \mathscr{S}'(\mathbb{R}^n)$:

$$\langle F_f, \varphi \rangle = \lim_{j\to\infty} \int_{\mathbb{R}^n} f(x)\varphi_j(x){\rm d}x. \tag{1.63}$$

Here $\{\varphi_j\}_{j=1}^\infty$ is any sequence of $C^\infty_{\rm c}(\mathbb{R}^n)$-functions convergent to $\varphi \in \mathscr{S}(\mathbb{R}^n)$ in the topology of $\mathscr{S}(\mathbb{R}^n)$ as $j \to \infty$. The set of all locally integrable functions is denoted by $\mathscr{S}'(\mathbb{R}^n) \cap L^1_{\rm loc}(\mathbb{R}^n) = L^1_{\rm loc}(\mathbb{R}^n) \cap \mathscr{S}'(\mathbb{R}^n)$. The element F_f is called the *distribution associated with f*.

We remark that the correspondence $f \mapsto F_f$ is injective thanks to the Lebesgue differentiation theorem.

Example 1.7

1. Denote by $\mathscr{P} = \mathscr{P}(\mathbb{R}^n)$ the set of all polynomials. Then using the mapping $f \mapsto F_f$ we can regard $\mathscr{P}(\mathbb{R}^n)$ as the subset of $\mathscr{S}'(\mathbb{R}^n)$. Denote by $\mathscr{P}_d(\mathbb{R}^n)$ the set of all polynomial functions with degree less than or equal to d, so that $\mathscr{P}(\mathbb{R}^n) \equiv \bigcup_{d=0}^\infty \mathscr{P}_d(\mathbb{R}^n)$. It is understood that $\mathscr{P}_{-1}(\mathbb{R}^n) = \{0\}$.
2. With ease we can check that the Dirac delta in Example 1.6 is not regular.

Despite Proposition 1.3, we do not have $L^p(\mathbb{R}^n) \hookrightarrow \mathscr{S}'(\mathbb{R}^n)$ for all $p \in (0, 1)$. In fact, we cannot define the coupling with $\mathscr{S}(\mathbb{R}^n)$ and any compactly supported nonnegative function which does not belong to $L^1(\mathbb{R}^n)$.

In the light of Theorem 1.22, we can say that a function $f \in L^1_{\rm loc}(\mathbb{R}^n)$ is a regular element in $\mathscr{S}'(\mathbb{R}^n)$ if and only if the limit $\lim\limits_{j\to\infty}\displaystyle\int_{\mathbb{R}^n} f(x)\varphi_j(x)\eta(x){\rm d}x$ exists, where φ is a prescribed $C^\infty_{\rm c}(\mathbb{R}^n)$-function such that $\chi_{B(1)} \le \varphi \le \chi_{B(2)}$ and $\varphi_j(x) = \varphi(2^{-j}\star)$ for all $j \in \mathbb{N}$.

Example 1.8 Let $1 \le p \le \infty$. Define the (harmonic) conjugate p' of p by

$$p' \equiv \begin{cases} \infty & (p=1), \\ \dfrac{p}{p-1} & (1 < p < \infty), \\ 1 & (p=\infty). \end{cases}$$

For $f \in L^p(\mathbb{R}^n)$, the integral $\displaystyle\int_{\mathbb{R}^n} \varphi(x)f(x){\rm d}x$ converges absolutely and

$$\left|\int_{\mathbb{R}^n} \varphi(x)f(x){\rm d}x\right| \le \int_{\mathbb{R}^n} |\varphi(x)f(x)|\,{\rm d}x \le \|f\|_p\|\varphi\|_{p'}. \tag{1.64}$$

We define $F_f \in \mathscr{S}'(\mathbb{R}^n)$ by

$$F_f(\varphi) = \langle F_f, \varphi\rangle \equiv \int_{\mathbb{R}^n} \varphi(x)f(x){\rm d}x \quad (\varphi \in \mathscr{S}(\mathbb{R}^n)).$$

We can prove that the mapping $f \in L^p(\mathbb{R}^n) \mapsto F_f \in \mathscr{S}'(\mathbb{R}^n)$ is injective, which together with (1.64) shows that $L^p(\mathbb{R}^n) \hookrightarrow \mathscr{S}'(\mathbb{R}^n)$; see Exercise 1.45.

Proof Let N be a sufficiently large integer, more precisely, let $N \equiv \left[\dfrac{n+1}{p'}\right] + 1$. Choose a test function $\varphi \in C^\infty_{\rm c}(\mathbb{R}^n)$. Then Hölder's inequality (Theorem 1.4) yields

$$|\langle F_f, \varphi\rangle| = \left|\int_{\mathbb{R}^n} f(x)\varphi(x){\rm d}x\right| \le \|f\|_p\|\varphi\|_{p'} \lesssim \|f\|_p p_N(\varphi).$$

Thus, we have $L^p(\mathbb{R}^n) \hookrightarrow \mathscr{S}'(\mathbb{R}^n)$.

Usually we identify f with F_f, so that $L^p(\mathbb{R}^n)$ is a subset of $\mathscr{S}'(\mathbb{R}^n)$ whenever $1 \le p \le \infty$.

1.2.2.3 $C^\infty(\mathbb{R}^n)$-Functions That Have at Most Polynomial Growth at Infinity

As more typical examples of F_f in Definition 1.12, we consider $C^\infty(\mathbb{R}^n)$-functions that have at most polynomial growth at infinity.

Definition 1.13 ($C^\infty(\mathbb{R}^n)$-function that has at most polynomial growth at infinity) A function $h \in C^\infty(\mathbb{R}^n)$ is said to *have at most polynomial growth at infinity*, if for all $\alpha \in \mathbb{N}_0{}^n$, there exists $N_\alpha > 0$ such that:

$$|\partial^\alpha h(x)| \lesssim_\alpha \langle x \rangle^{N_\alpha}, \quad x \in \mathbb{R}^n. \tag{1.65}$$

Denote by $\mathcal{O}_M(\mathbb{R}^n)$ the set of all $C^\infty(\mathbb{R}^n)$-functions that have at most polynomial growth at infinity.

As is seen from conditions (1.55) and (1.65), the sign of the power is reversed. As the example of $f(x) = \sin(|x|^{2m})$ shows, the number N_α can vary according to α. Keep in mind that the growth of the partial derivatives obeys a condition.

From the Leibniz rule, we have:

Lemma 1.14 *Let $h \in C^\infty(\mathbb{R}^n)$ have at most polynomial growth at infinity and let $\tau \in \mathcal{S}(\mathbb{R}^n)$. Then the pointwise product $h \cdot \tau$ belongs to $\mathcal{S}(\mathbb{R}^n)$. Furthermore, for each $C^\infty(\mathbb{R}^n)$-function that has at most polynomial growth at infinity h, the mapping $\tau \in \mathcal{S}(\mathbb{R}^n) \mapsto h \cdot \tau \in \mathcal{S}(\mathbb{R}^n)$ is continuous.*

We leave the proof to interested readers as Exercise 1.41. Motivated by Lemma 1.14 and Definition 1.12, we defined the *pointwise product of distributions and the functions in $\mathcal{O}_M(\mathbb{R}^n)$*.

Definition 1.14 (Product of distributions and the functions in $\mathcal{O}_M(\mathbb{R}^n)$) For $h \in \mathcal{O}_M(\mathbb{R}^n)$ and for $f \in \mathcal{S}'(\mathbb{R}^n)$, define the *pointwise product $h \cdot f \in \mathcal{S}'(\mathbb{R}^n)$* by $\langle h \cdot f, \varphi \rangle \equiv \langle f, h \cdot \varphi \rangle$ for $\varphi \in \mathcal{S}(\mathbb{R}^n)$.

We should keep in mind that $h \cdot \varphi \in \mathcal{S}(\mathbb{R}^n)$ if $\varphi \in \mathcal{S}(\mathbb{R}^n)$ and $h \in \mathcal{O}_M(\mathbb{R}^n)$.

Example 1.9 For $f \in \mathcal{S}'(\mathbb{R})$ and $m \in \mathbb{N}$, $a \in \mathbb{R}$, the elements x^m, $\sin x \cdot f$, $\langle x \rangle^a f$ make sense as elements in $\mathcal{S}'(\mathbb{R})$.

1.2.2.4 Topology of $\mathcal{S}'(\mathbb{R}^n)$

Equip the topology of $\mathcal{S}'(\mathbb{R}^n)$ with the weak-* topology.

Definition 1.15 (Topology of $\mathcal{S}'(\mathbb{R}^n)$) The topology of $\mathcal{S}'(\mathbb{R}^n)$ is the weakest topology among those such that the coupling $f \in \mathcal{S}'(\mathbb{R}^n) \mapsto \langle f, \varphi \rangle \in \mathbb{C}$ is continuous for all $\varphi \in \mathcal{S}(\mathbb{R}^n)$. Or equivalently, set

$$\mathcal{U}_{f,\varphi,\varepsilon} \equiv \{g \in \mathcal{S}'(\mathbb{R}^n) \,:\, |\langle f - g, \varphi \rangle| < \varepsilon\}, \quad (f \in \mathcal{S}'(\mathbb{R}^n), \varepsilon > 0, \varphi \in \mathcal{S}(\mathbb{R}^n))$$

and equip $\mathcal{S}'(\mathbb{R}^n)$ with the topology generated by $\{\mathcal{U}_{f,\varphi,\varepsilon}\}_{f \in \mathcal{S}'(\mathbb{R}^n), \varepsilon > 0, \varphi \in \mathcal{S}(\mathbb{R}^n)}$.

In short, $\{\mathcal{U}_{f,\varphi,\varepsilon}\}_{f \in \mathcal{S}'(\mathbb{R}^n), \varepsilon > 0, \varphi \in \mathcal{S}(\mathbb{R}^n)}$ plays the role of sub-basis of the topology. According to the definition of the topology, in $\mathcal{S}'(\mathbb{R}^n)$ an open set is a set U such that for each $f \in U$ there are $N = N_f$ elements $f_1, f_2, \ldots, f_N \in \mathscr{S}'(\mathbb{R}^n)$, $\varphi_1, \varphi_2, \ldots, \varphi_N \in \mathscr{S}(\mathbb{R}^n)$ and $\varepsilon_1, \varepsilon_2, \ldots, \varepsilon_N > 0$ such that $f \in \bigcap_{j=1}^{N} \mathscr{U}_{f_j,\varphi_j,\varepsilon_j}$.

Let $\{f_j\}_{j=1}^{\infty}$ be a sequence of $\mathscr{S}'(\mathbb{R}^n)$ which converges to $f \in \mathscr{S}'(\mathbb{R}^n)$. Then by definition of the topology $\langle f_j, \varphi\rangle \to \langle f, \varphi\rangle$ as $j \to \infty$ for all $\varphi \in \mathscr{S}(\mathbb{R}^n)$. But it matters that the converse is also true in the following sense: The next theorem is an important property of Schwartz distributions, which is akin to the uniformly bounded principle.

Theorem 1.22 (Weak sequential completeness of $\mathscr{S}'(\mathbb{R}^n)$) *Assume $\{F_j\}_{j=1}^{\infty} \subset \mathscr{S}'(\mathbb{R}^n)$ satisfies the following condition: For all $\varphi \in \mathscr{S}(\mathbb{R}^n)$, the limit $\lim_{j\to\infty} \langle F_j, \varphi\rangle$ exists for $\varphi \in \mathscr{S}(\mathbb{R}^n)$. Let $F : \varphi \mapsto \lim_{j\to\infty} \langle F_j, \varphi\rangle$. Then:*

1. *$F \in \mathscr{S}'(\mathbb{R}^n)$,*
2. *$F_j \to F$ in the topology of $\mathscr{S}'(\mathbb{R}^n)$ as $j \to \infty$,*
3. *there exists $N \in \mathbb{N}$ such that*

$$|\langle F_j, \varphi\rangle| \le N\, p_N(\varphi) \tag{1.66}$$

for all $\varphi \in \mathscr{S}(\mathbb{R}^n)$ and $j \in \mathbb{N}$.

Proof We note the heart of the matter is to find N satisfying (1.66), keeping in mind that $\mathscr{S}(\mathbb{R}^n)$ is a metric space. Other assertions follow immediately once (1.66) is shown. Define an increasing sequence $\{A_j\}_{j=1}^{\infty}$ of the subsets in $\mathscr{S}(\mathbb{R}^n)$ by

$$A_j \equiv \bigcap_{k=j}^{\infty} \left\{\varphi \in \mathscr{S}(\mathbb{R}^n) \,:\, |\langle F_j, \varphi\rangle - \langle F_k, \varphi\rangle| \le 1\right\}. \tag{1.67}$$

Then each A_j is a closed set, since $A_j = \bigcap_{k=j}^{\infty} G_{jk}{}^{-1}(\overline{\Delta}(1))$ for $G_{jk} \equiv F_j - F_k$ and a closed set $\overline{\Delta}(1) \equiv \{z \in \mathbb{C} \,:\, |z| \le 1\}$. Note that $\{\langle F_j, \varphi\rangle\}_{j=1}^{\infty}$ is a Cauchy sequence for any $\varphi \in \mathscr{S}(\mathbb{R}^n)$. A direct consequence of this fact is $\mathscr{S}(\mathbb{R}^n) = \bigcup_{j=1}^{\infty} A_j$. From the definition of A_j we have a symmetry;

$$\varphi \in A_j \iff -\varphi \in A_j. \tag{1.68}$$

The Baire category theorem shows that A_j has φ_0 as an interior point φ_0 as long as the nonnegative integer j is sufficiently large. Fix such a j.

Then by the definition of the interior point and the characterization of open sets in $\mathscr{S}(\mathbb{R}^n)$ (see Proposition 1.8), we can find $N \in \mathbb{N}$ so that

$$\{\varphi \in \mathscr{S}(\mathbb{R}^n) \,:\, N\, p_N(\varphi - \varphi_0) < 1\} \subset A_j. \tag{1.69}$$

Due to the symmetry (1.68), we have:

$$\{\varphi \in \mathscr{S}(\mathbb{R}^n) \,:\, N\, p_N(\varphi + \varphi_0) < 1\} \subset A_j. \tag{1.70}$$

Furthermore, according to the definition (1.67) of A_j, A_j is convex. Hence by virtue of (1.69) and (1.70), A_j contains the Minkovski sum of two neighborhoods of φ_0 and $-\varphi_0 \in \mathscr{S}'(\mathbb{R}^n)$:

$$\frac{1}{2}\{\varphi \in \mathscr{S}(\mathbb{R}^n) \,:\, N\, p_N(\varphi - \varphi_0) < 1\} + \frac{1}{2}\{\varphi \in \mathscr{S}(\mathbb{R}^n) \,:\, N\, p_N(\varphi + \varphi_0) < 1\} \subset A_j. \tag{1.71}$$

Here observe that:

$$\frac{1}{2}\{\varphi \in \mathscr{S}(\mathbb{R}^n) \,:\, N\, p_N(\varphi \pm \varphi_0) < 1\} = \mp\frac{1}{2}\varphi_0 + \{\varphi \in \mathscr{S}(\mathbb{R}^n) \,:\, 2N\, p_N(\varphi) < 1\}. \tag{1.72}$$

See Exercise 1.38. Hence

$$\begin{aligned} &\{\varphi \in \mathscr{S}(\mathbb{R}^n) \,:\, N\, p_N(\varphi) < 1\} \\ &= \frac{1}{2}\{\varphi \in \mathscr{S}(\mathbb{R}^n) \,:\, N\, p_N(\varphi - \varphi_0) < 1\} + \frac{1}{2}\{\varphi \in \mathscr{S}(\mathbb{R}^n) \,:\, N\, p_N(\varphi + \varphi_0) < 1\}, \end{aligned}$$

which yields

$$\{\varphi \in \mathscr{S}(\mathbb{R}^n) \,:\, N\, p_N(\varphi) < 1\} \subset A_j. \tag{1.73}$$

Equating this relation, we obtain $\big|\langle F_j, \varphi\rangle - \langle F_k, \varphi\rangle\big| \leq 1$ for all $\varphi \in \mathscr{S}(\mathbb{R}^n)$ and $k \geq j$ provided $N\, p_N(\varphi) = \dfrac{1}{2}$. A passage to the limit gives that $F - F_j$ is a continuous functional. Thus, it follows that F itself is continuous. Taking into account a finite number of functionals $F_1, F_2, \ldots, F_{j-1}$ and choosing N large enough again, we obtain the desired integer N.

We next want to differentiate elements in $\mathscr{S}'(\mathbb{R}^n)$. To this end for $f \in \mathscr{S}'(\mathbb{R}^n)$ and $\alpha \in \mathbb{N}_0{}^n$, we need to define $\partial^\alpha f(\varphi) = \langle \partial^\alpha, \varphi\rangle$ for any $\varphi \in \mathscr{S}(\mathbb{R}^n)$.

Definition 1.16 (Differentiation in $\mathscr{S}'(\mathbb{R}^n)$) For $f \in \mathscr{S}'(\mathbb{R}^n)$ and a multi-index $\alpha \in \mathbb{N}_0{}^n$, define $\partial^\alpha f \in \mathscr{S}'(\mathbb{R}^n)$ so that $\langle \partial^\alpha f, \varphi\rangle \equiv (-1)^{|\alpha|}\langle f, \partial^\alpha \varphi\rangle$ holds for all $\varphi \in \mathscr{S}(\mathbb{R}^n)$.

The element $\partial^\alpha f : \mathscr{S} \to \mathbb{C}$ is trivially $\mathbb{C}$-linear. It is not difficult to show that this linear mapping $\partial^\alpha f$ is continuous using Theorem 1.21. What matters about this definition is that we do not have to stick to the smoothness of f although we are considering the derivatives! One of the motivations of the above definition is the integration by parts formula: $\int_{\mathbb{R}^n} \partial_{x_j} f(x) \cdot g(x) \mathrm{d}x = -\int_{\mathbb{R}^n} f(x) \partial_{x_j} g(x) \mathrm{d}x$ for all $f, g \in \mathscr{S}(\mathbb{R}^n)$ and $j = 1, 2, \ldots, n$, so that $F_{\partial_{x_j} f} = \partial_{x_j} F_f$; see Exercise 1.50 for more. If $f \in \mathscr{S}'(\mathbb{R}^n)$ and $g \in \mathscr{O}_M(\mathbb{R}^n)$, then we can easily check $\partial_{x_j}(g \cdot f) = (\partial_{x_j} g) \cdot f + g \cdot \partial_{x_j} f$.

The differential of distributions differs largely from the usual one in that we can freely change the order of differentiation and the limit in $\mathscr{S}'(\mathbb{R}^n)$.

Theorem 1.23 (The order of the differentiation and the limit in $\mathscr{S}'(\mathbb{R}^n)$) *Let $\alpha, \beta \in {\mathbb{N}_0}^n$, and let $f \in \mathscr{S}'(\mathbb{R}^n)$.*

1. *If a sequence $\{f_j\}_{j=1}^\infty$ in $\mathscr{S}'(\mathbb{R}^n)$ converges to f, then $\lim\limits_{j \to \infty} \partial^\alpha f_j = \partial^\alpha f$.*
2. *$\partial^\alpha(\partial^\beta f) = \partial^\beta(\partial^\alpha f) = \partial^{\alpha+\beta} f$.*

We omit its simple proof; see Exercise 1.42 1.

Example 1.10 Denote by $H^1(\mathbb{R}^n)$ the set of all $L^2(\mathbb{R}^n)$-functions for which $\partial_{x_j} f \in L^2(\mathbb{R}^n)$. The norm of $f \in H^1(\mathbb{R}^n)$ is given by $\|f\|_{H^1} = \|f\|_2 + \sum_{j=1}^n \|\partial_{x_j} f\|_2$.

Define the *domain* Dom(∇) of ∇ by

$$\mathrm{Dom}(\nabla) \equiv \left\{ u \in L^2(\mathbb{R}^n) : \frac{\partial u}{\partial x_j} \in L^2(\mathbb{R}^n), \quad j = 1, 2, \ldots, n \right\},$$

and

$$\nabla u \equiv \left(\frac{\partial u}{\partial x_j} \right)_{j=1,2,\ldots,n} \in L^2(\mathbb{R}^n)^n.$$

Note that Dom(∇) $= H^1(\mathbb{R}^n)$. Since $\mathscr{S}(\mathbb{R}^n)$ is densely contained in $H^1(\mathbb{R}^n)$, ∇ is densely defined. Define div $\equiv -D^*$. We can easily check that $H^1(\mathbb{R}^n)^n \subset$ Dom(div).

We define $H^1(Q) \equiv \{f \in L^2(Q) : f = g|Q$ for some $g \in H^1(\mathbb{R}^n)\}$ for a cube Q. The following example shows the relation between the integral, $\|h\|_2$ and $\|\nabla h\|_{(L^2)^n}$.

Example 1.11 For $h \in H^1(Q)$, $\left| \int_Q \nabla h(x) \mathrm{d}x \right| \lesssim \ell(Q)^{\frac{n-1}{2}} \sqrt{\|h\|_{L^2(Q)} \|\nabla h\|_{L^2(Q)}}$. In fact, a translation and a dilation allow us to assume that Q is a unit cube. If $\|h\|_{L^2(Q)} \le \|\nabla h\|_{L^2(Q)}$ then the conclusion is clear. Thus, assume otherwise. Let $t \in (0, 1)$. Choose $\varphi \in C_c^\infty(Q)$ so that $\chi_{(1-t)Q} \le \varphi \le \chi_Q$ and that $\|\nabla \varphi\|_{(L^\infty)^n} \lesssim t^{-1}$. Then

$$\int_Q \nabla h(x) \mathrm{d}x = \int_Q (1 - \varphi(x)) \nabla h(x) \mathrm{d}x - \int_Q \varphi(x) \nabla h(x) \mathrm{d}x.$$

Thus, by the Cauchy–Schwartz inequality we have

$$\left|\int_Q \nabla h(x)\mathrm{d}x\right| \lesssim t^{\frac{1}{2}}\left(\int_Q |\nabla h(x)|^2\mathrm{d}x\right)^{\frac{1}{2}} + t^{-\frac{1}{2}}\left(\int_Q |h(x)|^2\mathrm{d}x\right)^{\frac{1}{2}}.$$

Although this inequality is proved for $t \in (0, 1)$, we can still incorporate the case $t \geq 1$ by the Hölder inequality. If we choose t suitably, we obtain the desired result.

1.2.2.5 Dense Subspace

We now consider a dense subspace in $\mathscr{S}(\mathbb{R}^n)$. Recall that for $j \in \mathbb{Z}$, $\Theta \in \mathscr{S}(\mathbb{R}^n)$, Θ_j is given by (26), see p. 5: $\Theta_j \equiv \Theta(2^{-j}\star)$.

Lemma 1.15 (Dyadic decomposition) *There exists a couple* $(\psi, \varphi) \in C_c^\infty(B(4)) \times C_c^\infty(B(8) \setminus B(2))$ *of nonnegative functions satisfying* $\psi + \sum_{j=1}^{\infty} \varphi_j \equiv 1$.

Although the proof is simple, since we need to use this construction many times, we exhibit how to construct ψ and φ. We leave the proof to the readers; for details see Exercise 1.44.

Proof Choose a $C^\infty(\mathbb{R}^n)$-function $\eta : \mathbb{R} \to [0, \infty)$ so that

$$\chi_{[-2,2]} \leq \eta \leq \chi_{[-3,3]}. \tag{1.74}$$

Then define $\psi, \varphi : \mathbb{R}^n \to [0, \infty)$ by $\psi \equiv \eta(|\star|)$, $\varphi \equiv \eta(|\star|) - \eta(2|\star|)$. A simple computation shows that $\psi, \varphi \in C_c^\infty(\mathbb{R}^n)$ enjoy the desired properties.

We frequently use the following result: The proof is simple but, for the sake of completeness, we supply the proof.

Theorem 1.24 *Let* ψ *and* φ *be functions in Lemma* 1.15. *Then for all* $\eta \in \mathscr{S}(\mathbb{R}^n)$,

$$\lim_{J\to\infty}\left(\psi + \sum_{j=1}^{J} \varphi_j\right) \cdot \eta = \eta \tag{1.75}$$

holds in the topology of $\mathscr{S}(\mathbb{R}^n)$. *In particular,* $C_c^\infty(\mathbb{R}^n)$ *is dense in* $\mathscr{S}(\mathbb{R}^n)$.

We have a "standard" way of approximate elements in $\mathscr{S}(\mathbb{R}^n)$ with those in $C_c^\infty(\mathbb{R}^n)$.

Proof Let $N \in \mathbb{N}$ be arbitrary. Our task is to show

$$\lim_{J\to\infty} p_N\left(\eta-\left(\psi+\sum_{j=1}^{J}\varphi_j\right)\cdot\eta\right)$$
$$=\lim_{J\to\infty}\left(\sup_{x\in\mathbb{R}^n}\sup_{|\alpha|\le N}\langle x\rangle^N\left|\partial^\alpha\left(\eta(x)-\psi(x)\eta(x)-\sum_{j=1}^{J}\varphi_j(x)\eta(x)\right)\right|\right)=0.$$

Let $x\in\mathbb{R}^n$. Note that

$$\eta(x)-\psi(x)\eta(x)-\sum_{j=1}^{J}\varphi_j(x)\eta(x)=\sum_{j=J+1}^{\infty}\varphi_j(x)\eta(x)=(1-\varphi(2^{-J}x))\eta(x),$$
$$\langle x\rangle^{N+1}|\partial^\alpha\eta(x)|\lesssim p_{N+1}(\eta)$$

for all $\alpha\in{\mathbb{N}_0}^n$ with length less than or equal to $N+1$. For such α we deduce from the size of support and the Leibniz rule that

$$\left|\partial^\alpha\left(\sum_{j=J+1}^{\infty}\varphi_j(x)\cdot\eta(x)\right)\right|\lesssim 2^{-J}p_{N+1}(\eta)\chi_{\mathbb{R}^n\setminus B(2^J)}(x)\,\langle x\rangle^{-N}.$$

Thus, the proof of (1.75) is complete.

In the proof we have actually shown that

$$p_N\left(\eta-\left(\psi+\sum_{j=1}^{J}\varphi_j\right)\cdot\eta\right)\lesssim 2^{-J}p_{N+1}(\eta).$$

Note that the semi-norms in both sides are different.

A direct corollary is that $C_c^\infty(\mathbb{R}^n)$ is dense in $\mathscr{S}(\mathbb{R}^n)$.

1.2.2.6 Support of Distributions

The support of functions is defined to be the closure on which the functions do not vanish. Defining the support of functions is possible because we can evaluate the functions. Since we cannot evaluate distributions, we need a trick to define the support of distributions; duality enables us this. Distributions having compact support are important because the Fourier transform converts smooth functions. In fact, this fact will be a starting point when we define our function spaces. In some sense, we are faced with the evaluation of distributions in the definition of the function spaces.

Definition 1.17 (Support of distributions) The *support of* $f \in \mathscr{S}'(\mathbb{R}^n)$ is the set of all points $x \in \mathbb{R}^n$ such that for all $r > 0$, there exists $\varphi \in \mathscr{S}(\mathbb{R}^n)$ such that $\text{supp}(\varphi) \subset B(x, r)$ and that $\langle f, \varphi \rangle \neq 0$. The support of $f \in \mathscr{S}'(\mathbb{R}^n)$ is denoted by $\text{supp}(f)$.

So the definition of the support is indirect like the one for the connectedness in topological spaces; we need to consider those points NOT in the support first. In fact, x is not in $\text{supp}(f)$ if there exists $r > 0$ such that $\langle f, \varphi \rangle = 0$ for all $\varphi \in \mathscr{S}(\mathbb{R}^n)$ supported in $B(x, r)$.

Example 1.12 We have $\text{supp}(\delta) = \{0\}$, where δ denotes the Dirac delta massed at the origin.

A passage to the complement shows that $\text{supp}(f)$ is a closed set. We justify something that is clear intuitively.

Proposition 1.10 *For $f \in \mathscr{S}'(\mathbb{R}^n)$ and $\varphi \in \mathscr{S}(\mathbb{R}^n)$ with disjoint support, $\langle f, \varphi \rangle =$ 0.*

Proof Choose a function $\eta \in \mathscr{S}(\mathbb{R}^n)$ so that $\chi_{B(1)} \leq \eta \leq \chi_{B(2)}$. According to (26), define $\eta_j \equiv \eta(2^{-j}\star)$ for $j \in \mathbb{N}$. Then from Theorem 1.24 we have $\lim\limits_{j\to\infty} \eta_j \cdot \varphi = \varphi$ in the $\mathscr{S}(\mathbb{R}^n)$-topology. Hence it can be assumed that φ is compactly supported.

Take $x \in \text{supp}(\varphi)$. Then by assumption, we have $x \notin \text{supp}(f)$. From the definition of $\text{supp}(f)$, there exists an open ball B_x whose double $2B_x$ does not intersect $\text{supp}(f)$ and which satisfies $\langle f, \psi \rangle = 0$ for all $\psi \in \mathscr{S}(\mathbb{R}^n)$ supported in B_x. Since $\text{supp}(\varphi)$ is compact, from the open covering $\{B_x\}_{x\in\text{supp}(\varphi)}$ of $\text{supp}(\varphi)$, we can find a finite sub-covering $B_1, B_2, \ldots, B_J$ of balls. Then $2B_x \cap \text{supp}(f) = \emptyset$, $x \in \text{supp}(\varphi)$ implies $\langle f, \psi \rangle = 0$ provided $\psi \in \mathscr{S}(\mathbb{R}^n)$ is supported on $2B_j$ for some $j = 1, 2, \ldots, J$.

We fix $\tau^{(j)} \in C^\infty_{\rm c}(\mathbb{R}^n)$ so that $\chi_{B_j} \leq \tau^{(j)} \leq \chi_{2B_j}$. Next, we define a function Θ by $\Theta \equiv \sum\limits_{k=1}^{J} \tau^{(k)} + \prod\limits_{k=1}^{J}(1-\tau^{(k)})$. In addition, we define $\kappa^{(j)} \equiv \Theta^{-1}\tau^{(j)} \in C^\infty_{\rm c}(\mathbb{R}^n)$ for $j = 1, 2, \ldots, J$. Since $\Theta > 0$, $\kappa^{(j)}$ is a smooth function. Also, on $\text{supp}(\varphi)$, since $\prod\limits_{j=1}^{J}(1-\tau^{(j)}) = 0$, we have $\sum\limits_{j=1}^{J} \kappa^{(j)} \equiv 1$. Hence $\langle f, \varphi \rangle = \sum\limits_{j=1}^{J} \langle f, \kappa^{(j)} \cdot \varphi \rangle = 0$. This is the desired result.

The next important theorem characterizes distributions supported only on a point: Recall that $\delta = \delta_0$ is the distribution given in Example 1.6 and it is clearly supported in $\{0\}$.

Theorem 1.25 (Characterization of distributions supported in one point) *A distribution $f \in \mathscr{S}'(\mathbb{R}^n)$ is zero or has support in a singleton $\{0\}$ if and only if there is a finite collection $\{a_\alpha\}_{\alpha\in\mathbb{N}_0{}^n, |\alpha|\leq L}$ of complex coefficients such that* $f = \sum\limits_{\alpha\in\mathbb{N}_0{}^n, |\alpha|\leq L} a_\alpha \, \partial^\alpha \delta$.

Proof Any distribution of the form $f = \sum_{\alpha \in \mathbb{N}_0{}^n, |\alpha| \le L} a_\alpha \, \partial^\alpha \delta$ is clearly supported in $\{0\}$. So, we concentrate on the proof of necessity. By virtue of Theorem 1.21, there exists a large integer N such that, for all $\varphi \in \mathscr{S}(\mathbb{R}^n)$,

$$|\langle f, \varphi \rangle| \le N \, p_N(\varphi). \tag{1.76}$$

Next, we set $E^\alpha(x) \equiv x^\alpha e^{-|x|^2}$, $\alpha \in \mathbb{N}_0{}^n$ for $x \in \mathbb{R}^n$ and then we define a set of coefficients $\{k_\alpha\}_{\alpha \in \mathbb{N}_0{}^n}$ independent of $\varphi \in \mathscr{S}(\mathbb{R}^n)$ so that $k_\alpha = 0$ for $|\alpha| \ge N+2$ and that

$$\varphi(x) - \sum_{\alpha \in \mathbb{N}_0{}^n} k_\alpha \cdot \partial^\alpha \varphi(0) \cdot E^\alpha(x) = \mathrm{O}(|x|^{N+2}) \tag{1.77}$$

as $x \to 0$; see Exercise 1.46. Also, we choose $\kappa \in \mathscr{S}(\mathbb{R}^n)$ so that $\chi_{B(1)} \le \kappa \le \chi_{B(2)}$. Then

$$\lim_{j \to \infty} p_N \left(\left(\varphi - \sum_{\alpha \in \mathbb{N}_0{}^n} k_\alpha \cdot \partial^\alpha \varphi(0) \cdot E^\alpha \right) (1 - \kappa(2^j \star)) \right) = 0 \tag{1.78}$$

thanks to (1.76). Therefore, since $\mathrm{supp}(f) \subset \{0\}$ and

$$\langle f, \varphi \rangle = \left\langle f, \sum_{\alpha \in \mathbb{N}_0{}^n} k_\alpha \cdot \partial^\alpha \varphi(0) \cdot E^\alpha \right\rangle + \left\langle f, \varphi - \sum_{\alpha \in \mathbb{N}_0{}^n} k_\alpha \cdot \partial^\alpha \varphi(0) \cdot E^\alpha \right\rangle,$$

we have

$$\begin{aligned}
\langle f, \varphi \rangle &= \left\langle f, \sum_{\alpha \in \mathbb{N}_0{}^n} k_\alpha \cdot \partial^\alpha \varphi(0) \cdot E^\alpha \right\rangle \\
&\quad + \lim_{j \to \infty} \left\langle f, \left(\varphi - \sum_{\alpha \in \mathbb{N}_0{}^n} k_\alpha \cdot \partial^\alpha \varphi(0) \cdot E^\alpha \right) \kappa(2^j \star) \right\rangle \\
&= \sum_{\alpha \in \mathbb{N}_0{}^n} \langle f, E^\alpha \rangle k_\alpha \partial^\alpha \varphi(0) = \left\langle \sum_{\alpha \in \mathbb{N}_0{}^n} (-1)^{|\alpha|} \langle f, E^\alpha \rangle k_\alpha \, \partial^\alpha \delta, \varphi \right\rangle
\end{aligned}$$

by (1.78).

Write $a_\alpha \equiv (-1)^{|\alpha|} \langle f, E^\alpha \rangle k_\alpha$ for each α and $L \equiv N+1$. Since $\varphi \in \mathscr{S}(\mathbb{R}^n)$ is chosen arbitrarily, we conclude that $f = \sum_{\alpha \in \mathbb{N}_0{}^n, |\alpha| \le L} a_\alpha \, \partial^\alpha \delta$ is the desired finite sum. Thus, the necessity is proved.

1.2.2.7 Convolution with Elements in $\mathscr{S}(\mathbb{R}^n)$ and $\mathscr{S}'(\mathbb{R}^n)$

Here we consider the convolution $f * g$ for $f \in \mathscr{S}(\mathbb{R}^n)$ and $g \in \mathscr{S}'(\mathbb{R}^n)$. If $g \in \mathscr{S}(\mathbb{R}^n)$, then

$$f * g(x) \equiv \int_{\mathbb{R}^n} f(x-y)g(y)\mathrm{d}y.$$

If $g \in \mathscr{S}'(\mathbb{R}^n)$, then the right-hand side does not make sense; we cannot integrate the element in $\mathscr{S}'(\mathbb{R}^n)$. So we begin with the simplest case where $f, g \in \mathscr{S}(\mathbb{R}^n)$.

Theorem 1.26 (Algebraic property of $*$ in $\mathscr{S}(\mathbb{R}^n)$) *The Schwartz space $\mathscr{S}(\mathbb{R}^n)$ is closed under the convolution $*$. That is, $f * g \in \mathscr{S}(\mathbb{R}^n)$ for $f, g \in \mathscr{S}(\mathbb{R}^n)$.*

Proof It is easy to check that $f * g \in C^\infty(\mathbb{R}^n)$. We omit the further details. To check that a $C^\infty(\mathbb{R}^n)$-function $f * g$ belongs to $\mathscr{S}(\mathbb{R}^n)$, we need to show that

$$p_N(f * g) = \sum_{\alpha \in \mathbb{N}_0{}^n,\ |\alpha| \le N} \left(\sup_{x \in \mathbb{R}^n} \langle x \rangle^N |\partial^\alpha [f * g](x)| \right)$$

is finite, where p_N is the semi-norm defined in (1.55). Choose $\alpha \in \mathbb{N}_0{}^n$ and $N \in \mathbb{N}$ so that $|\alpha| \le N$. Then

$$\begin{aligned} p_N(f * g) &= \sup_{x \in \mathbb{R}^n} \langle x \rangle^N \left| \int_{\mathbb{R}^n} \partial^\alpha f(x-y) \cdot g(y)\mathrm{d}y \right| \\ &\le \sup_{x \in \mathbb{R}^n} \langle x \rangle^N \int_{\mathbb{R}^n} |\partial^\alpha f(x-y)| \cdot |g(y)|\mathrm{d}y \\ &\le 2^N \sup_{x \in \mathbb{R}^n} \int_{\mathbb{R}^n} \langle x-y \rangle^N |\partial^\alpha f(x-y)| \cdot \langle y \rangle^N |g(y)|\mathrm{d}y \qquad (1.79) \\ &\le 2^N \sup_{x \in \mathbb{R}^n} \int_{\mathbb{R}^n} p_N(f) p_{N+n+1}(g) \langle y \rangle^{-n-1} \mathrm{d}y \simeq p_N(f) p_{N+n+1}(g) < \infty. \end{aligned}$$

We used Peetre's inequality to obtain the inequality (1.79). Thus, we conclude $f * g \in \mathscr{S}(\mathbb{R}^n)$.

Since the convolution is defined via an integral, it is plausible that the convolution can be expressed as the limit of the Riemannian sum. The next lemma shows that this is true. Although the formulation is easy, the proof is a burden.

Lemma 1.16 *Let $f, g \in \mathscr{S}(\mathbb{R}^n)$. Then in the topology of $\mathscr{S}(\mathbb{R}^n)$, we have*

$$\lim_{N \to \infty} \frac{1}{N^n} \sum_{m \in \mathbb{Z}^n} f\left(\star - \frac{m}{N}\right) g\left(\frac{m}{N}\right) = f * g.$$

A tacit understanding is that

$$\sum_{m\in\mathbb{Z}^n} f\left(\star-\frac{m}{N}\right)g\left(\frac{m}{N}\right) \in \mathscr{S}(\mathbb{R}^n). \tag{1.80}$$

We leave the proof to interested readers; see Exercise 1.47.

Proof Let $\alpha \in {\mathbb{N}_0}^n$, and let $N \in \mathbb{N}_0$ be fixed. It suffices to prove:

$$\lim_{N\to\infty}\left(\sup_{x\in\mathbb{R}^n}\langle x\rangle^m\left|\frac{1}{N^n}\sum_{m\in\mathbb{Z}^n}\partial^\alpha f\left(x-\frac{m}{N}\right)g\left(\frac{m}{N}\right)-(\partial^\alpha f)*g(x)\right|\right)=0. \tag{1.81}$$

Since $\partial^\alpha f \in \mathscr{S}(\mathbb{R}^n)$, we can assume that $\alpha = 0$.

We write $Q_N^{(m)} \equiv \left\{\frac{m}{N}+x : x\in\left[0,\frac{1}{N}\right]^n\right\} = \frac{m}{N}+\left[0,\frac{1}{N}\right]^n$. Then

$$\begin{aligned}
&f*g(x)-\frac{1}{N^n}\sum_{m\in\mathbb{Z}^n}f\left(x-\frac{m}{N}\right)g\left(\frac{m}{N}\right)\\
&=\sum_{m\in\mathbb{Z}^n}\int_{Q_N^{(m)}}\left(f(x-y)g(y)-f\left(x-\frac{m}{N}\right)g\left(\frac{m}{N}\right)\right)\mathrm{d}y\\
&=\sum_{m\in\mathbb{Z}^n}\int_{Q_N^{(m)}}\left\{\int_0^1\frac{\mathrm{d}}{\mathrm{d}t}\left(f\left(x-t\left(y-\frac{m}{N}\right)-\frac{m}{N}\right)g\left(t\left(y-\frac{m}{N}\right)+\frac{m}{N}\right)\right)\mathrm{d}t\right\}\mathrm{d}y
\end{aligned}$$

by the fundamental theorem of calculus. By the chain rule and Peetre's inequality, for $t \in [0, 1]$ and $y \in Q_N^{(m)}$, we have

$$\left|\langle x\rangle^m\frac{\mathrm{d}}{\mathrm{d}t}\left(f\left(x-t\left(y-\frac{m}{N}\right)-\frac{m}{N}\right)g\left(t\left(y-\frac{m}{N}\right)+\frac{m}{N}\right)\right)\right|\lesssim_{f,g}\frac{1}{N}\langle y\rangle^{-n-1}. \tag{1.82}$$

See Exercise 1.40 for details. If we insert (1.82) into the above expression and integrate it over $t \in [0, 1]$,

$$\sup_{x\in\mathbb{R}^n}\langle x\rangle^m\left|f*g(x)-\frac{1}{N^n}\sum_{m\in\mathbb{Z}^n}f\left(x-\frac{m}{N}\right)g\left(\frac{m}{N}\right)\right|\lesssim\frac{1}{N}.$$

Letting $N \to \infty$, we obtain (1.81).

Motivated by the definition of the convolution of two functions in $\mathscr{S}(\mathbb{R}^n)$, we define the convolution of $f \in \mathscr{S}(\mathbb{R}^n)$ and $g \in \mathscr{S}'(\mathbb{R}^n)$, as follows.

Definition 1.18 (Convolution of $f \in \mathscr{S}(\mathbb{R}^n)$ and $g \in \mathscr{S}'(\mathbb{R}^n)$) For $f \in \mathscr{S}(\mathbb{R}^n)$ and $g \in \mathscr{S}'(\mathbb{R}^n)$, define a **function** $f * g$ by

$$f*g(x)\equiv\langle g, f(x-\star)\rangle, x\in\mathbb{R}^n.$$

The function $f * g$ is called the *convolution of the functions f and g*.

In Definition 1.18, even when $g \in \mathscr{S}'(\mathbb{R}^n)$, we can evaluate $f * g$ at any point in $\mathbb{R}^n$. We summarize the property of this convolution. Recall that $\mathscr{O}_M$ is defined in Definition 1.13.

Theorem 1.27 (Elementary properties of convolution) *Let $f, g \in \mathscr{S}(\mathbb{R}^n)$ and $h \in \mathscr{S}'(\mathbb{R}^n)$.*

1. *The convolution $f * g$ of two $L^1(\mathbb{R}^n)$-functions:*
$$f * g(x) = \int_{\mathbb{R}^n} f(x-y)g(y)\mathrm{d}y$$
and that via Definition 1.18 *coincide. In terms of F_f in Definition* 1.12*, we have*
$$f * F_g(x) = \int_{\mathbb{R}^n} f(x-y)g(y)\mathrm{d}y. \tag{1.83}$$
So, two definitions are consistent.
2. *$f * h \in C^\infty(\mathbb{R}^n)$ and*
$$\partial^\alpha (f * h) = (\partial^\alpha f) * h = f * \partial^\alpha h \tag{1.84}$$
for $\alpha \in \mathbb{N}_0{}^n$.
Also, for all $\alpha \in \mathbb{N}_0{}^n$, we can find $N = N_\alpha \in \mathbb{N}_0$ such that
$$\sup_{x \in \mathbb{R}^n} |\partial^\alpha (f * h)(x)| \lesssim \langle x \rangle^N. \tag{1.85}$$
Thus, in particular, $f * h \in \mathscr{O}_M(\mathbb{R}^n)$; hence $f * h \in \mathscr{S}'(\mathbb{R}^n)$.
3. *The associative law $(f * g) * h = f * (g * h)$ holds.*

What matters in this proposition is that $f * h \in C^\infty(\mathbb{R}^n)$ and we can deal with the convolutions above as if they were for functions.

Proof From the definition of convolution for $f \in L^1(\mathbb{R}^n) \to F_f \in \mathscr{S}'(\mathbb{R}^n)$, (1.83) is clear. The left inequality (1.84) is clear from the definition of differential of distributions. We prove the right inequality (1.84) by mathematical induction. Once we prove the base case:
$$\partial^\alpha (f * h) = f * \partial^\alpha h, \quad |\alpha| = 1, \tag{1.86}$$
we can conclude the proof using (1.86) repeatedly. However, (1.86) is already included in Exercise 1.33 and we omit the details.

To prove (1.85), we fix a positive integer N such that, for all $\varphi \in \mathscr{S}(\mathbb{R}^n)$,
$$|\langle h, \varphi \rangle| \le N\, p_N(\varphi). \tag{1.87}$$

In (1.87) we choose $\varphi = f(x-\star)$. A simple calculation shows $p_N(\partial^\alpha f(x-\star)) \lesssim \langle x\rangle^N$; hence $\sup\limits_{x\in\mathbb{R}^n} |\partial^\alpha (f*h)(x)| \lesssim \langle x\rangle^N$.

Finally, we prove the associative law. Now it is time to apply Lemma 1.16, whose proof is very hard. By Lemma 1.16,

$$(f*g)*h = (g*f)*h \tag{1.88}$$

$$= \left(\lim_{N\to\infty} \frac{1}{N^n} \sum_{j\in\mathbb{Z}^n} g(\star - N^{-1}j) f(N^{-1}j)\right) * h \tag{1.89}$$

$$= \lim_{N\to\infty} \frac{1}{N^n} \sum_{j\in\mathbb{Z}^n} g(\star - N^{-1}j) * h \cdot f(N^{-1}j) \tag{1.90}$$

$$= \lim_{N\to\infty} \frac{1}{N^n} \sum_{j\in\mathbb{Z}^n} g*h(\star - N^{-1}j) \cdot f(N^{-1}j). \tag{1.91}$$

Since (1.85) is already proved, we have

$$(f*g)*h(x) = \int_{\mathbb{R}^n} g*h(x-y)f(y)\mathrm{d}y = (g*h)*f(x) = f*(g*h)(x) \tag{1.92}$$

by the Lebesgue convergence theorem. Therefore, the proof is complete.

In the course of the proof there are many $*$s. In (1.89), (1.90) and (1.91), $*$ stands for the convolution of elements in $\mathscr{S}(\mathbb{R}^n)$ and $\mathscr{S}'(\mathbb{R}^n)$. Meanwhile, in (1.88) and (1.92) we also consider convolution of two elements in $\mathscr{S}(\mathbb{R}^n)$.

1.2.2.8 Various Notions of Distributions

We list two important notions of distributions to conclude this section.

Definition 1.19 (Restricted at infinity, Vanishing weakly, Radial) Let $f \in \mathscr{S}'(\mathbb{R}^n)$.

1. A distribution f is said to be *restricted at infinity* if $\psi * f \in \bigcup\limits_{1\le r<\infty} L^r(\mathbb{R}^n)$ for all $\psi \in \mathscr{S}(\mathbb{R}^n)$.
2. A distribution $f \in \mathscr{S}'(\mathbb{R}^n)$ is said to *vanish weakly at infinity* if $\psi_j * f \to 0$ in $\mathscr{S}'(\mathbb{R}^n)$ as $j \to -\infty$ for all $\psi \in \mathscr{S}(\mathbb{R}^n)$. Denote by $\tilde{C}_0(\mathbb{R}^n)$ the distributions vanishing weakly at infinity.
3. Let $\mathrm{O}(n)$ denote the group of all orthonormal matrices, which acts on $\mathbb{R}^n$. A distribution f is said to be *radial* if it is invariant under the action of $\mathrm{O}(n)$.

The constant function $f \equiv 1 \in \mathscr{S}'(\mathbb{R}^n)$ does not vanish weakly at infinity, nor is it restricted at infinity.

Example 1.13 We present some examples of distributions vanishing weakly at infinity:

1. functions in $L^p(\mathbb{R}^n)$ with $1 \le p < \infty$,
2. derivative of functions in $L^\infty(\mathbb{R}^n)$,
3. derivative of functions in $\tilde{C}_0(\mathbb{R}^n)$.

Exercises

Exercise 1.38 Let $\psi \in \mathscr{S}(\mathbb{R}^n)$. Prove (1.72) by proving that the following are equivalent:

- $\psi \in \frac{1}{2}\{\varphi \in \mathscr{S}(\mathbb{R}^n) : Np_N(\varphi + \varphi_0) < 1\}$.
- There exists $\varphi \in \mathscr{S}(\mathbb{R}^n)$ such that $Np_N(\varphi + \varphi_0) < 1$ and that $\psi = \frac{1}{2}\varphi$.
- There exists $\overline{\varphi} \in \mathscr{S}(\mathbb{R}^n)$ such that $Np_N(\overline{\varphi}) < 1$ and that $\psi = -\frac{1}{2}\varphi_0 + \frac{1}{2}\overline{\varphi}$.
- There exists $\varphi^\dagger \in \mathscr{S}(\mathbb{R}^n)$ such that $Np_N(\varphi^\dagger) < \frac{1}{2}$ and that $\psi = -\frac{1}{2}\varphi_0 + \varphi^\dagger$.
- $\psi \in -\frac{1}{2}\varphi_0 + \{\varphi \in \mathscr{S}(\mathbb{R}^n) : 2Np_N(\varphi) < 1\}$.

Exercise 1.39 Let $f \in \mathscr{O}_M^\infty(\mathbb{R}^n)$ and $g \in \mathscr{S}(\mathbb{R}^n)$. Show that

$$\int_{\mathbb{R}^n} f(x)g(x)\mathrm{d}x = \lim_{N\to\infty} \frac{1}{N^n} \sum_{m\in\mathbb{Z}^n} f\left(\frac{m}{N}\right) g\left(\frac{m}{N}\right).$$

Hint: Reduce matters to the case of $f \equiv 1$. Then write the summation in the right-hand side by the use of an integral.

Exercise 1.40 Prove (1.82) using the Peetre inequality and paying attention to the position of y.

Exercise 1.41 Prove Lemma 1.14. Hint: The conclusion in Lemma 1.14 can be rephrased to the fact that, for all $\alpha, \beta \in \mathbb{N}_0{}^n$ to construct $K \in \mathbb{N}$ satisfying

$$p_{\alpha,\beta}(h \cdot \tau) \lesssim p_K(\tau) \quad (\tau \in \mathscr{S}(\mathbb{R}^n))$$

using (1.65).

Exercise 1.42

1. Prove Theorem 1.23.
2. For $f \in \mathscr{S}'(\mathbb{R}^n)$, show that the differential $\partial_{x_j} f$ extends a classical one naturally; if $\mathbf{e}_j$ is the j-th elementary vector, $\partial_{x_j} f = \lim_{h\to 0} \frac{1}{h}(f(\star + h\mathbf{e}_j) - f)$ in the topology of $\mathscr{S}'(\mathbb{R}^n)$. Hint: Exercise 1.33 is helpful.

Exercise 1.43 Show that a sequence $\{f_j\}_{j=1}^{\infty}$ of points (functions) in $\mathscr{S}'(\mathbb{R}^n)$ convergent to $f \in \mathscr{S}'(\mathbb{R}^n)$ if and only if $\lim_{j\to\infty} \langle f_j, \varphi \rangle = \langle f, \varphi \rangle$ for all $\varphi \in \mathscr{S}(\mathbb{R}^n)$.

Exercise 1.44 Here we verify the details of Lemma 1.15.

1. Construct a smooth function $\eta \in \mathscr{S}(\mathbb{R}^n)$ satisfying (1.74). Hint: How do you construct a nonzero nonnegative function in $C_c^{\infty}(\mathbb{R}^n)$? See [4, Theorem 2.28].
2. Check that the functions in Lemma 1.15 fulfill all the requirements.

Exercise 1.45 Let $1 \le p \le \infty$. Show that the mapping $f \in L^p(\mathbb{R}^n) \mapsto F_f \in \mathscr{S}'(\mathbb{R}^n)$ is injective using the equality condition of the Hölder inequality and the density of $\mathscr{S}(\mathbb{R}^n)$ in $L^{p'}(\mathbb{R}^n)$.

Exercise 1.46 Verify (1.77) using the Taylor expansion.

Exercise 1.47 Prove (1.80). Hint: For g, we solely use $|g(x)| \lesssim_N \langle x \rangle^{-N}$ for any $N \in \mathbb{N}$. Show that the series in the left-hand side converges uniformly first.

Exercise 1.48 [263, p. 46] Use Example 1.13 (repeatedly) and the induction on the degree of the polynomials to show that $\mathscr{P}(\mathbb{R}^n) \cap \tilde{C}_0(\mathbb{R}^n) = \{0\}$.

Exercise 1.49 Write $H \equiv \{z \in \mathbb{C} : \Re(z) > 0\}$. Let $f_z(t) \equiv \max(0, t)^z$ for $t \in \mathbb{R}$ and $z \in \mathbb{C}$ with positive real part. Fix $N \in \mathbb{N}$.

1. Show that $f_z \in \mathscr{S}'(\mathbb{R})$ for all $z \in \mathbb{C}$ with positive real part.
2. Write $U_N \equiv \{z \in \mathbb{C} : |\Re(z)| < N\}$. Using integration by parts, show that

$$z \in H \cap U_N \mapsto (z+1)(z+2)\cdots(z+N)\exp(-z^2)\langle f_z, \varphi \rangle \in \mathbb{C}$$

   extends to a bounded analytic function on U_N, which we write $G_{\star}(\varphi)$.
3. Prove that $G_z : \varphi \in \mathscr{S}(\mathbb{R}) \mapsto G_z(\varphi) \in \mathbb{C}$ is an element in $\mathscr{S}'(\mathbb{R})$ for all $z \in U_N$.
4. Show that $G_{-1} = c_N \delta_0$ for some constant $c_N \neq \{0\}$.

1.2.3 Definition of the Fourier Transform and Its Elementary Properties

1.2.3.1 Fourier Transform

We adopt the following definition of the Fourier transform among many other popular ones:

Definition 1.20 (Fourier transform, inverse Fourier transform) For $f \in L^1(\mathbb{R}^n)$, define the *Fourier transform* and the *inverse Fourier transform* by:

$$\mathscr{F} f(\xi) \equiv (2\pi)^{-\frac{n}{2}} \int_{\mathbb{R}^n} f(x) e^{-ix\cdot\xi} dx, \quad \mathscr{F}^{-1} f(x) \equiv (2\pi)^{-\frac{n}{2}} \int_{\mathbb{R}^n} f(\xi) e^{ix\cdot\xi} d\xi.$$

Some prefer to call the space on which $\mathscr{F} f$ is defined the *Fourier space* and the one on which f itself is defined the *physical space*.

Example 1.14 Let $m \in \mathbb{Z}^n$ and $\xi \in \mathbb{R}^n$. Then

$$\begin{aligned}\mathscr{F}\chi_{Q_{km}}(\xi) &= (2\pi)^{-\frac{n}{2}} \int_{\prod_{j=1}^n [2^{-k}m_j, 2^{-k}(m_j+1))} \exp(-ix\cdot\xi)\mathrm{d}x \\ &= (2\pi)^{-\frac{n}{2}} \prod_{j=1}^n \exp\left(-i\frac{m_j\xi_j}{2^k}\right)\frac{i}{\xi_j}\left(1-\exp\left(-i\frac{\xi_j}{2^k}\right)\right).\end{aligned}$$

Example 1.15 Set $E \equiv \exp\left(-\frac{1}{2}|\star|^2\right)$. Then for $\xi \in \mathbb{R}^n$ by completing the square we obtain

$$\begin{aligned}\mathscr{F}E(\xi) &= (2\pi)^{-\frac{n}{2}} \int_{\mathbb{R}^n} \exp\left(-\frac{1}{2}|x|^2 - ix\cdot\xi\right)\mathrm{d}x \\ &= (2\pi)^{-\frac{n}{2}} \int_{\mathbb{R}^n} \exp\left(-\frac{1}{2}(x+i\xi)\cdot(x+i\xi) - \frac{1}{2}|\xi|^2\right)\mathrm{d}x.\end{aligned}$$

Using the complex line integral, we obtain

$$\mathscr{F}E(\xi) = (2\pi)^{-\frac{n}{2}} \int_{\mathbb{R}^n} \exp\left(-\frac{1}{2}|x|^2 - \frac{1}{2}|\xi|^2\right)\mathrm{d}x = \exp\left(-\frac{1}{2}|\xi|^2\right) = E(\xi).$$

Theorem 1.28 (Transformation of operation by the Fourier transform) *We let $j = 1, 2, \ldots, n$. For $\varphi \in \mathscr{S}(\mathbb{R}^n)$, we have equalities between differentiation and the multiplication of monomials:*

$$\mathscr{F}\left[\frac{\partial\varphi}{\partial x_j}\right](\xi) = i\,\xi_j\mathscr{F}\varphi(\xi), \quad \mathscr{F}[x_j\varphi](\xi) = i\frac{\partial[\mathscr{F}\varphi]}{\partial\xi_j}(\xi).$$

In addition, for all $\alpha, \beta \in \mathbb{N}_0{}^n$, there exist constants $C_{\alpha,\beta} > 0$, $N_{\alpha,\beta} \in \mathbb{N}$ such that the pointwise estimate $|\xi^\alpha\partial^\beta\mathscr{F}\varphi(\xi)| \le C_{\alpha,\beta}\, p_{N_{\alpha,\beta}}(\varphi)$ holds.

In particular, the Fourier transform and its inverse Fourier transform map $\mathscr{S}(\mathbb{R}^n)$ to itself continuously.

The proof is left as Exercise 1.50.

We have the following properties on the $L^1(\mathbb{R}^n)$-functions:

Theorem 1.29 (Riemann–Lebesgue) *Let $f \in L^1(\mathbb{R}^n)$. Then*

$$\|\mathscr{F}f\|_\infty = \|\mathscr{F}^{-1}f\|_\infty \le (2\pi)^{-\frac{n}{2}}\|f\|_1 \tag{1.93}$$

$$\mathscr{F}f, \mathscr{F}^{-1}f \in \mathrm{BUC}(\mathbb{R}^n) \tag{1.94}$$

$$\lim_{|\xi|\to\infty} \mathscr{F}f(\xi) = \lim_{|x|\to\infty} \mathscr{F}^{-1}f(x) = 0. \tag{1.95}$$

Proof Observe that (1.93) follows by definition. To prove (1.94) and (1.95), we note that $\mathrm{BUC}(\mathbb{R}^n)$ is a closed set of $\mathrm{BC}(\mathbb{R}^n)$ and that $\mathscr{S}(\mathbb{R}^n)$ is dense in $L^1(\mathbb{R}^n)$. For $f \in \mathscr{S}(\mathbb{R}^n)$, we proved $\mathscr{F}f \in \mathscr{S}(\mathbb{R}^n)$ in Theorem 1.28. Thus, both (1.94) and (1.95) are clear. For $f \in L^1(\mathbb{R}^n)$, approximate with the functions in $\mathscr{S}(\mathbb{R}^n)$ in the topology of $L^1(\mathbb{R}^n)$.

Next, we investigate the inverse Fourier transform. Although it is simple, the next formula is the key to the inverse Fourier transform.

Theorem 1.30 (Elementary formula of $\mathscr{F}$) *For $\varphi, \psi \in \mathscr{S}(\mathbb{R}^n)$,*

$$\int_{\mathbb{R}^n} \varphi(x)\mathscr{F}\psi(x)\mathrm{d}x = \int_{\mathbb{R}^n} \mathscr{F}\varphi(\xi)\psi(\xi)\mathrm{d}\xi.$$

Proof If we write the both sides out in full, we notice that they coincide with

$$\frac{1}{\sqrt{(2\pi)^n}} \iint_{\mathbb{R}^n\times\mathbb{R}^n} \varphi(x)\psi(\xi)\mathrm{e}^{-i\,x\cdot\xi}\mathrm{d}x\mathrm{d}\xi$$

by virtue of Fubini's theorem; see Theorem 1.3.

Using Theorem 1.30 let us prove that $\mathscr{F}$ and $\mathscr{F}^{-1}$ are inverse to each other.

Theorem 1.31 (Inverse formula of the Fourier transform) *For all $\varphi \in \mathscr{S}(\mathbb{R}^n)$ $\mathscr{F}^{-1}[\mathscr{F}\varphi] = \mathscr{F}[\mathscr{F}^{-1}\varphi] = \varphi$.*

Proof The proof $\mathscr{F}[\mathscr{F}^{-1}\varphi] = \varphi$ being the same, we concentrate on the proof of $\mathscr{F}^{-1}[\mathscr{F}\varphi] = \varphi$. Then by Theorem 1.30,

$$\int_{\mathbb{R}^n} \mathscr{F}E_t(\xi)\varphi(\xi)\mathrm{d}\xi = \int_{\mathbb{R}^n} \mathscr{F}\varphi(x)E_t(x)\mathrm{d}x. \tag{1.96}$$

The Fourier transform of the Gauss kernel $E = \exp\left(-\frac{|\star|^2}{2}\right)$ is again the Gauss kernel E according to Example 1.15. Set $E_t \equiv E(t\star)$ for $t > 0$. A scaling yields

$$\mathscr{F}E_t = \frac{1}{(2\pi)^{\frac{n}{2}}t^n}E\left(\frac{\star}{t}\right). \tag{1.97}$$

If we insert this into (1.96), Then we obtain

$$\frac{1}{\sqrt{(2\pi)^n}}\int_{\mathbb{R}^n} t^{-n}E\left(\frac{\xi}{t}\right)\varphi(\xi)\mathrm{d}\xi = \int_{\mathbb{R}^n} \mathscr{F}\varphi(x)E(tx)\mathrm{d}x. \tag{1.98}$$

Subtracting $\varphi(0)$ from both sides, we have

$$\frac{1}{\sqrt{(2\pi)^n t^n}}\int_{\mathbb{R}^n} E\left(\frac{\xi}{t}\right)\varphi(\xi)\mathrm{d}\xi - \varphi(0) = \frac{1}{\sqrt{(2\pi)^n t^n}}\int_{\mathbb{R}^n} E\left(\frac{\xi}{t}\right)(\varphi(\xi)-\varphi(0))\mathrm{d}\xi$$
$$= \frac{1}{\sqrt{(2\pi)^n}}\int_{\mathbb{R}^n} E(\xi)(\varphi(t\xi)-\varphi(0))\mathrm{d}\xi.$$

By Lebesgue's convergence theorem, this converges to 0 as $t \downarrow 0$. Hence

$$\varphi(0) = \frac{1}{\sqrt{(2\pi)^n}}\int \mathscr{F}\varphi(\xi)\mathrm{d}\xi. \tag{1.99}$$

Note that (1.99) is an inversion formula for $x = 0$.

We use (1.99) to complete the proof of Theorem 1.31. Let $x \in \mathbb{R}^n$. First of all, by the translation and (1.99) we have $\varphi(x) = \dfrac{1}{\sqrt{(2\pi)^n}}\displaystyle\int_{\mathbb{R}^n} \mathscr{F}[\,\varphi(\star + x)\,](\xi)\mathrm{d}\xi$.

Using $\mathscr{F}[\,\varphi(\star + x)\,](\xi) = \mathscr{F}\varphi(\xi)\mathrm{e}^{ix\cdot\xi}$, we obtain

$$\varphi(x) = \frac{1}{\sqrt{(2\pi)^n}}\int_{\mathbb{R}^n} \mathscr{F}\varphi(\xi)\mathrm{e}^{ix\cdot\xi}\mathrm{d}\xi = \mathscr{F}^{-1}[\mathscr{F}\varphi](x),$$

which yields the inversion formula.

1.2.3.2 Fourier-Transform of $\mathscr{S}'(\mathbb{R}^n)$

Based upon Theorem 1.30, we define the Fourier transform for Schwartz distributions.

Definition 1.21 (Fourier transform for Schwartz distributions) Let $f \in \mathscr{S}'(\mathbb{R}^n)$. The *(distributional) Fourier transform* $\mathscr{F}f \in \mathscr{S}'(\mathbb{R}^n)$ is defined so that it satisfies

$$\langle \mathscr{F}f, \varphi\rangle \equiv \langle f, \mathscr{F}\varphi\rangle, \quad (\varphi \in \mathscr{S}(\mathbb{R}^n)).$$

Note that if $f \in L^1(\mathbb{R}^n)$, then this is compatible with F_f in Definition 1.12. By definition it is clear that the Fourier transform is an isomorphism from $\mathscr{S}'(\mathbb{R}^n)$ to itself.

Example 1.16 ([81, p. 253]*)* For $\alpha \in {\mathbb{N}_0}^n$, we calculate the Fourier transform $\mathscr{F}p_\alpha = \mathscr{F}[F_{p_\alpha}]$ of $p_\alpha(x) \equiv x^\alpha$ as an element in $\mathscr{S}'(\mathbb{R}^n)$, where F_{p_α} is defined in Definition 1.12 with f replaced by p_α. Let $\varphi \in \mathbb{R}^n$ be a test function and calculate $\langle \mathscr{F}p_\alpha, \varphi\rangle = \langle p_\alpha, \mathscr{F}\varphi\rangle$. If we write out the definition of p_α as a distribution, we have

$$\langle \mathscr{F}p_\alpha, \varphi\rangle = \int_{\mathbb{R}^n} x^\alpha \mathscr{F}\varphi(x)\mathrm{e}^{ix\cdot 0}\mathrm{d}x = \frac{\sqrt{(2\pi)^n}}{i^{|\alpha|}}\partial^\alpha[\mathscr{F}^{-1}[\mathscr{F}\varphi]](0) = \frac{\sqrt{(2\pi)^n}}{i^{|\alpha|}}\partial^\alpha\varphi(0).$$

Thus, $\mathscr{F} p_\alpha = i^{-|\alpha|}\sqrt{(2\pi)^n}\partial^\alpha \delta$.

1.2.3.3 Fourier-Transform of $L^2(\mathbb{R}^n)$: Plancherel Theorem

Recall now that a linear operator U on a Hilbert space $\mathscr{H}$ is said to be unitary, if U is surjective and $\|U x\|_{\mathscr{H}} = \|x\|_{\mathscr{H}}$ for all $x \in \mathscr{H}$.

We now prove Plancherel's theorem, asserting that the Fourier transform extends to an $L^2(\mathbb{R}^n)$-unitary transform. The next formula is a corollary of Theorems 1.30 and 1.31.

Theorem 1.32 (Plancherel) *Let $\varphi, \psi \in \mathscr{S}(\mathbb{R}^n)$. Then*

$$\int_{\mathbb{R}^n} \mathscr{F}\varphi(\xi)\overline{\mathscr{F}\psi(\xi)}\mathrm{d}\xi = \int_{\mathbb{R}^n} \varphi(x)\overline{\psi(x)}\mathrm{d}x.$$

Theorem 1.32 allows us to restrict the Fourier transform and its inverse, which are initially defined on $\mathscr{S}'(\mathbb{R}^n)(\supset L^2(\mathbb{R}^n))$, to have a unitary operator on $L^2(\mathbb{R}^n)$. This is also called *Plancherel's theorem.*

Proof Insert $\overline{\mathscr{F}\psi}$ to ψ in the right-hand side of the above formula. Observe that the Fourier transform and the conjugate operation are related by $\mathscr{F}[\overline{\mathscr{F}\psi}] = \mathscr{F}\mathscr{F}^{-1}[\overline{\psi}] = \overline{\psi}$, which together with Theorem 1.30, yields the desired result.

Recall that we defined $\mathscr{D}_k$ by Definition 1.1. We consider how smooth the indicator functions are by considering E_k defined below.

Example 1.17 Let $k \in \mathbb{Z}$, and let $0 < \alpha < 1/2$. Define the k-th dyadic average of $f \in L^1_{\mathrm{loc}}(\mathbb{R}^n)$ by

$$E_k f \equiv \sum_{Q \in \mathscr{D}_k} m_Q(f)\chi_Q. \tag{1.100}$$

Then we have $\| |\star|^\alpha \mathscr{F}[E_k f]\|_2 \lesssim 2^{k\alpha}\|f\|_2$ for $f \in L^2(\mathbb{R}^n)$.

Proof Let $m \in \mathbb{Z}^n$ and $\xi \in \mathbb{R}^n$. Thus, if we set

$$\Psi_k(\xi) \equiv |2^{-k}\xi|^\alpha \prod_{j=1}^n \frac{2^k}{\xi_j}\left(1 - \exp\left(-i\frac{\xi_j}{2^k}\right)\right) \quad (\xi \in \mathbb{R}^n),$$

then we have

$$\mathscr{F}[E_k f](\xi) = 2^{k(\alpha-n)} i^n \Psi_k(\xi) \sum_{m \in \mathbb{Z}^n} (2\pi)^{-\frac{n}{2}} \prod_{j=1}^n \exp\left(-i\frac{m_j \xi_j}{2^k}\right) m_{Q_{km}}(f)$$

by Example 1.14. Thus,

$$
\begin{aligned}
&\| |\star|^\alpha \mathscr{F}[E_k f] \|_2{}^2 \\
&\simeq 2^{2k(\alpha-n)} \int_{\mathbb{R}^n} |\Psi_0(\xi)|^2 \left| \sum_{m\in\mathbb{Z}^n} \prod_{j=1}^n \exp\left(-i\frac{m_j\xi_j}{2^k}\right) m_{Q_{km}}(f) \right|^2 d\xi \\
&\simeq 2^{2k(\alpha-n)} \sum_{l\in\mathbb{Z}^n} \int_{2^k(l+[0,2\pi]^n)} \langle l\rangle^{2\alpha-2} \left| \sum_{m\in\mathbb{Z}^n} \prod_{j=1}^n \exp\left(-i\frac{m_j\xi_j}{2^k}\right) m_{Q_{km}}(f) \right|^2 d\xi.
\end{aligned}
$$

In view of the periodicity and the Hölder inequality,

$$
\begin{aligned}
\| |\star|^\alpha \mathscr{F}[E_k f] \|_2{}^2 &\simeq 2^{2k(\alpha-n)} \int_{[0,2^{k+1}\pi]^n} \left| \sum_{m\in\mathbb{Z}^n} \prod_{j=1}^n \exp\left(-i\frac{m_j\xi_j}{2^k}\right) m_{Q_{km}}(f) \right|^2 d\xi \\
&\simeq 2^{2k(\alpha-n)} \sum_{m\in\mathbb{Z}^n} 2^{kn} |m_{Q_{km}}(f)|^2 \\
&\le (2^{k\alpha} \|f\|_2)^2,
\end{aligned}
$$

as was to be shown.

1.2.3.4 Convolution and the Fourier Transform

By Fubini's theorem, we obtain

$$
\mathscr{F}[\varphi * f] = (2\pi)^{\frac{n}{2}} \mathscr{F}\varphi \cdot \mathscr{F} f \tag{1.101}
$$

for $\varphi, f \in \mathscr{S}(\mathbb{R}^n)$. Even when $f \in \mathscr{S}'(\mathbb{R}^n)$, it is easy to prove (1.101). In this book we use Theorem 1.33 below many times. The proof is not trivial; we supply a detailed proof.

Theorem 1.33 (Convolution and the Fourier transform) *If $\varphi \in \mathscr{S}(\mathbb{R}^n)$ and $f \in \mathscr{S}'(\mathbb{R}^n)$, then*

$$
\mathscr{F}[\varphi * f] = (2\pi)^{\frac{n}{2}} \mathscr{F}\varphi \cdot \mathscr{F} f \tag{1.102}
$$

in the sense of $\mathscr{S}'(\mathbb{R}^n)$.

The first thing to observe is that Theorem 1.33 is not the one for functions; it is for distributions.

Proof Choose a test function $\psi \in \mathscr{S}(\mathbb{R}^n)$. Then by the definition of the Fourier transform for $\mathscr{S}'(\mathbb{R}^n)$, we have

$$\langle \mathscr{F}[\varphi * f], \psi \rangle = \langle \varphi * f, \mathscr{F}\psi \rangle.$$

We handle the right-hand side carefully. We define $\eta \equiv \mathscr{F}\psi(-\star)$. Then by Theorem 1.27,

$$\langle \mathscr{F}[\varphi * f], \psi \rangle = \langle \varphi * f, \eta(-\star) \rangle = \eta * (\varphi * f)(0) = (\eta * \varphi) * f(0) = \langle f, \eta * \varphi(-\star) \rangle. \tag{1.103}$$

Now we write out $\eta * \varphi(-x)$ for $x \in \mathbb{R}^n$ in full:

$$\eta * \varphi(-x) = \int_{\mathbb{R}^n} \eta(-x-y)\varphi(y)\mathrm{d}y.$$

Next, we change the variables

$$\eta * \varphi(-x) = \int_{\mathbb{R}^n} \mathscr{F}\psi(x+y)\varphi(y)\mathrm{d}y = \int_{\mathbb{R}^n} \mathscr{F}\psi(y)\varphi(y-x)\mathrm{d}y.$$

Finally, we write the Fourier transform out in full. Here we abbreviate $c_n = (2\pi)^{\frac{n}{2}}$. According to the related definitions,

$$\eta * \varphi(0-x) = \int_{\mathbb{R}^n} \psi(y)\mathscr{F}\varphi(y)\mathrm{e}^{-ix\cdot y}\mathrm{d}y = c_n \mathscr{F}[\psi \cdot \mathscr{F}\varphi](x). \tag{1.104}$$

If we insert (1.104) into (1.103), then

$$\langle \mathscr{F}[\varphi * f], \psi \rangle = c_n \langle f, \mathscr{F}[\psi \cdot \mathscr{F}\varphi] \rangle = c_n \langle \mathscr{F} f, \psi \cdot \mathscr{F}\varphi \rangle = c_n \langle \mathscr{F}\varphi \cdot \mathscr{F} f, \psi \rangle.$$

The function ψ being arbitrary, the proof is complete.

1.2.3.5 Band-Limited Distributions

A band-limited distribution is a distribution whose Fourier transform is compactly supported. The support of the Fourier transform of the distributions is called the *frequency support*. This class of distributions plays an important role in Littlewood–Paley theory and is dealt with in Sect. 3.1, as well as the theory of function spaces we develop because we cannot evaluate distribution. We can evaluate them after we use the Littlewood–Paley operator to make them band-limited.

Definition 1.22 (Band-limited distributions) A *band-limited distribution* is an element in $f \in \mathscr{S}'(\mathbb{R}^n)$ such that $\mathscr{F} f$ is compactly supported.

Example 1.18 The following functions are band-limited.

1. Any element in $\mathscr{D}(\mathbb{R}^n)$ is band-limited when it is embedded into $\mathscr{S}(\mathbb{R}^n)$.
2. If $\Phi \in C_c^\infty(\mathbb{R}^n)$, then $\Psi = \mathscr{F}^{-1}\Phi$ is band-limited.
3. Let $n = 1$ and $f(t) = \mathrm{sinc}(t) \equiv \dfrac{\sin t}{t}$ for $t \in \mathbb{R}$. In fact, consider the Fourier transform of the indicator function of intervals.

The following theorem explains why band-limited distributions are important.

Theorem 1.34 (Properties of band-limited distributions) *Band-limited distributions are represented by $C^\infty(\mathbb{R}^n)$-functions that have at most polynomial growth at infinity. That is, if $f \in \mathscr{S}'(\mathbb{R}^n)$ has the Fourier transform $\mathscr{F}f$ with compact support, then there exists a C^∞-function that has at most polynomial growth at infinity F such that*

$$\langle f, \varphi \rangle = \int_{\mathbb{R}^n} F(x)\varphi(x)\mathrm{d}x \quad (\varphi \in \mathscr{S}(\mathbb{R}^n));$$

namely, $f = F_f$, where F_f is from Definition 1.12.

Proof Let $f \in \mathscr{S}'(\mathbb{R}^n)$ be a band-limited distribution, and let $\psi \in \mathscr{S}(\mathbb{R}^n)$ be a function that assumes 1 on $\mathrm{supp}(\mathscr{F}f)$. Then

$$\langle f, \varphi \rangle = \langle \mathscr{F}f, \mathscr{F}^{-1}\varphi \rangle = \langle \psi \cdot \mathscr{F}f, \mathscr{F}^{-1}\varphi \rangle = \langle \mathscr{F}^{-1}[\psi \cdot \mathscr{F}f], \varphi \rangle$$

for any test function $\varphi \in \mathscr{S}(\mathbb{R}^n)$. Since $\varphi \in \mathscr{S}(\mathbb{R}^n)$ is arbitrary, with the help of Theorem 1.33, we have

$$f = \mathscr{F}^{-1}[\psi \cdot \mathscr{F}f] = (2\pi)^{-\frac{n}{2}}\mathscr{F}^{-1}\psi * f. \tag{1.105}$$

Therefore, the right-hand side being a function a $C^\infty(\mathbb{R}^n)$-function that has at most polynomial growth at infinity thanks to Theorem 1.27, the proof is complete.

Note that F in Theorem 1.34 is unique as is seen from

$$F(x_0)\|\varphi\|_1 = \lim_{r \downarrow 0} r^{-n} \int_{\mathbb{R}^n} F(x)\left|\varphi\left(\frac{x - x_0}{r}\right)\right| \mathrm{d}x$$

for all $\varphi \in C_c^\infty(\mathbb{R}^n)$. By Theorem 1.27, the most right-hand side of (1.105) is a smooth function. Thus, f is represented by a smooth function F, for which we write f again here and below. So the pointwise product $f \cdot g$ makes sense and this function belongs to $\mathscr{O}_M(\mathbb{R}^n)$, when $f, g \in \mathscr{S}'(\mathbb{R}^n)$ are band-limited distributions. Concerning this product, we have the following:

Proposition 1.11 *Let $f, g \in \mathscr{S}'(\mathbb{R}^n)$ be band-limited distributions. Then so is $f \cdot g \in \mathscr{S}'(\mathbb{R}^n)$ and satisfies* $\mathrm{supp}(\mathscr{F}[f \cdot g]) \subset \mathrm{supp}(\mathscr{F}f) + \mathrm{supp}(\mathscr{F}g)$. *Here in the right-hand side, we used the Minkovski sum; see* 29.

Proof When $f \in \mathscr{S}(\mathbb{R}^n)$, we can simply use Theorem 1.33. Choose $\varphi \in \mathscr{S}(\mathbb{R}^n)$ so that $\varphi(0) = 1$ and that $\mathrm{supp}(\mathscr{F}\varphi) \subset B(1)$. It suffices to prove $\xi \notin \mathrm{supp}(\mathscr{F}[f \cdot g])$ for any $\xi \notin \mathrm{supp}(\mathscr{F}f)+\mathrm{supp}(\mathscr{F}g)$. The additive operation $(a,b) \in \mathbb{R}^n \times \mathbb{R}^n \mapsto a+b \in \mathbb{R}^n$ being continuous, we see that $\mathrm{supp}(\mathscr{F}f)+\mathrm{supp}(\mathscr{F}g)$ is compact and hence closed. Hence there exists $r > 0$ such that $B(\xi, 2r) \cap (\mathrm{supp}(\mathscr{F}f)+\mathrm{supp}(\mathscr{F}g)) = \emptyset$. For such an r, as long as $t < r$, we have $\varphi(t\,\star)g \in \mathscr{S}(\mathbb{R}^n)$ and

$$\begin{aligned}\mathscr{F}[\varphi(t\,\star)f \cdot g] &= \mathscr{F}[f \cdot (\varphi(t\,\star)g)]\\ &\simeq_n \mathscr{F}f * [\mathscr{F}[(\varphi(t\,\star)g)]]\\ &\simeq_n \mathscr{F}f * \left(t^{-n}\mathscr{F}\varphi(t^{-1}\star) * \mathscr{F}g\right).\end{aligned}$$

For $\tau \in \mathscr{S}(\mathbb{R}^n)$ having its support in $B(\xi, r)$,

$$\begin{aligned}\langle \mathscr{F}[\varphi(t\,\star)f \cdot g], \tau\rangle &= \int_{\mathbb{R}^n} \mathscr{F}[\varphi(t\,\star)f \cdot g](x)\tau(x)\mathrm{d}x\\ &\simeq_n \frac{1}{t^n}\iiint_{\mathbb{R}^n\times\mathbb{R}^n\times\mathbb{R}^n} \mathscr{F}f(x-y-z)\mathscr{F}\varphi\left(\frac{y}{t}\right)\mathscr{F}g(z)\tau(x)\mathrm{d}x\mathrm{d}y\mathrm{d}z\\ &= 0.\end{aligned}$$

Letting $t \downarrow 0$, we obtain $\lim_{t\downarrow 0}\varphi(t\star)\tau = \tau$ in the topology of $\mathscr{S}'(\mathbb{R}^n)$; see Exercise 1.51. Consequently,

$$0 = \lim_{t\downarrow 0}\langle \mathscr{F}[\varphi(t\,\star)f \cdot g], \tau\rangle = \lim_{t\downarrow 0}\langle f \cdot g, \varphi(t\,\star)\mathscr{F}\tau\rangle = \langle f \cdot g, \mathscr{F}\tau\rangle = \langle \mathscr{F}[f \cdot g], \tau\rangle.$$

Since $\tau \in \mathscr{S}(\mathbb{R}^n)$ is an arbitrary function supported on a ball $B(\xi, r)$, we conclude $\xi \notin \mathrm{supp}(\mathscr{F}[f \cdot g])$.

It remains to show that the frequency support of $f \cdot g$ is compact. In view of Proposition 1.11 the support of $\mathscr{F}[f \cdot g]$ is bounded. Since the support of $\mathscr{F}[f \cdot g]$ is closed by definition, it follows that the support of $\mathscr{F}[f \cdot g]$ is compact. Thus $f \cdot g$ is band-limited.

Finally, we prove a counterpart to Theorem 1.24. Here and below we write $\psi(D)f \equiv \mathscr{F}^{-1}[\psi \cdot \mathscr{F}f]$ for $\psi \in \mathscr{S}(\mathbb{R}^n)$ and $f \in \mathscr{S}'(\mathbb{R}^n)$. The operator $\psi(D)$ is called the *Fourier multiplier* generated by ψ. It turns out that the following lemma motivates the definition of the function spaces; when we define the function spaces, it matters that we decompose the functions in the frequency side.

Theorem 1.35 (Calderón's reproducing formula) *Let ψ and φ be functions in Lemma* 1.15. *Then for all $f \in \mathscr{S}'(\mathbb{R}^n)$,*

$$\lim_{J\to\infty}\left(\psi(D)+\sum_{j=1}^{J}\varphi_j(D)\right)f=f \tag{1.106}$$

holds in the topology of $\mathscr{S}'(\mathbb{R}^n)$.

Proof Choose a test function $\eta\in\mathscr{S}(\mathbb{R}^n)$. Then by the linearity of the coupling,

$$\lim_{J\to\infty}\left\langle\psi(D)f+\sum_{j=1}^{J}\varphi_j(D)f,\eta\right\rangle=\langle\psi\cdot\mathscr{F}f,\mathscr{F}^{-1}\eta\rangle+\lim_{J\to\infty}\sum_{j=1}^{J}\left\langle\varphi_j\cdot\mathscr{F}f,\mathscr{F}^{-1}\eta\right\rangle.$$

From Theorem 1.24,

$$\lim_{J\to\infty}\left\langle\psi(D)f+\sum_{j=1}^{J}\varphi_j(D)f,\eta\right\rangle=\langle\mathscr{F}f,\mathscr{F}^{-1}\eta\rangle=\langle f,\eta\rangle.$$

Hence it follows that $\psi(D)f+\sum_{j=1}^{J}\varphi_j(D)f\to f$ in the $\mathscr{S}'(\mathbb{R}^n)$-topology.

Calderón's reproducing formula shows that any distribution can be estimated canonically by a sequence of smooth functions.

Exercises

Exercise 1.50 (Integration by parts) Let $j=1,2,\ldots,n$.

1. Let $f\in\mathscr{S}(\mathbb{R}^n)$. Show that $\int_{\mathbb{R}^n}\partial_{x_j}f(x)\mathrm{d}x=0$. Hint: Integrate first against x_j.
2. Assume that g is a $C^\infty(\mathbb{R}^n)$-function that has at most polynomial growth at infinity. Then show that $\int_{\mathbb{R}^n}g(x)\partial_{x_j}f(x)\mathrm{d}x=-\int_{\mathbb{R}^n}f(x)\partial_{x_j}g(x)\mathrm{d}x$ by using Lemma 1.14.
3. Prove Theorem 1.28.

Exercise 1.51 Prove $\lim_{t\downarrow 0}\varphi(t\star)\tau=\varphi(0)\tau$ for all $\varphi,\tau\in\mathscr{S}(\mathbb{R}^n)$. Hint: See Theorem 1.24.

Exercise 1.52 Let $t>0$. Define f_t by

$$f_t(x)\equiv(1+it)^{-n/2}\exp\left(-\frac{|x|^2}{1+it}\right)\quad(x\in\mathbb{R}^n).$$

1. Show that
$$|f_t(x)| = |1+it|^{-n/2} \exp\left(-\frac{|x|^2}{1+t^2}\right)$$
for all $x \in \mathbb{R}^n$.
2. Calculate $\|f_t\|_p$ for $1 \le p \le \infty$.
3. Use (1.97) and analytic extension to calculate $\mathscr{F} f_t$ to conclude that $\|\mathscr{F} f_t\|_p$ does not depend on t.
4. Let $2 < p \le \infty$. Show that $\mathscr{F}$ never maps $L^p(\mathbb{R}^n)$ to $L^{p'}(\mathbb{R}^n)$.

Exercise 1.53 Using Theorem 1.28, prove that a $C^\infty(\mathbb{R}^n)$-function φ is in $\mathscr{S}(\mathbb{R}^n)$ if and only if $\varphi_{(\alpha,\beta)}$ is in $L^1(\mathbb{R}^n)$, where $\varphi_{(\alpha,\beta)}$ is defined in Definition 1.8.

Exercise 1.54 Show that the function $E \equiv \exp\left(-\frac{1}{2}|\star|^2\right)$ has its Fourier transform $\mathscr{F}(E) = E$ using (1.97) and by means of the complex line integral.

Exercise 1.55 [805] A function $u : \mathbb{R}^n \to \mathbb{C}$ of a variable $x \in \mathbb{R}^n$ is said to be radially symmetric or radial if u depends only on $|x|$. For a smooth real-valued function u on $\mathbb{R}^n$, define
$$|\nabla_n^k u(x)|_{n^k} = \left(\sum_{i \in I_n^k} (D_i u(x))^2\right)^{1/2}$$
$$= \left(\sum_{i_1=1}^{n}\sum_{i_2=1}^{n}\cdots\sum_{i_k=1}^{n}\left(\frac{\partial}{\partial x_{i_1}}\frac{\partial}{\partial x_{i_2}}\cdots\frac{\partial}{\partial x_{i_k}}u(x)\right)^2\right)^{1/2} \quad \text{for } (x \in \mathbb{R}^n);$$
we make the agreement $\nabla_n^0 u(x) = u(x)$ and then $|\nabla_n^0 u(x)|_1 = |u(x)|$. If $u \in \mathscr{S}(\mathbb{R}^n)$ is real-valued and radially symmetric, then so is $|\nabla_n^k u|_{n^k}^2$ for $k \in \mathbb{Z}_+$. Hint: Use the Fourier transform.

1.2.4 *The Space $\mathscr{D}$–$\mathscr{D}(\mathbb{T}^n)$ and $\mathscr{D}(\Omega)$*

Up to now we have been considering $\mathscr{S}'(\mathbb{R}^n)$ in more depth. In the latter part of this book, we will work in the setting of an open set Ω in $\mathbb{R}^n$. So we deal with $\mathscr{D}$ and its dual $\mathscr{D}'(\mathbb{R}^n)$.

1.2.4.1 Function Spaces Made Up of 2π-Periodic Functions

Let us first recall the space of 2π-periodic functions. A function f is said to have period 2π if $f(x) = f(x+2\pi m)$ for almost every x for all $m \in \mathbb{Z}^n$. Here we write $\mathbb{T}^n \equiv [0, 2\pi)^n$ and investigate 2π-periodic functions.

Definition 1.23 ($\mathscr{D}(\mathbb{T}^n)$)

1. Denote by $\mathscr{D}(\mathbb{T}^n)$ the set of all 2π-periodic $C^\infty(\mathbb{R}^n)$-functions.
2. Equip $\mathscr{D}(\mathbb{T}^n)$ with a topology induced naturally by a family $\{p_\alpha\}_{\alpha \in \mathbb{N}_0{}^n}$ of functionals, where $p_\alpha(f) \equiv \sup\limits_{x \in \mathbb{T}^n} |\partial^\alpha f(x)|$ for $\alpha \in \mathbb{N}_0{}^n$.

As we did for $\mathscr{S}'(\mathbb{R}^n)$ starting from $\mathscr{S}(\mathbb{R}^n)$, we can consider $\mathscr{D}'(\mathbb{T}^n)$.

Definition 1.24 ($\mathscr{D}'(\mathbb{T}^n)$) The space $\mathscr{D}'(\mathbb{T}^n)$ denotes the set of all continuous linear functionals on $\mathscr{D}(\mathbb{T}^n)$.

We want to consider elements in $\mathscr{D}'(\mathbb{T}^n)$. One of the effective methods is to consider a mapping F_f as we did for $\mathscr{S}'(\mathbb{R}^n)$.

Definition 1.25 ($L^p(\mathbb{T}^n)$, $C(\mathbb{T}^n)$, $C^k(\mathbb{T}^n)$) Let $1 \le p \le \infty$. Then define

$$L^p(\mathbb{T}^n) \equiv \left\{ f : \mathbb{R}^n \to \mathbb{C} \; : \; f \text{ is } 2\pi\text{-periodic and } \|f\|_{L^p(\mathbb{T}^n)} \equiv \|\chi_{[0,2\pi]^n} f\|_p < \infty \right\}.$$

Set $C(\mathbb{T}^n) \equiv L^\infty(\mathbb{T}^n) \cap C(\mathbb{R}^n)$ to be the closed subspace of $L^\infty(\mathbb{R}^n)$. That is, define $\|f\|_{C(\mathbb{T}^n)} \equiv \|f\|_{L^\infty(\mathbb{T}^n)}$ for $f \in C(\mathbb{T}^n)$.

Furthermore, for $k \in \mathbb{N}_0$, define $C^k(\mathbb{T}^n) \equiv C^k(\mathbb{R}^n) \cap L^\infty(\mathbb{T}^n)$. The norm is given by $\|f\|_{C^k(\mathbb{T}^n)} \equiv \sum\limits_{\alpha \in \mathbb{N}_0{}^n, |\alpha| \le k} \|\partial^\alpha f\|_{C(\mathbb{T}^n)}$.

In this book, we use the following properties of periodic functions. We admit that (1.107) occurs in the topology of $L^2(\mathbb{T}^n)$, since it falls under the scope of elementary Fourier analysis: see Exercise 1.59.

Theorem 1.36 (Fourier series for periodic functions in $\mathscr{D}(\mathbb{T}^n)$) *Let $f \in \mathscr{D}(\mathbb{T}^n)$. Then*

$$f = (2\pi)^{-n} \sum_{m \in \mathbb{Z}^n} \langle f, \mathrm{e}^{-im \cdot \star} \rangle \mathrm{e}^{i\, m \cdot \star} \tag{1.107}$$

in the topology of $\mathscr{D}(\mathbb{T}^n)$.

Proof Let us see how rapidly the coefficients decay. By integration by parts,

$$\begin{aligned} \int_{[0,2\pi]^n} f(y)\, \mathrm{e}^{-i\, m \cdot y} \mathrm{d}y &= \frac{1}{\langle m \rangle^{2L}} \int_{[0,2\pi]^n} f(y) [\, (1-\Delta)^L\, \mathrm{e}^{-i\, m \cdot y}] \mathrm{d}y \\ &= \frac{1}{\langle m \rangle^{2L}} \int_{[0,2\pi]^n} [\, (1-\Delta)^L f(y)\,]\, \mathrm{e}^{-i\, m \cdot y} \mathrm{d}y. \end{aligned}$$

For all $L \in \mathbb{N}$, this decreases faster than the polynomial $\langle m \rangle^{2L}$ for all L. Consequently, the right-hand side of (1.107) converges uniformly and both sides of (1.107) are continuous. The same applies to the partial derivatives: we can change the order of differentiation and sum. Thus, (1.107) occurs in the topology of $\mathscr{D}(\mathbb{T}^n)$.

We dualize Theorem 1.36.

Theorem 1.37 (Fourier series for periodic functions in $\mathscr{D}'(\mathbb{T}^n)$) *Let $f \in \mathscr{D}'(\mathbb{T}^n)$.*

1. *Then there exists $N \in \mathbb{N}$ such that $|\langle f, \mathrm{e}^{-im\cdot\star}\rangle| \lesssim \langle m\rangle^N$ for all $m \in \mathbb{Z}^n$.*
2. *In the topology of $\mathscr{D}'(\mathbb{T}^n)$, $f = (2\pi)^{-n} \sum_{m\in\mathbb{Z}^n} \langle f, \mathrm{e}^{-im\cdot\star}\rangle \mathrm{e}^{i\,m\cdot\star}$.*

Proof We use an estimate similar to (1.61) for 1. As for 2, we resort to the duality using Theorem 1.36 as we did for the Calderón reproducing formula.

Remark 1.3 Let $f \in \mathscr{D}'(\mathbb{T}^n)$. Then expand f according to Theorem 1.37 2. Define the coupling $\langle f, \varphi\rangle$ for $\varphi \in \mathscr{S}(\mathbb{R}^n)$ by

$$\langle f, \varphi\rangle \equiv (2\pi)^{-n} \sum_{m\in\mathbb{Z}^n} \langle f, \mathrm{e}^{-im\cdot\star}\rangle\langle \mathrm{e}^{i\,m\cdot\star}, \varphi\rangle.$$

Then according to Theorem 1.37 1 and $\langle \mathrm{e}^{i\,m\cdot\star}, \varphi\rangle = (2\pi)^{\frac{n}{2}}\mathscr{F}\varphi(-m)$, we see that the mapping $\varphi \in \mathscr{S}(\mathbb{R}^n) \mapsto \langle f, \varphi\rangle \in \mathbb{C}$, defined above, belongs to $\mathscr{S}'(\mathbb{R}^n)$. In this sense, $\mathscr{D}'(\mathbb{R}^n)$ is a subset of $\mathscr{S}'(\mathbb{R}^n)$.

1.2.4.2 $\mathscr{D}(\Omega)$ and $\mathscr{D}'(\Omega)$

Let Ω be an open set here and below. To develop a theory of function spaces, we consider $\mathscr{D}(\Omega)$ and $\mathscr{D}'(\Omega)$. First of all, we develop a theory of $C_\mathrm{c}^\infty(\Omega)$-functions supported on a fixed compact subset K.

Definition 1.26 ($C_\mathrm{c}^\infty(\Omega)$ and $C_\mathrm{c}^\infty(\Omega; K)$) Let $\Omega \subset \mathbb{R}^n$ be an open set. Denote by $C_\mathrm{c}^\infty(\Omega)$ the set of all $C^\infty(\mathbb{R}^n)$-functions having a compact support in Ω.

1. By $\mathscr{K}(\Omega)$ denote the set of all compact sets in Ω.
2. Let $K \in \mathscr{K}(\Omega)$. Then define $C_\mathrm{c}^\infty(\Omega; K) \equiv \{\varphi \in C_\mathrm{c}^\infty(\Omega) : \mathrm{supp}(\varphi) \subset K\}$; that is, $C_\mathrm{c}^\infty(\Omega; K)$ denotes the set of all $C^\infty(\Omega)$-functions having support in K.
3. For $\varphi \in C_\mathrm{c}^\infty(\Omega)$ and a multi-index $\alpha \in \mathbb{N}_0{}^n$, define a functional p_α on $C_\mathrm{c}^\infty(\Omega)$ by $p_\alpha(\varphi) \equiv \|\partial^\alpha\varphi\|_{L^\infty(\Omega)}$.
4. Make $C_\mathrm{c}^\infty(\Omega; K)$ a locally convex topological space by means of the functionals $\{p_\alpha\}_{\alpha\in\mathbb{N}_0{}^n}$.
5. Equip $C_\mathrm{c}^\infty(\Omega; K)$ with the weakest topology under which each $p_\alpha, \alpha \in \mathbb{N}_0^n$ is continuous.

Example 1.19 Let Ω be an open set and $K \in \mathscr{K}(\Omega)$. Then for any $h \in L^\infty(\Omega)$, the functional $f \in C_\mathrm{c}^\infty(\Omega) \mapsto \|f \cdot h\|_{L^\infty(\Omega)} \in [0, \infty)$ is continuous.

Lemma 1.17 *Let Ω be an open set, and let K be a compact set contained in Ω. Then the space $C_\mathrm{c}^\infty(\Omega; K)$ is metrizable and $C_\mathrm{c}^\infty(\Omega; K)$ is complete with respect to the distance.*

Proof We content ourselves with defining the distance function which induces the topology:

$$d(f, g) \equiv \sum_{\alpha \in \mathbb{N}_0{}^n} \frac{1}{\alpha!} \min(1, p_\alpha(f - g)), \quad f, g \in C_c^\infty(\Omega; K). \tag{1.108}$$

The details are left to interested readers; see Exercise 1.62.

Keeping the definition of $C_c^\infty(\Omega; K)$ in mind, we induce a topology to $\mathscr{D}(\Omega)$.

Definition 1.27 (Topology of $C_c^\infty(\Omega)$) Let Ω be an open set in $\mathbb{R}^n$. The topology of $C_c^\infty(\Omega)$ is the strongest locally convex topology such that the inclusion mapping $C_c^\infty(\Omega; K) \subset C_c^\infty(\Omega)$ is continuous for all $K \in \mathscr{K}(\Omega)$. In particular, the symbol $\mathscr{D}(\Omega)$ stands for the topological linear space $C_c^\infty(\Omega)$.

In other words the open set system $\mathscr{O}_{C_c^\infty(\Omega)}$ of $C_c^\infty(\Omega)$ is given by

$$\mathscr{O}_{C_c^\infty(\Omega)} \equiv \bigcap_{K \in \mathscr{K}(\Omega)} \{U \in 2^{C_c^\infty(\Omega)} \, : \, U \cap C_c^\infty(\Omega; K) \text{ is open}\},$$

where $2^{C_c^\infty(\Omega)}$ denotes the family of all subsets in $C_c^\infty(\Omega)$.

1.2.4.3 Characterization of the Convergence of Sequences

With the definition of the topology in mind, we characterize the convergence of sequences.

Lemma 1.18 *Let $\{f_j\}_{j=1}^\infty$ be a sequence of functions in $\mathscr{D}(\Omega)$, and let $\{x_j\}_{j=1}^\infty$ be a sequence of points Ω without any accumulation point in Ω. We assume $f_j(x_j) \neq 0$ and $f_j(x_k) = 0$ for all $j, k \in \mathbb{N}$ such that $k < j$. Then $\{f_j\}_{j=1}^\infty$ never converges.*

Proof A passage to the subsequence allows us to assume that $\{x_j\}_{j=1}^\infty$ is made up of distinct points. Then by assumption, we can choose a sequence $\{r_j\}_{j=1}^\infty$ decreasing to 0 so that $x_k \notin B(x_j, 2r_j) \subset \Omega$ whenever $k \neq j$. Choose a sequence $\{\varphi_j\}_{j=1}^\infty$ of $C_c^\infty(\mathbb{R}^n)$-functions so that $\chi_{B(x_j, r_j)} \le \varphi_j \le \chi_{B(x_j, 2r_j)}$. With this setup, we define

$$p(f) \equiv \sum_{j=1}^\infty \frac{\|\varphi_j \cdot f\|_\infty}{|f_j(x_j)|}, \quad f \in \mathscr{D}(\Omega). \tag{1.109}$$

In order that $p : C_c^\infty(\Omega) \to [0, \infty)$ is a continuous functional, we need to prove that $\{ f \in C_c^\infty(\Omega) \, : \, p(f) < R\}$ is an open subset of $C_c^\infty(\Omega)$ for all $R > 0$. However, from the definition of the norm, this amounts to proving that $N(R) \equiv \{ f \in C_c^\infty(\Omega; K) \, : \, p(f) < R\}$ is an open set in $C_c^\infty(\Omega; K)$ for all $K \in \mathscr{K}(\Omega)$. If we restrict p to $C_c^\infty(\Omega; K)$, it is a finite sum. In fact, the sum at most involves those j belonging to

$$\rho_j \equiv \{j \in \mathbb{N} \, : \, B(x_j, 2r_j) \cap K \neq \emptyset\}.$$

Thus, $N(R)$ is an open set in $C_c^\infty(\Omega; K)$. Therefore, p is continuous. Meanwhile, unless j and k coincide, we have $p(f_j - f_k) \geq 1$. Thus, $\{f_j\}_{j=1}^\infty$ never converges.

Proposition 1.12 *A sequence $\{\varphi_k\}_{k=1}^\infty$ in $\mathscr{D}(\Omega)$ convergences to $\varphi \in \mathscr{D}(\Omega)$ if and only if we can find a compact set K such that:*

1. *for all $k \in \mathbb{N}$,*

$$\mathrm{supp}(\varphi_k) \subset K, \tag{1.110}$$

2. *for all $\alpha \in {\mathbb{N}_0}^n$,*

$$\lim_{k\to\infty} p_\alpha(\varphi_k - \varphi) = 0. \tag{1.111}$$

Note that (1.110) is a necessary condition for the convergence of $\{\varphi_k\}_{k=1}^\infty$ in $\mathscr{D}(\Omega)$.

Proof If we assume (1.110) and (1.111), it is easy to prove $\varphi_k \to \varphi$ as $k \to \infty$.

Assume that $\varphi_k \to \varphi$ as $k \to \infty$.

Let $\{A_k\}_{k=1}^\infty \in \mathscr{K}(\Omega)$ be an exhaustion of Ω; that is,

$$A_k \subset \mathrm{Int}(A_{k+1}), \quad \bigcup_{k=1}^\infty A_k = \Omega.$$

We claim that $\bigcup_{k=1}^\infty \mathrm{supp}(\varphi_k)$ is relatively compact in Ω. If not,

$$\bigcup_{k=1}^\infty \mathrm{supp}(\varphi_k) \subset A_L \cup \bigcup_{k=1}^L \mathrm{supp}(\varphi_k)$$

never happens for any $L \in \mathbb{N}$. This means that there exists $k(L) > L$ such that

$$\mathrm{supp}(\varphi_{k(L)}) \setminus \left(A_L \cup \bigcup_{k=1}^L \mathrm{supp}(\varphi_k) \right) \neq \emptyset.$$

A passage to a subsequence allows us to assume $k(L) = L + 1$, so that there exists a point x_L such that $\varphi_{L+1}(x_L) \neq 0$ and that $\varphi_l(x_L) = 0$ for $l \leq k(L)$.

Then by Lemma 1.18, $\{\varphi_l\}_{l=1}^\infty$ never converges. That is, (1.110) holds and hence (1.111) is clear from the definition of the convergence.

1.2.4.4 Bounded Sets in $\mathscr{D}(\Omega)$

We now define the bounded sets in $\mathscr{D}(\Omega)$, although $\mathscr{D}(\Omega)$ is not metrizable.

Definition 1.28 (Bounded sets in $\mathscr{D}(\Omega)$) A subset A in $\mathscr{D}(\Omega)$ is said to be *bounded*, if for any open neighborhood B of the origin, $A \subset k \cdot B \equiv \{k \cdot b : b \in B\}$ for some $k \in \mathbb{C}$.

Let us characterize bounded sets in $\mathscr{D}(\Omega)$, keeping the definition in mind.

Theorem 1.38 (A characterization of bounded sets in $\mathscr{D}(\Omega)$) *Let $A \subset \mathscr{D}(\Omega)$. Then A is bounded in $\mathscr{D}(\Omega)$ if and only if there exists a compact set $K \subset \mathbb{R}^n$ such that*

$$A \subset C_c^\infty(\Omega; K) \text{ and } \sup_{\varphi \in A} p_N(\varphi) < \infty \tag{1.112}$$

for each $N \in \mathbb{N}$.

Proof (Step 1: Assume that A satisfies (1.112)) Let $U \subset \mathscr{D}(\Omega)$ be an open set containing 0. Then $U \cap C_c^\infty(\Omega; K)$ is an open set in $C_c^\infty(\Omega; K)$ and U contains 0. Thus, there exists $\alpha \in \mathbb{C}$ such that $A \subset \alpha(U \cap C_c^\infty(\Omega; K)) \subset \alpha U$. Hence A is bounded in $\mathscr{D}(\Omega)$.

Proof (Step 2: Assume that A is bounded in $\mathscr{D}(\Omega)$) For any compact set $K \subset \Omega$, we assume for a contradiction that $A \subset C_c^\infty(\Omega; K)$ fails. Then for all nonnegative integers j, we can find a sequence $\{x_j\}_{j=1}^\infty$ of Ω without any accumulation point and a sequence $\{\varphi_j\}_{j=1}^\infty \subset A$ of $\varphi_j(x_k) = 0$ for all $j, k \in \mathbb{N}$ with $k < j$ and $\varphi_j(x_j) \neq 0$.

By discarding some points we can assume that $\{x_j\}_{j=1}^\infty$ is made up of distinct points. Hence we can find a sequence $\{B_j\}_{j=1}^\infty$ of balls such that each B_j is centered at x_j. Let ψ_j be a smooth function supported on B_j. Suppose in addition that each ψ_j assumes 1 near a neighborhood of x_j. Then define $p(\varphi) \equiv \sum_{j=1}^\infty j \dfrac{\|\psi_j \varphi\|_\infty}{|\varphi_j(x_j)|}$. In the same way as we did for the proof of Lemma 1.18, we can prove that p is a continuous functional. Hence there exists $M > 0$ such that $A \subset p^{-1}([0, M])$. However, for all $j > M$, we have $\varphi_j \in A \cap p^{-1}([j, \infty))$. This is a contradiction. Hence A is contained in $C_c^\infty(\Omega; K)$ for some compact set $K \subset \Omega$. This proves that the first condition of (1.112) is true. Once this is proved, the second one in (1.112) holds trivially.

1.2.4.5 Restriction of $\mathscr{S}'(\mathbb{R}^n)$ to Open Sets

So far, we have considered functions defined on $\mathbb{R}^n$. However, from the viewpoint of applications, this is not enough. Many examples from differential equations are staged in open sets. Motivated by this we will need the restriction mapping to define function spaces on open sets.

Definition 1.29 ($\mathscr{D}'(\Omega)$) Let Ω be an open set in $\mathbb{R}^n$. The space $\mathscr{D}'(\Omega)$ denotes continuous linear functionals on $\mathscr{D}(\Omega)$. The topology $\mathscr{D}'(\Omega)$ is the weakest topology such that $f \in \mathscr{D}'(\Omega) \mapsto f(\varphi) = \langle f, \varphi \rangle \in \mathbb{C}$ is continuous for all $\varphi \in \mathscr{D}(\Omega)$.

The space $\mathscr{D}'(\Omega)$ is fundamental but the following special elements are important in this book.

Definition 1.30 (Restriction of $\mathscr{S}'(\mathbb{R}^n)$ to open sets) Let Ω be an open set in $\mathbb{R}^n$. Extend each element in $\mathscr{D}(\Omega)$ by defining 0 outside Ω to have $\mathscr{D}(\Omega) \hookrightarrow \mathscr{S}(\mathbb{R}^n)$. This mapping being continuous, duality entails $\mathscr{S}'(\mathbb{R}^n) \hookrightarrow \mathscr{D}'(\Omega)$ in the sense of continuous embedding. Let $f \in \mathscr{S}'(\mathbb{R}^n)$. Write $f \mid \Omega$ for the element in $\mathscr{D}'(\Omega)$.

Exercises

Exercise 1.56 Show that $\mathscr{D}(\mathbb{T}^n)$ is metrizable and that it is complete with respect to the distance. Hint: Mimic (1.59).

Exercise 1.57 Let $k \in \mathbb{N}$.

1. Show that $C(\mathbb{T}^n)$ is a Banach space with its norm.
2. Show that $C^k(\mathbb{T}^n)$ is a Banach space with its norm.

Exercise 1.58 (Poisson summation) Let $f \in L^1(\mathbb{R}^n)$. Define $F \equiv \sum\limits_{m \in \mathbb{Z}^n} f(\star - 2\pi m)$. Show that the sum defining $F(x)$ is convergent for almost all $x \in \mathbb{R}^n$ and that $F \in L^1(\mathbb{T}^n)$.

Exercise 1.59 We admitted Theorem 1.36 for $L^2(\mathbb{T}^n)$. We refer to many books on the theory of Hilbert spaces; see [46] for example. For $L^2(\mathbb{T}^n)$, how did we prove Theorem 1.36? We summarize the outline as an exercise.

1. Use the mollifier to show that $C(\mathbb{T}^n)$ is dense in $L^2(\mathbb{T}^n)$.
2. Show that $\{e^{i\, m \cdot \star}\}_{m \in \mathbb{Z}^n}$ is dense in $C(\mathbb{T}^n)$ using the Weierstrass approximation theorem.
3. Show that $\{(2\pi)^{-\frac{n}{2}} e^{i\, m \cdot \star}\}_{m \in \mathbb{Z}^n}$ forms a complete orthonormal system in $L^2(\mathbb{T}^n)$.
4. Let $f \in L^2(\mathbb{T}^n)$. Show that (1.107) holds in $L^2(\mathbb{T}^n)$.

By a periodic lattice, we mean a subset Λ of $\mathbb{R}^n$ such that it is expressed:

$$\Lambda = \mathbb{Z}\, v_1 + \mathbb{Z}\, v_2 + \cdots + \mathbb{Z}\, v_n, \quad \mathbb{R}^n = \mathbb{R}\, v_1 + \mathbb{R}\, v_2 + \cdots + \mathbb{R}\, v_n$$

for some $v_1, v_2, \ldots, v_n \in \mathbb{R}^n$.

A function $f \in L^1_{\mathrm{loc}}(\mathbb{R}^n)$ is said to have the periodic lattice Λ if $f(x+\lambda) = f(x)$ holds for almost all $x \in \mathbb{R}^n$ and for all $\lambda \in \Lambda$.

Extend the above results to other lattices Λ. In particular, what happens when $\Lambda = A\mathbb{Z}\,\mathbf{e}_1 + A\,\mathbb{Z}\,\mathbf{e}_2 + \cdots + A\,\mathbb{Z}\,\mathbf{e}_n$ for $A > 0$?

Exercise 1.60 Show that

$$\left\{ f \in C_c^\infty(\Omega; K) : \sup_{x \in B(N)} |\partial^\alpha (f-g)(x)| < r \right\}$$

$$(r > 0, N \in \mathbb{N}, \alpha \in \mathbb{N}_0{}^n, g \in C_c^\infty(\Omega; K), K \in \mathcal{K}(\Omega))$$

generates the topology of $\mathcal{D}(\Omega)$. Hint: Uniform convergence.

Exercise 1.61 Equip $C^\infty(\mathbb{R}^n)$ with the distance; let $f, g \in C^\infty(\mathbb{R}^n)$ and define

$$d(f,g) \equiv \sum_{N=1}^{\infty} \sum_{\alpha \in \mathbb{N}_0{}^n} \frac{Z_{\alpha,N}(f,g)}{1 + 2^{N+|\alpha|} Z_{\alpha,N}(f,g)},$$

where $Z_{\alpha,N}(f,g) \equiv \|\partial^\alpha (f-g)\|_{L^\infty(B(N))}$. Show that $C^\infty(\mathbb{R}^n)$ is complete. Hint: Uniform convergence.

Exercise 1.62 Show that d given by (1.108) is a distance function and that it defines the topology of $C_c^\infty(\Omega; K)$. Hint: $\min(1, a+b) \le \min(1,a) + \min(1,b)$ for $a, b \ge 0$.

Exercise 1.63 Let $\varphi \in \mathcal{S}(\mathbb{R}^n)$ satisfy the moment condition of order ∞. and assume in addition that φ is compactly supported. Then show that $\varphi \equiv 0$. Hint: Consider

$$F(z) = \int_{\mathbb{R}^n} \mathrm{e}^{iz\cdot x} \varphi(x) \mathrm{d}x \quad (z \in \mathbb{C}^n)$$

and its Taylor expansion at $z = 0$.

Exercise 1.64 [510] Let $G \subset \mathbb{R}^n$ be an open set. Define $Z(G) \equiv \{\mathcal{F}\varphi : \varphi \in \mathcal{D}(G)\}$ and equip $Z(G)$ with the topology such that $\mathcal{F} : \mathcal{D}(G) \to Z(G)$ is a linear isomorphism. Denote by $Z'(G)$ the set of all continuous linear functionals on $Z(G)$. Show that the Fourier transform can be defined naturally so that $\mathcal{F} : Z'(G) \to \mathcal{D}'(G)$ is a linear isomorphism. Hint: Definition 1.21.

Exercise 1.65 Let $f \in \mathcal{D}'(\mathbb{R}^n)$.

1. Prove that there exists a finite set $A \subset \mathbb{N}_0{}^n$ such that $|\langle f, \varphi \rangle| \le C \sup_{\alpha \in A} p_\alpha(f)$ by mimicking the proof of (1.61).
2. Prove Theorem 1.37 1.
3. Prove Theorem 1.37 2 using Theorem 1.36.

Exercise 1.66 Show that $\mathcal{D}'(\mathbb{T}^n) \hookrightarrow \mathcal{S}'(\mathbb{R}^n)$ in the sense of continuous embedding by reexamining the estimate in Remark 1.3.

1.2.5 Some Functional Equations in $\mathscr{S}(\mathbb{R}^n)$

As we have stated, the elements in $\mathscr{S}(\mathbb{R}^n)$ are too good. In fact, they must decay very rapidly and they must be smooth. So, it is sometimes difficult to find nicer elements in $\mathscr{S}(\mathbb{R}^n)$. Here we collect some typical examples of nontrivial elements in $\mathscr{S}(\mathbb{R}^n)$.

1.2.5.1 Local Reproducing Formula

We want to consider the local version of Theorem 1.24. Unfortunately, if $\varphi \in C_c^\infty(\mathbb{R}^n)$ and $\mathscr{F}\varphi \in C_c^\infty(\mathbb{R}^n)$, then $\varphi = 0$ as is easily shown by the identity principle in $\mathbb{C}$. See Exercise 1.68. So, in Theorem 1.24 $f \mapsto \varphi_j(D)f \simeq \mathscr{F}\varphi_j * f$ is a convolution operator with noncompact functions. Sometimes, compactness of the function $\mathscr{F}\varphi_j$ will be helpful. So, we consider the case where the support of $\mathscr{F}\varphi_j$ is compact. The space $L^0(\mathbb{R}^n)$ denotes the set of all measurable functions considered modulo the difference on the set of measure zero. The value of the functions can be in $\mathbb{C}$ or $[-\infty, \infty]$.

Definition 1.31 (Moment condition, Lizorkin functions, $\mathscr{P}_L$, $L \geq -1$).

1. Let $L \in \mathbb{N}_0$. The set $\mathscr{P}_L(\mathbb{R}^n)^\perp$ denotes the set of all the measurable functions f for which $\langle\star\rangle^L f \in L^1(\mathbb{R}^n)$ and $\displaystyle\int_{\mathbb{R}^n} x^\alpha f(x)\mathrm{d}x = 0$ for all $\alpha \in \mathbb{R}^n$ with $|\alpha| \leq L$. Such a function f is said to satisfy the *moment condition of order* L. In this case, one also writes $f \perp \mathscr{P}_L(\mathbb{R}^n)$. If $f \perp \mathscr{P}_L(\mathbb{R}^n)$ for all $L \in \mathbb{N}$, one writes $f \perp \mathscr{P}(\mathbb{R}^n)$.
2. If $\varphi \in \mathscr{S}(\mathbb{R}^n)$ satisfies the moment condition of order L for any $L \in \mathbb{N}$, $\varphi \in \mathscr{S}(\mathbb{R}^n)$ is said to satisfy the moment condition of order ∞. Denote by $\mathscr{S}_\infty(\mathbb{R}^n)$ the set of all such functions. The functions in $\mathscr{S}_\infty(\mathbb{R}^n)$ are called *Lizorkin functions*.
3. One defines $\mathscr{P}_{-1}(\mathbb{R}^n)^\perp \equiv L^0(\mathbb{R}^n)$. The condition $a \perp \mathscr{P}_{-1}(\mathbb{R}^n)$ for $a \in L^1_{\mathrm{loc}}(\mathbb{R}^n)$ means that $a \in L^1_{\mathrm{loc}}(\mathbb{R}^n)$.

The next result is due to Rychkov. When $\varphi \in \mathscr{S}(\mathbb{R}^n)$, following (26), we write $\varphi^j \equiv 2^{jn}\varphi(2^j\star)$. By no means do ψ^{-1} and η^{-1} stand for the inverse mapping. They are given by a function in $\mathscr{S}(\mathbb{R}^n)$; see (26). In Sect. 1.2.5, we never use the symbol "$^{-1}$" to denote the mapping and we regard them as functions in $\mathscr{S}(\mathbb{R}^n)$ defined according to (26).

Theorem 1.39 (Local reproducing formula) *Let $L \in \mathbb{N}$ be fixed. Denote by $\delta \in \mathscr{S}'(\mathbb{R}^n)$ the Dirac delta at the origin. Suppose that $\psi \in \mathscr{S}(\mathbb{R}^n) \setminus \mathscr{P}_0(\mathbb{R}^n)^\perp$. Set $\varphi \equiv \psi - \psi^{-1} = \psi - 2^{-n}\psi(2^{-1}\star)$. Then there exist $\Psi \in \mathscr{S}(\mathbb{R}^n)$ and $\Phi \in \mathscr{S}(\mathbb{R}^n) \cap \mathscr{P}_L(\mathbb{R}^n)^\perp$ such that $\displaystyle\psi * \Psi + \sum_{j=1}^\infty \varphi^j * \Phi^j = \delta$ in the topology of $\mathscr{S}'(\mathbb{R}^n)$.*

If in addition $\psi \in C_c^\infty(\mathbb{R}^n)$, then we can arrange that Ψ and Φ be compactly supported.

We remark that ψ is *nondegenerate* since $\psi \in \mathscr{S}(\mathbb{R}^n) \setminus \mathscr{P}_0(\mathbb{R}^n)^\perp$ implies $\displaystyle\int_{\mathbb{R}^n} \psi(x)\mathrm{d}x \neq 0$.

Proof We abbreviate the l-fold convolution of $f \in \mathscr{S}(\mathbb{R}^n)$ to f^{*l}; $f^{*l} = f * f * \cdots * f$. For $j \in \mathbb{N}$, we set $\Theta \equiv \psi * \psi$ and $\kappa \equiv \Theta - \Theta^{-1} = \Theta - 2^n\Theta(2\star)$. Then by the nondegenerate condition, we can multiply a suitable constant to have $\displaystyle\Theta + \sum_{j=1}^{\infty} \kappa_j = \delta$ in the topology of $\mathscr{S}'(\mathbb{R}^n)$.

We consider the $(L+2)$-fold convolution of this formula to have

$$\left(\Theta + \sum_{j=1}^{\infty} \kappa^j\right)^{*(L+2)} = \delta.$$

We expand the left-hand side with respect to $\Theta, \kappa^1, \ldots, \kappa^j, \ldots$ to have

$$\sum_{l=1}^{L+2} \Theta^{*l} * h^{(0,l)} + \sum_{j=1}^{\infty}\sum_{l=1}^{L+2} (\kappa^j)^{*l} * h^{(j,l)} = \delta,$$

where

$$h^{(j,l)} \equiv {}_{L+2}C_l \left(\sum_{k=j+1}^{\infty} \kappa^k\right)^{*L+2-l} \qquad (j \in \mathbb{Z}, l = 1, 2, \ldots, L+2).$$

In view of the definition of $h^{(j,l)}$, we have

$$(h^{(j,l)})^k = 2^{kn} h^{(j,l)}(2^k\star) = h^{(j+k,l)} \quad (k \in \mathbb{Z}). \tag{1.113}$$

Define $\Psi, \Phi \in \mathscr{S}(\mathbb{R}^n)$ by

$$\Psi \equiv \sum_{l=1}^{L+2} \Theta^{*(l-1)} * h^{(0,l)} * \psi, \quad \Phi \equiv (\psi + \psi^{-1}) * \left(\sum_{l=1}^{L+2} \kappa^{*(l-1)} * h^{(0,l)}\right).$$

Then we obtain

$$\sum_{l=1}^{L+2} \Theta^{*l} * h^{(0,l)} = \Theta * \left(\sum_{l=1}^{L+2} \Theta^{*(l-1)} * h^{(0,l)}\right)$$

$$= \psi * \left(\sum_{l=1}^{L+2} \Theta^{*(l-1)} * h^{(0,l)} * \psi \right)$$
$$= \psi * \Psi$$

using the definition of φ, Θ and Ψ, as well as (1.113). Likewise, we can prove

$$\sum_{l=1}^{L+2} (\kappa^j)^{*l} * h^{(j,l)} = \kappa^j * \left(\sum_{l=1}^{L+2} (\kappa^j)^{*(l-1)} * h^{(j,l)} \right) = \varphi^j * \Phi^j$$

using the definition of Θ, κ and Ψ. Thus, we obtain $\psi * \Psi + \sum_{j=1}^{\infty} \varphi^j * \Phi^j = \delta$ in the topology of $\mathscr{S}'(\mathbb{R}^n)$.

By taking the Fourier transform and examining the order at 0, we see $\varphi \in \mathscr{P}_L(\mathbb{R}^n)$. Hence Ψ, Φ are desired functions.

1.2.5.2 Functional Equation $\Phi - 2^n \Phi(2\star) = \Psi$

One of the fundamental pieces of functions dealt with in this book is given by $\varphi \equiv \psi - \psi(2\star)$ for some $\psi \in \mathscr{S}(\mathbb{R}^n)$. Not only the function φ but also ψ is important. If we consider the Fourier transform of the relation $\varphi = \psi - \psi(2\star)$, then we are led to $\Phi - 2^{-n}\Phi(2^{-1}\star) = \Psi$, where $\Psi \equiv -2^{-n}\mathscr{F}\varphi(2^{-1}\star)$ and $\Phi \equiv \mathscr{F}\psi$. Regarding the transform $\varphi \in \mathscr{S}(\mathbb{R}^n) \mapsto \varphi - \varphi^{-1} = \varphi - 2^n\varphi(2\star) \in \mathscr{S}(\mathbb{R}^n)$, we note another important fact:

Theorem 1.40 *Fix $L \in \mathbb{N}$. Then there exist $\Phi^{(L)}, \Psi^{(L)} \in C^\infty_{\mathrm{c}}(\mathbb{R}^n)$ such that*

$$\int_{\mathbb{R}^n} \Phi^{(L)}(x)\mathrm{d}x = 1, \quad \Phi^{(L)} - (\Phi^{(L)})^{-1} = \Delta^L \Psi^{(L)}. \tag{1.114}$$

Recall that we defined $(\Phi^{(L)})^{-1}(x) = 2^{-n}\Phi^{(L)}(2^{-1}x)$ for $x \in \mathbb{R}^n$.

Proof Choose a function $\kappa^{(1)} \in C^\infty_{\mathrm{c}}(\mathbb{R})$ so that

$$\int_0^\infty \kappa^{(1)}(r) r^{n-1}\mathrm{d}r = 1, \quad \mathrm{supp}(\kappa^{(1)}) \subset (0,1). \tag{1.115}$$

We define $\kappa^{(2)}, \kappa^{(3)}, \ldots$ inductively. Suppose that we define $\kappa^{(L)}$. First we define $\kappa^{(L+1)}$ by

$$\kappa^{(L+1)} \equiv \mu_L \kappa^{(L)} + \lambda_L \kappa^{(L)}(2\star). \tag{1.116}$$

Here the coefficient μ_L and λ_L are unique solutions of a system of equations:

$$\mu_L + 2^{-n}\lambda_L = 1, \quad \mu_L + 2^{-n-2L}\lambda_L = 0. \tag{1.117}$$

Then as can be shown easily, we have

$$\mathrm{supp}(\kappa^{(L+1)}) \subset \mathrm{supp}(\kappa^{(L-1)}) \cup \frac{1}{2}\mathrm{supp}(\kappa^{(L-1)}) \subset \cdots \subset (0, 1). \tag{1.118}$$

In addition, by (1.115) and the recurrence formula (1.116) we have

$$\begin{aligned}
\int_0^\infty \kappa^{(L)}(r)r^{n-1}\mathrm{d}r &= \mu_{L-1}\int_0^\infty \kappa^{(L-1)}(r)r^{n-1}\mathrm{d}r + \lambda_{L-1}\int_0^\infty \kappa^{(L-1)}(2r)r^{n-1}\mathrm{d}r\\
&= \mu_{L-1}\int_0^\infty \kappa^{(L-1)}(r)r^{n-1}\mathrm{d}r + \frac{\lambda_{L-1}}{2^n}\int_0^\infty \kappa^{(L-1)}(r)r^{n-1}\mathrm{d}r\\
&= \left(\mu_{L-1} + \frac{\lambda_{L-1}}{2^n}\right)\int_0^\infty \kappa^{(L-1)}(r)r^{n-1}\mathrm{d}r\\
&= \int_0^\infty \kappa^{(L-1)}(r)r^{n-1}\mathrm{d}r = \cdots = 1.
\end{aligned}$$

For $r \in \mathbb{R}$, define $\eta^{(L)}(r) \equiv \kappa^{(L+1)}(r) - \frac{1}{2^n}\kappa^{(L+1)}\left(\frac{r}{2}\right)$ for $r > 0$. Then (1.117) yields

$$\begin{aligned}
&\mu_L\,\eta^{(L)}(r) + \lambda_L\,\eta^{(L)}(2r)\\
&= \mu_L\,\kappa^{(L+1)}(r) - \frac{\mu_L}{2^n}\kappa^{(L+1)}\left(\frac{r}{2}\right) + \lambda_L\kappa^{(L+1)}(2r) - \frac{\lambda_L}{2^n}\kappa^{(L+1)}(r)\\
&= \kappa^{(L+1)}(r) - \frac{1}{2^n}\kappa^{(L+1)}\left(\frac{r}{2}\right)\\
&= \eta^{(L+1)}(r).
\end{aligned}$$

Thus, we have: $\eta^{(L+1)} = \mu_L\,\eta^{(L)} + \lambda_L\,\eta^{(L)}(2\star)$.

Define a linear operator $T : C[0,\infty) \to C[0,\infty)$ by

$$(T\varphi)(r) \equiv \int_0^r \left(\int_0^t \left(\frac{s}{t}\right)^{n-1}\varphi(s)\mathrm{d}s\right)\mathrm{d}t \quad (r \geq 0).$$

We change the variables to have:

$$\begin{aligned}
(T\varphi)(2r) &= \int_0^{2r} \left(\int_0^t \left(\frac{s}{t}\right)^{n-1}\varphi(s)\mathrm{d}s\right)\mathrm{d}t\\
&= \int_0^r \left(\int_0^{2u} \left(\frac{s}{2u}\right)^{n-1}\varphi(s)\mathrm{d}s\right)2\mathrm{d}u
\end{aligned}$$

$$= 2^{-n+2} \int_0^r \left(\int_0^{2u} \left(\frac{s}{u}\right)^{n-1} \varphi(s) ds \right) du$$

$$= 4 \int_0^r \left(\int_0^u \left(\frac{v}{u}\right)^{n-1} \varphi(2v) dv \right) du$$

$$= 4 \cdot T[\varphi(2\star)](r) \quad (r \geq 0).$$

Denote by T^L the L-fold composition of T. By the induction on L and (1.118), we can easily verify that $T^L \eta^{(L)}$ vanishes in a neighborhood of $0 \in [0, \infty)$.

Next, in $[2, \infty)$, we prove that $T^L \eta^{(L)}$ agree with an even polynomial with degree $2L - 2$ inducting on L. Since it is trivial when $L = 0$, let $L \geq 1$. Then

$$T^{L+1} \eta^{(L+1)}(r) - T^{L+1} \eta^{(L+1)}(2)$$

$$= T[T^L \eta^{(L+1)}](r) - T[T^L \eta^{(L+1)}](2)$$

$$= \int_2^r \int_0^t \left(\frac{s}{t}\right)^{n-1} T^L \eta^{(L+1)}(s) ds dt$$

$$= \int_2^r \int_0^t \left(\frac{s}{t}\right)^{n-1} (\mu_L T^L \eta^{(L)}(s) + \lambda_L T^L[\eta^{(L)}(2\star)](s)) ds dt$$

$$= \int_2^r \int_0^t \left(\frac{s}{t}\right)^{n-1} (\mu_L T^L \eta^{(L)}(s) + \lambda_L 2^{-2L} T^L \eta^{(L)}(2s)) ds dt$$

for $r \geq 0$. Here for the last equality, we used $(T\varphi)(2r) = 4 \cdot T[\varphi(2\star)](r)$. Furthermore, using $\mu_L + 2^{-n-2L} \lambda_L = 0$, we have

$$T^{L+1} \eta^{(L+1)}(r) - T^{L+1} \eta^{(L+1)}(2)$$

$$= \int_2^r \left(\int_0^t \mu_L \left(\frac{s}{t}\right)^{n-1} T^L \eta^{(L)}(s) ds + \int_0^{2t} \lambda_L \left(\frac{s}{t}\right)^{n-1} \frac{T^L \eta^{(L)}(s)}{2^{2L+n}} ds \right) dt$$

$$= \int_2^r \left(\int_0^{2t} \lambda_L \left(\frac{s}{t}\right)^{n-1} \frac{T^L \eta^{(L)}(s)}{2^{2L+n}} ds - \int_0^t \left(\frac{s}{t}\right)^{n-1} \lambda_L T^L \eta^{(L)}(s) ds \right) dt$$

$$= \lambda_L 2^{-2L-n} \int_2^r \left(\int_t^{2t} \left(\frac{s}{t}\right)^{n-1} T^L \eta^{(L)}(s) ds \right) dt.$$

By virtue of this recurrence formula and

$$\int_2^r \left(\int_t^{2t} \left(\frac{s}{t}\right)^{n-1} s^{2k} ds \right) dt = c_{n,k}(r^{2k+2} - 2^{2k+2}), \quad k \in \mathbb{N}_0$$

we see that $T^L\eta^{(L)}$ agrees with an even polynomial of degree $2L-2$ on $[2,\infty)$. We denote this polynomial by $\sum_{l=0}^{L-1} a_l\, r^{2l}$.

Here we define

$$\psi^{(L)}(r) \equiv (T^L\eta^{(L)})(r) - \sum_{l=0}^{L-1} a_l\, r^{2l}, \quad 0 \le r < \infty.$$

By definition, $\psi^{(L)}$ agrees with an even polynomial of degree $2L-2$ on a neighborhood of $0 \in [0,\infty)$ and vanishes outside $B(2)$.

We let $S^{n-1} \equiv \{\omega \in \mathbb{R}^n : |\omega| = 1\}$. Then we obtain $\Phi^{(L)}, \Psi^{(L)}$ satisfying (1.114). That is, denoting by $\mathrm{vol}(S^{n-1})$ the surface area of S^{n-1}, define $\Phi^{(L)}, \Phi^{(L)} : \mathbb{R}^n \to [0,\infty)$ by

$$\Phi^{(L)} \equiv \mathrm{vol}(S^{n-1})^{-1}\kappa^{(L+1)}(|\star|), \quad \Psi^{(L)} \equiv \mathrm{vol}(S^{n-1})^{-1}\psi^{(L)}(|\star|).$$

A simple computation shows that $\Phi^{(L)}$ has integral 1.

Let $x \in \mathbb{R}^n$. In general, if a smooth function $\rho : [0,\infty) \to [0,\infty)$ vanishes near a neighborhood of the origin then, we have

$$\Delta T\rho(x) = \left[r^{1-n}\left(r^{n-1}(T\rho)'(r)\right)' \right]_{r=|x|} = \rho(|x|).$$

We write $r \equiv |x|$ and we have

$$\begin{aligned}
\Delta^L\psi^{(L)}(x) &= \mathrm{vol}(S^{n-1})^{-1}\Delta^L\left[(T^L\eta^{(L)})(r) - \sum_{l=0}^{L-1} a_l\, r^{2l} \right] \\
&= \mathrm{vol}(S^{n-1})^{-1}\Delta^L(T^L\eta^{(L)})(|x|) \\
&= \mathrm{vol}(S^{n-1})^{-1}\eta^{(L)}(|x|) \\
&= \Phi^{(L)}(x) - (\Phi^{(L)})^{-1}(x).
\end{aligned}$$

Hence $\Phi^{(L)}$ and $\Psi^{(L)}$ do satisfy (1.114).

Example 1.20 Let $d \in \mathbb{N}$. Then there exist $\psi_0, \Psi_0 \in C_c^\infty(\mathbb{R})$ such that $\int_{\mathbb{R}} \Psi_0(t)dt = 1$ and that $\Psi_0(t) - \frac{1}{2}\Psi_0\left(\frac{t}{2}\right) = \psi_0^{(d)}(t)$ for all $t \in \mathbb{R}$ as a special case of Theorem 1.40 with $n = 1$.

1.2.5.3 The Closed Subspace $\mathscr{S}_L(\mathbb{R}^n)$

The remander theorem on algebra shows that the residue of the polynomial $P(t)$ of $t \in \mathbb{R}$ divided by $(t-\alpha)$ is $P(\alpha)$. Therefore, $P(t)$ is a multiple of $(t-\alpha)$ if and only if $P(\alpha) = 0$. This theorem can be extended to higher dimensions and to $(t-\alpha)^k$ for $k \geq 2$. To formulate our results, set

$$\mathscr{S}_L(\mathbb{R}^n) \equiv \bigcap_{\alpha \in \mathbb{N}_0{}^n, |\alpha| \leq L-1} \{\varphi \in \mathscr{S}(\mathbb{R}^n) : \partial^\alpha \varphi(0) = 0\} \tag{1.119}$$

and equip $\mathscr{S}_L(\mathbb{R}^n)$ with the induced topology from $\mathscr{S}(\mathbb{R}^n)$. The following is a counterpart of this theorem to our theory of function spaces:

Theorem 1.41 *Let $L \in \mathbb{N}$. Then there exists a family $\{T_\gamma\}_{\gamma \in \mathbb{Z}^n, |\gamma|=L}$ of continuous mappings from $\mathscr{S}_L(\mathbb{R}^n)$ to $\mathscr{S}(\mathbb{R}^n)$ such that*

$$\varphi(x) = \sum_{\gamma \in \mathbb{N}_0{}^n, |\gamma|=L} x^\gamma T_\gamma \varphi(x) \quad (x \in \mathbb{R}^n) \tag{1.120}$$

for all $\varphi \in \mathscr{S}_L(\mathbb{R}^n)$.

Proof We induct on the dimension n. When $n = 1$, the result is clear from the fundamental theorem of calculus. Suppose that Theorem 1.41 holds for $n-1$. We write $x = (x_1, x_2, \ldots, x_n)$.

Choose $a \in C^\infty(\mathbb{R})$ so that $\chi_{[-1,1]} \leq a \leq \chi_{[-2,2]}$. For a function $\varphi \in \mathscr{S}(\mathbb{R}^n)_L$ and $x_1, x_2, \ldots, x_{n-1} \in \mathbb{R}$, $x_n \in \mathbb{R} \setminus \{0\}$, we define

$$R_\varphi(x) \equiv \frac{a(x_n)}{x_n{}^L}\left(\varphi(x) - \sum_{j=0}^{L} \frac{\partial_n{}^j \varphi(x', 0)}{j!} x_n{}^j\right).$$

Extend by continuity the function R_φ to $\mathbb{R}^n$. By the fundamental theorem of calculus, we see that $\varphi \mapsto R_\varphi$ is a continuous mapping from $\mathscr{S}(\mathbb{R}^n)_L$ to $\mathscr{S}(\mathbb{R}^n)$. In addition, define

$$T\varphi(x) \equiv R_\varphi(x) + \frac{1-a(x_n)}{x_n{}^L}\varphi(x_1, x_2, \ldots, x_n) \quad (x \in \mathbb{R}^n).$$

The function $\varphi \in \mathscr{S}(\mathbb{R}^n) \mapsto T\varphi \in \mathscr{S}(\mathbb{R}^n)$ is a continuous mapping. Note that $\varphi(x) = x_n{}^L T\varphi(x) + a(x_n)\sum_{j=0}^{L} \frac{\partial_n{}^j \varphi(x', 0)}{j!} x_n{}^j$. By the induction assumption, we obtain the conclusion (1.120) for n.

Set

$$\widetilde{\mathscr{S}}(\mathbb{R}^n)_L \equiv \mathscr{S}(\mathbb{R}^n) \cap \mathscr{P}_L(\mathbb{R}^n)^\perp$$

for $L \in \mathbb{N}$. By the Fourier transform, we have the following:

Corollary 1.1 *Let $L \in \mathbb{N}$. Then there exists $\{T_\gamma\}_{\gamma \in \mathbb{Z}^n}$ of continuous mappings from $\widetilde{\mathscr{S}}(\mathbb{R}^n)_L$ to $\mathscr{S}(\mathbb{R}^n)$ such that $\varphi = \sum_{\gamma \in {\mathbb{N}_0}^n, |\gamma| = L} \partial^\gamma T_\gamma \varphi$ for all $\varphi \in \mathscr{S}(\mathbb{R}^n)$.*

The proof is left to the interested readers as Exercise 1.69.

We present an application of Corollary 1.1.

Proposition 1.13 *Let $f_1, f_2, \ldots \in \mathscr{S}'(\mathbb{R}^n)$, $L \in \mathbb{N}$. For all $|\alpha| = L$, assume that $f_\alpha \equiv \lim_{j \to \infty} \partial^\alpha f_j$ exists in the topology of $\mathscr{S}'(\mathbb{R}^n)$. Then there exists a sequence $\{P_j\}_{j=1}^\infty \in \mathscr{P}_L(\mathbb{R}^n)$ such that $\{f_j + P_j\}_{j=1}^\infty$ converges in the topology of $\mathscr{S}'(\mathbb{R}^n)$.*

Proof We adopt the definitions of Corollary 1.1. As is seen from the induction on L, there exists a collection of polynomials P_α, $|\alpha| \le L-1$ with degree less than $L-1$ such that

$$\varphi_L = \varphi - \sum_{\alpha \in {\mathbb{N}_0}^n, |\alpha| \le L-1} \langle P_\alpha, \varphi \rangle \star^\alpha e^{-|\star|^2} \in \widetilde{\mathscr{S}}(\mathbb{R}^n)_L \tag{1.121}$$

for any test function $\varphi \in \mathscr{S}(\mathbb{R}^n)$. The mapping $\varphi \in \mathscr{S}(\mathbb{R}^n) \mapsto \varphi_L \in \mathscr{S}(\mathbb{R}^n)$ is continuous and $\varphi_L = \sum_{\gamma \in {\mathbb{N}_0}^n, |\gamma| = L} \partial^\gamma T_\gamma \varphi_L$. By virtue of (1.121), we have

$$\varphi(x) = \sum_{\alpha \in {\mathbb{N}_0}^n, |\alpha| \le L-1} \langle P_\alpha, \varphi \rangle x^\alpha e^{-|x|^2} + \sum_{\gamma \in {\mathbb{N}_0}^n, |\gamma| = L} \partial^\gamma T_\gamma \varphi_L(x) \quad (x \in \mathbb{R}^n).$$

Hence since

$$\langle f_j, \varphi \rangle = \sum_{\alpha \in {\mathbb{N}_0}^n, |\alpha| \le L-1} \langle P_\alpha, \varphi \rangle \langle f_j, x^\alpha e^{-|x|^2} \rangle + \sum_{\gamma \in {\mathbb{N}_0}^n, |\gamma| = L} (-1)^L \langle \partial^\gamma f_j, T_\gamma \varphi_L \rangle,$$

if we set $P_j \equiv - \sum_{\alpha \in {\mathbb{N}_0}^n, |\alpha| \le L-1} \langle f_j, x^\alpha e^{-|x|^2} \rangle P_\alpha$, then

$$\langle f_j + P_j, \varphi \rangle = \sum_{\gamma \in {\mathbb{N}_0}^n, |\gamma| = L} (-1)^L \langle \partial^\gamma f_j, T_\gamma \varphi_L \rangle.$$

Thus, we conclude

$$\lim_{j \to \infty} \langle f_j + P_j, \varphi \rangle = \sum_{\gamma \in {\mathbb{N}_0}^n, |\gamma| = L} (-1)^L \langle f_\gamma, T_\gamma \varphi_L \rangle$$

for any test function $\varphi \in \mathscr{S}(\mathbb{R}^n)$ from the assumption, which implies that the limit in the left-hand side exists. Thus, $\{f_j + P_j\}_{j=1}^\infty$ converges in the topology of $\mathscr{S}'(\mathbb{R}^n)$.

Finally, we present an application of Proposition 1.13. We end this section by investigating a little more about $\mathscr{S}'(\mathbb{R}^n)$; it is a sort of Poincare's lemma. We recall the following property:

Lemma 1.19 *Let $\Omega = \mathbb{R}^n$, or let $\mathbb{R}^n_+$. Suppose that we have $\{u_j\}_{j=1}^n \in C^\infty(\Omega)$, where $n \ge 2$. Assume $\partial_{x_j} u_k - \partial_{x_k} u_j$ for any $j, k \in \{1, 2, \dots, n\}$. Then there exists a function u such that $\partial_{x_j} u = u_j$ for all $j = 1, 2, \dots, k$.*

Proof This paraphrases $H^1_{\rm DR}(\Omega) = \{0\}$ or the well-known Poincaré lemma. See [114, Theorem 14.2], for example.

Example 1.21 Let $(f_1, f_2, \dots, f_n) \in \mathscr{S}'(\mathbb{R}^n)^n$ satisfy

$$\frac{\partial f_j}{\partial x_k} = \frac{\partial f_k}{\partial x_j} \quad (j, k = 1, \dots, n).$$

Let us show that there exists $f \in \mathscr{S}'(\mathbb{R}^n)$ such that $f_j = \dfrac{\partial f}{\partial x_j}$. Let $\phi \in C^\infty_{\rm c}(\mathbb{R}^n)$ be a nonnegative function such that $\|\phi\|_1 = 1$. Define $\phi_\varepsilon = \varepsilon^{-n}\phi(\varepsilon^{-1}\star)$ and $f_{j,\varepsilon} = \phi_\varepsilon * f$. Since $f_{j,\varepsilon} \in C^\infty(\mathbb{R}^n)$ and $\partial_{x_k} f_{j,\varepsilon} = \partial_{x_j} f_{k,\varepsilon}$ for all $j, k = 1, \dots, n$, there exists $f_\varepsilon \in C^\infty(\mathbb{R}^n)$ such that $f_{j,\varepsilon} = \partial_{x_j} f_\varepsilon$ thanks to Lemma 1.19. Since we know that $f_{j,\varepsilon} \to f_j$ in $\mathscr{S}'(\mathbb{R}^n)$ as $\varepsilon \downarrow 0$, we see that $f = \lim\limits_{\varepsilon \downarrow 0}(f_\varepsilon + c_\varepsilon)$ in $\mathscr{S}'(\mathbb{R}^n)$ thanks to Proposition 1.13. This function f is what we are looking for.

Exercises

Exercise 1.67 Let $L_1, L_2 \in \mathbb{N}_0$, and let $\varphi \in \mathscr{S}(\mathbb{R}^n) \cap \mathscr{P}_{L_1}(\mathbb{R}^n)^\perp$ and $\eta \in \mathscr{S}(\mathbb{R}^n) \cap \mathscr{P}_{L_2}(\mathbb{R}^n)^\perp$. Then show by binomial expansion that $\varphi * \eta \perp \mathscr{P}_{L_1+L_2+1}(\mathbb{R}^n)$.

Exercise 1.68 Let $\varphi \in C^\infty_{\rm c}(\mathbb{R}^n)$ be such that $\mathscr{F}\varphi \in C^\infty_{\rm c}(\mathbb{R}^n)$.

1. $\mathscr{F}\varphi$ extends to an analytic function on $\mathbb{C}^n$.
2. Show that $\varphi = 0$ inducting on n.

Exercise 1.69 Prove Corollary 1.1 using the Fourier transform.

Exercise 1.70 (Generalized Weyl's lemma) Let $L \in \mathbb{N}$. Assume that $f \in \mathscr{S}'(\mathbb{R}^n)$ satisfies $\Delta^L f = 0$.

1. Choose $\Phi \in \mathscr{S}(\mathbb{R}^n) \setminus \mathscr{P}_0(\mathbb{R}^n)^\perp$ and $\Psi \in \mathscr{S}(\mathbb{R}^n)$ so that $\Phi^1 - \Phi = \Delta^L \Psi$. Then for all positive integers $j \in \mathbb{N}$, show that $f * \Phi^j = f * \Phi^{j+1}$.
2. For some $N \in \mathbb{N}$, show that f is represented by a $C^\infty(\mathbb{R}^n)$-function F satisfying $|\partial^\alpha F(x)| \lesssim \langle x \rangle^N$ for all $\alpha \in {\mathbb{N}_0}^n$.

This is a well-known fact called the *Weyl lemma* for $L = 1$. We can modify the discussion so that we can work on domains.

Exercise 1.71 Replace $C_c^\infty(\mathbb{R}^n)$ with $\mathscr{S}(\mathbb{R}^n)$ in Theorem 1.40 and give an independent proof of Theorem 1.40. In fact, show that we can choose $\Phi^{(L)}$ of the form

$$\Phi^{(L)}(x) = \sum_{j=1}^{L} a_j e^{-j|x|^2} \quad (x \in \mathbb{R}^n)$$

for some $a_1, a_2, \ldots, a_L \in \mathbb{R}$.

Exercise 1.72 [185] Let $l \in \mathbb{N}$, $K \in C_c^\infty(\mathbb{R})$ supported on a compact interval in $(0, 1)$. Define

$$\Phi(t) \equiv \frac{\mathrm{d}^l}{\mathrm{d}t^l}\left(\frac{t^{l-1}}{(l-1)!}\int_0^t K(t)\mathrm{d}t\right) \quad (t \in \mathbb{R}).$$

Then show that $\Phi \in C_c^\infty(K)$ and $\displaystyle\int_0^1 \Phi(t)\mathrm{d}t = \int_0^1 K(t)\mathrm{d}t$.

Exercise 1.73

1. Choose a sequence $\{C_N\}_{N=1}^\infty$ of complex numbers so that

$$\lim_{N\to\infty}\left(C_N + \sum_{k=-N}^{N} \exp(2^k\pi i\star)\right)$$

in the topology of $\mathscr{S}'(\mathbb{R})$.

2. Choose a sequence $\{P_N\}_{N=1}^\infty \subset \mathscr{P}_1$ of complex linear functions so that

$$\lim_{N\to\infty}\left(P_N + \sum_{k=-N}^{N} 2^{-k}\exp(2^k\pi i\star)\right)$$

in the topology of $\mathscr{S}'(\mathbb{R})$.

Hint: There is no need to take care of $\displaystyle\sum_{k=0}^{N} \exp(2^k\pi i\star)$ since this converges in $\mathscr{S}'(\mathbb{R})$.

Exercise 1.74 [681, Lemma 4.3] Let $N \in \mathbb{N}$. Then construct a function $a \in C_c^\infty(\mathbb{R}^n) \cap \mathscr{P}_N(\mathbb{R}^n)^\perp$ satisfying $|a| \le \chi_{Q(1)}$.

1.2.6 Schwartz's Kernel Theorem

Finally, in this section, we deal with the Schwartz kernel theorem, which characterizes the continuous mapping from $\mathscr{S}(\mathbb{R}^n)$ to $\mathscr{S}'(\mathbb{R}^n)$.

1.2.6.1 Schwartz's Kernel Theorem

Let $\mathscr{S}(\mathbb{R}^n)$ be the topological linear space of all Schwartz functions. Furthermore, as a semi-norm of $\mathscr{S}(\mathbb{R}^n)$, we use $p_N(f) \equiv \sum_{\alpha \in \mathbb{N}_0{}^n, |\alpha| \le N} \sup_{x \in \mathbb{R}^n} \langle x \rangle^N |\partial^\alpha f(x)|$.

For $f, g \in \mathscr{S}(\mathbb{R}^n)$, we define the tensor product $f \otimes g \in \mathscr{S}(\mathbb{R}^n \times \mathbb{R}^n)$ of f, g by $f \otimes g(x, y) \equiv f(x)g(y)$. The following theorem claims that T and K have one-to-one correspondence.

Theorem 1.42 (Schwartz's kernel theorem)

1. *For $K \in \mathscr{S}'(\mathbb{R}^n \times \mathbb{R}^n)$, define the mapping $T : \mathscr{S}(\mathbb{R}^n) \to \mathscr{S}'(\mathbb{R}^n)$ by $\langle Tf, g \rangle \equiv K[f \otimes g]$ for $f, g \in \mathscr{S}(\mathbb{R}^n)$. Then $T : \mathscr{S}(\mathbb{R}^n) \to \mathscr{S}'(\mathbb{R}^n)$ is a continuous linear mapping. Furthermore, there exists $N \in \mathbb{N}$ such that*

$$|\langle Tf, g \rangle| \lesssim p_N(f) p_N(g) \tag{1.122}$$

   for all $f, g \in \mathscr{S}(\mathbb{R}^n)$.
2. *Conversely, let $T : \mathscr{S}(\mathbb{R}^n) \to \mathscr{S}'(\mathbb{R}^n)$ be a continuous linear mapping; that is, there exists $N \in \mathbb{N}$ satisfying* (1.122) *for all $f, g \in \mathscr{S}(\mathbb{R}^n)$. Then there uniquely exists $K : \mathscr{S}'(\mathbb{R}^n \times \mathbb{R}^n)$ such that $K[f \otimes g] = \langle Tf, g \rangle$, $f, g \in \mathscr{S}(\mathbb{R}^n)$.*

Proof Let $K \in \mathscr{S}'(\mathbb{R}^n \times \mathbb{R}^n)$. Then inequality (1.122) follows from the fact that there exists N that depends on T such that

$$|K[f \otimes g]| \lesssim p_N(f \otimes g) \lesssim p_N(f) p_N(g) \quad (f, g \in \mathscr{S}(\mathbb{R}^n)).$$

Let us prove that there exists K satisfying (1.122) for any continuous linear mapping $T : \mathscr{S}(\mathbb{R}^n) \to \mathscr{S}'(\mathbb{R}^n)$. First of all, choose a function $\kappa \in \mathscr{S}(\mathbb{R}^n)$ so that $\chi_{B(1)} \le \kappa \le \chi_{B(3/2)}$. We fix $F \in \mathscr{S}(\mathbb{R}^n \times \mathbb{R}^n)$. Then

$$F = \lim_{j \to \infty} F \cdot (\kappa_j \otimes \kappa_j) = F \cdot (\kappa_1 \otimes \kappa_1) + \sum_{j=2}^{\infty} \left\{ F \cdot (\kappa_j \otimes \kappa_j) - F \cdot (\kappa_{j-1} \otimes \kappa_{j-1}) \right\}$$

in the topology of $\mathscr{S}(\mathbb{R}^n \times \mathbb{R}^n)$. Let us define

$$\lambda_{k,m;j}(F) \equiv \frac{1}{(2^{j+1}\pi)^{2n}} \iint_{\mathbb{R}^{2n}} F(u, v) E_{j,k}(u) E_{j,l}(v) K_{(j)}(u, v) \mathrm{d}u \mathrm{d}v.$$

For $u, v \in \mathbb{R}^{2n}$, set

$$K_{(1)}(u, v) \equiv \kappa_1(u)\kappa_1(v)$$

$$K_{(j)}(u, v) \equiv \kappa_j(u)\kappa_j(v) - \kappa_{j-1}(u)\kappa_{j-1}(v), \quad j \ge 2$$

$$I_{(j)}(F) \equiv \frac{1}{(2^{j+1}\pi)^n} \sum_{k,m \in \mathbb{Z}^n} \lambda_{k,l;j}(F) \langle T E_{j+1,k}, E_{j+1,l} \rangle, \quad j \geq 2.$$

Now we define a function $E_{j,m}$ by

$$E_{j,m}(x) \equiv \kappa_{j+1}(x) \exp(2^{-j} m \cdot xi) \quad (x \in \mathbb{R}^n)$$

for $j \in \mathbb{N}$ and $m \in \mathbb{Z}^n$. We expand a $2\pi \cdot 2^j(\mathbb{Z}^n \times \mathbb{Z}^n)$-periodic function

$$\sum_{m_1, m_2 \in \mathbb{Z}^n} F \cdot (\kappa_j \otimes \kappa_j)(\star_1 - 2^{j+1}\pi m_1, \star_2 - 2^{j+1}\pi m_2)$$

into a Fourier series and then multiply by $E_{j+1,k} \otimes E_{j+1,l}$ to both sides to obtain

$$F \cdot (\kappa_j \otimes \kappa_j) = \frac{1}{(2^{j+1}\pi)^{2n}} \sum_{k,l \in \mathbb{Z}^n} \lambda_{k,l;j}(F) E_{j+1,k} \otimes E_{j+1,l}. \tag{1.123}$$

Using the expansion (1.123), we define $K \in \mathscr{S}'(\mathbb{R}^{2n})$ by

$$\langle K, F \rangle \equiv \sum_{j=1}^{\infty} I_{(j)}(F). \tag{1.124}$$

In the sums (1.123) and (1.124), we have an infinite sum, respectively; $\sum_{j=1}^{\infty}$ and $\sum_{k,m \in \mathbb{Z}^n}$. We have to be careful; does (1.124) makes sense? First of all, in view of the continuity of T, we can find $N \in \mathbb{N}$ such that $|\langle T E_{j+1,k}, E_{j+1,l} \rangle| \lesssim 2^{jN} \langle k \rangle^N \langle l \rangle^N$. Next, if we take a positive integer L sufficiently large, then a repeated integration by parts and the size of the support of K_j yields:

$$\left| \iint_{\mathbb{R}^{2n}} F(u,v) E_{j,k}(u) E_{j,l}(v) K_{(j)}(u,v) \mathrm{d}u\mathrm{d}v \right| \lesssim \frac{2^{-2j(N+n)}}{\langle k \rangle^{N+n+1} \langle l \rangle^{N+n+1}} p_L(F).$$

Note that $2^{-2j(N+n)}$ comes out since $F \in \mathscr{S}(\mathbb{R}^{2n})$ and $E_{j,k}, E_{j,l}$ are both supported on the dyadic annulus $B(2^{j+3}) \setminus B(2^{j-3})$. The implicit constants in $\lesssim$ do not depend on j, k, l. So, we see that the series (1.124) defining $\langle K, F \rangle$ is convergent and that $F \mapsto \langle K, F \rangle$ is continuous. The above observation shows that the expansion of F converges in the topology of $\mathscr{S}(\mathbb{R}^n \times \mathbb{R}^n)$, which forces K to be expressed as above. Furthermore, when $F = f \otimes g$, we can express the Fourier integral as a product:

$$\begin{aligned}&\iint_{\mathbb{R}^{2n}} F(u,v)E_{j,k}(u)E_{j,l}(v)\kappa_j(u)\kappa_j(v)\mathrm{d}u\mathrm{d}v\\&=\int_{\mathbb{R}^n} f(u)E_{j,k}(u)\kappa_j(u)\mathrm{d}u \times \int_{\mathbb{R}^n} g(v)E_{j,l}(v)\kappa_j(v)\mathrm{d}v.\end{aligned}$$

Using this identity, we obtain $K[f \otimes g] = \langle Tf, g\rangle$. The uniqueness is left as an exercise to interested readers; see Exercise 1.75.

We can define the dual using the above theorem.

Definition 1.32 (The dual T^*) Let $T : \mathscr{S}(\mathbb{R}^n) \to \mathscr{S}'(\mathbb{R}^n)$ be the same as in Theorem 1.42. Define a continuous linear mapping $T^* : \mathscr{S}(\mathbb{R}^n) \to \mathscr{S}'(\mathbb{R}^n)$ so that

$$\langle Tf, g\rangle = \langle T^*g, f\rangle, \quad (f, g \in \mathscr{S}(\mathbb{R}^n)).$$

The operator T^* is called the dual of T.

Exercises

Exercise 1.75 Let $m, n \in \mathbb{N}$. Define

$$\mathscr{S}(\mathbb{R}^m) \otimes \mathscr{S}(\mathbb{R}^n) \equiv \mathrm{Span}\{f \otimes g : f \in \mathscr{S}(\mathbb{R}^m), g \in \mathscr{S}(\mathbb{R}^n)\}.$$

1. Show that $\mathscr{S}(\mathbb{R}^m) \otimes \mathscr{S}(\mathbb{R}^n)$ is dense in $\mathscr{S}(\mathbb{R}^{m+n})$. Hint: Use first Theorem 1.24 to justify that approximation of compactly supported functions suffices and then we use the trigonometric expansion as we will do in the proof of Theorem 4.9.
2. Show the uniqueness of T in Theorem 1.42.

Exercise 1.76 Reexamine the proof of Theorem 1.42 to show that linear combinations of $C_c^\infty(\mathbb{R})$ -tensor functions are dense in $\mathscr{S}(\mathbb{R}^n)$.

Textbooks in Sect. 1.2

Fourier Transform

See [35, Chapter 1], [87, Chapter 5] and [91, Chapter 7].

Fourier Series

See [46, Chapters 1 and 2] and [121, Chapter 1].

Distributions

The original textbook is [81]; which is written in French. See [89, Chapter 3], [90, Chapter 1, Section 3] and [91, Chapter 1].

1.3 Difference/Oscillation Operators

Considering the difference of functions is a general method to measure their smoothness. In fact, a function $f : \mathbb{R}^n \to \mathbb{C}$ is continuous at a point $x \in \mathbb{R}^n$ if for all $\varepsilon > 0$ there exists $\delta > 0$ such that $|h| < \delta$ implies $|f(x) - f(x-h)| < \varepsilon$. Because the difference $f(x) - f(x-h)$ appears in this definition, we are convinced that considering the difference is the most fundamental method to measure the smoothness of functions.

Meanwhile, the Fourier transform converts the smoothness of functions into the size of functions. In this book, we use the Fourier transform to measure the smoothness of functions.

However, because the difference is a fundamental method to measure the smoothness of functions, we need to feed what is learned from the size of the Fourier transform back to the information on the differences.

From this viewpoint, we will collect some elementary formulas for the differences.

1.3.1 Elementary Formulas

One aspect of the studies on function spaces is to investigate the properties of the solutions of differential equations. Second order differential equations often appear in mathematics. Therefore, expressing functions with higher order smoothness in terms of the difference is useful. With this in mind, the *difference operator* with *step* $h \in \mathbb{R}^n$ of *order* $k \in \mathbb{N}$, which is denoted by Δ_h^k, is defined inductively by

$$\Delta_h^1 f \equiv \Delta_h f \equiv f(\star + h) - f, \quad \Delta_h^k \equiv \Delta_h^1 \circ \Delta_h^{k-1}, \quad k \geq 2 \tag{1.125}$$

for functions or distributions f.

Apart from the problem of how to use Δ_h^k to measure the smoothness, let us consider some important identities related to this difference.

1.3.1.1 An Elementary Formula for Binomial Coefficients

Because binomial coefficients appear naturally in definition (1.125), consider a fundamental formula, which is interesting in its own right, including its proof!

Lemma 1.20 *For all* $d \in \mathbb{N}$ $\sum_{m=1}^{d}(-1)^m m^d \binom{d}{m} = (-1)^d d!$.

Proof Set $f(t) \equiv \sum_{m=0}^{d}(-e^t)^m \binom{d}{m} = (1-e^t)^m$ for $t \in \mathbb{R}$. Then arithmetic shows: $\sum_{m=1}^{d} m^d(-1)^m \binom{d}{m} = f^{(d)}(0)$. Since $f(t) = (-1)^d t^d + \mathrm{O}(t^{d+1})$ as $t \to 0$ by Taylor expansion, it follows that $f^{(d)}(0) = (-1)^d d!$. Thus, the proof is complete.

1.3.1.2 Mollification and Difference

The mean-value theorem is probably one of the simplest ways to handle differences. Because expressing the differentiation by the difference seems daunting, we will discuss the relation between the mollification and the difference as an intermediate step. For $d = 1, 2, \dots$ and $\psi \in \mathscr{S}(\mathbb{R}^n)$, we set

$$\Psi \equiv \sum_{r=1}^{d}\sum_{m=1}^{d} \frac{(-1)^{r+m+d+1} m^{d-n}}{r^n d!} \binom{d}{m}\binom{d}{r} \psi\left(\frac{\star}{rm}\right). \tag{1.126}$$

By Lemma 1.20, we have

$$\begin{aligned}
\int_{\mathbb{R}^n} \Psi(x)\mathrm{d}x &= \sum_{r=1}^{d}\sum_{m=1}^{d} \frac{(-1)^{r+m+d+1} m^d}{d!} \binom{d}{m}\binom{d}{r} \int_{\mathbb{R}^n} \psi(x)\mathrm{d}x \\
&= \frac{(-1)^{d+1}}{d!}\left(\sum_{r=1}^{d}(-1)^r \binom{d}{r}\right)\left(\sum_{m=1}^{d}(-1)^m m^d \binom{d}{m}\right) \int_{\mathbb{R}^n} \psi(x)\mathrm{d}x \\
&= \frac{(-1)^{d+1}}{d!} \times (-1) \times (-1)^d d! \int_{\mathbb{R}^n} \psi(x)\mathrm{d}x = \int_{\mathbb{R}^n} \psi(x)\mathrm{d}x.
\end{aligned}$$

In view of the equality $\mathscr{F}\Psi(0) = (2\pi)^{-\frac{n}{2}} \int_{\mathbb{R}^n} \Psi(x)\mathrm{d}x$, we have the following:

Corollary 1.2 *If* $\mathscr{F}\psi$ *equals* 1 *on* $B(1)$, *then so does* $\mathscr{F}\Psi$ *on* $B(1)$.

If

$$f^j \equiv \Psi^j * f = 2^{jn}\Psi(2^j\star) * f, \tag{1.127}$$

where Ψ is given by (1.126), then $f^j - f$ can be expressed in terms of the difference operator.

Theorem 1.43 *Let $d \in \mathbb{N}$ be fixed. We define $\Theta \in \mathscr{S}(\mathbb{R}^n)$ by*

$$\Theta \equiv \sum_{m=1}^{d} \frac{(-1)^{m+d+1} m^d}{d!} \binom{d}{m} \Psi\left(\frac{\star}{m}\right), \tag{1.128}$$

where Ψ is further given by (1.126). *Then*

$$f^j(x) - f(x) = \int_{\mathbb{R}^n} \Theta(y) \Delta^d_{2^{-j}y} f(x) \mathrm{d}y \quad (j \in \mathbb{Z}, x \in \mathbb{R}^n)$$

for all $f \in \mathscr{S}'(\mathbb{R}^n)$. In particular, the conclusion is valid for $f \in L^1_{\mathrm{loc}}(\mathbb{R}^n)$ such that $\displaystyle\int_{\mathbb{R}^n} (1 + |x|)^{-N} |f(x)| \mathrm{d}x < \infty$ for some large N.

Proof We will suppose $f \in L^1_{\mathrm{loc}}(\mathbb{R}^n)$ and that $\displaystyle\int_{\mathbb{R}^n} (1 + |x|)^{-N} |f(x)| \mathrm{d}x < \infty$ for some large N. We can consider the case $f \in \mathscr{S}'(\mathbb{R}^n)$ similarly. Write $f^j - f$ fully. Then

$$\begin{aligned}
&f^j(x) - f(x) \\
&= \sum_{r=1}^{d} \sum_{m=1}^{d} \frac{2^{jn}(-1)^{r+m+d+1}}{m^{n-d} r^n d!} \binom{d}{m} \binom{d}{r} \int_{\mathbb{R}^n} \Psi\left(\frac{2^j y}{rm}\right) f(x-y) \mathrm{d}y - f(x) \\
&= \sum_{r=1}^{d} \sum_{m=1}^{d} \frac{(-1)^{r+m+d+1}}{m^{n-d} d!} \binom{d}{m} \binom{d}{r} \int_{\mathbb{R}^n} \Psi\left(\frac{y}{m}\right) f(x - 2^{-j} r y) \mathrm{d}y - f(x).
\end{aligned}$$

Thus, it follows from Lemma 1.20 that

$$\begin{aligned}
f^j(x) - f(x) &= \sum_{r=0}^{d} \sum_{m=1}^{d} \frac{(-1)^{r+m+d+1}}{m^{n-d} d!} \binom{d}{m} \binom{d}{r} \int_{\mathbb{R}^n} \Psi\left(\frac{y}{m}\right) f(x - 2^{-j} r y) \mathrm{d}y \\
&= \sum_{m=1}^{d} \frac{(-1)^{m+d+1}}{m^{n-d} d!} \binom{d}{m} \int_{\mathbb{R}^n} \Psi\left(\frac{y}{m}\right) \sum_{r=0}^{d} (-1)^r \binom{d}{r} f(x - 2^{-j} r y) \mathrm{d}y \\
&= \int_{\mathbb{R}^n} \left\{ \sum_{m=1}^{d} \frac{(-1)^{m+d+1} m^d}{d!} \binom{d}{m} \Psi\left(\frac{y}{m}\right) \right\} \Delta^d_{2^{-j}y} f(x) \mathrm{d}y \\
&= \int_{\mathbb{R}^n} \Theta(y) \Delta^d_{2^{-j}y} f(x) \mathrm{d}y.
\end{aligned}$$

Therefore, the proof is complete.

Now consider the expression of $f^{j+1} - f^j$.

When Ψ_0 is as in Example 1.20, the following expression holds:

Proposition 1.14 *Choose* $\Psi_0 \in \mathscr{S}(\mathbb{R})$ *as in Example* 1.20 *and set* $\Psi \equiv \otimes^n \Psi_0 = \Psi_0 \otimes \Psi_0 \otimes \ldots \otimes \Psi_0$. *Define* Θ *by* (1.128). *Then there exist* $\theta_1, \theta_2, \ldots, \theta_n \in \mathscr{S}(\mathbb{R}^n)$ *such that*

$$f^j(x) - f^{j+1}(x) = \int_{\mathbb{R}^n} \sum_{k=1}^n \partial_{x_k}{}^d \theta_k(y) \Delta^d_{2^{-j}y} f(x) \mathrm{d}y \quad (x \in \mathbb{R}^n) \tag{1.129}$$

for all $f \in L^1_{\mathrm{loc}}(\mathbb{R}^n)$.

Proof Note that $\Phi - \Phi^{-1} = \sum_{k=1}^n \partial_{x_k}{}^d \tau_k$ for some $\tau_1, \tau_2, \ldots, \tau_n \in \mathscr{S}(\mathbb{R}^n)$ and that $\Theta - \Theta^{-1} = \sum_{k=1}^n \partial_{x_k}{}^d \theta_k$ for some $\theta_1, \theta_2, \ldots, \theta_n \in \mathscr{S}(\mathbb{R}^n)$. It remains to use (1.128) to obtain the desired result.

Exercises

Exercise 1.77 Let $d \in \mathbb{N}$. Let $P^j \equiv \Psi^j * P = 2^{jn}\Psi(2^j \star) * P$ for all $j \in \mathbb{Z}$ and $P \in \mathscr{P}_{d-1}$, where Ψ is given by (1.126), Show that $P^j = P$ using Theorem 1.43.

Exercise 1.78 [187] Let $M \in \mathbb{N}$. Let $\eta \in L^1(\mathbb{R}^n)$, and define

$$\Psi(x) \equiv \sum_{k=1}^M \frac{(-1)^{M-k}}{k^{1+n}} \binom{M}{k} \eta\left(\frac{x}{k}\right) \quad (x \in \mathbb{R}^n).$$

Show that

$$\int_{\mathbb{R}^n} \Psi(x) \mathrm{d}x = \int_0^1 \frac{(t-1)^M - (-1)^M}{t} \mathrm{d}t \times \int_{\mathbb{R}^n} \eta(x) \mathrm{d}x.$$

Exercise 1.79 (Sobolev integral representation) Let ω be a continuous function on $[a, b]$ such that $\int_a^b \omega(t) \mathrm{d}t = 1$. Define

$$\Lambda(x, y) \equiv \chi_{(-\infty,x]}(y) \int_a^y \omega(t) \mathrm{d}t - \chi_{(x,\infty)}(y) \int_y^b \omega(t) \mathrm{d}t$$

for all $x, y \in (a, b)$. Then show that

$$f(x) = \int_a^b f(t)\omega(t)\mathrm{d}t + \int_a^b \Lambda(x,t)f'(t)\mathrm{d}t$$

for all $x \in (a, b)$ and all $f \in C^1[a, b]$. See [300, Teorema] for a related and generalized formula.

Exercise 1.80 [310] Let $m \in \mathbb{N}$, and let $b_k \in \mathbb{R}$, $k = 0, 1, 2, \ldots, m-1$. Define

$$a_{im} \equiv \sum_{j=i+1}^{m} \frac{(-1)^{j+1}}{j}\binom{m}{j} + \sum_{j=\max\{1,[2^{-1}(i+1)]\}}^{\min(i,m)} \frac{(-1)^j}{j}\binom{m}{j}, \quad i = 0, 1, \ldots, 2m-1$$

and define λ_{jm} for $j = 0, 1, \ldots, m-1$ and $k = 1, 2, \ldots, m-1$ by

$$\lambda_{0m} \equiv (-1)^m a_{0m}, \quad \lambda_{km} \equiv (-1)^m a_{km} + (-1)^{k+1}\sum_{j=0}^{k-1}(-1)^j\binom{k-j}{m}\lambda_{jm}.$$

Set $\Delta^m b_k \equiv \sum_{j=0}^{m}(-1)^{m+j}\binom{j}{m} b_{k+j}$. Then show that $\sum_{i=0}^{2m-1} a_{im} b_i = \sum_{k=0}^{m-1} \lambda_{km}\Delta^m b_k$ for $m \in \mathbb{N}$.

Exercise 1.81 [666, Lemma 22] Let $a, b > 0$, $M \in \mathbb{N}$, $h \in \mathbb{R}^n$ and let $f \in \mathscr{S}_{B(b)}(\mathbb{R}^n)$. Then show that

$$|\Delta_h^M f(x)| \lesssim \max(1, |bh|)^a \min(1, |bh|)^M \sup_{z \in \mathbb{R}^n} \frac{|f(x-z)|}{(1+|bz|)^a}$$

holds for every $x \in \mathbb{R}^n$.

1.3.2 Oscillation

It is possible that an integrable function satisfying some integral inequalities is continuous. More surprisingly, it may be Hölder continuous. Such a situation can be encountered when considering the weak solution of elliptic differential equations. In fact, the theory of function spaces allows the solution of the Laplace equation (and other important differential equations) to be grasped. However, the technique is indirect. First the weak solution (usually locally integrable) has to be considered, and the function must be deemed sufficiently smooth. Although information for derivatives is typically used, methods independent of derivatives do exist. Oscillation is one such technique. Here we use it to consider higher order smoothness.

1.3.2.1 Oscillation and the Best Polynomial

Let us define the oscillation. We defined $\mathscr{P}_d(\mathbb{R}^n)$ to be the set of all polynomials in $\mathbb{R}^n$ of order d. We write $\mathbb{R}^{n+1}_+ \equiv \mathbb{R}^n \times (0, \infty)$ and $\mathscr{P}_{-1}(\mathbb{R}^n) \equiv \{0\}$.

Definition 1.33 (Oscillation function) Let $d \in \mathbb{N}_0 \cup \{-1\}$, and let $u \in (0, \infty)$. For $f \in L^0(\mathbb{R}^n)$, the *oscillation function* $\mathrm{osc}^d_u f : \mathbb{R}^{n+1}_+ \to [0, \infty)$ is defined by

$$\mathrm{osc}^d_u f(x,t) \equiv \inf_{P \in \mathscr{P}_d(\mathbb{R}^n)} m^{(u)}_{B(x,t)}(f - P) \quad (x \in \mathbb{R}^n, t > 0).$$

From the above definition, we look for a polynomial of order d. Since x and t are given, the candidate should be quite familiar. Let $u \geq 1$, $d \in \mathbb{N}_0$, and let $\Psi \in C^\infty_c(B(1))$ satisfy $\int_{\mathbb{R}^n} \Psi(x)\mathrm{d}x = 1$. Choose Θ so that (1.128) holds for some $\Psi \in C^\infty_c(B(1))$ satisfying (1.126). Let $j \in \mathbb{Z}$ and $x \in \mathbb{R}^n$. For $f \in \mathscr{S}'(\mathbb{R}^n) \cap L^1_{\mathrm{loc}}(\mathbb{R}^n)$, define f^j by (1.127). Then by definition

$$\mathrm{osc}^{d-1}_u f(x, 2^{-j}) \leq m^{(u)}_{B(x,2^{-j})}\left(f - \sum_{\alpha \in \mathbb{N}_0{}^n, |\alpha| \leq d-1} \frac{1}{\alpha!}\frac{\partial^\alpha f^j}{\partial x^\alpha}(x)(\star - x)^\alpha\right) \quad (x \in \mathbb{R}^n).$$

Concerning the right-hand side of this formula, we have the following estimate:

Proposition 1.15 (Optimal polynomials) *Let $u \geq 1$, $d \in \mathbb{N}_0$, and let $\Psi \in C^\infty_c(B(1))$ satisfy $\int_{\mathbb{R}^n} \Psi(x)\mathrm{d}x = 1$. Choose Θ so that* (1.128) *holds for some $\Psi \in C^\infty_c(B(1))$ satisfying* (1.126). *Let $j \in \mathbb{Z}$ and $x \in \mathbb{R}^n$. For $f \in \mathscr{S}'(\mathbb{R}^n) \cap L^1_{\mathrm{loc}}(\mathbb{R}^n)$, define f^j by* (1.127). *Then*

$$\sup_{y \in B(x,2^{-j})} \left| f^j(y) - \sum_{\alpha \in \mathbb{N}_0{}^n, |\alpha| \leq d-1} \frac{1}{\alpha!}\frac{\partial^\alpha f^j}{\partial x}(x)(y-x)^\alpha \right| \lesssim \mathrm{osc}^{d-1}_u f(x, 2^{-j}(d+1)).$$

Proof The proof is made up of two parts: We prove

$$m^{(u)}_{B(x,2^{-j})}(f^j - f) \lesssim \mathrm{osc}^{d-1}_u f(x, 2^{-j}(d+1)) \tag{1.130}$$

and

$$\sup_{y \in B(x,2^{-j})} \left| f^j(y) - \sum_{\alpha \in \mathbb{N}_0{}^n, |\alpha| \leq d-1} \frac{1}{\alpha!}\frac{\partial^\alpha f^j}{\partial x}(x)(y-x)^\alpha \right| \lesssim \mathrm{osc}^{d-1}_u f(x, 2^{-j}d). \tag{1.131}$$

Since $\mathrm{supp}(\Theta) \subset B(d)$, we have

$$|f^j(x) - f(x)| \le |f^j(x)| + |f(x)| \lesssim |f(x)| + m_{B(x,2^{-j}d)}(|f|).$$

Let $P \in \mathscr{P}_{d-1}(\mathbb{R}^n)$. By considering $f - P$ instead of f, we have

$$|f^j(z) - f(z)| \lesssim |f(z) - P(z)| + 2^{jn} \int_{B(z,2^{-j}d)} |f(y) - P(y)| \mathrm{d}y$$

for all $z \in B(x, 2^{-j})$ from Exercise 1.77. Since $u \ge 1$, (1.130) follows. Additionally, note that

$$\begin{aligned} &\sup_{y \in B(x,2^{-j})} \left| f^j(y) - \sum_{\alpha \in \mathbb{N}_0{}^n, |\alpha| \le d-1} \frac{1}{\alpha!} \frac{\partial^\alpha f^j}{\partial x}(x)(y-x)^\alpha \right| \\ &\lesssim 2^{-jd} \sup\{|\partial^\alpha f^j(y)| \,:\, y \in B(x, 2^{-j}), \alpha \in \mathbb{N}_0{}^n, |\alpha| = d\} \\ &= 2^{-jd} \sup\{|\partial^\alpha [f - P]^j(y)| \,:\, y \in B(x, 2^{-j}), \alpha \in \mathbb{N}_0{}^n, |\alpha| = d\} \end{aligned}$$

for all $P \in \mathscr{P}_{d-1}(\mathbb{R}^n)$. Finally, recall $(f - P)^j = \Psi^j * (f - P)$, where Ψ is further given by (1.126). By the Hölder inequality, we have (1.131). Therefore, the proof is complete.

1.3.2.2 Additive and Multiplicative Inequalities

The additive and multiplicative inequalities are very useful. The following identity can be proved by the Fourier transform:

Lemma 1.21 *Let $d \in \mathbb{N}$. Then there exists a collection $\{a_0, a_1, \ldots, a_{d-1}\}$ of rational coefficients such that:*

$$\Delta_y^d f - 2^{-n} \Delta_{2y}^d f = \sum_{r=0}^{d-1} a_r \Delta_y^{d+1} f(\star + ry) \tag{1.132}$$

for all $y \in \mathbb{R}^n$ and $f \in L^1_{\mathrm{loc}}(\mathbb{R}^n)$ or $f \in \mathscr{S}'(\mathbb{R}^n)$.

The proof is left as Exercise 1.82.

We compare $\mathrm{osc}_u^{d-1} f$ and $\mathrm{osc}_u^d f$.

Proposition 1.16 (Additive inequality) *Let $1 \le u \le \infty$, $d \in \mathbb{N}_0$, and let $f \in \mathscr{S}'(\mathbb{R}^n) \cap L^1_{\mathrm{loc}}(\mathbb{R}^n)$. Then there exists $D \gg 1$ such that*

$$\mathrm{osc}_u^{d-1} f(x, 2^{-j}) \lesssim \sum_{l=0}^{j} 2^{-(j-l)d} \mathrm{osc}_u^d f(x, 2^{-l} D) + 2^{-jd} \|f\|_{L^1(B(x,1))}.$$

Proof Choose Θ so that (1.128) holds for some $\Psi \in C_c^\infty(B(1))$ satisfying (1.126). Define $f^j \equiv \Psi^j * f$. By Proposition 1.15

$$\mathrm{osc}_u^{d-1} f(x, 2^{-j}) \lesssim \mathrm{osc}_u^d f(x, 2^{-j}(d+1)) + 2^{-jd} \sum_{\alpha \in \mathbb{N}_0{}^n, |\alpha|=d} |\partial^\alpha f^j(x)|, \tag{1.133}$$

where f^j is given by (1.127). Let α be a multi-index with length d. Then from Theorem 1.43

$$\partial^\alpha (f^l - f)(x) = \int_{\mathbb{R}^n} \Theta(y) \Delta_{2^{-l}y}^d \partial^\alpha f(x) \mathrm{d}y = 2^{ld} \int_{\mathbb{R}^n} \partial^\alpha \Theta(y) \Delta_{2^{-l}y}^d f(x) \mathrm{d}y.$$

Thus, it follows from Lemma 1.21 that

$$\begin{aligned}
&\partial^\alpha (f^l - f^{l+1})(x) \\
&= 2^{ld} \int_{\mathbb{R}^n} \partial^\alpha \Theta(y) \Delta_{2^{-l}y}^d f(x) \mathrm{d}x - 2^{(l+1)d} \int_{\mathbb{R}^n} \partial^\alpha \Theta(y) \Delta_{2^{-l-1}y}^d f(x) \mathrm{d}x \\
&= 2^{ld} \int_{\mathbb{R}^n} \partial^\alpha \Theta(y) \left(\Delta_{2^{-l}y}^d f(x) - 2^d \Delta_{2^{-l-1}y}^d f(x) \right) \mathrm{d}x \\
&= 2^{(l+1)d} \int_{\mathbb{R}^n} \partial^\alpha \Theta(y) \sum_{r=0}^{d-1} a_r \Delta_{2^{-l-1}y}^{d+1} f(x + r2^{-l-1}y) \mathrm{d}x.
\end{aligned}$$

Hence

$$|\partial^\alpha f^{l+1}(x) - \partial^\alpha f^l(x)| \lesssim 2^{ld} \mathrm{osc}_u^d f(x, 2^{-l} D) \tag{1.134}$$

for some $D \gg 1$. A similar argument yields

$$|\partial^\alpha f^0(x)| \lesssim \|f\|_{L^1(B(x,1))}. \tag{1.135}$$

Therefore, combining (1.133), (1.134), and (1.135) completes the proof.

Now we consider the multiplicative inequality.

Theorem 1.44 (Multiplicative inequality) *Let $d \in \mathbb{N}_0$ be fixed, and let $0 < u < 1 < r \le \infty$. Define $\theta \in (0, 1)$ by the equation:*

$$\frac{1-\theta}{u} + \frac{\theta}{r} = 1.$$

We define f^j *by* $f^j \equiv \tau^j * f$*, where* $\psi \in C_c^\infty(B(1))$ *has integral* 1 *and*

$$\tau \equiv \sum_{\ell=1}^{d+1}\sum_{m=1}^{d+1} \frac{m^{d+1-n}(-1)^{\ell+m+d}}{\ell^n(d+1)!}\binom{d+1}{m}\binom{d+1}{\ell}\psi\left(\frac{\star}{\ell m}\right).$$

where ψ *is further given by* (1.126)*. Define* Θ *by*

$$\Theta \equiv \sum_{m=1}^{d+1} \frac{m^{d+1}(-1)^{m+d}}{(d+1)!}\binom{d+1}{m}\Psi\left(\frac{\star}{m}\right).$$

Then for all $f \in L^1_{\rm loc}(\mathbb{R}^n)$*,* $j \in \mathbb{Z}$ *and* $x \in \mathbb{R}^n$*,*

$$\begin{aligned}&{\rm osc}_1^d f(x,2^{-j})\\ &\lesssim (m^{(u)}_{B(x,2^{-j})}(f-f^j)+{\rm osc}_u^d f(x,2^{-j}))^{1-\theta}(m^{(r)}_{B(x,2^{-j})}(f-f^j)+{\rm osc}_r^d f(x,2^{-j}))^{\theta}.\end{aligned}$$

Proof We start from ${\rm osc}_1^d f(x,2^{-j}) \le m_{B(x,2^{-j})}(|f-f^j|)+{\rm osc}_1^d f_j(x,2^{-j})$. Note that Ψ is supported in $B(d+1)$. We write

$$P(z) \equiv \sum_{\alpha\in\mathbb{N}_0{}^n, |\alpha|\le d} \frac{1}{\alpha!}\frac{\partial^\alpha f^j}{\partial x^\alpha}(x)(z-x)^\alpha \quad (z \in \mathbb{R}^n).$$

Then

$$f^j(x)-f(x) = \int_{\mathbb{R}^n} \Theta(y)\Delta^{d+1}_{2^{-j}y}(f-P)(x){\rm d}y \quad (x \in \mathbb{R}^n)$$

according to Theorem 1.43. Consequently,

$$|f^j(x)-f(x)| \lesssim m^{(u)}_{B(x,2^{-j}(d+1))}(\Delta^{d+1}_{2^{-j}\star}(f-P))^{1-\theta} m^{(r)}_{B(x,2^{-j}(d+1))}(\Delta^{d+1}_{2^{-j}\star}(f-P))^{\theta}.$$

If we use Proposition 1.15, then we can handle $m_{B(x,2^{-j})}(|f-f^j|)$. Since

$$|f^j(z)-P(z)| \lesssim \sup_{|\beta|=d+1,\, w\in B(d+2)} \int_{\mathbb{R}^n} |\partial^\beta \tau^j(v)(f(w-v)-P(w-v))|{\rm d}v,$$

we can handle ${\rm osc}_1^d f_j(x,2^{-j})$.

Exercises

Exercise 1.82 Prove Lemma 1.21 using the Fourier transform.

Exercise 1.83 Let Q be a cube, and let $u \in C^1(Q)$. That is, u is a C^1-function defined in a neighborhood of Q.

1. Show that $m_Q(|u - m_Q(u)|) \lesssim \ell(Q) m_Q(|\nabla u|)$. [Poincaré–Wirtinger inequality] We sometimes call Poincaré–Wirtinger inequality Poincaré inequality.
2. Assume $\lim\limits_{j \to \infty} m_Q(u) = 0$. Then show that $|u(x)| \lesssim I_1(|\nabla u|)(x)$ for $x \in \mathbb{R}^n$ using the Lebesgue differentiation theorem. See Theorem 1.48 for the Lebesgue differentiation theorem.

Exercise 1.84 Let $f \in C^1(\mathbb{R}^n)$, and let $F : \mathbb{R}^n \to [0, \infty]$ be a measurable function.

1. Show that

$$|f(y) - f(x)| \le |x - y| \int_0^1 |\nabla f(x + t(y - x))| \mathrm{d}t$$

for $x, y \in \mathbb{R}^n$.
2. Show that $\displaystyle\int_{B(x,1)} \left(\int_{|x-z|}^1 F(z,t) \mathrm{d}t \right) \mathrm{d}z = \int_0^1 \left(\int_{B(x,t)} F(z,t) \mathrm{d}z \right) \mathrm{d}t$ for $x \in \mathbb{R}^n$.
3. Show that $|f(x) - m_{B(x,1)}(f)| \lesssim \displaystyle\int_{B(x,1)} \frac{|\nabla f(y)|}{|x - y|^{n-1}} \mathrm{d}y$ for $x \in \mathbb{R}^n$.

Exercise 1.85 [267, Lemma 6] Let $f : \mathbb{R}^n \to \mathbb{R}$ be a measurable function. Then show that the following are equivalent:

1. f is a sum of a bounded Borel function and a Lipschitz continuous function.
2. For all $x, y \in \mathbb{R}^n$, $|f(x) - f(y)| \lesssim \langle x - y \rangle$.
3. For all x, y with $|x - y| \le 1$, $|f(x) - f(y)| \lesssim 1$. Hint: Let ϕ be a $C^\infty(\mathbb{R}^n)$-function supported on $B(1)$. Define a function k by $k \equiv f - f * \phi$.

As a non-measurable set in $[0, 1]$ shows, the assumption of measurability is absolutely necessary.

Textbooks in Sect. 1.3

Difference

We refer to [100, Section 3.3] for differences.

Oscillations

We refer to [100, Section 3.4] for oscillations.

1.4 Boundedness of the Hardy–Littlewood Maximal Operator

Here we present fundamental results for the Hardy–Littlewood maximal operator.

In Sect. 1.4.1, we consider fundamental results of the boundedness of the Hardy–Littlewood maximal operator (Theorems 1.45, 1.46 and so on). In Sect. 1.4.2 we deal with the Fefferman–Stein vector-valued inequality (Theorem 1.49). The Fefferman–Stein vector-valued inequality is indispensable in the study of function spaces, in particular Triebel–Lizorkin spaces. In Sect. 1.4.3, we investigate the Plancherel–Polya–Nikolski'i inequality and related inequalities (Theorems 1.50, 1.51, 1.53 and so on). These play important roles in the study of Besov spaces and Triebel–Lizorkin spaces. Section 1.4.4 considers some integral inequalities for convolutions. We consider some duality results in Sect. 1.4.5. Section 1.4.5 is of a different nature from the remaining sections in Sect. 1.4.

1.4.1 Hardy–Littlewood Maximal Inequality

The history of the Hardy–Littlewood maximal operator dates back to 1932 (see [554]) and the original definition is, as follows:

$$Mf(t) \equiv \sup_{I \in \mathscr{I}} \frac{\chi_I(t)}{|I|} \int_I |f(s)| \mathrm{d}s \quad (t \in \mathbb{R})$$

for $f \in L^0(\mathbb{R})$. Here $\mathscr{I}$ denotes the set of all open intervals. The original aim was to apply it to the functions on the unit disk $\Delta(1)$ on the complex plane. So, originally the Hardy–Littlewood maximal operator was used to the analysis on the torus $\mathbb{T} = \mathbb{R}/2\pi\mathbb{Z}$. Here to study the function spaces on $\mathbb{R}$, we work in the setting of $\mathbb{R}$.

A natural passage to higher dimensions is

$$Mf(x) \equiv \sup_{B \in \mathscr{B}} \frac{\chi_B(x)}{|B|} \int_B |f(y)| \mathrm{d}y \quad (x \in \mathbb{R}^n)$$

for $f \in L^0(\mathbb{R}^n)$, which is due to N. Wiener [1112]. Now we observe that the ball comes into play because of the definition of the convergence of sequences in metric space $\mathbb{R}^n$. Here $\mathscr{B}$ denotes the set of all open balls. His purpose was to apply it to the ergodic theory.

However, the Hardy–Littlewood maximal operator appears in many mathematical contexts. Here let us see three examples after giving definitions.

1.4.1.1 Definition

The Hardy–Littlewood maximal operator is an elementary tool, and mostly we use it to study Besov spaces and Triebel–Lizorkin spaces in this book.

Definition 1.34 (Hardy–Littlewood maximal operator) For $f \in L^0(\mathbb{R}^n)$, dcfinc a function Mf by

$$Mf(x) \equiv \sup_{B\in\mathscr{B}} \frac{\chi_B(x)}{|B|}\int_B |f(y)|\mathrm{d}y \quad (x \in \mathbb{R}^n). \tag{1.136}$$

The mapping $M : f \mapsto Mf$ is called the *Hardy–Littlewood maximal operator*.

Note that M commutes dilation; $Mf(R\star) = M[f(\star)]$ for all $f \in L^0(\mathbb{R}^n)$ and $R > 0$. Denote by $\mathscr{Q}$ the set of all cubes whose edges are parallel to coordinate axes.

Remark 1.4 We replace $\mathscr{B}$ with $\mathscr{Q}$ in the definition of $Mf(x)$, which we sometimes do; we can define

$$Mf(x) \equiv \sup_{Q\in\mathscr{Q}} \frac{\chi_Q(x)}{|Q|}\int_Q |f(y)|\mathrm{d}y \quad (x \in \mathbb{R}^n).$$

But for the time being we study the Hardy–Littlewood maximal operator defined by $\mathscr{B}$ unless otherwise stated.

Let f be a measurable function and $\lambda \in \mathbb{R}$. According to the definition (1.136) of Mf, we note that $\{Mf > \lambda\}$ is an open set. Hence Mf is a measurable function. Let us explain why the Hardy–Littlewood maximal operator is important.

Example 1.22 The first example is closely related to PDE. Let I_α be the *fractional integral operator* given by

$$I_\alpha f(x) \equiv \int_{\mathbb{R}^n} \frac{f(y)}{|x-y|^{n-\alpha}}\mathrm{d}y \quad (x \in \mathbb{R}^n) \tag{1.137}$$

for a nonnegative measurable function $f : \mathbb{R}^n \to [0, \infty]$. Let $1 \le p < \frac{n}{\alpha}$ and define q so that $\frac{1}{q} = \frac{1}{p} - \frac{\alpha}{n}$. Then Hedberg showed in [580] that

$$I_\alpha f(x) \lesssim Mf(x)^{\frac{p}{q}} \|f\|_p{}^{1-\frac{p}{q}} \quad (x \in \mathbb{R}^n).$$

See Exercise 3.10 for more about this estimate.

Example 1.23 The next example is from Fourier analysis and theory of function spaces.

Let $R > 0$. Then the Plancherel–Polya–Nikolski'i theorem, which we formulate in Theorem 1.50, asserts that there exists a constant $C > 0$ that depends only on n and $\eta > 0$ such that

$$\sup_{y \in \mathbb{R}^n} \frac{|f(x-y)|}{(1+R|y|)^{n/\eta}} \lesssim M^{(\eta)}f(x) = M[|f|^\eta](x)^{1/\eta} \quad (x \in \mathbb{R}^n)$$

for all $f \in \mathscr{S}'(\mathbb{R}^n)$ whose frequency support is in $B(R)$; see Theoren 1.50.

Example 1.24 This example is closely related to the theory of the probability.

Let $f \in L^1_{\rm loc}(\mathbb{R}^n)$. Then consider E_j defined by (1.100). Note that $f_j \equiv E_j f$ is obtained by the averaging procedure of f with respect to the *grid* $\mathscr{D}_j = \{2^{-j}m + [0, 2^{-j})^n\}_{m \in \mathbb{Z}^n}$. Then it is easy to show that $|E_j f| \lesssim Mf$ with a constant independent of j and f, where M is the maximal operator generated by balls. This type of estimate is important when we consider the limit of $E_j f$ as $j \to \infty$.

Here we present examples of the calculation.

Example 1.25 Here we consider the case of $n = 1$. A simple calculation gives us $M\chi_{[-1,1]}(t) = \min\left\{1, 2(1+|t|)^{-1}\right\}$ for $t \in \mathbb{R}$. Thus, in particular, $M\chi_{[-1,1]}$ is not an $L^1(\mathbb{R})$-function:

$$\int_{-\infty}^{\infty} M\chi_{[-1,1]}(t){\rm d}t = \infty.$$

Mimicking this argument, we can show that $Mf \notin L^1(\mathbb{R}^n)$ when $f \in L^0(\mathbb{R}^n) \setminus \{0\}$.

Although we could find $M\chi_{[-1,1]}(t)$, $t \in \mathbb{R}$ in Example 1.25, it is in general hard to find $Mf(x)$, $x \in \mathbb{R}^n$ for a measurable function f on $\mathbb{R}^n$. The following crude estimate is useful on many occasions.

Example 1.26 Let Q be a cube, and let E be a measurable set containing x and contained by Q. Observe that $M\chi_E \geq \dfrac{|E|}{|Q|}$ on Q. Here M is the uncentered maximal operator generated by cubes. We can readily replace cubes by balls, in which case we need to replace the definition of M. Indeed, for $x \in Q$, we have

$$M\chi_E(x) = \sup\left\{\frac{|E \cap S|}{|S|} \,:\, S \text{ satisfies } x \in S \in \mathscr{Q}\right\} \geq \frac{|E|}{|Q|}.$$

In the definition of the Hardy–Littlewood maximal operator, Q moves over all cubes. The following technique is to used to localize the problem: $M[\chi_{\mathbb{R}^n \setminus 5Q} f](y)$, $y \in Q$ has a standard estimate and $M[\chi_{5Q} f](y)$, $y \in Q$ must be handled in a more careful manner.

Example 1.27 For all measurable functions f and cubes Q, we have

$$M[\chi_{\mathbb{R}^n\setminus 5Q}f](y) \lesssim \sup_{Q\subset R\in\mathcal{Q}} m_R(|f|) \quad (y \in Q). \tag{1.138}$$

To prove (1.138), we write out $M[\chi_{\mathbb{R}^n\setminus 5Q}f](y)$ in full:

$$M[\chi_{\mathbb{R}^n\setminus 5Q}f](y) = \sup_{R\in\mathcal{Q}} \frac{\chi_R(y)}{|R|}\|f\|_{L^1(R\setminus 5Q)},$$

where R moves over all cubes. In order that $\chi_R(y)\|f\|_{L^1(R\setminus 5Q)}$ be nonzero, we need to have $y \in R$ and $R \setminus 5Q \neq \emptyset$. Thus, R meets both Q and $\mathbb{R}^n \setminus 5Q$. If $R \in \mathcal{Q}$ is a cube that meets both Q and $\mathbb{R}^n \setminus 5Q$, then $\ell(R) \geq 2\ell(Q)$ and $2R \supset Q$. Thus, (1.138) follows.

1.4.1.2 Covering Lemma and the Weak-(1, 1) Boundedness

In view of Example 1.25 (as well as Exercise 1.87) Mf never belongs to $L^1(\mathbb{R}^n)$ even when $f \in L^1(\mathbb{R}^n)$. What can be said for $L^1(\mathbb{R}^n)$ functions? First of all, we state a geometrical lemma and then we investigate Mf for $L^1(\mathbb{R}^n)$-functions f.

We prove the following important estimate called the *weak-*(1, 1) *boundedness* or the L^1-$L^{1,\infty}$ boundedness of M.

Theorem 1.45 (Hardy–Littlewood maximal inequality) *For* $f \in L^1(\mathbb{R}^n)$, $\lambda > 0$,

$$\lambda|\{Mf > \lambda\}| \leq 3^n\|f\|_1.$$

Proof By the inner regularity (1.12) of the Lebesgue measure, it suffices to show

$$\lambda|K| \leq 3^n\|f\|_1 \tag{1.139}$$

for any compact set K contained in $\{Mf > \lambda\}$.

Let $x \in K$. By the definition of M, there exists a ball B_x centered at x such that

$$m_{B_x}(|f|) > \lambda. \tag{1.140}$$

By the compactness of K, there exists a finite collection $x_1, x_2, \ldots, x_J$ of points such that $K \subset \bigcup_{j=1}^{J} B_{x_j}$. To simplify the notation, let $B_j \equiv B_{x_j}$. By Theorem 1.7, if we relabel the B_j's, for some $L \leq J$, we have the following:

1. The ball $\{B_j\}_{j=1}^L$ is mutually disjoint, that is,

$$\sum_{j=1}^{L} \chi_{B_j} \le 1, \tag{1.141}$$

2.

$$K \subset \bigcup_{j=1}^{J} B_j \subset \bigcup_{j=1}^{L} 3\,B_j. \tag{1.142}$$

Hence if we use (1.142), (1.140) and (1.141) in that order, then

$$|K| \le \sum_{j=1}^{L} |3B_j| = 3^n \sum_{j=1}^{L} |B_j| \le \sum_{j=1}^{L} \frac{3^n}{\lambda} \int_{B_j} |f(x)| \mathrm{d}x \le \frac{3^n}{\lambda} \|f\|_1.$$

Thus, (1.139) and hence Theorem 1.45 are proved.

If we define the *weak L^p-space* $L^{p,\infty}(\mathbb{R}^n)$, $p \in [1, \infty)$ as the set of all the measurable functions f for which $\|f\|_{L^{p,\infty}} \equiv \sup\limits_{\lambda>0} \lambda |\{|f| > \lambda\}|^{1/p}$ is finite, then it follows that M is bounded from $L^1(\mathbb{R}^n)$ to $L^{1,\infty}(\mathbb{R}^n)$.

We can modify the argument above to have an estimate for the maximal function of the measure.

Example 1.28 Let μ be a nonnegative finite Radon measure and set (temporarily)

$$M\mu(x) \equiv \sup \left\{ \frac{\mu(Q)}{|Q|} \; : \; Q \in \mathcal{Q},\, c_Q = x \right\}. \tag{1.143}$$

Then modifying the argument above, we can show $\lambda |\{ M\mu > \lambda \}| \lesssim \mu(\mathbb{R}^n)$ for $\lambda > 0$. The proof is left as an exercise; see Exercise 1.97.

Example 1.29 (Kolmogorov inequality) Although we explained in Example 1.25 that the function Mf is never integrable, we still have the following substitute:

$$\int_E Mf(x)^p \mathrm{d}x \lesssim_p |E|^{1-p} \|f\|_1{}^p \tag{1.144}$$

for $0 < p < 1$, where E is a measurable set with finite measure and f is an integrable function. Note that (1.144) is a consequence of

$$\int_E Mf(x)^p \mathrm{d}x = \int_0^\infty p\lambda^{p-1} |\{x \in E \; : \; Mf(x) > \lambda\}| \mathrm{d}\lambda, \tag{1.145}$$

which follows from Theorem 1.51, and

$$|\{x \in E : Mf(x) > \lambda\}| \lesssim \min\{|E|, \lambda^{-1}\|f\|_1\}, \tag{1.146}$$

which in turn follows from Theorem 1.45. The idea that lies behind it is the use of the finiteness of the measure and the weak-(1, 1) boundedness of the operator. Inequality (1.144) is called the *Kolmogorov inequality*.

1.4.1.3 Dual Inequality of Stein-Type

We extend Theorem 1.45. We consider the uncentered maximal operator M generated by balls. For a nonnegative measurable function w and measurable set E, we define $w(E) \equiv \int_E w(x)\mathrm{d}x$.

Proposition 1.17 (Dual inequality of Stein-type) *Let w be a measurable function almost everywhere positive. Then $w\{Mf > \lambda\} \le \frac{3^n}{\lambda}\int_{\mathbb{R}^n} |f(x)|Mw(x)\mathrm{d}x$ for all measurable functions f. More precisely, $w\{Mf > \lambda\} \le \frac{3^n}{\lambda}\int_{\{Mf>\lambda\}} |f(x)|Mw(x)\mathrm{d}x$.*

We also call an inequality above the weighted weak (1, 1)-inequality. In Proposition 1.17, by letting $w \equiv 1$, we can recover Theorem 1.45.

Proof By the monotone convergence theorem, it can be assumed that w is bounded. For $\lambda > 0$, set $E_\lambda \equiv \{Mf > \lambda\}$. In view of the inner regularity of the measure $w(x)\mathrm{d}x$, it suffices to show

$$w(K) \le \frac{3^n}{\lambda}\int_{\mathbb{R}^n} |f(x)|Mw(x)\mathrm{d}x$$

for any compact set K contained in E_λ.

If we combine the compactness of K and Theorem 1.7, we can define a finite collection $B_1, B_2, \ldots, B_M$ of balls such that:

1. $B_1, B_2, \ldots, B_M$ are mutually disjoint:
$$\sum_{j=1}^{M} \chi_{B_j} \le 1. \tag{1.147}$$
2. The inequality
$$m_{B_j}(|f|) > \lambda \tag{1.148}$$
holds for $j = 1, 2, \ldots, M$.
3. K is covered by the triple of the balls $B_1, \ldots, B_M$; that is,

$$K \subset \bigcup_{j=1}^{M} 3\,B_j. \tag{1.149}$$

By the definition of the Hardy–Littlewood maximal operator M, we have

$$\frac{w(3B_j)}{|B_j|} \le 3^n \inf_{y \in B_j} Mw(y). \tag{1.150}$$

By (1.149), we have $w(K) \le \sum_{j=1}^{M} w(3\,B_j)$. If we use (1.148), (1.150), and (1.147), in that order, then

$$\begin{aligned} w(K) &\le \frac{1}{\lambda}\sum_{j=1}^{M}\int_{B_j}|f(x)|\mathrm{d}x \cdot \frac{w(3\,B_j)}{|B_j|} \\ &\le \frac{3^n}{\lambda}\sum_{j=1}^{M}\int_{B_j}|f(x)|Mw(x)\mathrm{d}x \\ &\le \frac{3^n}{\lambda}\int_{\mathbb{R}^n}|f(x)|Mw(x)\mathrm{d}x. \end{aligned}$$

Thus, the proof is complete.

Corollary 1.3 (Dual inequality of Stein-type) *Let f, w be measurable functions. Suppose in addition that $w \ge 0$ almost everywhere. Then for all $1 < p < \infty$,*

$$\int_{\mathbb{R}^n} Mf(x)^p \cdot w(x)\mathrm{d}x \le \frac{p2^p \cdot 3^n}{p-1}\int_{\mathbb{R}^n}|f(x)|^p \cdot Mw(x)\mathrm{d}x.$$

Proof First of all, whenever $\| f \|_\infty \le \lambda$, we have $Mf \le \lambda$. Together with the subadditivity $M[f+g] \le Mf + Mg$, we have $\{Mf > 2\lambda\} \subset \{M[\chi_{(\lambda,\infty)}(|f|)f] > \lambda\}$. By Proposition 1.17,

$$w\{Mf > 2\lambda\} \le w\{M[\chi_{(\lambda,\infty)}(|f|)f] > \lambda\} \le \frac{1}{\lambda}\int_{\{|f|>\lambda\}}|f(x)| \cdot Mw(x)\mathrm{d}x.$$

We use Theorem 1.51 for the function Mf and the measure $w(x)\mathrm{d}x$. If we change variables: $\sigma = 2\lambda$, then

$$\begin{aligned} \int_{\mathbb{R}^n} Mf(x)^p w(x)\mathrm{d}x &= p\int_{\mathbb{R}^n}\left(\int_0^{Mf(x)}\sigma^{p-1}\mathrm{d}\sigma\right)w(x)\mathrm{d}x \\ &= p \cdot 2^p \int_{\mathbb{R}^n}\left(\int_0^{2^{-1}Mf(x)}\lambda^{p-1}\mathrm{d}\lambda\right)w(x)\mathrm{d}x. \end{aligned} \tag{1.151}$$

By virtue of (1.151) and Proposition 1.17,

$$\int_{\mathbb{R}^n} Mf(x)^p w(x)\mathrm{d}x = p\,2^p \int_0^\infty \left(\int_{\mathbb{R}^n} \chi_{(2\lambda,\infty)}(Mf(x))w(x)\mathrm{d}x\right)\lambda^{p-1}\mathrm{d}\lambda$$
$$\leq p\,2^p\cdot 3^n \int_0^\infty \left(\int_{\mathbb{R}^n} \chi_{(\lambda,\infty)}(|f(x)|)|f(x)|\,Mw(x)\mathrm{d}x\right)\lambda^{p-2}\mathrm{d}\lambda.$$

By Fubini's theorem (Theorem 1.3), we calculate the 1-dimensional integral precisely to have:

$$\int_{\mathbb{R}^n} Mf(x)^p w(x)\mathrm{d}x \leq p\,2^p\cdot 3^n \int_0^\infty \left(\int_{\mathbb{R}^n} \chi_{(\lambda,\infty)}(|f(x)|)|f(x)|\,Mw(x)\mathrm{d}x\right)\lambda^{p-2}\mathrm{d}\lambda$$
$$= p\,2^p\cdot 3^n \int_{\mathbb{R}^n} \left(\int_0^\infty \chi_{(\lambda,\infty)}(|f(x)|)\lambda^{p-2}\mathrm{d}\lambda\right)|f(x)|\,Mw(x)\mathrm{d}x$$
$$= p\,2^p\cdot 3^n \int_{\mathbb{R}^n} \left(\int_0^{|f(x)|} \lambda^{p-2}\mathrm{d}\lambda\right)|f(x)|\,Mw(x)$$
$$= \frac{p\,2^p\cdot 3^n}{p-1}\int_{\mathbb{R}^n} |f(x)|^p Mw(x)\mathrm{d}x.$$

Therefore, the proof is complete.

Example 1.30 For all $f \in L^0(\mathbb{R}^n)$ and balls $B(x,r)$, we have

$$\int_{B(x,r)} Mf(y)^p\mathrm{d}y$$
$$= \int_{\mathbb{R}^n} Mf(y)^p \chi_{B(x,r)}(y)\mathrm{d}y \leq \frac{p\,2^p\cdot 3^n}{p-1}\int_{\mathbb{R}^n} |f(y)|^p M\chi_{B(x,r)}(y)\mathrm{d}y,$$

which can be regarded as a local norm estimate at all scales.

In particular, we can obtain the $L^p(\mathbb{R}^n)$-inequality of the Hardy–Littlewood maximal operator M as a corollary by letting $w \equiv 1$:

Theorem 1.46 ($L^p(\mathbb{R}^n)$-inequality of the Hardy–Littlewood maximal operator) *Let* $1 < p < \infty$. *Then* $\|Mf\|_p \leq \left(\frac{p\,2^p\cdot 3^n}{p-1}\right)^{\frac{1}{p}} \|f\|_p$ *for all* $f \in L^0(\mathbb{R}^n)$.

Remark 1.5 Theorem 1.46 is very important and we use it many times in this book. Usually, it is not necessary to learn the constant $\left(\frac{p\,2^p\cdot 3^n}{p-1}\right)^{\frac{1}{p}}$ by heart. However, it is not so difficult to keep track of the constants in the proof.

The proof relies upon the interpolation theorem due to Marcinkiewicz [763]. The next theorem is an example of estimates of linear operators by means of the Hardy–Littlewood maximal operator.

Theorem 1.47 *Let $\Phi : \mathbb{R}^n \to [0,\infty)$ be a continuous and integrable function satisfying the following condition: Φ is a radial and decreasing function. That is, there exists a decreasing function φ such that $\Phi(x) = \varphi(|x|)$ for all $x \in \mathbb{R}^n$. Then $|\Phi * f(x)| \le \|\Phi\|_1 Mf(x)$ holds for all nonnegative $f \in L^1_{\rm loc}(\mathbb{R}^n)$ and $x \in \mathbb{R}^n$.*

Proof Let $\varphi : [0,\infty) \to [0,\infty)$ be a function of the form $\varphi(r) = \sum_{j=1}^{k} a_j \, \chi_{[0,r_j]}(r)$ for $r \ge 0$. Then

$$|\Phi * f(x)| \le \sum_{j=1}^{k} a_j |B(x,r_j)| \frac{1}{|B(x,r_j)|} \int_{B(x,r_j)} |f(y)| {\rm d}y \le \|\Phi\|_1 Mf(x).$$

If we pass this to the limit, we see that this is valid for measurable functions Φ.

Here we list some examples of functions to which Theorem 1.47 is applicable.

Example 1.31

1. Gaussian in Example 1.4.
2. $\Phi = \chi_{B(1)}$.
3. Let $j \in \mathbb{N}_0$ and $m > n$. Define $\eta_{j,m}(x) \equiv 2^{jn}(1 + 2^j|x|)^{-m}$ for $x \in \mathbb{R}^n$. The function $\eta_{j,m}$ is called the *η-function*.

1.4.1.4 Applications to the Lebesgue Differentiation Theorem

Keeping the elementary inequalities in mind, we apply Theorem 1.46, the boundedness of the Hardy–Littlewood maximal operator to the Lebesgue differentiation theorem.

Theorem 1.48 (Lebesgue differentiation theorem) *Let $f \in L^1_{\rm loc}(\mathbb{R}^n)$. Then*

$$\lim_{r \downarrow 0} m_{B(x,r)}(|f - f(x)|) = 0 \tag{1.152}$$

for almost all $x \in \mathbb{R}^n$. In particular, for such a point x, we have $|f(x)| \le Mf(x)$.

Before we come to the proof, a helpful remark may be in order. The key estimate is:

$$m_{B(x,r)}(|f - f(x)|) \le |f(x)| + Mf(x). \tag{1.153}$$

It may be interesting to compare $|f| + Mf$ in (1.153) with the function g in Theorem 1.1, the dominated convergence theorem.

Proof The latter assertion follows immediately from the definition of the Hardy–Littlewood maximal operator. Thus, we concentrate on (1.152). To this end, we have only to show that the Lebesgue measure does not charge the set E of all points for which (1.152) fails. We decompose E; that is, we define

$$E_k \equiv \left\{ x \in \mathbb{R}^n \,:\, \limsup_{r \downarrow 0} m_{B(x,r)}(|f - f(x)|) > \frac{1}{k} \right\} \tag{1.154}$$

for $k \in \mathbb{N}$. Once we prove E_k has Lebesgue measure 0, then E will have Lebesgue measure 0. Condition (1.152) is local. So we can consider $\chi_{B(r)} f$ instead of f. We can assume that $f \in L^1(\mathbb{R}^n)$.

Furthermore, note that (1.152) is clear for $g \in L^1(\mathbb{R}^n) \cap \mathrm{BC}(\mathbb{R}^n)$.

Let $\varepsilon > 0$ be arbitrary. Let us conclude the proof of the theorem by showing that $|E_k| \le 2(3^n + 1)\, k\, \varepsilon$.

Choose $g \in L^1(\mathbb{R}^n) \cap \mathrm{BC}(\mathbb{R}^n)$ by density so that $\|f - g\|_1 \le \varepsilon$. We have:

$$\begin{aligned} E_k &= \left\{ x \in \mathbb{R}^n \,:\, \limsup_{r \downarrow 0} m_{B(x,r)}(|f - g + g(x) - f(x)|) > \frac{1}{k} \right\} \\ &\subset \left\{ x \in \mathbb{R}^n \,:\, M[f - g](x) + |f(x) - g(x)| > \frac{1}{k} \right\} \qquad (1.155) \\ &\subset \left\{ x \in \mathbb{R}^n \,:\, M[f - g](x) > \frac{1}{2k} \right\} \cup \left\{ x \in \mathbb{R}^n \,:\, |f(x) - g(x)| > \frac{1}{2k} \right\} \end{aligned}$$

thanks to (1.153); see also Exercise 1.88 for (1.155). If we combine (1.155) with the subadditivity, Chebychev's inequality (see Theorem 1.4) and the weak (1, 1)-maximal inequality, $|E_k| \le 2(3^n + 1)\, k\, \|f - g\|_1 \le 2(3^n + 1)\, k\, \varepsilon$. The number $\varepsilon > 0$ being arbitrary, we obtain (1.152).

Here we present an application of the Lebesgue differentiation theorem, which will be used at the end of this book.

Proposition 1.18 *Let $T \in B(L^1(\mathbb{R}^n))$, and let $K \in L^\infty(\mathbb{R}^{2n})$. Assume that*

$$Tf(x) \equiv \int_{\mathbb{R}^n} K(x, y) f(y) \mathrm{d}y \quad (f \in L^1(\mathbb{R}^n))$$

for almost every $x \in \mathbb{R}^n$. Then if we let

$$k(y) \equiv \int_{\mathbb{R}^n} |K(x, y)| \mathrm{d}x \quad (y \in \mathbb{R}^n),$$

then k is in $L^\infty(\mathbb{R}^n)$. Furthermore, $\|k\|_\infty = \|T\|_{B(L^1)}$.

Proof It is clear that $\|k\|_\infty \ge \|T\|_{B(L^1)}$. Thus, the heart of the matter is to prove the reverse inequality. Let $E_j \in B(L^1(\mathbb{R}^n))$ be an operator given by $E_j f \equiv \sum_{Q\in\mathscr{D}_j} m_Q(f)\chi_Q$ for $f \in L^1(\mathbb{R}^n)$. Then $E_j \circ T \circ E_j f \equiv \int_{\mathbb{R}^m} K_j(\star, y)f(y)\mathrm{d}y$, where

$$K_j(x, y) \equiv \sum_{Q,Q'\in\mathscr{D}_j} m_{Q\times Q'}(K_j)\chi_Q(x)\chi_{Q'}(y) \quad (x, y \in \mathbb{R}^n).$$

Since $E_j \circ T \circ E_j$ is bounded on $L^1(\mathbb{R}^n)$ with norm $\|T\|_{B(L^1)}$, we have

$$\left|\int_{\mathbb{R}^{2n}} \sum_{Q,Q'\in\mathscr{D}_j} m_{Q\times Q'}(K_j)\chi_Q(x)\chi_{Q'}(y)f(x)g(y)\mathrm{d}x\mathrm{d}y\right| \le \|T\|_{B(L^1)}\|f\|_\infty\|g\|_1$$

for all $f \in L^\infty(\mathbb{R}^n)$ and $g \in L^1(\mathbb{R}^n)$. Thus,

$$\int_{\mathbb{R}^n} \left|\sum_{Q\in\mathscr{D}_j} m_{Q\times Q'}(K_j)\chi_Q(x)\chi_{Q'}(y)f(x)\right| \mathrm{d}x \le \|T\|_{B(L^1)}\|f\|_\infty$$

for all cubes Q'. Another dualization yields

$$\int_{\mathbb{R}^n} \left|\sum_{Q\in\mathscr{D}_j} m_{Q\times Q'}(K_j)\chi_Q(x)\chi_{Q'}(y)\right| \mathrm{d}x \le \|T\|_{B(L^1)},$$

and thus, the proof is complete by the Lebesgue convergence theorem.

The following is an example of approximating the integral operator by means of the *triple dyadic average* $m_{3Q}(f)$, $Q \in \mathscr{D}$.

Example 1.32 Let f be a nonnegative measurable function. Then for $\eta_{\nu,m}$ defined in Example 1.31, $\nu \in \mathbb{N}_0$ and $m \in \mathbb{Z}^n$, we have

$$\eta_{\nu,m} * f(x) \lesssim \sum_{j=0}^{\infty} 2^{j(n-m)} \sum_{Q\in\mathscr{D}_{\nu-j}} \chi_{3Q}(x)m_Q(f) \quad (x \in \mathbb{R}^n).$$

In fact, we decompose

$$\begin{aligned}\eta_{\nu,m} * f(x) &= \int_{B(2^{-\nu})} \eta_{\nu,m}(y)f(x-y)\mathrm{d}y\\ &\quad+\sum_{l=1}^{\infty}\int_{B(2^{l-\nu})\setminus B(2^{l-1-\nu})} \eta_{\nu,m}(y)f(x-y)\mathrm{d}y.\end{aligned}$$

From the definition of $\eta_{\nu,m}$, we obtain

$$\begin{aligned}
&\eta_{\nu,m} * f(x)\\
&\lesssim 2^{\nu n}\int_{B(x,2^{-\nu})} f(y)\mathrm{d}y + \sum_{l=1}^{\infty} 2^{\nu n - ml}\int_{B(x,2^{l-\nu})} f(y)\mathrm{d}y\\
&= 2^{\nu n}\sum_{Q\in\mathscr{D}_\nu}\chi_Q(x)\int_{B(x,2^{-\nu})} f(y)\mathrm{d}y\\
&\quad + \sum_{l=1}^{\infty} 2^{\nu n - ml}\sum_{Q\in\mathscr{D}_{\nu-l}}\chi_Q(x)\int_{B(x,2^{l-\nu})} f(y)\mathrm{d}y.
\end{aligned}$$

A geometric observation shows that $x \in Q$ implies $B(x, \ell(Q)) \subset 3Q$. Thus

$$\begin{aligned}
\eta_{\nu,m} * f(x) &\lesssim 2^{\nu n}\sum_{Q\in\mathscr{D}_\nu}\chi_Q(x)\int_{3Q} f(y)\mathrm{d}y\\
&\quad + \sum_{l=1}^{\infty} 2^{\nu n - ml}\sum_{Q\in\mathscr{D}_{\nu-l}}\chi_Q(x)\int_{3Q} f(y)\mathrm{d}y\\
&= 2^{\nu n}\sum_{Q\in\mathscr{D}_\nu}\chi_{3Q}(x)\int_{Q} f(y)\mathrm{d}y\\
&\quad + \sum_{l=1}^{\infty} 2^{\nu n - ml}\sum_{Q\in\mathscr{D}_{\nu-l}}\chi_{3Q}(x)\int_{Q} f(y)\mathrm{d}y\\
&\simeq \sum_{j=0}^{\infty} 2^{j(n-m)}\sum_{Q\in\mathscr{D}_{\nu-j}}\chi_{3Q}(x) m_Q(f),
\end{aligned}$$

as required.

We considered the Lebesgue differentiation theorem (Theorem 1.48) as an application of the boundedness of the Hardy–Littlewood maximal operator. In retrospect, what lies behind this boundedness is the $5r$-covering lemma. So, covering lemmas are often used to study the smoothness of functions. Other purposes may be to describe the level set of the maximal operators as we witnessed in this chapter.

Exercises

Exercise 1.86 Show that $\{Mf > \lambda\} = \{M[f\chi_{(\lambda,\infty]}(Mf)] > \lambda\}$ for $f \in L^1_{\mathrm{loc}}(\mathbb{R}^n)$ and $\lambda > 0$. Hint: One inclusion is trivial since $|f| \ge |f\chi_{(\lambda,\infty]}(Mf)|$. To show the reverse inclusion, let $x \in \mathbb{R}^n$ be such that $Mf(x) > \lambda$. Then we can find a ball B_x such that (1.140) holds.

Exercise 1.87 Let $n = 1$. Prove that $M\chi_{[-1,1]}(t) = \min\left(1, 2(1+|t|)^{-1}\right)$ for $t \in \mathbb{R}$ by a direct calculation.

Exercise 1.88 By passing to the complement, verify the inclusion in (1.155).

Exercise 1.89 For $x \in \mathbb{R}^n$ and $\nu \in \mathbb{Z}$, denote by $m(x)$ a unique cube of the form $Q_{\nu m}$ such that $x \in Q_{\nu m}$ with $m \in \mathbb{Z}^n$. Then for $f \in L^1_{\rm loc}(\mathbb{R}^n)$, show that

$$\lim_{\nu\to\infty} m_{Q_{\nu m(x)}}(|f - f(x)|) = 0 \quad (\text{a.e.} x \in \mathbb{R}^n).$$

Hint: Mimic the proof of the Lebesgue differentiation theorem.

Exercise 1.90 Let $k \in C^\infty_{\rm c}(B(1))$ satisfy $\displaystyle\int_{\mathbb{R}^n} k(x){\rm d}x = 1$, and let $f \in L^p(\mathbb{R}^n)$.

1. If $1 \le p < \infty$, show that $f * k_\varepsilon \to f$ almost everywhere.
2. If $p = \infty$, show that $f * k_\varepsilon \to f$ almost everywhere. Hint: Reduce to the case of $p < \infty$.

Exercise 1.91 [350, Proposition 2.2] Let σ be a measure. Suppose that we have a positive sequence $\{\lambda_Q\}_{Q\in\mathscr{D}}$. Set

$$A_1(\{\lambda_Q\}_{Q\in\mathscr{D}}) \equiv \int_{\mathbb{R}^n}\left(\sum_{Q\in\mathscr{D}}\frac{\lambda_Q}{\sigma(Q)}\chi_Q(x)\right)^s {\rm d}\sigma(x)$$

$$A_2(\{\lambda_Q\}_{Q\in\mathscr{D}}) \equiv \sum_{Q\in\mathscr{D}}\lambda_Q\left(\frac{1}{\sigma(Q)}\sum_{Q'\subset Q}\lambda_{Q'}\right)^s$$

$$A_3(\{\lambda_Q\}_{Q\in\mathscr{D}}) \equiv \int_{\mathbb{R}^n}\sup_{Q\in\mathscr{D}}\left(\frac{1}{\sigma(Q)}\sum_{Q'\subset Q}\lambda_{Q'}\chi_Q(x)\right)^s {\rm d}\sigma(x).$$

Also define $M_\sigma f(x) \equiv \displaystyle\sup_{Q\in\mathscr{D}}\frac{\chi_Q(x)}{\sigma(Q)}\int_Q |f(y)|{\rm d}\sigma(y)$ for $x \in \mathbb{R}^n$. Let $1 < s < \infty$. We aim to show

$$A_1(\{\lambda_Q\}_{Q\in\mathscr{D}}) \le A_3(\{\lambda_Q\}_{Q\in\mathscr{D}}) \lesssim_s A_1(\{\lambda_Q\}_{Q\in\mathscr{D}}) \sim_s A_2(\{\lambda_Q\}_{Q\in\mathscr{D}}). \tag{1.156}$$

1. Show that $A_1(\{\lambda_Q\}_{Q\in\mathscr{D}}) \sim A_3(\{\lambda_Q\}_{Q\in\mathscr{D}})$ using Lemma 1.28.
2. Show that $A_2(\{\lambda_Q\}_{Q\in\mathscr{D}}) \lesssim A_1(\{\lambda_Q\}_{Q\in\mathscr{D}})^{1/s}A_3(\{\lambda_Q\}_{Q\in\mathscr{D}})^{1/s}$.
3. Using (1.7), show that $A_1(\{\lambda_Q\}_{Q\in\mathscr{D}}) \le sA_2(\{\lambda_Q\}_{Q\in\mathscr{D}})$ for $1 < s \le 2$.
4. Conclude the proof of (1.156) inducting on $m = -[-s]$ and using

$$\left(\sum_{Q'\in\mathscr{D}}\frac{\lambda_{Q'}\chi_{Q'}(x)}{\sigma(Q')}\left(\frac{1}{\sigma(Q')}\sum_{Q''\subset Q'}\lambda_{Q''}\right)^{s-2}\right)^{s-1}$$

$$\le\left(\sum_{Q'\in\mathscr{D}}\frac{\lambda_{Q'}\chi_{Q'}(x)}{\sigma(Q')}\right)\left(\sum_{Q'\in\mathscr{D}}\frac{\lambda_{Q'}\chi_{Q'}(x)}{\sigma(Q')}\left(\frac{1}{\sigma(Q')}\sum_{Q''\in\mathscr{D},Q''\subset Q'}\lambda_{Q''}\right)^{s-1}\right)^{s-2}.$$

Exercise 1.92 [517] Let $1 < p < q < \infty$, $\{g_j\}_{j=1}^{\infty} \subset L^q(\mathbb{R}^n)$, and let $\{Q_j\}_{j=1}^{\infty} \subset \mathscr{Q}$. Then via dualization of the left-hand side prove that

$$\left\|\sum_{j=1}^{\infty}|g_j|\chi_{Q_j}\right\|_p \lesssim_{p,q} \left\|\sum_{j=1}^{\infty}m_{Q_j}^{(q)}(g_j)\chi_{Q_j}\right\|_p.$$

1.4.2 Fefferman–Stein Vector-Valued Maximal Inequality

The Fefferman–Stein vector-valued maximal inequality, which we prove in this section, asserts that

$$\left\|\left(\sum_{j=1}^{\infty}Mf_j{}^q\right)^{\frac{1}{q}}\right\|_p \lesssim_{p,q} \left\|\left(\sum_{j=1}^{\infty}|f_j|^q\right)^{\frac{1}{q}}\right\|_p$$

for all sequences $\{f_j\}_{j=1}^{\infty} \subset L^0(\mathbb{R}^n)$. Note that this inequality differs from

$$\left\|M\left[\left(\sum_{j=1}^{\infty}|f_j|^q\right)^{\frac{1}{q}}\right]\right\|_p \lesssim_{p,q} \left\|\left(\sum_{j=1}^{\infty}|f_j|^q\right)^{\frac{1}{q}}\right\|_p,$$

which is a direct consequence of Theorem 1.46.

A basic idea of analysis is to decompose a function into a countable sum of elementary pieces of functions. The sum must be countable otherwise the sum may fail to be measurable. Because the sum is made up of functions which are not complicated, we pay attention to each summand. Therefore, we handle each term separately. The Hardy–Littlewood maximal operator controls in some sense functions whose structures are not clear. Therefore, the Fefferman–Stein vector-valued maximal inequality, whose left-hand side contains a countable sum of the Hardy–Littlewood maximal functions of functions, is useful.

1.4.2.1 Fefferman–Stein Vector-Valued Maximal Inequality: a Formulation

The Fefferman–Stein vector-valued maximal inequality is elementary and it is used in the study of Triebel–Lizorkin spaces. Theorem 1.49 is extremely important and is used hundreds of times in this book.

Theorem 1.49 (Fefferman–Stein vector-valued maximal inequality) *Let* $1 < p < \infty$ *and* $1 < q \le \infty$. *For all sequences* $\{f_j\}_{j=1}^\infty$ *of measurable functions,*

$$\left\| \left(\sum_{j=1}^\infty M f_j{}^q \right)^{\frac{1}{q}} \right\|_p \lesssim_{p,q} \left\| \left(\sum_{j=1}^\infty |f_j|^q \right)^{\frac{1}{q}} \right\|_p .$$

To formulate the vector-valued inequalities in general, we adopt the following notation; the index set J is usually taken as $J = \mathbb{N}, \mathbb{N}_0, \mathbb{Z}$. But here we present the definition generally.

Definition 1.35 (Vector-valued norm) Let $0 < p, q \le \infty$, and let $J \subset \mathbb{Z}$.

1. For a system $\{f_j\}_{j \in J} \subset L^0(\mathbb{R}^n)$ of functions, define

$$\|f_j\|_{\ell^q(L^p)} \equiv \|\{f_j\}_{j\in J}\|_{\ell^q(L^p)} = \left(\sum_{j \in J} \|f_j\|_p{}^q \right)^{\frac{1}{q}} .$$

   The space $\ell^q(L^p, \mathbb{R}^n)$ is the set of all collections $\{f_j\}_{j=1}^\infty$ for which the quantity $\|\{f_j\}_{j\in J}\|_{\ell^q(L^p)}$ is finite.
2. For a system $\{f_j\}_{j \in J} \subset L^0(\mathbb{R}^n)$ of functions, define

$$\|f_j\|_{L^p(\ell^q)} \equiv \|\{f_j\}_{j\in J}\|_{L^p(\ell^q)} = \left\| \left(\sum_{j \in J} |f_j|^q \right)^{\frac{1}{q}} \right\|_p .$$

   The space $L^p(\ell^q, \mathbb{R}^n)$ is the set of all collections $\{f_j\}_{j=1}^\infty$ for which the quantity $\|\{f_j\}_{j\in J}\|_{\ell^q(L^p)}$ is finite.

These two norms are called the *vector-valued norms*. A natural modification is made in the above when $q = \infty$.

We plan to prove Theorem 1.49 after proving a lemma.

Lemma 1.22 *Let* $1 \le p, q \le \infty$, *and let* $\{f_j\}_{j=1}^\infty$ *be a sequence of* $L^p(\mathbb{R}^n)$-*functions such that* $f_j = 0$ *a.e. if* j *is large enough. Then we can take* $\{g_j\}_{j=1}^\infty \in$

$L^{p'}(\ell^{q'})$ with norm 1 such that $\|f_j\|_{L^p(\ell^q)} = \sum_{j=1}^{\infty}\int_{\mathbb{R}^n} f_j(x)g_j(x)\mathrm{d}x$. If $\{f_j\}_{j=1}^{\infty}$ is nonnegative, then we can arrange that $\{g_j\}_{j=1}^{\infty}$ be nonnegative.

Proof We suppose $1 < q < \infty$ for simplicity. There is nothing to prove if $f_j(x) = 0$ for all nonnegative integers j and for almost all $x \in \mathbb{R}^n$; assume otherwise. In this case, we recall the construction of the duality $L^p(\mathbb{R}^n)$-$L^{p'}(\mathbb{R}^n)$; set

$$g_j(x) \equiv \frac{1}{A}\overline{\mathrm{sgn}(f_j)(x)}\,|f_j(x)|^{q-1}\left(\sum_{j=1}^{\infty}|f_j(x)|^q\right)^{\frac{p}{q}-1}.$$

Here A is a normalization constant: $A \equiv \|f_j\|_{L^p(\ell^q)}^{p-1}$. Since

$$\left(\sum_{j=1}^{\infty}|g_j(x)|^{q'}\right)^{\frac{p'}{q'}} = \left(\sum_{j=1}^{\infty}|f_j(x)|^q\right)^{\frac{p}{q}},$$

we have $\|g_j\|_{L^{p'}(\ell^{q'})} = 1$.

Furthermore, since

$$\sum_{j=1}^{\infty} f_j(x)g_j(x) = \frac{1}{A}\left(\sum_{j=1}^{\infty}|f_j(x)|^q\right)^{\frac{p}{q}},$$

we have $\|f_j\|_{L^p(\ell^q)} = \sum_{j=1}^{\infty}\int_{\mathbb{R}^n} f_j(x)g_j(x)\mathrm{d}x$.

1.4.2.2 Proof of Theorem 1.49

First we rephrase Theorem 1.49. We shall show

$$\int_{\mathbb{R}^n}\left(\sum_{j=1}^{\infty}Mf_j(x)^q\right)^{\frac{p}{q}}\mathrm{d}x \lesssim_{p,q} \int_{\mathbb{R}^n}\left(\sum_{j=1}^{\infty}|f_j(x)|^q\right)^{\frac{p}{q}}\mathrm{d}x.$$

By the monotone convergence theorem, we can assume that $f_j \equiv 0$ for large j.

Proof **A simple case 1: $q = p, \infty$.**

If $p = q$, Then Theorem 1.49 is clear by the $L^p(\mathbb{R}^n)$-boundedness of M and the monotone convergence theorem. If $q = \infty$, we invoke a trivial pointwise estimate:

$$\sup_{j \in \mathbb{N}} Mf_j(x) \le M \left[\sup_{j \in \mathbb{N}} |f_j| \right] (x)$$

for $x \in \mathbb{R}^n$.

Case 2: $p > q$. Keeping in mind that the left-hand side is at least finite, we resort to duality. Set $r \equiv \frac{p}{q}$. Then there exists a nonnegative measurable function g such that

$$\int_{\mathbb{R}^n} \left(\sum_{j=1}^{\infty} Mf_j(x)^q \right)^{\frac{p}{q}} dx = \int_{\mathbb{R}^n} \sum_{j=1}^{\infty} Mf_j(x)^q g(x) dx, \quad \|g\|_{r'} = 1$$

by virtue of the duality $L^r(\mathbb{R}^n)$-$L^{r'}(\mathbb{R}^n)$. By Corollary 1.3 and Hölder's inequality, we have

$$\begin{aligned} \int_{\mathbb{R}^n} \left(\sum_{j=1}^{\infty} Mf_j(x)^q \right)^{\frac{p}{q}} dx &\lesssim \int_{\mathbb{R}^n} \sum_{j=1}^{\infty} |f_j(x)|^q Mg(x) dx \\ &\le \|Mg\|_{r'}{}^p \int_{\mathbb{R}^n} \left(\sum_{j=1}^{\infty} |f_j(x)|^q \right)^{\frac{p}{q}} dx. \end{aligned}$$

Since we know that M is $L^{r'}(\mathbb{R}^n)$-bounded, $\|Mg\|_{r'} \lesssim 1$. Hence

$$\int_{\mathbb{R}^n} \left(\sum_{j=1}^{\infty} Mf_j(x)^q \right)^{\frac{p}{q}} dx \lesssim \int_{\mathbb{R}^n} \left(\sum_{j=1}^{\infty} |f_j(x)|^q \right)^{\frac{p}{q}} dx.$$

Therefore, the proof is complete.

Case 3: $p < q < \infty$.

Let $\theta \equiv \sqrt{p}$, $s \equiv p/\theta$, $t \equiv q/\theta$. Then

$$\int_{\mathbb{R}^n} \left(\sum_{j=1}^{\infty} Mf_j(x)^q \right)^{\frac{p}{q}} dx = \|Mf_j{}^\theta\|_{L^s(\ell^t)}^s.$$

By Lemma 1.22, there exists $\{g_j\}_{j=1}^{\infty} \subset L^{s'}(\ell^{t'}, \mathbb{R}^n)$ such that

$$\|Mf_j{}^\theta\|_{L^s(\ell^t)} = \sum_{j=1}^{\infty} \int_{\mathbb{R}^n} Mf_j(x)^\theta g_j(x) dx, \quad \|g_j\|_{L^{s'}(\ell^{t'})} = 1.$$

By Corollary 1.3, we can move M to the function g_j. By Hölder's inequality, we have

$$\sum_{j=1}^{\infty}\int_{\mathbb{R}^n} Mf_j(x)^\theta g_j(x)\mathrm{d}x \lesssim_{p,q} \sum_{j=1}^{\infty}\int_{\mathbb{R}^n} |f_j(x)|^\theta Mg_j(x)\mathrm{d}x$$

$$\leq \| \, |f_j|^\theta \, \|_{L^s(\ell^t)}^{\frac{s}{t}} \|Mg_j\|_{L^{s'}(\ell^{t'})}.$$

The relation for s, t is reversed if we pass to the conjugate index. Thus, by Step 2, $\|Mg_j\|_{L^{s'}(\ell^{t'})} \lesssim 1$. Consequently, we have

$$\int_{\mathbb{R}^n}\left(\sum_{j=1}^{\infty} Mf_j(x)^q \mathrm{d}x\right)^{\frac{p}{q}} \lesssim \int_{\mathbb{R}^n}\left(\sum_{j=1}^{\infty} |f_j(x)|^q \mathrm{d}x\right)^{\frac{p}{q}}.$$

Therefore, the proof is complete.

Exercises

Exercise 1.93 [902, Lemma 2] Let $\delta > 0$ be a positive constant. Suppose we have a sequence $\{g_j\}_{j=-\infty}^{\infty} \subset L^0(\mathbb{R}^n)$. Set $G_j \equiv \sum_{j=-\infty}^{\infty} 2^{-|k-j|\delta} g_k$. Let $0 < p, q \leq \infty$.

1. Show that $\{g_j\}_{j=-\infty}^{\infty} \mapsto \{G_j\}_{j=-\infty}^{\infty}$ is a bounded linear mapping on $\ell^q(L^p, \mathbb{R}^n)$.
2. Show that $\{g_j\}_{j=-\infty}^{\infty} \mapsto \{G_j\}_{j=-\infty}^{\infty}$ is a bounded linear mapping on $L^p(\ell^q, \mathbb{R}^n)$.

Exercise 1.94 Let $\{a_m\}_{m\in\mathbb{Z}^n}$ be a complex sequence. Then show that

$$\left(\int_{\mathbb{R}^n}\left(\sum_{m\in\mathbb{Z}^n}\frac{|a_m|^q}{1+|x-m|^{q(n+1)}}\right)^{\frac{p}{q}}\mathrm{d}x\right)^{\frac{1}{p}} \sim \left(\sum_{m\in\mathbb{Z}^n}|a_m|^p\right)^{\frac{1}{p}}$$

for $1 < p < \infty$ and $1 < q \leq \infty$. Hint: Use the Fefferman–Stein vector-valued maximal inequality and Example 1.25.

Exercise 1.95 Let $F \equiv \{f_j\}_{j=-\infty}^{\infty}$ be a sequence of measurable functions. Define

$$MF(x) \equiv \sup_{r>0} m_{B(x,r)}\left(\left(\sum_{j=-\infty}^{\infty}|f_j|^2\right)^{\frac{1}{2}}\right).$$

Mimic the proof of Theorem 1.46 to prove the following:

1. For all $\lambda > 0$, show that

$$|\{MF > \lambda\}| \leq \frac{3^n}{\lambda} \int_{\mathbb{R}^n} \left(\sum_{j=-\infty}^{\infty} |f_j(x)|^2 \right)^{\frac{1}{2}} dx.$$

2. Let $1 < p < \infty$. Show that

$$\int_{\mathbb{R}^n} MF(x)^p dx \leq \frac{2^n p}{p-1} \int_{\mathbb{R}^n} \left(\sum_{j=-\infty}^{\infty} |f_j(x)|^2 \right)^{\frac{p}{2}} dx.$$

Exercise 1.96 Here we check that the conclusion in Theorem 1.49 fails in the case when $p = \infty$. Let $f_j \equiv \chi_{B(2^j)\setminus B(2^{j-1})}$ for $j = 1, 2, \ldots$ and $0 < q < \infty$. Then show that

$$\left\| \left(\sum_{j=1}^{\infty} Mf_j{}^q \right)^{\frac{1}{q}} \right\|_{\infty} = \infty, \quad \left\| \left(\sum_{j=1}^{\infty} |f_j|^q \right)^{\frac{1}{q}} \right\|_{\infty} = 1$$

to disprove that the Fefferman–Stein vector-valued inequality fails for $p = \infty$. Hint: Show first that $Mf_j \gtrsim \chi_{B(2^j)}$ for each $j \in \mathbb{N}$ using Example 1.26.

Exercise 1.97 Using the $5r$-covering lemma and Theorem 1.45, prove Example 1.28.

1.4.3 Properties of Band-Limited Distributions

In this section among band-limited distributions we consider ones whose Fourier transforms are supported on a fixed compact set. In particular, we are interested in distributions and their inequality when the compact set is the unit ball. The *frequency support* of a distribution $f \in \mathscr{S}'(\mathbb{R}^n)$ is defined to be the support of $\mathscr{F}f$. To this end, we present the following definition:

Definition 1.36 ($\mathscr{S}'_\Omega(\mathbb{R}^n)$, $\mathscr{S}_\Omega(\mathbb{R}^n)$ and $L^p_\Omega(\mathbb{R}^n)$) Let Ω be a bounded set in $\mathbb{R}^n$.

1. Denote by $\mathscr{S}'_\Omega(\mathbb{R}^n)$ the set of all distributions whose Fourier transform is contained in the closure $\overline{\Omega}$. Define $\mathscr{S}_\Omega(\mathbb{R}^n) \equiv \mathscr{S}'_\Omega(\mathbb{R}^n) \cap \mathscr{S}(\mathbb{R}^n)$.
2. Let $0 < p < \infty$. Define $L^p_\Omega(\mathbb{R}^n) \equiv L^p(\mathbb{R}^n) \cap \mathscr{S}'_\Omega(\mathbb{R}^n)$.

Note that all these things are monotone in Ω; the smaller Ω is, the better functions the function spaces contain. It matters that we tolerate $0 < p < 1$ when we define

$L^p_\Omega(\mathbb{R}^n)$ and that $L^p_\Omega(\mathbb{R}^n) \hookrightarrow L^\infty_\Omega(\mathbb{R}^n)$ for all $0 < p < 1$. A remark about the definition may be in order.

Remark 1.6

1. The functions in $L^p_\Omega(\mathbb{R}^n)$ are band-limited functions. Hence the functions in $L^p_\Omega(\mathbb{R}^n)$ are continuous.
2. When $0 < p < 1$, there are elements in $L^p(\mathbb{R}^n)$ that do not belong to $\mathscr{S}'(\mathbb{R}^n)$. So one has to pay attention to the definition of $L^p_\Omega(\mathbb{R}^n)$.

1.4.3.1 Plancherel–Polya–Nikolski'i Inequality

The Plancherel–Polya–Nikolski'i inequality is an elementary tool to investigate the function spaces. It is convenient to present the following definition to formulate this inequality:

Definition 1.37 (Powered Hardy–Littlewood maximal operator $M^{(\eta)}$) Let $\eta > 0$. For $f \in L^0(\mathbb{R}^n)$, the *powered Hardy–Littlewood maximal operator* $M^{(\eta)}$ is defined by $M^{(\eta)} f \equiv [M[|f|^\eta]]^{\frac{1}{\eta}}$.

The Plancherel–Polya–Nikolski'i inequality is very elementary in this book. Recall that the band-limited functions are represented by $C^\infty(\mathbb{R}^n)$-functions. We have an extremely important quantitative estimate with respect to the size of the support.

Theorem 1.50 (Plancherel–Polya–Nikolski'i inequality) *Let $\eta > 0$, and let φ be a function in $\mathscr{S}_{B(1)}(\mathbb{R}^n)$. Then for $x \in \mathbb{R}^n$,*

$$\sup_{z\in\mathbb{R}^n} \langle z\rangle^{-\frac{n}{\eta}} |\nabla\varphi(x-z)| \lesssim \sup_{z\in\mathbb{R}^n} \langle z\rangle^{-\frac{n}{\eta}} |\varphi(x-z)|, \quad \sup_{z\in\mathbb{R}^n} \langle z\rangle^{-\frac{n}{\eta}} |\varphi(x-z)| \lesssim M^{(\eta)}\varphi(x).$$

The operator $f \mapsto \sup_{z\in\mathbb{R}^n} \langle z\rangle^{-\frac{n}{\eta}} |f(\star - z)|$ is called the *Peetre maximal operator.* Before we come to the proof of Theorem 1.50, we will need a general estimate for $C^1(\mathbb{R}^n)$-functions φ.

Lemma 1.23 *Let $\varphi : \mathbb{R}^n \to \mathbb{C}$ be a $C^1(\mathbb{R}^n)$-function, and let $x \in \mathbb{R}^n$. Then for all $\delta, \eta > 0$, there exist positive constants c_0 and c_δ such that c_0 is independent of δ and φ, c_δ is independent of φ and that $|\varphi(y)| \le c_\delta \|\varphi\|_{L^\eta(B(x,\delta))} + c_0\, \delta \sup_{w\in B(x,\delta)} |\nabla\varphi(w)|$ for all $y \in B(x, \delta)$.*

Proof (of Lemma 1.23) Choose $w_0 \in \overline{B(x,\delta)}$ so that $|\varphi|$ attains a minimum over $\overline{B(x,\delta)}$ at w_0. Then we have

$$|\varphi(y)| \le |\varphi(w_0)| + |\varphi(y) - \varphi(w_0)| = \min_{w\in\overline{B(x,\delta)}} |\varphi(w)| + |\varphi(y) - \varphi(w_0)|.$$

By the mean-value theorem, we estimate $|\varphi(y)|$:

$$|\varphi(y)| \leq \min_{w \in \overline{B(x,\delta)}} |\varphi(w)| + 2\delta \sup_{w \in B(x,\delta)} |\nabla \varphi(w)|. \tag{1.157}$$

Note that the minimum over B is less than the powered average of any power:

$$\min_{w \in \overline{B(x,\delta)}} |\varphi(w)| \leq c_\delta \|\varphi\|_{L^\eta(B(x,\delta))}. \tag{1.158}$$

Here we denote by c_δ various constants which depend on δ, where c_0 stands for the various constants which do not depend on δ. If we combine (1.157) and (1.158), then we obtain the desired estimate.

Proof (of Theorem 1.50*)* Let $x \in \mathbb{R}^n$, and let ψ be a compactly supported smooth function that assumes 1 on the unit ball. Then since $\mathscr{F}\varphi$ is supported on the closure of $B(1)$,

$$\varphi(x) \simeq_n \mathscr{F}^{-1}\psi * \varphi(x) = \int_{\mathbb{R}^n} \mathscr{F}^{-1}\psi(x-y)\varphi(y)\mathrm{d}y. \tag{1.159}$$

If we set $\tau \equiv \mathscr{F}^{-1}\psi$, we have $\varphi(x) \simeq_n \int_{\mathbb{R}^n} \tau(x-z)\varphi(z)\mathrm{d}z$ by virtue of (1.159).

If we differentiate (1.159), then $\partial_{x_j}\varphi(x) \simeq_n \int_{\mathbb{R}^n} \partial_{x_j}\tau(x-z)\varphi(z)\mathrm{d}z$. By the triangle inequality, we obtain

$$\langle x-y\rangle^{-\frac{n}{\eta}} |\partial_{x_j}\varphi(y)| \lesssim_n \int_{\mathbb{R}^n} \langle x-y\rangle^{-\frac{n}{\eta}} |\partial_{x_j}\tau(y-z)\varphi(z)|\mathrm{d}z.$$

By Peetre's inequality, we have

$$\begin{aligned}
\langle x-y\rangle^{-\frac{n}{\eta}} |\partial_{x_j}\varphi(y)| &\lesssim \int_{\mathbb{R}^n} \langle y-z\rangle^{\frac{n}{\eta}} |\partial_{x_j}\tau(y-z)| \langle x-z\rangle^{-\frac{n}{\eta}} |\varphi(z)|\mathrm{d}z \\
&\lesssim \left(\sup_{z \in \mathbb{R}^n} \langle x-z\rangle^{-\frac{n}{\eta}} |\varphi(z)| \right) \int_{\mathbb{R}^n} \langle y-z\rangle^{\frac{n}{\eta}} |\partial_{x_j}\tau(y-z)|\mathrm{d}z \\
&\lesssim \sup_{z \in \mathbb{R}^n} \langle x-z\rangle^{-\frac{n}{\eta}} |\varphi(z)|.
\end{aligned}$$

Thus, the left inequality is proved.

Next, we prove the right inequality. First of all, let $x, y \in \mathbb{R}^n$ satisfy $|x-y| \leq \delta \ll 1$. We use (1.158). Here we denote by c_0 various constants dependent of δ. Let $|z| \leq \delta$. In Lemma 1.23 we substitute $x-z$ to x. Then

$$
\begin{aligned}
|\varphi(x-z)| &\le c_\delta \left(\int_{B(\delta)} |\varphi(x-z-w)|^\eta \mathrm{d}w\right)^{\frac{1}{\eta}} + c_0\,\delta \sup_{y\in B(\delta)} |\nabla\varphi(x-z-y)| \\
&\le c_\delta \left(\int_{B(|z|+1)} |\varphi(x-u)|^\eta \mathrm{d}u\right)^{\frac{1}{\eta}} + c_0\,\delta \sup_{y\in B(1)} |\nabla\varphi(x-z-y)|.
\end{aligned}
$$

Dividing both sides by $\langle z\rangle^{\frac{n}{\eta}}$, we obtain

$$
\begin{aligned}
&\langle z\rangle^{-\frac{n}{\eta}} |\varphi(x-z)| \\
&\le c_\delta \left(\frac{1}{\langle z\rangle^n}\int_{B(|z|+1)} |\varphi(u)|^\eta \mathrm{d}u\right)^{\frac{1}{\eta}} + c_0\,\delta \sup_{y\in B(1)} \langle z\rangle^{-\frac{n}{\eta}} |\nabla\varphi(x-z-y)| \\
&\le c_\delta\, M^{(\eta)}\varphi(x) + c_0\,\delta \sup_{y\in B(1)} \langle z\rangle^{-\frac{n}{\eta}} |\nabla\varphi(x-z-y)|.
\end{aligned}
$$

When $y \in B(1)$, we have $\langle y+z\rangle \sim \langle z\rangle$ by Peetre's inequality. Consequently,

$$
\langle z\rangle^{-\frac{n}{\eta}} |\varphi(x-z)| \le c_\delta\, M^{(\eta)}\varphi(x) + c_0\,\delta \sup_{y\in B(1)} \langle y+z\rangle^{-\frac{n}{\eta}} |\nabla\varphi(x-y-z)|. \tag{1.160}
$$

As a result, we take sup in (1.160) over z to have

$$
\sup_{z\in\mathbb{R}^n} \langle z\rangle^{-\frac{n}{\eta}} |\varphi(x-z)| \le c_\delta\, M^{(\eta)}\varphi(x) + c_0\,\delta \sup_{z\in\mathbb{R}^n} \langle z\rangle^{-\frac{n}{\eta}} |\nabla\varphi(x-z)|. \tag{1.161}
$$

The left inequality is already proved. Inserting (1.161) into the left inequality, we obtain

$$
\sup_{z\in\mathbb{R}^n} \langle z\rangle^{-\frac{n}{\eta}} |\varphi(x-z)| \le c_\delta\, M^{(\eta)}\varphi(x) + c_0\,\delta \sup_{z\in\mathbb{R}^n} \langle z\rangle^{-\frac{n}{\eta}} |\varphi(x-z)|. \tag{1.162}
$$

For the constant c_0 in (1.162), we take $\delta \equiv \frac{1}{2c_0}$. Since we are assuming $\varphi \in \mathscr{S}(\mathbb{R}^n)$, all the terms in (1.162) are finite. Consequently, we can absorb the second term of the right-hand side of (1.162) to the left-hand side (1.162) to have

$$
\sup_{z\in\mathbb{R}^n} \langle z\rangle^{-\frac{n}{\eta}} |\varphi(x-z)| \lesssim M^{(\eta)}\varphi(x).
$$

Thus, the right inequality is proved.

We also scale the above inequality to have:

Corollary 1.4 *For $R, \eta > 0$, $x \in \mathbb{R}^n$ and a function $\varphi \in \mathscr{S}_{B(R)}(\mathbb{R}^n)$, we have*

$$
R^{-1} \sup_{z\in\mathbb{R}^n} \langle R\,z\rangle^{-\frac{n}{\eta}} |\,[\nabla\varphi](x-z)\,| \lesssim \sup_{z\in\mathbb{R}^n} \langle R\,z\rangle^{-\frac{n}{\eta}} |\varphi(x-z)| \lesssim_\eta M^{(\eta)}\varphi(x).
$$

The mapping $\varphi \mapsto \sup_{z\in\mathbb{R}^n} \langle R\, z\rangle^{-\frac{n}{\eta}}|\varphi(\star - z)|$ is also called the *Peetre maximal operator*.

Proof If we set $\varphi_R \equiv \varphi(R^{-1}\star)$, then $\text{supp}(\mathscr{F}(\varphi_R)) = \text{supp}(\mathscr{F}\varphi(R\star)) \subset \overline{B(1)}$. We thus obtain

$$\sup_{z\in\mathbb{R}^n} \langle z\rangle^{-\frac{n}{\eta}} |\,[\nabla\varphi_R\,](x-z)| \lesssim \sup_{z\in\mathbb{R}^n} \langle z\rangle^{-\frac{n}{\eta}} |\varphi_R(x-z)| \lesssim M^{(\eta)}\varphi_R(x).$$

Hence it follows that

$$\begin{aligned} R^{-1} \sup_{z\in\mathbb{R}^n} \langle z\rangle^{-\frac{n}{\eta}} |\,[\nabla\varphi\,](R^{-1}x - R^{-1}z)| &\lesssim \sup_{z\in\mathbb{R}^n} \langle z\rangle^{-\frac{n}{\eta}} |\varphi(R^{-1}x - R^{-1}z)| \\ &\lesssim M^{(\eta)}\varphi(R^{-1}x). \end{aligned}$$

Changing the variables gives

$$R^{-1} \sup_{z\in\mathbb{R}^n} \langle R\, z\rangle^{-\frac{n}{\eta}} |\,[\,\nabla\varphi\,](x-z)\,| \lesssim \sup_{z\in\mathbb{R}^n} \langle R\, z\rangle^{-\frac{n}{\eta}} |\varphi(x-z)| \lesssim M^{(\eta)}\varphi(x).$$

Therefore, the proof is complete.

If the readers are familiar with the Fourier series, then it may be interesting to compare Corollary 1.5 with the inequality $\|T_n'\|_{L^p(\mathbb{T})} \le n\|T_n\|_{L^p(\mathbb{T})}$ available for any trigonometric polynomial of order n. See [19, p. 104] for the case $0 < p \le 1$.

Corollary 1.5 (Bernstein's lemma) *Let* $0 < p \le \infty$. *Then* $\|\partial^\alpha f\|_p \lesssim_{p,\alpha} \|f\|_p$ *for all* $\alpha \in \mathbb{N}_0{}^n$ *and* $f \in \mathscr{S}_{B(1)}(\mathbb{R}^n)$.

Proof Let $\eta \equiv \frac{p}{2}$. Then by Theorem 1.50, we have $|\partial^\alpha f(x)| \lesssim_\alpha M^{(\eta)} f(x)$. If we use the Hardy–Littlewood maximal inequality, Then we obtain the desired result.

A scaling yields the following corollary, whose proof is left to readers as Exercise 1.98.

Corollary 1.6 (Bernstein's lemma) *For* $0 < p \le \infty$, $R > 0$, $\alpha \in \mathbb{N}_0{}^n$ *and* $f \in \mathscr{S}_{B(R)}(\mathbb{R}^n)$, $\|\partial^\alpha f\|_p \lesssim_{p,\alpha} R^{|\alpha|}\|f\|_p$.

Now we extend Theorem 1.50 from $\mathscr{S}(\mathbb{R}^n)$ to $L^p(\mathbb{R}^n)$ with $0 < p \le \infty$.

Proposition 1.19 *Let* $0 < p \le \infty$. *Then for* $f \in L^p_{B(1)}(\mathbb{R}^n)$, *the conclusion of Theorem* 1.50 *including Corollaries* 1.5 *and* 1.6 *still holds.*

Proof Let $\psi \in \mathscr{S}(\mathbb{R}^n)$ satisfy $\chi_{Q(1)} \le \psi \le \chi_{Q(2)}$. For $t \in (0,1]$, $f_t \equiv \mathscr{F}^{-1}\psi(t\star)\cdot f$ enjoys the following properties:

1. $\lim_{t\downarrow 0} f_t(x) = f(x)$.
2. $\text{supp}(\mathscr{F} f_t) \subset [-t-2, t+2]^n \subset [-3,3]^n$.
3. For each $j = 1, \ldots, n$ and $x \in \mathbb{R}^n$, $\lim_{t\downarrow 0} \partial_{x_j} f_t(x) = \partial_{x_j} f(x)$.

Then since f is a $C^\infty(\mathbb{R}^n)$-function that has at most polynomial growth at infinity and $\mathscr{F}^{-1}\psi(t\star)$ is a Schwartz function for each t, so is f_t. Hence

$$\sup_{z\in\mathbb{R}^n} \langle z\rangle^{-\frac{n}{\eta}} |\nabla f_t(x-z)| \lesssim \sup_{z\in\mathbb{R}^n} \langle z\rangle^{-\frac{n}{\eta}} |f_t(x-z)| \lesssim M^{(\eta)} f(x).$$

Since we have $|f_t(x)| \lesssim |f(x)|$ with a constant independent of t, a passage to the limit $t \downarrow 0$ shows that Theorem 1.50 is still valid for f.

Let us define

$$\sigma_p \equiv n\left(\frac{1}{p}-1\right)_+ = \max\left(\frac{n}{p}-n, 0\right).$$

Corollary 1.7 *Let $0 < p \le \infty$ and $R > 0$. Then for $f, g \in L^p_{B(R)}(\mathbb{R}^n)$ and for all $\alpha \in {\mathbb{N}_0}^n$, we have*

$$\|\partial^\alpha f\|_p \lesssim_{p,\alpha} R^{|\alpha|} \|f\|_p \tag{1.163}$$

and

$$\|f * g\|_p \lesssim_p R^{\sigma_p} \|f\|_p \|g\|_{\min(1,p)}. \tag{1.164}$$

Proof A scaling of Proposition 1.19 yields (1.163). Let us prove (1.164). When $p \ge 1$, the result is trivial by the Young inequality. Let $0 < p < 1$. We remark that Proposition 1.19 yields

$$|f*g(x)| \le \|f(\star)g(x-\star)\|_1 \lesssim R^{n\left(\frac{1}{p}-1\right)_+} \|f(\star)g(x-\star)\|_p, \quad x \in \mathbb{R}^n \tag{1.165}$$

since $f(\star)g(x-\star) \in L^p_{B(2R)}(\mathbb{R}^n)$. If we take the $L^p(\mathbb{R}^n)$-norm, then we obtain (1.164).

As an application, we can prove the sampling theorem.

Theorem 1.51 (Sampling theorem) *Let $f \in L^p_{B(1)}(\mathbb{R}^n)$ with $0 < p < \infty$. Then for sufficiently small $h > 0$, we have*

$$\left(\sum_{k\in\mathbb{Z}^n} \max_{x\in Q(hk,h)} |f(x)|^p\right)^{\frac{1}{p}} \lesssim h^{-\frac{n}{p}} \|f\|_p \lesssim \left(\sum_{k\in\mathbb{Z}^n} \min_{x\in Q(hk,h)} |f(x)|^p\right)^{\frac{1}{p}}.$$

Theorem 1.51 deserves its name, since we learn from this theorem that any value in $Q(hk, h)$ suffices to approximate $h^{-\frac{n}{p}} \|f\|_p$.

Proof Let us prove the left inequality. If we set $\eta \equiv 2^{-1}\min(1, p)$, then

$$\max_{y \in Q(hk,h)} |f(y)| \lesssim \sup_{z \in \mathbb{R}^n} \langle z \rangle^{-\frac{n}{\eta}} |f(x-z)| \lesssim M^{(\eta)} f(x)$$

for $x \in Q(hk, h)$.

Hence

$$h^{\frac{n}{p}} \left(\sum_{k \in \mathbb{Z}^n} \max_{x \in Q(hk,h)} |f(x)|^p \right)^{\frac{1}{p}} \lesssim \|M^{(\eta)} f\|_p \lesssim \|f\|_p,$$

which yields the left-hand inequality. We prove the right-hand inequality, as follows. Here we need to take h sufficiently small. Denote by c_0 the various constants independent of h. Let $x \in Q(hk, h)$. Then we have

$$\begin{aligned} |f(x)| &\le \min_{y \in Q(hk,h)} |f(y)| + c_0 h \max_{y \in Q(hk,h)} |\nabla f(y)| \\ &\le \min_{y \in Q(hk,h)} |f(y)| + c_0 h M^{(\eta)} f(x). \end{aligned}$$

Let $\eta \equiv \frac{1}{2}\min(p, 1)$. If we use the Hardy–Littlewood maximal inequality,

$$h^{-\frac{n}{p}} \|f\|_p \le c_0 \left(\sum_{k \in \mathbb{Z}^n} \min_{x \in Q(hk,h)} |f(x)|^p \right)^{\frac{1}{p}} + c_0 h \cdot h^{-\frac{n}{p}} \|f\|_p.$$

Hence if we take h sufficiently small, then we can absorb the second term of the right-hand side to the left-hand side. The result is:

$$\|f\|_p \lesssim h^{n/p} \left(\sum_{k \in \mathbb{Z}^n} \min_{x \in Q(hk,h)} |f(x)|^p \right)^{\frac{1}{p}}.$$

Proposition 1.20 *Let $f \in L^p_{B(1)}(\mathbb{R}^n)$, $p > 0$. Then $f \in L^\infty(\mathbb{R}^n)$ and $\|f\|_\infty \lesssim_p \|f\|_p$.*

Proof This is a direct corollary of the sampling theorem.

We consider the scaling argument and interpolation argument to have the following conclusion.

Corollary 1.8 *If $0 < p_0 < p_1 \le \infty$, then $\|f\|_{p_1} \lesssim R^{\frac{n}{p_0} - \frac{n}{p_1}} \|f\|_{p_0}$ for $f \in \mathscr{S}'_{B(R)}(\mathbb{R}^n)$.*

Proof By Hölder's inequality and the trivial case, $p_1 = p_0$, we may assume that $p_1 = \infty$. When $R = 1$, this is trivial from Proposition 1.20 and Hölder's inequality. For $R > 0$, to scale, we set $f_R \equiv f(R^{-1}\star)$. Then

$$f_R \in \mathscr{S}'_{B(1)}(\mathbb{R}^n), \quad \|f_R\|_{p_0} = R^{\frac{n}{p_0}} \|f\|_{p_0}, \quad \|f_R\|_\infty = \|f\|_\infty$$

and the result follows for any $R > 0$.

The following theorem is one of the fundamental ingredients for our theory of function spaces. Here and below we denote by $\mathrm{diam}(K)$ the diameter of K.

Theorem 1.52 *Suppose that the parameters p, q satisfy $0 < p < \infty$, $0 < q \le \infty$. Let $0 < \eta < \min(p, q)$. Let $\{K_j\}_{j=1}^\infty$ be a sequence of compact sets in $\mathbb{R}^n$. Write $d_j \equiv \mathrm{diam}(K_j)$. Suppose that we have $f_j \in L^p_{K_j}(\mathbb{R}^n)$ for each $j \in \mathbb{N}$. Then*

$$\left\| \sup_{z \in \mathbb{R}^n} \langle d_j z \rangle^{-\frac{n}{\eta}} |f_j(\star - z)| \right\|_{L^p(\ell^q)} \lesssim \|f_j\|_{L^p(\ell^q)}.$$

Proof By Theorem 1.50, $\sup_{z \in \mathbb{R}^n} \langle d_j z \rangle^{-\frac{n}{\eta}} |f_j(x - z)| \lesssim M^{(\eta)} f_j(x)$ for all $x \in \mathbb{R}^n$. Hence the left-hand side is bounded from above (modulo a multiplicative constant) by: $\|M^{(\eta)} f_j\|_{L^p(\ell^q)} = \left\|M[|f_j|^\eta]\right\|_{L^{p/\eta}(\ell^{q/\eta})}^{\frac{1}{\eta}}$. By Theorem (1.49), we obtain the desired result.

1.4.3.2 Multiplier Theorem

We are going to define function spaces whose definitions will involve a procedure of decomposing them by their frequency support. The decomposition amounts to the Fourier multipliers. Thus, the Fourier multipliers play the fundamental role in this book.

Definition 1.38 (Fourier multiplier) Let $f \in \mathscr{S}'(\mathbb{R}^n)$. Define $\tau(D)f \in \mathscr{S}'(\mathbb{R}^n)$ by $\tau(D)f \equiv \mathscr{F}^{-1}(\tau \cdot \mathscr{F} f)$ for $\tau \in \mathscr{S}(\mathbb{R}^n)$. The operator $\tau(D)$ is referred to as the *Fourier multiplier*.

The definition of $\tau(D)f$ requires three operations: the Fourier transform, the pointwise multiplication by τ and the inverse Fourier transform. It is clear that $\tau(D)f \in \mathscr{S}'(\mathbb{R}^n)$. But more precisely, since

$$\tau(D)f = (2\pi)^{-\frac{n}{2}} \mathscr{F}^{-1}\tau * f,$$

the evaluation of $\tau(D)f(x)$ at $x \in \mathbb{R}^n$ makes sense. Consequently, it follows that $\tau(D)f \in \mathscr{S}'(\mathbb{R}^n) \cap L^1_{\mathrm{loc}}(\mathbb{R}^n)$ and that $\tau(D)f(x) = (2\pi)^{-\frac{n}{2}} \langle f, \mathscr{F}^{-1}\tau(x - \star)\rangle$.

We seek to study the mapping property of $\tau(D)$ in $L^p(\mathbb{R}^n)$ or $L^p(\ell^q)$ with $0 < p, q \le \infty$. To this end, we need to postulate some condition on τ; that is, we need to measure some smoothness property of τ in a certain sense. We achieve this using the potential space $H^s_2(\mathbb{R}^n)$.

Let $s \in \mathbb{R}$. We denote by $H^s_2(\mathbb{R}^n) \equiv H^s(\mathbb{R}^n)$ the *potential space*

$$H^s(\mathbb{R}^n) \equiv \{H \in \mathscr{S}'(\mathbb{R}^n) : (1-\Delta)^{s/2} H \in L^2(\mathbb{R}^n)\},$$

whose norm is given by $\|H\|_{H^s} \equiv \|(1-\Delta)^{s/2} H\|_2$ for $H \in H^s(\mathbb{R}^n)$.

More generally, for $1 < p < \infty$ and $s \in \mathbb{R}$ the *potential space* $H^s_p(\mathbb{R}^n)$ is the set of all $f \in \mathscr{S}'(\mathbb{R}^n)$ for which $\mathscr{F}^{-1}[(1+|\star|^2)^{\frac{s}{2}}\mathscr{F} f] \in L^p(\mathbb{R}^n)$. For $f \in H^s_p(\mathbb{R}^n)$ the norm is given by

$$\|f\|_{H^s_p} \equiv \|\mathscr{F}^{-1}[(1+|\star|^2)^{\frac{s}{2}}\mathscr{F} f]\|_p. \tag{1.166}$$

We note that $H^s_2(\mathbb{R}^n)$ is made up of regular distributions by the Plancherel theorem.

The next theorem is indispensable for the study of function spaces.

Theorem 1.53 *Suppose that parameters p, q, s satisfy*

$$0 < p < \infty, \quad 0 < q \le \infty, \quad s > \frac{n}{\min(1, p, q)} + \frac{n}{2}.$$

Let $\{K_j\}_{j=1}^\infty$ be a sequence of compact sets in $\mathbb{R}^n$ and write $d_j \equiv \mathrm{diam}(K_j)$. Assume in addition that we have $f_j \in L^p_{K_j}(\mathbb{R}^n)$ and $H_{(j)} \in \mathscr{S}(\mathbb{R}^n)$ for $j \in \mathbb{N}$. Then $\|H_{(j)}(D) f_j\|_{L^p(\ell^q)} \lesssim \left(\sup_{j\in\mathbb{N}} \|H_{(j)}(d_j\star)\|_{H^s}\right) \|f_j\|_{L^p(\ell^q)}$.

Proof Once we prove Lemma 1.24 below, we are in the position of using a vector-valued inequality (Theorem 1.49) to conclude the proof of the theorem.

Lemma 1.24 *In addition to the assumption of Theorem 1.53, suppose that p, q, η satisfy*

$$0 < \eta < \min(p, q), \quad s > \frac{n}{\eta} + \frac{n}{2}.$$

Then $|H_{(k)}(D) f_k(x)| \lesssim \|H_{(k)}(d_k\star)\|_{H^s} M^{(\eta)} f_k(x)$ for all $x \in \mathbb{R}^n$.

Proof Choose $\delta > 0$ so that $s = \dfrac{n}{\eta} + \dfrac{n+\delta}{2}$. From the definition of $H_{(j)}(D) f$,

$$H_{(j)}(D) f_j(x) = \mathscr{F}^{-1}[H_{(j)} \cdot \mathscr{F} f_j] \simeq_n \int_{\mathbb{R}^n} \mathscr{F}^{-1} H_{(j)}(y) f_j(x-y)\mathrm{d}y. \tag{1.167}$$

By Corollary 1.4, $\langle d_j\, y\rangle^{-\frac{n}{\eta}} |f_j(x-y)| \lesssim M^{(\eta)} f_j(x)$, $x, y \in \mathbb{R}^n$. Hence

$$\left|\int_{\mathbb{R}^n} \mathscr{F}^{-1} H_{(j)}(y) f_j(x-y)\mathrm{d}y\right| \lesssim M^{(\eta)} f_j(x) \|\langle d_j\star\rangle^{\frac{n}{\eta}} \mathscr{F}^{-1} H_{(j)}\|_1. \tag{1.168}$$

If we apply the Cauchy–Schwartz inequality to $\|\langle d_j \star\rangle^{\frac{n}{\eta}} \mathscr{F}^{-1} H_{(j)}\|_1$ then

$$\|\langle d_j\star\rangle^{\frac{n}{\eta}} \mathscr{F}^{-1} H_{(j)}\|_1 \lesssim \left(\int_{\mathbb{R}^n} \langle d_k\, y\rangle^{2s} |\mathscr{F}^{-1} H_{(j)}(y)|^2 \mathrm{d}y\right)^{\frac{1}{2}} \left(\int_{\mathbb{R}^n} \frac{\mathrm{d}y}{\langle d_j\, z\rangle^{n+\delta}}\right)^{\frac{1}{2}}$$
$$\simeq \left(d_j{}^{-n} \int_{\mathbb{R}^n} \langle d_j\, y\rangle^{2s} |\mathscr{F}^{-1} H_{(j)}(y)|^2 \mathrm{d}y\right)^{\frac{1}{2}}.$$

We have

$$\int_{\mathbb{R}^n} \langle d_j\, y\rangle^{\frac{n}{\eta}} |\mathscr{F}^{-1} H_{(j)}(y)| \mathrm{d}y \lesssim \left(\int_{\mathbb{R}^n} |\mathscr{F}^{-1}[\, H_{(j)}(d_j\star)\,](y)|^2 \langle y\rangle^{2s} \mathrm{d}y\right)^{\frac{1}{2}}$$
$$= \|H_{(j)}(d_j\star)\|_{H^s} \tag{1.169}$$

from the property of the Fourier transform $\mathscr{F}$; $\mathscr{F} f(r\star) = r^{-n} \mathscr{F}[f(r^{-1}\star)]$, $r > 0$. If we combine (1.167), (1.168) and (1.169), then we obtain Lemma 1.24 and hence Theorem 1.53.

Exercises

Exercise 1.98 Mimic the proof of Corollary 1.5 and use Corollary 1.4 to prove Corollary 1.6.

Exercise 1.99 Let μ, ν be finite Radon measures on $\mathbb{R}^n$. Show that:

1. $\mathscr{F}\mu$ is represented by a bounded function given by

$$\xi \in \mathbb{R}^n \mapsto \frac{1}{\sqrt{(2\pi)^n}} \int_{\mathbb{R}^n} e^{-2\pi i x\cdot\xi} \mathrm{d}\mu(x) \in \mathbb{C}.$$

2. $\|\mathscr{F}\mu\|_\infty \le (2\pi)^{-n/2} \|\mu\|$, where $\|\mu\|$ denotes the total variation of μ.
3. if $\mathscr{F}\mu$ is in $L^2(\mathbb{R}^n)$, then μ is absolutely continuous.
4. $\mathscr{F}\mu$ is a $C^\infty(\mathbb{R}^n)$-function.
5. $\displaystyle\int_{\mathbb{R}^n} \mathscr{F}\mu(\xi)\mathrm{d}\nu(\xi) = \int_{\mathbb{R}^n} \mathscr{F}\nu(\xi)\mathrm{d}\mu(\xi)$.
6. If $\psi \in \mathscr{S}(\mathbb{R}^n)$, $\displaystyle\int_{\mathbb{R}^n} \mathscr{F}\psi(\xi)\mathrm{d}\nu(\xi) = \int_{\mathbb{R}^n} \psi(\xi)\mathscr{F}\nu(\xi)\mathrm{d}\xi$.
7. If μ is a supported on $B(R)$, $R > 0$, then $\|\partial^\alpha \mathscr{F}\mu\|_\infty \le (2\pi)^{-n/2} R^{|\alpha|} \|\mu\|$.

Exercise 1.100 [584, Lemma A.6] For $\nu \in \mathbb{Z}$ and $m > 0$. Define $\eta_{\nu,m}$ with $\nu \in \mathbb{N}_0$ and $m \in \mathbb{Z} \cap (n, \infty)$ as in Example 1.31. Then reexamine the proof of Theorem 1.50 to prove $|f|^r \lesssim \eta_{\nu,m} * [|f|^r]$ for all $r > 0$ and $f \in \mathscr{S}'_{B(2^\nu)}(\mathbb{R}^n)$.

1.4.4 *Some Integral Inequalities*

From the definition of $\mathscr{S}(\mathbb{R}^n)$, we frequently use an estimate of the form $|f(x)| \lesssim (1+|x|)^{-N}$ for some $N \in \mathbb{N}$. However, it is not so easy to calculate the integral containing a term similar to $(1+|x|)^{-N}$. So, we collect some estimates related to the integrals of the above type.

1.4.4.1 Convolution Estimates Without Moment Condition

As is guessed from an estimate of the form $|f(x)| \lesssim (1+|x|)^{-N}$ it is useful to collect estimates of convolution without a moment condition.

Theorem 1.54 (Estimates of convolutions of rapidly decreasing functions) *Let $\lambda > n$. Then for $0 < t \le 1$,*

$$\frac{1}{t^n}\int_{\mathbb{R}^n}\left\langle\frac{y}{t}\right\rangle^{-\lambda}\langle x-y\rangle^{-\lambda}\mathrm{d}y \sim_\lambda \langle x\rangle^{-\lambda} \quad (x \in \mathbb{R}^n) \tag{1.170}$$

with implicit constants independent of t.

Proof We decompose the integral in the left-hand side of (1.170) according to $B(t)$. On $B(t)$, we use Peetre's inequality $\langle x+y\rangle \le \sqrt{2}\langle x\rangle\langle y\rangle$ to have

$$\frac{1}{t^n}\int_{B(t)}\left\langle\frac{y}{t}\right\rangle^{-\lambda}\langle x-y\rangle^{-\lambda}\mathrm{d}y \sim \frac{1}{t^n}\int_{B(t)}\langle x-y\rangle^{-\lambda}\mathrm{d}y \sim \frac{1}{t^n}\int_{B(t)}\langle x\rangle^{-\lambda}\mathrm{d}y \simeq \langle x\rangle^{-\lambda}.$$

From this, we obtain estimate (1.170) from below.

Hence it suffices to prove

$$\frac{1}{t^n}\int_{\mathbb{R}^n\setminus B(t)}\left\langle\frac{y}{t}\right\rangle^{-\lambda}\langle x-y\rangle^{-\lambda}\mathrm{d}y \lesssim \langle x\rangle^{-\lambda} \tag{1.171}$$

to complete the proof of (1.170). When $|x| \le 10$, we use $\|f * g\|_1 \le \|f\|_\infty \|g\|_1$ for $f \in L^\infty(\mathbb{R}^n)$ and $g \in L^1(\mathbb{R}^n)$ and the result follows.

So it can be assumed that $|x| \ge 10$. Write $M(x;t) \equiv -1 + \log_2 \dfrac{|x|}{t}$ and

$$\mathrm{I} \equiv \sum_{1 \le j \le M(x;t)} \frac{1}{t^n}\int_{B(2^j t)} \frac{\langle x-y\rangle^{-\lambda}\mathrm{d}y}{2^{\lambda j}},$$

$$\mathrm{II} \equiv \sum_{j > M(x;t)} \frac{1}{t^n}\int_{B(2^j t)} \frac{\langle x-y\rangle^{-\lambda}\mathrm{d}y}{2^{\lambda j}}.$$

Then

$$\frac{1}{t^n}\int_{\mathbb{R}^n\setminus B(t)}\left\langle\frac{y}{t}\right\rangle^{-\lambda}\langle x-y\rangle^{-\lambda}\mathrm{d}y=\sum_{j=1}^{\infty}\frac{1}{t^n}\int_{B(2^jt)\setminus B(t)}\left\langle\frac{y}{t}\right\rangle^{-\lambda}\langle x-y\rangle^{-\lambda}\mathrm{d}y$$
$$\lesssim\sum_{j=1}^{\infty}\frac{1}{t^n}\int_{B(2^jt)}\frac{1}{2^{\lambda j}}\langle x-y\rangle^{-\lambda}\mathrm{d}y=\mathrm{I}+\mathrm{II}.$$

Note that $|x|\ge\max(10t,2^{j+1}t)$ as long as $1\le j\le M(x;t)=-1+\log_2\dfrac{|x|}{t}$.

Thus, $|x|\ge 10t$, $y\in B(2^jt)$ and $1\le j\le M(x;t)$, provided $2|x-y|\ge|x|$. Consequently,

$$\mathrm{I}\le\sum_{1\le j\le M(x;t)}\frac{1}{t^n}\int_{B(2^jt)}\left\langle\frac{x}{2}\right\rangle^{-\lambda}\frac{1}{2^{\lambda j}}\mathrm{d}y\lesssim\sum_{1\le j\le M(x;t)}\langle x\rangle^{-\lambda}\frac{1}{2^{j(\lambda-n)}}\lesssim\frac{1}{\langle x\rangle^{\lambda}}.$$

Meanwhile,

$$\mathrm{II}\le\sum_{j>M(x;t)}\frac{1}{t^n}\int_{\mathbb{R}^n}\frac{1}{2^{\lambda j}\langle x-y\rangle^{\lambda}}\mathrm{d}y\sim\sum_{j>M(x;t)}\frac{1}{2^{\lambda j}t^n}\lesssim\frac{1}{2^{\lambda M(x;t)}t^n}\lesssim\frac{t^{\lambda-n}}{\langle x\rangle^{\lambda}}.$$

If we combine I and II, then we obtain the desired inequality.

1.4.4.2 Convolution Estimates with Moment Condition

Using the triangle inequality for integrals, we are led to the estimates of convolution estimates without a moment condition. However, as is the case with the operator $\varphi_j(D)$, or equivalently, $f\in\mathscr{S}'(\mathbb{R}^n)\mapsto\mathscr{F}^{-1}\varphi_j*f$, we are fascinated with the moment condition the functions satisfy. The moment condition is an important condition which will lead us to a nontrivial estimate; the triangle inequality is beyond our reach. The following estimates are used many times in this book.

Theorem 1.55 (The relation between moment and the decay of the functions at infinity (1)) *Let $N\in\mathbb{N}$ be a constant and let $\lambda\in(n+N-1,\infty)$. Also assume that $a\in C^N(\mathbb{R}^n)$ satisfies the differential inequality:*

$$|\partial^\alpha a(x)|\le\langle x\rangle^{-\lambda}\quad(x\in\mathbb{R}^n)\tag{1.172}$$

for $|\alpha|\le N$ and that $\eta\perp\mathscr{P}_{N-1}$ satisfies the differential inequality:

$$|\eta(x)|\le\langle x\rangle^{-\lambda}\quad(x\in\mathbb{R}^n).\tag{1.173}$$

Set $\eta(\star; t) \equiv \frac{1}{t^n}\eta\left(\frac{\star}{t}\right)$ *for* $0 < t \le 1$. *Then*

$$\left|\int_{\mathbb{R}^n} a(y)\eta(x-y;t)\mathrm{d}y\right| \lesssim_\lambda t^N \langle x\rangle^{-\lambda}\int_t^2 r^{\lambda-N-n-1}\mathrm{d}r + t^{\lambda-n}\langle x\rangle^{-\lambda}.$$

In particular, when $\lambda > N+n$,

$$\left|\int_{\mathbb{R}^n} a(y)\eta(x-y;t)\mathrm{d}y\right| \lesssim t^N \langle x\rangle^{-\lambda}.$$

Proof When $\frac{1}{8} \le t \le 1$, Theorem 1.54 covers Theorem 1.55. So we assume $t \le \frac{1}{8}$. Change the variables of the integral in the left-hand side and then subtract the Taylor polynomial of a with degree $N-1$ from a;

$$\begin{aligned}
&\int_{\mathbb{R}^n} a(y)\eta(x-y;t)\mathrm{d}y \\
&= \int_{\mathbb{R}^n} a(x-t\,y)\eta(y)\mathrm{d}y \\
&= \int_{\mathbb{R}^n}\left(a(x-t\,y) - \sum_{\alpha\in\mathbb{N}_0{}^n, |\alpha|\le N-1}\frac{1}{\alpha!}\partial^\alpha a(x)(-ty)^\alpha\right)\eta(y)\mathrm{d}y. \qquad (1.174)
\end{aligned}$$

We estimate the integral in the most right-hand side of (1.174) by decomposing the integral of variable y according to $B(t^{-1})$ and obtaining (1.175) and (1.176) below.

Let us begin with working on $B(t^{-1})$. Here by assumption (1.172), the reminder term of the Taylor expansion is estimated by (1.173):

$$\left|a(x-t\,y) - \sum_{\alpha\in\mathbb{N}_0{}^n, |\alpha|\le N-1}\frac{1}{\alpha!}\partial^\alpha a(x)(-ty)^\alpha\right| \lesssim |t\,y|^N\langle x\rangle^{-\lambda} \qquad (1.175)$$

if $|y| \le t^{-1}$. We write $\min(a,b) \equiv a \wedge b$. Then

$$\int_{B(t^{-1})}\frac{|y|^N\mathrm{d}y}{\langle y\rangle^\lambda} \le \int_{B(t^{-1})}|y|^N \wedge |y|^{N-\lambda}\mathrm{d}y \simeq \int_0^{t^{-1}} r^{n+N-1}(1\wedge r^{-\lambda})\mathrm{d}r.$$

Changing the variables: $r \mapsto r^{-1}$ in the integral in the right-hand side yields

$$\int_0^{t^{-1}} r^{n+N-1}(1\wedge r^{-\lambda})\mathrm{d}r = \int_t^\infty r^{-n-N-1}(1\wedge r^\lambda)\mathrm{d}r.$$

Consequently,

$$\left|\int_{B(t^{-1})}\left(a(x-t\,y)-\sum_{\alpha\in\mathbb{N}_0{}^n,|\alpha|\le N-1}\frac{1}{\alpha!}\partial^\alpha a(x)(-ty)^\alpha\right)\eta(y)\mathrm{d}y\right|\lesssim\int_t^2\frac{\mathrm{d}r}{r^{N+n+1-\lambda}}.$$

Assume instead that $|y|\ge t^{-1}$. Then we obtain

$$\begin{aligned}&\left|a(x-t\,y)-\sum_{\alpha\in\mathbb{N}_0{}^n,|\alpha|\le N-1}\frac{1}{\alpha!}\partial^\alpha a(x)(-ty)^\alpha\right|\\&\le|a(x-t\,y)|+\sum_{\alpha\in\mathbb{N}_0{}^n,|\alpha|\le N-1}\left|\frac{1}{\alpha!}\partial^\alpha a(x)(-ty)^\alpha\right|\\&\lesssim\langle x-t\,y\rangle^{-\lambda}+|t\,y|^{N-1}\langle x\rangle^{-\lambda}\end{aligned}\tag{1.176}$$

by Peetre's inequality, $\langle x+y\rangle\le\sqrt{2}\langle x\rangle\langle y\rangle$ and (1.172). Hence

$$\left|\int_{\mathbb{R}^n\setminus B(t^{-1})}a(y)\eta(x-y;t)\mathrm{d}y\right|\lesssim\int_{\mathbb{R}^n\setminus B(t^{-1})}\left(\langle x-t\,y\rangle^{-\lambda}+|t\,y|^{N-1}\langle x\rangle^{-\lambda}\right)\langle y\rangle^{-\lambda}\mathrm{d}y.$$

We change the variables $y\mapsto z=ty$:

$$\begin{aligned}\int_{\mathbb{R}^n\setminus B(t^{-1})}\langle x-t\,y\rangle^{-\lambda}\langle y\rangle^{-\lambda}\mathrm{d}y&\le\int_{\mathbb{R}^n\setminus B(t^{-1})}\langle x-t\,y\rangle^{-\lambda}|y|^{-\lambda}\mathrm{d}y\\&=t^{\lambda-n}\int_{\mathbb{R}^n\setminus B(t^{-1})}\langle x-z\rangle^{-\lambda}|z|^{-\lambda}\mathrm{d}z\\&\lesssim t^{\lambda-n}\int_{\mathbb{R}^n\setminus B(1)}\langle x-z\rangle^{-\lambda}\langle z\rangle^{-\lambda}\mathrm{d}z\\&\le t^{\lambda-n}\int_{\mathbb{R}^n}\langle x-z\rangle^{-\lambda}\langle z\rangle^{-\lambda}\mathrm{d}z.\end{aligned}$$

From Theorem 1.54, we obtain

$$\int_{\mathbb{R}^n\setminus B(t^{-1})}\langle x-t\,y\rangle^{-\lambda}\langle y\rangle^{-\lambda}\mathrm{d}y\lesssim t^{\lambda-n}\langle x\rangle^{-\lambda}.$$

Recall also that $\lambda>N+n-1$. With this in mind we change the variables: $y\mapsto z=ty$ to have

$$\int_{\mathbb{R}^n\setminus B(t^{-1})}|y|^{N-1-\lambda}\mathrm{d}y=t^{\lambda-N-n+1}\int_{\mathbb{R}^n\setminus B(t^{-1})}|z|^{N-1-\lambda}\mathrm{d}z\sim t^{\lambda-N-n+1}.$$

Consequently,

$$\int_{\mathbb{R}^n \setminus B(t^{-1})} |t\,y|^{N-1} \langle y\rangle^{-\lambda} \mathrm{d}y < t^{N-1} \int_{\mathbb{R}^n \setminus B(t^{-1})} |y|^{N-1-\lambda} \mathrm{d}y$$
$$\simeq t^{\lambda-n}.$$

Hence

$$\left| \int_{\mathbb{R}^n \setminus B(t^{-1})} \left(a(x-t\,y) - \sum_{\alpha \in \mathbb{N}_0{}^n, |\alpha| \le N-1} \frac{1}{\alpha!} \partial^\alpha a(x)(-ty)^\alpha \right) \eta(y) \mathrm{d}y \right|$$
$$\lesssim t^N \langle x\rangle^{-\lambda} \int_t^2 r^{\lambda-N-n-1} \mathrm{d}r + t^{\lambda-n} \langle x\rangle^{-\lambda}.$$

The proof of Theorem 1.55 uses the mean-value theorem. We transform Theorem 1.55 so that we can easily handle it later.

Theorem 1.56 (The relation between moment and the decay of the functions at infinity (2)) *Maintain the same assumption on the constants* λ, N *and on the functions* a, η *as Theorem* 1.55. *For* $j \le \nu$ *we write* $a^j \equiv 2^{jn} a(2^j \star)$, $\eta^\nu \equiv 2^{\nu n} \eta(2^\nu \star)$. *Then* $|a^j * \eta^\nu(x)| \lesssim 2^{jn+(j-\nu)N} \langle 2^j x\rangle^{-\lambda} \int_{2^{j-\nu}}^2 r^{\lambda-N-n-1} \mathrm{d}r$ *for* $x \in \mathbb{R}^n$. *In particular, when* $\lambda > N+n$, $|a^j * \eta^\nu(x)| \lesssim 2^{jn+(j-\nu)N} \langle 2^j x\rangle^{-\lambda} + 2^{(j-\nu)(\lambda-n)} \langle 2^j x\rangle^{-\lambda}$.

Proof Since $a^j * \eta^\nu = (a * \eta^{\nu-j})^j$, simply use Theorem 1.55 with $t = 2^{j-\nu}$.

In the next example, it is worth checking where we used the cancellation condition.

Example 1.33 We set $\mathscr{D}(t) \equiv \mathscr{D}_{-[\log_2 t^{-1}]}$. Define the dyadic average operator by $S_t f \equiv \sum_{Q \in \mathscr{D}(t)} m_Q(f) \chi_Q$, $t > 0$. Fix $p \in C_c^\infty(B(1))$ with $p \ge 0$ and $\|p\|_1 = 1$. Define P_t by

$$P_t f \equiv p_t * f, \quad p_t \equiv \frac{1}{t^n} p\left(\frac{\star}{t}\right)$$

and $Z_t \equiv S_t - P_t$ for $t > 0$, so that Z_t has a cancellation property. Then we claim

$$\sqrt{\int_0^\infty \|Z_t f\|_2{}^2 \frac{\mathrm{d}t}{t}} \lesssim \|f\|_2 \quad (f \in L^2(\mathbb{R}^n)). \tag{1.177}$$

If we use the Fourier transform and $\|Z_t\|_{B(L^2)} \lesssim 1$, then we have

$$\sqrt{\int_0^\infty (\|Z_t(\mathrm{id}_{L^2} - P_t) f\|_2)^2 \frac{\mathrm{d}t}{t}} \tag{1.178}$$

$$\lesssim \sqrt{\int_0^\infty (\|(\mathrm{id}_{L^2} - P_t)f\|_2)^2 \frac{\mathrm{d}t}{t}} \lesssim \|f\|_2.$$

Thus, (1.177) reduces to:

$$\sqrt{\int_0^\infty \|Z_t P_t f\|_2{}^2 \frac{\mathrm{d}t}{t}} \lesssim \|f\|_2.$$

Let $t > 0$ be fixed. Denote by $Z_t(\star, \star)$ the integral kernel of Z_t. That is,

$$Z_t f(x) = \int_{\mathbb{R}^n} Z_t(x, y) f(y)\,\mathrm{d}y$$

for all $f \in L^2(\mathbb{R}^n)$. Then

$$|Z_t(x, y)| \lesssim t^{-n} \chi_{[0, 2^{n+2}t]}(|x - y|) \quad (x, y \in \mathbb{R}^n).$$

The integral kernel $\psi_t(\star, \star)$ of $W_t = Z_t P_t$ is given by

$$\psi_t(x, y) = \frac{1}{t^n} \int_{\mathbb{R}^n} Z_t(x, z) p\left(\frac{z - y}{t}\right) \mathrm{d}z \quad (x, y \in \mathbb{R}^n).$$

Since $Z_t(1) = 0$, the integral kernel $v_{s,t}(x, y)$ of $W_s^* W_t$ satisfies

$$v_{s,t}(x, y) \lesssim \frac{\min(s, t)}{\max(s, t)^{n+1}} \left(1 + \frac{|x - y|}{\max(s, t)}\right)^{-n-1} \quad (s, t > 0).$$

Thus, there exists $M > 0$ such that

$$\|W_t\|_{B(L^2)} \le M, \quad \|W_s^* W_t\|_{B(L^2)} \le M \frac{\min(s, t)}{\max(s, t)} \quad (s, t > 0).$$

From now on we use the technique by Cotlar. Fix $\delta > 0$. We define

$$Q_\delta \equiv \int_\delta^{\delta^{-1}} W_t^* W_t \frac{\mathrm{d}t}{t}.$$

Then

$$(Q_\delta)^N = \int_{[\delta, \delta^{-1}]^N} \prod_{j=1}^N W_{t_j}^* W_{t_j} \frac{\mathrm{d}t_1 \mathrm{d}t_2 \cdots \mathrm{d}t_N}{t_1 t_2 \cdots t_N}.$$

Thus, there exists $M^* \in (M, \infty)$ such that

$$\|(Q_\delta)^N\|_{B(L^2)} \le M^{N+1} \int_{[\delta, \delta^{-1}]^N} \prod_{j=2}^{N-1} \frac{\min(t_{j-1}, t_j)}{\max(t_{j-1}, t_j)} \frac{\mathrm{d}t_1 \mathrm{d}t_2 \cdots \mathrm{d}t_N}{t_1 t_2 \cdots t_N} \le M^{*N} \log \delta^{-1}.$$

Since Q_δ is self-adjoint, $\|(Q_\delta)^N\|_{B(L^2)} = \|Q_\delta\|_{B(L^2)}^N \le M^{*N} \log \delta^{-1}$. Since $N \in \mathbb{N}$ is arbitrary, it follows that $\|Q_\delta\|_{B(L^2)} \le M^*$. Since δ is also arbitrary,

$$\sqrt{\int_0^\infty \|W_t f\|_2{}^2 \frac{dt}{t}} = \sqrt{\int_0^\infty \langle W_t^* W_t f, f\rangle_2 \frac{dt}{t}} \le \sqrt{M^*}\|f\|_2.$$

Since $W_t = Z_t P_t$, putting together (1.178), we obtain (1.177).

Exercises

Exercise 1.101 Denote by $\ell^\infty(\mathbb{Z}^n)$ the set of all bounded sequences indexed by the set of lattice points $\mathbb{Z}^n$. Define a linear subspace $\mathscr{S}(\mathbb{Z}^n)$ of $\ell^\infty(\mathbb{Z}^n)$ by

$$\mathscr{S}(\mathbb{Z}^n) \equiv \bigcap_{N>0} \{\{a_j\}_{j\in\mathbb{Z}^n} \,:\, \{\langle j\rangle^N a_j\}_{j\in\mathbb{Z}^n} \in \ell^\infty(\mathbb{Z}^n)\}.$$

For $a = \{a_j\}_{j\in\mathbb{Z}^n}$ and $b = \{b_j\}_{j\in\mathbb{Z}^n}$, define a sequence $a * b$, the convolution of a and b, by $a * b \equiv \left\{\sum_{k\in\mathbb{Z}^n} a_k b_{j-k}\right\}_{j\in\mathbb{Z}^n}$. Show that the summand $\sum_{k\in\mathbb{Z}^n} a_k b_{j-k}$ of the infinite sum defining $a * b$ converges absolutely for each $j \in \mathbb{Z}^n$ and that $a * b \in \mathscr{S}(\mathbb{Z}^n)$. Hint: Mimic the proof of Theorem 1.26.

Exercise 1.102 Let $x, z \in \mathbb{R}^n$ and $a, b > 0$. Suppose $|x - z| \le \frac{1}{2}$. Set

$$I(a, b, x, z) \equiv \int_{B(x,1)} |x - y|^{a-n}|y - z|^{b-n} dy.$$

Show that:

1. $I(a, b, x, z) \lesssim 1$ if $a + b > n$.
2. $I(a, b, x, z) \lesssim \log|x - z|^{-1}$ if $a + b = n$.
3. $I(a, b, x, z) \lesssim |x - z|^{a+b-n}$ if $a + b < n$.
4. $\displaystyle\int_{B(x,1)} |x-y|^{1-n} \log|y-z|^{-1} dy \lesssim 1$. Hint: $\chi_{B(1)}(x)|x|^{-\alpha} \sim \sum_{l=1}^\infty 2^{l\alpha}\chi_{B(2^{-l})}(x)$ for $x \in \mathbb{R}^n$. In general, we may assume that $|x - y| \le \frac{1}{3}$ and consider the cases $2|x - y| \le |x - z|$, $2|x - z| \le |x - y|$ and $|x - y| \le 2|x - z| \le 4|x - y|$.

Exercise 1.103 [697, Lemma 6.1] Let $k \in \mathbb{N}_0 \cup \{-1\}$. Assume that $\eta \in L^1(\mathbb{R}^n)$ satisfies $\langle\star\rangle^{\max(0,k)}\eta \in L^1(\mathbb{R}^n)$ and assume $\eta \in \mathscr{P}_k(\mathbb{R}^n)^\perp$. Let $\chi_{B(4)\setminus B(2)} \le \varphi \le \chi_{B(8)\setminus B(1)}$. Then show that $\sum_{\nu=-\infty}^\infty |\varphi_\nu(D)\eta(x)| \lesssim (1 + |x|)^{-n-k-1}$ for all $x \in \mathbb{R}^n$.

Exercise 1.104 [791, Lemma 6.2] Let $0 < a, b < n$, and let $m, n > 0$ satisfy $m + n < a + b$. Then prove that

$$\sum_{\ell \in \mathbb{Z}^n \setminus \{0,k\}} (1 + |k - \ell| + |\ell|)^m |k - l|^{-a} |l|^{-b} \sim \langle k \rangle^{m+n-a-b}$$

for all $k \in \mathbb{Z}^n$.

Exercise 1.105 (Miyachi, private communication) Let $r_j(t) = \mathrm{sgn}[\sin(2^j \pi t)]$ for $t \in \mathbb{R}$ and $j \in \mathbb{N}$, and let $L \in \mathbb{N}$ be fixed. Define

$$f_j(t) \equiv 2r_j(t) - \sum_{\alpha=0}^{L} a_\alpha(j) t^\alpha \chi_{[0,1)}(t) \quad (t \in \mathbb{R}),$$

where $a_\alpha(j)$ is chosen so that $f_j \perp \mathscr{P}_L(\mathbb{R})$.

1. Give an explicit formula of $a_\alpha(j)$ to show that $a_\alpha(j) = \mathrm{o}(1)$ as $j \to \infty$.
2. Show that there exists $f \in L^\infty(\mathbb{R}) \cap \mathscr{P}_L(\mathbb{R})^\perp$ such that $\chi_{[0,1)} \le |f| \le 3\chi_{[0,1)}$.
3. Using the tensor product, conclude that there exists $f \in L^\infty(\mathbb{R}^n) \cap \mathscr{P}_L(\mathbb{R}^n)^\perp$ such that $\chi_{[0,1)^n} \le |f| \le 3\chi_{[0,1)^n}$.

1.4.5 Carleson Measure

We observe

$$\mathbb{R}^{n+1}_+ = \bigcup_{Q \in \mathscr{D}(\mathbb{R}^n)} Q \times [2^{-1}\ell(Q), \ell(Q)).$$

The product $Q \times [2^{-1}\ell(Q), \ell(Q))$ is called the *Whitney region* of Q. So, for $Q \in \mathscr{D} = \mathscr{D}(\mathbb{R}^n)$ its Whitney region $Q \times [2^{-1}\ell(Q), \ell(Q))$ and Q itself are closely related. Thus, it will be desirable to estimate $Q \times [2^{-1}\ell(Q), \ell(Q))$ in terms of Q. The Carleson measure will serve this role.

1.4.5.1 Carleson Measures

In connection with the dual inequality of Coifman, Meyer and Stein-type (see Theorem 3.23), we need the notion of the Carleson measures. These things are needed because we have to consider some substitute for $L^\infty(\mathbb{R}^n)$ spaces.

Definition 1.39 (Carleson measure) A Borel measure μ on $\mathbb{R}^{n+1}_+ \equiv \mathbb{R}^n \times (0, \infty)$ is said to be a *Carleson measure*, if $\|\mu\|_{\text{Carleson}} \equiv \sup\limits_{B \subset \mathbb{R}^n} \dfrac{\mu(\hat{B})}{|B|} < \infty$, where $\hat{B}$ is the Carleson tent of B defined in Definition 1.3 and B runs over all balls in $\mathbb{R}^n$.

By definition a Carleson measure is a measure such that the Carleson measure of the Carleson box of B is controlled by the Lebesgue measure of B. This property carries over to any open set. That is, we have:

Lemma 1.25 *If μ is a Carleson measure and O is an open set, then $\mu(\hat{O}) \leq 3^n \|\mu\|_{\text{Carleson}} |O|$.*

Proof Since $\mathbb{Q}$ is a countable set, $\hat{O}$ can be covered by a countable number of Carleson tents; see Exercise 1.23. Hence by the monotonicity of the measure, we can have only to prove

$$\mu\left(\bigcup_{j=1}^{N} \hat{B}_j\right) \leq 3^n \|\mu\|_{\text{Carleson}} |O|,$$

where O is made up of the union of a countable sequence $\{B_j\}_{j=1}^{\infty}$ of balls. In addition, by Theorem 1.7 we can find a nonnegative integer $M \leq N$ and a mapping $\iota : \{1, 2, \dots, N\} \to \{1, 2, \dots, M\}$ such that

$$\sum_{j=1}^{M} \chi_{B_j} \leq \chi_O, \quad B_j \subset 3B_{\iota(j)}.$$

Hence we have $\hat{O} \subset \bigcup\limits_{j=1}^{N} \hat{B}_j \subset \bigcup\limits_{j=1}^{M} \widehat{3B_{\iota(j)}}$. Altogether then,

$$\mu(\hat{O}) \leq \|\mu\|_{\text{Carleson}} \sum_{j=1}^{M} |3B_{\iota(j)}| = 3^n \|\mu\|_{\text{Carleson}} \sum_{j=1}^{M} |B_{\iota(j)}| \leq 3^n \|\mu\|_{\text{Carleson}} |O|.$$

Therefore, the proof is complete.

To state the Carleson embedding theorem, it is convenient to introduce the following definition:

Definition 1.40 (Cone, Non-tangential maximal function)

1. For $x \in \mathbb{R}^n$, define the *cone* $\Gamma(x)$ by $\Gamma(x) \equiv \{(y, t) \in \mathbb{R}^{n+1}_+ : |x - y| < t\}$.
2. For any measurable function ω on $\mathbb{R}^{n+1}_+$ and $x \in \mathbb{R}^n$, one defines its *nontangential maximal function* $N\omega(x)$ by setting

$$N\omega(x) \equiv \sup_{(y,t)\in\mathbb{R}^{n+1}_+, |y-x|<t} |\omega(y,t)| = \sup_{(y,t)\in\Gamma(x)} |\omega(y,t)|.$$

3. Let $\beta \ge 1$, and let $\omega : \mathbb{R}^{n+1}_+ \to \mathbb{C}$ be a function which is not always Borel measurable. Define

$$N_\beta\omega(x) \equiv \sup_{(y,\beta t)\in\Gamma(x)} \omega(y,t).$$

From the definition, we see $N\omega = N_1\omega$.

Although it is difficult to find the precise value of $N\omega(x)$ in general, we have an example of ω for which we can estimate $N\omega$.

Example 1.34 Let

$$\omega(x,t) \equiv \left(1 + \frac{a}{|x|+t}\right)^{-N} \quad ((x,t) \in \mathbb{R}^{n+1}_+)$$

for $N > 0$ and $a > 0$. Then

$$N\omega(x) = \sup_{(y,t)\in\Gamma(x)} \omega(y,t) \ge \sup_{t>0} \omega(x,t) = \left(1 + \frac{a}{|x|}\right)^{-N}.$$

Meanwhile, if $(y,t) \in \Gamma(x)$, then from the definition of the nontangential maximal function we deduce

$$\omega(y,t) = \left(1 + \frac{a}{|y|+t}\right)^{-N} \lesssim_N \left(1 + \frac{a}{|x|+t}\right)^{-N}.$$

Thus, taking the supremum over $\Gamma(x)$, we obtain

$$N\omega(x) = \sup_{(y,t)\in\Gamma(x)} \omega(y,t) \lesssim_N \sup_{t>0} \left(1 + \frac{a}{|x|+t}\right)^{-N} = \left(1 + \frac{a}{|x|}\right)^{-N}.$$

As a result, we obtain

$$N\omega(x) \sim \left(1 + \frac{a}{|x|}\right)^{-N},$$

where the implicit constant depends only on N.

Theorem 1.57 (Carleson's embedding theorem) *Let μ be a Carleson measure on $\mathbb{R}^{n+1}_{+}$, and let $F : \mathbb{R}^{n+1} \to [0, \infty)$ be a Borel measurable function. Then*

$$\int_{\mathbb{R}^{n+1}_{+}} F(x,t)\mathrm{d}\mu(x,t) \le 3^n \|\mu\|_{\mathrm{Carleson}} \int_{\mathbb{R}^n} \left(\sup_{(y,t)\in\Gamma(x)} F(y,t) \right) \mathrm{d}x.$$

Proof Set $F^*(x) \equiv \sup_{(y,t)\in\Gamma(x)} F(y,t)$. By the distribution function, we have

$$\int_{\mathbb{R}^{n+1}_{+}} F(x,t)\mathrm{d}\mu(x,t) = \int_0^\infty \mu\{(x,t) \in \mathbb{R}^{n+1}_{+} : F(x,t) > \lambda\}\mathrm{d}\lambda \tag{1.179}$$

from the layer cake representation; see Theorem 1.5. For $\lambda > 0$, we set

$$O_\lambda \equiv \{x \in \mathbb{R}^n : F^*(x) > \lambda\}. \tag{1.180}$$

Then O is an open set and O satisfies $\{(y,t) \in \mathbb{R}^{n+1}_{+} : F(y,t) > \lambda\} \subset \widehat{O_\lambda}$.

In fact, choose a point $(y,t) \in \mathbb{R}^{n+1}_{+}$ such that $F(y,t) > \lambda$. We see from the definition of $\widehat{O_\lambda}$ $B(y,t) \subset O$ if and only if $(y,t) \in \widehat{O_\lambda}$. Let $z \in B(y,t)$; that is, $(y,t) \in \Gamma(z)$. Then $F^*(z) = \sup_{(w,s)\in\Gamma(z)} F(w,s) > F(y,t)$. Hence we have $B(y,t) \subset O_\lambda$; that is, $(y,t) \in \widehat{O_\lambda}$. Thus,

$$\begin{aligned}\mu\{(y,t) \in \mathbb{R}^{n+1}_{+} : F(y,t) > \lambda\} &\le \mu\left\{x \in \mathbb{R}^n : \sup_{(y,t)\in\Gamma(x)} F(y,t) > \lambda\right\} \\ &\le 3^n \|\mu\|_{\mathrm{Carleson}} |O_\lambda|. \end{aligned} \tag{1.181}$$

If we combine (1.179), (1.180) and (1.181), then we obtain the desired result.

We discuss the dependency of the apertus and the Hausdorff capacity H^d.

Example 1.35 Let $0 < d \le n$, $\beta \ge 1$, and let $\omega : \mathbb{R}^{n+1}_{+} \to \mathbb{C}$ be a function which is not always Borel measurable. Let $O \equiv \{x \in \mathbb{R}^n : N\omega(x) > \mu\}$ for $\mu > 0$. According to Proposition 1.5, Then there exists a ball cover $\{B_\lambda\}_{\lambda\in\Lambda}$ of O such that

$$\sum_{\lambda\in\Lambda} |B_\lambda|^{\frac{d}{n}} \lesssim H^d(O), \quad \hat{O} \subset \bigcup_{\lambda\in\Lambda} \widehat{10^{10} B_\lambda}$$

and that for any ball B contained in O there exists $\lambda_B \in \Lambda$ such that $B \subset 10^{10} B_{\lambda_B}$. Thus, $\{N_\beta\omega > \mu\} \subset \bigcup_{\lambda\in\Lambda} 10^{10}\beta B_\lambda$, which implies $H^d(\{N_\beta\omega > \mu\}) \lesssim \beta^d H^d(\{x \in \mathbb{R}^n : N\omega(x) > \mu\})$.

Exercises

Exercise 1.106 Show that $N\omega$ is a measurable function no matter what $\omega : \mathbb{R}^{n+1}_+ \to \mathbb{C}$ is. (We tolerate the case where ω itself is not measurable.) Hint: Show that $\{N\omega > \lambda\}$ is an open set for all $\lambda \geq 0$.

Exercise 1.107 Let $0 < d \leq n$ and $\beta > 0$, and let $\omega : \mathbb{R}^{n+1}_+ \to \mathbb{C}$ be a measurable function such that $\displaystyle\int_0^\infty N\omega(x)\mathrm{d}H^d(x) < \infty$. Then show that $\displaystyle\int_0^\infty N_\beta\omega(x)\mathrm{d}H^d(x) < \infty$ using Example 1.35.

Textbooks in Sect. 1.4

Hardy–Littlewood Maximal Operator: Theorems 1.45 and 1.46

See [22, Chapter 2 Section 4], [31, Section 2.2], [32, Section 2.1] and [86, Chapter 1, Section 3].

Lebesgue Differentiation Theorem: Theorem 1.48

See [22, Corollary 2.13], [32, Corollary 2.1.16], [86, p. 13 Corollary] and [117, Theorems 6.3 and 6.4].

Fefferman–Stein Vector-Valued Inequalities: Theorem 1.49

See [86, Chapter 2, Section 1]. See also [31, Chapter 5, Theorem 4.2] for more about the vector-valued inequalities, which also includes the technique of Theorem 1.49 using the singular integral operators, which is considered in Sect. 1.5.

Plancherel–Polya–Nikolski'i Inequality

See [33, 99, 1069].

Duality of the Vector-Valued Spaces: Lemma 1.22

See [99, p. 177] for the dual space of $L^p(\ell^q)$ and $\ell^q(L^p)$ with $1 < p, q \leq \infty$ in connection with Lemma 1.22.

Others

We followed the proof in [71, p. 54, Lemma 1] for Corollary 1.8 with the help of the simplification pointed out by Noi; see Lemma 1.23. We referred to the textbooks [99, 100] for Theorems 1.50, 1.51, and 1.53 in the present form. I referred to the works by Ms. Mamiko Nagai and Ms. Mao Nakada in Tokyo Woman's Christian University for the statement and the proofs of Theorems 1.54, 1.55, and 1.56.

1.5 Singular Integral Operators

The singular integral operators, which are represented by the j-th Riesz transform given by,

$$R_j f(x) \equiv \lim_{\varepsilon \downarrow 0} \int_{\mathbb{R}^n \setminus B(x,\varepsilon)} \frac{x_j - y_j}{|x-y|^{n+1}} f(y) \mathrm{d}y, \tag{1.182}$$

are integral operators with singularity (mainly at the origin). As it turns out, this transform is important because it connects the Laplacian and the second pure derivative $\partial_{x_i}\partial_{x_j}$. Another example for $n = 2$ is the Ahlfors–Beuring transform in complex analysis which relates the ∂ operator with the $\bar{\partial}$ operator. They have a lot to do with $L^p(\mathbb{R}^n)$-spaces, Hardy spaces and Triebel–Lizorkin spaces. There is huge amount of literature of the theory of the singular integral operators. Our aim is rather modest; we seek to study them from the viewpoint of the function spaces and we are mainly interested in the singular integral operators of convolution type. In Sect. 1.5.1 we investigate the Calderón–Zygmund decomposition, which is needed for the proof of the (1, 1)-boundedness; see Theorem 1.59. By "the weak (1, 1)-boundedness", we mean an inequality of the form:

$$\lambda |\{|R_j f| > \lambda\}| \lesssim \|f\|_1 \quad (f \in L^1(\mathbb{R}^n), \lambda > 0).$$

We will deal with the singular integral operators and as an example, we consider the Riesz transform given by (1.182) and the boundedness in Sect. 1.5.2; see Theorems 1.62, 1.65, 1.66 and 1.67.

1.5.1 Dyadic Maximal Operator and the Calderón–Zygmund Decomposition

As we have mentioned, we are fascinated with the method of decomposition of functions into countable elementary pieces. That is, we will look into each function in much depth. But no matter how deeply we investigate the functions, we need to stop at a certain stage. So, we have to judge whether we stop or continue. To make such a decision, we need a tool; a collection of dyadic cubes.

1.5.1.1 Dyadic Maximal Operator and Calderón–Zygmund Cubes

With the definition of the dyadic cubes in mind, we move on to the definition of the dyadic maximal operators.

Definition 1.41 (Dyadic maximal operator) For $f \in L^1_{\rm loc}(\mathbb{R}^n)$, define

$$E_j f(x) \equiv E_j[f](x) \equiv \sum_{Q \in \mathcal{D}_j(\mathbb{R}^n)} \chi_Q(x) m_Q(f), \quad M_{\rm dyadic} f(x) \equiv \sup_{j \in \mathbb{Z}} E_j[\,|f|\,](x)$$

for $x \in \mathbb{R}^n$. The operator $M_{\rm dyadic}$ is called the *dyadic maximal operator*. The function $E_j f(x)$ stands for the average of f over a unique cube $Q \in \mathcal{D}_j(\mathbb{R}^n)$ containing x, or shortly, the *dyadic average operator of generation* j.

We have $|E_j f(x)| \le 2^{jn} \|f\|_1$ for all $j \in \mathbb{Z}$ and $x \in \mathbb{R}^n$. So the next lemma holds trivially.

Lemma 1.26 *We have* $\lim_{j \to -\infty} E_j f(x) = 0$ *for all* $f \in L^1(\mathbb{R}^n)$ *and* $x \in \mathbb{R}^n$.

The next theorem is the heart of the matter. To formulate this theorem, we define the *quadrant* as a set of the form

$$\{(x_1, x_2, \ldots, x_n) \in \mathbb{R}^n \, : \, (a_1 x_1, a_2 x_2, \ldots, a_n x_n) \in (0, \infty)^n\}$$

for some $(a_1, a_2, \ldots, a_n) \in \{-1, 1\}^n$.

Theorem 1.58 (Construction of Calderón–Zygmund cubes) *Let* $f \in L^1_{\rm loc}(\mathbb{R}^n)$, *and let* $\lambda > 0$. *Assume that* $\{x \in \mathbb{R}^n \, : \, M_{\rm dyadic} f(x) > \lambda\}$ *does not contain any quadrant. Then there exists a disjoint collection* $\{Q_j\}_{j \in J} \subset \mathcal{D}$ *such that*

$$\{x \in \mathbb{R}^n \, : \, M_{\rm dyadic} f(x) > \lambda\} = \bigcup_{j \in J} Q_j$$

and that $\lambda < m_{Q_j}(\,|f|\,) \le 2^n \lambda$ *for all* $j \in J$.

As it turns out from the weak $(1, 1)$-boundedness of M, if $f \in \bigcup_{1 \le p < \infty} L^p(\mathbb{R}^n)$, then the assumption in Theorem 1.58 is satisfied.

Proof To simplify the notation, we set $E_\lambda \equiv \{x \in \mathbb{R}^n \, : \, M_{\rm dyadic} f(x) > \lambda\}$. By the definition of the dyadic maximal operator, in particular by the property of "sup", we have $E_\lambda = \bigcup_{j \in \mathbb{Z}} \{E_j[|f|] > \lambda\}$. From Lemma 1.26, the set of j satisfying $x \in$

$\{E_j[|f|] > \lambda\}$ is bounded from below for each $x \in E_\lambda$. So there exists j^* such that $E_j[\,|f|\,](x) \le \lambda$ for all $j > j^*$. Hence we can partition E_λ:

$$E_\lambda = \bigcup_{j \in \mathbb{Z}} \bigcap_{k=-\infty}^{j-1} \{x \in \mathbb{R}^n \,:\, E_j[\,|f|\,](x) > \lambda,\, E_k[\,|f|\,](x) \le \lambda\}.$$

From Lemma 1.3, we can partition

$$\bigcap_{k \in \mathbb{Z} \cap (-\infty, j)} \{x \in \mathbb{R}^n \,:\, E_j[\,|f|\,](x) > \lambda,\, E_k[\,|f|\,](x) \le \lambda\} \tag{1.183}$$

into the sum of dyadic cubes $Q^j_m \in \mathscr{D}_j$ with $m = 1, 2, \ldots, j_M$. A rearrangement of Q^j_m's yields a partition of E_λ; $E_\lambda = \sum_{j \in J} Q_j$ and $\{Q_j\}_{j \in J} = \{Q^j_m\}_{m=1,\ldots,M_j,\, j \in \mathbb{Z}}$, where $M_j \in \mathbb{N} \cup \{\infty\}$.

Let $x \in Q_j =: Q_{j_0 m_0}$. From the choice of $Q_{j_0 m_0}$, we have $m_{Q_j}(|f|) = E_{j_0}[\,|f|\,](x) > \lambda$.

Let $Q^*_j \in \mathscr{D}_{j_0-1}$ be the dyadic parent of the dyadic cube Q_j; that is, the unique cube containing Q_j having volume as 2^n times as Q_j. From (1.183), the choice of $Q_{j_0 m_0}$, we have $m_{Q^*_j}(\,|f|\,) = E_{j_0-1}[\,|f|\,](x) \le \lambda$. From this formula, we conclude

$$m_{Q_j}(\,|f|\,) \le 2^n m_{Q^*_j}(\,|f|\,) = 2^n E_{j_0-1}[\,|f|\,](x) \le 2^n \lambda.$$

1.5.1.2 Calderón–Zygmund Decomposition

We consider the Calderón–Zygmund decomposition keeping the setup above in mind.

Theorem 1.59 (Calderón–Zygmund decomposition) *Let $f \in L^1(\mathbb{R}^n)$ and $\lambda > 0$. Then we can find a set $\{Q_j\}_{j \in J}$ of dyadic cubes and a collection $\{g\} \cup \{b_j\}_{j \in J}$ of countable collections of $L^1(\mathbb{R}^n)$-functions such that:*

1. *f admits the following decomposition:*

$$f = g + \sum_{j \in J} b_j. \tag{1.184}$$

2. *(The $L^1(\mathbb{R}^n)$-condition) The $L^1(\mathbb{R}^n)$-norm of g does not exceed that of f:*

$$\|g\|_1 \le \|f\|_1. \tag{1.185}$$

3. *(The $L^\infty(\mathbb{R}^n)$-condition) The function g is $L^\infty(\mathbb{R}^n)$-bounded: More quantitatively,*

$$\|g\|_\infty \leq 2^n \lambda. \tag{1.186}$$

4. (*The support condition*) *The function* b_j *is supported on the closure of* Q_j; *namely*

$$\text{supp}(b_j) \subset \overline{Q_j}. \tag{1.187}$$

5. (*The moment condition*) $b_j \perp \mathscr{P}_0(\mathbb{R}^n)$.
6. (*Partition of the set* $\{M_{\text{dyadic}} f > \lambda\}$) *The family* $\{Q_j\}_j$ *is disjoint and*

$$\{ M_{\text{dyadic}} f > \lambda \} = \sum_{j \in J} Q_j. \tag{1.188}$$

This decomposition is called the Calderón–Zygmund decomposition (at height λ).

Proof Choose a packing $\{Q_j\}_{j \in J}$ of dyadic cubes such that

$$\lambda \leq m_{Q_j}(|f|) \leq 2^n \lambda, \quad \left\{x \in \mathbb{R}^n : M_{\text{dyadic}} f(x) > \lambda\right\} = \bigcup_{j \in J} Q_j \tag{1.189}$$

according to Theorem 1.58 together with the remark there below. Define b_j, as follows:

$$b_j \equiv \chi_{Q_j}(f - m_{Q_j}(f)). \tag{1.190}$$

Based on b_j, we define g by $g \equiv f - \sum_{j \in J} b_j$, which yields a decomposition of f. Here note that the sum makes sense in the topology of $L^1(\mathbb{R}^n)$ and in the sense of almost all. Since the supports of b_j's are nonoverlapping, from the definition of b_j, the moment condition (1.185) follows. It remains to show (1.186).

Let $x \in Q_j$. Since $\{Q_j\}_{j \in J}$ are nonoverlapping, we have $g(x) = m_{Q_j}(f)$. Hence $|g(x)| \leq 2^n \lambda$ by (1.189).

Conversely, let $x \notin \bigcup_{j \in J} Q_j$. Inequality (1.186) follows from a trivial inequality which follows from the Lebesgue differentiation theorem:

$$\text{for almost all } x \in \mathbb{R}^n, \ |g(x)| = |f(x)| \leq M_{\text{dyadic}} f(x) \leq \lambda.$$

A helpful remark may be in order.

Remark 1.7 Let $f \in L^1(\mathbb{R}^n) \cap L^2(\mathbb{R}^n)$. If one reexamines the above proof, then one notices $b = \sum_{j \in J} b_j$ in the topology of $L^1(\mathbb{R}^n) \cap L^2(\mathbb{R}^n)$ and that $g \in L^1(\mathbb{R}^n) \cap L^2(\mathbb{R}^n)$.

Theorems 1.58 and 1.59 are widely known as the Calderón–Zygmund decomposition or the Calderón–Zygmund theory and these theorems have been applied to harmonic analysis.

1.5.1.3 Various Calderón–Zygmund Decompositions

So far, we have considered the decomposition for $f \in L^1_{\rm loc}(\mathbb{R}^n)$ based on the function $M_{\rm dyadic} f$. Here we consider decompositions based on the other maximal operators. The first one is a decomposition using $(M_{\rm dyadic}[|f|^p])^{1/p}$.

We have the following variant of Theorem 1.59:

Lemma 1.27 *Let $\lambda > 0$, $p_0 \in [1,\infty)$ and $f \in L^{p_0}(\mathbb{R}^n)$. Consider the decomposition of the set $\{M_{\rm dyadic}[|f|^{p_0}] > \lambda^{p_0}\}$ of maximal disjoint dyadic cubes satisfying:*

$$\{M_{\rm dyadic}[|f|^{p_0}] > \lambda^{p_0}\} = \bigcup_{j\in J} Q_j, \quad m_{Q_j}^{(p_0)}(f) > \lambda.$$

We set

$$g \equiv \sum_{j\in J} m_{Q_j}(f)\chi_{Q_j} + (1-\chi_{\bigcup_{j\in J} Q_j})f,$$

$$b_j \equiv \chi_{Q_j}(f - m_{Q_j}(f)) \in \mathscr{P}_0(\mathbb{R}^n)^\perp \tag{1.191}$$

for $j \in J$. Then we have

$$\|g\|_\infty \le \lambda, \tag{1.192}$$

$$\operatorname{supp}(b_j) \subset Q_j \quad (j \in J), \tag{1.193}$$

$$m_{Q_j}^{(p_0)}(b_j) \lesssim_{p_0} \lambda \quad (j \in J), \tag{1.194}$$

$$\sum_{j\in J} |Q_j| \le (\lambda^{-1}\|f\|_{p_0})^{p_0}. \tag{1.195}$$

We leave the proof of Lemma 1.27 to interested readers. See Exercise 1.108.

We consider another Calderón–Zygmund decomposition. We now move on to the Calderón–Zygmund decomposition of Sobolev functions. We need the following version of the Whitney covering lemma.

Theorem 1.60 *Let G be an open set in $\mathbb{R}^n$. Let $\mathscr{U}$ be the maximal family of dyadic cubes Q such that $9Q$ is included in G. Then if $Q_1, Q_2 \in \mathscr{U}$ are adjacent, namely $\partial Q_1 \cap \partial Q_2 \neq \emptyset$, then $\ell(Q_1) \le 4\ell(Q_2)$.*

Proof If $\ell(Q_1) > 4\ell(Q_2)$, then the dyadic parent S of Q_2 satisfies $9S \subset 9Q_1 \subset G$. This is a contradiction.

We need the following Calderón–Zygmund decomposition in the last chapter. Here $\dot{W}^{1,p}(\mathbb{R}^n)$ denotes the homogeneous Sobolev space. The homogeneous Sobolev norm is given by $\|f\|_{\dot{W}^{1,p}} \equiv \sum_{j=1}^{n} \|\partial_{x_j} f\|_p$.

Theorem 1.61 *Let $1 \le p < \infty$. For any $f \in L^1_{\rm loc}(\mathbb{R}^n) \cap \mathscr{S}'(\mathbb{R}^n)$ such that $\nabla f \in L^p(\mathbb{R}^n)^n$, there exist a locally integrable function g, a collection $\{Q_j\}_{j\in J}$ of dyadic cubes and a collection $\{b_j\}_{j\in J}$ of $C^1(\mathbb{R}^n)$-functions such that the following properties hold:*

1. *We have the size estimate*

$$\sum_{j\in J} |Q_j| \lesssim (\lambda^{-1}\|\nabla f\|_{(L^p)^n})^p \tag{1.196}$$

and overlapping property

$$\sum_{j\in J} \chi_{\frac{8}{7}Q_j} \lesssim 1. \tag{1.197}$$

2. *f has a decomposition:*

$$f = g + \sum_{j\in J} b_j. \tag{1.198}$$

3. *The good part g satisfies $\nabla g \in L^p(\mathbb{R}^n)^n \cap L^\infty(\mathbb{R}^n)^n$, more precisely, it satisfies the $\dot{W}^{1,p}(\mathbb{R}^n)$-condition*

$$\|\nabla g\|_{(L^p)^n} \lesssim \|\nabla f\|_{(L^p)^n}, \tag{1.199}$$

and the $\dot{W}^{1,\infty}(\mathbb{R}^n)$-condition

$$\|\nabla g\|_{(L^\infty)^n} \lesssim \lambda. \tag{1.200}$$

4. *Each bad part b_j satisfies*

$$\operatorname{supp}(b_j) \subset \frac{8}{7}Q_j, \tag{1.201}$$

$$m^{(p)}_{Q_j}(|\nabla b_j|) \lesssim \lambda. \tag{1.202}$$

Proof Let $G \equiv \{M^{(p)}[|\nabla f|] > \lambda\}$ and $F \equiv \mathbb{R}^n \setminus G$.

Let x be a point in F. Fix a dyadic cube Q containing x and let Q_k be the k-th dyadic child of Q containing x. Then by Poincaré's inequality (see Exercise 1.83)

$$|m_{Q_{k+1}}(f) - m_{Q_k}(f)| \lesssim m_{Q_{k+1}}(|f - m_{Q_{k+1}}(f)|) \lesssim \ell(Q_k) m^{(p)}_{Q_{k+1}}(|\nabla f|).$$

In view of the maximality of the cubes, we have

$$|m_{Q_{k+1}}(f) - m_{Q_k}(f)| \lesssim 2^k \ell(Q)\lambda, \tag{1.203}$$

since Q_{k+1} contains $x \in F$. Likewise, if Q and R are cubes containing x such that $\ell(Q) \sim \ell(R)$, then

$$|m_Q(f) - m_R(f)| \lesssim \ell(Q)\lambda.$$

It thus easily follows that $m_Q(f)$ has a limit $f^\dagger(x)$ as $|Q|$ tends to 0, where Q contains x. If, moreover, x is a Lebesgue point of f, then this limit is equal to $f(x)$.

Apply Theorem 1.60 to have a collection $\mathcal{U} = \{Q_j\}_{j \in J}$ of dyadic cubes. Note that $\mathcal{U}$ is a covering. According to Theorem 1.60, (1.197) is satisfied. Also by the weak-(1, 1) estimate for the Hardy–Littlewood maximal operator, (1.196) is satisfied.

Let $x_j \in 100Q_j \cap F$ be a reference point with respect to the covering $\mathcal{U}$. We form the partition of unity with respect to $\mathcal{U}$. Let $\Theta \in C^\infty_c(\mathbb{R}^n)$ be such that $\chi_{Q\left(\frac{1}{2}\right)} \le \Theta \le \chi_{Q\left(\frac{4}{7}\right)}$. Define $\Theta_j \equiv \Theta\left(\dfrac{\star - c(Q_j)}{\ell(Q_j)}\right)$, so that $\chi_{Q_j} \le \Theta_j \le \chi_{\frac{8}{7}Q_j}$. Write $K \equiv \sum_{j \in J} \Theta_j$ and $\kappa_j \equiv \dfrac{\Theta_j}{K}$ for $j \in J$. We define $b_j \equiv (f - f^\dagger(x_j))\kappa_j$ for $j \in J$. Then we have (1.201) from the support condition of κ_j.

We have $|m_{100Q_j}(f) - f^\dagger(x_j)| \lesssim \lambda \ell(Q_j)$ from (1.203). Hence

$$\int_{2Q_j} |m_{100Q_j}(f) - f^\dagger(x_j)|^p |\nabla \kappa_j(x)|^p \mathrm{d}x \lesssim \lambda^p |Q_j|. \tag{1.204}$$

Next, using the Poincaré inequality and the fact that $x_j \in 100Q_j \cap F$, we have

$$|m_{Q_j}(f) - m_{100Q_j}(f)| \lesssim m_{100Q_j}(|f - m_{100Q_j}(f)|) \lesssim m^{(p)}_{100Q_j}(|\nabla f|) \lesssim \lambda \ell(Q_j).$$

Hence

$$|m_{Q_j}(f) - m_{100Q_j}(f)|^p \int_{2Q_j} |\nabla \kappa_j(x)|^p \mathrm{d}x \lesssim \lambda^p |Q_j|. \tag{1.205}$$

Since $\nabla((f - m_{Q_j}(f))\kappa_j) = \kappa_j \nabla f + (f - m_{Q_j}(f))\nabla\kappa_j$, we have

$$\int_{\mathbb{R}^n} |\nabla[(f(x) - m_{Q_j}(f))\kappa_j(x)]|^p \mathrm{d}x \lesssim \lambda^p |Q_j| \tag{1.206}$$

thanks to the Poincaré inequality and the maximality of the cubes. Putting (1.204), (1.205), and (1.206) together, we obtain (1.202).

Set $p^* \equiv \frac{np}{n-p}$ if $p < n$ and $p^* \equiv \infty$ if $p \geq n$. Then for $r \in [p, p^*]$, we claim

$$\left\| \sum_{j \in J} \frac{1}{\ell(Q_j)} |b_j| \right\|_r \lesssim \lambda \left(\sum_{j \in J} |Q_j| \right)^{\frac{1}{r}}. \tag{1.207}$$

In fact, by the Poincaré inequality, the disjointness of cubes and (1.202), we have

$$\left(\left\| \sum_{j \in J} \frac{1}{\ell(Q_j)} |b_j| \right\|_r \right)^r \lesssim \sum_{j \in J} \ell(Q_j)^{-r} \int_{\mathbb{R}^n} |b_j(x)|^r \mathrm{d}r \lesssim \sum_{j \in J} \ell(Q_j)^{n - nr/p} \|\nabla b_j\|_{(L^p)^n}{}^r,$$

which proves (1.207).

We next claim that $b \equiv \sum_{j \in J} b_j$ is a well-defined distribution on $\mathbb{R}^n$ and that it converges in $L^1_{\mathrm{loc}}(\mathbb{R}^n)$. Indeed, for a test function $\varphi \in \mathscr{S}(\mathbb{R}^n)$, using the properties of Q_j, $j \in J$,

$$\begin{aligned} \sum_{j \in J} \|b_j \varphi\|_1 &\lesssim \int_{\mathbb{R}^n} \left(\sum_{j \in J} \ell(Q_j)^{-1} |b_j(x)| \right) |\varphi(x)| \mathrm{dist}(x, F) \mathrm{d}x \\ &\lesssim \int_{\mathbb{R}^n} \left(\sum_{j \in J} \ell(Q_j)^{-1} |b_j(x)| \right) |\varphi(x)| (1 + |x|) \mathrm{d}x \end{aligned}$$

and the last sum converges in $L^p(\mathbb{R}^n)$.

So far, information on the bad function b_j is ready. We want to consider

$$g \equiv f - \sum_{j \in J} b_j = f - b \in L^1_{\mathrm{loc}}(\mathbb{R}^n),$$

so that (1.198) holds automatically. Note that the sum defining g is locally finite, so that ∇g exists and belongs to $L^p_{\mathrm{loc}}(\mathbb{R}^n)^n$. We want to show that $\nabla g \in L^p(\mathbb{R}^n)^n \cap L^\infty(\mathbb{R}^n)^n$, which satisfies (1.199) and (1.200). Since we have proved (1.202), (1.199) follows from the estimate $|G| \lesssim \lambda^{-p} \|\nabla f\|_{(L^p)^n}$, which results from Theorem 1.8, the weak-(1, 1) inequality of M.

It remains to prove (1.200). To this end, we set $\mathbf{H} \equiv \sum_{j\in J} f^\dagger(x_j)\nabla\kappa_j$. Note that this sum is locally finite and $\mathbf{H}(x) = 0$ for $x \in F$. Note also that $\sum_{j\in J}\kappa_j = \chi_\Omega$ thanks to the property of the partition of unity. Since this sum is locally finite we have

$$\sum_{j\in J}\nabla\kappa_j = 0 \tag{1.208}$$

on Ω. We claim that

$$|\mathbf{H}(x)| \lesssim \lambda \quad (x \in \Omega). \tag{1.209}$$

Indeed, fix $x \in \Omega$. Let $j \in J$ be such that $x \in Q_j$, and let I_x be the set of indices i such that $x \in \frac{8}{7}Q_i$. We know that $\sharp I_x \lesssim 1$. Also for $i \in I_x$ we have that

$$\ell(Q_j) \sim \ell(Q_j), \quad |x_j - x_j| \lesssim \ell(Q_j).$$

Thanks to (1.203) and (1.208), we have

$$|\mathbf{H}(x)| = \left|\sum_{i\in I_x}(f(x) - f^\dagger(x_j))\nabla\kappa_j(x)\right| \lesssim \sum_{i\in I_x}|f(x) - f^\dagger(x_j)|\ell(Q_j)^{-1} \lesssim \lambda,$$

which proves (1.209).

Thus, from (1.209), $\nabla g = \mathbf{H} \in L^\infty(\mathbb{R}^n)^n$ and satisfies (1.200).

Remark 1.8 If $f \in W^{1,q}(\mathbb{R}^n)$ for some $1 \le q < \infty$, then the convergence takes place in $W^{1,q}(\mathbb{R}^n)$.

1.5.1.4 Dyadic Maximal Operator: Some Universal Estimates

As we have seen, dyadic cubes have some special geometric structure. In fact, among them some combinatric property play the key role on many occasions. Here we collect some of them, which are used later.

Denote by $M_{\mathscr{D}(Q)}$ the maximal operator generated by a cube Q; that is, for a measurable function f on Q we define

$$M_{\mathscr{D}(Q)}f(x) \equiv \sup_{R\in\mathscr{D}(Q)} \chi_R(x) m_R(|f|),$$

$$M_0 f(x) \equiv \sup_{R\in\mathscr{D}(Q)} \chi_R(x)\exp\left(m_R(\log|f|)\right).$$

We have the following universal estimates:

Lemma 1.28 (Universal estimates) *Let Q be a cube.*

1. *Let $f \in L^1(Q)$, $\lambda > 0$ and let $\Omega_\lambda \equiv \{x \in Q : M_{\mathscr{D}(Q)}f(x) > \lambda\}$. Then*

$$\lambda|\Omega_\lambda| \le \|f\chi_{\Omega_\lambda}\|_{L^1(Q)}. \tag{1.210}$$

2. *Let $f \in L^p(Q)$ with $1 < p < \infty$. Then*

$$\|M_{\mathscr{D}(Q)}f\|_{L^p(Q)} \le \frac{p}{p-1}\|f\|_{L^p(Q)}. \tag{1.211}$$

3. *Let $f \in L^1(Q)$. Then*

$$\|M_0 f\|_{L^1(Q)} \le e\|f\|_{L^1(Q)}. \tag{1.212}$$

Proof

1. Reexamine the argument before. Since the proof of (1.210) is similar, we omit the details.
2. Theorem 1.5 and (1.210) entail (1.211).
3. Let $u > 0$. Note that $M_0 f(x) \le M^{(u)}_{\mathscr{D}(Q)}f(x) \equiv \left(M_{\mathscr{D}(Q)}[|f|^u](x)\right)^{\frac{1}{u}}$ for all $x \in \mathbb{R}^n$. Thus,

$$\begin{aligned}\|M_0 f\|_{L^1(Q)} \le \|M^{(u)}_{\mathscr{D}(Q)}\|_{L^1(Q)} &= (\|M_{\mathscr{D}(Q)}[|f|^u]\|_{L^{\frac{1}{u}}(Q)})^{\frac{1}{u}}\\ &\le \left(\frac{1}{1-u}\right)^{\frac{1}{u}}\|f\|_{L^1(Q)}.\end{aligned}$$

If we let $u \downarrow 0$, then we obtain (1.212).

Exercises

Exercise 1.108 Prove Lemma 1.27 by mimicking the proof of Theorem 1.59.

Exercise 1.109 For a measurable function $f : \mathbb{R}^n \to \mathbb{C}$, define the *truncated maximal function* $f^\dagger$ by:

$$f^\dagger(x) \equiv \sup\left\{m_{B(y,r)}(|f|) : x \in B(y,r), \quad r < 1\right\}. \tag{1.213}$$

Then show that $\{f^\dagger > \alpha\}$ is an open set for all $\alpha > 0$. Hint: Mimic (1.140).

Exercise 1.110 [772, p. 718, Theorem] Let $0 < \alpha \le 1$. Suppose that a function u is integrable on Q and satisfies $m_R(|u - m_R(u)|) \le \ell(R)^\alpha$ for all cubes R. Denote by E the set of all Lebesgue points of u.

1. Let $x, y \in E$, and let $Q_{x,y}$ be the smallest cube containing x and y. Establish that $|u(x) - u(y)| \lesssim \ell(Q_{x,y})^\alpha \sim |x - y|^\alpha$.
2. Show that the limit $v(x) \equiv \lim\limits_{y \to x, y \in E} u(y)$ exists for all $x \in Q$ and that $v(x) = u(x)$ for all $x \in E$.
3. Show that v satisfies $|v(x) - v(y)| \lesssim |x - y|^\alpha$ for all $x, y \in Q$.

1.5.2 Singular Integral Operators

We are now oriented to the integral operators which are continuous versions of the matrix multiplication operators. Here we are considering the case where we cannot control well near the diagonal. The integral operators having singularity play important roles in many branches of analysis such as Fourier analysis and partial differential equations. These operators are called the singular integral operators. Therefore, the study of the singular integral operators is and will be a big branch of mathematics. In this book, we are not oriented to such a direction; we aim to investigate the properties of functions. So, we minimize the content; the results will be related directly to the study of function spaces.

1.5.2.1 Singular Integral Operators

Here we are interested in the following type of the singular integral operators:

Definition 1.42 (Singular integral operator) A *singular integral operator* is an $L^2(\mathbb{R}^n)$-bounded linear operator T that comes with a function $K \in C^1(\mathbb{R}^n \setminus \{0\})$ satisfying the following conditions:

1. (*Size condition*) For all $x \in \mathbb{R}^n$,

$$|K(x)| \lesssim |x|^{-n}. \tag{1.214}$$

2. (*Gradient condition*) For all $x \in \mathbb{R}^n$,

$$|\nabla K(x)| \lesssim |x|^{-n-1}. \tag{1.215}$$

3. Let f be an $L^2(\mathbb{R}^n)$-function. For almost all $x \notin \operatorname{supp}(f)$,

$$Tf(x) = \int_{\mathbb{R}^n} K(x - y) f(y) \mathrm{d}y. \tag{1.216}$$

The function K is called the *integral kernel* of T.

In (1.216) the support of f does not get into the singularity. In addition to the size condition, we need the gradient condition. Sometimes we can relax the gradient condition in the sense that we can reduce the order of smoothness; even the fractional smoothness is sometimes enough. The basic difficulty of the singular integral operators is the size condition. In fact, the mapping $y \mapsto |x-y|^{-n}$ fails to be (locally and globally) integrable, so that we cannot use the Young inequality. To overcome this problem, when we handle the Riesz transform, we need to use the cancellation property to show that the Riesz transform is $L^2(\mathbb{R}^n)$-bounded.

The most prominent example is the (j-th) Riesz transform, generated by the Fourier multiplier

$$m(\xi) \equiv \frac{\xi_j}{|\xi|} \quad (j = 1, 2, \ldots, n)$$

for $\xi \in \mathbb{R}^n$. Let $R_j f \equiv \mathscr{F}^{-1}[m\mathscr{F} f]$. After multiplying the nonzero constant, this operator has the kernel expression:

$$R_j f(x) = \int_{\mathbb{R}^n} \frac{x_j - y_j}{|x-y|^{n+1}} f(y)\mathrm{d}y \quad (x \in \mathbb{R}^n). \tag{1.217}$$

When $n = 1$, the Riesz transform is called the Hilbert transform. See Exercise 1.111. We can say that the definition is given so that it captures the essential properties of the kernel of these operators.

In general, for a bounded linear operator $T \in B(L^2(\mathbb{R}^n))$, on the Hilbert space $L^2(\mathbb{R}^n)$, the adjoint operator T^* is a unique bounded linear operator T^* satisfying:

$$\langle Tf, g\rangle_2 = \langle f, T^*g\rangle_2 \quad (f, g \in L^2(\mathbb{R}^n)).$$

If T is a singular integral operator, so is the adjoint operator.

Lemma 1.29 *Let T be a singular integral operator with an integral kernel K. Then so is T^* and the kernel is K^*, where $K^*(x, y) \equiv \overline{K(y, x)}$, $x, y \in \mathbb{R}^n$.*

Proof It suffices to check (1.216). To this end, we take $f, g \in L^2(\mathbb{R}^n)$ with disjoint support. Use

$$\int_{\mathbb{R}^n} T^* f(x)\overline{g(x)}\mathrm{d}x = \int_{\mathbb{R}^n} f(x)\overline{Tg(x)}\mathrm{d}x = \int_{\mathbb{R}^n} \left(\int_{\mathbb{R}^n} \overline{K(y, x)} f(y)\mathrm{d}y\right) g(x)\mathrm{d}x.$$

Therefore, the proof is complete.

1.5.2.2 Weak-$L^1(\mathbb{R}^n)$ Boundedness

Unfortunately, the singular integral operators are not $L^1(\mathbb{R}^n)$-bounded in general; there is a counterexample of the Hilbert transform and $f = \chi_{[-1,1]}$. So, we are interested in its substitute. The following theorem is an answer:

Theorem 1.62 *Let T be a singular integral operator. Then for all $\lambda > 0$ and $f \in L^1(\mathbb{R}^n) \cap L^2(\mathbb{R}^n)$,*

$$\lambda |\{ |Tf| > \lambda \}| \lesssim \|f\|_1.$$

A direct consequence of Theorem 1.62 is that we can define Tf for $f \in L^1(\mathbb{R}^n)$.

Proof We form the Calderón–Zygmund decomposition of f at height λ. We decompose $f = g + \sum_{j \in J} b_j = g + b$. From Remark 1.7, we have $b, g \in L^2(\mathbb{R}^n)$ and $Tf = Tg + Tb$. If we consider the complement, we have the following inclusion:

$$\{ |Tf| > \lambda \} \subset \left\{ |Tg| > \frac{\lambda}{2} \right\} \cup \left\{ |Tb| > \frac{\lambda}{2} \right\}. \tag{1.218}$$

See Exercise 1.88 for more details. Hence it suffices to show

$$\left| \left\{ |Tg| > \frac{\lambda}{2} \right\} \right| \lesssim \lambda^{-1} \|f\|_1, \tag{1.219}$$

$$\left| \left\{ |Tb| > \frac{\lambda}{2} \right\} \right| \lesssim \lambda^{-1} \|f\|_1. \tag{1.220}$$

The proof of (1.219) is easy. If we use Chebychev's inequality (Theorem 1.4), $L^2(\mathbb{R}^n)$-boundedness of T, (1.186) and (1.185), we obtain

$$|\{ |Tg| > \lambda \}| \le \frac{\|Tg\|_2{}^2}{\lambda^2} \lesssim \frac{\|g\|_2{}^2}{\lambda^2} \lesssim \frac{\|g\|_1}{\lambda} \le \frac{\|f\|_1}{\lambda}.$$

It is harder to prove (1.220). First of all, from (1.188), we have

$$\left| \bigcup_{j \in J} 2Q_j \right| \le 2^n \sum_{j \in J} |Q_j| = 2^n |\{ M_{\text{dyadic}} f > \lambda \}| \le \frac{2^n}{\lambda} \|f\|_1. \tag{1.221}$$

Hence we may concentrate on proving

$$\left| \{ |Tb| > \lambda \} \setminus \bigcup_{j \in J} 2Q_j \right| \lesssim \lambda^{-1} \|f\|_1. \tag{1.222}$$

The left-hand side of (1.222) is controlled by virtue of Chebychev's inequality (Theorem 1.4):

$$\left| \{ |Tb| > \lambda \} \setminus \bigcup_{j \in J} 2Q_j \right| \le \frac{1}{\lambda} \sum_{j \in J} \|Tb_j\|_{L^1(\mathbb{R}^n \setminus 2Q_j)}.$$

Thus, (1.220) is reduced to: $\|Tb_j\|_{L^1(\mathbb{R}^n \setminus 2Q_j)} \lesssim \|b_j\|_{L^1(Q_j)}$ for $j \in J$. For $x \in \mathbb{R}^n \setminus 2Q_j$, we use the support condition of b_j, (1.187) and the moment condition; $b_j \perp \mathscr{P}_0(\mathbb{R}^n)$ to have

$$|Tb_j(x)| \le \int_{Q_j} |K(x-y) - K(x - c(Q_j))| \cdot |b_j(y)| \mathrm{d}y \lesssim \frac{\ell(Q_j)^{n+1} m_{Q_j}(|b_j|)}{|x - c(Q_j)|^{n+1}}$$

since

$$Tb_j(x) = \int_{Q_j} K(x-y) b_j(y) \mathrm{d}y = \int_{Q_j} (K(x-y) - K(x - c(Q_j))) b_j(y) \mathrm{d}y. \tag{1.223}$$

Integrate this inequality over $\mathbb{R}^n \setminus 2Q_j$ and to conclude (1.222).

1.5.2.3 Marcinkiewicz Interpolation Theorem

We are oriented to the boundedness of the singular integral operators and we obtained a variant of the $L^1(\mathbb{R}^n)$-boundedness, as well as the $L^2(\mathbb{R}^n)$-boundedness. Once we obtain these two boundednesses, we can resort to a general framework. The next theorem is the Marcinkiewicz interpolation theorem, which yields the $L^p(\mathbb{R}^n)$-boundedness of the singular integral operators for many p.

Theorem 1.63 (Interpolation of the weak $(1,1)$ and $L^p(\mathbb{R}^n)$-boundedness) *Let $1 < p < \infty$. Assume that the mapping $S : L^p(\mathbb{R}^n) \to L^p(\mathbb{R}^n)$ satisfies the following conditions:*

1. (homogeneity) *Let $a \in \mathbb{C}$ and $f \in L^p(\mathbb{R}^n)$. Then for almost all $x \in \mathbb{R}^n$, $S[af](x) = |a| Sf(x)$.*
2. (subadditivity) *Let $f, g \in L^p(\mathbb{R}^n)$. Then for almost all $x \in \mathbb{R}^n$, $S[f+g](x) \le Sf(x) + Sg(x)$.*
3. (weak $(1,1)$-boundedness) $\lambda |\{ |Sf| > \lambda \}| \le c_1 \|f\|_1$ *for all $\lambda > 0$ and $f \in L^1(\mathbb{R}^n) \cap L^p(\mathbb{R}^n)$.*
4. ($L^p(\mathbb{R}^n)$-boundedness) $\|Sf\|_p \le c_p \|f\|_p$ *for $f \in L^p(\mathbb{R}^n)$.*

Then for $1 < q < p$ and $f \in L^p(\mathbb{R}^n)$, $\|Sf\|_q \lesssim_{c_1, c_p, q} \|f\|_q$.

Proof Using the layer cake representation, Theorem 1.5, we express the integral:

$$\|Sf\|_q{}^q = q\,2^q \int_0^\infty \lambda^{q-1}|\{\,|Sf| > 2\lambda\,\}|\mathrm{d}\lambda.$$

Here for $\lambda > 0$, by virtue of subadditivity, the weak $(1, 1)$-boundedness, the Chebyshev inequality and the $L^p(\mathbb{R}^n)$-boundedness, we have

$$\begin{aligned}
|\{\,|Sf| > 2\lambda\,\}| &\le |\{\,|S[\chi_{(\lambda,\infty)}(|f|)f]| > \lambda\,\}| + |\{\,|S[\chi_{(0,\lambda]}(|f|)f]| > \lambda\,\}| \\
&\lesssim \frac{1}{\lambda}\int_{\{|f|>\lambda\}} |f(x)|\mathrm{d}x + \frac{1}{\lambda^p}\int_{\{|f|\le\lambda\}} |f(x)|^p \mathrm{d}x \\
&= \int_{\mathbb{R}^n}\left(\frac{\chi_{(\lambda,\infty)}(|f(x)|)|f(x)|}{\lambda} + \frac{\chi_{(0,\lambda]}(|f(x)|)|f(x)|^p}{\lambda^p}\right)\mathrm{d}x.
\end{aligned}$$

Hence

$$\|Sf\|_q{}^q \lesssim \int_0^\infty \left\{\int_{\mathbb{R}^n}\left(\frac{\chi_{(\lambda,\infty)}(|f(x)|)|f(x)|}{\lambda} + \frac{\chi_{(0,\lambda]}(|f(x)|)|f(x)|^p}{\lambda^p}\right)\mathrm{d}x\right\}\mathrm{d}\lambda.$$

By Fubini's theorem (Theorem 1.3) we change the order of integrals; if we integrate against λ, we have $\|Sf\|_q \lesssim \|f\|_q$.

We have the following generalization, which is also called the Marcinkiewicz interpolation theorem.

Theorem 1.64 (Interpolation of the weak $(\tilde{p}, \tilde{p})$ and weak (p, p)-boundedness) *Let $1 \le \tilde{p} < p < \infty$. Assume that the mapping $S : L^p(\mathbb{R}^n) \to L^p(\mathbb{R}^n)$ satisfies the following conditions:*

1. (homogeneity) *Let $a \in \mathbb{C}$, and let $f \in L^p(\mathbb{R}^n)$. For almost all $x \in \mathbb{R}^n$, $S[a\,f\,](x) = |a|Sf(x)$.*
2. (subadditivity) *Let $f, g \in L^p(\mathbb{R}^n)$. For almost all $x \in \mathbb{R}^n$, $S[\,f + g\,](x) \le Sf(x) + Sg(x)$.*
3. (weak $(\tilde{p}, \tilde{p})$-boundedness) *For all $f \in L^{\tilde{p}}(\mathbb{R}^n) \cap L^p(\mathbb{R}^n)$ and $\lambda > 0$ we have $\lambda|\{\,|Sf| > \lambda\,\}|^{\frac{1}{\tilde{p}}} \le c_{\tilde{p}}\|f\|_{\tilde{p}}$.*
4. ($L^p(\mathbb{R}^n)$-boundedness) *$\|Sf\|_p \le c_p\|f\|_p$ for $f \in L^p(\mathbb{R}^n)$.*

Then for $\tilde{p} < q < p$ and $f \in L^p(\mathbb{R}^n) \cap L^q(\mathbb{R}^n)$, $\|Sf\|_q \lesssim_{c_{\tilde{p}},c_p,q} \|f\|_q$.

We omit the proof since the proof is the same as Theorem 1.63.

1.5.2.4 $L^p(\mathbb{R}^n)$-Boundedness of the Singular Integral Operators

By the interpolation and the duality we proceed to the case of $L^p(\mathbb{R}^n)$.

Theorem 1.65 ($L^p(\mathbb{R}^n)$-boundedness of the singular integral operators) *Let T be a singular integral operator, and let $1 < p < \infty$. Then there exists a constant*

$c_p > 0$ *such that* $\|Tf\|_p \le c_p \|f\|_p$ *for all* $f \in L^2(\mathbb{R}^n) \cap L^p(\mathbb{R}^n)$. *Furthermore, we can take* c_p, $1 < p < \infty$ *so that* $c_p = c_{p'}$.

Proof When $p = 2$, this follows from Definition 1.42. We have to consider two cases: $1 < p < 2$ and $2 < p < \infty$. When $1 < p < 2$, we use Theorem 1.63 to interpolate Theorem 1.62 and the $L^2(\mathbb{R}^n)$-boundedness. When $2 < p < \infty$, we invoke duality. We proceed as follows. Fix $\varepsilon > 0$ and choose $g \in L^2(\mathbb{R}^n) \cap L^{p'}(\mathbb{R}^n)$ so that

$$\|Tf\|_p \le (1+\varepsilon) \left| \int_{\mathbb{R}^n} Tf(x) \cdot \overline{g}(x) \mathrm{d}x \right|, \quad \|g\|_{p'} = 1. \tag{1.224}$$

Since $f, g \in L^2(\mathbb{R}^n)$, T and T^* satisfy

$$\int_{\mathbb{R}^n} Tf(x)\overline{g(x)}\mathrm{d}x = \int_{\mathbb{R}^n} f(x)\overline{T^*g(x)}\mathrm{d}x.$$

By Hölder's inequality, $\|Tf\|_p \le (1+\varepsilon)\|f\|_p\|T^*g\|_{p'}$.

Since $1 < p' < 2$ and T^* is also a Calderón–Zygmund operator, we are in the position of using the $L^{p'}$-boundedness of T^* to have $\|Tf\|_p \le c_{p'}(1+\varepsilon)\|f\|_p$. Since $\varepsilon > 0$ is arbitrary, $\|Tf\|_p \le c_{p'} \|f\|_p$.

Thus, the proof of Theorem 1.65 is complete.

We summarize and transform the above observation of the theory of the singular integral operators above. To avoid the problem of convergence of the integrals, we consider the case where the kernel is in $\mathscr{S}(\mathbb{R}^n)$.

Theorem 1.66 ($L^p(\mathbb{R}^n)$-boundedness of the singular integral operators of convolution type) *Let* $1 < p < \infty$, *and let* $k \in \mathscr{S}(\mathbb{R}^n)$. *Set*

$$A \equiv \sup_{x \in \mathbb{R}^n} |x|^n |k(x)|, \quad B \equiv \sup_{x \in \mathbb{R}^n} |x|^{n+1} |\nabla k(x)|, \quad C \equiv \sup_{\xi \in \mathbb{R}^n} |\mathscr{F}k(\xi)|.$$

Then $\|k * f\|_p \lesssim_{A,B,C} \|f\|_p$ *for all* $f \in L^p(\mathbb{R}^n)$.

Proof Note that A yields the size condition and that B yields the Hörmander condition. By Plancherel's theorem, the norm of the operator $f \in L^2(\mathbb{R}^n) \mapsto k*f \in L^2(\mathbb{R}^n)$ is a constant multiple of C. We can rephrase Theorem 1.65 in terms of the constants A, B, C.

Finally, we present an example. This is the Hölmander–Michlin multiplier theorem.

Theorem 1.67 (Hölmander–Michlin multiplier theorem) *Let* $K \in \mathbb{N}$. *Choose* $\psi \in C_c^\infty(B(4) \setminus B(1))$ *so that* $\sum_{j=-\infty}^{\infty} \psi_j \equiv \chi_{\mathbb{R}^n \setminus \{0\}}$. *Assume that* $m \in C^K(\mathbb{R}^n \setminus \{0\}) \cap L^\infty(\mathbb{R}^n)$ *satisfies*

$$M_\alpha \equiv \sup_{\xi \in \mathbb{R}^n \setminus \{0\}} |\xi|^{|\alpha|} |\partial^\alpha m(\xi)| < \infty \tag{1.225}$$

for all $|\alpha| \leq K$.

1. *The function* $\mathscr{K} \equiv \sum_{j=-\infty}^{\infty} \mathscr{F}^{-1}[\psi(2^{-j}\star)m]$ *is independent of the choice of* ψ *and all the partial derivatives up to order* $K - n - 1$ *converge uniformly over any compact sets* $\mathbb{R}^n \setminus \{0\}$. *Furthermore, if* $\beta \in {\mathbb{N}_0}^n$ *satisfies* $K \geq |\beta| + n + 1$, *then*

$$|\partial^\beta \mathscr{K}(x)| \lesssim_{\{M_\alpha\}_{\alpha \in {\mathbb{N}_0}^n, |\alpha| \leq K}} |x|^{-n-|\beta|} \quad (x \in \mathbb{R}^n \setminus \{0\}) \tag{1.226}$$

and

$$\int_{\mathbb{R}^n} \mathscr{K}(x)\varphi(x)\mathrm{d}x = \langle \mathscr{F}^{-1}m, \varphi \rangle \tag{1.227}$$

for all $\varphi \in C_c^\infty(\mathbb{R}^n \setminus \{0\})$.

2. *If* $K \geq n+2$ *and* $1 < p < \infty$, *then the* $L^2(\mathbb{R}^n)$*-bounded operator*

$$f \mapsto m(D)f \equiv \mathscr{F}^{-1}[m \cdot \mathscr{F}f] \quad (f \in L^2(\mathbb{R}^n))$$

extends to an $L^p(\mathbb{R}^n)$*-bounded linear operator naturally. More precisely, if* $f \in L^2(\mathbb{R}^n)$ *is compactly supported, then*

$$m(D)f(x) = (2\pi)^{\frac{n}{2}} \int_{\mathbb{R}^n} \mathscr{K}(y)f(x-y)\mathrm{d}y \tag{1.228}$$

for almost all $x \notin \mathrm{supp}(f)$.

Consequently, we are in the position of applying the Calderón–Zygmund theory; see Theorem 1.65.

Proof Let $j \in \mathbb{Z}$. Ignoring the general remark 26 the notation (see p. 5), we set $m_j \equiv \psi(2^{-j}\star) \cdot m$.

For $\varphi \in \mathscr{S}(\mathbb{R}^n)$, Lebesgue's convergence theorem yields

$$\begin{aligned} \int_{\mathbb{R}^n} m(\xi)\mathscr{F}^{-1}\varphi(\xi)\mathrm{d}\xi &= \sum_{j=-\infty}^{\infty} \int_{\mathbb{R}^n} \psi(2^{-j}\xi)m(\xi)\mathscr{F}^{-1}\varphi(\xi)\mathrm{d}\xi \\ &= \sum_{j=-\infty}^{\infty} \langle \psi(2^{-j}\star)m, \mathscr{F}^{-1}\varphi \rangle; \end{aligned}$$

hence

$$\langle \mathscr{F}^{-1}m, \varphi\rangle = \langle m, \mathscr{F}^{-1}\varphi\rangle = \sum_{j=-\infty}^{\infty} \langle \mathscr{F}^{-1}[\psi(2^{-j}\star)m], \varphi\rangle = \sum_{j=-\infty}^{\infty} \langle m_j, \varphi\rangle. \tag{1.229}$$

We estimate $|x^\alpha \partial^\beta \mathscr{F}^{-1} m_j(x)|$ in two ways. Without integration by parts, we use the triangle inequality directly:

$$|x^\alpha \partial^\beta \mathscr{F}^{-1} m_j(x)| \le |x|^{|\alpha|} \int_{\mathbb{R}^n} |m_j(\xi)||\xi|^{|\beta|} \mathrm{d}\xi \lesssim_{M_0} 2^{j(n+|\beta|)}|x|^{|\alpha|}. \tag{1.230}$$

If we use integration by parts, then

$$|x^\alpha \partial^\beta \mathscr{F}^{-1} m_j(x)| = \left| x^\alpha \int_{\mathbb{R}^n} m_j(\xi)\partial_x^\beta[\mathrm{e}^{ix\cdot\xi}]\mathrm{d}\xi \right| = \left| \int_{\mathbb{R}^n} m_j(\xi)\xi^\beta \partial_\xi^\alpha[\mathrm{e}^{ix\cdot\xi}]\mathrm{d}\xi \right|.$$

By the triangle inequality, we obtain

$$|x^\alpha \partial^\beta \mathscr{F}^{-1} m_j(x)| \le \int_{\mathbb{R}^n} |\partial_\xi^\alpha[m_j(\xi)\xi^\beta]|\mathrm{d}\xi.$$

By the differential inequality $|\partial_\xi^\alpha[m_j(\xi)\xi^\beta]| \lesssim_{M_\gamma, \gamma\le\alpha} |\xi|^{-|\alpha|+|\beta|}\chi_{B(2^{j+2})\setminus B(2^j)}$ (ξ) we have

$$|x^\alpha \partial^\beta \mathscr{F}^{-1} m_j(x)| \lesssim 2^{j(n-|\alpha|+|\beta|)}. \tag{1.231}$$

If we combine (1.230) and (1.231), we obtain

$$|\partial^\beta \mathscr{F}^{-1} m_j(x)| \lesssim \min(2^{j(n+|\beta|)}, |x|^{-K}2^{j(n-K+|\beta|)}). \tag{1.232}$$

If we sum (1.232) over $j \in \mathbb{Z}$, we have

$$\begin{aligned}
\sum_{j=-\infty}^{\infty} |\partial^\beta \mathscr{F}^{-1} m_j(x)| &\lesssim \sum_{j=-\infty}^{\infty} \min(2^{j(n+|\beta|)}, |x|^{-K}2^{j(n-K+|\beta|)}) \\
&\lesssim \int_0^\infty \min(t^{n+|\beta|}, |x|^{-K}t^{n-K+|\beta|}) \frac{\mathrm{d}t}{t} \\
&\lesssim \int_0^\infty \min(|x|^{-n-|\beta|}s^{n+|\beta|}, |x|^{-n-|\beta|}s^{n-K+|\beta|}) \frac{\mathrm{d}s}{s} \\
&\simeq |x|^{-n-|\beta|}.
\end{aligned}$$

See Exercise 1.9. From this estimate, we see that (1.226), and (1.227) hold and that the definition of $\mathscr{K}$ do not depend on ψ.

We consider the boundedness of $m(D)$. Let $f \in L^2(\mathbb{R}^n)$ be a compactly supported function, and let $x \notin \mathrm{supp}(f)$. In view of the definition of $\mathrm{supp}(f)$, we can choose $r > 0$ so that $\overline{B(x,r)} \cap \mathrm{supp}(f) = \emptyset$.

Let $f_j \in C_{\mathrm{c}}^\infty(\mathbb{R}^n)$ be such that $\|f - f_j\|_2 \le j^{-1}$ and that $\overline{B(x,r)} \cap \mathrm{supp}(f_j) = \emptyset$. Since $m(D)f_j$ is convergent to $m(D)f$ in the $L^2(\mathbb{R}^n)$-topology, a passage to the subsequence allows us to assume that $m(D)f_j(x) \to m(D)f(x)$ for almost all $x \in \mathbb{R}^n$.

Since we have $\overline{B(x,r)} \cap \mathrm{supp}(f_j) = \emptyset$, if we fix $x \notin \mathrm{supp}(f)$, then

$$\lim_{j\to\infty} \int_{\mathbb{R}^n} \mathscr{F}^{-1}m(y) f_j(x-y)\mathrm{d}y = \int_{\mathbb{R}^n} \mathscr{F}^{-1}m(y) f(x-y)\mathrm{d}y.$$

Thus, we can justify the assumption that $f \in C_{\mathrm{c}}^\infty(\mathbb{R}^n)$.

When $f \in C_{\mathrm{c}}^\infty(\mathbb{R}^n)$, we have

$$\begin{aligned} m(D)f(x) &= (2\pi)^{-\frac{n}{2}} \lim_{R\to\infty} \int_{B(R)} m(\xi)\mathscr{F}f(\xi)\mathrm{e}^{ix\cdot\xi}\mathrm{d}\xi \\ &= (2\pi)^{-\frac{n}{2}} \int_{\mathbb{R}^n} m(\xi)\mathscr{F}f(\xi)\mathrm{e}^{ix\cdot\xi}\mathrm{d}\xi \end{aligned}$$

in $L^2(\mathbb{R}^n)$ by the definition of the Fourier transform in $L^2(\mathbb{R}^n)$. Since $\mathscr{F}f \in L^1(\mathbb{R}^n)$, we have

$$\begin{aligned} m(D)f(x) &= (2\pi)^{-\frac{n}{2}} \sum_{j=-\infty}^{\infty} \int_{\mathbb{R}^n} m_j(\xi)\mathscr{F}f(\xi)\mathrm{e}^{ix\cdot\xi}\mathrm{d}\xi \\ &= (2\pi)^{\frac{n}{2}} \sum_{j=-\infty}^{\infty} \langle \mathscr{F}^{-1}m_j * f(x-\star)\rangle \end{aligned}$$

by Lebesgue's convergence theorem. Since $\mathrm{supp}(f(x-\star))$ is a compact set that does not contain 0, again by Lebesgue's convergence theorem we have

$$\begin{aligned} m(D)f(x) &= (2\pi)^{\frac{n}{2}} \int_{\mathbb{R}^n} \left(\sum_{j=-\infty}^{\infty} \mathscr{F}^{-1}m_j(y) \right) f(x-y)\mathrm{d}y \\ &= (2\pi)^{\frac{n}{2}} \int_{\mathbb{R}^n} \mathscr{K}(y) f(x-y)\mathrm{d}y. \end{aligned}$$

Therefore, the proof is complete.

Example 1.36 The most important example of Theorem 1.67 is the *Riesz transform* defined in (1.182).

1.5.2.5 Good λ-Inequality and Application to the Proof of the Boundedness of the Operators

Here we present some other methods to prove the weak $L^p(\mathbb{R}^n)$-boundedness of the operators in Chap. 6. We can say that the methods that follow refine the proof of the weak $L^1(\mathbb{R}^n)$-estimate for the singular integral operators. First we consider the case of $1 \le p < 2$. Here by "sublinear", we mean that T is a linear mapping from the domain where T is defined into the set of all nonnegative measurable functions and that T satisfies $T[af] = |a|Tf$ and $T[f+g] \le Tf + Tg$.

Theorem 1.68 *Let $p_0 \in [1,2)$ and $a > 0$. Suppose that T is an $L^2(\mathbb{R}^n)$-bounded linear or sublinear operator and let $\{A_r\}_{r>0} \subset B(L^2(\mathbb{R}^n))$. Assume for $j \ge 2$*

$$m^{(2)}_{2^{j+1}Q\setminus 2^jQ}(T \circ (\mathrm{id}_{L^2} - A_{\ell(Q)})[\chi_Q f]) \lesssim \exp(-a2^j) m^{(p_0)}_Q(f) \tag{1.233}$$

$$m^{(2)}_{4Q}(A_{\ell(Q)}[\chi_Q f]) \lesssim m^{(p_0)}_Q(f) \tag{1.234}$$

and

$$m^{(2)}_{2^{j+1}Q\setminus 2^jQ}(A_{\ell(Q)}[\chi_Q f]) \lesssim \exp(-a2^j) m^{(p_0)}_Q(f) \tag{1.235}$$

for all cubes Q and $f \in L^2(\mathbb{R}^n)$. Then T satisfies

$$|\{|Tf| > \lambda\}| \lesssim (\lambda^{-1}\|f\|_{p_0})^{p_0} \quad (f \in L^2(\mathbb{R}^n) \cap L^{p_0}(\mathbb{R}^n), \lambda > 0). \tag{1.236}$$

In particular, T extends to a bounded linear operator on $L^p(\mathbb{R}^n)$ for all $p \in (p_0, 2]$.

Proof Once we show (1.236), the boundedness of T on $L^p(\mathbb{R}^n)$ for $p \in (p_0, 2]$ results from Theorem 1.64, the Marcinkiewicz interpolation theorem. We form the Calderón–Zygmund decomposition of $|f|^{p_0}$ at height λ^{p_0}; see Lemma 1.27. Since T is assumed $L^2(\mathbb{R}^n)$-bounded, we can handle the good part as we did for the singular integral operators. We need to handle the bad part. Let b_j be one of the bad parts, and let Q_j be the corresponding cube. Since $f \in L^2(\mathbb{R}^n)$, the sum in the right-hand side of (1.191) converges in $L^2(\mathbb{R}^n)$. This means that there exists an increasing sequence $\{J_l\}_{l=1}^\infty$ such that

$$T\left(\sum_{j\in J} b_j\right)(x) = \lim_{l\to\infty} \sum_{j \in J_l} Tb_j(x), \quad J = \bigcup_{l=1}^\infty J_l \tag{1.237}$$

for almost all $x \in \mathbb{R}^n$. Thus, we can assume that J is a finite set. In this case

$$\left|T\left(\sum_{j\in J} b_j\right)\right| \le \sum_{j\in J} |T \circ (\mathrm{id}_{L^2} - A_{\ell(Q_j)})b_j| + \left|T\left(\sum_{j\in J} A_{\ell(Q_j)} b_j\right)\right|.$$

As we did for the singular integral operators in (1.220), let us work within $E \equiv \mathbb{R}^n \setminus \bigcup_{j\in J} 4Q_j$, so that we have

$$\chi_E \left| T\left(\sum_{j\in J} b_j\right)\right| \le \sum_{j\in J} \chi_{\mathbb{R}^n\setminus 4Q_j} |T\circ(\mathrm{id}_{L^2} - A_{\ell(Q_j)})b_j| + \left| T\left(\sum_{j\in J} A_{\ell(Q_j)} b_j\right)\right|.$$

We use the Chebyshev inequality (see Theorem 1.4) to have

$$\begin{aligned}
&\left|\left\{\sum_{j\in J} \chi_{\mathbb{R}^n\setminus 4Q_j} |T\circ(\mathrm{id}_{L^2} - A_{\ell(Q_j)})b_j| > \lambda\right\} \cap E\right| \\
&\le \frac{1}{\lambda^2}\int_E \left(\sum_{j\in J} \chi_{\mathbb{R}^n\setminus 4Q_j}(x) |T\circ(\mathrm{id}_{L^2} - A_{\ell(Q_j)})b_j(x)|\right)^2 \mathrm{d}x.
\end{aligned}$$

To dualize the right-hand side, we choose a nonnegative function $h \in L^2(\mathbb{R}^n)$ arbitrarily. Then we decompose

$$\begin{aligned}
&\int_E \sum_{j\in J} h(x)\chi_{\mathbb{R}^n\setminus 4Q_j}(x)|T\circ(\mathrm{id}_{L^2} - A_{\ell(Q_j)})b_j(x)|\mathrm{d}x \\
&= \sum_{j\in J}\int_{\mathbb{R}^n\setminus 4Q_j} h(x)|T\circ(\mathrm{id}_{L^2} - A_{\ell(Q_j)})b_j(x)|\mathrm{d}x \\
&= \sum_{j\in J}\sum_{l=2}^{\infty}\int_{2^{l+1}Q_j\setminus 2^l Q_j} h(x)|T\circ(\mathrm{id}_{L^2} - A_{\ell(Q_j)})b_j(x)|\mathrm{d}x.
\end{aligned}$$

By the Cauchy–Schwarz inequality and (1.233), we have

$$\begin{aligned}
&\int_E \sum_{j\in J} h(x)\chi_{\mathbb{R}^n\setminus 4Q_j}(x)|T\circ(\mathrm{id}_{L^2} - A_{\ell(Q_j)})b_j(x)|\mathrm{d}x \\
&\le \sum_{j\in J}\sum_{l=2}^{\infty}\left(\int_{2^{l+1}Q_j\setminus 2^l Q_j} |h(x)|^2\mathrm{d}x \int_{2^{l+1}Q_j\setminus 2^l Q_j} |T\circ(\mathrm{id}_{L^2} - A_{\ell(Q_j)})b_j(x)|^2\mathrm{d}x\right)^{\frac{1}{2}} \\
&\lesssim \sum_{j\in J}\sum_{l=2}^{\infty} 2^{ln}\exp(-a2^l)\inf_{y\in Q_j} M^{(2)}h(y) m_{Q_j}^{(p_0)}(b_j)|Q_j|.
\end{aligned}$$

We recall that we formed the Calderón–Zygmund decomposition of $|f|^{p_0}$ at height λ^{p_0} using Lemma 1.27. Thus, we can estimate $m_{Q_j}^{(p_0)}(b_j)$ to have

$$
\begin{aligned}
&\int_E \sum_{j \in J} h(x)\chi_{\mathbb{R}^n \setminus 4Q_j}(x)|T \circ (\mathrm{id}_{L^2} - A_{\ell(Q_j)})b_j(x)|\mathrm{d}x \\
&\lesssim \lambda \sum_{l=2}^{\infty} 2^{ln} \exp(-a2^l) \int_{\bigcup_{j \in J} Q_j} M^{(2)}h(y)\mathrm{d}y \simeq \lambda \int_{\bigcup_{j \in J} Q_j} M^{(2)}h(y)\mathrm{d}y \\
&\lesssim \lambda \left| \bigcup_{j \in J} Q_j \right|^{1/2} (\|h^2\|_{L^1(E)})^{1/2} = \lambda \left| \bigcup_{j \in J} Q_j \right|^{1/2} \|h\|_{L^2(E)}.
\end{aligned}
$$

thanks to Example 1.29 and the $L^2(\mathbb{R}^n)$-boundedness of M. Using the above estimate and taking the supremum over h, we obtain

$$
\left| \left\{ \sum_{j \in J} \chi_{\mathbb{R}^n \setminus 4Q_j} |T \circ (\mathrm{id}_{L^2} - A_{\ell(Q_j)})b_j| > \lambda \right\} \cap E \right| \lesssim (\lambda^{-1}\|f\|_{p_0})^{p_0}.
$$

Meanwhile, by the $L^2(\mathbb{R}^n)$-boundedness of T and the Chebyshev inequality,

$$
\left| \left\{ \left| T\left(\sum_{j \in J} A_{\ell(Q_j)}b_j \right) \right| > \lambda \right\} \cap E \right| \lesssim \frac{1}{\lambda^2} \int_{\mathbb{R}^n} \left| \sum_{j \in J} A_{\ell(Q_j)}b_j(x) \right|^2 \mathrm{d}x.
$$

Thus, if we go through a dualization argument similar to that above using (1.234) and (1.235), we obtain

$$
\left| \left\{ \left| T\left(\sum_{j \in J} A_{\ell(Q_j)}b_j \right) \right| > \lambda \right\} \cap E \right| \lesssim (\lambda^{-1}\|f\|_{p_0})^{p_0}.
$$

The proof is complete, since λ is arbitrary and we have the desired bound for $|E|$.

We consider the counterpart for $p > 2$.

Theorem 1.69 *Let $2 < p < \infty$. Suppose that T is an $L^2(\mathbb{R}^n)$-bounded linear or sublinear operator and let $\{A_r\}_{r>0} \subset B(L^2(\mathbb{R}^n))$. Assume*

$$
m_Q^{(2)}(T \circ (\mathrm{id}_{L^2} - A_{\ell(Q)})f) \lesssim \inf_{y \in Q} M^{(2)}f(y) \tag{1.238}
$$

$$
m_Q^{(p)}(T \circ A_{\ell(Q)}f) \lesssim \inf_{y \in Q} M^{(2)} \circ Tf(y) \tag{1.239}
$$

for all $f \in L^2(\mathbb{R}^n)$ and all cubes Q. Then $\|Tf\|_p \lesssim_{p,n} \|f\|_p$ for all $f \in L^p(\mathbb{R}^n) \cap L^2(\mathbb{R}^n)$.

To prove Theorem 1.69, we need a lemma.

Lemma 1.30 *Let $a, q \in [1, \infty)$. Suppose that F, G are nonnegative measurable functions and that $\{G_Q\}_{Q \in \mathcal{D}}$ and $\{H_Q\}_{Q \in \mathcal{D}}$ are collections of nonnegative measurable functions with the following properties for all $Q \in \mathcal{D}$:*

$$\chi_Q F \le G_Q + H_Q, \tag{1.240}$$

$$m_Q(G_Q)\chi_Q \le G, \tag{1.241}$$

$$m_Q^{(q)}(H_Q)\chi_Q \le a M_{\rm dyadic} F. \tag{1.242}$$

Denote by c_q the weak (q, q)-constant of $M_{\rm dyadic}$:

$$\lambda^q |\{M_{\rm dyadic} F > \lambda\}| \le {c_q}^q \|F\|_q{}^q \quad (F \in L^q(\mathbb{R}^n), \lambda > 0).$$

1. *For all $\lambda > 0$, $K \gg 1$ and $\gamma < 1$, letting $D \equiv 2^n K^{-q} a^q c_q + 2K^{-1}\gamma$, we have*

$$|\{M_{\rm dyadic} F > K\lambda, G \le \gamma\lambda\}| \le D|\{M_{\rm dyadic} F > \lambda\}|. \tag{1.243}$$

2. *Let $1 < r < \infty$. If*

$$\min(1, M_{\rm dyadic} F) \in L^r(\mathbb{R}^n), \tag{1.244}$$

then

$$\|M_{\rm dyadic} F\|_r \lesssim_{q,r,a} \|G\|_r. \tag{1.245}$$

We can say that (1.243) above is a relative distributional inequality.

Proof

1. Let $E_\lambda \equiv \{M_{\rm dyadic} F > \lambda\}$. We may assume that $|E_\lambda| < \infty$; otherwise there is nothing to prove. We consider the maximal collection of dyadic cubes $\{Q_j\}_{j \in J}$ of E_λ:

$$E_\lambda = \sum_{j \in J} Q_j.$$

Since the family $\{Q_j\}_{j \in J}$ is disjoint,

$$|\{M_{\rm dyadic} F > K\lambda, G \le \gamma\lambda\}| = \sum_{j \in J} |Q_j \cap \{M_{\rm dyadic} F > K\lambda, G \le \gamma\lambda\}|.$$

Let $j \in J$ satisfy $Q_j \cap \{M_{\rm dyadic}F > K\lambda, G \le \gamma\lambda\} \ne \emptyset$. Choose a point $y_j \in Q_j$ so that $M_{\rm dyadic}F(y_j) > K\lambda$ and that $G(y_j) \le \gamma\lambda$. We note that

$$Q_j \cap \{M_{\rm dyadic}F > K\lambda\} \subset \{M_{\rm dyadic}[\chi_{Q_j}F] > K\lambda\}$$

by the maximality of Q_j. We obtain

$$\begin{aligned}&\{M_{\rm dyadic}[\chi_{Q_j}F] > K\lambda\}\\ &\subset \left\{M_{\rm dyadic}[\chi_{Q_j}G_{Q_j}] > \frac{K\lambda}{2}\right\} \cup \left\{M_{\rm dyadic}[\chi_{Q_j}H_{Q_j}] > \frac{K\lambda}{2}\right\}\end{aligned}$$

thanks to (1.240). Using the universal estimate of $M_{\rm dyadic}$ (see Lemma 1.28), we obtain

$$\left|\left\{M_{\rm dyadic}[\chi_{Q_j}G_{Q_j}] > \frac{K\lambda}{2}\right\}\right| \le \frac{2\|G_{Q_j}\|_{L^1(Q_j)}}{K\lambda}.$$

Using y_j and (1.242), we have

$$\left|\left\{M_{\rm dyadic}[\chi_{Q_j}G_{Q_j}] > \frac{K\lambda}{2}\right\}\right| \le \frac{2|Q_j|G(y_j)}{K\lambda} \le \frac{2|Q_j|\gamma}{K}.$$

Meanwhile, choosing a point $z_j \in Q_j^* \setminus E_\lambda$, where Q_j^* denotes the dyadic parent of Q_j, from (1.242) we obtain

$$\begin{aligned}\left|\left\{M_{\rm dyadic}[\chi_{Q_j}H_{Q_j}] > \frac{K\lambda}{2}\right\}\right| &\le \frac{c_q}{K^q\lambda^q}\int_{Q_j^*} H_{Q_j}(x)^q{\rm d}x\\ &\le \frac{2^n c_q}{K^q\lambda^q}|Q_j|a^q M_{\rm dyadic}F(z_j)^q\\ &\le \frac{2^n c_q}{K^q}|Q_j|a^q.\end{aligned}$$

Thus,

$$|Q_j \cap \{M_{\rm dyadic}F > K\lambda, G \le \gamma\lambda\}| \le D|Q_j|. \tag{1.246}$$

Adding (1.246) over $j \in J$, we obtain (1.243).

2. We remark that (1.245) is a consequence of the layer cake representation and (1.243). Let $R > 0$. Note that

$$|\{\min(R, M_{\rm dyadic}F) > K\lambda, G \le \gamma\lambda\}| \le D|\{\min(R, M_{\rm dyadic}F) > \lambda\}|$$

from (1.243) and $K > 1$. Thus,

$$\begin{aligned}
&\int_{\mathbb{R}^n} \min(R, M_{\text{dyadic}}F(x))^r \mathrm{d}x \\
&= K^r \int_0^\infty r\lambda^{r-1}|\{\min(R, M_{\text{dyadic}}F) > K\lambda\}|\mathrm{d}\lambda \\
&\le C_1 \int_0^\infty \lambda^{r-1}|\{G > \gamma\lambda\}|\mathrm{d}\lambda + K^r D \int_0^\infty r\lambda^{r-1}|\{\min(R, M_{\text{dyadic}}F) > \lambda\}|\mathrm{d}\lambda \\
&= C_2 \int_{\mathbb{R}^n} G(x)^r \mathrm{d}x + K^r D \int_{\mathbb{R}^n} \min(R, M_{\text{dyadic}}F(x))^r \mathrm{d}x.
\end{aligned}$$

Here C_1 and C_2 are constants independent of R and f. If we choose γ, K, a so that $2K^r D < 1$ then we have $\|\min(R, M_{\text{dyadic}}F)\|_r \lesssim_{K,r,q,a,\gamma} \|G\|_r$ from (1.244). Letting $R \to \infty$, we obtain (1.245).

Proof (of Theorem 1.69*)* Let $f \in L^2(\mathbb{R}^n) \cap L^p(\mathbb{R}^n)$, and let $\tilde{a} \gg 1$ and $q = \frac{p}{2}$. Define

$$F \equiv |Tf|^2, \; G \equiv \tilde{a}M[|f|^2], \; G_Q \equiv 2|T\circ(\mathrm{id}_{L^2} - A_{\ell(Q)})f|^2, \; H_Q \equiv 2|T\circ A_{\ell(Q)}f|^2$$

for $Q \in \mathcal{Q}$, so that (1.240) holds. If $\tilde{a} \gg 1$, then we have (1.242) thanks to (1.239). Finally, we have (1.241) thanks to (1.238) if $\tilde{a} \gg 1$. Note also that $M_{\text{dyadic}}F \in L^{1,\infty}(\mathbb{R}^n)$, since $|Tf|^2 \in L^1(\mathbb{R}^n)$. Thus (1.244) is satisfied thanks to Theorem 1.5. Consequently, from (1.245), we have $\|Tf\|_p^2 \le \|M_{\text{dyadic}}F\|_{p/2} \lesssim \|G\|_{p/2} \lesssim \|f\|_p^2$, as desired.

Exercises

Exercise 1.111 Prove (1.217) by reexamining the proof of Theorem 1.67.

Exercise 1.112 Let $b \in \mathbb{R}$. Let us define $(1 - \Delta)^{ib}$ in analogy with $(1 - \Delta)^b$. Then show that $(1 - \Delta)^{ib} \in B(L^p(\mathbb{R}^n))$ by reexamining the proof of Theorem 1.67.

Exercise 1.113 (Interpolation of $L^\infty(\mathbb{R}^n)$-norms) Suppose a function $f \in C^2(\mathbb{R}^n)$ is such that $f, \Delta f \in L^\infty(\mathbb{R}^n)$. Let $j = 1, 2, \ldots, n$.

1. Show that $\partial_{x_j}(1 - \Delta)^{-1}$ has an integral kernel by reexamining the proof of Theorem 1.67.
2. Show that $\partial_{x_j} f \in L^\infty(\mathbb{R}^n)$.

Exercise 1.114 Let $K \gg 1$ and $\rho \in C_c^\infty(\mathbb{R}^n)$ satisfy $0 \le \rho \le \chi_{Q(2)\setminus Q(1)}$. Assume that $m \in C^K(\mathbb{R}^n \setminus \{0\}) \cap L^\infty(\mathbb{R}^n)$ satisfies (1.225). Then show that the function $\mathcal{K}_t \equiv \sum_{j=-\infty}^{\infty} \mathcal{F}^{-1}[\rho(t^{-1}\star)m]$ satisfies that all the partial derivatives up to order

$K - n - 1$ converge uniformly over any compact sets $\mathbb{R}^n \setminus \{0\}$. Furthermore,

$$|\partial^\beta \mathscr{K}_t(x)| \lesssim_{\{M_\alpha\}_{\alpha \in \mathbb{N}_0{}^n,\, |\alpha| \le K}} |x|^{-n-|\beta|} \quad (x \in \mathbb{R}^n \setminus \{0\}), \tag{1.247}$$

where the implicit constant is independent of $t > 0$. Hint: See Theorem 1.67 for (1.247).

Textbooks in Sect. 1.5

Calderón–Zygmund Theory: Theorems 1.58 and 1.59

See [86, Chapter 1, Section 4] and [117, Theorem 6.9] for Theorems 1.58 and 1.59. Lemma 1.30 and Theorem 1.61 are comparatively new. See [146, Proposition 1.5] for Lemma 1.30 and [146, Proposition 1.5] from Theorem 1.61, respectively.

Boundedness of the Singular Integral Operators: Theorems 1.62, 1.65 and 1.66

See [22, Chapters 3 and 4] and [92, Chapter 2].

The Marcinkiewcz Interpolation Theorem: Theorems 1.64, 1.68 and 1.69

See [22, Chapter 2 Section 3], [30, Chapter 10], [31, Section 2.5], [32, Section 1.3.1] and [117, Theorem 6.7] for Theorem 1.64. See [146, Theorems 1 and 2] for Theorems 1.68 and 1.69.

1.6 Harmonic Functions

A smooth function u is said to be harmonic if $\Delta u = 0$. Because harmonic functions play key roles in complex analysis and potential theory, harmonic functions constitute a huge branch of mathematics. This book does not go into detail, but provides a minimum knowledge of harmonic functions. Section 1.6.1 considers the polynomial harmonic functions, which are the most elementary form. Harmonic functions defined on $\mathbb{R}^n$ are nice. But we need to know well the harmonic functions defined on elementary domains. We consider the harmonic functions defined on balls and the half space in Sect. 1.6.2. Harmonic functions are nice for many reasons. One of the reasons is that they are smooth. Another reason is that they satisfy the mean-value property. However, it is not closed under taking the maximum. To accommodate such a property, we study subharmonic functions in Sect. 1.6.3, which will supplement some properties of harmonic functions considered in Sects. 1.6.1 and 1.6.2.

1.6.1 Harmonic Polynomials

1.6.1.1 Applications of the Polynomial Harmonic Functions to Inequalities

We aim to prove an algebraic inequality via polynomial harmonic functions in this section. We write $m = (m', m_l) = (m'', m_{l-1}, m_l)$ for $m \in \mathbb{Z}^l$, $m' \in \mathbb{Z}^{l-1}$, $m'' \in \mathbb{Z}^{l-2}$ and $m_{l-1}, m_l \in \mathbb{Z}$. Denote by S_n the set of all the bijections from $\{1, 2, \ldots, n\}$ to itself. We write $a_{\sigma(m)} \equiv \{a_{(\sigma(m_1),\sigma(m_2),\ldots,\sigma(m_l))}\}_{m\in\{1,2,\ldots,n\}^l}$ when we are given a sequence $\{a_m\}_{m\in\{1,2,\ldots,n\}^l} \in \mathbb{C}^{n^l}$.

Theorem 1.70 *Let $l, n \in \{2, 3, \ldots\}$, and let $\{a_m\}_{m\in\{1,2,\ldots,n\}^l} \in \mathbb{C}^{n^l}$ satisfy*

$$a_m = a_{\sigma(m)} \tag{1.248}$$

for all $\sigma \in S_n$ and

$$\sum_{\mu=1}^{n} a_{(m'',\mu,\mu)} = 0 \tag{1.249}$$

for all $m'' \in \{1, 2, \ldots, n\}^{l-2}$. Then for $\mu = 1, 2, \ldots, n$

$$\sum_{m'\in\{1,2,\ldots,n\}^{l-1}} |a_{(m',\mu)}|^2 \le \frac{l+n-3}{2l+n-4} \sum_{m\in\{1,2,\ldots,n\}^l} |a_m|^2. \tag{1.250}$$

If $n = 2$, we have equality in (1.250). *In particular,*

$$\sum_{m_l=1}^{n} \left| \sum_{m'\in\{1,2,\ldots,n\}^{l-1}} a_m b_{m'} \right|^2 \le \frac{l+n-3}{2l+n-4} \sum_{m\in\{1,2\ldots,n\}^l} |a_m|^2 \sum_{m'\in\{1,2,\ldots,n\}^{l-1}} |b_{m'}|^2 \tag{1.251}$$

for any sequences $\{a_m\}_{m\in\{1,2,\ldots,n\}^l} \in \mathbb{C}^{n^l}$ and $\{b_{m'}\}_{m'\in\{1,2,\ldots,n\}^{l-1}} \in \mathbb{C}^{n^{l-1}}$,

For the proof of (1.250), we may assume $\mu = 1$. The case of $n = 2$ is simple. Indeeed, since $a_{(m'',1,1)} + a_{(m'',2,2)} = 0$ and $a_{(m'',1,2)} = a_{(m'',2,1)}$ for all $m'' \in \{1, 2\}^{l-2}$,

$$\sum_{m'\in\{1,2\}^{l-1}} |a_{(m',1)}|^2 = \sum_{m'\in\{1,2\}^{l-1}} |a_{(m',2)}|^2.$$

Hence

$$\sum_{m'\in\{1,2\}^{l-1}} |a_{(m',1)}|^2 = \frac{1}{2}\sum_{m\in\{1,2\}^l} |a_m|^2.$$

Thus, it is the case where $n \geq 3$ that matters.

Let us remark that, once (1.250) is proved, (1.251) is easy to prove by the fact that $\ell^2(\{1,2,\dots,n\}^{l-1})$ is a Hilbert space.

1.6.1.2 The Space $\mathcal{H}_k$

Let $k \in \mathbb{N}_0$. Recall that $\mathcal{P}_k(\mathbb{R}^n)$ denotes the set of all polynomials with a degree less than or equal to k. Define $\mathcal{H}_k(\mathbb{R}^n) \equiv \{P \in \mathcal{P}_k(\mathbb{R}^n) : \Delta P = 0, P(x) = \mathrm{O}(|x|^k) \text{ as } x \to 0\}$. Denote by $\dfrac{\partial}{\partial n}$ the normal derivative of $S^{n-1} \equiv \{x \in \mathbb{R}^n : |x| = 1\}$, so that we require homogeneity for elements in $\mathcal{H}_k$. Denote by σ the *surface measure* of S^{n-1}. It is easy to verify that the Euler formula

$$\frac{\partial P}{\partial n}(x) = \sum_{j=1}^{n} x_j \frac{\partial P}{\partial x_j}(x) = kP(x) \tag{1.252}$$

holds on S^{n-1} and that

$$\int_{B(0,1)} P(x)\mathrm{d}x = \frac{1}{n+k}\int_{S^{n-1}} P(x)\mathrm{d}\sigma(x) \tag{1.253}$$

for all homogeneous $P \in \mathcal{P}$ of degree k.

We equip $L^2(S^{n-1}, d\sigma)$ with the inner product given by

$$\langle P, Q\rangle_{L^2(S^{n-1},\mathrm{d}\sigma)} \equiv \int_{S^{n-1}} P(x)\overline{Q(x)}\mathrm{d}\sigma(x) \quad (P, Q \in L^2(S^{n-1}, d\sigma)).$$

Now we write $\|P\|_{L^2(S^{n-1},\mathrm{d}\sigma)} \equiv \sqrt{\langle P, P\rangle_{L^2(S^{n-1},\mathrm{d}\sigma)}}$ to make $L^2(S^{n-1}, \mathrm{d}\sigma)$ into a Hilbert space under this norm. Then

$$\mathcal{H}_k(\mathbb{R}^n) \perp \mathcal{H}_{k'}(\mathbb{R}^n) \tag{1.254}$$

for all distinct $k, k' \in \mathbb{N}_0$. We leave the proof of (1.252), (1.253), and (1.254) as an exercise to interested readers. See Exercises 1.115, 1.116, and 1.117.

We observe that the $\ell^2(\{1,2,\dots,n\}^l)$-norm is described in terms of the inner product in $L^2(S^{n-1}, \mathrm{d}\sigma)$.

Lemma 1.31 *Let $l \in \mathbb{N}$ and $n \in \{2, 3, \dots\}$. Assume $\{a_m\}_{m \in \{1,2,\dots,n\}^l} \in \mathbb{C}^{n^l}$ satisfies* (1.248) *and* (1.249). *For $x \in \mathbb{R}^n$ define*

$$P(x) \equiv \frac{1}{l!} \sum_{m \in \{1,2,\dots,n\}^l} a_m x^m. \tag{1.255}$$

Here it will be understood that (1.248) *and* (1.249) *with $l = 1$ mean no condition.*

1. *The polynomial P is harmonic; that is,*

$$P \in \mathscr{H}_l(\mathbb{R}^n). \tag{1.256}$$

2. *We have*

$$\langle P, P \rangle_{L^2(S^{n-1}, \mathrm{d}\sigma)} = \Gamma\left(\frac{n}{2}\right) \frac{\sigma_{n-1}(S^{n-1})}{2^l l!} \Gamma\left(l + \frac{n}{2}\right)^{-1} \sum_{m \in \{1,2,\dots,n\}^l} |a_m|^2. \tag{1.257}$$

Proof It is straightforward to check (1.256). See Exercise 1.118.

We prove (1.257) inducting on l. If $l = 1$, then (1.257) is clear since for the monomial x_j, $n\langle x_j, x_j \rangle_{L^2(S^{n-1}, \mathrm{d}\sigma)} = \langle 1, 1 \rangle_{L^2(S^{n-1}, \mathrm{d}\sigma)} = \sigma_{n-1}(S^{n-1})$. Assume that (1.257) is true of $l - 1$. We observe $\|P\|_{L^2(S^{n-1}, \mathrm{d}\sigma)}{}^2 = \langle |P|^2, 1 \rangle_{L^2(S^{n-1}, \mathrm{d}\sigma)}$. Since $|P|^2$ is homogeneous of degree $2l$, we have

$$\|P\|_{L^2(S^{n-1}, \mathrm{d}\sigma)}{}^2 = \int_{S^{n-1}} |P(x)|^2 \, \mathrm{d}\sigma(x) = \frac{1}{2l} \left\langle \frac{\partial (P\bar{P})}{\partial n}, 1 \right\rangle_{L^2(S^{n-1}, \mathrm{d}\sigma)}$$

by the Euler formula (1.252). Using the Stokes formula and $\Delta P = 0$, we obtain

$$\|P\|_{L^2(S^{n-1}, \mathrm{d}\sigma)}{}^2 = \frac{1}{2l} \int_{B(1)} \Delta(P(x)\overline{P(x)}) \mathrm{d}x = \frac{1}{l} \int_{B(1)} \sum_{j=1}^{n} |\partial_{x_j} P(x)|^2 \mathrm{d}x,$$

since $\partial_{x_j} P$ is homogeneous of degree $l - 1$. If we use (1.253) with $k = 2l - 2$, then we have

$$\|P\|_{L^2(S^{n-1}, \mathrm{d}\sigma)}{}^2 = \frac{1}{l(2l + n - 2)} \sum_{j=1}^{n} \|\partial_{x_j} P\|_{L^2(S^{n-1}, \mathrm{d}\sigma)}{}^2.$$

It remains to use the induction assumption.

1.6.1.3 Generalized Legendre Polynomials

We seek to find an orthonormal system in $L^2([-1,1],(1-s^2)^{\nu-\frac{1}{2}})$, where

$$\langle f, g\rangle_{L^2([-1,1],(1-s^2)^{\nu-\frac{1}{2}})} = \int_{-1}^{1} f(s)\overline{g(s)}(1-s^2)^{\nu-\frac{1}{2}}\mathrm{d}s$$

for $f, g \in L^2([-1,1],(1-s^2)^{\nu-\frac{1}{2}})$. We claim that the following system is the one we are looking for:

Definition 1.43 (Generalized Legendre polynomial) Let $D \equiv \{(s,t) \in \mathbb{R}^2 : 1-2st+t^2>0\}$. For $\nu \in (0,\infty)$ satisfying $2\nu \in \mathbb{N}$ and $k \in \mathbb{N}_0$, define $C_k^\nu \in \mathscr{P}_k$, $k = 1, 2, \ldots$, which is called the *generalized Legendre polynomial* of degree k, so that for any $s, t \in D$

$$(1-2st+t^2)^{-\nu} = \sum_{k=0}^{\infty} C_k^\nu(s)t^k. \tag{1.258}$$

A couple of helpful remarks may be in order.

Remark 1.9 Let k and ν be the same as in Definition 1.43.

1. Since $(s,t) \in \left[-\frac{1}{2},\frac{1}{2}\right]^2 \mapsto (1-2st+t^2)^{-\nu}$ is an even function, C_k^ν is odd if k is odd and C_k^ν is even if k is even. So the function $f(x) \equiv |x|^k C_k^\nu(|x|^{-1}x_n)$, which is defined for $x \in \mathbb{R}^n \setminus \{0\}$, is a restriction of a polynomial function.
2. By differentiating (1.258) k times in s, one can check

$$\frac{\mathrm{d}^k C_k^\nu}{\mathrm{d}s^k}(0) = \frac{2^k\Gamma(\nu+k)}{\Gamma(\nu)}. \tag{1.259}$$

   By expanding $(1-t)^{-2\nu}$, one can also check

$$C_k^\nu(1) = \frac{\Gamma(2\nu+k)}{k!\Gamma(2\nu)}. \tag{1.260}$$

3. Let $D = \{(s,t) \in \mathbb{R}^2 : 1-2st+t^2>0\}$. If $F(s,t) \equiv (1-2st+t^2)^{-\nu}$, $s, t \in D$, then

$$(1-s^2)F_{ss}(s,t) - (2\nu+1)sF_s(s,t) + t^2F_{tt}(s,t) + (2\nu+1)tF_t(s,t) = 0. \tag{1.261}$$

   As a direct consequence of (1.261),

$$(1-s^2)(C_k^\nu)''(s) - (2\nu+1)s(C_k^\nu)'(s) + k(2\nu+k)C_k^\nu(s) = 0. \tag{1.262}$$

See Exercise 1.121 for (1.261) and (1.262).

4. Let $y(s) \equiv (1-s^2)^{-\nu+\frac{1}{2}} \dfrac{d^k}{ds^k}(1-s^2)^{k+\nu-\frac{1}{2}}$ for $s \in \mathbb{R}$. Then

$$(1-s^2)y''(s) - (2\nu+1)sy'(s) + k(2\nu+k)y(s) = 0$$

with initial condition

$$y^{(k)}(0) = (-1)^k \frac{k!\Gamma(2k+2\nu)}{\Gamma(k+2\nu)}. \tag{1.263}$$

Since y and C_k^ν solve the same differential equation and both are polynomials with degree k such that

$$y^{(k-1)}(0) = (C_k^\nu)^{(k-1)}(0) = 0. \tag{1.264}$$

This implies that C_k^ν and y are linearly dependent. Consequently, from (1.259) and (1.263),

$$C_k^\nu(s) = \frac{(-2)^k\Gamma(\nu+k)\Gamma(k+2\nu)}{\Gamma(k+1)\Gamma(\nu)\Gamma(2k+2\nu)}(1-s^2)^{-\nu+\frac{1}{2}} \frac{d^k}{ds^k}(1-s^2)^{k+\nu-\frac{1}{2}}. \tag{1.265}$$

Furthermore, we can show that

$$(1-s^2)(C_k^\nu)'(s) + ksC_k^\nu(s) - (k+2\nu-1)C_{k-1}^\nu(s) = 0 \tag{1.266}$$

for s with $|s| < 1$; see Exercise 1.121.

We will show that $\{C_k^\nu\}_{k=1}^\infty$ forms an orthogonal family.

Lemma 1.32 *For all $k, k' \in \mathbb{N}_0$ and ν as in Definition* 1.43, *we have*

$$a_{kk'}^\nu \equiv \int_{-1}^1 C_k^\nu(s)C_{k'}^\nu(s)(1-s^2)^{\nu-\frac{1}{2}}ds = \delta_{kk'}\frac{2^{1-2\nu}\pi\Gamma(k+2\nu)}{k!(k+\nu)\Gamma(\nu)^2}. \tag{1.267}$$

In particular if $k \in \mathbb{N}$, letting $a_k^\nu = a_{kk}^\nu$, we have

$$a_k^\nu = \frac{(k-1+\nu)(k-1+2\nu)}{k(k+\nu)}a_{k-1}^\nu. \tag{1.268}$$

Proof If $k \neq k'$, then the result is clear from (1.265) and integration by parts. Let us assume $k = k'$. Then

$$\int_{-1}^1 |C_k^\nu(s)|^2(1-s^2)^{\nu-\frac{1}{2}}ds = \frac{2^k\Gamma(\nu+k)}{\Gamma(k+1)\Gamma(\nu)}\int_{-1}^1 s^kC_k^\nu(s)(1-s^2)^{\nu-\frac{1}{2}}ds.$$

If we insert (1.265) into the expression of the right-hand side and integrate by parts k-times, then we obtain the desired result.

The next lemma will be useful when we decompose $\mathcal{H}_\mu(\mathbb{R}^n)$. Note that a smooth function H belongs to $\ker(\partial_{x_n})$ if and only if H is independent of x_n.

Lemma 1.33 *Let $n \in \mathbb{N} \cap [2, \infty)$, $l, \mu \in \mathbb{N}_0$, and let $H \in \mathcal{H}_\mu(\mathbb{R}^n) \cap \ker(\partial_{x_n})$. Assume that*

$$\mu + \frac{n-2}{2} > 0, \quad \mu \le l.$$

Then

$$P(x) \equiv |x|^{l-\mu} C_{l-\mu}^{\mu-1+n/2}(|x|^{-1}x_n)H(x),$$

which is initially defined on $\mathbb{R}^n \setminus \{0\}$, extends to a harmonic function $P \in \mathcal{H}_\mu(\mathbb{R}^n)$.

Proof Let $\mathbf{e}_n$ be the unit vector in the x_n direction. We learn from the Euler formula that $f(x) \equiv |x|^{-2\mu-n+2}H(x)$ is harmonic on $\mathbb{R}^n \setminus \{0\}$. For $t > 0$ let $Q_t \equiv \dfrac{H}{|\mathbf{e}_n - t \star|^{2\mu+n-2}}$. Since H is homogeneous of degree μ and $H \in \ker(\partial_{x_n})$, for any $t > 0$

$$Q_t(x) = \frac{H(tx)}{t^\mu |\mathbf{e}_n - tx|^{2\mu+n-2}} = \frac{H(tx - \mathbf{e}_n)}{t^\mu |\mathbf{e}_n - tx|^{2\mu+n-2}} \quad (x \in \mathbb{R}^n \setminus \{t\mathbf{e}_n\}),$$

so Q_t is harmonic on $\mathbb{R}^n \setminus \{t\mathbf{e}_n\}$. Let $k \equiv 2\mu + n - 2$. Since $P(x) = \lim_{t \downarrow 0} \partial_t^k Q_t(x)$ for all $x \in \mathbb{R}^n$, we see that P is also harmonic.

Based on Lemma 1.33, define $P_{l,\mu;H} \in \mathcal{P}$ so that

$$P_{l,\mu;H}(x) = |x|^{l-\mu} C_{l-\mu}^{\mu-1+n/2}(|x|^{-1}x_n)H(x) \quad (x \in \mathbb{R}^n \setminus \{0\}),$$

when we have H, l, μ as in Lemma 1.33. From Lemma 1.33 $P_{l,\mu;H} \in \mathcal{H}_l(\mathbb{R}^n)$. Define a linear mapping $I_{l,\mu} : \mathcal{H}_k(\mathbb{R}^n) \cap \ker(\partial_{x_n}) \to \mathcal{H}_l(\mathbb{R}^n)$ by $I_{l,\mu}(H) \equiv P_{l,\mu;H}$.

We have the following decomposition of $\mathcal{H}_l(\mathbb{R}^n)$ as a sum of linear spaces:

Lemma 1.34 *Let $l \in \mathbb{N}_0$ and $n \ge 3$. Then*

$$\mathcal{H}_l(\mathbb{R}^n) = \sum_{\mu=0}^{l} \mathrm{Im}(I_{l,\mu} : \mathcal{H}_\mu(\mathbb{R}^n) \cap \mathrm{Ker}(\partial_{x_n}) \to \mathcal{H}_l(\mathbb{R}^n)), \tag{1.269}$$

where for a linear mapping $f : V_1 \to V_2$ from a linear space V_1 to another linear space V_2, $\mathrm{Im}(f)$ stands for the image.

Proof We have shown that

$$\mathscr{H}_l(\mathbb{R}^n) \subset \sum_{\mu=0}^{l} \operatorname{Im}(I_{l,\mu} : \mathscr{H}_\mu(\mathbb{R}^n) \cap \operatorname{Ker}(\partial_{x_n}) \to \mathscr{H}_l(\mathbb{R}^n))$$

in Lemma 1.33. We induct on n. If $n = 1$, it is easy to check (1.269). Thus, we need to show the reverse inclusion for $n \geq 2$. To this end, let $P \in \mathscr{H}_l(\mathbb{R}^n)$. Let μ be the degree of P as a polynomial with respect to x_n. If $\mu = 0$, then we can use the induction assumption to have $P \in \operatorname{Im}(I_{l,0} : \mathscr{H}_0(\mathbb{R}^n) \cap \operatorname{Ker}(\partial_{x_n}) \to \mathscr{H}_l(\mathbb{R}^n))$.

Let $P \in \mathscr{H}_l(\mathbb{R}^n)$. Assuming $\mu \in \mathbb{N}$, we can define $Q(x')$ so that it satisfies

$$Q(x') = \frac{1}{\mu!} \frac{\mathrm{d}^\mu P}{\mathrm{d}{x_n}^\mu}(x) \quad (x = (x', x_n) \in \mathbb{R}^n).$$

This definition makes sense since the right-hand side does not depend on x_n. Thus,

$$\sum_{j=1}^{n-1} \frac{\partial^2 Q}{\partial {x_j}^2}(x') = \frac{1}{\mu!} \sum_{j=1}^{n} \frac{\partial^{2+\mu} P}{\partial {x_n}^\mu \partial {x_j}^2}(x) = 0 \quad (x \in \mathbb{R}^n).$$

Consequently, $P - C_{l-\mu}^{\mu-1+n/2}(1)^{-1} I_{l,l-\mu} Q \in \mathscr{H}_l(\mathbb{R}^n)$ and its degree as a polynomial of x_n is strictly less than to that of P. If we continue this procedure, then we obtain the desired result.

Along with the decomposition above, we investigate the structure of $\mathscr{H}_\mu(\mathbb{R}^n)$.

Lemma 1.35 *Let $n \geq 3$, $l \in \mathbb{N}$, $1 \leq \mu, \nu \leq l$, and let $H_\mu \in \mathscr{H}_\mu(\mathbb{R}^n) \cap \ker(\partial_{x_n})$ and $H_\nu \in \mathscr{H}_\nu(\mathbb{R}^n) \cap \ker(\partial_{x_n})$.*

1. *If $\mu \neq \nu$, then*

$$\langle I_{l,\mu}(H_\mu), I_{l,\nu}(H_\nu) \rangle_{L^2(S^{n-1},\mathrm{d}\sigma)} = 0. \tag{1.270}$$

2. *We have*

$$\begin{aligned} &\langle \partial_{x_n} I_{l,\mu}(H_\mu), \partial_{x_n} I_{l,\nu}(H_\nu) \rangle_{L^2(S^{n-1},\mathrm{d}\sigma)} \\ &= \delta_{\mu,\nu} \frac{(l+\mu+n-3)(l-\mu)(2l+n-2)}{2l+n-4} \langle I_{l,\mu}(H_\mu), I_{l,\nu}(H_\nu) \rangle_{L^2(S^{n-1},\mathrm{d}\sigma)}. \end{aligned} \tag{1.271}$$

In particular,

$$\sup_{P \in \mathscr{H}_l(\mathbb{R}^n) \setminus \{0\}} \frac{\|\partial_{x_n} P\|_{L^2(S^{n-1},\mathrm{d}\sigma)}}{\|P\|_{L^2(S^{n-1},\mathrm{d}\sigma)}} = \sqrt{\frac{l(l+n-3)(2l+n-2)}{2l+n-4}}. \tag{1.272}$$

Proof We verify that (1.272) is a consequence of (1.270) and (1.271). Expand P according to Lemma 1.34 to have $P = \sum_{\mu=0}^{l} I_{l,\mu}(H_\mu)$ for some sequence

$$(H_0, H_1, \ldots, H_l) \in \prod_{j=0}^{l} \mathscr{H}_j(\mathbb{R}^n).$$

Then

$$\begin{aligned}
\|\partial_{x_n} P\|_{L^2(S^{n-1},\mathrm{d}\sigma)}{}^2 &= \sum_{\mu=0}^{l-1} \|\partial_{x_n} I_{l,\mu}(H_\mu)\|_{L^2(S^{n-1},\mathrm{d}\sigma)}{}^2 \\
&\le \frac{l(l+n-3)(2l+n-2)}{(2l+n-4)} \sum_{\mu=0}^{l} \|I_{l,\mu}(H_\mu)\|_{L^2(S^{n-1},\mathrm{d}\sigma)}{}^2 \\
&= \frac{l(l+n-3)(2l+n-2)}{(2l+n-4)} \|P\|_{L^2(S^{n-1},\mathrm{d}\sigma)}{}^2
\end{aligned}$$

from (1.270) and (1.271). Using (1.271) with $\mu = 0$, we can check that we have equality in (1.272).

1. Write $P(x_n) \equiv C_{l-\mu}^{\mu-1+n/2}(x_n) C_{l-\nu}^{\nu-1+n/2}(x_n)$ for $x_n \in \mathbb{R}$ and $H(x') \equiv H_\mu(x') H_\nu(x')$ for $x' \in \mathbb{R}^{n-1}$. Due to the structure of $I_{l,\mu}$ and $I_{l,\nu}$,

$$\langle I_{l,\mu}(H_\mu), I_{l,\nu}(H_\nu)\rangle_{L^2(S^{n-1},\mathrm{d}\sigma)} = \int_{S^n} H(x') P(x_n) \mathrm{d}\sigma(x).$$

Since $H(x')P(x_n)$ is a homogeneous polynomial of degree $2l$, we obtain

$$\begin{aligned}
&\langle I_{l,\mu}(H_\mu), I_{l,\nu}(H_\nu)\rangle_{L^2(S^{n-1},\mathrm{d}\sigma)} \\
&= (2l+n) \int_{B(1)} H(x') P(x_n) \mathrm{d}x \\
&= (2l+n) \int_{-1}^{1} \left(\int_{|x'|^2 \le 1 - x_n{}^2} H(x') P(x_n) \mathrm{d}x' \right) \mathrm{d}x_n \\
&= (2l+n) \int_{-1}^{1} \left(\int_{B(1)} P(x_n)(1-x_n)^{\frac{\mu+\nu+n-1}{2}} H(x') \mathrm{d}x' \right) \mathrm{d}x_n
\end{aligned}$$

from (1.253). Equality (1.270) is clear from (1.254).

2. From (1.266) with $k = l - \mu$ and $\nu = \mu - 1 + \frac{n}{2}$,

$$\partial_{x_n} I_{l,\mu}(H_\mu)(x) \div H_\mu(x)$$
$$= |x|^{l-\mu-1}\left((l-\mu)\frac{x_n}{|x|}C_{l-\mu}^{\mu-1+n/2}\left(\frac{x_n}{|x|}\right) + \left(1 - \frac{{x_n}^2}{|x|^2}\right)\frac{d\,C_{l-\mu}^{\mu-1+n/2}}{ds}\left(\frac{x_n}{|x|}\right)\right)$$
$$= |x|^{l-\mu-1}(l+\mu+n-3)C_{l-\mu-1}^{\mu-1+n/2}\left(\frac{x_n}{|x|}\right). \tag{1.273}$$

If $\nu \neq \mu$, then (1.271) is clear from (1.267) and (1.273). So, let us assume $\mu = \nu$ below. If we define a_k^ν by (1.267) with $k = k'$, then from (1.268),

$$\frac{\langle \partial_{x_n} I_{l,\mu}(H_\mu), \partial_{x_n} I_{l,\mu}(H_\mu)\rangle_{L^2(S^{n-1},\mathrm{d}\sigma)}}{\langle I_{l,\mu}(H_\mu), I_{l,\mu}(H_\mu)\rangle_{L^2(S^{n-1},\mathrm{d}\sigma)}} = \frac{(l+n+\mu-3)^2 a_{l-\mu-1}^{\mu-1+n/2}}{a_{l-\mu}^{\mu-1+n/2}}$$
$$= \frac{(l+n+\mu-3)(l-\mu)(2l+n-2)}{(2l+n-4)},$$

which yields (1.271).

We return to the proof of Theorem 1.70. We note that the precise value of $|S^{n-1}|$ can be determined by

$$\pi^{\frac{n}{2}} = \int_{\mathbb{R}^n} e^{-|x|^2}\mathrm{d}x = |S^{n-1}|\int_0^\infty r^{n-1}e^{-r^2}\mathrm{d}r = \frac{|S^{n-1}|}{2}\Gamma\left(\frac{n}{2}\right). \tag{1.274}$$

For the proof of (1.250) we have to consider $n \geq 3$, since the case of $n = 2$ was settled down already. Let P be defined by (1.255). Then consider

$$\frac{{\|\partial_{x_n} P\|_{L^2(S^{n-1},\mathrm{d}\sigma)}}^2}{{\|P\|_{L^2(S^{n-1},\mathrm{d}\sigma)}}^2}\left(\leq \frac{l(l+n-3)(2l+n-2)}{2l+n-4}\right). \tag{1.275}$$

Note that

$$\partial_{x_n} P(x) = \sum_{m\in\{1,2,\dots,n\}^l} \frac{m_l a_m}{m!} x^{(m',m_l-1)}.$$

Hence

$${\|\partial_{x_n} P\|_{L^2(S^{n-1},\mathrm{d}\sigma)}}^2 = \Gamma\left(\frac{n}{2}\right)\frac{\sigma_{n-1}(S^{n-1})}{2^{l-1}(l-1)!}\Gamma\left(l-1+\frac{n}{2}\right)^{-1}\sum_{m'\in\{1,2,\dots,n\}^{l-1}}|a_{(m',n)}|^2$$

from (1.257). Thus, it follows from (1.257) once again that

$$\frac{\|\partial_{x_n} P\|_{L^2(S^{n-1},d\sigma)}{}^2}{\|P\|_{L^2(S^{n-1},d\sigma)}{}^2} \sum_{m\in\{1,2,\dots,n\}^l} |a_{(m',m_l)}|^2 = l(2l+n-2) \sum_{m'\in\{1,2,\dots,n\}^{l-1}} |a_{1,m'}|^2,$$

which together with (1.275) yields (1.250).

Exercises

Exercise 1.115 Verify (1.252) using the chain rule. Equality (1.252) is called the *Euler relation* or the *Euler formula.*

Exercise 1.116 Verify (1.253) using the polar coordinate.

Exercise 1.117 Verify (1.254) using the polar coordinate, (1.252) and the Stokes theorem.

Exercise 1.118 Prove (1.256) by induction. If $n = 2$, this is clear from (1.249). Use

$$f(x) = \frac{1}{l!}\sum_{k=1}^{n} x_k \sum_{(m_2,m_3,\dots,m_l)\in\{1,2,\dots,n\}^{l-1}} a_{(k,m_2,m_3,\dots,m_l)} x_{m_2} x_{m_3} \cdots x_{m_l}.$$

for $x \in \mathbb{R}^n$.

Exercise 1.119

1. Check that $C_k^\nu \in \mathscr{P}_k$.
2. Verify (1.259).

Exercise 1.120 Verify (1.260) using $C_k^\nu(1) = \dfrac{1}{k!}\lim_{t\to 0}\dfrac{\partial^k}{\partial t^k}(1-t)^{-2-2\nu}$.

Exercise 1.121 Let $G(s,t) = 1 - 2st + t^2$ and $F(s,t) = G(s,t)^{-\nu}$ for $\nu \in \mathbb{R}$. Here $(s,t) \in D = \{(s,t) \in \mathbb{R}^2 : 1 - 2st + t^2 > 0\}$. Verify (1.261) and (1.262) using

$$F_s(s,t) = 2\nu t G(s,t)^{-\nu-1}$$

$$F_{ss}(s,t) = 4\nu(\nu+1)t^2 G(s,t)^{-\nu-2}$$

$$F_t(s,t) = 2\nu(s-t)G(s,t)^{-\nu-1}$$

$$F_{tt}(s,t) = 4\nu(\nu+1)(s-t)^2 G(s,t)^{-\nu-2} - 2\nu G(s,t)^{-\nu-1}.$$

Furthermore, show that

$$(1-s^2)F_s(s,t) + t(s-t)F_t(s,t) - 2\nu t F(s,t) = 0$$

and that

$$(1 - s^2)(C_k^\nu)'(s) + ksC_k^\nu(s) - (k + 2\nu - 1)C_{k-1}^\nu(s) = 0$$

for all $s, t \in D$.

1.6.2 Harmonic Functions on the Unit Ball and the Half-Plane

In this section, let $n \geq 2$. Denote by σ the surface measure.

1.6.2.1 Harmonic Functions on the Unit Ball

Let U be an open set. We start by investigating harmonic functions on U.

Theorem 1.71 *Let u be a harmonic function defined on U. Then for all balls $B(x, r)$ whose closure is included in U,*

$$u(x) = \frac{1}{\sigma(\partial B(x, r))} \int_{\partial B(x,r)} u(y) d\sigma(y).$$

Proof Observe that

$$\frac{1}{\sigma(\partial B(x, r))} \int_{\partial B(x,r)} u(y) d\sigma(y) = \frac{1}{\sigma(\partial B(1))} \int_{\partial B(1)} u(x + ry) d\sigma(y).$$

If this is differentiated with respect to r and the Stokes theorem is applied, then

$$\begin{aligned}
&\frac{\partial}{\partial r}\left(\frac{1}{\sigma(\partial B(x, r))} \int_{\partial B(x,r)} u(y) d\sigma(y)\right) \\
&= \frac{1}{\sigma(\partial B(1))} \int_{\partial B(1)} y \cdot \nabla u(x + ry) d\sigma(y) \\
&= \frac{1}{\sigma(\partial B(x, r))} \int_{B(x,r)} \Delta u(y) dy = 0.
\end{aligned}$$

As a result,

$$\frac{1}{\sigma(\partial B(x, r))} \int_{\partial B(x,r)} u(y) d\sigma(y) = \lim_{r' \to 0} \frac{1}{\sigma(\partial B(x, r'))} \int_{\partial B(x,r')} u(y) d\sigma(y) = u(x).$$

A topological argument yields the following maximum principle:

Corollary 1.9 *If $u \in C^2(U)$ is a nonconstant harmonic function defined on a connected open set U, then u never attains its maximum.*

Using Corollary 1.9 we can further obtain *Poisson's formula for balls.*

Corollary 1.10 *Let $g : \partial B(x_0, r) \to \mathbb{R}$ be a continuous function. Define*

$$u(x) = \frac{r^2 - |x - x_0|^2}{n|B(1)|r} \int_{B(x_0,r)} \frac{g(y)}{|x - y|^n} d\sigma(y)$$

for all $x \in B(x_0, r)$. Then u is harmonic on $B(x_0, r)$ and

$$\lim_{B(x_0,r) \ni x \to z} u(x) = g(z) \quad (z \in \partial B(x_0, r)).$$

The proof is left as Exercise 1.122.

Next, consider the harmonic functions on the half-plane.

Definition 1.44 (Poisson kernel) The *Poisson kernel* $P = P_t$ over $\mathbb{R}^{n+1}_+$ is given by

$$P(x) = P_t(x) \equiv \frac{1}{\pi^{\frac{n+1}{2}}} \Gamma\left(\frac{n+1}{2}\right) \frac{t}{(t^2 + |x|^2)^{\frac{n+1}{2}}} \quad (x, t) \in \mathbb{R}^{n+1}_+.$$

A direct calculation shows that P is harmonic on $\mathbb{R}^{n+1}_+$. Here and in the rest of this section let $P_t : \mathbb{R}^n \to [0, \infty)$ to be the Poisson kernel.

The next lemma shows that P_t is normalized.

Lemma 1.36 *For any $t > 0$, $\|P_t\|_1 = 1$.*

Proof By changing the variables, we can assume $t = 1$. Changing the polar coordinate gives

$$\int_{\mathbb{R}^n} \langle x \rangle^{-n-1} dx = |S^{n-1}| \int_0^\infty \frac{r^{n-1}}{(1 + r^2)^{\frac{n+1}{2}}} dr = \frac{|S^{n-1}|}{2} \int_0^\infty \frac{r^{\frac{n}{2}-1}}{(1 + r)^{\frac{n+1}{2}}} dr.$$

Changing variables: $r \mapsto v = (1 + r)^{-1}$ yields

$$\int_{\mathbb{R}^n} \langle x \rangle^{-n-1} dx = \frac{|S^{n-1}|}{2} \int_0^1 v^{-\frac{1}{2}}(1-v)^{\frac{n}{2}-1} dv = \frac{|S^{n-1}|}{2} \Gamma\left(\frac{n}{2}\right) \pi^{\frac{1}{2}} \Gamma\left(\frac{n+1}{2}\right)^{-1}.$$

From (1.274) with n replaced by $n + 1$, we conclude

$$\|P_t\|_1 = \frac{1}{\pi^{\frac{n+1}{2}}} \Gamma\left(\frac{n+1}{2}\right) \int_{\mathbb{R}^n} \langle x \rangle^{-n-1} dx = 1. \tag{1.276}$$

Thus, the proof is complete.

Proposition 1.21 *Let $t > 0$. The Fourier transform of P_t is given by:*

$$\mathscr{F} P_t(\xi) = \frac{1}{\sqrt{(2\pi)^n}} \exp(-t|\xi|) \quad (\xi \in \mathbb{R}^n).$$

Proof In fact, we can check that

$$\frac{\partial^2}{\partial t^2}\mathscr{F} P_t(\xi) = \mathscr{F}\left[\frac{\partial^2 P_t}{\partial t^2}\right](\xi) = -\mathscr{F}[\Delta P_t](\xi) = |\xi|^2 \mathscr{F} P_t(\xi).$$

Thus, since $\mathscr{F} P_t(\xi)$ is radial in ξ and $\mathscr{F} P_t(\mathbf{e}_1) = \mathscr{F} P_1(t\mathbf{e}_1) \to 0$ as $t \to \infty$,

$$\mathscr{F} P_t(\xi) = c(\xi)\exp(-t|\xi|)$$

for some function $c(\xi)$. Since $\mathscr{F} P_t(0) = (2\pi)^{-\frac{n}{2}}$, $\mathscr{F} P_t(\xi) = \mathscr{F} P_1(t\xi)$ and $\mathscr{F} P_t(\xi) = \mathscr{F} P_t(|\xi|\mathbf{e}_1)$, the proof is complete.

Next, consider the *reflection principle*.

Lemma 1.37 (Reflection principle) *Let $u : \mathbb{R}^{n+1}_+ \to \mathbb{C}$ be a harmonic function such that u can be extended to a continuous function U on $\overline{\mathbb{R}^{n+1}_+}$ which satisfies $U(x,0) = 0$ for all $x \in \mathbb{R}^n$. Then if we define*

$$v(x,t) \equiv \begin{cases} U(x,t) & (x,t) \in \overline{\mathbb{R}^{n+1}_+}, \\ -U(x,-t) & (x,t) \in \mathbb{R}^n \setminus \overline{\mathbb{R}^{n+1}_+} \end{cases}$$

for $(x,t) \in \mathbb{R}^{n+1}$, then v is a harmonic function.

Proof Note that v is trivially harmonic on $\mathbb{R}^{n+1} \setminus \partial\mathbb{R}^{n+1}_+$. Therefore, it suffices to show that v is harmonic in

$$W \equiv \bigcup_{x_0 \in \mathbb{R}^n} B((x_0,0),1) = \{(x,t) \in \mathbb{R}^{n+1} \,:\, -1 < t < 1\}.$$

To this end, let $x_0 \in \mathbb{R}^n$ be fixed and solve the Poisson problem

$$-\Delta w = 0, \quad w|\partial B((x_0,0),1) = v$$

using the Poisson kernel. Then $w = 0$ on $\partial\mathbb{R}^{n+1}_+ \cap B((x_0,0),1)$. As a result, $w - v$ is zero on the boundary of $\mathbb{R}^{n+1}_+ \cap B((x_0,0),1)$. By the maximum principle, $w = v$ on $\mathbb{R}^{n+1}_+ \cap B((x_0,0),1)$. The same applies to $(\mathbb{R}^n \setminus \overline{\mathbb{R}^{n+1}_+}) \cap B((x_0,0),1)$. Therefore, it follows that $w = v$ on $B((x_0,0),1)$. Hence v is harmonic on $B((x_0,0),1)$.

Corollary 1.11 (Identity principle for the harmonic functions on the half space) *Let $u : \mathbb{R}^{n+1}_+ \to \mathbb{C}$ be a bounded harmonic function that admits a continuous extension U to the boundary $\partial\mathbb{R}^{n+1}_+$. Also assume that $U|_{\partial\mathbb{R}^{n+1}_+} = 0$. Then $U \equiv 0$.*

Proof According to Lemma 1.37, it can be assumed that u is a bounded harmonic function defined on $\mathbb{R}^{n+1}$ and that $u|_{\partial\mathbb{R}^{n+1}_+} = 0$. This implies that u is constant by virtue of the Lioueville theorem for harmonic functions, whose proof is the same as that for entire functions on $\mathbb{C}$. Thus, it follows that $U \equiv 0$.

Corollary 1.12 (Semi-group property of the harmonic extension) *Let $u : \mathbb{R}^{n+1}_+ \to \mathbb{C}$ be a bounded harmonic function. Then*

$$u(x, t+s) = \int_{\mathbb{R}^n} P_t(x-y)u(y,s)\mathrm{d}y$$

for all $s, t > 0$ and $x \in \mathbb{R}^n$.

Proof Fix $s > 0$. Apply Corollary 1.11 to the function

$$v_s(x,t) \equiv u(x, t+s) - \int_{\mathbb{R}^n} P_t(x-y)u(y,s)\mathrm{d}y,$$

defined for all $t > 0$ and $x \in \mathbb{R}^n$.

Later, the following lemma will be needed in Chap. 3.

Lemma 1.38 *For all $\eta > 0$, $0 < R \le 1$ and all harmonic functions $u : B(2) \to \mathbb{R}$ $|u(0)| \lesssim_\eta m^{(\eta)}_{B(R)}(u)$.*

Proof Assume that $0 < \eta < 1$; otherwise, this is trivial from the Hölder inequality and Theorem 1.71. By the scaling argument, we can assume

$$\int_{B(1)} |u(y)|^\eta \mathrm{d}y = 1 \tag{1.277}$$

and $R = 1$. Write $m(r) \equiv \|u\|_{L^\infty(B(r))}$ for $r > 0$. If $U \equiv m(1/2) \le 1$, there is nothing to prove; $m(r) \le U + 2^n$ from (1.277). Thus, suppose that $m(1/2) \ge 1$. Then thanks to Theorem 1.71 we have $m(s) \lesssim (1 - r^{-1}s)^{-n} m_{B(r)}(|u|)$ for all $0 < s < r < 1$, since $B(x, r-s) \subset B(r)$ for all $x \in B(s)$, which yields for $a > 1$,

$$m(r^a) \le C(1 - r^{a-1})^{-n} m(r)^{1-\eta} \times m_{B(r)}(|u|^\eta)$$

for all $r \in (1/2, 1)$. Hence

$$\log m(r^a) \le C + n \log \frac{1}{1 - r^{a-1}} + (1-\eta)\log m(r)$$

for all $r \in (1/2, 1)$. If this is integrated over $r \in (1/2, 1)$ against $\frac{dr}{r}$, the Haar measure on $(0, \infty)$, and (1.277) is used, then

$$\int_{1/2}^{1} \log m(r^a) \frac{dr}{r} \leq A + (1-\eta) \int_{1/2}^{1} \log m(r) \frac{dr}{r}$$

for some constant $A > 0$. By a change of variables, and the fact that $U \geq 1$, we obtain

$$\left(\frac{1}{a} - 1 + \eta\right) \int_{1/2}^{1} \log m(r) \frac{dr}{r} \leq A.$$

Choosing $a \in (1, 2)$ so that $a^{-1} - 1 + \eta > 0$, we obtain U is bounded by an absolute constant. Thus, the maximum principle can be applied once again, yielding $|u(0)| \leq U \lesssim_\eta 1$.

1.6.2.2 A Generalized Cauchy–Riemann Equation

Consider a generalized Cauchy–Riemann equation on

$$\mathbb{R}^{n+1}_+ \equiv \{(x_1, x_2, \ldots, x_n, t) \,:\, t \in (0, \infty), (x_1, x_2, \ldots, x_n) \in \mathbb{R}^n\}.$$

We consider that a vector field $\mathbf{F} \equiv (u_1, \ldots, u_n, u_{n+1})$ of harmonic functions on $\mathbb{R}^{n+1}_+$ is said to satisfy a generalized Cauchy Riemann equation if

$$\frac{\partial u_0}{\partial t} + \sum_{l=1}^{n} \frac{\partial u_l}{\partial x_l} = 0, \quad \frac{\partial u_0}{\partial x_k} = \frac{\partial u_k}{\partial t}, \quad \frac{\partial u_j}{\partial x_k} = \frac{\partial u_k}{\partial x_j}$$

for all $j, k = 1, 2, \ldots, n$. The generalized Cauchy–Riemann equation has a lot to do with complex analysis but we do not go into any detail in this direction. Instead, we need to generalize of this notion to the tensor-valued case.

Let $m \in \mathbb{N}$ and $\{\mathbf{e}_1, \mathbf{e}_2, \ldots, \mathbf{e}_n, \mathbf{e}_0\}$ be an orthonormal basis in $\mathbb{R}^{n+1} \equiv \{(x, t) \in \mathbb{R}^n \times \mathbb{R}\}$. In particular, $\mathbf{e}_0$ is the unit vector in the $t = x_0$ direction. The *tensor product* of m copies of $\mathbb{R}^{n+1}$ is defined as

$$\otimes^m \mathbb{R}^{n+1} \equiv \text{Span}\left\{\mathbf{e}_{j_1} \otimes \mathbf{e}_{j_2} \otimes \cdots \otimes \mathbf{e}_{j_m} \,:\, 0 \leq j_1, j_2, \ldots, j_m \leq n\right\}.$$

Let $G : \mathbb{R}^{n+1} \to \otimes^m \mathbb{R}^{n+1}$ be a *tensor-valued function* of rank m in the form that, for all $(x, t) \in \mathbb{R}^{n+1}_+$,

$$G(x, t) = \sum_{j_1, j_2, \ldots, j_m = 0}^{n} G_{j_1, j_2, \ldots, j_m}(x, t) \mathbf{e}_{j_1} \otimes \mathbf{e}_{j_2} \otimes \cdots \otimes \mathbf{e}_{j_m}, \tag{1.278}$$

where each $G_{j_1,j_2,\ldots,j_m}(x,t)$ is a complex number. Define its *gradient* by

$$\nabla G(x,t) \equiv \sum_{j=0}^{n} \sum_{j_1,j_2,\ldots,j_m=0}^{n} \frac{\partial G_{j_1,j_2,\ldots,j_m}}{\partial x_j}(x,t)\mathbf{e}_{j_1} \otimes \mathbf{e}_{j_2} \otimes \cdots \otimes \mathbf{e}_{j_m} \otimes \mathbf{e}_j$$

for $0 \le j_1, j_2, \ldots, j_m \le n$ as long as the definition makes sense. Then the tensor-valued function G of rank m is said to be *symmetric* if

$$G_{j_1,j_2,\ldots,j_m}(x,t) = G_{j_{\sigma(1)},j_{\sigma(2)},\ldots,j_{\sigma(m)}}(x,t)$$

for all bijections $\sigma : \{1,2,\ldots,m\} \to \{1,2,\ldots,m\}$. For such a tensor, we say that G is *trace free* or *of trace zero*, if

$$\sum_{j=0}^{n} G_{j,j,j_3,\ldots,j_m}(x,t) = 0.$$

A tensor-valued function G is said to satisfy the generalized Cauchy–Riemann equation if both G and ∇G are symmetric and of trace zero.

Equip $\otimes^m \mathbb{R}^{n+1}$ with an inner product so that $\{\mathbf{e}_{j_1} \otimes \mathbf{e}_{j_2} \otimes \cdots \otimes \mathbf{e}_{j_m}\}_{0 \le j_1,j_2,\ldots,j_m \le n}$ forms a complete orthonormal system.

Theorem 1.72 [111, Theorem 14.3] *Let $m \in \mathbb{N} \cap [2,\infty)$, and let $G : \mathbb{R}^{n+1}_+ \to \otimes^m \mathbb{R}$ be a C^∞-tensor-valued function of rank m such that both G and ∇G are symmetric and G is of trace zero. Then there exists a harmonic function u on $\mathbb{R}^{n+1}_+$ such that $\nabla^m u = G$. In particular, $|G|^\eta$ is subharmonic on $\mathbb{R}^{n+1}_+$ for all $\eta \ge \dfrac{n-1}{n+m-1}$.*

Proof We induct on m. If $m = 1$, then this is Lemma 1.19 itself. Suppose that the conclusion is correct for $m-1$ with $m \ge 2$. We aim to prove Theorem 1.72 for m also. To this end, assume that the tensor-valued function G is given so that (1.278) is true.

According to Lemma 1.19, $v_{j_2,j_3,\ldots,j_m}$, we can find $j_2, j_3, \ldots, j_m = 1,2,\ldots,n$ so that

$$G_{j_1,j_2,\ldots,j_m} = \frac{\partial v_{j_2,j_3,\ldots,j_m}}{\partial x_{j_1}}.$$

Observe that we can compute $\Delta v_{j_2,j_3,\ldots,j_m} = \displaystyle\sum_{j=1}^{n} \frac{\partial^2 v_{j_2,j_3,\ldots,j_m}}{\partial^2 x_j}$, as follows:

$$\Delta v_{j_2,j_3,\ldots,j_m} = \sum_{j=1}^{n} \frac{\partial G_{j,j_2,\ldots,j_m}}{\partial x_j} = \sum_{j=1}^{n} \frac{\partial G_{j_2,j,j_3,j_4,\ldots,j_m}}{\partial x_j} = \frac{\partial}{\partial x_{j_2}} \sum_{j=1}^{n} G_{j,j,j_3,j_4,\ldots,j_m} = 0.$$

Note also that

$$\frac{\partial v_{j_2,j_3,\dots,j_m}}{\partial x_{j_1}} = G_{j_1,j_2,\dots,j_m} = G_{\sigma(j_1),\sigma(j_2),\dots,\sigma(j_m)} = \frac{\partial v_{\sigma(j_2),\sigma(j_3),\dots,\sigma(j_m)}}{\partial x_{\sigma(j_1)}}.$$

Thus, we can use the induction assumption.

1.6.2.3 Bounded Distributions and Harmonic Extensions

We want to solve the Poisson equation $\Delta u = 0$ on $\mathbb{R}^{n+1}_+$ subject to the boundary condition $u(\star, 0) = f \in \mathscr{S}'(\mathbb{R}^n)$. As is well known, we have the expression formula using the Poisson kernel P_t. However, the Poisson kernel does not belong to $\mathscr{S}(\mathbb{R}^n)$, so that $P_t * f$ does not make sense. So we have to postulate an assumption on f.

Definition 1.45 (Bounded distribution) A distribution $f \in \mathscr{S}'(\mathbb{R}^n)$ is said to be bounded, if $\varphi * f \in L^\infty(\mathbb{R}^n)$.

Below are examples of bounded and unbounded distributions.

Example 1.37

1. Let $1 \le p \le \infty$. Then any function $f \in L^p(\mathbb{R}^n)$ is bounded.
2. Let $j = 1, 2, \dots, n$. Then the distribution x_j is not bounded.

Definition 1.46 (Poisson extension) Let $f \in \mathscr{S}'(\mathbb{R}^n)$ be a bounded distribution. Choose $\varphi \in \mathscr{S}(\mathbb{R}^n)$ so that $\chi_{B(1)} \le \varphi \le \chi_{B(2)}$. Then the *Poisson extension* of f is given by

$$e^{-t\sqrt{-\Delta}} f(x) \equiv \varphi(D) f * P_t(x) + \mathscr{F}^{-1}[\{\mathscr{F} P_t \cdot (1-\varphi)\} \cdot \mathscr{F} f](x) \quad ((x,t) \in \mathbb{R}^{n+1}_+).$$

Based on the definition of $e^{-t\sqrt{-\Delta}} f(x)$ a couple of remarks may be in order.

Remark 1.10 Let $(x, t) \in \mathbb{R}^{n+1}_+$.

1. The first term $\varphi(D) f * P_t(x)$ makes sense because $\varphi(D) f \in L^\infty(\mathbb{R}^n)$ and $P_t \in L^1(\mathbb{R}^n)$.
2. The second term $\mathscr{F}^{-1}[\{\mathscr{F} P_t \cdot (1-\varphi)\} \cdot \mathscr{F} f](x)$ also makes sense because $\mathscr{F} P_t \cdot (1-\varphi) \in \mathscr{S}(\mathbb{R}^n)$ from Proposition 1.21.
3. The function $(x, t) \in \mathbb{R}^{n+1}_+ \mapsto e^{-t\sqrt{-\Delta}} f(x) \in \mathbb{R}^n$ is a harmonic function on $\mathbb{R}^{n+1}_+$.
4. Since $\tau * e^{-t\Delta} f = e^{-t\Delta}[\tau * f]$ for all $\tau \in \mathscr{S}(\mathbb{R}^n) \subset L^1(\mathbb{R}^n)$, $t > 0$ and bounded distributions f, the definition of $e^{-t\Delta} f$ does not depend on φ.

As a corollary of Corollary 1.12, the following semi-group property of $e^{-t\sqrt{-\Delta}}$ holds.

Corollary 1.13 (Semi-group property of $e^{-t\sqrt{-\Delta}}$) *Let f be a bounded distribution, and let $s, t > 0$. Then $e^{-s\sqrt{-\Delta}}[e^{-t\sqrt{-\Delta}} f] = e^{(s+t)\sqrt{-\Delta}} f$.*

The following property is necessary later:

Lemma 1.39 *If $\{f_j\}_{j=1}^{\infty}$ is a bounded sequence of $L^\infty(\mathbb{R}^n)$-functions, which converges in $\mathscr{S}'(\mathbb{R}^n)$ to $f \in L^\infty(\mathbb{R}^n)$, then $\lim_{j\to\infty} \mathrm{e}^{-t\sqrt{-\Delta}} f_j(x) = \mathrm{e}^{-t\sqrt{-\Delta}} f(x)$ for all $x \in \mathbb{R}^n$ and $t > 0$.*

The proof is left as an exercise; see Exercise 1.123.

Exercises

Exercise 1.122 Prove Corollary 1.10 by a change of variables.

Exercise 1.123

1. Show that the integral kernel of $\exp(t\sqrt{-\Delta})$ is integrable for all $t > 0$.
2. Prove Lemma 1.39 using the duality $L^1(\mathbb{R}^n)$-$L^\infty(\mathbb{R}^n)$. Hint: The Banach Alaoglu theorem allows us to assume $\{f_j\}_{j=1}^{\infty}$ is a bounded sequence of $L^\infty(\mathbb{R}^n)$-functions, which converges in the weak$*$ topology of $L^\infty(\mathbb{R}^n)$ to $f \in L^\infty(\mathbb{R}^n)$.

1.6.3 Subharmonic Functions

Consider subharmonic functions. Let Ω be an open set. Recall that in general an upper semi-continuous function defined on a topological space X is a function $u : X \to [-\infty, \infty)$ such that $\{x \in X : u(x) < \lambda\}$ is an open set in X for all $\lambda \in \mathbb{R}$. An upper semi-continuous function $u : \Omega \to [-\infty, \infty)$ is said to be subharmonic if $u(x) \le m_{B(x,r)}(u)$ for all balls $B(x,r)$ satisfying $\overline{B}(x,r) \subset \Omega$. Note that u is bounded above on any compact set in Ω due to upper semi-continuity, so that $m_{B(x,r)}(u) = \sup\limits_{B(x,r)} u - m_{B(x,r)}\left(\sup\limits_{B(x,r)} u - u\right)$ makes sense.

1.6.3.1 An Example of Subharmonic Functions

The rest of this chapter will prove the following theorem:

Theorem 1.73 *Let $m \in \mathbb{N}$, $n \ge 2$, and let u be a harmonic function on $\mathbb{R}^n_+$. Define*

$$\nabla^m u(x) \equiv \{\partial^\alpha u(x)\}_{\alpha \in {\mathbb{N}_0}^{n+1}, |\alpha|=m} \quad (x \in \mathbb{R}^n_+)$$

and

$$|\nabla^m u(x)| \equiv \sqrt{\sum_{\alpha \in {\mathbb{N}_0}^{n+1}, |\alpha|=m} |\partial^\alpha u(x)|^2} \quad (x \in \mathbb{R}^n_+).$$

Then $|\nabla^m u|^\eta$ *is subharmonic for all* $\eta \in (0, \infty) \cap [(n+m-2)^{-1}(n-2), \infty)$.

Proof Let $\varepsilon > 0$ be arbitrary. Write

$$|\nabla^m u(x)|_\varepsilon^\eta \equiv \sqrt{\left(\varepsilon + \sum_{\alpha \in {\mathbb N_0}^{n+1}, |\alpha|=m} |\partial^\alpha u(x)|^2\right)^\eta} \quad (x \in \mathbb{R}^n_+).$$

We calculate the Laplacian of $\Delta(|\nabla^m u(x)|_\varepsilon)^\eta$ to show that $\Delta(|\nabla^m u(x)|_\varepsilon^\eta) \geq 0$. The result is

$$\begin{aligned}\Delta(|\nabla^m u(x)|_\varepsilon^\eta) &= \eta \sum_{j=1}^n \sum_{\beta \in {\mathbb N_0}^2, |\beta|=m} \frac{\partial}{\partial x_j}\left[\partial^\beta u(x) \partial^{\beta+\mathbf{e}_j} u(x) |\nabla^m u(x)|_\varepsilon^{\eta-2}\right] \\ &= \eta(\eta-2) \sum_{j=1}^n \left(\sum_{\beta \in {\mathbb N_0}^2, |\beta|=m} \partial^\beta u(x) \partial^{\beta+\mathbf{e}_j} u(x)\right)^2 |\nabla^m u(x)|_\varepsilon^{\eta-4} \\ &\quad + \eta \sum_{j=1}^n \sum_{\beta \in {\mathbb N_0}^2, |\beta|=m} |\partial^{\beta+\mathbf{e}_j} u(x)|^2 |\nabla^m u(x)|_\varepsilon^{\eta-2}.\end{aligned}$$

By Theorem 1.70, we obtain the desired result.

Corollary 1.14 *Let* $f \in L^u(\mathbb{R}^n)$ *with* $1 < u < \infty$ *have compact frequency support in* $\mathbb{R}^n \setminus \{0\}$. *Define*

$$u_\beta(x,t) \equiv {R_{j_1}}^{\beta_1} {R_{j_2}}^{\beta_2} \cdots {R_{j_n}}^{\beta_n} e^{-t\sqrt{-\Delta}} f(x) \quad ((x,t) \in \mathbb{R}^{n+1}_+)$$

for $\beta = (\beta_1, \beta_2, \ldots, \beta_{n+1}) \in {\mathbb N_0}^{n+1}$. *Then*

$$u_\beta(x,t) = c_\beta \partial_x^{(\beta_1, \beta_2, \ldots, \beta_n)} \partial_t^{\beta_{n+1}} \mathscr{F}^{-1}\left(\frac{e^{-t|\star|}}{|\star|^{|\beta|}} \mathscr{F} f\right)(x) \tag{1.279}$$

for all $\beta = (\beta_1, \beta_2, \ldots, \beta_{n+1}) \in {\mathbb N_0}^{n+1}$, *where* c_β *is a complex constant independent of* u, f, x *and* t *with modulus of* 1. *In particular,*

$$V(x,t) \equiv \left(\sum_{|\beta|=m} \frac{1}{\beta!} |u_\beta(x,t)|^2\right)^{\frac{\eta}{2}} \quad ((x,t) \in \mathbb{R}^{n+1}_+)$$

is subharmonic.

Proof The relation (1.279) can readily be checked in view of the size of the frequency support. Theorem 1.73 guarantees that V is subharmonic. Note also that

$\beta!$ comes into play here because the summation is acquired in a manner different from Theorem 1.73.

1.6.3.2 Poisson Integral and Subharmonic Functions

Here we aim to deduce some important inequalities related to subharmonic functions.

Lemma 1.40 *Let f be a $C^2(\mathbb{R}^n)$-function defined on an open set containing $\overline{\mathbb{R}^n_+}$. Assume in addition that $\lim_{|x|\to\infty,\, x_n\geq 0} f(x) = 0$, $f|\partial\mathbb{R}^n_+ = 0$ and that $\Delta f = f$ on $\mathbb{R}^n_+$. Then $f \equiv 0$.*

Proof We can assume that f is real-valued. If $f(x_0) > 0$ for some $x_0 \in \mathbb{R}^n$, then by assumption we can assume that $f(x_0)$ is a maximum in $\mathbb{R}^n_+$. In this case $f_{x_j x_j}(x_0) \leq 0$ for $j = 1, 2, \ldots$. Thus $0 < f(x_0) - \sum_{j=1}^{n} f_{x_j x_j}(x_0) = (1-\Delta)f(x_0) \leq 0$. This is a contradiction.

Lemma 1.41 *Suppose that $v : \mathbb{R}^{n+1}_+ \to \mathbb{R}$ is a bounded $C^2(\mathbb{R}^n)$-function such that $v_t \in L^\infty(\mathbb{R}^{n+1}_+)$ and that $\Delta v \geq 0$ Then for all $a > 0$ and $(x,t) \in \mathbb{R}^{n+1}_+$, we have $v(x, t+a) \leq v(\star, a) * P_t(x)$.*

Proof Set $V(x,t) \equiv v(x, t+a) - v(\star, a) * P_t(x)$. We can use Corollary 1.11.

Next, we collect some information on the Poisson integral on the disk. We define

$$\mu_0(t) \equiv \frac{\sin \pi\theta}{2[\cosh(\pi t) - \cos \pi\theta]} \text{ and } \mu_1(t) \equiv \frac{\sin \pi\theta}{2[\cosh(\pi t) + \cos \pi\theta]}. \tag{1.280}$$

Note that $\|\mu_0\|_{L^1(\mathbb{R})} = 1 - \theta$ and $\|\mu_1\|_{L^1(\mathbb{R})} = \theta$.

We need a conformal mapping from $S \equiv \{z \in \mathbb{C} : 0 < \Re(z) < 1\}$ to $\Delta(1) \equiv \{z \in \mathbb{C} : |z| < 1\}$.

Lemma 1.42 *The mapping $g(z) \equiv \dfrac{e^{i\pi z} - i}{e^{i\pi z} + i}$ maps S to $\Delta(1)$ conformally.*

Proof Once we fix the branch of $\log : \mathbb{C} \setminus (-\infty, 0]$ suitably, we learn that $g(z) \in \Delta(1)$ when $z \in S$ and that $g^{-1}(z) = \dfrac{1}{\pi i} \log\left(\dfrac{i(1+z)}{1-z}\right)$. Notice that $\Im(\frac{i(1+z)}{1-z}) > 0$ for every $z \in \Delta(1)$, so $z \in \Delta(1) \mapsto \log\left(\frac{i(1+z)}{1-z}\right)$ is a well-defined holomorphic function on $\Delta(1)$. Thus, g^{-1} maps $\Delta(1)$ conformally to S.

Theorem 1.74 *Let F be a function which is analytic on the open strip S and continuous on its closure such that it grows mildly in the sense that*

$$\sup_{z \in \overline{S}} e^{-a|\Im(z)|} \log|F(z)| < \infty \tag{1.281}$$

for some $a \in (0, \pi)$. *Then for all* $0 < \theta < 1$, *we have*

$$\log|F(\theta)| \le \int_{\mathbb{R}} \left[\mu_0(t) \log|F(it)| + \mu_1(t) \log|F(1+it)|\right] dt. \tag{1.282}$$

As the example of $f(z) = \exp(i \exp(2\pi i z))$ together with the fact that $|f(x+iy)| = \exp(-\exp(-2\pi y) \sin 2\pi x)$ implies, we need assumption (1.281).

Proof It suffices to deal with the case where f satisfies $\lim\limits_{y \to \pm\infty} \sup\limits_{x \in [0,1]} |f(x+iy)| = 0$ by replacing $f(z)$, $z \in \bar{S}$ by $f(z)e^{-\varepsilon z^2}$, $z \in \bar{S}$ if necessary, where $\varepsilon > 0$. We use the function g in Lemma 1.42. For each $z \in \overline{S}$, define $H(z) \equiv \log|F(z)|$. For each $z \in \Delta(1)$, define $G(z) \equiv H(g^{-1}(z))$. Since $H(z)$ is subharmonic on S, we see that $G(z)$ is subharmonic on $\Delta(1)$. Then for every $r \in (0, \rho)$ with $\rho < 1$ and $0 \le s \le 2\pi$, we have

$$G(re^{is}) \le \frac{1}{2\pi} \int_0^{2\pi} \frac{\rho^2 - r^2}{\rho^2 - 2\rho r \cos(t-s) + r^2} G(\rho e^{it}) dt.$$

For every $\rho \in (r/2, 1)$, we have

$$\begin{aligned}
\frac{\rho^2 - r^2}{\rho^2 - 2\rho r \cos(t-s) + r^2} \Re(G(\rho e^{it})) &\le \sup_{z \in \overline{S}} H(z) \frac{\rho^2 - r^2}{\rho^2 - 2\rho r + r^2} \\
&= \sup_{z \in \overline{S}} H(z) \frac{\rho + r}{\rho - r} \\
&\le \sup_{z \in \overline{S}} H(z) \frac{1+r}{\frac{r+1}{2} - r} \\
&= \sup_{z \in \overline{S}} H(z) \frac{2+2r}{1-r}.
\end{aligned}$$

By the Fatou lemma and continuity of G, we get

$$\begin{aligned}
G(re^{is}) &\le \limsup_{\rho \uparrow 1} \frac{1}{2\pi} \int_0^{2\pi} \frac{\rho^2 - r^2}{\rho^2 - 2\rho r \cos(t-s) + r^2} G(\rho e^{it}) dt \\
&= \frac{1}{2\pi} \int_0^{2\pi} \frac{1 - r^2}{1 - 2r \cos(t-s) + r^2} G(e^{it}) dt.
\end{aligned}$$

For $\theta \in (0, 1)$, we have $g(\theta) = \frac{e^{i\pi\theta} - i}{e^{i\pi\theta} + i} = -i \frac{\cos \pi\theta}{1 + \sin \pi\theta}$, so, the solution of

$$re^{is} = g(\theta) \quad (r \in (0, 1),\, s \in (0, 2\pi))$$

is

$$(r,s)=\begin{cases}\left(\frac{\cos\pi\theta}{1+\sin\pi\theta},\frac{3\pi}{2}\right) & \text{if } \theta\in(0,1/2]\\ \left(-\frac{\cos\pi\theta}{1+\sin\pi\theta},\frac{\pi}{2}\right) & \text{if } \theta\in(1/2,1).\end{cases}\tag{1.283}$$

For (r,s) in (1.283), we have

$$H(\theta)=G(g(\theta))\le\frac{1}{2\pi}\int_0^{2\pi}\frac{1-r^2}{1-2r\cos(t-s)+r^2}H\left(g^{-1}(\mathrm{e}^{it})\right)\mathrm{d}t$$

and

$$\frac{1-r^2}{1-2r\cos(t-s)+r^2}=\frac{\sin\pi\theta}{1+\sin t\cos\pi\theta},$$

so

$$H(\theta)\le\frac{1}{2\pi}\int_0^{\pi}\frac{H(g^{-1}(\mathrm{e}^{it}))\sin\pi\theta}{1+\sin t\cos\pi\theta}\mathrm{d}t+\frac{1}{2\pi}\int_\pi^{2\pi}\frac{H(g^{-1}(\mathrm{e}^{it}))\sin\pi\theta}{1+\sin t\cos\pi\theta}\mathrm{d}t.$$

For $t\in[0,\pi]$, let $1+iy=g^{-1}(\mathrm{e}^{it})$. Then arithmetic shows: $\mathrm{e}^{it}=g(1+iy)=-\tanh\pi y+i\mathrm{sech}\pi y$. See Exercise 1.124. Consequently,

$$\begin{aligned}&\frac{1}{2\pi}\int_0^{\pi}\frac{\sin\pi\theta}{1+\sin t\cos\pi\theta}H(g^{-1}(\mathrm{e}^{it}))\mathrm{d}t\\ &\quad=\frac{1}{2\pi}\int_{-\infty}^{\infty}\frac{\sin\pi\theta}{1+\mathrm{sech}\pi y\cos\pi\theta}H(1+iy)\,\pi\ \mathrm{sech}\pi y\mathrm{d}y\\ &\quad=\frac{1}{2}\int_{-\infty}^{\infty}\frac{\sin\pi\theta}{\cosh\pi y+\cos\pi\theta}H(1+iy)\mathrm{d}y.\end{aligned}\tag{1.284}$$

For $t\in[\pi,2\pi]$, let $iy=g^{-1}(\mathrm{e}^{it})$. Then $\mathrm{e}^{it}=g(iy)=-\tanh\pi y-i\mathrm{sech}\pi y$. See Exercise 1.124 again. Therefore,

$$\begin{aligned}&\frac{1}{2\pi}\int_\pi^{2\pi}\frac{\sin\pi\theta}{1+\sin t\cos\pi\theta}H(g^{-1}(\mathrm{e}^{it}))\mathrm{d}t\\ &\quad=\frac{1}{2\pi}\int_{\infty}^{-\infty}\frac{\sin\pi\theta}{1-\mathrm{sech}\pi y\cos\pi\theta}H(iy)\,(-\pi\mathrm{sech}\pi y)\mathrm{d}y\\ &\quad=\frac{1}{2}\int_{-\infty}^{\infty}\frac{\sin\pi\theta}{\cosh\pi y-\cos\pi\theta}H(iy)\mathrm{d}y.\end{aligned}\tag{1.285}$$

By combining (1.284) and (1.285), and $H(z)=\log|F(z)|$, we get the desired inequality.

Theorem 1.75 *Let $\theta \in (0,1)$. Then for all functions F analytic on the open strip S and continuous on its closure satisfying* (1.281)*, we have*

$$|F(\theta)| \le \left(\int_{\mathbb{R}} |F(it)|\frac{\mu_0(t)}{1-\theta}\mathrm{d}t\right)^{1-\theta} \left(\int_{\mathbb{R}} |F(1+it)|\frac{\mu_1(t)}{\theta}\mathrm{d}t\right)^{\theta},$$

where μ_0 and μ_1 are defined by (1.280)*. In particular,*

$$|F(\theta)| \le \sup_{t\in\mathbb{R}} |F(it)|^{1-\theta} \sup_{t\in\mathbb{R}} |F(1+it)|^{\theta}. \tag{1.286}$$

Inequality (1.286) *is called Doetsch's three-line lemma.*

The three-line theorem is a variant of the three-circle theorem by Hadamard [528]. Since it initially appeared explicitly in Doetsch [426], it is sometimes called the Doetsch three-line theorem.

Proof Let $j \in \{0,1\}$. Equip $S_j \equiv \{(j,t) : t \in \mathbb{R}\}$ with a probability measure P_j given by

$$P_j(\{j\}\times E) \equiv \int_E \frac{\mu_j(t)}{\|\mu_j\|_{L^1(\mathbb{R})}}\mathrm{d}t$$

for any measurable set $E \subseteq \mathbb{R}$. We use (1.282) and the Jensen inequality [630] to get

$$\begin{aligned}
|F(\theta)| &\le \left[\exp\left(\int_{\mathbb{R}} \frac{\mu_0(t)}{1-\theta}\log|F(it)|\mathrm{d}t\right)\right]^{1-\theta} \left[\exp\left(\int_{\mathbb{R}} \frac{\mu_1(t)}{\theta}\log|F(1+it)|\mathrm{d}t\right)\right]^{\theta} \\
&\le \left[\exp\left(\int_{S_0} \log|F(iT(\omega))|\,\mathrm{d}P_0(\omega)\right)\right]^{1-\theta} \\
&\quad\times \left[\exp\left(\int_{S_1} \log|F(1+iT(\omega))|\,\mathrm{d}P_1(\omega)\right)\right]^{\theta} \\
&\le \left(\int_{S_0} |F(iT(\omega))|\,\mathrm{d}P_0(\omega)\right)^{1-\theta} \left(\int_{S_1} |F(1+iT(\omega))|\,\mathrm{d}P_1(\omega)\right)^{\theta} \\
&= \left(\int_{\mathbb{R}} |F(it)|\frac{\mu_0(t)}{1-\theta}\mathrm{d}t\right)^{1-\theta} \left(\int_{\mathbb{R}} |F(1+it)|\frac{\mu_1(t)}{\theta}\mathrm{d}t\right)^{\theta}.
\end{aligned}$$

Thus, Theorem 1.75 is proved.

Exercises

Exercise 1.124 For $t \in [0,\pi]$, let $1+iy = g^{-1}(\mathrm{e}^{it})$, where $g(\theta) = \frac{\mathrm{e}^{i\pi\theta}-i}{\mathrm{e}^{i\pi\theta}+i}$.

1. Verify $e^{it} = g(1 + iy) = -\tanh \pi y + i\text{sech}\pi y$ for $t \in [0, \pi]$.
2. Verify $e^{it} = g(1 + iy) = -\tanh \pi y + i\text{sech}\pi y$ for $t \in [\pi, 2\pi]$.

Exercise 1.125 Verify $\|\mu_0\|_{L^1(\mathbb{R})} = 1 - \theta$ and $\|\mu_1\|_{L^1(\mathbb{R})} = \theta$, where μ_0, μ_1 are given by (1.280).

Textbooks in Sect. 1.6

Poisson Integral Over the Half Space

See [31, Section 2.4], [85, Chapter 2, Section 2] and [90, Section 2.1].

Harmonic Functions on the Unit Ball

See [25, Chapter 2, Section 2.2] and [31, Section 1.1].

Harmonic Polynomials

See [111] for example.

Subharmonic Functions

See [31, Section 1.1], [112, Chapter 3] and [90, Chapter 2].

1.7 Notes for Chap. 1

Section 1.1

Section 1.1.1

Inequalities for Sequences: Lemmas 1.1 and 1.2

See [853, Lemma 2.17] for the proof of Lemma 1.1, which slightly improves [99, p. 68]. See [902, Lemma 3] for Lemma 1.2.

Quasi-Banach Spaces and $L^p(\mu)$ with $0 < p \le \infty$

The space $L^p([0,1])$ dates back to 1940. Day considered the space $L^p([0,1])$. See the paper by Day [14], where a property of the dual spaces is discussed. Although $L^p(\mathbb{R}^n)$ is not a metric space with its norm when $0 < p \le 1$, the Lebesgue space $L^p(\mathbb{R}^n)$ is metrizable for any $0 < p \le \infty$. See the works [137, p. 593 (2)] and [893, p. 472] for a technique that is applicable for general quasi-Banach spaces.

Layer Cake Representation: Theorem 1.5

It seems difficult to decide who found out this result. As is written in this book, this theorem is used to prove the $L^p(\mathbb{R}^n)$-boundedness of the Hardy–Littlewood maximal operator. The author feels that Theorem 1.5 is due to Hardy and Littlewood [554, III, maximal problem for integrals].

Section 1.1.2

5r-Covering Lemma: Theorems 1.7 and 1.8

Theorems 1.7 and 1.8 are due to Vitali [1097].

Whitney Covering Lemma: Theorem 1.9

Theorem 1.9 dates back to 1934; see the work by Whitney [1111]. Originally, Whitney proved Theorem 1.9 to extend functions smoothly. Generally speaking, covering lemmas are used to investigate the smoothness of functions.

A Partition of Unity Associated with Whitney Covering

We refer to [301, p. 44], as well as [302, 1111] for Proposition 1.4.

Section 1.1.3

The Dyadic Hausdorff Capacity–$\tilde{H}_0^d$

Yang and Yuan defined $\tilde{H}_0^d$ in [1151, Definition2.1], which asserts that $\tilde{H}_0^d$ is strongly subadditive; see [1151, Proposition 2.4]. Proposition 1.5, the canonical ball cover with respect to Hausdorff capacity, is from [401, Lemma 4.1]. Proposition 1.6, which asserts $H^d \lesssim_n \tilde{H}_0^d \sim H^d$, is from [1152, Proposition 2.3]. Theorem 1.10, the strong subadditivity of the Hausdorff capacity, is [1152, Proposition 2.1]. See [1152, Theorem 2.1] for Lemma 1.5. The monotone property of $\tilde{H}_0^d$ can be found in [1152, Theorem 2.1]. See [1152, Proposition 2.4] for the subadditivity of the integral.

Section 1.1.4

Choquet Integral

See [363].

Others

The authors are indebted to Professor Akihiko Miyachi for the proof of (1.51).

Section 1.1.5

Helly's Theorem: Theorem 1.14

It was discovered by Eduard Helly in 1913 and published in 1923 [586].

A Lemma Related to the Closed Graph Theorem–Lemma 1.9

Lemma 1.9 attributes to Mitrea [781, Lemma 2.4].

Section 1.2

Section 1.2.1

The Definition of $\mathscr{S}(\mathbb{R}^n)$

The definition of $\mathscr{S}(\mathbb{R}^n)$ dates back to 1951 (see [81, p. 233, Section 3]). It seems that Tosio Kato introduced the notion of distributions earlier than this book; see [45].

Some Special Functions: Lemma 1.12

Lemma 1.12 is due to Rychkov [900]. See [900, p. 254] for the function G defined by (1.56).

Fourier Transform in $\mathscr{S}(\mathbb{R}^n)$: Theorem 1.31

We refer to [81, p. 249, Théorème XII].

Section 1.2.2

The Definition of $\mathscr{S}'(\mathbb{R}^n)$

The definition of $\mathscr{S}'(\mathbb{R}^n)$ also dates back to 1951 (see [81, p. 237, Section 4]).

Multiplication in $\mathscr{S}'(\mathbb{R}^n)$

See [81, p. 245, La multiplication dans $\mathscr{S}'$] for Lemma 1.14 and Definition 1.14.

Distributions Supported in a Point: Theorem 1.25

An expression equivalent to Theorem 1.25 using $\mathscr{D}'(\mathbb{R}^n)$ can be found in [81, p. 100, Théorème XXXV].

Regular Elements

The definition of regular elements as in Definition 1.12 is based on [81, p. 238, Théorème IV].

Calderón-Reproducing Formula for $\mathscr{S}'(\mathbb{R}^n)$: Theorems 1.24 and 1.35

This formula goes back to [325]. See [81, p. 237] for Theorem 1.24.

Compactness of the Closed Bounded Sets: Theorem 1.20

See [514, Theorem 2.2].

Convolution in $\mathscr{S}'(\mathbb{R}^n)$

See [81, p. 246, La convolusion dans $\mathscr{S}'$].

Section 1.2.3

Fourier Transform from $\mathscr{S}'(\mathbb{R}^n)$ to Itself

We refer to [81, p. 251, Théorème XIII].

Fourier Transform and Differentiation in $\mathscr{S}'(\mathbb{R}^n)$: Theorems 1.28

See [82, p. 109, Section 7. Examples].

Section 1.2.4

The Spaces $\mathscr{D}(\mathbb{R}^n)$ and $\mathscr{D}'(\mathbb{R}^n)$

See [1001].

Section 1.2.5

Theorem 1.39: Local Reproducing Formula

Theorem 1.39 is [904, Theorem 1.6].

The Functional Equation $\Phi^{(L)} - (\Phi^{(L)})^{-1} = \Delta^L \Psi^{(L)}$: Theorem 1.40

See [940, Proposition].

Taylor Type Expansion of $\mathscr{S}(\mathbb{R}^n)$: Theorem 1.41

See [282, Lemma 6.1].

Section 1.2.6

Schwartz Kernel Theorem: Theorem 1.42

See [82].

Section 1.3

Section 1.3.1

Expression of the Difference of Higher Order: Theorem 1.43

Many people have studied equations similar to the one in Theorem 1.43. For example, Besov obtained some formulas similar to Theorem 1.43 in his work [186,

Section 6]. We refer to the paper [962, Lemma 10] for more about a pointwise estimate involving the difference operator and $\varphi_j(D)f$.

Sobolev's Representation Formula

Besov pointed out the tight connection of Sobolev's integral representation to the multidimensional Taylor's formula [179, 181].

Section 1.3.2

Oscillation

There are several ways of expressing continuity of functions. For example, it can be expressed in terms of the oscillation. See the papers [772, Theorem] and [826, Section 8].

Section 1.4

Section 1.4.1

Hardy Littlewood Maximal Operator: Theorems 1.45 and 1.46

The Hardy–Littlewood maximal operator is used to investigate the Fourier series on the torus [554]. Hardy and Littlewood studied the 1-dimensional maximal operator and later Wiener generalized the boundedness of the Hardy–Littlewood maximal operator to higher dimensions; see [1112], where Wiener used the covering lemma for the proof of the boundedness of the maximal operator.

The Origin of the Weak $(1, 1)$-Estimates

Kolmogorov proved the weak $(1, 1)$-estimate for the Hilbert transform in [678], where the notion of the weak $(1, 1)$-estimate emerged. The person who named the weak $(1, 1)$-estimate is Wolff; see the mathereview of [1191].

The space $L^{p,\infty}$

The definition is due to Lorentz [750, 751]. The word "Lorentz space" is due to Calderón [324, 326].

Lebesgue Differential Theorem: Theorem 1.48

This theorem dates back to 1904; consult the book [44] by Lebesgue himself when $n = 1$. If $n = 2$, Theorem 1.48 goes back 1908; see the paper by Vitali [1097]. For $n \geq 3$, Lebesgue proved Theorem 1.48 in 1910 [704]. Banach proposed the method of the proof of Theorem 1.48 [160].

Band-limited Distributions: Theorems 1.50, 1.51, 1.52 and 1.53

Theorem 1.50 can be found in Triebel's textbook [96, Section 1.3.2]. Once Theorem 1.50 is proved, Theorem 1.53 is proved systematically by means of the inequalities of the Hardy–Littlewood maximal operator. See [99, p. 31, Theorem] for this systematic approach. We refer to [99, Section 1.4.1] for Theorem 1.52 and [99, Section 1.4.1] for Theorem 1.53. We consult [99, Section 1.5.3] including Remark 2 there and [71, p. 234, Lemma 1] for Corollary 1.7.

Furthermore, we refer to the textbook [40] for Theorems 1.50, 1.51, and 1.53.

Corollary 1.7 can be found in the textbooks [71] and [99, p. 28].

Integral Inequalities: Theorems 1.54, 1.55 and 1.56

We refer to the textbook [32, Appendix] for more generalizations.

Control of η-Function: Example 1.32

See [422, Lemma 5.2] and [877].

Section 1.4.2

Fefferman–Stein Vector-Valued Maximal Inequality: Theorem 1.49

Fefferman and Stein proved Theorem 1.49 [459, Theorem 1]. The proof here is based upon the method employed in the proof of the author's paper [913, Theorem 1.3].

Section 1.4.3

Sampling Theorem: Theorem 1.51

The sampling theorem has a long history. It dates back to 1920 [857] as is described in [288].

Theorem 1.51 dates back to 1937 (see the work of Plancherel and Polya [880, Théorème III]).

Band-Limited Distributions: Theorems 1.50, 1.51, 1.52 and Corollary 1.7

When $1 \le p \le q \le \infty$, inequality (1.163) is due to Nilol'skii [69, 845]. See [367, Theorem 1.2] for what happens if η is smaller than the critical value in Theorem 1.52, where Christ and Seeger presented us a counterexample. See [855] for the case of compact Lie groups.

A Convolution Estimate: Corollary 1.7

See [1051, Theorem].

Section 1.4.4

Integral Inequalities: Theorems 1.54, 1.55 and 1.56

Due to their importance, Theorems 1.54, 1.55 and 1.56 appear in many forms. For example, see [482, Lemma 3.3] for Theorem 1.56.

Section 1.4.5

Carleson Measure

The notion of the Carleson measure goes back to [347, 348].

Section 1.5

Section 1.5.1

Calderón–Zygmund Theory: Theorems 1.58 and 1.59

This theory dates back to 1952 (see the fundamental papers [330, 334]) and has been expanded in various directions.

Section 1.5.2

Interpolation Theorem of Marcinkiewicz Type: Theorem 1.64

The origin is Marcinkiewicz, who proved the interpolation theorem. Theorem 1.64 first appeared without proof in his brief note [763]. After his death in World War II, nobody seemed to be aware of Theorem 1.64 until Zygmund reintroduced it in [121, 1191]. See the biography [761] for more about this interpolation theorem.

Fourier Multiplier

Many mathematicians considered the boundedness of the Fourier multipliers. Theorem 1.53 is one of the standard forms in this field. Hölmander and Michlin conducted pioneering works [606, 779, 780]. Theorem 1.53 played an important rule in defining Besov spaces and Triebel–Lizorkin spaces in the 1960s. We refer to the works [201, 314, 742, 868, 1035] for more generalization.

Singular Integral Operators on Lebesgue Spaces: Theorems 1.65, 1.66 and 1.67

The theory of the singular integral operators is due to Calderón, Zygmund, Babenko and Stein and many other mathematicians. It was refined in the 1950s. Due to the immerse volume of research, we are content mentioning some fundamental works [156, 330–332, 334, 1008]. Singular integral operators are investigated in connection with the Fourier transform on $\mathbb{R}^n$ under various assumptions; see the works [606, 690, 778] due to Hörmander, Kree and Michlin, which are related to this book.

Theorem 1.67 is due to Marcinkiewicz [763]. See the Russian papers [247, Teorema 1] and [248] for the result related to Theorem 1.66.

Many people have applied Theorems 1.65 and 1.66, as well as Theorem 1.67 to partial differential equations. Due to the immense volume of research, we are content just mentioning some fundamental works; [326, 331, 332, 334].

Section 1.6

Section 1.6.1

Harmonic Polynomials

See the pioneering paper [333, Theorem 1] for Theorem 1.70 where a variational approach is taken. Uchiyama replaced this variational argument with an approach using the Hilbert space $\mathcal{H}_k(\mathbb{R}^n)$ in his monograph [114].

Section 1.6.2

Mean-Value Property of Harmonic Functions: Lemma 1.38

See Hardy's work [553], as well as [856] for Lemma 1.38.

Section 1.6.3

Subharmonic Functions Generated by a Quadratic Form: Theorem 1.73 and Corollary 1.14

Theorem 1.72 is from [114, Theorem 14.3], while Theorem 1.73 is from [333, Theorem 1]. See [333, Chapter 1, Theorem 1]. We can remove the assumption of compactness of the frequency support of f in Corollary 1.14 using the spherical harmonic similar to the above paper [333]. See the works [333, Chapter 3, Theorem 1] and [333, Chapter 2, Theorem 2] for the extension of Corollary 1.14.

Chapter 2
Besov Spaces, Triebel–Lizorkin Spaces and Modulation Spaces

Having set down elementary facts in the previous chapter, we take a detailed look at Besov spaces and Triebel–Lizorkin spaces, which are the main theme of this book. Chapter 2 is devoted to the introduction of elementary definitions together with some fundamental properties. First we define the Besov space $B^s_{pq}(\mathbb{R}^n)$ with $1 \le p, q \le \infty$ and $s \in \mathbb{R}$ in the spirit of Peetre, although Besov introduced Besov spaces in [171, 172]. After the Besov space $B^s_{pq}(\mathbb{R}^n)$ for such a restricted case we define $A^s_{pq}(\mathbb{R}^n)$, which unifies the Besov space $B^s_{pq}(\mathbb{R}^n)$ with $0 < p, q \le \infty$ and $s \in \mathbb{R}$ and the Triebel–Lizorkin space $F^s_{pq}(\mathbb{R}^n)$ with $0 < p < \infty$, $0 < q \le \infty$ and $s \in \mathbb{R}$. We will see that we can replace Sobolev spaces with Besov spaces; many of the important theorems on Sobolev spaces will carry over to Besov spaces and Triebel–Lizorkin spaces. For $m \in \mathbb{N}$ and $1 \le p \le \infty$, we recall that the *Sobolev space* $W^{m,p}(\mathbb{R}^n)$ is defined by

$$W^{m,p}(\mathbb{R}^n) \equiv \bigcap_{\alpha \in \mathbb{N}_0{}^n, \, |\alpha| \le m} \{f \in L^p(\mathbb{R}^n) \, : \, \partial^\alpha f \in L^p(\mathbb{R}^n)\}. \tag{2.1}$$

The norm is given by

$$\|f\|_{W^{m,p}} \equiv \sum_{\alpha \in \mathbb{N}_0{}^n, \, |\alpha| \le m} \|\partial^\alpha f\|_p. \tag{2.2}$$

The norm (2.2) is called the *Sobolev norm*.

In Sect. 2.1.1 we define the Besov space $B^s_{pq}(\mathbb{R}^n)$ with $s \in \mathbb{R}$, $1 \le p, q \le \infty$, see Definitions 2.1 and Theorem 2.1. In Sect. 2.1.2 using the Young inequality, we investigate the lifting property (Theorem 2.3), completeness (Theorem 2.4), and the Sobolev inequality (Theorem 2.5); see also Propositions 2.1, 2.2 and 2.3. Section 2.1 considers the Besov space $B^s_{pq}(\mathbb{R}^n)$ with $1 \le p, q \le \infty$ and $s \in \mathbb{R}$. In Sect. 2.3 we plan to define the Besov space $B^s_{pq}(\mathbb{R}^n)$ with $0 < p, q \le \infty$ and $s \in \mathbb{R}$ and the Triebel–Lizorkin space $F^s_{pq}(\mathbb{R}^n)$ with $0 < p < \infty$, $0 < q \le \infty$ and $s \in \mathbb{R}$.

Y. Sawano, *Theory of Besov Spaces*, Developments in Mathematics 56,
https://doi.org/10.1007/978-981-13-0836-9_2

However, although the definition for the Besov space $B^s_{pq}(\mathbb{R}^n)$ with $1 \le p, q \le \infty$ and $s \in \mathbb{R}$ and the function spaces for other parameters is the same, there is a gap between them. To justify the definition is one of the big problems. So we will consider these cases separately. Section 2.2 provides us with the application of Besov spaces defined in Sect. 2.1. After considering the simplest case – the Besov space $B^s_{pq}(\mathbb{R}^n)$ with $1 \le p, q \le \infty$ and $s \in \mathbb{R}$, in Sect. 2.3 we consider the Besov space $B^s_{pq}(\mathbb{R}^n)$ with $0 < p, q \le \infty$ and $s \in \mathbb{R}$ and the Triebel–Lizorkin space $F^s_{pq}(\mathbb{R}^n)$ with $0 < p < \infty$, $0 < q \le \infty$ and $s \in \mathbb{R}$, as well as the modulation space $M^s_{pq}(\mathbb{R}^n)$ for $0 < p, q \le \infty$ and $s \in \mathbb{R}$. Section 2.4 is oriented to another direction. In some sense, the Besov norm $\| \star \|_{B^s_{pq}}$ and the Triebel–Lizorkin norm $\| \star \|_{F^s_{pq}}$ are made up of two parts: the low-frequency part and the high-frequency part. Therefore, these norms do not have symmetry. In fact, as it turns out, they are not dilation invariant; they lack the scaling property. We are interested in the norms having the scaling property. Section 2.4 presents us with such norms. The Besov space $B^s_{pq}(\mathbb{R}^n)$ and the Triebel–Lizorkin space $F^s_{pq}(\mathbb{R}^n)$, defined in Sects. 2.1 and 2.3, are called nonhomogeneous, while the Besov space $\dot{B}^s_{pq}(\mathbb{R}^n)$ and the Triebel–Lizorkin space $\dot{F}^s_{pq}(\mathbb{R}^n)$, defined in Sect. 2.4, are called homogeneous. Finally, Sect. 2.5 looks for some equivalent norms that will help us understand our function spaces better.

Besov spaces are sometimes called Nikolski'i–Besov spaces. Although we frequently use the term "Nikolski'i–Besov spaces", we simply write Besov spaces in this book.

2.1 Definition of the Nikolskii–Besov Space $B^s_{pq}(\mathbb{R}^n)$ with $1 \le p, q \le \infty$ and $s \in \mathbb{R}$

The definition of Nikolskii–Besov spaces is difficult and many mathematicians dislike the space. However, it is useful in many branches in mathematics such as partial differential equations, potential theory and Fourier analysis. We give a brief view of the spaces. Section 2.1.1 defines the Besov space $B^s_{pq}(\mathbb{R}^n)$ with $1 \le p, q \le \infty$ and $s \in \mathbb{R}$ and Sect. 2.1.2 investigates its property.

2.1.1 Definition of Nikolskii–Besov Spaces

2.1.1.1 Nikolskii–Besov Spaces $B^s_{pq}(\mathbb{R}^n)$ with $1 \le p, q \le \infty, s \in \mathbb{R}$

Let us define the Besov space $B^s_{pq}(\mathbb{R}^n)$ for $1 \le p, q \le \infty$ and $s \in \mathbb{R}$. To this end, we need notation. For $\tau \in \mathscr{S}(\mathbb{R}^n)$ and $f \in \mathscr{S}'(\mathbb{R}^n)$, we defined

$$\tau(D)f(x) \equiv (2\pi)^{-\frac{n}{2}}\langle f, \mathscr{F}^{-1}\tau(x - \star)\rangle = \mathscr{F}^{-1}[\tau \cdot \mathscr{F} f](x) \quad (x \in \mathbb{R}^n).$$

See Definition 1.38. Thus, although we cannot evaluate f itself at a point x, we can do so for $\tau(D)f(x)$.

We will define Nikolskii–Besov spaces.

Definition 2.1 (Besov norm $\| \star \|_{B^s_{pq}}$, $1 \le p, q \le \infty, s \in \mathbb{R}$) Suppose that the parameters p, q, s satisfy $1 \le p, q \le \infty, s \in \mathbb{R}$. Furthermore, choose $\psi, \varphi \in \mathscr{S}(\mathbb{R}^n)$ so that

$$\psi(\xi) = \begin{cases} 1 & (|\xi| \le 4) \\ 0 & (|\xi| \ge 8) \end{cases} \quad \varphi(\xi) = \begin{cases} 1 & (2 \le |\xi| \le 4), \\ 0 & (|\xi| \le 1 \text{ or } |\xi| \ge 8). \end{cases}$$

For $j \in \mathbb{N}$, let $\varphi_j \equiv \varphi(2^{-j}\star)$. For $f \in \mathscr{S}'(\mathbb{R}^n)$, define

$$\|f\|_{B^s_{pq}} \equiv \begin{cases} \|\psi(D)f\|_p + \left(\sum_{j=1}^{\infty} 2^{jqs} \|\varphi_j(D)f\|_p{}^q \right)^{\frac{1}{q}} & (1 \le q < \infty) \\ \|\psi(D)f\|_p + \sup_{j \in \mathbb{N}} 2^{js} \|\varphi_j(D)f\|_p & (q = \infty). \end{cases} \tag{2.3}$$

Here and below we prefer to say Besov spaces instead of Nikolskii–Besov spaces. Defining Besov spaces consists of the following six steps:

- Transform the function f on the time domain to $\mathscr{F}f$ on the frequency domain.
- Multiply ψ and φ_j and localize $\mathscr{F}f$.
- Consider the functions $\mathscr{F}^{-1}[\psi\mathscr{F}f]$ and $\mathscr{F}^{-1}[\varphi_j\mathscr{F}f]$ by the use of the inverse Fourier transform. The operators $\mathscr{F}^{-1}[\psi\mathscr{F}f]$ and $\mathscr{F}^{-1}[\varphi_j\mathscr{F}f]$ are called the Littlewood–Paley operators.
- Multiply the weight 2^{ks} according to the size of localization.
- Consider the $L^p(\mathbb{R}^n)$-norm of $\mathscr{F}^{-1}[\psi\mathscr{F}f]$ and each $\mathscr{F}^{-1}[\varphi_j\mathscr{F}f]$.
- Finally, consider the $\ell^q(\mathbb{N})$-norm with respect to j and add $\|\psi(D)f\|_p$.

As is seen from the definition, Besov spaces use the dyadic partition of the Fourier space.

For the notation we have the following remark.

Remark 2.1 Some prefer to write Δ_k instead of $\varphi_k(D)$.

In Chap. 2, we will define $\| \star \|_{B^s_{pq}}$ for $0 < p, q \le \infty, s \in \mathbb{R}$; the definition of $\| \star \|_{B^s_{pq}}$ for $1 \le p, q \le \infty, s \in \mathbb{R}$ is easier to grasp and in this chapter we define $B^s_{pq}(\mathbb{R}^n)$ only for $1 \le p, q \le \infty, s \in \mathbb{R}$.

Example 2.1 We draw the graph of ψ and φ. The horizontal axis is $|\xi|$ and the vertical axis is the value of the functions.

1. The graph of ψ: It assumes 1 near the origin.
2. The graph of φ: It vanishes near the origin.

For a sequence $\{a_j\}_{j=1}^{\infty}$, define

$$\|a_j\|_{\ell^p} = \|\{a_j\}_{j=1}^{\infty}\|_{\ell^p} \equiv \begin{cases} \left(\sum_{j=1}^{\infty} |a_j|^p\right)^{\frac{1}{p}} & \text{for } 0 < p < \infty, \\ \sup_{j \in \mathbb{N}} |a_j| & \text{for } p = \infty. \end{cases}$$

Then the Besov norm can be expressed:

$$\|f\|_{B^s_{pq}} = \|\psi(D)f\|_p + \|\{\|2^{js}\varphi_j(D)f\|_p\}_{j=1}^{\infty}\|_{\ell^q}.$$

2.1.1.2 Validity of the Definition of Besov Spaces

One of the main concerns in the theory of function spaces is whether the definition can be justified. The Besov norm $B^s_{pq}(\mathbb{R}^n)$ does depend on ψ and φ. However, for different admissible φ and ψ the definitions of $B^s_{pq}(\mathbb{R}^n)$ are the same. Let us formulate this.

Theorem 2.1 (The validity of $B^s_{pq}(\mathbb{R}^n)$) *In Definition 2.1, we obtain the equivalent norms for the admissible choice of ψ and φ. That is, suppose $\widetilde{\psi}, \widetilde{\varphi} \in \mathscr{S}(\mathbb{R}^n)$ satisfy*

$$\chi_{B(4)} \leq \widetilde{\psi} \leq \chi_{B(8)}, \quad \chi_{B(4)\setminus B(2)} \leq \widetilde{\varphi} \leq \chi_{B(8)\setminus B(1)}. \tag{2.4}$$

For $j \in \mathbb{N}$, we write $\widetilde{\varphi}_j \equiv \widetilde{\varphi}(2^{-j}\star)$. For $f \in \mathscr{S}'(\mathbb{R}^n)$, define

$$\|f\|^*_{B^s_{pq}} \equiv \begin{cases} \|\widetilde{\psi}(D)f\|_p + \left(\sum_{j=1}^{\infty} 2^{jqs}\|\widetilde{\varphi}_j(D)f\|_p{}^q\right)^{\frac{1}{q}} & (1 \leq q < \infty), \\ \|\widetilde{\psi}(D)f\|_p + \sup_{j \in \mathbb{N}} 2^{js}\|\widetilde{\varphi}_j(D)f\|_p & (q = \infty). \end{cases} \tag{2.5}$$

Then the norm equivalence

$$\|f\|_{B^s_{pq}} \sim \|f\|^*_{B^s_{pq}}, \quad f \in \mathscr{S}'(\mathbb{R}^n) \tag{2.6}$$

holds.

Theorem 2.1 says that the definition of $B^s_{pq}(\mathbb{R}^n)$ makes sense. In mathematics, it sometimes happens that being able to define something is more important than proving something. Theorem 2.1 is a sort of this thing.

Proof We assume $q < \infty$. When $q = \infty$, we simply replace the summation with the $\ell^\infty(\mathbb{N})$-norm; a natural modification will be made. Choose compactly supported smooth functions Φ, Ψ, ζ so that

$$\Psi(\psi + \varphi_1) = \widetilde{\psi}, \quad \Phi(\psi + \varphi_1 + \varphi_2) = \widetilde{\varphi}_1, \quad \zeta(\varphi_{-1} + \varphi + \varphi_1) = \widetilde{\varphi}.$$

Define a function ζ_j by $\zeta_j \equiv \zeta(2^{-j}\star)$ for $j \in \mathbb{N}$. Then

$$\widetilde{\psi} = \Psi(\psi + \varphi_1), \quad \widetilde{\varphi}_1 = \Phi(\psi + \varphi_1 + \varphi_2), \quad \widetilde{\varphi}_j = \zeta_j(\varphi_{j-1} + \varphi_j + \varphi_{j+1})$$

for $j \ge 2$. To use this relation, we rewrite $\|f\|^*_{B^s_{pq}}$ and then we estimate it using the triangle inequality for $\ell^q(\mathbb{N})$:

$$\begin{aligned}
\|f\|^*_{B^s_{pq}} &= \|\widetilde{\psi}(D)f\|_p + \left(\sum_{j=1}^\infty 2^{jqs}\|\widetilde{\varphi}_j(D)f\|_p{}^q\right)^{\frac{1}{q}} \\
&\le \|\widetilde{\psi}(D)f\|_p + 2^s\|\widetilde{\varphi}_1(D)f\|_p + \left(\sum_{j=2}^\infty 2^{jqs}\|\widetilde{\varphi}_j(D)f\|_p{}^q\right)^{\frac{1}{q}}.
\end{aligned}$$

Using the relation between the convolution of $\mathscr{S}(\mathbb{R}^n)$ and $\mathscr{S}(\mathbb{R}^n)$ and the Fourier transform (see Theorem 1.27), we obtain

$$\begin{aligned}
\|f\|^*_{B^s_{pq}} &\le \|\Psi(D)(\psi + \varphi_1)(D)f\|_p + 2^s\|\Phi(D)(\psi + \varphi_1 + \varphi_2)(D)f\|_p \\
&\quad + \left(\sum_{j=2}^\infty 2^{jqs}\|[\zeta_j(\varphi_{j-1} + \varphi_j + \varphi_{j+1})](D)f\|_p{}^q\right)^{\frac{1}{q}} \\
&\simeq_n \|\mathscr{F}^{-1}\Psi * (\psi + \varphi_1)(D)f\|_p + 2^s\|\mathscr{F}^{-1}\Phi * (\psi + \varphi_1 + \varphi_2)(D)f\|_p \\
&\quad + \left(\sum_{j=2}^\infty 2^{jqs}\|\mathscr{F}^{-1}[\zeta_j] * (\varphi_{j-1} + \varphi_j + \varphi_{j+1})(D)f\|_p{}^q\right)^{\frac{1}{q}}.
\end{aligned}$$

Here note that $\|\mathscr{F}^{-1}[\zeta_j]\|_1 = 2^{jn}\|\mathscr{F}^{-1}\zeta(2^j\star)\|_1 = \|\mathscr{F}^{-1}\zeta\|_1 < \infty$. Hence we have

$$\begin{aligned}
\|f\|^*_{B^s_{pq}} &\lesssim \|(\psi + \varphi_1)(D)f\|_p + \|(\psi + \varphi_1 + \varphi_2)(D)f\|_p \\
&\quad + \left(\sum_{j=2}^\infty 2^{jqs}\|(\varphi_{j-1} + \varphi_j + \varphi_{j+1})(D)f\|_p{}^q\right)^{\frac{1}{q}}
\end{aligned}$$

by the Young inequality (see Theorem 1.4). If we use the triangle inequality, then we have

$$\|f\|^*_{B^s_{pq}} \lesssim \|\psi(D)f\|_p + \|\varphi_1(D)f\|_p + \|\varphi_2(D)f\|_p + \left(\sum_{j=2}^{\infty} 2^{jqs}\|\varphi_{j-1}(D)f\|_p{}^q\right)^{\frac{1}{q}}$$

$$+ \left(\sum_{j=2}^{\infty} 2^{jqs}\|\varphi_j(D)f\|_p{}^q\right)^{\frac{1}{q}} + \left(\sum_{j=2}^{\infty} 2^{jqs}\|\varphi_{j+1}(D)f\|_p{}^q\right)^{\frac{1}{q}}.$$

If we change the index, then we have

$$\|f\|^*_{B^s_{pq}} \lesssim \|\psi(D)f\|_p + \left(\sum_{j=1}^{\infty} 2^{jqs}\|\varphi_j(D)f\|_p{}^q\right)^{\frac{1}{q}} = \|f\|_{B^s_{pq}}.$$

A symmetry yields the reverse inequality of (2.6). Thus, we obtain (2.6).

Here and below, we fix ψ and φ to consider the Besov norm $\| \star \|_{B^s_{pq}}$. However, we still need to consider another definition as the following remark shows:

Remark 2.2 In Theorem 2.1, we can assume that $\tilde{\varphi}$ and $\tilde{\psi}$ satisfy

$$\tilde{\psi} = \tilde{\varphi} - \tilde{\varphi}_{-1}, \quad \tilde{\varphi}(\xi) = \begin{cases} 1 & \text{if } |\xi| \le 1, \\ 0 & \text{if } |\xi| \ge 2. \end{cases} \tag{2.7}$$

The proof is the same as Theorem 2.1. Sometimes, (2.7) is easier to handle than the norm defined by (2.3), so in Sect. 2.3 we use (2.7) without notice. We refer to Lemma 1.15 for the existence of the function $\tilde{\varphi}$ satisfying (2.7).

Definition 2.2 (Besov space $B^s_{pq}(\mathbb{R}^n)$, $1 \le p, q \le \infty, s \in \mathbb{R}$) Take functions $\tilde{\psi}, \tilde{\varphi} \in \mathscr{S}(\mathbb{R}^n)$ satisfying (2.4) or (2.7). For $1 \le p, q \le \infty$, $s \in \mathbb{R}$, define the norm by (2.3) and (2.5) accordingly. The *Besov space* $B^s_{pq}(\mathbb{R}^n)$ is the set given by

$$B^s_{pq}(\mathbb{R}^n) \equiv \{f \in \mathscr{S}'(\mathbb{R}^n) : \|f\|_{B^s_{pq}} < \infty\}.$$

We will obtain a different norm if we define the Besov norm $\| \star \|_{B^s_{pq}}$ by (2.5) using the different functions $\tilde{\psi}, \tilde{\varphi}$ satisfying (2.4) or (2.7). However, by virtue of Theorem 2.1, this will result in the same set of $\mathscr{S}'(\mathbb{R}^n)$ as the one in Definition 2.2.

Since we are considering the dyadic partition, we can say that the weight 2^{js} grows moderately. So we simply need to investigate every frequency using the natural cut-off.

Exercises

Exercise 2.1 Define $\mathscr{F}M(\mathbb{R}^n)$ as the set of all continuous functions f on $\mathbb{R}^n$ such that there exists a finite Radon measure μ such that

$$f(x) = \frac{1}{(2\pi)^{n/2}} \int_{\mathbb{R}^n} \mathrm{e}^{-ix\cdot\xi} \mathrm{d}\mu(\xi)$$

for all $x \in \mathbb{R}^n$. For such an f, the norm is defined by setting $\|f\|_{\mathscr{F}M} = \|\mu\|$, where $\|\mu\|$ denotes the total variation of μ. Then show that $\mathscr{F}M(\mathbb{R}^n)$ is embedded into $B^0_{\infty\infty}(\mathbb{R}^n)$. Hint: Choose an even function $\psi \in \mathscr{S}(\mathbb{R}^n)$. Then we have

$$\psi(D)f(x) = c_n \int_{\mathbb{R}^n} \mathscr{F}^{-1}\psi(\xi - x)\mathrm{d}\mu(\xi).$$

Exercise 2.2 Let $1 \le p \le \infty$, $1 \le q < \infty$ and $s \in \mathbb{R}$. Let $a \in \mathbb{R}^n$. Let δ_a be the Dirac delta massed at a.

1. Show that $\delta_a \in B^s_{pq}(\mathbb{R}^n)$ if and only if $s < \dfrac{n}{p}$.
2. Show that $\delta_a \in B^s_{p\infty}(\mathbb{R}^n)$ if and only if $s \le \dfrac{n}{p}$.

2.1.2 Elementary Properties of the Besov Space $B^s_{pq}(\mathbb{R}^n)$ with $1 \le p, q \le \infty$ and $s \in \mathbb{R}$

Having defined Besov spaces as above, we are faced with a natural question: what properties do Besov spaces have? First of all, let us investigate the elementary embeddings. A Banach space X is said to be embedded continuously into a Banach space Y, if X is a subset of Y and there exists a constant such that $\|x\|_Y \le C\|x\|_X$. If a Banach space X is embedded into a Banach space Y and T is a bounded mapping from Z to X, then we can say that T is a bounded mapping from Z to Y. Note that a linear operator T is not always bounded from Z to X when T is a bounded linear operator from Z to Y. Thus, the embedding can measure the "quality" of the results.

2.1.2.1 $L^p(\mathbb{R}^n)$ Space and Inclusions

Here we first consider the role of p and check that $B^0_{pq}(\mathbb{R}^n)$ is close to $L^p(\mathbb{R}^n)$. The following proposition shows that $B^0_{pq}(\mathbb{R}^n)$, $1 \le p, q \le \infty$ is close to $L^p(\mathbb{R}^n)$:

Proposition 2.1 *Let $1 \le p \le \infty$. Then $B^0_{p1}(\mathbb{R}^n) \hookrightarrow L^p(\mathbb{R}^n) \hookrightarrow B^0_{p\infty}(\mathbb{R}^n)$.*

Here a tacit understanding is that $L^p(\mathbb{R}^n)$ is embedded into $\mathscr{S}'(\mathbb{R}^n)$ via the mapping $f \in L^p(\mathbb{R}^n) \mapsto F_f \in \mathscr{S}'(\mathbb{R}^n)$; see Definition 1.12. In some cases (but not always), the second parameter q in $B^s_{pq}(\mathbb{R}^n)$ plays the role of smoothness as is seen from Proposition 2.1.

Suppose that $\psi, \varphi \in \mathscr{S}(\mathbb{R}^n)$ satisfy (2.7). For the proof of Proposition 2.1, we use ψ and φ to define the norms $\| \star \|_{B^0_{p1}}$ and $\| \star \|_{B^0_{p\infty}}$.

Proof Let us prove $B^0_{p1}(\mathbb{R}^n) \hookrightarrow L^p(\mathbb{R}^n)$. Choose $f \in B^0_{p1}(\mathbb{R}^n)$ arbitrarily. By the definition of the norm,

$$\|f\|_{B^0_{p1}} = \|\psi(D)f\|_p + \sum_{j=1}^{\infty} \|\varphi_j(D)f\|_p < \infty.$$

From (2.7), we deduce

$$\|f\|_p = \left\| \psi(D)f + \sum_{j=1}^{\infty} \varphi_j(D)f \right\|_p .$$

So if we start from the functions ψ and φ satisfying (2.7), then we have $\|f\|_p \leq \|f\|_{B^0_{p1}}$ by the triangle inequality for $L^p(\mathbb{R}^n)$-spaces. Hence we obtain $B^0_{p1}(\mathbb{R}^n) \hookrightarrow L^p(\mathbb{R}^n)$.

Let us prove $L^p(\mathbb{R}^n) \hookrightarrow B^0_{p\infty}(\mathbb{R}^n)$. Let $f \in L^p(\mathbb{R}^n)$. First of all, we write the norm $B^0_{p\infty}(\mathbb{R}^n)$ out in full to have $\|f\|_{B^0_{p\infty}} = \|\psi(D)f\|_p + \sup_{j\in\mathbb{N}} \|\varphi_j(D)f\|_p$. Since $\mathscr{F}^{-1}\varphi_j = 2^{jn}\mathscr{F}^{-1}\varphi(2^j\star)$, $j \geq 1$,

$$\|\mathscr{F}^{-1}\varphi_j\|_1 = \|2^{jn}\mathscr{F}^{-1}\varphi(2^j\star)\|_1 = \|\mathscr{F}^{-1}\varphi\|_1 \tag{2.8}$$

for all $j \in \mathbb{N}$. If we combine (2.8) with the Young inequality and $\varphi_j(D)f \simeq_n \mathscr{F}^{-1}\varphi_j * f$, then

$$\|\psi(D)f\|_p + \sup_{j\in\mathbb{N}} \|\varphi_j(D)f\|_p \leq (\|\mathscr{F}^{-1}\psi\|_1 + \|\mathscr{F}^{-1}\varphi\|_1)\|f\|_p \simeq \|f\|_p < \infty.$$

Hence we obtain the right-hand side inclusion $L^p(\mathbb{R}^n) \hookrightarrow B^0_{p\infty}(\mathbb{R}^n)$.

2.1.2.2 Elementary Embeddings Between Besov Spaces

It is important to compare two Besov spaces $B^{s_0}_{p_0q_0}(\mathbb{R}^n)$ and $B^{s_1}_{p_1q_1}(\mathbb{R}^n)$ for given parameters $p_0, p_1, q_0, q_1, s_0, s_1$ because we sometimes obtain some results in terms of various Besov spaces.

It clearly follows from the definition of the sequence space $\ell^q(\mathbb{N})$, $1 \le q \le \infty$ that we have the following assertion which explains the role of q:

Proposition 2.2 (Monotonicity of the Besov norm in q**)** *Let $1 \le p \le \infty$, $1 \le q_0 \le q_1 \le \infty$, and $s \in \mathbb{R}$. Then $B^s_{pq_0}(\mathbb{R}^n) \hookrightarrow B^s_{pq_1}(\mathbb{R}^n)$ in the sense of the continuous embedding. More precisely, using the common functions $\psi, \varphi \in \mathscr{S}(\mathbb{R}^n)$ in the definition of the norms for $B^s_{pq_0}(\mathbb{R}^n)$ and $B^s_{pq_1}(\mathbb{R}^n)$, we have $\|f\|_{B^s_{pq_1}} \le \|f\|_{B^s_{pq_0}}$ for all $f \in B^s_{pq_0}(\mathbb{R}^n)$.*

The proof is left to the interested readers as Exercise 2.6.

We deduce from the structure of the underlying sequence space which defines the norms of the Besov space. This proposition shows that q is free at the (small) cost of s.

Proposition 2.3 *Let $1 \le p, q_1, q_2 \le \infty$, $s \in \mathbb{R}$, $\delta > 0$. Then $B^{s+\delta}_{pq_1}(\mathbb{R}^n) \hookrightarrow B^s_{pq_2}(\mathbb{R}^n)$ in the sense of continuous embedding.*

Proof By writing the definition of the norms out in full, Proposition 2.3 reduces to Proposition 1.21.

2.1.2.3 Meaning of the Parameter s

We now aim to verify that the parameter s stands for the smoothness. To this end, we use a piece of information for the differential of quotient functions.

Lemma 2.1 *Let $a \in \mathbb{R}$, and let $\alpha = (\alpha_1, \alpha_2, \dots, \alpha_n) \in \mathbb{N}_0{}^n$. Then there exists a polynomial $P_{a,\alpha}(x)$ with $x_1, x_2, \dots, x_n$ of degree $|\alpha|(= \alpha_1 + \alpha_2 + \cdots + \alpha_n)$ such that $\partial^\alpha \langle x \rangle^a$ can be expressed: $\partial^\alpha \langle x \rangle^a = P_{a,\alpha}(x) \langle x \rangle^{a-2|\alpha|}$. In particular, for $1/4 \le |x| \le 4$, we have $|\partial^\alpha \langle 2^k x \rangle^a| \lesssim_{a,\alpha} 2^{ak}$.*

Proof By the mathematical induction, we can prove this easily; see Exercise 2.4.

We will prove the next theorem which concerns the differentiability of functions using Lemma 2.1. We will investigate the differentiability of functions in more depth in Sect. 2.2.2.

Theorem 2.2 (Lift operator) *Let $1 \le p, q \le \infty$, $s \in \mathbb{R}$.*

1. *For $k = 1, 2, \dots, n$, the partial differential operator $\partial_k = \partial_{x_k} : B^s_{pq}(\mathbb{R}^n) \to B^{s-1}_{pq}(\mathbb{R}^n)$ is continuous. That is,*

$$\|\partial_{x_k} f\|_{B^{s-1}_{pq}} \lesssim \|f\|_{B^s_{pq}}, \quad f \in B^s_{pq}(\mathbb{R}^n). \tag{2.9}$$

2. *For $\sigma \in \mathbb{R}$ the lift operator $(1-\Delta)^\sigma : B^s_{pq}(\mathbb{R}^n) \approx B^{s-2\sigma}_{pq}(\mathbb{R}^n)$, whose Fourier multiplier is generated by $(1+|\xi|^2)^\sigma$, is an isomorphism; that is,*

$$\|(1-\Delta)^\sigma f\|_{B^{s-2\sigma}_{pq}} \sim \|f\|_{B^s_{pq}}, \quad f \in B^s_{pq}(\mathbb{R}^n). \tag{2.10}$$

3. *For $m \in \mathbb{N}$ the operator $1+(-\Delta)^m : B^s_{pq}(\mathbb{R}^n) \approx B^{s-2m}_{pq}(\mathbb{R}^n)$ is an isomorphism; that is,*

$$\|(1+(-\Delta)^m)f\|_{B^{s-2m}_{pq}} \sim \|f\|_{B^s_{pq}}, \quad f \in B^s_{pq}(\mathbb{R}^n). \tag{2.11}$$

4. *For $m \in \mathbb{N}$ the operator $(1+\partial_1{}^{4m}+\cdots+\partial_n{}^{4m}) : B^s_{pq}(\mathbb{R}^n) \approx B^{s-4m}_{pq}(\mathbb{R}^n)$ is an isomorphism; that is,*

$$\|(1+\partial_1{}^{4m}+\cdots+\partial_n{}^{4m})f\|_{B^{s-4m}_{pq}} \sim \|f\|_{B^s_{pq}}, \quad f \in B^s_{pq}(\mathbb{R}^n). \tag{2.12}$$

Proof We content ourselves with (2.9) and the $\lesssim$ part of (2.10).

We can prove the $\gtrsim$ part of (2.10), (2.11), and (2.12) similarly.

Also, a natural modification is made when $q = \infty$. Replace the sum over j with the $\ell^\infty(\mathbb{N})$-norm. We thus suppose that $q < \infty$.

We start with the proof of (2.9), which is not so hard.

For ψ and $j \in \mathbb{Z}$ satisfying (2.7), we write $\psi_j \equiv \psi(2^{-j}\star)$. Furthermore, for $k = 1, 2, \ldots, n$, denote simply by ξ_k the coordinate function $\xi \in \mathbb{R}^n \mapsto \xi_k \in \mathbb{R}$. In view of the size of supports, we have

$$\begin{aligned}
&\|\partial_{x_k} f\|_{B^{s-1}_{pq}} \\
&= \|\mathscr{F}^{-1}[\xi_k \cdot \mathscr{F} f]\|_{B^{s-1}_{pq}} \\
&= \|\mathscr{F}^{-1}[\psi \cdot \xi_k \cdot \mathscr{F} f]\|_p + \left(\sum_{j=1}^{\infty} 2^{jq(s-1)} \|\mathscr{F}^{-1}[\varphi_j \cdot \xi_k \cdot \mathscr{F} f]\|_p{}^q\right)^{\frac{1}{q}} \\
&= \|\mathscr{F}^{-1}[\psi \cdot \psi_1 \cdot \xi_k \cdot \mathscr{F} f]\|_p + \left(\sum_{j=1}^{\infty} 2^{jq(s-1)} \|\mathscr{F}^{-1}[\varphi_j \cdot \psi_{j+1} \cdot \xi_k \cdot \mathscr{F} f]\|_p{}^q\right)^{\frac{1}{q}}.
\end{aligned}$$

Let $\psi^{(j,k)}(\xi) \equiv (2^{-j}\xi_k)\psi_j(\xi)$ for $\xi \in \mathbb{R}^n$. By the Young inequality and Theorem 1.27, we have

$$\begin{aligned}
&\|\partial_{x_k} f\|_{B^{s-1}_{pq}} \\
&\simeq_n \|\mathscr{F}^{-1}[\psi_1 \cdot \xi_k] * \psi(D) f\|_p + \left(\sum_{j=1}^{\infty} 2^{jq(s-1)} \|\mathscr{F}^{-1}[\psi_{j+1} \cdot \xi_k] * \varphi_j(D) f\|_p{}^q\right)^{\frac{1}{q}} \\
&\lesssim \|\mathscr{F}^{-1}[\psi_1 \cdot \xi_k]\|_1 \|\psi(D) f\|_p + \left(\sum_{j=1}^{\infty} 2^{jqs} \|\mathscr{F}^{-1}\psi^{(j,k)}\|_1{}^q \|\varphi_j(D) f\|_p{}^q\right)^{\frac{1}{q}}
\end{aligned}$$

$$\lesssim \|\psi(D)f\|_p + \left(\sum_{j=1}^{\infty} 2^{jqs} \|\varphi_j(D)f\|_p{}^q\right)^{\frac{1}{q}}$$

$$= \|f\|_{B^s_{pq}}.$$

As for (2.10), we argue in the same way to have

$$\|(1-\Delta)^\sigma f\|_{B^{s-\sigma}_{pq}} \le D\|f\|_{B^s_{pq}}. \tag{2.13}$$

Here

$$D \equiv \|\mathscr{F}^{-1}[\psi(1+|\star|^2)^\sigma]\|_1 + \sup_{j\in\mathbb{N}} 2^{-2j\sigma}\|\mathscr{F}^{-1}[(1+|\star|^2)^\sigma(\psi_{j+1}-\psi_{j-3})]\|_1.$$

Hence it remains to establish that $D < \infty$. The first term is clearly finite.

Note that Lemma 2.1 yields

$$\|\partial^\beta[(1+|\star|^2)^\sigma(\psi_{j+3}-\psi_{j-1})]\|_1 \lesssim 2^{j(2\sigma+n-|\beta|)}.$$

Thus, by the Riemann–Lebesgue theorem we have

$$|\mathscr{F}^{-1}[(1+|\star|^2)^\sigma(\psi_{j+3}-\psi_{j-1})](x)| \lesssim_M 2^{j(2\sigma+n-M)}|x|^{-M} \quad (x \in \mathbb{R}^n)$$

for all $M \ge 0$. If we integrate this inequality with $M = 0$ and $M = n+1$, then we obtain the desired result.

The next corollary is a kind of interpolation inequality; in analysis, a significant motto is that we can control mixed derivatives $\partial^\alpha f$, $|\alpha| \le m$ using pure derivative $\partial_{x_j}{}^m f$, $j = 1, 2, \ldots, n$ and the original function f.

We illustrate this situation. Let $n = 2$. Suppose that we have information on the points $\alpha = (0,0)$, $(m,0)$, $(0,m)$. Then by (2.9), for any lattice point $\mathrm{P}(m_1, m_2)$ inside the triangle $\triangle$ABO or on each vertex of the triangle $\triangle$ABO, we have

$$\|\partial_1{}^{m_1}\partial_2{}^{m_2} f\|_{B^s_{pq}} \lesssim \|f\|_{B^{s+m}_{pq}}. \tag{2.14}$$

So we have information on the norm for the differential operator $\partial_1{}^{m_1}\partial_2{}^{m_2}$.

The following theorem can be located as a corollary of Theorem 2.2. However, it is an important theorem in its own right. Let $f \in L^2(\mathbb{R}^n)$. Then we know that the weak solution to the equation $(1-\Delta)u = f$ exists and belongs to the Sobolev space $W^{2,2}(\mathbb{R}^n)$. This fact can be derived from Theorem 2.2 and the fact that $B^0_{22}(\mathbb{R}^n)$ is isomorphic to $L^2(\mathbb{R}^n)$. This fact is also deduced from the fundamental technique in the elliptic differential equations. Now let us suppose that we obtained the existence result from the theory of elliptic partial equations and that $f \in W^{m,2}(\mathbb{R}^n)$. See (2.1). To show that $u \in W^{m+2,2}(\mathbb{R}^n)$, we use the fact that $v = \partial^\alpha u$ is the solution of

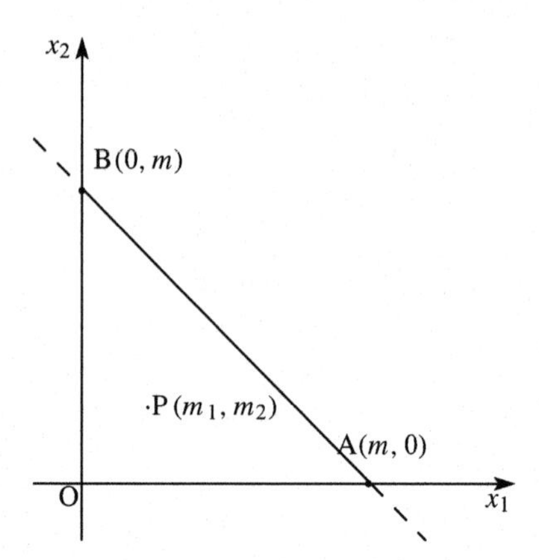

$(1-\Delta)v = \partial^\alpha f$ for all $\alpha \in \mathbb{N}_0{}^n$ with $|\alpha| \le m$. According to the fundamental theory of elliptic differential equations, it follows that $\partial^\alpha u \in W^{2,2}(\mathbb{R}^n)$ for all $\alpha \in \mathbb{N}_0{}^n$ with $|\alpha| \le m$. For Sobolev spaces, it follows directly from the definition of the Sobolev norm $\| \star \|_{W^{2+m,2}}$ that $u \in W^{2+m,2}(\mathbb{R}^n)$. Note that this argument does not directly work for Besov spaces. However, the next theorem guarantees that an analogy for Besov spaces is also available.

Theorem 2.3 *Let* $1 \le p, q \le \infty$, $m \in \mathbb{N}$ *and* $s \in \mathbb{R}$. *Then the norm equivalence*

$$\|f\|_{B^{s+m}_{pq}} \sim \|f\|_{B^s_{pq}} + \sum_{j=1}^n \|\partial_{x_j}{}^m f\|_{B^s_{pq}} \quad (f \in \mathscr{S}'(\mathbb{R}^n))$$

holds.

From this theorem, we learn that $B^s_{pq}(\mathbb{R}^n)$ can describe the fractional smoothness.

Proof By Theorem 2.2, $B^{s+m}_{pq}(\mathbb{R}^n) \hookrightarrow B^s_{pq}(\mathbb{R}^n)$ and the differential operator $\partial_{x_k} : B^s_{pq}(\mathbb{R}^n) \to B^{s-1}_{pq}(\mathbb{R}^n)$ is continuous. Thus, $\|f\|_{B^s_{pq}} + \sum_{j=1}^n \|\partial_{x_j}{}^m f\|_{B^s_{pq}} \lesssim \|f\|_{B^{s+m}_{pq}}$ is trivial. By (2.9) and (2.12), we also have

$$\begin{aligned}
\|f\|_{B^{s+m}_{pq}} &\sim \|(1 + \partial_1{}^{4m} + \cdots + \partial_n{}^{4m})f\|_{B^{s-3m}_{pq}} \\
&\lesssim \|f\|_{B^{s-3m}_{pq}} + \sum_{j=1}^n \|\partial_{x_j}{}^{4m} f\|_{B^{s-3m}_{pq}} \\
&\lesssim \|f\|_{B^s_{pq}} + \sum_{j=1}^n \|\partial_{x_j}{}^m f\|_{B^s_{pq}}.
\end{aligned}$$

Hence the proof is complete.

2.1.2.4 An Embedding Into BUC($\mathbb{R}^n$)

It is important that we know the relations of function spaces. One of the ways to do this is to investigate the embedding relation between them. Here we explain how to understand Besov spaces including those between Besov spaces using the function space BUC($\mathbb{R}^n$), the set of all bounded uniformly continuous functions.

Definition 2.3 ($\mathscr{B}^m$) For $m = 0, 1, 2, \ldots$, define $\mathscr{B}^m(\mathbb{R}^n)$ to be the set of all $f \in C^m(\mathbb{R}^n)$ having bounded partial derivatives up to order m.

Recall that BUC($\mathbb{R}^n$) denotes the Banach space of all uniformly continuous bounded functions.

Proposition 2.4 (Embedding $B^m_{\infty 1}(\mathbb{R}^n) \hookrightarrow \mathscr{B}^m(\mathbb{R}^n)$) *For any $m \in \mathbb{N}_0$, in the sense of continuous embedding, $B^m_{\infty 1}(\mathbb{R}^n) \hookrightarrow \mathscr{B}^m(\mathbb{R}^n)$ and all the partial derivatives up to order m of any element in $B^m_{\infty 1}(\mathbb{R}^n)$ is uniformly continuous. In particular, $B^0_{\infty 1}(\mathbb{R}^n) \hookrightarrow$ BUC($\mathbb{R}^n$).*

Proof By Theorem 2.3, the lifting property, it can be assumed that $m = 0$. Let $f \in B^0_{\infty 1}(\mathbb{R}^n)$. We use ψ and φ satisfying (2.7) to define the norms. By Theorem 1.24,

$$\lim_{N \to \infty} \left(\psi(D)f + \sum_{j=1}^{N} \varphi_j(D)f \right) = f \tag{2.15}$$

in the topology of $\mathscr{S}'(\mathbb{R}^n)$.

The condition $f \in B^0_{\infty 1}(\mathbb{R}^n)$ means $\|f\|_{B^0_{\infty 1}} = \|\psi(D)f\|_\infty + \sum_{j=1}^{\infty} \|\varphi_j(D)f\|_\infty < \infty$. We have (2.15) in the topology of $L^\infty(\mathbb{R}^n)$. Since $\psi(D)f + \sum_{j=1}^{N} \varphi_j(D)f \in \mathscr{B}^1(\mathbb{R}^n)$, in particular, it is uniformly continuous. Since f is realized as the $L^\infty(\mathbb{R}^n)$-limit of such a uniformly continuous function, f itself is uniformly continuous.

2.1.2.5 Completeness, Density and Sobolev's Embedding

Once the readers become familiar with Sobolev spaces, which date back to 1935–1938; see Sobolev's works [1000–1002], they will agree that completeness and density are elementary notions for the theory of the function spaces. The main problem we encounter in the proof of the completeness is to find a candidate of the limit for a given Cauchy sequence. This is not so easy to prove as we did for Lebesgue spaces. To prove the completeness of our function spaces we need the one for $L^\infty(\mathbb{R}^n)$; see [4, Theorem 2. 16] for example. The reader will see how we manage to do this from now on.

Theorem 2.4 (Elementary inclusions, completeness and density) *Let* $1 \le p, q \le \infty$ *and* $s \in \mathbb{R}$.

1. *In the sense of continuous embedding,*

$$\mathscr{S}(\mathbb{R}^n) \hookrightarrow B^s_{pq}(\mathbb{R}^n) \hookrightarrow \mathscr{S}'(\mathbb{R}^n). \tag{2.16}$$

2. *The space* $B^s_{pq}(\mathbb{R}^n)$ *is a Banach space. That is, the normed space* $B^s_{pq}(\mathbb{R}^n)$ *is complete under the norm.*
3. *Let* $1 \le p, q < \infty$. *Then* $C^\infty_{\mathrm{c}}(\mathbb{R}^n)$ *is dense in* $B^s_{pq}(\mathbb{R}^n)$.

Embedding (2.16) is important because it shows that $B^s_{pq}(\mathbb{R}^n)$ is *nontrivial*; namely, $B^s_{pq}(\mathbb{R}^n)$ is not the zero linear space for any $p, q \in [1, \infty]$ and $s \in \mathbb{R}$.

Proof

1. Proposition 2.1 yields $\mathscr{S}(\mathbb{R}^n) \hookrightarrow L^p(\mathbb{R}^n) \hookrightarrow B^0_{p\infty}(\mathbb{R}^n) \hookrightarrow B^{-1}_{pq}(\mathbb{R}^n)$. Hence using $\mathscr{S}(\mathbb{R}^n) \hookrightarrow B^{-1}_{pq}(\mathbb{R}^n)$ and the lift operator $(1-\Delta)^{\frac{s+1}{2}}$, we obtain $\mathscr{S}(\mathbb{R}^n) \hookrightarrow B^s_{pq}(\mathbb{R}^n)$. Likewise, $B^1_{pq}(\mathbb{R}^n) \hookrightarrow B^0_{p1}(\mathbb{R}^n) \hookrightarrow L^p(\mathbb{R}^n) \hookrightarrow \mathscr{S}'(\mathbb{R}^n)$ follows. We thus obtain $B^s_{pq}(\mathbb{R}^n) \hookrightarrow \mathscr{S}'(\mathbb{R}^n)$ using the embedding $B^1_{pq}(\mathbb{R}^n) \hookrightarrow \mathscr{S}'(\mathbb{R}^n)$ and the lift operator $(1-\Delta)^{-\frac{s+1}{2}} : B^s_{pq}(\mathbb{R}^n) \to B^{-1}_{pq}(\mathbb{R}^n)$.
2. The lift operator allows us to assume that $s > 0$. Then $B^s_{pq}(\mathbb{R}^n) \hookrightarrow B^s_{p1}(\mathbb{R}^n) \hookrightarrow L^p(\mathbb{R}^n)$. Hence if we have a $B^s_{pq}(\mathbb{R}^n)$-Cauchy sequence $\{f_j\}_{j=1}^\infty$, it is convergent to f in $L^p(\mathbb{R}^n)$ thanks to the completeness of $L^p(\mathbb{R}^n)$. By the Fatou lemma, we have $\|f - f_j\|_{B^s_{pq}} \le \liminf_{k\to\infty} \|f_k - f_j\|_{B^s_{pq}}$. Thus,

$$\limsup_{j\to\infty} \|f - f_j\|_{B^s_{pq}} \le \limsup_{j\to\infty} \left(\liminf_{k\to\infty} \|f_k - f_j\|_{B^s_{pq}} \right) = 0.$$

   Thus, any $B^s_{pq}(\mathbb{R}^n)$-Cauchy sequence $\{f_j\}_{j=1}^\infty$ is convergent to f in $B^s_{pq}(\mathbb{R}^n)$.
3. Define the norm using the functions ψ and φ satisfying (2.7). Let $k \ge 3$. By virtue of the size of the support, we have

$$\|f - \mathscr{F}^{-1}[\psi_{k+3} \cdot \mathscr{F} f]\|_{B^s_{pq}} = \left(\sum_{j=k}^{\infty} 2^{jqs} \|\mathscr{F}^{-1}[(1-\psi_{k+3})\varphi_j \cdot \mathscr{F} f]\|_p{}^q \right)^{\frac{1}{q}}.$$

   For $k \le j \le k+5$ we use the Young inequality to have

$$\|f - \mathscr{F}^{-1}[\psi_{k+3} \cdot \mathscr{F} f]\|_{B^s_{pq}} \lesssim \left(\sum_{j=k}^{\infty} 2^{jqs} \|\varphi_j(D) f\|_p{}^q \right)^{\frac{1}{q}}.$$

Hence since $q < \infty$, we have

$$\lim_{k\to\infty} \mathscr{F}^{-1}[\psi_{k+3} \cdot \mathscr{F} f] = f$$

in the topology of $B^s_{pq}(\mathbb{R}^n)$. So, instead of $f \in B^s_{pq}(\mathbb{R}^n)$, we have only to approximate g of the form: $\mathscr{F}^{-1}[\kappa \cdot \mathscr{F} f]$, $f \in B^s_{pq}(\mathbb{R}^n)$, $\kappa \in C^\infty_c(\mathbb{R}^n)$. Here we have

$$\mathscr{F}^{-1}\big[\varphi_j \cdot \mathscr{F}[g - \psi_k g]\big] = \mathscr{F}^{-1}[\varphi_j \cdot \mathscr{F} g] - c_n 2^{kn}[\mathscr{F}^{-1}\psi](2^k\star) \\ *\mathscr{F}^{-1}[\varphi_j \cdot \mathscr{F} g]$$

for $j, k \in \mathbb{N}$. Since $1 \le p < \infty$, $\|g - \psi_k g\|_{B^s_{pq}}$ converges to 0 as $k \to \infty$. Since $\psi_k g \in C^\infty_c(\mathbb{R}^n)$, $C^\infty_c(\mathbb{R}^n)$ is dense in $B^s_{pq}(\mathbb{R}^n)$.

Note that we could manage to prove the completeness of the function spaces. This was a long journey after we define Besov spaces at the beginning of this chapter.

Let $1 \le p \le \infty$. The Young inequality $\|f * g\|_p \le \|f\|_p \|g\|_1$ is a special case of the usual Young inequality

$$\|f * g\|_r \le \|f\|_p \|g\|_q, \quad 1 \le p, q, r \le \infty, \quad \frac{1}{r} = \frac{1}{p} + \frac{1}{q} - 1. \tag{2.17}$$

By means of the classical Young inequality, we will prove the following Sobolev embedding theorem:

Theorem 2.5 (Sobolev's embedding theorem for Besov spaces) *Suppose that the parameters p_0, p_1, q, s_0, s_1 satisfy*

$$1 \le p_0 < p_1 \le \infty, \quad 1 \le q \le \infty, \quad -\infty < s_1 < s_0 < \infty$$

and

$$s_0 - \frac{n}{p_0} = s_1 - \frac{n}{p_1}. \tag{2.18}$$

Then $B^{s_0}_{p_0 q}(\mathbb{R}^n) \hookrightarrow B^{s_1}_{p_1 q}(\mathbb{R}^n)$ in the sense of continuous embedding.

As (2.18) implies, $s - \frac{n}{p}$ is an important quantity in the theory of function spaces, which is called the *differential index*.

The same can be said for the classical Sobolev embedding theorem. This will be taken up once again in Sect. 2.3.

Proof Let $q < \infty$. A minor modification works for the case where $q = \infty$. We define (2.7) from ψ and φ to define the norm. Let $f \in B^{s_0}_{p_0q}(\mathbb{R}^n)$. Define $r > 0$ by $\dfrac{1}{p_1} = \dfrac{1}{p_0} + \dfrac{1}{r} - 1$. By the definition of the norm, we have

$$\|f\|_{B^{s_1}_{p_1q}} = \|\mathscr{F}^{-1}[\psi \cdot \mathscr{F} f]\|_{p_1} + \left(\sum_{j=1}^{\infty} 2^{jqs_1}\|\mathscr{F}^{-1}[\varphi_j \cdot \mathscr{F} f]\|_{p_1}{}^q\right)^{\frac{1}{q}}$$

$$= \|\mathscr{F}^{-1}[\psi_1 \cdot \psi \mathscr{F} f]\|_{p_1} + \left(\sum_{j=1}^{\infty} 2^{jqs_1}\|\mathscr{F}^{-1}[\psi_{j+1} \cdot \varphi_j \cdot \mathscr{F} f]\|_{p_1}{}^q\right)^{\frac{1}{q}}.$$

By the Young inequality, we have

$$\|f\|_{B^{s_1}_{p_1q}}$$

$$\simeq_n \|\mathscr{F}^{-1}\psi_1 * \mathscr{F}^{-1}[\psi \mathscr{F} f]\|_{p_1} + \left(\sum_{j=1}^{\infty} 2^{jqs_1}\|\mathscr{F}^{-1}\psi_{j+1} * \mathscr{F}^{-1}[\varphi_j \cdot \mathscr{F} f]\|_{p_1}{}^q\right)^{\frac{1}{q}}$$

$$\lesssim \|\mathscr{F}^{-1}[\psi \mathscr{F} f]\|_{p_0} + \left(\sum_{j=1}^{\infty} \|\mathscr{F}^{-1}\psi_{j+1}\|_r{}^q 2^{jqs_1}\|\mathscr{F}^{-1}[\varphi_j \cdot \mathscr{F} f]\|_{p_0}{}^q\right)^{\frac{1}{q}}.$$

Here by (2.18), $2^{js_1}\|\mathscr{F}^{-1}\psi_{j+1}\|_r \simeq 2^{j(s_1+n-\frac{n}{r})} = 2^{j\left(s_1+\frac{n}{p_0}-\frac{n}{p_1}\right)} = 2^{js_0}$, we obtain $\|f\|_{B^{s_1}_{p_1q}} \lesssim \|f\|_{B^{s_0}_{p_0q}}$. Hence we have $B^{s_0}_{p_0q}(\mathbb{R}^n) \hookrightarrow B^{s_1}_{p_1q}(\mathbb{R}^n)$.

Exercises

Exercise 2.3 Let $s \geq 0$, and let $P_t, t > 0$ be the Poisson kernel defined in Definition 1.44. The norm

$$\|f\|_{B^s_{11}} = \|f\|_1 + \sum_{j=1}^{\infty} 2^{js}\|\varphi_j(D)f\|_1,$$

where $\varphi \in C^\infty_c(\mathbb{R}^n)$ is chosen so that $\chi_{B(4)\setminus B(2)} \leq \varphi \leq \chi_{B(8)\setminus B(1)}$.

1. Show that $B^s_{11}(\mathbb{R}^n)$ is embedded into $L^1(\mathbb{R}^n)$ using Proposition 2.1.
2. Show that $P_t \in B^s_{11}(\mathbb{R}^n)$.

Exercise 2.4 Prove Lemma 2.1. Hint: Induct on $|\alpha|$.

Exercise 2.5 Let $m(\xi) \equiv \langle\xi\rangle^{-2}$ for $\xi \in \mathbb{R}^n$. Using $m(\xi) = \sum_{j=1}^{\infty}(-1)^{j+1}\langle\xi\rangle^{-2j}$ for $|\xi| \geq 2$, reprove $(1-\Delta) : B^{s+2}_{pq}(\mathbb{R}^n) \approx B^s_{pq}(\mathbb{R}^n)$. Here we admit that $(1-\Delta) : B^{s+2}_{pq}(\mathbb{R}^n) \approx B^s_{pq}(\mathbb{R}^n)$ is at least continuous.

Textbooks in Sect. 2.1

Sobolev Spaces

See [3], [2, Section 1.2.1], [4, Chapter 3], [6], [25, Chapter 5] and [92, Chapter 3]. In particular, [92, Chapter 3] includes interpolation inequalities of the function spaces. See also [8].

$\mathscr{C}^\gamma(\mathbb{R}^n) \approx B^\gamma_{\infty\infty}(\mathbb{R}^n)$: Theorems 2.7 and 2.8

See [99, p. 110] and [99, p. 113] for Theorems 2.7 and 2.8, respectively.

Brownian Motion

See [89, Chapter 6] for more about the Brownian motion including its existence.

The Besov Space $B^s_{pq}(\mathbb{R}^n)$ with $1 \le p, q \le \infty$ and $s \in \mathbb{R}$

See [7, Chapter 6] and [71, p. 48, Definition 1] for example.

Introductory Textbooks of Besov Spaces

See [2, 32, 71] and [71, p. 232, Definition 1] for the property of Besov spaces.

Introductory Description of Besov Spaces

[100, Chapter 1] overviews the theory of function spaces including the history.

2.2 Besov Spaces in Analysis

Section 2.2.1 considers the Besov space $B^0_{22}(\mathbb{R}^n)$. In fact, we show $B^0_{22}(\mathbb{R}^n) \approx L^2(\mathbb{R}^n)$ in Theorem 2.6. In Sect. 2.2.2 we define Hölder–Zygmund spaces, as well as the notion of a fractional differential. In the theory of Hölder–Zygmund spaces the differential of integer order is less natural than that of fractional order. Furthermore, for $s > 0$, we establish $\mathscr{C}^s(\mathbb{R}^n) \approx B^s_{\infty\infty}(\mathbb{R}^n)$, where $\mathscr{C}^s(\mathbb{R}^n)$ denotes the Hölder–Zygmund space. In Sect. 2.2.3 we present an example of fractal analysis and function spaces and an application of Besov spaces to the Fourier transform.

2.2.1 Sobolev Spaces and Besov Spaces

Compared with the definition of Sobolev spaces, that for Besov spaces is complicated. However, due to the complicated but good definition, the proof becomes systematic.

When we compare Sobolev spaces and Besov spaces, two possibilities are likely:

- Besov spaces completely coincide with Sobolev spaces.
- Besov spaces are suitably replaced by Sobolev spaces; Besov spaces are taken to be close to Sobolev spaces.

More precisely, as for the second case, $W^{m,p}(\mathbb{R}^n)$ are close to $B^m_{p2}(\mathbb{R}^n)$. For $m \in \mathbb{N}$, we define the Sobolev space $W^{m,p}(\mathbb{R}^n)$ by replacing $L^2(\mathbb{R}^n)$-norm with $L^p(\mathbb{R}^n)$-norm in (2.1). Here we do not given any further account. This is just a matter of what we feel. Here let us consider the first case.

2.2.1.1 The Lebesgue Space $L^2(\mathbb{R}^n)$ and the Sobolev Space $W^{2,m}(\mathbb{R}^n)$ with $m \in \mathbb{N}$

By Plancherel's theorem (see Theorem 1.32), we have

$$\mathscr{F} f(\xi) = \lim_{R\to\infty} \frac{1}{\sqrt{(2\pi)^n}} \int_{B(R)} f(x) \mathrm{e}^{-ix\xi} dx \quad (f \in L^2(\mathbb{R}^n))$$

in the topology of $L^2(\mathbb{R}^n)$ and

$$\|\mathscr{F} f\|_2 = \|f\|_2 \quad (f \in L^2(\mathbb{R}^n)).$$

We will give the proof in Theorem 1.32. This yields the following result:

Theorem 2.6 **($B^0_{22}(\mathbb{R}^n) \sim L^2(\mathbb{R}^n)$)** *In the following sense the spaces $B^0_{22}(\mathbb{R}^n)$ and $L^2(\mathbb{R}^n)$ are isomorphic:*

1. *If f belongs to $B^0_{22}(\mathbb{R}^n)$, then it is represented by an $L^2(\mathbb{R}^n)$-function.*
2. $L^2(\mathbb{R}^n) \hookrightarrow B^0_{22}(\mathbb{R}^n)$.
3. *The norm equivalence*

$$\|f\|_{B^0_{22}} \sim \|f\|_2 \tag{2.19}$$

holds for all $f \in L^2(\mathbb{R}^n)$.

Proof Since $\mathscr{S}(\mathbb{R}^n)$ is dense both in $B^0_{22}(\mathbb{R}^n)$ (see Theorem 2.4) and in $L^2(\mathbb{R}^n)$, we have only to show (2.19) for $f \in \mathscr{S}(\mathbb{R}^n)$; three statements will follow automatically from (2.19) with $f \in \mathscr{S}(\mathbb{R}^n)$. We have

$$\|f\|_{B^0_{22}} \simeq_n \|\psi \cdot \mathscr{F}f\|_2 + \left(\sum_{j=1}^{\infty} \|\varphi_j \cdot \mathscr{F}f\|_2{}^2\right)^{\frac{1}{2}}$$

thanks to the definition of the norm and Plancherel's theorem; see Theorem 1.32. Now that $\sqrt{a+b} \le \sqrt{a} + \sqrt{b} \le \sqrt{2(a+b)}$ for $a, b > 0$, we have

$$\begin{aligned}
\|f\|_{B^0_{22}} &\simeq \left(\int_{\mathbb{R}^n} \left[|\psi(\xi)|^2|\mathscr{F}f(\xi)|^2 + \sum_{j=1}^{\infty} |\varphi_j(\xi)|^2|\mathscr{F}f(\xi)|^2\right] \mathrm{d}\xi\right)^{\frac{1}{2}} \\
&\simeq \left(\int_{\mathbb{R}^n} |\mathscr{F}f(\xi)|^2 \mathrm{d}\xi\right)^{\frac{1}{2}} \\
&= \|f\|_2.
\end{aligned}$$

Hence we obtain (2.19).

From this, we obtain the following corollary:

Corollary 2.1 *For $m \in \mathbb{N}$, we have $B^m_{22}(\mathbb{R}^n) \approx W^{m,2}(\mathbb{R}^n)$ in the sense similar to Theorem* 2.6.

Proof Combine Theorems 2.3 and 2.6.

Exercises

Exercise 2.6 Let $1 \le p \le \infty$, $1 \le q_0 \le q_1 \le \infty$, and $s \in \mathbb{R}$. Prove Proposition 2.2. Hint: Look for a similar relation to ℓ^p-spaces.

Exercise 2.7

1. Let $1 \le p, u \le \infty$, $s \in \mathbb{R}$ and $\tau \in C^\infty_{\mathrm{c}}(\mathbb{R}^n)$. Show that $\tau(D)$ maps $L^p(\mathbb{R}^n)$ to $B^s_{pu}(\mathbb{R}^n)$ and that $\tau(D)$ maps $B^s_{pu}(\mathbb{R}^n)$ to $L^p(\mathbb{R}^n)$.
2. Let $1 \le p \le 2 \le q \le \infty$. Suppose that $T : B^0_{p2}(\mathbb{R}^n) \to B^0_{q2}(\mathbb{R}^n)$ is a bounded linear operator. Then using Proposition 2.1 show that $\|Tf\|_q \lesssim \|f\|_p$ for all $f \in L^p(\mathbb{R}^n)(\hookrightarrow B^0_{p2}(\mathbb{R}^n))$.
3. Let $1 \le p \le 2 \le q \le \infty$ and $M \gg 1$. Suppose that $T : B^0_{p\infty}(\mathbb{R}^n) \to B^{-M}_{q\infty}(\mathbb{R}^n)$ is a bounded linear operator. Assume in addition that $\|\varphi_j(D)T\varphi_k(D)\|_{L^p \to L^{p'}} \lesssim 2^{-\delta|j-k|}$ for some $\delta > 0$. Then show that T extends to a bounded linear operator from $B^0_{p2}(\mathbb{R}^n)$ to $B^0_{q2}(\mathbb{R}^n)$ and hence from $L^p(\mathbb{R}^n)$ to $L^q(\mathbb{R}^n)$. In this sense, we are occasionally interested in the local estimate: $\|\varphi_j(D)T\varphi_k(D)\|_{L^p \to L^q} \lesssim 2^{-\delta|j-k|}$.

Exercise 2.8 Let $n \ge 3$ and $1 \le q \le \dfrac{2n}{n-2}$. Disprove that $W^{2,1}(\mathbb{R}^n) \hookrightarrow L^q(\mathbb{R}^n)$ is compact. Hint: Consider translation.

Exercise 2.9 [110, Proposition 6.5] Let $1 \le p, q \le \infty$ and $s \in \mathbb{R}$. Find a necessary and sufficient condition for $\chi_{[0,1]^n}$ to be a member in $B^s_{pq}(\mathbb{R}^n)$. Hint: Reduce matters to the case where $n = 1$.

2.2.2 Hölder–Zygmund Spaces and Besov Spaces

As we have discussed in Sect. 1.3, it is fundamental that we use the ε-δ method when we consider the continuity of functions. However, we have a very bad memory for this. In many practical cases, we encounter Lipschitz functions f satisfying the Lipschitz condition $|f(x) - f(y)| \lesssim |x - y|$ and we prove such an estimate by means of the mean-value theorem. However, the path of Brownian motion fails to satisfy this property. Instead such a path f almost surely satisfies $|f(x) - f(y)| \le C_\alpha |x - y|^\alpha$ for any $\alpha \in (0, 1/2)$. How do we guarantee the continuity of functions in general cases? It is more than enough to have $|f(x) - f(y)| \le C|x - y|^\alpha$ for some $\alpha \in (0, 1]$. We will here see that Besov spaces cover such a class of functions.

2.2.2.1 Hölder–Zygmund Spaces of Order $\gamma \in (0, 1)$

When we consider function spaces having γ smoothness for $\gamma \in (0, 1)$, matters are simple; simply consider the difference of the first order. However, as the example of the a priori estimates of the elliptic differential equations shows, we sometimes need to consider function spaces having higher smoothness. We need to go step by step. First of all, we define the Hölder–Zygmund space $\mathscr{C}^\gamma(\mathbb{R}^n)$ for $0 < \gamma < 1$.

Definition 2.4 (Hölder–Zygmund space $\mathscr{C}^\gamma(\mathbb{R}^n)$, $0 < \gamma < 1$) For $0 < \gamma < 1$ and a bounded continuous function $f : \mathbb{R}^n \to \mathbb{C}$, define

$$\|f\|_{\mathscr{C}^\gamma} \equiv \|f\|_\infty + \sup_{x,y \in \mathbb{R}^n, x \neq y} \frac{|f(x) - f(y)|}{|x - y|^\gamma}.$$

The *Hölder–Zygmund space* $\mathscr{C}^\gamma(\mathbb{R}^n)$ is the set of all bounded continuous functions f for which the norm $\|f\|_{\mathscr{C}^\gamma}$ is finite.

One can consider the space $\mathscr{C}^\gamma(\mathbb{R}^n)$ as the generalization of $\mathrm{Lip}(\mathbb{R}^n)$ made up of all the functions f satisfying

$$\|f\|_{\mathrm{Lip}} \equiv \sup_{x,y \in \mathbb{R}^n, x \neq y} \frac{|f(x) - f(y)|}{|x - y|} < \infty.$$

Note that we omitted $\|f\|_\infty$ in the above definition in order to make $\mathrm{Lip}(\mathbb{R}^n)$ a Banach space of functions modulo constants.

We have the following function $\mathscr{C}^\gamma(\mathbb{R}^n)$ as an example. A *linear spline* on I is a piecewise linear function which is not differentiable at finitely many points called *nodes*. If I is a compact interval, then the boundary points are by definition nodes.

Example 2.2 (Cantor function) Starting from the closed line segment

$$L = \{(t, a\,t + b) \,:\, t_0 \le t \le t_1\}$$

in $\mathbb{R}^2$, we define L_1, L_2, L_3, as follows:

$$L_1 \equiv \left\{\left(t, \frac{3}{2}a\,(t - t_0) + a\,t_0 + b\right) \in \mathbb{R}^2 \,:\, t_0 \le t \le \frac{2t_0 + t_1}{3}\right\}$$

$$L_2 \equiv \left\{\left(t, a\,\frac{t_0 + t_1}{2} + b\right) \in \mathbb{R}^2 \,:\, \frac{2t_0 + t_1}{3} \le t \le \frac{t_0 + 2t_1}{3}\right\}$$

$$L_3 \equiv \left\{\left(t, \frac{3}{2}a\,(t - t_1) + a\,t_1 + b\right) \in \mathbb{R}^2 \,:\, \frac{t_0 + 2t_1}{3} \le t \le t_1\right\}.$$

Define inductively $L_{12} \equiv (L_1)_2$ and so on. For $\star_1, \star_2, \ldots, \star_k \in \{1, 2, 3\}$, define $L_{\star_1\star_2\ldots\star_k}$ likewise. We start from $L = \{(t, t) \,:\, 0 \le t \le 1\}$ we can define 3^k closed segments $A_k \equiv \{L_{\star_1\star_2\ldots\star_k} \,:\, \star_1, \star_2, \ldots, \star_k \in \{1, 2, 3\}\}$ and connect them to have a linear spline $y = f_k(x)$, so that $f_k : [0, 1] \to [0, 1]$ has nodes at $3^{-k}\mathbb{Z} \cap [-1, 1]$. Extend continuously on $[1, \infty)$ and $(-\infty, 0]$ so that f_k is a constant. Then:

1. $\{f_k\}_{k=1}^\infty$ is convergent uniformly to a function f.
2. If $\gamma = \log_3 2$, then $\|f_k\|_{\mathscr{C}^\gamma} = \|f\|_{\mathscr{C}^\gamma} = 2$.

The limit function f is said to be the *Cantor function.*.

2.2.2.2 Hölder–Zygmund Spaces and Besov Spaces

As the following theorem shows, Besov spaces cover Hölder–Zygmund spaces.

Theorem 2.7 (Norm equivalence for Hölder–Zygmund spaces) *Let* $0 < \gamma < 1$. *Then* $\mathscr{C}^\gamma(\mathbb{R}^n) \approx B^\gamma_{\infty\infty}(\mathbb{R}^n)$ *with equivalence of norms.*

Proof We fix ψ and φ so that they satisfy (2.7). Let $f \in \mathscr{C}^\gamma(\mathbb{R}^n)$. Namely, we suppose that f is a bounded function satisfying the inequality: $|f(x - y) - f(x)| \le |y|^\gamma \|f\|_{\mathscr{C}^\gamma}$.

Then since $\mathscr{F}^{-1}\varphi_j \perp \mathscr{P}_0(\mathbb{R}^n)$, we have

$$(2\pi)^{\frac{n}{2}}\varphi_j(D)f(x) = f * \mathscr{F}^{-1}\varphi_j(x) = \int_{\mathbb{R}^n} (f(x - y) - f(x))\mathscr{F}^{-1}\varphi_j(y)dy.$$

Note that $\mathscr{F}[f(a\star)] = a^{-n}\mathscr{F}f(a^{-1}\star)$ for any $a > 0$. Hence a change of variables yields

$$
\begin{aligned}
\|\varphi_j(D)f\|_\infty &\lesssim \left(\int_{\mathbb{R}^n} |y|^\gamma \cdot |\mathscr{F}^{-1}\varphi_j(y)|dy\right) \|f\|_{\mathscr{C}^\gamma} \\
&= 2^{jn}\left(\int_{\mathbb{R}^n} |y|^\gamma \cdot |\mathscr{F}^{-1}\varphi(2^j y)|dy\right) \|f\|_{\mathscr{C}^\gamma} \\
&= 2^{-j\gamma}\left(\int_{\mathbb{R}^n} |z|^\gamma \cdot |\mathscr{F}^{-1}\varphi(z)|dz\right) \|f\|_{\mathscr{C}^\gamma} \\
&\simeq 2^{-j\gamma}\|f\|_{\mathscr{C}^\gamma}.
\end{aligned}
$$

By the Young inequality $\|\psi(D)f\|_\infty \lesssim \|f\|_{\mathscr{C}^\gamma}$ follows trivially, which shows that $f \in B^\gamma_{\infty\infty}(\mathbb{R}^n)$.

If $f \in B^\gamma_{\infty\infty}(\mathbb{R}^n)$, we have

$$
f = \psi(D)f + \sum_{j=1}^{\infty} \varphi_j(D)f
$$

as we did in the proof of Proposition 2.1. As we have seen in Proposition 2.4, we have the $L^\infty(\mathbb{R}^n)$-norm convergence: We are left with the task of proving Lipschitz continuity. To this end, since we know that f is bounded, it suffices to show that

$$
\sup_{x\in\mathbb{R}^n} |f(x-y) - f(y)| \lesssim |y|^\gamma \|f\|_{B^\gamma_{\infty\infty}} \tag{2.20}
$$

for $y \in B(1) \setminus \{0\}$. Let $x \in \mathbb{R}^n$ be fixed. Write

$$
\mathrm{I} \equiv \psi(D)f(x-y) - \psi(D)f(x), \quad \mathrm{II}_j \equiv \varphi_j(D)f(x-y) - \varphi_j(D)f(x).
$$

We decompose $f(x-y) - f(x)$ into three parts:

$$
f(x-y) - f(x) = \mathrm{I} + \sum_{j=1}^{[\log_2 |y|^{-1}]} \mathrm{II}_j + \sum_{j=[1+\log_2 |y|^{-1}]}^{\infty} \mathrm{II}_j.
$$

By the integral form of the mean-value theorem, II_j satisfies:

$$
|\mathrm{II}_j| \le \int_0^1 |y \cdot \nabla[\varphi_j(D)f](x-ty)|dt \le \sum_{l=1}^{n} |y| \cdot \|\partial_l[\varphi_j(D)f]\|_\infty.
$$

Take an auxiliary function $\kappa \in \mathscr{S}(\mathbb{R}^n)$ so that $\chi_{B(8)} \le \kappa \le \chi_{B(16)}$. If we set

$$\kappa_{(l)}(\xi) \equiv \xi_l \kappa(\xi)$$

for $\xi \in \mathbb{R}^n$, then a Fourier transform of both sides yields

$$\partial_l \varphi_j(D) f \simeq (\kappa_{(l)}\varphi)_j(D) f \simeq 2^{jn}\mathscr{F}^{-1}\kappa_{(l)}(2^j\star) * \varphi_j(D) f. \tag{2.21}$$

Hence we have

$$\begin{aligned}|\mathrm{II}_j| &\lesssim 2^{jn}\sum_{l=1}^{n}|y|\cdot\|\mathscr{F}^{-1}\kappa_{(l)}(2^j\star) * \varphi_j(D) f\|_\infty \\ &= 2^{jn}\sum_{l=1}^{n}|y|\cdot\|\mathscr{F}^{-1}\kappa_{(l)}(2^j\star)\|_1\|\varphi_j(D) f\|_\infty \\ &\lesssim |y|\cdot\|\varphi_j(D) f\|_\infty.\end{aligned}$$

Since $\gamma < 1$, we obtain

$$\sum_{j=1}^{[\log_2|y|^{-1}]}|\mathrm{II}_j| \lesssim |y|\sum_{j=1}^{[\log_2|y|^{-1}]}2^{j(1-\gamma)}\|f\|_{B^\gamma_{\infty\infty}} \lesssim |y|^\gamma\|f\|_{B^\gamma_{\infty\infty}}. \tag{2.22}$$

We deal with $\sum_{j=[1+\log_2|y|^{-1}]}^{\infty}\mathrm{II}_j$. By the fact $|\mathrm{II}_j| \le 2\|\varphi_j(D) f\|_\infty \le 2^{1-j\gamma}\|f\|_{B^\gamma_{\infty\infty}}$, we have

$$\left|\sum_{j=[1+\log_2|y|^{-1}]}^{\infty}\mathrm{II}_j\right| \lesssim |y|^\gamma\|f\|_{B^\gamma_{\infty\infty}}.$$

Since we have a similar estimate for I, we obtain (2.20).

2.2.2.3 Hölder–Zygmund Spaces of Order $\gamma \in (0, 2)$

We define $\mathscr{C}^\gamma(\mathbb{R}^n)$ with $0 < \gamma < 2$. Definition 2.5 below overlaps Definition 2.4, when $0 < \gamma < 1$. We will check that the two definitions are consistent.

Definition 2.5 (Hölder–Zygmund space $\mathscr{C}^\gamma(\mathbb{R}^n)$, $0 < \gamma < 2$) For $0 < \gamma < 2$ and a bounded continuous function f, define

$$\|f\|_{\mathscr{C}^\gamma} \equiv \|f\|_\infty + \sup_{x,y\in\mathbb{R}^n, y\neq 0}|y|^{-\gamma}|f(x+y) - 2f(x) + f(x-y)|.$$

The *Hölder–Zygmund* space $\mathscr{C}^\gamma(\mathbb{R}^n)$ is the set of all bounded continuous functions f for which the norm $\|f\|_{\mathscr{C}^\gamma}$ is finite.

For $0 < \gamma < 1$, two different definitions for $\mathscr{C}^\gamma(\mathbb{R}^n)$ are equivalent, as the next theorem shows; two different definitions turn out to be compatible.

Theorem 2.8 (Norm equivalence for Hölder–Zygmund spaces) *For $0 < \gamma < 2$, define the norm of $\mathscr{C}^\gamma(\mathbb{R}^n)$ via Definition 2.5. Then $\mathscr{C}^\gamma(\mathbb{R}^n) \approx B^\gamma_{\infty\infty}(\mathbb{R}^n)$ with equivalence of norms.*

Proof Let $f \in \mathscr{C}^\gamma(\mathbb{R}^n)$. That is, f is bounded and satisfies the inequality

$$\sup_{x\in\mathbb{R}^n} |f(x+y) - 2f(x) + f(x-y)| \le |y|^\gamma \|f\|_{\mathscr{C}^\gamma}.$$

The definition of the norms does not depend upon the admissible choice of ψ and φ. So it can be assumed that ψ and φ are even. Then

$$\int_{\mathbb{R}^n} \mathscr{F}^{-1}\varphi_j(y)\mathrm{d}y = (2\pi)^{\frac{n}{2}}\mathscr{F}[\mathscr{F}^{-1}\varphi_j](0) = (2\pi)^{\frac{n}{2}}\varphi_j(0) = 0. \tag{2.23}$$

Hence we have

$$\varphi_j(D)f(x) \simeq f * \mathscr{F}^{-1}\varphi_j(x) = \frac{1}{2}\int_{\mathbb{R}^n}(f(x+y) - 2f(x) + f(x-y))\mathscr{F}^{-1}\varphi_j(y)\mathrm{d}y.$$

As a consequence, again by $\mathscr{F}[f(a\star)] = a^{-n}\mathscr{F}f(a^{-1}\star), a > 0$ and change variables, we obtain

$$\|\varphi_j(D)f\|_\infty \lesssim \int_{\mathbb{R}^n} |y|^\gamma \cdot |\mathscr{F}^{-1}\varphi_j(y)|\mathrm{d}y = 2^{jn}\int_{\mathbb{R}^n} |y|^\gamma \cdot |\mathscr{F}^{-1}\varphi(2^j y)|\mathrm{d}y \simeq 2^{-j\gamma}.$$

Here if we use the Young inequality, then

$$\|\psi(D)f\|_\infty \simeq_n \|\mathscr{F}^{-1}\psi * f\|_\infty \le \|\mathscr{F}^{-1}\psi\|_1 \cdot \|f\|_\infty \lesssim \|f\|_{\mathscr{C}^\gamma},$$

which shows that $f \in B^\gamma_{\infty\infty}(\mathbb{R}^n)$.

Conversely, if $f \in B^\gamma_{\infty\infty}(\mathbb{R}^n)$, we can use the same argument as in Theorem 2.7 to prove $f \in \mathscr{C}^\gamma(\mathbb{R}^n)$.

We end this section with a couple of examples.

Example 2.3 Let $0 < \gamma < 2$ and $m \in \mathbb{N}$. Then define

$$\mathscr{C}^{m+\gamma}(\mathbb{R}^n) \equiv \bigcap_{\alpha\in\mathbb{N}_0{}^n,\, |\alpha|\le m} \{f \in \mathscr{C}^\gamma(\mathbb{R}^n) : \partial^\alpha f \in \mathscr{C}^\gamma(\mathbb{R}^n)\}.$$

Then:

- If $m \ge 1$ we have an equivalent definition of $\mathscr{C}^{m+\gamma}(\mathbb{R}^n)$:

$$\mathscr{C}^{m+\gamma}(\mathbb{R}^n) = \bigcap_{\alpha \in \mathbb{N}_0{}^n,\ |\alpha| \le m-1} \{f \in \mathscr{B}^{m-1}(\mathbb{R}^n) \,:\, \partial^\alpha f \in \mathscr{C}^{1+\gamma}(\mathbb{R}^n)\}.$$

Here $\mathscr{B}^K(\mathbb{R}^n)$ stands for the set of all $C^K(\mathbb{R}^n)$ functions whose derivatives up to order K are bounded.

- The equivalence $\|f\|_{\mathscr{C}^{m+\gamma}} \sim \|f\|_{\mathscr{C}^\gamma} + \sum_{j=1}^{n} \|\partial_{x_j}{}^m f\|_{\mathscr{C}^\gamma}$ holds for $f \in \mathscr{C}^{m+\gamma}(\mathbb{R}^n)$.

For the proof, we use Theorem 2.3, the lifting property and $B^\gamma_{\infty\infty}(\mathbb{R}^n) \approx \mathscr{C}^\gamma(\mathbb{R}^n)$.

Examples 2.4 and 2.5 show that $\mathrm{Lip}(\mathbb{R}^n)$ and $\mathscr{C}^1(\mathbb{R}^n)$ are different function spaces.

Example 2.4 Let $n = 1$. Define $f(t) \equiv \int_0^t \log|s|\,ds$, $t \in \mathbb{R}$. Take a smooth cut-off function $\rho : \mathbb{R} \to \mathbb{R}$ so that $\chi_{(-2,2)} \le \rho \le \chi_{(-3,3)}$. Note that $g = f \cdot \rho$ is not a Lipschitz continuous function. Let us show that $f \in \mathscr{C}^1(\mathbb{R}^n)$. Write $I_1 = (-\infty, 1]$, $I_2 = [-2, 2]$, $I_3 \in [1, \infty)$. Let $x \in \mathbb{R}$ and $y > 0$.We consider three model cases: other cases are similar to these cases. Assume first that $x + xy, x - xy \in I_1$. Then since g' is bounded on I_1, by the use of the mean-value theorem we obtain the desired result. Assume next that $x + xy, x - xy \in I_2$. Since $g = f$ on I_2, we have

$$\begin{aligned} g(x+xy) - 2g(x) + g(x-xy) &= x \int_0^y \big(g'(x+xu) - g'(x-xu)\big)\,du \\ &= x \int_0^y \log\left|\frac{1+u}{1-u}\right| du. \end{aligned}$$

Thus $|g(x+xy) - 2g(x) + g(x-xy)| \lesssim |xy|$. Assume finally that $x + xy \in I_1 \setminus I_2$ and that $x - xy \in I_2 \setminus I_1$. Then $|xy| \ge 1$. Since g is bounded, we still have $|g(x + xy) - 2g(x) + g(x-xy)| \lesssim |xy|$. Thus in total $|g(x+z) - 2g(x) + g(x-z)| \lesssim |z|$ if $x \ne 0$. Continuity of f allows us to include the case of $x = 0$. Hence $g \in \mathscr{C}^1(\mathbb{R}^n)$.

Example 2.5 In Example 2.4 by letting $F(x_1, x_2, \ldots, x_n) \equiv g(x_n)$, we can construct an example in $\mathbb{R}^n$ similar to Example 2.4. We leave these matters to interested readers; see Exercise 2.13.

Exercises

Exercise 2.10 Define $f : \mathbb{R} \to \mathbb{R}$ by $f(t) \equiv e^t \sin(e^t) = (-\cos e^t)'$ for $t \in \mathbb{R}$.

1. Show that $f \in \mathscr{S}'(\mathbb{R}^n)$ using Definition 1.16.
2. Show that $f \in B^{-1}_{\infty\infty}(\mathbb{R}^n)$ using Theorem 2.2 and Proposition 2.1.

See [81, p. 245, Example] for another example of f; in fact Schwartz considered $f(t) = \exp(i\pi t^2)$ there.

Exercise 2.11

1. Let $f : \mathbb{R} \to \mathbb{R}$ be continuous. Show that the solution u to $-u'' = f$ is $C^2(\mathbb{R})$.
2. Define $u(x_1, x_2) \equiv \sqrt{{x_1}^2 + {x_2}^2} \cdot \log({x_1}^2 + {x_2}^2)$ for $(x_1, x_2) \in \mathbb{R}^2$. Calculate $f \equiv -\Delta u$ in the sense of $\mathscr{S}'(\mathbb{R}^2)$ by using the Stokes theorem and check that Δf is continuous but that f is not $C^2(\mathbb{R}^2)$. The result shows that the space $C^2(\mathbb{R}^2)$ is not sufficient when we consider the Laplace equation.

Exercise 2.12 Show that $g \in \mathscr{C}^1(\mathbb{R})\setminus\mathrm{Lip}(\mathbb{R})$ in Example 2.4 using the Rademacher theorem that asserts that any function in $\mathrm{Lip}(\mathbb{R})$ is differentiable almost everywhere and the derivative is in $L^\infty(\mathbb{R})$.

Exercise 2.13 Take a $C^\infty(\mathbb{R})$-function $\kappa_0 : \mathbb{R} \to \mathbb{R}$ so that $\chi_{[-1,1]} \le \kappa_0 \le \chi_{[-2,2]}$. Let $\kappa(x) \equiv \kappa_0 \otimes \kappa_0 \otimes \cdots \otimes \kappa_0(x) \equiv \kappa_0(x_1)\kappa_0(x_2)\cdots\kappa_0(x_n)$ for $x = (x_1, x_2, \dots, x_n) \in \mathbb{R}^n$. Use $\psi \equiv \kappa$, $\varphi = \kappa - \kappa_{-1}$ to define the Besov norm $\| \star \|_{B^s_{pq}}$ by (2.3). Using this Besov norm, prove the assertion of Example 2.5.

Exercise 2.14 [1114, Lemma 3] Let $0 < \alpha \le 1$ and $\varepsilon > 0$. The *Uchiyama class* $\mathscr{U}_{(\alpha,\varepsilon)}$ is the set of all the functions $\phi : \mathbb{R}^n \to \mathbb{R}$ such that

$$|\phi(x)| \le \frac{1}{(1+|x|)^{n+\varepsilon}}, \quad |\phi(x) - \phi(x')| \le \frac{|x-x'|^\alpha}{(1+|x|)^{n+\varepsilon+\alpha}} + \frac{|x-x'|^\alpha}{(1+|x'|)^{n+\varepsilon+\alpha}}$$

for all $x, x' \in \mathbb{R}^n$ and that $\mathscr{F}\phi(0) = 0$.

1. Let $\psi \in \mathscr{U}_{(\alpha,\varepsilon)}$, and let $\rho \in C^\infty_c(B(1))$. Define
$$C_k \equiv -\frac{1}{\|\rho_k\|_1}\left(\int_{\mathbb{R}^n} \sum_{l=k+1}^{\infty} \rho_l(y)\psi(y)\mathrm{d}y\right).$$
Show that $|C_k| \lesssim 2^{-k(n+\varepsilon)}$.
2. Let $\mu = 1 - \rho_1 - \rho_2 - \cdots$. We let $g_k \equiv C_k\rho_k$ for $k \in \mathbb{N}$ and define $g_0 = C_0\mu$. Show that
$$\int_{\mathbb{R}^n} (g_k(x) - g_{k-1}(x))\mathrm{d}x = \int_{\mathbb{R}^n} \psi(x)\rho_k(x)\mathrm{d}x$$
for all $k \in \mathbb{N}$ and that
$$\int_{\mathbb{R}^n} g_0(x)\mathrm{d}x = \int_{\mathbb{R}^n} \psi(x)\mu(x)\mathrm{d}x.$$
3. Set $\phi_0 \equiv \psi\mu - g_0$ and $\phi_k \equiv \psi\rho_k - g_k + g_{k-1}$ for $k \in \mathbb{N}$. For any $k \in \mathbb{N}_0$, show that $\phi_k \perp \mathscr{P}_0(\mathbb{R}^n)$, that $\mathrm{supp}(\phi_k) \subset B(2^k)$ and that $\|\phi_k\|_{\mathrm{Lip}} \lesssim 2^{-k(n+\alpha)}$.

Exercise 2.15 [1114, p. 788, Free lunch lemma] Let $0 < \alpha \le 1$ and $\varepsilon > 0$. Suppose $\alpha' \in (0, \alpha] \cap (0, \varepsilon)$. Set $\varepsilon' \equiv \varepsilon - \alpha'$. Assume that a continuous function ϕ satisfies

$$|\phi(x)| \le \frac{1}{(1+|x|)^{n+\varepsilon}}, \quad |\phi(x) - \phi(x')| \le \frac{|x-x'|^{\alpha}}{(1+|x|)^{n+\varepsilon}} + \frac{|x-x'|^{\alpha}}{(1+|x'|)^{n+\varepsilon}}$$

for all $x, x' \in \mathbb{R}^n$. Then show that

$$|\phi(x) - \phi(x')| \lesssim |x-x'|^{\alpha'} \langle x \rangle^{-n-\varepsilon'-\alpha'} + |x-x'|^{\alpha'} \langle x' \rangle^{-n-\varepsilon'-\alpha'}$$

for all $x, x' \in \mathbb{R}^n$.

Exercise 2.16 [806] For $0 < \gamma < 2$ and a function f, define

$$\|f\|_{\dot{\mathscr{C}}^{\gamma}} \equiv \sup_{x,y \in \mathbb{R}^n, y \ne 0} |y|^{-\gamma} |f(x+y) - 2f(x) + f(x-y)|.$$

Construct a discontinuous function f for which $\|f\|_{\dot{\mathscr{C}}^{\gamma}} = 0$.

2.2.3 Applications to Fractals and the Fourier Transform

The fractal arises naturally in many branches of science. However, due to its complicated structure, it is very hard to describe what is inside the fractals. Here we will take up two examples showing that Besov spaces are useful when we describe the complexity. Besov spaces will let us know what is more than continuity.

2.2.3.1 Weierstrass Function and Besov Spaces

Let us see the notion of fractional differentiability by way of the Weierstrass function as an example. Recall that the Weierstrass function is nowhere differentiable.

Example 2.6 Let $1 < a \le 2$ and we will work in the setting of real line $\mathbb{R}$. Let

$$f(t) \equiv \sum_{j=2}^{\infty} a^{-j} \cos(2^j t) \quad (t \in \mathbb{R}). \tag{2.24}$$

The function f is a continuous function called the *Weierstrass function* but $1 < a \le 2$ this is known to be nowhere differentiable like the well-known Takagi function. See [34, 46] for more about the Takagi function. Let us investigate how continuous/discontinuous f is. When we consider such a subtle continuity, the Hölder–Zygmund norm plays a key role. In this example we can calculate the Fourier transform of f:

$$\mathscr{F} f \simeq_n \sum_{j=2}^{\infty} a^{-j}(\delta_{2^j} + \delta_{-2^j}). \tag{2.25}$$

To measure the Besov norm skillfully, we choose $\psi, \varphi \in \mathscr{S}(\mathbb{R}^n)$ so that

$$\chi_{B(2)} \le \psi \le \chi_{B(3)}, \quad \varphi = \psi - \psi_{-1} = \psi - \psi(2\star). \tag{2.26}$$

By virtue of (2.25) and (2.26),

$$\psi(D)f(t) = 0, \quad \varphi_j(D)f(t) = a^{-j}\cos(2^{j+1}t), \quad j \in \mathbb{N}, \quad t \in \mathbb{R}. \tag{2.27}$$

Hence from Theorem 2.7, we deduce

$$\|f\|_{\mathscr{C}^{\log_2 a}} \sim \|f\|_{B^{\log_2 a}_{\infty\infty}} = \|\psi(D)f\|_\infty + \sup_{j\in\mathbb{N}} 2^{j\log_2 a}\|\varphi_j(D)f\|_\infty < \infty$$

$$\|f\|_{\mathscr{C}^{\varepsilon+\log_2 a}} \sim \|f\|_{B^{\varepsilon+\log_2 a}_{\infty\infty}} = \|\psi(D)f\|_\infty + \sup_{j\in\mathbb{N}} 2^{j(\varepsilon+\log_2 a)}\|\varphi_j(D)f\|_\infty = \infty$$

for all $\varepsilon > 0$. Hence we conclude $f \in \mathscr{C}^{\log_2 a}(\mathbb{R}^n) \setminus \mathscr{C}^{\varepsilon+\log_2 a}(\mathbb{R}^n)$ for $\varepsilon > 0$.

2.2.3.2 Brownian Motion and Besov Spaces

A probability space is a measure space with the total measure 1. The Brownian motion is one of the most attractive objects in many branches of science, not only in mathematics. Mathematically, the definition is given as follows.

Definition 2.6 Let $(\Omega, \mathscr{F}, P)$ be a probability space. A mapping $X : \Omega \times [0, \infty) \to \mathbb{R}$ is said to be a Brownian motion starting from 0, if:

(A) $X(0, \star) = 0$.
(B) $X(\star, \omega)$ is a continuous function for almost sure $\omega \in \Omega$.
(C) If $s > t$, then the distribution of $X(s, \star) - X(t, \star)$ is the normal distribution of mean zero and variance $s - t$:

$$P(\omega \in \Omega : X(s,\omega) - X(t,\omega) \ge a) = \int_a^\infty \frac{1}{\sqrt{2\pi(s-t)}} \exp\left(-\frac{v^2}{2(s-t)}\right) dv.$$

(D) If $t_1 < t_2 < \cdots < t_d$ is an increasing sequence, then

$$X(t_d, \star) - X(t_{d-1}, \star), X(t_{d-1}, \star) - X(t_{d-2}, \star), \ldots, X(t_2, \star) - X(t_1, \star)$$

is independent.

It is known that a Brownian motion does exist for a suitably chosen probability space $(\Omega, \mathscr{F}, P)$; see Exercise 2.20. Recall also that a probability space $(\Omega, \mathscr{F}, P)$ is said to be complete if $A \in \mathscr{F}$ for all $A, B \subset \Omega$ whenever $B \in \mathscr{F}$, $P(B) = 0$ and $A \subset B$.

2.2.3.3 Refinements of the Riemann–Lebesgue Theorem

One of the elementary mathematical problems that have been tormenting us is the properties of the mathematical transform. This applies both to the theoretical point of view and to that of applications. Among them we want to investigated more about the Fourier series, the Fourier transform and the Laplace transform, all of which are quite fundamental mathematical transforms. For example, there are many problems for the Fourier transform in several variables that need to be settled. Here let us see how Besov spaces describe the boundedness property of the Fourier transform. Recall that the classical Riemann–Lebesgue theorem asserts that the Fourier transform maps $L^1(\mathbb{R}^n)$-functions to the set of uniformly continuous bounded functions that decay to 0 as $|x| \to \infty$; see [22, Chapter 1] It seems that no continuity property can be expected beyond this estimate. However, under the framework of Besov spaces, the result improves.

Theorem 2.9 (A refinement of the Riemann–Lebesgue theorem) *The Fourier transform $\mathscr{F}$ sends $L^1(\mathbb{R}^n)$-functions to the space $B^0_{\infty 1}(\mathbb{R}^n)$ continuously.*

If we combine Proposition 2.4 with Theorem 2.9, we learn that Theorem 2.9 turns out to be stronger than the Riemann–Lebesgue theorem, the well-known fact that $\mathscr{F}$ maps $L^1(\mathbb{R}^n)$ to $\mathrm{BUC}(\mathbb{R}^n)$ continuously.

Furthermore, the assertion if $f \in L^1(\mathbb{R}^n)$, then, $\mathscr{F}f$ converges to 0 as $|x| \to \infty$ follows immediately from the density of $C^\infty_{\mathrm{c}}(\mathbb{R}^n)$ in $L^1(\mathbb{R}^n)$. As is seen from Theorem 2.9 or the corollary to the Riemann–Lebesgue theorem, this theorem is interesting in that this is a self-improvement of the Riemann–Lebesgue theorem.

Proof Choose $\psi, \varphi \in \mathscr{S}(\mathbb{R}^n)$ so that $\chi_{B(1)} \le \psi \le \chi_{B(2)}$ Set $\varphi \equiv \psi - \psi(2\star)$. Since $\mathscr{F}f = \mathscr{F}^{-1}f(-\star)$, we can concentrate on $\mathscr{F}^{-1}$ instead of $\mathscr{F}$. From the definition we have

$$\|\mathscr{F}^{-1}f\|_{B^0_{\infty 1}} = \|\mathscr{F}^{-1}[\psi \cdot f]\|_\infty + \sum_{j=1}^{\infty} \|\mathscr{F}^{-1}[\varphi_j \cdot f]\|_\infty \lesssim \|\psi \cdot f\|_1 + \sum_{j=1}^{\infty} \|\varphi_j \cdot f\|_1.$$

The most right-hand side is smaller than $\|f\|_1$. The result follows. Here for the first inequality, we used the Riemann–Lebesgue theorem.

We have the following version: Note that $L^1(\mathbb{R}^n) \hookrightarrow B^0_{1\infty}(\mathbb{R}^n)$; see Proposition 2.1.

Theorem 2.10 (A sufficient condition for the image of the Fourier transform to be bounded) *The Fourier transform $\mathscr{F}$ sends $B^0_{1\infty}(\mathbb{R}^n)$-functions to $L^\infty(\mathbb{R}^n)$ continuously.*

Proof We content ourselves with proving $\|\mathscr{F} f\|_\infty \lesssim \|f\|_{B^0_{1\infty}}$ for $f \in \mathscr{S}(\mathbb{R}^n)$. The general case can be handled by the use of the limiting argument together with the Banach–Alaoglu theorem. As usual we will choose the functions $\psi, \varphi \in \mathscr{S}(\mathbb{R}^n)$ satisfying $\chi_{B(1)} \le \psi \le \chi_{B(2)}$ and that $\varphi = \psi - \psi(2\star)$, so that

$$\|\mathscr{F} f\|_\infty \le \|\psi \cdot \mathscr{F} f\|_\infty + \sup_{j \in \mathbb{N}} \|\varphi_j \cdot \mathscr{F} f\|_\infty.$$

By the classical Riemann–Lebesgue theorem, we have

$$\begin{aligned}
\|\mathscr{F} f\|_\infty &\le \left\|\mathscr{F}[\mathscr{F}^{-1}[\psi \cdot \mathscr{F} f]]\right\|_\infty + \sup_{j \in \mathbb{N}} \left\|\mathscr{F}[\mathscr{F}^{-1}[\varphi_j \cdot \mathscr{F} f]]\right\|_\infty \\
&\lesssim \|\mathscr{F}^{-1}[\psi \cdot \mathscr{F} f]\|_1 + \sup_{j \in \mathbb{N}} \|\mathscr{F}^{-1}[\varphi_j \cdot \mathscr{F} f]\|_1 \\
&= \|f\|_{B^0_{1\infty}}.
\end{aligned}$$

Therefore, the proof is complete.

On the probability space, a random variable is a measurable function. As we mentioned, Besov spaces are useful when we study the path of the Brownian motion. See Exercises 2.20 and 2.21 below.

Exercises

Exercise 2.17 For $j \in \mathbb{Z}^n$ and a continuous function $f \in C(\mathbb{T}^n)$, we set

$$a_j \equiv \frac{1}{(2\pi i)^n} \int_{[0,2\pi)^n} f(y) e^{ijy} dy.$$

1. Show $|a_j| \le \|f\|_{\mathrm{BC}}$.
2. Show $|a_j| \lesssim (1 + |j|)^{-m} \|f\|_{\mathscr{B}^m}$.
3. Let $\theta \ge 0$. Show $|a_j| \lesssim (1 + |j|)^{-\theta} \|f\|_{B^\theta_{\infty\infty}}$.

4. (Bernstein's theorem) Let $n = 1$. Let also $\frac{1}{2} < \theta \le 1$. Using $\sum_{j=-\infty}^{\infty} |a_j|^2 < \infty$, if f satisfies the Lipschitz continuity of order θ: $\|f\|_{\mathrm{Lip}(\theta)} \equiv \sup_{-\infty<s<t<\infty} \frac{|f(s)-f(t)|}{(t-s)^\theta} < \infty$, show that $\sum_{j=-\infty}^{\infty} a_j e^{ij\star}$ converges uniformly.

Exercise 2.18 Show that $f * g \in B^0_{\infty 1}(\mathbb{R}^n)$ for all $f, g \in L^2(\mathbb{R}^n)$.

Exercise 2.19 Show that the Fourier transform $\mathscr{F}$ maps $B^{\frac{n}{2}}_{21}(\mathbb{R}^n)$ to $L^1(\mathbb{R}^n)$ continuously by the use of the Plancherel theorem: $\mathscr{F} : B^{\frac{n}{2}}_{21}(\mathbb{R}^n) \to L^1(\mathbb{R}^n)$.

Exercise 2.20 (Smoothness of the paths of Brownian motion) Let $(\Omega, \mathscr{F}, P)$ be a probability space. Suppose that we have an independent sequence $\{X_n\}_{n=1}^{\infty}$ of random variables such that the distribution of $X_n : \Omega \to \mathbb{R}$ is the standard normal distribution. Define the Brownian motion by

$$B(t, \omega) = a \sum_{n=1}^{\infty} \frac{X_n(\omega)}{n} \sin nt \tag{2.28}$$

for $(t, \omega) \in (0, \infty) \times \Omega$, where $a > 0$ is a normalization constant. Choose $\psi \in \mathscr{S}(\mathbb{R}^n)$ so that $\chi_{B(1)} \le \mathscr{F}\psi \le \chi_{B(2)}$. Let $0 < \varepsilon < 1$.

1. For any $t > 0$, prove that the series defining $B(t, \omega)$ converges in $L^2(P)$.
2. Choose a so that the variance of $B(t, \star)$ is t for any $0 < t < 1$.
3. Let a be chosen as above. Show that the distribution of $t^{-\frac{1}{2}} B(t, \star)$ is the standard normal distribution.
4. Let $0 = t_0 \le t_1 < t_2 < \cdots < t_N = 1$. Show that $\{B(t_j, \star) - B(t_{j-1}, \star)\}_{j=1}^N$ is independent.
5. Using the Borel–Cantelli theorem, prove that $P(X_n > \log n$ infinitely often$) = 0$.
6. Here and below we extend B to $\mathbb{R} \times \Omega$ by (2.28). Let $0 < \varepsilon < 1$. Show that $\psi \cdot B(\star, \omega) \in B^{1-\varepsilon}_{22}(\mathbb{R})$ for almost sure $\omega \in \Omega$.
7. Show that $\psi \cdot B(\star, \omega) \in \mathscr{C}^{1/2-\varepsilon}(\mathbb{R})$ for almost sure $\omega \in \Omega$. Hint: Combine Theorems 2.5 and 2.7.

Exercise 2.21 Let $(\Omega, \mathscr{F}, P)$ be a complete probability space and let $X : \Omega \times [0, \infty) \to \mathbb{R}$ be a Brownian motion starting from 0. The aim of this problem is to prove that the function $X(\star, \omega)$ is nondifferentiable almost surely on $[0, \infty)$; that is, there exists a measurable set Ω_0 such that $X(\star, \omega)$ is nondifferentiable for all $\omega \in \Omega_0$. Here we content ourselves with proving that the set of all $\omega \in \Omega$ such that

$$\limsup_{t \to t_0} \frac{X(t, \omega) - X(t_0, \omega)}{t - t_0} = \liminf_{t \to t_0} \frac{X(t, \omega) - X(t_0, \omega)}{t - t_0} \le 1$$

for some $t_0 \in [0, 1)$ forms a set of measure zero. If we generalize what we are going to show, it is not so hard to show that Ω is nondifferentiable everywhere with respect to the second variable for almost all the first variables.

1. Let $N, k \in \mathbb{N}$. Define

$$Y_{N,k} \equiv \max_{j=0,1,2} \left| X\left(\star, \frac{k+j}{N}\right) - X\left(\star, \frac{k+j-1}{N}\right) \right|.$$

   Show that

$$P\left(Y_{N,0} \leq \frac{1}{N}\right) \leq \frac{100M}{N}, \quad P\left(Y_{N,k} \leq \frac{1}{N}\right) \leq \frac{1000}{N\sqrt{N}}.$$

2. Let $\omega \in \Omega$ be such that $X(\omega, t)$ is differentiable at $t = t_0 \in [0, 1)$. Then show that there exists $s \in [0, 1 - N^{-1})$ such that $|X(t, \omega) - X(s, \omega)| \leq |X_t(\omega, t_0)| \cdot |t - s|$. Thus, $Y_{N,k}(\omega) \leq [|X_t(\omega, t_0)| + 1]$.
3. Show that such events ω form a null set.

Exercise 2.22 Let us investigate Example 2.6 in more depth. A strictly increasing sequence $\{n_k\}_{k=1}^{\infty}$ of positive integers is said to have the *Hadamard gap* if there exists $q > 1$ such that

$$n_{k+1} \geq q\, n_k. \tag{2.29}$$

1. Let $0 < p, q \leq \infty$ and $s \in \mathbb{R}$. Then using (2.27) prove that $f \in B^s_{pq}(\mathbb{R})$ if and only if

$$p = \infty, \quad q \in \mathbb{R}, \quad s < \log_2 a \text{ or } p = q = \infty, \quad s = \log_2 a. \tag{2.30}$$

2. In (2.24), replace cos with sin. What can we say?
3. For $q \neq 2$, we have an analogy to the following result: For the sake of simplicity let $q = 2$ in (2.29) and consider a sequence $\{n_k\}_{k=1}^{\infty}$ of nonnegative integers. Let $s > 0$. We define $g(t) \equiv \sum_{k=2}^{\infty} \frac{1}{{n_k}^s} \cos(n_k\, t)$ for $t \in \mathbb{R}$. Show that $g \in B^s_{\infty\infty}(\mathbb{R})$.
4. Let $s > 1$. We define $h(t) \equiv \sum_{k=2}^{\infty} \frac{1}{k^s} \cos(2^k\, t)$ for $t \in \mathbb{R}$. Show that $h \in B^0_{\infty 1}(\mathbb{R})$.

This fact is due to Paley, Wiener and Zygmund [863] and the proof given here is due to Dvoretzky, Erdös and Kakutani [443].

Textbooks in Sect. 2.2

Hölder–Zygmund Spaces

See [99, p. 90, Theorem].

Fourier Transform and Besov Spaces: Theorem 2.9

See [71, p. 116 (3)] for the fact that the Fourier transform maps $\dot{B}^0_{1\infty}(\mathbb{R}^n)$ into $L^\infty(\mathbb{R}^n)$. See [71, Chapter 6] for various extensions of Theorem 2.9.

Others

The author is indebted to Dr. Naoko Ogata for Exercises 2.22 3 and 4.

2.3 Besov Spaces, Triebel–Lizorkin Spaces and Modulation Spaces with $0 < p, q \leq \infty$ and $s \in \mathbb{R}$

In this section, we define Besov spaces and Triebel–Lizorkin spaces using our harmonic analytic tool. Unlike other function spaces such as the well-known Sobolev spaces, it is a big burden to define them and justify the definition. As a result, many mathematicians dislike Besov spaces and Triebel–Lizorkin spaces. However, due to the complicated definition, the function has rich and nice properties and they are applied to initial problems of partial differential equations. Based on Chap. 1, we define Triebel–Lizorkin spaces and investigate elementary properties.

First of all, in Sect. 2.3.1 we define these spaces and justify the definition. In Sect. 2.3.2 we study function spaces. The completeness and density are quite elementary but their proofs are both complicated. So we start with a proposition which can be proved easily. The easiest thing is, maybe, that we have trivial embeddings. Once we investigate embeddings, we can prove completeness and density easily. We can also define modulation spaces in Sect. 2.3.3. Section 2.3.3 parallels Sects. 2.3.1 and 2.3.2 to a large extent. Let us start with what can be seen easily.

2.3.1 *Definition of* $A^s_{pq}(\mathbb{R}^n)$ *with* $0 < p, q \le \infty$ *and* $s > 0$

When $p < 1$ and/or $q < 1$, the study of these function spaces is difficult; $L^p(\mathbb{R}^n)$ and ℓ^q is not a Banach space and the Hardy–Littlewood maximal operator is not $L^p(\mathbb{R}^n)$-bounded if $p \le 1$ according to Example 1.25.

2.3.1.1 Definition of Besov Spaces, Triebel–Lizorkin Spaces and Modulation Spaces

We define Besov spaces and Triebel–Lizorkin spaces. As we will state later, $A^s_{pq}(\mathbb{R}^n)$ is a symbol used to denote either the Besov space $B^s_{pq}(\mathbb{R}^n)$ or the Triebel–Lizorkin space $F^s_{pq}(\mathbb{R}^n)$.

Definition 2.7 (ψ and φ for the definition of function spaces) Let $\psi, \varphi \in \mathscr{S}(\mathbb{R}^n)$ satisfy

$$\chi_{B(4)} \le \psi \le \chi_{B(8)}, \quad \chi_{B(4)\setminus B(2)} \le \varphi \le \chi_{B(8)\setminus B(1)}. \tag{2.31}$$

Write $\varphi_j \equiv \varphi(2^{-j}\star)$ for $\varphi \in \mathscr{S}(\mathbb{R}^n)$ and $j \in \mathbb{Z}$.

Remark 2.3 We may postulate the following conditions:

$$\mathrm{supp}(\varphi) \subset B(4) \setminus B(1), \quad \mathrm{supp}(\psi) \subset B(4), \quad \psi + \sum_{j=1}^{\infty} \varphi_j \equiv 1. \tag{2.32}$$

Condition (2.32) seems better but it is not always necessary. We assume conditions in Definition 2.7 to define function space.

After understanding the definition quite well, we consider many conditions on φ.

Here to describe the norms used in Sect. 2.3 we define the vector-valued norms; let $\{f_j\}_{j=1}^{\infty}$ be a sequence of measurable functions. Then define

$$\|f_j\|_{\ell^q(L^p)} (= \|\{f_j\}_{j=1}^{\infty}\|_{\ell^q(L^p)}) \equiv \left(\sum_{j=1}^{\infty} \|f_j\|_p{}^q\right)^{\frac{1}{q}} \tag{2.33}$$

$$\|f_j\|_{L^p(\ell^q)} (= \|\{f_j\}_{j=1}^{\infty}\|_{L^p(\ell^q)}) \equiv \left\| \left(\sum_{j=1}^{\infty} |f_j|^q\right)^{\frac{1}{q}} \right\|_p . \tag{2.34}$$

Definition 2.8 (Besov/Triebel–Lizorkin norm) Let $0 < p, q \le \infty$, $s \in \mathbb{R}$, and let $f \in \mathscr{S}'(\mathbb{R}^n)$.

1. Define the quasi-norm $\| \star \|_{B^s_{pq}(\varphi,\psi)}$ by

$$\|f\|_{B^s_{pq}} \equiv \|f\|_{B^s_{pq}(\varphi,\psi)} \equiv \|\psi(D)f\|_p + \|2^{js}\varphi_j(D)f\|_{\ell^q(L^p)}, \tag{2.35}$$

where $\ell^q(L^p)$ stands for the space given by (2.33). The space $B^s_{pq}(\mathbb{R}^n, \varphi, \psi)$ is the set of all $f \in \mathscr{S}'(\mathbb{R}^n)$ for which the quasi-norm $\|f\|_{B^s_{pq}(\varphi,\psi)}$ is finite.
2. Furthermore, let $p < \infty$. The quasi-norm $\| \star \|_{F^s_{pq}(\varphi,\psi)}$ is defined by

$$\|f\|_{F^s_{pq}} \equiv \|f\|_{F^s_{pq}(\varphi,\psi)} \equiv \|\psi(D)f\|_p + \|2^{js}\varphi_j(D)f\|_{L^p(\ell^q)}, \tag{2.36}$$

where $L^p(\ell^q)$ means the space given by (2.34). The function space $F^s_{pq}(\mathbb{R}^n, \varphi, \psi)$ is the set of all $f \in \mathscr{S}'(\mathbb{R}^n)$ for which the quasi-norm $\|f\|_{F^s_{pq}(\varphi,\psi)}$ is finite.

Defining Triebel–Lizorkin spaces essentially consists of six steps:

- Transform the function f on the time domain to $\mathscr{F}f$ on the frequency domain.
- Multiply ψ and φ_j and localize $\mathscr{F}f$.
- Consider the functions $\mathscr{F}^{-1}[\psi\mathscr{F}f]$ and $\mathscr{F}^{-1}[\varphi_j\mathscr{F}f]$ by the use of the inverse Fourier transform.
- Multiply the weight 2^{ks} according to the size of localization.
- Consider the $\ell^q(\mathbb{N})$-norm of $\mathscr{F}^{-1}[\varphi_j\mathscr{F}f]$ with respect to j and then add $\mathscr{F}^{-1}[\psi\mathscr{F}f]$.
- Finally, consider the $L^p(\mathbb{R}^n)$-norm.

In the definition of $F^s_{pq}(\mathbb{R}^n, \varphi, \psi)$, we assume $p < \infty$: we need to use the Fefferman–Stein vector-valued inequality; see Theorem 1.49.

We consider $F^s_{\infty q}(\mathbb{R}^n)$ mainly only in Sect. 3.3.3: We do not consider $F^s_{\infty q}(\mathbb{R}^n)$ elsewhere.

With the definition, a natural question arises:

Problem 2.1 Do $B^s_{pq}(\mathbb{R}^n, \varphi, \psi)$ and $F^s_{pq}(\mathbb{R}^n, \varphi, \psi)$ depend on functions ψ and φ?

This is where the first issue arises. Since the definition is complicated, we want to look for an affirmative answer to this problem. We do not know from where to start, but a clue is that $\varphi_j(D)f$ has compact frequency support.

Theorem 2.11 (Validity of Besov spaces and Triebel–Lizorkin spaces) *Suppose that the parameters p, q, s satisfy $0 < p, q \le \infty$ and $s \in \mathbb{R}$. The function spaces $B^s_{pq}(\mathbb{R}^n, \varphi, \psi)$ and $F^s_{pq}(\mathbb{R}^n, \varphi, \psi)$ do not depend on the admissible choice of functions ψ and φ.*

Proof Suppose that the two functions $\widetilde{\varphi}$, $\widetilde{\psi}$ enjoy the same properties (2.31) as ψ and φ, respectively. From the symmetry, it suffices to prove that

$$\|f\|_{B^s_{pq}(\widetilde{\varphi},\widetilde{\psi})} \lesssim \|f\|_{B^s_{pq}(\varphi,\psi)}, \tag{2.37}$$

$$\|f\|_{F^s_{pq}(\widetilde{\varphi},\widetilde{\psi})} \lesssim \|f\|_{F^s_{pq}(\varphi,\psi)}. \tag{2.38}$$

Here in particular,

$$\|2^{(j+3)s}\widetilde{\varphi}_{j+3}(D)f\|_{L^p(\ell^q)} \lesssim \|2^{js}\varphi_j(D)f\|_{L^p(\ell^q)}. \tag{2.39}$$

Of course, (2.39) itself is not sufficient for the proof of (2.37) and (2.38). We need to supplement it by proving;

$$\|2^{(j+3)s}\widetilde{\varphi}_{j+3}(D)f\|_{\ell^q(L^p)} \lesssim \|2^{js}\varphi_j(D)f\|_{\ell^q(L^p)}, \tag{2.40}$$

$$\|\widetilde{\psi}_0(D)f\|_p+\sum_{j=1}^{3}\|2^{js}\widetilde{\psi}_j(D)f\|_p \lesssim \|\psi(D)f\|_p + \|2^{js}\varphi_j(D)f\|_{L^p(\ell^q)}, \tag{2.41}$$

$$\|\widetilde{\psi}_0(D)f\|_p+\sum_{j=1}^{3}\|2^{js}\widetilde{\psi}_j(D)f\|_p \lesssim \|\psi(D)f\|_p+\|2^{js}\varphi_j(D)f\|_{\ell^q(L^p)}. \tag{2.42}$$

But due to similarity, we omit the details here. See Exercise 2.27. For $j \in \{4, 5, 6\ldots\}$, we have

$$\begin{aligned}
\widetilde{\varphi}_j(D)f &= \mathscr{F}^{-1}(\varphi_j\mathscr{F}f)\\
&= \mathscr{F}^{-1}\left[(\widetilde{\varphi}_{j-1}+\widetilde{\varphi}_j+\widetilde{\varphi}_{j+1})\frac{\varphi_j}{\widetilde{\varphi}_{j-1}+\widetilde{\varphi}_j+\widetilde{\varphi}_{j+1}}\mathscr{F}f\right]\\
&= \left(\frac{\varphi_j}{\widetilde{\varphi}_{j-1}+\widetilde{\varphi}_j+\widetilde{\varphi}_{j+1}}\right)(D)\mathscr{F}^{-1}[\,(\widetilde{\varphi}_{j-1}+\widetilde{\varphi}_j+\widetilde{\varphi}_{j+1})\mathscr{F}f\,]\\
&= \left(\frac{\varphi}{\widetilde{\varphi}_{-1}+\widetilde{\varphi}+\widetilde{\varphi}_1}\right)_j(D)\mathscr{F}^{-1}[\,(\widetilde{\varphi}_{j-1}+\widetilde{\varphi}_j+\widetilde{\varphi}_{j+1})\mathscr{F}f\,].
\end{aligned}$$

Here $\widetilde{\varphi}_j$ is expressed as the pointwise product of Φ_j and a linear combination of φ_j's. So, we are in the position of using Theorem 1.53. Thus, we obtain

$$\|2^{js}\widetilde{\varphi}_{j+3}(D)f\|_{L^p(\ell^q)} \lesssim \sum_{l=2}^{4}\|2^{js}\varphi_{j+l}(D)f\|_{L^p(\ell^q)} \lesssim \|2^{js}\varphi_j(D)f\|_{L^p(\ell^q)}$$

by Theorem 1.53.

The same as Sect. 2.1 applies to the choice of the functions ψ and φ in the definition of $A^s_{pq}(\mathbb{R}^n)$.

In the above, only Triebel–Lizorkin spaces were considered. However, the above proof sees no difference between Besov spaces and Triebel–Lizorkin spaces except in that we use $\ell^q(L^p)$ and $L^p(\ell^q)$ for Besov spaces and Triebel–Lizorkin spaces, respectively. The same applies to many other proofs of the propositions in this book.

2.3.1.2 Elementary Properties

Having proved a extremely significant property that the definition of function spaces do not depend on ψ and φ, here and below, we omit ψ and φ in the notation of the norms and the function spaces. Hence we will write $B^s_{pq}(\mathbb{R}^n)$, $F^s_{pq}(\mathbb{R}^n)$, $\| \star \|_{B^s_{pq}}$ and $\| \star \|_{F^s_{pq}}$ for $B^s_{pq}(\mathbb{R}^n, \psi, \varphi)$, $F^s_{pq}(\mathbb{R}^n, \psi, \varphi)$, $\| \star \|_{B^s_{pq}(\psi,\varphi)}$ and $\| \star \|_{F^s_{pq}(\psi,\varphi)}$.

Definition 2.9 ($A^s_{pq}(\mathbb{R}^n)$) Let $0 < p, q \le \infty$, $s > 0$. Denote by $A^s_{pq}(\mathbb{R}^n)$ either $B^s_{pq}(\mathbb{R}^n)$ or $F^s_{pq}(\mathbb{R}^n)$. Assume tacitly that $p < \infty$ when $A^s_{pq}(\mathbb{R}^n) = F^s_{pq}(\mathbb{R}^n)$.

An advantage of this notation is that we can formulate many propositions in a unified manner.

We give examples of the elements in $A^s_{pq}(\mathbb{R}^n)$.

Example 2.7 Let us consider what function space $f \equiv 1 \in \mathscr{S}'(\mathbb{R}^n)$ belongs to. Since $\mathscr{F} f = (2\pi)^{-\frac{n}{2}}\delta$, where δ is the Dirac delta massed at the origin, we have $\varphi_j(D)f = 0$ and $\psi(D)f = f$. Hence it follows that:

1. Let $0 < p, q \le \infty$ and $s \in \mathbb{R}$. Then $1 \in B^s_{pq}(\mathbb{R}^n)$ if and only if $p = \infty$.
2. Let $0 < p < \infty$, $0 < q \le \infty$ and $s \in \mathbb{R}$. Then $1 \in F^s_{pq}(\mathbb{R}^n)$ never happens.

The same can be said for $f \equiv \sin(a \cdot \star)$, $\cos(a \cdot \star)$ with $a \in \mathbb{R}^n$.

Example 2.8 Let $0 < q \le \infty$. We will show that $L^\infty(\mathbb{R}^n)$ and $B^0_{\infty q}(\mathbb{R}^n)$ are different spaces. When $0 < q \le 1$, we have $B^0_{\infty q}(\mathbb{R}^n) \hookrightarrow B^0_{\infty 1}(\mathbb{R}^n) \hookrightarrow \mathrm{BUC}(\mathbb{R}^n) \hookrightarrow L^\infty(\mathbb{R}^n)$ and the last embedding is proper. Thus, our assertion is true for $0 < q \le 1$.

Let $1 < q \le \infty$. We will show $B^0_{\infty q}(\mathbb{R}^n) \setminus L^\infty(\mathbb{R}^n) \neq \emptyset$. In fact, we will show

$$B^{n/p}_{pq}(\mathbb{R}^n) \setminus L^\infty(\mathbb{R}^n) \neq \emptyset$$

for $0 < p \le \infty$. Choose $\Psi \in \mathscr{S}(\mathbb{R}^n) \setminus \{0\}$ so that $\chi_{B(3.9)\setminus B(3.1)} \le \Psi \le \chi_{B(4)\setminus B(3)}$. Choose $\psi, \varphi \in \mathscr{S}(\mathbb{R}^n)$ so that $\chi_{B(2)} \le \psi \le \chi_{B(3)}$ and $\varphi = \psi - \psi_{-1}$, so that we can calculate the Besov norm with ease.

Choose a positive sequence $a \equiv \{a_k\}_{k=1}^\infty$ so that $a \in \ell^q(\mathbb{N}) \setminus \ell^1(\mathbb{N})$. Define

$$f \equiv \sum_{k=1}^{\infty} a_k \mathscr{F}^{-1}\Psi_k.$$

Let $0 < p \le \infty$. Then since $\varphi_j(D)f = a_j \mathscr{F}^{-1}[\varphi_j \cdot \Psi_{j+1}]$, we have $f \in B^{n/p}_{pq}(\mathbb{R}^n)$. Since $\mathscr{F} f \ge 0$ by the inverse Fourier transform,

$$\|\psi_M(D)f\|_\infty \simeq_n \int_{\mathbb{R}^n} \psi_M(\xi)\mathscr{F}f(\xi)\mathrm{d}\xi = \sum_{k=1}^{\infty} a_k \int_{\mathbb{R}^n} \psi_M(\xi)\Psi_k(\xi)\mathrm{d}\xi \to \infty \tag{2.43}$$

as $M \to \infty$. Thus, $f \notin L^\infty(\mathbb{R}^n)$.

Proposition 2.5 will be used for two different purposes. One is when the functions in $A^s_{pq}(\mathbb{R}^n)$ are continuous and the other is to prove the completeness as we did in Proposition 2.4.

Proposition 2.5 *Suppose that the positive parameters p, q, s satisfy*

$$0 < p, q \leq \infty, \; s > \frac{n}{p}.$$

Then $A^s_{pq}(\mathbb{R}^n) \hookrightarrow \mathrm{BUC}(\mathbb{R}^n)$.

Proof By Corollary 1.8, we have $A^s_{pq}(\mathbb{R}^n) \hookrightarrow B^0_{\infty 1}(\mathbb{R}^n) \hookrightarrow \mathrm{BUC}(\mathbb{R}^n)$.

We remark that we did not define $F^s_{\infty q}(\mathbb{R}^n)$. We cannot replace the $L^p(\mathbb{R}^n)$-norm with the $L^\infty(\mathbb{R}^n)$-norm in the definition of the $F^s_{pq}(\mathbb{R}^n)$-norm. As in [97], the definition does depend on ψ and φ, if we formally extend our definition to the case when $p = \infty$. In Sect. 3.3 we will define the function space $F^s_{\infty q}(\mathbb{R}^n)$.

Exercises

Exercise 2.23 (Vector-valued maximal inequality for powered maximal operators) Let $0 < p, q \leq \infty$. Recall that the powered Hardy–Littlewood maximal operator $M^{(\eta)}$ is defined in Definition 1.37 for $\eta > 0$; $M^{(\eta)}f(x) \equiv M[|f|^\eta](x)^{\frac{1}{\eta}}$ for $x \in \mathbb{R}^n$ and a measurable function $f : \mathbb{R}^n \to \mathbb{C}$. We transform the Hardy–Littlewood maximal inequality and the Fefferman–Stein vector-valued inequality into the form in which we use in this book:

1. Let $0 < \eta < \min(1, p)$. Show that $\|M^{(\eta)}f_j\|_{\ell^q(L^p)} \lesssim \|f_j\|_{\ell^q(L^p)}$ for all sequences $\{f_j\}_{j=1}^\infty \subset L^0(\mathbb{R}^n)$.
2. Let $p < \infty$, $0 < \eta < \min(1, p, q)$. Show that $\|M^{(\eta)}f_j\|_{L^p(\ell^q)} \lesssim \|f_j\|_{L^p(\ell^q)}$ for all sequences $\{f_j\}_{j=1}^\infty \subset L^0(\mathbb{R}^n)$.

Exercise 2.24 In this exercise, we reconsider the definition of Besov spaces and Triebel–Lizorkin spaces. Prove that the norms $\| \star \|^{(i)}_{A^s_{pq}}$ with $i = 1, 2, 3, 4$ below are equivalent to the norm $\|f\|_{A^s_{pq}}$ defined in this section. Furthermore, show that:

$$A^s_{pq}(\mathbb{R}^n) = \{f \in \mathscr{S}'(\mathbb{R}^n) : \|f\|^{(j)}_{A^s_{pq}} < \infty\}, \quad j = 1, 2, 3, 4. \tag{2.44}$$

1. Choose $\psi, \varphi \in \mathscr{S}(\mathbb{R}^n)$ so that

$$\chi_{B(1)} \le \psi \le \chi_{B(2)}, \quad \varphi \equiv \psi - \psi(2\star). \tag{2.45}$$

Set

$$\begin{cases} \|f\|^{(1)}_{B^s_{pq}} \equiv \|\psi(D)f\|_p + \|2^{js}\varphi_j(D)f\|_{\ell^q(L^p)}, \\ \|f\|^{(1)}_{F^s_{pq}} \equiv \|\psi(D)f\|_p + \|2^{js}\varphi_j(D)f\|_{L^p(\ell^q)}. \end{cases} \tag{2.46}$$

2. Choose $\psi, \varphi \in \mathscr{S}(\mathbb{R}^n)$ so that $\mathrm{supp}(\psi) \subset B(4)$, $\mathrm{supp}(\varphi) \subset B(4) \setminus B(2)$ and that

$$\psi(\xi)^2 + \sum_{j=1}^{\infty} \varphi_j(\xi)^2 \equiv 1 \quad (\xi \in \mathbb{R}^n). \tag{2.47}$$

Define

$$\begin{cases} \|f\|^{(2)}_{B^s_{pq}} \equiv \|\psi(D)f\|_p + \|2^{js}\varphi_j(D)f\|_{\ell^q(L^p)}, \\ \|f\|^{(2)}_{F^s_{pq}} \equiv \|\psi(D)f\|_p + \|2^{js}\varphi_j(D)f\|_{L^p(\ell^q)}. \end{cases} \tag{2.48}$$

3. Choose again $\psi \in \mathscr{S}(\mathbb{R}^n)$ so that $\chi_{B(1)} \le \psi \le \chi_{B(3)}$. Set $\varphi \equiv \psi - \psi(3\star)$ for $\xi \in \mathbb{R}^n$. Then define

$$\begin{cases} \|f\|^{(3)}_{B^s_{pq}} \equiv \|\psi(D)f\|_p + \|3^{js}\mathscr{F}^{-1}[\varphi(3^{-j}\star)\mathscr{F}f]\|_{\ell^q(L^p)}, \\ \|f\|^{(3)}_{F^s_{pq}} \equiv \|\psi(D)f\|_p + \|3^{js}\mathscr{F}^{-1}[\varphi(3^{-j}\star)\mathscr{F}f]\|_{L^p(\ell^q)}. \end{cases} \tag{2.49}$$

4. Choose $\psi \in \mathscr{S}(\mathbb{R}^n)$ so that $\chi_{B(1)} \le \psi \le \chi_{B(2)}$. Set $\varphi \equiv \psi - \psi(2\star)$ for $\xi \in \mathbb{R}^n$. In this book, the definition of the Fourier transform $\mathscr{F}$ and its inverse $\mathscr{F}^{-1}$ is supposed to be:

$$\mathscr{F}f(\xi) \equiv \int_{\mathbb{R}^n} f(x)\mathrm{e}^{-ix\cdot\xi}\frac{\mathrm{d}x}{\sqrt{(2\pi)^n}}, \quad \mathscr{F}^{-1}f(x) \equiv \int_{\mathbb{R}^n} f(\xi)\mathrm{e}^{ix\cdot\xi}\frac{\mathrm{d}\xi}{\sqrt{(2\pi)^n}}.$$

But here to define $\|\star\|^{(4)}_{A^s_{pq}}$ we redefine the Fourier transform and its inverse by

$$\mathscr{F}f(\xi) \equiv \int_{\mathbb{R}^n} f(x)e^{-2\pi ix\cdot\xi}\mathrm{d}x, \quad \mathscr{F}^{-1}f(x) \equiv \int_{\mathbb{R}^n} f(\xi)e^{2\pi ix\cdot\xi}\mathrm{d}\xi. \tag{2.50}$$

With the definition (2.50) in mind, we define

$$\begin{cases} \|f\|^{(4)}_{B^s_{pq}} \equiv \|\psi(D)f\|_p + \|2^{js}\mathscr{F}^{-1}[\varphi_j \cdot \mathscr{F}f]\|_{\ell^q(L^p)}, \\ \|f\|^{(4)}_{F^s_{pq}} \equiv \|\psi(D)f\|_p + \|2^{js}\mathscr{F}^{-1}[\varphi_j \cdot \mathscr{F}f]\|_{L^p(\ell^q)}. \end{cases} \tag{2.51}$$

The norms $\|f\|^{(j)}_{A^s_{pq}}$, $j = 1, 2$ defined in Exercise 2.24 above are also denoted by $\|f\|_{A^s_{pq}}$. Choose appropriate definitions for each problem.

Theorem 1.53, used in the proof of Theorem 2.11, is extremely important and later in this book we use it very frequently without a detailed explanation. To be able to use Theorem 1.53, we add more exercises.

Exercise 2.25 [850, Lemma 4.9] Let $0 < p, q \le \infty$, $s \in \mathbb{R}$, $R, r > 0$, and let $\kappa \in \mathscr{S}(\mathbb{R}^n)$, $\tau \in \mathscr{S}(\mathbb{R}^n)$ satisfy $\chi_{B(R)} \le \kappa \le \chi_{B(4R)}$ and $\chi_{B(8r)\setminus B(4r)} \le \tau \le \chi_{B(32r)\setminus B(r)}$. Then show that

$$f \in A^s_{pq}(\mathbb{R}^n) \mapsto \kappa(D)f + \sum_{j=1}^{\infty} \tau_j(D)f \in A^s_{pq}(\mathbb{R}^n)$$

is bounded linear operator.

Exercise 2.26 (Peetre maximal characterization) Let $0 < p, q \le \infty$, $s \in \mathbb{R}$, and let $f \in \mathscr{S}'(\mathbb{R}^n)$. We let $\psi, \varphi \in C^\infty_c(\mathbb{R}^n)$ be the same function as in the definition of Besov spaces. Let $0 < \eta < \min(p, q, 1)$.

1. Show that the following semi-norms are equivalent to the original Besov norm $\| \star \|_{B^s_{pq}}$:

$$\left\| \sup_{z \in \mathbb{R}^n} \langle z \rangle^{-\frac{n}{\eta}} |\psi(D)f(\star - z)| \right\|_p + \left\| 2^{js} \sup_{z \in \mathbb{R}^n} \langle 2^j z \rangle^{-\frac{n}{\eta}} |\varphi_j(D)f(\star - z)| \right\|_{\ell^q(L^p)} .$$

2. Let $p < \infty$. Then show that the following semi-norm is equivalent to the original Triebel–Lizorkin norm $\| \star \|_{F^s_{pq}}$:

$$\left\| \sup_{z \in \mathbb{R}^n} \langle z \rangle^{-\frac{n}{\eta}} |\psi(D)f(\star - z)| \right\|_p + \left\| 2^{js} \sup_{z \in \mathbb{R}^n} \langle 2^j z \rangle^{-\frac{n}{\eta}} |\varphi_j(D)f(\star - z)| \right\|_{L^p(\ell^q)} .$$

Exercise 2.27 Mimic the proof of (2.39) to prove (2.40), (2.41), and (2.42).

2.3.2 Fundamental Properties of Function Spaces

Most of the fundamental results obtained for Besov spaces in the previous chapter carry over to Triebel–Lizorkin spaces. We can extend them to the case of $0 < p, q \le \infty$. However, we have to look for another method of the proof; the Young inequality is not available in the case of $0 < p, q \le \infty$ and in the case of Triebel–Lizorkin spaces.

2.3.2.1 Trivial Inclusions

Here we collect some trivial inclusions. The proof parallels the previous case.

Proposition 2.6 (Monotonicity of $A^s_{pq}(\mathbb{R}^n)$ with respect to q) *Let $0 < p \le \infty$, $0 < q_1 \le q_2 \le \infty$ and $s \in \mathbb{R}$. Then $A^s_{pq_1}(\mathbb{R}^n) \hookrightarrow A^s_{pq_2}(\mathbb{R}^n)$.*

Proof The same argument as that for Besov spaces works for Triebel–Lizorkin spaces; see the proof of Proposition 2.2. Even when $p, q < 1$, the same argument works.

We can describe the relation between $F^s_{pq}(\mathbb{R}^n)$ and $B^s_{pq}(\mathbb{R}^n)$ in terms of q.

Proposition 2.7 (The relation between $F^s_{pq}(\mathbb{R}^n)$ and $B^s_{pq}(\mathbb{R}^n)$ for fixed p, q) *Let $0 < p < \infty$, $0 < q \le \infty$ and $s \in \mathbb{R}$. Then $B^s_{p\,\min(p,q)}(\mathbb{R}^n) \hookrightarrow F^s_{pq}(\mathbb{R}^n) \hookrightarrow B^s_{p\,\max(p,q)}(\mathbb{R}^n)$. In particular, we have $B^s_{pp}(\mathbb{R}^n) = F^s_{pp}(\mathbb{R}^n)$ with norm coincidence.*

Proof From the definition, it is trivial that $B^s_{pp}(\mathbb{R}^n) = F^s_{pp}(\mathbb{R}^n)$. We will show $B^s_{p\,\min(p,q)}(\mathbb{R}^n) \hookrightarrow F^s_{pq}(\mathbb{R}^n) \hookrightarrow B^s_{p\,\max(p,q)}(\mathbb{R}^n)$.

Let $0 < q \le p < \infty$. By what we have been gathering, we have $F^s_{pq}(\mathbb{R}^n) \hookrightarrow F^s_{pp}(\mathbb{R}^n) = B^s_{pp}(\mathbb{R}^n) = B^s_{p\,\max(p,q)}(\mathbb{R}^n)$. Thus it remains to show $B^s_{pq}(\mathbb{R}^n) \hookrightarrow F^s_{pq}(\mathbb{R}^n)$. To this end, it suffices to show $\| f_k \|_{\ell^q(L^p)} \ge \| f_k \|_{L^p(\ell^q)}$. But this is trivial from the Minkovski inequality.

Assume $q \ge p$. As we did in the case $q \le p$, we reformulate what we need to prove. Using the Minkovski inequality, we complete the proof.

Proposition 2.8 (The relation between parameters s and q) *Let $0 < p < \infty$, $0 < q_1, q_2 \le \infty$, $s \in \mathbb{R}$ and $\delta > 0$. Then $A^{s+\delta}_{pq_1}(\mathbb{R}^n) \hookrightarrow A^s_{pq_2}(\mathbb{R}^n)$. In particular $A^{s+\delta}_{pq_1}(\mathbb{R}^n) \hookrightarrow A^s_{pq_1}(\mathbb{R}^n)$.*

Proof For Besov spaces the same proof as Proposition 2.3 works even when $0 < p, q_1, q_2 \le \infty$. For Triebel–Lizorkin spaces, we combine what we obtained for Besov spaces and Proposition 2.7 to have

$$F^{s+\delta}_{pq_1}(\mathbb{R}^n) \hookrightarrow B^{s+\delta}_{p\,\max(p,q_1)}(\mathbb{R}^n) \hookrightarrow B^s_{p\,\min(p,q_2)}(\mathbb{R}^n) \hookrightarrow F^s_{pq_2}(\mathbb{R}^n).$$

Therefore, the proof is complete.

2.3.2.2 Lift Operator

Let us establish that the parameter s stands for the smoothness of functions and that the lift operator controls s freely, which enables us to prove many things considerably easily. The operators (2.53)–(2.55) are called lift operators. Among others, we frequently use (2.53).

Theorem 2.12 (Lift operator) *Let $0 < p < \infty$, $0 < q \le \infty$, $s \in \mathbb{R}$, $\sigma \in \mathbb{R}$, $k = 1, 2, \ldots, n$ and $m \in \mathbb{N}$. The differentiation mapping $\partial_{x_k} : A^s_{pq}(\mathbb{R}^n) \to A^{s-1}_{pq}(\mathbb{R}^n)$ is a continuous mapping. That is,*

$$\|\partial_{x_k} f\|_{A^{s-1}_{pq}} \lesssim \|f\|_{A^s_{pq}}. \tag{2.52}$$

Furthermore, the following linear mappings are all isomorphisms:

$$(1-\Delta)^\sigma : A^s_{pq}(\mathbb{R}^n) \approx A^{s-2\sigma}_{pq}(\mathbb{R}^n), \tag{2.53}$$

$$(1+(-\Delta)^m) : A^s_{pq}(\mathbb{R}^n) \approx A^{s-2m}_{pq}(\mathbb{R}^n), \tag{2.54}$$

$$(1+\partial_1{}^{4m}+\cdots+\partial_n{}^{4m}) : A^s_{pq}(\mathbb{R}^n) \approx A^{s-4m}_{pq}(\mathbb{R}^n). \tag{2.55}$$

We want to use Theorem 1.53. So Lemma 2.1 will be of use again.

Proof We prove (2.52). To this end, it suffices to show that

$$\|\psi(D)[\partial_{x_k} f]\|_p \lesssim \|f\|_{A^s_{pq}}, \tag{2.56}$$

$$\|2^{(s-1)j}\varphi_j(D)[\partial_{x_k} f]\|_{\ell^q(L^p)} \lesssim \|f\|_{B^s_{pq}}, \tag{2.57}$$

$$\|2^{(s-1)j}\varphi_j(D)[\partial_{x_k} f]\|_{L^p(\ell^q)} \lesssim \|f\|_{F^s_{pq}}. \tag{2.58}$$

Let us now prove the most complicated inequality (2.58). We leave the proof of other inequalities as Exercise 2.32 to interested readers.

By taking the Fourier transform, $\varphi_j(D)[\partial_{x_k} f] = \psi_{j+2}(D)\partial_{x_k}[\varphi_j(D)f]$. If we set $\Psi^{(k)}(\xi) \equiv \xi_k\psi(\xi)$ for $\xi \in \mathbb{R}^n$, we have

$$\varphi_j(D)[\partial_{x_k} f] \simeq_n 2^j\Psi^{(k)}_{j+2}(D)[\varphi_j(D)f]$$

by the properties of the Fourier transform. Hence if $M \gg 1$, then

$$\begin{aligned}\|2^{(s-1)j}\varphi_j(D)[\partial_{x_k} f]\|_{L^p(\ell^q)} &\simeq_n \|2^{js}\Psi_{j+2}(D)[\varphi_j(D)f]\|_{L^p(\ell^q)}\\ &\lesssim \sup_{j\in\mathbb{N}}\|\Psi_{j+2}(2^{j+2}\star)\|_{H^M}\cdot\|2^{js}\varphi_j(D)f\|_{L^p(\ell^q)}\\ &\lesssim \|2^{js}\varphi_j(D)f\|_{L^p(\ell^q)}\end{aligned}$$

by virtue of Theorem 1.53. Thus, we obtain (2.58).

Next, we show that the three mappings are isomorphisms; namely, we will show

$$(1-\Delta)^\sigma : A^s_{pq}(\mathbb{R}^n) \approx A^{s-2\sigma}_{pq}(\mathbb{R}^n),$$

$$(1+(-\Delta)^m) : A^s_{pq}(\mathbb{R}^n) \approx A^{s-2m}_{pq}(\mathbb{R}^n),$$

$$(1 + \partial_1{}^{4m} + \cdots + \partial_n{}^{4m}) : A^s_{pq}(\mathbb{R}^n) \approx A^{s-4m}_{pq}(\mathbb{R}^n).$$

Due to similarity and the fact that the first mapping is the most standard, we concentrate on the first one.

To prove this, we note that the inverse is given by the mapping $(1 - \Delta)^{-\sigma}$ and the proof of the continuity of the inverse is completely the same as that of $(1 - \Delta)^{\sigma}$. We prove the continuity of $(1 - \Delta)^{\sigma}$.

To this end, we need to show that

$$\|\psi(D)[\,(1-\Delta)^{\sigma} f\,]\|_p \lesssim \|f\|_{A^s_{pq}} \tag{2.59}$$

and that

$$\|2^{j(s-2\sigma)}\varphi_j(D)[\,(1-\Delta)^{\sigma} f\,]\|_{\ell^q(L^p)} \lesssim \|f\|_{B^s_{pq}}, \tag{2.60}$$

$$\|2^{j(s-2\sigma)}\varphi_j(D)[\,(1-\Delta)^{\sigma} f\,]\|_{L^p(\ell^q)} \lesssim \|f\|_{F^s_{pq}} \tag{2.61}$$

for $j \geq 1$. Here we content ourselves with proving (2.61); we leave the proof of other assertions as Exercise 2.32 to interested readers. We have

$$\begin{aligned}&\|2^{j(s-2\sigma)}\varphi_j(D)[\,(1-\Delta)^{\sigma} f\,]\|_{L^p(\ell^q)}\\ &\lesssim \left(\sup_{j\in\mathbb{N}} 2^{-2j\sigma}\|(\psi_{j+2}(2^{j+2}\star) - \psi_{j-3}(2^{j+2}\star))\langle 2^{j+2}\star\rangle^{2\sigma}\|_{H^M}\right)\cdot\|f\|_{F^s_{pq}}\end{aligned}$$

by Theorem 1.53. Hence the heart of the matter is to show

$$\sup_{j\in\mathbb{N}} 2^{-2j\sigma}\|(\psi_{j+2}(2^{j+2}\star) - \psi_{j-3}(2^{j+2}\star))\langle 2^{j+2}\star\rangle^{2\sigma}\|_{H^M} < \infty. \tag{2.62}$$

By the Leibniz rule and Lemma 2.1,

$$\left|\partial^{\alpha}\left[2^{-2j\sigma}(\psi_{j+2}(2^{j+2}x) - \psi_{j-3}(2^{j+2}x))\langle 2^{j+2}x\rangle^{2\sigma}\right]\right| \lesssim_{\alpha} \chi_{B(8)}(x)$$

for each $\alpha \in \mathbb{N}_0{}^n$ and $x \in \mathbb{R}^n$; hence (2.62) holds. Thus, we obtain (2.61)

Corollary 2.2 *Let $m \in \mathbb{N}$, $0 < p < \infty$, $0 < q \leq \infty$ and $s \in \mathbb{R}$. Then one has a norm equivalence:* $\|f\|_{A^s_{pq}} + \sum_{j=1}^{n} \|\partial_{x_j}{}^m f\|_{A^s_{pq}} \sim \|f\|_{A^{s+m}_{pq}}$ *for* $f \in \mathscr{S}'(\mathbb{R}^n)$.

Proof Mimic the proof of Theorem 2.3, where we will use Theorem 2.12, Proposition 2.8 and (2.52).

Example 2.9 Choose $\rho \in C^{\infty}_{c}(\mathbb{R})$ so that $\chi_{(2,\infty)} \leq \rho \leq \chi_{(1,\infty)}$. We prove that $\rho \in B^s_{\infty\infty}(\mathbb{R}^n)$, $s > 0$: since $L^{\infty}(\mathbb{R}^n) \hookrightarrow B^0_{\infty\infty}(\mathbb{R}^n) \hookrightarrow B^{s-m}_{\infty\infty}(\mathbb{R}^n)$, where $m \equiv \max(1, [s+1])$, we have

$$\|\rho\|_{B^s_{\infty\infty}} \sim \|\rho\|_{B^{s-m}_{\infty\infty}} + \|\partial_t{}^m \rho\|_{B^{s-m}_{\infty\infty}} \lesssim \|\rho\|_\infty + \|\partial_t{}^m \rho\|_\infty \lesssim 1.$$

We have an interpolation inequality describing the smoothness of functions, which we leave as an exercise; see Exercise 2.30.

2.3.2.3 Sobolev Index

Recall first that $H^s_p(\mathbb{R}^n) = (1-\Delta)^{-\frac{s}{2}} L^p(\mathbb{R}^n)$ for $s > 0$ and $p > 1$ as in (1.166).

Sobolev's embedding theorem asserts:

Theorem 2.13 (Sobolev's embedding theorem for Sobolev spaces) *Suppose that the positive parameters p_0, p_1, s_0, s_1 satisfy*

$$p_0 > p_1 > 1, \quad s_0 > s_1, \quad s_0 - \frac{n}{p_0} = s_1 - \frac{n}{p_1},$$

then $H^{s_0}_{p_0}(\mathbb{R}^n) \hookrightarrow H^{s_1}_{p_1}(\mathbb{R}^n)$.

The quantity $s - n/p$ is called the *Sobolev index* or the *differential index.* Theorem 2.13 is well known and we do not recall the proof. See [2], for example. However, Theorem 2.13 can be proved using what is proved in this section and Sect. 3.1. We have the following counterparts for Besov spaces and Triebel–Lizorkin spaces:

Theorem 2.14 (Sobolev's embedding theorem for Besov and Triebel–Lizorkin spaces) *Suppose that the parameters p_0, p_1, s_0, s_1 satisfy*

$$0 < p_0 < p_1 \le \infty, \quad s_0 > s_1, \quad s_0 - \frac{n}{p_0} = s_1 - \frac{n}{p_1},$$

Then

$$B^{s_0}_{p_0q}(\mathbb{R}^n) \hookrightarrow B^{s_1}_{p_1q}(\mathbb{R}^n), \tag{2.63}$$

$$F^{s_0}_{p_0\infty}(\mathbb{R}^n) \hookrightarrow F^{s_1}_{p_1q}(\mathbb{R}^n). \tag{2.64}$$

This theorem extends the classical Sobolev embedding theorem and recaptures the results for Sobolev space nicely. Since Corollary 1.8 is already proved, the proof of Theorem 2.14 is considerably easy.

Proof Use Corollary 1.8 with $R = 2^{j+3}$. Then

$$\|\varphi_j(D)f\|_{p_1} \lesssim 2^{\frac{jn}{p_0} - \frac{jn}{p_1}} \|\varphi_j(D)f\|_{p_0}. \tag{2.65}$$

Here and below, we will prove (2.63) and (2.64) separately.

First of all, let us prove (2.63). From (2.65), we have

$$2^{js_1}\|\varphi_j(D)f\|_{p_1} \lesssim 2^{j\left(s_1-\frac{n}{p_1}+\frac{n}{p_0}\right)}\|\varphi_j(D)f\|_{p_0} = 2^{js_0}\|\varphi_j(D)f\|_{p_0}. \quad (2.66)$$

Thus, we write out two norms of the function spaces $B^{s_0}_{p_0q}(\mathbb{R}^n)$ and $B^{s_1}_{p_1q}(\mathbb{R}^n)$ to have

$$\|f\|_{B^{s_0}_{p_0q}} = \|\psi(D)f\|_{p_0} + \|2^{js_0}\varphi_j(D)f\|_{\ell^q(L^{p_0})},$$

$$\|f\|_{B^{s_1}_{p_1q}} = \|\psi(D)f\|_{p_1} + \|2^{js_1}\varphi_j(D)f\|_{\ell^q(L^{p_1})}.$$

If we use (2.66), we obtain (2.63).

We prove (2.64), which is delicate. To simplify the notation, we write

$$F(x) \equiv |\psi(D)f(x)| + \sup_{j\in\mathbb{N}} 2^{js_0}|\varphi_j(D)f(x)|,$$

$$G(x) \equiv |\psi(D)f(x)| + \left(\sum_{j=1}^{\infty}|2^{js_1}\varphi_j(D)f(x)|^q\right)^{\frac{1}{q}}$$

for $x \in \mathbb{R}^n$. We estimate the $L^{p_0}(\mathbb{R}^n)$-norm of F and the $L^{p_1}(\mathbb{R}^n)$-norm of G. By (2.65) and the definition of F, we have

$$|2^{js_1}\varphi_j(D)f(x)| \lesssim \min\left(2^{(s_1-s_0)j}F(x), 2^{\frac{jn}{p_1}}\|f\|_{F^{s_0}_{p_0\infty}}\right). \quad (2.67)$$

Hence

$$G(x) \lesssim \left(\sum_{j=1}^{\infty}\min\left(2^{(s_1-s_0)j}F(x), 2^{\frac{jn}{p_1}}\|f\|_{F^{s_0}_{p_0\infty}}\right)^q\right)^{\frac{1}{q}}$$

$$\leq \left(\sum_{j=-\infty}^{\infty}\min\left(2^{(s_1-s_0)j}F(x), 2^{\frac{jn}{p_1}}\|f\|_{F^{s_0}_{p_0\infty}}\right)^q\right)^{\frac{1}{q}}.$$

A passage to the continuous variable t yields

$$G(x) \lesssim \left(\sum_{j=-\infty}^{\infty}\int_{2^j}^{2^{j+1}}\min\left(t^{(s_1-s_0)}F(x), t^{\frac{jn}{p_1}}\|f\|_{F^{s_0}_{p_0\infty}}\right)^q\frac{dt}{t}\right)^{\frac{1}{q}}$$

$$\lesssim \left(\sum_{j=-\infty}^{\infty}\int_{0}^{\infty}\min\left(t^{(s_1-s_0)}F(x), t^{\frac{jn}{p_1}}\|f\|_{F^{s_0}_{p_0\infty}}\right)^q\frac{dt}{t}\right)^{\frac{1}{q}}$$

$$\sim F(x)^{\frac{p_0}{p_1}} (\|f\|_{F^{s_0}_{p_0\infty}})^{1-\frac{p_0}{p_1}}.$$

See Exercise 1.9. Since the $L^{p_0}(\mathbb{R}^n)$-norm of F equals $F^{s_0}_{p_0\infty}(\mathbb{R}^n)$-norm of f, if we consider the p_1-th power and integrate the inequality, we obtain the desired result.

Before we go further, a helpful remark may be in order.

Remark 2.4 We proved an embedding $A^{s_0}_{p_0q_0}(\mathbb{R}^n) \hookrightarrow A^{s_1}_{p_1q_1}(\mathbb{R}^n)$ in Theorem 2.14. We give a remark on the parameters q_0, q_1.

1. The inclusion $F^{s_0}_{p_0\infty}(\mathbb{R}^n) \hookrightarrow F^{s_1}_{p_1q}(\mathbb{R}^n)$ in (2.64) follows from Proposition 2.6. From $F^{s_0}_{p_0q}(\mathbb{R}^n) \hookrightarrow F^{s_0}_{p_0\infty}(\mathbb{R}^n)$ we deduce $F^{s_0}_{p_0q}(\mathbb{R}^n) \hookrightarrow F^{s_1}_{p_1q}(\mathbb{R}^n)$. Hence (2.64) covers (2.63) with B replaced by F.
2. Conversely, if $q' > q$, then we can disprove that $B^{s_0}_{p_0q'}(\mathbb{R}^n) \hookrightarrow B^{s_1}_{p_1q}(\mathbb{R}^n)$; see Exercise 2.31.

One of the interesting things after proving the theorems in the section is to compare the Sobolev embeddings in many textbooks such as [4, 51, 57].

We will need the following *Fatou property* of the function spaces for the proof of completeness of the function spaces.

Proposition 2.9 (Fatou property for $A^s_{pq}(\mathbb{R}^n)$) *Let $0 < p, q \le \infty$ and $s \in \mathbb{R}$. Suppose that a bounded sequence $\{f_j\}_{j=1}^\infty$ in $A^s_{pq}(\mathbb{R}^n)$ converges to $f \in \mathscr{S}'(\mathbb{R}^n)$ in the topology of $\mathscr{S}'(\mathbb{R}^n)$. Then $f \in A^s_{pq}(\mathbb{R}^n)$. More precisely, fix the functions ψ and φ defining the $A^s_{pq}(\mathbb{R}^n)$-norm. Then $\|f\|_{A^s_{pq}} \le \liminf_{j\to\infty} \|f_j\|_{A^s_{pq}}$.*

Proof Concentrate on Triebel–Lizorkin spaces. First of all, remark that

$$\Phi * f(x) = \langle f, \Phi(x - \star)\rangle = \lim_{k\to\infty} \langle f_k, \Phi(x - \star)\rangle = \lim_{k\to\infty} \Phi * f_k(x) \tag{2.68}$$

for all $x \in \mathbb{R}^n$ and $\Phi \in \mathscr{S}(\mathbb{R}^n)$. We apply (2.68) to $\Phi = \mathscr{F}^{-1}\psi$ and to $\Phi = \mathscr{F}^{-1}\varphi_j$, $j \in \mathbb{N}$. The result is:

$$\|f\|_{F^s_{pq}} = \|\psi(D)f\|_p + \left\| \left(\sum_{j=1}^\infty |2^{js}\varphi_j(D)f|^q \right)^{\frac{1}{q}} \right\|_p$$

$$= \left\| \liminf_{k\to\infty} |\psi_j(D)f_k| \right\|_p + \left\| \left(\sum_{j=1}^\infty \liminf_{k\to\infty} |2^{js}\varphi_j(D)f_k|^q \right)^{\frac{1}{q}} \right\|_p.$$

By the Fatou lemma, we have

$$\|f\|_{F^s_{pq}} \le \liminf_{k\to\infty}\left(\|\psi_j(D)f_k\|_p + \left\|\left(\sum_{j=1}^{\infty}|2^{js}\varphi_j(D)f_k|^q\right)^{\frac{1}{q}}\right\|_p\right) \le \liminf_{k\to\infty}\|f_k\|_{F^s_{pq}}.$$

Hence we obtain the desired result.

Theorem 2.15 (Quasi-Banach space property of $A^s_{pq}(\mathbb{R}^n)$) *Let $0 < p, q \le \infty$ and $s \in \mathbb{R}$. Then $A^s_{pq}(\mathbb{R}^n)$ is a Banach space continuously embedded into $\mathscr{S}'(\mathbb{R}^n)$. That is, $A^s_{pq}(\mathbb{R}^n)$ is a complete quasi-normed space in the following sense: If $\{f_j\}_{j=1}^{\infty} \subset A^s_{pq}(\mathbb{R}^n)$ satisfies*

$$\lim_{j,k\to\infty}\|f_j - f_k\|_{A^s_{pq}} = 0, \tag{2.69}$$

then there exists $f \in A^s_{pq}(\mathbb{R}^n)$ such that

$$\lim_{j\to\infty}\|f - f_j\|_{A^s_{pq}} = 0. \tag{2.70}$$

If $p, q \ge 1$, then $A^s_{pq}(\mathbb{R}^n)$ is Banach space.

Proof It is easy to show that $A^s_{pq}(\mathbb{R}^n)$ is a quasi-normed space. Next, if we use Propositions 2.7 and 2.6, and Theorem 2.14 in that order, then

$$A^s_{pq}(\mathbb{R}^n) \hookrightarrow B^s_{p\max(p,q)}(\mathbb{R}^n) \hookrightarrow B^s_{p\infty}(\mathbb{R}^n) \hookrightarrow B^{s-n/p}_{\infty\infty}(\mathbb{R}^n).$$

Next, take a positive integer m so that $2m + s > \frac{n}{p}$. From Theorem 2.12, we deduce

$$(1-\Delta)^{-m} : B^{s-n/p}_{\infty\infty}(\mathbb{R}^n) \approx B^{2m+s-n/p}_{\infty\infty}(\mathbb{R}^n) \tag{2.71}$$

Finally, if we combine Propositions 2.4 and 2.8, we obtain

$$B^{2m+s-n/p}_{\infty\infty}(\mathbb{R}^n) \hookrightarrow B^0_{\infty 1}(\mathbb{R}^n) \hookrightarrow \mathrm{BUC}(\mathbb{R}^n). \tag{2.72}$$

We obtain

$$(1-\Delta)^{-m} : A^s_{pq}(\mathbb{R}^n) \to \mathrm{BUC}(\mathbb{R}^n) \hookrightarrow \mathscr{S}'(\mathbb{R}^n) \tag{2.73}$$

using (2.71) and (2.72). Since $(1-\Delta)^m : \mathscr{S}'(\mathbb{R}^n) \approx \mathscr{S}'(\mathbb{R}^n)$ is an isomorphism, we see $A^s_{pq}(\mathbb{R}^n) \hookrightarrow \mathscr{S}'(\mathbb{R}^n)$ is continuous.

We prove the completeness. Let $f_j \in A^s_{pq}(\mathbb{R}^n)$, $j = 1, 2, \ldots$. Suppose that $\{f_j\}_{j=1}^{\infty}$ is a Cauchy sequence in the sense of (2.69). From (2.73), $(1-\Delta)^{-m} f_j$ converges to $F \in \mathrm{BUC}(\mathbb{R}^n)$ in $\mathrm{BUC}(\mathbb{R}^n)$. Hence we have $f_j \to f \in \mathscr{S}'(\mathbb{R}^n)$. By

Proposition 2.9, we have $f \in A^s_{pq}(\mathbb{R}^n)$ and $\|f - f_j\|_{A^s_{pq}} \le \liminf_{k\to\infty} \|f_k - f_j\|_{A^s_{pq}}$. Hence we obtain (2.70).

We now prove the density. We have to come up with another method unlike the case of $B^s_{pq}(\mathbb{R}^n)$ with $1 \le p, q < \infty$. Indeed, the Young inequality played the key role before.

Proposition 2.10 (Density of $\mathscr{S}(\mathbb{R}^n)$ in $A^s_{pq}(\mathbb{R}^n)$) *Let $0 < p, q < \infty$, $s \in \mathbb{R}$. Then $\mathscr{S}(\mathbb{R}^n)$ is dense in $A^s_{pq}(\mathbb{R}^n)$.*

As is the case with $L^p(\mathbb{R}^n)$, the proof consists of convolution and truncation.

Proof We assume that ψ and φ satisfy (2.32). The proof for the Besov space $B^s_{pq}(\mathbb{R}^n)$ is somewhat easier than that for the Triebel–Lizorkin space $F^s_{pq}(\mathbb{R}^n)$. Concentrate on the Triebel–Lizorkin space $F^s_{pq}(\mathbb{R}^n)$. We have

$$\psi(D)f + \sum_{j=1}^{\infty} \varphi_j(D)f = f \tag{2.74}$$

in the topology of $F^s_{pq}(\mathbb{R}^n)$. In fact, choose M sufficiently large according to p, q, s. Then by Theorem 1.53 and (2.32), for $N \ge 3$,

$$\begin{aligned}
\left\| f - \psi(D)f - \sum_{j=1}^{N} \varphi_j(D)f \right\|_{F^s_{pq}} &= \left\| 2^{js} \sum_{k=N+1}^{\infty} \varphi_k(D)\varphi_j(D)f \right\|_{L^p(\ell^q)} \\
&= \left\| 2^{js} \chi_{[\max(N,k-1),\infty)}(j)\varphi_k(D)\varphi_j(D)f \right\|_{L^p(\ell^q)} \\
&\lesssim \left\| \sum_{k=j-1}^{j+1} \varphi_k(2^k\star) \right\|_{H^M_2} \left\| \{2^{js}\varphi_j(D)f\}_{j=N}^{\infty} \right\|_{L^p(\ell^q)} \\
&\lesssim \left\| \{2^{js}\varphi_j(D)f\}_{j=N}^{\infty} \right\|_{L^p(\ell^q)}.
\end{aligned}$$

Since $0 < p, q < \infty$, $\psi(D)f + \sum_{j=1}^{\infty} \varphi_j(D)f = f$ in the topology of $F^s_{pq}(\mathbb{R}^n)$ as $N \to \infty$. Hence, setting $f_N \equiv \psi(D)f + \sum_{j=1}^{N} \varphi_j(D)f$, we can approximate each f_N. Choose $\psi \in \mathscr{S}(\mathbb{R}^n)$ so that $\chi_{B(2)} \le \psi \le \chi_{B(1)}$. If we set $G_N \equiv \dfrac{f_N \mathscr{F}\psi(\delta\star)}{\mathscr{F}\psi(0)}$ and take δ sufficiently small, Lebesgue's convergence theorem yields

$$\operatorname{supp}(\mathscr{F}(f_N - G_N)) \subset B(2^{N+4}), \quad \|f_N - G_N\|_p = o(1) \quad \delta \downarrow 0. \tag{2.75}$$

Since $\psi \in \mathscr{S}(\mathbb{R}^n)$, $G_N \in \mathscr{S}(\mathbb{R}^n)$. Hence $\mathscr{S}(\mathbb{R}^n)$ is dense in $F^s_{pq}(\mathbb{R}^n)$ from Corollary 1.7.

Now we explain the meaning of three parameters p, q, s and the "fourth" parameter $s - \frac{n}{p}$, which the function spaces carry. It is hard to grasp the meaning of the q.

Remark 2.5

1. The meaning of p: If $1 < p < \infty$, we will prove $L^p(\mathbb{R}^n) \approx F^0_{p2}(\mathbb{R}^n)$ with equivalence of norms in Sect. 3.1. So we can say that p stands for integrability of the functions.
2. The meaning of q: Let $0 < p \le \infty$ and $0 < q_1 \le q_2 \le \infty$. For $s \in \mathbb{R}$, we have $A^s_{pq_1}(\mathbb{R}^n) \hookrightarrow A^s_{pq_2}(\mathbb{R}^n)$. Let $0 < p \le \infty$ and $0 < q_1, q_2 \le \infty$. For $\delta > 0$ and $s \in \mathbb{R}$, we have shown that $A^{s+\delta}_{pq_1}(\mathbb{R}^n) \hookrightarrow A^s_{pq_2}(\mathbb{R}^n)$ holds. Hence the parameter q seems to describe the property of function spaces. But compared with the smoothness parameter s, the parameter q does not affect the properties of functions so much. In fact, it is natural to consider q after p and s are fixed. The smaller q is, the better the properties the functions in $A^s_{pq}(\mathbb{R}^n)$ enjoy. Also, q has another meaning; in Sect. 4.2 this has a lot to do with the parameters from interpolation theory.
3. The meaning of s: As we have seen that $\partial_{x_k} : A^s_{pq}(\mathbb{R}^n) \to A^{s-1}_{pq}(\mathbb{R}^n)$ and that
$$B^s_{\infty\infty}(\mathbb{R}^n) \approx \mathscr{C}^s(\mathbb{R}^n) \quad \text{(Hölder–Zygmund space)}$$
with equivalence of norms in Sect. 2.3, s can be regarded as the differential properties of the functions.
4. The meaning of $s-n/p$: As we have seen in Theorem 2.14, this can be seen as the Sobolev index. Let us think this over more deeply. For each $p \in (0, \infty]$, we say that the functions can be s-times differentiable in $L^p(\mathbb{R}^n)$. But this is an abstract sense. However, when $p = \infty$, as is seen from the property of $\mathscr{C}^s(\mathbb{R}^n)$, the properties that the functions are s times differentiable is the same as the classical sense. From Theorem 2.14 we have $A^s_{pq}(\mathbb{R}^n) \subset B^{s-n/p}_{\infty\infty}(\mathbb{R}^n)$. Hence we can say $s - n/p$ stands for the differentiability translated into the classical sense.

2.3.2.4 Diversity of Function Spaces

When we define function spaces, it is very important to check that they contain $C^\infty_c(\mathbb{R}^n)$ and that they are embedded into $\mathscr{S}'(\mathbb{R}^n)$ or $L^1_{\mathrm{loc}}(\mathbb{R}^n)$. However, it is also important to check that the "new" function spaces are actually new. Even if the spaces are isomorphic to the existing function spaces, we are happy enough: The existing function spaces turn out to have an equivalent expression. However, as for Besov spaces and Triebel–Lizorkin spaces, there is no new coincidence except some exceptional cases.

Theorem 2.16 (Diversity of function spaces) *The function spaces $F^s_{pq}(\mathbb{R}^n)$ and $B^s_{pq}(\mathbb{R}^n)$ differ except in the trivial cases listed below:*

1. *Suppose that the parameters $p_0, p_1, q_0, q_1, s_0, s_1$ satisfy the following conditions:*

$$0 < p_0, p_1, q_0, q_1 \le \infty, \quad s_0, s_1 \in \mathbb{R}.$$

   Then if $B^{s_0}_{p_0q_0}(\mathbb{R}^n) = B^{s_1}_{p_1q_1}(\mathbb{R}^n)$ as a set, then $p_0 = p_1, q_0 = q_1, s_0 = s_1$.
2. *Suppose that the parameters $p_0, p_1, q_0, q_1, s_0, s_1$ satisfy the following conditions:*

$$0 < p_0, p_1 < \infty, \quad 0 < q_0, q_1 \le \infty, \quad s_0, s_1 \in \mathbb{R}.$$

   Then if $F^{s_0}_{p_0q_0}(\mathbb{R}^n) = F^{s_1}_{p_1q_1}(\mathbb{R}^n)$ as a set, then $p_0 = p_1, q_0 = q_1, s_0 = s_1$.
3. *Suppose that the parameters $p_0, p_1, q_0, q_1, s_0, s_1$ satisfy the following conditions:*

$$0 < p_0 < \infty, \quad 0 < p_1, q_0, q_1 \le \infty, \quad s_0, s_1 \in \mathbb{R}. \tag{2.76}$$

   Then if $F^{s_0}_{p_0q_0}(\mathbb{R}^n) = B^{s_1}_{p_1q_1}(\mathbb{R}^n)$ as a set, then $p_0 = p_1 = q_0 = q_1, s_0 = s_1$.

To prove the theorem, it suffices to prove Lemma 2.2 below.

Lemma 2.2 *Let $0 < p, q \le \infty$ and $s \in \mathbb{R}$, and let $\eta \in \mathscr{S}(\mathbb{R}^n)$ satisfy*

$$\chi_{B(2.1)\setminus B(1.9)} \le \eta_1 = \eta(2^{-1}\star) \le \chi_{B(2.2)\setminus B(1.8)}. \tag{2.77}$$

Fix a sequence $a = \{a_k\}_{k=1}^\infty$. Let $\boldsymbol{e}_1$ be the unit vector in the x_1 direction. Set

$$\alpha^{(k)} \equiv (\mathscr{F}^{-1}\eta)^k, \quad \beta^{(k)} \equiv \mathscr{F}^{-1}[\eta(\star - 2^k\boldsymbol{e}_1)] \tag{2.78}$$

and

$$\gamma^{(k)} \equiv \sum_{l=1}^k \mathscr{F}^{-1}[\eta(\star - 2^l\boldsymbol{e}_1)](\star - 2^l\boldsymbol{e}_1), \quad \delta^{(k)} \equiv \sum_{l=1}^k a_l\,\mathscr{F}^{-1}[\eta(\star - 2^l\boldsymbol{e}_1)] \tag{2.79}$$

for $k \in \mathbb{N}$. Then

$$\|\alpha^{(k)}\|_{A^s_{pq}} \sim \|2^{ks}\alpha^{(k)}\|_p \simeq 2^{(s+n)k-nk/p}, \tag{2.80}$$

$$\|\beta^{(k)}\|_{A^s_{pq}} \sim \|2^{ks}\beta^{(k)}\|_p \simeq 2^{ks}, \tag{2.81}$$

$$\|\gamma^{(k)}\|_{B^0_{pq}} \sim k^{\frac{1}{q}}, \tag{2.82}$$

$$\|\mathscr{F}^{-1}(\gamma^{(k)})\|_{F^0_{pq}} \sim k^{\frac{1}{p}}. \tag{2.83}$$

We have a uniform estimate of the sequence $\{a_k\}_{k=1}^{\infty}$*:*

$$\|\delta^{(k)}\|_{A^0_{pq}} \sim \|a\|_{\ell^q}. \tag{2.84}$$

Admitting Lemma 2.2, let us complete the proof of Theorem 2.16. If the assumption of the theorem is satisfied, then from (2.80) and (2.81) we obtain $p_0 = p_1, s_0 = s_1$. Once this is proved, it can be assumed that $s_0 = s_1 = 0$ by the use of the lift operator (see Theorem 2.12). Now we can use (2.82), (2.83) and (2.84) to complete the proof of the theorem.

Proof (of Lemma 2.2) Choose $\psi \in \mathscr{S}(\mathbb{R}^n)$ so that

$$\chi_{Q(1.1)} \le \psi \le \chi_{Q(1.5)}, \quad \varphi = \psi - \psi(2\star). \tag{2.85}$$

From (2.77), (2.78) and (2.85), we have

$$\psi(D)\alpha^{(k)} = \psi(D)\beta^{(k)} = 0, \quad \varphi_j(D)\alpha^{(k)} = \delta_{jk}\alpha^{(k)}, \quad \varphi_j(D)\beta^{(k)} = \delta_{jk}\beta^{(k)} \tag{2.86}$$

for $j, k \in \mathbb{N}$. Thus, (2.80) and (2.81) are proved. However, we need to handle (2.82), (2.83) and (2.84) with care.

For (2.82), we have

$$\begin{aligned}\|\gamma^{(k)}\|_{B^0_{pq}} &= \left(\sum_{l=1}^{k} \|\mathscr{F}^{-1}[\eta(\star - 2^l \mathbf{e}_1)](\star - 2^l \mathbf{e}_1)\|_p{}^q\right)^{\frac{1}{q}} \\ &= \left(\sum_{l=1}^{k} \|\mathscr{F}^{-1}\eta\|_p{}^q\right)^{\frac{1}{q}} \sim k^{\frac{1}{q}}.\end{aligned} \tag{2.87}$$

Define

$$\mathrm{I} \equiv \left\| \left(\sum_{l=0}^{k} |\mathscr{F}^{-1}\eta(\star - 2^l \mathbf{e}_1)|^q\right)^{\frac{1}{q}} \right\|_p.$$

As for (2.83), we have

$$\|\gamma^{(k)}\|_{F^0_{pq}} \sim \left\| \left(\sum_{l=0}^{k} |\mathscr{F}^{-1}[\eta(\star - 2^l \mathbf{e}_1)](\star - 2^l \mathbf{e}_1)|^q\right)^{\frac{1}{q}} \right\|_p \sim \mathrm{I}.$$

Here by the continuity of $\mathscr{F}^{-1}\eta(0) \neq 0$ and continuity of $\mathscr{F}^{-1}\eta$, we have

$$\chi_{B(2^l \mathbf{e}_1, a)} \lesssim |\mathscr{F}^{-1}\eta(\star - 2^l \mathbf{e}_1)|$$

as long as $0 < a \ll 1$.

Hence

$$\mathrm{I} \gtrsim \left(\sum_{l=0}^{k} |\chi_{B(2^l \mathbf{e}_1, a)}|^q \right)^{\frac{1}{q}}. \tag{2.88}$$

Let $\rho \in (0, \min(1, p, q))$.

$$\left(\sum_{l=0}^{k} |\mathscr{F}^{-1}\eta(\star - 2^l \mathbf{e}_1)|^q \right)^{\frac{1}{q}} \lesssim \left(\sum_{l=0}^{k} M^{(\rho)}[\chi_{B(2^l \mathbf{e}_1, a)}]^q \right)^{\frac{1}{q}}. \tag{2.89}$$

By (2.89) and Theorem 1.49, we obtain

$$\mathrm{I} \sim \left\| \left(\sum_{l=0}^{k} |\chi_{B(2^l \mathbf{e}_1, a)}|^q \right)^{\frac{1}{q}} \right\|_p = \left\| \sum_{l=0}^{k} \chi_{B(2^l \mathbf{e}_1, a)} \right\|_p = k^{\frac{1}{p}}. \tag{2.90}$$

By virtue of (2.87) and (2.90), (2.83) holds.

Finally, by (2.84), the definition of $\delta^{(k)}$, we obtain

$$\|\delta^{(k)}\|_{B^0_{pq}} \sim \left(\sum_{l=0}^{k} \left\| a_l \, \mathscr{F}^{-1}[\eta(\star - 2^l \mathbf{e}_1)] \right\|_p{}^q \right)^{\frac{1}{q}} \sim \|a\|_{\ell^q},$$

$$\|\delta^{(k)}\|_{F^0_{pq}} \sim \left\| \left(\sum_{l=0}^{k} |a_l \, \mathscr{F}^{-1}[\eta(\star - 2^l \mathbf{e}_1)]|^q \right)^{\frac{1}{q}} \right\|_p \sim \|a\|_{\ell^q}.$$

Therefore, the proof is complete.

2.3.2.5 Duality

Finally, we investigate duality. We will make use of the following lemma for our purpose of considering the duality:

Lemma 2.3 *Let $1 \le p < \infty$, and let $\{F^{(j)}\}_{j=1}^{\infty} \in L^p(\ell^1, \mathbb{R}^n)$. Then*

$$\|\{\varphi_j(D)F^{(j)}\}_{j=1}^{\infty}\|_{L^p(\ell^1)} \lesssim \|\{F^{(j)}\}_{j=1}^{\infty}\|_{L^p(\ell^1)}.$$

Proof We express the left-hand side by means of the duality:

$$
\begin{aligned}
&\|\{\varphi_j(D)F^{(j)}\}_{j=1}^{\infty}\|_{L^p(\ell^1)}\\
&=\left\|\sum_{j=1}^{\infty}|\varphi_j(D)F^{(j)}|\right\|_p\\
&=\sup_{\substack{G\in L^{p'}(\mathbb{R}^n)\\ \|G\|_{p'}=1}}\left\|\sum_{j=1}^{\infty}|\varphi_j(D)F^{(j)}|G\right\|_1\\
&=\sup_{\substack{G\in L^{p'}(\mathbb{R}^n)\\ \|G\|_{p'}=1}}\sup_{\substack{\{a_j\}_{j=1}^{\infty}\in L^{\infty}(\ell^{\infty},\mathbb{R}^n)\\ \|a_j\|_{L^{\infty}(\ell^{\infty})}=1}}\left|\int_{\mathbb{R}^n}\sum_{j=1}^{\infty}\varphi_j(D)F^{(j)}(x)a_j(x)G(x)\mathrm{d}x\right|.
\end{aligned}
$$

By the definition of $\varphi_j(D)$, we have

$$
\begin{aligned}
&\|\{\varphi_j(D)F^{(j)}\}_{j=1}^{\infty}\|_{L^p(\ell^1)}\\
&=\sup_{\substack{G\in L^{p'}(\mathbb{R}^n)\\ \|G\|_{p'}=1}}\sup_{\substack{\{a_j\}_{j=1}^{\infty}\in L^{\infty}(\ell^{\infty},\mathbb{R}^n)\\ \|a_j\|_{L^{\infty}(\ell^{\infty})}=1}}\left|\int_{\mathbb{R}^n}\sum_{j=1}^{\infty}F^{(j)}(x)\varphi_j(D)[a_jG](x)\mathrm{d}x\right|.
\end{aligned}
$$

By the Hardy–Littlewood maximal operator, we obtain

$$
\begin{aligned}
\|\{\varphi_j(D)F^{(j)}\}_{j=1}^{\infty}\|_{L^p(\ell^1)}&\lesssim\sup_{\substack{G\in L^{p'}(\mathbb{R}^n)\\ \|G\|_{p'}=1}}\sup_{\substack{\{a_j\}_{j=1}^{\infty}\in L^{\infty}(\ell^{\infty},\mathbb{R}^n)\\ \|a_j\|_{L^{\infty}(\ell^{\infty})}=1}}\int_{\mathbb{R}^n}\sum_{j=1}^{\infty}|F^{(j)}(x)|M[a_jG](x)\mathrm{d}x\\
&\le\sup_{\substack{G\in L^{p'}(\mathbb{R}^n)\\ \|G\|_{p'}=1}}\sup_{\substack{\{a_j\}_{j=1}^{\infty}\in L^{\infty}(\ell^{\infty},\mathbb{R}^n)\\ \|a_j\|_{L^{\infty}(\ell^{\infty})}=1}}\int_{\mathbb{R}^n}\sum_{j=1}^{\infty}|F^{(j)}(x)|MG(x)\mathrm{d}x\\
&\lesssim\|\{F^{(j)}\}_{j=1}^{\infty}\|_{L^p(\ell^1)}.
\end{aligned}
$$

Therefore, the proof is complete.

We can describe duality of Besov spaces and Triebel–Lizorkin spaces using Theorem 1.15 together with duality $L^p(\mathbb{R}^n)$-$L^{p'}(\mathbb{R}^n)$ and $\ell^q(\mathbb{N}_0)$-$\ell^{q'}(\mathbb{N}_0)$. See [4, Theorems 2.44 and 2.45] for the duality $L^p(\mathbb{R}^n)$-$L^{p'}(\mathbb{R}^n)$, which is known as the Riesz representation theorem. Based on the duality $L^p(\mathbb{R}^n)$-$L^{p'}(\mathbb{R}^n)$, we can describe duality of Besov spaces and Triebel–Lizorkin spaces.

Theorem 2.17 (Duality of Besov spaces and Triebel–Lizorkin spaces) *Suppose that the parameters p, q, s satisfy $1 \le p < \infty$, $1 \le q < \infty$, $s \in \mathbb{R}$ for Besov spaces and $1 < p < \infty$, $1 \le q < \infty$, $s \in \mathbb{R}$ for Triebel–Lizorkin spaces.*

1. *Let* $g \in A^{-s}_{p'q'}(\mathbb{R}^n)$. *Then*

$$|\langle g, \eta \rangle| \lesssim \|g\|_{A^{-s}_{p'q'}} \|\eta\|_{A^s_{pq}} \tag{2.91}$$

for all $\eta \in \mathscr{S}(\mathbb{R}^n)$. *Hence* $\eta \in \mathscr{S}(\mathbb{R}^n) \mapsto \langle g, \eta \rangle \in \mathbb{C}$ *extends to a continuous linear functional on* $A^s_{pq}(\mathbb{R}^n)$.

2. *Let* L_0 *be a bounded linear functional on* $A^s_{pq}(\mathbb{R}^n)$. *Then there exists* $g \in A^{-s}_{p'q'}(\mathbb{R}^n)$ *such that*

$$\|g\|_{A^{-s}_{p'q'}} \lesssim \|L_0\|_{(A^s_{pq}(\mathbb{R}^n))^*}, \quad L_0(\eta) = \langle g, \eta \rangle \tag{2.92}$$

for all $\eta \in \mathscr{S}(\mathbb{R}^n)$.

Proof Concentrate on Triebel–Lizorkin spaces. For the time being, we suppose that $q > 1$. We indicate later the modification of the proof when $q = 1$ in Remark 2.6. Choose even functions $\varphi, \psi \in \mathscr{S}(\mathbb{R}^n)$ so that

$$0 \le \psi \le \chi_{B(4)}, \quad 0 \le \varphi \le \chi_{B(2)\setminus B(1)}$$

and that

$$\psi^2 + \sum_{j=1}^{\infty} \varphi_j{}^2 \equiv 1. \tag{2.93}$$

Now that ψ and φ are even and that they satisfy (2.93), we have:

$$\begin{aligned}
\langle g, \eta \rangle &= \langle \psi(D)^2 g, \eta \rangle + \sum_{j=1}^{\infty} \langle \varphi_j(D)^2 g, \eta \rangle \\
&= \langle \psi(D) g, \psi(D)\eta \rangle + \sum_{j=1}^{\infty} \langle \varphi_j(D) g, \varphi_j(D)\eta \rangle.
\end{aligned}$$

We use Hölder's inequality twice. The result is:

$$\begin{aligned}
|\langle g, \eta \rangle| &\le \int_{\mathbb{R}^n} |\psi(D)g(x)\psi(D)\eta(x)|\mathrm{d}x + \sum_{j=1}^{\infty} \int_{\mathbb{R}^n} |\varphi_j(D)g(x)\varphi_j(D)\eta(x)|\mathrm{d}x \\
&\le \|g\|_{F^{-s}_{p'q'}} \|\eta\|_{F^s_{pq}} + \|2^{-js}\varphi_j(D)g\|_{L^{p'}(\ell^{q'})} \|2^{js}\varphi_j(D)\eta\|_{L^p(\ell^q)} \\
&\le 2\|g\|_{F^{-s}_{p'q'}} \|\eta\|_{F^s_{pq}}.
\end{aligned}$$

Thus, (2.92) is proved.

Now let $L_0 \in (A^s_{pq}(\mathbb{R}^n))^*$. Choose $\psi, \varphi \in \mathscr{S}(\mathbb{R}^n)$ so that

$$\chi_{B(1)} \le \psi \le \chi_{B(4)}, \quad 0 \le \varphi \le \chi_{B(4)\setminus B(1)}, \quad \psi + \sum_{j=1}^{\infty} \varphi_j \equiv 1. \tag{2.94}$$

Define a linear functional L on $L^p(\ell^q)$ by

$$L(\{F^{(j)}\}_{j=0}^{\infty}) \equiv L_0\left(\sum_{j=0}^{\infty} 2^{-js}\varphi_j(D)F^{(j)}\right), \quad \{F^{(j)}\}_{j=0}^{\infty} \in L^p(\ell^q). \tag{2.95}$$

The mapping L actually defines a bounded linear functional on $L^p(\ell^q)$. Indeed, by Theorems 1.49 and 1.53, we have

$$\begin{aligned}
&\left|L_0\left(\psi(D)F^{(0)} + \sum_{j=1}^{\infty} 2^{-js}\varphi_j(D)F^{(j)}\right)\right| \\
&\lesssim \left\|\psi(D)F^{(0)} + \sum_{j=1}^{\infty} 2^{-js}\varphi_j(D)F^{(j)}\right\|_{F^s_{pq}} \|L_0\|_{(F^s_{pq}(\mathbb{R}^n))^*} \\
&\lesssim \left(\|\psi(D)F^{(0)}\|_p + \|\{\varphi_j(D)F^{(j)}\}_{j=1}^{\infty}\|_{L^p(\ell^q)}\right) \|L_0\|_{(F^s_{pq}(\mathbb{R}^n))^*} &(2.96)\\
&\lesssim \|\{MF^{(j)}\}_{j=0}^{\infty}\|_{L^p(\ell^q)} \|L_0\|_{(F^s_{pq}(\mathbb{R}^n))^*} &(2.97)\\
&\lesssim \|\{F^{(j)}\}_{j=0}^{\infty}\|_{L^p(\ell^q)} \|L_0\|_{(F^s_{pq}(\mathbb{R}^n))^*}. &(2.98)
\end{aligned}$$

Hence there exists a sequence $\{h^{(j)}\}_{j=1}^{\infty} \in L^{p'}(\ell^{q'})$ that realizes L. That is, we can find $\{h^{(j)}\}_{j=1}^{\infty} \in L^{p'}(\ell^{q'})$ such that:

$$L(\{F^{(j)}\}_{j=0}^{\infty}) = \sum_{j=0}^{\infty} \int_{\mathbb{R}^n} F^{(j)}(x)h^{(j)}(x)\mathrm{d}x \tag{2.99}$$

and that $\|\{h^{(j)}\}_{j=0}^{\infty}\|_{L^{p'}(\ell^{q'})} \lesssim \|L_0\|_{(F^s_{pq}(\mathbb{R}^n))^*}$. We define

$$g \equiv \psi(D)h^{(0)} + \sum_{j=1}^{\infty} 2^{js}\varphi_j(D)h^{(j)}. \tag{2.100}$$

We ignore the problem of the convergence in the right-hand side. We leave the matter to interested readers; see Exercise 2.29. By Theorem 1.53,

$$\|g\|_{F^{-s}_{p'q'}} \lesssim \|\{h^{(j)}\}_{j=0}^{\infty}\|_{L^{p'}(\ell^{q'})} \lesssim \|L_0\|_{(F^s_{pq}(\mathbb{R}^n))^*} < \infty.$$

Here let $F^{(0)} \equiv \psi(D)h^{(0)}$, $F^{(j)} \equiv 2^{js}\varphi_j(D)h^{(j)}$ for $j \in \mathbb{N}$. Then by (2.93) and (2.99), we have

$$L_0(\eta) = L_0\left(\psi(D)[\psi(D)\eta] + \sum_{j=1}^{\infty}\varphi_j(D)[\varphi_j(D)\eta]\right) = L(\{F^{(j)}\}_{j=0}^{\infty}).$$

for $\eta \in \mathscr{S}(\mathbb{R}^n)$. We have

$$\begin{aligned} L_0(\eta) &= \int_{\mathbb{R}^n}\sum_{j=0}^{\infty}h^{(j)}(x)F^{(j)}(x)\mathrm{d}x \\ &= \int_{\mathbb{R}^n}\left(h^{(0)}(x)\psi(D)\eta(x) + \sum_{j=1}^{\infty}2^{js}h^{(j)}(x)\varphi_j(D)\eta(x)\right)\mathrm{d}x \end{aligned}$$

by (2.99). Now that ψ and φ are even,

$$L_0(\eta) = \left\langle \psi(D)h^{(0)} + \sum_{j=0}^{\infty}2^{js}\varphi_j(D)h^{(j)}, \eta\right\rangle = \langle g, \eta\rangle.$$

Hence g is the desired function satisfying (2.92).

Three helpful remarks about the proof may be in order.

Remark 2.6

1. We invoke Lemma 2.3 to go from (2.97) to (2.98) when $q = 1$.
2. We can not go from (2.97) to (2.98) when $A = B$, $p = 1$. When $A = B$, $p = 1$, we invoke (2.96) to prove (2.98) directly using the Young inequality.
3. Let $A = B$, and let $1 \le p, q \le \infty$, $s \in \mathbb{R}$. From the proof of the theorem, for $f \in B^s_{pq}(\mathbb{R}^n)$ and $g \in B^{-s}_{p'q'}(\mathbb{R}^n)$, we can define

$$\langle f, g\rangle = \int_{\mathbb{R}^n}\psi(D)f(x)\psi(D)g(x)\mathrm{d}x + \sum_{j=1}^{\infty}\int_{\mathbb{R}^n}\varphi_j(D)f(x)\varphi_j(D)g(x)\mathrm{d}x.$$

A calculation similar to Theorem 2.17 works to have $B^{-s}_{p'q'}(\mathbb{R}^n) \hookrightarrow (B^s_{pq}(\mathbb{R}^n))^*$ for $1 \le p, q, \le \infty$ and $s \in \mathbb{R}$.

Exercises

Exercise 2.28 (min(p, q, 1)-triangle inequality) Let $0 < p, q \le \infty$ and $s \in \mathbb{R}$. Define $\eta \equiv \min(p, q, 1)$. Then show that $\|f + g\|_{A^s_{pq}}^{\eta} \le \|f\|_{A^s_{pq}}^{\eta} + \|g\|_{A^s_{pq}}^{\eta}$ for $f, g \in A^s_{pq}(\mathbb{R}^n)$, which is called the *$\eta$-triangle inequality*. Hint: See Sect. 1.1.1.

Exercise 2.29 Show that the series in the right-hand side of (2.100) defining g converges in $\mathscr{S}'(\mathbb{R}^n)$. Hint: Use the Fatou property.

Exercise 2.30 (Interpolation of the norms) Let $\varepsilon > 0$. Then show that there exists $c_\varepsilon > 0$ which depends on p, q, s and ε such that $\|f\|_{A^s_{pq}} \le \varepsilon \|f\|_{A^{s+1}_{pq}} + c_\varepsilon \|f\|_{A^{s-1}_{pq}}$. for all $f \in A^{s+1}_{pq}(\mathbb{R}^n)$. Hint: Refine the proof of Theorem 2.12, where we considered $(1 - \Delta)^{-1}$ instead of $(\varepsilon^a - \Delta)^{-1}$ for some $a \in \mathbb{R}$.

Exercise 2.31 Choose $\psi, \varphi \in \mathscr{S}(\mathbb{R}^n)$ so that

$$\chi_{B(1.6)} \le \psi \le \chi_{B(1.7)}, \quad \varphi = \psi - \psi(2\star). \tag{2.101}$$

In this exercise, the Besov norm $\|f\|_{B^s_{pq}}$ is specified by

$$\|f\|_{B^s_{pq}} \equiv \|\psi(D)f\|_p + \left(\sum_{j=1}^{\infty} 2^{jqs} \|\varphi_j(D)f\|_p{}^q\right)^{\frac{1}{q}}$$

using ψ and φ satisfying (2.101).

Furthermore, we choose $\kappa \in \mathscr{S}(\mathbb{R}^n)$ so that $\chi_{B(2.1)\setminus B(1.9)} \le \kappa \le \chi_{B(2.2)\setminus B(1.8)}$. For $a = \{a_j\}_{j=1}^{\infty} \in \ell^q(\mathbb{N})$, let $f_a \equiv \sum_{j=1}^{\infty} 2^{-j(n+s-n/p)} a_j \mathscr{F}^{-1}[\kappa_j]$.

1. Calculate $\|f_a\|_{B^s_{pq}}$ to show

$$\|f_a\|_{B^s_{pq}} \sim \left(\sum_{j=1}^{\infty} |a_j|^q\right)^{\frac{1}{q}}. \tag{2.102}$$

2. Let $\tilde{q} > q$. Disprove that $B^{s_0}_{p_0\tilde{q}}(\mathbb{R}^n) \hookrightarrow B^{s_1}_{p_1q}(\mathbb{R}^n)$ using (2.102) in Theorem 2.14.

Exercise 2.32 Show (2.56), (2.57), (2.59), and (2.60). Furthermore, show that

$$1 + (-\Delta)^m : A^s_{pq}(\mathbb{R}^n) \approx A^{s-2m}_{pq}(\mathbb{R}^n)$$

and that

$$1 + \partial_1{}^{4m} + \cdots + \partial_n{}^{4m} : A^s_{pq}(\mathbb{R}^n) \approx A^{s-4m}_{pq}(\mathbb{R}^n).$$

Hint: Mimic the proof of Theorem 2.12 or use Theorem 2.12 itself.

Exercise 2.33 Let $A = \{a_{ij}\}_{i,j=1,\ldots,n}$ be a positive definite matrix. Then we define the second elliptic differential operator L by

$$L \equiv 1 - \sum_{i,j=1}^{n} a_{ij} \partial_{x_i} \partial_{x_j}.$$

Show that $L : A^s_{pq}(\mathbb{R}^n) \approx A^{s-2}_{pq}(\mathbb{R}^n)$ is an isomorphism by using Thoerem 2.12 and by reexamining its proof.

Exercise 2.34 Use Fatou's property to prove that the three spaces X, Y, Z below cannot be realized as $A^s_{pq}(\mathbb{R}^n)$ no matter how we choose p, q, s.

1. $X = L^1(\mathbb{R}^n)$, Hint: Consider the sequence $f_j = j^n \chi_{(0,j^{-1})^n}$ for $j \in \mathbb{N}$.
2. $Y = \mathrm{BC}(\mathbb{R}^n)$, Hint: Consider the sequence $g_j = \min(1, \max(-1, j\star_1))$ for $j \in \mathbb{N}$.
3. $Z = \mathrm{BUC}(\mathbb{R}^n)$. Hint: Consider the sequence g_j above for $j \in \mathbb{N}$.

Exercise 2.35 (Diversity of the spaces $L^p(\mathbb{R}^n)$ and ℓ^q) Let $0 < p_0, p_1, q_0, q_1 \leq \infty$. If $L^{p_0}(\mathbb{R}^n) = L^{p_1}(\mathbb{R}^n)$ and $\ell^{q_0}(\mathbb{N}) = \ell^{q_1}(\mathbb{N})$ as sets, show that $p_0 = p_1, q_0 = q_1$. Hint: Start with sequence spaces.

Exercise 2.36 Define $A^{s+}_{pq}(\mathbb{R}^n) \equiv \bigcup_{\varepsilon>0} A^{s+\varepsilon}_{pq}(\mathbb{R}^n)$ for $0 < p, q \leq \infty$ and $s \in \mathbb{R}$. Exclude the case where $p = \infty$ if $A = F$.

1. Show that $A^{s+}_{pq}(\mathbb{R}^n)$ is a linear space.
2. Show that $A^{s+}_{pq}(\mathbb{R}^n) = \bigcup_{(\varepsilon,r)\in(0,\infty)\times(0,\infty]} A^{s+\varepsilon}_{pr}(\mathbb{R}^n)$, so that $A^{s+}_{pq}(\mathbb{R}^n)$ is independent of q as a set.
3. Show that $A^{s+}_{pq}(\mathbb{R}^n) \subsetneq A^s_{pq}(\mathbb{R}^n)$. Hint: Refine the example of the functions used in the proof of Theorem 2.16.

2.3.3 Modulation Spaces

To conclude this section, we investigate modulation spaces, which are defined in a similar way to Besov spaces. The history of modulation spaces is shorter than that of Besov spaces, but these spaces shed light on the boundedness of pseudo-differential operators in Sect. 5.2.

2.3.3.1 Definition of Modulation Spaces

The definition of modulation spaces and that of Besov spaces look alike. If we use Theorem 1.53, which is a key theorem to the study of Besov spaces, we can easily define and investigate modulation spaces.

Definition 2.10 (Modulation space) Let $0 < p, q \leq \infty$ and $s \in \mathbb{R}$. Suppose that a real-valued function $\tau \in C_{\mathrm{c}}^{\infty}(\mathbb{R}^n)$ satisfies

$$\inf_{x \in \mathbb{R}^n} \left(\sum_{k \in \mathbb{Z}^n} \tau(x-k) \right) > 0. \tag{2.103}$$

Then define

$$\begin{aligned} \|f\|_{M^s_{pq}} &\equiv \left(\sum_{k \in \mathbb{Z}^n} (\langle k \rangle^s \, \|\tau(D-k)f\|_p)^q \right)^{\frac{1}{q}} \\ &\simeq_n \left(\sum_{k \in \mathbb{Z}^n} (\langle k \rangle^s \, \|\mathscr{F}^{-1}[\tau(\star - k)] * \mathscr{F}f\|_p)^q \right)^{\frac{1}{q}}. \end{aligned}$$

Denote by $M^s_{pq}(\mathbb{R}^n)$ the set of all $f \in \mathscr{S}'(\mathbb{R}^n)$ for which $\|f\|_{M^s_{pq}} < \infty$. The space $M^s_{pq}(\mathbb{R}^n)$ is called the *modulation space*.

Unlike Besov spaces, modulation spaces use the uniform partition.

We make a notational remark.

Remark 2.7 Some prefer to write $\square_k \equiv \psi(D-k)$.

Among the spaces $M^s_{pq}(\mathbb{R}^n)$, the following spaces have special names.

Example 2.10

1. A *Banach algebra* is a Banach space equipped with the multiplication $* : X \times X \to X$ such that $\|x * y\|_X \leq \|x\|_X \|y\|_X$ for all $x, y \in X$. and that $(x+y)*z = x*z+y*z$, $x*(y+z) = x*y+x*z$, $(x*y)*z = x*(y*z)$ for all $x, y, z \in X$. A *Segal algebra* on $\mathbb{R}^n$, defined in [76, Chapter 6], is a Banach algebra of the dense subspace $L^1(\mathbb{R}^n)$ whose norm is translation invariant. The space $M^0_{11}(\mathbb{R}^n)$ is a *Segal algebra*. This space deserves its name because it is closed under pointwise multiplication; see Theorem 4.39 for a similar assertion.
2. The space $M^0_{\infty 1}(\mathbb{R}^n)$ is called the *Sjöstrand class* [973, 974]. Sjöstrand's class contains the *Hörmander class* S^0_{00}, defined in Definition 5.10, and also includes nonsmooth symbols. He proved some fundamental results about the L^2-boundedness and the algebra property.

We have an analogy with Besov spaces for modulation spaces.

Theorem 2.18 *Let* $0 < p, q \leq \infty$ *and* $s \in \mathbb{R}$.

1. *The definition of* $M^s_{pq}(\mathbb{R}^n)$ *does not depend on the choice of* τ *satisfying* (2.103).
2. $M^s_{pq}(\mathbb{R}^n)$ *is a quasi-Banach space.*
3. $\mathscr{S}(\mathbb{R}^n) \hookrightarrow M^s_{pq}(\mathbb{R}^n) \hookrightarrow \mathscr{S}'(\mathbb{R}^n)$.
4. *When* $0 < p, q < \infty$, $\mathscr{S}(\mathbb{R}^n)$ *is dense in* $M^s_{pq}(\mathbb{R}^n)$.
5. *When* $1 \le p \le \infty$, $M^0_{p1}(\mathbb{R}^n) \hookrightarrow L^p(\mathbb{R}^n) \hookrightarrow M^0_{p\infty}(\mathbb{R}^n)$.

We have the following embedding: Unlike Besov spaces and Triebel–Lizorkin spaces, the index behaves like p in $\ell^p(\mathbb{N})$.

Theorem 2.19 *Let* $0 < p_0 \le p_1 \le \infty$, $0 < q_0 \le q_1 \le \infty$, $-\infty < s_1 \le s_0 < \infty$. *Then* $M^{s_0}_{p_0q_0}(\mathbb{R}^n) \hookrightarrow M^{s_1}_{p_1q_1}(\mathbb{R}^n)$.

Proof We may assume that $s = s_0 = s_1$. We will prove $M^s_{p_0q_0}(\mathbb{R}^n) \hookrightarrow M^s_{p_0q_1}(\mathbb{R}^n)$ and $M^s_{p_0q_1}(\mathbb{R}^n) \hookrightarrow M^s_{p_1q_1}(\mathbb{R}^n)$. The first embedding is obtained from Corollary 1.8. Meanwhile, the second embedding follows from the embedding of the underlying sequence spaces $\ell^{q_0}(\mathbb{Z}^n) \hookrightarrow \ell^{q_1}(\mathbb{Z}^n)$.

Exercises

Exercise 2.37 Prove that $M^0_{\infty 1}(\mathbb{R}^n)$ is continuously embedded into $\mathrm{BC}(\mathbb{R}^n)$ by choosing $\psi \in \mathscr{S}(\mathbb{R}^n)$ satisfying $\sum\limits_{k\in\mathbb{Z}^n} \psi(\star - k) \equiv 1$.

Exercise 2.38 Show $\chi_{[-1,1]} \in L^1(\mathbb{R}) \setminus M^0_{11}(\mathbb{R})$. Hint: To disprove $\chi_{[-1,1]} \in M^0_{11}(\mathbb{R})$, choose a suitable partition of unity.

Exercise 2.39 Let $f(t) \equiv \max(0, 1 - |t|)$ for $t \in \mathbb{R}$. Then show that $f \in M^0_{\infty 1}(\mathbb{R})$. Hint: Let $g = \chi_{(-1/2,1/2)}$. Then $f = g * g$.

Exercise 2.40 Show that $\mathscr{F} : L^1(\mathbb{R}^n) \mapsto M^0_{\infty 1}(\mathbb{R}^n)$ is bounded. Hint: Reexamine the proof of Theorem 2.9.

Exercise 2.41 [681, Lemmas 4.1 and 4.2] Let $0 < p \le 1$ and $q \in (0, 2] \cup \{\infty\}$. Assume that $a \in L^\infty_{\mathrm{c}}(\mathbb{R}^n) \cap \mathscr{P}_L(\mathbb{R}^n)^\perp$ is supported on a cube $Q \equiv Q(x, r)$ with $0 < r \le 1$ and that $\|a\|_\infty \le |Q|^{-n/p}$. If $L \gg 1$, then show that a is contained in a bounded set in $M^{n-n/p-n/q}_{pq}(\mathbb{R}^n)$.

Exercise 2.42 [1021, Lemma 3.2] Let $1 < p, q < \infty$, and let $\varphi \in \mathscr{S}(\mathbb{R}^n)$ satisfy $\chi_{B(1/4)} \le \psi \le \chi_{B(1/2)}$. Define

$$f_{j,\varepsilon,\delta} \equiv \sum_{k' \in B(2^{j\delta})\setminus\{0\}} |k'|^{-n/q-\varepsilon} \mathrm{e}^{ik'(\star - k')} \mathscr{F}^{-1}\psi(\star - k'),$$

where $0 < \delta < 1$ and $j_0 \in \mathbb{N}_0$ is chosen to satisfy

$$\sqrt[4]{2}(1 - 2^{j_0(\delta-1)+1}) \ge 1, \quad 2^{3-j_0\delta/2}\sqrt{n} \le 1.$$

Then show that $\|f_{j,\varepsilon,\delta}(2^{j\delta}\star)\|_{M^0_{pq}} \lesssim 2^{j\delta n(\frac{1}{q}-1)}$.

Exercise 2.43 [1021, Lemma 3.6] Let $1 < p, q < \infty$, and let $\varphi, \psi \in \mathscr{S}(\mathbb{R}^n)$ satisfy

$$\chi_{B(1/8)} \le \varphi \le \chi_{B(1/4)}, \quad \psi \equiv \varphi_1.$$

Moreover, we assume that φ is radial. Define

$$g_{j,\varepsilon,\delta} \equiv \sum_{k,k' \in B(2^{j\delta})\setminus\{0\}} |k'|^{-n/q-\varepsilon} e^{ik'(\star-k')-ik(\star-k)} \mathscr{F}^{-1}\varphi(\star-k)\mathscr{F}^{-1}\psi(\star-k'),$$

where $0 < \delta < 1$ and $j_0 \in \mathbb{N}_0$ is chosen to satisfy

$$\sqrt[4]{2}(1-2^{j_0(\delta-1)+1}) \ge 1, \quad 2^{3-j_0\delta/2}\sqrt{n} \le 1.$$

Also define

$$B(t_1, t_2, \dots, t_n) \equiv \prod_{j=1}^{n} \chi_{[-1,0]} * \chi_{[0,1]}(t_i) \quad (t_1, t_2, \dots, t_n) \in \mathbb{R}^n.$$

1. Show that $B \in M^0_{pq}(\mathbb{R}^n)$.
2. Show that $\|g_{j,\varepsilon,\delta}(2^{j\delta}\star)\|_{M^0_{pq}} \gtrsim 2^{-j\delta(n/q+\varepsilon)}$.

Exercise 2.44 [1020, Lemma 3.8] Let $1 < p, q < \infty$ and $\varepsilon > 0$. Set

$$f \equiv \sum_{k \in \mathbb{Z}^n\setminus\{0\}} |k|^{-n/q-\varepsilon} e^{ik\cdot\star} \exp(-|\star|^2).$$

1. Show that the sum defining f converges in $\mathscr{S}'(\mathbb{R}^n)$.
2. Show that $f \in M^0_{pq}(\mathbb{R}^n)$.
3. Show that $\varphi \equiv \exp(-|\star|^2) \in M^0_{p'q'}(\mathbb{R}^n)$.
4. Show that $\|f(\lambda\star)\|_{M^0_{pq}} \gtrsim \lambda^{n(\frac{1}{q}-1)+\varepsilon}$ for all $\lambda \in (0,1)$.

Exercise 2.45 [1020, Lemma 3.9] Let $1 < p, q < \infty$ and $\varepsilon > 0$. Suppose that $\psi \in C^\infty(\mathbb{R}^n)$ satisfy $\chi_{Q(1/4)} \le \psi \le \chi_{Q(1/2)}$. Define $f \equiv \sum_{k \in \mathbb{Z}^n\setminus\{0\}} |k|^{-n/q-\varepsilon}$ $e^{ik\cdot\star}\psi(\star-k)$.

1. Show that the sum defining f converges in $\mathscr{S}'(\mathbb{R}^n)$.
2. Show that $f \in M^0_{pq}(\mathbb{R}^n)$.

3. Show that $\|f(\lambda\star)\|_{M^0_{pq}} \gtrsim \lambda^{-n(\frac{2}{p}-\frac{1}{q})+\varepsilon}$ for all $\lambda \in (0, 1)$.

Exercise 2.46 [1020, Lemma 3.10] Let $1 \le p \le \infty$ and $\varepsilon > 0$. Suppose that $\psi \in C^\infty(\mathbb{R}^n)$ satisfy $\chi_{Q(1/4)} \le \psi \le \chi_{Q(1/2)}$. Define $f \equiv \sum\limits_{k\in\mathbb{Z}^n\setminus\{0\}} \mathrm{e}^{ik\cdot\star}\psi(\star - k)$.

1. Show that the sum defining f converges in $\mathscr{S}'(\mathbb{R}^n)$.
2. Show that $f \in M^0_{p\infty}(\mathbb{R}^n)$.
3. Let $1 \le p \le 2$. Show that $\|f(\lambda\star)\|_{M^0_{p\infty}} \sim \lambda^{-n(\frac{2}{p}-\frac{1}{q})+\varepsilon}$ for all $\lambda \in (0, 1)$.

Exercise 2.47 [1020, Lemmas 4.1–4.3] Let $1 < p, q < \infty$ and $\varepsilon > 0$. Suppose that $\psi \in \mathscr{S}(\mathbb{R}^n)$ satisfies $\chi_{Q(1/2)} \le \psi \le \chi_{Q(1)}$. Define

$$f \equiv \sum_{\ell\in\mathbb{Z}^n\setminus\{0\}} \ell|^{-n/p-\varepsilon}\mathscr{F}^{-1}\psi(\star - \ell), \quad g \equiv e^{8\star_1 i} \sum_{\ell\in\mathbb{Z}^n\setminus\{0\}} \ell|^{-n/p-\varepsilon}\mathscr{F}^{-1}\psi(\star - \ell).$$

1. Show that the sum defining f converges in $\mathscr{S}'(\mathbb{R}^n)$.
2. Show that $f \in M^0_{p\infty}(\mathbb{R}^n)$.
3. Show that $\|f(\lambda\star)\|_{M^0_{p\infty}} \gtrsim \lambda^{-n/p-\varepsilon}$ for all $\lambda \in (1, \infty)$.
4. Show that $\|g(\lambda\star)\|_{M^0_{p\infty}} \gtrsim \lambda^{-n/p-\varepsilon}$ for all $\lambda \in (1, \infty)$.
5. Show that $\|g(2^k\star)\|_{B^s_{pq}} \lesssim 2^{k(s-n/p)}$.

Exercise 2.48 [1020, Lemmas 4.5 and 4.6] Let $1 < p, q < \infty$ and $\varepsilon, s > 0$. Suppose that $\psi \in \mathscr{S}(\mathbb{R}^n)$ satisfy $\chi_{Q(1/4)} \le \psi \le \chi_{Q(1/2)}$. Set

$$f_j \equiv 2^{-jn/p} \sum_{k\in([-2^j,2^j]\cap\mathbb{Z}\setminus\{0\})^n} |k|^{-\frac{n}{p}-\varepsilon}\mathrm{e}^{ikt}\mathscr{F}^{-1}\psi(\star - k) \quad (j \in \mathbb{N}).$$

1. Show that the sum defining f_j converges in $\mathscr{S}'(\mathbb{R}^n)$.
2. Show that $f_j \in M^0_{pq}(\mathbb{R}^n)$ for all $j \in \mathbb{N}$.
3. Show that $\|f_j\|_{M^0_{pq}} \gtrsim 2^{-jn(\frac{2}{p}-\frac{1}{q})-j\varepsilon}$ for all $j \in \mathbb{N}$.
4. Show that $\|f_j\|_{M^0_{pq}} \gtrsim 2^{-jn(\frac{2}{p}-\frac{1}{q})-j\varepsilon}$ for all $j \in \mathbb{N}$.
5. Show that $\|f_j\|_{B^s_{pq}} \lesssim 2^{j(s-\frac{n}{p})}$ for all $j \in \mathbb{N}$.

Exercise 2.49 Let $0 < p, q \le \infty$. Let $E = \exp(-|\star|^2)$ be the Gaussian defined in Example 1.4. One defines

$$V_g f(x, \xi) \equiv \langle f, E(\star - x)\mathrm{e}^{-i\cdot\xi}\rangle \quad ((x, \xi) \in \mathbb{R}^{2n}).$$

1. Show that $\|f\|_{M^0_{pq}}$ and $\|V_g f\|_{L^q_\xi L^p_x}$ are equivalent for $f \in \mathscr{S}'(\mathbb{R}^n)$.
2. Define the Wiener amalgam norm by $\|f\|_{W^0_{pq}} = \|V_g f\|_{L^p_x L^q_\xi}$ for $f \in \mathscr{S}'(\mathbb{R}^n)$. The *Wiener amalgam* space $W^0_{pq}(\mathbb{R}^n)$ is the set of all $f \in \mathscr{S}'(\mathbb{R}^n)$ for which $\|f\|_{W^0_{pq}} < \infty$. Show that $\mathscr{F} : M^0_{pq}(\mathbb{R}^n) \to W^0_{pq}(\mathbb{R}^n)$ is an isomorphism.

The *Wiener amalgam space* $W^0_{pq}(\mathbb{R}^n)$ was defined in 1983 by Feichtinger [461–463].

Textbooks in Sect. 2.3

Nonhomogeneous Besov Spaces

For fundamental facts on function spaces, we refer to [99]. Indeed, most fundamental facts are originally covered in the full range of the parameters. See [2, Chapter 4], [7, Section 6.2] (only Besov spaces) and [33, Section 2.2.1] for nonhomogeneous Besov spaces and Triebel–Lizorkin spaces. The textbook [99] is an exhaustive account of nonhomogeneous Besov spaces. We followed [99, p. 46, Proposition 1] for our proof of Theorem 2.11. See also Grafakos [33, Definition 2.2.1] for a similar approach of defining the norm of $A^s_{pq}(\mathbb{R}^n)$. We refer to [23, 2.3.1] and [262, p. 1081] for the dilation properties of Besov spaces.

Nonhomogeneous Triebel–Lizorkin Spaces

See [99, p. 85, 2.5.4] for a detailed discussion of Lizorkin's approach. The spaces $\Lambda^r_{p,\theta}(\mathbb{R}^n)$ and $F^r_{p\theta}(\mathbb{R}^n)$ are isomorphic with norm equivalence. The key tool is the vector-valued inequality of the Hilbert transform; see [31, Corollary 2.13] for more details about this vector-valued inequality. See [37, p. 49, Proposition 2.63] for Theorem 2.12.

Modulation Spaces

We refer to [35, Chapters 11 and 12] for modulation spaces.

2.4 Homogeneous Besov Spaces and Homogeneous Triebel–Lizorkin Spaces

In Sect. 2.3, we have been dealing with Besov spaces and Triebel–Lizorkin spaces. Strictly speaking, these spaces are called nonhomogeneous Besov spaces and nonhomogeneous Triebel–Lizorkin spaces. The norms of these function spaces were defined as follows. That is, choose $\psi, \varphi \in \mathscr{S}(\mathbb{R}^n)$ so that

$$\chi_{B(4)} \le \psi \le \chi_{B(8)}, \quad \chi_{B(4)\setminus B(2)} \le \varphi \le \chi_{B(8)\setminus B(1)}$$

and define

$$\|f\|_{B^s_{pq}} \equiv \|\psi(D)f\|_p + \|\{2^{js}\varphi_j(D)f\}_{j=1}^\infty\|_{\ell^q(L^p)},$$

$$\|f\|_{F^s_{pq}} \equiv \|\psi(D)f\|_p + \|\{2^{js}\varphi_j(D)f\}_{j=1}^\infty\|_{L^p(\ell^q)}$$

for $f \in \mathscr{S}'(\mathbb{R}^n)$. As a simpler definition, we have homogeneous Besov spaces and homogeneous Triebel–Lizorkin spaces. We widely use homogeneous Besov spaces and homogeneous Triebel–Lizorkin spaces in partial differential equations. From the viewpoint of Fourier analysis, the action of Δ on f equals $\mathscr{F}^{-1}[|\xi|^2\mathscr{F}f]$ modulo a multiplicative constant. In nonhomogeneous spaces, we consider norms that kill the behavior of the function $|\xi|^2$ at $\xi = 0$. But in the homogeneous spaces, we can consider this behavior neatly.

As is stated later, for $f \in \mathscr{S}'(\mathbb{R}^n)$, the homogeneous Besov norm $\|\star\|_{\dot{B}^s_{pq}}$ and the homogeneous Triebel–Lizorkin norm $\|\star\|_{\dot{F}^s_{pq}}$ are defined to be

$$\|f\|_{\dot{B}^s_{pq}} \equiv \|\{2^{js}\varphi_j(D)f\}_{j=-\infty}^\infty\|_{\ell^q(L^p)}, \quad 0 < p, q \le \infty, \quad s \in \mathbb{R},$$

$$\|f\|_{\dot{F}^s_{pq}} \equiv \|\{2^{js}\varphi_j(D)f\}_{j=-\infty}^\infty\|_{L^p(\ell^q)}, \quad 0 < p < \infty, \quad 0 < q \le \infty, \quad s \in \mathbb{R}.$$

In homogeneous Besov spaces, and homogeneous Triebel–Lizorkin spaces, as we will show in Theorem 2.28, we can consider the power $(-\Delta)^\alpha$ of the positive Laplacian $-\Delta$. However, the "norm" of the function spaces of homogeneous type $\dot{B}^s_{pq}(\mathbb{R}^n)$ and $\dot{F}^s_{pq}(\mathbb{R}^n)$ does not satisfy the condition of the norms; they annihilate polynomial functions as distributions. Instead of $\mathscr{S}'(\mathbb{R}^n)$ itself, we need to consider the quotient space $\mathscr{S}'(\mathbb{R}^n)/\mathscr{P}(\mathbb{R}^n)$. We want to equip $\mathscr{S}'(\mathbb{R}^n)/\mathscr{P}(\mathbb{R}^n)$ with a topology. Here we consider a topological vector space isomorphic to $\mathscr{S}'(\mathbb{R}^n)/\mathscr{P}(\mathbb{R}^n)$ in Sect. 2.4.1. Here by a topological vector space we mean a linear space equipped with a topology under which the addition and the scalar multiplication are both continuous. We investigate function spaces of homogeneous type (or for short, homogeneous spaces) in Sect. 2.4.2. We recall the following notation of the vector-valued norm in Sect. 2.4: Let $\{f_j\}_{j=-\infty}^\infty$ be a sequence of measurable functions on $\mathbb{R}^n$. Then define

$$\|f_j\|_{\ell^q(L^p)} \equiv \left(\sum_{j=-\infty}^\infty \|f_j\|_p{}^q\right)^{\frac{1}{q}}, \tag{2.104}$$

$$\|f_j\|_{L^p(\ell^q)} \equiv \left\|\left(\sum_{j=-\infty}^\infty |f_j|^q\right)^{\frac{1}{q}}\right\|_p. \tag{2.105}$$

Meanwhile, $\tau_j \equiv \tau(2^{-j}\star)$ for $\tau \in \mathscr{S}(\mathbb{R}^n)$ and $j \in \mathbb{Z}$ as we did in Sect. 2.3. As we will see, $\dot{A}^s_{pq}(\mathbb{R}^n)$ is not a subset of $\mathscr{S}'(\mathbb{R}^n)$. This is not pleasant because the subset of $\mathscr{S}'(\mathbb{R}^n)$ can be applied to many cases. Here we provide a way to make $\dot{A}^s_{pq}(\mathbb{R}^n)$ a subset of $\mathscr{S}'(\mathbb{R}^n)$ in Sect. 2.4.3.

2.4.1 $\mathscr{S}_\infty(\mathbb{R}^n)$ and its Dual $\mathscr{S}'_\infty(\mathbb{R}^n)$

Our purpose for the time being is to define the homogeneous space $\dot{A}^s_{pq}(\mathbb{R}^n)$. To this end, we need to get familiar with the quotient space $\mathscr{S}'(\mathbb{R}^n)/\mathscr{P}(\mathbb{R}^n)$. This section is a preparatory step for this purpose.

2.4.1.1 Definition of $\mathscr{S}_\infty(\mathbb{R}^n)$ and $\mathscr{S}'_\infty(\mathbb{R}^n)$

As we have explained, it is not good to focus upon $\mathscr{S}'(\mathbb{R}^n)$ when we consider the homogeneous spaces. We need to work on $\mathscr{S}'(\mathbb{R}^n)/\mathscr{P}(\mathbb{R}^n)$. Generally speaking, it is a bit nasty to consider the quotient space; handling the representative is not so intuitive. Therefore, we seek to find an expression of the quotient $\mathscr{S}'(\mathbb{R}^n)/\mathscr{P}(\mathbb{R}^n)$. From this standpoint, the following definition of $\mathscr{S}'_\infty(\mathbb{R}^n)$ is natural. We recall the definition of $\mathscr{S}_\infty(\mathbb{R}^n)$ see Definition 1.31 and expand our theory.

Definition 2.11 ($\mathscr{S}_\infty(\mathbb{R}^n)$ **and** $\mathscr{S}'_\infty(\mathbb{R}^n)$) Define the *Lizorkin function space* $\mathscr{S}_\infty(\mathbb{R}^n)$ by $\mathscr{S}_\infty(\mathbb{R}^n) \equiv \mathscr{S}(\mathbb{R}^n) \cap \mathscr{P}(\mathbb{R}^n)^\perp$. Equip $\mathscr{S}_\infty(\mathbb{R}^n)$ with the topology induced by $\mathscr{S}(\mathbb{R}^n)$. The *Lizorkin distribution space* $\mathscr{S}'_\infty(\mathbb{R}^n)$ is the topological dual space $\mathscr{S}_\infty(\mathbb{R}^n)$. That is, define

$$\begin{aligned}\mathscr{S}'_\infty(\mathbb{R}^n) &\equiv \{F \in \mathrm{Hom}_{\mathbb{C}}(\mathscr{S}_\infty(\mathbb{R}^n), \mathbb{C}) \,:\, F : \mathscr{S}_\infty(\mathbb{R}^n) \to \mathbb{C} \text{ is continuous}\}\\ &= \{F : \mathscr{S}_\infty(\mathbb{R}^n) \to \mathbb{C} \,:\, F \text{ is continuous and } \mathbb{C}\text{-linear}\}.\end{aligned}$$

Equip $\mathscr{S}'_\infty(\mathbb{R}^n)$ with the topology later.

Various other notation is also used for $\mathscr{S}'_\infty(\mathbb{R}^n)$.

Remark 2.8 Some prefer to write $\mathscr{Z}(\mathbb{R}^n)$ (see [99]) and $\mathscr{S}_0(\mathbb{R}^n)$ instead of $\mathscr{S}_\infty(\mathbb{R}^n)$ and $\mathscr{Z}'(\mathbb{R}^n)$ and $\mathscr{S}'_0(\mathbb{R}^n)$ instead of $\mathscr{S}'_\infty(\mathbb{R}^n)$.

2.4.1.2 Homogeneous Phase Decomposition

We now consider an analogue of Theorem 1.35 adapted to $\mathscr{S}'_\infty(\mathbb{R}^n)$. We write $\langle f, \varphi\rangle \equiv f(\varphi)$ for $\varphi \in \mathscr{S}_\infty(\mathbb{R}^n)$ and $f \in \mathscr{S}'_\infty(\mathbb{R}^n)$.

Proposition 2.11 *Choose smooth functions $\varphi : \mathbb{R}^n \to \mathbb{R}$ so that*

$$\mathrm{supp}(\varphi) \subset B(8) \setminus B(1), \qquad \sum_{j=-\infty}^{\infty} \varphi_j \equiv \chi_{\mathbb{R}^n \setminus \{0\}}. \tag{2.106}$$

We have:

1. *For all $\tau \in \mathscr{S}_\infty(\mathbb{R}^n)$, $\lim_{J\to\infty} \sum_{j=-J}^{J} \varphi_j(D)\tau = \tau$ in the topology of $\mathscr{S}_\infty(\mathbb{R}^n)$.*
2. *For all $F \in \mathscr{S}'_\infty(\mathbb{R}^n)$ and $\tau \in \mathscr{S}_\infty(\mathbb{R}^n)$,*

$$\lim_{J\to\infty} \left\langle \sum_{j=-J}^{J} \varphi_j(D)F, \tau \right\rangle = \langle F, \tau \rangle. \tag{2.107}$$

Note that (2.107) turns out to be equivalent to saying that $\lim_{J\to\infty} \sum_{j=-J}^{J} \varphi_j(D)F = F$ holds in the topology of $\mathscr{S}'_\infty(\mathbb{R}^n)$. We induce the topology so that this is the case.

Proof As we did for $\mathscr{S}(\mathbb{R}^n)$, it suffices to prove (2.106). Then duality entails (2.107). Let us estimate $\mathrm{I} \equiv p_{\alpha,\beta}\left(\mathscr{F}\left\{\tau - \sum_{j=-J}^{J} \varphi_j(D)\tau\right\}\right)$. We write out I in full:

$$\mathrm{I} = \sup_{x\in\mathbb{R}^n} \left| x^\alpha \partial^\beta \left(\mathscr{F}\tau(x) - \sum_{j=-J}^{J} \varphi_j(x)\mathscr{F}\tau(x) \right) \right|.$$

First of all, choose a test function $\tau \in \mathscr{S}_\infty(\mathbb{R}^n)$ arbitrarily. Furthermore, set X_J by $X_J \equiv B(2^{3-J}) \cup (\mathbb{R}^n \setminus B(2^J))$. Then

$$\begin{aligned}
\mathrm{I} &\lesssim_\beta \sup_{x\in\mathbb{R}^n} \left| \left(1 - \sum_{j=-J}^{J} \varphi_j(x) \right) x^\alpha \partial^\beta \mathscr{F}\tau(x) \right| \\
&\quad + \sup_{x\in\mathbb{R}^n} \sum_{\gamma,\delta\in\mathbb{N}_0{}^n, |\gamma|,|\delta|\le|\beta|} \left| x^\alpha \partial^\gamma \mathscr{F}\tau(x) \partial_x^\delta \left[\sum_{j=-J}^{J} \varphi(2^{-j}x) \right] \right| \\
&\lesssim_\beta \sup_{x\in X_J} \left| x^\alpha \partial^\beta \mathscr{F}\tau(x) \right| + \sum_{\gamma,\delta\in\mathbb{N}_0{}^n, |\gamma|,|\delta|\le|\beta|} \sup_{x\in X_J} \left| x^\alpha \partial^\gamma \mathscr{F}\tau(x)|x|^{-|\delta|} \right|.
\end{aligned}$$

Assuming that $\tau \in \mathscr{S}_\infty(\mathbb{R}^n)$, we learn I converges to 0 as $J \to \infty$. Thus,

$$\lim_{J\to\infty}\sum_{j=-J}^{J}\varphi_j(D)\tau=\tau. \tag{2.108}$$

Therefore, the proof is complete.

Proposition 2.11 is a discrete version of the reproducing formula. We also have a continuous version. We collect an application of Proposition 2.11 and the subsequent remark.

Proposition 2.12 *Let $\varphi \in C^\infty_{\rm c}(\mathbb{R}^n \setminus \{0\})$ be a function satisfying*

$$\int_0^\infty \varphi(t\xi)^2\frac{{\rm d}t}{t}=1 \quad (\xi \in \mathbb{R}^n \setminus \{0\}).$$

Assume that we have a family $\{R_s\}_{s>0}$ of bounded operators on $L^2(\mathbb{R}^n)$ such that for all $s,t>0$ $\|R_s\varphi(tD)\|_{B(L^2)} \le \min\left(\frac{t}{s},\frac{s}{t}\right)$. Then $\sqrt{\int_0^\infty \|R_s f\|_2{}^2\frac{{\rm d}s}{s}} \lesssim \|f\|_2$ for all $f \in L^2(\mathbb{R}^n)$.

Proof We observe

$$\|R_s f\|_2{}^2=\int_{\mathbb{R}^n}\lim_{R\to\infty}\left|R_s\left(\int_{R^{-1}}^R \varphi(tD)^2 f(x)\frac{{\rm d}t}{t}\right)\right|^2{\rm d}x,$$

since

$$\lim_{R\to\infty}\int_{R^{-1}}^R \varphi(tD)^2 f\frac{{\rm d}t}{t}=f$$

in $L^2(\mathbb{R}^n)$. Here and below in the proof we fix $R>0$. Since R_s is a bounded operator on $L^2(\mathbb{R}^n)$, by the Fatou lemma, we have

$$\|R_s f\|_2{}^2 \le \liminf_{R\to\infty}\int_{\mathbb{R}^n}\left|\int_{R^{-1}}^R R_s\varphi(tD)^2 f(x)\frac{{\rm d}t}{t}\right|^2{\rm d}x.$$

We dualize the right-hand side: choose $g \in L^2(\mathbb{R}^n)$ with norm 1 such that

$$\int_{\mathbb{R}^n}\left|\int_{R^{-1}}^R R_s\varphi(tD)^2 f(x)\frac{{\rm d}t}{t}\right|^2{\rm d}x=\left(\int_{\mathbb{R}^n}g(x)\left(\int_{R^{-1}}^R R_s\varphi(tD)^2 f(x)\frac{{\rm d}t}{t}\right){\rm d}x\right)^2.$$

Using the adjoint, we obtain

$$\int_{\mathbb{R}^n}g(x)\left(\int_{R^{-1}}^R R_s\varphi(tD)^2 f(x)\frac{{\rm d}t}{t}\right){\rm d}x$$

$$= \int_{\mathbb{R}^n} (R_s\varphi(tD))^* g(x) \left(\int_{R^{-1}}^{R} \varphi(tD) f(x) \frac{\mathrm{d}t}{t} \right) \mathrm{d}x$$

$$= \int_{\mathbb{R}^n \times (R^{-1},R)} (R_s\varphi(tD))^* g(x) \varphi(tD) f(x) \mathrm{d}x \frac{\mathrm{d}t}{t}.$$

If we use the Cauchy–Schwarz inequality and our assumption, then we have

$$\begin{aligned}
&\int_{\mathbb{R}^n} \left| \int_{R^{-1}}^{R} R_s\varphi(tD)^2 f(x) \frac{\mathrm{d}t}{t} \right|^2 \mathrm{d}x \\
&\le \int_{\mathbb{R}^n \times (R^{-1},R)} |(R_s\varphi(tD))^* g(x)|^2 \max\left(\frac{t}{s}, \frac{s}{t}\right) \mathrm{d}x \frac{\mathrm{d}t}{t} \\
&\quad \times \int_{\mathbb{R}^n \times (R^{-1},R)} |\varphi(tD) f(x)|^2 \min\left(\frac{t}{s}, \frac{s}{t}\right) \mathrm{d}x \frac{\mathrm{d}t}{t} \\
&\le \int_{\mathbb{R}^n \times (R^{-1},R)} |g(x)|^2 \min\left(\frac{t}{s}, \frac{s}{t}\right) \mathrm{d}x \frac{\mathrm{d}t}{t} \\
&\quad \times \int_{\mathbb{R}^n \times (R^{-1},R)} |\varphi(tD) f(x)|^2 \min\left(\frac{t}{s}, \frac{s}{t}\right) \mathrm{d}x \frac{\mathrm{d}t}{t} \\
&\simeq \int_{\mathbb{R}^n} |g(x)|^2 \mathrm{d}x \times \int_{\mathbb{R}^n \times (R^{-1},R)} |\varphi(tD) f(x)|^2 \min\left(\frac{t}{s}, \frac{s}{t}\right) \mathrm{d}x \frac{\mathrm{d}t}{t}.
\end{aligned}$$

If we integrate against $\dfrac{\mathrm{d}s}{s}$ over $(0, \infty)$, we obtain

$$\begin{aligned}
&\int_{\mathbb{R}^n} \left| \int_{R^{-1}}^{R} R_s\varphi(tD)^2 f(x) \frac{\mathrm{d}t}{t} \right|^2 \mathrm{d}x \\
&\lesssim \int_{\mathbb{R}^n \times (R^{-1},R)} |\varphi(tD) f(x)|^2 \min\left(\frac{t}{s}, \frac{s}{t}\right) \mathrm{d}x \frac{\mathrm{d}s}{s} \lesssim \|f\|_2{}^2
\end{aligned}$$

by the Plancherel theorem. If we let $R \to \infty$, then we obtain the desired result.

2.4.1.3 $\mathscr{S}'(\mathbb{R}^n)/\mathscr{P}(\mathbb{R}^n)$ and $\mathscr{S}'_\infty(\mathbb{R}^n)$ as Linear Spaces

Next, we show that $\mathscr{S}'(\mathbb{R}^n)/\mathscr{P}(\mathbb{R}^n)$ is isomorphic to $\mathscr{S}'_\infty(\mathbb{R}^n)$. We need the Hahn–Banach theorem; see Theorem 1.18.

An argument similar to Proposition 1.8 works and we have the following:

Proposition 2.13 *Let* $\{F_j\}_{j=1}^\infty \subset \mathscr{S}'_\infty(\mathbb{R}^n)$ *be a convergent sequence in* $\mathscr{S}'_\infty(\mathbb{R}^n)$. *Then there exists* $N \in \mathbb{N}$ *such that* $|\langle F_j, \varphi \rangle| \le N\, p_N(\varphi) \quad (\varphi \in \mathscr{S}_\infty(\mathbb{R}^n))$.

We leave the proof to interested readers; see Exercise 2.54.

We can consider the quotient topology linear space $\mathscr{S}'(\mathbb{R}^n)/\mathscr{P}(\mathbb{R}^n)$. See Definition 2.15 for its definition. As the first step to this purpose, we now review the quotient linear spaces. The quotient linear space $\mathscr{S}'(\mathbb{R}^n)/\mathscr{P}(\mathbb{R}^n)$ is given by the set of all the elements of

$$[f] \equiv f + \mathscr{P}(\mathbb{R}^n) = \{f + P \, : \, P \in \mathscr{P}(\mathbb{R}^n)\} \subset \mathscr{S}'(\mathbb{R}^n).$$

That is, as a set

$$\mathscr{S}'(\mathbb{R}^n)/\mathscr{P}(\mathbb{R}^n) \equiv \{[f] \, : \, f \in \mathscr{S}'(\mathbb{R}^n)\} \subset 2^{\mathscr{S}'(\mathbb{R}^n)}.$$

For $[f] \in \mathscr{S}'(\mathbb{R}^n)/\mathscr{P}(\mathbb{R}^n)$, we say that $[f]$ is a representative of $f \in \mathscr{S}'(\mathbb{R}^n)$. For $f, g \in \mathscr{S}'(\mathbb{R}^n)$, since $\mathscr{P}(\mathbb{R}^n)$ is a linear space, we have

$$[f] = [g] \Longleftrightarrow f - g \in \mathscr{P}(\mathbb{R}^n) \Longleftrightarrow f - g \text{ is (represented by) a polynomial.} \tag{2.109}$$

See Exercise 2.58 1.

Define the *addition* and the *scalar multiplication* by

$$[f] + [g] \equiv [f + g], \quad k \cdot [f] \equiv [k \cdot f], \quad f, g \in \mathscr{S}'(\mathbb{R}^n), \quad k \in \mathbb{C}. \tag{2.110}$$

See Exercise 2.58 2. Note that the multiplication and the addition, given by (2.110), do not depend on f, g; see Exercise 2.58.

Lemma 2.4 *Let $g \in \mathscr{S}'_\infty(\mathbb{R}^n)$. Then there exists $N \in \mathbb{N}$ such that*

$$|\langle g, \varphi \rangle| \le N p_N(\varphi) \tag{2.111}$$

for all $\varphi \in \mathscr{S}_\infty(\mathbb{R}^n)$.

Proof Suppose that $g : \mathscr{S}_\infty(\mathbb{R}^n) \to \mathbb{C}$ is continuous; our task is to find $N \in \mathbb{N}$ such that (2.111) holds. Let $\Delta(1) = \{z \in \mathbb{C} \, : \, |z| < 1\}$ as before. By the continuity of g, the set

$$g^{-1}(\Delta(1)) = \{\varphi \in \mathscr{S}_\infty(\mathbb{R}^n) \, : \, |g(\varphi)| < 1\}$$

is an open set of $\mathscr{S}_\infty(\mathbb{R}^n)$ that contains 0.

Therefore, if $L \in \mathbb{N}$ is sufficiently large, we conclude

$$\{\varphi \in \mathscr{S}_\infty(\mathbb{R}^n) \, : \, L\, p_L(\varphi) < 1\} \subset \{\varphi \in \mathscr{S}_\infty(\mathbb{R}^n) \, : \, |g(\varphi)| < 1\}. \tag{2.112}$$

Hence $|g(\varphi)| \le 1$ as long as $\varphi \in \mathscr{S}_\infty(\mathbb{R}^n)$ satisfies $2L\, p_{2L}(\varphi) = 1$.

Now we suppose that we have $\varphi \in \mathscr{S}_\infty(\mathbb{R}^n) \setminus \{0\}$. Then the function $\psi \equiv \dfrac{1}{2L\, p_{2L}(\varphi)}\varphi$ satisfies $2L\, p_{2L}(\psi) = 1$. Thus, $|g(\psi)| \le 1$. In view of the definition of ψ, we have $|g(\varphi)| \le 2L\, p_{2L}(\varphi)$, $\varphi \in \mathscr{S}_\infty(\mathbb{R}^n)\setminus\{0\}$. The case where $\varphi = 0$ can be readily incorporated. Thus, by letting $N = 2L$, we can choose N satisfying (2.111).

The following theorem is a consequence of elementary algebra.

Theorem 2.20 *As a linear space, we have* $\mathscr{S}'(\mathbb{R}^n)/\mathscr{P}(\mathbb{R}^n) \approx \mathscr{S}'_\infty(\mathbb{R}^n)$.

We construct a linear mapping, $\Phi : \mathscr{S}'(\mathbb{R}^n)/\mathscr{P}(\mathbb{R}^n) \to \mathscr{S}'_\infty(\mathbb{R}^n)$ and its inverse $\Psi : \mathscr{S}'_\infty(\mathbb{R}^n) \to \mathscr{S}'(\mathbb{R}^n)/\mathscr{P}(\mathbb{R}^n)$.

Proof (Step 1: The construction of the mapping $\Phi : \mathscr{S}'(\mathbb{R}^n)/\mathscr{P}(\mathbb{R}^n) \to \mathscr{S}'_\infty(\mathbb{R}^n)$*)* Let $P \in \mathscr{P}(\mathbb{R}^n)$; that is, P be a polynomial function. From the definition of $\mathscr{S}_\infty(\mathbb{R}^n)$ we have $\displaystyle\int_{\mathbb{R}^n} P(x)\tau(x)\mathrm{d}x = 0$ for all $\tau \in \mathscr{S}_\infty(\mathbb{R}^n)$.

Consequently, the restriction mapping $f \in \mathscr{S}'(\mathbb{R}^n) \mapsto f|_{\mathscr{S}_\infty(\mathbb{R}^n)} \in \mathscr{S}'_\infty(\mathbb{R}^n)$ factors through the projection mapping from $\mathscr{S}'(\mathbb{R}^n)$ to $\mathscr{S}'(\mathbb{R}^n)/\mathscr{P}(\mathbb{R}^n)$.

Proof (Step 2: The construction of the mapping $\Psi : \mathscr{S}'_\infty(\mathbb{R}^n) \to \mathscr{S}'(\mathbb{R}^n)/\mathscr{P}(\mathbb{R}^n)$*)* Let $f \in \mathscr{S}'_\infty(\mathbb{R}^n)$. Then there exists $N \in \mathbb{N}$ such that

$$|\langle f, \tau\rangle| \le N\, p_N(\tau) \tag{2.113}$$

for all $\tau \in \mathscr{S}_\infty(\mathbb{R}^n)$.

By the Hahn–Banach theorem (Theorem 1.18), f extends to a continuous linear functional F on $\mathscr{S}(\mathbb{R}^n)$ satisfying (2.113) for all $\tau \in \mathscr{S}(\mathbb{R}^n)$. Since $F \in \mathscr{S}'(\mathbb{R}^n)$, $[F] \in \mathscr{S}'(\mathbb{R}^n)/\mathscr{P}(\mathbb{R}^n)$ makes sense. We obtain F by the Hahn–Banach theorem, so that F is not uniquely determined. However, we claim

$$f \in \mathscr{S}'_\infty(\mathbb{R}^n) \mapsto [F] \in \mathscr{S}'(\mathbb{R}^n)/\mathscr{P}(\mathbb{R}^n) \tag{2.114}$$

does not depend on the choice of F. Once this is achieved, the mapping is linear by definition.

Let G be another extension of $f : \mathscr{S}_\infty(\mathbb{R}^n) \to \mathbb{C}$. Then if restricted to $\mathscr{S}_\infty(\mathbb{R}^n)$, F and G agree with f. So $F-G$ vanishes on $\mathscr{S}_\infty(\mathbb{R}^n)$. As a result, $\langle F-G, \mathscr{F}\varphi\rangle = 0$ for all $\varphi \in \mathscr{S}(\mathbb{R}^n)$ whose support is a compact set in $\mathbb{R}^n \setminus \{0\}$.

From Theorem 1.25, we can express $\mathscr{F}(F-G)$ as $\displaystyle\sum_{\alpha\in\mathbb{N}_0{}^n, |\alpha|\le L} a_\alpha \partial^\alpha \delta_0$. As a result, F and G differ by a polynomial.

Let us return to the proof of the theorem. From the definition of the mapping Φ, Ψ, these mappings are inverse to each other. So, there is a linear isomorphism giving $\mathscr{S}'_\infty(\mathbb{R}^n) \approx \mathscr{S}'(\mathbb{R}^n)/\mathscr{P}(\mathbb{R}^n)$.

Via the isomorphism $\mathscr{S}'_\infty(\mathbb{R}^n) \approx \mathscr{S}'(\mathbb{R}^n)/\mathscr{P}(\mathbb{R}^n)$ we temporarily define the topology of $\mathscr{S}'(\mathbb{R}^n)/\mathscr{P}(\mathbb{R}^n)$.

Definition 2.12 (The topology of $\mathscr{S}'(\mathbb{R}^n)/\mathscr{P}(\mathbb{R}^n)$) The topology of the quotient space $\mathscr{S}'(\mathbb{R}^n)/\mathscr{P}(\mathbb{R}^n)$ is induced by the isomorphism in Theorem 2.20.

From the property of the topology, for $\{[f_j]\}_{j=1}^{\infty} \in \mathscr{S}'(\mathbb{R}^n)/\mathscr{P}(\mathbb{R}^n)$, $\lim_{j\to\infty} [f_j] = [f]$ if and only if $\lim_{j\to\infty} \langle f_j, \varphi\rangle = \langle f, \varphi\rangle$ for all $\varphi \in \mathscr{S}_\infty(\mathbb{R}^n)$.

Definition 2.13 (Abbreviation of the elements $\mathscr{S}'(\mathbb{R}^n)/\mathscr{P}(\mathbb{R}^n)$) Unless possible confusion is not likely, for $f \in \mathscr{S}'(\mathbb{R}^n)$, identify $[f] \in \mathscr{S}'(\mathbb{R}^n)/\mathscr{P}(\mathbb{R}^n)$ with $f \in \mathscr{S}'(\mathbb{R}^n)/\mathscr{P}(\mathbb{R}^n)$.

We remark that this abuse of notation is done in many cases in this book or in the actual application.

2.4.1.4 $\mathscr{S}'(\mathbb{R}^n)/\mathscr{P}(\mathbb{R}^n)$ and $\mathscr{S}_\infty'(\mathbb{R}^n)$ as Topological Spaces

Since $\mathscr{S}_\infty(\mathbb{R}^n)$ is continuously embedded into $\mathscr{S}(\mathbb{R}^n)$, the dual operator R, called the restriction, is continuous from $\mathscr{S}'(\mathbb{R}^n)$ to $\mathscr{S}_\infty'(\mathbb{R}^n)$. We prove the following theorem:

Theorem 2.21 *The restriction mapping $R : f \in \mathscr{S}'(\mathbb{R}^n) \mapsto f|\mathscr{S}_\infty(\mathbb{R}^n) \in \mathscr{S}_\infty'(\mathbb{R}^n)$ is open; namely the image $R(U)$ is open in $\mathscr{S}_\infty'(\mathbb{R}^n)$ for any open set U in $\mathscr{S}'(\mathbb{R}^n)$.*

Next, we prove the following theorem, which is the heart of the matter:

Theorem 2.22 *Let $K, L \in \mathbb{N}$, and let $\Phi_1, \Phi_2, \dots, \Phi_L \in \mathscr{S}(\mathbb{R}^n) \setminus \mathscr{S}_\infty(\mathbb{R}^n)$ satisfy that $[\Phi_1], [\Phi_2], \dots, [\Phi_L] \in \mathscr{S}(\mathbb{R}^n)/\mathscr{S}_\infty(\mathbb{R}^n)$ is linearly independent in $\mathscr{S}(\mathbb{R}^n)/\mathscr{S}_\infty(\mathbb{R}^n)$ and let also $\varphi_1, \varphi_2, \dots, \varphi_K \in \mathscr{S}_\infty(\mathbb{R}^n)$. Then the image of*

$$\mathscr{U} \equiv \left(\bigcap_{k=1}^{K}\{F \in \mathscr{S}'(\mathbb{R}^n) : |\langle F, \varphi_k\rangle| < 1\}\right)\cap\left(\bigcap_{l=1}^{L}\{F \in \mathscr{S}'(\mathbb{R}^n) : |\langle F, \Phi_l\rangle| < 1\}\right)$$

by R is exactly

$$U \equiv \bigcap_{k=1}^{K}\{f \in \mathscr{S}_\infty'(\mathbb{R}^n) : |\langle f, \varphi_k\rangle| < 1\}.$$

Proof Note that the family of the mappings $\Theta_l : f \in \mathscr{P}(\mathbb{R}^n) \mapsto \langle f, \Phi_l\rangle \in \mathbb{C}$ is linearly independent for $l = 1, 2, \dots, L$. This means that $\bigcap_{l'\in\{1,2,\dots,L\}\setminus\{l\}} \ker(\Theta_{l'})$ is not contained in $\ker(\Theta_l)$ for all $l = 1, 2, \dots, L$ by the Helly theorem. Thus for all $l = 1, 2, \dots, L$, there exists $p_l \in \mathscr{P}(\mathbb{R}^n)$ such that $\langle p_l, \Phi_{l'}\rangle = \delta_{l'l}$ for all $l' = 1, 2, \dots, L$.

Let $\tilde{F} \in \mathscr{S}'(\mathbb{R}^n)$ be any extension of $f \in U$. Set

$$F \equiv \tilde{F} - \sum_{l=1}^{L}\langle \tilde{F}, \Phi_l\rangle p_l.$$

Then $\langle F, \Phi_l\rangle = 0$, implying that $F \in \mathscr{U}$.

Theorem 2.23 *Let $K, L, L^* \in \mathbb{N}_0$, let $\varphi_1, \varphi_2, \dots, \varphi_K \in \mathscr{S}_\infty(\mathbb{R}^n)$, and let*

$$\Phi_1, \Phi_2, \dots, \Phi_L, \dots, \Phi_{L^*} \in \mathscr{S}(\mathbb{R}^n) \setminus \mathscr{S}_\infty(\mathbb{R}^n).$$

Assume

$$[\Phi_1], [\Phi_2], \dots, [\Phi_L] \in \mathscr{S}(\mathbb{R}^n) \setminus \mathscr{S}_\infty(\mathbb{R}^n)$$

is linearly independent in $\mathscr{S}(\mathbb{R}^n)/\mathscr{S}_\infty(\mathbb{R}^n)$ and that $\mathscr{S}_\infty(\mathbb{R}^n)$ and $\Phi_1, \Phi_2, \dots, \Phi_L \in \mathscr{S}(\mathbb{R}^n) \setminus \mathscr{S}_\infty(\mathbb{R}^n)$ span $\Phi_{L+1}, \dots, \Phi_{L^*}$. More precisely, we assume

$$\Phi_l = \varphi_l^* + \sum_{l=1}^{L} \beta_{l,k} \Phi_l \quad (l = L+1, \dots, L^*) \tag{2.115}$$

for some $\varphi_{L+1}^*, \dots, \varphi_{L^*}^* \in \mathscr{S}_\infty(\mathbb{R}^n)$. Then the image of

$$\mathscr{U} \equiv \left(\bigcap_{k=1}^{K} \{F \in \mathscr{S}'(\mathbb{R}^n) \,:\, |\langle F, \varphi_k \rangle| < 1\} \right) \cap \left(\bigcap_{l=1}^{L^*} \{F \in \mathscr{S}'(\mathbb{R}^n) \,:\, |\langle F, \Phi_l \rangle| < 1\} \right)$$

by R contains

$$\left(\bigcap_{k=1}^{K} \{f \in \mathscr{S}'_\infty(\mathbb{R}^n) \,:\, |\langle f, \varphi_k \rangle| < 1\} \right) \cap \left(\bigcap_{k=L+1}^{L^*} \{f \in \mathscr{S}'_\infty(\mathbb{R}^n) \,:\, |\langle f, \varphi_k^* \rangle| < 1\} \right).$$

Proof We know that the image U of

$$\tilde{\mathscr{U}} \equiv \left(\bigcap_{k=1}^{K} \{F \in \mathscr{S}'(\mathbb{R}^n) \,:\, |\langle F, \varphi_k \rangle| < 1\} \right) \cap \left(\bigcap_{l=1}^{L} \{F \in \mathscr{S}'(\mathbb{R}^n) \,:\, |\langle F, \Phi_l \rangle| < 1\} \right)$$
$$\cap \left(\bigcap_{k=L+1}^{L^*} \{F \in \mathscr{S}'(\mathbb{R}^n) \,:\, |\langle F, \varphi_k^* \rangle| < 1\} \right)$$

by R is exactly

$$\left(\bigcap_{k=1}^{K} \{f \in \mathscr{S}'_\infty(\mathbb{R}^n) \,:\, |\langle f, \varphi_k \rangle| < 1\} \right) \cap \left(\bigcap_{k=L+1}^{L^*} \{f \in \mathscr{S}'_\infty(\mathbb{R}^n) \,:\, |\langle f, \varphi_k^* \rangle| < 1\} \right)$$

thanks to Theorem 2.22. According to (2.115), we can say that the image of

$$\begin{aligned}
&\tilde{\mathscr{U}}^* \\
&\equiv \left(\bigcap_{k=1}^{K}\{F \in \mathscr{S}'(\mathbb{R}^n) : |\langle F, \varphi_k\rangle| < 1\}\right) \cap \left(\bigcap_{l=1}^{L^*}\{F \in \mathscr{S}'(\mathbb{R}^n) : |\langle F, \Phi_l\rangle| < 1\}\right) \\
&\quad \cap \left(\bigcap_{l=L+1}^{L^*}\{F \in \mathscr{S}'(\mathbb{R}^n) : |\langle F, \varphi_l^*\rangle| < 1\}\right)
\end{aligned}$$

is U. Since $\mathscr{U}$ contains $\tilde{\mathscr{U}}^*$, the image of $\mathscr{U}$ by R contains U. Therefore, the proof is complete.

Now the proof of Theorem 2.21 is easy. In fact, let $\mathscr{U}_0$ be a neighborhood of 0. Then according to Theorem 2.23, $\mathscr{U}_0$ contains a set of the form $\mathscr{U}$ described in Theorem 2.23. According to Theorem 2.23, we know that $0 \in U = R(\mathscr{U}) \subset R(\mathscr{U}_0)$. Thus, 0 is an interior point of $R(\mathscr{U}_0)$. By the translation, we can show that any point $R(f)$ with $f \in R(\mathscr{U}_0)$ can be proved to be an interior point of $R(\mathscr{U}_0)$.

Let us rephrase Theorem 2.21 in terms of the quotient topology. To begin with let us recall some elementary facts on general topology.

Definition 2.14 (Equivalence relation) An equivalence relation of a set X is a subset R of $X \times X$ satisfying the following. Below, for $x, y \in X$, $x \sim y$ means that $(x, y) \in R$.

1. $x \sim x$ for all $x \in X$ (Reflexivity).
2. Let $x, y \in X$. Then $x \sim y$ implies $y \sim x$ (Symmetry).
3. Let $x, y, z \in X$. Then $x \sim y$ and $y \sim z$ implies $x \sim z$ (Transivity).

In this case $\sim$ is an equivalence relation of X. Given an equivalence relation of X, we write $[x] \equiv \{y \in X : x \sim y\} \in 2^X$ for $x \in X$ and $X/\sim \equiv \{[x] : x \in X\} \subset 2^X$.

Definition 2.15 (Quotient topology) Let $\sim$ be an equivalence relation of a topological space X. Then the quotient topology of X with respect to $\sim$ is the weakest topology such that the natural projection $p : X \to X/\sim, x \mapsto [x]$ is continuous. The mapping p is said to be the (canonical/natural) projection.

According to the definition, we see that $U \subset X/\sim$ is open, if and only if $p^{-1}(U)$ is open.

As for this topology, the following is elementary.

Theorem 2.24 *Let X and Y be topological spaces and $\sim$ an equivalence relation of X. A mapping $f : X/\sim \to Y$ is continuous, if and only if $f \circ p : X \to Y$ is continuous.*

Equip $\mathscr{S}'(\mathbb{R}^n)/\mathscr{P}(\mathbb{R}^n)$ with the quotient topology.

Theorem 2.25 *The spaces $\mathscr{S}'(\mathbb{R}^n)/\mathscr{P}(\mathbb{R}^n)$, endowed with the quotient topology, and $\mathscr{S}'_\infty(\mathbb{R}^n)$, endowed with the weak-$*$ topology, are homeomorphic.*

Proof According to Theorem 2.24, the mapping $\Phi : [f] \in \mathscr{S}'(\mathbb{R}^n)/\mathscr{P}(\mathbb{R}^n) \mapsto R(f) \in \mathscr{S}'_\infty(\mathbb{R}^n)$ is continuous. Let O be an open set in $\mathscr{S}'(\mathbb{R}^n)/\mathscr{P}(\mathbb{R}^n)$. Then $O = p(\mathscr{O}+\mathscr{P}(\mathbb{R}^n))$ for some open set $\mathscr{O}$ in $\mathscr{S}'(\mathbb{R}^n)$. Thus, $\Phi(O) = R(\mathscr{O})$ is open according to Theorem 2.21.

Exercises

Exercise 2.50

1. Show that $\mathscr{S}_\infty(\mathbb{R}^n)$ is a closed subspace of $\mathscr{S}(\mathbb{R}^n)$. Hint: Let

$$\ell_\alpha(f) \equiv \int_{\mathbb{R}^n} x^\alpha f(x)\mathrm{d}x$$

for $\alpha \in \mathbb{N}_0{}^n$. Then $\mathscr{S}(\mathbb{R}^n) = \bigcap_{\alpha\in\mathbb{N}_0{}^n} \ker(\ell_\alpha)$.

2. If $\varphi \in \mathscr{S}(\mathbb{R}^n)$ does not contain 0 as its support, show that $\mathscr{F}\varphi \in \mathscr{S}_\infty(\mathbb{R}^n)$.

Exercise 2.51 [697, Proposition 6.1] Let $k \in \mathbb{N}_0 \cap \{-1\}$, and let $f \in L^1_{\mathrm{loc}}(\mathbb{R}^n)$. Also assume that

$$\int_{\mathbb{R}^n} |f(x)|\langle x\rangle^{-n-k-1}\mathrm{d}x < \infty.$$

Let $\psi \in \mathscr{S}(\mathbb{R}^n)$ be such that $\chi_{B(1)} \le \psi \le \chi_{B(2)}$. Define $\varphi \equiv \psi - \psi(2\star)$. Then show that

$$\sum_{j=-\infty}^{\infty} \partial^\alpha[\varphi_j(D)f] = \partial^\alpha f$$

in $\mathscr{S}'(\mathbb{R}^n)$ for all α with $|\alpha| = k+1$.

Exercise 2.52 [697, Proposition 6.2] Let $k \in \mathbb{N}_0 \cap \{-1\}$, let $f \in L^1_{\mathrm{loc}}(\mathbb{R}^n) \cap \mathscr{S}'(\mathbb{R}^n)$ satisfy

$$|\langle f, \eta\rangle| \le p_k(\eta) \quad (\eta \in \mathscr{S}(\mathbb{R}^n)),$$

and let $\psi \in \mathscr{S}(\mathbb{R}^n)$ satisfy $\chi_{B(1)} \le \psi \le \chi_{B(2)}$. Define $\varphi \equiv \psi - \psi(2\star)$. Then show that

$$\sum_{j=-\infty}^{\infty} \partial^\alpha[\varphi_j(D)f] = \partial^\alpha f$$

in $\mathscr{S}'(\mathbb{R}^n)$ for all α with $|\alpha| = k+1$.

Exercise 2.53 [697, Section 2] Let $L \in \mathbb{N}_0$. Define $\mathscr{S}(\mathbb{R}^n)_L \equiv \mathscr{S}(\mathbb{R}^n) \cap \mathscr{P}_L(\mathbb{R}^n)^\perp$. Endow $\mathscr{S}(\mathbb{R}^n)_L$ with the topology induced by $\mathscr{S}(\mathbb{R}^n)$. Denote by $(\mathscr{S}(\mathbb{R}^n)_L)'$ the topological dual of $\mathscr{S}(\mathbb{R}^n)_L$. Let $\psi \in \mathscr{S}(\mathbb{R}^n)$ be chosen so that $\chi_{B(1)} \le \psi \le \chi_{B(2)}$ and define $\varphi \equiv \psi(2^{-1}\star) - \psi$.

1. Show that $\mathscr{S}(\mathbb{R}^n)_L$ is a closed subspace of $\mathscr{S}(\mathbb{R}^n)$.
2. Let $f \in (\mathscr{S}(\mathbb{R}^n)_L)'$. Then

$$f = \sum_{j=-\infty}^{\infty} \varphi_j(D)f$$

   in $\mathscr{S}'(\mathbb{R}^n)/\mathscr{P}_L(\mathbb{R}^n)$, which is equipped with the quotient topology.
3. Show that $\mathscr{S}'(\mathbb{R}^n)/\mathscr{P}_L(\mathbb{R}^n)$ and $(\mathscr{S}(\mathbb{R}^n)_L)'$ are isomorphic.

2.4.2 *Function Spaces of Homogeneous Type*

2.4.2.1 Homogeneous Besov/Triebel–Lizorkin Spaces

Having set down the fundamental properties $\mathscr{S}_\infty(\mathbb{R}^n)$ and $\mathscr{S}'_\infty(\mathbb{R}^n)$ in the detailed manner, we define $\dot{F}^s_{pq}(\mathbb{R}^n)$ and $\dot{B}^s_{pq}(\mathbb{R}^n)$.

Definition 2.16 (Homogeneous Besov/Triebel–Lizorkin spaces) Let $0 < p, q \le \infty$ and $s \in \mathbb{R}$. Choose a function $\varphi \in \mathscr{S}(\mathbb{R}^n)$ so that $\chi_{B(4)\setminus B(2)} \le \varphi \le \chi_{B(8)\setminus B(1)}$.

1. Define the quasi-norm $\|\star\|_{\dot{B}^s_{pq}}$ by $\|f\|_{\dot{B}^s_{pq}} \equiv \|\{2^{js}\varphi_j(D)f\}_{j=-\infty}^{\infty}\|_{\ell^q(L^p)}$ for $f \in \mathscr{S}'(\mathbb{R}^n)/\mathscr{P}(\mathbb{R}^n)$. The space $\dot{B}^s_{pq}(\mathbb{R}^n)$ is the set of all $f \in \mathscr{S}'(\mathbb{R}^n)/\mathscr{P}(\mathbb{R}^n)$ for which the norm $\|f\|_{\dot{B}^s_{pq}}$ is finite.
2. Assume in addition $p < \infty$. For $f \in \mathscr{S}'(\mathbb{R}^n)/\mathscr{P}(\mathbb{R}^n)$, define the quasi-norm $\|\star\|_{\dot{F}^s_{pq}}$ by $\|f\|_{\dot{F}^s_{pq}} \equiv \|\{2^{js}\varphi_j(D)f\}_{j=-\infty}^{\infty}\|_{L^p(\ell^q)}$. The space $\dot{F}^s_{pq}(\mathbb{R}^n)$ is the set of all $f \in \mathscr{S}'(\mathbb{R}^n)/\mathscr{P}(\mathbb{R}^n)$ for which the quasi-norm $\|f\|_{\dot{F}^s_{pq}}$ is finite.
3. The space $\dot{A}^s_{pq}(\mathbb{R}^n)$ stands for either $\dot{B}^s_{pq}(\mathbb{R}^n)$ or $\dot{F}^s_{pq}(\mathbb{R}^n)$. When $\dot{A}^s_{pq}(\mathbb{R}^n)$ stands for $\dot{F}^s_{pq}(\mathbb{R}^n)$, exclude the case $p = \infty$ tacitly.

Observe that the definition makes sense in view of Example 1.16. In fact, for $f \in \mathscr{P}(\mathbb{R}^n) \subset \mathscr{S}'(\mathbb{R}^n)$, we have $\varphi_j(D)f = 0$ using the fact that

$$\mathscr{F} f \in \operatorname{Span}\left\{\partial^\alpha \delta\right\}_{\alpha \in \mathbb{N}_0{}^n};$$

hence $\varphi_j \cdot \mathscr{F} f = 0$. This discussion can be reversed so that we can check that $\dot{A}^s_{pq}(\mathbb{R}^n)$ is a quasi-normed space.

2.4.2.2 Fundamental Properties of Homogeneous Besov/Triebel–Lizorkin Spaces

As is the case for nonhomogeneous function spaces, we can prove the following theorems: Here and below in this section, the proof is omitted if the proof is similar to the one for nonhomogeneous function spaces.

Theorem 2.26 *Let p, q, s satisfy $0 < p, q \le \infty$ and $s \in \mathbb{R}$.*

1. *For different admissible choices of φ one gets equivalent quasi-norms in Definition 2.16. Hence the set $\dot{A}^s_{pq}(\mathbb{R}^n)$ does not depend on φ.*
2. *Assume $p < \infty$. Then $\dot{B}^s_{p\,\min(p,q)}(\mathbb{R}^n) \hookrightarrow \dot{F}^s_{pq}(\mathbb{R}^n) \hookrightarrow \dot{B}^s_{p\,\max(p,q)}(\mathbb{R}^n)$ in the sense of continuous embedding.*

Proof Mimic the proof for the corresponding assertions in the nonhomogeneous function spaces. See Proposition 2.7 and Theorem 2.11.

We have a counterpart of the lifting property for the nonhomogeneous spaces.

Theorem 2.27 (The boundedness of lift operators in $\mathscr{S}_\infty(\mathbb{R}^n)$) *Let $\alpha \in \mathbb{R}$. Choose a smooth function $\varphi \in \mathscr{S}(\mathbb{R}^n)$ so that*

$$\mathrm{supp}(\varphi) \subset B(8) \setminus B(1), \quad \sum_{j=-\infty}^{\infty} \varphi_j \equiv \chi_{\mathbb{R}^n \setminus \{0\}}. \tag{2.116}$$

Then for $f \in \mathscr{S}_\infty(\mathbb{R}^n)$, define

$$(-\Delta)^\alpha f \equiv \mathscr{F}^{-1}[|\star|^{2\alpha} \mathscr{F} f] = \lim_{J \to \infty} \sum_{j=-J}^{\infty} \mathscr{F}^{-1}[|\star|^{2\alpha} \varphi_j \cdot \mathscr{F} f]. \tag{2.117}$$

By duality and the isomorphism $\mathscr{S}'_\infty(\mathbb{R}^n) \simeq \mathscr{S}'(\mathbb{R}^n)/\mathscr{P}(\mathbb{R}^n)$, we extend the definition of $\mathscr{S}'_\infty(\mathbb{R}^n)$ and $\mathscr{S}'(\mathbb{R}^n)/\mathscr{P}(\mathbb{R}^n)$. Then we have the following:

1. *For $f \in \mathscr{S}_\infty(\mathbb{R}^n)$,* (2.117) *the convergence in the topology of $\mathscr{S}_\infty(\mathbb{R}^n)$ and the limit does not depend on the choice of $\varphi \in \mathscr{S}(\mathbb{R}^n)$ satisfying* (2.116). *Furthermore, $(-\Delta)^\alpha$ induces the isomorphism from $\mathscr{S}_\infty(\mathbb{R}^n)$ to itself.*
2. *Duality entails that $(-\Delta)^\alpha$ is the isomorphism from $\mathscr{S}'(\mathbb{R}^n)/\mathscr{P}(\mathbb{R}^n)$.*

The operator $(-\Delta)^\alpha$ is called the lift operator in the setting of homogeneous spaces.

Proof We write $\mathscr{F}\mathscr{S}_\infty(\mathbb{R}^n)$ for the image of $\mathscr{S}_\infty(\mathbb{R}^n)$ by the Fourier transform. Then

$$\mathscr{F}\mathscr{S}_\infty(\mathbb{R}^n) = \bigcap_{\alpha \in \mathbb{N}_0{}^n} \{\varphi \in \mathscr{S}(\mathbb{R}^n) : \partial^\alpha \varphi(0) = 0\}.$$

Hence the mapping

$$\varphi \in \mathscr{F}\mathscr{S}_\infty(\mathbb{R}^n) \mapsto |\star|^\alpha \varphi \in \mathscr{F}\mathscr{S}_\infty(\mathbb{R}^n)$$

is a linear isomorphism. Here for $\alpha < 0$, we need to check that the mapping is well defined; see Exercise 2.56. 1. is obtained by the Fourier transform of (2.117). Finally, 2. follows by duality. Thus, the proof is complete.

Theorem 2.28 (Lift operator in homogeneous function spaces) *Suppose that the parameters p, q, s and α satisfy $0 < p, q \le \infty, s, \alpha \in \mathbb{R}$. Then $(-\Delta)^\alpha : \dot{A}^s_{pq}(\mathbb{R}^n) \approx \dot{A}^{s-2\alpha}_{pq}(\mathbb{R}^n)$ is an isomorphism.*

Proof Mimic the proof of the propositions for nonhomogeneous function spaces; see Theorem 2.12.

Theorem 2.29 (Inclusions between homogeneous function spaces) *In the sense of continuous embedding, we have $\mathscr{S}_\infty(\mathbb{R}^n) \hookrightarrow \dot{A}^s_{pq}(\mathbb{R}^n) \hookrightarrow \mathscr{S}'(\mathbb{R}^n)/\mathscr{P}(\mathbb{R}^n)$. More precisely:*

1. $p_N(\Phi) \gtrsim \|\Phi\|_{\dot{A}^s_{pq}}$ *for all $\Phi \in \mathscr{S}_\infty(\mathbb{R}^n)$.*
2. *There exists $N = N_{p,q,s} \in \mathbb{N}$ such that*

$$|\langle f, \varphi\rangle| \lesssim \|f\|_{\dot{A}^s_{pq}} p_N(\varphi) \tag{2.118}$$

for all $\varphi \in \mathscr{S}_\infty(\mathbb{R}^n)$ and $f \in \dot{A}^s_{pq}(\mathbb{R}^n)$.

Note that we identify $f \in \mathscr{S}'(\mathbb{R}^n)$ with $[f] \in \mathscr{S}'(\mathbb{R}^n)/\mathscr{P}(\mathbb{R}^n)$ when we consider $\|f\|_{\dot{A}^s_{pq}}$ in (2.118).

Proof To prove $\mathscr{S}_\infty(\mathbb{R}^n) \hookrightarrow \dot{A}^s_{pq}(\mathbb{R}^n)$, we take $\Phi \in \mathscr{S}_\infty(\mathbb{R}^n)$ arbitrarily. Choose $\psi \in \mathscr{S}(\mathbb{R}^n)$ so that $\chi_{B(1)} \le \psi \le \chi_{B(2)}$. We decompose $\Phi = \psi(D)\Phi + (1 - \psi(D))\Phi$. Here $\psi(D)$ plays the role of the *low pass filter*, that is, $\psi(D)$ kills the high frequency, while $(1 - \psi(D))$ is the *high pass filter* that complements the low pass filter. So $\psi(D)f$ is called the low-frequency part while $(1 - \psi(D))$ is called the high-frequency part. According to the results for nonhomogeneous spaces, we have $\Phi \in \mathscr{S}(\mathbb{R}^n) \subset A^s_{pq}(\mathbb{R}^n)$ and hence $(1-\psi(D))\Phi \in \dot{A}^s_{pq}(\mathbb{R}^n)$ is trivial. Hence it remains to show $\psi(D)\Phi \in \dot{A}^s_{pq}(\mathbb{R}^n)$. From Theorem 1.55, there exists $N \gg 1$ such that

$$\|\varphi_j(D)\Phi\|_p \lesssim p_N(\Phi), \quad j \le 0. \tag{2.119}$$

According to Theorem 2.28, it can be assumed that $s > 0$. Assuming $s > 0$, we can easily show that $\varphi_0(D)\Phi \in \dot{A}^s_{pq}(\mathbb{R}^n)$ from (2.119). Hence it follows that $\mathscr{S}_\infty(\mathbb{R}^n) \hookrightarrow \dot{A}^s_{pq}(\mathbb{R}^n)$.

Let us prove $\dot{A}^s_{pq}(\mathbb{R}^n) \hookrightarrow \mathcal{S}'(\mathbb{R}^n)/\mathcal{P}(\mathbb{R}^n)$. From $\dot{F}^s_{pq}(\mathbb{R}^n) \hookrightarrow \dot{B}^s_{p\max(p,q)}(\mathbb{R}^n)$, we may concentrate on Besov spaces. In fact, similar to Proposition 2.7, we have

$$\dot{B}^s_{p\min(p,q)}(\mathbb{R}^n) \hookrightarrow \dot{F}^s_{pq}(\mathbb{R}^n) \hookrightarrow \dot{B}^s_{p\max(p,q)}(\mathbb{R}^n). \tag{2.120}$$

See Exercise 2.55. From Theorem 2.14, we have $\dot{B}^s_{pq}(\mathbb{R}^n) \hookrightarrow \dot{B}^{s-n/p}_{\infty\infty}(\mathbb{R}^n)$. Thus, the heart of the matter is to define $\Phi([f]) \in \mathcal{S}'_\infty(\mathbb{R}^n)$ by

$$\langle \Phi([f]), \tau \rangle \equiv \sum_{j=-\infty}^{\infty} \langle \varphi_j(D)f, \tau \rangle, \tau \in \mathcal{S}_\infty(\mathbb{R}^n) \tag{2.121}$$

and to show that $\Phi : \dot{B}^{-n-1}_{\infty\infty}(\mathbb{R}^n) \to \mathcal{S}'_\infty(\mathbb{R}^n)$ is continuous. When $j \le 0$,

$$|\langle \varphi_j(D)f, \tau \rangle| \le \|\varphi_j(D)f\|_\infty \cdot \|\tau\|_1 \le 2^{(n+1)j}\|f\|_{\dot{B}^{-n-1}_{\infty\infty}} \cdot \|\tau\|_1 \tag{2.122}$$

and, when $j \ge 0$,

$$\begin{aligned}
|\langle \varphi_j(D)f, \tau \rangle| &= 2^{-2(n+1)j}|\langle ((|2^{-j} \star|^{-2(n+1)}\varphi_j)(D)f, \Delta^{n+1}\tau \rangle| \\
&\le 2^{-2(n+1)j}\|((|2^{-j} \star|^{-2(n+1)}\varphi_j)(D)f\|_\infty \cdot \|\Delta^{n+1}\tau\|_1 \\
&\lesssim 2^{-2(n+1)j}\|\varphi_j(D)f\|_\infty\|\Delta^{n+1}\tau\|_1 \\
&\lesssim 2^{-(n+1)j}\|f\|_{\dot{B}^{-n-1}_{\infty\infty}}\|\Delta^{n+1}\tau\|_1.
\end{aligned}$$

Hence

$$|\langle \Phi([f]), \tau \rangle| \lesssim \|f\|_{\dot{B}^{-n-1}_{\infty\infty}} p_{2n+2}(\tau), \quad \tau \in \mathcal{S}_\infty(\mathbb{R}^n).$$

Thus, Theorem 2.29 is proved.

Corollary 2.3 (Completeness of $\dot{A}^s_{pq}(\mathbb{R}^n)$) *If $0 < p, q \le \infty, s \in \mathbb{R}$, then $\dot{A}^s_{pq}(\mathbb{R}^n)$ is a quasi-Banach space.*

Proof The proof is the same as that of Theorem 2.15.

Theorem 2.30 (Density of $\mathcal{S}_\infty(\mathbb{R}^n)$ in $\dot{A}^s_{pq}(\mathbb{R}^n)$) *If $0 < p, q < \infty$ and $s \in \mathbb{R}$, $\mathcal{S}_\infty(\mathbb{R}^n)$ is dense in $\dot{A}^s_{pq}(\mathbb{R}^n)$.*

Proof Since $p, q < \infty$, $\sum_{k=-l}^{l} \varphi_k(D)f$ converges to f in $\dot{A}^s_{pq}(\mathbb{R}^n)$ as $l \to \infty$. As a result, it can be assumed that f is expressed as $f = \sum_{k=-l}^{l} \varphi_k(D)g$ for some $g \in \dot{A}^s_{pq}(\mathbb{R}^n)$ and $l \in \mathbb{N}$. Assuming that f takes the form above, we approximate such an f. With a minor modification of the proof for $A^s_{pq}(\mathbb{R}^n)$, we obtain the results for $\dot{A}^s_{pq}(\mathbb{R}^n)$.

Exercises

Exercise 2.54 Prove Proposition 2.13 by reexamining the proof of Theorem 1.21.

Exercise 2.55 Show (2.120) by reexamining the case of nonhomogeneous spaces.

Exercise 2.56 Let $F : \mathbb{R} \to \mathbb{C}$ be a continuous function differentiable at any point in $\mathbb{R} \setminus \{0\}$. Furthermore, assume that $a = \lim_{t \to 0, t \neq 0} F'(t)$ exists. Then show also that F is differentiable at 0 and that $F'(0) = a$ using the mean-value theorem.

Exercise 2.57 (The operator norm of the dilation operators in homogeneous spaces, [6, Remark 2.19], [99, p. 239, Remark 4]) Let $0 < p, q \leq \infty$ and $s \in \mathbb{R}$. For $\lambda > 0$, show that

$$\|f(\lambda\star)\|_{\dot{A}^s_{pq}} \sim \lambda^{s-n/p}\|f\|_{\dot{A}^s_{pq}} \quad (f \in \dot{A}^s_{pq}(\mathbb{R}^n)). \tag{2.123}$$

Hint: $\mathscr{F}[f(\lambda\star)] = \lambda^{-n}\mathscr{F}f(\lambda^{-1}\star)$.

Exercise 2.58

1. Prove (2.109) using the definition of $\mathscr{P}$.
2. Show that the multiplication and addition do not depend on f, g in (2.110). That is, for $f_1, f_2, g_1, g_2 \in \mathscr{S}'(\mathbb{R}^n)$, assume $[f_1] = [f_2]$ and $[g_1] = [g_2]$. Then prove $[f_1 + g_1] = [f_2 + g_2]$ and $[k \cdot f_1] = [k \cdot f_2]$ for all $k \in \mathbb{C}$.

2.4.3 Realization of $\dot{A}^s_{pq}(\mathbb{R}^n)$

The definition of $\dot{A}^s_{pq}(\mathbb{R}^n)$ is simpler than that of the nonhomogeneous counterpart. However, it has a disadvantage for partial differential equations. For example, $\dot{A}^s_{pq}(\mathbb{R}^n)$ is defined modulo polynomials. It does not make sense to consider the graph of f in general; $\{(x, f(x)) \in \mathbb{R}^n \times \mathbb{C} : x \in \mathbb{R}^n\}$. Nevertheless, in some cases, there is no need to consider the space modulo polynomials. Our aim here is to consider the conditions for which such a thing happens.

2.4.3.1 Realization in $\mathscr{S}'(\mathbb{R}^n)/\mathscr{P}_L(\mathbb{R}^n)$

We recall that $\mathscr{P}_L(\mathbb{R}^n)$ for the set of all polynomials with a degree less than or equal to L. We start with the general case together with the following notion:

Definition 2.17 (Banach distribution space) Let $L \in \mathbb{N}_0 \cup \{-1\}$. One says that a *Banach distribution space* (BDS) in the quotient space $\mathscr{S}'(\mathbb{R}^n)/\mathscr{P}_L(\mathbb{R}^n)$ or a Banach space of distributions modulo $\mathscr{P}_L(\mathbb{R}^n)$ is a vector subspace of E of $\mathscr{S}'(\mathbb{R}^n)/\mathscr{P}_L(\mathbb{R}^n)$ endowed with a complete norm which renders continuous the natural mapping $E \hookrightarrow \mathscr{S}'(\mathbb{R}^n)/\mathscr{P}_L(\mathbb{R}^n)$.

We are especially interested in the case of $L = -1$ because in this case we can consider $\dot{A}^s_{pq}(\mathbb{R}^n)$ as a subset of $\mathscr{S}'(\mathbb{R}^n)$.

Theorem 2.31 (Realization of $\dot{A}^s_{pq}(\mathbb{R}^n)$) *Let $0 < p, q \le \infty$ and $s \in \mathbb{R}$. Choose two functions $\psi, \varphi \in \mathscr{S}(\mathbb{R}^n)$ so that $\chi_{B(1)} \le \psi \le \chi_{B(2)}$, $\varphi = \psi - \psi(2\star) = \psi - \psi_{-1}$. Define a mapping $\Phi : \mathscr{S}'(\mathbb{R}^n) \to \mathscr{S}'(\mathbb{R}^n)/\mathscr{P}(\mathbb{R}^n)$ by*

$$\Phi(f) \equiv \sum_{j=-\infty}^{\infty} [\varphi_j(D)f] \in \mathscr{S}'(\mathbb{R}^n)/\mathscr{P}(\mathbb{R}^n), \quad f \in \mathscr{S}'(\mathbb{R}^n). \tag{2.124}$$

Let also $L \in \mathbb{N}_0$ satisfy

$$L > s - \frac{n}{p}. \tag{2.125}$$

1. *Let L be fixed. For any multi-index α with length L and $f \in \dot{A}^s_{pq}(\mathbb{R}^n)$,*

$$\sum_{j=-\infty}^{-1} \partial^\alpha \varphi_j(D)f \tag{2.126}$$

converges in the topology of $\mathscr{S}'(\mathbb{R}^n)$.

2. *For all $f \in \dot{A}^s_{pq}(\mathbb{R}^n)$, there exists a sequence $\{P_{-j}\}_{j=1}^{\infty} \subset \mathscr{P}_{L-1}(\mathbb{R}^n)$ such that*

$$\left\{\left(\sum_{k=-j}^{\infty} \varphi_k(D)f\right) + P_{-j}\right\}_{j=1}^{\infty} \tag{2.127}$$

converges in the topology of $\mathscr{S}'(\mathbb{R}^n)$.

Theorem 2.27 ensures that the limit in (2.124) exists.

Proof Concentrate on the convergence of (2.126): (2.127) follows from Proposition 1.13 and (2.126).

Assume (2.125). From $\dot{A}^s_{pq}(\mathbb{R}^n) \hookrightarrow \dot{B}^{s-n/p}_{\infty\infty}(\mathbb{R}^n)$, we have only to establish $\sum_{j=-\infty}^{-1} \partial^\alpha \varphi_j(D)f$ converges in the topology of $\mathscr{S}'(\mathbb{R}^n)$ as long as

$$f \in \dot{B}^s_{\infty\infty}(\mathbb{R}^n), \quad L > s. \tag{2.128}$$

Choose an auxiliary even function $\kappa \in \mathscr{S}(\mathbb{R}^n)$, as well as a test function $\tau \in \mathscr{S}(\mathbb{R}^n)$ arbitrarily so that $\chi_{B(2)\setminus B(1/2)} \le \kappa \le \chi_{B(4)\setminus B(1/4)}$. Then writing $\tau^j \equiv 2^{jn}\tau(2^j\star)$ for $\tau \in \mathscr{S}(\mathbb{R}^n)$, we have

$$|\langle \partial^\alpha \varphi_j(D)f, \tau\rangle| \simeq_n |\langle \partial^\alpha[(\mathscr{F}\kappa)^j * \varphi_j(D)f], \tau\rangle| = |\langle \varphi_j(D)f, (\mathscr{F}\kappa)^j * \partial^\alpha\tau\rangle| \tag{2.129}$$

Here by Theorem 1.56, we have a pointwise estimate:

$$|(\mathscr{F}\kappa)^j * \partial^\alpha\tau(x)| \lesssim 2^{(n+|\alpha|)j}\langle 2^j x\rangle^{-n-1}. \tag{2.130}$$

Hence if we insert (2.130) into (2.129) and then use Hölder's inequality, we have

$$|\langle \partial^\alpha \varphi_j(D)f, \tau\rangle| \lesssim 2^{(n+|\alpha|)j}\|\varphi_j(D)f\|_\infty \int_{\mathbb{R}^n} \langle 2^j x\rangle^{-n-1}\mathrm{d}x. \tag{2.131}$$

Change variables in the integral of the right-hand side of (2.131). Then (2.131) yields

$$|\langle \partial^\alpha \varphi_j(D)f, \tau\rangle| \lesssim 2^{|\alpha|j}\|\varphi_j(D)f\|_\infty. \tag{2.132}$$

Assuming $L > s$, we can sum (2.132) over $j \in \mathbb{N}_0 \cap \{-1\}$. Since $|\alpha| = L$, we have

$$\sum_{j=-\infty}^{-1} |\langle \partial^\alpha \varphi_j(D)f, \tau\rangle| \lesssim \sum_{j=-\infty}^{-1} 2^{(|\alpha|-s)j}\|f\|_{\dot{B}^s_{\infty\infty}} \lesssim_{s,L} \|f\|_{\dot{B}^s_{\infty\infty}}.$$

Therefore, the proof is complete.

If we reexamine the proof of Theorem 2.31, we can sum (2.132) over nonnegative integers j. For example, we have the following:

Theorem 2.32 *Let $0 < p, q \le \infty$ and $L \in \mathbb{N}_0$ satisfy*

$$L \ge s - \frac{n}{p}, \quad q \le 1. \tag{2.133}$$

Then for any multi-index α with length L and $f \in \dot{B}^s_{pq}(\mathbb{R}^n)$,

$$\sum_{j=-\infty}^{-1} \partial^\alpha \varphi_j(D)f \tag{2.134}$$

converges in the topology of $\mathscr{S}'(\mathbb{R}^n)$.

The proof is left to the interested readers as Exercise 2.60.

Note that (2.134) converges but that (2.127) converges with the help of differentiation. So, we can say that the differentiation makes things better in the world of the band-limited distributions. This is a contrast to the proposition that, if $\{f_j\}_{j=1}^\infty \subset \mathrm{BC}([a, b])$ converges to $f \in \mathrm{BC}([a, b])$ uniformly, then

$$\lim_{j\to\infty}\int_a^b f_j(t)\mathrm{d}t = \int_a^b f(t)\mathrm{d}t$$

holds but generally $\lim_{j\to\infty} f_j'(t) = f'(t)$ fails.

2.4.3.2 Realization in $\mathscr{S}'(\mathbb{R}^n)$

Let us consider the case where $L = 0$ in Theorems 2.31 and 2.32. Since $L = 0$, $\sum_{j=-\infty}^{\infty} \varphi_j(D)f$ converges in $\mathscr{S}'(\mathbb{R}^n)$, so that we can say $\Phi(f) \in \mathscr{S}'(\mathbb{R}^n)$. In this case we redefine

$$\Phi(f) \equiv \sum_{j=-\infty}^{\infty} \varphi_j(D)f.$$

The operation $f \mapsto \Phi(f)$ is to add a filter on f.

Theorem 2.33 *Let $0 < p, q \le \infty$ and $s \in \mathbb{R}$. Choose $\psi, \varphi \in \mathscr{S}(\mathbb{R}^n)$ so that*

$$\chi_{B(1)} \le \psi \le \chi_{B(2)}, \quad \varphi = \psi - \psi(2\star) = \psi - \psi_{-1}. \tag{2.135}$$

Furthermore, suppose that p and s satisfy

$$s < \frac{n}{p}. \tag{2.136}$$

Then we define the quasi-normed space $\dot{A}^{s*}_{pq}(\mathbb{R}^n)$ by

$$\dot{A}^{s*}_{pq}(\mathbb{R}^n) \equiv \{f \in \mathscr{S}'(\mathbb{R}^n) : \Phi(f) = f, \|f\|_{\dot{A}^{s*}_{pq}} \equiv \|[f]\|_{\dot{A}^{s}_{pq}} < \infty\}.$$

$\Psi : [f] \in \dot{A}^{s}_{pq}(\mathbb{R}^n) \mapsto \Phi(f) \in \dot{A}^{s*}_{pq}(\mathbb{R}^n)$ is an isomorphism between homogeneous Besov spaces. The same can be said for $\dot{B}^{s}_{pq}(\mathbb{R}^n)$ with (2.136) or $s = \frac{n}{p}$ and $q \le 1$.

We plan to prove that Ψ is surjective and that Ψ is injective. After that, we prove that Ψ is a topological isomorphism.

Proof (of the surjectivity of Ψ) Let $f \in \dot{A}^{s*}_{pq}(\mathbb{R}^n)$. Then since $\Psi([f]) = \Phi(f) = f$, we see that Ψ is surjective.

Proof (of the injectivity of Ψ) Let $[f] \in \dot{A}^{s}_{pq}(\mathbb{R}^n)$ satisfy $\Psi([f]) = 0$. This means that

$$\sum_{j=-\infty}^{\infty} \varphi_j(D)f = \Phi(f) = \Psi([f]) = 0.$$

Hence we deduce $\operatorname{supp}(\mathscr{F}f) \subset \{0\}$ from this equality. By Theorem 1.25, we have $f \in \mathscr{P}(\mathbb{R}^n)$; hence $[f] = 0$.

Proof (of the fact that Ψ is a homeomorphism) To prove this, for $f \in \dot{A}^{s*}_{pq}(\mathbb{R}^n)$, we need to check $\| [f] \|_{\dot{A}^s_{pq}} = \|f\|_{\dot{A}^{s*}_{pq}}$. But this is trivial by definition.

Definition 2.18 (Realization in $\dot{A}^{s*}_{pq}(\mathbb{R}^n)$) In Theorem 2.33 $f \in \dot{A}^{s*}_{pq}(\mathbb{R}^n)$, satisfying

$$f = \sum_{j=-\infty}^{\infty} \varphi_j(D)f$$

in the topology of $\mathscr{S}'(\mathbb{R}^n)$, is called the *realization of* $[f]$.

A passage to the realizations allows us to embed $\dot{A}^s_{pq}(\mathbb{R}^n)$ into $\mathscr{S}'(\mathbb{R}^n)$.

Exercises

Exercise 2.59 Establish that $\dot{A}^s_{pq}(\mathbb{R}^n)$ is a Banach distribution space (BDS) in $\mathscr{S}'(\mathbb{R}^n)/\mathscr{P}_m(\mathbb{R}^n)$ when $s < m + \dfrac{n}{p}$. Hint: Use Corollary 1.5.

Exercise 2.60 Prove Theorem 2.32 by reexamining the proof of Theorem 2.31.

Exercise 2.61 [1096] Let $f \in \mathrm{BC}^1(\mathbb{R}^n)$ satisfy $\|\nabla f\|_{(L^p)^n} < \infty$ with $1 \le p < n$, so that $f \in \dot{L}^{1,p}(\mathbb{R}^n)$. Let $\psi \in C^\infty_{\mathrm{c}}(\mathbb{R}^n)$ satisfy $\chi_{B(1)} \le \psi \le \chi_{B(2)}$. Define $\varphi \equiv \psi - \psi^{-1}$.

1. Show that $g \equiv \lim\limits_{j\to\infty} (\psi_j(D)f - \psi_{-j}(D)f)$ exists in $\mathscr{S}'(\mathbb{R}^n)$.
2. Show that $g - f$ is a constant function.

Exercise 2.62 [810, Theorem 1.2] Let $0 < p, q \le \infty$ and $s \in \mathbb{R}$. Suppose that $m \in \mathbb{N}$ is a fixed integer specified later. We define

$$\begin{aligned}
&\mathscr{A}^s_{pq}(\mathbb{R}^n)\\
&\equiv \left\{ f \in \mathscr{S}'(\mathbb{R}^n) \,:\, [f] \in \dot{A}^s_{pq}(\mathbb{R}^n) \right\}\\
&\quad \cup \bigcap_{\substack{\alpha,\beta\in\mathbb{N}_0{}^n\\ |\alpha|=|\beta|+1=m}} \left\{ f \in \mathscr{S}'(\mathbb{R}^n) \,:\, f \in C^{m-1}(\mathbb{R}^n), \frac{\partial^\beta f}{\partial x^\beta}(0) = 0, \frac{\partial^\alpha f}{\partial x^\alpha} \in \tilde{C}_0(\mathbb{R}^n) \right\}.
\end{aligned}$$

Let $\varphi \in \mathscr{S}(\mathbb{R}^n)$ be such that $\varphi = \phi_1 - \phi$ for some $\phi \in \mathscr{S}(\mathbb{R}^n)$ satisfying $\chi_{B(1)} \leq \phi \leq \chi_{B(2)}$.

1. Assume that $s - \frac{n}{p} \in (0, \infty) \setminus \mathbb{N}$ or that $0 < q \leq 1$, $s - \frac{n}{p} \in (0, \infty)\mathbb{N}$. If $u \in \dot{B}^s_{pq}(\mathbb{R}^n)$, then show that

$$f = \sum_{j=0}^{\infty} \varphi_j(D)u + \sum_{j=-\infty}^{\infty} \varphi_j(D)f - \sum_{|\alpha| \leq m} \frac{1}{\alpha!} \frac{\partial^\alpha [\varphi_j(D)u]}{\partial x^\alpha}(0)x^\alpha$$

is the unique element in $\mathscr{B}^s_{pq}(\mathbb{R}^n)$ such that $[f] = u$.

2. Assume that $s - \frac{n}{p} \in (0, \infty) \setminus \mathbb{N}$ or that $0 < p \leq 1$, $s - \frac{n}{p} \in (0, \infty)\mathbb{N}$. If $u \in \dot{F}^s_{pq}(\mathbb{R}^n)$, then show that

$$f = \sum_{j=0}^{\infty} \varphi_j(D)u + \sum_{j=-\infty}^{\infty} \varphi_j(D)f - \sum_{|\alpha| \leq m} \frac{1}{\alpha!} \frac{\partial^\alpha [\varphi_j(D)u]}{\partial x^\alpha}(0)x^\alpha$$

is the unique element in $\mathscr{F}^s_{pq}(\mathbb{R}^n)$ such that $[f] = u$. If necessary, resort to the embedding in Theorem 4.11.

In each case note that the mapping $u \mapsto f$ is dilation commuting.

Exercise 2.63 Let $0 < q \leq \infty$. Show that $\dot{B}^{-1}_{\infty q}(\mathbb{R}^n) \hookrightarrow \dot{B}^{-1}_{\infty\infty}(\mathbb{R}^n) \hookrightarrow B^{-1}_{\infty\infty}(\mathbb{R}^n)$ using Theorem 2.33, where the right embedding is understood via the realization.

Textbooks in Sect. 2.4

The Spaces $\mathscr{S}_\infty(\mathbb{R}^n)$ and $\mathscr{S}'_\infty(\mathbb{R}^n)$: Theorems 2.20 and 2.21

For Theorems 2.20 and 2.21 we refer to [95, Propositions 35.4 and 35.5]. Note also that Holschneider considered Theorems 2.20 and 2.21 in the context of wavelet analysis in [41, Theorem 24.0.4], where he applied Treves's general result to this special setting.

The Isomorphism $(-\Delta)^\alpha$: Theorem 2.27

See [99, Section 5.2.3], as well as the recent paper [368]. Note that Cho further investigated the Fourier multiplier in [368].

Homogeneous Besov Spaces

We reconstructed Sect. 2.4.2 based on [99, Chapter 5] and [71]. See also [7, Section 6.3] and [33, Section 2.2.1] for homogeneous Besov spaces. The paper [922] is a self-contained survey.

2.5 Local Means

In this section we consider the local means.

We continue to write $\tau_j \equiv \tau(2^{-j}\star)$ and $\tau^j \equiv 2^{jn}\tau(2^j\star)$ for $j \in \mathbb{Z}$ for $\tau \in \mathscr{S}(\mathbb{R}^n)$. To describe the Besov norm and the Triebel–Lizorkin norm we choose $\psi, \varphi \in \mathscr{S}(\mathbb{R}^n)$ so that $\chi_{B(4)} \le \psi \le \chi_{B(8)}$ and that $\chi_{B(4)\setminus B(2)} \le \varphi \le \chi_{B(8)\setminus B(1)}$. In Sect. 2.3 we defined the Besov norm and the Triebel–Lizorkin norm by

$$\|f\|_{B^s_{pq}} \equiv \|\psi(D)f\|_p + \|\{2^{js}\varphi_j(D)f\}_{j=1}^{\infty}\|_{\ell^q(L^p)}, \tag{2.137}$$

$$\|f\|_{F^s_{pq}} \equiv \|\psi(D)f\|_p + \|\{2^{js}\varphi_j(D)f\}_{j=1}^{\infty}\|_{L^p(\ell^q)}. \tag{2.138}$$

If we write the Fourier transform fully and change variables, then

$$\mathscr{F}[\varphi_j] = [\mathscr{F}\varphi]^j \tag{2.139}$$

for all $j \in \mathbb{Z}$. See Exercise 2.66. A question now arises: do we absolutely need to assume that φ has a compact support when we consider $\varphi_j(D)f = \mathscr{F}^{-1}[\varphi_j \cdot \mathscr{F}f] = c_n[\mathscr{F}^{-1}\varphi]^j * f$ in the definition?

The Laplacian is transformed into the multiplier via the Fourier transform; that is,

$$\mathscr{F}[\,2^{js}(\Delta^L\psi)^j\,](\xi) = 2^{js}(-|2^{-j}\xi|^2)^L\mathscr{F}\psi(2^{-j}\xi) \tag{2.140}$$

for $\xi \in \mathbb{R}^n$. See Exercise 2.64. So $2^{j(s+n)}(\Delta^L\psi)_{-j} * f$ and $2^{js}\varphi_j(D)f$ look alike. Thus, φ_j should concentrate on "$\{\,|\xi| \sim 2^j\,\}$" in a certain sense. To formulate this observation, we define

$$\|f\|^*_{B^s_{pq}} \equiv \|\psi * f\|_p + \|\{2^{js}(\Delta^L\psi)^j * f\}_{j=1}^{\infty}\|_{\ell^q(L^p)},$$

$$\|f\|^*_{F^s_{pq}} \equiv \|\psi * f\|_p + \|\{2^{js}(\Delta^L\psi)^j * f\}_{j=1}^{\infty}\|_{L^p(\ell^q)}.$$

The norms are called the local means. We ask ourselves whether the local means are equivalent to the original norms. Here we have $(\mathscr{F}^{-1}\varphi)^j \perp \mathscr{P}(\mathbb{R}^n)$. We need some assumption on φ^j when we consider the convolution.

As a counterpart to the Plancherel–Polya–Nikolski'i inequality, we deal with a maximal inequality of the Peetre maximal functions in Sect. 2.5.1.

In Sect. 2.5.2 we consider the problem addressed above; that is, whether the norm equivalences $\|f\|_{B^s_{pq}} \sim \|f\|^*_{B^s_{pq}}$ and $\|f\|_{F^s_{pq}} \sim \|f\|^*_{F^s_{pq}}$ hold or not for any $f \in \mathscr{S}'(\mathbb{R}^n)$. Furthermore, we will take up other norm equivalences by means of differences and oscillation in Sect. 2.5.3.

2.5.1 *Maximal Inequality Adapted to the Local Means*

Let $L \in \mathbb{N}$ be sufficiently large. When we consider the convolutions for compactly supported functions, we choose $\psi, k, K \in \mathscr{S}(\mathbb{R}^n)$ so that

$$\chi_{B(2)} \le \psi \le \chi_{B(4)}, \quad K = \psi, \quad k = \Delta^L \psi. \tag{2.141}$$

In (2.137) and (2.138), we will establish that k and K correspond to $\mathscr{F}^{-1}\psi$ and to $\mathscr{F}^{-1}\varphi$, respectively, as long as $L \gg 1$. $\mathscr{F}^{-1}\varphi \perp \mathscr{P}(\mathbb{R}^n)$. When we consider the parameters p, q, s, we notice that being perpendicular to $\mathscr{P}(\mathbb{R}^n)$ is too strong. In the definition of k, we have the L-fold composition of Laplacian Δ. So integration by parts shows that $k \perp \mathscr{P}_{2L-1}(\mathbb{R}^n)$. Then assuming $0 \in \operatorname{supp}(\psi)$, we can choose $\zeta, \theta \in \mathscr{D}$ so that

$$\zeta \cdot \mathscr{F}K + \sum_{l=1}^{\infty} \theta_l \cdot (\mathscr{F}k)_l \equiv 1, \quad 0 \notin \operatorname{supp}(\theta). \tag{2.142}$$

See Exercise 2.67.

2.5.1.1 Peetre's Maximal Operator for Local Means

Recall that we wrote $\tau^j = 2^{jn}\tau(2^j\star)$ for $j \in \mathbb{Z}$ and $\tau \in \mathscr{S}(\mathbb{R}^n)$. We define the Peetre maximal operator, as follows.

Definition 2.19 (Peetre's maximal operator) Fix $K, k \in \mathscr{S}(\mathbb{R}^n)$ satisfying (2.141). For $j \in \mathbb{N}$, $A > 0$ and $f \in \mathscr{S}'(\mathbb{R}^n)$, define the maximal operator $\mathscr{M}_{A,j}f$ by

$$\mathscr{M}_{A,j}f(x) \equiv \sup_{y \in \mathbb{R}^n, l \ge j} 2^{A(j-l)}\langle 2^j y\rangle^{-A}|k^l * f(x-y)| \tag{2.143}$$

$$\mathscr{M}_{A,0}f(x) \equiv \mathscr{M}_{A,1}f(x) + \sup_{y \in \mathbb{R}^n} \langle y\rangle^{-A}|K * f(x-y)| \tag{2.144}$$

for $x \in \mathbb{R}^n$. The operators

$$f \mapsto \sup_{y \in \mathbb{R}^n} \langle 2^j y \rangle^{-A} |k^j * f(\star - y)|, \quad f \mapsto \sup_{y \in \mathbb{R}^n} \langle y \rangle^{-A} |K * f(\star - y)|$$

are called *Peetre's maximal operators*.

Let $x \in \mathbb{R}^n$. As (2.143) and (2.144) show, we have

$$\mathcal{M}_{A,j} f(x) \geq |k^j * f(x)|. \tag{2.145}$$

See Exercise 2.65. Furthermore, $\mathcal{M}_{A,j} f(x)$ contains some information on $|k^j * f(x - y)|$ for any $y \in \mathbb{R}^n$ but due to the factor $\langle y \rangle^{-A}$, the influence is weaker as $|y|$ gets larger.

2.5.1.2 Maximal Inequality of Local Means

Proposition 2.14 (Maximal inequality of local means) *Let* $0 < p, q \leq \infty$ *and* $s \in \mathbb{R}$.

1. *Assume that* A, p, s *satisfy*

$$A > \max\left(\frac{n}{\min(p, 1)}, \frac{n}{\min(p, 1)} - s\right).$$

Then

$$\left\| \left\{ 2^{js} \mathcal{M}_{A,j} f \right\}_{j=0}^{\infty} \right\|_{\ell^q(L^p)} \sim \|K * f\|_p + \|\{2^{js} k^j * f\}_{j=1}^{\infty}\|_{\ell^q(L^p)} \tag{2.146}$$

for all $f \in \mathcal{S}'(\mathbb{R}^n)$.

2. *Assume that* A, p, q, s *satisfy*

$$p < \infty, \quad A > \max\left(\frac{n}{\min(p, q, 1)}, \frac{n}{\min(p, q, 1)} - s\right).$$

Then

$$\left\| \left\{ 2^{js} \mathcal{M}_{A,j} f \right\}_{j=0}^{\infty} \right\|_{L^p(\ell^q)} \sim \|K * f\|_p + \|\{2^{js} k^j * f\}_{j=1}^{\infty}\|_{L^p(\ell^q)} \tag{2.147}$$

for all $f \in \mathcal{S}'(\mathbb{R}^n)$.

Proof The inequality $\gtrsim$ follows trivially from the definition of $\mathcal{M}_{A,j} f$. Hence we have only to prove $\lesssim$. The proof shows that $\mathcal{M}_{A,0} f$ can be handled like $\mathcal{M}_{A,j} f$ with $j \in \mathbb{N}$. So we concentrate on $\mathcal{M}_{A,j} f$ for $j \in \mathbb{N}$.

We can mimic the proof of (2.147) for the proof (2.146). So it is enough to prove (2.147). Let $0 < \eta < \min(1, p, q)$ be fixed so that $A\eta > n$.

Choose $\zeta, \theta \in \mathscr{D}$ satisfying (2.142). Then by (2.139) and (2.142)

$$k^j * f \simeq_n K^j * (\mathscr{F}^{-1}\zeta)^j * k^j * f + \sum_{l=j+1}^{\infty} k^j * (\mathscr{F}^{-1}\theta)^l * k^l * f. \quad (2.148)$$

Let $x \in \mathbb{R}^n$ and $j, l \in \mathbb{N}$ with $j < l$. We estimate the convolution $k^j * (\mathscr{F}^{-1}\theta)^l$. We invoke Theorem 1.56 with

$$a = k, \quad \eta = \mathscr{F}^{-1}\theta, \quad L \geq 2[A+1], \quad \lambda = A$$

to have

$$|k^j * (\mathscr{F}^{-1}\theta)^l(x)| \lesssim 2^{(j-l)(L+1)+jn}\langle 2^j x\rangle^{-A}. \quad (2.149)$$

If we insert (2.149) into each term of (2.148), then

$$\begin{aligned} |k^j * (\mathscr{F}^{-1}\theta)^l * k^l * f(x)| &\leq \int_{\mathbb{R}^n} |k^j * (\mathscr{F}^{-1}\theta)^l(y)| \cdot |k^l * f(x-y)| \mathrm{d}y \\ &\lesssim 2^{(j-l)(L+1)+jn} \\ &\quad \int_{\mathbb{R}^n} \langle 2^j y\rangle^{-A} |k^l * f(x-y)| \mathrm{d}y. \end{aligned} \quad (2.150)$$

We apply Theorem 1.54 to the first term of the right-hand side of (2.148). We obtain an estimate of (2.150) with $j = l$;

$$|K^j * (\mathscr{F}^{-1}\zeta)^j * k^j * f(x)| \lesssim 2^{jn} \int_{\mathbb{R}^n} \langle 2^j y\rangle^{-A} |k^j * f(x-y)| \mathrm{d}y; \quad (2.151)$$

interchange ζ and θ. If we insert (2.150) and (2.151) into (2.148), then

$$|k^j * f(x)| \lesssim \sum_{l=j}^{\infty} 2^{(j-l)(L+1)+jn} \int_{\mathbb{R}^n} \langle 2^j y\rangle^{-A} |k^l * f(x-y)| \mathrm{d}y. \quad (2.152)$$

Hence for all $x, z \in \mathbb{R}^n$

$$\langle 2^j z\rangle^{-A} |k^j * f(x-z)| \lesssim \sum_{l=j}^{\infty} 2^{(j-l)(L+1)+jn} \int_{\mathbb{R}^n} \langle 2^j y\rangle^{-A} \langle 2^j z\rangle^{-A} |k^l * f(x-y-z)| \mathrm{d}y.$$

By the Peetre inequality,

$$\langle 2^j z\rangle^{-A}|k^j*f(x-z)|\lesssim\sum_{l=j}^{\infty}2^{(j-l)(L+1)+jn}\int_{\mathbb{R}^n}\langle 2^j(y+z)\rangle^{-A}|k^l*f(x-y-z)|\mathrm{d}y.$$

If we change variables in the right-hand side, then we obtain

$$\langle 2^j z\rangle^{-A}|k^j*f(x-z)|\lesssim\sum_{l=j}^{\infty}2^{(j-l)(L+1)+jn}\int_{\mathbb{R}^n}\langle 2^j y\rangle^{-A}|k^l*f(x-y)|\mathrm{d}y.$$

If we use $\langle 2^j y\rangle\geq 2^{j-l}\langle 2^l y\rangle$,

$$\langle 2^j z\rangle^{-A}|k^j*f(x-z)|\lesssim\sum_{l=j}^{\infty}2^{(j-l)(L+1-A+n)+ln}\int_{\mathbb{R}^n}\langle 2^l y\rangle^{-A}|k^l*f(x-y)|\mathrm{d}y.$$

As a result,

$$\begin{aligned}&\sup_{z\in\mathbb{R}^n}\langle 2^j z\rangle^{-A}|k^j*f(x-z)|\\&\lesssim\sum_{l=j}^{\infty}2^{(j-l)(L+1-A+n)+ln}\int_{\mathbb{R}^n}\frac{|k^l*f(x-y)|^\eta}{\langle 2^l y\rangle^{A\eta}}\mathrm{d}y\left(\sup_{z\in\mathbb{R}^n}\langle 2^j z\rangle^{-A}|k^j*f(x-z)|\right)^{1-\eta}.\end{aligned}$$

Since $f\in\mathscr{S}'(\mathbb{R}^n)$, $\mathscr{M}_{A,j}f(x)\lesssim_{x,f}B^j$, where B also depends on A and f. Thus, Lemma 1.2 yields

$$\left(\sup_{z\in\mathbb{R}^n}\langle 2^j z\rangle^{-A}|k^j*f(x-z)|\right)^{\eta}\lesssim\sum_{l=j}^{\infty}2^{(j-l)(L+1-A+n)+ln}\int_{\mathbb{R}^n}\frac{|k^l*f(x-y)|^\eta}{\langle 2^l y\rangle^{A\eta}}\mathrm{d}y; \tag{2.153}$$

hence

$$\mathscr{M}_{A,j}f(x)^\eta\lesssim\sum_{m=j}^{\infty}2^{(j-m)A\eta}M[|k^m*f|^\eta](x).$$

Hence the Fefferman–Stein vector-valued inequality (Theorem 1.49) yields (2.147).

Corollary 2.4 (Maximal inequality for local means) *Maintain the same notation as Proposition* 2.14.

1. *Inequality*

$$\left\|\sup_{z\in\mathbb{R}^n}\langle z\rangle^{-A}|K*f(\star-z)|\right\|_p\lesssim\|K*f\|_p+\|\{2^{js}k^j*f\}_{j=1}^{\infty}\|_{\ell^q(L^p)}$$

holds for $f \in \mathscr{S}'(\mathbb{R}^n)$.

2. *Inequality*

$$\left\| \left\{ 2^{js} \sup_{z \in \mathbb{R}^n} \langle 2^j z \rangle^{-A} |k^j * f(\star - z)| \right\}_{j=1}^{\infty} \right\|_{\ell^q(L^p)} \lesssim \|K * f\|_p + \|\{2^{js} k^j * f\}_{j=1}^{\infty}\|_{\ell^q(L^p)}$$

holds for $f \in \mathscr{S}'(\mathbb{R}^n)$.

3. *Let* $p < \infty$. *Then*

$$\left\| \sup_{z \in \mathbb{R}^n} \langle z \rangle^{-A} |K * f(\star - z)| \right\|_p \lesssim \|K * f\|_p + \|\{2^{js} k^j * f\}_{j=1}^{\infty}\|_{L^p(\ell^q)}$$

holds for $f \in \mathscr{S}'(\mathbb{R}^n)$.

4. *Let* $p < \infty$. *Then*

$$\left\| \left\{ 2^{js} \sup_{z \in \mathbb{R}^n} \langle 2^j z \rangle^{-A} |k^j * f(\star - z)| \right\}_{j=1}^{\infty} \right\|_{L^p(\ell^q)} \lesssim \|K * f\|_p + \|\{2^{js} k^j * f\}_{j=1}^{\infty}\|_{L^p(\ell^q)}$$

holds for $f \in \mathscr{S}'(\mathbb{R}^n)$.

Exercises

Exercise 2.64 Verify (2.140) using Theorem 1.28.

Exercise 2.65 Verify (2.145) using the definition (2.143) and (2.144) of $\mathscr{M}_{A,j} f(x)$.

2.5.2 *Local Means*

We describe the original Besov norm and the original Triebel–Lizorkin norm for a setup. Choose a function $\Psi \in \mathscr{S}(\mathbb{R}^n)$ so that $\chi_{B(1)} \leq \Psi \leq \chi_{B(2)}$. Define $\varphi, \psi \in \mathscr{S}(\mathbb{R}^n)$ by $\psi \equiv \Psi$ and $\varphi \equiv \Psi - 2^n \Psi(2\star)$. Let $0 < p, q \leq \infty$, $s \in \mathbb{R}$, and let $f \in \mathscr{S}'(\mathbb{R}^n)$. The Besov norm and the Triebel–Lizorkin norm were defined as follows:

$$\|f\|_{B^s_{pq}} \equiv \|\psi(D)f\|_p + \|\{2^{js}\varphi_j(D)f\}_{j=1}^{\infty}\|_{\ell^q(L^p)},$$

$$\|f\|_{F^s_{pq}} \equiv \|\psi(D)f\|_p + \|\{2^{js}\varphi_j(D)f\}_{j=1}^{\infty}\|_{L^p(\ell^q)}.$$

Of course we exclude the case $p = \infty$ in the above. Here we rewrite the norms.

2.5.2.1 Local Means

Fix L so that $2L > s$. Define $K \equiv \Psi$ and $k \equiv \Delta^L\Psi$. We have the following equivalent expression of the function spaces:

Theorem 2.34 *Let $L \in \mathbb{N}_0$, $0 < p, q \le \infty$, let $s \in \mathbb{R}$ satisfy $2L > s$, and let $f \in \mathscr{S}'(\mathbb{R}^n)$.*

1. *We have $\|f\|_{B^s_{pq}} \sim \|K * f\|_p + \|\{2^{js}k^j * f\}_{j=1}^{\infty}\|_{\ell^q(L^p)}$.*
2. *Also assume that $p < \infty$. Then $\|f\|_{F^s_{pq}} \sim \|K * f\|_p + \|\{2^{js}k^j * f\}_{j=1}^{\infty}\|_{L^p(\ell^q)}$.*

We call the right-hand side the *local means*.

Proof We prove *1* since we can prove 2 similarly. To investigate the relation between the norm in the right-hand side and the original norm, choose $\zeta, \eta \in \mathscr{S}(\mathbb{R}^n)$ so that

$$\mathscr{F}\zeta, \mathscr{F}\eta \in C^{\infty}_{\mathrm{c}}(\mathbb{R}^n), \quad 0 \notin \mathrm{supp}(\mathscr{F}\eta), \quad \zeta * K + \sum_{m=1}^{\infty} \eta^m * k^m = \delta, \tag{2.154}$$

where the convergence of the left-hand side of the last formula takes place in $\mathscr{S}'(\mathbb{R}^n)$. Fix $j \in \mathbb{N}$. Then by (2.154), we have

$$\begin{aligned}\varphi_j(D)f &= \varphi_j(D)[\zeta * K * f] + \sum_{m=1}^{\infty} \varphi_j(D)[\eta^m * k^m * f] \\ &\simeq_n \{(\mathscr{F}^{-1}\varphi)^j * \zeta\} * K * f + \sum_{m=1}^{\infty} \{(\mathscr{F}^{-1}\varphi)^j * \eta^m\} * k^m * f.\end{aligned} \tag{2.155}$$

Let $A \gg 1$, and let $x \in \mathbb{R}^n$. Theorem 1.54 yields

$$|\,(\mathscr{F}^{-1}\varphi)^j * \zeta(x)\,| \lesssim_A 2^{-2jL}\langle x\rangle^{-A-n-1}. \tag{2.156}$$

We will choose $N \in \mathbb{N}$ so that $\theta \equiv \min(2L - s, N - A + s) > 0$. For $m, j \ge 1$, Theorem 1.56 yields

$$|\,(\mathscr{F}^{-1}\varphi)^j * \eta^m(x)\,| \lesssim_N \begin{cases} 2^{2(m-j)L+mn}\langle 2^m x\rangle^{-A-n-1} & j \ge m, \\ 2^{(j-m)N+jn}\langle 2^j x\rangle^{-A-n-1} & m \ge j. \end{cases} \tag{2.157}$$

We insert (2.156) and (2.157) into (2.155). Then

$$
\begin{aligned}
&|2^{js}\varphi_j(D)f(x)| \\
&\lesssim 2^{j(s-2L)}\int_{\mathbb{R}^n}\frac{|K*f(x-y)|}{\langle y\rangle^{A+n+1}}\mathrm{d}y \\
&\quad +\sum_{m=1}^{j}2^{(j-m)s-2(j-m)L+mn}\int_{\mathbb{R}^n}\frac{|2^{ms}k^m*f(x-y)|}{\langle 2^m y\rangle^{A+n+1}}\mathrm{d}y \\
&\quad +\sum_{m=j+1}^{\infty}2^{(j-m)s-(m-j)N+jn}\int_{\mathbb{R}^n}\frac{|2^{ms}k^m*f(x-y)|}{\langle 2^j y\rangle^{A+n+1}}\mathrm{d}y \\
&\lesssim 2^{-j\theta}\sup_{y\in\mathbb{R}^n}\langle y\rangle^{-A}|K*f(x-y)|+\sum_{m=1}^{\infty}2^{sm-|j-m|\theta}\sup_{y\in\mathbb{R}^n}\langle 2^m y\rangle^{-A}|k^m*f(x-y)|.
\end{aligned}
$$

Since $\theta > 0$, we have

$$
\begin{aligned}
&\left(\sum_{j=1}^{\infty}|2^{js}\varphi_j(D)f(x)|^q\right)^{\frac{1}{q}} \\
&\lesssim \sup_{y\in\mathbb{R}^n}\langle y\rangle^{-A}|K*f(x-y)|+\left(\sum_{j=1}^{\infty}\left(2^{js}\sup_{y\in\mathbb{R}^n}\langle 2^j y\rangle^{-A}|k^j*f(x-y)|\right)^q\right)^{\frac{1}{q}}
\end{aligned}
\tag{2.158}
$$

by the Hölder inequality.

Likewise, we have a counterpart to (2.158):

$$
\begin{aligned}
|\psi(D)f(x)| &\lesssim \sup_{y\in\mathbb{R}^n}\langle y\rangle^{-A}|K*f(x-y)| \\
&\quad +\left(\sum_{j=1}^{\infty}\sup_{y\in\mathbb{R}^n}2^{jsq}\langle 2^j y\rangle^{-Aq}|k^j*f(x-y)|^q\right)^{\frac{1}{q}}.
\end{aligned}
$$

By Corollary 2.4, we have

$$
\|f\|_{F^s_{pq}} \lesssim \|K*f\|_p+\|\{2^{js}k^j*f\}_{j=1}^{\infty}\|_{L^p(\ell^q)}. \tag{2.159}
$$

We can prove the reverse inequality of (2.159) similarly.

Example 2.11 Although the heat kernel $E(x) = \exp(-|x|^2)$ is not compactly supported, we can use it instead of k. In this case, we have $K*f = E*f = \mathrm{e}^{-\Delta}f$. Thus, if we choose $L \in \mathbb{N}_0$ so that $2L > s$, then we have the following norm equivalence for all $f \in \mathscr{S}'(\mathbb{R}^n)$:

$$\|f\|_{B^s_{pq}} \sim \|e^{-\Delta} f\|_p + \|\{\{2^{j(s-2L)} \Delta^L e^{-\Delta} f\}_{j=1}^\infty\|_{\ell^q(L^p)}$$

$$\|f\|_{F^s_{pq}} \sim \|e^{-\Delta} f\|_p + \|\{\{2^{j(s-2L)} \Delta^L e^{-\Delta} f\}_{j=1}^\infty\|_{L^p(\ell^q)}.$$

We used the Laplacian for the definition of k. It is not the Laplacian that counts. We need to rely upon the fact that $k \in \mathscr{P}_{2L-1}(\mathbb{R}^n)$. Hence we have the following theorem, whose proof is analogous to that of Theorem 2.34:

Theorem 2.35 *Let $0 < p, q \le \infty$, $s \in \mathbb{R}$, and let L be an integer.*

1. *Assume $L > \sigma_p + s$. Then $\|f\|_{B^s_{pq}} \sim \|K * f\|_p + \sum_{l=1}^n \|\{2^{js}[\partial_l{}^L k]^j * f\}_{j=1}^\infty\|_{\ell^q(L^p)}$ for all $f \in \mathscr{S}'(\mathbb{R}^n)$.*
2. *Assume $L > \sigma_{pq} + s$ and $p < \infty$. Then $\|f\|_{F^s_{pq}} \sim \|K * f\|_p + \sum_{l=1}^n \|\{2^{js}[\partial_l{}^L k]^j * f\}_{j=1}^\infty\|_{L^p(\ell^q)}$ for all $f \in \mathscr{S}'(\mathbb{R}^n)$.*

We present a couple of applications. As an application of local means, we have the following:

Proposition 2.15 *Let $0 < p \le \infty$ and $K \in \mathbb{N}_0$. Assume that A is a compactly supported $C^K(\mathbb{R}^n)$-function with some $K \in \mathbb{N}_0$. Then $A \in B^K_{p\infty}(\mathbb{R}^n)$.*

Proof By Theorem 2.12, we have norm equivalence:

$$\|A\|_{B^K_{p\infty}} \sim \|A\|_{B^0_{p\infty}} + \sum_{j=1}^n \|\partial_{x_j}{}^K A\|_{B^0_{p\infty}}.$$

So it can be assumed that $K = 0$. Consider the local means in Theorem 2.34. Since any function in the family $\{k^j * A\}_{j=1}^\infty$ vanishes outside a compact set, we can assume $p \ge 1$.

In this case by Proposition 2.1, $A \in L^p(\mathbb{R}^n) \hookrightarrow B^0_{p\infty}(\mathbb{R}^n)$, as was to be shown.

Next, we generalize $L^p(B(R)) \hookrightarrow L^q(B(R))$ with $0 < q < p < \infty$ and $R > 0$, as follows.

Proposition 2.16 *Let $0 < p_0 < p \le \infty$, $0 < q \le \infty$, $s \in \mathbb{R}$. If $f \in A^s_{pq}(\mathbb{R}^n)$ vanishes outside a ball with radius R, we have*

$$\|f\|_{A^s_{p_0 q}} \lesssim_R \|f\|_{A^s_{pq}}. \tag{2.160}$$

Proof Concentrate on Triebel–Lizorkin spaces. By Theorem 2.34, we have

$$\|f\|_{F^s_{p_0 q}} \sim \|K * f\|_{p_0} + \|\{2^{js} k^j * f\}_{j=1}^\infty\|_{L^{p_0}(\ell^q)}.$$

Here for any $j \in \mathbb{N}$, $K * f$ and $k^j * f$ are supported on a ball of radius $R+2$; by Hölder's inequality we have

$$\|K * f\|_{p_0} + \|\{2^{js}k^j * f\}_{j=1}^{\infty}\|_{L^{p_0}(\ell^q)} \lesssim_R \|K * f\|_p + \|\{2^{js}k^j * f\}_{j=1}^{\infty}\|_{L^p(\ell^q)}.$$

Hence again by Theorem 2.34, (2.160) holds.

2.5.2.2 Structure Theorem of $\mathscr{S}'(\mathbb{R}^n)$

Theorem 2.34 and Corollary 2.4 show a property of the Peetre maximal operator and it also yields another important result.

Theorem 2.36 (The structure theorem of $\mathscr{S}'(\mathbb{R}^n)$) *Let $f \in \mathscr{S}'(\mathbb{R}^n)$. Then for sufficiently large $M, N \in \mathbb{N}$, we have $(1-\Delta)^{-M}[(1+|x|^2)^{-N}f] \in L^\infty(\mathbb{R}^n)$.*

Proof By Theorem 1.21, there exists $N \gg 1$ such that

$$|\langle f, \varphi \rangle| \lesssim_f p_N(\varphi) = \sum_{\substack{\alpha \in \mathbb{N}_0{}^n \\ |\alpha| \le N}} \sup_{x \in \mathbb{R}^n} \langle x \rangle^N |\partial^\alpha \varphi(x)|, \quad \varphi \in \mathscr{S}(\mathbb{R}^n). \tag{2.161}$$

Hence

$$|\langle (1+|x|^2)^{-N} f, \varphi \rangle| \lesssim_f \sum_{\substack{\alpha \in \mathbb{N}_0{}^n \\ |\alpha| \le N}} \sup_{x \in \mathbb{R}^n} |\partial^\alpha \varphi(x)|, \quad \varphi \in \mathscr{S}(\mathbb{R}^n). \tag{2.162}$$

By replacing N with a larger number we use $\mathscr{F} : \mathscr{S}(\mathbb{R}^n) \simeq \mathscr{S}(\mathbb{R}^n)$ to have

$$|\langle (1+|x|^2)^{-N} f, \varphi \rangle| \lesssim_f p_N(\mathscr{F}\varphi) = \sum_{\substack{\alpha \in \mathbb{N}_0{}^n \\ |\alpha| \le N}} \sup_{x \in \mathbb{R}^n} |\partial^\alpha \mathscr{F}\varphi(x)|, \quad \varphi \in \mathscr{S}(\mathbb{R}^n). \tag{2.163}$$

Hence

$$\begin{aligned} |\langle (1-\Delta)^{-N}[(1+|x|^2)^{-N} f], \varphi \rangle| &= |\langle (1+|x|^2)^{-N} f, (1-\Delta)^{-N}\varphi \rangle| \\ &\lesssim_f \sum_{\substack{\alpha \in \mathbb{N}_0{}^n \\ |\alpha| \le N}} \sup_{y \in \mathbb{R}^n} |\partial^\alpha [\mathscr{F}(1-\Delta)^{-N}\varphi(y)]|. \end{aligned}$$

We calculate

$$|\langle (1-\Delta)^{-N}[(1+|x|^2)^{-N} f], \varphi \rangle| \lesssim_f \sum_{\substack{\alpha \in \mathbb{N}_0{}^n \\ |\alpha| \le N}} \sup_{y \in \mathbb{R}^n} \left| \partial^\alpha [\langle y \rangle^{-2N} \mathscr{F}\varphi(y)] \right|.$$

We fix $x \in \mathbb{R}^n$ and $j \in \mathbb{N}$. We will take $K(x - \star)$ or $2^{jn}k(2^j x - 2^j\star) = k^j(x - \star)$. Then

$$|(1 - \Delta)^{-N}[(1 + |x|^2)^{-N} f] * k^j(x)| \lesssim_f 2^{(N+n+1)j},$$
$$|(1 - \Delta)^{-N}[(1 + |x|^2)^{-N} f] * K(x)| \lesssim_f 1.$$

Hence by Theorem 2.34 and Corollary 2.4, we have

$$(1 - \Delta)^{-N}[(1 + |x|^2)^{-N} f] \in B^{-N-n-1}_{\infty\infty}(\mathbb{R}^n).$$

Hence $(1-\Delta)^{-2N-n-1}[(1+|x|^2)^{-N} f] \in B^{N+n+1}_{\infty\infty}(\mathbb{R}^n) \hookrightarrow L^\infty(\mathbb{R}^n)$. Therefore, the proof is complete.

2.5.2.3 Local Means for Homogeneous Spaces

We have the homogeneous counterparts of the results above. However, we need to formulate carefully: we have to take care of polynomials. Choose $\psi \in \mathscr{S}(\mathbb{R}^n)$ so that $\chi_{B(1)} \le \psi \le \chi_{B(2)}$ and define $\varphi \equiv \psi - \psi_{-1} = \psi - \psi(2\star)$ as usual.

Lemma 2.5 *Let $f \in \dot{A}^s_{pq}(\mathbb{R}^n)$. If $L_0 \in \mathbb{Z}$ satisfies $L_0 > \dfrac{s - \sigma_p}{2}$, then, there exists a sequence $\{P_j\}_{j=1}^\infty \subset \mathscr{P}_{2L_0-1}(\mathbb{R}^n)$ such that $P_K + \lim\limits_{K\to\infty} \sum\limits_{j=-K}^{\infty} \varphi_j(D) f$ converges in $\mathscr{S}'(\mathbb{R}^n)$. In particular,*

$$\Delta^{L_0} f \equiv \lim_{K\to\infty} \sum_{j=-K}^{\infty} \Delta^{L_0}(\varphi_j(D) f) \tag{2.164}$$

converges in $\mathscr{S}'(\mathbb{R}^n)$.

Proof This follows clearly from Theorem 2.31.

To consider a counterpart of Theorem 2.34, we need to take care of the difference of polynomials. Lemma 2.5 will serve this purpose.

Theorem 2.37 (Local means for homogeneous spaces) *Suppose that the parameters p, q, s satisfy $0 < p, q \le \infty$, $s \in \mathbb{R}$. Choose $L \in \mathbb{N}$ much larger than L_0 in Lemma 2.5.*

1. *The norm equivalence $\|f\|_{\dot{B}^s_{pq}} \sim \|2^{j(s-2L)} k_j * \Delta^L f\|_{\ell^q(L^p)}$ holds for $f \in \dot{B}^s_{pq}(\mathbb{R}^n)$.*
2. *Assume $p < \infty$ in addition. Then the norm equivalence*

$$\|f\|_{\dot{F}^s_{pq}} \sim \|2^{j(s-2L)} k_j * \Delta^L f\|_{L^p(\ell^q)}$$

holds for $f \in \dot{F}^s_{pq}(\mathbb{R}^n)$.

Here $\Delta^L f$, defined by (2.164), *makes sense despite ambiguity of the representative of* $[f]$.

We omit the proof of the theorem since it resembles that of Theorem 2.34.

Exercises

Exercise 2.66 Prove (2.139) by change of variables.

Exercise 2.67 Let $K \in C_c^\infty(\mathbb{R}^n)$ satisfy $\chi_{B(2)} \le K \le \chi_{B(4)}$. Show that there exist $\zeta, \theta \in \mathscr{D}$ satisfying (2.142) by estimating the size of $\mathscr{F}K$ carefully.

Exercise 2.68 Show that x^α, $|x| \notin B^\rho_{\infty\infty}(\mathbb{R}^n)$ for $\alpha \in \mathbb{N}_0{}^n \setminus \{0\}$, $\rho \in \mathbb{R}$ using Theorem 2.34.

Exercise 2.69 Let $\psi \in \mathscr{S}(\mathbb{R}^n)$. If $0 < r < \infty$, $A > 0$, $j \in \mathbb{N}$, $f \in \mathscr{S}'(\mathbb{R}^n)$ and $x_0 \in \mathbb{R}^n$ satisfy $2^{jn}\displaystyle\int_{\mathbb{R}^n} \sup_{l \ge j} \langle 2^j y\rangle^{-Ar}|\psi^l * f(x_0 - y)|^r \mathrm{d}y < \infty$, then show that for all $x \in \mathbb{R}^n$ $2^{jn}\displaystyle\int_{\mathbb{R}^n} \sup_{l \ge j} \langle 2^j y\rangle^{-Ar}|\psi^l * f(x - y)|^r \mathrm{d}y < \infty$.

2.5.3 Characterizations by Means of the Difference and the Oscillation

Besov spaces were first considered in 1959–1961. At that time, the Fourier transform was not used.

Let us recall that we can define the Triebel–Lizorkin norm $\|\star\|_{F^s_{pq}}$ with $0 < p < \infty$, $0 < q \le \infty$ and $s \in \mathbb{R}$ by (2.36) via (2.45).

Let $1 \le p < \infty$, $0 < q \le \infty$ and $s > 0$. Then using Propositions 2.1, 2.7 and 2.8, we can show that $F^s_{pq}(\mathbb{R}^n) \hookrightarrow L^p(\mathbb{R}^n)$; see Exercise 2.70. So here and below, we mainly consider the case where the functions are embedded into some $L^u(\mathbb{R}^n)$ space for some $1 \le u \le \infty$. Using Proposition 1.2, we can show that

$$\|f\|_{F^s_{pq}} \sim \|f\|_p + \|\{2^{js}(f - \psi_j(D)f)\}_{j=1}^\infty\|_{L^p(\ell^q)} \tag{2.165}$$

for all $f \in L^p(\mathbb{R}^n)$, where $\psi_j = \psi(2^{-j}\star)$. See Exercise 2.70. We are convinced that $\psi_j(D)$ is a good operator since $\psi \in C_c^\infty(\mathbb{R}^n)$. However, the condition $\psi \in C_c^\infty(\mathbb{R}^n)$ being too strong we are fascinated with a weaker condition. In fact, the functions $C_c^\infty(\mathbb{R}^n)$ are in general too artificial. So we aim to consider some "natural" candidate of $\psi_j(D)f$. The difference played a key role. In this section, we seek an equivalent definition by the use of the difference. We also characterize the Besov norm using the oscillation. The same idea applies to the Triebel–Lizorkin norm.

2.5.3.1 Ball Means of Differences

For $f \in L^1_{\rm loc}(\mathbb{R}^n)$, $S \in \mathbb{N}$ and $t > 0$, define the *ball means of differences* by

$$d_t^S f(x) \equiv t^{-n} \int_{B(t)} |\Delta_h^S f(x)| \mathrm{d}x.$$

Define the *Besov norm of the ball means of differences* by

$$\|f\|_{B^s_{pq}:\text{ B. M. D.}} \equiv \|f\|_p + \left(\int_0^1 (t^{-s}\|d_t^S f\|_p)^q \frac{\mathrm{d}t}{t}\right)^{\frac{1}{q}}$$

for $f \in L^1_{\rm loc}(\mathbb{R}^n)$.

Likewise the *Triebel–Lizorkin norm of ball means of differences* can be defined by

$$\|f\|_{F^s_{pq}:\text{ B. M. D.}} \equiv \|f\|_p + \left\|\left(\int_0^1 (t^{-s} d_t^S f)^q \frac{\mathrm{d}t}{t}\right)^{\frac{1}{q}}\right\|_p$$

for $f \in L^1_{\rm loc}(\mathbb{R}^n)$.

Theorem 2.38 (Characterization of Besov spaces by the ball means of differences) *Let $0 < p \le \infty$, and let $f \in L^{\max(1,p)}(\mathbb{R}^n)$.*

1. *Let $0 < q \le \infty$ and $s > \sigma_p$. Assume that an integer S satisfies $S > [s]$. Then the norm equivalence $\|f\|_{B^s_{pq}} \sim \|f\|_{B^s_{pq}:\text{ B. M. D.}}$ holds.*
2. *Let $p < \infty$, $0 < q \le \infty$ and $s > \sigma_{pq}$. Assume that an integer S satisfies $S > [s]$. Then the norm equivalence $\|f\|_{F^s_{pq}} \sim \|f\|_{F^s_{pq}:\text{ B. M. D.}}$ holds.*

Proof Use Exercise 1.81 to have $\|f\|_{A^s_{pq}} \gtrsim \|f\|_{A^s_{pq}:\text{difference}}$. For the reverse inequality use Proposition 1.14.

2.5.3.2 Classical Definition of Besov Spaces

We can define Besov spaces by means of the difference operator. This is a classical definition of Besov spaces. For $h \in \mathbb{R}^n$ and a function or a distribution f, the difference operator is defined by $\Delta_h f = \Delta_h^1 f \equiv f(\star + h) - f$. For $k \ge 2$, it is defined inductively as $\Delta_h^k f = (\Delta_h)^k f \equiv \Delta_h[(\Delta_h)^{k-1} f]$.

Consider the Besov space $B^s_{pq}(\mathbb{R}^n)$ with $1 \le p, q \le \infty$ and $s > 0$ in the field of partial differential equations. Some people prefer to define the Besov space $B^s_{pq}(\mathbb{R}^n)$ with $1 \le p, q \le \infty$ and $s > 0$ by

$$\|f\|_{B^s_{pq}:\text{difference}} \equiv \|f\|_p + \left(\int_{B(1)} \left(|h|^{-s}\|\Delta_h^S f\|_p\right)^q \frac{\mathrm{d}h}{|h|^n}\right)^{\frac{1}{q}} \tag{2.166}$$

for $f \in L^p(\mathbb{R}^n)$, where S is an integer satisfying $S > [s]$. A natural modification is made via the $\ell^\infty(\mathbb{N})$-norm when $q = \infty$. This is the classical definition of the Besov norm.

Likewise, we can consider the Triebel–Lizorkin norm. Let $1 \le p < \infty$, $1 \le q \le \infty$ and $s > 0$. For $f \in L^p(\mathbb{R}^n)$, we define

$$\|f\|_{F^s_{pq}:\text{difference}} \equiv \|f\|_p + \left\| \left(\int_{B(1)} \left(|h|^{-s} |\Delta_h^S f| \right)^q \frac{dh}{|h|^n} \right)^{\frac{1}{q}} \right\|_p . \tag{2.167}$$

To unify $\| \star \|_{B^s_{pq}:\text{difference}}$ and $\| \star \|_{F^s_{pq}:\text{difference}}$ we write $\| \star \|_{A^s_{pq}:\text{difference}}$. Let us show that this norm is equivalent to the one in Definition 2.1. Let $s > 0$ and $1 \le p, q \le \infty$. Then recall that we defined

$$\|f\|_{B^s_{pq}} \equiv \|\psi(D)f\|_p + \|\{2^{js}\varphi_j(D)f\}_{j=1}^\infty\|_{\ell^q(L^p)}$$

for $f \in \mathscr{S}'(\mathbb{R}^n)$. We note that if $s > 0$ and $f \in L^p(\mathbb{R}^n)$, then we have

$$\|f\|_{B^s_{pq}} \sim \|f\|_p + \|\{2^{js}\varphi_j(D)f\}_{j=1}^\infty\|_{\ell^q(L^p)}.$$

A natural modification can be made using the $\ell^\infty(\mathbb{N})$-norm when $q = \infty$. If $f \in B^s_{pq}(\mathbb{R}^n)$, then $f \in L^p(\mathbb{R}^n)$ by Propositions 2.1 and 2.3, We can prove:

Theorem 2.39 (Besov spaces by means of differences) *Let $s > 0$, $1 \le q \le \infty$, let $S \in ([s], \infty) \cap \mathbb{N}$, and let $f \in L^p(\mathbb{R}^n)$.*

1. *Let $1 \le p \le \infty$. Then the norm equivalence $\|f\|_{B^s_{pq}} \sim \|f\|_{B^s_{pq}:\text{difference}}$ holds.*
2. *Let $1 < p < \infty$. Then the norm equivalence $\|f\|_{F^s_{pq}} \sim \|f\|_{F^s_{pq}:\text{difference}}$ holds.*

Proof Mimic the proof of Theorem 2.35 to have $\|f\|_{A^s_{pq}} \gtrsim \|f\|_{A^s_{pq}:\text{difference}}$. Let us prove the reverse inequality. Using Theorem 2.38, we see that it is just a matter of removing the average, which can be done by the Minkowski inequality or by the vector-valued boundedness of the Hardy–Littlewood maximal operator. See Exercise 2.71.

2.5.3.3 Characterization of Function Spaces by Means of Oscillation

Function spaces can be characterized by means of oscillation. Let $f \in L^1_{\text{loc}}(\mathbb{R}^n)$, and let $M \in \mathbb{N}$. Recall that its oscillation is defined as follows:

$$\text{osc}_1^M f(x,r) \equiv \inf_{P \in \mathscr{P}_M(\mathbb{R}^n)} m_{B(x,r)}(|f - P|) \quad (x \in \mathbb{R}^n, r > 0).$$

The function $\text{osc}_1^M f$ is called the *oscillation of f*. See Definition 1.33.

With the above definition, we will prove that the function spaces are characterized by means of the oscillation. It should be noted that since s is sufficiently large, the distributions are automatically regular.

Theorem 2.40 *Let* $0 < p, q \le \infty$, $s > 0$, *and let* $f \in L^{\max(1,p)}(\mathbb{R}^n)$.

1. *For* $s > \sigma_p = n\left(\dfrac{1}{p} - 1\right)_+$ *and a positive integer* $M > [s]$,

$$\|f\|_{B^s_{pq}} \sim \|f\|_p + \|\{2^{js}\mathrm{osc}_1^M f(\star, 2^{-j})\}_{j=0}^\infty\|_{\ell^q(L^p)}.$$

2. *For* $s > \sigma_{pq} = \max(\sigma_p, \sigma_q)$, *and positive integer* $M > [s]$,

$$\|f\|_{F^s_{pq}} \sim \|f\|_p + \|\{2^{js}\mathrm{osc}_1^M f(\star, 2^{-j})\}_{j=0}^\infty\|_{L^p(\ell^q)}.$$

Proof Concentrate on $F^s_{pq}(\mathbb{R}^n)$. For $L \gg M$, we choose $\Psi \in C^\infty_c(B(1)) \setminus \mathscr{P}_0(\mathbb{R}^n)^\perp$ and $\Phi \in C^\infty_c(B(1))$ so that $\Psi - \Psi^1 = \Delta^L\Phi$ using Theorem 1.40. Hence

$$\|f\|_{F^s_{pq}} \sim \|\Psi * f\|_p + \left\|\left(\sum_{j=0}^\infty 2^{jsq}|(\Delta^L\Phi)^j * f|^q\right)^{\frac{1}{q}}\right\|$$

with equivalence of norms by Theorem 2.34.

By assumption, we can choose $0 < \eta < \min(1, p, q)$ and $A > 0$ so that

$$A > \frac{n}{\min(1, p, q)}, \quad A(1-\eta) < s.$$

Since $f = \Psi * f + \sum_{j=0}^\infty (\Psi^{j+1} * f - \Psi^j * f)$ and $s > 0$, $\|f\|_p \lesssim \|f\|_{F^s_{pq}}$. Thanks to Proposition 1.15

$$\begin{aligned}\mathrm{osc} f(x, 2^{-j}) &\lesssim m_{B(x,2^{-j})}(|f - \Psi^j * f|)\\ &\quad + m_{B(x,2^{-j})}\left(\left|\Psi^j * f - \sum_{\alpha\in\mathbb{N}_0{}^n, |\alpha|\le M}\frac{(\star - x)^\alpha}{\alpha!}\partial^\alpha\Psi^j * f(x)\right|\right).\end{aligned}$$

Then

$$\begin{aligned}&m_{B(x,2^{-j})}(|f - \Psi^j * f|)\\ &\le \sum_{l=j}^\infty m_{B(x,2^{-j})}(|(\Delta^L\Phi)^l * f|)\\ &\lesssim \sum_{l=j}^\infty 2^{A(l-j)(1-\eta)} M[|(\Delta^L\Phi)^l * f|^\eta](x)\left(\sup_{y\in\mathbb{R}^n}\frac{|(\Delta^L\Phi)^l * f(y)|}{(1+2^l|x-y|)^A}\right)^{1-\eta}\end{aligned}$$

$$\lesssim \sum_{l=j}^{\infty} 2^{A(l-j)(1-\eta)} M^{(\eta)}[(\Delta^L \Phi)^l * f](x)$$

$$+ \sum_{l=j}^{\infty} 2^{A(l-j)(1-\eta)} \left(\sup_{y \in \mathbb{R}^n} \frac{|(\Delta^L \Phi)^l * f(y)|}{(1+2^l|x-y|)^A} \right)$$

and

$$m_{B(x,2^{-j})} \left(\left| \Psi^j * f - \sum_{\alpha \in \mathbb{N}_0{}^n, |\alpha| \le M} \frac{(\star - x)^\alpha}{\alpha!} \partial^\alpha \Psi^j * f(x) \right| \right)$$

$$\lesssim_N \sup_{\alpha \in \mathbb{N}_0{}^n, |\alpha| = M+1} \sup_{z \in \mathbb{R}^n} \frac{|(\partial^\alpha \Psi)^j * f(x-y)|}{(1+2^j|y|)^A}$$

with a constant that depends on $A \gg 1$. Hence

$$\|\{2^{js} \mathrm{osc}_1^M f(\star, 2^{-j})\}_{j=0}^\infty\|_{L^p(\ell^q)} \lesssim \|f\|_{F^s_{pq}}$$

using Proposition 2.14. Next, suppose that $f \in L^{\max(1,p)}(\mathbb{R}^n)$ satisfies

$$\|f\|_p + \|\{2^{js} \mathrm{osc}_1^M f(\star, 2^{-j})\}_{j=0}^\infty\|_{L^p(\ell^q)} < \infty.$$

Then

$$|(\Delta^L \Phi)^j * f(x)| = |(\Delta^L \Phi)^j * (f - P)(x)| \lesssim m_{B(x,2^{-j})}(|f - P|)$$

for any polynomial $P \in \mathscr{P}_{M-1}(\mathbb{R}^n)$. Hence $|k^j * f(x)| \lesssim \mathrm{osc}_1^M f(x, 2^{-j})$. Meanwhile,

$$\Psi * f(x) = f - \sum_{j=0}^\infty (\Psi^{j+1} * f - \Psi^j * f) = f + \sum_{j=0}^\infty (\Delta^L \Phi)^j * f.$$

Thus,

$$|\Psi * f(x)| \lesssim |f(x)| + \sum_{j=0}^\infty |(\Delta^L \Phi)^j * f(x)| \lesssim |f(x)| + \sum_{j=0}^\infty \mathrm{osc}_1^M f(x, 2^{-j}).$$

Since $s > 0$, $\|\Psi * f\|_p \lesssim \|f\|_p + \|\{2^{js} \mathrm{osc}_1^M f(\star, 2^{-j})\}_{j=0}^\infty\|_{L^p(\ell^q)}$. It is easy to see $|\Psi^{j+1} * f(x) - \Psi^j * f(x)| \lesssim \mathrm{osc}_1^M f(x, 2^{-j})$ from $\Psi^{j+1} - \Psi^j = -(\Delta^L \Psi)^j$. Thus, it follows that $\|f\|_{F^s_{pq}} \lesssim \|f\|_p + \|\{2^{js} \mathrm{osc}_1^M f(\star, 2^{-j})\}_{j=0}^\infty\|_{L^p(\ell^q)}$. Therefore, the proof is complete.

2.5.3.4 Characterization of Function Spaces by the Average Over Balls

Let $1 < p < \infty$, $1 < q \le \infty$ and $0 < s < 2$. Here we aim to refine the local means in some sense to have the following equivalent norms.

Theorem 2.41 *Let* $1 < q \le \infty$ *and* $0 < s < 2$.

1. *Let* $1 < p \le \infty$. *Then* $f \in B^s_{pq}(\mathbb{R}^n)$ *if and only if* $f \in L^p(\mathbb{R}^n)$ *and*

$$\|f\|_p + \|\{2^{ks}(m_{B(\star,2^{-k})}(f) - f)\}_{k=-\infty}^{\infty}\|_{\ell^q(L^p)} < \infty.$$

2. *Let* $1 < p < \infty$. *Then* $f \in F^s_{pq}(\mathbb{R}^n)$ *if and only if* $f \in L^p(\mathbb{R}^n)$ *and*

$$\|f\|_p + \|\{2^{ks}(m_{B(\star,2^{-k})}(f) - f)\}_{k=-\infty}^{\infty}\|_{L^p(\ell^q)} < \infty. \tag{2.168}$$

Proof We concentrate on Triebel–Lizorkin spaces. Let $f \in F^s_{pq}(\mathbb{R}^n)$. Since $s > 0$, we have

$$\begin{aligned} &\|f\|_p + \|\{2^{ks}(m_{B(\star,2^{-k})}(f) - f)\}_{k=-\infty}^{\infty}\|_{L^p(\ell^q)} \\ &\lesssim \|f\|_p + \|\{2^{ks}(m_{B(\star,2^{-k})}(f) - m_{B(\star,2^{-k+1})}(f))\}_{k=-\infty}^{\infty}\|_{L^p(\ell^q)}. \end{aligned}$$

Let $\Psi \equiv \dfrac{1}{|B(1)|}\chi_{B(1)}$. Define $\Phi \equiv \Psi - \Psi^{-1}$. Then we have

$$\|\{2^{ks}(m_{B(\star,2^{-k})}(f) - m_{B(\star,2^{-k+1})}(f))\}_{k=-\infty}^{\infty}\|_{L^p(\ell^q)} = \|\{2^{ks}\Phi^k * f\}_{k=-\infty}^{\infty}\|_{L^p(\ell^q)}.$$

Choose $\varphi \in C^\infty_{\rm c}(\mathbb{R}^n)$ so that

$$\sum_{j=-\infty}^{\infty} \varphi(2^{-k}\star)^2 \equiv \chi_{\mathbb{R}^n\setminus\{0\}}.$$

Let $x \in \mathbb{R}^n$ be fixed. We note that $|\Phi^k * \mathscr{F}^{-1}\varphi_{k+l} * F(x)| \lesssim MF(x)$. Since we have a key estimate

$$\mathscr{F}^{-1}\Psi(\xi) = (2\pi)^{-\frac{n}{2}} + {\rm O}(|\xi|^2) \quad (\xi \to 0),$$

we also have $|\Phi^k * \mathscr{F}^{-1}\varphi_{k+l} * F(x)| \lesssim 2^{2l}MF(x)$. Thus,

$$|\Phi^k * \mathscr{F}^{-1}\varphi_{k+l} * F(x)| \lesssim \min(1, 2^{2l})MF(x).$$

As a result, since $0 < s < 2$,

$$\left(\sum_{k=-\infty}^{\infty} 2^{ksq}|\Phi^k * f(x)|^q\right)^{\frac{1}{q}} \lesssim \left(\sum_{k=-\infty}^{\infty} 2^{ksq} M[\varphi_k(D)f](x)^q\right)^{\frac{1}{q}}.$$

Taking the $L^p(\mathbb{R}^n)$-norm, we obtain

$$\|\{2^{ks}\Phi^k * f\}_{k=-\infty}^{\infty}\|_{L^p(\ell^q)} \lesssim \|f\|_{F^s_{pq}} < \infty.$$

Conversely, let $f \in L^p(\mathbb{R}^n)$ satisfy (2.168). Then we have

$$|\varphi_{k+k_0}(D)f(x)| \leq M[f - \Phi^k * f](x) \quad (x \in \mathbb{R}^n)$$

for some $k_0 \in \mathbb{Z}$, which is independent of x and f. Thus, by the Fefferman–Stein vector-valued inequality, we obtain the desired result.

Exercises

Exercise 2.70 Let $1 \leq p < \infty$, $0 < q \leq \infty$ and $s > 0$.

1. Prove $F^s_{pq}(\mathbb{R}^n) \hookrightarrow L^p(\mathbb{R}^n)$ using Propositions 2.1, 2.7, and 2.8. See also [2, Theorem 4.1.3 (b)].
2. Prove (2.165) using Proposition 1.2.

Exercise 2.71 Prove Theorem 2.39.

Textbooks in Sect. 2.5

Various Characterizations by Difference

In the textbook [99, Section 2.5.12] we can find the characterization by way of differences. In the textbook [99, Section 2.5.11] we can find the characterization by way of ball means of differences. See [99, p. 101, Theorem] and [99, p. 110, Theorem] for the characterization of Triebel–Lizorkin spaces and Besov spaces by means of differences, respectively.

In the textbook [100], Theorem 2.40 is explained very precisely.

Various Characterization Similar to Local Means

In the textbook [99, Section 2.12.2], we can find the characterization by means of the heat kernel, Gauss–Weierstrass semi-group and the Poisson kernel, Cauchy–Poisson semi-group, both of which are variants of local means. See also [101, Chapter 2].

Others

We referred to the textbook [100] for this section.

2.6 Notes for Chap. 2

Section 2.1

Section 2.1.1

Sobolev Space $W^m_p(\mathbb{R}^n)$ with $m \in \mathbb{N}$ and $1 \le p < \infty$

The Sobolev space $W^m_p(\mathbb{R}^n)$ dates back to 1936 [1001].

Sobolev Space $W^s_p(\mathbb{R}^n)$ with $s \in (0,\infty) \setminus \mathbb{N}$ and $1 \le p < \infty$

The space $W^s_p(\mathbb{R}^n)$ with $s \in (0,\infty) \setminus \mathbb{N}$ and $1 \le p \le \infty$ dates back to 1955–1958 (see [139, 493, 998]).

Potential Space $H^p_s(\mathbb{R}^n)$

It is defined and investigated by Aronszajn, Smith, and Calderón [140, 323].

Definition of $B^s_{pq}(\mathbb{R}^n)$ with $1 \le p, q \le \infty$ and $s \in \mathbb{R}$ by Fourier analysis

The definition of the Besov space $B^s_{pq}(\mathbb{R}^n)$ with $1 \le p \le \infty$, $1 \le q \le \infty$ and $s \in \mathbb{R}$ in the form in this book dates back to 1967. See [870, p. 281] and see the textbook [7, Definition 6.2.2] for the case of $1 \le p \le \infty$, $1 \le q \le \infty$ and $s \in \mathbb{R}$. Triebel used the Fourier multipliers systematically in [1039, Theorem 3.5] to define $B^s_{pq}(\mathbb{R}^n)$ with $1 < p, q < \infty$ and $s \in \mathbb{R}$ by means of the Littlewood–Paley decomposition. See [862] for more information.

Inclusion $B^0_{p1}(\mathbb{R}^n) \hookrightarrow L^p(\mathbb{R}^n) \hookrightarrow B^0_{p\infty}(\mathbb{R}^n)$: Proposition 2.1

Proposition 2.1, or more precisely $B^0_{p\,\min(p,2)}(\mathbb{R}^n) \hookrightarrow L^p(\mathbb{R}^n) \hookrightarrow B^0_{p\,\max(p,2)}(\mathbb{R}^n)$, was originally investigated in [1040, 5.2.5 Remark] for the case of $1 < p, q \le \infty$. See also [99, p. 89] for Propositions 2.1 and 2.4.

Section 2.1.2

Embedding $B^s_{pq_0}(\mathbb{R}^n) \hookrightarrow B^s_{pq_1}(\mathbb{R}^n)$ with $1 \le p, q_0, q_1 \le \infty$ and $s \in \mathbb{R}$: Proposition 2.2

See [1039, Theorem 5.21, (5.7)] if $1 \le p, q_0, q_1 \le \infty$ and $s \in \mathbb{R}$.

Lifting Operators for $B^s_{pq}(\mathbb{R}^n)$ with $1 \le p, q \le \infty$ and $s \in \mathbb{R}$: Theorem 2.2

Theorem 2.2 dates back to 1965 [848, p. 132, Corollary 1] for potential spaces. See [1039, Theorem 5.1.1]. See also [71, p. 67, Theorem 9] and [71, p. 67, Theorem 8] for (2.9) and (2.10), respectively.

Embedding $\mathscr{S}(\mathbb{R}^n) \hookrightarrow B^s_{pq}(\mathbb{R}^n) \hookrightarrow \mathscr{S}'(\mathbb{R}^n)$ with $1 \le p, q \le \infty$ and $s \in \mathbb{R}$: Theorem 2.4 *1*

See [71, p. 61, Theorem 2] for $\mathscr{S}(\mathbb{R}^n) \hookrightarrow B^s_{pq}(\mathbb{R}^n)$ and [71, p. 61, Theorem 2] for $B^s_{pq}(\mathbb{R}^n) \hookrightarrow \mathscr{S}'(\mathbb{R}^n)$ including the case of $1 \le p \le \infty$ and $0 < q \le \infty$.

Completeness of $B^s_{pq}(\mathbb{R}^n)$ with $1 \le p, q \le \infty$ and $s \in \mathbb{R}$: Theorem 2.4 *2*

See [71, p. 58, Theorem 1] including the case of $1 \le p \le \infty$ and $0 < q \le \infty$.

Density of $C^\infty_c(\mathbb{R}^n)$ in $B^s_{pq}(\mathbb{R}^n)$ with $1 \le p, q \le \infty$ and $s \in \mathbb{R}$: Theorem 2.4 *3*

See [1039, Theorem 6.2.1] and [71, p. 61, Theorem 2].

Sobolev Embedding $B^{s_0}_{p_0q}(\mathbb{R}^n) \hookrightarrow B^{s_1}_{p_1q}(\mathbb{R}^n)$: Theorem 2.5

Theorem 2.5 dates back to 1976 (see the textbook [7, Theorem 6.5.1]) when $1 \le p_0, p_1, q \le \infty$.

Section 2.2.1

$L^2(\mathbb{R}^n)$ and Besov Spaces: Corollary 2.1

For the space $B^s_{22}(\mathbb{R}^n)$ we refer to the papers [865, 1100]. Corollary 2.1 can be found in the book [71, p. 76, Theorem 1], where Peetre claimed that $B^s_{22}(\mathbb{R}^n) \approx W^s_2(\mathbb{R}^n)$ with equivalence of norms.

Section 2.2.2

Weierstrass Function

See [641] for an interesting approach to the Weierstrass function.

Hölder–Zygmund Spaces

See [1190] for the definition of Hölder–Zygmund spaces, where Zygmund also pointed out that Δ_h^2 is useful.

$\mathscr{C}^\gamma(\mathbb{R}^n) \approx B^\gamma_{\infty\infty}(\mathbb{R}^n)$: Theorems 2.7 and 2.8

Theorems 2.7 and 2.8 go back to Hardy and Littlewood [555]. Taibleson refined [1026, Theorem 4]. We refer to the papers [519, 871, 1050] for further details. Equivalence $\mathscr{C}^\gamma(\mathbb{R}^n) \approx B^\gamma_{\infty\infty}(\mathbb{R}^n)$ is due to Triebel [1050, Theorem 4 (40)], while the homogeneous counterpart is due to Grevholm [519]. See also Triebel [1050, Theorem 4 (41)] for the homogeneous counterpart. See [419] for an approach using some maximal operators, which expands the technique of defining BMO. For $s > 2$, $\mathscr{C}^s(\mathbb{R}^n)$ is a function space applied to elliptic differential equations in the paper [124]. See [124, 999] for the explicit expression of the solution operator.

Others

The author is indebted to Professor Bin Xu for Example 2.8.

Section 2.2

Section 2.2.3

Applications of Function Spaces to Fractals: Example 2.6

Example 2.6 is a simple application of Besov spaces to fractals and is itself an example of fractal analysis. Saka used oscillations to investigate the local property of functions [906]. The function defined by formula (2.24) is called a lacunary (Fourier) series. Example 2.6 demonstrates that Besov spaces are applicable to fractals. We refer to the paper by Triebel [1075] for examples of fractal analysis, where Triebel considered the elliptic Laplacian on fractals. Currently, let us just say that the atomic decomposition appearing in the latter half of this book is useful for fractal analysis. See also [322, 695].

Applications of Besov Spaces to the Fourier Transform: Theorem 2.9 and Exercise 2.19

Theorem 2.9 is due to Rivière and Sagher [891, Theorem III]. Gabisoniya proved $\|f\|_1 \lesssim \|f\|_{B^{\frac{n}{2}}_{21}}$ for all $f \in L^2(\mathbb{R}^n) \cap B^{\frac{n}{2}}_{21}(\mathbb{R}^n)$ [490]. Exercise 2.19 is pointed out in [7, p. 164, Exercise 16]. On the torus, Bernstein considered a counterpart to Example 2.19. Golovkin and Solonnikov proved the results assuming that both sides are finite in their paper [513, Section 3, Theorem 7]. Madych gave a sharper version of Exercise 2.19 in his paper [759, Theorems 2–4].

Applications of Besov Spaces to the Fourier Series

We use Besov spaces and Triebel–Lizorkin spaces to investigate the convergence of the Fourier series. For example, Schmeisser and Sickel developed the theory of function spaces on an n-dimensional torus [936–938]. Later, based on these papers, Sautbekova and Sickel considered Besov–Morrey spaces and Triebel–Lizorkin–Morrey spaces to consider the convergence of the Fourier series [911]. See the series of works [203, 225, 228, 899, 952, 956] for more details.

Applications of Besov Spaces to Convolution

Exercise 2.18 demonstrates that Besov spaces can describe what is gained from convolution more precisely. Burenkov obtained similar results [303, Teorema 3].

Other Applications of Besov Spaces

Investigation of the Riemannian integral by means of Besov spaces can be found in [192, 193].

Section 2.3

Section 2.3.1

Definition of $B^s_{pq}(\mathbb{R}^n)$ with $0 < p, q \le \infty$ and $s \in \mathbb{R}$ by the Fourier Transform

There is a long history in Theorem 2.11 for Besov spaces, which dates back to 1959 [171]. In fact, Besov defined the function spaces by the use of the difference. He also considered the trace operators in [171, Teorema 4] and obtained the characterization by means of the oscillation [171, Teorema 1].

Let us move on to the definition of the Besov space $B^s_{pq}(\mathbb{R}^n)$ with $0 < p, q \le \infty$ and $s \in \mathbb{R}$ by way of the Fourier series. The definition dates back to 1973. See the paper by Peetre [872], who was inspired by Flett [471].

We refer to [71, pp. 225–227] for the motivation of $B^s_{pq}(\mathbb{R}^n)$ with $0 < p, q < \infty$. In his earlier paper [872, p. 947, Lemme], using Corollary 1.8 in this book, Peetre got to the definition of Besov spaces presented in this book. In [872, p. 947] he also defined homogeneous Besov spaces. He clearly claimed that the homogeneous Besov space $\dot{B}^s_{pq}(\mathbb{R}^n)$ is a subset of $\mathscr{S}'(\mathbb{R}^n)/\mathscr{P}(\mathbb{R}^n)$. He compared his definitions with those by Flett [471]; see [872, p. 948]. In his 1976 book [71] Peetre showed that the definition of $\dot{B}^s_{pq}(\mathbb{R}^n)$ is independent of the choice of φ using (1.164) whose proof in that book is similar to the one in this book. See also [2, Definition 4.1.1] for the definition of the Besov space $B^s_{pq}(\mathbb{R}^n)$ for $0 < p, q \le \infty$ and $s \in \mathbb{R}$.

Definition of $F^s_{pq}(\mathbb{R}^n)$ with $1 < p < \infty$, $1 < q \le \infty$ and $s \in \mathbb{R}$: Lizorkin's Contribution

Lizorkin [745] obtained the following equivalent norm for $F^s_{p2}(\mathbb{R}^n)$ with $1 < p < \infty$, $s \in \mathbb{R}$:

$$\|\chi_{Q(1)}(D)f\|_p + \|2^{js}\chi_{[-2^j,2^j]^n\setminus[-2^{j-1},2^{j-1}]^n}(D)f\|_{L^p(\ell^2)}.$$

By the Hilbert transform, that is, the Riesz transform when $n = 1$, we can show that this norm is equivalent to the norm of $F^s_{p2}(\mathbb{R}^n)$ defined in this book. Lizorkin considered an operator of the form

$$f \mapsto \varphi(x,t) \equiv \int_t^{2t} (f(x) - f(x-h))h^{-1-r}\mathrm{d}h$$

to define the space $L^{(r)}_{p,\theta}(\mathbb{R})$, which is connected to the Triebel–Lizorkin space $F^r_{p\theta}(\mathbb{R}^n)$ [746]. Lizorkin also considered the space $\Lambda^r_{p,\theta}(\mathbb{R}^n)$ whose norm is given by

$$\|f\|_{\Lambda^r_{p,\theta}} = \|\chi_{[-1,1]^n}(D)f\|_p + \|2^{jr}\chi_{[-2^j,2^j]^n\setminus[-2^{j-1},2^{j-1}]^n}(D)f\|_{\ell^\theta(L^p)}$$

for $1 < p < \infty$, $1 \le \theta \le \infty$ and $r \in \mathbb{R}$ in [748].

Definition of $F^s_{pq}(\mathbb{R}^n)$ with $0 < p < \infty, 0 < q \le \infty$ and $s \in \mathbb{R}$

The Triebel–Lizorkin space $F^s_{pq}(\mathbb{R}^n)$ with $0 < p < \infty, 0 < q \le \infty$ and $s \in \mathbb{R}$ dates back to 1975 via a couple of important steps mentioned above and [1039, 1040].

The definition of the Triebel–Lizorkin space $F^s_{pq}(\mathbb{R}^n)$ with $0 < p < \infty$ and $0 < q \le \infty$ dates back to 1975 (see the paper by Peetre [874, Definition 1.1]). Here it counts that the Fefferman–Stein vector-valued inequality is used there. Theorem 2.1 corresponds to Theorem 2.11 with $p \ge 1$ for Besov spaces and we can use the Young inequality for the proof of Theorem 2.1. So the proof is a little simpler.

Triebel considered $F^s_{pq}(\mathbb{R}^n)$ with $1 < p, q < \infty$ and $s \in \mathbb{R}$ in [1039]. We refer to [655] for more related to (Triebel–)Lizokin spaces. Later, Peetre defined $F^s_{pq}(\mathbb{R}^n)$ with $0 < p < \infty$ and $0 < q \le \infty$ in 1975 [874, Definition 1.1]. Although Peetre defined $F^s_{\infty q}(\mathbb{R}^n)$ and $0 < q \le \infty$, its definition did not use the Carleson box. The Peetre maximal operator is used; see [874, Lemma 2.1] and [874, Theorem 3.1]. Based on the Peetre maximal operator, Peetre proved Theorem 2.11 in [874, Corollary 3.1] and the multiplier theorem in [874, Theorem 5.1]. See also [875].

We can use Theorem 2.11 for the proof of the convolution inequality as in the paper [1051]. In Theorem 2.11, we have to take into account the possibility other than $p \ge 1$. So we need to use Theorem 1.53 or the (weighted) convolution inequality in [1051, Theorem 1].

Triebel used the notation $A^s_{pq}(\mathbb{R}^n)$ with $0 < p, q \le \infty$ and $s \in \mathbb{R}$ [99, p. 45], which is in this book.

Section 2.3.2

Embedding $B^s_{p\,\min(p,q)}(\mathbb{R}^n) \hookrightarrow F^s_{pq}(\mathbb{R}^n) \hookrightarrow B^s_{p\,\max(p,q)}(\mathbb{R}^n)$: Proposition 2.7

See [1039, Theorem 5.2.3].

Embedding $A^{s+\delta}_{pq_1}(\mathbb{R}^n) \hookrightarrow A^s_{pq_2}(\mathbb{R}^n)$: Proposition 2.8

See [1039, Theorem 5.2.1].

Density of $\mathscr{S}(\mathbb{R}^n)$ and $C_c^\infty(\mathbb{R}^n)$ in $A^s_{pq}(\mathbb{R}^n)$: Proposition 2.10 Including Remark 4.2

Triebel proved that $C_c^\infty(\mathbb{R}^n)$ is dense in $A^s_{pq}(\mathbb{R}^n)$ [1039, Theorem 6.1.1], if $1 < p, q < \infty$ and $s \in \mathbb{R}$. Proposition 2.10 is extended to the case $0 < p, q \le \infty$ and $s \in \mathbb{R}$ in [99, p. 48, Theorem]. See [6, Proposition 2.27] for the case of $A = B$, where the authors treated the density of $\mathscr{S}_\infty(\mathbb{R}^n)$ in $\dot{A}^s_{pq}(\mathbb{R}^n)$.

Lift Operator: Theorem 2.12

Let $1 < p < \infty$, $1 \le q \le \infty$ and $s \in \mathbb{R}$ when $A = B$ and $1 < p, q < \infty$ and $s \in \mathbb{R}$ when $A = F$. Theorem 2.12 dates back to 1973 [1039, Theorem 5.1.1].

We refer to [71, p. 244, Theorem 9] and [71, p. 244, Theorem 8] for the lifting properties (2.52) and (2.53) of the Besov space $B^s_{pq}(\mathbb{R}^n)$ with $0 < p, q \le \infty$ and $s \in \mathbb{R}$, respectively. (2.63) is [71, p. 242, Theorem 6].

We refer to [99, p. 58, Theorem] for Theorem 2.12 in full generality.

Sobolev Embedding Theorems: Theorem 2.14 Including Theorem 4.11 and Related Sharpness

Theorem 2.14 is due to Jawerth [628, Theorem 2.1] and, when $1 < p, q < \infty$, Theorem 2.14 is due to Triebel [1039, (7) and (8)]. See [352, Theorem 1] and subsequent results for more technique using (2.67).

Theorem 4.11 is due to Jawerth and Frank. In [628, Theorem 2.1 (iii)] Jawerth proved $F^s_{pq}(\mathbb{R}^n) \hookrightarrow B^{s_1}_{p_1 p}(\mathbb{R}^n)$, the first part of Theorem 4.11, while Frank proved $B^{s_0}_{p_0 p}(\mathbb{R}^n) \hookrightarrow F^s_{pq}(\mathbb{R}^n)$ [477, Theorem 1]. Our proof of $F^s_{pq}(\mathbb{R}^n) \hookrightarrow B^{s_1}_{p_1 p}(\mathbb{R}^n)$ is based largely on the reproof by Vybíral [1104].

Theorem 2.14 for $B^s_{pq}(\mathbb{R}^n)$ dates back to [1040, Theorem], which is a supplemental paper of [1039], while Theorem 2.14 for $F^s_{pq}(\mathbb{R}^n)$ is [1039, Theorem 6.1.1].

We followed [648, Proposition 1] for our method of the proof of $F^{s_0}_{p_0\infty}(\mathbb{R}^n) \hookrightarrow F^{s_1}_{p_1 q}(\mathbb{R}^n)$ (2.64). The paper [648] contains a historical account of this embedding.

As for the sharpness results, see [477, Theorem 1] for Theorem 4.16.

Completeness of the Spaces: Theorem 2.15

Theorem 2.15, which deals with the completeness of $A^s_{pq}(\mathbb{R}^n)$ is proved in [1039] for $1 < p, q < \infty$. Proposition 2.1, which ensures that the function spaces are embedded into $\mathrm{BUC}(\mathbb{R}^n)$, can be used for the proof of completeness of $A^s_{pq}(\mathbb{R}^n)$; see [904, Lemma 2.15] for the proof of the completeness by means of this fact.

Diversity: Theorem 2.16

Theorem 2.16 in full generality dates back to 1983 (see [99, p. 61, Theorem]).

Duality of $B^s_{pq}(\mathbb{R}^n)$ with $0 < p, q < \infty$ and $s \in \mathbb{R}$: Theorem 2.17 for $B^s_{pq}(\mathbb{R}^n)$

Theorem 2.17 is investigated in [1039, Theorem 7.1.7] for Triebel-Lizorkin spaces and [1039, Theorem 7.2.2] for Besov spaces when $1 < p, q < \infty$. Theorem 2.17 for $F^s_{pq}(\mathbb{R}^n)$ can be found in [1039, Theorem 7.1.1]. See also [1050, Theorem 2]. See [99, p. 178] for the extension to the case $1 \le p < \infty$ and $0 < q \le \infty$ for Besov spaces. See [818, Theorems 1.2–1.4] for the case of the Besov space $B^s_{pq}(\mathbb{R}^n)$ with $1 \le p, q \le \infty$ and $s \in \mathbb{R}$.

Peetre considered the dual space of $B^s_{pq}(\mathbb{R}^n)$ with $0 < p < 1$, $0 < q \le 1$ and $s \in \mathbb{R}$ in [71, p. 249, Theorem 13]. See [873, (3.3)] in the case of $1 < p < \infty$ and $0 < q < 1$, where Peetre used the real interpolation. The paper [249] follows this spirit. See also [867] and [71, p. 74, Theorem 12] for the dual space of $B^s_{pq}(\mathbb{R}^n)$ including $(B^s_{pq}(\mathbb{R}^n))^* \approx B^{-s}_{p\infty}(\mathbb{R}^n)$ with equivalence of norms for $1 \le p < \infty$, $q \in (0, 1]$ and $s \in \mathbb{R}$ and [2, Theorem 4.1.3(d)] for the dual space of $B^s_{pq}(\mathbb{R}^n)$ with $1 \le p, q < \infty$.

Furthermore, in the case $0 < p, q \le \infty$ and $s \in \mathbb{R}$ there are many results in [99] concerning the duality, which we omitted. For $0 < p < 1$, we refer to [628, Theorem 4.2].

Another Meaning of q in $B^s_{pq}(\mathbb{R}^n)$

Triebel showed that $\lim_{s \uparrow m}(m - s)^{\frac{1}{q}} \|f\|_{B^{s**}_{pq}} \sim_{p,m} q^{-\frac{1}{q}} \sum_{|\alpha| \le m} \|\partial^\alpha f\|_p$ in [1082, Theorem 1.2].

Boundedness of the Singular Integral Operators on $\dot{A}^s_{pq}(\mathbb{R}^n)$

Frazier, Torres and Weiss proved $\dot{F}^s_{pq}(\mathbb{R}^n)$-boundedness of the singular integral operators dealt with in Sect. 1.5.2 [481].

See [542] for further information.

Dilation Properties of Besov Spaces

We refer to [1019, Proposition 1.1].

Section 2.3.3

Definition of Modulation Spaces

The modulation space $M^s_{pq}(\mathbb{R}^n)$, $1 \le p, q \le \infty$, $s \in \mathbb{R}$ is a function space defined by Feichtinger [463] and it is widely investigated by Feichtinger and Gröchenig; see [464, 468, 469], where an approach different from Theorem 1.53 is taken to investigate modulation spaces with $M^s_{pq}(\mathbb{R}^n)$, $0 < p, q \le \infty$, $s \in \mathbb{R}$ [676, Lemma 2.6]; Kobayashi considered the convolution estimate as in Corollary 1.7.

Singular Integral Operators on Modulation Spaces

See the paper by Sugimoto and Tomita [1022], whose counterexample is based on the pseudo-differential operator with symbol S^0_{11}; see Definition 5.10 for S^0_{11}.

Dilation Properties of Modulation Spaces

We have a surprising property.

Proposition 2.17 *Let $1 \le p, q \le \infty$ and $s \in \mathbb{R}$. Then there exists a couple (θ_1, θ_2) of real numbers with $\theta_1 < \theta_2$ and functions $\varphi_1, \varphi_2 \in M^s_{pq}(\mathbb{R}^n)$ such that*

$$\|\varphi_j(t\star)\|_{M^s_{pq}} \sim t^{\theta_j} \tag{2.169}$$

for all $j = 1, 2$ and $0 < t \le 1$.

Observe (2.169) cannot happen in $L^p(\mathbb{R}^n)$ and this is characteristic to modulation spaces; see the paper by Sugimoto and Tomita [1020] for the proof of (2.169).

Relations Between Modulation Spaces and Besov Spaces

Sugimoto and Tomita investigated embedding theorems between Besov spaces and modulation spaces [1021]. Furthermore, Cunanan, Kato Kobayashi, Miyachi, Sugimoto and Tomita showed the inclusion relationships between weighted $L^2(\mathbb{R}^n)$-spaces, local Hardy spaces and modulation spaces [396, 662, 681].

Brownian Motion and Modulation Spaces

We refer to [168] for a discussion of how smooth the path of the Brownian motion is in terms of modulation spaces.

Section 2.4

Section 2.4.1

The Spaces $\mathscr{S}_\infty(\mathbb{R}^n)$ and $\mathscr{S}'_\infty(\mathbb{R}^n)$ and Their Topologies: Theorems 2.20 and 2.21

It seems that the space $\mathscr{S}'(\mathbb{R}^n)/\mathbb{P}(\mathbb{R}^n)$ emerged in [71] and was developed in [482]. Meanwhile, the symbols $\mathscr{Z}(\mathbb{R}^n)$ and $\mathscr{Z}'(\mathbb{R}^n)$ appeared in [99].

We refer to [830, Section 6] for the proof of Theorems 2.20 and 2.21, which asserts that $\mathscr{S}_\infty(\mathbb{R}^n)/\mathscr{P}(\mathbb{R}^n) \approx \mathscr{S}'_\infty(\mathbb{R}^n)$.

Section 2.4.2

Definition of Homogeneous Besov Spaces

The definition of the homogeneous Besov space $\dot{B}^s_{pq}(\mathbb{R}^n)$ with $1 \le p \le \infty$ and $0 < q \le \infty$ dates back to 1976 (see [71, p. 51, Definition 1]).

Embedding $\mathscr{S}_\infty(\mathbb{R}^n) \hookrightarrow \dot{A}^s_{pq}(\mathbb{R}^n) \hookrightarrow \mathscr{S}'(\mathbb{R}^n)/\mathscr{P}(\mathbb{R}^n)$: Theorem 2.29

We remark that Theorem 2.29 is in [99, p. 240, Theorem] for admissible parameters.

See also [2, Definition 4.1.1] for the definition of the Besov space $B^s_{pq}(\mathbb{R}^n)$ for $0 < p, q \le \infty$ and $s \in \mathbb{R}$.

Dilation Property of Homogeneous Besov Spaces: Exercise 2.57

Concerning (2.123) in Exercise 2.57, we have

$$\|f(\lambda\star)\|_{A^s_{pq}} \sim \lambda^{s-n/p}\|f\|_{A^s_{pq}} \tag{2.170}$$

as long as $\lambda \in (0,1]$, $f \in A^s_{pq}(\mathbb{R}^n)$ and $\mathrm{supp}(f) \subset B(\lambda)$. See [320, Theorem 6] for the case of Besov spaces and [103, Corollary 5.16] for the case of Triebel–Lizorkin spaces.

Section 2.4.3

Realizations

An example in [697, Example p. 140–141 (2.5)] presents the case where the realization cannot be considered. See [697, Lemma 6.2, Proposition 6.3] for Proposition 1.13. Definition 2.17 is [266, Definition 1].

Realization in $\dot{F}^2_{p_0 2}(\mathbb{R}^n)$

Masaki constructed a realization in $\dot{F}^2_{p_0 2}(\mathbb{R}^n)$ [764, Theorem 2.1] for $1 < p_0 < 2$.

Realization in $\dot{L}^{1,p}(\mathbb{R}^n)$

There are many extensions of Exercise 2.61; see [181, 1003].

Translation Invariant Realization

Realization can be considered not only in $\mathscr{S}'(\mathbb{R}^n)$; we can extend the notion to the space $\mathscr{S}'(\mathbb{R}^n)/\mathscr{P}_k(\mathbb{R}^n)$. We refer to [266] for realization which is commutative with transition or dilation.

Realization of $\dot{W}^{m,p}(\mathbb{R}^n)$

See [266, Proposition 5], [266, Theorem 2] and [266, Theorem 3] for the realization of $\dot{W}^{m,\infty}(\mathbb{R}^n)$, $\dot{W}^{m,1}(\mathbb{R}^n)$ with $m < n$ and $\dot{W}^{n,1}(\mathbb{R}^n)$, respectively.

Realization of $\dot{A}^s_{pq}(\mathbb{R}^n)$

We refer to [264] for more of the above realizations to various function spaces: Theorems 2.31 and 2.32 correspond to [264, Le cas $s < n/p$, page 47] and [264, Le cas $s = \dfrac{n}{p}$, $p < \infty$, $q = 1$, Le cas $s = 0$, $p = \infty$, $q = 1$], respectively.

Section 2.5

Section 2.5.1

Peetre's Maximal Operator

See [1092, (2.29)], where Ullrich showed that we can use the Peetre maximal operator in various ways to define the function spaces.

Section 2.5.2

Assumptions on the Local Means

We followed [902] in Sect. 2.5.2. In [100] we can find the theory of local means, but the assumption is somehow weaker. Via [294, 295], Rychikov proved Theorem 2.34 [902, Theorem 3.2]. Theorem 2.36 seems to have been known earlier, but we used the idea of [902, Proposition 3.1] for the proof.

Local Means: Theorems 2.34 and 2.35

As in Theorems 2.34 and 2.35, equivalent norms of function spaces enable us to obtain the properties of the function spaces. The proof has a history. According to [100, Remark 1], we can find the original form of Theorems 2.34 and 2.35 in [1063]. In [1058, Theorems 1–3], the prototype of the result is obtained. However, the assumption was strong; from the start we need to assume $f \in A^s_{pq}(\mathbb{R}^n)$ instead of $f \in \mathscr{S}'(\mathbb{R}^n)$. We try to obtain equivalent norms for the function spaces. We use the key estimate (2.153) when we want to get rid of the Peetre maximal operator. See [666, Proof of Theorem 13, (25)] and [1092, (2.29)] for the equivalence of the norms by means of the Peetre maximal operator.

Examples of Local Means

An inspection of the proof shows that k and K need not be compactly supported. Based on this idea, we can obtain many characterization related to PDEs. For example, we have the heat characterization and the harmonic characterization; see [1063, Theorem 9], [1058, Corollary 3] for the heat characterization and the Poisson characterization, as well as [100, 101, 1058]. See [727, 728] for a more general approach.

Section 2.5.3

Characterizations via the Ball Means of Differences

See the papers [410–412, 579, 910, 1158, 1159] which are motivated by the paper [127]. Alabern, Mateu and Verdera characterized $\dot{W}^{s,p}(\mathbb{R}^n)$ via the ball means of the difference in [127]. Yang, Yuan and Zhou extended their results to $\dot{F}^s_{pq}(\mathbb{R}^n)$ where $1 < p < \infty$, $1 < q \le \infty$ and $s \in (0, 2)$ [1159]. Furthermore, they extended the results to $B^s_{pq}(\mathbb{R}^n)$ with $1 < p \le \infty$, $0 < q \le \infty$ and $s \in (0, \infty) \setminus 2\mathbb{N}$ and $F^s_{pq}(\mathbb{R}^n)$ with $1 < p < \infty$, $1 < q \le \infty$ and $s \in (0, \infty) \setminus 2\mathbb{N}$. He, Yang and Yuan characterized $W^{2,p}(\mathbb{R}^n)$ for $1 < p < \infty$ [579]. See [721] for the case of anisotropic weighted function spaces. and [411] for the characterization of $A^s_{pq}(\mathbb{R}^n)$ for $s > 0$ via averages on balls.

See [358] for the Littlewood–Paley characterizations of Hajłasz-Sobolev and Triebel–Lizorkin spaces using averages on balls. This idea is expanded to the general differential operators [1180].

Characterizations by Means of the Oscillation: Theorem 2.40

We refer to the papers [429, 430, 943, 1067, 1068], as well as [285], which explains the hidden results in the early 1960's for Theorem 2.40. There are many approaches to local means; see [656, Teorema 6] for example.

Characterizations by Means of the Difference Including the Pointwise Characterization

Via a pointwise inequality involving the higher order differences, Haroske and Triebel [561, 562, 1081, 1082] obtained some pointwise characterizations, in the spirit of Hajłasz [533].

The pointwise characterizations of Besov and Triebel–Lizorkin spaces play important and key roles in the study for the invariance of these function spaces under quasi-conformal mappings; see [257, 504, 505, 587, 675, 686] for example.

Classical Definition of the Besov Norm: Theorem 2.39

Theorem 2.39 dates back to the late 1950's. In fact, The Besov space $B^s_{pq}(\mathbb{R}^n)$ with $1 \le p, q \le \infty$ and $s \in \mathbb{R}$ in the form of difference goes back to 1959 [171, p. 1164]. The space $\Lambda^s_{p,\infty}(\mathbb{R}^n)$ is defined by Besov in [171, 172]. Let $f \in L^p(\mathbb{R}^n)$ with $1 \le p < \infty$. When $0 < s < 1$, the norm $\|f\|_{\Lambda^s_{p,\infty}}$ is given by

$$\|f\|_{\Lambda^s_{p,\infty}} \equiv \|f\|_p + \sup_{h \in \mathbb{R}^n \setminus \{0\}} |h|^{-s} \|\Delta_h^2 f\|_p.$$

When $s = 1$, the norm $\|f\|_{\Lambda^s_{p,\infty}}$ is given by

$$\|f\|_{\Lambda^s_{p,\infty}} \equiv \|f\|_{W^1_p} + \sup_{h \in \mathbb{R}^n \setminus \{0\}} \|\nabla \Delta_h^2 f\|_{(L^p)^n}.$$

When $1 < s < 2$, the norm $\|f\|_{\Lambda^s_{p,\infty}}$ is given by

$$\|f\|_{\Lambda^s_{p,\infty}} \equiv \|f\|_{W^1_p} + \sup_{h \in \mathbb{R}^n \setminus \{0\}} |h|^{-s+1} \|\nabla \Delta_h^2 f\|_{(L^p)^n}.$$

Let $1 \le p \le \infty$ and $s > 0$. Theorem 2.39, dealing with $\| \star \|_{B^s_{pq}} *$, is called the classical Besov norm. The definition of the standard norm $B^s_{p\infty}(\mathbb{R}^n)$ is given by Nikolski'i [845]. The definition of $B^s_{pq}(\mathbb{R}^n)$ is due to Aronszajn [139], Slobodeckij [998] and Gagliardo [493]. See [7, p. 162]. We refer to [654, Teorema 1] and [1063, Theorems 5 and 6] for the case of Triebel–Lizorkin spaces. We remark that the work by Nikolski'i including [845] is explained in [216]. See [71, p. 8] for further details together with a motivation of the definition of Besov spaces.

Not only (2.166) but also

$$\|f\|_{B^{s**}_{pq}} \equiv \|f\|_p + \left(\int_0^1 \left(t^{-s} \sup_{|h| \le t} \|\Delta_h^S f\|_p \right)^q \frac{dt}{t} \right)^{\frac{1}{q}} \tag{2.171}$$

is called the *classical Besov norm.*

Chapter 3
Relation with Other Function Spaces

Although the definition of function spaces is complicated, we are still fascinated with Besov spaces and Triebel–Lizorkin spaces. One of the reasons, as we have alluded to in the preface, is that these scales realize many important function spaces. We already witnessed that the Besov space $B^{\sigma}_{\infty\infty}(\mathbb{R}^n)$ is the Hölder–Zygmund space $\mathscr{C}^{\sigma}(\mathbb{R}^n)$ for $\sigma > 0$; see Theorem 2.8. We still have many similar phenomena. To describe them, we need to know other function spaces as we did for Hölder–Zygmund spaces to establish that the Besov space $B^{\sigma}_{\infty\infty}(\mathbb{R}^n)$ and the Hölder–Zygmund space $\mathscr{C}^{\sigma}(\mathbb{R}^n)$ are isomorphic for $\sigma > 0$.

This chapter proceeds as follows. After investigating $L^p(\mathbb{R}^n)$ with $1 < p < \infty$ in Sect. 3.1, we consider Hardy spaces $H^p(\mathbb{R}^n)$ with $0 < p \le 1$ in Sect. 3.2 and establish that $H^p(\mathbb{R}^n)$ and $\dot{F}^0_{p2}(\mathbb{R}^n)$ are isomorphic in Theorem 3.17. Section 3.3 is dual to Sect. 3.2. So far, we did not define $\dot{F}^0_{\infty q}(\mathbb{R}^n)$ and $F^0_{\infty q}(\mathbb{R}^n)$ for $0 < q \le \infty$ due to a serious reason; the corresponding Fefferman–Stein vector-valued maximal inequality does not hold as was shown in Exercise 1.96. The function space $\mathrm{BMO}(\mathbb{R}^n)$ is the starting point for Sect. 3.3. After studying $\mathrm{BMO}(\mathbb{R}^n)$, we will show that we can characterize $\mathrm{BMO}(\mathbb{R}^n)$ by means of the Carleson measure. This characterization motivates us to define $\dot{F}^0_{\infty q}(\mathbb{R}^n)$ and $F^0_{\infty q}(\mathbb{R}^n)$. After defining these spaces, we establish that $\dot{F}^0_{\infty 2}(\mathbb{R}^n)$ is isomorphic to $\mathrm{BMO}(\mathbb{R}^n)$ in Theorem 3.26.

3.1 $L^p(\mathbb{R}^n)$ Spaces and Sobolev Spaces

In this section, we prove

$$\dot{F}^0_{p2}(\mathbb{R}^n) \approx F^0_{p2}(\mathbb{R}^n) \approx L^p(\mathbb{R}^n), \quad 1 < p < \infty \tag{3.1}$$

with equivalence of norms. The readers may wonder how to cope with the ℓ^2-norm appearing in the definition of the norms of $\dot{F}^0_{p2}(\mathbb{R}^n)$ and $F^0_{p2}(\mathbb{R}^n)$. However, we have

Y. Sawano, *Theory of Besov Spaces*, Developments in Mathematics 56,
https://doi.org/10.1007/978-981-13-0836-9_3

a good tool for this. In Sect. 3.1.1, we deal with the Rademacher sequence. This sequence is a strong weapon with which to analyze the space $\ell^2(\mathbb{N})$. We prove (3.1) in Sect. 3.1.2.

3.1.1 Rademacher Sequence

3.1.1.1 Rademacher Sequence

The Rademacher sequence, which appears as an elementary tool in probability theory, is also a very strong tool in harmonic analysis. Now we investigate the properties of the Rademacher sequence and then apply it to function spaces. We need some terminology of probability in Sect. 3.1.1, but since this is a book of analysis, we minimize probability theory. Here we deal only with the probability space arising naturally from the Lebesgue measure and we denote by $\mathscr{B}([0,1))$ the Borel σ-field over $[0,1)$.

Definition 3.1 (Rademacher sequence) For $j = 1, 2, \ldots$ let r_j on the probability space $(\Omega, \mathscr{F}, P) = ([0,1), \mathscr{B}([0,1)), dx|[0,1))$ be defined by

$$r_j(t) \equiv \sum_{k=1}^{2^j} (-1)^{k-1} \chi_{[(k-1)2^{-j}, k2^{-j})}(t) \quad (t \in [0,1)).$$

The sequence $\{r_j\}_{j=1}^{\infty}$ is called the *Rademacher sequence*.

We sometimes regard r_j as a function on $\mathbb{R}$ defined by this formula. Note that $r_j(t) = \mathrm{sgn}[\sin(2^j \pi t)]$ for almost every $t \in [0,1)$ and that a sequence $\{r_j\}_{j=1}^{\infty}$ of functions is a family of independent random variables; see Exercise 3.1 for the definition of independent random variables.

3.1.1.2 Khintchine's Inequality

In general it is difficult to handle the $L^p(\ell^q)$-norm and the $\ell^q(L^p)$-norm for $0 < p, q \le \infty$. However, when $p = q$, matters become simple dramatically. In fact, thanks to the monotone convergence theorem we learn that the $L^p(\ell^q)$-norm and the $\ell^q(L^p)$-norm turn out to be the same. As we have announced, in this section we are faced with the $L^p(\ell^2)$-norm. So, everything is difficult when $p \neq 2$. Khintchine's inequality is a magical tool that allows us to transform the ℓ^2-norm into the L^p-norm.

Theorem 3.1 (Khintchine's inequality) *Let* $0 < p < \infty$ *and* $\{a_j\}_{j=1}^{\infty} \in \ell^2(\mathbb{N})$. *Then* $\sum_{j=1}^{\infty} a_j r_j$ *converges in* $L^p(\mathbb{R})$ *and*

$$\left\| \sum_{j=1}^{\infty} a_j r_j \right\|_p \sim_p \|a_j\|_{\ell^2}, \tag{3.2}$$

where the implicit constant does not depend on $\{a_j\}_{j=1}^{\infty} \in \ell^2(\mathbb{N})$.

Proof First of all, we note $\{r_j\}_{j=1}^{\infty}$ is a complete orthonormal system in $L^2[0,1)$. Thus, we can replace $\sim$ with the equality when $p = 2$.

By the monotone convergence theorem and the Fatou lemma, we can assume $a_j = 0$ with finite number of exception. Next, we prove the "$\lesssim$" part of (3.2). A normalization allows us to assume $\|a_j\|_{\ell^2} = 1$.

We write $X \equiv \sum_{j=1}^{\infty} a_j r_j$ to prove the assertion based on important notions in probability theory, for example we use the notion of independent random variables.

By the Taylor expansion, we have

$$|t|^p \lesssim_p \cosh t = \frac{\mathrm{e}^t + \mathrm{e}^{-t}}{2} \le \exp(t^2) \tag{3.3}$$

(see Exercise 3.2 1).

By virtue of the left-hand inequality of (3.3) and the fact that $\{r_j\}_{j=1}^{\infty}$ is an independent sequence, we have $\|X\|_p{}^p \lesssim \int_0^1 \cosh\left(\sum_{j=1}^{\infty} a_j r_j(t)\right) \mathrm{d}t \lesssim \prod_{j=1}^{\infty} \int_0^1 \exp(a_j r_j(t))\mathrm{d}t$. Furthermore,

$$\|X\|_p{}^p \lesssim \prod_{j=1}^{\infty} \cosh a_j \lesssim \prod_{j=1}^{\infty} \exp(a_j{}^2) = \mathrm{e} \tag{3.4}$$

by the right-hand inequality in (3.3) and the symmetry of the Rademacher sequence (see Exercise 3.2 1). Thus, we obtain the "$\lesssim$" part of (3.2).

To prove the "$\gtrsim$" part of (3.2), it can be assumed that $p < 2$ in view of the monotonicity of the $L^p[0,1)$-norm. Recall that we have an equality when $p = 2$. Choose $0 < \theta < 1$ so that

$$\frac{1}{2} = \frac{1-\theta}{p} + \frac{\theta}{4}.$$

Then $\|a_j\|_{\ell^2} = \|X\|_2$ and hence $\|a_j\|_{\ell^2} \lesssim \|X\|_p{}^{1-\theta} \|X\|_4{}^{\theta} \lesssim \|X\|_p{}^{1-\theta} \|a_j\|_{\ell^2}^{\theta}$. If we rearrange this inequality, we obtain the "$\gtrsim$" part of (3.2).

Exercises

Exercise 3.1 A sequence $\{X_j\}_{j=1}^{\infty}$ of measurable functions on $[0, 1)$ is said to be independent, if

$$\left|\bigcap_{j=1}^{J} \{t \in [0,1) : X_j(t) \in A_j\}\right| = \prod_{j=1}^{J} |\{t \in [0,1) : X_j(t) \in A_j\}|$$

for any **finite** positive integer $J \in \mathbb{N}$ and for Borel measurable sets $A_1, A_2, \ldots, A_J$ on $\mathbb{R}$. Based on this definition, let us show that the Rademacher sequence $\{r_j\}_{j=1}^{\infty}$ is a family of independent random variables. Hint: Consider $\{r_j\}_{j=1}^{J}$ after we fix $J \in \mathbb{N}$.

Exercise 3.2

1. Prove (3.3) using the Taylor series.
2. As for (3.4), let $f_1, f_2, \ldots, f_J$ be a function from the set $\{-1, 0, 1\}$ to $\mathbb{R}$. Show that $\displaystyle\int_0^1 \prod_{j=1}^{J} f_j(r_j(t))\mathrm{d}t = \prod_{j=1}^{J} \frac{f_j(-1) + f_j(1)}{2}$.
3. Justify the claim that we can assume $a_j = 0$ for $j \gg 1$ in the proof of Theorem 3.1.

Exercise 3.3 [737] Let $N \in \mathbb{N}$ be fixed. Let $\{a_{m,n}\}_{n=1}^{\infty}$ be a doubly indexed complex sequence such that $a_{m,n} = 0$ whenever $|m| + |n| \geq N$. Show that

$$\sum_{n=1}^{\infty} \left(\sum_{m=1}^{\infty} |a_{m,n}|^2\right)^{1/2} \lesssim \left\| \sum_{m,n=1}^{\infty} a_{m,n} r_m \otimes r_n \right\|_{L^{\infty}([0,1]^2)},$$

where the implicit constant is independent of N.

Exercise 3.4 [138] We aim here to generalize what we did in this section. Let N be fixed. Expand $t \in [0, 1]$ into the infinite N-adic series: $t = \dfrac{n_1(t)}{N} + \dfrac{n_2(t)}{N^2} + \cdots + \dfrac{n_k(t)}{N^k} + \cdots$, where $n_1(t), n_2(t), \ldots \in \{0, 1, 2, \ldots, N-1\}$ and $\{n_k(t)\}_{k=1}^{\infty} \notin \ell^1(\mathbb{N})$. Define $s_k(t) \equiv \exp\left(\dfrac{2\pi i}{N} n_k(t)\right)$ for $t \in [0, 1)$.

1. Let $1 \leq j_1 < j_2 < \cdots < j_k$ be an increasing sequence of integers. Let also $m_1, m_2, \ldots, m_k \in \mathbb{N}$. Show that $\displaystyle\int_0^1 s_{j_1}(t)^{m_1} s_{j_2}(t)^{m_2} \cdots s_{j_k}(t)^{m_k}\mathrm{d}t = 1$ if and only if N divides all $m_1, m_2, \ldots, m_k$.
2. Let $j_1, j_2, \ldots, j_N \in \mathbb{N}$. Show that $\displaystyle\int_0^1 s_{j_1}(t) s_{j_2}(t) \cdots s_{j_N}(t)\mathrm{d}t = 1$ if and only if $j_1 = j_2 = \cdots = j_N$.

Exercise 3.5 [111, 134] Let $\{a_j\}_{j=1}^{\infty}$ be a real sequence. If there exists a set $E \subset [0, 1)$ with positive measure such that $\left|\sum_{j=1}^{N} a_j r_j(t)\right| \le 1$ for all $t \in E$, then show that $\{a_j\}_{j=1}^{\infty} \in \ell^2(\mathbb{N})$.

Exercise 3.6 [762, Lemma 1] Show that $\|a\cos\star + b\sin\star\|_{L^p(\mathbb{T})} \simeq_p \sqrt{a^2+b^2}$ for all $a, b \in \mathbb{R}$.

Exercise 3.7 We have been dealing with the ℓ^2-norm of infinite sequences. Here, we propose a substitute for finite sequences. Let $v \in \mathbb{R}^n$, and let $\mathbf{e}_1$ be the unit vector in the x_1 direction. Show that $\left(\int_{S^{n-1}} |v \cdot x|^p d\sigma(x)\right)^{\frac{1}{p}} = |v| \left(\int_{S^{n-1}} |\mathbf{e}_1 \cdot x|^p d\sigma(x)\right)^{\frac{1}{p}}$ by the use of an orthogonal matrix A_x satisfying $A_x e = \mathbf{e}_1$.

3.1.2 *The $L^p(\mathbb{R}^n)$ Space and the Triebel–Lizorkin Spaces $F^0_{p2}(\mathbb{R}^n)$ and $\dot{F}^0_{p2}(\mathbb{R}^n)$ with $1 < p < \infty$*

Pointwise control of the singular integral operators by means of a sensible geometric observation is by no means clear. One thing we might do is to try to replace "pointwise control" by some sort of "local norm control". The Littlewood–Paley decomposition is one such example.

3.1.2.1 $L^p(\mathbb{R}^n)$ and $F^0_{p2}(\mathbb{R}^n)$ with $1 < p < \infty$

One of the reasons why we are interested in Triebel–Lizorkin spaces is that we can grasp $L^p(\mathbb{R}^n)$ systematically with a wider framework. In the proof we are convinced that the Calderón–Zygmund operators appear in disguise.

Theorem 3.2 *Let $1 < p < \infty$. Then in the sense of equivalence of norms, we have $L^p(\mathbb{R}^n) \approx F^0_{p2}(\mathbb{R}^n)$; any element f in $F^0_{p2}(\mathbb{R}^n)$ is realized by an $L^p(\mathbb{R}^n)$-function g and*

$$\|g\|_p \lesssim \|f\|_{F^0_{p2}}. \tag{3.5}$$

Conversely, if $f \in L^p(\mathbb{R}^n)$, then

$$\|f\|_{F^0_{p2}} \lesssim \|f\|_p. \tag{3.6}$$

This inequality is only a part of the Littelwood-Paley theory, which is a big world. In this book, we can encounter such a big world through some examples.

Proof Since $\mathscr{S}(\mathbb{R}^n)$ is dense both in $L^p(\mathbb{R}^n)$ and in $F^0_{p2}(\mathbb{R}^n)$, we have only to prove

$$\|f\|_p \sim \|f\|_{F^0_{p2}} \tag{3.7}$$

for all $f \in \mathscr{S}(\mathbb{R}^n)$.

In the proof, ψ and φ are assumed to satisfy

$$0 \le \psi \le \chi_{B(4)}, \quad 0 \le \varphi \le \chi_{B(2)\setminus B(1)}, \quad \psi^2 + \sum_{j=1}^{\infty} \varphi_j{}^2 \equiv 1. \tag{3.8}$$

Use the functions ψ and φ satisfying (3.8). As the definition of the function space $F^0_{p2}(\mathbb{R}^n)$, we adopt the following norm:

$$\|f\|_{F^0_{p2}} = \left\| \left(|\psi(D)f|^2 + \sum_{j=1}^{\infty} |\varphi_j(D)f|^2 \right)^{\frac{1}{2}} \right\|_p . \tag{3.9}$$

Of course, we have other expressions of this space but (3.9) simplifies the proof when $p = 2$.

If $p = 2$, we have $\|f\|_{F^0_{22}} = \|f\|_{B^0_{22}} = \|f\|_2$ by (3.8) and Plancherel's theorem; see the proof of Theorem 2.6. We thus have (3.5) and (3.6).

For the proof of the inequality for $1 < p < \infty$, we will prove:

$$\|\psi(D)f\|_p + \left\| \left(\sum_{j=1}^{N} |\varphi_j(D)f|^2 \right)^{\frac{1}{2}} \right\|_p \lesssim_p \|f\|_p, \tag{3.10}$$

where the implicit constants are independent of $N \in \mathbb{N}$.

By Hölder's inequality, we can incorporate $\psi(D)f$. So we do not deal with $\psi(D)f$. Choose a Rademacher sequence $\{r_j\}_{j=1}^{\infty}$ and prove

$$\left\| \sum_{j=1}^{N} r_j(t)\varphi_j(D)f \right\|_p \lesssim \|f\|_p \tag{3.11}$$

with the implicit constant independent of $t \in [0, 1]$ and $N \in \mathbb{N}$. Once (3.11) is proved, we can take the $L^p[0, 1)$-norm to have

$$\left\| \left(\sum_{j=1}^{N} |\varphi_j(D)f|^2 \right)^{\frac{1}{2}} \right\|_p \simeq \left\| \left(\int_0^1 \left| \sum_{j=1}^{N} r_j(t)\varphi_j(D)f \right|^p dt \right)^{\frac{1}{p}} \right\|_p$$
$$= \left(\int_0^1 \left(\left\| \sum_{j=1}^{N} r_j(t)\varphi_j(D)f \right\|_p \right)^p dt \right)^{\frac{1}{p}}$$
$$\lesssim \left(\int_0^1 \|f\|_p{}^p dt \right)^{\frac{1}{p}} = \|f\|_p$$

by Theorem 3.1 and we can prove (3.10). The integral kernel of the operator

$$T_{N,t} : f \in L^p(\mathbb{R}^n) \mapsto \sum_{j=1}^{N} r_j(t)\varphi_j(D)f \in L^p(\mathbb{R}^n) \tag{3.12}$$

is denoted by $K_N(t, x)$. That is, we set $K_N(t, x) \equiv \sum_{j=1}^{N} r_j(t)\mathscr{F}^{-1}\varphi_j(x)$ for $t \in [0, 1]$ and $x \in \mathbb{R}^n$. When $p = 2$, Plancherel's theorem yields $\{T_{N,t}\}_{N\in\mathbb{N}}$ is uniformly $L^p(\ell^2, \mathbb{R}^n)$-bounded. Also, since

$$|x|^{n+|\alpha|}|\partial^\alpha \mathscr{F}^{-1}\varphi_j(x)| = 2^{j|\alpha|}|x|^{n+|\alpha|}|\mathscr{F}^{-1}[(2^{-j}\xi)^\alpha \varphi_j](x)|$$
$$= |2^j x|^{n+|\alpha|}|\mathscr{F}^{-1}[\xi^\alpha \varphi](2^j x)|$$
$$\lesssim |2^j x|^{n+|\alpha|}\langle 2^j x\rangle^{-n-|\alpha|-1},$$

for $\alpha \in \mathbb{N}_0{}^n$, we have

$$|x|^{n+|\alpha|}|\partial^\alpha K_N(t, x)| \lesssim \sum_{j=-\infty}^{\infty} \frac{|2^j x|^{n+|\alpha|}}{\langle 2^j x\rangle^{n+|\alpha|+1}} \sim \sum_{j=-\infty}^{\infty} \frac{2^{j(n+|\alpha|)}}{(1+2^j)^{n+|\alpha|+1}} \simeq 1,$$

for $t \in [0, 1]$, $x \in \mathbb{R}^n$ and $N \in \mathbb{N}$. By the Calderón–Zygmund theory (see Theorem 1.65) $\|T_{N,t}f\|_p \lesssim_p \|f\|_p$; hence we have (3.10). Thus, the proof of (3.6) is complete.

To prove (3.5), we can assume that $f \in \mathscr{S}(\mathbb{R}^n)$. Let $g \in \mathscr{S}(\mathbb{R}^n)$ and estimate $\mathrm{I} \equiv \int_{\mathbb{R}^n} f(x)\overline{g(x)}dx$. By Plancherel's theorem and Lebesgue's convergence theorem, we have

$$\mathrm{I} = \int_{\mathbb{R}^n} \left(\psi(\xi)\mathscr{F}f(\xi)\overline{(\psi(\xi)\mathscr{F}g(\xi))} + \sum_{j=1}^{\infty} \varphi_j(\xi)\mathscr{F}f(\xi)\overline{\varphi_j(\xi)\mathscr{F}g(\xi)} \right) d\xi$$

$$= \int_{\mathbb{R}^n} \psi(\xi)\mathscr{F}f(\xi)\overline{\psi(\xi)\mathscr{F}g(\xi)}\mathrm{d}\xi + \sum_{j=1}^{\infty}\int_{\mathbb{R}^n} \varphi_j(\xi)\mathscr{F}f(\xi)\overline{\varphi_j(\xi)\mathscr{F}g(\xi)}\mathrm{d}\xi.$$

By Plancherel's theorem and Lebesgue's convergence theorem again, we have

$$\int_{\mathbb{R}^n} f(x)\overline{g(x)}\mathrm{d}x = \int_{\mathbb{R}^n} \psi(D)f(x)\overline{\psi(D)g(x)}\mathrm{d}x + \sum_{j=1}^{\infty}\int_{\mathbb{R}^n} \varphi_j(D)f(x)\overline{\varphi_j(D)g(x)}\mathrm{d}x$$
$$= \int_{\mathbb{R}^n}\left(\psi(D)f(x)\overline{\psi(D)g(x)} + \sum_{j=1}^{\infty}\varphi_j(D)f(x)\cdot\overline{\varphi_j(D)g(x)}\right)\mathrm{d}x.$$

By Hölder's inequality, $|\mathrm{I}| \le \|f\|_{F^0_{p2}}\|g\|_{F^0_{p'2}}$. Since we already have (3.6), it follows that $|\mathrm{I}| \lesssim \|f\|_{F^0_{p2}}\|g\|_{p'}$. The function $g \in \mathscr{S}(\mathbb{R}^n)$ being arbitrary, we have

$$\|f\|_p = \sup\left\{\left|\int_{\mathbb{R}^n} f(x)\overline{g(x)}\mathrm{d}x\right| : g \in \mathscr{S}(\mathbb{R}^n),\ \|g\|_{p'} = 1\right\} \lesssim \|f\|_{F^0_{p2}}.$$

Therefore, the proof is complete.

$L^p(\mathbb{R}^n)$ and $\dot{F}^0_{p2}(\mathbb{R}^n)$ with $1 < p < \infty$

Furthermore, we can prove $\dot{F}^0_{p2}(\mathbb{R}^n) \approx L^p(\mathbb{R}^n)$. The proof is similar to that of $F^0_{p2}(\mathbb{R}^n) \approx L^p(\mathbb{R}^n)$. However, due to its importance, we dare to formulate it:

Theorem 3.3 *Let $1 < p < \infty$. Denote by $[f]$ the image by the natural mapping $\mathscr{S}'(\mathbb{R}^n) \to \mathscr{S}'(\mathbb{R}^n)/\mathscr{P}(\mathbb{R}^n)$ for $f \in \mathscr{S}'(\mathbb{R}^n)$. Then:*

1. *If $f \in \mathscr{S}'(\mathbb{R}^n)$ satisfies $[f] \in \dot{F}^0_{p2}(\mathbb{R}^n)$, then the limit $g \equiv \lim\limits_{J\to\infty}\sum\limits_{j=-J}^{\infty}\varphi_j(D)f$ exists in $L^p(\mathbb{R}^n)$ and coincides with f modulo $\mathscr{P}(\mathbb{R}^n)$. Furthermore, $\|g\|_p \lesssim \|[f]\|_{\dot{F}^0_{p2}}$.*
2. *Let $f \in L^p(\mathbb{R}^n)$. Then $[f] \in \dot{F}^0_{p2}(\mathbb{R}^n)$ and $\|[f]\|_{\dot{F}^0_{p2}} \lesssim \|f\|_p$.*

Thus, $\dot{F}^0_{p2}(\mathbb{R}^n) \approx L^p(\mathbb{R}^n)$ with equivalence of norms.

Unlike Theorem 3.2, Theorem 3.3 allows us to assume that f has frequency support in a dyadic annulus. Theorems 3.2 and 3.3 are key theorems in this section. Once we prove Theorem 3.2, we can characterize the Sobolev space $W^m_p(\mathbb{R}^n)$ with $m \in \mathbb{N}_0$ and $1 < p < \infty$.

In (1.166), we defined the space $H^s_p(\mathbb{R}^n)$. The spaces $W^s_p(\mathbb{R}^n)$ and $H^s_p(\mathbb{R}^n)$ are related as follows. The proof is basically dependent on the Calderón–Zygmund theory. We can prove this with Theorem 2.12 and $F^0_{p2}(\mathbb{R}^n) \approx L^p(\mathbb{R}^n)$. The following corollary is the main theorem of the paper with extremely short title $H = W$ [773]:

Corollary 3.1 ($H = W$) *Let $m \in \mathbb{N}_0$, $1 < p < \infty$. Then with equivalence of norms $H^m_p(\mathbb{R}^n) \approx W^m_p(\mathbb{R}^n)$. That is,*

$$\|(1-\Delta)^{m/2} f\|_p \sim \sum_{\substack{\alpha \in \mathbb{N}_0{}^n \\ |\alpha| \le m}} \|\partial^\alpha f\|_p \left(\sim \|f\|_p + \sum_{j=1}^n \|\partial^m_{x_j} f\|_p \right)$$

for all $f \in H^m_p(\mathbb{R}^n) \cup W^m_p(\mathbb{R}^n)$. In particular, $H^m_p(\mathbb{R}^n) \approx W^m_p(\mathbb{R}^n) \approx F^m_{p2}(\mathbb{R}^n)$ with equivalence of norms.

Proof Note that the norms of both spaces are equivalent to $F^m_{p2}(\mathbb{R}^n)$.

It may be interesting to prove Corollary 3.1 directly using the Calderón–Zygmund decomposition.

Exercises

Exercise 3.8 [7, Theorem 6.4.4] By combining Corollary 3.1 with Proposition 2.7, show that $B^s_{p\,\min(p,2)}(\mathbb{R}^n) \hookrightarrow H^s_p(\mathbb{R}^n) \hookrightarrow B^s_{p\,\max(p,2)}(\mathbb{R}^n)$ for $1 < p < \infty$.

Exercise 3.9 [29, (5.3)]

1. Let $s > 0$, $1 \le p \le \infty$, and $0 < q \le \infty$, or let $s = 0$, $1 < p < \infty$, and $0 < q \le \min(p, 2)$. Show that

$$\|f\|_{B^s_{pq}} \sim \|f\|_p + \|\{2^{js}\varphi_j(D)f\}_{j=1}^\infty\|_{\ell^q(L^p)} \sim \|f\|_p + \|\{2^{js}\varphi_j(D)f\}_{j=-\infty}^\infty\|_{\ell^q(L^p)}.$$

2. Let $s > 0$, $1 \le p < \infty$, and $0 < q \le \infty$ or $s = 0$, $1 < p < \infty$, and $0 < q \le 2$. Show that

$$\|f\|_{F^s_{pq}} \sim \|f\|_p + \|\{2^{js}\varphi_j(D)f\}_{j=1}^\infty\|_{L^p(\ell^q)} \sim \|f\|_p + \|\{2^{js}\varphi_j(D)f\}_{j=-\infty}^\infty\|_{L^p(\ell^q)}.$$

Exercise 3.10 Let $0 < \alpha < n$, and let f be a nonnegative measurable function defined on $\mathbb{R}^n$. Define the fractional integral operator I_α by (1.137).

1. Show that $\displaystyle\int_0^\infty \chi_{B(x,l)}(y) \frac{dl}{l^{n+1-\alpha}} = \frac{|x-y|^{-n+\alpha}}{n-\alpha}$ for $x, y \in \mathbb{R}^n$.
2. Show that $\displaystyle I_\alpha f(x) = (n-\alpha)\int_0^\infty \left(\frac{1}{l^{n+1-\alpha}} \int_{B(x,l)} f(y)dy \right) dl$ for $x, y \in \mathbb{R}^n$.
3. [580] Let $1 \le p < \dfrac{n}{\alpha}$ and define q so that $\dfrac{1}{q} = \dfrac{1}{p} - \dfrac{\alpha}{n}$. Then show that

$$I_\alpha f(x) \lesssim Mf(x)^{\frac{p}{q}} \|f\|_p{}^{1-\frac{p}{q}} \quad (x \in \mathbb{R}^n).$$

4. Let $k = 1, 2, \ldots, n$. Regarding the inverse Fourier transform of the functions $m_k(\xi) \equiv \xi_k|\xi|^{-2}$, show that $|\mathscr{F}^{-1}m_k(x)| \lesssim |x|^{-n+1}$.
5. For $f \in C_c^\infty(\mathbb{R}^n)$, show that $|f(x)| \lesssim I_1[|\nabla f|](x)$. Hint: Scale Exercise 1.83.
6. Denote by $\dot{L}^{1,2}(\mathbb{R}^n)$ the completion of $\mathscr{D}$ by the Dirichlet norm $\|\nabla f\|_{(L^2)^n}$. Show that $\dot{L}^{1,2}(\mathbb{R}^n) \hookrightarrow L^{n/(n-2)}(\mathbb{R}^n)$ for $n \geq 3$ using a counterpart of (2.64) to homogeneous Triebel–Lizorkin spaces.
7. Show that $\dot{L}^{1,2}(\mathbb{R}^n) \approx \dot{F}^1_{22}(\mathbb{R}^n)$ with equivalence of norms for $n \geq 3$.

Exercise 3.11 (Helmholtz projection) Let $1 < p < \infty$.

1. Show that $\mathscr{S}_\infty(\mathbb{R}^n)$ is dense in $L^p(\mathbb{R}^n)$. Hint: Use Theorems 3.3 and 2.30.
2. Given $u \in \mathscr{S}_\infty(\mathbb{R}^n)^n$, show that there exists a unique element $\pi \in \mathscr{S}_\infty(\mathbb{R}^n)$ such that $\Delta\pi = \nabla \cdot u$. Hint: Use that $\Delta : \mathscr{S}_\infty(\mathbb{R}^n) \to \mathscr{S}_\infty(\mathbb{R}^n)$ is an isomorphism.
3. Let $u \in \mathscr{S}_\infty(\mathbb{R}^n)^n$. Then define $\mathrm{Pr}(u) \equiv u - \nabla\pi \in \mathscr{S}_\infty(\mathbb{R}^n)^n$, where $\pi \in \mathscr{S}_\infty(\mathbb{R}^n)$ satisfies $\Delta\pi = \nabla \cdot u$. Then show that $\|\mathrm{Pr}(u)\|_{L^p(\mathbb{R}^n)^n} \lesssim \|u\|_{L^p(\mathbb{R}^n)^n}$ using Theorem 1.65.

Exercise 3.12 (Lizorkin) Let $1 < p < \infty$, $s \in \mathbb{R}$, and let $f \in L^p(\mathbb{R}^n)$. For $m \in L^\infty(\mathbb{R}^n)$, we write $m(D)f \equiv \mathscr{F}^{-1}[m \cdot \mathscr{F}f]$ as long as the right-hand side makes sense.

1. Distinguish three cases: $p = 2$, $1 < p < 2$, and $2 < p < \infty$ to prove that

$$\begin{aligned}&\|\chi_{Q(1)}(D)f_j\|_p + \left\|\{2^{js}\chi_{Q(2^j)\setminus Q(2^{j-1})}(D)f_j\}_{j=1}^\infty\right\|_{L^p(\ell^2)}\\&\lesssim \|\psi(D)f_j\|_p + \left\|\{2^{js}\varphi_j(D)f_j\}_{j=1}^\infty\right\|_{L^p(\ell^2)}.\end{aligned}$$

2. Show that $F^s_{p2}(\mathbb{R}^n)$ has the following equivalent norm:

$$\|\chi_{Q(1)}(D)f\|_p + \left\|\{2^{js}\chi_{Q(2^j)\setminus Q(2^{j-1})}(D)f_j\}_{j=1}^\infty\right\|_{L^p(\ell^2)}.$$

Hint: We need a vector-valued maximal inequality for the Hilbert transform. See (1.217) for the definition of the Hilbert transform.

Exercise 3.13 Let $1 < p < \infty$. Assume that $\varphi \in C^\infty(\mathbb{R}^n)$ satisfies $\chi_{\mathbb{R}^{n-1}\times[2,4]} \leq \varphi \leq \chi_{\mathbb{R}^{n-1}\times[1,8]}$. Then show that $\|f\|_p \sim \|\{\varphi_j(D)f\}_{j\in\mathbb{Z}}\|_{L^p(\ell^2)}$ for all $f \in L^p(\mathbb{R}^n)$.

Section 3.1

Textbooks in Sect. 3.1

See [22, Chapter 8, Section 2], [86, Chapter 3, Section 4],

Khintchine's Inequality: Theorem 3.1

See [31, p. 483, Lemma 2.4], [30, Section 12.3] and [117, Proposition 7.8].

$L^p(\mathbb{R}^n) \approx F^0_{p2}(\mathbb{R}^n) \approx \dot{F}^0_{p2}(\mathbb{R}^n)$: Theorems 3.2 and 3.3

We refer to the textbooks [22, 24, 29], [31, Section 5.5] and [86] for the Littlewood–Paley theory, of which Theorems 3.2 and 3.3 are small parts.

We can find (3.5) and (3.6) in [71, p. 80, (3)] and [71, p. 80, (4)].

Others

The author referred to García-Cuerva and Rubio de Francia [31] for the proof of Theorem 3.1, where García-Cuerva and Rubio de Francia actually took an approach via the normal distributions [31, Lemma 2.4].

3.2 Hardy Spaces

The Hardy space $H^1(\mathbb{R}^n)$ often substitutes for the space $L^1(\mathbb{R}^n)$ in real analysis. As is the case in stochastic analysis, it plays an important role. Here we will investigate real-analytic Hardy spaces. We prove that the homogeneous Triebel–Lizorkin space $\dot{F}^0_{p2}(\mathbb{R}^n)$ is isomorphic to $H^p(\mathbb{R}^n)$ for all $0 < p \le 1$.

In Sect. 3.2 we use the semi-norm p_N on $\mathscr{S}(\mathbb{R}^n)$ as we did in (1.55):

$$p_N(\psi) \equiv \sum_{\substack{\alpha \in \mathbb{N}_0{}^n \\ |\alpha| \le N}} \sup_{x \in \mathbb{R}^n} \langle x \rangle^N |\partial^\alpha \psi(x)|. \tag{3.13}$$

Furthermore, we define the unit ball $\mathscr{F}_N$ with respect to p_N by

$$\mathscr{F}_N \equiv \{\psi \in \mathscr{S}(\mathbb{R}^n) \,:\, p_N(\psi) \le 1\}. \tag{3.14}$$

For $j \in \mathbb{Z}$ and $\psi \in \mathscr{S}(\mathbb{R}^n)$, we write

$$\psi^j \equiv 2^{jn}\psi(2^j \star). \tag{3.15}$$

In Sect. 3.2.1, we define the Hardy space $H^p(\mathbb{R}^n)$ by means of a vertical maximal operator:

$$\|f\|_{H^p} \equiv \left\| \sup_{\psi \in \mathscr{F}_N, j \in \mathbb{Z}} |\psi^j * f| \right\|_p, \quad 0 < p < \infty. \tag{3.16}$$

To connect Hardy spaces with Triebel–Lizorkin spaces we prove the boundedness of the singular integral operators in Sect. 3.2.2. Meanwhile, as will be stated in Sect. 3.2.3, we can define Hardy spaces by means of atoms. These two definitions will be unified in Sect. 3.2.6 via some auxiliary observations in Sects. 3.2.4 and 3.2.5. Hardy spaces are characterized by the Riesz transform as is proved in Sect. 3.2.7. To connect the nonhomogeneous Triebel–Lizorkin space $F^0_{p2}(\mathbb{R}^n)$, we define the local Hardy space $h^p(\mathbb{R}^n)$ in Sect. 3.2.8. In Sect. 3.2.9, we will prove that the homogeneous Triebel–Lizorkin space $\dot{F}^0_{p2}(\mathbb{R}^n)$ and the nonhomogeneous Triebel–Lizorkin space $F^0_{p2}(\mathbb{R}^n)$ are isomorphic to $H^p(\mathbb{R}^n)$ and $h^p(\mathbb{R}^n)$, respectively.

3.2.1 *Definition of Hardy Spaces*

3.2.1.1 Hardy Spaces Generated by a Single Function

Let $\psi \in \mathscr{S}(\mathbb{R}^n) \setminus \mathscr{P}_0(\mathbb{R}^n)^\perp$. Using this function, we define the Hardy norm by:

$$\|f\|^\psi_{H^p} \equiv \left\| \sup_{j \in \mathbb{Z}} |\psi^j * f| \right\|_p \quad (0 < p < \infty, \ f \in \mathscr{S}'(\mathbb{R}^n)). \tag{3.17}$$

In this section, we will show that the norm (3.17) is equivalent to the norm (3.16). Let $\varphi \equiv \psi - 2^{-n}\psi(2^{-1}\star)$. By the Fourier transform, there exist $\tau, \zeta \in \mathscr{S}(\mathbb{R}^n)$ such that

$$\delta = \psi * \tau + \sum_{k=1}^{\infty} \varphi^k * \zeta^k. \tag{3.18}$$

Note that the Fourier transform of ζ vanishes up to order infinity.

Example 3.1 Let $F \in \mathscr{S}'(\mathbb{R}^n)$. Then the *heat semigroup* $\{e^{t\Delta}\}_{t>0}$ is given by:

$$e^{t\Delta}F(x) \equiv \left\langle F, \frac{1}{\sqrt{(4\pi t)^n}} \exp\left(-\frac{|x-\star|^2}{4t}\right)\right\rangle \quad (x \in \mathbb{R}^n).$$

So it follows that we can define the Hardy space $H^p(\mathbb{R}^n)$ with $0 < p < \infty$ as the set of all $f \in \mathscr{S}'(\mathbb{R}^n)$ satisfying

$$\left\| \sup_{t>0} |e^{t\Delta} f| \right\|_p < \infty.$$

Let $f \in \mathscr{S}'(\mathbb{R}^n)$, and let $B > 0$. Define the maximal operator M_B by

$$M_B f(x; j) \equiv \sup \left\{ 2^{(j-k)B} \langle 2^k y \rangle^{-B} |\kappa^k * f(x-y)| \, : \, \kappa \in \mathscr{F}_N, k \geq j, y \in \mathbb{R}^n \right\}. \tag{3.19}$$

3.2.1.2 A Maximal Inequality

One of the convenient tools is the Peetre maximal operator when we analyze Besov spaces and Triebel–Lizorkin spaces. We look for the counterpart for Hardy spaces. We first prove the inequality which is used in the proof of Lemma 3.2.

Lemma 3.1 *Let $f \in \mathscr{S}'(\mathbb{R}^n)$, and let $r > 0$. Suppose that the constants A and B satisfy $Ar > n$, $Br > n$. Furthermore, assume that*

$$M_B f(x, j)^r \lesssim_f 2^{jn} \int_{\mathbb{R}^n} \sup_{l \geq j} \langle 2^j y \rangle^{-Ar} |\psi^l * f(x-y)|^r \mathrm{d}y < \infty \tag{3.20}$$

for all $j \in \mathbb{Z}$ and $x \in \mathbb{R}^n$ with a constant dependent on f. Then $M_A f(x; j) < \infty$.

Proof Let $\kappa \in \mathscr{F}_N$. From the definition (3.19) of M_B and assumption (3.20), we deduce

$$\begin{aligned} |\kappa^k * f(x-y)|^r &\leq M_B f(x-y, k)^r \\ &\lesssim_f 2^{kn} \int_{\mathbb{R}^n} \sup_{l \geq k} \langle 2^k z \rangle^{-Ar} |\psi^l * f(x-y-z)|^r \mathrm{d}z \\ &< \infty. \end{aligned}$$

By the relation between j, k and l in the summand and Peetre's inequality $\langle y+z \rangle \leq \sqrt{2} \langle y \rangle \langle z \rangle$, we have

$$\begin{aligned} M_A f(x; j)^r &= \sup_{\kappa \in \mathscr{F}_N, k \geq j, y \in \mathbb{R}^n} 2^{(j-k)Ar} \frac{|\kappa^k * f(x-y)|^r}{\langle 2^k y \rangle^{Ar}} \\ &\lesssim_f \sup_{k \geq j, y \in \mathbb{R}^n} 2^{(j-k)Ar+kn} \int_{\mathbb{R}^n} \left(\sup_{l \geq k} (\langle 2^j y \rangle \langle 2^j z \rangle)^{-Ar} |\psi^l * f(x-y-z)|^r \right) \mathrm{d}z \\ &\lesssim_f \sup_{k \geq j, y \in \mathbb{R}^n} 2^{jn} \int_{\mathbb{R}^n} \left(\sup_{l \geq k} \langle 2^j (y+z) \rangle^{-Ar} |\psi^l * f(x-y-z)|^r \right) \mathrm{d}z \\ &= 2^{jn} \int_{\mathbb{R}^n} \left(\sup_{l \geq j} \langle 2^j z \rangle^{-Ar} |\psi^l * f(x-z)|^r \right) \mathrm{d}z < \infty. \end{aligned}$$

Therefore, the proof is complete.

The next lemma corresponds to Proposition 2.14, which is of great importance.

Lemma 3.2 *Let* $j \in \mathbb{N}_0$, $x \in \mathbb{R}^n$, *and let* $0 < A, r < \infty$ *satisfy* $A > \max(n, n/r)$. *Then*

$$M_A f(x; j)^r \lesssim_{A,r} 2^{jn} \int_{\mathbb{R}^n} \sup_{l \geq j} \langle 2^j y \rangle^{-Ar} |\psi^l * f(x-y)|^r \mathrm{d}y. \tag{3.21}$$

Proof Let $L \gg A$. Suppose that the moment of ζ vanishes up to order L. From (3.18),

$$\kappa^j * f(x) = \kappa^j * \tau^j * \psi^j * f(x) + \sum_{k=j+1}^{\infty} \kappa^j * \zeta^k * \varphi^k * f(x). \tag{3.22}$$

Fix $M \in \mathbb{N}$. If we let $A \ll M \ll L$, we have

$$|\kappa^j * \zeta^k(x)| \lesssim 2^{(j-k)L+jn} \langle 2^j x \rangle^{-M} \tag{3.23}$$

from Theorem 1.56 whenever $j \leq k$. Since $\kappa * \tau \in \mathscr{S}(\mathbb{R}^n)$, we obtain

$$|\kappa^j * \tau^j(x)| = |(\kappa * \tau)^j(x)| = 2^{jn} |\kappa * \tau(2^j x)| \lesssim 2^{jn} \langle 2^j x \rangle^{-M}. \tag{3.24}$$

from Theorem 1.56.

If we insert (3.23) and (3.24) into (3.22), then

$$\begin{aligned} &|\kappa^j * f(x)| \\ &\lesssim 2^{jn} \int_{\mathbb{R}^n} \langle 2^j y \rangle^{-A} |\psi^j * f(x-y)| \mathrm{d}y + \sum_{k=j+1}^{\infty} 2^{(j-k)A+jn} \int_{\mathbb{R}^n} \langle 2^j y \rangle^{-A} |\varphi^k * f(x-y)| \mathrm{d}y. \end{aligned} \tag{3.25}$$

In (3.25), if we replace z with y and x with $x - y$, respectively, then

$$\begin{aligned} |\kappa^j * f(x)| &\lesssim 2^{jn} \int_{\mathbb{R}^n} \langle 2^j z \rangle^{-A} |\psi^j * f(x-y-z)| \mathrm{d}z \\ &\quad + \sum_{k=j+1}^{\infty} 2^{(j-k)A+jn} \int_{\mathbb{R}^n} \langle 2^j z \rangle^{-A} |\varphi^k * f(x-y-z)| \mathrm{d}z. \end{aligned} \tag{3.26}$$

In (3.26), if we replace j with k and k with l respectively, then

$$|\kappa^k * f(x)| \lesssim 2^{kn} \int_{\mathbb{R}^n} \langle 2^k z \rangle^{-A} |\psi^k * f(x-y-z)| \mathrm{d}z$$

$$+\sum_{l=k+1}^{\infty} 2^{(k-l)A+kn}\int_{\mathbb{R}^n}\langle 2^k z\rangle^{-A}|\varphi^l * f(x-y-z)|\mathrm{d}z. \quad (3.27)$$

If we insert $\varphi = \psi - 2^{-n}\psi(2^{-1}\star)$ into (3.27),

$$|\kappa^k * f(x-y)| \lesssim 2^{kn}\int_{\mathbb{R}^n}\left(\sup_{l\geq k}\langle 2^k z\rangle^{-A}|\psi^l * f(x-y-z)|\right)\mathrm{d}z. \quad (3.28)$$

Let $k \geq j$. By (3.28) and Peetre's inequality $\langle x+y\rangle \leq \sqrt{2}\langle x\rangle\langle y\rangle$, we have

$$\begin{aligned}
&2^{(j-k)A}\langle 2^k y\rangle^{-A}|\kappa^k * f(x-y)| \\
&\lesssim 2^{(j-k)A+kn}\int_{\mathbb{R}^n}\sup_{l\geq k}\langle 2^k y\rangle^{-A}\langle 2^k z\rangle^{-A}|\psi^l * f(x-y-z)|\mathrm{d}z \\
&\lesssim 2^{(j-k)A+kn}\int_{\mathbb{R}^n}\sup_{l\geq k}\langle 2^k (y+z)\rangle^{-A}|\psi^l * f(x-y-z)|\mathrm{d}z.
\end{aligned}$$

Changing variables: $y+z \mapsto z$ gives an integral in the last formula that does not depend on y. That is,

$$\begin{aligned}
2^{(j-k)A}\langle 2^k y\rangle^{-A}|\kappa^k * f(x-y)| &\lesssim 2^{(j-k)A+kn}\int_{\mathbb{R}^n}\sup_{l\geq k}\langle 2^k z\rangle^{-A}|\psi^l * f(x-z)|\mathrm{d}z \\
&= 2^{(j-k)(A-n)+jn}\int_{\mathbb{R}^n}\sup_{l\geq k}\langle 2^k z\rangle^{-A}|\psi^l * f(x-z)|\mathrm{d}z.
\end{aligned}$$

Recall the definition of the maximal operator M_A:

$$M_A f(x;j) = \sup_{\kappa\in\mathscr{F}_N, k\geq j}\ \sup_{y\in\mathbb{R}^n} 2^{(j-k)A}\langle 2^k y\rangle^{-A}|\kappa^k * f(x-y)|.$$

Since $A > n$ and $k \geq j$, we have

$$\begin{aligned}
&2^{(j-k)A}\langle 2^k y\rangle^{-A}|\kappa^k * f(x-y)| \\
&\lesssim 2^{(j-k)(A-n)+jn}\int_{\mathbb{R}^n}\sup_{l\geq k}\langle 2^k z\rangle^{-A}|\psi^l * f(x-z)|\mathrm{d}z \\
&\lesssim 2^{(j-k)(A-n)(1-r)+jn}\int_{\mathbb{R}^n}\sup_{l\geq k}\langle 2^k z\rangle^{-A}|\psi^l * f(x-z)|\mathrm{d}z \\
&\lesssim M_A f(x;j)^{1-r}\times 2^{jn}\int_{\mathbb{R}^n}\sup_{l\geq j}\langle 2^j z\rangle^{-Ar}|\psi^l * f(x-z)|^r\mathrm{d}z.
\end{aligned}$$

Here if we take the supremum over $\kappa \in \mathscr{F}_N$, $k \geq j$, $y \in \mathbb{R}^n$, we obtain

$$M_A f(x;\, j) \lesssim M_A f(x;\, j)^{1-r} \times 2^{jn} \int_{\mathbb{R}^n} \sup_{l \geq j} \langle 2^j z \rangle^{-Ar} |\psi^l * f(x-z)|^r \mathrm{d}z. \quad (3.29)$$

In (3.29) it seems that it is enough to divide both sides by $M_A f(x;\, j)$. But this is not that easy, since it may happen that $M_A f(x;\, j) = \infty$. We need a trick. For $f \in \mathscr{S}'(\mathbb{R}^n)$, we have an $A_f > 0$ that depends on f such that $M_{A_f} f(x;\, j) < \infty$. When $A \geq A_f$, we have

$$M_A f(x;\, j)^r \lesssim_f 2^{jn} \int_{\mathbb{R}^n} \sup_{l \geq j} \langle 2^j z \rangle^{-Ar} |\psi^l * f(x-z)|^r \mathrm{d}z \quad (3.30)$$

with constants independent of f. For all $A > n/r$, we set $B \equiv \max(A, A_f)$. By the monotonicity of the right-hand side with respect to A, we have

$$M_B f(x;\, j)^r \lesssim_f 2^{jn} \int_{\mathbb{R}^n} \sup_{l \geq j} \langle 2^j z \rangle^{-Ar} |\psi^l * f(x-z)|^r \mathrm{d}z.$$

Hence Lemma 3.1 yields $M_A f(x;\, j) < \infty$ if we assume

$$2^{jn} \int_{\mathbb{R}^n} \sup_{l \geq j} \langle 2^j z \rangle^{-Ar} |\psi^l * f(x-z)|^r \mathrm{d}z < \infty. \quad (3.31)$$

Hence no matter whether (3.31) holds or not, from (3.29) we obtain (3.21).

Using Lemma 3.2 we have control of the maximal operator M_A, which yields equivalence of norms together with control of the powered Hardy–Littlewood maximal operator $M^{(r)}$.

Corollary 3.2 *Let $r \in (0, \infty)$, $j \in \mathbb{Z}$, and let $f \in \mathscr{S}'(\mathbb{R}^n)$. For $M^{(r)}$, one has*

$$M_A f(x;\, j) \lesssim_r M^{(r)} \left[\sup_{l \geq j} |\psi^l * f| \right](x), \quad x \in \mathbb{R}^n.$$

3.2.1.3 Hardy Spaces by Means of the Grand Maximal Operator

When we analyze Hardy spaces, the grand maximal operator plays an important role.

Definition 3.2 (Vertical maximal operator) Let $f \in \mathscr{S}'(\mathbb{R}^n)$, and let $N \gg 1$. Define the (*discrete*) *vertical maximal operator* by

$$\mathscr{M} f(x) = \mathscr{M}_{(N)} f(x) \equiv \sup_{\kappa \in \mathscr{F}_N,\, j \in \mathbb{Z}} |\kappa^j * f(x)| \quad (x \in \mathbb{R}^n).$$

The next theorem justifies the definition of Hardy spaces.

Theorem 3.4 *Let $0 < p < \infty$. Suppose that $\psi \in \mathscr{S}(\mathbb{R}^n) \setminus \mathscr{P}_0(\mathbb{R}^n)^\perp$. If $N \gg 1$, then*

$$\|\mathscr{M} f\|_p = \|\mathscr{M}_{(N)} f\|_p \sim_{N,\psi} \left\| \sup_{j\in\mathbb{Z}} |\psi^j * f| \right\|_p \quad (f \in \mathscr{S}'(\mathbb{R}^n)).$$

Proof It is trivial that

$$\|\mathscr{M} f\|_p = \|\mathscr{M}_{(N)} f\|_p \gtrsim_{N,\psi} \left\| \sup_{j\in\mathbb{Z}} |\psi^j * f| \right\|_p.$$

Choose positive constants r, A so that $\frac{n}{A} < r < \min(1, p)$. Let $x \in \mathbb{R}^n$. We deduce

$$\sup_{j\in\mathbb{Z}} |\psi^j * f(x)| \lesssim \sup_{\kappa\in\mathscr{F}_N, j\in\mathbb{Z}} |\kappa^j * f(x)| \le \sup_{j\in\mathbb{Z}} M_A f(x; j) \lesssim M^{(r)} \left[\sup_{j\in\mathbb{Z}} |\psi^j * f| \right](x)$$

from the definition of $M_A f(x; j)$, Definition 3.2 and Corollary 3.2.

Since $\frac{p}{r} > 1$, we can use the maximal inequality to conclude the proof of Theorem 3.4.

With the property of the vertical maximal operator in mind, we define Hardy spaces.

Definition 3.3 (Hardy space $H^p(\mathbb{R}^n)$) Let $0 < p < \infty$. Suppose that $\psi \in \mathscr{S}(\mathbb{R}^n)$ satisfies the *nondegenerate condition* $\int_{\mathbb{R}^n} \psi(x) \mathrm{d}x \neq 0$, or equivalently, $\psi \in \mathscr{S}(\mathbb{R}^n) \setminus \mathscr{P}_0(\mathbb{R}^n)^\perp$. The Hardy space $H^p(\mathbb{R}^n)$ is the set of all $f \in \mathscr{S}'(\mathbb{R}^n)$ which satisfies

$$\|f\|_{H^p} \equiv \left\| \sup_{j\in\mathbb{Z}} |\psi^j * f| \right\|_p < \infty.$$

Although we should have written $H^{p;\psi}(\mathbb{R}^n)$ as we did for Besov spaces and Triebel–Lizorkin spaces, we see from Theorem 3.4 that the definition of $H^p(\mathbb{R}^n)$ is independent of the admissible choice of ψ. Furthermore, the norm equivalence $\|\mathscr{M} f\|_p \sim_p \|f\|_{H^p}$ with $0 < p < \infty$ holds.

Remark 3.1 For example, assume that $\psi \in \mathscr{S}(\mathbb{R}^n)$ satisfies $\chi_{B(1)} \le \psi \le \chi_{B(2)}$. Set $\varphi \equiv \psi - \psi^{-1}$. Then we have the norm of the homogeneous Triebel–Lizorkin space $\dot{F}^0_{\infty 2}(\mathbb{R}^n)$: $\|f\|_{\dot{F}^0_{p\infty}} \equiv \left\| \sup_{j\in\mathbb{Z}} |\varphi^j * f| \right\|_p$. This function space differs from the

Hardy space $H^p(\mathbb{R}^n)$. As is seen from Theorem 3.17, $H^p(\mathbb{R}^n)$ is isomorphic to $\dot{F}^0_{p2}(\mathbb{R}^n)$.

Recall that the distribution f is bounded if there exists $r \in (1, \infty)$ such that $\psi * f \in L^r(\mathbb{R}^n)$ for all $\psi \in \mathscr{S}(\mathbb{R}^n)$; see Definition 1.45. Also recall the notion of the distributions restricted at infinity; see Definition 1.19.

Proposition 3.1 *Let $0 < p \le 1$. Then $f * \psi \in L^1(\mathbb{R}^n) \cap L^\infty(\mathbb{R}^n)$ whenever $f \in H^p(\mathbb{R}^n)$ and $\psi \in \mathscr{S}(\mathbb{R}^n)$. In particular, f is bounded and restricted at infinity.*

Proof Let $\tau \in \mathscr{S}(\mathbb{R}^n)$ be a function such that $\chi_{B(1)} \le \tau \le \chi_{B(2)}$. Set

$$\psi^{(0)} \equiv \psi\tau, \quad \psi^{(j)} \equiv \psi(\tau_j - \tau_{j-1}) \quad (j \in \mathbb{N}).$$

Then $\psi^{(0)} * f \in L^p(\mathbb{R}^n)$, since $|\psi^{(0)} * f| \lesssim \mathscr{M} f$. Since $\tau_j - \tau_{j-1}$ is supported in $B(2^j) \setminus B(2^{j-1})$, we see that $|\psi^{(j)} * f| \lesssim_k 2^{-jk}\mathscr{M} f$ for all $k \in \mathbb{N}$. Since

$$\|f\|_\infty \lesssim R^{\frac{n}{p}}\|f\|_p, \quad \|f\|_1 \lesssim R^{\frac{n}{p}-n}\|f\|_p \quad (f \in L^p_{B(R)}(\mathbb{R}^n)),$$

we conclude $\psi * f = \sum_{j=0}^{\infty} \psi^{(0)} * f \in L^1(\mathbb{R}^n) \cap L^\infty(\mathbb{R}^n)$, as was to be shown.

As we did for Besov spaces and Triebel–Lizorkin spaces, we want to prove that the choice of $\psi \in \mathscr{S}(\mathbb{R}^n) \setminus \mathscr{P}_0(\mathbb{R}^n)^\perp$ does not affect the definition of $H^p(\mathbb{R}^n)$, where $H^p(\mathbb{R}^n)$ is defined by the norm $\| \star \|^{\psi}_{H^p}$ in (3.17).

Theorem 3.5 *Let $0 < p \le 1$. Let $f \in \mathscr{S}'(\mathbb{R}^n)$ be a bounded distribution. Then $f \in H^p(\mathbb{R}^n)$ if and only if $\sup_{t>0} |\mathrm{e}^{-t\sqrt{-\Delta}} f| \in L^p(\mathbb{R}^n)$.*

Proof Let $f \in H^p(\mathbb{R}^n)$. Fix $x \in \mathbb{R}^n$. Then

$$\begin{aligned}
&\mathrm{e}^{-t\sqrt{-\Delta}} f(x)\\
&= \mathscr{F}^{-1}[(1-\tau(t\star))\mathrm{e}^{-t|\star|}\mathscr{F}](x) + \frac{1}{\pi^{\frac{n+1}{2}}}\Gamma\left(\frac{n+1}{2}\right)\int_{\mathbb{R}^n} \frac{t\tau(tD)f(y)}{(t^2+|x-y|^2)^{\frac{n+1}{2}}}\mathrm{d}y.
\end{aligned}$$

Here $\tau \in \mathscr{S}(\mathbb{R}^n)$ is taken so that $\chi_{B(1)} \le \tau \le \chi_{B(2)}$. Write $\rho(\xi) \equiv (1-\tau(\xi))e^{-|\xi|}$ for $\xi \in \mathbb{R}^n$. Let $x \in \mathbb{R}^n$. Since $\rho \in \mathscr{S}(\mathbb{R}^n)$, it follows that

$$\sup_{t>0} |\mathscr{F}^{-1}[(1-\tau(t\star))\mathrm{e}^{-t|\star|}\mathscr{F}](x)| \lesssim \mathscr{M} f(x).$$

We next define

$$\rho^\dagger(x) \equiv \frac{\tau(x)}{(1+|x|^2)^{\frac{n+1}{2}}}, \quad \rho^{\dagger,k}(x) \equiv \frac{\tau(x)-\tau(2x)}{(2^{-k}+|x|^2)^{\frac{n+1}{2}}} \quad (x \in \mathbb{R}^n)$$

for $k \in \mathbb{N}$. Then it is not difficult to see that $p_N(\rho^\dagger) + \sup_{k\in\mathbb{N}} p_N(\rho^{\dagger,k}) < \infty$. Meanwhile, we decompose

$$\begin{aligned}
&\int_{\mathbb{R}^n} \frac{t}{(t^2+|x-y|^2)^{\frac{n+1}{2}}}\tau(tD)f(y)\mathrm{d}y \\
&=\int_{\mathbb{R}^n} \frac{t}{(t^2+|x-y|^2)^{\frac{n+1}{2}}}\tau(t^{-1}(x-y))\tau(tD)f(y)\mathrm{d}y \\
&\quad+\sum_{k=1}^{\infty}\int_{\mathbb{R}^n} \frac{t}{(t^2+|x-y|^2)^{\frac{n+1}{2}}}(\tau(t^{-1}2^{-k}(x-y))-\tau(t^{-1}2^{1-k}(x-y)))\tau(tD)f(y)\mathrm{d}y \\
&=t^{-n}\rho^\dagger\left(\frac{\star}{t}\right)*f(x)+\sum_{k=1}^{\infty}2^{-k}(2^{-kn}t^{-n})*\rho^{\dagger,k}\left(\frac{\star}{2^k t}\right)*f(x).
\end{aligned}$$

With this decomposition, we obtain

$$\sup_{t>0}\left|\int_{\mathbb{R}^n}\frac{t\tau(tD)f(y)}{(t^2+|x-y|^2)^{\frac{n+1}{2}}}\mathrm{d}y\right| \lesssim \mathscr{M}[\tau(D)f](x) \lesssim \mathscr{M}f(x).$$

As a result, we see that $\sup_{t>0}|\mathrm{e}^{-t\sqrt{-\Delta}}f| \in L^p(\mathbb{R}^n)$.

Conversely, assume $\sup_{t>0}|\mathrm{e}^{-t\sqrt{-\Delta}}f| \in L^p(\mathbb{R}^n)$. We define $\Phi \equiv \int_1^\infty \eta(s)P_s(\star)\mathrm{d}s$, where η is a function in Lemma 1.12. Then $\Phi \in \mathscr{S}(\mathbb{R}^n)$ because Φ is smooth in a neighborhood of the origin. Observe also that $t^{-n}\Phi\left(\frac{\star}{t}\right) = \int_1^\infty \eta(s)P_{st}(\star)\mathrm{d}s$.

Let $\psi \in \mathscr{S}(\mathbb{R}^n)$ be such that $\chi_{B(1)} \le \psi \le \chi_{B(2)}$. Fix $x \in \mathbb{R}^n$. Then

$$\begin{aligned}
t^{-n}\Phi\left(\frac{\star}{t}\right)*f(x) &= t^{-n}\Phi\left(\frac{\star}{t}\right)*\psi(D)f(x)+t^{-n}\Phi\left(\frac{\star}{t}\right)*(1-\psi(D))f(x) \\
&=\int_1^\infty\left(\int_{\mathbb{R}^n}\eta(s)P_{st}(y)\psi(D)f(x-y)\mathrm{d}y\right)\mathrm{d}s \\
&\quad+\left\langle(1-\psi(D))f,\int_1^\infty\eta(s)P_{st}(x-\star)\mathrm{d}s\right\rangle \\
&=\int_1^\infty\left(\int_{\mathbb{R}^n}\eta(s)P_{st}(y)\psi(D)f(x-y)\mathrm{d}y\right)\mathrm{d}s \\
&\quad+\left\langle(1-\psi)\mathscr{F}f,\int_1^\infty\eta(s)\mathrm{e}^{-ix\xi}\exp(-st|\star|)\mathrm{d}s\right\rangle \\
&=\int_1^\infty\eta(s)\left(\int_{\mathbb{R}^n}P_{st}(y)\psi(D)f(x-y)\mathrm{d}y\right)\mathrm{d}s
\end{aligned}$$

$$+\left\langle \mathscr{F} f, \int_1^\infty \eta(s) \mathrm{e}^{-ix\xi}(1-\psi)\exp(-st|\star|)\mathrm{d}s \right\rangle.$$

Since

$$\int_1^\infty \eta(s)\mathrm{e}^{-ix\xi}(1-\psi)\exp(-st|\star|)\mathrm{d}s = \lim_{L\to\infty}\int_1^L \eta(s)\mathrm{e}^{-ix\xi}(1-\psi)\exp(-st|\star|)\mathrm{d}s$$

in $\mathscr{S}(\mathbb{R}^n)$, we obtain

$$\begin{aligned} t^{-n}\Phi\left(\frac{\star}{t}\right) * f(x) &= \int_1^\infty \left(\int_{\mathbb{R}^n} \eta(s) P_{st}(y)\psi(D) f(x-y)\mathrm{d}y\right)\mathrm{d}s \\ &\quad + \int_1^\infty \eta(s)\left\langle \mathscr{F} f, \mathrm{e}^{-ix\xi}(1-\psi)\exp(-st|\star|)\right\rangle \mathrm{d}s \\ &= \int_1^\infty \eta(s)\mathrm{e}^{-st\sqrt{-\Delta}} f(x)\mathrm{d}s. \end{aligned}$$

As a result, we obtain $\sup\limits_{t>0}|t^{-n}\Phi(t^{-1}\star) * f| \le \sup\limits_{t>0}|\mathrm{e}^{-t\sqrt{-\Delta}} f| \in L^p(\mathbb{R}^n)$; hence $f \in H^p(\mathbb{R}^n)$.

Exercises

Exercise 3.14 Let $M_B f(x; j)$ be the maximal operator given by (3.19). If $B > 0$, $j \in \mathbb{N}$, $f \in \mathscr{S}'(\mathbb{R}^n)$ and $x_0 \in \mathbb{R}^n$ satisfy $M_B f(x_0; j) < \infty$, then show that $M_B f(x; j) < \infty$ for all $x \in \mathbb{R}^n$.

Exercise 3.15 Show that $\liminf\limits_{j\to\infty} \mathscr{M} f_j(x) \ge \mathscr{M} f(x)$ when $f_j \to f$ in $\mathscr{S}'(\mathbb{R}^n)$. Hint: Go back to the definition of the grand maximal operator.

Exercise 3.16 Let $f \in H^1(\mathbb{R}^n)$, and let $\psi \in C^\infty(\mathbb{R}^n)$ satisfy $\chi_{B(1)} \le \psi \le \chi_{B(2)}$.

1. Show that $|f * \psi^j| \lesssim \mathscr{M} f$ for $j \in \mathbb{Z}$.
2. The dual of $\mathrm{BC}(\mathbb{R}^n)$ is known as the set of all finite Radon measures. Combine the Banach–Alaoglu theorem and this fact to find an increasing sequence $\{j_k\}_{k=1}^\infty$ and a finite Radon measure μ such that

$$\lim_{k\to\infty}\int_{\mathbb{R}^n} f * \psi^{j_k}(x) g(x)\mathrm{d}x = \int_{\mathbb{R}^n} g(x)\mathrm{d}\mu(x), \quad g \in \mathrm{BC}(\mathbb{R}^n). \tag{3.32}$$

3. The measure μ satisfying (3.32) is absolutely continuous with respect to the Lebesgue measure.

4. Denote by F the Radon–Nykodim derivative of μ. Then show that f is represented by F and that $\|F\|_1 \lesssim \|f\|_{H^1}$.

Altogether then, $H^1(\mathbb{R}^n) \hookrightarrow L^1(\mathbb{R}^n)$.

Exercise 3.17 Let $1 < p < \infty$. Show that $L^p(\mathbb{R}^n) \approx H^p(\mathbb{R}^n)$ with equivalence of norms in the following sense:

1. If $f \in L^p(\mathbb{R}^n)$, show that $\|f\|_{H^p} \lesssim \|f\|_p$ using Theorem 1.46.
2. If $f \in H^p(\mathbb{R}^n)$, show that f is represented by an $L^p(\mathbb{R}^n)$-function g and that $\|g\|_p \lesssim \|f\|_{H^p}$.

We can skip Exercise 3.17 because we do not use $H^p(\mathbb{R}^n) \approx L^p(\mathbb{R}^n)$ for $1 < p < \infty$.

Exercise 3.18 (S. Nakamura, private communication) Let $t > 0$, $\kappa \in \mathscr{S}(\mathbb{R}^n)$ and $f \in \mathscr{S}'(\mathbb{R}^n)$.

1. Prove $\left|t^{-n}\kappa(t^{-1}\star) * f\right| \le p_N(\kappa)\mathscr{M}f$.
2. Prove $|\kappa * f| \le p_N(t^{-n}\kappa(t^{-1}\star))\mathscr{M}f$.

Hint: Go back to the definition of the grand maximal operator.

Exercise 3.19 [1114] Let $\mathscr{C}_\alpha$ be the family of functions $\phi : \mathbb{R}^n \to \mathbb{R}$ in $C_c(B(1)) \cap \mathscr{P}_0(\mathbb{R}^n)^\perp \cap \mathrm{Lip}^\alpha(\mathbb{R}^n)$. Set

$$A_\alpha(f)(x,t) \equiv \sup_{\phi \in \mathscr{C}_\alpha} t^{-n}\left|f * \phi(t^{-1}\star)(x)\right| \quad ((x,t) \in \mathbb{R}^{n+1}_+).$$

Show that $A_\alpha(f)$ is continuous on $\mathbb{R}^{n+1}_+$.

Exercise 3.20 Let $\psi \in \mathscr{S}(\mathbb{R}^n)$ be fixed.

1. Prove $\mathscr{M}[\psi * f] \lesssim p_N(\psi)\mathscr{M}f$ for all $f \in \mathscr{S}'(\mathbb{R}^n)$, where the numbers N in the grand maximal operator $\mathscr{M}$ appearing in both sides are the same.
2. For $0 < p \le 1$, show that $f \in H^p(\mathbb{R}^n) \mapsto \psi * f \in H^p(\mathbb{R}^n)$ is a bounded linear operator.

3.2.2 Singular Integral Operators on Hardy Spaces

Keeping Exercise 3.17 in mind, we study $H^p(\mathbb{R}^n)$ with $0 < p \le 1$. When $1 < p < \infty$, Exercise 3.17 shows that $H^p(\mathbb{R}^n) \approx L^p(\mathbb{R}^n)$ with equivalence of norms. So we concentrate on the case where $0 < p \le 1$, although the equivalence $H^p(\mathbb{R}^n) \approx L^p(\mathbb{R}^n)$ with $1 < p < \infty$ is interesting and important in its own right.

3.2.2.1 Whitney Decomposition and the Calderón–Zygmund Decomposition

We study the relation between the singular integral operators and Hardy spaces. We can list the Hilbert transform as an example of the singular integral operators. Since it is difficult to define the image of the singular integral operators for the functions in Hardy spaces, we content ourselves with investigating the boundedness of the operator $f \in H^p(\mathbb{R}^n) \mapsto m(D)f \in L^p(\mathbb{R}^n)$ with $0 < p \le 1$ and $m \in C_c^\infty(\mathbb{R}^n)$.

3.2.2.2 Boundedness of the Singular Integral Operators on $H^p(\mathbb{R}^n)$

Set

$$\Omega = \Omega_\lambda \equiv \{\mathcal{M} f > \lambda\} \subsetneq \mathbb{R}^n, \quad \lambda > 0. \tag{3.33}$$

Form the Whitney decomposition (Theorem 1.9) of Ω. There exists a packing $\{Q_j\}_{j=1}^\infty$ such that (1.17) and (1.19) are fulfilled. With this Whitney covering $\{Q_j\}_{j=1}^\infty$, we will show the boundedness of the singular integral operators on Hardy spaces. Although it is not difficult to elaborate the statement, we assume $m \in C_c^\infty(\mathbb{R}^n)$ to ignore the problem of the validity of the definition.

Theorem 3.6 (Singular integral operators on $H^p(\mathbb{R}^n)$) *Let $m \in C_c^\infty(\mathbb{R}^n)$. Write*

$$M_\alpha \equiv \sup_{\xi \in \mathbb{R}^n} |\xi|^{|\alpha|} |\partial^\alpha m(\xi)|. \tag{3.34}$$

Then there exists a constant $N \gg 1$ such that

$$\|m(D)f\|_{H^p} \lesssim \|f\|_{H^p} \quad (f \in H^p(\mathbb{R}^n)) \tag{3.35}$$

with the implicit constant that depends only on M_α, $|\alpha| \le N$.

Proof Choose an auxiliary function $\psi \in \mathcal{S}(\mathbb{R}^n)$ so that $\chi_{B(1)} \le \psi \le \chi_{B(2)}$. Let $R \gg 1$. Since $m(D)f = m(D)\psi(R^{-1}D)f$, we have

$$\|\psi(R^{-1}D)f\|_{H^p} = \left\| \sup_{j \in \mathbb{Z}} |\psi(2^{-j}D)\psi(R^{-1}D)f| \right\|_p .$$

Since

$$\sup_{j \in \mathbb{Z}} |\psi(2^{-j}D)\psi(R^{-1}D)f| \lesssim \mathcal{M} f, \tag{3.36}$$

we have an estimate independent of R:

$$\|\psi(R^{-1}D)f\|_{H^p} \lesssim \|\mathcal{M} f\|_p = \|f\|_{H^p}.$$

Thus, by considering $\psi(R^{-1})f$ instead of f, we can assume that f is a band-limited distribution. In this case, in particular, f is smooth. For each $\lambda > 0$, we estimate the measure of the level set $\{ |m(D)f| > \lambda \}$. Let $\{\varphi^{(j)}\}_{j=1}^{\infty}$ be a partition of unity subordinate to these cubes. That is, we take a sequence $\{\varphi^{(j)}\}_{j=1}^{\infty}$ of $C_c^{\infty}(\mathbb{R}^n)$-functions as in Proposition 1.4. Define $I^{(j)}, a^{(j)} \in \mathbb{C}$ by

$$I^{(j)} \equiv \int_{\mathbb{R}^n} \varphi^{(j)}(y)\mathrm{d}y, \quad a^{(j)} \equiv \frac{1}{I^{(j)}} \int_{\mathbb{R}^n} f(y)\varphi^{(j)}(y)\mathrm{d}y. \tag{3.37}$$

We need the following estimate:

Lemma 3.3 *Let $k \in C_c^{\infty}(\mathbb{R}^n)$, $j = 1, 2, \ldots$ and $x \in \mathbb{R}^n \setminus \Omega$. Assume*

$$M_\alpha \equiv \sup_{w \in \mathbb{R}^n} |w|^{n+|\alpha|} |\partial^\alpha k(w)| < \infty.$$

Define

$$\Theta_x^{(j)} \equiv \ell(Q_j)^n \left(k(x - z_j + \ell(Q_j)\star) - \frac{1}{I^{(j)}} \int_{\mathbb{R}^n} k(x-z)\varphi^{(j)}(z)\mathrm{d}z \right) \varphi^{(j)}(z_j - \ell(Q_j)\star)$$

for $y \in \mathbb{R}^n$. Then for all $N \in \mathbb{N}$, there exists a finite set A of ${\mathbb{N}_0}^n$ such that

$$p_N(\Theta_x^{(j)}) \lesssim \sup_{\alpha \in A} M_\alpha \times \frac{\ell(Q_j)^{n+1}}{|x - c(Q_j)|^{n+1}},$$

where the implicit constant does not depend on x, j.

We admit Lemma 3.3 for the time being. Note that $f = f\chi_{\mathbb{R}^n \setminus \Omega} + \sum_{j=1}^{\infty} f\varphi^{(j)}$ is further decomposed as follows:

$$f = \left\{ f\chi_{\mathbb{R}^n \setminus \Omega} + \sum_{j=1}^{\infty} a^{(j)}\varphi^{(j)} \right\} + \sum_{j=1}^{\infty} (f - a^{(j)})\varphi^{(j)}. \tag{3.38}$$

Set $\widehat{\varphi^{(j)}} \equiv \varphi^{(j)}(-\star)$. By (1.19) and the definition (3.33) of Ω, we have

$$|a^{(j)}| \lesssim \frac{1}{\ell(Q_j)^n} \left| \int_{\mathbb{R}^n} f(y)\widehat{\varphi^{(j)}}(z_j - y - z_j)\mathrm{d}y \right| = \frac{1}{\ell(Q_j)^n} \left| f * [\widehat{\varphi^{(j)}}(\star - z_j)](z_j) \right|$$

for the reference point z_j. By the definition of $\mathcal{M}f$, we have $|a^{(j)}| \lesssim \mathcal{M}f(z_j) \lesssim \lambda$. Hence

$$\tilde{f} \equiv \chi_{\mathbb{R}^n \setminus \Omega} f + \sum_{j=1}^{\infty} a^{(j)} \cdot \varphi^{(j)} \tag{3.39}$$

does not exceed a constant multiple of λ in the absolute value in Ω. Furthermore, the same argument works to have

$$|\tilde{f}(x)| \lesssim \mathcal{M} f(x) \quad x \in \mathbb{R}^n. \tag{3.40}$$

According to Theorem 1.5, we have

$$\begin{aligned}\left(\left\|\sup_{t>0} |e^{t\Delta} m(D) f|\right\|_p\right)^p &= \int_0^\infty p\,\lambda^{p-1} \left|\left\{\sup_{t>0} |e^{t\Delta} m(D) f| > \lambda\right\}\right| d\lambda \\ &= p\,2^p \int_0^\infty \lambda^{p-1} \left|\left\{\sup_{t>0} |e^{t\Delta} m(D) f| > 2\lambda\right\}\right| d\lambda.\end{aligned}$$

By Chebychev's inequality (see Theorem 1.4), we have

$$\left|\left\{\sup_{t>0} |e^{t\Delta} m(D) \tilde{f}| > \lambda\right\}\right| \le \frac{1}{\lambda^2} \int_{\mathbb{R}^n} |m(D)\tilde{f}(x)|^2 dx.$$

By Theorem 1.47, (3.39), (3.40) and the $L^2(\mathbb{R}^n)$-boundedness of $m(D)$, we have

$$\left|\left\{\sup_{t>0} |e^{t\Delta} m(D) \tilde{f}| > \lambda\right\}\right| \lesssim \frac{1}{\lambda^2} \int_{\mathbb{R}^n} |\tilde{f}(x)|^2 dx \lesssim \frac{1}{\lambda^2} \int_{\mathbb{R}^n \setminus \Omega} |\mathcal{M} f(x)|^2 dx + |\Omega|.$$

Next, we estimate the second term: we have

$$\begin{aligned}&m(D)[f\varphi^{(j)} - a^{(j)}\varphi^{(j)}](x) \\ &\simeq_n \int_{\mathbb{R}^n} \mathcal{F}^{-1} m(x-y) \left(f(y)\varphi^{(j)}(y) - a^{(j)}\varphi^{(j)}(y)\right) dy \\ &= \int_{\mathbb{R}^n} \mathcal{F}^{-1} m(x-y) f(y)\varphi^{(j)}(y) dy \\ &\quad - \frac{1}{I^{(j)}} \iint_{\mathbb{R}^{2n}} \mathcal{F}^{-1} m(x-y) f(z)\varphi^{(j)}(z)\varphi^{(j)}(y) dy dz \\ &= \int_{\mathbb{R}^n} \mathcal{F}^{-1} m(x-y) f(y)\varphi^{(j)}(y) dy \\ &\quad - \frac{1}{I^{(j)}} \iint_{\mathbb{R}^{2n}} \mathcal{F}^{-1} m(x-z) f(y)\varphi^{(j)}(z)\varphi^{(j)}(y) dy dz \\ &= \int_{\mathbb{R}^n} \left(\mathcal{F}^{-1} m(x-y) - \frac{1}{I^{(j)}} \int_{\mathbb{R}^n} \mathcal{F}^{-1} m(x-z)\varphi^{(j)}(z) dz\right) f(y)\varphi^{(j)}(y) dy\end{aligned}$$

for each $j \in \mathbb{N}$.

Here we fix $x \notin \Omega$. Set

$$\Psi_x^{(j)}(y) \equiv \left(\mathscr{F}^{-1}m(x-z_j+y) - \frac{1}{I^{(j)}}\int_{\mathbb{R}^n}\mathscr{F}^{-1}m(x-z)\varphi^{(j)}(z)\mathrm{d}z\right)\varphi^{(j)}(z_j-y)$$

for $x \in \mathbb{R}^n$. Then the above calculation shows:

$$m(D)[f\varphi^{(j)} - a^{(j)}\varphi^{(j)}](x) = \Psi_x^{(j)} * f(z_j). \tag{3.41}$$

Now we rewrite

$$\begin{aligned}\Psi_x^{(j)}(y) &= \frac{\varphi^{(j)}(z_j-y)}{I^{(j)}}\left(\int_{\mathbb{R}^n}\left[\mathscr{F}^{-1}m(x-z_j+y) - \mathscr{F}^{-1}m(x-z)\right]\varphi^{(j)}(z)\mathrm{d}z\right)\\ &= \frac{\varphi^{(j)}(z_j-y)}{I^{(j)}}\left(\int_{\mathbb{R}^n}\left[\mathscr{F}^{-1}m(x-z_j+y) - \mathscr{F}^{-1}m(z)\right]\varphi^{(j)}(x-z)\mathrm{d}z\right)\end{aligned}$$

again. Theorem 1.67 yields $|\partial^\alpha \mathscr{F}^{-1}m(D)(w)| \lesssim_{M_\alpha,\alpha} |w|^{-n-|\alpha|}$ for each $\alpha \in \mathbb{N}_0{}^n$ and $w \in \mathbb{R}^n$.

Hence it follows that $|y|, |z_j| \gtrsim |x - c(Q_j)| \gtrsim \ell(Q_j)$. As a result,

$$p_N(\ell(Q_j)^n \Psi_x^{(j)}(\ell(Q_j)\star)) \lesssim \frac{\ell(Q_j)^{n+1}}{|x-c(Q_j)|^{n+1}}.$$

Hence

$$|\Psi_x^{(j)} * f(z_j)| \lesssim \frac{\ell(Q_j)^{n+1}}{|x-c(Q_j)|^{n+1}}\mathscr{M}f(z_j) \lesssim \frac{\lambda\ell(Q_j)^{n+1}}{|x-c(Q_j)|^{n+1}}.$$

We have

$$\begin{aligned}\chi_{\mathbb{R}^n\setminus\Omega}(x)|m(D)[f-\tilde{f}](x)| &\lesssim \sum_{j=1}^{\infty}|m(D)[f\varphi^{(j)} - a^{(j)}\varphi^{(j)}](x)| \\ &\lesssim \lambda\sum_{j=1}^{\infty}\frac{\ell(Q_j)^{n+1}}{|x-c(Q_j)|^{n+1}}\end{aligned} \tag{3.42}$$

for $x \in \mathbb{R}^n$. In (3.42) we can use $m_t \equiv m\exp(-t|\star|^2)$ for $t > 0$ instead of m. Thus,

$$|m_t(D)[f-\tilde{f}](x)| \lesssim \lambda\sum_{j=1}^{\infty}\frac{\ell(Q_j)^{n+1}}{|x-c(Q_j)|^{n+1}}$$

and, by considering the supremum over $t > 0$, we obtain

$$\sup_{t>0} |e^{t\Delta}m(D)[f - \tilde{f}](x)| \lesssim \lambda \sum_{j=1}^{\infty} \frac{\ell(Q_j)^{n+1}}{|x - c(Q_j)|^{n+1}}. \tag{3.43}$$

If we integrate the most right-hand side over Ω^c, then we see the value is estimated from the above:

$$\int_{\Omega^c} \sum_{j=1}^{\infty} \frac{\ell(Q_j)^{n+1}dx}{|x - c(Q_j)|^{n+1}} \le \sum_{j=1}^{\infty} \int_{\mathbb{R}^n \setminus Q_j} \frac{\ell(Q_j)^{n+1}dx}{|x - c(Q_j)|^{n+1}} \lesssim \sum_{j=1}^{\infty} \ell(Q_j)^n \lesssim |\Omega|. \tag{3.44}$$

Chebychev's inequality (Theorem 1.4), (3.42) and (3.44) give:

$$\begin{aligned}
\left| \left\{ \sup_{t>0} |e^{t\Delta}m(D)[f - \tilde{f}]| > \lambda \right\} \right| &\le |\Omega| + \left| \left\{ \sup_{t>0} |e^{t\Delta}m(D)[f - \tilde{f}]| > \lambda \right\} \cap \Omega^c \right| \\
&\le |\Omega| + \frac{1}{\lambda} \int_{\mathbb{R}^n \setminus \Omega} \sup_{t>0} |e^{t\Delta}m(D)[f - \tilde{f}]|(x)dx \\
&\lesssim |\Omega|.
\end{aligned}$$

From the definition (3.33) of Ω, we have

$$\begin{aligned}
&\left| \left\{ \sup_{t>0} |e^{t\Delta}m(D)f| > 2\lambda \right\} \right| \\
&\le |\{ \sup_{t>0} |e^{t\Delta}m(D)\tilde{f}| > \lambda \}| + |\{ \sup_{t>0} |e^{t\Delta}m(D)[f - \tilde{f}]| > \lambda \}| \\
&\lesssim \frac{1}{\lambda^2} \int_{\{\mathcal{M}f \le \lambda\}} |\tilde{f}(x)|^2 dx + |\{ \mathcal{M}f > \lambda \}| \\
&\lesssim \frac{1}{\lambda^2} \int_{\{\mathcal{M}f \le \lambda\}} \mathcal{M}f(x)^2 dx + |\{ \mathcal{M}f > \lambda \}|.
\end{aligned}$$

Multiply this chain of inequalities by $p\, 2^p\, \lambda^{p-1}$ and then integrate it over $0 < \lambda < \infty$:

$$\begin{aligned}
&\left(\left\| \sup_{t>0} |e^{t\Delta}m(D)f| \right\|_p \right)^p \\
&\lesssim \int_0^{\infty} \lambda^{p-3} \left(\int_{\{\mathcal{M}f \le \lambda\}} \mathcal{M}f(x)^2 dx \right) d\lambda + \int_0^{\infty} \lambda^{p-1} |\{ \mathcal{M}f > \lambda \}| d\lambda \\
&\lesssim \int_{\mathbb{R}^n} \mathcal{M}f(x)^2 \left(\int_{\mathcal{M}f(x)}^{\infty} \lambda^{p-3} d\lambda \right) dx + \int_{\mathbb{R}^n} \mathcal{M}f(x)^p dx \\
&\sim \int_{\mathbb{R}^n} \mathcal{M}f(x)^p dx \sim \|f\|_{H^p}{}^p.
\end{aligned}$$

Therefore, the proof is complete.

Proof (of Lemma 3.3) Let $y \in \mathbb{R}^n$. Note that $\Theta_x^{(j)}(y) \neq 0$ only if $z_j - \ell(Q_j)y \in 50Q_j$. This implies $\Theta_x^{(j)}(y) \neq 0$ if and only if $\|y\|_\infty \leq 10050$. Therefore, we have

$$p_N(\Theta_x^{(j)}) \lesssim \sup_{|\alpha|\leq N} \sup_{y\in\mathbb{R}^n} |\partial^\alpha \Theta_x^{(j)}(y)|.$$

By the mean-value theorem, we have

$$\Theta_x^{(j)}(y) \equiv \frac{\ell(Q_j)^n}{I^{(j)}}\varphi^{(j)}(z_j - \ell(Q_j)y) \times \iint_{\mathbb{R}^n\times[0,1]} (z_j+\ell(Q_j)y-z)\cdot\nabla k(x-z-t(z_j+\ell(Q_j)y-z))\varphi^{(j)}(z)\mathrm{d}z\mathrm{d}t;$$

hence, as long as y is in the support of $\Theta^{(j)}$,

$$\begin{aligned}
&|\Theta_x^{(j)}(y)| \\
&\lesssim \iint_{\mathbb{R}^n\times[0,1]} |(z_j + \ell(Q_j)y - z)\cdot\nabla k(x - z - t(z_j + \ell(Q_j)y - z))\varphi^{(j)}(z)|\mathrm{d}z\mathrm{d}t \\
&\lesssim \sup_{\alpha'\in A} M_{\alpha'} \times \frac{\ell(Q_j)}{|x - c(Q_j)|^{n+1}} \iint_{\mathbb{R}^n\times[0,1]} |\varphi^{(j)}(z)|\mathrm{d}z\mathrm{d}t \\
&\sim \sup_{\alpha'\in A} M_{\alpha'} \times \frac{\ell(Q_j)}{|x - c(Q_j)|^{n+1}}.
\end{aligned}$$

So, we have

$$\sup_{y\in\mathbb{R}^n} |\partial^\alpha \Theta_x^{(j)}(y)| \lesssim \sup_{\alpha'\in A} M_{\alpha'} \times \frac{\ell(Q_j)^{n+1}}{|x - c(Q_j)|^{n+1}} \tag{3.45}$$

when $\alpha = 0$. If $\alpha \neq 0$, we can go through a similar argument to have (3.45). Thus, the proof of Lemma 3.3 is complete.

3.2.2.3 Vector-Valued Function Space

We are oriented to showing that $H^p(\mathbb{R}^n) \approx \dot{F}^0_{p2}(\mathbb{R}^n)$ here. Let us define vector-valued Hardy spaces to this end.

Definition 3.4 (Matrix notation) For $F \equiv \{f_j\}_{j=-\infty}^\infty \subset \mathscr{S}'(\mathbb{R}^n)$ use the following notation: Suppose we have an $\mathscr{S}(\mathbb{R}^n)$-valued $\mathbb{Z}\times\mathbb{Z}$-matrix $\Phi = \{\Phi(\star; j,k)\}_{j,k\in\mathbb{Z}} \subset \mathscr{S}(\mathbb{R}^n)$. Here we assume that $\Phi(\star; j,k)$ is chosen arbitrarily. By no means for any $j_1, j_2, k_1, k_2 \in \mathbb{Z}$, do $\Phi(\star; j_1,k_1)$ and $\Phi(\star; j_2,k_2)$ have any special relation.

Following the standard rule of the matrix calculation, we define

$$\Phi * F \equiv \left\{ \sum_{k=-\infty}^{\infty} \Phi(\star; j,k) * f_k \right\}_{j=-\infty}^{\infty}. \tag{3.46}$$

In order to avoid a problem of the convergence of the infinite sum, we deal with $\Phi(\star; j,k)$, $j,k \in \mathbb{Z}$ such that $\Phi(\star; j,k) = 0$ with a finite number of exceptions. Likewise, define $\partial^\alpha \Phi \equiv \{\partial^\alpha \Phi(\star; j,k)\}_{j,k\in\mathbb{Z}}$ for $\alpha \in \mathbb{N}_0{}^n$ and $\Phi = \{\Phi(\star; j,k)\}_{j,k\in\mathbb{Z}} \subset \mathscr{S}(\mathbb{R}^n)$. Denote by $\widetilde{\mathscr{S}}(\mathbb{R}^n;\mathbb{Z})$ the set of all collections $\Phi = \{\Phi(\star; j,k)\}_{j,k\in\mathbb{Z}} \subset \mathscr{S}(\mathbb{R}^n)$ such that $\Phi(\star; j,k) \equiv 0$ with a finite number of exceptions. Furthermore, define

$$p_N(\Phi) \equiv \sum_{\substack{\alpha\in\mathbb{N}_0{}^n \\ |\alpha|\le N}} \sup_{x\in\mathbb{R}^n} \langle x\rangle^{|\alpha|} \|\partial^\alpha \Phi(x)\|_{B(\ell^2)} \tag{3.47}$$

for $N \in \mathbb{N}_0$ and $\Phi \in \widetilde{\mathscr{S}}(\mathbb{R}^n;\mathbb{Z})$. Here one defines $\|A\|_{B(\ell^2)}$ for the operator norm of A: $\|A\|_{B(\ell^2)} \equiv \sup\{\|Ax\|_{\ell^2} : x \in \ell^2(\mathbb{Z}), \|x\|_{\ell^2(\mathbb{Z})} = 1\}$ for $A \in M(\mathbb{Z},\mathbb{C}) \equiv \{\{a_{ij}\}_{i,j\in\mathbb{Z}} : a_{ij} \in \mathbb{C}\}$. Finally, define $\widetilde{\mathscr{F}_N} \equiv \{\Phi \in \widetilde{\mathscr{S}}(\mathbb{R}^n;\mathbb{Z}) : p_N(\Phi) < 1\}$.

We can prove the next theorem in the same way as the scalar-value case.

Theorem 3.7 *Set* $\mathbb{G}^l \equiv \{2^{nl}\delta_{jk}\exp(-|2^l \star|^2)\}_{j,k\in\mathbb{Z}}$ *and* $\Psi(\star; j,k)^l \equiv 2^l\Psi(2^l\star; j,k)$ *for* $j,k,l \in \mathbb{Z}$. *Let* $0 < p < \infty$. *Then*

$$\left\| \sup_{l\in\mathbb{Z}, \{\Psi(\star;j,k)\}_{j,k\in\mathbb{Z}}\in\widetilde{\mathscr{F}_N}} \|\{\Psi(\star; j,k)^l\}_{j,k\in\mathbb{Z}} * F(\star)\|_{\ell^2} \right\|_p \lesssim \left\| \sup_{l\in\mathbb{Z}} \|\mathbb{G}^l * F(\star)\|_{\ell^2} \right\|_p$$

for all $F \in \mathscr{S}'(\mathbb{R}^n;\mathbb{Z})$.

Let $g \equiv \exp(-|\star|^2)$ be the Gaussian and $\mathbb{G}^j \equiv \{\delta_{jk} g^j\}_{j,k\in\mathbb{Z}}$. Define the *vector-valued Hardy space* $H^p(\ell^2)$ by

$$\|F\|_{H^p(\ell^2)} \equiv \left\| \sup_{j\in\mathbb{Z}} \|\{\mathbb{G}^j * f_k\}_{k\in\mathbb{Z}}\|_{\ell^2} \right\|_p \tag{3.48}$$

for $F \equiv \{f_k\}_{k\in\mathbb{Z}} \in \mathscr{S}'(\mathbb{R}^n;\mathbb{Z})$. We have a vector-valued counterpart of Theorem 3.6, the boundedness of the Fourier multiplier.

Theorem 3.8 *Let* $\Psi \in \widetilde{\mathscr{S}}(\mathbb{R}^n)$. *Define a finite subset* A_N *of* $\mathbb{R}$ *by*

$$A_N \equiv \left\{ \sup_{x\in\mathbb{R}^n} |x|^{|\alpha|} \|\partial^\alpha \Psi(x)\|_{B(\ell^2)} : \alpha \in \mathbb{N}_0{}^n, \quad |\alpha| \le N \right\}.$$

Then $\|\Psi * F\|_{H^p(\ell^2)} \lesssim_{A_N} \|F\|_{H^p(\ell^2)}$ *if* $N \gg 1$.

The proof is the adaptation of the scalar case.

Exercises

Exercise 3.21 Check (3.36). Hint: We distinguish two cases: R and 2^j are close and otherwise. In the former case, we can show that $2^{jn}R^{-n}\psi(2^jR^{-1}\star)\psi(\star)$ belongs to $\mathscr{F}_N$ modulo an unimportant constant. In the latter case, either $\psi(2^{-j}D)$ or $\psi(R^{-1}D)$ is absorbed in the other.

Exercise 3.22 Let $0 < \alpha < n$. Assume $\dfrac{1}{q} = 1 - \dfrac{\alpha}{n} > 0$. Show that I_α maps $H^1(\mathbb{R}^n)$ continuously to $L^p(\mathbb{R}^n)$ using $H^1(\mathbb{R}^n) \hookrightarrow \dot{F}^0_{1\infty}(\mathbb{R}^n)$.

3.2.3 *Atoms for Hardy Spaces*

3.2.3.1 Atoms for Hardy Spaces

The Hardy space $H^p(\mathbb{R}^n)$, $0 < p \le 1$ serves to substitute the Lebesgue space $L^p(\mathbb{R}^n)$. As we will see later, the singular integral operators behave well. However, the ambiguity of the choice of $\psi \in \mathscr{S}(\mathbb{R}^n) \setminus \mathscr{P}_0(\mathbb{R}^n)^\perp$ makes the definition of the norm more complicated. So, we want to investigate Hardy spaces from a different side. We are oriented to the structure of the function spaces.

Define $\sigma_p \equiv n\left(\frac{1}{p} - 1\right)_+ = n\left(\frac{1}{p} - 1\right)$ for $0 < p \le 1$. This index σ_p is important in the definition. Furthermore, recall that $[\star]$ stands for the integer part.

Definition 3.5 (Atoms for Hardy spaces) Let $0 < p \le 1 < q \le \infty$. Fix $L \ge L_0 \equiv [\sigma_p]$. A *$(p,q)$-atom centered at a cube Q* is a function $A \in L^q(\mathbb{R}^n) \cap \mathscr{P}_L(\mathbb{R}^n)^\perp$ satisfying $\text{supp}(A) \subset Q$ and $\|A\|_q \le |Q|^{-\frac{1}{p}+\frac{1}{q}}$.

We will see why a must belong to $\mathscr{P}_L(\mathbb{R}^n)$ in Exercise 4.16.

Let a be a (p,q)-atom with $0 < p \le 1 < q \le \infty$. Then $\|a\|_p \le 1$ by Hölder's inequality and the fact $\|a\|_q \le |Q|^{-\frac{1}{p}+\frac{1}{q}}$.

3.2.3.2 Synthesis of Atoms

We are oriented to the synthesis part of the atomic decomposition; that is, when we have a sum

$$f = \sum_{j=1}^{\infty} \lambda_j a_j,$$

we are interested in the convergence of the sum defining f and in how to estimate the norm of f. We are mainly concerned with the case where $q = 2, \infty$. The value of q will do as long as $1 < q \le \infty$.

Proposition 3.2 *Let $0 < p \le 1$. Suppose that we have a sequence $\{a_j\}_{j=1}^{\infty}$ of $(p, 2)$-atoms and $\{\lambda_j\}_{j=1}^{\infty} \in \ell^p(\mathbb{N})$. Then $\sum_{j=1}^{\infty} \lambda_j a_j$ is convergent in $L^p(\mathbb{R}^n)$ and $\mathscr{S}'(\mathbb{R}^n)$.*

Proof Let a be a $(p, 2)$-atom. Since the $L^p(\mathbb{R}^n)$-norm of $(p, 2)$-atom is less than 1,

$$\left(\left\| \sum_{j=1}^{\infty} |\lambda_j a_j| \right\|_p \right)^p \le \sum_{j=1}^{\infty} \| \lambda_j a_j \|_p{}^p \le \sum_{j=1}^{\infty} |\lambda_j|^p \le 1$$

by (1.3). Thus, we obtain the $L^p(\mathbb{R}^n)$-convergence.

Let us prove the $\mathscr{S}'(\mathbb{R}^n)$-convergence. From Theorem 1.56,

$$\left| \int_{\mathbb{R}^n} a(x)\varphi(x)\mathrm{d}x \right| \lesssim p_{L+1}(\varphi)\ell(Q)^{L+1} \int_Q |a(x)|\mathrm{d}x \lesssim p_{L+1}(\varphi)\ell(Q)^{n+L+1-\frac{n}{p}} p_{L+1}(\varphi). \tag{3.49}$$

Meanwhile, by the definition of the $(p, 2)$-atom, $\|a\|_2 \le |Q|^{\frac{1}{2}-\frac{1}{p}}$ and Hölder's inequality, we have

$$\left| \int_{\mathbb{R}^n} a(x)\varphi(x)\mathrm{d}x \right| \le p_0(\varphi) \int_Q |a(x)|\mathrm{d}x \le \ell(Q)^{n-\frac{n}{p}} p_0(\varphi) \tag{3.50}$$

without using the moment condition. Since $L \ge L_0 = [\sigma_p]$, if we combine (3.49) and (3.50), then

$$\left| \int_{\mathbb{R}^n} a(x)\varphi(x)\mathrm{d}x \right| \lesssim p_{n+L+1}(\varphi) \tag{3.51}$$

for any $(p, 2)$-atom a.

Suppose that we have $\{a_j\}_{j=1}^{\infty}$ and $\{\lambda_j\}_{j=1}^{\infty} \in \ell^p(\mathbb{N})$. Since $0 < p \le 1$, we have

$$\sum_{j=1}^{\infty} \left| \lambda_j \int_{\mathbb{R}^n} a_j(x)\varphi(x)\mathrm{d}x \right| \lesssim p_{n+L+1}(\varphi) \sum_{j=1}^{\infty} |\lambda_j| \lesssim p_{n+L+1}(\varphi)\|\lambda\|_{\ell^p} \tag{3.52}$$

by (3.51). Consequently, (3.52) shows that $\sum_{j=1}^{\infty} \lambda_j a_j$ is convergent in $\mathscr{S}'(\mathbb{R}^n)$.

The next lemma shows that $(p, 2)$-atoms are contained in the bounded set of $H^p(\mathbb{R}^n)$.

Lemma 3.4 *For a $(p, 2)$-atom A, we have $\|\mathcal{M}A\|_p \lesssim 1$.*

Proof Let Q be a cube in Definition 3.5. A translation allows us to assume that Q is centered at the origin. Fix $\varphi \in \mathcal{S}(\mathbb{R}^n)$ and $t > 0$. Then since α is arbitrary as long as $|\alpha| = n+L+1$, we have $\mathcal{M}A(x) \lesssim \ell(Q)^{n+L+1-\frac{n}{p}}|x|^{-n-L-1}$ for all $x \in \mathbb{R}^n \setminus 2Q$ thanks to Theorem 1.56. Meanwhile, $\mathcal{M}A(x) \lesssim MA(x)$ for all $x \in \mathbb{R}^n$.

Hence we have

$$\mathcal{M}A(x) \lesssim \chi_{2Q}(x)MA(x) + \ell(Q)^{n+L+1-\frac{n}{p}}|x|^{-n-L-1}\chi_{\mathbb{R}^n\setminus Q}(x) \tag{3.53}$$

for all $x \in \mathbb{R}^n$. If we integrate (3.53) over $\mathbb{R}^n$, we have

$$\|\mathcal{M}A\|_p \lesssim \|\chi_{2Q}MA\|_p + 1 \lesssim |Q|^{\frac{1}{p}-\frac{1}{2}}\|MA\|_2 + 1$$

by the assumption $n + L + 1 > \frac{n}{p}$. Since the Hardy–Littlewood maximal operator M is $L^2(\mathbb{R}^n)$-bounded, we obtain $\|\mathcal{M}A\|_p \lesssim |Q|^{\frac{1}{p}-\frac{1}{2}}\|A\|_2 + 1 \lesssim 1$. Lemma 3.4 is thus proved.

Proposition 3.3 *Let $0 < p \le 1$. Suppose that we have a collection $\{a_j\}_{j=1}^\infty$ of $(p, 2)$-atoms and $\lambda = \{\lambda_j\}_{j=1}^\infty \in \ell^p(\mathbb{N})$. Then*

$$\left\|\sum_{j=1}^\infty \lambda_j a_j\right\|_{H^p} = \left\|\mathcal{M}\left[\sum_{j=1}^\infty \lambda_j a_j\right]\right\|_p \lesssim_p \|\lambda\|_{\ell^p}.$$

Proof By Lemma 3.4 and the Fatou lemma, we have

$$\left\|\mathcal{M}\left[\sum_{j=1}^\infty \lambda_j a_j\right]\right\|_p \le \liminf_{J\to\infty}\left\|\mathcal{M}\left[\sum_{j=1}^J \lambda_j a_j\right]\right\|_p.$$

Here since $0 < p \le 1$,

$$\left\|\mathcal{M}\left[\sum_{j=1}^\infty \lambda_j a_j\right]\right\|_p \le \liminf_{J\to\infty}\left\{\sum_{j=1}^J \|\mathcal{M}[\lambda_j a_j]\|_p{}^p\right\}^{\frac{1}{p}} \lesssim \|\lambda\|_{\ell^p}.$$

Therefore, the proof is complete.

Finite Atomic Decomposition of $H^1(\mathbb{R}^n)$

Here we discuss a variant of the finite atomic decomposition in $H^1(\mathbb{R}^n)$. Let $1 < q \leq \infty$, so that we consider $(1, q)$-atoms. One defines $X^1(\mathbb{R}^n) = X^{1,q}(\mathbb{R}^n) \equiv \mathrm{Span}(\{(1, q)\text{-atom}\}) \subset H^1(\mathbb{R}^n)$. For $f \in X^1(\mathbb{R}^n)$, one defines

$$|||f||| \equiv \inf \left\{ \|\lambda\|_{\ell^1(F)} \,:\, f = \sum_{j \in F} \lambda_j a_j, \quad \sharp F < \infty \text{ and each } a_j \text{ is a } (1, \infty)\text{-atom} \right\},$$

$$||f|| \equiv \inf \left\{ \|\lambda\|_{\ell^1(\mathbb{N})} \,:\, f = \sum_{j=1}^{\infty} \lambda_j a_j, \quad \text{each } a_j \text{ is a } (1, \infty)\text{-atom} \right\}.$$

For a linear mapping T defined on $X^1(\mathbb{R}^n)$ and assuming its value in $\mathbb{C}$, we define two norms by

$$|||T||| \equiv \sup \left\{ \frac{|T(f)|}{|||f|||} \,:\, f \in X^1(\mathbb{R}^n) \setminus \{0\} \right\},$$

$$||T|| \equiv \sup \left\{ \frac{|T(f)|}{||f||} \,:\, f \in X^1(\mathbb{R}^n) \setminus \{0\} \right\}.$$

It is easy to see

$$|||T||| \equiv \sup\{|T(a)| \,:\, a \text{ is a } (1, \infty)\text{-atom}\}, \quad |||T||| \leq ||T||, \quad |||f||| \leq ||f|| \tag{3.54}$$

for all $f \in X^1(\mathbb{R}^n)$ and all operator T.

Proposition 3.4

1. *The norms* $\| \cdot \|$ *and* $||| \cdot |||$ *on* $X^1(\mathbb{R}^n)$ *are not equivalent.*
2. *There exists a linear operator* $T : X^1(\mathbb{R}^n) \to \mathbb{C}$ *such that* $|||T||| < ||T|| = \infty$.

Proof

1. Let $\delta > 0$ be fixed. We consider a disjoint sequence $\{I_j\}_{j=1}^{\infty}$ of cubes in $[0, 1]^n$ such that $\sum_{j=1}^{\infty} |I_j| \leq \delta$ and that $\overline{\bigcup_{j=1}^{\infty} I_j} = [0, 1]^n$. Let b_{I_j} be a function such that $b_{I_j} = \chi_{E_j} - \chi_{F_j}$, where $|E_j| = |F_j| = \frac{1}{2}|I_j|$ and $E_j \cup F_j = I_j$. We let $f = \sum_{j=1}^{\infty} b_{I_j}$. Then it is easy to see that $|f| \leq \chi_{[0,1]^n}$. Note that $\|f\| \leq \sum_{j=1}^{\infty} |I_j| \leq \delta$. We prove $|||f||| \geq 1$. To this end, we take a finite decomposition

$f = \sum_{j=1}^{J} \lambda_j a_j$ arbitrarily, where a_j is a $(1, \infty)$ supported near Q_j. Then we have $|f| \le \sum_{j=1}^{J} \frac{|\lambda_j|}{|Q_j|} \chi_{Q_j}$. Since this holds almost everywhere, the right-hand side is continuous almost everywhere and $|f|$ is almost everywhere 1, it follows that $1 \le \sum_{j=1}^{J} \frac{|\lambda_j|}{|Q_j|} \chi_{Q_j}$. If we integrate over $[0, 1]^n$, then $1 \le \sum_{j=1}^{J} |\lambda_j|$, showing that $|||f||| \ge 1$.

2. We now consider the following functions $\{f_k\}_{k=1}^{\infty} \subset X^1(\mathbb{R}^n)$ and disjoint cubes $\{Q_k\}_{k=1}^{\infty}$: $|f_k| = \chi_{Q_k}$ and that $||f_k|| \le 2^{-k}|Q_k|$. Consider $f \in \mathrm{Span}(\{f_j\}_{j=1}^{\infty})$. If $\lambda_1, \ldots, \lambda_K$ satisfies $f = \sum_{k=1}^{K} \lambda_k f_k$, then define $T(f) = \sum_{k=1}^{K} \lambda_k |Q_k|$. Note that this mapping is well defined since the Q_k's are disjoint. As we did above, we can show that

$$|T(f)| \le |||f|||$$

using the continuity argument. Thus, if we use the Hahn-Banach theorem, then we can extend it to a bounded linear functional on $X^1(\mathbb{R}^n)$. Meanwhile, T is not bounded from $(X^1(\mathbb{R}^n), || \cdot ||)$ to $\mathbb{C}$, since $||f_k|| = 2^{-k}|Q_k|$ and $T(f_k) = |Q_k|$.

Exercises

Exercise 3.23 Let $0 < p < 1 = q \le \infty$. Fix $L \ge L_0 \equiv [\sigma_p]$. Assume that A is an $L^1(\mathbb{R}^n)$-function such that A is supported in B and that $\|A\|_q \le |B|^{-\frac{1}{p}+1}$ for some ball B. Then show that $\|A\|_{H^p} \lesssim 1$. Hint: Show that $\|\chi_B MA\|_p \lesssim 1$ using Theorem 1.45.

Exercise 3.24 [596, p. 370, proof of Theorem 9] Let $L \in \mathbb{N}_0$ and $1 \le q \le \infty$. Suppose that we have $h \in L^q(\mathbb{R}^n)$ and $\{a_\gamma\}_{\gamma \in {\mathbb{N}_0}^n, |\gamma| \le L} \subset C_c^\infty(\mathbb{R}^n)$ such that

$$\mathrm{supp}(h) \subset B(2\mathbf{e}_1, 1), \quad \int_{\mathbb{R}^n} x^\beta a_\gamma(x) \mathrm{d}x = \delta_{\beta\gamma}$$

for all multi-indexes β with $|\beta| \le L$. Define

$$A(x) \equiv h(x) - \sum_{|\gamma| \le L} \left(\int_{\mathbb{R}^n} y^\gamma h(y) \mathrm{d}y \right) \varphi_\gamma(x) \quad (x \in \mathbb{R}^n).$$

Then show that $\mathrm{supp}(A) \subset B(3)$ and that $A \in \mathscr{P}_L(\mathbb{R}^n)^\perp$.

Exercise 3.25 Let $0 < p < 1$.

1. Let A be a nonzero function satisfying $|A| \le \chi_{[1,2)^n}$ and A has a moment of order $[n/p - n]$. Then show that

$$f = \sum_{j=1}^{\infty} \frac{2^{jn/p}}{j^{2/p}} A(2^j \star) \tag{3.55}$$

belongs to $H^p(\mathbb{R}^n)$. Hint: Show $\|2^{jn/p} A(2^j \star)\|_{H^p} \lesssim 1$.

2. Present an example of the distribution $f \in H^p(\mathbb{R}^n)$ such that f is not represented by an $L^1_{\rm loc}(\mathbb{R}^n)$-function. Hint: Define f by (3.55). Note that f also converges almost everywhere. Assume to the contrary that f is represented by an $L^1_{\rm loc}(\mathbb{R}^n)$-function g. Then it is easy to see that

$$\int_{\mathbb{R}^n} \varphi(x) \sum_{j=1}^{\infty} \frac{2^{jn/p}}{j^{2/p}} A(2^j x) {\rm d}x = \int_{\mathbb{R}^n} \varphi(x) g(x) {\rm d}x$$

for any $\varphi \in C^\infty_{\rm c}(\mathbb{R}^n \setminus \{0\})$. Thus, thanks to the Lebesgue differentiation theorem,

$$g(x) = \sum_{j=1}^{\infty} \frac{2^{jn/p}}{j^{2/p}} A(2^j x)$$

for almost every $x \in \mathbb{R}^n$.

Exercise 3.26 Let $0 < p < \infty$. Denote by $X^p(\mathbb{R}^n)$ the finite linear span of (p, ∞)-atoms. Let $\eta(x) = \exp(-|x|^2)$, $x \in \mathbb{R}^n$.

1. Show that X^p and $\{{\rm e}^{im\cdot\star}\eta\}_{m \in \mathbb{Z}^n}$ is linearly independent.
2. Using the fact that every linear space has a basis algebraically, show that there exists a subset $\mathscr{A}$ of $H^p(\mathbb{R}^n)$ such that any $f \in H^p(\mathbb{R}^n)$ admits the finite decomposition:

$$f = \sum_{g \in \mathscr{A}} c_g \cdot g + \sum_{l \in \mathbb{Z}^n} a_l {\rm e}^{il\cdot\star} \eta + h,$$

where $\sharp\{c_g \ne 0\} + \sharp\{l \in \mathbb{Z}^n : a_l \ne 0\} < \infty$ and $h \in X^p(\mathbb{R}^n)$.

3. Show that the mapping

$$T : \sum_{g \in \mathscr{A}} c_g \cdot g + \sum_{l \in \mathbb{Z}^n} a_l {\rm e}^{il\cdot\star} \eta + h \to \sum_{g \in \mathscr{A}} c_g \cdot g + \sum_{l \in \mathbb{Z}^n} |l| a_l {\rm e}^{il\cdot\star} \eta + h$$

is an unbounded linear mapping on $H^p(\mathbb{R}^n)$ such that T is bounded if we restrict it to $X^p(\mathbb{R}^n)$.

3.2.4 Hilbert Space $\mathcal{H}_j$

Let $0 < p \le 1$. Furthermore, let $f \in H^p(\mathbb{R}^n)$ in the sense of Definition 3.3. Let us prove the converse of Proposition 3.3; we formulate the result fully in Theorem 3.9. When the functions are complex-valued, we decompose them into the real part and the imaginary part. Here and below, the functions are assumed to be real-valued.

3.2.4.1 A Setup

For $\lambda > 0$, define an open set Ω by $\Omega \equiv \{\mathcal{M}f > \lambda\}$. Let us consider the Whitney decomposition of Ω (Theorem 1.9). Let $\{Q_j\}_{j=1}^\infty$ be a sequence of cubes satisfying (1.17) and (1.19). Let $\psi \in \mathcal{S}(\mathbb{R}^n)$ satisfy $\chi_{Q(10)} \le \psi \le \chi_{Q(100)}$. We define

$$\psi^{(j)} \equiv \psi\left(\frac{\star - c(Q_j)}{\ell(Q_j)}\right), \quad \varphi^{(j)} \equiv \frac{\psi^{(j)}}{\chi_{\mathbb{R}^n\setminus\Omega} + \sum_{k=1}^\infty \psi^{(k)}}, \quad I^{(j)} \equiv \|\varphi^{(j)}\|_1 \tag{3.56}$$

for each $j \in \mathbb{N}$ by keeping this setup in mind. Since $\mathrm{supp}(\psi^{(j)}) \subset \Omega$, we have $\varphi^{(j)} \in C_c^\infty(\mathbb{R}^n)$. We remark that

$$I^{(j)} \sim \ell(Q_j)^n. \tag{3.57}$$

We have the following quantitative information:

Lemma 3.5 *The function $\varphi^{(j)}$ satisfies the differential inequality:*

$$|\partial^\alpha \varphi^{(j)}(x)| \lesssim_\alpha \ell(Q_j)^{-|\alpha|}\chi_{100Q_j}(x) \quad (x \in \mathbb{R}^n) \tag{3.58}$$

for each $j \in \mathbb{N}$ and $\alpha \in {\mathbb{N}_0}^n$.

Lemma 3.5 shows that $\psi^{(j)}$ and $\varphi^{(j)}$ behave alike.

Proof Simply use the differential inequalities below:

$$\sum_{k=1}^\infty \psi^{(k)}(x) \gtrsim \chi_\Omega(x), \quad \sum_{k=1}^\infty |\partial^\alpha \psi^{(k)}(x)| \lesssim_\alpha \mathrm{dist}(x, \mathbb{R}^n \setminus \Omega)^{-|\alpha|} \quad (x \in \mathbb{R}^n).$$

Take $L \in \mathbb{N}$ so that $L \ge L_0 \equiv [\sigma_p]$ here and below. Next, we consider a "projection" of $f \in \mathcal{S}'(\mathbb{R}^n)$. The function $p^{(j)} \in \mathcal{P}_{L+n}(\mathbb{R}^n)$ will play its role.

Lemma 3.6 *Let $f \in \mathcal{S}'(\mathbb{R}^n)$ be such that $\mathcal{M}f$ is finite almost everywhere. Then there uniquely exists a polynomial $p^{(j)} \in \mathcal{P}_{L+n}(\mathbb{R}^n)$ such that*

$$\langle f - p^{(j)}, \varphi^{(j)}q\rangle = 0 \tag{3.59}$$

for all $q \in \mathscr{P}_{L+n}(\mathbb{R}^n)$.

Proof This is a matter of linear algebra. We express

$$p^{(j)}(x) \equiv \sum_{\alpha \in \mathbb{N}_0{}^n, |\alpha| \le L+n} \lambda_\alpha \, x^\alpha$$

via coefficients λ_α. We write out condition (3.59) in full. Then (3.59) holds if and only if

$$\sum_{\alpha \in \mathbb{N}_0{}^n, |\alpha| \le L+n} \lambda_\alpha \int_{\mathbb{R}^n} x^{\alpha+\beta} \varphi^{(j)}(x) \mathrm{d}x = \langle f, \varphi^{(j)} \cdot x^\beta \rangle \tag{3.60}$$

for all multi-indexes β with length less than $L+n$.

To show that (3.60) has a unique solution, we consider the symmetric matrix $A = \{a_{\alpha,\beta}\}_{|\alpha|,|\beta| \le L+n}$ whose entry is

$$a_{\alpha,\beta} \equiv \int_{\mathbb{R}^n} x^{\alpha+\beta} \varphi^{(j)}(x) \mathrm{d}x, \quad |\alpha|, |\beta| \le L+n.$$

In fact, this matrix is positive definite.

Let $\{k_\alpha\}_{\alpha \in \mathbb{N}_0{}^n, |\alpha| \le L+n} \subset \mathbb{R}$ be a real vector. Arithmetic shows that

$$\sum_{\alpha,\beta \in \mathbb{N}_0{}^n, |\alpha|,|\beta| \le L+n} a_{\alpha,\beta} k_\alpha k_\beta = \int_{\mathbb{R}^n} \left| \sum_{\alpha \in \mathbb{N}_0{}^n, |\alpha| \le L+n} k_\alpha x^\alpha \right|^2 \varphi^{(j)}(x) \mathrm{d}x.$$

This quadratic form vanishes if and only if all coefficients k_α are 0. Hence it follows that A is nondegenerate.

3.2.4.2 Hilbert Space $\mathscr{H}_j$

Denote by $\mathscr{H}_j$ the Hilbert space of measurable functions on $100Q_j$ such that

$$\|g\|_{\mathscr{H}_j} \equiv \sqrt{\frac{1}{I^{(j)}} \int_{100Q_j} |g(x)|^2 \varphi^{(j)}(x) \mathrm{d}x} \tag{3.61}$$

is finite. The space $\mathscr{H}_{j,L}$ is defined to be a closed subspace $\mathscr{H}_j \cap \mathscr{P}_{L+n}(\mathbb{R}^n)$. By the definition of the inner product (3.61), we have

$$\langle g_1, g_2 \rangle_{\mathscr{H}_j} = \frac{1}{I^{(j)}} \int_{100Q_j} g_1(x) g_2(x) \varphi^{(j)}(x) \mathrm{d}x. \tag{3.62}$$

We now invoke a fundamental fact on functional analysis that any linear mapping from a normed space V_1 to another normed space V_2 is bounded as long as the dimension of V_1 is finite. We use this fact, as follows.

Lemma 3.7 *Fix a multi-index $\alpha \in \mathbb{N}_0{}^n$. Then*

$$\|\partial^\alpha q\|_{\mathscr{B}^0(100Q_j)} \lesssim_{\alpha,L} \ell(Q_j)^{-|\alpha|}\|q\|_{\mathscr{H}_j}$$

for all $q \in \mathscr{H}_{j,L}$.

Proof We obtain an equivalent norm, if we replace $\varphi^{(j)}$ in (3.61) with $\psi^{(j)}$. Since $\mathscr{H}_{j,L}$ is a finite-dimensional Hilbert space, a scaling together with the differential inequality allows us to assume $\ell(Q) = 1$. In this case the conclusion holds trivially.

Proposition 3.5 *Let $f \in \mathscr{S}'(\mathbb{R}^n)$ be such that $\mathscr{M}f$ is finite almost everywhere. Using a finite index set K_j, we express a complete orthonormal system $\{e^{(j,k)}\}_{k\in K_j}$ of $\mathscr{H}_{j,L}$. Then $e^{(j,k)}\varphi^{(j)} \in \mathscr{S}(\mathbb{R}^n)$ and*

$$p^{(j)} = \sum_{k\in K_j} \frac{\langle f, e^{(j,k)}\varphi^{(j)}\rangle}{I^{(j)}} \cdot e^{(j,k)}. \tag{3.63}$$

Proof Since $\varphi^{(j)}$ is compactly supported, $e^{(j,k)}\varphi^{(j)} \in \mathscr{S}(\mathbb{R}^n)$. We write $d^{(j)}$ for the function in the right-hand side. We compare this with the conditions that $p^{(j)}$ satisfies. Let $q \in \mathscr{P}_{L+n}(\mathbb{R}^n)$.

Since $\langle f, g\rangle_{\mathscr{H}} = \dfrac{1}{I^{(j)}}\langle f, g\rangle$ for $f \in \mathscr{H}_j$ and $g \in \mathscr{S}(\mathbb{R}^n)$, we have

$$\begin{aligned}
\langle f - d^{(j)}, \varphi^{(j)}q\rangle &= \langle f, \varphi^{(j)}q\rangle - \sum_{k\in K_j} \frac{\langle f, e^{(j,k)}\varphi^{(j)}\rangle \cdot \langle e^{(j,k)}, \varphi^{(j)}q\rangle}{I^{(j)}} \\
&= \langle f, \varphi^{(j)}q\rangle - \left\langle f, \sum_{k\in K_j} \frac{\langle e^{(j,k)}, \varphi^{(j)}q\rangle}{I^{(j)}} e^{(j,k)}\varphi^{(j)}\right\rangle \\
&= \langle f, \varphi^{(j)}q\rangle - \left\langle f, \left(\sum_{k\in K_j} \langle e^{(j,k)}, q\rangle_{\mathscr{H}_j} \cdot e^{(j,k)}\right)\varphi^{(j)}\right\rangle \\
&= \langle f, \varphi^{(j)}q\rangle - \langle f, q\varphi^{(j)}\rangle = 0,
\end{aligned}$$

as required.

From the next lemma, we see that $p^{(j)}\varphi^{(j)}$ behaves like $\lambda\chi_{100Q_j}$.

Lemma 3.8 *Let $f \in \mathscr{S}'(\mathbb{R}^n)$ be such that $\mathscr{M}f$ is finite almost everywhere. Then*

$$|p^{(j)}(x)\,\varphi^{(j)}(x)| \lesssim \lambda\chi_{100Q_j}(x)$$

for all j *and* $x \in \mathbb{R}^n$.

Proof Since K_j is a finite set, it suffices to show

$$\left| \frac{\langle f, e^{(j,k)} \varphi^{(j)} \rangle}{I^{(j)}} \cdot e^{(j,k)}(x) \varphi^{(j)}(x) \right| \lesssim \lambda \tag{3.64}$$

for each $k \in K_j$ and $x \in 100Q_j$.

By Lemma 3.7, we have

$$|\partial^\alpha e^{(j,k)}(x)| \lesssim_\alpha \ell(Q_j)^{-|\alpha|} \tag{3.65}$$

for $\alpha \in \mathbb{N}_0{}^n$.

By Lemma 3.5 and Proposition 3.5 we can control $|e^{(j,k)}(x)\varphi^{(j)}(x)|$:

$$|e^{(j,k)}(x)\varphi^{(j)}(x)| \lesssim 1. \tag{3.66}$$

Now we control $\langle f, p^{(j)}\varphi^{(j)} \rangle$. Firstly, we have

$$\mathcal{M} f(z_j) \lesssim \lambda. \tag{3.67}$$

Keeping in mind that $\langle f, p^{(j)} \cdot \varphi^{(j)} \rangle = f * [\, p^{(j)}(z_j - \star)\varphi^{(j)}(z_j - \star)\,](z_j)$, we set

$$\tau^{(j)} \equiv p^{(j)}(z_j - \ell(Q_j)\star)\varphi^{(j)}(z_j - \ell(Q_j)\star).$$

By virtue of Lemma 3.5 and (3.65), we have

$$\begin{aligned}
p_N(\tau^{(j)}) &= \sum_{\substack{\alpha \in \mathbb{N}_0{}^n \\ |\alpha| \le N}} \sup_{x \in \mathbb{R}^n} \langle x \rangle^N |\, \partial_x^\alpha [p^{(j)}(z_j - \ell(Q_j)x)\varphi^{(j)}(z_j - \ell(Q_j)x)\,]\,| \\
&= \sum_{\substack{\alpha \in \mathbb{N}_0{}^n \\ |\alpha| \le N}} \sup_{x \in B\left(\frac{z_j - c(Q_j)}{\ell(Q_j)}, 1\right)} \langle x \rangle^N |\, \partial_x^\alpha [p^{(j)}(z_j - \ell(Q_j)x)\varphi^{(j)}(z_j - \ell(Q_j)x)]| \\
&= \sum_{\substack{\alpha \in \mathbb{N}_0{}^n \\ |\alpha| \le N}} \sup_{1 \le |x| \le 2000} \langle x \rangle^N |\, \partial_x^\alpha [p^{(j)}(z_j - \ell(Q_j)x)\varphi^{(j)}(z_j - \ell(Q_j)x)]| \lesssim 1.
\end{aligned}$$

Hence

$$|\langle f, p^{(j)} \cdot \varphi^{(j)} \rangle| = \left| \left\langle f, \tau^{(j)}\left(\frac{\star}{\ell(Q_j)}\right) \right\rangle \right| \lesssim I^{(j)} \mathcal{M} f(z_j) \lesssim \lambda\, I^{(j)}. \tag{3.68}$$

If we combine (3.66) and (3.68), we obtain the desired result.

In this section, we present some examples of the elements in $\mathscr{F}_N$. Here we assume that

$$\varphi \in \mathscr{F}_N \tag{3.69}$$

and that $\tilde{\varphi} \in C_c^\infty(\mathbb{R}^n)$ satisfies

$$\chi_{Q(100)} \le \tilde{\varphi} \le \chi_{Q(200)}. \tag{3.70}$$

Define

$$\tilde{\varphi}^{(j)} \equiv \tilde{\varphi}\left(\frac{\star - c(Q_j)}{\ell(Q_j)}\right). \tag{3.71}$$

Let $f \in \mathscr{S}'(\mathbb{R}^n)$ be such that $\mathscr{M} f$ is finite almost everywhere in this section, so that we have the desired Whitney covering.

Example 3.2 Let $f \in \mathscr{S}'(\mathbb{R}^n)$. Thanks to Lemma 3.7, we have

$$W \equiv \sup_{j\in\mathbb{N}} p_N([e^{(j,k)}(x_0 - \ell(Q_j)\star)\varphi^{(j)}(x_0 - \ell(Q_j)\star)]) < \infty. \tag{3.72}$$

Thus, defining Ψ_j by

$$\Psi_j = \frac{1}{W} e^{(j,k)}(x_0 - \ell(Q_j)\star)\varphi^{(j)}(x_0 - \ell(Q_j)\star), \tag{3.73}$$

we obtain $p_n(\Psi_j) \lesssim 1$.

Example 3.3 Let $x_0 \in 10^6 Q_j \setminus 10^4 Q_j$ and $0 < t \le \ell(Q_j)$. Define

$$\Psi_{j,t}(x) = \varphi^{(j)}(z_j - tx)\varphi\left(\frac{x_0 - z_j}{t} + x\right) \quad (x \in \mathbb{R}^n). \tag{3.74}$$

Then from (1.21), we learn that $\Psi_{j,t}(x) = 0$ unless $z_j - tx \in 50Q_j$. Since $z_j \in 10000Q_j$, we have only to consider the cases

$$|x| \le \frac{20000n\ell(Q_j)}{t}, \text{ and } |x_0 - z_j + tx| \ge 10^3\ell(Q_j).$$

Thus, for all α with $|\alpha| \le N$, we obtain

$$\begin{aligned}
\langle x\rangle^N \left|\frac{\partial^\alpha \varphi}{\partial x^\alpha}\left(\frac{x_0 - z_j}{t} + x\right)\right| &\le \langle x\rangle^N \left(1 + \frac{|x_0 - z_j + tx|}{t}\right)^{-N} \\
&\le \left(1 + \frac{20000n\ell(Q_j)}{t}\right)^N \left(1 + \frac{1000\ell(Q_j)}{t}\right)^{-N} \\
&\le (20n)^N.
\end{aligned}$$

Thus, by Lemma 3.8 and the Leibniz rule we obtain

$$p_N(\Psi_{j,t}) \lesssim_N \sum_{|\beta|\le N} t^{|\beta|} \sup_{v\in\mathbb{R}^n} \left|\frac{\partial^\beta \varphi^{(j)}}{\partial x^\beta}(v)\right| \lesssim 1.$$

Example 3.4 Let $x_0 \in 10^6 Q_j \setminus 10^4 Q_j$ and $t > \ell(Q_j)$. Define

$$\Psi_{j,t} \equiv \frac{\ell(Q_j)^n}{t^n}\varphi^{(j)}(z_j - \ell(Q_j)\star)\varphi\left(\frac{x_0 - z_j}{t} + \frac{\ell(Q_j)}{t}\star\right). \tag{3.75}$$

Note that $\mathrm{supp}(\Psi_{j,t}) \subset Q(10050)$. Thus, by Lemma 3.8 and the Leibniz rule we obtain $p_N(\Psi_{j,t}) \lesssim_N 1$.

Example 3.5 We suppose $N \ge n + L + 1$ here. Let $x_0 \notin 10^6 Q_j$ and $t \ge \ell(Q_j)$. Denote by $P(y)$ the Taylor polynomial of order $n + L$ of the function $y \mapsto \varphi\left(\frac{x_0 - y}{t}\right)$ around the point $y = c(Q_j)$. Define:

$$\Psi_j \equiv \varphi^{(j)}(z_j - \ell(Q_j)\star)\left(\varphi\left(\frac{x_0 - z_j}{t} + \frac{\ell(Q_j)}{t}\star\right) - P(z_j - \ell(Q_j)\star)\right), \tag{3.76}$$

$$\tilde{\Psi}_j \equiv \tilde{\varphi}^{(j)}(z_j - \ell(Q_j)\star)\left(\varphi\left(\frac{x_0 - z_j}{t} + \frac{\ell(Q_j)}{t}\star\right) - P(z_j - \ell(Q_j)\star)\right),$$

where $\varphi^{(j)}$ is given by (3.71). Note that

$$p_N(\Psi_j), p_N(\tilde{\Psi}_j) \lesssim \frac{\ell(Q_j)^{n+L+1}}{|x_0 - c(Q_j)|^{n+L+1}}. \tag{3.77}$$

Observe first that $\mathrm{supp}(\Psi_j) \cup \mathrm{supp}(\tilde{\Psi}_j) \subset Q(10200)$. When $|z| \le 10200n$ and $|\alpha| \le n + L$, we have

$$\begin{aligned}&\left|\frac{\partial^\alpha}{\partial z^\alpha}\left(\varphi\left(\frac{x_0 - z_j + \ell(Q_j)z}{t}\right) - P(z_j - \ell(Q_j)z)\right)\right| \\ &\lesssim p_N(\varphi) \sup_{w\in B(20200n)} \left(1 + \frac{|x_0 - z_j + \ell(Q_j)w|}{t}\right)^{-n-L-1} \frac{\ell(Q_j)^{n+L+1}}{t^{n+L+1}}.\end{aligned}$$

Since $t \le \ell(Q_j)$,

$$\begin{aligned}&\left|\frac{\partial^\alpha}{\partial z^\alpha}\left(\varphi\left(\frac{x_0 - z_j + \ell(Q_j)z}{t}\right) - P(z_j - \ell(Q_j)z)\right)\right| \\ &\lesssim p_N(\varphi)\left(1 + \frac{|x_0 - c(Q_j)|}{t}\right)^{-n-L-1} \frac{\ell(Q_j)^{n+L+1}}{t^{n+L+1}}\end{aligned}$$

$$\lesssim \left(1+\frac{|x_0-c(Q_j)|}{t}\right)^{-n-L-1}\frac{\ell(Q_j)^{n+L+1}}{t^{n+L+1}}$$
$$\lesssim \frac{\ell(Q_j)^{n+L+1}}{|x_0-c(Q_j)|^{n+L+1}}$$

thanks to conditions $N \geq n+L+1$, (1.21) and (3.70). When $n+L+1 \leq |\alpha| \leq N$, we use $\dfrac{\partial^\alpha}{\partial z^\alpha}P(z_j-\ell(Q_j)z)=0$ to obtain

$$\left|\frac{\partial^\alpha}{\partial z^\alpha}\left(\varphi\left(\frac{x_0-z_j+\ell(Q_j)z}{t}\right)-P(z_j-\ell(Q_j)z)\right)\right|$$
$$\lesssim p_N(\varphi)\sup_{w\in B(20200n)}\left(1+\frac{|x_0-w|}{t}\right)^{-n-L-1}\frac{\ell(Q_j)^{|\alpha|}}{t^{|\alpha|}}$$
$$\lesssim p_N(\varphi)\left(1+\frac{|x_0-c(Q_j)|}{t}\right)^{-n-L-1}\frac{\ell(Q_j)^{n+L+1}}{t^{n+L+1}}$$
$$\lesssim \frac{\ell(Q_j)^{n+L+1}}{|x_0-c(Q_j)|^{n+L+1}}.$$

Thus, as long as $|\alpha| \leq N$, we obtain

$$\left|\frac{\partial^\alpha}{\partial z^\alpha}\left(\varphi\left(\frac{x_0-z_j+\ell(Q_j)z}{t}\right)-P(z_j-\ell(Q_j)z)\right)\right|\lesssim \frac{\ell(Q_j)^{n+L+1}}{|x_0-c(Q_j)|^{n+L+1}}.$$

From (3.59), we conclude (3.77).

Example 3.6 Let $f \in \mathscr{S}'(\mathbb{R}^n)$ be such that $\mathscr{M}f$ is finite almost everywhere. Let $x_0 \notin 10^6 Q_j$ and $0 < t \leq \ell(Q_j)$. Define

$$\Psi_{j,t} \equiv \varphi\left(\frac{x_0-z_j}{t}+\star\right)\varphi^{(j)}(z_j-t\star), \tag{3.78}$$
$$\tilde{\Psi}_{j,t} \equiv \varphi\left(\frac{x_0-z_j}{t}+\star\right)\tilde{\varphi}^{(j)}(z_j-t\star).$$

Let us prove

$$p_N(\Psi_j) \lesssim \frac{\ell(Q_j)^N}{|x_0-c(Q_j)|^N} \tag{3.79}$$

and

$$p_N(\tilde{\Psi}_j) \lesssim \frac{\ell(Q_j)^N}{|x_0 - c(Q_j)|^N}. \tag{3.80}$$

If $\Psi_j(y) \neq 0$, then $z_j - ty \in 200Q_j$. So, we obtain

$$|y| \leq \frac{20000n\ell(Q_j)}{t}, \quad |x_0 - c(Q_j)| \gtrsim |z_j - ty + c(Q_j)|.$$

Since $0 < t \leq \ell(Q_j)$, we obtain

$$\begin{aligned}
p_N(\Psi_j) &= \sum_{|\alpha| \leq N} \sup_{y \in \mathbb{R}^n} (1 + |y|)^N |\partial^\alpha \Psi_j(y)| \\
&\lesssim p_N(\varphi) \left(\frac{\ell(Q_j)}{t} \right)^N \left| \frac{x_0 - z_j}{t} + y \right|^{-N} \\
&= p_N(\varphi) \ell(Q_j)^N |x_0 - z_j + ty|^{-N} \\
&\lesssim p_N(\varphi) \ell(Q_j)^N |x_0 - c(Q_j)|^{-N} \\
&\lesssim \frac{\ell(Q_j)^N}{|x_0 - c(Q_j)|^N}.
\end{aligned}$$

Thus, we are led to (3.79). Likewise we can prove (3.80).

We conclude this section with a property of the polynomial $p^{(j)}$.

Proposition 3.6 *Let $f \in \mathscr{S}'(\mathbb{R}^n)$ be such that $\mathscr{M} f$ is finite almost everywhere. Let $N \geq n + L + 1$ and $x_0, x_1 \in 10000Q_j$. Then*

$$\mathscr{M}[e^{(j,k)} \varphi^{(j)}](x_0) \lesssim 1, \tag{3.81}$$

$$|\langle f, e^{(j,k)} \varphi^{(j)} \rangle| \lesssim \ell(Q_j)^n \mathscr{M} f(x_0) \tag{3.82}$$

and

$$\mathscr{M}[p^{(j)} \varphi^{(j)}](x_1) \lesssim \mathscr{M} f(x_0). \tag{3.83}$$

Proof Define Ψ_j by (3.73) and W by (3.72). Then

$$|\langle f, e^{(j,k)} \varphi^{(j)} \rangle| = W \ell(Q_j)^n \left| f * \left(\frac{1}{\ell(Q_j)^n} \Psi_j \left(\frac{\star}{\ell(Q_j)} \right) \right) (x_0) \right| \leq W \ell(Q_j)^n \mathscr{M} f(x_0),$$

which proves (3.82). By the use of Lemma 3.7 for $q = e^{(j,k)}, \alpha = 0$, we have $W' \equiv \sup_{j \in \mathbb{N}} \|e^{(j,k)} \varphi^{(j)}\|_\infty < \infty$. Thus,

$$\mathcal{M}[e^{(j,k)}\varphi^{(j)}](x_0) \lesssim W' \sup_{t>0} t^{-n} \sup_{\kappa \in \mathcal{F}_N} \int_{\mathbb{R}^n} \left|\kappa\left(\frac{w}{t}\right)\right| dw \lesssim 1,$$

which proves (3.81). As for (3.83), we use (3.37), Proposition 3.5 and

$$\mathcal{M}[e^{(j,k)}\varphi^{(j)}](x_1) \lesssim 1, \quad |\langle f, e^{(j,k)}\varphi^{(j)}\rangle| \lesssim \ell(Q_j)^n \mathcal{M} f(x_0).$$

By these inequalities and (3.72), we have

$$\begin{aligned}\mathcal{M}[p^{(j)}\varphi^{(j)}](x_1) &\lesssim \sum_{k\in K_j} \frac{|\langle f, e^{(j,k)}\varphi^{(j)}\rangle|}{I^{(j)}} \mathcal{M}[e^{(j,k)}\varphi^{(j)}](x_1) \\ &\lesssim \sum_{k\in K_j} \frac{|\langle f, e^{(j,k)}\varphi^{(j)}\rangle|}{I^{(j)}} \lesssim \mathcal{M} f(x_0),\end{aligned}$$

which proves (3.83).

Exercises

Exercise 3.27 Let $\mathcal{H}_{j,L}$ be a closed subspace $\mathcal{H}_j \cap \mathcal{P}_{L+n}(\mathbb{R}^n)$. Find $\dim(\mathcal{H}_{j,L}) = \sharp K_j$ by counting the number of the solutions $\alpha_1 + \alpha_2 + \cdots + \alpha_n \le L + n$, where $\alpha_1, \alpha_2, \ldots, \alpha_n$ are nonnegative integers.

Exercise 3.28 Let $f \in \mathcal{S}'(\mathbb{R}^n)$ be such that $\mathcal{M} f$ is finite almost everywhere.

1. Using the polarization formula, prove (3.62).
2. Show that we will obtain an equivalent inner product, if we replace $\varphi^{(j)}$ with $\psi^{(j)}$ in the definition of $\langle \star_1, \star_2 \rangle_{\mathcal{H}_j}$.

3.2.5 *Calderón–Zygmund Decomposition for Distributions*

We consider the Calderón–Zygmund decomposition for $f \in \mathcal{S}'(\mathbb{R}^n)$ be such that $\mathcal{M} f$ is finite almost everywhere.

3.2.5.1 Good Part and Bad Part

Let $f \in \mathcal{S}'(\mathbb{R}^n)$. Let us define

$$b^{(j)} \equiv (f - p^{(j)})\varphi^{(j)}, \quad j \in \mathbb{N}, \quad g \equiv f - \sum_{j=1}^{\infty} b^{(j)}. \tag{3.84}$$

This is the Calderón–Zygmund decomposition for distributions. In this Calderón–Zygmund decomposition, $b^{(j)}$ is the bad part and g is the good part.

We are oriented to the pointwise estimates of their grand maximal function. So we fix $x \in \mathbb{R}^n$ and denote it by x_0. Begin with the estimate for the bad part $\mathscr{M}b^{(j)}(x)$.

The first lemma is an auxiliary one.

Lemma 3.9 *Let $j \in \mathbb{N}$.*

1. *For $x_0 \in 100Q_j$, $\mathscr{M}[\varphi^{(j)} f](x_0) \lesssim \mathscr{M}f(x_0)$.*
2. *For $x_0 \in 100Q_j$, $\mathscr{M}[\varphi^{(j)} p^{(j)}](x_0) \lesssim \mathscr{M}f(x_0)$.*

Proof

1. By the definition of the grand maximal operator $\mathscr{M}$, we have

$$\begin{aligned}\mathscr{M}[\varphi^{(j)} f](x_0) &= \sup\left\{\left|\left\langle \varphi^{(j)} f, \frac{1}{t^n}\kappa\left(\frac{x_0 - \star}{t}\right)\right\rangle\right| : \kappa \in \mathscr{F}_N, t > 0\right\} \\ &= \sup\left\{\left|\left\langle f, \varphi^{(j)} \cdot \frac{1}{t^n}\kappa\left(\frac{x_0 - \star}{t}\right)\right\rangle\right| : \kappa \in \mathscr{F}_N, t > 0\right\}.\end{aligned}$$

Let us fix $\kappa \in \mathscr{F}_N$ and $t > 0$.

First of all, let $t \le \ell(Q_j)$. Then consider $\Phi_{j,t} \equiv \varphi^{(j)}(x_0 - t\star)\kappa$. By Lemma 3.7, we have

$$\sup\{p_N(\Phi_{j,t}) : j \in \mathbb{N}, \quad 0 < t \le \ell(Q_j)\} < \infty.$$

Hence it follows that

$$\begin{aligned}\left|\left\langle f, \varphi^{(j)} \cdot \frac{1}{t^n}\kappa\left(\frac{x_0 - \star}{t}\right)\right\rangle\right| &= \left|\left\langle f, \frac{1}{t^n}\Phi_{j,t}\left(\frac{x_0 - \star}{t}\right)\right\rangle\right| \\ &= \left|f * \left[\frac{1}{t^n}\Phi_{j,t}\left(\frac{\star}{t}\right)\right](x_0)\right| \\ &\lesssim \mathscr{M}f(x_0).\end{aligned} \tag{3.85}$$

Let $t > \ell(Q_j)$. Consider a function

$$\Psi_{j,t} \equiv \frac{\ell(Q_j)^n}{t^n}\varphi^{(j)}(x_0 - \ell(Q_j)\star)\kappa\left(\frac{\ell(Q_j)}{t}\star\right),$$

supported on $Q(100)$. In view of the size of the supports and the assumption $t > \ell(Q_j)$, we obtain $\sup\limits_{j \in \mathbb{N}} p_N(\Psi_{j,t}) < \infty$. Hence

$$
\begin{aligned}
\left|\left\langle f, \varphi^{(j)} \cdot \frac{1}{t^n}\kappa\left(\frac{x_0-\star}{t}\right)\right\rangle\right| &= \left|\left\langle f, \frac{1}{\ell(Q_j)^n}\Psi_{j,t}\left(\frac{x_0-\star}{\ell(Q_j)}\right)\right\rangle\right| \\
&= \left| f * \left[\frac{1}{\ell(Q_j)^n}\Psi_{j,t}\left(\frac{\star}{\ell(Q_j)}\right)\right](x_0)\right| \\
&\lesssim \mathcal{M} f(x_0). \qquad (3.86)
\end{aligned}
$$

Thus, we obtain the desired result.

2. By Proposition 3.5, we have

$$
\mathcal{M}[\, p^{(j)}\varphi^{(j)}\,](x_0) \le \frac{1}{I^{(j)}} \sum_{k\in K_j} |\langle f, e^{(j,k)}\varphi^{(j)}\rangle| \cdot \mathcal{M}[\, e^{(j,k)}\varphi^{(j)}\,](x_0). \qquad (3.87)
$$

Lemma 3.7 yields $\sup_{j\in\mathbb{N}} p_N[\, e^{(j,k)}(x_0-\ell(Q_j)\star)\cdot\varphi^{(j)}(x_0-\ell(Q_j)\star)\,] < \infty$. By the definition of the grand maximal operator $\mathcal{M}$, we have

$$
\mathcal{M}[e^{(j,k)}\varphi^{(j)}](x_0) \lesssim \ell(Q_j)^n, \quad |\langle f, e^{(j,k)}\varphi^{(j)}\rangle| \lesssim \mathcal{M} f(x_0). \qquad (3.88)
$$

Hence we conclude $\mathcal{M}[p^{(j)}\varphi^{(j)}](x_0) \lesssim \mathcal{M} f(x_0)$ from (3.87) and (3.88). Thus, *2* is proved.

The estimate for the bad part is the following:

Proposition 3.7 *Let $N \ge n+L+1$ and $x_0 \in \mathbb{R}^n$. Then*

$$
\mathcal{M} b^{(j)}(x_0) \lesssim \mathcal{M} f(x_0)\chi_{1000Q_j}(x_0) + \frac{\lambda \ell(Q_j)^{n+L+1}}{\ell(Q_j)^{n+L+1} + |x_0 - c(Q_j)|^{n+L+1}}.
$$

Proof We already handled the case where $x_0 \in 1000Q_j$ in Lemma 3.9. We can suppose $x_0 \in \mathbb{R}^n \setminus 1000Q_j$. Since $\mathcal{M} f(z_j) \le \lambda$ thanks to the definition of Ω, to prove Proposition 3.7, we have only to establish

$$
\mathcal{M} b^{(j)}(x_0) \lesssim \frac{\lambda \ell(Q_j)^{n+L+1}}{\ell(Q_j)^{n+L+1} + |x_0 - c(Q_j)|^{n+L+1}},
$$

or equivalently,

$$
\begin{aligned}
\frac{1}{t^n}\left| b_j * \varphi\left(\frac{\star}{t}\right)(x_0)\right| &= \frac{1}{t^n}\left|\left\langle f - p^{(j)}, \varphi^{(j)}\varphi\left(\frac{x_0-\star}{t}\right)\right\rangle\right| \\
&\lesssim \frac{\ell(Q_j)^{n+L+1}\mathcal{M} f(z_j)}{\ell(Q_j)^{n+L+1} + |x_0 - c(Q_j)|^{n+L+1}}
\end{aligned}
$$

with the implicit constant above independent of $\varphi \in \mathscr{F}_N$ and $t > 0$. We have to consider four cases.

1. First suppose $x_0 \in 10^6 Q_j \setminus 10^4 Q_j$ and $0 < t \le \ell(Q_j)$. Define $\Psi_{j,t}$ by (3.74). Then

$$\Psi_{j,t}\left(\frac{z_j - y}{t}\right) = \varphi^{(j)}(y)\varphi\left(\frac{x_0 - y}{t}\right) \quad (y \in \mathbb{R}^n);$$

hence

$$\frac{1}{t^n}\left|b_j * \varphi\left(\frac{\star}{t}\right)(x_0)\right| \le \frac{1}{t^n}\left|\Psi_{j,t}\left(\frac{\star}{t}\right) * f(z_j)\right| + \frac{1}{t^n}\left|\varphi\left(\frac{\star}{t}\right) * (p^{(j)}\varphi^{(j)})(x_0)\right|.$$

Since $p_N(\Psi_{j,t}) \lesssim 1$, we have

$$\frac{1}{t^n}\left|b_j * \varphi\left(\frac{\star}{t}\right)(x_0)\right| \lesssim \mathscr{M}f(z_j) + \mathscr{M}[p^{(j)}\varphi^{(j)}](x_0) \lesssim \mathscr{M}f(z_j).$$

2. Let $x_0 \in 10^6 Q_j \setminus 10^4 Q_j$ and $\ell(Q_j) \le t < \infty$. Define $\Psi_{j,t}$ by (3.75). Then

$$\Psi_{j,t}\left(\frac{z_j - y}{\ell(Q_j)}\right) = \frac{\ell(Q_j)^n}{t^n}\varphi^{(j)}(y)\varphi\left(\frac{x_0 - y}{t}\right) \quad (y \in \mathbb{R}^n).$$

Since $p_N(\Psi_{j,t}) \lesssim 1$, we obtain

$$\frac{1}{t^n}\left|b_j * \varphi\left(\frac{\star}{t}\right)(x_0)\right| \lesssim \left|\frac{1}{\ell(Q_j)^n}\left\langle \Psi_{j,t}\left(\frac{z_j - \star}{\ell(Q_j)}\right), f\right\rangle\right| + \mathscr{M}f(z_j) \lesssim \mathscr{M}f(z_j).$$

3. Let $x_0 \notin 10^6 Q_j$ and $t \ge \ell(Q_j)$. Define $\Psi_{j,t}, \tilde{\Psi}_{j,t}$ by (3.76). Then

$$\Psi_{j,t}\left(\frac{z_j - y}{\ell(Q_j)}\right) = \tilde{\varphi}^{(j)}(y)\left(\varphi\left(\frac{x_0 - y}{t}\right) - P(y)\right) \quad (y \in \mathbb{R}^n),$$

$$\tilde{\Psi}_{j,t}\left(\frac{z_j - y}{\ell(Q_j)}\right) = \tilde{\varphi}^{(j)}(y)\left(\varphi\left(\frac{x_0 - y}{t}\right) - P(y)\right) \quad (y \in \mathbb{R}^n);$$

hence

$$\begin{aligned}
&\frac{1}{t^n}\left|b_j * \varphi\left(\frac{\star}{t}\right)(x_0)\right| \\
&= \frac{\ell(Q_j)^n}{t^n}\left|\left\langle (f - p^{(j)})\varphi^{(j)}, \frac{1}{\ell(Q_j)^n}\tilde{\Psi}_{j,t}\left(\frac{z_j - \star}{\ell(Q_j)}\right)\right\rangle\right| \\
&\le \left|\left\langle (f - p^{(j)})\varphi^{(j)}, \frac{1}{\ell(Q_j)^n}\tilde{\Psi}_{j,t}\left(\frac{z_j - \star}{\ell(Q_j)}\right)\right\rangle\right|
\end{aligned}$$

$$\leq \left|\left\langle (f, \frac{1}{\ell(Q_j)^n}\varphi^{(j)}\tilde{\Psi}_{j,t}\left(\frac{z_j - \star}{\ell(Q_j)}\right)\right\rangle\right|$$
$$+ \left|[p^{(j)}\varphi^{(j)}] * \left[\frac{1}{\ell(Q_j)^n}\Psi_{j,t}\left(\frac{\star}{\ell(Q_j)}\right)\right](z_j)\right|$$
$$\leq \left|f * \left[\frac{1}{\ell(Q_j)^n}\Psi_{j,t}\left(\frac{\star}{\ell(Q_j)}\right)\right](z_j)\right|$$
$$+ \left|[p^{(j)}\varphi^{(j)}] * \left[\frac{1}{\ell(Q_j)^n}\Psi_{j,t}\left(\frac{\star}{\ell(Q_j)}\right)\right](z_j)\right|.$$

Since $p_N(\Psi_{j,t}), p_N(\tilde{\Psi}_{j,t}) \lesssim_N \dfrac{\ell(Q_j)^{n+L+1}}{|x_0 - c(Q_j)|^{n+L+1}}$, we have

$$\frac{1}{t^n}\left|b_j * \varphi\left(\frac{\star}{t}\right)(x_0)\right|$$
$$\lesssim \frac{\ell(Q_j)^{n+L+1}}{\ell(Q_j)^{n+L+1} + |x - c(Q_j)|^{n+L+1}}(\mathscr{M}f(z_j) + \mathscr{M}[p^{(j)}\varphi^{(j)}](z_j))$$
$$\lesssim \frac{\ell(Q_j)^{n+L+1}}{\ell(Q_j)^{n+L+1} + |x - c(Q_j)|^{n+L+1}}\mathscr{M}f(z_j).$$

4. Let $x_0 \notin 10^6 Q_j$ and $0 < t \leq \ell(Q_j)$. Then

$$\frac{1}{t^n}\left|b_j * \varphi\left(\frac{\star}{t}\right)(x_0)\right| = \frac{1}{t^n}\left|\left\langle f - p^{(j)}, \varphi\left(\frac{x_0 - \star}{t}\right)\varphi^{(j)}\right\rangle\right|.$$

Define $\Psi_{j,t}, \tilde{\Psi}_{j,t}$ by (3.78). Then since

$$\Psi_{j,t}\left(\frac{z_j - \star}{t}\right) = \varphi\left(\frac{x_0 - \star}{t}\right)\varphi^{(j)}, \quad \tilde{\Psi}_{j,t}\left(\frac{z_j - \star}{t}\right) = \varphi\left(\frac{x_0 - \star}{t}\right)\tilde{\varphi}^{(j)}$$

and $p_N(\Psi_{j,t}), p_N(\tilde{\Psi}_{j,t}) \lesssim \dfrac{\ell(Q_j)^N}{|x_0 - c(Q_j)|^N}$, we obtain

$$\frac{1}{t^n}\left|b_j * \varphi\left(\frac{\star}{t}\right)(x_0)\right| \leq \left|f * \left[\frac{1}{t^n}\Psi_{j,t}\left(\frac{\star}{t}\right)\right](z_j)\right| + \left|[p^{(j)}\varphi^{(j)}] * \left[\frac{1}{t^n}\Psi_{j,t}\left(\frac{\star}{t}\right)\right](z_j)\right|$$
$$\lesssim \frac{\ell(Q_j)^N}{|x_0 - c(Q_j)|^N}\mathscr{M}f(z_j)$$
$$\lesssim \frac{\ell(Q_j)^{n+L+1}}{\ell(Q_j)^{n+L+1} + |x_0 - c(Q_j)|^{n+L+1}}\mathscr{M}f(z_j).$$

Therefore, the proof is complete.

The estimate for the good part is the following one:

Proposition 3.8 *Let $f \in \mathscr{S}'(\mathbb{R}^n)$ be such that $\mathscr{M} f$ is finite almost everywhere. Then for all $x \in \mathbb{R}^n$,*

$$\mathscr{M} g(x) \lesssim \chi_{\mathbb{R}^n \setminus \Omega}(x) \mathscr{M} f(x) + \lambda \sum_{j=1}^{\infty} \frac{\ell(Q_j)^{n+L+1}}{\ell(Q_l)^{n+L+1} + |x - c(Q_l)|^{n+L+1}}.$$

Proof When $x \notin \Omega$, there is nothing to prove thanks to Proposition 3.7. Thus, suppose $x \in \Omega$. Note that we can find $j \in J$ so that $x \in 10Q_j$. Let

$$\mathscr{J}_{j,0} \equiv \{l \in \mathbb{N} : 100Q_l \cap 100Q_j \neq \emptyset\}, \quad \mathscr{J}_{j,1} \equiv \mathbb{N} \setminus \mathscr{J}_{j,0}.$$

Then $\mathscr{J}_{j,0}$ and $\mathscr{J}_{j,1}$ to partition $\mathbb{N}$.

Decompose g: $g = \left(f - \sum_{l \in \mathscr{J}_{j,0}} b^{(l)} \right) - \sum_{l \in \mathscr{J}_{j,1}} b^{(l)} =: \mathrm{I} + \mathrm{II}$. The second term II can be estimated by the triangle inequality $\mathscr{M}[F + G] \leq \mathscr{M} F + \mathscr{M} G$; for $x \in \mathbb{R}^n$:

$$\mathscr{M}\left[\sum_{l \in \mathscr{J}_{j,1}} b^{(l)}\right](x) \leq \sum_{l \in \mathscr{J}_{j,1}} \mathscr{M} b^{(l)}(x) \lesssim \sum_{l \in \mathscr{J}_{j,1}} \frac{\lambda \ell(Q_l)^{n+L+1}}{\ell(Q_l)^{n+L+1} + |x - c(Q_l)|^{n+L+1}}.$$

The estimate for I is harder. We decompose the grand maximal function of the first term I; for $x \in \mathbb{R}^n$

$$\mathscr{M}\left[f - \sum_{l \in \mathscr{J}_{j,0}} b^{(l)}\right](x) \leq \mathscr{M}\left[f - \sum_{l \in \mathscr{J}_{j,0}} \varphi^{(l)} \cdot f\right](x) + \sum_{l \in \mathscr{J}_{j,0}} \mathscr{M}\left[p^{(l)} \cdot \varphi^{(l)}\right](x).$$

The second term can be handled similarly: Consider $\mathscr{M} f(z_j)$. We have

$$\sum_{l \in \mathscr{J}_{j,0}} \mathscr{M}\left[p^{(l)} \cdot \varphi^{(l)}\right](x) \lesssim \frac{\lambda\, \ell(Q_j)^{n+L+1}}{\ell(Q_l)^{n+L+1} + |x - c(Q_l)|^{n+L+1}}$$

for $x \in \mathbb{R}^n$.

We will estimate the remaining term by taking care of the size of the support of $f - \sum_{l \in \mathscr{J}_{j,0}} \varphi^{(l)} \cdot f$. That is, $f - \sum_{l \in \mathscr{J}_{j,0}} \varphi^{(l)} \cdot f$ vanishes outside $100Q_j$. So, let us prove

$$\mathscr{M}\left[f - \sum_{l \in \mathscr{J}_{j,0}} \varphi^{(l)} \cdot f\right](x) \lesssim \lambda. \tag{3.89}$$

For $\varphi \in \mathscr{S}(\mathbb{R}^n)$ and $y \in \mathbb{R}^n$, we set

$$\widetilde{\Psi}_{j,t}(y) \equiv \frac{1}{t^n}\varphi\left(\frac{x-y}{t}\right)\left(1-\sum_{l\in\mathscr{J}_{j,0}}\varphi^{(l)}(y)\right)$$

$$\Psi_{j,t}(y) \equiv \widetilde{\Psi}_{j,t}(z_j-y) = \frac{1}{t^n}\varphi\left(\frac{x-z_j+y}{t}\right)\left(1-\sum_{l\in\mathscr{J}_{j,0}}\varphi^{(l)}(z_j-y)\right).$$

Once we prove

$$\left|\left\langle f-\sum_{l\in\mathscr{J}_{j,0}}\varphi^{(l)}\cdot f, \frac{1}{t^n}\varphi\left(\frac{z_j-\star}{t}\right)\right\rangle\right| = |\Psi_{j,t} * f(z_j)| \lesssim \mathscr{M}f(z_j) \lesssim \lambda \tag{3.90}$$

uniformly over $t>0$, we will have been proving (3.89) and hence the lemma.

Assume $t > \ell(Q_j)$. Fix $x \in \mathbb{R}^n$. Let $\varphi \in \mathscr{F}_N$. Set

$$\varphi_{t,j}(w) \equiv \varphi\left(\frac{x-z_j}{t}+w\right),$$

$$\psi_{t,j}(w) \equiv \sum_{l\in\mathscr{J}_{j,0}}\varphi^{(l)}(z_j-\ell(Q_j)w)\cdot\frac{\ell(Q_j)^n}{t^n}\varphi\left(\frac{x-z_j+\ell(Q_j)w}{t}\right),$$

for $w \in \mathbb{R}^n$ and we have $p_N(\varphi_{t,j}), p_N(\psi_{t,j}) \lesssim 1$. Hence

$$\begin{aligned}
&\left|\left\langle f-\sum_{l\in\mathscr{J}_{j,0}}\varphi^{(l)}\cdot f, \frac{1}{t^n}\varphi\left(\frac{x-\star}{t}\right)\right\rangle\right| \\
&\le \left|\left\langle f, \frac{1}{t^n}\varphi\left(\frac{x-\star}{t}\right)\right\rangle\right| + \left|\left\langle \sum_{l\in\mathscr{J}_{j,0}}\varphi^{(l)}\cdot f, \frac{1}{t^n}\varphi\left(\frac{x-\star}{t}\right)\right\rangle\right| \\
&\le \left|f*\left(\frac{1}{t^n}\varphi_{t,j}\left(\frac{\star}{t}\right)\right)(z_j)\right| + \left|f*\left(\frac{1}{\ell(Q_j)^n}\psi_{t,j}\left(\frac{\star}{\ell(Q_j)}\right)\right)(z_j)\right| \\
&\lesssim \mathscr{M}f(z_j) \lesssim \lambda.
\end{aligned}$$

Conversely, assume $t \le \ell(Q_j)$. Then observe that $|y| \ge \frac{1}{100}|z_j-x|$ whenever $|z_j-y-x| \ge \frac{1}{100}|z_j-x|$. Thus,

$$p_N(t^n\Psi_{j,t}(t\star))$$

$$= \sum_{\substack{\alpha \in \mathbb{N}_0{}^n \\ |\alpha| \le N}} \sup_{v \in \mathbb{R}^n} \langle v \rangle^N \left| \partial^\alpha \left[\varphi\left(\frac{x - z_j + t\,v}{t}\right)\left(1 - \sum_{l \in \mathscr{J}_{j,0}} \varphi^{(l)}(z_j - t\,v)\right)\right]\right| \lesssim 1.$$

Hence we obtain (3.90).

3.2.5.2 Parametrized Calderón–Zygmund Decomposition

So far, we have been considering general estimates for $f \in \mathscr{S}'(\mathbb{R}^n)$. Now we consider what happens for $f \in H^p(\mathbb{R}^n)$. Here and below, we write g_λ instead of g in order to stress that g is dependent on λ. The same can be said for $b^{(j)}$ and Q_j.

So we have a collection of a Hilbert space $\mathscr{H}_{k,\lambda}$, $k \in \mathbb{N}$, a Whitney covering $\{Q_{j,\lambda}\}_{j=1}^\infty$ of $\{\mathscr{M} f > \lambda\}$ and pointwise estimates:

$$\mathscr{M} g_\lambda \lesssim \chi_{\mathbb{R}^n \setminus \Omega_\lambda} \mathscr{M} f + \sum_{k=1}^\infty \frac{\lambda\, \ell(Q_{k,\lambda})^{n+L+1}}{\ell(Q_{k,\lambda})^{n+L+1} + |\star - c(Q_{k,\lambda})|^{n+L+1}}, \tag{3.91}$$

$$\mathscr{M} b_\lambda^{(j)} \lesssim \chi_{100 Q_{j,\lambda}} \mathscr{M} f + \frac{\lambda\, \ell(Q_{j,\lambda})^{n+L+1}}{\ell(Q_{j,\lambda})^{n+L+1} + |\star - c(Q_{j,\lambda})|^{n+L+1}}, \tag{3.92}$$

for $j \in \mathbb{N}$. Furthermore, let us set $b_\lambda \equiv \sum_{j=1}^\infty b_\lambda^{(j)}$.

By integrating Propositions 3.7 and 3.8, we obtain the following important estimates:

Proposition 3.9 *Let* $\lambda > 0$ *be fixed, and let* $0 < p \le 1$.

1. *Let* g_λ *as before. Then* $\displaystyle\int_{\mathbb{R}^n} \mathscr{M} g_\lambda(x)\mathrm{d}x \lesssim_p \lambda^{1-p} \int_{\mathbb{R}^n} \mathscr{M} f(x)^p \mathrm{d}x$.
2. *Let* b_λ *as before. Then* $\displaystyle\int_{\mathbb{R}^n} \mathscr{M} b_\lambda(x)^p \mathrm{d}x \lesssim_p \int_{\{\mathscr{M} f > \lambda\}} \mathscr{M} f(x)^p \mathrm{d}x$.

Proof

1. Integrate the inequality in Proposition 3.8 over $\mathbb{R}^n$. Then

$$\int_{\mathbb{R}^n} \mathscr{M} g_\lambda(x)\mathrm{d}x \lesssim \int_{\{\mathscr{M} f \le \lambda\}} \mathscr{M} f(x)\mathrm{d}x + \lambda \sum_{j=1}^\infty \ell(Q_j)^n,$$

$$\lesssim \lambda^{1-p} \int_{\mathbb{R}^n} \mathscr{M} f(x)^p \mathrm{d}x + \lambda \sum_{j=1}^\infty \ell(Q_j)^n.$$

Since $\sum_{j=1}^{\infty} \ell(Q_j)^n = \sum_{j=1}^{\infty} |Q_j| \le |\{\mathcal{M} f > \lambda\}| < \infty$, we have only to use Proposition 3.8.

2. Sum the estimate over $j \in \mathbb{N}$ to have

$$\mathcal{M} b_\lambda \lesssim \chi_\Omega \mathcal{M} f + \sum_{j=1}^{\infty} \frac{\lambda \ell(Q_j)^{n+L+1}}{\ell(Q_l)^{n+L+1} + |\star - c(Q_l)|^{n+L+1}}. \tag{3.93}$$

Since $0 < p \le 1$, if we raise both sides of (3.93) to the p-th power,

$$\mathcal{M} b_\lambda{}^p \lesssim \chi_\Omega \mathcal{M} f^p + \sum_{j=1}^{\infty} \frac{\lambda^p \ell(Q_j)^{p(n+L+1)}}{\ell(Q_j)^{p(n+L+1)} + |\star - c(Q_j)|^{p(n+L+1)}}. \tag{3.94}$$

Since $L \ge [\sigma_p]$, we can integrate (3.94) over $\mathbb{R}^n$ to have

$$\int_{\mathbb{R}^n} \mathcal{M} b_\lambda(x)^p \mathrm{d}x \le \int_{\{\mathcal{M} f > \lambda\}} \mathcal{M} f(x)^p \mathrm{d}x + \lambda^p \sum_{j=1}^{\infty} \ell(Q_j)^n.$$

The same argument as before works:

$$\lambda^p \sum_{j=1}^{\infty} \ell(Q_j)^n \le \int_{\{\mathcal{M} f > \lambda\}} \lambda^p \mathrm{d}x \le \int_{\{\mathcal{M} f > \lambda\}} \mathcal{M} f(x)^p \mathrm{d}x.$$

Therefore, the proof is complete.

From Proposition 3.9, we have:

Proposition 3.10 *Let $0 < p \le 1$. Then for all $f \in \mathcal{S}'(\mathbb{R}^n)$ such that $\mathcal{M} f \in L^p(\mathbb{R}^n)$, we have $g_\lambda \in H^1(\mathbb{R}^n) \cap H^p(\mathbb{R}^n)$ and $g_\lambda \to f$ in $H^p(\mathbb{R}^n)$.*

The function g_λ has a pointwise estimate; g_λ deserves its name of the good part.

Lemma 3.10 *Let $f \in L^1_{\mathrm{loc}}(\mathbb{R}^n) \cap \mathcal{S}'(\mathbb{R}^n)$. Then $\|g_\lambda\|_\infty \lesssim \lambda$ for all $\lambda > 0$.*

Proof When $x \notin \Omega_\lambda$, we have

$$b_\lambda^{(j)}(x) = \varphi^{(j)}(x) \cdot (f(x) - p_\lambda^{(j)}(x)), \quad g_\lambda(x) = f(x) - \sum_{j=1}^{\infty} b_\lambda^{(j)}(x)$$

by the definition of g_λ. Hence $|g_\lambda(x)| = |f(x)| \lesssim \mathcal{M} f(x) \lesssim \lambda$ follows trivially. If instead $x \in \Omega_\lambda$, then we can choose j such that $x \in 10 Q_{j,\lambda}$. We then invoke a pointwise estimate $|p_\lambda^{(j)}(x) \varphi_\lambda^{(j)}(x)| \lesssim \lambda$.

Exercises

Exercise 3.29 Let $f \in L^1_{\rm loc}(\mathbb{R}^n) \cap H^p(\mathbb{R}^n)$ with $0 < p \leq 1$, and let g_λ be a function in Lemma 3.10. Show that $\mathscr{M} g_\lambda \in L^q(\mathbb{R}^n)$ for all $q \in (p, \infty)$.

Exercise 3.30 Let $0 < p \leq 1$, and let $f \in \mathscr{S}'(\mathbb{R}^n)$ satisfy $\mathscr{M} f \in L^{q,\infty}(\mathbb{R}^n)$ for some $q \in (p, \infty)$. Then use Proposition 3.9 to show that $b_\lambda \to 0$ in $H^p(\mathbb{R}^n)$ as $\lambda \to \infty$.

3.2.6 Atomic Decomposition of Hardy Spaces

This is a culmination of what we have been doing. We will show that the Hardy spaces defined by means of the atomic decomposition agree with the ones by means of the grand maximal operator. Before we go further, let us summarize what we have been doing.

3.2.6.1 Summary of Our Observations

We are oriented to the converse of the atomic decomposition. Here we will rearrange what we have obtained:

Lemma 3.11 *Let D and Y be linear subspaces of a linear space X such that $D \subset Y \subset X$. Let X and Y be equipped with quasi-norms $\| \star \|_X$ and $\| \star \|_Y$. Assume the following:*

1. *D is dense in X.*
2. *$(Y, \| \star \|_Y)$ is embedded continuously into $(X, \| \star \|_X)$.*
3. *For all $y \in D$,*

$$\|y\|_X \sim \|y\|_Y. \tag{3.95}$$

4. *The space $(Y, \| \star \|_Y)$ is a quasi-Banach space.*

Then $X = Y$ as a set and their norms are equivalent.

Proof Let $x \in X$. Then we can find a sequence $\{d_j\}_{j=1}^\infty \subset D$ convergent to x in the topology of X. Since two quasi-norms are equivalent in D,

$$\lim_{j,k \to \infty} \|d_j - d_k\|_Y = 0. \tag{3.96}$$

Since $(Y, \| \star \|_Y)$ is a quasi-Banach space, we have $\lim\limits_{j \to \infty} d_j =: y \in Y$ in the topology of Y. Since Y is embedded continuously into X, we still have $\lim\limits_{j \to \infty} d_j = y$ in the topology of X. Since X is a quasi-Banach space; hence a Hausdorff space,

$x = y \in Y$. Since $x \in X$ is chosen arbitrarily, it follows that $X = Y$. Since any element in X can be approximated in $\| \star \|_X$ and $\| \star \|_Y$, a passage of (3.95) to the limit shows that these two norms are equivalent.

We apply Lemma 3.11 for the following pairs:

$$X = H^p(\mathbb{R}^n) = \{f \in \mathscr{S}'(\mathbb{R}^n) \,:\, \|\mathscr{M} f\|_p < \infty\},$$
$$Y = Y^p = \{f \in \mathscr{S}'(\mathbb{R}^n) \,:\, f \text{ admits atomic decomposition}\},$$
$$D = \{f \in H^p(\mathbb{R}^n) \,:\, \|\mathscr{M} f\|_1 < \infty\}.$$

The norm of Y is given by

$$\|f\|_Y \equiv \inf \|\lambda\|_{\ell^p}. \tag{3.97}$$

Here the infimum of the right-hand side of (3.97) is taken over all the elements $\lambda = \{\lambda_j\}_{j=1}^{\infty} \in \ell^p(\mathbb{N})$ and a sequence $\{a_j\}_{j=1}^{\infty}$ of atoms satisfying $f = \sum_{j=1}^{\infty} \lambda_j a_j$.

So far, we have the following:

1. We showed that D is dense in $H^p(\mathbb{R}^n)$ in Proposition 3.10 and Lemma 3.10.
2. Proposition 3.3 shows that $Y^p \hookrightarrow H^p(\mathbb{R}^n)$ in the sense of the continuous embedding.
3. As is shown easily, the space Y^p is a quasi-Banach space; see Exercise 3.33.

3.2.6.2 Two Definitions of Hardy Spaces: Via Atoms and via the Grand Maximal Operators

Hence it remains to prove the following proposition to establish $Y^p \approx H^p(\mathbb{R}^n)$:

Proposition 3.11 *For* $f \in D$, $\|f\|_{Y^p} \lesssim \|f\|_{H^p}$.

Proof From (3.91), we have

$$\begin{aligned}\int_{\mathbb{R}^n} \mathscr{M} g_\lambda(x)^p \mathrm{d}x &\lesssim \int_{\{\mathscr{M} f \le \lambda\}} \mathscr{M} f(x)^p \mathrm{d}x + \lambda^p \sum_{j=1}^{\infty} |Q_{j,\lambda}| \\ &\lesssim \int_{\mathbb{R}^n} \min(\mathscr{M} f(x), \lambda)^p \mathrm{d}x,\end{aligned}$$

which shows that $g_\lambda \to 0$ as $\lambda \to 0$. Hence

$$f = \lim_{J \to \infty} (g_{2^J} - g_{2^{-J-1}}) \tag{3.98}$$

in the topology of $H^p(\mathbb{R}^n)$.

For $j \in \mathbb{Z}$ and $k \in \mathbb{N}$, we denote by P_k^j the projection onto the Hilbert space $\mathscr{H}_{k,2^j}$. So, when $f \in \mathscr{S}'(\mathbb{R}^n)$ satisfies $\|\mathscr{M} f\|_p < \infty$, the polynomial $P_k^j(f)$ is the unique polynomial satisfying

$$\langle f, q \cdot \eta_{2^j}^k \rangle = \langle P_k^j(f), q \cdot \eta_{2^j}^k \rangle \tag{3.99}$$

for all $q \in \mathscr{P}_L(\mathbb{R}^n)$.

Let $p^{(j,k,l)} \equiv P_l^j[(f - p_{2^j}^{(l)})\eta_{2^{j-1}}^{(k)}]$. If $100Q_{l,2^j} \cap 100Q_{k,2^{j-1}} = \emptyset$, (3.99) yields $p^{(j,k,l)} = 0$.

Since $\Omega_{2^j} \subset \Omega_{2^{j-1}}$,

$$\ell(Q_{l,2^j}) \le 2\ell(Q_{k,2^{j-1}}) \tag{3.100}$$

or $Q_{l,2^j} \cap Q_{k,2^{j-1}} = \emptyset$ all j, k and l. See Exercise 3.31. This together with Proposition 3.5 yields

$$|p^{(j,k,l)}(x)\eta_{2^j}^{(l)}(x)| \lesssim 2^j \chi_{500Q_k^{2^{j-1}}}(x) \quad (x \in \mathbb{R}^n). \tag{3.101}$$

See Exercise 3.32. Here and below, unless otherwise stated, the equality makes sense in $\mathscr{S}'(\mathbb{R}^n)$. Since $\sum_{k=1}^{\infty}(f - p_{2^j}^{(l)})\eta_{2^{j-1}}^{(k)} = (f - p_{2^j}^{(l)})\chi_{\Omega_{2^{j-1}}}$ in the topology of $L^2(\mathbb{R}^n)$,

$$\sum_{k,l=1}^{\infty} p^{(j,k,l)}\eta_{2^j}^{(l)} = \sum_{l=1}^{\infty} P_l^j[(f - p_{2^j}^{(l)})\chi_{\Omega_{2^{j-1}}}]\eta_{2^j}^{(l)} = \sum_{l=1}^{\infty}(p_{2^j}^{(l)} - p_{2^j}^{(l)})\eta_{2^j}^{(l)} = 0 \tag{3.102}$$

by virtue of the definition of P_l^j.

Define a function $A_{j,k}$ by

$$A_{j,k} \equiv (f - p_{2^{j-1}}^{(k)})\eta_{2^{j-1}}^{(k)} - \sum_{l=1}^{\infty}(f - p_{2^j}^{(l)})\eta_{2^j}^{(l)}\eta_{2^{j-1}}^{(k)} + \sum_{l=1}^{\infty} p^{(j,k,l)}\eta_{2^j}^{(l)}. \tag{3.103}$$

The above geometric observation shows that the support of $A_{j,k}$ is contained in $500Q_{k,2^{j-1}}$.

Next, we rewrite (3.103), the definition of $A_{j,k}$,

$$A_{j,k} = \left(1 - \sum_{l=1}^{\infty}\eta_{2^j}^{(l)}\right)\eta_{2^{j-1}}^{(k)} f - p_{2^{j-1}}^{(k)}\eta_{2^{j-1}}^{(k)} + \sum_{l=1}^{\infty} p_{2^j}^{(l)}\eta_{2^j}^{(l)}\eta_{2^{j-1}}^{(k)} + \sum_{l=1}^{\infty} p^{(j,k,l)}\eta_{2^j}^{(l)}.$$

Then by virtue of Lemma 3.8,

$$|A_{j,k}| \lesssim \chi_{500Q_{k,2^{j-1}}} \left\{ \left(1 - \sum_{l=1}^{\infty} \eta_{2^j}^{(l)}\right) \eta_{2^{j-1}}^{(k)} |f| + 2^j \right\} \lesssim 2^j \chi_{500Q_{k,2^{j-1}}}. \tag{3.104}$$

Furthermore, since $\{Q_{k,2^{j-1}}\}_{k=1}^{\infty}$ is a Whitney covering of $\{\mathcal{M} f > 2^{j-1}\}$, we have

$$\sum_{j=-\infty}^{\infty} \sum_{k=1}^{\infty} 2^{jp} |Q_{k,2^{j-1}}| \lesssim \sum_{j} 2^{jp} |\{\mathcal{M} f > 2^{j-1}\}| \lesssim \int_{\mathbb{R}^n} \mathcal{M} f(x)^p \mathrm{d}x < \infty. \tag{3.105}$$

See also Exercise 1.8.

Now we verify that the functions satisfy the moment condition. Let $q \in \mathcal{P}_L(\mathbb{R}^n)$. From the definition of $A_{j,k}$, we have

$$\begin{aligned}
&\int_{\mathbb{R}^n} A_{j,k}(x) q(x) \mathrm{d}x \\
&= \left\langle (f - p_{2^{j-1}}^{(k)}) - \sum_{l=1}^{\infty} (f - p_{2^j}^{(l)}) \eta_{2^j}^{(l)}, q \cdot \eta_{2^{j-1}}^{(k)} \right\rangle + \sum_{l=1}^{\infty} \langle p^{(j,k,l)}, q \cdot \eta_{2^j}^{(l)} \rangle.
\end{aligned}$$

Note that $f - p_{2^{j-1}}^{(k)}$ vanishes because of $p_{2^{j-1}}^{(k)}$. Furthermore, since k is fixed, we have

$$\int_{\mathbb{R}^n} A_{j,k}(x) q(x) \mathrm{d}x = -\sum_{l=1}^{\infty} \langle (f - p_{2^j}^{(l)}) \eta_{2^j}^{(l)}, q \cdot \eta_{2^{j-1}}^{(k)} \rangle + \sum_{l=1}^{\infty} \langle p^{(j,k,l)}, q \cdot \eta_{2^j}^{(l)} \rangle$$

by Lebesgue's convergence theorem. Thus, $A_{j,k} \perp \mathcal{P}_L(\mathbb{R}^n)$ in view of the property of P_l^j.

If we combine (3.104) and (3.105) with $A_{j,k} \perp \mathcal{P}_L(\mathbb{R}^n)$, we see that $h \equiv \sum_{j=-\infty}^{\infty} \sum_{k=1}^{\infty} A_{j,k}$ satisfies the condition of the atomic decomposition. So, it remains to prove $h = f$. So we can change the order of j, k in the above sum. Hence

$$\begin{aligned}
h &= \lim_{J \to \infty} \sum_{j=-J}^{J} \sum_{k=1}^{\infty} \left((f - p_{2^{j-1}}^{(k)}) \eta_{2^{j-1}}^{(k)} - \sum_{l=1}^{\infty} (f - p_{2^j}^{(l)}) \eta_{2^j}^{(l)} \eta_{2^{j-1}}^{(k)} + \sum_{l=1}^{\infty} p^{(j,k,l)} \eta_{2^j}^{(l)} \right) \\
&= \lim_{J \to \infty} \sum_{j=-J}^{J} \left(\sum_{k=1}^{\infty} (f - p_{2^{j-1}}^{(k)}) \eta_{2^{j-1}}^{(k)} - \sum_{l=1}^{\infty} (f - p_{2^j}^{(l)}) \eta_{2^j}^{(l)} \right).
\end{aligned}$$

Here we invoke (3.98) for the second equality. If we use the decomposition, we have

$$h = \lim_{J\to\infty}\left(\sum_{j=-J}^{J}(b_{2^{j-1}} - b_{2^j})\right) = \lim_{J\to\infty}\left(\sum_{j=-J}^{J}(g_{2^j} - g_{2^{j-1}})\right).$$

If we use (3.98), then we see that $h = \lim_{J\to\infty}(g_{2^J} - g_{2^{-J-1}}) = f$.

Here again note that the equality holds in the sense of $\mathscr{S}'(\mathbb{R}^n)$.

In view of estimates (3.104)–(3.105) and the above calculation, we conclude that $\sum_{j=-\infty}^{\infty}\sum_{k=1}^{\infty} A_{j,k} = f$ is the desired atomic decomposition of f and that $\|f\|_{H^p} \lesssim \|f\|_{Y^p}$. Thus, the proposition is proved.

Let us summarize what we have obtained.

Theorem 3.9 *Let $0 < p \le 1$, and let $f \in \mathscr{S}'(\mathbb{R}^n)$. Set $L_0 \equiv [\sigma_p]$. Choose a positive integer L with $L \ge L_0$.*

1. *The following conditions are all equivalent:*

(H1) *If we define the grand maximal function $\mathscr{M} f$ by*

$$\mathscr{M} f(x) \equiv \sup_{\substack{\kappa\in\mathscr{F} \\ t>0}} \left|\frac{1}{t^n}\kappa\left(\frac{\star}{t}\right) * f(x)\right|,$$

then $\|\mathscr{M} f\|_p < \infty$.

(H2) *Let $\psi \in \mathscr{S}(\mathbb{R}^n) \setminus \mathscr{P}_0(\mathbb{R}^n)^{\perp}$; that is, $\int_{\mathbb{R}^n}\psi(x)\mathrm{d}x \neq 0$. Then*

$$\left\|\sup_{j\in\mathbb{Z}} 2^{jn}|\psi(2^j\star) * f|\right\|_p < \infty. \tag{3.106}$$

(H3) *The distribution f admits the following decomposition:*

$$f = \sum_{j=1}^{\infty}\lambda_j a_j. \tag{3.107}$$

Here each a_j is a $(p, 2)$-atom, and the coefficient $\{\lambda_j\}_{j=1}^{\infty}$ is in $\ell^p(\mathbb{N})$.

(H4) *The distribution f admits the following decomposition:*

$$f = \sum_{j=1}^{\infty}\lambda_j a_j. \tag{3.108}$$

Here each a_j is a (p, ∞)-atom and the coefficient $\{\lambda_j\}_{j=1}^{\infty}$ is in $\ell^p(\mathbb{N})$. Furthermore, if f satisfies (H1)–(H4)*, we have the following norm equivalence:*

$$\|\mathcal{M}f\|_p \sim \left\| \sup_{j\in\mathbb{Z}} 2^{jn} |\psi(2^j\star) * f| \right\|_p \sim \inf_{(3.107)} \|\lambda\|_p \sim \inf_{(3.108)} \|\lambda\|_p, \tag{3.109}$$

Here the infimum is over all elements satisfying (3.107) *and* (3.108).

2. *The space* $H^1(\mathbb{R}^n) \cap H^p(\mathbb{R}^n)$ *is dense in* $H^p(\mathbb{R}^n)$.

Proof The above argument shows the implications (H3)$\Longrightarrow$(H2)$\Longrightarrow$(H1)$\Longrightarrow$(H4), as well as 2. Observe that (H4) trivially implies (H3) because of the definition of atoms.

The atomic decomposition in this theorem is called the nonsmooth atomic decomposition.

Finally, using the atomic decomposition we show that the singular integral operators (Calderón–Zygmund or CZ-operators) are bounded.

Theorem 3.10 *Let* $T \in B(L^2(\mathbb{R}^n))$ *be a CZ operator, which automatically extends to a bounded linear operator from* $L^1(\mathbb{R}^n)$ *to* $L^{1,\infty}(\mathbb{R}^n)$. *If we restrict* T *to* $H^1(\mathbb{R}^n) \subset L^1(\mathbb{R}^n)$, T *is a bounded operator from* $H^1(\mathbb{R}^n)$ *to* $L^1(\mathbb{R}^n)$.

Proof Let A be $(1, \infty)$-atoms centered at a cube Q. By (1.215), we have

$$|TA(x)| \lesssim |TA(x)|\chi_{2Q}(x) + \frac{\ell(Q)}{(\ell(Q) + |x - c(Q)|)^{n+1}} \quad (x \in \mathbb{R}^n).$$

We have $\|TA\|_1 \lesssim 1$. Since T is weakly $L^1(\mathbb{R}^n)$-bounded, by the atomic decomposition, we can extend T to a bounded operator from $H^1(\mathbb{R}^n)$ to $L^1(\mathbb{R}^n)$. We proceed as follows. First we take $f \in H^1(\mathbb{R}^n)$. Then we can find a sequence $\{a_j\}_{j=1}^{\infty}$ of (p, ∞)-atoms and a complex sequence $\{\lambda_j\}_{j=1}^{\infty}$ such that

$$f = \sum_{j=1}^{\infty} \lambda_j a_j, \quad \sum_{j=1}^{\infty} |\lambda_j| \le 2\|f\|_{H^1}.$$

Since $H^1(\mathbb{R}^n) \hookrightarrow L^1(\mathbb{R}^n)$, we readily have $f = \sum_{j=1}^{\infty} \lambda_j a_j$ in the topology of $L^1(\mathbb{R}^n)$. Then there exists a strictly increasing sequence $\{j_k\}_{k=1}^{\infty}$ such that $j_1 = 1$ and that

$$Tf(x) = \lim_{N\to\infty} \sum_{k=1}^{N} \left(\sum_{j=j_k}^{j_{k+1}-1} \lambda_j Ta_j(x) \right)$$

for almost all $x \in \mathbb{R}^n$. As we verified earlier, $\|Ta_j\|_1 \lesssim 1$. Thus by the Fatou lemma,

$$\|Tf\|_1 \le \sum_{k=1}^{\infty} \left\| \sum_{j=j_k}^{j_{k+1}-1} \lambda_j T a_j \right\|_1 \lesssim \sum_{k=1}^{\infty} \sum_{j=j_k}^{j_{k+1}-1} |\lambda_j| = \sum_{j=1}^{\infty} |\lambda_j| \lesssim \|f\|_{H^1}.$$

Hence T is an $H^1(\mathbb{R}^n)$-$L^1(\mathbb{R}^n)$ bounded linear operator.

3.2.6.3 Harmonic Functions and Hardy Spaces

The next theorem describes the relation between Hardy spaces and harmonic functions.

Theorem 3.11 *The proof uses the fact that $B^s_{11}(\mathbb{R}^n)$ is separable for any $s \in \mathbb{R}$, that is, there exists a countable set in $B^s_{11}(\mathbb{R}^n)$ whose closure is $B^s_{11}(\mathbb{R}^n)$. See Exercise 3.39. This will allow us to use the Banach-Alaoglu theorem. Let $u : \mathbb{R}^{n+1}_+ \to \mathbb{C}$ be a harmonic function. Write*

$$u^*(x) \equiv \sup_{(y,t)\in\Gamma(x)} |u(y,t)| = \|u\|_{L^\infty(\Gamma(x))} \quad (x \in \mathbb{R}^n).$$

Let $0 < p \le 1$. Then $u^* \in L^p(\mathbb{R}^n)$ if and only if there exists $f \in H^p(\mathbb{R}^n)$ such that

$$u(x,t) = \mathrm{e}^{-t\sqrt{-\Delta}} f(x)$$

for all $(x,t) \in \mathbb{R}^{n+1}_+$. Moreover, in this case we have $\|f\|_{H^p} \sim \|u^*\|_p$.

Proof Let us suppose that $u(x,t) = \mathrm{e}^{-t\sqrt{-\Delta}} f(x)$ for some $f \in H^p(\mathbb{R}^n)$. We use the atomic decomposition of f:

$$f = \sum_{j=1}^{\infty} \lambda_j a_j, \quad \left(\sum_{j=1}^{\infty} |\lambda_j|^p\right)^{\frac{1}{p}} \lesssim \|f\|_{H^p},$$

where each a_j is a (p,∞)-atom such that $a_j \perp \mathscr{P}_d$ for some fixed integer $d \in (\sigma_p, \infty)$. Then we can check that

$$u^*(x) = \sup_{t>0} |\mathrm{e}^{-t\sqrt{-\Delta}} f(x)| \lesssim \sum_{j=1}^{\infty} |\lambda_j| \cdot |Q_j|^{-\frac{1}{p}} \left(1 + \frac{|x - c(Q_j)|}{\ell(Q_j)}\right)^{-n-d-1}.$$

As a result, $u^* \in L^p(\mathbb{R}^n)$ and $\|u^*\|_p \lesssim \|f\|_{H^p}$.

If $u^* \in L^p(\mathbb{R}^n)$, then $u(\star, t)$ is bounded for all $t > 0$ according to Lemma 1.38. Since $\mathrm{e}^{-s\sqrt{-\Delta}}[u(\star,t)] = u(\star, t+s)$ for all $s > 0$, it follows that $u(\star,t) \in H^p(\mathbb{R}^n)$ according to Theorem 3.5. Since $H^p(\mathbb{R}^n)$ is embedded into $B^{-n/p}_{\infty\infty}(\mathbb{R}^n)$ and $B^{-n/p}_{\infty\infty}(\mathbb{R}^n)$ admits the predual $B^{n/p}_{11}(\mathbb{R}^n)$ according to Theorem 2.17, by the

Banach–Alaoglu theorem, there exists a sequence $\{t_j\}_{j=1}^{\infty}$ decreasing to 0 such that $u(\star, t_j)$ tends to a limit f in the weak-* topology of $B_{\infty\infty}^{-n/p}(\mathbb{R}^n)$. Since $P_t \in B_{11}^{n/p}(\mathbb{R}^n)$ (see Exercise 2.3), by the Fatou property of $H^p(\mathbb{R}^n)$ we obtain $f \in H^p(\mathbb{R}^n)$. It remains to notice

$$e^{-t\sqrt{-\Delta}} f(x) = \lim_{j\to\infty} e^{-t\sqrt{-\Delta}}[u(\star, t_j)](x) = \lim_{j\to\infty} u(x, t_j+t) = u(x,t) \quad (x,t) \in \mathbb{R}_+^{n+1}$$

according to Lemma 1.39 and Corollary 1.12. Thus, the proof is complete.

We need the following fact to prove Theorem 3.13:

Theorem 3.12 *Let $u : \mathbb{R}_+^{n+1} \to \mathbb{C}$ be a harmonic function. Write $u^*(x) \equiv \|u\|_{L^\infty(\Gamma(x))}$ for $x \in \mathbb{R}^n$.*

Let $0 < p \leq 1$. Assume that $u^* \in L^p(\mathbb{R}^n)$. Then there exists a function $f \in H^p(\mathbb{R}^n)$ such that $u = f * P_t$.

Proof Assume $u^* \in L^p(\mathbb{R}^n)$. Let $f_\varepsilon \equiv u(\star, \varepsilon) \in \mathscr{S}'(\mathbb{R}^n)$. Then

$$f_\varepsilon * P_t = u(\star, \varepsilon + t)$$

according to Corollary 1.12. Hence if we define

$$u_\varepsilon^*(x) \equiv \sup_{(y,t)\in\Gamma(x)} |f_\varepsilon * P_t(y)| \quad (x \in \mathbb{R}^n),$$

then $u_\varepsilon^* \leq u^* \in L^p(\mathbb{R}^n)$. Thus, $\{f_\varepsilon\}_{\varepsilon>0}$ forms a bounded set in $H^p(\mathbb{R}^n)$. Since $H^p(\mathbb{R}^n) \hookrightarrow \dot{B}_{\infty\infty}^{-n/p}(\mathbb{R}^n) \simeq (\dot{B}_{11}^{n/p}(\mathbb{R}^n))^*$, a subsequence $\{f_{\varepsilon(k)}\}_{k=1}^{\infty}$ of $\{f_\varepsilon\}_{\varepsilon>0}$ converges to an element f in $\mathscr{S}'(\mathbb{R}^n)$. By a Fatou property of $H^p(\mathbb{R}^n) \simeq \dot{F}_{p2}^0(\mathbb{R}^n)$, we see that $f \in \dot{F}_{p2}^0(\mathbb{R}^n) \simeq H^p(\mathbb{R}^n)$.

Exercises

Exercise 3.31 Let $f \in \mathscr{S}'(\mathbb{R}^n)$ be such that $\mathscr{M} f$ is finite almost everywhere. Prove (3.100) assuming $Q_{l,2^j} \cap Q_{k,2^{j-1}} \neq \emptyset$. Hint: Use the maximality of the covering.

Exercise 3.32 Prove (3.101) using Proposition 3.5.

Exercise 3.33 Show that Y^p is a quasi-Banach space for $p \in (0, 1]$.

Exercise 3.34 Fix $0 < p \leq 1$, $0 < r < \infty$ and $L \in \mathbb{N}_0$. Let $f \in H^p(\mathbb{R}^n)$. Then show that we can find $\{a_j\}_{j=1}^{\infty} \subset L^\infty(\mathbb{R}^n) \cap \mathscr{P}_L(\mathbb{R}^n)^\perp$ and a sequence $\{Q_j\}_{j=1}^{\infty}$ of cubes:

1. $\text{supp}(a_j) \subset Q_j$,
2. $f = \sum_{j=1}^{\infty} a_j$ in $\mathscr{S}'(\mathbb{R}^n)$,

3. $\left\{\sum_{j=1}^{\infty}\left(\|a_j\|_\infty \chi_{Q_j}\right)^r\right\}^{1/r} \lesssim \mathcal{M}f.$

Reexamine the proof of Proposition 3.11. In particular, what did we need for (3.105)?

Exercise 3.35 Let $f \in L^p(\mathbb{R}) \cap \mathcal{P}_0(\mathbb{R})^\perp$ be supported on $[-1, 1]$.

1. Let $f^\dagger$ be the function defined by (1.213). For each $\alpha \geq 1$, write $E_\alpha = \{f^\dagger > \alpha\}$. Decompose E_α to the connected component, so that we choose disjoint open intervals $\{I_j^\alpha\}_{j \in J_\alpha}$ that decompose E_α. Define

$$g^\alpha \equiv f\chi_{\mathbb{R}^n \setminus E_\alpha} + \sum_{j \in J_\alpha} m_j^\alpha \chi_{I_j^\alpha}, \quad b_j^\alpha \equiv (f - m_j^\alpha)\chi_{I_j^\alpha}$$

for each $j \in J_\alpha$, where $m_j^\alpha \equiv m_{I_j^\alpha}(f)$. Fix $j \in J_\alpha$.

(a) Show that $|m_j^\alpha| \leq \alpha$. Hint: If $|I_j^\alpha| \geq 1$ use $f \in \mathcal{P}_0(\mathbb{R})^\perp$.
(b) Prove that $|g(x)| \leq \alpha$ for almost every $x \in \mathbb{R}^n$.
(c) Show $b_j^\alpha(x) = 0$ for almost every $x \in \mathbb{R}^n \setminus E_\alpha$.
(d) Show $b_j^\alpha \in \mathcal{P}_0(\mathbb{R}^n)^\perp$.

2. For each $\alpha = 2^0, 2^1, \ldots, 2^k, \ldots$, we consider the decomposition $f = g^\alpha + \sum_{j \in J_\alpha} b_j^\alpha$ as above. For each k and each $j \in J_{2^k}$, we define

$$A_j^k \equiv b_j^k - \sum_{i' \in J_{2^{k+1}} : I_{i'}^{k+1} \subset I_j^k} b_{i'}^{k+1}.$$

(a) Prove $\lim_{k \to \infty} g^k(x) = f(x)$ for almost every $x \in \mathbb{R}^n$.
(b) Prove $|A_j^k| \leq 3 \cdot 2^k \chi_{I_j^k}$.
(c) Show that $A_j^k \perp \mathcal{P}_0(\mathbb{R}^n)$.

3. Set $\lambda_j^k \equiv 3 \cdot 2^k |I_j^k|$ and $a_j^k \equiv (\lambda_j^k)^{-1} A_j^k$.

(a) Show that $\sum_{k=0}^{\infty} \sum_{j \in J_{2^k}} \lambda_j^k \leq 3 \int_{\mathbb{R}^n} f^\dagger(x) \mathrm{d}x$.
(b) Prove that $|a_j^k| \leq |I_j^k|^{-1} \chi_{I_j^k}$ and that $a_j^k \perp \mathcal{P}_0(\mathbb{R}^n)$.
(c) Show that $f = g^0 + \sum_{k=0}^{\infty} \sum_{j \in J_{2^k}} \lambda_j^k a_j^k$.

Exercise 3.36 Use a similar technique to Exercise 3.35 to state and prove an analogue for $L^p(\mathbb{R}^n)$.

Exercise 3.37 [784] Let $0 < p \le 1$ and $M \in \mathbb{Z} \cap [\sigma_p, \infty)$. Define

$$\mathscr{A}_{p,M}(r) \equiv \bigcap_{|\alpha| \le M} \{f \in L^2(\mathbb{R}^n) : \mathrm{supp}(\mathscr{F} f) \subset \mathbb{R}^n \setminus B(r^{-1}), \|\partial^\alpha \mathscr{F} f\|_2 \le r^{|\alpha| - \frac{n}{p} + \frac{n}{2}}\}$$

for $r > 0$ and

$$\mathscr{A}_{p,M} \equiv \bigcup_{0<r<\infty} \mathscr{A}_{p,M}(r).$$

Let $\psi \in C_c^\infty(\mathbb{R}^n)$ satisfy $\chi_{B(1)} \le \psi \le \chi_{B(2)}$. Write $\varphi_j \equiv \psi_{j+1} - \psi_j$ for $j \in \mathbb{Z}$.

1. If $a \in \mathscr{M}_{p,M}(r)$, then show that $\sup\limits_{j \in \mathbb{Z}} |\psi_j(D) f| \lesssim |Q(r)|^{-\frac{1}{p}} M^{(N)} \chi_{Q(r)}$ for some $N \in \mathbb{N}$.
2. If a is a (p, ∞)-atom centered at $Q(2^{-k})$, then show that there exist constants $\mu, C > 0$ such that $C 2^{-\mu|j-k|} \varphi_j(D) a \in \mathscr{A}_{p,M}(2^{-j})$.
3. Show that we can replace atoms with functions of the form $a(\star - x)$ for some $a \in \mathscr{A}_{p,M}$.

Exercise 3.38 Let f be a measurable function such that $\displaystyle\int_{\mathbb{R}^n} \frac{|f(y)|}{|y|^{n+1} + 1} \mathrm{d}y < \infty$.

- Show that f is a bounded distribution, so that $u(x,t) \equiv \exp(-t\sqrt{-\Delta}) f(x)$ makes sense for $(x,t) \in \mathbb{R}^{n+1}_+$.
- Let $\Gamma_\beta(x) \equiv \{(y,t) \in \mathbb{R}^{n+1}_+ : |x - y| < \beta t\}$. One defines the *Luzin* function $S(f)$ by $S_\beta(f) \equiv \int_{\Gamma_\beta(\star)} |\nabla u(y,t)|^2 \frac{\mathrm{d}y\mathrm{d}t}{t^{n-1}}$.
- One also defines $\Gamma(x) \equiv \Gamma_1(x)$ and $S(f) \equiv S_1(f)$. One defines the *Littlewood–Paley g-function* by $g(f) \equiv \left(\int_0^\infty t|\nabla u(\star, t)|^2 \mathrm{d}t\right)^{\frac{1}{2}}$.
- One defines the g_λ^*-*function* by $g_\lambda(f) \equiv \left(\displaystyle\int_{\mathbb{R}^{n+1}_+} \frac{|\nabla u(y,t)|^2}{(t + |\star - y|)^\lambda} \frac{\mathrm{d}y\mathrm{d}t}{t^{n-\lambda-1}}\right)^{1/2}$ for $\lambda > 0$.

1. For $\beta < \beta'$, show that $S_\beta(f) \le S_{\beta'}(f)$.
2. For $\lambda < \lambda'$, show that $g_\lambda^*(f) \ge g_{\lambda'}^*(f)$.
3. For $\beta > 0$ and $\lambda > 0$, show that $S_\beta(f) \lesssim_{\lambda,\beta} g_\lambda^*(f)$.
4. By using the submean inequality for $|\nabla u|^2$, show that $g(f) \lesssim S_\beta(f)$.

Exercise 3.39 Let $s \in \mathbb{R}$. Let $\psi \in C_c^\infty(\mathbb{R}^n)$ satisfy $\chi_{B(1)} \le \psi \le \chi_{B(2)}$.

1. Show that $\psi_j(D) f \to f$ in $B^s_{11}(\mathbb{R}^n)$ for all $f \in B^s_{11}(\mathbb{R}^n)$.
2. Show that $B^s_{11}(\mathbb{R}^n)$ is separable.

3.2.7 Characterization of Hardy Spaces via Riesz Transforms

So far, we have obtained two different characterizations of Hardy spaces; one is by the grand maximal operator and the other is by the atomic decomposition. Here we aim to obtain another characterization. We use the Riesz transform to characterize Hardy spaces. In this section we assume that $0 < p \leq 1$.

3.2.7.1 Characterization of Hardy Spaces via Riesz Transforms

Recall that the distribution f is bounded if $\psi * f \in L^\infty(\mathbb{R}^n)$ for all $\psi \in \mathscr{S}(\mathbb{R}^n)$; see Definition 1.45. Also, recall the notion of a distribution restricted at infinity; see Definition 1.19.

We aim here to prove the following theorems:

Theorem 3.13

1. *Let $f \in L^1(\mathbb{R}^n)$. Then*

$$\sum_{j=1}^{n} \|R_j f\|_1 < \infty \tag{3.110}$$

if and only if $f \in H^1(\mathbb{R}^n)$.

2. *Let $\dfrac{n-1}{n} < p \leq 1$, and let $f \in \mathscr{S}'(\mathbb{R}^n)$ be a distribution restricted at infinity. Let $\psi \in \mathscr{S}_{B(2)}(\mathbb{R}^n)$ satisfy the nondegenerate condition*

$$\chi_{B(1)} \leq \mathscr{F}\psi \leq \chi_{B(2)}. \tag{3.111}$$

Then the estimate

$$\sup_{j\in\mathbb{Z}} \left(\|f * \psi^j\|_p + \sum_{k=1}^{n} \|R_k[f * \psi^j]\|_p \right) < \infty \tag{3.112}$$

holds if and only if $f \in H^p(\mathbb{R}^n)$.

Next, consider the general case. In this case, for any $0 < p \leq 1$ there exists $m \in \mathbb{N}$ so that

$$\frac{n-1}{n+m-1} < p \leq 1.$$

We aim here to prove the following theorem in this case using Theorems 1.72 and 1.73:

Theorem 3.14 *Let $m \in \mathbb{N}$, $\dfrac{n-1}{n+m-1} < p \leq 1$, and let $f \in \mathscr{S}'(\mathbb{R}^n)$ be a distribution restricted at infinity. Let $\psi \in \mathscr{S}_{B(2)}(\mathbb{R}^n)$ satisfy* (3.111). *Then the estimate*

$$\sup_{j\in\mathbb{Z}} \sum_{\alpha=(\alpha_1,\alpha_2,\ldots,\alpha_n)\in\mathbb{N}^n, |\alpha|\le m} \|R_1{}^{\alpha_1} R_2{}^{\alpha_2} \cdots R_n{}^{\alpha_n}[f * \psi^j]\|_p < \infty \tag{3.113}$$

holds if and only if $f \in H^p(\mathbb{R}^n)$.

For the proof of Theorem 3.13, a couple of remarks may be in order.

Remark 3.2 The heart of the proof of Theorem 3.13 is to deduce $f \in H^p(\mathbb{R}^n)$ from the assumptions (3.110) with $p = 1$ or (3.112) with $0 < p \le 1$:

1. Note that $R_k[f * \Psi^j]$ and $R_1{}^{\alpha_1} R_2{}^{\alpha_2} \cdots R_n{}^{\alpha_n}[f * \psi^j]$ makes sense in the assumptions (3.112) and (3.113) because $f * \Psi^j \in L^v(\mathbb{R}^n)$ for some $1 < v < \infty$.
2. Let $f \in H^1(\mathbb{R}^n)$. Then
$$\|f\|_1 + \sum_{j=1}^n \|R_j f\|_1 \lesssim \|f\|_{H^1} + \sum_{j=1}^n \|R_j f\|_{H^1} \lesssim \|f\|_{H^1} < \infty$$
Thus, one already has the "if" part thanks to Theorem 3.6.
3. Let $\dfrac{n-1}{n+m-1} < p \le 1$, and let $f \in H^p(\mathbb{R}^n)$. Then in a similar manner (3.112) can be proven.
4. When $p = 1$, the assumptions (3.110) and (3.112) are equivalent by virtue of the Young inequality.

Proposition 3.12 *Let* $p \in \left(\dfrac{n-1}{n+m-1}, 1\right]$, *and let* $u : \mathbb{R}^{n+1}_+ \to \mathbb{C}$ *be a harmonic function. Define* $u^*(x) \equiv \|u\|_{L^\infty(\Gamma(x))}$ *for* $x \in \mathbb{R}^n$. *Then* $u^* \in L^p(\mathbb{R}^n)$ *if and only if there exists a tensor-valued function*

$$G(x,t) = \sum_{j_1,j_2,\ldots,j_m=0}^n G_{j_1,j_2,\ldots,j_m}(x,t)\boldsymbol{e}_{j_1} \otimes \boldsymbol{e}_{j_2} \otimes \cdots \otimes \boldsymbol{e}_{j_m} \quad ((x,t) \in \mathbb{R}^{n+1}_+)$$

satisfying the generalized Cauchy–Riemann equation,

$$u = G_{0,0,\ldots,0}$$

and

$$\sup_{t>0} \sum_{j_1,j_2,\ldots,j_m=0}^n \|G_{j_1,j_2,\ldots,j_m}(\star, t)\|_p < \infty. \tag{3.114}$$

Proof We prove the sufficiency. Set

$$F(x,t) \equiv \left(\sum_{j_1,j_2,\dots,j_m=0}^{n} |G_{j_1,j_2,\dots,j_m}(x,t)|^2 \right)^{\frac{1}{2}} \quad ((x,t) \in \mathbb{R}^{n+1}_+).$$

Then the assumption (3.114) reads as: $\{F(\star,t)^{\frac{n-1}{n+m-1}}\}_{t>0}$ forms a bounded set in $L^{\frac{(n+m-1)p}{n-1}}(\mathbb{R}^n)$. Note that $\dfrac{(n+m-1)p}{n-1} > 1$. Thus, by the Banach–Alaoglu theorem, a subsequence $\{t_k\}_{k=1}^{\infty}$ can be found so that $\{F(\star,t_k)^{\frac{n-1}{n+m-1}}\}_{k=1}^{\infty}$ converges weakly to a nonnegative function h in $L^{\frac{(n+m-1)p}{n-1}}(\mathbb{R}^n)$. This implies

$$\lim_{k\to\infty} [F(\star,t_k)^{\frac{n-1}{n+m-1}}] * P_t(x) = h * P_t(x) \tag{3.115}$$

for all $x \in \mathbb{R}^n$. Meanwhile, $F^{\frac{n-1}{n+m-1}}$ is subharmonic according to Theorem 1.73. Hence

$$F(x,t)^{\frac{n-1}{n+m-1}} = \lim_{k\to\infty} F(x,t+t_k)^{\frac{n-1}{n+m-1}} \le \lim_{k\to\infty} [F(\star,t_k)^{\frac{n-1}{n+m-1}}] * P_t(x) \tag{3.116}$$

by Lemma 1.41. Combining equality (3.115) and inequality (3.116), we obtain

$$F(x,t)^{\frac{n-1}{n+m-1}} \le h * P_t(x) = \mathrm{e}^{-t\sqrt{-\Delta}}h(x).$$

As a result,

$$\|u^*\|_p \le \|F^*\|_p \le \left(\left\| \sup_{(y,t)\in\Gamma(\star)} |\mathrm{e}^{-t\sqrt{-\Delta}}h(y)| \right\|_{\frac{(n+m-1)p}{n-1}} \right)^{\frac{n-1}{n+m-1}},$$

where $F^* \equiv \sup_{(y,t)\in\Gamma(\star)} |F(y,t)|$. Assuming that $(n+m-1)p > n-1$, we learn that

$$\begin{aligned}
\|u^*\|_p &\lesssim \left(\|h\|_{\frac{(n+m-1)p}{n-1}} \right)^{\frac{n-1}{n+m-1}} \\
&\lesssim \sup_{t>0} \left(\|F(\star,t)^{\frac{n-1}{n+m-1}}\|_{\frac{(n+m-1)p}{n-1}} \right)^{\frac{n-1}{n+m-1}} \\
&= \sup_{t>0} \|F(\star,t)\|_p < \infty
\end{aligned}$$

according to Theorem 1.47, implying that $u^* \in L^p(\mathbb{R}^n)$.

If $u^* \in L^p(\mathbb{R}^n)$, then a function $f \in H^p(\mathbb{R}^n)$ can be found so that for all $x \in \mathbb{R}^n$ and $t > 0$, $u(x,t) = \exp(-t\sqrt{-\Delta})f(x)$ and that $\|f\|_{H^p} \lesssim \|u^*\|_p$ using Theorem 3.11. Since $\mathscr{S}_\infty(\mathbb{R}^n)$ is dense in $H^p(\mathbb{R}^n)$ according to Corollary 3.3,

$\{f_k\}_{k=1}^{\infty} \subset \mathscr{S}_\infty(\mathbb{R}^n)$ is convergent to f in $H^p(\mathbb{R}^n)$. If necessary, by passing to a subsequence $\|f_k\|_{H^p} \le 2\|f\|_{H^p}$ can be assumed for all $k \in \mathbb{N}$.

We claim that

$$G(x,t) = \sum_{j_1,j_2,\dots,j_m=0}^{n} G_{j_1,j_2,\dots,j_m}(x,t)\mathbf{e}_{j_1} \otimes \mathbf{e}_{j_2} \otimes \cdots \otimes \mathbf{e}_{j_m} \quad ((x,t) \in \mathbb{R}^{n+1}_+) \tag{3.117}$$

is the desired tensor-valued function, where

$$G_{j_1,j_2,\dots,j_m}(x,t) \equiv R_{j_1}R_{j_2}\cdots R_{j_m}[\exp(-t\sqrt{-\Delta})f](x).$$

We define $u_k^{(\alpha)}(x,t) \equiv {R_1}^{\alpha_1}{R_2}^{\alpha_2}\cdots{R_n}^{\alpha_n}[\exp(-t\sqrt{-\Delta})f_k](x)$ and $u^{(\alpha)}(x,t) \equiv {R_1}^{\alpha_1}{R_2}^{\alpha_2}\cdots{R_n}^{\alpha_n}[\exp(-t\sqrt{-\Delta})f](x)$ for $\alpha = (\alpha_0,\alpha_1,\dots,\alpha_n) \in {\mathbb{N}_0}^{n+1}$. Note that

$$\sup_{\substack{\alpha\in{\mathbb{N}_0}^{n+1}\\|\alpha|=m}}\sup_{t>0}\|u_k^{(\alpha)}(\star,t)\|_p \lesssim \sup_{\substack{\alpha\in{\mathbb{N}_0}^{n+1}\\|\alpha|=m}}\sup_{t>0}\|u_k^{(\alpha)}(\star,t)\|_{H^p}$$

for each $\alpha \in {\mathbb{N}_0}^{n+1}$, since each $u_k^{(\alpha)} \in \mathscr{S}_\infty(\mathbb{R}^n)$. Using the boundedness of the Riesz transform on $H^p(\mathbb{R}^n)$ gives

$$\sup_{\substack{\alpha\in{\mathbb{N}_0}^{n+1}\\|\alpha|=m}}\sup_{t>0}\|u_k^{(\alpha)}(\star,t)\|_p \lesssim \sup_{\substack{\alpha\in{\mathbb{N}_0}^{n+1}\\|\alpha|=m}}\sup_{t>0}\|u_k^{(\alpha)}(\star,t)\|_{H^p} \lesssim \|f_k\|_{H^p} \lesssim \|f\|_{H^p}.$$

Consequently,

$$\sup_{t>0}\sum_{\alpha\in{\mathbb{N}_0}^{n+1},|\alpha|=m}\|u_k^{(\alpha)}(\star,t)\|_p \lesssim \|f\|_{H^p},$$

where the implicit constant does not depend on k. Fix $t > 0$ and $(\alpha_1,\alpha_2,\dots,\alpha_n) \in {\mathbb{N}_0}^n$. Note that $\mathrm{e}^{-t\sqrt{-\Delta}}f \in L^2(\mathbb{R}^n)$ for all $t > 0$; see Exercise 3.41. Once $t > 0$ is fixed, by passing to a subsequence and using

$$R_1^{\alpha_1}R_2^{\alpha_2}\cdots R_n^{\alpha_n}[\exp(-t\sqrt{-\Delta})f] = \lim_{k\to\infty}\exp(-t\sqrt{-\Delta})[R_1^{\alpha_1}R_2^{\alpha_2}\cdots R_n^{\alpha_n}f_k]$$

in $L^2(\mathbb{R}^n)$, it can be assumed that

$$R_1^{\alpha_1}R_2^{\alpha_2}\cdots R_n^{\alpha_n}[\exp(-t\sqrt{-\Delta})f](x) = \lim_{k\to\infty}R_1^{\alpha_1}R_2^{\alpha_2}\cdots R_n^{\alpha_n}[\exp(-t\sqrt{-\Delta})f_k](x)$$

for almost every $x \in \mathbb{R}^n$. Thus, we are in the position of using the Fatou lemma to have

$$\sup_{t>0}\sum_{\alpha\in{\mathbb{N}_0}^{n+1},|\alpha|=m}\|u^{(\alpha)}(\star,t)\|_p \lesssim \|f\|_{H^p}.$$

Thus, G given by the formula (3.117) is the desired tensor-valued function.

3.2.7.2 Proof of Theorem 3.13

Due to similarity, we will prove only that (3.113) implies $f \in H^p(\mathbb{R}^n)$ for any bounded distributions f. Set

$$u_{j,\alpha}(x,t) \equiv \exp(-t\sqrt{-\Delta})[{R_1}^{\alpha_1}{R_2}^{\alpha_2}\cdots{R_n}^{\alpha_n}[f*\psi^j]](x)$$

for $(x,t) \in \mathbb{R}^{n+1}_+$ and $\alpha = (\alpha_1, \alpha_2, \ldots, \alpha_n) \in {\mathbb{N}_0}^n$ by keeping in mind that $f*\psi^j \in L^r(\mathbb{R}^n)$ for some $1 < r < \infty$. Let us also set

$$u_{j,\alpha}(0,t) \equiv {R_1}^{\alpha_1}{R_2}^{\alpha_2}\cdots{R_n}^{\alpha_n}[f*\psi^j](x)$$

for $\alpha = (\alpha_1, \alpha_2, \ldots, \alpha_n) \in {\mathbb{N}_0}^n$ and $t > 0$. Observe that

$$u_{j,\alpha}(x,t) \equiv \exp(-t\sqrt{-\Delta})[\psi^{j+1}*[{R_1}^{\alpha_1}{R_2}^{\alpha_2}\cdots{R_n}^{\alpha_n}[f*\psi^j]]](x)$$

and that $\psi^{j+1}*[{R_1}^{\alpha_1}{R_2}^{\alpha_2}\cdots{R_n}^{\alpha_n}[f*\psi^j]] \in L^\infty(\mathbb{R}^n)$. Thus $u_{j,\alpha} \in L^\infty(\mathbb{R}^n)$ for all $j = 0, 1, \ldots, n$ and $(\alpha_1, \alpha_2, \ldots, \alpha_n) \in {\mathbb{N}_0}^n$. Consequently,

$$\left(\sum_{|\alpha|\le m}\frac{|u_{j,\alpha}(x,t)|^2}{\alpha!}\right)^{\frac{n-1}{2(n+m-1)}} \le \exp(-t\sqrt{-\Delta})\left[\left(\sum_{|\alpha|\le m}\frac{|u_{j,\alpha}(\star,0)|^2}{\alpha!}\right)^{\frac{n-1}{2(n+m-1)}}\right](x)$$

from Lemma 1.41. Thus,

$$\left(\sum_{|\alpha|\le m}\frac{|u_{j,\alpha}(x,t)|^2}{\alpha!}\right) \le M^{\left(\frac{n-1}{n+m-1}\right)}\left[\left(\sum_{|\alpha|\le m}\frac{|u_{j,\alpha}(\star,0)|^2}{\alpha!}\right)^{\frac{1}{2}}\right](x)$$

for all $(x,t) \in \mathbb{R}^{n+1}_+$. Therefore,

$$\sup_{j\in\mathbb{N}}\sup_{t>0}\int_{\mathbb{R}^n}\left(\sum_{|\alpha|\le m}\frac{1}{\alpha!}|u_{j,\alpha}(x,t)|^2\right)^{\frac{p}{2}}\mathrm{d}x < \infty.$$

Thus, $\{f*\psi^j\}_{j=1}^\infty$ forms a bounded set in $H^p(\mathbb{R}^n)$ according to Proposition 3.12. Due to the Fatou property of $H^p(\mathbb{R}^n)$ (see Exercise 3.42), $f \in H^p(\mathbb{R}^n)$.

Exercises

Exercise 3.40 Let $f \in \bigcup_{1<u<\infty} L^u(\mathbb{R}^n)$, and let $1 < p < \infty$. Then using Theorem 1.65, show that the estimate $\|f\|_p + \sum_{j=1}^{n} \|R_j f\|_p < \infty$ holds if and only if $f \in L^p(\mathbb{R}^n)$.

Exercise 3.41 Let $f \in H^p(\mathbb{R}^n)$ with $0 < p \le 1$. Choose $\psi \in \mathscr{S}(\mathbb{R}^n)$ so that $\chi_{B(1)} \le \psi \le \chi_{B(2)}$.

1. Show that $\psi(D)f \in L^1(\mathbb{R}^n)$ using Corollary 1.8.
2. Show that $e^{-t\sqrt{-\Delta}} f \in L^2(\mathbb{R}^n)$ for all $t > 0$ using Corollary 1.8 and $e^{-t\sqrt{-\Delta}} f = e^{-t\sqrt{-\Delta}}\psi(D)f + e^{-t\sqrt{-\Delta}}(1 - \psi(D))f$.

Exercise 3.42 Let $0 < p \le 1$ and $\{f_j\}_{j=1}^{\infty}$ be a bounded sequence in $H^p(\mathbb{R}^n)$. If $\{f_j\}_{j=1}^{\infty}$ converges to $f \in \mathscr{S}'(\mathbb{R}^n)$ in $\mathscr{S}'(\mathbb{R}^n)$, then show that $f \in H^p(\mathbb{R}^n)$. Hint: Consider establishing the Fatou property for Hardy spaces.

3.2.8 *Local Hardy Spaces*

There are two types of Triebel–Lizorkin spaces: nonhomogeneous Triebel–Lizorkin spaces and homogeneous Triebel–Lizorkin spaces. So, we are led to thinking that there should be "homogeneous" Hardy spaces and "nonhomogeneous" Hardy spaces. The local Hardy spaces correspond to the Hardy spaces in this sense. So far, it is not clear which Hardy space is the homogeneous one, although from the definition we can have a guess. We clarify this point in Sect. 3.2.9. We omit the proof of some results on local Hardy spaces because they are similar to the assertions corresponding to Hardy spaces.

3.2.8.1 Local Hardy Spaces

The proof of the next theorem is omitted to an extent because it is often similar to that for Hardy spaces.

Theorem 3.15 *Maintain the same assumption as Theorem 3.4. Then*

$$\left\| \sup_{\tilde{\psi} \in \mathscr{F}_N, 0<t\le 1} \frac{1}{t^n} \left| \tilde{\psi}\left(\frac{\star}{t}\right) * f \right| \right\|_p \lesssim \left\| \sup_{0<t\le 1} \frac{1}{t^n} \left| \psi\left(\frac{\star}{t}\right) * f \right| \right\|_p$$

for all $f \in \mathscr{S}'(\mathbb{R}^n)$.

This theorem justifies the following definition:

Definition 3.6 (Local Hardy spaces) Let $0 < p < \infty$. The *local Hardy space* $h^p(\mathbb{R}^n)$ is the set of all $f \in \mathscr{S}'(\mathbb{R}^n)$ such that

$$\|f\|_{h^p} \equiv \left\| \sup_{0<t\leq 1} \frac{1}{t^n} \left| \psi\left(\frac{\star}{t}\right) * f \right| \right\|_p < \infty, \tag{3.118}$$

where $\psi \in \mathscr{S}(\mathbb{R}^n) \setminus \mathscr{P}_0(\mathbb{R}^n)^{\perp}$ is a fixed function.

The following definition yields an equivalent norm:

$$\|f\|_{h^p}^* \equiv \left\| \sup_{\psi \in \mathscr{F}_N, 0<t\leq 1} \frac{1}{t^n} \left| \psi\left(\frac{\star}{t}\right) * f \right| \right\|_p \quad (f \in \mathscr{S}'(\mathbb{R}^n)). \tag{3.119}$$

3.2.8.2 Equivalent Norms

Sometimes we choose $\widetilde{\psi} \in \mathscr{S}(\mathbb{R}^n)$ so that $\widetilde{\psi}(0) \neq 0$. For $f \in \mathscr{S}'(\mathbb{R}^n)$, define

$$\|f\|_{h^p}^{**} \equiv \left\| \sup_{0<t\leq 1} |\widetilde{\psi}(t D) * f| \right\|_p, \quad \|f\|_{h^p}^{***} \equiv \left\| \sup_{\kappa \in \mathscr{F}_N, 0<t\leq 1} |\kappa(t D) * f| \right\|_p. \tag{3.120}$$

This may be more convenient. See Exercise 3.46 for (3.118), (3.119), (3.120).

3.2.8.3 Hardy Spaces and Local Hardy Spaces

By definition, it is easy to check that Hardy spaces are embedded into local Hardy spaces continuously. Let us quantify this embedding more precisely. Note that local Hardy spaces can be characterized by means of Hardy spaces, as the following lemma shows:

Lemma 3.12 *Let $\psi \in \mathscr{S}(\mathbb{R}^n)$ satisfy $\chi_{B(4)} \leq \psi \leq \chi_{B(8)}$. Then we have the following norms equivalence for $f \in \mathscr{S}'(\mathbb{R}^n)$: $\|f\|_{h^p} \sim \|\psi(D)f\|_p + \|(1 - \psi(D))f\|_{H^p}$.*

Proof Since $\psi(t D)(1 - \psi(D))f = 0$ for $t \geq 2$, we have

$$\|(1 - \psi(D))f\|_{H^p} \lesssim \|(1 - \psi(D))f\|_{h^p} \lesssim \|\psi(D)f\|_{h^p} + \|f\|_{h^p}.$$

Since $\sup\limits_{0<t\leq 1} p_N(\psi \cdot \psi(t\star)) < \infty$, we have $\sup\limits_{0<t\leq 1} |\psi(D)\psi(t D)f(x)| \lesssim \mathscr{M} f(x)$. Hence using the norm $\| \star \|_{h^p}^{***}$ given by (3.120), we have

$$\|\psi(D)f\|_{h^p} \lesssim \|f\|_{h^p}^{***} \lesssim \|f\|_{h^p}.$$

Furthermore, by definition of the $h^p(\mathbb{R}^n)$-norm, we have $\|\psi(D)f\|_p \lesssim \|f\|_{h^p}$. Hence it remains to show the reverse inequality $\lesssim$. By the quasi-triangle inequality and Theorem 1.53, we have

$$\|f\|_{h^p} \lesssim \|\psi(D)f\|_{h^p} + \|(1-\psi(D))f\|_{h^p} \lesssim \|\psi(D)f\|_{h^p} + \|(1-\psi(D))f\|_{H^p}.$$

So, to prove the reverse inequality $\lesssim$, we have only to prove

$$\|\psi(D)f\|_{h^p} \lesssim \|\psi(D)f\|_p. \tag{3.121}$$

Let $0 < t \le 1$. For $0 < \eta < \min(1, p)$, we have

$$\begin{aligned}|\psi(t\,D)\psi(D)f(x)| &\lesssim t^{-n}\int_{\mathbb{R}^n} |\mathscr{F}^{-1}\psi(t^{-1}y)| \cdot |\psi(D)f(x-y)|\mathrm{d}y \\ &\lesssim t^{-n} M^{(\eta)}[\,\psi(D)f\,](x)\int_{\mathbb{R}^n} |\mathscr{F}^{-1}\psi(t^{-1}y)| \cdot \langle y\rangle^{\frac{n}{\eta}}\mathrm{d}y \\ &\lesssim t^{-n} M^{(\eta)}[\,\psi(D)f\,](x)\int_{\mathbb{R}^n} |\mathscr{F}^{-1}\psi(t^{-1}y)| \cdot \langle t^{-1}y\rangle^{\frac{n}{\eta}}\mathrm{d}y \\ &\simeq M^{(\eta)}[\,\psi(D)f\,](x).\end{aligned}$$

Here $M^{(\eta)}$ denotes the powered Hardy–Littlewood maximal operator. Taking the supremum over $0 < t \le 1$, we conclude

$$\sup_{t>0} |\psi(t\,D)\psi(D)f(x)| \lesssim M^{(\eta)}[\,\psi(D)f\,](x).$$

By the maximal inequality we obtain (3.121).

3.2.8.4 Atomic Decomposition for $h^p(\mathbb{R}^n)$

Note that $H^p(\mathbb{R}^n)$ is smaller than $h^p(\mathbb{R}^n)$. This means that we have to expand the notion of atoms; we also need to use blocks.

Definition 3.7 (**(p,q)-block centered at Q**) Let $0 < p \le 1 < q \le \infty$, and let $Q \in \mathscr{Q}$. A function $A \in L^q(\mathbb{R}^n)$ is said to be a *(p,q)-block centered at Q*, if $\mathrm{supp}(A)$ is contained in a cube Q and A satisfies $\|A\|_q \le |Q|^{\frac{1}{q}-\frac{1}{p}}$. Call (p,q)-blocks centered at Q $h^p(\mathbb{R}^n)$-blocks.

Note that we postulated atoms on the moment condition but that the blocks do not necessarily satisfy the moment condition.

Lemma 3.13 *Let $0 < p \le 1 < q \le \infty$. If the sidelength of Q exceeds* 1, *we have $\|A\|_{h^p} \lesssim_{p,q} 1$ for all (p,q)-blocks centered at Q.*

Proof Choose $\psi \in \mathscr{S}(\mathbb{R}^n) \setminus \{0\}$ so that $0 \le \psi \le \chi_{Q(1)}$. Since $|Q| \ge 1$, we have

$$\sup_{0<t\le 1} \left| \frac{1}{t^n}\psi\left(\frac{\star}{t}\right) * A(x) \right| \lesssim \chi_{2Q}(x) MA(x). \tag{3.122}$$

Hence

$$\begin{aligned}\left\| \sup_{0<t\le 1} \left| \frac{1}{t^n}\psi\left(\frac{\star}{t}\right) * A \right| \right\|_p &\lesssim \left(\int_{2Q} MA(x)^p \mathrm{d}x \right)^{\frac{1}{p}} \\ &\lesssim |Q|^{\frac{1}{p}-\frac{1}{q}} \left(\int_{2Q} MA(x)^q \mathrm{d}x \right)^{\frac{1}{q}} \\ &\lesssim 1\end{aligned}$$

by the $L^q(\mathbb{R}^n)$-boundedness of the Hardy–Littlewood maximal operator M, as required.

We have the atomic decomposition for $h^p(\mathbb{R}^n)$. Unlike that for $H^p(\mathbb{R}^n)$, the proof is considerably short. This is because we can resort to that for $H^p(\mathbb{R}^n)$.

Theorem 3.16 (Atomic decomposition for $h^p(\mathbb{R}^n)$) *Let $0 < p \le 1 < q \le \infty$.*

1. *Let $\{A_j\}_{j=1}^\infty$ be a collection of $H^p(\mathbb{R}^n)$-atoms, and let $\{B_j\}_{j=1}^\infty$ be a collection of $h^p(\mathbb{R}^n)$-blocks with the volume of the corresponding cubes larger than* 1. *Let $\lambda = \{\lambda_j\}_{j=1}^\infty$, $\mu = \{\mu_j\}_{j=1}^\infty$ be sequences in $\ell^p(\mathbb{N})$. Then*

$$f \equiv \sum_{j=1}^\infty \lambda_j A_j + \sum_{j=1}^\infty \mu_j B_j \tag{3.123}$$

converges in the topology of $h^p(\mathbb{R}^n)$ and the norm estimate

$$\|f\|_{h^p} \lesssim_p \|\lambda\|_{\ell^p} + \|\mu\|_{\ell^p} \tag{3.124}$$

holds.

2. *Any $f \in h^p(\mathbb{R}^n)$ admits a decomposition:*

$$f = \sum_{j=1}^\infty \lambda_j A_j + \sum_{j=1}^\infty \mu_j B_j, \tag{3.125}$$

where $\{A_j\}_{j=1}^\infty$ is a collection of $H^p(\mathbb{R}^n)$-atoms, $\{B_j\}_{j=1}^\infty$ is a collection of $h^p(\mathbb{R}^n)$-blocks with the volume of the corresponding cubes larger than 1 *and $\lambda = \{\lambda_j\}_{j=1}^\infty$, $\mu = \{\mu_j\}_{j=1}^\infty$ are sequences in $\ell^p(\mathbb{N})$. Furthermore, the sequences λ and μ satisfy*

$$\|\lambda\|_{\ell^p} + \|\mu\|_{\ell^p} \lesssim_p \|f\|_{h^p}. \tag{3.126}$$

Proof We already proved (3.123) and (3.124) in Theorem 3.9 and Lemma 3.13, respectively. We will establish (3.125) and (3.126). Let $f \in h^p(\mathbb{R}^n)$. Choose $\psi \in \mathscr{S}(\mathbb{R}^n)$ so that $\chi_{B(1)} \le \psi \le \chi_{B(2)}$. By Lemma 3.12, $f \in h^p(\mathbb{R}^n) \mapsto (1-\psi(D))f \in H^p(\mathbb{R}^n)$ is continuous. Since $H^p(\mathbb{R}^n)$ admits decomposition without blocks (see Sect. 3.2.6), we concentrate on $\varphi(D)f$. Furthermore, by the monotonicity q, it can be assumed that $q = \infty$. Here define a function V_m by $V_m \equiv \chi_{m+[0,1]^n}\psi(D)f$. By the sampling theorem (Theorem 1.51), we have

$$\left(\sum_{m\in\mathbb{Z}^n}\sup_{x\in\mathbb{R}^n}|V_m(x)|^p\right)^{\frac{1}{p}} \lesssim \|\psi(D)f\|_p. \tag{3.127}$$

Hence we can decompose $\psi(D)f$ into blocks: $\psi(D)f = \sum_{m\in\mathbb{Z}^n}\chi_{m+[0,1]^n}\psi(D)f$. If we use (3.127) to estimate coefficients, then we have (3.126).

Exercises

Exercise 3.43 Let $p > 1$. Then show that $h^p(\mathbb{R}^n) \approx L^p(\mathbb{R}^n)$ with equivalence of norms using the reflexivity and the $L^p(\mathbb{R}^n)$-boundedness of the Hardy–Littlewood maximal operator.

Exercise 3.44 Show that the norms given by (3.120) are equivalent to the norm of $h^p(\mathbb{R}^n)$ by mimicking the proof of the similar assertion for Hardy spaces.

Exercise 3.45 [681, Lemma 3.2] Let $0 < p \le 1$, $0 < q \le \infty$, $s \in \mathbb{R}$, and let $\eta \in \mathscr{S}(\mathbb{R}^n)$ satisfy $\chi_{Q(1/4)} \le \eta \le \chi_{Q(1/2)}$. For a finitely supported sequence $\{c_l\}_{l\in\mathbb{Z}^n}$, define

$$f(x) = \sum_{l\in\mathbb{Z}^n} c_l e^{2\pi i l x}\eta(x-l) \quad (x \in \mathbb{R}^n).$$

1. Calculate $\|f\|_p$.
2. Assume that p, q, s is such that $M^s_{pq}(\mathbb{R}^n) \hookrightarrow h^p(\mathbb{R}^n)$. Use $\|f\|_p \lesssim \|f\|_{h^p}$ with the implicit constant independent of $\{c_l\}_{l\in\mathbb{Z}^n}$ to show that

$$\left(\sum_{k\in\mathbb{Z}^n}|c_k|^p\right)^{\frac{1}{p}} \lesssim \left(\sum_{k\in\mathbb{Z}^n}|c_k|^q(1+|k|)^{sq}\right)^{\frac{1}{q}}.$$

3.2.9 The Hardy Space $H^p(\mathbb{R}^n)$ and the Triebel–Lizorkin Space $\dot{F}^0_{p2}(\mathbb{R}^n)$

Let $0 < p \le 1$. We investigate the relation between the Hardy space $H^p(\mathbb{R}^n)$ and the homogeneous Triebel–Lizorkin space $\dot{F}^0_{p2}(\mathbb{R}^n)$ and, as a corollary, that between the local Hardy space $h^p(\mathbb{R}^n)$ and the nonhomogeneous Triebel–Lizorkin space $F^0_{p2}(\mathbb{R}^n)$. In this section, we always assume that $\psi \in \mathscr{S}(\mathbb{R}^n)$ satisfies $\chi_{B(1)} \le \psi \le \chi_{B(2)}$. Furthermore, $\varphi \in \mathscr{S}(\mathbb{R}^n)$ satisfies $\text{supp}(\varphi) \subset B(8) \setminus B(1)$ and $\sum_{j=-\infty}^{\infty} {\varphi_j}^2 \equiv \chi_{\mathbb{R}^n \setminus \{0\}}$.

3.2.9.1 Hardy Spaces and Homogeneous Triebel–Lizorkin Spaces

Recall that $\dot{F}^0_{p2}(\mathbb{R}^n)$ with $0 < p \le 1$ admits the realization. So for $f \in \mathscr{S}'(\mathbb{R}^n)$, the Hardy space $H^p(\mathbb{R}^n)$ and the Triebel–Lizorkin space $\dot{F}^0_{p2}(\mathbb{R}^n)$ are given by

$$\|f\|_{H^p} = \left\| \sup_{j \in \mathbb{Z}} |\psi_j * f| \right\|_p, \quad \|f\|_{\dot{F}^0_{p2}} = \left\| \varphi_j(D) f \right\|_{L^p(\ell^2)},$$

respectively. We aim to show that these spaces are the same.

Let us check:

Lemma 3.14 *Let $0 < p \le 1$, and let $f \in \mathscr{S}'(\mathbb{R}^n)$.*

1. *If $f \in H^p(\mathbb{R}^n)$, then*

$$f = \sum_{j=-\infty}^{\infty} \varphi_j(D)^2 f \tag{3.128}$$

holds in the topology of $\mathscr{S}'(\mathbb{R}^n)$.

2. *If $f \in \dot{F}^0_{p2}(\mathbb{R}^n)$ (precisely $[f] \in \dot{F}^0_{p2}(\mathbb{R}^n)$), then*

$$\sum_{j=-\infty}^{\infty} \varphi_j(D)^2 f \tag{3.129}$$

converges in $\mathscr{S}'(\mathbb{R}^n)$ to an element whose difference from f is a polynomial.

Proof We can use the atomic decomposition for the proof of (3.128). If f is a (p, ∞)-atom, then (3.128) holds trivially. The convergence (3.129) can be obtained from Theorem 2.33.

When $f \in \dot{F}^0_{p2}(\mathbb{R}^n)$, recall that $\sum_{j=-\infty}^{\infty} \varphi_j(D)^2 f$ is called a realization. Modulo polynomials, we can choose realizations, and $H^p(\mathbb{R}^n)$ and $\dot{F}^0_{p2}(\mathbb{R}^n)$ are Banach spaces embedded into $\mathscr{S}'(\mathbb{R}^n)$. If f is a realization of these function spaces, $f = \sum_{j=-\infty}^{\infty} \varphi_j(D)^2 f$ holds in the topology of $\mathscr{S}'(\mathbb{R}^n)$.

3.2.9.2 Hardy Spaces and Triebel–Lizorkin Spaces

As we have announced, Hardy spaces are realized by way of Triebel–Lizorkin spaces.

Theorem 3.17 *Let* $0 < p \leq 1$. *Let also* $f \in H^p(\mathbb{R}^n)$, *or let* $f \in \dot{F}^0_{p2}(\mathbb{R}^n)$ *be its realization. Then* $\|f\|_{H^p} \sim \|f\|_{\dot{F}^0_{p2}}$.

This theorem in turn can be regarded as a natural introduction of Triebel–Lizorkin spaces.

Proof We recall the definition of the norm;

$$\|\{f_j\}_{j=-\infty}^{\infty}\|_{H^p(\ell^2)} = \left\| \sup_{l \in \mathbb{Z}} \left(\sum_{j=-\infty}^{\infty} |2^{ln} \exp(-4^l |\star|^2) * f_j|^2 \right)^{1/2} \right\|_p .$$

Thus,

$$\|\{\varphi_j(D) f\}_{j=-\infty}^{\infty}\|_{H^p(\ell^2)} = \left\| \sup_{l \in \mathbb{Z}} \left(\sum_{j=-\infty}^{\infty} |2^{ln} \exp(-4^l |\star|^2) * \varphi_j(D) f|^2 \right)^{1/2} \right\|_p .$$

Define $\varphi \equiv \psi(2^{-1}\star) - \psi$ by way of ψ. By the size of supports, we have

$$2^{ln} \exp(-4^l |\star|^2) * \varphi_j(D) f = \sum_{j'=-2}^{2} 2^{ln} \exp(-4^l |\star|^2) * [\varphi_{j+j'}(D) \varphi_j(D) f].$$

Now we have an estimate independent of l and j, j':

$$|2^{ln} \exp(-4^l |\star|^2) * [\varphi_{j+j'}(D) \varphi_j(D) f]| \lesssim_\eta M^{(\eta)}[\varphi_j(D) f] \quad (\eta > 0).$$

Since $j' = -2, -1, 0, 1, 2$, we consider the special case $j' = 0$. Let us show

$$|2^{ln} \exp(-4^l | \star |^2) * [\varphi_j(D)\varphi_j(D)f]| \lesssim_\eta M^{(\eta)}[\varphi_j(D)f] \quad (\eta > 0)$$

with constants independent of l and j, j'.

To this end, we use Lemma 1.24. By this lemma, this amounts to establishing $\| \exp(-4^{j-l} | \star |^2)\varphi\|_{H^s} \lesssim 1$, where $s > \dfrac{n}{\eta} + \dfrac{n}{2}$. We define $M \equiv 2\left[\dfrac{n}{\eta} + \dfrac{n}{2} + 1\right]$. Thus, it suffices to show $\| \exp(-4^{j-l} | \star |^2)\varphi\|_{H^M} \lesssim 1$. Now

$$\begin{aligned}
&\sup_{|\beta|=m} \|\partial^\beta [\exp(-4^{j-l} | \star |^2)\varphi]\|_\infty \\
&\lesssim_m \sum_{\alpha,\beta \in \mathbb{N}_0{}^n, |\alpha|+|\beta| \le m} \left(\sup_{x \in \mathbb{R}^n} |\partial^\alpha [\exp(-4^{j-l} | \star |^2)](x) \partial^\beta \varphi(x)| \right)
\end{aligned}$$

for $m = 0, 1, 2, \ldots$ and the support condition of φ. If we define P_α, $\alpha \in \mathbb{N}_0{}^n$ by

$$P_\alpha(x) \equiv \exp(|x|^2)\partial^\alpha \exp(-|x|^2) \quad (x \in \mathbb{R}^n),$$

since $0 \notin \operatorname{supp}(\varphi)$,

$$\begin{aligned}
&|\partial^\alpha [\exp(-4^{j-l} | \star |^2)](x) \partial^\beta \varphi(x)| \\
&= 2^{|\alpha|(j-l)} |P_\alpha(2^{j-l}x) \exp(-4^{j-l}|x|^2) \partial^\beta \varphi(x)| \\
&= |2^{j-l}x|^{|\alpha|} |P_\alpha(2^{j-l}x) \exp(-4^{j-l}|x|^2)| \times |x|^{-|\alpha|} |\partial^\beta \varphi(x)| \\
&\le \chi_{B(8)}(x) \sup_{y \in \mathbb{R}^n} |y|^{|\alpha|} |P_\alpha(y)| \exp(-|y|^2) \times \sup_{y \in \mathbb{R}^n} |y|^{-|\alpha|} |\partial^\beta \varphi(y)| \lesssim_{\alpha,\beta} \chi_{B(8)}(x).
\end{aligned}$$

Thus, for all $x \in \mathbb{R}^n$,

$$\| \exp(-4^{j-l} | \star |^2)\varphi\|_{H^M} \sim \sup_{|\beta|=m} \|\partial^\beta [\exp(-4^{j-l} | \star |^2)\varphi]\|_p \lesssim \|\chi_{B(8)}\|_2 \sim 1.$$

Let us prove the inequality $\gtrsim$. Recall that we defined the space $H^p(\ell^2, \mathbb{R}^n)$ by (3.48) before we formulate Theorem 3.8. Let $N \in \mathbb{N}$. Define a linear operator T_N from $H^p(\ell^2, \mathbb{R}^n)$ to itself by

$$T_N(\{f_j\}_{j=-\infty}^\infty) \equiv \left\{ \sum_{l=-N}^N \delta_{jl} \varphi_l(D) f_0 \right\}_{j=-\infty}^\infty .$$

Thanks to Theorem 3.8, T_N is a bounded linear operator with the norm bounded by a constant independent of N. Letting $N \to \infty$, we have $\|\varphi_j(D)f\|_{H^p(\ell^2)} \lesssim \|f\|_{H^p}$. Next, we deduce $\|f\|_{\dot{F}^0_{p2}} \lesssim \|\varphi_j(D)f\|_{H^p(\ell^2)}$ from the definition of $H^p(\ell^2, \mathbb{R}^n)$.

Let us prove the inequality $\lesssim$. Let $N \in \mathbb{N}$.

Let S_N be a linear operator from $H^p(\ell^2, \mathbb{R}^n)$ to $H^p(\mathbb{R}^n)$, which is given by

$$S_N(\{f_j\}_{j=-\infty}^{\infty}) \equiv \sum_{l=-N}^{N} \varphi_l(D) f_l.$$

Then from Theorem 3.8, we have

$$\|S_N(\{f_j\}_{j=-\infty}^{\infty})\|_{H^p} = \left\| \sum_{l=-N}^{N} \varphi_l(D) f_l \right\|_{H^p} \lesssim \|\{f_l\}_{l\in\mathbb{Z}}\|_{H^p(\ell^2)}. \tag{3.130}$$

In (3.130), let $f_l \equiv \varphi_l(D)f$ for $l \in \mathbb{Z}$, and we obtain $\|f\|_{H^p} \lesssim \|\{\varphi_l(D)f\}_{l\in\mathbb{Z}}\|_{H^p(\ell^2)}$ by letting $N \to \infty$. Theorem 1.53 yields

$$\|\{\varphi_l(D)f\}_{l\in\mathbb{Z}}\|_{H^p(\ell^2)} \lesssim \|\{\varphi_l(D)f\}_{l\in\mathbb{Z}}\|_{L^p(\ell^2)} = \|f\|_{\dot{F}^0_{p2}}.$$

Thus the left-hand inequality is proved.

Although we have not considered density in $H^p(\mathbb{R}^n)$ yet, we now can propose the following natural dense subspace:

Corollary 3.3 *Let* $0 < p \le 1$. *Then* $\mathscr{S}_\infty(\mathbb{R}^n)$ *is dense in* $H^p(\mathbb{R}^n)$.

Proof Use $H^p(\mathbb{R}^n) \approx \dot{F}^0_{p2}(\mathbb{R}^n)$ with equivalence of norms and the fact that $\mathscr{S}_\infty(\mathbb{R}^n)$ is dense in $\dot{F}^0_{p2}(\mathbb{R}^n)$.

Having set down elementary facts on homogeneous spaces, we move on to nonhomogeneous spaces. We have an analogy for local Hardy spaces:

Theorem 3.18 *Let* $0 < p \le 1$. *Then* $h^p(\mathbb{R}^n) \approx F^0_{p2}(\mathbb{R}^n)$ *with equivalence of norms.*

Proof Since $\dot{F}^0_{p2}(\mathbb{R}^n) \approx H^p(\mathbb{R}^n)$ with equivalence of norms, we have

$$\|f\|_{h^p} \sim \|\psi(D)f\|_p + \|(1-\psi(D))f\|_{H^p} \sim \|\psi(D)f\|_p + \|(1-\psi(D))f\|_{\dot{F}^0_{p2}}.$$

Meanwhile, Theorem 1.53 also yields

$$\|(1-\psi(D))f\|_{\dot{F}^0_{p2}} \lesssim \|f\|_{F^0_{p2}} \lesssim \|\psi(D)f\|_p + \|(1-\psi(D))f\|_{\dot{F}^0_{p2}}.$$

Thus, we obtain $\|f\|_{h^p} \sim \|f\|_{F^0_{p2}}$ with equivalence of norms.

The next corollary is a nonhomogeneous counterpart of Corollary 3.3.

Corollary 3.4 *Let* $0 < p \le 1$. *Then* $\mathscr{S}(\mathbb{R}^n)$ *is dense in* $h^p(\mathbb{R}^n)$.

Proof Note that $h^p(\mathbb{R}^n) \approx F^0_{p2}(\mathbb{R}^n)$ with equivalence of norms and that $\mathscr{S}(\mathbb{R}^n)$ is dense in $F^0_{p2}(\mathbb{R}^n)$ according to Proposition 2.10.

Exercises

Exercise 3.46 Show that the norms (3.118), (3.119), (3.120) are equivalent by mimicking the proof of the corresponding assertion for Hardy spaces.

Exercise 3.47 [382, Proposition 4 when $1 < p < \infty$] Let $0 < p < \infty$. Suppose that a collection $\{f_j\}_{j=1}^{\infty} \subset L^p(\mathbb{R}^n) \cap L^1_{\rm loc}(\mathbb{R}^n) \cap \mathscr{S}'(\mathbb{R}^n)$ satisfies $\mathscr{F} f_j \subset B(2^{j+1}) \setminus B(2^{j-1})$ and

$$\left\| \left(\sum_{j=1}^{\infty} |f_j|^2 \right)^{\frac{1}{2}} \right\|_p < \infty.$$

Then show that

$$\left\| \sum_{j=1}^{\infty} f_j \right\|_p \lesssim \left\| \left(\sum_{j=1}^{\infty} |f_j|^2 \right)^{\frac{1}{2}} \right\|_p .$$

Hint: Since p is finite, we can reduce matters to proving

$$\left\| \sum_{j=1}^{J} f_j \right\|_p \lesssim \left\| \left(\sum_{j=1}^{J} |f_j|^2 \right)^{\frac{1}{2}} \right\|_p$$

with the implicit constant independent of J. When $1 < p < \infty$, use Theorem 3.3. When $0 < p \le 1$, we use

$$\left\| \sum_{j=1}^{J} f_j \right\|_p \lesssim \left\| \sum_{j=1}^{J} f_j \right\|_{H^p}$$

and Theorem 3.17.

Textbooks in Sect. 3.2

The Hardy Space $H^1(\mathbb{R}^n)$

See [22, Chapter 6], [40, Definition 5.5] and [117, Section 6.2].

The Hardy Space $H^p(\mathbb{R}^n)$ for $0 < p \leq 1$

In addition to the above references, see [31, Chapter 3], [33, Section 2.1], [86, Chapter 3] and [90, Chapter 3] for the Hardy space $H^p(\mathbb{R}^n)$.

The Wavelet Decomposition of $H^p(\mathbb{R}^n)$

See [40, Theorem 5.8] for the wavelet decomposition.

Vector-Valued Hardy Space: Theorem 3.7

See [97, Section 3.2.1].

Equivalence $H^p(\mathbb{R}^n) \approx \dot{F}^0_{p2}(\mathbb{R}^n)$

[86, Chapter 1, Section 6] contains an idea of the proof of $H^s_p(\mathbb{R}^n) \approx F^s_{p2}(\mathbb{R}^n)$ employed in this book.

Characterization of Hardy Spaces via Riesz Transforms: Theorem 3.13

See [86, p. 123 Proposition 3 and p. 133, 5.16] for Theorem 3.13.

A Characterization of Local Hardy Spaces: Theorem 3.17

See [99, p. 93, Remark].

Isomorphism $h^1(\mathbb{R}^n) \approx F^0_{p2}(\mathbb{R}^n)$: Theorem 3.18

See [99, p. 92, Theorem].

Isomorphism $F^0_{p2}(\mathbb{R}^n) \approx H^p(\mathbb{R}^n)$: Theorem 3.21

See [70].

Others

The author referred to the textbook [99] for the proof of Theorems 3.7 and 3.8, and is indebted to Professor Akihiko Miyachi for Exercise 3.25. In writing 3.2.7, we followed largely the paper [1167].

Isomorphism $h^1(\mathbb{R}^n) \approx F^0_{p2}(\mathbb{R}^n)$: Theorem 3.18

See [99, p. 92, Theorem 1].

3.3 BMO($\mathbb{R}^n$)

As we mentioned earlier, the singular integral operators do not map $L^\infty(\mathbb{R}^n)$ to itself. So we need to look for a good substitute of $L^\infty(\mathbb{R}^n)$. As we will see, they map $L^\infty(\mathbb{R}^n)$ to the space BMO($\mathbb{R}^n$). This fact is a dual to the one that the singular integral operators map $H^1(\mathbb{R}^n)$ to $L^1(\mathbb{R}^n)$. BMO is an abbreviation of "bounded mean oscillation".

In Sect. 3.3.1, we study the space BMO($\mathbb{R}^n$). Our main aim is to prove the duality $H^1(\mathbb{R}^n)$-BMO($\mathbb{R}^n$) and the property of BMO($\mathbb{R}^n$) itself. In Sect. 3.3.2, we define and investigate bmo, the local BMO($\mathbb{R}^n$) space. Here we are concerned with the duality $h^1(\mathbb{R}^n)$-bmo($\mathbb{R}^n$). In Sect. 3.3.3 we define $F^s_{\infty q}(\mathbb{R}^n)$ and investigate the relation between bmo($\mathbb{R}^n$) and $F^0_{\infty 2}(\mathbb{R}^n)$.

3.3.1 BMO($\mathbb{R}^n$)*: Definition and Fundamental Properties*

Let us start with the definition of BMO($\mathbb{R}^n$).

3.3.1.1 BMO($\mathbb{R}^n$)

We aim to specify the dual of $H^1(\mathbb{R}^n)$. We first give the answer to this problem by defining BMO($\mathbb{R}^n$) Recall that $m_Q(f)$ denotes the average of the locally integrable function f over a cube Q.

Definition 3.8 (BMO($\mathbb{R}^n$) **(space))** Define

$$\|f\|_{\mathrm{BMO}} \equiv \sup_{Q\in\mathcal{Q}} \frac{1}{|Q|}\int_Q |f(y)-m_Q(f)|\mathrm{d}y = \sup_{Q\in\mathcal{Q}} m_Q(|f-m_Q(f)|)$$

for $f \in L^1_{\mathrm{loc}}(\mathbb{R}^n)$. The "norm" $\|\star\|_{\mathrm{BMO}}$ is called the BMO($\mathbb{R}^n$) *norm*.

Concerning terminology, a clarifying remark may be in order.

Remark 3.3

1. Strictly speaking, the BMO($\mathbb{R}^n$)-norm is not a norm appearing in function analysis: in fact, the "BMO($\mathbb{R}^n$) norm" annihilates nonzero constants. Conversely, if a function f has 0 "BMO($\mathbb{R}^n$)-norm", then f is a constant almost everywhere. So, consider the quotient linear space, and redefine the "BMO($\mathbb{R}^n$) norm" there. Then BMO($\mathbb{R}^n$) as a quotient linear space is a Banach space. Nevertheless, in this book, the BMO($\mathbb{R}^n$) space is not a quotient space; so there are nonzero functions whose BMO($\mathbb{R}^n$) norm is 0.
2. Nevertheless, one can make BMO($\mathbb{R}^n$) into a Banach space by the norm given by

$$\|f\|_{\mathrm{BMO}^+} \equiv \|f\|_{L^1(Q(1))} + \|f\|_{\mathrm{BMO}}. \tag{3.131}$$

Although the definition and the fundamental property is complicated at first glance, it has a lot to do with the embedding result.

Example 3.7 Let $f \in W^{1,n}(\mathbb{R}^n)$, so that $f \in L^n(\mathbb{R}^n)$, $\nabla f \in L^n(\mathbb{R}^n)^n$. Then for any cube Q,

$$m_Q(|f-m_Q(f)|) \lesssim |Q|^{\frac{1}{n}-1}\int_Q |\nabla f(x)|\mathrm{d}x$$

according to the Poincarë–Wirtinger inequality. See Exercise 1.83. So $f \in$ BMO($\mathbb{R}^n$). Let $f(x) \equiv \kappa(4x)\,|\log|x||^\theta$ for $x \in \mathbb{R}^n$, where $\kappa \in \mathscr{S}(\mathbb{R}^n)$ satisfies $\chi_{B(1)} \le \kappa \le \chi_{B(2)}$ and $0 < \theta \ll 1$. Then $f \in W^{1,n}(\mathbb{R}^n) \setminus L^\infty(\mathbb{R}^n)$.

3.3.1.2 John–Nirenberg Inequality

We go into the detailed discussion of the property of BMO($\mathbb{R}^n$). For the time being, we are oriented to showing that BMO($\mathbb{R}^n$) is close to $L^\infty(\mathbb{R}^n)$. The next theorem describes such an aspect.

Theorem 3.19 (John–Nirenberg inequality) *For any* $\lambda > 0$*, a cube* Q *and a nonconstant* BMO($\mathbb{R}^n$)*-function* b*,*

$$\left|\{x\in Q : |b(x)-m_Q(b)| > \lambda\}\right| \lesssim_n \exp\left(-\frac{D\lambda}{\|b\|_{\mathrm{BMO}}}\right),$$

where D depends only on the dimension.

Proof If λ is small, then the assertion is trivial; the right-hand side is more than the constant multiple of $|Q|$ and the left-hand side never exceeds $|Q|$. With this in mind, let us prove

$$\left|\left\{x \in Q \,:\, |b(x) - m_Q(b)| > k\,2^{n+2}\,\|b\|_{\rm BMO}\right\}\right| \le 2^{1-k}|Q|$$

by induction on k. Once this is achieved, let $\lambda \equiv k\,2^{n+2}\,\|b\|_{\rm BMO}$ and then pass to the continuous variable; we obtain the inequality.

When $k = 1$, then this is trivial. Assume that

$$\left|\left\{x \in R \,:\, |b(x) - m_R(b)| > k\,2^{n+1}\,\|b\|_{\rm BMO}\right\}\right| \le 2^{1-k}|R| \tag{3.132}$$

for all cubes R. We aim to prove the inequality

$$\left|\left\{x \in Q \,:\, |b(x) - m_Q(b)| > (k+1)\,2^{n+1}\,\|b\|_{\rm BMO}\right\}\right| \le 2^{-k}|Q|. \tag{3.133}$$

We can assume the set in the left-hand side in (3.133) is not empty.

A dyadic cube with respect to Q is a cube obtained by bisecting all the edges of Q several times. By the Lebesgue differentiation theorem (Theorem 1.48), there exists a proper dyadic cube with respect to Q such that

$$S \subset Q, \quad m_S(|b - m_Q(b)|) > 2\|b\|_{\rm BMO}$$

as long as $x \in Q$ satisfies $|b(x) - m_Q(b)| > (k+1)\,2^{n+1}\,\|b\|_{\rm BMO}$. Choose a maximal cube $S(x)$ among such cubes. By the maximality, we have

$$m_{S(x)}(|b - m_Q(b)|) \le 2^{n+1}\|b\|_{\rm BMO};$$

hence

$$|m_{S(x)}(b) - m_Q(b)| \le m_{S(x)}(|b - m_Q(b)|) \le 2^{n+1}\|b\|_{\rm BMO}.$$

If $S(x)$ and $S(y)$ have an interior point in common, maximality forces $S(x)$ and $S(y)$ to be identical. Thus, we have a countable collection $\{Q_\lambda\}_{\lambda\in\Lambda}$ of nonoverlapping cubes for Q:

$$2\|b\|_{\rm BMO} \le m_{Q_\lambda}(|b - m_Q(b)|), \tag{3.134}$$

$$|m_{Q_\lambda}(b) - m_Q(b)| \le 2^{n+1}\|b\|_{\rm BMO}, \tag{3.135}$$

$$\{x \in Q \,:\, |b(x) - m_Q(b)| > (k+1)\,2^{n+1}\,\|b\|_{\rm BMO}\} \subset \bigcup_{\lambda\in\Lambda} Q_\lambda. \tag{3.136}$$

Since $\{Q_\lambda\}_{\lambda\in\Lambda}$ is a family of nonoverlapping dyadic cubes with respect to Q, and inequality (3.134) holds, we have

$$\begin{aligned}\sum_{\lambda\in\Lambda}|Q_\lambda| &\le \frac{1}{2\|b\|_{\mathrm{BMO}}}\sum_{\lambda\in\Lambda}\int_{Q_\lambda}|b(x)-m_Q(b)|\mathrm{d}x\\ &\le \frac{1}{2\|b\|_{\mathrm{BMO}}}\int_Q|b(x)-m_Q(b)|\mathrm{d}x\\ &\le \frac{|Q|}{2}.\end{aligned} \tag{3.137}$$

Hence

$$\begin{aligned}&|\{x\in Q : |b(x)-m_Q(b)|>(k+1)\,2^{n+1}\,\|b\|_{\mathrm{BMO}}\}|\\ &=\sum_{\lambda\in\Lambda}|\{x\in Q_\lambda : |b(x)-m_Q(b)|>(k+1)\,2^{n+1}\,\|b\|_{\mathrm{BMO}}\}|\\ &\qquad\text{(from (3.136) and the fact that }\{Q_\lambda\}_{\lambda\in\Lambda}\text{ is nonoverlapping)}\\ &\le\sum_{\lambda\in\Lambda}|\{x\in Q_\lambda : |b(x)-m_{Q_\lambda}(b)|>k\,2^{n+1}\,\|b\|_{\mathrm{BMO}}\}|\quad\text{(from (3.135))}\\ &\le\sum_{\lambda\in\Lambda}2^{1-k}|Q_\lambda|\quad\text{(from the induction assumption (3.132) on }k)\\ &\le 2^{-k}|Q|\quad\text{(from (3.137)).}\end{aligned}$$

Thus, (3.133) is proved.

The John–Nirenberg inequality yields a corollary. This shows that any BMO function is $L^p(\mathbb{R}^n)$-integrable for all $1\le p<\infty$, and as a result we can say that BMO is close to $L^\infty(\mathbb{R}^n)$.

Corollary 3.5 *Let* $1\le p<\infty$. *Then for all* $b\in\mathrm{BMO}(\mathbb{R}^n)$ *and cubes* Q,

$$m_Q^{(p)}(b-m_Q(b))\lesssim_p\|b\|_{\mathrm{BMO}}.$$

More precisely, there exists a constant M *independent of* $1\le p<\infty$ *such that*

$$m_Q^{(p)}(b-m_Q(b))\le Mp\,\|b\|_{\mathrm{BMO}}. \tag{3.138}$$

Proof Denote by Γ the Gamma function. We calculate the right-hand side using the distribution function of b:

$$\int_Q|b(x)-m_Q(b)|^p\mathrm{d}x=p\int_0^\infty\lambda^{p-1}|\{x\in Q : |b(x)-m_Q(b)|>\lambda\}|\mathrm{d}\lambda.$$

By Theorem 3.19, we have

$$\begin{aligned}\int_Q |b(x)-m_Q(b)|^p \mathrm{d}x &\lesssim p\int_0^\infty \lambda^{p-1}\exp\left(-\frac{c_1\lambda}{\|b\|_{\mathrm{BMO}}}\right)\mathrm{d}\lambda\\ &\simeq p\Gamma(p)\|b\|_{\mathrm{BMO}}{}^p\\ &=\Gamma(p+1)\|b\|_{\mathrm{BMO}}{}^p.\end{aligned}$$

Since we know $\sqrt[p]{\Gamma(p+1)} \sim p$ for $1 \le p < \infty$ by the Stirling formula we obtain the desired result (3.138) with the size of constants.

As the following corollary shows, the integrability of $\mathrm{BMO}(\mathbb{R}^n)$ is high.

Corollary 3.6 *There exists a constant $\theta > 0$ that depends on n such that the following inequality:*

$$m_Q^{(\theta)}\left(\exp\left(\frac{|b-m_Q(b)|}{\|b\|_{\mathrm{BMO}}}\right)\right) \lesssim 1$$

holds for all nonconstant $\mathrm{BMO}(\mathbb{R}^n)$*-functions b.*

Proof Use the constant M in Corollary 3.5. Set $\theta \equiv \dfrac{1}{3M}$. Then

$$\lim_{j\to\infty}\frac{j}{\sqrt[j]{j!}} = \lim_{j\to\infty}\frac{j!}{j^j}\frac{(j+1)^{j+1}}{(j+1)!} = e < 3.$$

Hence

$$\begin{aligned}m_Q\left(\exp\left(\theta\frac{|b-m_Q(b)|}{\|b\|_{\mathrm{BMO}}}\right)\right) &= \sum_{j=0}^\infty \frac{1}{j!}m_Q\left(\theta^j\frac{|b-m_Q(b)|^j}{\|b\|_{\mathrm{BMO}}^j}\right)\\ &\le 1+\sum_{j=1}^\infty\frac{(M\theta)^j j^j}{j!} = 1+\sum_{j=1}^\infty\frac{j^j}{j!3^j} \simeq 1.\end{aligned}$$

Therefore, the proof is complete.

We call Theorem 3.19 including Corollaries 3.5 and 3.6 the *John–Nirenberg inequality*.

3.3.1.3 Singular Integral Operators on $L^\infty(\mathbb{R}^n)$

We consider the boundedness of the singular integral operators.

Theorem 3.20 *Let $T \in B(L^2(\mathbb{R}^n))$ be a singular integral operator. Then there exists a linear operator $S : L^\infty(\mathbb{R}^n) \to \mathrm{BMO}(\mathbb{R}^n)$ such that $Sf = Tf$ for $f \in L^2(\mathbb{R}^n) \cap L^\infty(\mathbb{R}^n)$.*

In general, for a singular integral operator T, we conventionally write T for S. For the sake of preciseness, here in the proof of this theorem, we distinguish S and T.

We remark that BMO($\mathbb{R}^n$) is taken as the set of all the measurable functions modulo additive constants.

Proof Since T^* is also a singular integral operator, we can define $S : L^\infty(\mathbb{R}^n) \to \mathrm{BMO}(\mathbb{R}^n)$ as the dual of $T^* : H^1(\mathbb{R}^n) \to L^1(\mathbb{R}^n)$.

3.3.1.4 Duality $H^1(\mathbb{R}^n)$ and BMO($\mathbb{R}^n$)

Now we investigate duality to show that the dual of $H^1(\mathbb{R}^n)$ is isomorphic to BMO($\mathbb{R}^n$). To this end, we need a lemma, which strongly motivates our duality results.

Lemma 3.15 *Let A be a $(1, \infty)$-atom, and let $b \in \mathrm{BMO}(\mathbb{R}^n)$. Then*

$$\left| \int_{\mathbb{R}^n} A(x)b(x)\mathrm{d}x \right| \le \|b\|_{\mathrm{BMO}}.$$

Proof By the moment condition of A, we have

$$\int_{\mathbb{R}^n} A(x)b(x)\mathrm{d}x = \int_{\mathbb{R}^n} (b(x) - m_Q(b))A(x)\mathrm{d}x.$$

Thus,

$$\left| \int_{\mathbb{R}^n} A(x)b(x)\mathrm{d}x \right| \le \int_{\mathbb{R}^n} |b(x)-m_Q(b)|\cdot|A(x)|\mathrm{d}x \le m_Q(|b-m_Q(b)|) \le \|b\|_{\mathrm{BMO}}.$$

Therefore, the proof is complete.

The dual of $H^1(\mathbb{R}^n)$ is BMO($\mathbb{R}^n$) as the following theorem shows:

Theorem 3.21 (Duality $H^1(\mathbb{R}^n)$-BMO($\mathbb{R}^n$)) *The space BMO($\mathbb{R}^n$) is canonically identified with the dual space of $H^1(\mathbb{R}^n)$ in the following sense:*

1. *For any $b \in \mathrm{BMO}(\mathbb{R}^n)$ there uniquely exists a continuous linear functional $L_b \in (H^1(\mathbb{R}^n))^*$ such that $L_b(A) = \int_{\mathbb{R}^n} b(x)A(x)\mathrm{d}x$ for all $(1, \infty)$-atoms A.*
2. *Conversely, any $L \in (H^1(\mathbb{R}^n))^*$ is realized as $L = L_b$ for some $b \in \mathrm{BMO}(\mathbb{R}^n)$. Furthermore, if $b, b' \in \mathrm{BMO}(\mathbb{R}^n)$ satisfies $L_b = L_{b'}$, then $b - b'$ is constant almost everywhere.*

In the proof, we suppose that the functions are real-valued. There are three stages to the proof: the definition of L_b, $H^1(\mathbb{R}^n)^* \hookrightarrow \mathrm{BMO}(\mathbb{R}^n)$, and the uniqueness of b for a given functional.

Proof (Step 1: The definition of L_b) First of all, we *do not* consider $\mathrm{BMO}(\mathbb{R}^n)$ modulo additive constants. Furthermore, any function here assumes its value in $\mathbb{R}$. When $b \in \mathrm{BMO}(\mathbb{R}^n)$, then it is easy to see that $|b| \in \mathrm{BMO}(\mathbb{R}^n)$ and that

$$\| \, |b| \, \|_{\mathrm{BMO}} \le 2\|b\|_{\mathrm{BMO}}. \tag{3.139}$$

See Exercise 3.48 for more details. Hence if we set $b_R \equiv \max(-R, \min(R, b))$, then $\|b_R\|_{\mathrm{BMO}} \le 10\|b\|_{\mathrm{BMO}}$. With this in mind, define

$$L_{b_R}(f) \equiv \int_{\mathbb{R}^n} f(x) b_R(x) \mathrm{d}x$$

for $f \in L^1(\mathbb{R}^n)$. Note that thanks to the John–Nirenberg inequality

$$\lim_{R\to\infty} L_{b_R}(f) = \int_{\mathbb{R}^n} f(x) b(x) \mathrm{d}x$$

for all $(1, \infty)$-atoms f. For b_R, we can make use of the embedding $H^1(\mathbb{R}^n) \hookrightarrow L^1(\mathbb{R}^n)$ and the duality $L^1(\mathbb{R}^n)$-$L^\infty(\mathbb{R}^n)$, so that $L_{b_R} \in (H^1(\mathbb{R}^n))^*$. Once we can show that $\|b_R\|_{(H^1)^*} \lesssim \|b\|_{\mathrm{BMO}}$, then there exists an increasing sequence $R(1) < R(2) < \cdots$ such that $L_{b_{R(j)}} \to L$ takes place in the weak-* topology of $(H^1(\mathbb{R}^n))^*$ due to the Banach–Alaoglu theorem. However, as we have discussed, L agrees with the mapping

$$f \mapsto \int_{\mathbb{R}^n} f(x) b(x) \mathrm{d}x$$

defined for $(1, \infty)$-atoms f. Thus, it follows that $L_b = L$ once we prove $\|b_R\|_{(H^1)^*} \lesssim \|b\|_{\mathrm{BMO}}$.

For $f \in H^1(\mathbb{R}^n)$, we can find $\{\lambda_j\}_{j=1}^\infty \in \ell^1(\mathbb{N})$ and a collection $\{a_j\}_{j=1}^\infty$ of $(1, \infty)$-atoms such that $f = \sum_{j=1}^\infty \lambda_j a_j$ in $L^1(\mathbb{R}^n)$. Hence we have $L_b(f) = \sum_{j=1}^\infty \lambda_j L_b(a_j)$ since $b \in L^\infty(\mathbb{R}^n)$.

Consequently, thanks to Lemma 3.15 and the fact that $\{\lambda_j\}_{j=1}^\infty \in \ell^1(\mathbb{N})$, we conclude that L_b is a continuous linear mapping from $H^1(\mathbb{R}^n)$ to $\mathbb{C}$.

Proof (Step 2: $H^1(\mathbb{R}^n)^* \hookrightarrow \mathrm{BMO}(\mathbb{R}^n)$) Let $l \in (H^1(\mathbb{R}^n))^*$. For a compact set A in $\mathbb{R}^n$, we define $L^2(A) \subset L^2(\mathbb{R}^n)$ to be the subspace of all $f \in L^2(\mathbb{R}^n)$ for which $f = \chi_A f$. We also define $L^2(A)_0 = L^2(A) \cap \mathscr{P}_0(\mathbb{R}^n)^\perp \subset L^2(\mathbb{R}^n)$. Any element in $f \in L^2(Q(j))_0$ is an atom modulo a multiplicative constant. Thus, $f \in L^2(Q(j))_0 \mapsto l(f) \in \mathbb{C}$ is a bounded linear functional; indeed, for $f \in L^2(Q(j))_0$

$$|l(f)| \le \|l\|_{(H^1)^*}\|f\|_{H^1} \lesssim |Q(j)|^{\frac{1}{2}} \cdot \|l\|_{(H^1)^*}\|f\|_{L^2(Q(j))}.$$

Since $L^2(Q(j))_0$ is a Hilbert space, we can uniquely find $g_j \in L^2(Q(j))_0$ such that

$$\int_{\mathbb{R}^n} g_j(x)f(x)\mathrm{d}x = l(f), \quad f \in L^2(Q(j))_0, \quad \|g_j\|_2 \lesssim |Q(j)|^{\frac{1}{2}}\|l\|_{(H^1)^*} \tag{3.140}$$

by the Riesz representation theorem asserting that any bounded linear functional on Hilbert spaces can be realized by the inner product. Let $j < k$. Then

$$\int_{\mathbb{R}^n} (g_k(x) - m_{Q(j)}(g_k))\chi_{Q(j)}(x)f(x)\mathrm{d}x = \int_{\mathbb{R}^n} g_k(x)f(x)\mathrm{d}x = l(f)$$

for $f \in L^2(Q(j))_0$; hence

$$\int_{\mathbb{R}^n} g_k(x)f(x)\mathrm{d}x = l(f) = \int_{\mathbb{R}^n} g_j(x)f(x)\mathrm{d}x$$

for $f \in L^2(Q(j))_0$. Hence by the uniqueness of (3.140), $g_j = (g_k - m_{Q(j)}(g_k))\chi_{Q(j)}$. Let $h_j \equiv g_j - m_{Q(1)}(g_j)$. For $1 \le j < k$, we have

$$\begin{aligned} h_k(x) - h_j(x) &= g_k(x) - g_j(x) + m_{Q(1)}(g_j) - m_{Q(1)}(g_k) \\ &= m_{Q(j)}(g_k) + m_{Q(1)}(g_j) - m_{Q(1)}(g_j + m_{Q(j)}(g_k)) \\ &= 0 \end{aligned}$$

for almost all $x \in Q(j)$. Hence if we define

$$g(x) \equiv \limsup_{j\to\infty} h_j(x) \quad (x \in \mathbb{R}^n),$$

then g is an $L^2(\mathbb{R}^n)$-locally integrable function.

Let us show that $g \in \mathrm{BMO}(\mathbb{R}^n)$. To this end choose a cube Q arbitrarily. Let j be chosen large enough as to have $Q \subset Q(j)$. Then

$$m_Q(|g_j - m_Q(g_j)|^2) = m_Q(g_j \cdot (\overline{g_j} - m_Q(\overline{g_j}))) = \frac{1}{|Q|}l((\overline{g_j} - m_Q(\overline{g_j}))\chi_Q).$$

Here by the property of $(2, \infty)$-atoms,

$$m_Q(|g_j - m_Q(g_j)|^2) \lesssim \|l\|_{(H^1)^*} \cdot \sqrt{m_Q(|g_j - m_Q(g_j)|^2)}.$$

This shows that $g \in \mathrm{BMO}(\mathbb{R}^n)$.

Now let A be a $(1, \infty)$-atom centered at Q. If we choose j_0 so that $Q \subset Q(j_0)$, then

$$l(A) = \int_{\mathbb{R}^n} g_j(x)A(x)\mathrm{d}x = \int_{\mathbb{R}^n} h_j(x)A(x)\mathrm{d}x$$

for $j \geq j_0$ and $h_j(x) = h_{j_0}(x)$ for almost all $x \in Q(j_0)$. Thus, we obtain

$$l(A) = \int_{\mathbb{R}^n} g(x)A(x)\mathrm{d}x.$$

Therefore, the proof is complete.

Proof (Step 3: The uniqueness of b for a given functional) We will consider atoms in $H^1(\mathbb{R}^n)$. Let $b \in \mathrm{BMO}(\mathbb{R}^n)$. It suffices to show that b is a constant as long as $A \cdot b \perp \mathscr{P}_0(\mathbb{R}^n)$ for all $(1, \infty)$-atoms A.

We have $b(x) = \lim\limits_{t \downarrow 0} t^n \int_{\mathbb{R}^n} \chi_{[0,t]^n}(x - y)b(y)\mathrm{d}y$ for almost all $x \in \mathbb{R}^n$ by virtue of the Lebesgue differentiation theorem; see Theorem 1.48. Here $\chi_{[0,t]^n}(x - \star) - \chi_{[0,t]^n}$ is a $(1, \infty)$-atom whenever $0 < t \ll c_x$. Thus, $b(x) = \lim\limits_{t \downarrow 0} t^n \int_{\mathbb{R}^n} \chi_{[0,t]^n}(y)b(y)\mathrm{d}y$ holds for almost all $x \in \mathbb{R}^n$. The right-hand side being a constant, we conclude that b equals a constant almost everywhere.

Remark 3.4 It is not the case that $f \cdot g \in L^1(\mathbb{R}^n)$ when $f \in H^1(\mathbb{R}^n)$ and g is a representative of a function in $\mathrm{BMO}(\mathbb{R}^n)$; see Exercise 3.51 for a counterexample.

Exercises

Exercise 3.48 Prove (3.139) using the triangle inequality.

We state a proposition showing how integrable $\mathrm{BMO}(\mathbb{R}^n)$ functions are.

Exercise 3.49 Let $b \in \mathrm{BMO}(\mathbb{R}^n)$, and let Q be a cube of sidelength 1. Then prove that

$$\int_{\mathbb{R}^n} \frac{|b(x) - m_Q(b)|\mathrm{d}x}{|x - c(Q)|^{n+1} + 1} \lesssim \|b\|_{\mathrm{BMO}}, \quad Q \in \mathscr{Q}. \tag{3.141}$$

Hint: We decompose $\mathbb{R}^n$ dyadically.

Exercise 3.50 The space $\mathrm{VMO}(\mathbb{R}^n)$, vanishing mean oscillation, is defined to be a subset of $\mathrm{BMO}(\mathbb{R}^n)$ made up of all the elements f such that

$$\lim_{\varepsilon \downarrow 0} \left(\sup_{Q : |Q| \le \varepsilon} \frac{1}{|Q|^2} \iint_{Q \times Q} |f(x) - f(y)| \mathrm{d}x \mathrm{d}y \right) = 0.$$

Show that VMO($\mathbb{R}^n$) is a closed subspace of BMO($\mathbb{R}^n$).

Exercise 3.51 Let $Q_{jm} \equiv [2^{-j}m, 2^{-j}(m+1))$ for $j, m \in \mathbb{Z}$.

1. Define

$$f = \sum_{j=1}^{\infty} j^{-2} (\chi_{Q_{j2}} - \chi_{Q_{j3}}).$$

   Then show that $f \in H^1(\mathbb{R})$ using the atomic decomposition.
2. Let $g(t) \equiv \mathrm{sgn}(t) \log |t| = (2\chi_{(0,\infty)}(t) - 1) \log |t|$ for $t \in \mathbb{R}$. Then show that $f \cdot g \notin L^1(\mathbb{R})$.

Exercise 3.52 (Modified singular integral operator) Let T be a singular integral operator defined in Definition 1.42. For $f \in L^\infty(\mathbb{R}^n)$, we define

$$Uf(x) \equiv \int_{\mathbb{R}^n} \big(K(x, y) - \chi_{Q(1)^c}(y) K(0, y)\big) f(y) \mathrm{d}y.$$

Then modulo an additive constant show that $Uf = Sf$ using (1.216), where S is defined in Theorem 3.20.

3.3.2 *Local* bmo($\mathbb{R}^n$) *Space*

In analogy with duality $H^1(\mathbb{R}^n)$-BMO($\mathbb{R}^n$), we consider the dual space of $h^1(\mathbb{R}^n)$; the answer is given by the local bmo space.

3.3.2.1 The Local bmo Space

In analogy with BMO($\mathbb{R}^n$), we define bmo($\mathbb{R}^n$), as follows.

Definition 3.9 (Local bmo($\mathbb{R}^n$) **space,** bmo($\mathbb{R}^n$)**)** Define

$$\|f\|_{\mathrm{bmo}} \equiv \|f\|_{\mathrm{BMO}} + \sup \left\{ \|f\|_{L^1(Q)} : Q \in \mathcal{Q}, \quad \ell(Q) = 1 \right\}$$

for $f \in L^1_{\mathrm{loc}}(\mathbb{R}^n)$.

We remark that $h^1(\mathbb{R}^n) \hookleftarrow H^1(\mathbb{R}^n)$ but that bmo($\mathbb{R}^n$) $\hookrightarrow$ BMO($\mathbb{R}^n$).

3.3.2.2 Duality $h^1(\mathbb{R}^n)$ and $\mathrm{bmo}(\mathbb{R}^n)$

With what we have gathered, we can easily specify the dual space of $h^1(\mathbb{R}^n)$.

Theorem 3.22 *The space* $\mathrm{bmo}(\mathbb{R}^n)$ *is canonically identified with the dual space of* $h^1(\mathbb{R}^n)$.

In the course of the proof of Theorem 3.22 we describe how to identify $\mathrm{bmo}(\mathbb{R}^n)$ with the dual space of $h^1(\mathbb{R}^n)$.

Proof (Step 1: $\mathrm{bmo}(\mathbb{R}^n) \hookrightarrow (h^1(\mathbb{R}^n))^*$*)* Choose an even function $\psi \in \mathscr{S}(\mathbb{R}^n)$ so that $\chi_{B(1)} \le \psi \le \chi_{B(2)}$. Let $g \in \mathrm{bmo}(\mathbb{R}^n)$. Then for $f \in h^1(\mathbb{R}^n)$, we decompose

$$L_g(f) = \int_{\mathbb{R}^n} \psi(D)g(x)f(x)\mathrm{d}x + \int_{\mathbb{R}^n} g(x)(1-\psi(D))f(x)\mathrm{d}x.$$

Since $f \in h^1(\mathbb{R}^n) \hookrightarrow L^1(\mathbb{R}^n)$, the first term makes sense. Meanwhile, as for the second term, we use $(1-\psi(D))f \in H^1(\mathbb{R}^n)$ and $g \in \mathrm{bmo}(\mathbb{R}^n) \subset \mathrm{BMO}(\mathbb{R}^n)$. If we denote by $\mathscr{L}$ temporarily the mapping $\mathrm{BMO}(\mathbb{R}^n) \mapsto H^1(\mathbb{R}^n)$ obtained in Theorem 3.21, then the second term will be understood as the quantity $\mathscr{L}(g)((1-\psi(D))f)$. Therefore, the second term also makes sense. Furthermore, it is clear that this mapping is one-to-one. Thus, $(h^1(\mathbb{R}^n))^* \approx \mathrm{bmo}(\mathbb{R}^n)$ with equivalence of norms.

Proof (Step 2: The dual of $h^1(\mathbb{R}^n)$ *is realized by* $\mathrm{bmo}(\mathbb{R}^n)$*)* Conversely, suppose that we have a bounded linear functional $l \in (h^1(\mathbb{R}^n))^*$. Choose an even function $\psi \in \mathscr{S}(\mathbb{R}^n)$ so that $\chi_{B(1)} \le \psi \le \chi_{B(2)}$. Consider two functionals $l_1 \in (L^1(\mathbb{R}^n))^*$ and $l_2 \in (H^1(\mathbb{R}^n))^*$ by

$$l_1(f) \equiv l(\psi(D)f) \quad (f \in L^1(\mathbb{R}^n)), \qquad l_2(f) \equiv l(f) \quad (f \in H^1(\mathbb{R}^n)).$$

Then since $(L^1(\mathbb{R}^n))^* \approx L^\infty(\mathbb{R}^n)$ with coincidence of norms and $(H^1(\mathbb{R}^n))^* \approx \mathrm{BMO}(\mathbb{R}^n)$ by Theorem 3.21, we can find $g \in L^\infty(\mathbb{R}^n)$ and $h \in \mathrm{BMO}(\mathbb{R}^n)$ such that

$$l_1(f) = \int_{\mathbb{R}^n} g(x)f(x)\mathrm{d}x \quad (f \in L^1(\mathbb{R}^n)),$$

and that

$$l_2(f) = \int_{\mathbb{R}^n} h(x)f(x)\mathrm{d}x \quad (f \in H^1(\mathbb{R}^n)).$$

Hence we have

$$\begin{aligned} l(f) &= l(\,\psi(D)f\,) + l(\,(1-\psi(D))f\,) \\ &= l_1(f) + l_2(\,(1-\psi(D))f\,) \end{aligned}$$

$$= \int_{\mathbb{R}^n} f(x)g(x)\mathrm{d}x + \mathscr{L}_h((1-\psi(D))f)$$

for all $f \in h^1(\mathbb{R}^n)$.

In particular, if f is a $(1,\infty)$-atom, then

$$l(f) = \int_{\mathbb{R}^n} f(x)\cdot(g(x)+h(x)-(1-\psi(D))h(x))\mathrm{d}x. \tag{3.142}$$

So our candidate for the bmo($\mathbb{R}^n$) function b should be $b \equiv g + h - (1-\psi(D))h$. Let us prove that $b \in \mathrm{bmo}(\mathbb{R}^n)$. Since $g \in L^\infty(\mathbb{R}^n) \hookrightarrow \mathrm{bmo}(\mathbb{R}^n)$, once we prove $h - \psi(D)h \in \mathrm{bmo}(\mathbb{R}^n)$, then $(h^1(\mathbb{R}^n))^* \hookrightarrow \mathrm{bmo}(\mathbb{R}^n)$.

Since $h \in \mathrm{BMO}(\mathbb{R}^n)$, it is trivial to prove $h - \psi(D)h \in \mathrm{BMO}(\mathbb{R}^n)$ using the translation invariant property. Choose a cube Q with sidelength 1. Then we have

$$\begin{aligned}
\|h-\psi(D)h\|_{L^1(Q)} &= \int_Q \left| h(x) - \int_{\mathbb{R}^n} (2\pi)^{-\frac{n}{2}} \mathscr{F}^{-1}\psi(x-y)h(y)\mathrm{d}y \right| \mathrm{d}x \\
&\simeq_n \int_Q \left| \int_{\mathbb{R}^n} \mathscr{F}^{-1}\psi(x-y)(h(y)-h(x))\mathrm{d}y \right| \mathrm{d}x \\
&\lesssim_n \int_{Q\times\mathbb{R}^n} |\mathscr{F}^{-1}\psi(x-y)(h(y)-h(x))|\mathrm{d}y\mathrm{d}x.
\end{aligned}$$

Here we decompose the integral into a number of dyadic annulus:

$$\int_{Q\times\mathbb{R}^n} = \int_{Q\times Q} + \sum_{j=1}^{\infty} \int_{Q\times(2^j Q\setminus 2^{j-1}Q)}.$$

Since $\psi \in \mathscr{S}(\mathbb{R}^n)$, we have

$$\int_{Q\times\mathbb{R}^n} |\mathscr{F}^{-1}\psi(x-y)(h(y)-h(x))|\mathrm{d}y\mathrm{d}x \lesssim \sum_{j=0}^{\infty} \frac{1}{2^{j(n+1)}} \int_{Q\times 2^j Q} |h(x)-h(y)|\mathrm{d}y\mathrm{d}x.$$

Note that $\displaystyle\int_{Q\times 2^j Q} |h(x)-h(y)|\mathrm{d}y\mathrm{d}x \lesssim j\cdot 2^{jn}|Q|^2\|h\|_{\mathrm{BMO}} = j\cdot 2^{jn}\,\|h\|_{\mathrm{BMO}}$; hence we conclude $\displaystyle\int_Q |h(x)-\psi(D)h(x)|\mathrm{d}x \lesssim \|h\|_{\mathrm{BMO}}$. Therefore, the proof is complete.

Exercises

Exercise 3.53 Let $\psi \in C_c^\infty(\mathbb{R}^n)$ be a bump function that equals 1 near a neighborhood of the origin. Show by a direct calculation that $\log|x| \in \mathrm{BMO}(\mathbb{R}^n)$ and that $\psi(x)\log|x| \in \mathrm{bmo}(\mathbb{R}^n)$.

Exercise 3.54 [267, Lemma 8] Let $u \in C^\infty(\mathbb{R})$ be such that $\chi_{(-\infty,-1)} \le u \le \chi_{(-\infty,0)}$. Define

$$\theta(x) \equiv u(\log_2|x|), \quad \psi(x) \equiv u(\log_2|x|-2) \quad (x \in \mathbb{R}^n).$$

Then prove that $\theta_j \to 0$ in $\mathrm{bmo}(\mathbb{R}^n)$ and that $\psi_j \to 0$ in $\mathrm{BMO}(\mathbb{R}^n)$ using the fact that $v(x) \equiv \log|x| \in \mathrm{BMO}(\mathbb{R}^n)$.

3.3.3 Function Spaces $F^s_{\infty q}(\mathbb{R}^n)$ and $\dot{F}^s_{\infty q}(\mathbb{R}^n)$

The spaces $\mathrm{BMO}(\mathbb{R}^n)$ and $\mathrm{bmo}(\mathbb{R}^n)$ are realized by Triebel–Lizorkin spaces. To this end, we need to define and investigate $F^s_{\infty q}(\mathbb{R}^n)$ and $\dot{F}^s_{\infty q}(\mathbb{R}^n)$, which have been left untouched. Note that we do not consider $F^s_{\infty q}(\mathbb{R}^n)$ and $\dot{F}^s_{\infty q}(\mathbb{R}^n)$ so much.

3.3.3.1 The Definition of $F^s_{\infty q}(\mathbb{R}^n)$ and $\dot{F}^s_{\infty q}(\mathbb{R}^n)$ and Their Validity as Function Spaces

Let us define the spaces $F^s_{\infty q}(\mathbb{R}^n)$ and $\dot{F}^s_{\infty q}(\mathbb{R}^n)$ for $0 < q \le \infty$ and $s \in \mathbb{R}$. Later our definition turns out to be natural. Recall that we denote by $\mathscr{D}$ the set of all dyadic cubes.

Definition 3.10 ($F^s_{\infty q}(\mathbb{R}^n)$, $\dot{F}^s_{\infty q}(\mathbb{R}^n)$) Choose $\psi, \varphi \in \mathscr{S}(\mathbb{R}^n)$ so that (2.31) holds. Let $s \in \mathbb{R}$ and $0 < q \le \infty$.

1. For $f \in \mathscr{S}'(\mathbb{R}^n)$, define

$$\|f\|_{F^s_{\infty q}} \equiv \|\psi(D)f\|_\infty + \sup_{P \in \mathscr{D}, |P| \le 1} \left(\sum_{j=-\log_2 \ell(P)}^\infty \frac{2^{jqs}}{|P|} \int_P |\varphi_j(D)f(x)|^q \mathrm{d}x \right)^{\frac{1}{q}}.$$

   Denote by $F^s_{\infty q}(\mathbb{R}^n)$ the set of all the elements in $\mathscr{S}'(\mathbb{R}^n)$ for which the norm $\|f\|_{F^s_{\infty q}}$ is finite.
2. For $f \in \mathscr{S}'(\mathbb{R}^n)/\mathscr{P}(\mathbb{R}^n)$,

$$\|f\|_{\dot{F}^s_{\infty q}} \equiv \sup_{P \in \mathscr{D}} \left(m_P \left(\sum_{j=-\log_2 \ell(P)}^\infty 2^{jqs} |\varphi_j(D)f|^q \right) \right)^{\frac{1}{q}}.$$

Denote by $\dot{F}^s_{\infty q}(\mathbb{R}^n)$ the set of all the elements in $\mathscr{S}'(\mathbb{R}^n)/\mathscr{P}(\mathbb{R}^n)$ for which the norm $\| \star \|_{\dot{F}^s_{\infty q}}$ is finite.

As before, we need to prove that the definition of the sets $F^s_{\infty q}(\mathbb{R}^n)$ and $\dot{F}^s_{\infty q}(\mathbb{R}^n)$ are independent of the choice of φ as we did in Sect. 2.3.

To this end, we prove the following lemma: To investigate function spaces $F^s_{\infty q}(\mathbb{R}^n)$ and $\dot{F}^s_{\infty q}(\mathbb{R}^n)$, the classical maximal inequality (Theorem 1.50) is not enough. So we need to refine this maximal inequality:

Lemma 3.16 *Let $P \in \mathscr{D}$, $0 < q \le \infty$, $0 < \eta < \min(1, q)$, and let j be an integer such that $j \ge -\log_2 \ell(P)$. Then we have*

$$\|\langle 2^j(x-\star)\rangle^{-\frac{n}{\eta}} f\|_\infty \lesssim_{\eta,q} M^{(\eta)}[\chi_{10nP} f](x) + 2^{jn\left(\frac{1}{q}-\frac{1}{\eta}\right)} \ell(P)^{-\frac{n}{\eta}} \sup_{k\in\mathbb{Z}^n} \|f\|_{L^q(\ell(P)k+P)}$$

for all $f \in \mathscr{S}'_{B(2^j)}(\mathbb{R}^n)$ and $x \in P$.

Proof As we did in Theorem 1.50, we can assume that $f \in \mathscr{S}(\mathbb{R}^n)$. A scaling allows us to assume $j = 0$. In this case, the sidelength of P exceeds 1. From (1.157), we have

$$|f(y)| \le \min_{w\in \overline{B(y,\delta)}} |f(w)| + 2\delta \sup_{w\in B(y,\delta)} |\nabla f(w)|, \quad \delta \ll 1. \tag{3.143}$$

Hence

$$\begin{aligned}
&\sup_{y\in\mathbb{R}^n} \langle x-y\rangle^{-\frac{n}{\eta}} |f(y)| \\
&\le \sup_{y\in\mathbb{R}^n} \langle x-y\rangle^{-\frac{n}{\eta}} \left(\min_{w\in \overline{B(y,\delta)}} |f(w)| \right) + \delta \sup_{y\in\mathbb{R}^n} \langle x-y\rangle^{-\frac{n}{\eta}} \left(\sup_{w\in B(y,\delta)} |\nabla f(w)| \right) \\
&\lesssim_\eta \sup_{y\in\mathbb{R}^n} \langle x-y\rangle^{-\frac{n}{\eta}} \left(\min_{w\in \overline{B(y,\delta)}} |f(w)| \right) + \delta \sup_{w\in\mathbb{R}^n} \langle x-w\rangle^{-\frac{n}{\eta}} |\nabla f(w)|,
\end{aligned}$$

with the constant independent of $\delta > 0$. According to Theorem 1.50,

$$\sup_{w\in\mathbb{R}^n} \langle x-w\rangle^{-\frac{n}{\eta}} |\nabla f(w)| \lesssim_\eta \sup_{w\in\mathbb{R}^n} \langle x-w\rangle^{-\frac{n}{\eta}} |f(w)|.$$

If we take $\delta \ll 1$, then

$$\sup_{y\in\mathbb{R}^n} \langle x-y\rangle^{-\frac{n}{\eta}} |f(y)| \lesssim_\eta \sup_{y\in\mathbb{R}^n} \langle x-y\rangle^{-\frac{n}{\eta}} \left(\min_{w\in \overline{B(y,\delta)}} |f(w)| \right).$$

In the right-hand side, we have to consider two cases according to whether $y \in 7nP$ or not. That is, we write and deal with

$$\mathrm{I} \equiv \sup_{y \in 7nP} \left(\min_{w \in B(y,\delta)} \frac{|f(w)|}{\langle x-y \rangle^{\frac{n}{\eta}}} \right), \quad \mathrm{II} \equiv \sup_{y \in \mathbb{R}^n \setminus 7nP} \left(\min_{w \in B(y,\delta)} \frac{|f(w)|}{\langle x-y \rangle^{\frac{n}{\eta}}} \right).$$

As for I, we proceed as in Theorem 1.50. A change of variables yields

$$\begin{aligned} \mathrm{I} &\lesssim \sup_{y \in 7nP} \left(\langle x-y \rangle^{-n} \int_{B(y,\delta)} \chi_{10nP}(w)|f(w)|^\eta \mathrm{d}w \right)^{\frac{1}{\eta}} \\ &= \sup_{y \in 7nP} \left(\langle x-y \rangle^{-n} \int_{B(y-x,\delta)} \chi_{10nP}(x+w)|f(x+w)|^\eta \mathrm{d}w \right)^{\frac{1}{\eta}}, \end{aligned}$$

since P has volume greater than 1. A geometric observation shows:

$$\begin{aligned} \mathrm{I} &\le \sup_{y \in 7nP} \left(\langle x-y \rangle^{-n} \int_{B(1+|x-y|)} \chi_{10nP}(x+w)|f(x+w)|^\eta \mathrm{d}w \right)^{\frac{1}{\eta}} \\ &\sim \sup_{y \in 7nP} \left(\frac{1}{(1+|x-y|)^n} \int_{B(1+|x-y|)} \chi_{10nP}(x+w)|f(x+w)|^\eta \mathrm{d}w \right)^{\frac{1}{\eta}} \\ &\lesssim_\eta M^{(\eta)}[\chi_{10Q} f](x). \end{aligned}$$

Hence the proof of I is complete.

As for II, keeping in mind that the length of $x-y$ exceeds $4n$, we estimate

$$\mathrm{II} \le \sup_{y \in \mathbb{R}^n \setminus 7nP} |x-y|^{-\frac{n}{\eta}} \left(\inf_{w \in B(y,\delta)} |f(w)| \right).$$

Observe that the infimum of any nonnegative measurable function F over $B(y,\delta)$ is less than the constant times its integral. This then yields

$$\mathrm{II} \lesssim_{q,\eta} \sup_{y \in \mathbb{R}^n \setminus 7nP} |x-y|^{-\frac{n}{\eta}} \|f\|_{L^q(B(y,\delta))}. \tag{3.144}$$

Since $\{Q_{0m}\}_{m \in \mathbb{Z}^n}$ partitions $\mathbb{R}^n$, for each $y \in \mathbb{R}^n \setminus 7nP$ and $\delta \le 1$, we can find $m \in \mathbb{Z}^n$ such that $y \in Q_{0m}$. Then $B(y,\delta) \subset 3Q_{0m}$. Furthermore, since

$$|x-y| \ge |c(P)-m| - |y-m| - |x-c(P)| \ge |c(P)-m| - \frac{|x-y|}{4} - \frac{|x-y|}{4},$$

we have $|c(P)-m| \le 2|x-y|$. Hence it follows that

$$\mathrm{II} \lesssim_{q,\eta} \sup_{m\in\mathbb{Z}^n, Q_{0m}\in\mathbb{R}^n\setminus 7nP} |c(P)-m|^{-\frac{n}{\eta}} \|f\|_{L^q(3Q_{jm})}.$$

Since we can assume that Q_{0m} lies outside $7nP$ and that $\ell(P) \geq 1$, we have

$$\begin{aligned}\mathrm{II} &\lesssim_{q,\eta} \sum_{k\in\mathbb{Z}^n} \sup_{m\in\mathbb{Z}^n : Q_{0m}\in(\ell(P)k+P)\setminus 7nP} |c(P)-m|^{-\frac{n}{\eta}} \left(\int_{3Q_{0m}} |f(w)|^q \mathrm{d}w\right)^{\frac{1}{q}} \\ &\lesssim_{q,\eta} \sum_{k\in\mathbb{Z}^n} \sup_{m\in\mathbb{Z}^n : Q_{jm}\in(\ell(P)k+P)\setminus P} |\ell(P)k|^{-\frac{n}{\eta}} \|f\|_{L^q(Q_{0m})}.\end{aligned}$$

We do not have to consider the term of the infinite sum corresponding to $k=0$; this amounts to investigating m satisfying $Q_{0m} \in P \setminus P$, which never happens. So, we obtain

$$\mathrm{II} \lesssim_{q,\eta} \sup_{k\in\mathbb{Z}^n\setminus\{0\}} |\ell(P)k|^{-\frac{n}{\eta}} \|f\|_{L^q(\ell(P)k+P)}. \tag{3.145}$$

Since $0<\eta<1$, we have $\mathrm{II} \lesssim_{q,\eta} \ell(P)^{-\frac{n}{\eta}} \sup_{k\in\mathbb{Z}^n} \|f\|_{L^q(\ell(P)k+P)}$. Hence the estimate of II is now valid.

Proposition 3.13 (ψ and φ do not affect $F^s_{\infty q}(\mathbb{R}^n)$ and $\dot{F}^s_{\infty q}(\mathbb{R}^n)$) *Let $s\in\mathbb{R}$ and $0<q\leq\infty$. Then the definition of the spaces $F^s_{\infty q}(\mathbb{R}^n)$ and $\dot{F}^s_{\infty q}(\mathbb{R}^n)$ does not depend on ψ and φ satisfying* (2.31).

Proof We handle $\dot{F}^s_{\infty q}(\mathbb{R}^n)$ for simplicity. Let ψ^* and φ^* satisfy the same condition as ψ and φ, respectively. Set $\eta \equiv 2^{-1}\min(1,q)$. To consider two types of norms, we write

$$\|f\|_{\dot{F}^s_{\infty q}(\varphi,\psi)} \equiv \sup_{P\in\mathscr{D}} \left(m_P\left(\sum_{j=-\log_2\ell(P)}^{\infty} |2^{js}\varphi_j(D)f|^q\right)\right)^{\frac{1}{q}},$$

$$\|f\|_{\dot{F}^s_{\infty q}(\varphi^*,\psi)} \equiv \sup_{P\in\mathscr{D}} \left(m_P\left(\sum_{j=-\log_2\ell(P)}^{\infty} |2^{js}\varphi^*_j(D)f|^q\right)\right)^{\frac{1}{q}}.$$

According to Lemma 3.16, we have

$$\begin{aligned}\sup_{y\in\mathbb{R}^n} \langle 2^j(x-y)\rangle^{-\frac{n}{\eta}} |2^{js}\varphi^*_j(D)f(y)| &\lesssim M^{(\eta)}[\chi_{10nP}2^{js}\varphi^*_j(D)f](x) \\ &+2^{jn\left(\frac{1}{q}-\frac{1}{\eta}\right)}\ell(P)^{-\frac{n}{\eta}} \sup_{k\in\mathbb{Z}^n}\left(\int_{(\ell(P)k+P)} |2^{js}\varphi^*_j(D)f(x)|^q\mathrm{d}x\right)^{\frac{1}{q}}\end{aligned}$$

$$\lesssim M^{(\eta)}[\chi_{10nP}2^{js}\varphi_j^*(D)f](x)+2^{jn\left(\frac{1}{q}-\frac{1}{\eta}\right)}\ell(P)^{n\left(\frac{1}{q}-\frac{1}{\eta}\right)}\|f\|_{\dot{F}^s_{\infty q}(\varphi^*,\psi^*)}.$$

Observe that there exists $\psi \in \mathscr{S}$ such that $\varphi_j = \psi_j(\varphi_{j-1}^* + \varphi_j^* + \varphi_{j+1}^*)$. If we use this identity, then

$$\begin{aligned}
&\sum_{j=-\log_2 \ell(P)}^{\infty} |2^{js}\varphi_j(D)f(x)|^q \\
&= \sum_{j=-\log_2 \ell(P)}^{\infty} |2^{js}\psi_j(D)(\varphi_{j-1}^*(D)+\varphi_j^*(D)+\varphi_{j+1}^*(D))f(x)|^q \\
&\lesssim \sum_{j=-\log_2 \ell(P)-1}^{\infty} M^{(\eta)}[\chi_{10nP}2^{js}\varphi_j^*(D)f](x)^q \\
&\quad + \sum_{j=-\log_2 \ell(P)-1}^{\infty} 2^{jn\left(1-\frac{q}{\eta}\right)}\ell(P)^{n-\frac{qn}{\eta}}\|f\|^q_{\dot{F}^s_{\infty q}(\varphi^*,\psi^*)} \\
&\lesssim \sum_{j=-\log_2 \ell(P)-1}^{\infty} M^{(\eta)}[\chi_{10nP}2^{js}\varphi_j^*(D)f](x)^q + \|f\|^q_{\dot{F}^s_{\infty q}(\varphi^*,\psi^*)}.
\end{aligned}$$

If we take the average over P and then take the q-th root of both sides, then

$$\|f\|_{\dot{F}^s_{\infty q}(\varphi,\psi)} \lesssim \|f\|_{\dot{F}^s_{\infty q}(\varphi^*,\psi^*)}.$$

By symmetry, we have the desired result.

3.3.3.2 Properties of $F^s_{\infty q}(\mathbb{R}^n)$

Having set down the definition of $F^s_{\infty q}(\mathbb{R}^n)$, we consider their properties as we did for $F^s_{pq}(\mathbb{R}^n)$. We do not investigate all the properties; here we content ourselves with typical ones.

The next proposition corresponds to Proposition 2.7 with $p = \infty$. This proposition can be used to show that $F^s_{\infty q}(\mathbb{R}^n)$ is complete.

Proposition 3.14 (Inclusion between $B^s_{\infty q}(\mathbb{R}^n)$ and $F^s_{\infty q}(\mathbb{R}^n)$) *Let $s \in \mathbb{R}$ and $0 < q \le \infty$. Then*

$$B^s_{\infty q}(\mathbb{R}^n) \hookrightarrow F^s_{\infty q}(\mathbb{R}^n) \hookrightarrow B^s_{\infty\infty}(\mathbb{R}^n), \quad \dot{B}^s_{\infty q}(\mathbb{R}^n) \hookrightarrow \dot{F}^s_{\infty q}(\mathbb{R}^n) \hookrightarrow \dot{B}^s_{\infty\infty}(\mathbb{R}^n)$$

in the sense of continuous embedding.

Proof If we write out the related norms, we see that $B^s_{\infty q}(\mathbb{R}^n) \hookrightarrow F^s_{\infty q}(\mathbb{R}^n)$ and $\dot{B}^s_{\infty q}(\mathbb{R}^n) \hookrightarrow \dot{F}^s_{\infty q}(\mathbb{R}^n)$ holds in the sense of continuous embedding; see Exercise 3.55. Lemma 3.16 yields $F^s_{\infty q}(\mathbb{R}^n) \hookrightarrow B^s_{\infty\infty}(\mathbb{R}^n)$ and $\dot{F}^s_{\infty q}(\mathbb{R}^n) \hookrightarrow \dot{B}^s_{\infty\infty}(\mathbb{R}^n)$.

Proposition 3.15 (Completeness of $F^s_{\infty q}(\mathbb{R}^n)$ and $\dot{F}^s_{\infty q}(\mathbb{R}^n)$) *Let $s \in \mathbb{R}$ and $0 < q \le \infty$. Then $F^s_{\infty q}(\mathbb{R}^n)$ and $\dot{F}^s_{\infty q}(\mathbb{R}^n)$ are complete (quasi-)Banach spaces.*

Proof We content ourselves with outlining the proof, for the proof is similar to that for $F^s_{pq}(\mathbb{R}^n)$, $0 < p < \infty$, $0 < q \le \infty$ and $s \in \mathbb{R}$. Since we have $F^s_{\infty q}(\mathbb{R}^n) \hookrightarrow B^s_{\infty\infty}(\mathbb{R}^n)$, we see that any Cauchy sequence in $F^s_{\infty q}(\mathbb{R}^n)$ is convergent in $B^s_{\infty\infty}(\mathbb{R}^n)$. By the Fatou lemma, we see that it is also convergent in the topology of $F^s_{\infty q}(\mathbb{R}^n)$.

3.3.3.3 Dual Inequality of Coifman, Meyer and Stein-Type

We need an important dual inequality to consider duality. Here it will be understood that the definition of $\mathcal{T}_q(\{F_j\}_{j\in A})(x)$ and $\mathcal{G}_q(\{G_j\}_{j\in A})(x)$ below is modified naturally when $q = \infty$.

Theorem 3.23 (Dual inequality of Coifman, Meyer and Stein-type) *Let $1 \le q \le \infty$. For each $j \in \mathbb{Z}$, we suppose that we have nonnegative measurable functions F_j, G_j. Let $A \subset \mathbb{Z}$ be a finite set. For $x \in \mathbb{R}^n$, we set*

$$\mathcal{T}_q(\{F_j\}_{j\in A})(x) \equiv \left\{ \sum_{j\in A} \sum_{m\in\mathbb{Z}^n} \chi_{Q_{jm}}(x) m^{(q)}_{Q_{jm}}(F_j)^q \right\}^{\frac{1}{q}},$$

$$\mathcal{G}_{q'}(\{G_j\}_{j\in A})(x) \equiv \sup_{j\in\mathbb{Z}, m\in\mathbb{Z}^n} \left\{ \chi_{Q_{jm}}(x) \sum_{k\in A, k\ge j} m^{(q')}_{Q_{jm}}(G_k)^{q'} \right\}^{\frac{1}{q'}}.$$

Then

$$\sum_{j\in A} \int_{\mathbb{R}^n} F_j(x) G_j(x) \mathrm{d}x \lesssim \int_{\mathbb{R}^n} \mathcal{T}_q(\{F_j\}_{j=-\infty}^{\infty})(x) \mathcal{G}_{q'}(\{G_j\}_{j=-\infty}^{\infty})(x) \mathrm{d}x.$$

For the proof, since F_j, G_j are compactly supported bounded functions, it can be assumed that $F_j, G_j \equiv 0$ with finite exception. Under this assumption, we prove the following sublemma: let

$$\mathcal{G}_q(\{F_j\}_{j\in A})(x) \equiv \sup_{j\in\mathbb{Z}, m\in\mathbb{Z}^n} \left\{ m_{Q_{jm}} \left(\sum_{k\in A, k\ge j} F_k{}^q \right) \chi_{Q_{jm}}(x) \right\}^{\frac{1}{q}} \quad (x \in \mathbb{R}^n).$$

Lemma 3.17 *We have*

$$|Q_{jm} \cap \{\mathcal{T}_{q'}(\{G_k\}_{k=j}^{\infty}) > 2^{\frac{1}{q'}}\mathcal{G}_{q'}(\{G_k\}_{k=j}^{\infty})\}| \le \frac{1}{2}|Q_{jm}|$$

for each compactly supported bounded nonnegative measurable function G_k as long as $G_k \equiv 0$ for k whose absolute value is sufficiently large.

Proof Abbreviate $\mathbb{G}_j \equiv \{G_k\}_{k=j}^{\infty}$ and $\mathbb{G} \equiv \{G_k\}_{k=-\infty}^{\infty}$. By the definitions, we have

$$\begin{aligned}\int_{Q_{jm}} \mathcal{T}_{q'}(\mathbb{G}_j)(x)^{q'}\mathrm{d}x &= \int_{Q_{jm}} \left(\sum_{k=j}^{\infty} \sum_{m' \in \mathbb{Z}^n} \chi_{Q_{jm'}}(x) m_{Q_{km'}}(G_k{}^{q'}) \right) \chi_{Q_{km'}}(x)\mathrm{d}x \\ &\le \int_{Q_{jm}} \left(\sum_{k=j}^{\infty} G_k(x)^{q'} \right) \mathrm{d}x \\ &\le 2^{-jn} \inf_{x \in Q_{jm}} \mathcal{G}_{q'}(\mathbb{G}_j)(x)^{q'}.\end{aligned}$$

Hence by the Chebychev inequality,

$$\begin{aligned}|Q_{jm} \cap \{\mathcal{T}_{q'}(\mathbb{G}_j) > 2^{\frac{1}{q'}}\mathcal{G}_{q'}(\mathbb{G}_j)\}| &\le \frac{1}{2}\int_{Q_{jm}} \frac{\mathcal{T}_{q'}(\mathbb{G}_j)(x)^{q}}{\mathcal{G}_{q'}(\mathbb{G}_j)(x)^{q'}}\mathrm{d}x \\ &\le \frac{\int_{Q_{jm}} \mathcal{T}_q(\mathbb{G}_j)(x)^{q'}\mathrm{d}x}{2\inf\{\mathcal{G}_{q'}(\mathbb{G}_j)(x)^{q'} \,:\, x \in Q_{jm}\}} \\ &\le \frac{1}{2}|Q_{jm}|.\end{aligned}$$

Therefore, the proof is complete.

Now we go back to the proof of Theorem 3.23.

Proof We define $E_j \subset \mathbb{R}^n$ by

$$E_j \equiv \{x \in \mathbb{R}^n \,:\, \mathcal{T}_q(\{F_k\}_{k=j}^{\infty})(x) \le 2^{\frac{1}{q'}}\mathcal{G}_q(\{F_k\}_{k=-\infty}^{\infty})(x)\}.$$

Lemma 3.17 yields

$$\begin{aligned}\|F_j G_j\|_{L^1(l^1)} &= 2\sum_{j=-\infty}^{\infty} \int_{\mathbb{R}^n} \frac{2^{jn}F_j(x)G_j(x)}{2^{jn+1}}\mathrm{d}x \\ &\le 2\sum_{(j,m)\in\mathbb{Z}^{n+1}} \int_{Q_{jm}} |Q_{jm} \cap E_j| 2^{jn} F_j(x)G_j(x)\mathrm{d}x\end{aligned}$$

$$= 2 \sum_{(j,m)\in\mathbb{Z}^{n+1}} \iint_{Q_{jm}\times(Q_{jm}\cap E_j)} 2^{jn} F_j(x)G_j(x)\mathrm{d}y\mathrm{d}x$$

$$= \iint_{\mathbb{R}^{2n}} \sum_{(j,m)\in\mathbb{Z}^{n+1}} \chi_{Q_{jm}}(y)\chi_{Q_{jm}\cap E_j}(x)2^{jn+1}F_j(y)G_j(y)\mathrm{d}y\mathrm{d}x.$$

By the Hölder inequality,

$$\|F_jG_j\|_{L^1(l^1)}$$

$$\lesssim \int_{\mathbb{R}^n} \left(\sum_{(j,m)\in\mathbb{Z}^{n+1}} m^{(q)}_{Q_{jm}}(F_j)^q \right)^{\frac{1}{q}} \left(\sum_{(j,m)\in\mathbb{Z}^{n+1}} \chi_{Q_{jm}\cap E_j}(x) m^{(q')}_{Q_{jm}}(G_j)^{q'} \right)^{\frac{1}{q'}} \mathrm{d}x$$

$$\lesssim \int_{\mathbb{R}^n} \mathscr{T}_q(\{F_j\}_{j=-\infty}^{\infty})(x) \left(\sum_{(j,m)\in\mathbb{Z}^{n+1}} \chi_{Q_{jm}\cap E_j}(x) m^{(q')}_{Q_{jm}}(G_j)^{q'} \right)^{\frac{1}{q'}} \mathrm{d}x.$$

We now resort to the so-called stopping time argument. Let $x \in \mathbb{R}^n$. Set

$$j(x) \equiv \inf(\{\infty\} \cup \{j \in \mathbb{Z} : \mathscr{T}_q(\{G_k\}_{k=j}^{\infty})(x) \le 2^{\frac{1}{q'}} \mathscr{G}_q(\{F_k\}_{k=-\infty}^{\infty})\}).$$

Then we have $\mathscr{T}_q(\{G_k\}_{k=j(x)}^{\infty})(x) \le 2^{\frac{1}{q'}} \mathscr{G}_q(\{F_k\}_{k=-\infty}^{\infty})\})$. Then

$$\|F_jG_j\|_{L^1(l^1)} \lesssim \int_{\mathbb{R}^n} \mathscr{T}_q(\{F_j\}_{j=-\infty}^{\infty})(x) \left(\sum_{m\in\mathbb{Z}^n} \sum_{j=j(x)}^{\infty} m^{(q')}_{Q_{j(x)m}}(G_j)^{q'} \right)^{\frac{1}{q'}} \mathrm{d}x$$

$$= \int_{\mathbb{R}^n} \mathscr{T}_q(\{F_j\}_{j=-\infty}^{\infty})(x) \mathscr{T}_{q'}(\{G_j\}_{j=j(x)}^{\infty})(x)\mathrm{d}x$$

$$\lesssim \int_{\mathbb{R}^n} \mathscr{T}_q(\{F_j\}_{j=-\infty}^{\infty})(x) \mathscr{G}_{q'}(\{G_j\}_{j=-\infty}^{\infty})(x)\mathrm{d}x,$$

as was to be shown.

3.3.3.4 Duality

Now we aim to prove $(\dot{F}^{-s}_{1q}(\mathbb{R}^n))^* \sim \dot{F}^{s}_{\infty q'}(\mathbb{R}^n)$ for all $1 \le q < \infty$ and $s \in \mathbb{R}$.

We want to prove a duality result, which requires the following lemma:

Lemma 3.18 *Let $1 \le q < \infty$, $s \in \mathbb{R}$, $J \in \mathbb{N}$, and let $P \in \mathscr{D}$. Suppose that we have $\{a_j\}_{j=-\log_2 \ell(P)}^{\infty} \subset C^{\infty}_{\mathrm{c}}(\mathrm{Int}(P))$ satisfying*

$$\int_{\mathbb{R}^n} \sum_{k=-\log_2 \ell(P)}^{\infty} |a_k(x)|^q \mathrm{d}x = 1, \quad a_j = 0 \quad j \geq J.$$

Let $\varphi \in \mathscr{S}_\infty(\mathbb{R}^n)$. If we set

$$\psi_P \equiv \frac{1}{|P|} \sum_{j=-\log_2 \ell(P)}^{\infty} 2^{-js} \varphi_j(D) a_j, \tag{3.146}$$

then $\psi_P \in \mathscr{S}_\infty(\mathbb{R}^n)$ and $\|\psi_P\|_{\dot{F}^s_{1q}} \lesssim 1$, where the implicit constant does not depend on J.

Proof For the sake of simplicity, let $1 < q < \infty$. We content ourselves with indicating where to modify when $q = 1$; see Remark 3.5.

Note that

$$\sum_{j=-\log_2 \ell(P)}^{\infty} |\varphi_j(D) a_j(x)|^q \lesssim \sum_{j=-\log_2 \ell(P)}^{\infty} M a_j(x)^q. \tag{3.147}$$

Let $L > n$ be fixed. If $k \in \mathbb{N}$ and $x \in 2^{k+1}P \setminus 2^k P$, then

$$|\varphi_j(D) a_j(x)| \leq 2^{jn} \int_P |a_j(y) \mathscr{F}^{-1}\varphi(2^j(x-y))| \mathrm{d}y \lesssim \frac{2^{jn}}{(2^{k+j}\ell(P))^L} \int_P |a_j(y)| \mathrm{d}y;$$

hence

$$\begin{aligned} \left(\sum_{j=-\log_2 \ell(P)}^{\infty} |\varphi_j(D) a_j(x)|^q \right)^{\frac{1}{q}} &\lesssim \frac{2^{-kqL}}{\ell(P)^L} \int_P \left(\sum_{j=-\log_2 \ell(P)}^{\infty} 2^{j(n-L)q} |a_j(y)|^q \right)^{\frac{1}{q}} \mathrm{d}y \\ &\lesssim \frac{2^{-kqL}}{\ell(P)^n} \int_P \left(\sum_{j=-\log_2 \ell(P)}^{\infty} |a_j(y)|^q \right)^{\frac{1}{q}} \mathrm{d}y. \end{aligned}$$

Thus it follows that

$$\left(\sum_{j=-\log_2 \ell(P)}^{\infty} |\varphi_j(D) a_j(x)|^q \right)^{\frac{1}{q}} \lesssim \frac{1}{2^{kqL}|P|} \int_P \left(\sum_{j=-\log_2 \ell(P)}^{\infty} |a_j(y)|^q \right)^{\frac{1}{q}} \mathrm{d}y \tag{3.148}$$

for all $x \in 2^{k+1}P \setminus 2^k P$. Combining (3.147) and (3.148), we obtain $\|\psi_P\|_{\dot{F}^s_{1q}} \lesssim 1$.

Remark 3.5 When $q = 1$, we can go through a similar argument without using the Hardy–Littlewood maximal operator in (3.147).

We have the following analogue of the duality $L^1(\mathbb{R}^n)$-$L^\infty(\mathbb{R}^n)$:

Theorem 3.24 (Duality $\dot{F}^{-s}_{1q}(\mathbb{R}^n)$-$\dot{F}^{s}_{\infty q'}(\mathbb{R}^n)$) *Let $s \in \mathbb{R}$ and $1 \le q < \infty$. Then with norm equivalence of norms, $(\dot{F}^{s}_{1q}(\mathbb{R}^n))^* \approx \dot{F}^{-s}_{\infty q'}(\mathbb{R}^n)$. More precisely, if we have $f \in \dot{F}^{-s}_{\infty q'}(\mathbb{R}^n)$, then the functional $\psi \in \mathscr{S}_\infty(\mathbb{R}^n) \mapsto \langle f, \psi \rangle \in \mathbb{C}$ satisfies*

$$|\langle f, \psi \rangle| \lesssim \|f\|_{\dot{F}^{-s}_{\infty q'}} \|\psi\|_{\dot{F}^{s}_{1q}} \quad (\psi \in \mathscr{S}_\infty(\mathbb{R}^n)).$$

Using Theorem 1.15 and the density of $\mathscr{S}_\infty(\mathbb{R}^n)$ in $\dot{F}^{s}_{1q}(\mathbb{R}^n)$, we can define a continuous linear functional on $\dot{F}^{s}_{1q}(\mathbb{R}^n)$. Conversely, any continuous linear functional in $\dot{F}^{s}_{1q}(\mathbb{R}^n)$ can be realized in this way. We have an analogy for $F^{-s}_{1q}(\mathbb{R}^n)$ and $F^{s}_{\infty q'}(\mathbb{R}^n)$.

Proof First suppose that we have $f \in \dot{F}^{-s}_{\infty q'}(\mathbb{R}^n)$. Choose an even function $\varphi \in \mathscr{S}(\mathbb{R}^n)$ so that

$$\operatorname{supp}(\varphi) \in B(4) \setminus B(1), \qquad \sum_{j=-\infty}^{\infty} \varphi_j \equiv \chi_{\mathbb{R}^n \setminus \{0\}}. \tag{3.149}$$

Then

$$\langle f, \psi \rangle = \sum_{j=-\infty}^{\infty} \int_{\mathbb{R}^n} \varphi_j(D) f(x) \varphi_j(D)\psi(x) \mathrm{d}x$$

for $\psi \in \mathscr{S}_\infty(\mathbb{R}^n)$. We write

$$\mathrm{I} \equiv \int_{\mathbb{R}^n} \left(\sum_{j=-\infty}^{\infty} (m^{(q)}_{B(x,2^{-j})}(2^{-js}\varphi_j(D)\psi))^q \right)^{\frac{1}{q}} \mathrm{d}x,$$

$$\mathrm{II} \equiv \left(\sup_{j \in \mathbb{Z},\, y \in \mathbb{R}^n,\, x \in B(y,2^{-j})} 2^{jn} \int_{B(x,2^{n-j})} |2^{js}\varphi_j(D) f(z)|^{q'} \mathrm{d}z \right)^{\frac{1}{q'}}.$$

Hence by Theorem 3.23,

$$|\langle f, \psi \rangle| \le \sum_{j=-\infty}^{\infty} \int_{\mathbb{R}^n} |\varphi_j(D) f(x) \varphi_j(D)\psi(x)| \mathrm{d}x \lesssim \mathrm{I} \times \mathrm{II}.$$

If we set $\eta \equiv \dfrac{1}{2}\min(1, q)$, then

$$
\begin{aligned}
\mathrm{I} &\lesssim \int_{\mathbb{R}^n} \left(\sum_{j=-\infty}^{\infty} 2^{jn} \int_{B(x,2^{-j})} M[|2^{-js}\varphi_j(D)\psi|^\eta](x)^{\frac{q}{\eta}} \mathrm{d}y \right)^{\frac{1}{q}} \mathrm{d}x \\
&\lesssim \int_{\mathbb{R}^n} \left(\sum_{j=-\infty}^{\infty} M[|2^{-js}\varphi_j(D)\psi|^\eta](x)^{\frac{q}{\eta}} \right)^{\frac{1}{q}} \mathrm{d}x \\
&\lesssim \int_{\mathbb{R}^n} \left(\sum_{j=-\infty}^{\infty} |2^{-js}\varphi_j(D)\psi(x)|^q \right)^{\frac{1}{q}} \mathrm{d}x \\
&\lesssim \|\psi\|_{\dot{F}^{-s}_{1q}}
\end{aligned}
$$

by Theorem 1.50. As for II, we have

$$
\mathrm{II} \lesssim \|f\|_{\dot{F}^{s}_{\infty q'}} \tag{3.150}
$$

by the definition of the norm $\dot{F}^{s}_{\infty q'}(\mathbb{R}^n)$. Hence $\dot{F}^{-s}_{\infty q'}(\mathbb{R}^n)$ defines a continuous linear functional by way of Theorem 3.24.

Conversely, let $L \in (\dot{F}^{s}_{1q}(\mathbb{R}^n))^*$. Then since $\mathscr{S}_\infty(\mathbb{R}^n) \hookrightarrow \dot{F}^{s}_{1q}(\mathbb{R}^n)$, L, restricted to $\mathscr{S}_\infty(\mathbb{R}^n)$, can be regarded as an element f in $\mathscr{S}'_\infty(\mathbb{R}^n) \sim \mathscr{S}'(\mathbb{R}^n)/\mathscr{P}(\mathbb{R}^n)$. Let us estimate the $\dot{F}^{s}_{\infty q'}(\mathbb{R}^n)$-norm of f. Once we can show that this is finite, then we can say that $L_f = L$ on $\mathscr{S}_\infty(\mathbb{R}^n)$. We fix $P \in \mathscr{D}$. Then we need to estimate

$$
\left\{ m_P \left(\sum_{j=-\log_2 \ell(P)}^{\infty} |2^{-js}\varphi_j(D)f|^{q'} \right) \right\}^{\frac{1}{q'}}.
$$

By the duality $L^q(\ell^q)$-$L^{q'}(\ell^{q'})$, we have

$$
\begin{aligned}
&\left\{ m_P \left(\sum_{j=-\log_2 \ell(P)}^{\infty} |2^{-js}\varphi_j(D)f|^{q'} \right) \right\}^{\frac{1}{q'}} \\
&\le \Re \left[\frac{1}{|P|} \int_{\mathbb{R}^n} \left(\sum_{j=-\log_2 \ell(P)}^{\infty} 2^{1-js} a_j(x)\varphi_j(D)f(x) \right) \mathrm{d}x \right]
\end{aligned}
$$

where a sequence $\{a_j\}_{j=-\log_2 \ell(P)}^{\infty} \subset C^\infty_{\mathrm{c}}(\mathrm{Int}(P))$ satisfies

$$
\int_P \sum_{k=-\log_2 \ell(P)}^{\infty} |a_k(x)|^q \mathrm{d}x = |P|, \quad a_j(x) = 0, \quad j \ge J
$$

almost everywhere for some large J.

Set

$$\psi_P \equiv \frac{1}{|P|} \sum_{j=-\log_2 \ell(P)}^{\infty} 2^{-js} \varphi_j(D) a_j.$$

Then $\|\psi_P\|_{\dot{F}^s_{1q}} \lesssim 1$ thanks to Lemma 3.18.

Recall that φ is an even function. Now that

$$\left\{ \frac{1}{|P|} \int_P \sum_{j=-\log_2 \ell(P)}^{\infty} |2^{-js} \varphi_j(D) f(x)|^{q'} \mathrm{d}x \right\}^{\frac{1}{q'}} \leq 2|\langle f, \psi_P \rangle| = 2|L(\psi_P)|,$$

we have

$$\left\{ \frac{1}{|P|} \int_P \sum_{j=-\log_2 \ell(P)}^{\infty} |2^{-js} \varphi_j(D) f(x)|^{q'} \mathrm{d}x \right\}^{\frac{1}{q'}} \lesssim \|L\|_{(\dot{F}^s_{\infty q})^*}.$$

Thus, the proof is complete.

Now we describe the dual space of the nonhomogeneous space $F^{-s}_{1q}(\mathbb{R}^n)$.

Theorem 3.25 (Duality $F^{-s}_{1q}(\mathbb{R}^n)$-$F^s_{\infty q'}(\mathbb{R}^n)$) *We let $1 \leq q < \infty$ and $s \in \mathbb{R}$. Then $(F^s_{1q}(\mathbb{R}^n))^* \approx F^{-s}_{\infty q'}(\mathbb{R}^n)$ with equivalence of norms.*

Proof Since the argument for $F^s_{1q}(\mathbb{R}^n)$ parallels the one for the homogeneous counterpart, we omit the proof.

Theorem 3.26 *We have $F^0_{\infty 2}(\mathbb{R}^n) \approx \mathrm{bmo}(\mathbb{R}^n)$ and $\dot{F}^0_{\infty 2}(\mathbb{R}^n) \approx \mathrm{BMO}(\mathbb{R}^n)$ with equivalence of norms.*

Proof Combine what we proved: $\dot{F}^0_{\infty 2}(\mathbb{R}^n) \approx (\dot{F}^0_{12}(\mathbb{R}^n))^* \approx (H^1(\mathbb{R}^n))^* \approx \mathrm{BMO}(\mathbb{R}^n)$. We can prove the first assertion for $\mathrm{bmo}(\mathbb{R}^n)$ similarly.

We conclude this section by characterizing $\mathrm{BMO}(\mathbb{R}^n)$ in terms of Carleson measures. We can rephrase Theorem 3.26 as follows:

Theorem 3.27 *Choose $\psi \in \mathscr{S}(\mathbb{R}^n)$ so that $\chi_{B(1)} \leq \psi \leq \chi_{B(2)}$. Set $\varphi \equiv \psi - \psi_{-1}$. Let a be a measurable function satisfying $\displaystyle\int_{\mathbb{R}^n} \frac{|a(x)|}{|x|^{n+1}+1} \mathrm{d}x < \infty$. Denote by $\mathscr{B}$ the set of all Borel measurable sets. Let μ be a measure over $\mathbb{R}^n \times (0, \infty)$ satisfying $\mu(A \times (0,t)) = \displaystyle\sum_{j \in \mathbb{Z} \cap (\log_2 t^{-1}, \infty)} \int_A |\varphi_j(D) a(x)|^2 \mathrm{d}x$ for $t > 0$ and $A \in \mathscr{B}$. Then $a \in \mathrm{BMO}(\mathbb{R}^n)$ if and only if μ is a Carleson measure. Furthermore, if this is the case, we have $\|a\|_{\mathrm{BMO}} \sim \sqrt{\|\mu\|_{\mathrm{Carleson}}}$.*

If we combine Theorems 1.57 and 3.27, then:

Theorem 3.28 *Choose $\psi \in \mathscr{S}(\mathbb{R}^n)$ so that $\chi_{B(1)} \le \psi \le \chi_{B(2)}$. Set $\varphi \equiv \psi - \psi_{-1}$. Furthermore, let $\kappa \in \mathscr{S}(\mathbb{R}^n)$. Then*

$$\sum_{j=-\infty}^{\infty} \int_{\mathbb{R}^n} |\varphi_j(D)a(x)\kappa_j(D)f(x)|^2 \mathrm{d}x \lesssim \|a\|_{\mathrm{BMO}}{}^2 \|f\|_2{}^2.$$

Proof If we set $F(x,t) \equiv \kappa_{[-\log_2 t]}(D)f(x)$ for $(x,t) \in \mathbb{R}^{n+1}_+$, then we have

$$\sum_{j=-\infty}^{\infty} \int_{\mathbb{R}^n} |\varphi_j(D)a(x)\kappa_j(D)f(x)|^2 \mathrm{d}x \lesssim \|a\|_{\mathrm{BMO}}^2 \int_{\mathbb{R}^n} \sup_{(y,t)\in\Gamma(x)} |\kappa_{[-\log_2 t]}(D)f(y)|^2 \mathrm{d}x$$

by virtue of Theorems 1.57 and 3.27. Meanwhile, Theorem 1.47 yields

$$\sup_{(y,t)\in\Gamma(x)} |\kappa(tD)f(y)| \lesssim Mf(x) \quad (x \in \mathbb{R}^n),$$

where M denotes the Hardy–Littlewood maximal operator. Hence we obtain the desired result from the $L^2(\mathbb{R}^n)$-boundedness of M.

Exercises

Exercise 3.55 Let $0 < q < \infty$. Using Lemma 3.16, prove $\|f\|_\infty \lesssim_q \sup_{Q\in\mathscr{D}_0} \|f\chi_Q\|_q$ for all $f \in \mathscr{S}'_{B(1)}(\mathbb{R}^n)$.

Exercise 3.56 [483, Corollary 5.6] Let $0 < q \le \infty$, and let $\{\lambda_{\nu m}\}_{(\nu,m)\in\mathbb{Z}^{n+1}}$ be a complex sequence. Show that

$$\sup_{Q\in\mathscr{D}} m_Q^{(q)}\left(\sum_{(\nu,m)\in\mathbb{Z}^{n+1}} \chi_{[-\log_2 \ell(Q),\infty)}(j)\lambda_{\nu m}\chi_{Q_{\nu m}}\right)$$
$$\sim \inf_{\mathfrak{E}=\{E_{\nu m}\}_{(\nu,m)\in\mathbb{Z}^{n+1}}} \|\{\lambda_{\nu m}\chi_{E_{\nu m}}\}_{(\nu,m)\in\mathbb{Z}^{n+1}}\|_{L^\infty(\ell^q)},$$

where $\mathfrak{E}$ moves over all collections $\{E_{\nu m}\}_{(\nu,m)\in\mathbb{Z}^{n+1}}$ such that $E_{\nu m} \subset Q_{\nu m}$ and $2|E_{\nu m}| \ge |Q_{\nu m}|$ for all $(\nu,m) \in \mathbb{Z}^{n+1}$.

Exercise 3.57 Let $f \in \mathscr{S}'(\mathbb{R}^n)$. Show that the following are equivalent:

(*A*) $f = \sum_{j=-\infty}^{\infty} \varphi_j(D)f$ holds in the topology of $\mathscr{S}'(\mathbb{R}^n)$ and $[f] \in \dot{F}^{-1}_{\infty 2}(\mathbb{R}^n)$.

(*B*) There exists $g \in \mathrm{BMO}(\mathbb{R}^n)$ such that $f = \sqrt{-\Delta}g$.

(C) There exist $g_1, g_2, \ldots, g_n \in \mathrm{BMO}(\mathbb{R}^n)$ such that $f = \sum_{j=1}^{n} \partial_{x_j} g_j$.

$$(D) \quad \sup_{(x_0,t)\in\mathbb{R}^{n+1}_+} t^{-\frac{n}{2}} \int_0^t \left(\int_{|x-x_0|<\sqrt{t}} |e^{s\Delta} f(x)|^2 \mathrm{d}x \right) \mathrm{d}s < \infty.$$

Hint: It is not difficult to show (A), (B), (C) are equivalent using Theorem 3.21 and the lifting property for $\dot{F}^s_{pq}(\mathbb{R}^n)$ for $0 < p, q \le \infty$ and $s \in \mathbb{R}$. (D) is its continuous version.

Textbooks in Sect. 3.3

BMO($\mathbb{R}^n$)

See [31, Section 2.3], [22, Chapter 6, Section 2], [33, Section 3.1], [40, Section 5.5] [86, Chapter 4] and [117, Section 6.2] for BMO($\mathbb{R}^n$).

$F^0_{\infty 2}(\mathbb{R}^n) \approx \mathrm{bmo}(\mathbb{R}^n)$ and $\dot{F}^0_{\infty 2}(\mathbb{R}^n) \approx \mathrm{BMO}(\mathbb{R}^n)$

See [99, p. 93] for $F^0_{\infty 2}(\mathbb{R}^n) \approx \mathrm{bmo}(\mathbb{R}^n)$, the first half of Theorem 3.26. See [97, Section 3.2.2] for $\dot{F}^0_{\infty 2}(\mathbb{R}^n) \approx \mathrm{BMO}(\mathbb{R}^n)$, the second half of Theorem 3.26.

3.4 Notes for Chap. 3

Section 3.1.1

Rademacher Sequence

Theorem 3.1 is known as Khintchine's theorem [670, Lemma 2]. We refer to [143] for an exhaustive account of this direction.

Section 3.1.2

See Littlewood and Paley [738–740] for $\mathbb{R}$ and Stein [1009] for $\mathbb{R}^n$.

Benedik, Calderón and Panzone proved the vector-valued boundedness of the singular integral operators in the framework of the Bochner integral. Seemingly, this is a generalization for its own sake but as the proofs of Theorems 3.2 and 3.3 show, this is meaningful. See also [71, p. 81, Theorem 3] for the Hilbert space-valued Calderón–Zygmund operators.

Equivalence $F^0_{p2}(\mathbb{R}^n) \approx \dot{F}^0_{p2}(\mathbb{R}^n) \approx L^p(\mathbb{R}^n)$: Theorems 3.2 and 3.3

Theorem 3.2 is due to Peetre [70, 872] and Theorem 3.3 is due to Bui [289]. As is seen from the names, the Littlewood and Paley theory stemmed from [738–740] and it is investigated in complex analysis in [1009]. Triebel proved $H^s_p(\mathbb{R}^n) \approx F^s_{p2}(\mathbb{R}^n)$ with equivalence of norms in [1039, Theorem 5.2.3]. The main result $H^m_p(\mathbb{R}^n) = W^m_p(\mathbb{R}^n)$ in Corollary 3.1 dates back to 1964 [773], whose short title is "$H = W$". In the case of non-integers, we refer to [746, 747, 1012]. See also [99, Section 2.5.6]. See [4, Theorem 3.17] for the case of domains for example. See [99, p. 88, Theorem] for Corollary 3.1. See also [2, Theorem 4.2.2].

See the textbooks [63, 97] for the expressions $\dot{F}^0_{p2}(\mathbb{R}^n)$ and $F^0_{p2}(\mathbb{R}^n)$.

See [144] for the case of BMO.

Section 3.2

Section 3.2.1

Definition of Hardy Spaces

The origin of Hardy spaces is the paper on complex analysis [553], where Hardy proved that the mean-value of the p-th power of the modulus of an analytic function on the unit disk is an increasing function of the radius and its logarithm is a convex function of the logarithm of the radius. The theory of Hardy spaces $H^p(\mathbb{R}^n)$ for $p \in (0, 1]$ was originally initiated by Stein and Weiss [1010]. After that, via several complex analytic definitions, in the paper [460] Fefferman and Stein defined Hardy spaces as is described in this book. Theorem 3.4 is due to Fefferman and Stein [460, Theorem 11].

Section 3.2.2

Singular Integral Operators on Hardy Spaces: Theorems 3.6 and 3.10

Theorem 3.6 is due to Fefferman and Stein [460, Theorem 6]. The vector-valued boundedness of the singular integral operators considered in [166] lies behind the theory. As is seen from Theorem 3.10, Hardy spaces are good tools with which to describe the boundedness of the singular integral operators appearing in partial differential equations.

Hardy Spaces and PDE

The theory of Hardy spaces has been widely used in various fields of analysis and partial differential equations; see, for example, [380, 381, 385, 814, 944], as well as the textbooks [33, 86].

Section 3.2.3

Atomic Hardy Spaces: Theorems 3.15 and 3.16

Coifman investigated atomic decomposition in contrast to Hardy spaces via harmonic extension [380] when $n = 1$, while Latter presented a method of decomposing $f \in \mathscr{S}'(\mathbb{R}^n)$ in [703, Theorem A]. Furthermore, Wilson gave a different proof in [1113]. See also [384, 385, 824, 888, 1027] for the atomic decomposition of $H^p(\mathbb{R}^n)$.

We followed [86] to prove Theorems 3.15 and 3.16, the atomic decomposition via $\mathscr{H}_j$.

Finite Decomposition: Proposition 3.4

The author is grateful to Professor Akihiko Miyachi for letting me know about Proposition 3.4, which is from Miyachi's note of 1992.

Section 3.2.4

The Space $\mathscr{H}_j$

See [380] for its prototype idea.

Section 3.2.5

Various Applications of the Calderón–Zygmund Decomposition

The conclusion obtained in Sect. 3.2.5 can be applied to many other function spaces because we have only to assume $f \in \mathscr{S}'(\mathbb{R}^n)$. Actually, many authors applied this conclusion to Hardy–Morrey spaces [604], Morrey spaces [622], Musielak–Orlicz spaces [714], Orlicz spaces [827] and Orlicz–Morrey spaces [923] starting from Hardy spaces with variable exponents [394, 826]. See Chap. 6 for the related definition. See [927] for more general results.

Section 3.2.6

Boundedness of the Operators and Atomic Decomposition: Theorem 3.10

The idea by Miyachi was summarized in Proposition 3.4, which Yabuta noted in his paper [1138]. Later Meda, P. Sjögren and M. Vallarino solidified their idea in [770]. It seems sufficient in the proof of Theorem 3.10, if we show that the image of atoms is contained in a bounded set. However, in [277, Theorem 2], Bownik constructed a counterexample showing that this does not suffice. See also [1164].

Section 3.2.7

Characterization of Hardy Spaces via Riesz Transforms: Theorem 3.13

Theorem 3.13 dates back to 1972 [460, Chapter 8].

Section 3.2.8

Local Hardy Spaces: Theorems 3.17 and 3.18

Theorems 3.17 and 3.18 dates back to 1979 (see the papers by Goldberg [507, 508]).

Section 3.2.9

Isomorphism $F^0_{p2}(\mathbb{R}^n) \approx H^p(\mathbb{R}^n)$: Theorem 3.21

See the original work by Peetre [872].

Isomorphism $h^p(\mathbb{R}^n) \approx F^0_{p2}(\mathbb{R}^n)$: Theorem 3.22

Theorem 3.22, which asserts $h^p(\mathbb{R}^n) \approx F^0_{p2}(\mathbb{R}^n)$, is due to Bui and Triebel [289, 1054]. This equivalence motivates the definition of the space $F^s_{pq}(\mathbb{R}^n)$.

Section 3.3

Section 3.3.1

John–Nirenberg Inequality and John–Nirenberg Space: Theorem 3.19 and Corollaries 3.5, 3.6

In [639], the John–Nirenberg inequality (Theorem 3.19) is proved (for functions defined on cubes), but the terminology BMO($\mathbb{R}^n$) does not appear.

Corollary 3.5 is extended to various directions; see [596, Theorem 8], as well as [525, 599, 601, 619, 622].

Duality $H^1(\mathbb{R}^n)$-BMO($\mathbb{R}^n$): Theorem 3.21

Theorem 3.21 is [460, Theorem 2]; see also [455].

Akihito Uchiyama's Great Contribution–the Decomposition of the BMO Functions

The space BMO($\mathbb{R}^n$) admits decomposition. We refer to the fundamental paper by Uchiyama [1091, Main Lemma]. Uchiyama utilized Carleson measures to obtain the decomposition theory of the space BMO($\mathbb{R}^n$). This technique influenced the successors, who obtained the decomposition theory of Besov spaces and Triebel–Lizorkin spaces. Uchiyama's great contribution to Hardy spaces and BMO($\mathbb{R}^n$) is described in the monograph [114], which is edited by Fujii, Miyachi and Yabuta. The description of Hardy spaces and BMO($\mathbb{R}^n$) in [114, Chapters 3 and 4] spaces will be helpful.

Section 3.3.2

Local BMO

See the papers by Goldberg [507, 508].

The Dual of $h^1(\mathbb{R}^n)$: Theorem 3.22

Theorem 3.22 is due to [507, p. 36].

Section 3.3.3

The Space $F^s_{\infty q}(\mathbb{R}^n)$

The definition of $F^s_{\infty q}(\mathbb{R}^n)$ dates back to 1980 (see the paper [1054], as well as the textbook [99, Section 2.3.4]). Later Frazier and Jawerth expanded this idea [483, Section 5]. See also [99, p. 50, Definition] for another expression of the space $F^s_{\infty q}(\mathbb{R}^n)$ and [99, Section 2.3.4] for Theorem 3.24. The inequality in Theorem 3.23 is a parametrized version of [383, Theorem 1(a)].

Note that we cannot define $F^s_{\infty q}(\mathbb{R}^n)$ analogously to $F^s_{pq}(\mathbb{R}^n)$ with $0 < p < \infty$; see [97, 2.1.4].

Duality, $(\dot{F}^s_{1q}(\mathbb{R}^n))^* \approx \dot{F}^{-s}_{\infty q'}(\mathbb{R}^n)$: Theorem 3.24

See [99, p. 239] for Theorem 3.24.

The Characterization of $F^s_{\infty q}(\mathbb{R}^n)$ by Means of the Peetre Maximal Function

See [902, Lemma 5].

Chapter 4
Decomposition of Function Spaces and Its Applications

The field of harmonic analysis dates back to the nineteenth century; the branch of function spaces dates back to the first half of twentieth century, and has its roots in the study of the decomposition of functions using Fourier series and the Fourier transform. This aspect will be stressed for Besov spaces and Triebel–Lizorkin spaces. One of the elementary tools in harmonic analysis is to decompose functions or distributions into linear sums of elementary units. Such theorems are called decomposition theorems. In general, when the functions are compactly supported, the decomposition is usually called the atomic decomposition; otherwise it is usually called the molecular decomposition. In the atomic decomposition, the elementary unit is called the atom; in the molecular decomposition, it is called the molecule. The wavelet decomposition is a special case of these two decompositions. We also deal with the quarkonial decomposition. The quarkonial decomposition is another kind of atomic decomposition. With a new parameter, it is more complicated. This in turn overcomes the disadvantages. In general, when we are able to express functions by means of the infinite sum of good functions, this can be used to prove the boundedness property of the function spaces, as we will show later in this book. Such a decomposition is very interesting. However, if we postulate too many conditions, it can happen that atoms satisfying all the conditions do not exist; see Exercise 4.3. Another example is the wavelet expansion in [40, 117]. If we assume that the functions are compactly supported, then the function is not $C^\infty(\mathbb{R}^n)$. Furthermore, note that a function f is zero if the Fourier transform and f itself are compactly supported. As these examples show, we cannot make too many good conditions compatible.

Generally speaking, the decomposition theorem for function spaces is made up of two parts: one is analysis, that is, the property that any function in the function space admits a decomposition into the linear sum of "nice" functions; the other is synthesis, that is, the property that any linear sum of functions belongs to the function space. These two aspects of decompositions correspond to specify the basis in linear algebra.

Y. Sawano, *Theory of Besov Spaces*, Developments in Mathematics 56,
https://doi.org/10.1007/978-981-13-0836-9_4

Each decomposition theorem has advantages and disadvantages. For example, in the decomposition theorem, it is not always the case that the nice functions form a linearly independent system. The wavelet decomposition ensures that the system is linearly independent. But in general, such a wavelet system is not mapped into some wavelet system by integral operators; this is a disadvantage of wavelet decomposition. We discuss advantages and disadvantages of other decompositions after we formulate them in Sect. 4.1.

The results in Sect. 4.1 can be applied to various directions. In Sect. 4.2 we consider interpolation. In Sect. 4.3 we consider the pointwise product of distributions which belong to Besov spaces or Triebel–Lizorkin spaces. Finally, in Sect. 4.4 we discuss some important results when we develop a theory of function spaces on domains. Overall, the results in Sects. 4.3 and 4.4 are oriented to applications to partial differential equations.

4.1 Decomposition of Function Spaces

In Sect. 4.1.1, we consider the atomic decomposition and the molecular decomposition. Section 4.1.2 is dedicated to wavelet decomposition and Sect. 4.1.3 is devoted to the quarkonial decomposition. As prominent applications, we consider the properties of function spaces in Sect. 4.1.4. Finally, we investigate the decompositions in this book. At least, since $H^p(\mathbb{R}^n)$ is isomorphic to $\dot{F}^0_{p2}(\mathbb{R}^n)$ for $0 < p \le 1$, we will have two different decompositions for $H^p(\mathbb{R}^n)$. Due to the fact that the definition of Besov spaces and Triebel–Lizorkin spaces is complicated, it is hard to investigate their properties directly. But decompositions of function spaces enable us to do this rather easily. Furthermore, with the atomic decomposition and the molecular decomposition, we will have a better understanding of the parameters in the function spaces. For example, in Definition 4.2, the definition of atoms and molecules, the quantity $s - \frac{n}{p}$ comes about. So, we will be convinced again that $s - \frac{n}{p}$ is an important quantity in the function space $A^s_{pq}(\mathbb{R}^n)$. Here we concentrate on the nonhomogeneous spaces. We can formulate the homogeneous counterpart for overall results in this chapter, most of which we omit.

4.1.1 Atomic Decomposition and Molecular Decomposition

By the "smooth atomic decomposition", we mean that f is decomposed into a sum of the form

$$f = \sum_{\nu=0}^{\infty} \sum_{m \in \mathbb{Z}^n} \lambda_{\nu m} a_{\nu m}$$

with nice properties and that such a sum belongs to the function spaces. The parameter m in the second layer describes the position of atoms and the parameter ν in the first sum describes the smoothness of the functions.

4.1.1.1 Sequence Spaces and Atoms

Let us investigate atomic decomposition. Let $\nu \in \mathbb{Z}$ and $m = (m_1, m_2, \ldots, m_n) \in \mathbb{Z}^n$. Recall that a dyadic cube $Q_{\nu m}$ is defined by $Q_{\nu m} \equiv \prod_{j=1}^{n} \left[\frac{m_j}{2^\nu}, \frac{m_j+1}{2^\nu}\right)$ in Definition 1.1.

Definition 4.1 (Sequence spaces $\mathbf{b}_{pq}(\mathbb{R}^n)$, $\mathbf{f}_{pq}(\mathbb{R}^n)$, $\mathbf{a}_{pq}(\mathbb{R}^n)$, $\dot{\mathbf{b}}_{pq}(\mathbb{R}^n)$, $\dot{\mathbf{f}}_{pq}(\mathbb{R}^n)$, $\dot{\mathbf{a}}_{pq}(\mathbb{R}^n)$) Let $0 < p, q \le \infty$.

1. Let $\nu \in \mathbb{Z}$ and $m \in \mathbb{Z}^n$. The *$L^p(\mathbb{R}^n)$-normalized indicator function* $\chi_{\nu m}^{(p)}$ is defined by $\chi_{\nu m}^{(p)} \equiv 2^{\nu n/p} \chi_{Q_{\nu m}}$.
2. Let $\lambda = \{\lambda_{\nu m}\}_{\nu \in \mathbb{N}_0,\, m \in \mathbb{Z}^n}$ be a doubly indexed complex sequence. Then define

$$\|\lambda\|_{\mathbf{b}_{pq}} \equiv \|\Lambda_\nu\|_{\ell^q(L^p)} = \left\| \sum_{m \in \mathbb{Z}^n} \lambda_{\nu m} \chi_{\nu m}^{(p)} \right\|_{\ell^q(L^p)},$$

$$\|\lambda\|_{\mathbf{f}_{pq}} \equiv \|\Lambda_\nu\|_{L^p(\ell^q)} = \left\| \sum_{m \in \mathbb{Z}^n} \lambda_{\nu m} \chi_{\nu m}^{(p)} \right\|_{L^p(\ell^q)},$$

   where $\Lambda_\nu \equiv \sum_{m \in \mathbb{Z}^n} \lambda_{\nu m} \chi_{\nu m}^{(p)}$, $\nu \in \mathbb{N}_0$. Here we exclude the case where $p < \infty$ for $\|\lambda\|_{\mathbf{f}_{pq}}$.
3. The space $\mathbf{b}_{pq}(\mathbb{R}^n)$ consists of all complex sequences $\lambda = \{\lambda_{\nu m}\}_{\nu \in \mathbb{N}_0,\, m \in \mathbb{Z}^n}$ for which $\|\lambda\|_{\mathbf{b}_{pq}} < \infty$. Likewise if $p < \infty$ the space $\mathbf{f}_{pq}(\mathbb{R}^n)$ consists of all complex sequences $\lambda = \{\lambda_{\nu m}\}_{\nu \in \mathbb{N}_0,\, m \in \mathbb{Z}^n}$ for which $\|\lambda\|_{\mathbf{f}_{pq}} < \infty$. As before, use $\mathbf{a}_{pq}(\mathbb{R}^n)$ to unify $\mathbf{b}_{pq}(\mathbb{R}^n)$ and $\mathbf{f}_{pq}(\mathbb{R}^n)$.

The spaces $\dot{\mathbf{b}}_{pq}(\mathbb{R}^n)$, $\dot{\mathbf{f}}_{pq}(\mathbb{R}^n)$, $\dot{\mathbf{a}}_{pq}(\mathbb{R}^n)$ are defined similar to $\mathbf{b}_{pq}(\mathbb{R}^n)$, $\mathbf{f}_{pq}(\mathbb{R}^n)$, $\mathbf{a}_{pq}(\mathbb{R}^n)$ except that the index ν moves over $\mathbb{Z}$ everywhere in the definition.

It may be convenient to note that

$$\mathbf{b}_{p\,\min(p,q)}(\mathbb{R}^n) \hookrightarrow \mathbf{f}_{pq}(\mathbb{R}^n) \hookrightarrow \mathbf{b}_{p\,\max(p,q)}(\mathbb{R}^n) \tag{4.1}$$

similar to Proposition 2.7. To formulate the results for the Besov space $B^s_{pq}(\mathbb{R}^n)$ and the Triebel–Lizorkin space $F^s_{pq}(\mathbb{R}^n)$ in a unified manner, denote as usual by $\mathbf{a}_{pq}(\mathbb{R}^n)$ either $\mathbf{b}_{pq}(\mathbb{R}^n)$ or $\mathbf{f}_{pq}(\mathbb{R}^n)$. Exclude the case where $p = \infty$ in the notation $\mathbf{f}_{pq}(\mathbb{R}^n)$.

Let $\lambda = \{\lambda_{\nu m}\}_{(\nu,m)\in\mathbb{N}_0\times\mathbb{Z}^n}$. If we write the norm out in full, we have

$$\|\lambda\|_{\mathbf{b}_{\infty\infty}} = \sup\{|\lambda_{\nu m}| \,:\, (\nu, m) \in \mathbb{N}_0 \times \mathbb{Z}^n\} \tag{4.2}$$

and

$$\|\lambda\|_{\mathbf{b}_{pp}} = \left(\sum_{\nu=0}^{\infty}\sum_{m\in\mathbb{Z}^n} |\lambda_{\nu m}|^p\right)^{\frac{1}{p}},$$

for example.

Definition 4.2 (Atoms for Besov spaces and Triebel–Lizorkin spaces) Let $0 < p \le \infty$, $s \in \mathbb{R}$, $\nu \in \mathbb{N}$, and let $m \in \mathbb{Z}^n$. Suppose that the integers $K, L \in \mathbb{Z}$ satisfy $K \ge 0$ and $L \ge -1$.

1. A function $a \in C^K(\mathbb{R}^n)$ is said to be an *atom centered at* Q_{0m} if the following differential inequality holds:

$$|\partial^\alpha a| \le \chi_{3Q_{0m}}, \quad |\alpha| \le K. \tag{4.3}$$

2. A function $a \in C^K(\mathbb{R}^n) \cap \mathscr{P}_L(\mathbb{R}^n)^\perp$ is said to be an *atom centered at* $Q_{\nu m}$, if it satisfies the following differential inequality:

$$|\partial^\alpha a| \le 2^{-\nu\left(s-\frac{n}{p}\right)+\nu|\alpha|}\chi_{3Q_{\nu m}}, \quad |\alpha| \le K. \tag{4.4}$$

One important observation is that $\psi(D)f$ does not need to satisfy the moment condition but that $\varphi_j(D)f \perp \mathscr{P}(\mathbb{R}^n)$. Therefore, it is natural that a_{0m} does not need to satisfy the moment condition, while $a_{\nu m}$ with $\nu \ge 1$ satisfies the moment condition to some extent.

We required that the atoms are compactly supported. However, from the definitions of various important operators such as the singular integral operators and the pseudo-differential operators, we want to relax this condition. So, the following definition seems natural.

Definition 4.3 (Molecules for Besov spaces and Triebel–Lizorkin spaces) Let $0 < p \le \infty$, $s \in \mathbb{R}$, $\nu \in \mathbb{N}$ and $m \in \mathbb{Z}^n$. Suppose that the integers $K, L \in \mathbb{Z}$ satisfy $K \ge 0$ and $L \ge -1$. Furthermore, $M \gg 1$, say, $M \ge \frac{10n}{\min(1,p,q)} + 2$.

1. A function $\Psi \in C^K(\mathbb{R}^n)$ is said to be a *molecule centered at* Q_{0m}, if it satisfies the differential inequality:

$$|\partial^\alpha \Psi(x)| \le \langle x - m\rangle^{-M} \quad x \in \mathbb{R}^n \tag{4.5}$$

for $|\alpha| \le K$.

2. A function $\Psi \in C^K(\mathbb{R}^n) \cap \mathscr{P}_L(\mathbb{R}^n)^\perp$ is said to be a molecule centered at $Q_{\nu m}$, if it satisfies the differential inequality:

$$|\partial^\alpha \Psi(x)| \leq 2^{-\nu\left(s-\frac{n}{p}\right)+\nu|\alpha|} \langle 2^\nu x - m \rangle^{-M} \quad (|\alpha| \leq K, x \in \mathbb{R}^n). \tag{4.6}$$

We remark that the moment condition is postulated only when $\nu \geq 1$.

4.1.1.2 Norm Estimates of Atomic Decomposition

In this section, we are oriented to the norm estimates of the atomic decomposition and the molecular decomposition.

The atomic decomposition and the molecular decomposition decompose the function f:

$$f = \sum_{\nu=0}^{\infty} \left(\sum_{m \in \mathbb{Z}^n} \lambda_{\nu m} a_{\nu m} \right), \quad \sum_{\nu=0}^{\infty} \left(\sum_{m \in \mathbb{Z}^n} \lambda_{\nu m} m_{\nu m} \right),$$

respectively. Let us check that the above two sums actually converge in $\mathscr{S}'(\mathbb{R}^n)$. It suffices to deal with the molecular decomposition.

Lemma 4.1 *Let $0 < p, q \leq \infty$ and $s \in \mathbb{R}$. Suppose that the integers $K, L, M \in \mathbb{Z}$ satisfy*

$$K \geq [1+s]_+, \quad L \geq \max(-1, [\sigma_p - s]), \quad M \geq \max\left(L, \frac{10n}{\min(1, p, q)} + 2\right)$$

in Definition 4.3. Let $\lambda = \{\lambda_{\nu m}\}_{(\nu,m) \in \mathbb{N}_0 \times \mathbb{Z}^n} \in \boldsymbol{b}_{pq}(\mathbb{R}^n)$. Or more generally, let $\lambda = \{\lambda_{\nu m}\}_{(\nu,m) \in \mathbb{N}_0 \times \mathbb{Z}^n} \in \boldsymbol{b}_{\infty\infty}(\mathbb{R}^n)$. Furthermore, suppose that we have a molecule $m_{\nu m}$ centered at $Q_{\nu m}$ for each $(\nu, m) \in \mathbb{N}_0 \times \mathbb{Z}^n$. Then

$$f = \sum_{\nu=0}^{\infty} \left(\sum_{m \in \mathbb{Z}^n} \lambda_{\nu m} m_{\nu m} \right)$$

converges in the topology of $\mathscr{S}'(\mathbb{R}^n)$.

Proof The infinite sum $f_\nu \equiv \sum_{m \in \mathbb{Z}^n} \lambda_{\nu m} m_{\nu m}$ with respect to m trivially converges once we fix $\nu \in \mathbb{N}_0$. Hence the heart of the matter is to consider the convergence of the sum with respect to ν. Let $\varphi \in \mathscr{S}(\mathbb{R}^n)$. Then

$$\left| \int_{\mathbb{R}^n} m_{\nu m}(x) \varphi(x) \mathrm{d}x \right| \lesssim_M \frac{p_N(\varphi)}{2^{\nu(s+n+L+1)} \langle 2^{-\nu} m \rangle^M} \tag{4.7}$$

by virtue of Theorem 1.55.

We use (4.7) to estimate the norm of each f_ν:

$$
\begin{aligned}
&\left|\left\langle \sum_{m\in\mathbb{Z}^n} \lambda_{\nu m} m_{\nu m}, \varphi \right\rangle\right|^{\min(1,p)} \\
&\lesssim_M \sum_{m\in\mathbb{Z}^n} \frac{|\lambda_{\nu m}|^{\min(1,p)} p_N(\varphi)^{\min(1,p)}}{2^{\nu(s+n+L+1)\min(1,p)} \langle 2^{-\nu} m\rangle^{M\min(1,p)}} \\
&\lesssim \left(\sup_{m\in\mathbb{Z}^n} |\lambda_{\nu m}|\right)^{\min(1,p)} 2^{\{n-(s+n+L+1)\min(1,p)\}\nu} p_N(\varphi)^{\min(1,p)}.
\end{aligned}
$$

Next, let $\theta \equiv s+n+L+1-\dfrac{n}{\min(1,p)}$. Then

$$
\left|\left\langle \sum_{m\in\mathbb{Z}^n} \lambda_{\nu m} m_{\nu m}, \varphi \right\rangle\right| \lesssim 2^{-\nu\theta} p_N(\varphi) \|\lambda\|_{\mathbf{b}_{\infty\infty}} \tag{4.8}
$$

by virtue of equality (4.2).

Since we assume $L \geq \min(-1, [\sigma_p - s])$, arithmetic shows:

$$
\theta \geq s+n+[\sigma_p - s]+1-\frac{n}{\min(1,p)} > n+\sigma_p - \frac{n}{\min(1,p)} = 0.
$$

If we sum (4.8) over j, then

$$
\sum_{\nu=0}^{\infty} |\langle f_\nu, \varphi\rangle| \lesssim p_N(\varphi)\|\lambda\|_{\mathbf{b}_{\infty\infty}}, \quad \varphi \in \mathscr{S}(\mathbb{R}^n).
$$

That is, $f = \sum\limits_{\nu=0}^{\infty} f_\nu$ converges in the topology of $\mathscr{S}'(\mathbb{R}^n)$.

We propose a way with which to control functions $\langle\star\rangle^{-n-\varepsilon}$, $\varepsilon > 0$ by means of the indicator function, which is decreasing sufficiently rapidly.

Lemma 4.2 *Let $\kappa \geq n$ and $\varepsilon > 0$. Fix $\nu \in \mathbb{N}_0$. Then for any complex sequence $\{\lambda_{\nu m}\}_{m\in\mathbb{Z}^n}$*

$$
\left|\sum_{m\in\mathbb{Z}^n} \lambda_{\nu m} \langle 2^\nu \star - m\rangle^{-\kappa-\varepsilon}\right| \lesssim_\varepsilon M^{(\kappa^{-1}n)}\left[\sum_{m\in\mathbb{Z}^n} \lambda_{\nu m}\chi_{Q_{\nu m}}\right].
$$

Here $M^{(\kappa^{-1}n)}$ denotes the powered Hardy–Littlewood maximal operator in Definition 1.37 with $\eta \equiv \kappa^{-1}n$.

Proof We calculate the left-hand side. Fix $x \in \mathbb{R}^n$. We decompose

$$\left|\sum_{m\in\mathbb{Z}^n} \lambda_{\nu m}\langle 2^\nu x - m\rangle^{-\kappa-\varepsilon}\right| \le \sum_{j=0}^{\infty}\left(\sum_{m\in\mathbb{Z}^n\cap Q(2^\nu x,2^j)} 2^{-j(\kappa+\varepsilon)}|\lambda_{\nu m}|\right)$$
$$\le \sum_{j=0}^{\infty} 2^{-j\varepsilon}\left(2^{-jn}\sum_{m\in\mathbb{Z}^n\cap Q(2^\nu x,2^j)} |\lambda_{\nu m}|^{\kappa^{-1}n}\right)^{\frac{\kappa}{n}}. \tag{4.9}$$

If $m \in \mathbb{Z}^n \cap Q(2^\nu x, 2^j)$, then

$$\int_{Q(x,2^{j-\nu})} \chi_{Q_{\nu m}}(y)\mathrm{d}y \le \int_{Q(2^{-\nu}m,2^{j-\nu+1})} \chi_{Q_{\nu m}}(y)\mathrm{d}y = 2^{-\nu n}.$$

Hence it follows that

$$\sum_{m\in\mathbb{Z}^n\cap Q(2^\nu x,2^j)} |\lambda_{\nu m}|^{\kappa^{-1}n} \le 2^{\nu n}\int_{Q(x,2^{j-\nu+1})}\sum_{m\in\mathbb{Z}^n}|\lambda_{\nu m}|^{\kappa^{-1}n}\chi_{Q_{\nu m}}(y)\mathrm{d}y$$
$$= 2^{\nu n}\int_{Q(x,2^{j-\nu+1})}\left|\sum_{m\in\mathbb{Z}^n}\lambda_{\nu m}\chi_{Q_{\nu m}}(y)\right|^{\kappa^{-1}n}\mathrm{d}y$$
$$\lesssim 2^{jn}M^{(\kappa^{-1}n)}\left[\sum_{m\in\mathbb{Z}^n}\lambda_{\nu m}\chi_{Q_{\nu m}}\right](x).$$

If we insert this inequality into (4.9), then we obtain Lemma 4.2.

Now let us formulate the atomic decomposition and the molecular decomposition.

Theorem 4.1 (Synthesis of molecules and atoms) *Suppose that the parameters p, q, s satisfy*

$$0 < p \le \infty,\ 0 < q \le \infty,\ s \in \mathbb{R}.$$

Let K be an integer satisfying $K \ge [1+s]_+$. Furthermore, let $L, N \in \mathbb{Z}$ satisfy

$$L \ge \max(-1, [\sigma_p - s]) \tag{4.10}$$

and

$$N \ge \max(L+1, 10\sigma_p + 10n)$$

for Besov spaces, and

$$L \geq \max(-1, [\sigma_{pq} - s]) \tag{4.11}$$

and

$$N \geq \max\left(L+1, 10\sigma_{p,q} + 10n\right)$$

for Triebel–Lizorkin spaces.

Furthermore, let $\lambda = \{\lambda_{\nu m}\}_{\nu \in \mathbb{N}_0,\ m \in \mathbb{Z}^n} \in \boldsymbol{a}_{pq}(\mathbb{R}^n)$ *and* $m_{\nu m}$ *be a molecule centered at* $Q_{\nu m}$ *for each* ν, m. *Then*

$$\left\| \sum_{\nu=0}^{\infty} \sum_{m \in \mathbb{Z}^n} \lambda_{\nu m} m_{\nu m} \right\|_{A^s_{pq}} \lesssim \|\lambda\|_{\boldsymbol{a}_{pq}}.$$

In particular, we have the same conclusions when each $m_{\nu m}$ *is an atom.*

Proof We use the norm in Theorem 2.34. Some symbols are used in a different manner in Theorems 2.34 and 4.1; we follow the notation in Theorem 4.1. With this in mind, let us reformulate Theorem 2.34. Choose a function $\Psi \in \mathscr{S}(\mathbb{R}^n)$ so that

$$\chi_{B(2)} \leq \Psi \leq \chi_{B(4)}.$$

Choose the integers K, L, N as above and a smooth function Θ by $\Theta \equiv \Delta^J \Psi$, where $J \equiv [K + L + N + |s| + 1]$. According to Theorem 2.34, the norms of $B^s_{pq}(\mathbb{R}^n)$ and $F^s_{pq}(\mathbb{R}^n)$ are equivalent to

$$\|f\|_{B^s_{pq}} = \|\Psi * f\|_p + \|\{2^{js}\Theta^j * f\}_{j=1}^{\infty}\|_{\ell^q(L^p)},$$

$$\|f\|_{F^s_{pq}} = \|\Psi * f\|_p + \|\{2^{js}\Theta^j * f\}_{j=1}^{\infty}\|_{L^p(\ell^q)},$$

respectively. Here and below, we concentrate on $F^s_{pq}(\mathbb{R}^n)$ as usual. Let $j \geq 1$; we can readily incorporate the case where $j = 0$. We apply Theorem 1.56, as follows.

1. When $j \leq \nu$, write N for the maximum of the order of the moment condition plus 1 and differential order. We let N, λ, a, η be $L, N, \Theta, M_{\nu m}$, respectively.
2. When $j \geq \nu$, write N for the maximum of the order of the moment condition plus 1 and differential order. We let N, λ, a, η be $K, N, M_{\nu m}, \Theta$, respectively.

Let $x \in \mathbb{R}^n$. We invoke Theorem 1.56 for each pair to obtain:

$$|2^{js}\Theta^j * m_{\nu m}(x)| \lesssim \begin{cases} 2^{\frac{\nu n}{p} - (\nu - j)(L+s+n+1)} \langle 2^j(x - 2^{-\nu}m)\rangle^{-N}, & j \leq \nu, \\ 2^{\frac{\nu n}{p} - (j-\nu)(K-s)} \langle 2^\nu x - m\rangle^{-N}, & j \geq \nu. \end{cases} \tag{4.12}$$

Here, by the conditions of the parameters p, q, s, we consider the relation of K, L, N. By conditions (4.10) and (4.11) of K, L,

$$K - s \geq [1+s]_+ - s = \max([s+1]-s, -s) > 0, \quad L \geq [\sigma_{pq} - s] > \sigma_{pq} - s - 1.$$

Set $\delta \equiv \min(L + s + 1 + n - \kappa - \varepsilon, K - s - \varepsilon)$. If we take $\kappa > 0$ and $\varepsilon > 0$ sufficiently small, then

$$\frac{n}{\min(1, p, q)} < \kappa < N, \quad \delta > 0. \tag{4.13}$$

Here and below, we postulate (4.13) on κ and δ. From (4.12) and the condition $\kappa < N$, when $j \leq \nu$, we have $|2^{js}\Theta^j * m_{\nu m}(x)| \lesssim 2^{\frac{\nu n}{p}-(\nu-j)(L+s+n+1)}\langle 2^j(x - 2^{-\nu}m)\rangle^{-\kappa}$ and when $j \geq \nu$, $|2^{js}\Theta^j * m_{\nu m}(x)| \lesssim 2^{\frac{\nu n}{p}-(j-\nu)(K-s)}\langle 2^\nu x - m\rangle^{-\kappa}$. Hence it follows that

$$|2^{js}\Theta^j * m_{\nu m}(x)| \lesssim 2^{\frac{\nu n}{p}-|\nu-j|\delta}\langle 2^\nu x - m\rangle^{-\kappa} \tag{4.14}$$

for $\kappa > 0$, since $\langle 2^{j-\nu}x\rangle \geq 2^{j-\nu}\langle x\rangle$.

By virtue of Lemma 4.2, we have

$$\left|2^{js}\Theta^j * \sum_{m\in\mathbb{Z}^n} \lambda_{\nu m} m_{\nu m}(x)\right| \lesssim 2^{-\delta|j-\nu|} M^{(\kappa^{-1}n)}\left[\sum_{m\in\mathbb{Z}^n} \lambda_{\nu m} \chi_{\nu m}^{(p)}\right](x). \tag{4.15}$$

Note that (4.15) is valid even for $j = 0$. Since $\delta > 0$, Proposition 1.2 yields

$$\sum_{j=0}^{\infty}\left|2^{js}\Theta^j * \sum_{m\in\mathbb{Z}^n} \lambda_{\nu m} m_{\nu m}(x)\right|^q \lesssim \sum_{\nu=0}^{\infty}\left(M^{(\kappa^{-1}n)}\left[\sum_{m\in\mathbb{Z}^n} \lambda_{\nu m} \chi_{\nu m}^{(p)}\right](x)\right)^q.$$

Finally, if we take the $L^p(\mathbb{R}^n)$-norm over this inequality, then

$$\left\|\left\{2^{js}\Theta^j * \sum_{m\in\mathbb{Z}^n} \lambda_{\nu m} m_{\nu m}\right\}_{j=0}^{\infty}\right\|_{L^p(\ell^q)} \lesssim \|\lambda\|_{\mathbf{f}_{pq}}.$$

by virtue of the fact that $\kappa > \frac{n}{\min(1,p,q)}$ and the Fefferman–Stein vector-valued maximal inequality, Theorem 1.49. This is the desired result.

4.1.1.3 Atomic Decomposition and Molecular Decomposition

We consider atomic decomposition and molecular decomposition for the function spaces.

Theorem 4.2 (Atomic decomposition and molecular decomposition) *Suppose that the parameters p, q, s satisfy $0 < p, q \le \infty$, $s \in \mathbb{R}$. Let an integer K satisfy $K \ge [1+s]_+$. Furthermore, suppose that $L \in \mathbb{Z}$ satisfies* (4.10) *for Besov spaces and* (4.11) *for Triebel–Lizorkin spaces. Then $f \in A^s_{pq}(\mathbb{R}^n)$ admits a decomposition:*

$$f = \sum_{\nu=0}^{\infty} \left(\sum_{m \in \mathbb{Z}^n} \lambda_{\nu m} a_{\nu m} \right) \tag{4.16}$$

in the topology of $\mathscr{S}'(\mathbb{R}^n)$. Here each $a_{\nu m}$ is an atom centered at $Q_{\nu m}$ and the coefficient $\lambda \equiv \{\lambda_{\nu m}\}_{\nu \in \mathbb{N}_0,\, m \in \mathbb{Z}^n}$ satisfies $\|\lambda\|_{\boldsymbol{a}_{pq}} \lesssim \|f\|_{A^s_{pq}}$. We can replace atoms with molecules.

Proof Concentrate on Triebel–Lizorkin spaces. Let $f \in F^s_{pq}(\mathbb{R}^n)$. Choose functions $\psi, \varphi, \widetilde{\psi}, \widetilde{\varphi} \in C^\infty_{\rm c}(\mathbb{R}^n)$ so that φ satisfies the moment condition of order L from Theorem 1.39. A scaling allows us to assume that these four functions are supported in a cube $Q(1/10)$. Following the rule 26, we define φ^j. We decompose f:

$$\begin{aligned} f &= \widetilde{\psi} * \psi * f + \sum_{j=1}^{\infty} \widetilde{\varphi}^j * \varphi^j * f \\ &= \sum_{m \in \mathbb{Z}^n} \int_{Q_{0m}} \widetilde{\psi}(\star - y)\psi * f(y){\rm d}y + \sum_{j=1}^{\infty} \sum_{m \in \mathbb{Z}^n} \int_{Q_{jm}} \widetilde{\varphi}^j(\star - y)\varphi^j * f(y){\rm d}y \end{aligned}$$

in $\mathscr{S}'(\mathbb{R}^n)$. For $j \in \mathbb{N}_0$, $m \in \mathbb{Z}^n$, we set

$$B_{0m} \equiv \int_{Q_{0m}} \widetilde{\psi}(\star - y)\psi * f(y){\rm d}y, \quad \mathscr{V}_0 f \equiv \sup_{y \in \mathbb{R}^n} \langle y \rangle^{-N} |\psi * f(\star - y)|$$

and for $j \ge 1$

$$B_{jm} \equiv \int_{Q_{jm}} \widetilde{\varphi}^j(\star - y)\varphi^j * f(y){\rm d}y, \quad \mathscr{V}_j f \equiv \sup_{y \in \mathbb{R}^n} \langle 2^j y \rangle^{-N} |\varphi^j * f(\star - y)|.$$

We notice that B_{jm} is supported on $3Q_{jm}$. Let $x \in 3Q_{jm}$. There exists a constant $M_K > 0$ that depends only on K such that

$$|\partial^\alpha B_{jm}(x)| \le M_K\, 2^{j|\alpha|} \left(\inf_{y \in Q_{jm}} \mathscr{V}_j f(y) \right)$$

for all multi-indexes α with length less than K. We write

$$\lambda_{jm} \equiv 2^{\nu\left(s - \frac{n}{p}\right)} M_K \left(\inf_{y \in Q_{jm}} \mathscr{V}_j f(y) \right), \quad A_{jm} \equiv \begin{cases} \dfrac{1}{\lambda_{jm}} B_{jm} & \text{if } \lambda_{jm} \neq 0, \\ 0 & \text{otherwise.} \end{cases}$$

The function A_{jm} is an atom, since $\widetilde{\varphi} \perp \mathscr{P}_L(\mathbb{R}^n)$. We have

$$\|\lambda\|_{\mathbf{f}_{pq}} = M_K \left\| \sum_{m\in\mathbb{Z}^n} 2^{js} \left(\inf_{y\in Q_{jm}} \mathscr{V}_j f(y) \right) \chi_{Q_{jm}} \right\|_{L^p(\ell^q)} \le M_K \, \|2^{js}\mathscr{V}_j f\|_{L^p(\ell^q)}$$

for the coefficient $\lambda = \{\lambda_{jm}\}_{j\in\mathbb{N}_0, m\in\mathbb{Z}^n}$. Meanwhile, Proposition 2.14 yields

$$\|\lambda\|_{\mathbf{f}_{pq}} \le M_K \, \|2^{js}\mathscr{V}_j f\|_{L^p(\ell^q)} \lesssim \|f\|_{F^s_{pq}}.$$

Hence the proof is complete.

Note that the coefficient mapping $f \mapsto \lambda_{\nu m}$ is not linear.

We give applications.

Example 4.1 Let $n = 1$ and $s = \frac{1}{p}$. Let us prove $\chi_{[0,2]} \in B^s_{p\infty}(\mathbb{R})$ for $0 < p \le \infty$. According to Theorem 2.14, we have $B^1_{1\infty}(\mathbb{R}) \hookrightarrow B^0_{\infty\infty}(\mathbb{R})$. So it can be assumed that $p < \infty$. Let $\kappa \in C^\infty_c((1,3)) \subset \mathscr{S}(\mathbb{R})$ satisfy $\sum_{j=1}^{\infty} \kappa(2^j t) = 1$ for $t \in (0,1]$. Set $k \equiv \sum_{j=1}^{\infty} \kappa(2^j \star)$. Singularity lies in the points $t = 0, 1$ and their properties are the same. So we can replace f with k and it suffices to show that $B^s_{p\infty}(\mathbb{R})$. Set $a_j(t) \equiv \kappa(2^j t)$ for $t \in \mathbb{R}$. Then we have a differential inequality: $|a_j^{(m)}(t)| \lesssim_m 2^{jm}$ for $t \in \mathbb{R}$.

There is no need to consider the moment condition when we consider the atomic decomposition for $B^s_{p\infty}(\mathbb{R})$ since $s > \left(\frac{1}{p} - 1\right)_+$. Hence we can regard a_j as the atom centered at Q_{j0}. We define $\{\lambda_{jm}\}_{j\in\mathbb{N}_0, m\in\mathbb{Z}^n}$ by $\lambda_{jm} \equiv \delta_{m0}$ for $j \in \mathbb{N}_0, m \in \mathbb{Z}^n$. Note that

$$\|k\|_{B^s_{p\infty}(\mathbb{R})} \lesssim \|\lambda\|_{\mathbf{b}_{p\infty}(\mathbb{R})} \lesssim 1. \tag{4.17}$$

Hence it follows that k is decomposed into the sum of atoms in $B^s_{p\infty}(\mathbb{R})$ and that $k \in B^s_{p\infty}(\mathbb{R})$.

Example 4.2 Let $n = 1$ again. Let $\rho \in C^\infty(\mathbb{R})$ so that $\chi_{(-\infty,0)} \le \rho \le \chi_{(-\infty,2)}$. Set $R_{(j)} \equiv \rho(2^j \star)$. If we go through the same argument as Example 4.1, we see that $\{R_{(j)} - \rho\}_{j=1}^\infty$ forms a bounded set of $B^{1/p}_{p\infty}(\mathbb{R})$. Hence so does $\{R_{(j)}\}_{j=1}^\infty$.

Example 4.3 Choose $\psi \in \mathscr{S}(\mathbb{R}^n)$ so that $\chi_{B(1)} \le \psi \le \chi_{B(2)}$. Let $\alpha > 0$. If we set $f(x) \equiv |x|^\alpha \psi(x)$ for $x \in \mathbb{R}^n$, then an argument similar to the above examples shows that $f \in B^{\frac{n}{p}+\alpha}_{p\infty}(\mathbb{R}^n)$. We refer to [6, Proposition 2.21] for a counterpart to $\dot{B}^{\frac{n}{p}+\alpha}_{p\infty}(\mathbb{R}^n)$.

We remark that the theory of homogenous spaces parallels what we have done for nonhomogeneous spaces.

4.1.1.4 A Refinement

The function space $L^2(\mathbb{R}^n)$ is a fundamental space. Sometimes, we need to refine the result obtained in this section. The following is an example of this attempt.

Proposition 4.1 *Let $s > 0$ and $\lambda = \{\lambda_{\nu m}\}_{(\nu,m)\in\mathbb{Z}^{n+1}} \in \dot{f}_{22}(\mathbb{R}^n)$. That is,*

$$\|\lambda\|_{\dot{f}_{22}} = \sqrt{\sum_{(\nu,m)\in\mathbb{Z}^{n+1}} |\lambda_{\nu m}|^2} < \infty.$$

Let $\{M_{\nu m}\}_{(\nu,m)\in\mathbb{Z}^{n+1}} \subset \mathrm{Lip}^s(\mathbb{R}^n) \cap \mathscr{P}_0(\mathbb{R}^n)^\perp$ be a set of functions satisfying;

$$\text{(pointwise estimate) } |M_{\nu m}(x)| \le 2^{\frac{\nu n}{2}} \langle 2^\nu x - m\rangle^{-n-s},$$

$$\text{(Lipschitz continuity) } |M_{\nu m}(x) - M_{\nu m}(y)| \le 2^{\frac{\nu n}{2}} 2^{\nu s}|x-y|^s \langle 2^\nu x - m\rangle^{-n-s}.$$

Then $F \equiv \sum_{\nu=-\infty}^{\infty}\left(\sum_{m\in\mathbb{Z}^n} \lambda_{\nu m} M_{\nu m}\right)$ converges in the topology of $L^2(\mathbb{R}^n)$ and $\|F\|_2 \lesssim_s \|\lambda\|_{\dot{f}_{22}}$.

Proposition 4.1 modifies Theorem 4.1 a little. We reexamine the proof of Theorem 4.1.

Proof We suppose $\lambda_{\nu m} = 0$ with a finite number of exceptions, so that we ignore the problem of convergence of the sum defining F. Let $\varphi, \psi \in \mathscr{S}(\mathbb{R}^n)$ satisfy $\chi_{B(1)} \le \psi \le \chi_{B(2)}$ and $\varphi = \psi - \psi^{-1}$. Then define

$$F_{\nu j} \equiv \sum_{m\in\mathbb{Z}^n} \lambda_{\nu m} \varphi_j(D) M_{\nu m}, \quad \nu, j \in \mathbb{Z}. \tag{4.18}$$

Fix $x \in \mathbb{R}^n$. Let $j \ge \nu$. Then we have

$$\begin{aligned}F_{\nu j}(x) &\simeq_n \sum_{m\in\mathbb{Z}^n} \lambda_{\nu m} \int_{\mathbb{R}^n} \mathscr{F}^{-1}\varphi_j(x-y) M_{\nu m}(y)\mathrm{d}y \\ &\simeq_n \sum_{m\in\mathbb{Z}^n} \lambda_{\nu m} \int_{\mathbb{R}^n} \mathscr{F}^{-1}\varphi_j(x-y)(M_{\nu m}(y) - M_{\nu m}(x))\mathrm{d}y.\end{aligned}$$

Thus, by the Lipschitz continuity and $|\mathscr{F}^{-1}\varphi_j(z)| \lesssim \langle z\rangle^{-n-2s}$,

$$|F_{\nu j}(x)| \lesssim 2^{(\nu-j)s+jn+\frac{\nu n}{2}} \sum_{m\in\mathbb{Z}^n} |\lambda_{\nu m}| \int_{\mathbb{R}^n} \langle 2^j(x-y)\rangle^{-n-s} \langle 2^\nu y - m\rangle^{-n-s} \mathrm{d}y$$
$$= 2^{(\nu-j)s+jn+\frac{\nu n}{2}} \int_{\mathbb{R}^n} \langle 2^j(x-y)\rangle^{-n-s} \sum_{m\in\mathbb{Z}^n} |\lambda_{\nu m}| \langle 2^\nu y - m\rangle^{-n-s} \mathrm{d}y.$$

By the definition of the Hardy–Littlewood maximal operator,

$$\sum_{m\in\mathbb{Z}^n} \frac{|\lambda_{\nu m}|}{\langle 2^\nu y - m\rangle^{n+s}} \lesssim \sum_{k=1}^{\infty} \frac{1}{2^{k(s+n)}} \sum_{m\in\mathbb{Z}^n, |2^\nu y - m|\le 2^k} |\lambda_{\nu m}|$$
$$\lesssim \sum_{k=1}^{\infty} \frac{2^{\frac{\nu n}{2}}}{2^{k(s+n)}} \int_{B(y, 2^{k-\nu+n+1})} \sum_{m\in\mathbb{Z}^n} |\lambda_{\nu m}| \chi_{\nu m}^{(2)}(z) \mathrm{d}z$$
$$\lesssim 2^{-\frac{\nu n}{2}} M\left[\sum_{m\in\mathbb{Z}^n} \lambda_{\nu m}\chi_{\nu m}^{(2)}\right](y).$$

We let $M^2[F] \equiv M[M[F]]$, $F \in L^0(\mathbb{R}^n)$ stand for the double composition of M. Then it follows that

$$|F_{\nu j}(x)| \lesssim 2^{(\nu-j)s} M^2\left[\sum_{m\in\mathbb{Z}^n} \lambda_{\nu m}\chi_{\nu m}^{(2)}\right](x). \tag{4.19}$$

We can go through a similar argument in the case where $\nu \ge j$. Thus,

$$|F_{\nu j}(x)| \lesssim 2^{-|\nu-j|s} M^2\left[\sum_{m\in\mathbb{Z}^n} \lambda_{\nu m}\chi_{\nu m}^{(2)}\right](x) \tag{4.20}$$

for all $\nu, j \in \mathbb{Z}$. Hence (4.20) is summable over $j \in \mathbb{Z}$:

$$\sum_{j=-\infty}^{\infty} |F_{\nu j}(x)| \le M^2\left[\sum_{m\in\mathbb{Z}^n} \lambda_{\nu m}\chi_{\nu m}^{(2)}\right](x).$$

If we insert this pointwise estimate into the definition of the $L^2(\mathbb{R}^n)$-norm, then

$$\|F\|_2{}^2 \sim \|F\|_{\dot{F}^0_{22}}^2 \lesssim \int_{\mathbb{R}^n} \sum_{\nu=-\infty}^{\infty} \left(\sum_{j=-\infty}^{\infty} |F_{\nu j}(x)|\right)^2 \mathrm{d}x$$
$$\lesssim \int_{\mathbb{R}^n} \sum_{\nu=-\infty}^{\infty} M^2\left[\sum_{m\in\mathbb{Z}^n} \lambda_{\nu m}\chi_{\nu m}^{(2)}\right](x)^2 \mathrm{d}x.$$

By the $L^2(\mathbb{R}^n)$-boundedness of the Hardy–Littlewood maximal operator, we have

$$\|F\|_2 \lesssim \left\{ \int \sum_{\nu=-\infty}^{\infty} \left| \sum_{m \in \mathbb{Z}^n} \lambda_{\nu m} \chi_{\nu m}^{(2)}(x) \right|^2 \mathrm{d}x \right\}^{\frac{1}{2}} = \|\lambda\|_{\dot{\mathbf{f}}_{22}},$$

as was to be shown.

4.1.1.5 Decomposition of Sequences

We have used the sequence spaces to describe the coefficients. Here we consider a method to decompose sequences. Let $0 < q \leq \infty$, and let $\lambda \equiv \{\lambda_{\nu m}\}_{\nu \in \mathbb{N}_0, m \in \mathbb{Z}^n}$ be a complex sequence be such that

$$\left(\sum_{\nu=0}^{\infty} \left| \sum_{m \in \mathbb{Z}^n} \lambda_{\nu m} \chi_{\nu m}^{(p)}(x) \right|^q \right)^{\frac{1}{q}} < \infty \tag{4.21}$$

for almost all $x \in \mathbb{R}^n$.

Let $k \in \mathbb{Z}^n$. We set

$$\mathscr{A}_k \equiv \left\{ (\tilde{\nu}, \tilde{m}) \in \mathbb{N}_0 \times \mathbb{Z}^n \, : \, \inf_{Q_{\tilde{\nu} m}} \left(\sum_{\nu=0}^{\tilde{\nu}} \left| \sum_{m \in \mathbb{Z}^n} \chi_{Q_{\nu m}}(2^{-\tilde{\nu}} \tilde{m}) \lambda_{\nu m} \chi_{\nu m}^{(p)} \right|^q \right)^{\frac{1}{q}} > 2^k \right\}, \tag{4.22}$$

so that we take into account (ν, m) satisfying $Q_{\tilde{\nu}\tilde{m}} \subset Q_{\nu m}$ in the summation.

Lemma 4.3 *Let $k \in \mathbb{Z}$. Suppose we have $\lambda \equiv \{\lambda_{\nu m}\}_{\nu \in \mathbb{N}_0, m \in \mathbb{Z}^n}$ satisfying* (4.21).

1. $\bigcap_{k=-\infty}^{\infty} \mathscr{A}_k = \emptyset$.
2. *If $(\tilde{\nu}, \tilde{m}) \in \mathbb{N}_0 \times \mathbb{Z}^n$ satisfies $\lambda_{\tilde{\nu}\tilde{m}} \neq 0$, then $(\tilde{\nu}, \tilde{m}) \in \bigcup_{k=-\infty}^{\infty} \mathscr{A}_k$.*

Proof

1. This follows from (4.22).
2. If $k \in \mathbb{Z}$ satisfies $2^{\tilde{\nu}n/p}|\lambda_{\tilde{\nu}\tilde{m}}| > 2^k$, then $(\tilde{\nu}, \tilde{m}) \in \mathscr{A}_k$.

We set

$$\mathscr{B}_k \equiv \bigcap_{(\nu^\dagger, m^\dagger) \in \mathscr{A}_k \setminus \mathscr{A}_{k+1}} \{(\tilde{\nu}, \tilde{m}) \in \mathscr{A}_k \, : \, Q_{\tilde{\nu}\tilde{m}} \supset Q_{\nu^\dagger m^\dagger} \text{ or } Q_{\tilde{\nu}\tilde{m}} \cap Q_{\nu^\dagger m^\dagger} = \emptyset\}. \tag{4.23}$$

Lemma 4.4 *For all $(\nu, m) \in \mathbb{N}_0 \times \mathbb{Z}^n$ such that $\lambda_{\nu m} \neq 0$ there uniquely exist $k \in \mathbb{Z}$ and $(\tilde{\nu}, \tilde{m}) \in \mathscr{B}_k$ such that $Q_{\tilde{\nu}\tilde{m}} \supset Q_{\nu m}$ and that $(\nu, m) \in \mathscr{A}_k \setminus \mathscr{A}_{k+1}$.*

Proof This is clear from Lemma 4.3 and (4.23).

Lemma 4.5 *For each $k \in \mathbb{Z}$ and $(\tilde{\nu}, \tilde{m}) \in \mathscr{B}_k$,*

$$\left(\sum_{\nu=\tilde{\nu}}^{\infty}\left|\sum_{m\in\mathbb{Z}^n}\chi_{Q_{\tilde{\nu}\tilde{m}}}(2^{-\nu}m)\chi_{\mathscr{A}_k\setminus\mathscr{A}_{k+1}}(\nu,m)\lambda_{\nu m}\chi_{\nu m}^{(p)}\right|^q\right)^{\frac{1}{q}} \le 2^{k+1}\chi_{Q_{\tilde{\nu}\tilde{m}}}.$$

Proof If $x \in \mathbb{R}^n \setminus Q_{\tilde{\nu}\tilde{m}}$, then both sides equal 0. If

$$\left(\sum_{\nu=0}^{\infty}\left|\sum_{m\in\mathbb{Z}^n}\lambda_{\nu m}\chi_{\nu m}^{(p)}(x)\right|^q\right)^{\frac{1}{q}} \le 2^{k+1},$$

then there is nothing to prove. If $x \in Q_{\tilde{\nu}\tilde{m}}$ satisfies

$$\left(\left|\sum_{m\in\mathbb{Z}^n}\lambda_{0m}\chi_{0m}^{(p)}(x)\right|^q\right)^{\frac{1}{q}} > 2^{k+1},$$

then $(\nu, M_\nu(x)) \in \mathscr{A}_{k+1}$ for all $\nu \in \mathbb{N}_0$, where $M_\nu(x) \in \mathbb{Z}^n$ satisfies $Q_{\nu M_\nu(x)} \ni x$. Thus, the right-hand side is zero.

Let us suppose

$$\left(\sum_{\nu=0}^{\infty}\left|\sum_{m\in\mathbb{Z}^n}\lambda_{\nu m}\chi_{\nu m}^{(p)}(x)\right|^q\right)^{\frac{1}{q}} > 2^{k+1} \ge \left(\left|\sum_{m\in\mathbb{Z}^n}\lambda_{0m}\chi_{0m}^{(p)}(x)\right|^q\right)^{\frac{1}{q}}.$$

Let $x \in Q_{\tilde{\nu}\tilde{m}}$ and $N(x) \in \mathbb{Z}^n$ satisfy

$$\left(\sum_{\nu=0}^{N(x)}\left|\sum_{m\in\mathbb{Z}^n}\lambda_{\nu m}\chi_{\nu m}^{(p)}(x)\right|^q\right)^{\frac{1}{q}} \le 2^{k+1} < \left(\sum_{\nu=0}^{N(x)+1}\left|\sum_{m\in\mathbb{Z}^n}\lambda_{\nu m}\chi_{\nu m}^{(p)}(x)\right|^q\right)^{\frac{1}{q}}.$$

Let $N^*(x) \equiv N(x) + 1$. We choose $M(x), M^*(x) \in \mathbb{Z}^n$ so that $x \in Q_{N^*(x)M^*(x)} \subset Q_{N(x)M(x)}$. Then we have $(N(x), M(x)) \notin \mathscr{A}_{k+1}$. If $x \in Q_{\nu m}$, then either $Q_{\nu m} \subset Q_{N^*(x)M^*(x)}$ or $Q_{\nu m} \supset Q_{N(x)M(x)}$ happens. If the former happens, then $(\nu, m) \in \mathscr{A}_{k+1}$. Thus, we have only to consider the cubes such that $Q_{\nu m} \supset Q_{N(x)M(x)}$ Thus,

$$\left(\sum_{\nu=\tilde{\nu}}^{\infty}\left|\sum_{m\in\mathbb{Z}^n}\chi_{Q_{\tilde{\nu}\tilde{m}}}(2^{-\nu}m)\chi_{\mathscr{A}_k\setminus\mathscr{A}_{k+1}}(\nu,m)\lambda_{\nu m}\chi_{\nu m}^{(p)}(x)\right|^q\right)^{\frac{1}{q}}$$

$$\leq \left(\sum_{\nu=0}^{N(x)} \left| \sum_{m \in \mathbb{Z}^n} \chi_{Q_{\tilde{\nu}\tilde{m}}}(2^{-\nu}m)\chi_{\mathscr{A}_k \setminus \mathscr{A}_{k+1}}(\nu, m)\lambda_{\nu m}\chi_{\nu m}^{(p)}(x) \right|^q \right)^{\frac{1}{q}} \leq 2^{k+1},$$

as required.

Based on these preparatory observation, we obtain the following conclusion:

Theorem 4.3 *Let $0 < q \leq \infty$ and let $\{\lambda_{\nu m}\}_{\nu \in \mathbb{N}_0, m \in \mathbb{Z}^n}$ be a complex sequence satisfying* (4.21). *If we set*

$$\lambda_{\nu m}^{k,\tilde{\nu},\tilde{m}} \equiv \begin{cases} \lambda_{\nu m} & (\nu, m) \in \mathscr{A}_k \setminus \mathscr{A}_{k+1} \text{ and } Q_{\nu m} \subset Q_{\tilde{\nu}\tilde{m}}, \\ 0 & \text{otherwise} \end{cases}$$

for $k \in \mathbb{Z}$, $\tilde{\nu} \in \mathbb{N}_0$ and $m \in \mathbb{Z}^n$ and $\lambda^{k,\tilde{\nu},\tilde{m}} \equiv \{\lambda_{\nu m}^{k,\tilde{\nu},\tilde{m}}\}_{\nu \in \mathbb{N}_0, m \in \mathbb{Z}^n}$, then

$$\left(\sum_{\nu=0}^{\infty} \left| \sum_{m \in \mathbb{Z}^n} \lambda_{\nu m}^{k,\tilde{\nu},\tilde{m}} \chi_{\nu m}^{(p)} \right|^q \right)^{\frac{1}{q}} \leq 2^{k+1} \chi_{Q_{\tilde{\nu}\tilde{m}}}$$

and

$$\lambda_{\nu m} = \sum_{k=-\infty}^{\infty} \sum_{\tilde{\nu}=0}^{\infty} \sum_{m \in \mathbb{Z}^n} \lambda_{\nu m}^{k,\tilde{\nu},\tilde{m}}$$

for all $\nu \in \mathbb{N}$ and $m \in \mathbb{Z}^n$.

Note that the last sum contains at most only one nonzero term. Using Theorem 4.3, we obtain nonsmooth decomposition for function spaces. We omit the detail.

Example 4.4 Let us consider the $B^s_{pq}(\mathbb{R}^2)$-norm of the functions when $0 < p, q \leq \infty$ and s satisfy $1 \geq [\sigma_p - s]$. We let $\Phi \in C^\infty_{\rm c}(\mathbb{R}^2)$. Define

$$\eta_k(x_1, x_2) \equiv \sum_{j_1, j_2=1}^{2} (-1)^{j_1+j_2} \Phi(2^k x_1 + (-1)^{j_1}, 2^k x_2 + (-1)^{j_2}) \quad (x_1, x_2 \in \mathbb{R}, k \in \mathbb{N}).$$

Note that modulo the trivial constant $2^{-sk+2k/p}\eta_k$ is an atom centered at Q_{k0}. Thus,

$$\left\| \sum_{k=1}^{\infty} \lambda_k \eta_k \right\|_{B^s_{pq}(\mathbb{R}^2)} \lesssim \left(\sum_{k=1}^{\infty} (2^{sk-2k/p}|\lambda_k|)^q \right)^{\frac{1}{q}}.$$

Using this example, Bourgain and Li controlled the $H^1(\mathbb{R}^2) = B^1_{22}(\mathbb{R}^2)$ and proved the strong ill-posedness of the incompressible Euler equation in borderline Sobolev spaces in [283].

Exercises

Exercise 4.1 [317, Theorem 2.5] Let $0 < p, q < \infty$ and $s \in \mathbb{R}$.

1. Show that the convergence of (4.16) takes place in $A^s_{pq}(\mathbb{R}^n)$.
2. Show that the sum defining f converges in $A^s_{pq}(\mathbb{R}^n)$ in Theorem 4.2.

Since p and q are finite, we can use the Lebesgue convergence theorem.

Exercise 4.2 Let $\alpha > 0$ and $0 < p \le 1$. Set $f_\alpha(x) \equiv \exp(-|x|^\alpha)$ for $x \in \mathbb{R}^n$ and $\beta = \alpha + \frac{n}{p}$.

1. Prove that $f_\alpha \in B^\beta_{p\infty}(\mathbb{R}^n)$ by using Theorem 4.1.
2. Use $(1-\Delta)^{\frac{\beta}{2}} f_\alpha \in B^0_{p\infty}(\mathbb{R}^n)$ and Theorem 2.10 to show $|\mathscr{F} f_\alpha(\xi)| \lesssim \langle \xi \rangle^{-\alpha-n}$.

Exercise 4.3 Let $Q \in \mathscr{D}$. Assume that $a \in C^\infty(\mathbb{R}^n)$ satisfies $|\partial^\alpha a| \le \ell(Q)^{-|\alpha|}\chi_{3Q}$ for **all multi-indexes** α. Then show that $a \equiv 0$. Hint: Use the identity theorem for holomorphic functions in the complex plane.

Exercise 4.4 Let $\kappa > 1$. Show that in the definition of atoms in Definition 4.2 we can replace 3 with κ in order to develop a theory of decompositions by reexamining the proof of Theorem 4.2.

Exercise 4.5 Let $\Theta \in C^\infty_c(\mathbb{R}^n)$ be such that $\chi_{[-0.1,0.1]^n} \le \Theta \le \chi_{[-0.2,0.2]^n}$. Define

$$\Psi_j(x) \equiv 2^{jn/p}\mathscr{F}^{-1}\Theta(2^j x - \mathbf{e}_1) \quad (x \in \mathbb{R}^n).$$

Let $2 < p < \infty$. Then show that $\left\| \sum_{j=1}^N \Psi_j \right\|_p \sim_p N^{\frac{1}{p}}$ and that $\mathscr{F}$ never sends $L^p(\mathbb{R}^n)$ to $L^{p'}(\mathbb{R}^n)$.

Exercise 4.6 [697, Lemma 7.4] Let $M > n$, $1 < p < \infty$ and $j, \nu \in \mathbb{Z}$ with $\nu \ge j$. If $\lambda = \{\lambda_Q\}_{Q \in \mathscr{D}_\nu}$ is a sequence of complex numbers, then

$$\left\{ \sum_{R \in \mathscr{D}_j} \left(\sum_{Q \in \mathscr{D}_\nu} \left(1 + \frac{|c(R) - c(Q)|}{\ell(R)} \right)^{-M} \lambda_R \right)^p \right\}^{1/p} \lesssim 2^{(\nu-j)d/p'} \|\lambda\|_{\ell^p}.$$

Hint: Use the Fefferman–Stein vector-valued inequality (Theorem 1.49).

Exercise 4.7 (Minimality of $\dot{B}^0_{11}(\mathbb{R}^n)$)

1. [29, Theorem (3.11)] Write out Theorems 4.1 and 4.2 fully for $\dot{B}^0_{11}(\mathbb{R}^n)$.
2. [29, Corollary (3.13)] If B is a Banach space embedded into $\mathscr{S}'_\infty(\mathbb{R}^n)$ and containing $\mathscr{S}_\infty(\mathbb{R}^n)$ such that $\|f(t\star)\|_B \sim t^{-n}\|f\|_B$ and that $\|f(\star - y)\|_B \sim \|f\|_B$ for all $t > 0$ and $y \in \mathbb{R}^n$, then show that $\dot{B}^0_{1,1}(\mathbb{R}^n)$ is embedded into B. Hint: Resort to the atomic decomposition; Theorem 4.2. Observe also that we can consider the realization in $\dot{B}^0_{11}(\mathbb{R}^n)$; see Theorem 2.32.

Exercise 4.8 [1114, p. 778] Let $\mathbb{R}^{n+1}_+ = \sum_{j=1}^{\infty} E_j$ be a measurable partition, and let $\Phi : \mathbb{R}^{n+1}_+ \mapsto \mathscr{C}_\alpha$ be a mapping such that $\Phi|E_j$ is constant, where the set $\mathscr{C}_\alpha$ is given in Exercise 3.19. Let $L^2(\mathbb{R}^{n+1}_+, \mathrm{d}x\mathrm{d}t/t)$ be the set of all the measurable functions $g : \mathbb{R}^{n+1}_+ \to \mathbb{R}$ such that

$$\|g\|_{L^2(\mathbb{R}^{n+1}_+,\mathrm{d}x\mathrm{d}t/t)} \equiv \sqrt{\int_{\mathbb{R}^{n+1}_+} |g(x,t)|^2 \frac{\mathrm{d}x\mathrm{d}t}{t}}$$

is finite. Let $K \subset \mathbb{R}^{n+1}_+$ be a compact set. Define

$$Tg(x) \equiv \int_K t^{-n-1} g(y,t)\Phi^{(y,t)}\left(\frac{x-y}{t}\right)\mathrm{d}y\mathrm{d}t \quad (g \in L^2(\mathbb{R}^{n+1}_+, \mathrm{d}x\mathrm{d}t/t))$$

and for $Q \in \mathscr{D}$,

$$T_Q g(x)$$
$$\equiv \int_{K\cap(Q\times[\ell(Q)/2,\ell(Q)))} t^{-n-1} g(y,t)\Phi^{(y,t)}\left(\frac{x-y}{t}\right)\mathrm{d}y\mathrm{d}t \quad (g \in L^2(\mathbb{R}^{n+1}_+, \mathrm{d}x\mathrm{d}t/t)).$$

1. Show that $\mathrm{supp}(T_Q g) \subset 3\overline{Q}$.
2. Show that $T_Q g \perp \mathscr{P}_0(\mathbb{R}^n)$ for all $g \in L^2(\mathbb{R}^{n+1}_+, \mathrm{d}x\mathrm{d}t/t)$.
3. Use Proposition 4.1 to conclude that T is bounded from $L^2(\mathbb{R}^{n+1}_+, \mathrm{d}x\mathrm{d}t/t)$ to $L^2(\mathbb{R}^n)$.

Exercise 4.9 [1070, Theorem 1] Let $0 < p, q \le \infty$ and $s \in \mathbb{R}$. Suppose that we have $f \in C^M_{\mathrm{c}}(Q(0.1))$ with $M \in \mathbb{N} \cap (s, \infty)$. Let $\{a_k\}_{k\in\mathbb{Z}^n} \in \ell^p(\mathbb{Z}^n)$. Then show that

$$\left\|\sum_{k\in\mathbb{Z}^n} a_k f(\star - k)\right\|_{A^s_{pq}} \lesssim \left(\sum_{k\in\mathbb{Z}^n} |a_k|^p\right)^{\frac{1}{p}} \|f\|_{A^s_{pq}}.$$

Hint: When $A = F$, use Theorem 4.38. When $B = F$, use the local means.

4.1.2 *Wavelet Decomposition*

As a further application of the atomic decomposition, we consider wavelet expansion. Let $0 < p, q \leq \infty$ and $s \in \mathbb{R}$ as usual. Denote by $\langle \star_1, \star_2 \rangle$ the complex L^2-inner product in $L^2(\mathbb{R}^n)$. We are now interested in wavelets. By a wavelet we mean a function Ψ such that the set $\{2^{\nu n/2}\Psi(2^\nu \star - m)\}_{(\nu,m)\in\mathbb{Z}^{n+1}}$ is a complete orthonormal basis in $L^2(\mathbb{R}^n)$. We aim to show that this function generates a "basis" in $\dot{A}^s_{pq}(\mathbb{R}^n)$.

4.1.2.1 Wavelet Decomposition

We can generalize E below to an extent, but we content ourselves with as simple an argument as possible.

Theorem 4.4 *There exists a function $\varphi \in \mathscr{S}_\infty(\mathbb{R})$ such that the system $\{2^{j/2}\varphi(2^j \star - m)\}_{j\in\mathbb{Z},m\in\mathbb{Z}}$ forms a complete orthonormal system in $L^2(\mathbb{R})$.*

Proof Start with an odd function $\kappa \in C^\infty(\mathbb{R})$ such that $\chi_{[1/3,\infty)} - \chi_{[-\infty,1/3)} \leq \kappa \leq \chi_{[-1/3,\infty)} - \chi_{[-\infty,-1/3)}$.

We define an even function Θ by

$$\Theta \equiv \sqrt{\frac{1}{2}(1+\kappa(\star-\pi))\left(1+\kappa\left(\pi-\frac{\star}{2}\right)\right)+\frac{1}{2}(1-\kappa(\star+\pi))\left(1+\kappa\left(\pi+\frac{\star}{2}\right)\right)}.$$

Let φ be a function given by $\varphi \equiv \mathscr{F}^{-1}\Theta\left(\star - \frac{1}{2}\right)$.

Let us check that $\{\varphi_{j,k}\}_{j\in\mathbb{Z},k\in\mathbb{Z}} \equiv \{2^{j/2}\varphi(2^j\star-k)\}_{j\in\mathbb{Z},k\in\mathbb{Z}}$ forms an orthonormal system. To this end, we have to show that

$$\langle \varphi_{j,k}, \varphi_{j',k'} \rangle_{L^2(\mathbb{R})} = \delta_{jj'}\delta_{kk'}. \tag{4.24}$$

In view of the size of the support, we may assume that $|j - j'| \leq 1$; otherwise (4.24) is trivial by Plancherel's theorem. Assume first that $j = j'$. In this case, we can assume $k' = 0$ by translation. We calculate

$$\begin{aligned}
\langle \varphi_{j,k}, \varphi_{j',k'} \rangle_{L^2(\mathbb{R})} &= \int_{\mathbb{R}} 2^j\varphi(2^j t - k)\overline{\varphi(2^j t)}\mathrm{d}t \\
&= \int_{\mathbb{R}} \varphi(t-k)\overline{\varphi(t)}\mathrm{d}t \\
&= \int_{\mathbb{R}} e^{k\tau i}|\Theta(\tau)|^2\mathrm{d}\tau \\
&= 2\int_0^\infty |\Theta(\tau)|^2\cos(k\tau)\mathrm{d}\tau
\end{aligned}$$

for the last line, where we used the parity of φ. If we write out the definition of Θ, then we have

$$\begin{aligned}
&\langle \varphi_{j,k}, \varphi_{j',k'} \rangle_{L^2(\mathbb{R})} \\
&= \int_{2\pi/3}^{4\pi/3} (1+\kappa(\tau-\pi))\cos(k\tau)\mathrm{d}\tau + \int_{4\pi/3}^{8\pi/3} \left(1+\kappa\left(\pi-\frac{\tau}{2}\right)\right)\cos(k\tau)\mathrm{d}\tau \\
&= \int_{-\pi/3}^{\pi/3} (1+\kappa(\tau))\cos(k(\tau+\pi))\mathrm{d}\tau + \int_{-\pi/3}^{\pi/3} (1+\kappa(\tau))\cos(2k(\pi-\tau))\mathrm{d}\tau \\
&= \delta_{jk}.
\end{aligned}$$

Thus, it remains to handle the case where $j = j' \pm 1$. By symmetry we let $j' = j+1$. Then

$$\begin{aligned}
\langle \varphi_{j,k}, \varphi_{j',k'} \rangle_{L^2(\mathbb{R})} &= \sqrt{2}\int_{\mathbb{R}} 2^j \varphi(2^j t - k)\overline{\varphi(2^{j+1}t - k')}\mathrm{d}t \\
&= \sqrt{2}\int_{\mathbb{R}} \varphi(t-k)\overline{\varphi(2t-k')}\mathrm{d}t \\
&= \sqrt{2}\int_{\mathbb{R}} \Theta(\tau)\mathrm{e}^{-ik\tau}e^{-\tau/2}\overline{\Theta(2^{-1}\tau)\mathrm{e}^{-i2^{-1}k'\tau}e^{-\tau/4}}\mathrm{d}\tau \\
&= \sqrt{2}\int_{\mathbb{R}} \Theta(\tau)\mathrm{e}^{-iK\tau/4}\overline{\Theta(2^{-1}\tau)}\mathrm{d}\tau,
\end{aligned}$$

where K is an odd integer. Again by parity of κ, we have

$$\langle \varphi_{j,k}, \varphi_{j',k'} \rangle_{L^2(\mathbb{R})} = \sqrt{2}\int_{4\pi/3}^{8\pi/3} \Theta(\tau)\Theta(2^{-1}\tau)\cos\left(\frac{K}{4}\tau\right)\mathrm{d}\tau.$$

If we change variables: $\tau' = \tau - 2\pi$, we obtain $\langle \varphi_{j,k}, \varphi_{j',k'} \rangle_{L^2(\mathbb{R})} = 0$.

We next need to show that $\{\varphi_{j,k}\}_{j,k\in\mathbb{Z}}$ is complete. To this end, we assume that $f \in L^2(\mathbb{R})$ is perpendicular to this system. If we use the general formulas

$$\sum_{j=0}^{\infty} \mathscr{F}\varphi(2^j\star)\overline{\mathscr{F}\varphi(2^j\star + 2^{j+1}k\pi)} = 0, \tag{4.25}$$

$$\sum_{j=-\infty}^{\infty} |\mathscr{F}\varphi(2^j\star)|^2 = \chi_{\mathbb{R}\setminus\{0\}} \tag{4.26}$$

and

$$\mathscr{F}\left(\sum_{m=-\infty}^{\infty} \langle f, \varphi_{jm} \rangle_{L^2(\mathbb{R})} \varphi_{jm}\right) \tag{4.27}$$

$$= \mathscr{F}\varphi(2^{-j}\star) \sum_{m=-\infty}^{\infty} \mathscr{F} f(\star + 2^{j+1}k\pi)\overline{\mathscr{F}\varphi(2^{-j}\star + 2k\pi)},$$

then we see that $\mathscr{F} f$ vanishes everywhere; hence $f = 0$. See Exercise 4.12 for (4.25) and (4.26).

We modify the idea of Theorem 4.4 to discuss inhomogeneous decomposition.

Theorem 4.5 *There exist functions $\psi \in \mathscr{S}(\mathbb{R})$, $\varphi \in \mathscr{S}_\infty(\mathbb{R})$ such that the system $\{\psi(\star - m)\}_{m\in\mathbb{Z}} \cup \{2^{j/2}\varphi(2^j \star - m)\}_{j\in\mathbb{N}_0, m\in\mathbb{Z}}$ forms a complete orthonormal system in $L^2(\mathbb{R})$.*

Proof We modify the proof of Theorem 4.4. We content ourselves with the orthonormality. We omit the proof of the completeness of the wavelet system. For φ, use the same function as Theorem 4.4. We define

$$\Theta^* \equiv \sqrt{\frac{1}{4}\left(1 + \kappa\left(\pi - \frac{t}{2}\right)\right)\left(1 + \kappa\left(\pi + \frac{t}{2}\right)\right)}.$$

Let ψ be a function given by

$$\psi \equiv \mathscr{F}^{-1}\Theta^*\left(\star - \frac{1}{2}\right).$$

We go through almost the same argument as Theorem 4.4 to prove that the system

$$\{\psi(\star - m)\}_{m\in\mathbb{Z}} \cup \{2^{j/2}\varphi(2^j \star - m)\}_{j\in\mathbb{N}, m\in\mathbb{Z}}$$

is complete. Let us check the orthogonality:

$$\begin{aligned}
&\langle \psi(\star - m), \psi(\star - m') \rangle_{L^2(\mathbb{R})} \\
&= \frac{1}{4}\int_{\mathbb{R}} \left(1 + \kappa\left(\pi - \frac{t}{2}\right)\right)\left(1 + \kappa\left(\pi + \frac{t}{2}\right)\right) e^{-\frac{m-m'}{2}it} dt \\
&= \frac{1}{2}\int_0^\infty \left(1 + \kappa\left(\pi - \frac{t}{2}\right)\right) \cos\left(\frac{m - m'}{2}t\right) dt \\
&= \frac{1}{2}\int_0^{4\pi} \left(1 + \kappa\left(\pi - \frac{t}{2}\right)\right) \cos\left(\frac{m - m'}{2}t\right) dt.
\end{aligned}$$

If we change variables: $s = 4\pi - t$, then we have

$$\begin{aligned}
&\langle \psi(\star - m), \psi(\star - m')\rangle_{L^2(\mathbb{R})} \\
&= \frac{1}{4}\int_0^{4\pi} \left(2 + \kappa\left(\pi - \frac{t}{2}\right) + \kappa\left(\pi + \frac{t}{2}\right)\right) \cos\left(\frac{m - m'}{2} t\right) dt \\
&= \frac{1}{2}\int_0^{4\pi} \cos\left(\frac{m - m'}{2} t\right) dt = 0.
\end{aligned}$$

Likewise, we can show that

$$\langle \psi(\star - m), \varphi(2\star - m')\rangle_{L^2(\mathbb{R})} = \langle \varphi(\star - m), \varphi(2\star - m')\rangle_{L^2(\mathbb{R})} = 0.$$

Thus, the proof is complete.

Based on this construction, we go back to $\mathbb{R}^n$.

Theorem 4.6 *Set*

$$E \equiv \{\varepsilon = (\varepsilon_1, \varepsilon_2, \ldots, \varepsilon_n) \in \{0, 1\}^n : \varepsilon_1 + \varepsilon_2 + \cdots + \varepsilon_n > 0\} = \{0, 1\}^n \setminus \{(0, 0, \ldots, 0)\}.$$

Then we can choose real-valued functions $\Psi \in \mathscr{S}(\mathbb{R}^n)$ and $\{\Phi^\varepsilon\}_{\varepsilon \in E} \subset \mathscr{S}_\infty(\mathbb{R}^n)$ satisfies the following conditions:

1. *The system $\{\Psi(\star - m)\}_{m \in \mathbb{Z}^n} \cup \{2^{jn/2}\Phi^\varepsilon(2^j \star - m)\}_{j \in \mathbb{N}_0, m \in \mathbb{Z}^n, \varepsilon \in E}$ forms a complete orthonormal system in $L^2(\mathbb{R}^n)$.*
2. *Ψ and Φ^ε, $\varepsilon \in E$ can be expressed:*

$$\Psi(x) = \prod_{j=1}^n \varphi^{(0)}(x_j), \quad \Phi^\varepsilon(x) = \prod_{j=1}^n \varphi^{(\varepsilon_j)}(x_j), \quad x = (x_1, x_2, \ldots, x_n) \in \mathbb{R}^n$$

for some $\varphi^{(0)}, \varphi^{(1)} \in C^\infty(\mathbb{R})$.

Proof Let $(\varphi^{(0)}, \varphi^{(1)}) = (\psi, \varphi)$, where ψ and φ are defined in Theorem 4.5. Then it is easy to see that the system is orthonormal. To show that the system is complete we use a multidimensional version of (4.27) and mimic the proof of Theorem 4.4.

4.1.2.2 Wavelet Expansion

Here and below, we use the notation in Theorem 4.6.

To use Theorem 4.2, we set

$$\Psi_m \equiv \Psi(\star - m), \quad \Phi^\varepsilon_{j,m} \equiv 2^{-j\left(s - \frac{n}{p}\right)}\Phi^\varepsilon(2^j \star - m), \quad m \in \mathbb{Z}^n, \quad j \in \mathbb{N}_0, \quad \varepsilon \in E.$$

Any $f \in L^2(\mathbb{R}^n)$ can be expanded into an orthonormal series:

$$f = \sum_{\varepsilon \in E} \sum_{m \in \mathbb{Z}^n} \langle f, \Psi_m \rangle_2 \Psi_m + \sum_{\varepsilon \in E} \sum_{j=0}^{\infty} \sum_{m \in \mathbb{Z}^n} 2^{j\left(2s - \frac{2n}{p} + n\right)} \langle f, \Phi^{\varepsilon}_{j,m} \rangle_2 \Phi^{\varepsilon}_{j,m},$$

where the right-hand side converges in $L^2(\mathbb{R}^n)$ by Theorem 4.6.

We formulate the wavelet decomposition for $A^s_{pq}(\mathbb{R}^n)$.

Theorem 4.7 *Let $0 < p, q \le \infty$ and $s \in \mathbb{R}$. Then for $f \in A^s_{pq}(\mathbb{R}^n)$,*

$$\|f\|_{A^s_{pq}} \sim \left(\sum_{m \in \mathbb{Z}^n} |\langle f, \Phi(\star - m) \rangle_2|^p \right)^{\frac{1}{p}} + \sum_{\varepsilon \in E} \left\| \{2^{j\left(s - \frac{n}{p} + n\right)} \langle f, \Psi^{\varepsilon}(2^j \star - m) \rangle_2 \}_{j \in \mathbb{N}_0, m \in \mathbb{Z}^n} \right\|_{\boldsymbol{a}_{pq}}.$$

Note that this time the coefficient mapping is linear.

Proof By the molecular decomposition, Theorem 4.2, we have

$$\left(\sum_{m \in \mathbb{Z}^n} |\langle f, \Phi(\star - m) \rangle_2|^p \right)^{\frac{1}{p}} + \sum_{\varepsilon \in E} \left\| \left\{ 2^{j\left(s - \frac{n}{p} + n\right)} \langle f, \Psi^{\varepsilon}(2^j \star - m) \rangle_2 \right\}_{j \in \mathbb{N}_0, m \in \mathbb{Z}^n} \right\|_{\mathbf{a}_{pq}} \lesssim \|f\|_{A^s_{pq}}.$$

Let us prove the reverse inequality. Set

$$f_{\le J} \equiv \sum_{m \in \mathbb{Z}^n \cap B(J)} \langle f, \varphi(\star - m) \rangle \varphi(\star - m) + \sum_{j=0}^{J} \sum_{m \in \mathbb{Z}^n \cap B(J)} \sum_{\varepsilon \in E} \langle f, 2^{jn/2} \psi^{\varepsilon}(2^j \star - m) \rangle 2^{jn/2} \psi^{\varepsilon}(2^j \star - m)$$

for $f \in A^s_{pq}(\mathbb{R}^n)$. We claim that $f_{\le J}$ converges to f in the topology of $\mathscr{S}'(\mathbb{R}^n)$. Choose $1 < u, v < \infty$, $\delta > 0$ so that $K > \sigma_u - s + \delta$ and that $L \ge \max(-1, [\sigma_u - s + \delta])$. The embedding $A^s_{pq}(\mathbb{R}^n) \hookrightarrow B^{s-\sigma_u-\delta}_{uv}(\mathbb{R}^n)$ allows us to assume $p > 1$ and $1 < q < \infty$; see Theorem 2.14.

Let $g \in \mathscr{S}(\mathbb{R}^n)$. Theorem 4.6 shows that $g_J \to g$ in the $L^2(\mathbb{R}^n)$-topology. Furthermore, Φ, Ψ^{ε}, $\varepsilon \in E$ are real-valued; hence $\langle f_{\le J}, g \rangle = \langle f, g_{\le J} \rangle$ by the definition of the coupling.

To estimate $g_{\le J}$, we use the atomic decomposition (see Theorem 4.2),

$$\|g_{\le J} - g_{\le J'}\|_{A^{-s}_{p'q'}} \sim \left(\sum_{\substack{m\in\mathbb{Z}^n \\ J'<|m|\le J}} |\langle g, \Phi(\star - m)\rangle_2|^{p'} \right)^{\frac{1}{p'}}$$

$$+\sum_{\varepsilon\in E}\sum_{m\in\mathbb{Z}^n\cap B(J)} \left\| \{2^{j\left(s-\frac{n}{p}\right)}\chi_{[0,J]}(j)\langle g, 2^{jn}\Psi^\varepsilon(2^j \star - m)\rangle_2\}_{j\in\mathbb{N}_0, m\in\mathbb{Z}^n} \right\|_{\mathbf{a}_{p'q'}}$$

$$-\sum_{\varepsilon\in E}\sum_{m\in\mathbb{Z}^n\cap B(J)'} \left\| \{2^{j\left(s-\frac{n}{p}\right)}\chi_{[0,J']}(j)\langle g, 2^{jn}\Psi^\varepsilon(2^j \star - m)\rangle_2\}_{j\in\mathbb{N}_0, m\in\mathbb{Z}^n} \right\|_{\mathbf{a}_{p'q'}},$$

whenever the integers J and J' satisfy $J' < J$.

Hence $\{g_{\le J}\}_{J\in\mathbb{N}}$ is a Cauchy sequence in $A^{-s}_{p'q'}(\mathbb{R}^n)$. Since $g_{\le J} \to g$ in $L^2(\mathbb{R}^n)$, $g_{\le J}$ converges to g in $A^{-s}_{p'q'}(\mathbb{R}^n)$, implying $\lim_{J\to\infty} \langle f_{\le J}, g\rangle = \lim_{J\to\infty} \langle f, g_{\le J}\rangle = \langle f, g\rangle$. Thus,

$$\|f_{\le J}\|_{A^s_{pq}}$$

$$\lesssim \left(\sum_{m\in\mathbb{Z}^n\cap B(J)} |\langle f, \Phi(\star - m)\rangle_2|^{p} \right)^{\frac{1}{p}}$$

$$+\sum_{\varepsilon\in E}\sum_{m\in\mathbb{Z}^n\cap B(J)} \left\| \{2^{j\left(s-\frac{n}{p}\right)}\chi_{[0,J]}(j)\langle f, 2^{jn}\Psi^\varepsilon(2^j \star - m)\rangle_2\}_{j\in\mathbb{N}_0, m\in\mathbb{Z}^n} \right\|_{\mathbf{a}_{pq}}$$

by the atomic decomposition; Theorem 4.2. Finally, the Fatou lemma yields the reverse inequality.

Since $p = \infty$ and/or $q = \infty$, the convergence of the wavelet expansion does not always take place in $f \in A^s_{pq}(\mathbb{R}^n)$. Using the atomic decomposition, we can say that this takes place at least in $\mathscr{S}'(\mathbb{R}^n)$. We summarize this observation.

Corollary 4.1 *In Theorem* 4.7, *we have*

$$f = \lim_{J\to\infty} \sum_{m\in\mathbb{Z}^n\cap B(J)} \langle f, \varphi(\star - m)\rangle \varphi(\star - m)$$

$$+ \lim_{J\to\infty} \left(\sum_{j=0}^{J} \sum_{m\in\mathbb{Z}^n\cap B(J)} \sum_{\varepsilon\in E} \langle f, 2^{jn/2}\Psi^\varepsilon(2^j \star - m)\rangle 2^{jn/2}\Psi^\varepsilon(2^j \star - m) \right)$$

in the topology of $\mathscr{S}'(\mathbb{R}^n)$ *for all* $f \in A^s_{pq}(\mathbb{R}^n)$.

Here we list some criteria with which to measure that the approximation is valid.

Definition 4.4 (Unconditional convergence, Schauder basis, Greedy) Let $0 < p, q < \infty$ and $s \in \mathbb{R}$.

1. For a sequence $\{f_k\}_{k=1}^{\infty} \subset A^s_{pq}(\mathbb{R}^n)$, a convergent series $f = \sum_{k=1}^{\infty} f_k$ is said to converge *unconditionally* if $f = \sum_{k=1}^{\infty} f_{\sigma(k)}$ in $A^s_{pq}(\mathbb{R}^n)$ for all bijections $\sigma : \mathbb{N} \to \mathbb{N}$.
2. A sequence $\{f_k\}_{k=1}^{\infty} \subset A^s_{pq}(\mathbb{R}^n)$ is said to be a Schauder basis if there exists a unique sequence $\{c_k(f)\}_{k=1}^{\infty} \subset \mathbb{C}$ such that $x = \sum_{k=1}^{\infty} c_k(f) f_k$ in $A^s_{pq}(\mathbb{R}^n)$ for all $f \in A^s_{pq}(\mathbb{R}^n)$. Furthermore, if the convergence above is always unconditional, then the basis is said to be *unconditional.*
3. An unconditional basis $\{\psi_k\}_{k=1}^{\infty}$ is said to be *greedy* in $A^s_{pq}(\mathbb{R}^n)$, if the norm of the finite sum $\sum_{k \in F} \psi_k$ is equivalent to $N(\sharp F)$, where N is a certain function.

Exercises

Exercise 4.10 Let $0 < p, q \le \infty$ and $s \in \mathbb{R}$. Using the wavelet decomposition, show that $A^s_{pq}(\mathbb{R}^n)$ is separable if and only if $p + q < \infty$.

Exercise 4.11 [317, Theorem 2.5] Let $0 < p, q < \infty$ and $s \in \mathbb{R}$. Show that

$$f = \sum_{m \in \mathbb{Z}^n} \langle f, \varphi(\star - m)\rangle \varphi(\star - m) + \sum_{j=0}^{\infty} \sum_{m \in \mathbb{Z}^n} \langle f, 2^{jn/2}\varphi(2^j \star - m)\rangle 2^{jn/2}\varphi(2^j \star - m),$$

in $A^s_{pq}(\mathbb{R}^n)$ in Theorem 4.7. Hint: Since p, q are finite, we can use the Lebesgue convergence theorem.

Exercise 4.12

1. Using the property of κ, prove (4.25) and (4.26).
2. Using the Fourier series, prove (4.27).

Exercise 4.13 Let $0 < p, q \le \infty$ and $s \in \mathbb{R}$. For $L \in \mathbb{N}$ let $\mathscr{S}_L(\mathbb{R}^n)$ be the closed subspace of $\mathscr{S}(\mathbb{R}^n)$ given by (1.119). Denote by $\mathscr{S}'_L(\mathbb{R}^n)$ be its topological dual.

1. Use Proposition 1.13 to show that $\dot{A}^s_{pq}(\mathbb{R}^n)$ is embedded into $\mathscr{S}'_L(\mathbb{R}^n)$ for $L > \frac{n}{p}$.
2. Let $L > \frac{n}{p}$. Choose $\varphi \in C_c^{[|s|+1]}(\mathbb{R}^n) \cap \mathscr{P}_L^{\perp}(\mathbb{R}^n)$ so that $\{\varphi(2^\nu \star - m)\}_{j \in \mathbb{Z}_0, m \in \mathbb{Z}^n}$ is a complete orthonormal basis. Using this function φ, develop a theory of wavelet expansion in $\dot{A}^s_{pq}(\mathbb{R}^n)$.

4.1.3 Quarkonial Decomposition

The atomic decomposition is a technique with which to express functions f into the double sum: $\sum_{j=1}^{\infty}\sum_{m\in\mathbb{Z}^n}\lambda_{jm}a_{jm}$. But, as the proof shows, we see that the coefficients do not depend linearly on f. We have a hint to overcome this difficulty. In view of the wavelet decomposition, it is attractive to add another parameter to $j \in \mathbb{N}_0$ and $m \in \mathbb{Z}^n$. So, we are led to a triple sum: $\sum_{\beta\in{\mathbb{N}_0}^n}\sum_{j=1}^{\infty}\sum_{m\in\mathbb{Z}^n}\cdots$. Based on this idea, we consider the quarkonial decomposition.

4.1.3.1 Quarks

To formulate the quarkonial decomposition, we need to fix some notation.

Definition 4.5 (ψ for the quarkonial decomposition) Throughout this section, the function $\psi \in \mathscr{S}(\mathbb{R}^n)$ is fixed so that $\{\psi(\star - m)\}_{m\in\mathbb{Z}^n}$ forms a *partition of unity*; $\sum_{m\in\mathbb{Z}^n}\psi(\star - m) \equiv 1$. Accordingly, choose $r > 0$ so that $\mathrm{supp}(\psi) \subset B(2^r)$.

With ψ specified as above, we define quarks. As before, the parameters p, q, s satisfy the conditions for the function space $A^s_{pq}(\mathbb{R}^n)$. Here the symbol qu stands for "quark" as in [103].

Definition 4.6 (Regular quark) Let $0 < p \le \infty$, $s \in \mathbb{R}$, $\beta \in {\mathbb{N}_0}^n$, $\nu \in \mathbb{N}_0$, and let $m \in \mathbb{Z}^n$. Then define ψ^β and $(\beta\mathrm{qu})_{\nu m}$ by

$$\psi^\beta(x) \equiv x^\beta\psi(x), \quad (\beta\mathrm{qu})_{\nu m}(x) \equiv 2^{-\nu\left(s-\frac{n}{p}\right)}\psi^\beta(2^\nu x - m) \tag{4.28}$$

for $x \in \mathbb{R}^n$. Each $(\beta\mathrm{qu})_{\nu m}$ is called the *quark*.

In the setting of quarkonial decomposition, we use the following notation of sequence spaces:

Definition 4.7 (Sequence spaces $\mathbf{a}_{pq,\rho}(\mathbb{R}^n)$ for quarkonial decomposition) Let r be as in Definition 4.5. For $\rho > r$, define functions recursively. For a triply parametrized sequence $\lambda = \{\lambda^\beta_{\nu m}\}_{\beta\in{\mathbb{N}_0}^n,\ \nu\in\mathbb{N}_0,\ m\in\mathbb{Z}^n}$, define

$$\lambda^\beta \equiv \{\lambda^\beta_{\nu m}\}_{\nu\in\mathbb{N}_0,\ m\in\mathbb{Z}^n}, \quad \|\lambda\|_{\mathbf{a}_{pq},\rho} \equiv \sup_{\beta\in{\mathbb{N}_0}^n} 2^{\rho|\beta|}\|\lambda^\beta\|_{\mathbf{a}_{pq}}.$$

As usual, when the Triebel–Lizorkin space comes into play, exclude the case where $p = \infty$. The space $\mathbf{a}_{pq,\rho}(\mathbb{R}^n)$ collects all triply parametrized sequence λ for which $\|\lambda\|_{\mathbf{a}_{pq,\rho}}$ is finite.

4.1.3.2 Quarkonial Decomposition for the Regular Case

Here we deal with the quarkonial decomposition for the regular case:

$$0 < p \leq \infty, \ 0 < q < \infty, \ s > \sigma_p \tag{4.29}$$

for Besov spaces and

$$0 < p < \infty, \ 0 < q \leq \infty, \ s > \sigma_{pq} \tag{4.30}$$

for Triebel–Lizorkin spaces. Also assume that

$$\rho > r. \tag{4.31}$$

Here r is a constant in Definition 4.5.

Theorem 4.8 (Quarkonial decomposition for the regular case) *Assume that ρ satisfies* (4.31). *Also assume that the parameters p, q, s satisfy* (4.29) *for Besov spaces and* (4.30) *for Triebel–Lizorkin spaces. Then a distribution $f \in \mathscr{S}'(\mathbb{R}^n)$ belongs to $A^s_{pq}(\mathbb{R}^n)$ if and only if f has an expression:*

$$f = \sum_{\beta \in \mathbb{N}_0{}^n} \sum_{\nu=0}^{\infty} \sum_{m \in \mathbb{Z}^n} \lambda^{\beta}_{\nu m} (\beta \mathrm{qu})_{\nu m} \tag{4.32}$$

in the topology of $\mathscr{S}'(\mathbb{R}^n)$, where $\lambda \equiv \{\lambda^{\beta}_{\nu m}\}_{\beta \in \mathbb{N}_0{}^n, \ \nu \in \mathbb{N}_0, \ m \in \mathbb{Z}^n}$ is a triply indexed sequence such that

$$\|\lambda\|_{\boldsymbol{a}_{pq},\rho} < \infty. \tag{4.33}$$

If this is the case, we can choose λ so that

$$\|\lambda\|_{\boldsymbol{a}_{pq},\rho} \sim \|f\|_{A^s_{pq}}. \tag{4.34}$$

The “if” part is the synthesis part, while the “only if” part is the analysis part.

The proof will be long. However, the atomic decomposition being proved, we can easily prove the sufficiency: Assuming (4.31), we choose $\varepsilon > 0$ so that $0 < \varepsilon < \rho - r$ to conclude that $2^{-(r+\varepsilon)|\beta|}(\beta \mathrm{qu})_{\nu m}$ is an atom centered at $Q_{\nu m}$ modulo a multiplicative constant. Regarding the support condition, we may not be able to realize that $\mathrm{supp}((\beta \mathrm{qu})_{\nu m}) \subset d\, Q_{\nu m}$ with d less than or equal to 3. However, this is immaterial. With this in mind, we set

$$f^{\beta} \equiv \sum_{\nu=0}^{\infty} \sum_{m \in \mathbb{Z}^n} \lambda^{\beta}_{\nu m} (\beta \mathrm{qu})_{\nu m} \tag{4.35}$$

and invoke Theorem 4.1 to conclude that

$$\|f^{\beta}\|_{A^s_{pq}} \lesssim 2^{-(\rho-r-\varepsilon)|\beta|}\|\lambda\|_{\mathbf{a}_{pq},\rho}. \tag{4.36}$$

For $f_1, f_2 \in A^s_{pq}(\mathbb{R}^n)$, we have the $\min(1, p, q)$-triangle inequality

$$\|f_1 + f_2\|_{A^s_{pq}}{}^{\min(1,p,q)} \le \|f_1\|_{A^s_{pq}}{}^{\min(1,p,q)} + \|f_2\|_{A^s_{pq}}{}^{\min(1,p,q)}.$$

Thus, $f \in A^s_{pq}(\mathbb{R}^n)$ and we have the norm estimate (4.33). Since $\mathrm{supp}(\psi) \subset B(2^r)$, there exists $r' \in (0, r)$ such that $\mathrm{supp}(\psi) \subset B(2^{r'})$. Thus, it can be assumed that $r = \rho$.

The heart of the matter is to prove the necessity; that is, we are requested to expand $f \in A^s_{pq}(\mathbb{R}^n)$ in a desired manner. To this end, we need to prove some other facts.

Theorem 4.9 (The Frazier–Jawerth φ-transform) *Let $R > 0$. Assume that $\kappa \in \mathscr{S}(\mathbb{R}^n)$ satisfies $\chi_{Q(3)} \le \kappa \le \chi_{Q(3+1/100)}$. Then any $f \in \mathscr{S}'_{Q(3R)}(\mathbb{R}^n)$ admits an expansion:*

$$f = (2\pi)^{-\frac{n}{2}} \sum_{m\in\mathbb{Z}^n} f(R^{-1}m)\mathscr{F}^{-1}\kappa(R\star - m),$$

where the convergence takes place in $\mathscr{S}'(\mathbb{R}^n)$.

This theorem does not contain φ in the statement. However, in the original papers [482, 483], instead of the symbol κ the authors used the symbol φ. This is why Theorem 4.9 deserves its name.

It is surprising that $f(x)$ is realized by means of the discrete data $\{f(R^{-1}m)\}_{m\in\mathbb{Z}^n}$.

The larger R is, the stronger the assumption is, implying that we need more discrete data.

Proof Let $\kappa_R(\star) \equiv \kappa(R^{-1}\star)$ for $R > 0$. Choose a test function $\tau \in \mathscr{S}(\mathbb{R}^n)$ arbitrarily. Then

$$\langle \mathscr{F} f, \tau\rangle = \langle \mathscr{F} f, \kappa_R \cdot \kappa_R \cdot \tau\rangle \tag{4.37}$$

in view of the size of support $\mathscr{F} f$. Define a $2\pi R$-periodic function by $\tau^* \equiv \sum_{l\in\mathbb{Z}^n} \kappa_R(\star - 2\pi Rl)\tau(\star - 2\pi Rl)$. If we expand τ^* into the Fourier series, then

$$\tau^* = \sum_{m\in\mathbb{Z}^n} a_m \exp\left(\frac{\star \cdot m}{R} i\right), \tag{4.38}$$

where the coefficient satisfies

$$\begin{aligned}
a_m &= \frac{1}{(2\pi R)^n} \int_{Q(\pi R)} \tau^*(x) \exp\left(-\frac{x \cdot m}{R} i\right) \mathrm{d}x \\
&= \frac{1}{(2\pi R)^n} \int_{Q(\pi R)} \left(\sum_{l \in \mathbb{Z}^n} (\kappa_R \cdot \tau)(x - 2\pi R l)\right) \exp\left(-\frac{x \cdot m}{R} i\right) \mathrm{d}x \\
&= \frac{1}{(2\pi R)^n} \int_{\mathbb{R}^n} \kappa_R(x) \tau(x) \exp\left(-\frac{x \cdot m}{R} i\right) \mathrm{d}x.
\end{aligned}$$

We have

$$\kappa_R(x)\kappa_R(x)\tau(x) = \kappa_R(x)\kappa_R(x)\tau^*(x) = \sum_{m \in \mathbb{Z}^n} a_m \, \kappa_R(x) \exp\left(\frac{x \cdot m}{R} i\right) \tag{4.39}$$

for $x \in \mathbb{R}^n$ in view of the size of the support of functions.

If we write (4.37) using (4.38) and (4.39), then

$$\begin{aligned}
\langle \mathscr{F} f, \tau \rangle &= \sum_{m \in \mathbb{Z}^n} a_m \left\langle \mathscr{F} f, \kappa_R \exp\left(\frac{\star \cdot m}{R} i\right) \right\rangle \\
&= \sum_{m \in \mathbb{Z}^n} \frac{1}{(2\pi R)^n} \left\langle \kappa_R \exp\left(-\frac{\star \cdot m}{R} i\right), \tau \right\rangle \cdot \left\langle \mathscr{F} f, \kappa_R \exp\left(\frac{\star \cdot m}{R} i\right) \right\rangle \\
&= \left\langle \left\{ \sum_{m \in \mathbb{Z}^n} \frac{1}{(2\pi R)^n} \left\langle \mathscr{F} f, \kappa_R \exp\left(\frac{\star \cdot m}{R} i\right) \right\rangle \cdot \kappa_R \exp\left(-\frac{\star \cdot m}{R} i\right) \right\}, \tau \right\rangle.
\end{aligned}$$

Finally, $\left\langle \mathscr{F} f, \kappa_R \cdot \exp\left(\frac{\star \cdot m}{R} i\right) \right\rangle = (2\pi)^{\frac{n}{2}} f(R^{-1} m)$ in view of the size of the support of $\mathscr{F}$.

Since the test function τ is arbitrary, we have

$$\mathscr{F} f = (2\pi)^{\frac{n}{2}} \sum_{m \in \mathbb{Z}^n} \frac{1}{(2\pi R)^n} f(R^{-1} m) \cdot \kappa_R \exp\left(-\frac{\star \cdot m}{R} i\right).$$

If we take the inverse Fourier transform, then we obtain the desired result.

Here and below, $\tau, \varphi \in \mathscr{S}(\mathbb{R}^n)$ satisfy

$$\chi_{B(2)} \le \tau \le \chi_{B(3)}, \ \varphi_j = \tau(2^{-j} \star) - \tau(2^{-j+1} \star), \ j \in \mathbb{N}.$$

Since $f = \tau(D) f + \sum_{\nu=1}^{\infty} \varphi_\nu(D) f$ for $f \in \mathscr{S}'(\mathbb{R}^n)$, we have the following corollary:

Corollary 4.2 *Let κ be the same as Theorem 4.9. Any $f \in \mathscr{S}'(\mathbb{R}^n)$ can be expanded:*

$$f = (2\pi)^{-\frac{n}{2}} \sum_{m \in \mathbb{Z}^n} \tau(D) f(m) \mathscr{F}^{-1}\kappa(\star - m)$$
$$+ (2\pi)^{-\frac{n}{2}} \sum_{\nu=1}^{\infty} \left(\sum_{m \in \mathbb{Z}^n} \varphi_\nu(D) f(2^{-\nu} m) \mathscr{F}^{-1}\kappa(2^\nu \star - m) \right).$$

The next lemma concerns the translation by $l \in \mathbb{Z}^n$:

$$\{\lambda_{\nu m}\}_{\nu \in \mathbb{N}_0,\, m \in \mathbb{Z}^n} \mapsto \lambda^l \equiv \{\lambda_{\nu\, m+l}\}_{\nu \in \mathbb{N}_0,\, m \in \mathbb{Z}^n}. \tag{4.40}$$

Lemma 4.6 *Whenever $l \in \mathbb{Z}^n$ and $0 < \eta < \min(p, q)$, $\|\lambda^l\|_{\mathbf{a}_{pq}} \lesssim \langle l \rangle^{\frac{n}{\eta}} \|\lambda\|_{\mathbf{a}_{pq}}$ for any complex sequence $\{\lambda_{\nu m}\}_{\nu \in \mathbb{N}_0,\, m \in \mathbb{Z}^n}$, where $\lambda^l \equiv \{\lambda_{\nu m+l}\}_{\nu \in \mathbb{N}_0,\, m \in \mathbb{Z}^n}$.*

Proof Let $x \in \mathbb{R}^n$. Fix ν and then choose m so that $x \in Q_{\nu m}$. A geometric observation shows:

$$|\lambda_{\nu\, m+l}| \lesssim \langle l \rangle^{\frac{n}{\eta}} m^{(\eta)}_{B(x, c\, 2^{-\nu}\langle l \rangle)} \left(\sum_{m \in \mathbb{Z}^n} \lambda_{\nu m} \chi_{Q_{\nu m}} \right) \lesssim \langle l \rangle^{\frac{n}{\eta}} M^{(\eta)} \left[\sum_{m \in \mathbb{Z}^n} \lambda_{\nu m} \chi_{Q_{\nu m}} \right](x).$$

So, we are in the position of using the Fefferman–Stein vector-valued inequality of the Hardy–Littlewood maximal operator.

The next lemma concerns how fast the Fourier transform grows with respect to the derivatives.

Lemma 4.7 *Let $\kappa \in \mathscr{S}(\mathbb{R}^n)$ satisfy $\chi_{Q(3)} \le \kappa \le \chi_{Q(3+1/100)}$. Then*

$$|\partial^\alpha \mathscr{F}^{-1}\kappa(x)| \lesssim_N \langle \alpha \rangle^{2N} \langle x \rangle^{-2N} \quad (x \in \mathbb{R}^n, \alpha \in {\mathbb{N}_0}^n) \tag{4.41}$$

for $N \gg 1$.

Proof We integrate by parts in $\partial^\alpha \mathscr{F}^{-1}\kappa(x) \simeq_n \int_{\mathbb{R}^n} (iz)^\alpha \kappa(z) \exp(iz \cdot x) dz$ to have

$$\partial^\alpha \mathscr{F}^{-1}\kappa(x) \simeq_n \langle x \rangle^{-2N} \int_{\mathbb{R}^n} (iz)^\alpha \kappa(z) \left((1 - \Delta_z)^N \exp(iz \cdot x) \right) dz$$
$$\simeq_n \langle x \rangle^{-2N} \int_{\mathbb{R}^n} \left((1 - \Delta_z)^N (iz)^\alpha \kappa(z) \right) \exp(iz \cdot x) dz.$$

Hence (4.41) follows.

We discuss the analysis part of Theorem 4.8.

Proof Let $f \in A^s_{pq}(\mathbb{R}^n)$. Then

$$f = (2\pi)^{-\frac{n}{2}} \sum_{m \in \mathbb{Z}^n} \tau(D) f(m) \mathscr{F}^{-1}\kappa(\star - m)$$
$$+ (2\pi)^{-\frac{n}{2}} \sum_{\nu=1}^{\infty} \left(\sum_{m \in \mathbb{Z}^n} \varphi_\nu(D) f(2^{-\nu} m) \mathscr{F}^{-1}\kappa(2^\nu \star - m) \right)$$

by Corollary 4.2. For $(\nu, m) \in \mathbb{N}_0 \times \mathbb{Z}^n$, we set

$$\Lambda_{\nu m} = \begin{cases} \tau(D) f(m) & (\nu = 0), \\ 2^{\nu\left(s - \frac{n}{p}\right)} \varphi_\nu(D) f(2^{-\nu} m) & (\nu \geq 1). \end{cases}$$

Then (4.42) can be rephrased:

$$f \simeq_n \sum_{\nu=0}^{\infty} \sum_{m \in \mathbb{Z}^n} 2^{-\nu\left(s - \frac{n}{p}\right)} \Lambda_{\nu m} \mathscr{F}^{-1}\kappa(2^\nu \star - m). \tag{4.42}$$

Here and below, we do not consider the case where $\nu = 0$, for we can readily incorporate the term for $\nu = 0$.

To prove Theorem 4.8, we can assume that ρ is a large integer by replacing ρ with $[\rho + 1]$. Let $x \in \mathbb{R}^n$. By the Taylor expansion, we have

$$\psi(2^{\nu+\rho} x - l) \mathscr{F}^{-1}\kappa(2^\nu x - m)$$
$$= \sum_{\beta \in \mathbb{N}_0{}^n} \frac{\partial^\beta \mathscr{F}^{-1}\kappa(2^{-\rho} l - m)(2^\nu x - 2^{-\rho} l)^\beta \psi(2^{\nu+\rho} x - l)}{\beta!}$$
$$= \sum_{\beta \in \mathbb{N}_0{}^n} \frac{2^{-\rho|\beta|} \partial^\beta \mathscr{F}^{-1}\kappa(2^{-\rho} l - m) \psi^\beta(2^{\nu+\rho} x - l)}{\beta!}.$$

Since $\sum_{m \in \mathbb{Z}^n} \psi(\star - m) \equiv 1$, we can expand $\varphi_\nu(D) f$ further:

$$\varphi_\nu(D) f \simeq_n 2^{-\nu\left(s - \frac{n}{p}\right)} \sum_{m \in \mathbb{Z}^n} \sum_{l \in \mathbb{Z}^n} \sum_{\beta \in \mathbb{N}_0{}^n} \frac{\Lambda_{\nu m}}{\beta! 2^{\rho|\beta|}} \partial^\beta \mathscr{F}^{-1}\kappa(2^{-\rho} l - m) \psi^\beta(2^{\nu+\rho} \star - l).$$

Recall that $L^\infty(\mathbb{R}^n)$ is the dual of $L^1(\mathbb{R}^n)$. Since the convergence takes place unconditionally in the weak-* topology of $L^\infty(\mathbb{R}^n)$, we have

$$\varphi_\nu(D) f(x) \simeq_n \sum_{l \in \mathbb{Z}^n} \sum_{\beta \in \mathbb{N}_0{}^n} \sum_{m \in \mathbb{Z}^n} \frac{1}{\beta! 2^{\rho|\beta|}} \Lambda_{\nu m}\, \partial^\beta \mathscr{F}^{-1}\kappa(2^{-\rho} l - m) (\beta\mathrm{qu})_{\nu+\rho\, l}(x)$$

using quarks. Let

$$\lambda^{\beta}_{\nu+\rho\, l} \equiv \frac{1}{\beta! 2^{\rho|\beta|}} \sum_{m\in\mathbb{Z}^n} \Lambda_{\nu m}\, \partial^{\beta}\mathscr{F}^{-1}\kappa(2^{-\rho}l - m).$$

If we use this expression, then

$$f = \sum_{\nu=0}^{\infty} \varphi_\nu(D) f \simeq_n \sum_{\nu=0}^{\infty} \sum_{l\in\mathbb{Z}^n} \sum_{\beta\in\mathbb{N}_0{}^n} \lambda^{\beta}_{\nu+\rho\, l} (\beta \mathrm{qu})_{\nu+\rho\, l}. \tag{4.43}$$

Next, we investigate the size of coefficients. Let $l \in \mathbb{Z}^n$ and $x \in Q_{\nu+\rho\ 2^{\rho}l+l_0}$. We use (4.41) to have

$$|\lambda^{\beta}_{\nu+\rho\ 2^{\rho}l+l_0}| \lesssim 2^{-\rho|\beta|} \sum_{m\in\mathbb{Z}^n} \langle l-m\rangle^{-N} |\Lambda_{\nu m}| = 2^{-\rho|\beta|} \sum_{m\in\mathbb{Z}^n} \langle m\rangle^{-N} |\Lambda_{\nu\ m+l}|,$$

where l_0 is a lattice point in $[0, 2^{\rho})^n$. We fix m and we define

$$\eta_0 \equiv \min(p,q), \quad \Lambda^m \equiv \{|\Lambda_{\nu\ m+l}|\}_{\nu\in\mathbb{N}_0,\, l\in\mathbb{Z}^n}. \tag{4.44}$$

Then

$$\|\lambda^{\beta}\|_{\mathbf{a}_{pq}} \lesssim 2^{-\rho|\beta|} \left\| \sum_{m\in\mathbb{Z}^n} \langle m\rangle^{-N} \Lambda^m \right\|_{\mathbf{a}_{pq}} \lesssim 2^{-\rho|\beta|} \left(\sum_{m\in\mathbb{Z}^n} \langle m\rangle^{-N\eta_0} \|\Lambda^m\|_{\mathbf{a}_{pq}}^{\eta_0} \right)^{\frac{1}{\eta_0}}.$$

Since N can be chosen sufficiently large, we have

$$\|\lambda^{\beta}\|_{\mathbf{a}_{pq},\rho} \lesssim 2^{-\rho|\beta|} \left(\sum_{m\in\mathbb{Z}^n} \langle m\rangle^{(2n/\min(1,p,q)-N)\eta_0} \|\Lambda\|_{\mathbf{a}_{pq}}^{\eta_0} \right)^{\frac{1}{\eta_0}} \lesssim \|\Lambda\|_{\mathbf{a}_{pq}} \tag{4.45}$$

using Lemma 4.6. Since $\rho > r$, we have $\|\lambda\|_{\mathbf{a}_{pq},\rho} \lesssim \|\Lambda\|_{\mathbf{a}_{pq}}$.

Theorem 1.50 yields $|\Lambda_{\nu m}| \lesssim 2^{\nu\left(s-\frac{n}{p}\right)} \inf_{y\in Q_{\nu m}} M^{\left(\frac{\eta_0}{2}\right)}[\varphi_\nu(D)f](y)$ for $\nu > 0$ and

$$|\Lambda_{0m}| \lesssim \inf_{y\in Q_{0m}} M^{\left(\frac{\eta_0}{2}\right)}[\tau(D)f](y).$$

When $\nu \geq 1$, choose $y \in Q_{\nu m}$ and apply Theorem 1.50 to have

$$|\Lambda_{\nu m}| = 2^{\nu\left(s-\frac{n}{p}\right)} \left|\varphi_\nu(D)f(2^{-\nu}m)\right|$$

$$
\begin{aligned}
&= (1+2^{\nu}|y-2^{-\nu}m|)^{2n/\eta_0} \frac{2^{\nu\left(s-\frac{n}{p}\right)}}{(1+2^{\nu}|y-2^{-\nu}m|)^{2n/\eta_0}} \left|\varphi_\nu(D)f(2^{-\nu}m)\right| \\
&\le (1+n)^{2n/\eta_0} \frac{2^{\nu\left(s-\frac{n}{p}\right)}}{(1+2^{\nu}|y-2^{-\nu}m|)^{2n/\eta_0}} \left|\varphi_\nu(D)f(2^{-\nu}m)\right| \\
&\lesssim 2^{\nu\left(s-\frac{n}{p}\right)} M^{\left(\frac{\eta_0}{2}\right)}[\varphi_\nu(D)f](y),
\end{aligned}
$$

where we used

$$
\frac{1}{(1+2^{\nu}|y-2^{-\nu}m|)^{2n/\eta_0}} \left|\varphi_\nu(D)f(2^{-\nu}m)\right| \lesssim M^{\left(\frac{\eta_0}{2}\right)}[\varphi_\nu(D)f](y).
$$

If we take the infimum over y, we obtain

$$
|\Lambda_{\nu m}| \lesssim 2^{\nu\left(s-\frac{n}{p}\right)} \inf_{y\in Q_{\nu m}} M^{\left(\frac{\eta_0}{2}\right)}[\varphi_\nu(D)f](y).
$$

A similar estimate for $\nu = 0$ is readily available; hence we have

$$
\|\Lambda\|_{\mathbf{a}_{pq}} \lesssim \|f\|_{A^s_{pq}}.
$$

The proof of quarkonial decomposition is complete.

4.1.3.3 Application of φ-Transform: General Principle

We remark that the φ-transform is useful. Here we prove a general fact on functional analysis. Note that we can suppose that the φ-transform is a special case.

Proposition 4.2 *Let X_0, X_1, Y_0, Y_1 be quasi-Banach spaces. Suppose that we have bounded linear operators $A_0 : X_0 \to Y_0$, $B_0 : Y_0 \to X_0$, $A_1 : X_1 \to Y_1$, $B_1 : Y_1 \to X_1$, $\Phi : X_0 \to X_1$ and $\Psi : Y_0 \to Y_1$ satisfying*

$$
B_0 \circ A_0 = \mathrm{id}_{X_0}, \quad B_1 \circ A_1 = \mathrm{id}_{X_1}, \quad A_1 \circ \Phi = \Psi \circ A_0, \quad B_1 \circ \Psi = \Phi \circ B_0.
$$

If Ψ is an isomorphism, then so is Φ.

Proof Let us show that Φ is injective. To this end, assume that $x_0 \in \ker(\Phi)$. Then $\Psi \circ A_0(x_0) = A_1 \circ \Phi(x_0) = 0$. Since Ψ is an isomorphism, $A_0(x_0) = 0$. Thus, $x_0 = B_0 \circ A_0(x_0) = 0$.

Let us show that Φ is surjective. To this end, take $x_1 \in X_1$. Then $A_1(x_1) = \Psi(y_0)$ for some $y_0 \in Y_0$. Thus, $x_1 = B_1 \circ A_1(x_1) = B_1 \circ \Psi(y_0) = \Phi(B_0(y_0))$. Thus, x_1 is in the image of Φ.

4.1.3.4 Quarkonial Decomposition for General Case

We consider quarkonial decomposition for the general case.

Definition 4.8 (Quark for general case) Let $L \in \{-1, 1, 3, 5, \dots\}$ be an odd integer and $(\nu, m) \in \mathbb{Z}^{n+1}$, define a function $(\beta\mathrm{qu})_{\nu m}$ by

$$(\beta\mathrm{qu})^{(L)}_{\nu m}(x) \equiv 2^{-\nu\left(s-\frac{n}{p}\right)}\left((-\Delta)^{\frac{L+1}{2}}\psi^{\beta}\right)(2^{\nu}x - m) \tag{4.46}$$

for $x \in \mathbb{R}^n$, where p and s are parameters in the function space $A^s_{pq}(\mathbb{R}^n)$.

Let us formulate the result.

Theorem 4.10 (Quarkonial decomposition for general case) *Suppose that the parameters p, q, s, ρ, r satisfy $0 < p, q \le \infty$, $s \in \mathbb{R}$ and $\rho > r > 0$. Furthermore, let the parameter σ satisfy $\sigma > \max(\sigma_p, s)$ for Besov spaces and $\sigma > \max(\sigma_{pq}, s)$ for Triebel–Lizorkin spaces, and let an odd integer $L \in \mathbb{Z}$ satisfy* (4.10) *for Besov spaces and* (4.11) *for Triebel–Lizorkin spaces. Furthermore, define $(\beta\mathrm{qu})_{\nu m}$ and $(\beta\mathrm{qu})^{(L)}_{\nu m}$ by* (4.28) *with s replaced by σ and* (4.46)*, respectively.*

Then $f \in A^s_{pq}(\mathbb{R}^n)$ if and only if there exist triply indexed sequences

$$\eta = \{\eta^{\beta}_{\nu m}\}_{\beta \in \mathbb{N}_0{}^n,\ \nu \in \mathbb{N}_0,\ m \in \mathbb{Z}^n}, \quad \lambda = \{\lambda^{\beta}_{\nu m}\}_{\beta \in \mathbb{N}_0{}^n,\ \nu \in \mathbb{N}_0,\ m \in \mathbb{Z}^n}$$

such that f is represented as

$$f = \sum_{\beta \in \mathbb{N}_0{}^n}\sum_{\nu=0}^{\infty}\sum_{m \in \mathbb{Z}^n}\eta^{\beta}_{\nu m}(\beta\mathrm{qu})_{\nu m} + \sum_{\beta \in \mathbb{N}_0{}^n}\sum_{\nu=0}^{\infty}\sum_{m \in \mathbb{Z}^n}\lambda^{\beta}_{\nu m}(\beta\mathrm{qu})^{(L)}_{\nu m}$$

in the topology of $\mathscr{S}'(\mathbb{R}^n)$ and the coefficients η and λ satisfy

$$\|\eta\|_{\boldsymbol{a}_{pq},\rho} + \|\lambda\|_{\boldsymbol{a}_{pq},\rho} < \infty.$$

Then in this case, λ, η can be chosen to depend linearly on f and they satisfy

$$\|\eta\|_{\boldsymbol{a}_{pq},\rho} + \|\lambda\|_{\boldsymbol{a}_{pq},\rho} \sim \|f\|_{A^s_{pq}}. \tag{4.47}$$

Proof If f is decomposed as in the theorem, it is clear that $f \in A^s_{pq}(\mathbb{R}^n)$. We can prove sufficiency using Theorem 4.1 as we did for the regular quarkonial decomposition. We use Theorem 2.12, the boundedness of the lift operator to prove a necessary condition. Let M be a sufficiently large odd number, say, $M > \max(L, \sigma - s)$. Then we set

$$g_1 \equiv (1 + (-\Delta)^{\frac{M+1}{2}})^{-1}f \in A^{s+M+1}_{pq}(\mathbb{R}^n) \hookrightarrow A^{\sigma}_{pq}(\mathbb{R}^n),$$

$$g_2 \equiv (-\Delta)^{\frac{M-L}{2}}(1+(-\Delta)^{\frac{M+1}{2}})^{-1} f \in A_{pq}^{s+L+1}(\mathbb{R}^n).$$

By Theorem 2.12, we have $f = g_1 + (-\Delta)^{\frac{L+1}{2}} g_2$, with $g_1 \in A_{pq}^{\sigma}(\mathbb{R}^n)$ and $g_2 \in A_{pq}^{s+L+1}(\mathbb{R}^n)$. We apply Theorem 4.8 to g_1 and g_2. Then we obtain the quarkonial decomposition for g_1, g_2:

$$g_1 = \sum_{\beta \in \mathbb{N}_0{}^n} \sum_{\nu=0}^{\infty} \sum_{m \in \mathbb{Z}^n} \eta_{\nu m}^{\beta} (\beta \mathrm{qu})_{\nu m}$$

$$g_2 = \sum_{\beta \in \mathbb{N}_0{}^n} \sum_{\nu=0}^{\infty} \sum_{m \in \mathbb{Z}^n} \lambda_{\nu m}^{\beta} \left\{ 2^{-\nu\left(s+L+1-\frac{n}{p}\right)} \psi^{\beta}(2^{\nu} * -m) \right\},$$

where the coefficient satisfies $\|\eta\|_{\mathbf{a}_{pq},\rho} + \|\lambda\|_{\mathbf{a}_{pq},\rho} \lesssim \|g_2\|_{A_{pq}^{s+L+1}} + \|g_1\|_{A_{pq}^{\sigma}} \lesssim \|f\|_{A_{pq}^s}$. Also, since $(-\Delta)^{\frac{L+1}{2}} g_2 = \sum_{\beta \in \mathbb{N}_0{}^n} \sum_{\nu=0}^{\infty} \sum_{m \in \mathbb{Z}^n} \lambda_{\nu m}^{\beta} (\beta \mathrm{qu})_{\nu m}^{(L)}$, we have

$$f = \sum_{\beta \in \mathbb{N}_0{}^n} \sum_{\nu=0}^{\infty} \sum_{m \in \mathbb{Z}^n} \eta_{\nu m}^{\beta} (\beta \mathrm{qu})_{\nu m} + \sum_{\beta \in \mathbb{N}_0{}^n} \sum_{\nu=0}^{\infty} \sum_{m \in \mathbb{Z}^n} \lambda_{\nu m}^{\beta} (\beta \mathrm{qu})_{\nu m}^{(L)} \tag{4.48}$$

and the quarkonial decomposition of f is obtained.

We end this section with a couple of remarks.

Remark 4.1 Decompositions for homogeneous function spaces are also available. Due to similarity we omit the details.

Remark 4.2 A direct corollary of Theorems 4.8 and 4.10 is that $C_c^{\infty}(\mathbb{R}^n)$ is dense in $A_{pq}^s(\mathbb{R}^n)$ if $0 < p, q < \infty$ and $s \in \mathbb{R}$.

Exercises

Exercise 4.14 Formulate and establish the results of quarkonial decomposition for the regular case of homogeneous spaces.

Exercise 4.15 [99, p. 21] If $\varphi \in \mathscr{S}_{[-b,b]^n}(\mathbb{R}^n)$ with $b > 0$, then show that

$$\varphi(x) = \sum_{k \in \mathbb{Z}^n} \varphi\left(\frac{\pi}{b} k\right) \prod_{j=1}^{n} \frac{\sin(b x_j - \pi k_j)}{b x_j - \pi k_j}$$

for all $x = (x_1, x_2, \ldots, x_n) \in \mathbb{R}^n$ such that $\pi^{-1} b x \notin \mathbb{Z}^n$.

4.1.4 Applications of the Atomic Decomposition to the Embedding Theorems

In general the function spaces are more difficult to handle than the sequence spaces. So far, we have obtained a language to translate the function spaces into the sequence spaces. Therefore, we expect the language to facilitate the analysis of the function spaces. Here we present a typical application of the decompositions.

4.1.4.1 Applications of the Atomic Decomposition to the Embedding Theorems

As another application of atomic decomposition, we supplement Theorem 2.14.

Theorem 4.11 (Sobolev embedding of Frank–Jawerth-type) *Let the real parameters p, p_0, p_1, q, s_0, s_1, s satisfy $0 < p_0 < p < p_1 \le \infty$, $0 < q \le \infty$, $-\infty < s_1 < s < s_0 < \infty$ and*

$$s_0 - \frac{n}{p_0} = s - \frac{n}{p} = s_1 - \frac{n}{p_1}. \tag{4.49}$$

Then $B^{s_0}_{p_0 p}(\mathbb{R}^n) \hookrightarrow F^s_{pq}(\mathbb{R}^n) \hookrightarrow B^{s_1}_{p_1 p}(\mathbb{R}^n)$.

For the proof we need a lemma on sequence spaces.

Lemma 4.8 *Let $0 < p < p_1 \le \infty$ and $\nu_0 \in \mathbb{N}$. Then*

$$\left(\sum_{\nu=0}^{\infty}\left(\sum_{m\in\mathbb{Z}^n} |\lambda_{\nu m}|^{p_1}\right)^{\frac{p}{p_1}}\right)^{\frac{1}{p}} \lesssim \left(\sum_{m\in\mathbb{Z}^n} |\lambda_{\nu_0 m}|^{p}\right)^{\frac{1}{p}}$$

for any doubly indexed sequence such that $\lambda_{\nu m} = 0$ for $(\nu, m) \in (\mathbb{Z}\cap(\nu_0,\infty))\times\mathbb{Z}^n$, and that

$$\sup_{m\in\mathbb{Z}^n} 2^{\nu_0 n/p}|\lambda_{\nu_0 m}|\chi_{Q_{\nu_0 m}} \ge \sup_{m\in\mathbb{Z}^n} 2^{\nu n/p}|\lambda_{\nu m}|\chi_{Q_{\nu m}}, \quad \nu = 0, 1, \dots, \nu_0.$$

Proof By the monotone convergence theorem we may assume that

$$\sharp\{(\nu, m) \in \mathbb{N}_0 \times \mathbb{Z} : \lambda_{\nu m} \neq 0\} < \infty.$$

We sort $\{\lambda_{\nu m}\}_{m\in\mathbb{Z}^n}$ in decreasing order for $\nu = 0, 1, \dots, \nu_0 - 1$; let $\{\lambda_{\nu_0 m}\}_{m\in\mathbb{Z}^n} = \{a_j\}_{j=1}^{\infty}$ satisfy $|a_1| \ge |a_2| \ge \cdots$ and $\{\lambda_{\nu m}\}_{m\in\mathbb{Z}^n} = \{a_{j,\nu}\}_{j=1}^{\infty}$ satisfy $|a_{1,\nu}| \ge |a_{2,\nu}| \ge \cdots$ for $\nu = 0, 1, \dots, \nu_0 - 1$. Then

$$\sum_{m\in\mathbb{Z}^n}|\lambda_{\nu m}|^{p_1}=\sum_{j=1}^{\infty}|a_{j,\nu}|^{p_1}\le 2^{(\nu_0-\nu)np_1/p}\sum_{j=1}^{\infty}|a_{2^{(\nu_0-\nu)n}j}|^{p_1},$$

since $|a_{j,\nu}|\le|a_{2^{(\nu_0-\nu)n}j}|$ for $\nu=0,1,\ldots,\nu_0-1$. Thus, we have

$$\sum_{m\in\mathbb{Z}^n}|\lambda_{\nu m}|^{p_1}\lesssim\sum_{k=0}^{\infty}2^{(\nu_0-\nu)np_1/p+nk}|a_{2^{(\nu_0-\nu+k)n}}|^{p_1}.$$

Since $p<p_1$, we have

$$\left(\sum_{m\in\mathbb{Z}^n}|\lambda_{\nu m}|^{p_1}\right)^{\frac{p}{p_1}}\lesssim\sum_{k=0}^{\infty}2^{(\nu_0-\nu)n+knp/p_1}|a_{2^{(\nu_0-\nu+k)n}}|^{p}.$$

Thanks to the monotonicity of the sequence $\{a_j\}_{j=1}^{\infty}$,

$$\begin{aligned}\sum_{\nu=0}^{\nu_0}2^{\nu n}|a_{2^{\nu n}}|^p&\le|a_1|^p+2^n|a_{2^n}|^p+4^n|a_{4^n}|^p+\cdots\\&\lesssim|a_1|^p+(2^n-1)|a_{2^n}|^p+(4^n-2^n)|a_{4^n}|^p+\cdots\\&\le|a_1|^p+|a_2|^p+\cdots.\end{aligned}$$

We add this over $\nu=0,1,\ldots$:

$$\begin{aligned}\sum_{\nu=0}^{\infty}\left(\sum_{m\in\mathbb{Z}^n}|\lambda_{\nu m}|^{p_1}\right)^{\frac{p}{p_1}}&\lesssim\sum_{\nu=0}^{\nu_0}\sum_{k=0}^{\infty}2^{(\nu_0-\nu)n+knp/p_1}|a_{2^{(\nu_0-\nu+k)n}}|^p\\&=\sum_{\nu=0}^{\nu_0}\sum_{k=0}^{\infty}2^{\nu n+knp/p_1}|a_{2^{(\nu+k)n}}|^p\\&\lesssim\sum_{j=1}^{\infty}|a_j|^p+\sum_{\nu=0}^{\nu_0}\sum_{k=1}^{\infty}2^{\nu n+knp/p_1}|a_{2^{(\nu+k)n}}|^p.\end{aligned}$$

Thus since $p<p_1$, we obtain

$$\begin{aligned}\sum_{\nu=0}^{\infty}\left(\sum_{m\in\mathbb{Z}^n}|\lambda_{\nu m}|^{p_1}\right)^{\frac{p}{p_1}}&\lesssim\sum_{j=1}^{\infty}|a_j|^p+\sum_{\nu=0}^{\nu_0}\sum_{k=1}^{\infty}\sum_{j=2^{(\nu+k-1)n}+1}^{2^{(\nu+k)n}}2^{knp/p_1-kn}|a_j|^p\\&=\sum_{j=1}^{\infty}|a_j|^p+\sum_{k=1}^{\infty}\sum_{j=2^{(k-1)n}+1}^{2^{(\nu_0+k)n}}2^{knp/p_1-kn}|a_j|^p\end{aligned}$$

$$\leq \sum_{j=1}^{\infty} |a_j|^p + \sum_{j=1}^{\infty}\sum_{k=1}^{\infty} 2^{knp/p_1-kn}|a_j|^p$$

$$\simeq \sum_{j=1}^{\infty} |a_j|^p = \sum_{m\in\mathbb{Z}^n} |\lambda_{\nu_0 m}|^p.$$

With Lemma 4.8 in mind, we prove Theorem 4.11.

Proof (Step 1: $F^s_{pq}(\mathbb{R}^n) \hookrightarrow B^{s_1}_{p_1 p}(\mathbb{R}^n)$) Theorem 2.14 allows us to assume $p_1 < \infty$. In view of the embedding $F^s_{pq}(\mathbb{R}^n) \hookrightarrow F^s_{p\infty}(\mathbb{R}^n)$, we can also assume that $q = \infty$. Furthermore, by Theorem 2.12 we can assume that $s = 0$. The atomic decomposition reduces the matter to the estimates of the coefficients; it is enough to see

$$\left[\sum_{\nu=0}^{\infty}\left(\sum_{m\in\mathbb{Z}^n} 2^{\nu p_1\left(s_1-\frac{n}{p_1}\right)}|\lambda_{\nu m}|^{p_1}\right)^{\frac{p}{p_1}}\right]^{\frac{1}{p}} \lesssim \left\| \sup_{\nu\in\mathbb{N}_0{}^n} |\lambda_{\nu m}|\chi_{Q_{\nu m}}\right\|_p. \tag{4.50}$$

Once the matter is reduced to that of sequence spaces, we can assume the positivity: $\lambda_{\nu m} \geq 0$, $(\nu, m) \in \mathbb{N}_0 \times \mathbb{Z}^n$. In addition, by the monotone convergence theorem, we can even assume that $\lambda = \{\lambda_{\nu m}\}_{(\nu,m)\in\mathbb{N}_0\times\mathbb{Z}^n}$ has only a finite number of nonzero entries. Let

$$H_\lambda(x) \equiv \sup_{\nu\in\mathbb{N}_0{}^n} \lambda_{\nu m}\chi_{Q_{\nu m}}(x) \quad (x \in \mathbb{R}^n).$$

Then $\lambda_{\nu m} \leq \inf_{x\in Q_{\nu m}} H_\lambda(x)$ for $(\nu, m) \in \mathbb{N}_0 \times \mathbb{Z}^n$ and (4.49) show that

$$\sum_{\nu=0}^{\infty}\left(\sum_{m\in\mathbb{Z}^n} 2^{\nu p_1\left(s_1-\frac{n}{p_1}\right)}\lambda_{\nu m}{}^{p_1}\right)^{\frac{p}{p_1}} \leq \sum_{\nu=0}^{\infty} 2^{-\nu n}\left(\sum_{m\in\mathbb{Z}^n}\inf_{x\in Q_{\nu m}} H_\lambda(x)^{p_1}\right)^{\frac{p}{p_1}}.$$

Let

$$N \equiv \sup_{m\in\mathbb{Z}^n} \max\{\{0\}\cup\{\nu \in \mathbb{N}_0 : \lambda_{\nu m} \neq 0\}\}.$$

For $(\nu, m) \in \mathbb{N}_0 \times \mathbb{Z}^n$, define $\lambda' \equiv \{\lambda'_{\nu m}\}_{(\nu,m)\in\mathbb{N}_0\times\mathbb{Z}^n}$, where

$$\lambda'_{\nu m} \equiv \delta_{\nu N} \sup\{\lambda_{\mu m_1} : \mu \in \mathbb{N}_0,\ m_1 \in \mathbb{Z}^n,\ Q_{\mu m_1} \supset Q_{Nm}\}.$$

We can assume that $\lambda_{\nu m} = 0$ if $\nu \neq N$ by replacing λ with λ' if necessary. In this case, since we have $H_\lambda(x) = \sum_{m\in\mathbb{Z}^n} \lambda_{\nu m}\chi_{Q_{\nu m}}(x)$, it follows that

$$\left(\sum_{m\in\mathbb{Z}^n} \inf_{x\in Q_{\nu m}} H_\lambda(x)^{p_1}\right)^{\frac{p}{p_1}} \le \left(\sum_{m\in\mathbb{Z}^n} 2^{(\nu-N)_+ n}\lambda_{Nm}{}^{p_1}\right)^{\frac{p}{p_1}} \le \sum_{m\in\mathbb{Z}^n} 2^{(\nu-N)_+\frac{np}{p_1}}\lambda_{Nm}{}^{p}.$$

Since $p_1 > p$, we obtain

$$\begin{aligned}\sum_{\nu=0}^{\infty} 2^{-\nu n}\left(\sum_{m\in\mathbb{Z}^n} \inf_{x\in Q_{\nu m}} H_\lambda(x)^{p_1}\right)^{\frac{p}{p_1}} &\lesssim \sum_{\nu=0}^{\infty} 2^{-\nu n}\sum_{m\in\mathbb{Z}^n} 2^{(\nu-N)_+\frac{np}{p_1}}\lambda_{Nm}{}^{p}\\ &\sim 2^{-\nu N}\sum_{m\in\mathbb{Z}^n}\lambda_{Nm}{}^{p}\\ &\sim \|H_\lambda\|_p^p.\end{aligned}$$

The proof of $F^s_{pq}(\mathbb{R}^n) \hookrightarrow B^{s_1}_{p_1 p}(\mathbb{R}^n)$ is therefore complete.

Proof (Step 2: $B^{s_0}_{p_0 p}(\mathbb{R}^n) \hookrightarrow F^s_{pq}(\mathbb{R}^n)$) As we did in Step 1, we can reduce matters to the estimate of sequence spaces. In the sequence space, we can take the power freely for nonnegative sequences, which allows us to assume $p_0, p, q > 1$. Furthermore, by the trivial embedding $F^s_{pq}(\mathbb{R}^n) \hookrightarrow F^s_{p\infty}(\mathbb{R}^n)$, it can be assumed that $q < \infty$. As we proved earlier, $F^s_{pq}(\mathbb{R}^n) \hookrightarrow B^{s_1}_{p_1 p}(\mathbb{R}^n)$ and duality shows

$$F^s_{pq}(\mathbb{R}^n) \approx (F^{-s}_{p'q'}(\mathbb{R}^n))^* \hookleftarrow (B^{-s_0}_{p_0' p'}(\mathbb{R}^n))^* \approx B^{s_0}_{p_0 p}(\mathbb{R}^n). \tag{4.51}$$

The proof of $B^{s_0}_{p_0 p}(\mathbb{R}^n) \hookrightarrow F^s_{pq}(\mathbb{R}^n)$ is therefore complete.

4.1.4.2 Embeddings: Complements

We now survey the embedding theorems of Lebesgue spaces, Besov spaces and Triebel–Lizorkin spaces. We fully state the results; all the results will be necessary and sufficient conditions on the parameters. Some of the parts of implications in Theorem 4.12–4.19 and the sufficiency are already proved from what we have been gathering. We need to present some examples in order to prove other necessary conditions or the remaining implication. But we omit the details, which are covered by some exercises.

Theorem 4.12 (Inclusions between $B^s_{pq}(\mathbb{R}^n)$, $F^s_{pq}(\mathbb{R}^n)$, $L^p(\mathbb{R}^n)$) *Suppose that the positive parameters p, q, s, u, v satisfy $0 < p < \infty$, and $0 < q, u, v \le \infty$.*

1. $B^s_{pu}(\mathbb{R}^n) \hookrightarrow F^s_{pq}(\mathbb{R}^n)$ *if and only if* $u \le \min(p, q)$.
2. $F^s_{pq}(\mathbb{R}^n) \hookrightarrow B^s_{pv}(\mathbb{R}^n)$ *if and only if* $v \ge \max(p, q)$.
3. $B^0_{1u}(\mathbb{R}^n) \hookrightarrow L^1(\mathbb{R}^n)$ *if and only if* $u \le 1$.
4. $L^1(\mathbb{R}^n) \hookrightarrow B^0_{1v}(\mathbb{R}^n)$ *if and only if* $v = \infty$.
5. $F^0_{1u}(\mathbb{R}^n) \hookrightarrow L^1(\mathbb{R}^n)$ *if and only if* $u \le 2$.

6. $L^1(\mathbb{R}^n) \hookrightarrow F^0_{1\infty}(\mathbb{R}^n)$ *fails.*
7. $B^0_{\infty u}(\mathbb{R}^n) \hookrightarrow L^\infty(\mathbb{R}^n)$ *if and only if* $u \le 1$.
8. $L^\infty(\mathbb{R}^n) \hookrightarrow B^0_{\infty v}(\mathbb{R}^n)$ *if and only if* $v = \infty$.

Theorem 4.13 (Sobolev's embedding theorem for $B^s_{pq}(\mathbb{R}^n)$ and $F^s_{pq}(\mathbb{R}^n)$) *Suppose that we have the positive parameters* $p_0, p_1, s_0, s_1, s, q, u, v$ *satisfying* $0 < p_0, p_1, p, q, \le \infty$

$$0 < p_0 < p < p_1 \le \infty, \quad s_0 - \frac{n}{p_0} = s_1 - \frac{n}{p_1} = s - \frac{n}{p}.$$

1. $B^{s_0}_{p_0 u}(\mathbb{R}^n) \hookrightarrow F^s_{pq}(\mathbb{R}^n)$ *if and only if* $u \le p$.
2. $F^s_{pq}(\mathbb{R}^n) \hookrightarrow B^{s_1}_{p_1 v}(\mathbb{R}^n)$ *if and only if* $v \ge p$.
3. $F^s_{p\infty}(\mathbb{R}^n) \hookrightarrow F^{s_1}_{p_1 q}(\mathbb{R}^n)$.
4. $B^{s_0}_{p_0 u}(\mathbb{R}^n) \hookrightarrow B^{s_1}_{p_1 v}(\mathbb{R}^n)$ *if and only if* $u \le v$.

Theorem 4.14 (Sobolev's embedding theorem from $A^s_{pq}(\mathbb{R}^n)$ to $L^r(\mathbb{R}^n)$) *Suppose that the positive parameters* p, q, r *satisfy* $p < r$, *and* $r \ge 1$.

1. $B^{\frac{n}{p}-\frac{n}{r}}_{pq}(\mathbb{R}^n) \hookrightarrow L^r(\mathbb{R}^n)$ *if and only if* $0 < q \le r$.
2. $B^{\frac{n}{p}-\frac{n}{r}}_{p\infty}(\mathbb{R}^n) \hookrightarrow L^r(\mathbb{R}^n)$ *fails.*
3. $F^{\frac{n}{p}-\frac{n}{r}}_{p\infty}(\mathbb{R}^n) \hookrightarrow L^r(\mathbb{R}^n)$.

Theorem 4.15 (Embedding of $B^s_{pq}(\mathbb{R}^n)$ into $L^\infty(\mathbb{R}^n)$, BUC($\mathbb{R}^n$)) *Let* $p \in (0, \infty)$, $q \in (0, \infty]$ *and* $s \in \mathbb{R}$. *Then the following are equivalent:*

1. $B^s_{pq}(\mathbb{R}^n) \hookrightarrow L^\infty(\mathbb{R}^n)$.
2. $B^s_{pq}(\mathbb{R}^n) \hookrightarrow \mathrm{BUC}(\mathbb{R}^n)$.
3. $s > \frac{n}{p}$, *or* $s = \frac{n}{p}$ *and* $0 < q \le 1$.

Theorem 4.16 (Embedding of $F^s_{pq}(\mathbb{R}^n)$ into $L^\infty(\mathbb{R}^n)$, BUC($\mathbb{R}^n$)) *Let* $p \in (0, \infty)$, $q \in (0, \infty]$ *and* $s \in \mathbb{R}$. *Then the following are equivalent:*

1. $F^s_{pq}(\mathbb{R}^n) \hookrightarrow L^\infty(\mathbb{R}^n)$.
2. $F^s_{pq}(\mathbb{R}^n) \hookrightarrow \mathrm{BUC}(\mathbb{R}^n)$.
3. $s > \frac{n}{p}$, *or* $s = \frac{n}{p}$ *and* $0 < p \le 1$.

Theorem 4.17 (Embedding of $B^s_{pq}(\mathbb{R}^n)$ into $L^1_{\mathrm{loc}}(\mathbb{R}^n)$) *Let* $p \in (0, \infty)$, $q \in (0, \infty]$ *and* $s \in \mathbb{R}$. *Then the following are equivalent:*

1. $B^s_{pq}(\mathbb{R}^n) \hookrightarrow L^1_{\mathrm{loc}}(\mathbb{R}^n)(\cap \mathscr{S}'(\mathbb{R}^n))$; *that is, for all* $f \in B^s_{pq}(\mathbb{R}^n)$ *there exists* $g \in L^1_{\mathrm{loc}}(\mathbb{R}^n)$ *such that* $\langle f, \varphi \rangle = \int_{\mathbb{R}^n} g(x)\varphi(x)\mathrm{d}x$ *for all* $\varphi \in C^\infty_{\mathrm{c}}(\mathbb{R}^n)$.
2. $B^s_{pq}(\mathbb{R}^n) \hookrightarrow L^{\max(1,p)}(\mathbb{R}^n)$.
3. $s > \sigma_p$, *or* $s = \sigma_p$ *and* $0 < q \le \min(\max(1, p), 2)$.

Theorem 4.18 (Embedding of $B^s_{\infty q}(\mathbb{R}^n)$ into $L^1_{\mathrm{loc}}(\mathbb{R}^n)$) *Let* $s \in \mathbb{R}$ *and* $0 < q \le \infty$. *Then the following are equivalent:*

1. $B^s_{\infty q}(\mathbb{R}^n) \hookrightarrow L^1_{loc}(\mathbb{R}^n)(\cap \mathscr{S}'(\mathbb{R}^n))$.
2. $B^s_{\infty q}(\mathbb{R}^n) \hookrightarrow \mathrm{b}mo(\mathbb{R}^n)$.
3. $s > 0$, *or* $s = 0$ *and* $0 < q \le 2$.

Theorem 4.19 (Embedding of $F^s_{pq}(\mathbb{R}^n)$ into $L^1_{\mathrm{loc}}(\mathbb{R}^n)$) *Suppose that the real parameters* p, q, s *satisfy* $0 < p < \infty$, $0 < q \le \infty$. *Then the following are equivalent:*

1. $F^s_{pq}(\mathbb{R}^n) \hookrightarrow L^1_{\mathrm{loc}}(\mathbb{R}^n)(\cap \mathscr{S}'(\mathbb{R}^n))$.
2. $F^s_{pq}(\mathbb{R}^n) \hookrightarrow L^{\max(1,p)}(\mathbb{R}^n)$.
3. $s > \sigma_p$, $s = \sigma_p$ *and* $0 < p < 1$, *or* $s = \sigma_p$, $1 \le p < \infty$ *and* $0 < q \le 2$.

By the embedding theorem, we can prove the Hardy inequality.

Theorem 4.20 (Hardy inequality) *Let* $0 < p \le 1$. *For* $f \in H^p(\mathbb{R}^n)$, *the image by the Fourier transform* $\mathscr{F} f$ *is an* $L^1_{\mathrm{loc}}(\mathbb{R}^n)$*-function and satisfies*

$$\left(\int_{\mathbb{R}^n} |\mathscr{F} f(\xi)|^p |\xi|^{(p-2)n} \mathrm{d}\xi \right)^{\frac{1}{p}} \lesssim \|f\|_{H^p}. \tag{4.52}$$

We leave the proof as Exercise 4.16 to interested readers.

Exercises

Exercise 4.16 Let $0 < p \le 1$. Prove the Hardy inequality (4.52) and its related fact as follows:

1. Reduce matters to the case where $f \in \mathscr{S}_\infty(\mathbb{R}^n)$.
2. Let $k \in \mathbb{Z}$. Using the Hölder inequality, show that

$$\left(\int_{B(2^k)\setminus B(2^{k-1})} |\mathscr{F} f(\xi)|^p \mathrm{d}\xi \right)^{\frac{1}{p}} \lesssim 2^{kn\left(\frac{1}{p}-\frac{1}{2}\right)} \left(\int_{B(2^k)\setminus B(2^{k-1})} |\mathscr{F} f(\xi)|^2 \mathrm{d}\xi \right)^{\frac{1}{2}}.$$

3. Check the embedding $H^p(\mathbb{R}^n) \approx \dot{F}^0_{p2}(\mathbb{R}^n) \hookrightarrow \dot{B}^{n/2-n/p}_{2p}(\mathbb{R}^n)$.
4. Conclude the proof of the Hardy inequality.
5. Taking care of the singularity of the integrand in the left-hand side of (4.52), show that $H^p(\mathbb{R}^n) \cap \mathscr{S}(\mathbb{R}^n) \subset \mathscr{P}_{[\sigma_p]}(\mathbb{R}^n)^\perp$ if $0 < p \le 1$, which explains why we need the moment condition for the definition of atoms for Hardy spaces.

Exercise 4.17

1. Let $0 < v \le \infty$. Prove that $L^\infty(\mathbb{R}^n) \hookrightarrow B^0_{\infty v}(\mathbb{R}^n)$ only if $v = \infty$.
2. Let $1 < q < \infty$ and $\{\lambda_j\}_{j=1}^\infty \in \ell^q(\mathbb{N}) \setminus \ell^1(\mathbb{N})$ be a positive sequence. Show that

$$f \equiv \sum_{j=1}^{\infty} \lambda_j e^{2^j \star 1 i} \in B^0_{\infty q}(\mathbb{R}^n) \setminus L^\infty(\mathbb{R}^n).$$

3. Let $0 < v \le \infty$. Prove that $L^1(\mathbb{R}^n) \hookrightarrow B^0_{1v}(\mathbb{R}^n)$ only if $v = \infty$. Hint: If $L^1(\mathbb{R}^n) \hookrightarrow B^0_{1v}(\mathbb{R}^n)$ were true with $1 \le v < \infty$, then we would have $B^0_{\infty v'}(\mathbb{R}^n) \hookrightarrow L^\infty(\mathbb{R}^n)$.

Exercise 4.18

1. Disprove that $\chi_{[-1,1]^n} \in B^0_{\infty q}(\mathbb{R}^n)$ for $0 < q < \infty$. Hint: We may assume that $n = 1$. Denote by H the Hilbert transform. Estimate from below the quantity:

$$H\varphi_j(D)\chi_{[-1,1]}(1) = \frac{1}{2\pi}\int_0^\infty e^{i\xi}\varphi_j(\xi)\frac{\sin\xi}{\xi}d\xi.$$

2. Prove that $B^0_{1u}(\mathbb{R}^n) \hookrightarrow L^1(\mathbb{R}^n)$ only if $u \le 1$. Hint: If this were true with $1 < u < \infty$, then we would have $L^\infty(\mathbb{R}^n) \hookrightarrow B^0_{\infty u'}(\mathbb{R}^n)$ by duality.

Exercise 4.19

1. Let $\mathscr{F}^0_{\infty 1}(\mathbb{R}^n)$ be the closure of $\mathscr{S}(\mathbb{R}^n)$ in $F^0_{1\infty}(\mathbb{R}^n)$. Then show that the dual of $\mathscr{F}^0_{\infty 1}(\mathbb{R}^n)$ is $F^0_{\infty 1}(\mathbb{R}^n)$.
2. Let $\psi \in \mathscr{S}$ satisfy $\chi_{Q(1)} \le \psi \le \chi_{Q(2)}$. Then show that

$$f \equiv \sum_{j=1}^{\infty} \psi(2^{-j}\star) \in B^n_{1\infty}(\mathbb{R}^n)$$

using the atomic decomposition.
3. Show that $B^n_{1\infty}(\mathbb{R}^n)$ is not contained in $L^\infty(\mathbb{R}^n)$.
4. Disprove $L^1(\mathbb{R}^n) \hookrightarrow F^0_{1\infty}(\mathbb{R}^n)$. Hint: if $L^1(\mathbb{R}^n) \hookrightarrow F^0_{1\infty}(\mathbb{R}^n)$ were true, then we would have $L^1(\mathbb{R}^n) \hookrightarrow \mathscr{F}^0_{1\infty}(\mathbb{R}^n)$. By duality, we in turn would have $F^0_{\infty 1}(\mathbb{R}^n) \hookrightarrow L^\infty(\mathbb{R}^n)$. What can we say if we combine this with $B^{2n}_{\frac{1}{2}\infty}(\mathbb{R}^n) \hookrightarrow F^0_{\infty 1}(\mathbb{R}^n)$?

Exercise 4.20 Let $0 < p < \infty$, $0 < q \le \infty$ and $s \in \mathbb{R}$.

1. Let $\{a_k\}_{k=1}^\infty \in \ell^q(\mathbb{R}^n)$ and $\psi \in C^\infty_c(\mathbb{R}^n)$ be such that $\chi_{B(1)} \le \psi \le \chi_{B(2)}$. Then define

$$f \equiv \sum_{k=1}^{\infty} a_k \mathscr{F}^{-1}[\psi(\star - 2^k \mathbf{e}_1)].$$

Let $0 < r < 1$. Mimic Example 2.6 to show that the sum defining f converges in $\mathscr{S}'(\mathbb{R}^n)$ and that $\|f\|_{F^0_{rq}} \simeq \|f\|_{B^0_{\infty q}} \simeq \|\{a_k\}_{k=1}^\infty\|_{\ell^q}$.

2. Prove

$$\left| t^n \psi\left(\frac{x}{t}\right) \left\{ \sum_{k=1}^{\infty} a_k \mathscr{F}^{-1}[\psi(\star - 2^k \mathbf{e}_1)](x) \right\} \right|$$
$$\lesssim \langle x \rangle^{-N} M \left[\sum_{k=1}^{\infty} a_k \exp(2^k i \star_1) \right](x) \quad (x \in \mathbb{R}^n)$$

for $t > 0$ and $N \gg 1$. Hint: Mimic Example 3.5.
3. Let $0 < r < 1$. Prove that $\|M[\chi_{[0,2\pi]^n} F]\|_{L^r[0,2\pi]^n} \lesssim_r \|F\|_1$ for all $F \in L^1(\mathbb{R}^n)$ using Theorems 1.5 and 1.45.
4. Show that

$$\|f\|_{F^0_{r2}} \simeq \|\{a_k\}_{k=1}^{\infty}\|_{\ell^2} \lesssim \left\| \sum_{k=1}^{\infty} a_k \exp(2^k i \star_1) \right\|_{L^1([0,2\pi]^n)}$$

using Theorem 3.18.
5. Let $p, q \geq 1$ and $s \in \mathbb{R}$. Assume that $A^s_{pq}(\mathbb{R}^n)$ is included in $L^1_{\mathrm{loc}}(\mathbb{R}^n)$ as a set. Then use the closed graph theorem to show that $g \in A^s_{pq}(\mathbb{R}^n) \mapsto \psi(4^{-1}\star) \cdot g \in L^1(\mathbb{R}^n)$ is bounded.
6. Prove that $F^s_{pq}(\mathbb{R}^n) \hookrightarrow L^1_{\mathrm{loc}}(\mathbb{R}^n)$ only if $s > \sigma_p$, or $s = \sigma_p$ and $0 < p < 1$, or $s = \sigma_p$, $1 \leq p < \infty$ and $0 < q \leq 2$. Hint: Putting our observation together when $q > 2$, $f \in F^0_{pq}(\mathbb{R}^n) \cap F^0_{1/2q}(\mathbb{R}^n) \setminus L^1_{\mathrm{loc}}(\mathbb{R}^n)$.
7. Prove that $B^s_{\infty q}(\mathbb{R}^n) \hookrightarrow L^1_{\mathrm{loc}}(\mathbb{R}^n)$ only if $s > 0$, or $s = 0$ and $0 < q \leq 2$.
8. Prove that $F^0_{1u}(\mathbb{R}^n) \hookrightarrow L^1(\mathbb{R}^n)$ only if $u \leq 2$.

Textbooks in Sect. 4.1

Atomic Decomposition

Theorems 4.1 and 4.2 are stated in many concrete cases; see [29, Theorem (1.4)], [29, Theorem (2.4)], and [29, Theorem (3.5)] for $L^2(\mathbb{R}^n)$, $\dot{\mathscr{C}}^\alpha(\mathbb{R}^n)$ with $0 < \alpha < 1$ and $\dot{B}^0_{1,1}(\mathbb{R}^n)$, respectively.

We refer to [2, Section 4.6], [33, Chapter 2], [100, Chapter 3] and [101, Section 13] for various decompositions including the atomic decomposition of $A^s_{pq}(\mathbb{R}^n)$.

The book [99, Section 2.5.5] seems to contain the prototype of the atomic decomposition, where the extra parameter $t = 1, 2, \ldots, T(\equiv 2^n - 1)$ is added.

We note that the embedding (4.1) is [103, Proposition 2.3].

Wavelet Expansion of $L^2(\mathbb{R})$

For wavelet expansion used in this book, we refer to [17, Chapters 1 and 3], [35, Chapter 10], [117, Chapters 1–5], [40, Chapters 1–4], [41, Chapter 1], [43] and [64, Chapter 2]. In particular, we can find a vivid description of the motivations for wavelets in [17, Chapter 1].

These books contain the wavelet expansion and it is developed independently of the atomic decomposition.

Wavelet Expansion of $A^s_{pq}(\mathbb{R}^n)$

See [105, Remark 1. 14] for a discussion to justify the coupling for $f \in A^s_{pq}(\mathbb{R}^n)$ and $\varphi \in C^K_{\rm c}(\mathbb{R}^n) = C_{\rm c}(\mathbb{R}^n) \cap C^K(\mathbb{R}^n)$ with $K \gg 1$.

Wavelet Expansion of $H^1(\mathbb{R})$ and $L^p(\mathbb{R}^n)$ with $1 < p < \infty$

See [117, Chapter 8], where Wojtaszczyk discussed the type of convergence.

φ-Transform: Theorem 4.9

We refer to [29, Chapter 6] for more on the φ-transform. See also [99, p. 21] for a similar equality. See [500] for the consideration at the level of $\mathscr{S}'_\infty(\mathbb{R}^n)$.

Quarkonial Decomposition: Theorems 4.8 and 4.10

The notion of quarks is introduced in [101] and it is stated very precisely in [103].

Others

We depended on the idea of Georgiadis, Johnsen and Nielsen [500] to prove Theorem 4.4.

4.2 Interpolation Theory

Interpolation theory reveals and studies many situations of the following kind. Suppose that X_0, X_1 are Banach spaces both contained continuously in some bigger space X, and Y_0, Y_1 are Banach spaces both contained continuously in some (other) bigger space Y. Interpolation theory gives us various ways to construct and describe Banach spaces X and Y such that $T : X \to Y$ is bounded if $T|X_0 : X_0 \to Y_0$ and $T|X_1 : X_1 \to Y_1$ are bounded. An interpolation theory deals with function spaces

X_0 and X_1 as if they are points in the plane; we consider a point which separates "the line segment X_0X_1" into $1 - \theta : \theta$.

Roughly there are real interpolation and complex interpolation in the theory of interpolation of Banach spaces. Section 4.2.1 defines what the interpolation spaces are in general. In Sect. 4.2.2 we consider some elementary facts on real interpolation and then consider the real interpolation for Besov spaces and Triebel–Lizorkin spaces. Furthermore, Sect. 4.2.3 considers some elementary facts on complex interpolation and then considers the complex interpolation of Besov spaces and Triebel–Lizorkin spaces.

4.2.1 Topological Vector Spaces and Compatible Couple

Before we go into the detail of the interpolation of Besov spaces, we start with the definition of the interpolation.

4.2.1.1 Topological Vector Spaces

We start with some elementary facts that are used for real interpolation and complex interpolation.

Definition 4.9 (Topological vector space) A *topological vector space* is a complex linear space X equipped with the topology $\mathcal{O}_X$ under which the scalar multiplication and the addition

$$(\alpha, x) \in \mathbb{C} \times X \mapsto \alpha x \in X, \quad (x, y) \in X \times X \mapsto x + y \in X$$

are continuous.

Likewise we can define the real topological vector spaces. However, we are mainly interested in complex topological vector spaces.

The function spaces dealt with in this book are all topological vector spaces.

Definition 4.10 (Compatible couple) Let X_0, X_1 be complex quasi-Banach spaces. Then (X_0, X_1) is said to be a compatible couple, if there exists a topological vector space $(X, \mathcal{O}_X)$ into which $(X_0, \| \star \|_{X_0})$ and $(X_1, \| \star \|_{X_1})$ are continuously embedded. When one needs to specify X, (X_0, X_1) is said to be a compatible couple embedded into a topological vector space X. The space X is sometimes called the *containing space*.

In Definition 4.10, if X_0 and X_1 are Banach spaces, then we say that (X_0, X_1) is a compatible couple of Banach spaces.

Example 4.5 Let $0 < p_0, p_1, q_0, q_1 \le \infty$ and $s_0, s_1 \in \mathbb{R}$. Then $(B^{s_0}_{p_0q_0}(\mathbb{R}^n), B^{s_1}_{p_1q_1}(\mathbb{R}^n))$ is a compatible couple. In fact, they are continuously embedded into $\mathscr{S}'(\mathbb{R}^n)$.

Definition 4.11 (Sum space, Intersection subspace) Let (X_0, X_1) be a compatible couple of quasi-Banach spaces embedded into a topological vector space X.

1. The *sum quasi-Banach space* $X_0 + X_1$ is defined to be the algebraic sum of X_0 and X_1 as a linear subspace of X. That is, define

$$X_0 + X_1 \equiv \{x \in X : x_0 \in X_0,\ x_1 \in X_1,\ x = x_0 + x_1\}.$$

The norm of $X_0 + X_1$ is defined by

$$\|x\|_{X_0+X_1} \equiv \inf\{\|x_0\|_{X_0} + \|x_1\|_{X_1} : x_0 \in X_0, x_1 \in X_1, x = x_0 + x_1\} \quad (x \in X_0 + X_1).$$

2. Define the *intersection quasi-Banach space* $X_0 \cap X_1$ to be the intersection subspace $X_0 \cap X_1$ of X_0 and X_1 as a set and define the norm of $X_0 \cap X_1$ by $\|x\|_{X_0 \cap X_1} \equiv \max(\|x\|_{X_0}, \|x\|_{X_1})$ for $x \in X_0 \cap X_1$.

We remark that $X_0 + X_1 \hookrightarrow X$ and we are led to subspaces of $X_0 + X_1$ when we consider interpolation spaces of spaces X_0, X_1. Among other properties, we use the following duality fact:

Theorem 4.21 (Duality theorem) *Suppose that (X_0, X_1) is a compatible couple of Banach spaces. Assume in addition that $X_0 \cap X_1$ is dense in X_0 and X_1. Then*

$$(X_0 \cap X_1)^* \approx {X_0}^* + {X_1}^*,\ (X_0 + X_1)^* \approx {X_0}^* \cap {X_1}^*$$

with coincidence of norms; namely,

$$\|x^*\|_{{X_0}^*+{X_1}^*} = \sup_{x \in X_0 \cap X_1} \frac{|\langle x^*, x\rangle|}{\|x\|_{X_0 \cap X_1}} \text{ for all } x^* \in {X_0}^* + {X_1}^*,$$

$$\|x^*\|_{{X_0}^* \cap {X_1}^*} = \sup_{x \in X_0 + X_1} \frac{|\langle x^*, x\rangle|}{\|x\|_{X_0 + X_1}} \text{ for all } x^* \in {X_0}^* \cap {X_1}^*.$$

Proof It is straightforward to prove that

$$(X_0 \cap X_1)^* \hookleftarrow {X_0}^* + {X_1}^*,\ (X_0 + X_1)^* \approx {X_0}^* \cap {X_1}^* \tag{4.53}$$

and that

$$\|x^*\|_{{X_0}^*+{X_1}^*} \ge \sup_{x \in X_0 \cap X_1} \frac{|\langle x^*, x\rangle|}{\|x\|_{X_0 \cap X_1}}, \tag{4.54}$$

$$\|x^*\|_{X_0{}^* \cap X_1{}^*} = \sup_{x \in X_0+X_1} \frac{|\langle x^*, x\rangle|}{\|x\|_{X_0+X_1}}. \tag{4.55}$$

We leave the proof of (4.53), (4.54) and (4.55) as an exercise (see Exercise 4.26).

Let us prove the reverse inclusion of (4.53). To this end, we take $x^* \in (X_0 \cap X_1)^*$. Define the norm of $X_0 \oplus X_1$ by

$$\|(x_0, x_1)\|_{X_0 \oplus X_1} = \max(\|x_0\|_{X_0}, \|x_1\|_{X_1}) \quad ((x_0, x_1) \in X_0 \oplus X_1),$$

which immediately makes $E \equiv \{(x_0, x_1) \in X_0 \oplus X_1 \,:\, x_0 = x_1 \in X_0 \cap X_1\}$ into a closed subspace of $X_0 \oplus X_1$. Furthermore, the dual of $X_0 \oplus X_1$ is canonically identified with $X_0{}^* \oplus X_1{}^*$, whose norm is given by

$$\|(x_0{}^*, x_1{}^*)\|_{X_0 \oplus X_1} = \|x_0{}^*\|_{X_0} + \|x_1{}^*\|_{X_1}.$$

Then $l \,:\, (x_0, x_1) \in E \mapsto \dfrac{1}{2} x^*(x_0 + x_1)$ is a continuous functional which is dominated by the norm of $X_0 \oplus X_1$. Therefore, l extends to a continuous linear functional L on $X_0 \oplus X_1$ thanks to the Hahn–Banach theorem in such a way that $\|L\|_{(X_0 \oplus X_1)^*} \le \|x^*\|_{(X_0 \cap X_1)^*}$ As a result we obtain $x_0{}^*$ and $x_1{}^*$ such that

$$\|x_0{}^*\|_{X_0{}^*} + \|x_1{}^*\|_{X_1{}^*} = \|L\|_{(X_0 \oplus X_1)^*} \le \|x^*\|_{(X_0 \cap X_1)^*} \tag{4.56}$$

and that $L(x_0, x_1) = \langle x_0{}^*, x_0\rangle + \langle x_1{}^*, x_1\rangle$ for all $(x_0, x_1) \in X_0 \oplus X_1$. Letting $x_0 = x_1 = x$, we obtain $\langle x^*, x\rangle = L(x, x) = \langle x_0{}^*, x\rangle + \langle x_1{}^*, x\rangle$. Thus, $x^* = x_0{}^*|X_0 \cap X_1 + x_1{}^*|X_0 \cap X_1$ and we deduce from (4.56) that $\|x^*\|_{X_0{}^*+X_1{}^*} \le \|x_0{}^*\|_{X_0{}^*} + \|x_1{}^*\|_{X_1{}^*} \le \|x^*\|_{(X_0 \cap X_1)^*}$. This is the desired converse inequality.

Exercises

Exercise 4.21 Let $0 < p_0 < \infty$, $0 < p_1, q_0, q_1 \le \infty$ and $s_0, s_1 \in \mathbb{R}$. Then show that $(F^{s_0}_{p_0 q_0}(\mathbb{R}^n), B^{s_1}_{p_1 q_1}(\mathbb{R}^n))$ is a compatible couple. Hint: Use Proposition 2.7.

Exercise 4.22 Let (X, μ) be a σ-finite space. $0 < p_0 \le p \le p_1 \le \infty$, $f \in L^p(\mu)$, and let $\{f_j\}_{j=1}^{\infty}$ be a sequence of measurable functions such that $|f_j| \le |f|$ for all $j \in \mathbb{N}$. Set $f_j^{(0)} \equiv f_j \chi_{[1,\infty]}(|f|)$ and $f_j^{(1)} \equiv f_j \chi_{[0,1)}(|f|)$. Then show that $f_j^{(0)} \to f \chi_{[1,\infty]}(|f|)$ and that $f_j^{(1)} \to f \chi_{[0,1)}(|f|)$ in $L^p(\mu)$ as $j \to \infty$. Show also that $L^p(\mu) \subset L^{p_0}(\mu) + L^{p_1}(\mu)$ using this decomposition.

Exercise 4.23 Let (X_0, X_1) be a compatible couple of quasi-Banach spaces embedded into a topological vector space X.

1. Show that $X_0 \cap X_1 \hookrightarrow X_0, X_1 \hookrightarrow X_0 + X_1 \hookrightarrow X$.

2. Suppose that we have quasi-Banach spaces X_0, X_1 such that $X_0 \hookrightarrow X_1$. If we understand that X_1 is embedded into X_1 itself, (X_0, X_1) is a compatible couple. Then show that $X_0 \approx X_0 \cap X_1$ and $X_0 + X_1 \approx X_1$.
3. When X_0 and X_1 are both Banach spaces, then show that $X_0 + X_1$ and $X_0 \cap X_1$ are Banach spaces.

See [7, Chapter 3] for some elementary facts on real interpolation.

Exercise 4.24 Let (X_0, X_1) be a compatible couple of quasi-Banach spaces and let $\{x_j\}_{j=1}^{\infty} \subset X_0 + X_1$. Then show that $\{x_j\}_{j=1}^{\infty}$ converges in $X_0 + X_1$ if and only if there exist $\{y_j\}_{j=1}^{\infty} \subset X_0$ and $\{z_j\}_{j=1}^{\infty} \subset X_1$ such that $x = y_j + z_j$, that y_j is convergent to 0 in X_0 and that z_j is convergent to 0 in X_1. Hint: Choose any y_j and z_j such that $\|y_j\|_{X_0} + \|z_j\|_{X_1} \le 2\|x_j\|_{X_0+X_1}$. Why is this possible?

Exercise 4.25 Let $0 < \theta < 1 \le p < \infty$, and let (X_0, X_1) be a compatible couple of Banach spaces. Show that $(X_0, X_1)_{\theta,p}$ is a Banach space. Hint: It is trivial that $(X_0, X_1)_{\theta,p}$ is a normed space. So, completeness matters. Assume $p = \infty$ first.

Exercise 4.26 Prove (4.53), (4.54), and (4.55) from the definition of the sum spaces and the intersection spaces.

4.2.2 Real Interpolation

There are several methods of interpolation. The real interpolation is one of the most fundamental methods. Let (X_0, X_1) be a compatible couple of Banach spaces.

4.2.2.1 Real Interpolation Functor

Now we deal with real interpolation. We give a brief review of the fundamental theory,

We start with the definition of the K-functional.

Definition 4.12 (K-functional) For $x \in X_0 + X_1$, $t > 0$ and a compatible couple (X_0, X_1) of quasi-Banach spaces, define the K-functional by

$$K(t,x) = K(t,x; X_0, X_1) \equiv \inf\{\|x_0\|_{X_0} + t\,\|x_1\|_{X_1} \ :\ x_0 \in X_0,\ x_1 \in X_1,\ x = x_0 + x_1\}.$$

The K-functional gives us an easy way to construct interpolation spaces: Based on the K-functional, we define the real interpolation functor.

Definition 4.13 (Real interpolation functor) Let (X_0, X_1) be a compatible couple of quasi-Banach spaces, and let θ, p satisfy $0 < \theta < 1, 0 < p \le \infty$.

1. For $x \in X_0 + X_1$, define

$$\|x\|_{(X_0,X_1)_{\theta,p}} \equiv \left(\int_0^\infty (t^{-\theta}K(t,x))^p \frac{\mathrm{d}t}{t}\right)^{\frac{1}{p}}$$

for $0 < p < \infty$ and

$$\|x\|_{(X_0,X_1)_{\theta,\infty}} \equiv \sup_{t>0} t^{-\theta}K(t,x).$$

2. The real interpolation quasi-Banach space $((X_0,X_1)_{\theta,p}, \| \star \|_{(X_0,X_1)_{\theta,p}})$ is the subspace of $X_1 + X_0$ given by

$$(X_0,X_1)_{\theta,p} \equiv \{x \in X_0 + X_1 \, : \, \|x\|_{(X_0,X_1)_{\theta,p}} < \infty\}.$$

The correspondence $(X_0,X_1) \mapsto (X_0,X_1)_{\theta,p}$ is called the *real interpolation functor* for each θ and p. One sometimes abbreviate $((X_0,X_1)_{\theta,p}, \|\star\|_{(X_0,X_1)_{\theta,p}})$ to $(X_0,X_1)_{\theta,p}$.

Let us consider the meaning of the parameter p of the real interpolation functor. As is seen from the next theorem, the space $(X_0,X_1)_{\theta,p}$ is monotone with respect to p.

Theorem 4.22 (Relations between the real interpolation spaces) *Let (X_0,X_1) be a compatible couple of quasi-Banach spaces. Let $0 < \theta < 1$ and $0 < p \le q < \infty$. Then*

$$X_0 \cap X_1 \hookrightarrow (X_0,X_1)_{\theta,p} \hookrightarrow (X_0,X_1)_{\theta,q} \hookrightarrow (X_0,X_1)_{\theta,\infty} \hookrightarrow X_0 + X_1$$

holds in the sense of continuous embedding for a compatible couple (X_0,X_1) of quasi-Banach spaces.

The proof is left to the interested readers as Exercise 4.28. The inclusion

$$(X_0,X_1)_{\theta,p} \hookrightarrow (X_0,X_1)_{\theta,q} \hookrightarrow (X_0,X_1)_{\theta,\infty}$$

is referred to as the *comparison theorem*.

Next, we show the completeness of the real interpolation spaces.

Theorem 4.23 (Completeness of the real interpolation spaces) *Let $0 < \theta < 1, 0 < p < \infty$ and let (X_0,X_1) be a compatible couple of quasi-Banach spaces. Then $(X_0,X_1)_{\theta,p}$ is complete.*

Proof Any Cauchy sequence of $(X_0,X_1)_{\theta,p}$ is convergent in the topology of $X_0 + X_1$. The Fatou lemma shows that it is convergent in the topology of $(X_0,X_1)_{\theta,p}$.

The next theorem estimates the operator norm of interpolation spaces. This is one of the elementary theorems in the theory of interpolation spaces.

Theorem 4.24 (Real interpolation of the operator norms) *Let* $0 < \theta < 1, 0 < p \le \infty$ *and let* X_0, X_1 *and* Y_0, Y_1 *be compatible couples of quasi-Banach spaces. Suppose that* $T : X_0 + X_1 \to Y_0 + Y_1$ *is a bounded linear mapping. Assume if* T*, restricted to* X_l *for each* $l = 0, 1$*, is a bounded operator from* X_l *to* Y_l*. That is,*

$$M_l > 0, \quad T(X_l) \subset Y_l, \quad \|Tx\|_{Y_l} \le M_l \|x\|_{X_l}, \quad x \in X_l, \quad l = 0, 1.$$

Then $\|Tx\|_{(Y_0,Y_1)_{\theta,q}} \le {M_0}^{1-\theta} {M_1}^{\theta} \|x\|_{(X_0,X_1)_{\theta,q}}$ *for all* $x \in (X_0, X_1)_{\theta,q}$.

Thus if we redefine $M_0 \equiv \|T\|_{X_0 \to Y_0}$ and $M_1 \equiv \|T\|_{X_1 \to Y_1}$, then the operator norm $M_\theta = \|T\|_{(X_0,X_1)_{\theta,q} \to (Y_0,Y_1)_{\theta,q}}$ with $0 < \theta < 1$ obeys the *logarithmic convexity formula* $M_\theta \le {M_0}^{1-\theta} {M_1}^{\theta}$ for $0 < \theta < 1$.

The proof is left to the interested readers as Exercise 4.29.

As is guessed, the underlying idea of interpolation is that we want to use two different properties of a linear or sublinear operator T to deduce that it has yet other different properties.

4.2.2.2 Real Interpolation of Function Spaces

Having set down elementary facts on the real interpolation, we consider the interpolation of function space; a simple description of the pair $(A^{s_0}_{pq_0}(\mathbb{R}^n), A^{s_1}_{pq_1}(\mathbb{R}^n))_{\theta,q}$. Denote by ℓ^q_s, $0 < q \le \infty$, $s \in \mathbb{R}$ the space of all sequences $\{a_j\}_{j=1}^\infty$ for which $\|\{a_j\}_{j=1}^\infty\|_{\ell^q_s} = \|\{2^{js} a_j\}_{j=1}^\infty\|_{\ell^q_s}$ is finite. Let $q_0, q_1, q \in (0, \infty]$, $s_0, s_1, s \in \mathbb{R}$ and $\theta \in (0, 1)$. Assume that $s_0 \ne s_1$ and that $s = (1 - \theta)s_0 + \theta s_1$. Then we can show that $[\ell^{q_0}_{s_0}, \ell^{q_1}_{s_1}]_{\theta,q} = \ell^q_s$. See Exercise 4.35.

Theorem 4.25 (Real interpolation of function spaces with p **fixed)** *Suppose that* $p, q, q_0, q_1 \in (0, \infty]$ *and the real parameters* s_0, s_1 *satisfy* $s_0 \ne s_1$*. For* $\theta \in (0, 1)$*, define the intermediate index* $s \in (s_0, s_1)$ *by* $s = (1 - \theta)s_0 + \theta s_1$*. Then with equivalence of norms,*

$$(B^{s_0}_{pq_0}(\mathbb{R}^n), B^{s_1}_{pq_1}(\mathbb{R}^n))_{\theta,q} \approx (F^{s_0}_{pq_0}(\mathbb{R}^n), F^{s_1}_{pq_1}(\mathbb{R}^n))_{\theta,q} \approx B^s_{pq}(\mathbb{R}^n).$$

In particular, when $p > 1$ *with equivalence of norms,*

$$(H^{s_0}_p(\mathbb{R}^n), H^{s_1}_p(\mathbb{R}^n))_{\theta,q} \approx B^s_{pq}(\mathbb{R}^n). \tag{4.57}$$

Before the proof of the theorem, let us consider what this theorem means.

Remark 4.3

1. Observe that the minimal assumption is postulated on the parameters q, q_0, q_1. Hence there is almost no relation between these parameters. Hence one can say that the parameter q in the Besov space $B^s_{pq}(\mathbb{R}^n)$ comes from this theorem.
2. For parameters, let us assume the same conditions as Theorem 4.25. Also assume that $p > 1$. For the definition of the fractional Sobolev space $W^s_p(\mathbb{R}^n)$, one can adopt the definition: $W^s_p(\mathbb{R}^n) \equiv (W^{s_0}_p(\mathbb{R}^n), W^{s_1}_p(\mathbb{R}^n))_{\theta,p}$, when $s_0 \ne s_1$, $s_0, s_1 \in$

$\mathbb{N}_0$, $s = (1-\theta)s_0 + \theta s_1 \in \mathbb{R} \setminus \mathbb{N}_0$. Note that this function space is the Besov space $B^s_{pp}(\mathbb{R}^n)$ as Theorem 4.25 asserts.

Proof We remark that (4.57) is a special case where $q_0 = q_1 = 2$. We assume $q < \infty$; otherwise modify the proof in a natural manner. Recall that

$$B^s_{pr}(\mathbb{R}^n) \hookrightarrow B^s_{p\,\min(p,q)}(\mathbb{R}^n) \hookrightarrow A^s_{pq}(\mathbb{R}^n) \hookrightarrow B^s_{p\,\max(p,q)}(\mathbb{R}^n) \hookrightarrow B^s_{p\infty}(\mathbb{R}^n)$$

for $0 < p, q, r \le \infty$ and $0 < r \le \min(p,q)$, so that

$$\begin{aligned}(B^{s_0}_{pr}(\mathbb{R}^n), B^{s_1}_{pr}(\mathbb{R}^n))_{\theta,q} &\hookrightarrow (B^{s_0}_{p\,\min(p,q)}(\mathbb{R}^n), B^{s_1}_{p\,\min(p,q)}(\mathbb{R}^n))_{\theta,q}\\ &\hookrightarrow (A^{s_0}_{pq_0}(\mathbb{R}^n), A^{s_1}_{pq_1}(\mathbb{R}^n))_{\theta,q}\\ &\hookrightarrow (B^{s_0}_{p\infty}(\mathbb{R}^n), B^{s_1}_{p\infty}(\mathbb{R}^n))_{\theta,q}\end{aligned}$$

when $r \le \min(p,q)$. Hence it suffices to show

$$(B^{s_0}_{p\infty}(\mathbb{R}^n), B^{s_1}_{p\infty}(\mathbb{R}^n))_{\theta,q} \hookrightarrow B^s_{pq}(\mathbb{R}^n) \hookrightarrow (B^{s_0}_{pr}(\mathbb{R}^n), B^{s_1}_{pr}(\mathbb{R}^n))_{\theta,q} \tag{4.58}$$

for $r \in (0, \min(p_0, q_0, p_1, q_1, 1))$. When $s_0 < s_1$, we can swap the roles of 0 and 1. So we assume that $s_0 > s_1$. Let $\delta(s) \equiv s_0 - s_1 > 0$. In the proof we choose $\psi \in \mathscr{S}(\mathbb{R}^n)$ so that $\chi_{B(1)} \le \psi \le \chi_{B(2)}$ and then define $\varphi \equiv \psi - \psi_{-1}$. As usual, we define the norm of the function spaces by way of $\{\psi\} \cup \{\varphi_j\}_{j=1}^\infty$.

Let $f \in (B^{s_0}_{p\infty}(\mathbb{R}^n), B^{s_1}_{p\infty}(\mathbb{R}^n))_{\theta,q}$. The definition of the K-functional yields a decomposition: $f = f_0^{[t]} + f_1^{[t]}$, $f_0^{[t]} \in B^{s_0}_{p\infty}(\mathbb{R}^n)$, $f_1^{[t]} \in B^{s_1}_{p\infty}(\mathbb{R}^n)$ and

$$\|f_0^{[t]}\|_{B^{s_0}_{p\infty}} + t\,\|f_1^{[t]}\|_{B^{s_1}_{p\infty}} \le 2K(t, f; B^{s_0}_{p\infty}(\mathbb{R}^n), B^{s_1}_{p\infty}(\mathbb{R}^n)) \tag{4.59}$$

for each $t > 0$. From the definition of the norm in $B^{s_0}_{p\infty}(\mathbb{R}^n)$ and $B^{s_1}_{p\infty}(\mathbb{R}^n)$, we have

$$\begin{aligned}2^{js_0}\|\varphi_j(D)f_0^{[t]}\|_p &\le 2K(t, f; B^{s_0}_{p\infty}(\mathbb{R}^n), B^{s_1}_{p\infty}(\mathbb{R}^n)),\\ t\,2^{js_1}\|\varphi_j(D)f_1^{[t]}\|_p &\le 2K(t, f; B^{s_0}_{p\infty}(\mathbb{R}^n), B^{s_1}_{p\infty}(\mathbb{R}^n)).\end{aligned}$$

Hence we have

$$2^{js}\|\varphi_j(D)f_0^{[t]}\|_p \le 2^{1+j(s-s_0)}K(t, f; B^{s_0}_{p\infty}(\mathbb{R}^n), B^{s_1}_{p\infty}(\mathbb{R}^n)), \tag{4.60}$$

$$2^{js}\|\varphi_j(D)f_1^{[t]}\|_p \le 2^{1+j(s-s_1)}t^{-1}K(t, f; B^{s_0}_{p\infty}(\mathbb{R}^n), B^{s_1}_{p\infty}(\mathbb{R}^n)). \tag{4.61}$$

Let $t \equiv 2^{\delta(s)j}$ so as to optimize (4.60) and (4.61) in t. Inequalities (4.60) and (4.61) yield

$$\begin{aligned}2^{js}\|\varphi_j(D)f\|_p &\le 2^{js}\|\varphi_j(D)f_0^{[2^{\delta(s)j}]}\|_p + 2^{js}\|\varphi_j(D)f_1^{[2^{\delta(s)j}]}\|_p\\ &\le 2^{-j\theta\delta(s)+2}K(2^{\delta(s)j}, f; B^{s_0}_{p\infty}(\mathbb{R}^n), B^{s_1}_{p\infty}(\mathbb{R}^n)),\end{aligned}$$

since $s = (1-\theta)s_0 + \theta s_1$. If we insert this into the definition of the norm, then

$$\|(1-\psi(D))f\|_{B^s_{pq}} \lesssim \left[\sum_{j=1}^{\infty}\left(2^{-j\theta\delta(s)}K(2^{\delta(s)j}, f; B^{s_0}_{p\infty}(\mathbb{R}^n), B^{s_1}_{p\infty}(\mathbb{R}^n))\right)^q\right]^{\frac{1}{q}}$$
$$\lesssim \|f\|_{(B^{s_0}_{p\infty}(\mathbb{R}^n), B^{s_1}_{p\infty}(\mathbb{R}^n))_{\theta,q}}. \tag{4.62}$$

Meanwhile, for any decomposition $\psi(D)f = g_0 + g_1$, $g_0 \in B^{s_0}_{p\infty}(\mathbb{R}^n)$, $g_1 \in B^{s_1}_{p\infty}(\mathbb{R}^n)$ of $\psi(D)f$,

$$\|g_0\|_{B^{s_0}_{p\infty}} + \|g_1\|_{B^{s_1}_{p\infty}} \gtrsim \|g_0+g_1\|_{B^{s_1}_{p\infty}} = \|\psi(D)f\|_{B^{s_1}_{p\infty}} \simeq \|\psi(D)f\|_{B^s_{p\infty}}.$$

Thus, $K(t, \psi(D)f; B^{s_0}_{p\infty}(\mathbb{R}^n), B^{s_1}_{p\infty}(\mathbb{R}^n)) \gtrsim \|\psi(D)f\|_{B^s_{pq}}$ for $t > 1$.

Hence it follows that

$$\|\psi(D)f\|_{B^s_{pq}} \lesssim \left[\sum_{j=1}^{\infty}\left(2^{-j\theta\delta(s)}K(2^{\delta(s)j}, f; B^{s_0}_{p\infty}(\mathbb{R}^n), B^{s_1}_{p\infty}(\mathbb{R}^n))\right)^q\right]^{\frac{1}{q}}$$
$$\lesssim \|f\|_{(B^{s_0}_{p\infty}(\mathbb{R}^n), B^{s_1}_{p\infty}(\mathbb{R}^n))_{\theta,q}}. \tag{4.63}$$

From (4.62) and (4.63), we conclude $(B^{s_0}_{pr}(\mathbb{R}^n), B^{s_1}_{pr}(\mathbb{R}^n))_{\theta,q} \hookrightarrow B^s_{pq}(\mathbb{R}^n)$.

Let $f \in B^s_{pq}(\mathbb{R}^n)$. Use a trivial decomposition $f = f + 0 \in B^{s_0}_{pr}(\mathbb{R}^n) + B^{s_1}_{pr}(\mathbb{R}^n)$ to conclude that $K(t, f; B^{s_0}_{pr}(\mathbb{R}^n), B^{s_1}_{pr}(\mathbb{R}^n)) \lesssim t\,\|f\|_{B^s_{pq}}$. Thus,

$$\left(\int_0^1 (t^{-\theta}K(t, f; B^{s_0}_{pr}(\mathbb{R}^n), B^{s_1}_{pr}(\mathbb{R}^n)))^q \frac{\mathrm{d}t}{t}\right)^{\frac{1}{q}}$$
$$\lesssim \left(\int_0^1 (t^{1-\theta})^q \frac{\mathrm{d}t}{t}\right)^{\frac{1}{q}} \|f\|_{B^s_{pq}} \simeq \|f\|_{B^s_{pq}}.$$

Next, we estimate $A \equiv \left(\int_1^\infty (t^{-\theta}K(t, f; B^{s_0}_{pr}(\mathbb{R}^n), B^{s_1}_{pr}(\mathbb{R}^n)))^q \frac{\mathrm{d}t}{t}\right)^{\frac{1}{q}}$. Among the decompositions $f = f_0 + f_1$ of f, we choose $f_0 = \psi(D)f + \sum_{k=1}^{j}\varphi_k(D)f$ and $f_1 = \sum_{k=j+1}^{\infty}\varphi_k(D)f$. Estimate the K-functional by

$$K(2^{\delta(s)j}, f; B^{s_0}_{pr}(\mathbb{R}^n), B^{s_1}_{pr}(\mathbb{R}^n)) \le \|f_0\|_{B^{s_0}_{pr}} + 2^{\delta(s)j}\|f_1\|_{B^{s_1}_{pr}}.$$

By Theorem 1.53, we have

$$\|f_0\|_{B^{s_0}_{pr}} \sim \|\psi(D)f\|_p + \left(\sum_{k=1}^{j}(2^{ks_0}\|\varphi_k(D)f\|_p)^r\right)^{\frac{1}{r}},$$

$$\|f_1\|_{B^{s_1}_{pr}} \sim \left(\sum_{k=j+1}^{\infty}(2^{ks_1}\|\varphi_k(D)f\|_p)^r\right)^{\frac{1}{r}}.$$

Hence

$$K(2^{\delta(s)j}, f; B^{s_0}_{pr}(\mathbb{R}^n), B^{s_1}_{pr}(\mathbb{R}^n))$$

$$\lesssim \|\psi(D)f\|_p + \left(\sum_{k=1}^{j}(2^{ks_0}\|\varphi_k(D)f\|_p)^r\right)^{\frac{1}{r}} + \delta(s)\left(\sum_{k=j+1}^{\infty}(2^{ks_1}\|\varphi_k(D)f\|_p)^r\right)^{\frac{1}{r}}.$$

Since $s = (1-\theta)s_0 + \theta s_1$, we have

$$A \lesssim \left(\sum_{j=1}^{\infty}(2^{-j\theta\delta(s)}K(2^{\delta(s)j}, f; B^{s_0}_{pr}, B^{s_1}_{pr}))^q\right)^{\frac{1}{q}}$$

$$\lesssim \|\psi(D)f\|_p + \left(\sum_{j=1}^{\infty}\left(\sum_{k=1}^{j}(2^{ks_0-j\theta\delta(s)}\|\varphi_k(D)f\|_p)^r\right)^{\frac{q}{r}}\right)^{\frac{1}{q}}$$

$$+ \left(\sum_{j=1}^{\infty}\left(\sum_{k=j+1}^{\infty}(2^{ks_1+j(1-\theta)\delta(s)}\|\varphi_k(D)f\|_p)^r\right)^{\frac{q}{r}}\right)^{\frac{1}{q}}.$$

Arithmetic shows that

$$A \lesssim \|\psi(D)f\|_p + \left(\sum_{j=1}^{\infty}\left(\sum_{k=1}^{j}(2^{(k-j)\theta\delta(s)}2^{ks}\|\varphi_k(D)f\|_p)^r\right)^{\frac{q}{r}}\right)^{\frac{1}{q}}$$

$$+ \left(\sum_{j=1}^{\infty}\left(\sum_{k=j+1}^{\infty}(2^{(j-k)(1-\theta)\delta(s)}2^{ks}\|\varphi_k(D)f\|_p)^r\right)^{\frac{q}{r}}\right)^{\frac{1}{q}}$$

$$\lesssim \|f\|_{B^s_{pq}}.$$

Thus, the proof is complete.

Exercises

Exercise 4.27 Let $1 < p < \infty$. Define the modulus of continuity by

$$\omega(f, t) = \sup_{h \in B(t)} \|f(\cdot + h) - f\|_p.$$

Then show that

$$K(t, f; L^p, W^{1,p}) \sim \min(1, t)\|f\|_p + \omega(f, t)$$

for all $t > 0$ and $f \in L^p(\mathbb{R}^n)$.

Exercise 4.28 Prove Theorem 4.22. Hint: $K(t, f; X_0, X_1)$ is monotone increasing in $t > 0$ for any $f \in X_0 + X_1$.

Exercise 4.29 Complete the proof of Theorem 4.24 by showing

$$K(t, Tx; Y_0, Y_1) \le M_0\, K(M_1\, t/M_0, x; X_0, X_1), \quad (x \in X_0 + X_1).$$

Exercise 4.30 Let $0 < \theta < 1$, $0 < p \le \infty$. For a compatible couple (X_0, X_1), prove that $(X_0, X_1)_{\theta,p} = (X_1, X_0)_{1-\theta,p}$ with coincidence of norms by the transform $s = t^{-1}$.

Exercise 4.31 [71, p. 30, Example 4], [98, p. 189] Let $0 < s_0 < s_1 < \infty$ and $0 < \theta < 1$. Then show that $(\mathscr{C}^{s_0}(\mathbb{R}^n), \mathscr{C}^{s_1}(\mathbb{R}^n))_{\theta,\infty} \approx \mathscr{C}^{(1-\theta)s_0+\theta s_1}(\mathbb{R}^n)$. It is a dramatic contrast to the real interpolation of $(\mathrm{BC}(\mathbb{R}^n), \mathrm{BC}(\mathbb{R}^n) \cap \mathrm{Lip}(\mathbb{R}^n))_{\theta,\infty}$, which remains open. Hint: Combine Theorems 2.2, 2.8 and 4.25.

Exercise 4.32 (Real interpolation of function space with $p > 0$ fixed) We apply Theorem 4.25 to obtain a simple but important result. Prove that

$$(B^{s_0}_{pq_0}(\mathbb{R}^n), F^{s_1}_{pq_1}(\mathbb{R}^n))_{\theta,q} \approx (F^{s_0}_{pq_0}(\mathbb{R}^n), B^{s_1}_{pq_1}(\mathbb{R}^n))_{\theta,q} \approx B^{s}_{pq}(\mathbb{R}^n)$$

with equivalence of norms under the same assumption of Theorem 4.25. Hint: Combine Proposition 2.7 and Theorem 4.25.

Exercise 4.33 The numbers K and L defining the atoms and the molecules must be integers. However, we can consider the case where K and L are nonnegative real numbers. Let $\nu \in \mathbb{N}$ and $m \in \mathbb{Z}^n$. Let us say that $a \in B^K_{\infty\infty}(\mathbb{R}^n)$ is a $[K, L]$-*atom* centered at $Q_{\nu m}$, if it is supported on $3\,Q_{\nu m}$, satisfies $\|a(2^{-\nu}\star)\|_{B^K_{\infty\infty}} \le 1$ and

$$\left|\int_{\mathbb{R}^n} \psi(x)a(x)\mathrm{d}x\right| \le 2^{-\nu(L+n)}\|\psi\|_{B^L_{\infty\infty}}.$$

See [506, Definition 3.1].

1. [506, Lemma 3.6] Let $\psi \in C_c^\infty((0,1)^n)$ satisfy $\psi(x) = O(|x|^R)$ as $x \to 0$, where an integer R satisfies $R \geq L+1$. Then show that $a_{\nu m} \equiv 2^{jn/p}\psi(2^j \star - m)$ is a $[K, L]$-atom centered at $Q_{\nu m}$.
2. [506, Lemma 3.7] By using the real interpolation

$$[C^{m_1}(\mathbb{R}^n), C^{m_2}(\mathbb{R}^n)]_{\theta,\infty} \approx B^{m(1-\theta)+m_2\theta}_{\infty\infty}(\mathbb{R}^n),$$

extend Theorem 1.56.
3. [506, Theorem 3.14] Extend Theorems 4.1 and 4.2 using this generalized notion.

Exercise 4.34 [99, Section 2.4.3] Let $0 < p_1 < p_0 \leq \infty$ and $s_0, s_1 \in \mathbb{R}$. For $0 < \theta < 1$, define

$$\frac{1}{p} = \frac{1-\theta}{p_0} + \frac{\theta}{p_1}, \qquad s = (1-\theta)s_0 + \theta s_1.$$

Then use Proposition 4.2 to show $(B^{s_0}_{p_0 p_0}(\mathbb{R}^n), B^{s_1}_{p_1 p_1}(\mathbb{R}^n))_{\theta,p} \approx B^s_{pp}(\mathbb{R}^n)$ with equivalence of norms.

Exercise 4.35 [7] Let $q_0, q_1, q \in (0,\infty]$, $s_0, s_1, s \in \mathbb{R}$ and $\theta \in (0,1)$. Assume that $s_0 \neq s_1$ and that $s = (1-\theta)s_0 + \theta s_1$. Then reexamine the proof of Theorem 4.25 to show that $[\ell^{q_0}_{s_0}, \ell^{q_1}_{s_1}]_{\theta,q} = \ell^q_s$.

4.2.3 Complex Interpolation

The complex interpolation functors are originally from a famous theorem on complex analysis: Theorem 1.75, Doetsch's three-line lemma, is a source of this theory. This lemma is based on the maximum principle of holomorphic functions. As a special case of domains, we consider $\{0 < \Re(z) < 1\}$ and its boundary $\{\Re(z) = 0, 1\}$. In this special case, we know that any continuous function on the closed set $\{0 \leq \Re(z) \leq 1\}$ which is holomorphic in the interior $\{0 < \Re(z) < 1\}$ and which grows sufficiently mildly is bounded on $\{0 \leq \Re(z) \leq 1\}$ whenever it is bounded on $\{\Re(z) = 0, 1\}$. From this theorem, we learn that the property of the boundary is transferred to the interior with some appropriate sense. The complex interpolation functor is a tool to convey information on the boundary to the interior, where complex analysis in one variable plays a key role. Information on the boundary can be described in terms of linear spaces; they will give us some qualitative and quantitative information. One of the strongest and most fundamental tools in complex analysis in one variable is the Cauchy integral formula. So, we want to consider line integrals of functions. By the atomic decomposition, we do not have to work within the Banach space setting when we consider Besov spaces and Triebel–Lizorkin spaces.

4.2.3.1 Complex Interpolation of $L^p(\mathbb{R}^n)$

We start with the following definition by Calderón. One of the fundamental results in complex interpolation of Banach spaces is the use of the following complex interpolation functors defined by Calderón [325].

Definition 4.14 ((Open) strip domain S, closed strip domain $\overline{S}$, continuous functions over $\overline{S}$, holomorphic functions over S)

1. Define the *(open) strip domain* $S \equiv \{z \in \mathbb{C} : 0 < \Re(z) < 1\}$ and the *closed strip domain* $\overline{S} \equiv \{z \in \mathbb{C} : 0 \le \Re(z) \le 1\}$.
2. A function $F : \overline{S} \to \mathscr{S}'(\mathbb{R}^n)$ is said to be (weakly) continuous, if the mapping $z \in \overline{S} \mapsto \langle F(z), \varphi\rangle \in \mathbb{C}$ is continuous for all $\varphi \in \mathscr{S}(\mathbb{R}^n)$.
3. A function $F : S \to \mathscr{S}'(\mathbb{R}^n)$ is said to be (weakly) holomorphic, if the mapping $z \in S \mapsto \langle F(z), \varphi\rangle \in \mathbb{C}$ is holomorphic for all $\varphi \in \mathscr{S}(\mathbb{R}^n)$.

This definition of holomorphic is a special case where the containing space is $\mathscr{S}'(\mathbb{R}^n)$. When we interpolate Banach spaces, we have the following natural definition.

Definition 4.15 (Calderón's first complex interpolation space) Let $\overline{X} = (X_0, X_1)$ be a compatible couple of Banach spaces.

1. Define $\mathscr{F}(X_0, X_1)$ as the set of all the functions $F : \bar{S} \to X_0 + X_1$ such that:
   (a) F is bounded and continuous on $\bar{S}$; that is, $\sup\limits_{z\in\bar{S}} \|F(z)\|_{X_0+X_1} < \infty$,
   (b) F is holomorphic on S,
   (c) the functions $t \in \mathbb{R} \mapsto F(j + it) \in X_j$ are bounded and continuous on $\mathbb{R}$ for $j = 0, 1$.

   The space $\mathscr{F}(X_0, X_1)$ is equipped with the norm

$$\|F\|_{\mathscr{F}(X_0,X_1)} \equiv \max\left\{\sup_{t\in\mathbb{R}} \|F(it)\|_{X_0}, \sup_{t\in\mathbb{R}} \|F(1 + it)\|_{X_1}\right\}.$$

2. Let $\theta \in (0, 1)$. Define the complex interpolation space $[X_0, X_1]_\theta$ with respect to (X_0, X_1) to be the set of all the elements $x \in X_0 + X_1$ such that $x = F(\theta)$ for some $F \in \mathscr{F}(X_0, X_1)$. The norm on $[X_0, X_1]_\theta$ is defined by

$$\|x\|_{[X_0,X_1]_\theta} \equiv \inf\{\|F\|_{\mathscr{F}(X_0,X_1)} : x = F(\theta) \text{ for some } F \in \mathscr{F}(X_0, X_1)\}.$$

   The space $[X_0, X_1]_\theta$ is called the *Calderón first complex interpolation space* and the operation $(X_0, X_1) \mapsto [X_0, X_1]_\theta$ is called the (Calderón) *first complex interpolation functor*.

In these definitions we want to pay attention to what it means to say that a function $f : S \to E$ is holomorphic for various containing spaces E.

We have the following result for complex interpolation of $L^p(\mathbb{R}^n)$ spaces. This result convinces us that the definition above works well.

Theorem 4.26 *Let $1 \le p_0 < p_1 < \infty$ and $0 < \theta < 1$. Define p by*

$$\frac{1}{p} = \frac{1-\theta}{p_0} + \frac{\theta}{p_1}. \tag{4.64}$$

Then

$$[L^{p_0}(\mathbb{R}^n), L^{p_1}(\mathbb{R}^n)]_\theta = L^p(\mathbb{R}^n) \tag{4.65}$$

with coincidence of norms. In particular, for any linear operator T defined on $L^{p_0}(\mathbb{R}^n) + L^{p_1}(\mathbb{R}^n)$, we have the logarithmic convexity formula

$$\|T|L^p(\mathbb{R}^n)\|_{B(L^p)} \le (\|T|L^{p_0}(\mathbb{R}^n)\|_{B(L^{p_0})})^{1-\theta}(\|T|L^{p_1}(\mathbb{R}^n)\|_{B(L^{p_1})})^{\theta}. \tag{4.66}$$

We will see the idea of the three-line lemma lurking in the background.

Proof Let $f \in L^p(\mathbb{R}^n)$. We complexity (4.64); define $p(z)$ and $p'(z)$ by

$$\frac{1}{p(z)} = \frac{1-z}{p_0} + \frac{z}{p_1}, \quad \frac{1}{p'(z)} = \frac{1-z}{p'_0} + \frac{z}{p'_1} \tag{4.67}$$

for $z \in \bar{S}$. Let $z \in \bar{S}$ and $x \in \mathbb{R}^n$. Put

$$F(z;x) \equiv \frac{f(x)}{|f(x)|}|f(x)|^{\frac{p}{p(z)}}\chi_{\{f\neq 0\}}(x) = \lim_{\varepsilon\downarrow 0}\frac{f(x)}{|f(x)|+\varepsilon}(|f(x)|+\varepsilon)^{\frac{p}{p(z)}}. \tag{4.68}$$

Observe that $F \in \mathscr{F}(L^{p_0}(\mathbb{R}^n), L^{p_1}(\mathbb{R}^n))$. In fact,

$$|F(it;x)| = |f(x)|^{\frac{p}{p_0}} \quad (t \in \mathbb{R}, x \in \mathbb{R}^n)$$

implies $\|F(it;\star)\|_{p_0} \le \|f\|_p$. Similarly $\|F(1+it;\star)\|_{p_1} \le \|f\|_p$. Therefore

$$\|F\|_{\mathscr{F}(L^{p_0},L^{p_1})} \le \|f\|_p$$

for all $\varepsilon > 0$. Hence $\|f\|_{[L^{p_0},L^{p_1}]_\theta} \le \|f\|_p$.

To prove the converse inequality, we choose $f \in [L^{p_0}(\mathbb{R}^n), L^{p_1}(\mathbb{R}^n)]_\theta$. Then we have

$$\|f\|_p = \sup\left\{\left|\int_{\mathbb{R}^n} f(x)\cdot g(x)\mathrm{d}x\right| : g \in L^\infty_{\mathrm{c}}(\mathbb{R}^n), \quad \|g\|_{p'} = 1\right\}.$$

Let $\varepsilon > 0$ be arbitrary and choose $F \in \mathscr{F}(L^{p_0}(\mathbb{R}^n), L^{p_1}(\mathbb{R}^n))$ with

$$\|f\|_{[L^{p_0},L^{p_1}]_\theta} \le (1+\varepsilon)\|F\|_{\mathscr{F}(L^{p_0},L^{p_1})}.$$

Fix $g \in L^\infty_{\rm c}(\mathbb{R}^n)$ with $\|g\|_{p'} = 1$. Set

$$G(z;x) \equiv |g(x)|^{\frac{p'}{p'(z)}} \operatorname{sgn}(g)(x) \quad (x \in \mathbb{R}^n, z \in \bar{S}).$$

Then $H(z) \equiv \int_{\mathbb{R}^n} f(z)(x)G(z;x){\rm d}x$ defines a holomorphic function on S that satisfies

$$|H(it)|,\ |H(1+it)| \le \|F\|_{\mathscr{F}(L^{p_0},L^{p_1})}.$$

Therefore, the three-line lemma (1.286) yields $\|F\|_p \le \|F\|_{\mathscr{F}(L^{p_0},L^{p_1})}$. Hence (4.65) is proved.

The proof of (4.66) amounts to proving

$$\|f\|_{[L^{p_0},L^{p_1}]_\theta} = \inf_{F \in \mathscr{F}(L^{p_0}(\mathbb{R}^n),L^{p_1}(\mathbb{R}^n))} \left\{ \sup_{t \in \mathbb{R}} \|F(i\,t)\|_{p_0}^{1-\theta} \sup_{t\in\mathbb{R}} \|F(1+i\,t)\|_{p_1}^{\theta} \right\} \tag{4.69}$$

for $f \in [\mathscr{F}(L^{p_0}(\mathbb{R}^n), L^{p_1}(\mathbb{R}^n))]$. In fact, let $f \in L^p(\mathbb{R}^n)$ be arbitrary. We define $F \in \mathscr{F}(L^{p_0}(\mathbb{R}^n), L^{p_1}(\mathbb{R}^n))$ by (4.68). Then we have

$$\begin{aligned}
&\|Tf\|_p = \|Tf\|_{[\,L^{p_0},L^{p_1}\,]_\theta} = \|T[F(\theta)]\|_{[\,L^{p_0},L^{p_1}\,]_\theta}\\
&\le \sup_{t\in\mathbb{R}} \|TF(i\,t)\|_{p_0}^{1-\theta} \sup_{t\in\mathbb{R}} \|TF(1+i\,t)\|_{p_1}^{\theta}\\
&\le \sup_{t\in\mathbb{R}}(\|T|L^{p_0}(\mathbb{R}^n)\|_{B(L^{p_0})}\|F(i\,t)\|_{p_0})^{1-\theta} \sup_{t\in\mathbb{R}}(\|T|L^{p_1}(\mathbb{R}^n)\|_{B(L^{p_1})}\|F(1+i\,t)\|_{p_1})^{\theta}\\
&= (\|T|L^{p_0}(\mathbb{R}^n)\|_{B(L^{p_0})})^{1-\theta}(\|T|L^{p_1}(\mathbb{R}^n)\|_{B(L^{p_1})})^{\theta}\|f\|_p,
\end{aligned}$$

since $TF \in \mathscr{F}(L^{p_0}(\mathbb{R}^n), L^{p_1}(\mathbb{R}^n))$.

By definition, the inequality $\le$ of (4.69) is trivial:

$$\begin{aligned}
\|f\|_{[\,L^{p_0},L^{p_1}\,]_\theta} &= \inf_F \left[\max\left(\sup_{t\in\mathbb{R}} \|F(i\,t)\|_{p_0}, \sup_{t\in\mathbb{R}} \|F(1+i\,t)\|_{p_1} \right)\right]\\
&\ge \inf_F \left[\sup_{t\in\mathbb{R}} \|F(i\,t)\|_{p_0}^{1-\theta} \sup_{t\in\mathbb{R}} \|F(1+i\,t)\|_{p_1}^{\theta} \right].
\end{aligned}$$

Let us prove the reverse inequality. Let $F \in [\,L^{p_0}(\mathbb{R}^n), L^{p_1}(\mathbb{R}^n)\,]_\theta$, and let $\delta > 0$ be fixed. Consider a function $G_F \in [\,L^{p_0}(\mathbb{R}^n), L^{p_1}(\mathbb{R}^n)\,]_\theta$ over $\overline{S}$ given by

$$G_F(z) \equiv e^{K(z-\theta)+\delta z^2-\delta\theta^2} F(z), \quad z \in \overline{S},$$

where $K \in \mathbb{R}$ satisfies the equation

$$\sup_{t\in\mathbb{R}} \|e^{-K\theta} F(i\,t)\|_{p_0} = \sup_{t\in\mathbb{R}} \|e^{K(1-\theta)} F(1+i\,t)\|_{p_1}.$$

We write $M_\delta \equiv \sup_{z\in S} |e^{\delta z^2-\delta\theta^2}|$. Choose θ, G_F as above. Then

$$\begin{aligned}\|f\|_{[L^{p_0},L^{p_1}]_\theta} &\le \inf_F \left[\max\left(\sup_{t\in\mathbb{R}} \|G_F(i\,t)\|_{p_0}, \sup_{t\in\mathbb{R}} \|G_F(1+i\,t)\|_{p_1}\right)\right] \\ &\le M_\delta \inf_F \left[\sup_{t\in\mathbb{R}} \|F(i\,t)\|_{p_0}^{1-\theta} \sup_{t\in\mathbb{R}} \|F(1+i\,t)\|_{p_1}^{\theta}\right].\end{aligned}$$

Since $\lim_{\delta\downarrow 0} M_\delta = 1$, we obtain the reverse inequality of (4.69).

Here we present an application of the complex interpolation functors.

Theorem 4.27 *Let (X_0, X_1) and (Y_0, Y_1) be a compatible couple of Banach spaces, and let $T : (X_0, X_1) \to (Y_0, Y_1)$ be a bounded linear mapping. Assume that $T : (X_0, X_1)_{\theta^\dagger} \to (Y_0, Y_1)_{\theta^\dagger}$ is isomorphic for some $\theta^\dagger \in (0, 1)$. Then there exists an open neighborhood $U \subset (0, 1)$ of $\theta^\dagger$ such that $T : (X_0, X_1)_\theta \to (Y_0, Y_1)_\theta$ is isomorphic for any $\theta \in U$.*

We first prove the surjectivity of T by using Lemma 1.9 and then after proving a lemma we prove that T is injective.

Proof (of surjectivity of T) Let $y \in (Y_0, Y_1)_\theta$. Then $y = G(\theta)$ for some $G \in \mathscr{F}(Y_0, Y_1)$ such that $2\|y\|_{(Y_0,Y_1)_\theta} \ge \|G\|_{\mathscr{F}(Y_0,Y_1)}$. Since $T : (X_0, X_1)_{\theta^\dagger} \to (Y_0, Y_1)_{\theta^\dagger}$ is surjective, we can find $x_\dagger \in (X_0, X_1)_{\theta^\dagger}$ such that $Tx_\dagger = G(\theta^\dagger)$. We have

$$\|T^{-1}\|_{(Y_0,Y_1)_{\theta^\dagger}\to(X_0,X_1)_{\theta^\dagger}} \|G(\theta^\dagger)\|_{(Y_0,Y_1)_{\theta^\dagger}} \ge \|x_\dagger\|_{(X_0,X_1)_{\theta^\dagger}},$$

since $T^{-1} : (Y_0, Y_1)_{\theta^\dagger} \to (X_0, X_1)_{\theta^\dagger}$ is bounded. Let $F \in \mathscr{F}(X_0, X_1)$ be such that $x_\dagger = F(\theta^\dagger)$ and that $2\|x_\dagger\|_{(X_0,X_1)_\theta} \ge \|F\|_{\mathscr{F}(Y_0,Y_1)}$. We consider $\varphi(z) = \langle H'(z), TF(z) - G(z)\rangle$, where $H \in \mathscr{G}(Y_0^*, Y_1^*)$ and $z \in \mathbb{C}$ satisfy $0 < \Re(z) < 1$. Then $\varphi(\theta^\dagger) = 0$ and $|\varphi(z)| \le \|H\|_{\mathscr{G}(Y_0^*,Y_1^*)}(\|F\|_{\mathscr{F}(X_0,X_1)} + \|G\|_{\mathscr{F}(Y_0,Y_1)})$. By the Schwartz lemma,

$$|\varphi(z)| \lesssim \|H\|_{\mathscr{G}(Y_0^*,Y_1^*)}(\|F\|_{\mathscr{F}(X_0,X_1)} + \|G\|_{\mathscr{F}(Y_0,Y_1)})|z - \theta|.$$

Since H is arbitrary and $G(\theta^\dagger) = y$,

$$\|TF(\theta^\dagger) - y\|_{(Y_0,Y_1)_\theta} \lesssim (\|F\|_{\mathscr{F}(X_0,X_1)} + \|G\|_{\mathscr{F}(Y_0,Y_1)})|\theta^\dagger - \theta|$$

and hence $\|TF(\theta^\dagger)-y\|_{(Y_0,Y_1)_\theta} \lesssim \|y\|_{(Y_0,Y_1)_\theta}|\theta^\dagger-\theta|$. Thus, choosing θ sufficiently close to 0, we obtain the surjectivity of $T : (X_0, X_1)_\theta \to (Y_0, Y_1)_\theta$ by Lemma 1.9.

Lemma 4.9 *Let $\theta^\dagger \in (0, 1)$. There exists $D > 0$ such that we have*

$$\|F(z)\|_{(X_0,X_1)_{\Re(z)}} \geq \frac{1}{2}\|F(\theta^\dagger)\|_{(X_0,X_1)_{\theta^\dagger}} - D|z-\theta^\dagger| \cdot \|F\|_{\mathscr{F}(X_0,X_1)}$$

as long as $|z-\theta^\dagger| \ll 1$ and $F \in \mathscr{F}(X_0, X_1)$.

Proof Choose $G \in \mathscr{F}(X_0, X_1)$ such that $G(\theta^\dagger) = F(\theta^\dagger)$ and that

$$\|G\|_{\mathscr{F}(X_0,X_1)} \leq 2\|F(\theta^\dagger)\|_{(X_0,X_1)_{\theta^\dagger}}.$$

Set

$$H(z) \equiv \frac{G(z)-F(z)}{z-\theta^\dagger}$$

for $z \in \mathbb{C}$ with $0 \leq \Re(z) \leq 1$, so that $H \in \mathscr{F}(X_0, X_1)$ with

$$\begin{aligned}\|H\|_{\mathscr{F}(X_0,X_1)} &\leq C(\theta^\dagger)(\|F\|_{\mathscr{F}(X_0,X_1)} + \|G\|_{\mathscr{F}(X_0,X_1)})\\ &\leq C(\theta^\dagger)(\|F\|_{\mathscr{F}(X_0,X_1)} + 2\|F(\theta^\dagger)\|_{(X_0,X_1)_{\theta^\dagger}})\\ &\leq 5C(\theta^\dagger)\|F\|_{\mathscr{F}(X_0,X_1)}.\end{aligned} \tag{4.70}$$

Consequently, by Schwartz's lemma in complex analysis

$$\|F(z)-G(z)\|_{(X_0,X_1)_{\Re(z)}} \leq \frac{|z-\theta^\dagger|}{\min(1-\theta^\dagger, \theta^\dagger)}\|H\|_{\mathscr{F}(X_0,X_1)}.$$

This implies

$$\begin{aligned}2\|F(\theta^\dagger)\|_{(X_0,X_1)_{\theta^\dagger}} &\geq \|G(z)\|_{(X_0,X_1)_{\Re(z)}}\\ &\geq \|F(z)\|_{(X_0,X_1)_{\Re(z)}} - \frac{|z-\theta^\dagger|}{\min(1-\theta^\dagger, \theta^\dagger)}\|H\|_{\mathscr{F}(X_0,X_1)}.\end{aligned} \tag{4.71}$$

Consequently, if $|z-\theta^\dagger| \ll 1$, we deduce from (4.70) and (4.71),

$$\|F(\theta^\dagger)\|_{(X_0,X_1)_{\theta^\dagger}} \gtrsim \|F(z)\|_{(X_0,X_1)_{\Re(z)}}.$$

Proof (of injectivity) Let $x \in (X_0, X_1)_\theta$ be such that $Tx = 0$. Then we can choose $F \in \mathscr{F}(X_0, X_1)$ so that $x = F(\theta)$ and that $2\|F(\theta)\|_{(X_0,X_1)_\theta} \geq \|F\|_{\mathscr{F}(X_0,X_1)}$. We set $\varphi(z) \equiv TF(z)$. Denote by C_1, C_2 and C_3 some important constants for later consideration. Then thanks to Lemma 1.9 we have

$$0 = \|\varphi(\theta)\|_{(Y_0,Y_1)_\theta} \geq \frac{1}{2}\|\varphi(\theta^\dagger)\|_{(Y_0,Y_1)_{\theta^\dagger}} - D\|\varphi\|_{\mathscr{F}(X_0,X_1)},$$

so that using the injectivity of T, we have

$$\|T^{-1}\|_{(Y_0,Y_1)_\theta \to (X_0,X_1)_\theta}\|\varphi(\theta^\dagger)\|_{(Y_0,Y_1)_{\theta^\dagger}} \geq \|x\|_{(X_0,X_1)_{\theta^\dagger}}.$$

Once again using the property of F, we have

$$\|F(\theta^\dagger)\|_{(X_0,X_1)_{\theta^\dagger}} \geq \frac{1}{2}\|F(\theta)\|_{(X_0,X_1)_\theta} - D|\theta^\dagger - \theta|\|F\|_{\mathscr{F}(X_0,X_1)}.$$

All together then,

$$2|z-\theta^\dagger| \cdot \|F(\theta)\|_{(X_0,X_1)_\theta} \geq |z-\theta^\dagger| \cdot \|F\|_{\mathscr{F}(X_0,X_1)} \geq D'\|F(\theta)\|_{(X_0,X_1)_\theta}$$

for some D' as long as $|\theta - \theta^\dagger| \ll 1$. As a consequence if $|\theta - \theta^\dagger| \ll 1$, then $F(\theta) = 0$.

4.2.3.2 Special Complex Interpolation Functor

We defined the interpolation functor as above. However, we want to deal with a quasi-Banach space $A^s_{pq}(\mathbb{R}^n)$ $(0 < p, q \leq \infty, s \in \mathbb{R})$. So by "mimicking" the definition of complex interpolation in [7], we define the complex interpolation space $[\,A^{s_0}_{p_0q_0}(\mathbb{R}^n), A^{s_1}_{p_1q_1}(\mathbb{R}^n)\,]_\theta$ as follows. However, we will distort the definition slightly even for the case of Banach spaces.

We remark that the following definition differs a little from those by Peetre or Calderón:

Definition 4.16 (Complex interpolation functor for $A^s_{pq}(\mathbb{R}^n)$) Suppose that the parameters $p_0, p_1, q_0, q_1, s_0, s_1, \theta$ satisfy

$$0 < p_0, p_1, q_0, q_1 \leq \infty, \quad s_0, s_1 \in \mathbb{R}, \quad 0 < \theta < 1.$$

1. The set $\mathscr{F}(A^{s_0}_{p_0q_0}(\mathbb{R}^n), A^{s_1}_{p_1q_1}(\mathbb{R}^n))$ is defined to be the set of all continuous functions $F : \overline{S} \to \mathscr{S}'(\mathbb{R}^n)$ satisfying:

   (a) $F|S : S \to \mathscr{S}'(\mathbb{R}^n)$ is a holomorphic function,
   (b) $F(j+it) \in A^{s_j}_{p_jq_j}(\mathbb{R}^n)$ for $j = 0, 1, t \in \mathbb{R}$,
   (c) the quasi-norm $\|F\|_{\mathscr{F}(A^{s_0}_{p_0q_0}, A^{s_1}_{p_1q_1})} \equiv \sup\{\|F(j+it)\|_{A^{s_j}_{p_jq_j}} : t \in \mathbb{R}, j = 0, 1\}$ is finite.

2. The set $[\, A^{s_0}_{p_0q_0}(\mathbb{R}^n),\, A^{s_1}_{p_1q_1}(\mathbb{R}^n)\,]_\theta$ of *complex interpolation space of level* θ is defined to be

$$[\, A^{s_0}_{p_0q_0}(\mathbb{R}^n),\, A^{s_1}_{p_1q_1}(\mathbb{R}^n)\,]_\theta \equiv \{\, F(\theta) \in \mathscr{S}'(\mathbb{R}^n) \,:\, F \in \mathscr{F}(A^{s_0}_{p_0q_0}(\mathbb{R}^n),\, A^{s_1}_{p_1q_1}(\mathbb{R}^n))\,\}.$$

The quasi-norm $\|f\|_{[\, A^{s_0}_{p_0q_0}, A^{s_1}_{p_1q_1}\,]_\theta}$ over $[\, A^{s_0}_{p_0q_0}(\mathbb{R}^n),\, A^{s_1}_{p_1q_1}(\mathbb{R}^n)\,]_\theta$ is defined by setting

$$\|f\|_{[\, A^{s_0}_{p_0q_0}, A^{s_1}_{p_1q_1}\,]_\theta} \equiv \inf\left\{ \|F\|_{\mathscr{F}(A^{s_0}_{p_0q_0}, A^{s_1}_{p_1q_1})} \,:\, F \in \mathscr{F}(A^{s_0}_{p_0q_0}(\mathbb{R}^n),\, A^{s_1}_{p_1q_1}(\mathbb{R}^n)),\ F(\theta) = f \right\}$$

for $f \in [\, A^{s_0}_{p_0q_0}(\mathbb{R}^n),\, A^{s_1}_{p_1q_1}(\mathbb{R}^n)\,]_\theta$.

The definition of $[\, A^{s_0}_{p_0q_0}(\mathbb{R}^n),\, A^{s_1}_{p_1q_1}(\mathbb{R}^n)\,]_\theta$ with $1 \le p_0, p_1, q_0, q_1 \le \infty$ here differs from Definition 4.15 slightly. We discuss their difference at the end of this section.

For the norm of the complex interpolation defined in Definition 4.16, we have the following:

Theorem 4.28 (Complex interpolation of the operator norms) *Suppose that we have seven parameters* $p_0, p_1, q_0, q_1, s_0, s_1, \theta$ *satisfying*

$$0 < p_0, p_1, q_0, q_1 \le \infty, \quad s_0, s_1 \in \mathbb{R}, \quad 0 < \theta < 1. \tag{4.72}$$

Then

$$\|f\|_{[\, A^{s_0}_{p_0q_0}, A^{s_1}_{p_1q_1}\,]_\theta} = \inf_F \left[\sup_{t\in\mathbb{R}} \|F(i\,t)\|^{1-\theta}_{A^{s_0}_{p_0q_0}} \sup_{t\in\mathbb{R}} \|F(1+i\,t)\|^{\theta}_{A^{s_1}_{p_1q_1}} \right].$$

Here F *moves over all* $F \in \mathscr{F}(\, A^{s_0}_{p_0q_0}(\mathbb{R}^n),\, A^{s_1}_{p_1q_1}(\mathbb{R}^n)\,)$ *such that* $f = F(\theta)$.

Proof Similar to (4.69).

By the following corollary, we can make a complex interpolation of the boundedness of operators:

Corollary 4.3 *Suppose that the parameters*

$$p_0, p_1, \widetilde{p_0}, \widetilde{p_1}, q_0, q_1, \widetilde{q_0}, \widetilde{q_1}, s_0, s_1, \widetilde{s_0}, \widetilde{s_1}, \theta$$

satisfy

$$0 < p_0, p_1, \widetilde{p_0}, \widetilde{p_1}, q_0, q_1, \widetilde{q_0}, \widetilde{q_1} \le \infty, \quad s_0, s_1, \widetilde{s_0}, \widetilde{s_1} \in \mathbb{R}, \quad 0 < \theta < 1. \tag{4.73}$$

Define $p, \widetilde{p}, q, \widetilde{q}, s, \widetilde{s}$ *by*

$$\frac{1}{p} = \frac{1-\theta}{p_0} + \frac{\theta}{p_1}, \quad \frac{1}{\widetilde{p}} = \frac{1-\theta}{\widetilde{p_0}} + \frac{\theta}{\widetilde{p_1}}, \quad \frac{1}{q} = \frac{1-\theta}{q_0} + \frac{\theta}{q_1}, \quad \frac{1}{\widetilde{q}} = \frac{1-\theta}{\widetilde{q_0}} + \frac{\theta}{\widetilde{q_1}},$$

as well as $s = (1-\theta)s_0 + \theta s_1$ *and* $\widetilde{s} = (1-\theta)\widetilde{s_0} + \theta\widetilde{s_1}$. *If a continuous linear mapping* $T : \mathscr{S}'(\mathbb{R}^n) \to \mathscr{S}'(\mathbb{R}^n)$ *satisfies* $\|Tf\|_{A^{\widetilde{s_l}}_{\widetilde{p_l}\widetilde{q_l}}} \le M_l \|f\|_{A^{s_l}_{p_l q_l}}$ *for* $l = 0, 1$, *then* $\|Tf\|_{[A^{\widetilde{s_0}}_{\widetilde{p_0}\widetilde{q_0}}, A^{\widetilde{s_1}}_{\widetilde{p_1}\widetilde{q_1}}]_\theta} \le {M_0}^{1-\theta}{M_1}^{\theta}\|f\|_{[A^{s_0}_{p_0q_0}, A^{s_1}_{p_1q_1}]_\theta}$.

Proof We use Theorem 4.28.

Let $F \in \mathscr{F}(A^{s_0}_{p_0q_0}(\mathbb{R}^n), A^{s_1}_{p_1q_1}(\mathbb{R}^n))$ be arbitrary. Then the $\mathscr{S}'(\mathbb{R}^n)$-valued mapping $z \in \overline{S} \mapsto T[F(z)] \in \mathscr{S}'(\mathbb{R}^n)$ belongs to $\mathscr{F}(A^{\widetilde{s_0}}_{\widetilde{p_0}\widetilde{q_0}}, A^{\widetilde{s_1}}_{\widetilde{p_1}\widetilde{q_1}})$. Thus,

$$\begin{aligned}
\|Tf\|_{[A^{\widetilde{s_0}}_{\widetilde{p_0}\widetilde{q_0}}, A^{\widetilde{s_1}}_{\widetilde{p_1}\widetilde{q_1}}]_\theta} &\le \inf_F \left[\sup_{t\in\mathbb{R}} \|T[F(it)]\|^{1-\theta}_{A^{\widetilde{s_0}}_{\widetilde{p_0}\widetilde{q_0}}} \sup_{t\in\mathbb{R}} \|T[F(1+it)]\|^{\theta}_{A^{\widetilde{s_1}}_{\widetilde{p_1}\widetilde{q_1}}}\right] \\
&\le \inf_F \left(M_0 \sup_{t\in\mathbb{R}} \|F(it)\|_{A^{s_0}_{p_0q_0}}\right)^{1-\theta} \left(M_1 \sup_{t\in\mathbb{R}} \|F(1+it)\|_{A^{s_1}_{p_1q_1}}\right)^{\theta} \\
&= {M_0}^{1-\theta}{M_1}^{\theta}\|f\|_{[A^{s_0}_{p_0q_0}, A^{s_1}_{p_1q_1}]_\theta}.
\end{aligned}$$

Thus, the proof is complete.

It is convenient to transform the three-line lemma (1.286) into the one adapted to Triebel–Lizorkin spaces.

Lemma 4.10 (Generalized three-line theorem) *Let* $\{f_{\nu m}\}_{\nu\in\mathbb{N}_0, m\in\mathbb{Z}^n}$ *be a family of bounded continuous functions on* $\overline{S}$ *such that, restricted to* S, *each* $f_{\nu m}$ *is holomorphic. Suppose also that we have seven parameters* $p, p_0, p_1, q, q_0, q_1, \theta$ *satisfying*

$$0 < p, p_0, p_1, q, q_0, q_1 \le \infty, \quad 0 < \theta < 1, \quad \frac{1}{p} = \frac{1-\theta}{p_0} + \frac{\theta}{p_1}, \quad \frac{1}{q} = \frac{1-\theta}{q_0} + \frac{\theta}{q_1}.$$

Define $\lambda(z) \equiv \{f_{\nu m}(z)\}_{(\nu,m)\in\mathbb{N}_0\times\mathbb{Z}^n}$. *Then:*

$$\|\lambda(\theta)\|_{\boldsymbol{a}_{pq}} \le \sup_{t\in\mathbb{R}} \|\lambda(it)\|^{1-\theta}_{\boldsymbol{a}_{p_0q_0}} \sup_{t\in\mathbb{R}} \|\lambda(1+it)\|^{\theta}_{\boldsymbol{a}_{p_1q_1}}.$$

Proof Let $r \equiv 2^{-1}\min(1, p_0, p_1, q_0, q_1) < 1$. By Theorem 1.75 and the Jensen inequality [630],

$$|f(\theta)|^r \le \left(\int_{\mathbb{R}} |f(i\tau)|^r \frac{\mu_0(\tau)}{1-\theta} \mathrm{d}\tau\right)^{1-\theta} \left(\int_{\mathbb{R}} |f(1+i\tau)|^r \frac{\mu_1(\tau)}{\theta} \mathrm{d}\tau\right)^{\theta} \tag{4.74}$$

for $0 < r < \infty$, where μ_0 and μ_1 are defined by (1.280). To simplify the notation, for $\nu \in \mathbb{N}_0$ and $m \in \mathbb{Z}^n$, we write

$$g_{\nu m} \equiv \left(\int_{\mathbb{R}} |f_{\nu m}(i\tau)|^r \frac{\mu_0(\tau)}{1-\theta} d\tau\right)^{1-\theta}, \quad h_{\nu m} \equiv \left(\int_{\mathbb{R}} |f_{\nu m}(1+i\tau)|^r \frac{\mu_1(\tau)}{\theta} d\tau\right)^{\theta}.$$

Then from (4.74) we deduce

$$|f_{\nu m}(\theta)|^r \le g_{\nu m} h_{\nu m}. \tag{4.75}$$

Since $\frac{p_0}{r}, \frac{q_0}{r}, \frac{p_1}{r}, \frac{q_1}{r} > 1$, we obtain

$$\begin{aligned}\|\lambda(\theta)\|_{\mathbf{a}_{pq}} &\le \|\{g_{\nu m} h_{\nu m}\}_{(\nu,m)\in\mathbb{N}_0\times\mathbb{Z}^n}\|_{\mathbf{a}_{\frac{p}{r}\frac{q}{r}}}^{\frac{1}{r}} \\ &\le \|\{g_{\nu m}\}_{(\nu,m)\in\mathbb{N}_0\times\mathbb{Z}^n}\|_{\mathbf{a}_{\frac{p_0}{r}\frac{q_0}{r}}}^{\frac{1}{r}} \|\{h_{\nu m}\}_{(\nu,m)\in\mathbb{N}_0\times\mathbb{Z}^n}\|_{\mathbf{a}_{\frac{p_1}{r}\frac{q_1}{r}}}^{\frac{1}{r}}\end{aligned}$$

by Hölder's inequality and (4.75). We have

$$\begin{aligned}\left\|\{g_{\nu m}\}_{(\nu,m)\in\mathbb{N}_0\times\mathbb{Z}^n}\right\|_{\mathbf{a}_{\frac{p_0}{r}\frac{q_0}{r}}} &\le \left(\int_{\mathbb{R}} \left\|\{|f_{\nu m}(i\tau)|^r\}_{(\nu,m)\in\mathbb{N}_0\times\mathbb{Z}^n}\right\|_{\mathbf{a}_{\frac{p_0}{r}\frac{q_0}{r}}} \frac{\mu_0(\tau)d\tau}{1-\theta}\right)^{1-\theta} \\ &= \left(\int_{\mathbb{R}} \|\lambda(i\tau)\|_{\mathbf{a}_{p_0q_0}}^r \frac{\mu_0(\tau)d\tau}{1-\theta}\right)^{1-\theta} \\ &\le \sup_{t\in\mathbb{R}} \|\lambda(it)\|_{\mathbf{a}_{p_0q_0}}^{r(1-\theta)}\end{aligned}$$

by the Minkovski inequality.

Likewise by the Minkovski inequality, we obtain

$$\left\|\{h_{\nu m}\}_{(\nu,m)\in\mathbb{N}_0\times\mathbb{Z}^n}\right\|_{\mathbf{a}_{\frac{p_1}{r}\frac{q_1}{r}}} \le \sup_{t\in\mathbb{R}} \|\lambda(1+it)\|_{\mathbf{a}_{p_1q_1}}^{r\theta}.$$

If we combine the above inequalities, then we obtain Lemma 4.10.

4.2.3.3 Complex Interpolation of Function Spaces

We calculate the complex interpolation in the sense of Definition 4.16. Unlike the interpolation $[L^{p_0}(\mathbb{R}^n), L^{p_1}(\mathbb{R}^n)]_\theta$ we cannot consider the modulus of distribution. So, it is a different matter to consider the interpolation $[A^{s_0}_{p_0q_0}(\mathbb{R}^n), A^{s_1}_{p_1q_1}(\mathbb{R}^n)]_\theta$. However, we have the atomic decomposition of function spaces which reduces the matter to sequence. The result hinges on the property of the underlying sequence spaces.

Theorem 4.29 (Complex interpolation of function spaces) *Suppose that we have seven parameters $p_0, p_1, q_0, q_1, s_0, s_1$ satisfying*

$$0 < p_0, p_1, q_0, q_1 \le \infty, \quad s_0, s_1 \in \mathbb{R}.$$

For $\theta \in (0, 1)$, define p, q, s by

$$\frac{1}{p} = \frac{1-\theta}{p_0} + \frac{\theta}{p_1}, \quad \frac{1}{q} = \frac{1-\theta}{q_0} + \frac{\theta}{q_1}, \quad s = (1-\theta)s_0 + \theta s_1. \tag{4.76}$$

Then $[A^{s_0}_{p_0q_0}(\mathbb{R}^n), A^{s_1}_{p_1q_1}(\mathbb{R}^n)]_\theta \approx A^s_{pq}(\mathbb{R}^n)$ with equivalence of norms, or equivalently we have $[F^{s_0}_{p_0q_0}(\mathbb{R}^n), F^{s_1}_{p_1q_1}(\mathbb{R}^n)]_\theta \approx F^s_{pq}(\mathbb{R}^n)$ and $[B^{s_0}_{p_0q_0}(\mathbb{R}^n), B^{s_1}_{p_1q_1}(\mathbb{R}^n)]_\theta \approx B^s_{pq}(\mathbb{R}^n)$ with equivalence of norms.

Proof By the lift operator, it can be assumed that $s_l > \sigma_{p_lq_l}$ for $l = 0, 1$. Since the proof for Besov spaces is somewhat easier, we concentrate on Triebel–Lizorkin spaces. Let $f \in F^s_{pq}(\mathbb{R}^n)$. We aim to prove $f \in [F^{s_0}_{p_0q_0}(\mathbb{R}^n), F^{s_1}_{p_1q_1}(\mathbb{R}^n)]_\theta$. We can readily assume $f \neq 0$. Since s is sufficiently large, we have the following atomic decomposition of f: $f = \sum_{\nu=0}^{\infty} \sum_{m\in\mathbb{Z}^n} \lambda_{\nu m} a_{\nu m}$ in the topology of $\mathscr{S}'(\mathbb{R}^n)$, where each $a_{\nu m}$ satisfies $|\partial^\alpha a_{\nu m}| \le 2^{|\alpha|\nu} \chi_{3Q_{\nu m}}$. The value s being large, there is no need to postulate the moment condition. Furthermore, the coefficient $\lambda = \{\lambda_{\nu m}\}_{\nu\in\mathbb{N}_0, m\in\mathbb{Z}^n}$ satisfies

$$\left\| \left\{ 2^{\nu s} \sum_{m\in\mathbb{Z}^n} \lambda_{\nu m} \chi_{Q_{\nu m}} \right\}_{\nu=0}^{\infty} \right\|_{L^p(\ell^q)} \lesssim \|f\|_{F^s_{pq}}.$$

In fact, letting $f \in F^s_{pq}(\mathbb{R}^n)$, we write $\lambda^* \equiv \{\lambda^*_{\nu m}\}_{\nu\in\mathbb{N}_0}$ using the atomic decomposition:

$$f = \sum_{\nu=0}^{\infty} \sum_{m\in\mathbb{Z}^n} \lambda^*_{\nu m} a^*_{\nu m},$$

where the convergence takes place in the topology of $\mathscr{S}'(\mathbb{R}^n)$ and λ^* and each $a_{\nu m}$ satisfies

$$\|\lambda^*\|_{\mathbf{f}_{pq}} \lesssim \|f\|_{F^s_{pq}}, \quad |\partial^\alpha a^*_{\nu m}| \le 2^{-\nu(s-n/p)+|\alpha|\nu} \chi_{3Q_{\nu m}}.$$

Set $a_{\nu m} \equiv 2^{\nu(s-n/p)} a^*_{\nu m}$ and $\lambda_{\nu m} \equiv 2^{-\nu(s-n/p)} \lambda^*_{\nu m}$. Then

$$f = \sum_{\nu=0}^{\infty} \sum_{m\in\mathbb{Z}^n} \lambda_{\nu m} a_{\nu m},$$

where the convergence takes place in the topology of $\mathscr{S}'(\mathbb{R}^n)$. Furthermore,

$$\|\lambda^*\|_{\mathbf{f}_{pq}} = \left\| \left\{ \sum_{m\in\mathbb{Z}} 2^{\nu n/p} \lambda^*_{\nu m} \chi_{Q_{\nu m}} \right\}_{\nu=0}^{\infty} \right\|_{L^p(\ell^q)} = \left\| \left\{ \sum_{m\in\mathbb{Z}} 2^{\nu s} \lambda_{\nu m} \chi_{Q_{\nu m}} \right\}_{\nu=0}^{\infty} \right\|_{L^p(\ell^q)}.$$

For $\nu \in \mathbb{N}_0$, we define

$$\Lambda_\nu \equiv \sum_{m\in\mathbb{Z}^n} |\lambda_{\nu m}| \chi_{Q_{\nu m}}, \tag{4.77}$$

so that $\|\lambda^*\|_{\mathbf{f}_{pq}} = \|\{2^{\nu s} \Lambda_\nu\}_{\nu=0}^{\infty}\|_{L^p(\ell^q)}$.

Define linear functions $\rho_1, \rho_2, \rho_3, \rho_4$ of the variable $z \in \mathbb{C}$ uniquely by

$$\rho_1(l) = s\frac{q}{q_l} - s_l, \quad \rho_2(l) = \frac{p}{p_l} - \frac{q}{q_l}, \quad \rho_3(l) = 1 - \frac{p}{p_l}, \quad \rho_4(l) = \frac{q}{q_l}, \quad l = 0, 1.$$

Since $\rho_k(\theta) = (1-\theta)\rho_k(0) + \theta\, \rho_k(1)$, $\rho_k(\theta) = 0$ for $k = 1, 2, 3$ and $\rho_4(\theta) = 1$.

Let $\Lambda = \{2^{js} \Lambda_j\}_{j=0}^{\infty}$. Then define complex numbers $\lambda_{\nu m}(z)$, $\nu \in \mathbb{N}_0$, $m \in \mathbb{Z}^n$ by

$$2^{\nu\rho_1(z)} \left(\sum_{j=0}^{\nu} |2^{js} \Lambda_j(x)|^q \right)^{\frac{\rho_2(z)}{q}} \|\Lambda\|_{L^p(\ell^q)}^{\rho_3(z)} |\Lambda_\nu(x)|^{\rho_4(z)} = \sum_{m\in\mathbb{Z}^n} \lambda_{\nu m}(z) \chi_{Q_{\nu m}}(x) \tag{4.78}$$

uniquely. We claim that

$$\left\| \left\{ 2^{\nu s} \sum_{m\in\mathbb{Z}^n} \lambda_{\nu m} \chi_{Q_{\nu m}} \right\}_{\nu\in\mathbb{N}_0} \right\|_{L^p(\ell^q)} \sim \left\| \left\{ 2^{\nu s} \sum_{m\in\mathbb{Z}^n} \lambda_{\nu m}(l+it) \chi_{Q_{\nu m}} \right\}_{\nu\in\mathbb{N}_0} \right\|_{L^{p_l}(\ell^{q_l})} \tag{4.79}$$

for $t \in \mathbb{R}$. In fact, a direct calculation shows that

$$\left\{ \sum_{\nu=0}^{\infty} \left(2^{\nu s} \sum_{m\in\mathbb{Z}^n} |\lambda_{\nu m}(l+it)| \chi_{Q_{\nu m}}(x) \right)^{q_l} \right\}^{\frac{1}{q_l}}$$

$$= \left(\sum_{j=0}^{\infty} \left(2^{sk_l} 2^{\rho_1(l)} \left(\sum_{j=0}^{\nu} |2^{js} \Lambda_j(x)|^q \right)^{\frac{\rho_2(l)}{q}} \|\Lambda\|_{L^p(\ell^q)}^{\rho_3(l)} \Lambda_\nu(x)^{\rho_4(l)} \right)^{q_l} \right)^{\frac{1}{q_l}}$$

$$= \|\Lambda\|_{L^p(\ell^q)}^{\rho_3(l)} \left(\sum_{j=0}^{\infty} |2^{js} \Lambda_j(x)|^q \left(\sum_{j=0}^{\nu} |2^{js} \Lambda_j(x)|^q \right)^{\frac{q_l}{q}\frac{p}{p_l} - 1} \right)^{\frac{1}{q_l}}$$

for $t \in \mathbb{R}$ and $x \in \mathbb{R}^n$. By Lemma 1.1, we have

$$\left\{\sum_{\nu=0}^{\infty}\left(2^{\nu s}\sum_{m\in\mathbb{Z}^n}|\lambda_{\nu m}(l+it)|\chi_{Q_{\nu m}}(x)\right)^{q_l}\right\}^{\frac{1}{q_l}} \sim \|\Lambda\|_{L^p(\ell^q)}^{\rho_3(l)}\left(\sum_{j=0}^{\infty}|2^{js}\Lambda_j(x)|^q\right)^{\frac{p}{q\,p_l}}$$

for $l = 0, 1$, $t \in \mathbb{R}$ and $x \in \mathbb{R}^n$. If we use (4.78), then

$$\left\|\left\{2^{\nu s}\sum_{m\in\mathbb{Z}^n}\lambda_{\nu m}(l+it)\chi_{Q_{\nu m}}\right\}_{\nu\in\mathbb{N}_0}\right\|_{L^{p_l}(\ell^{q_l})} \sim \|\Lambda\|_{L^p(\ell^q)}^{\rho_3(l)}\|\Lambda\|_{L^p(\ell^q)}^{\frac{p}{p_l}} = \|\Lambda\|_{L^p(\ell^q)},$$

which yields (4.79). Hence

$$F(z) \equiv \sum_{\nu=0}^{\infty}\sum_{m\in\mathbb{Z}^n}\lambda_{\nu m}(z)a_{\nu m} \tag{4.80}$$

satisfies:

1. For $l = 0, 1$, $F(l+i\,t) \in F^{s_l}_{p_l q_l}$ and $\|F(l+it)\|_{F^{s_l}_{p_l q_l}} \lesssim \|f\|_{F^s_{pq}}$.
2. $f = F(\theta)$.

Thus, it follows that $f \in [F^{s_0}_{p_0q_0}(\mathbb{R}^n), F^{s_1}_{p_1q_1}(\mathbb{R}^n)]_\theta$. Conversely, suppose that a distribution $f \in [F^{s_0}_{p_0q_0}(\mathbb{R}^n), F^{s_1}_{p_1q_1}(\mathbb{R}^n)]_\theta$ is expressed as $f = F(\theta)$ using a function $F \in \mathscr{F}(F^{s_0}_{p_0q_0}(\mathbb{R}^n), F^{s_1}_{p_1q_1}(\mathbb{R}^n))$. Then F satisfies $F(l+i\,t) \in F^{s_l}_{p_lq_l}(\mathbb{R}^n)$ and $\|F(l+it)\|_{F^{s_l}_{p_lq_l}} \lesssim \|f\|_{F^s_{pq}}$ for $l = 0, 1$. By the use of the molecular decomposition, for each ν, m we can find a holomorphic function $\lambda_{\nu m} : S \to \mathbb{C}$ such that λ can be extended continuously to $\bar{S}$ and that $F(z) = \sum_{\nu=0}^{\infty}\sum_{m\in\mathbb{Z}^n}\lambda_{\nu m}(z)m_{\nu m}$ where the convergence takes place in the topology of $\mathscr{S}'(\mathbb{R}^n)$. Finally, note that $\|f\|_{F^s_{pq}} \lesssim \|F\|_{[F^{s_0}_{p_0q_0}, F^{s_1}_{p_1q_1}]_\theta}$ by Lemma 4.10. Thus, the proof is complete.

4.2.3.4 Comparision of Two Complex Interpolation Functors

We defined two complex interpolation functors in Definitions 4.15 and 4.16, so far. Here we investigate what happens in the common framework: we suppose $1 \le p_0, p_1, q_0, q_1 \le \infty$, $s_0, s_1 \in \mathbb{R}$. Their difference is the requirement of the continuity of functions. Definition 4.15 requires that $t \in \mathbb{R} \mapsto F(l+it) \in A^{s_l}_{p_lq_l}(\mathbb{R}^n)$ is continuous for $l = 0, 1$ but the continuity of $t \in \mathbb{R} \mapsto F(l+it) \in \mathscr{S}'(\mathbb{R}^n)$ suffices in Definition 4.16. Comparing two functors in this way, we see that Definition 4.15 requires much more than Definition 4.16. The following theorem explains that the gap can be closed when $1 \le p_0, p_1, q_0, q_1 < \infty$.

Theorem 4.30 (Complex interpolation of function spaces) *Suppose we have six parameters $p_0, p_1, q_0, q_1, s_0, s_1$ satisfying $1 \le p_0, p_1, q_0, q_1 < \infty$. For $\theta \in (0, 1)$, define p, q, s by* (4.76). *Then the two complex interpolation spaces, considered in Definitions* 4.15 *and* 4.16, $[A^{s_0}_{p_0q_0}(\mathbb{R}^n), A^{s_1}_{p_1q_1}(\mathbb{R}^n)]_\theta$ *coincide.*

Proof To fill the gap between Definitions 4.15 and 4.16, we have only to show that $t \in \mathbb{R} \mapsto F(l+it) \in A^{s_l}_{p_lq_l}(\mathbb{R}^n)$ is continuous and bounded for F defined by (4.80) via $f \in A^s_{pq}(\mathbb{R}^n)$. Once this is shown, we can show that the mapping $z \in S \mapsto F(z) \in A^{s_0}_{p_0q_0}(\mathbb{R}^n) + A^{s_1}_{p_1q_1}(\mathbb{R}^n)$ is holomorphic by reducing the matter to the case where the sum (4.80) is a finite sum. We reexamine the proof of Theorem 4.29. For $t_1, t_2 \in \mathbb{R}$, $l = 0, 1$ and $x \in \mathbb{R}^n$, we obtain

$$\sum_{\nu=0}^{\infty}\left(2^{\nu s}\sum_{m\in\mathbb{Z}^n}|\lambda_{\nu m}(l+it_1)-\lambda_{\nu m}(l+it_2)|\chi_{\nu m}\right)^{q_l} \lesssim \|\Lambda\|^{\rho_3(l)}_{L^p(\ell^q)}\left(\sum_{j=0}^{\infty}|2^{js}\Lambda_j|^q\right)^{\frac{pq_l}{qp_l}}$$

similar to (4.78). Thus we conclude $t \in \mathbb{R} \mapsto F(j+it) \in A^s_{pq}(\mathbb{R}^n)$ is continuous using the Lebesgue convergence theorem together with the finiteness of p_0, p_1, q_0, q_1.

We left untouched the case where $\infty \in \{p_0, p_1, q_0, q_1\}$. Even in this case, it is not difficult to prove that $F(j+i\star) \in A^{s_j}_{p_jq_j}(\mathbb{R}^n)$ is a bounded function and that

$$[A^{s_0}_{p_0q_0}(\mathbb{R}^n), A^{s_1}_{p_1q_1}(\mathbb{R}^n)]_\theta \hookrightarrow A^s_{pq}(\mathbb{R}^n), \tag{4.81}$$

where $[A^{s_0}_{p_0q_0}(\mathbb{R}^n), A^{s_1}_{p_1q_1}(\mathbb{R}^n)]_\theta$ denotes the interpolation space given by Definition *4.15* with $1 \le p_0, p_1, q_0, q_1 \le \infty$. In this case, the two complex interpolation spaces $[A^{s_0}_{p_0q_0}(\mathbb{R}^n), A^{s_1}_{p_1q_1}(\mathbb{R}^n)]_\theta$ considered in Definitions 4.15 and 4.16 are not the same. To obtain a result similar to Theorem 4.29, we need the second complex interpolation functor.

Definition 4.17 (Calderón's second complex interpolation space) Suppose that $\overline{X} = (X_0, X_1)$ is a compatible couple of Banach spaces.

1. Define $\mathscr{G}(X_0, X_1)$ as the set of all the functions $G : \overline{S} \to X_0 + X_1$ such that
   (a) G is continuous on $\overline{S}$ and $\sup_{z\in\overline{S}}\left\|\frac{G(z)}{1+|z|}\right\|_{X_0+X_1} < \infty$,
   (b) G is holomorphic on S,
   (c) the functions $t \in \mathbb{R} \mapsto G(j+it) - G(j) \in X_j$ are Lipschitz continuous on $\mathbb{R}$ for $j = 0, 1$.

   The space $\mathscr{G}(X_0, X_1)$ is equipped with the mapping

$$\|G\|_{\mathscr{G}(X_0,X_1)} \equiv \max\left\{\|G(i\star)\|_{\mathrm{Lip}(\mathbb{R},X_0)},\ \|G(1+i\star)\|_{\mathrm{Lip}(\mathbb{R},X_1)}\right\}. \tag{4.82}$$

2. Let $\theta \in (0, 1)$. Define the complex interpolation space $[X_0, X_1]^\theta$ with respect to (X_0, X_1) to be the set of all the elements $x \in X_0 + X_1$ such that $x = G'(\theta)$ for some $G \in \mathcal{G}(X_0, X_1)$. The norm on $[X_0, X_1]^\theta$ is defined by

$$\|x\|_{[X_0,X_1]^\theta} \equiv \inf\{\|G\|_{\mathcal{G}(X_0,X_1)} : x = G'(\theta) \text{ for some } G \in \mathcal{G}(X_0, X_1)\}.$$

The space $[X_0, X_1]^\theta$ is called the *Calderón second complex interpolation space* and the operation $(X_0, X_1) \mapsto [X_0, X_1]^\theta$ is called the *(Calderón) second complex interpolation functor*.

We have the following interpolation result:

Theorem 4.31 *Suppose that (X_0, X_1) is a compatible couple such that $X_0 \cap X_1$ are dense in X_0 and X_1 respectively. Then $[X_0, X_1]_\theta{}^* = [X_0{}^*, X_1{}^*]^\theta$ with coincidence of norms.*

Proof $\boxed{[X_0, X_1]_\theta{}^* \hookleftarrow [X_0{}^*, X_1{}^*]^\theta.}$ Let $x^* \in [X_0{}^*, X_1{}^*]^\theta$ and $\varepsilon > 0$. Then there exists $f^* \in \mathcal{G}(X_0{}^*; X_1{}^*)$ such that $f^{*\prime}(\theta) = x^*$ with

$$\|f^*\|_{\mathcal{G}(X_0{}^*;X_1{}^*)} \le (1+\varepsilon)\, \|x^*\|_{(X_0{}^*,X_1{}^*)_\theta}. \tag{4.83}$$

Let $x \in [X_0, X_1]_\theta$. Then x can be realized as $x = g(\theta)$ with some $g \in \mathcal{F}(X_0; X_1)$ such that $\|g\|_{\mathcal{F}(X_0;X_1)} \le (1+\varepsilon)\, \|x\|_{[X_0,X_1]_\theta}$. Therefore, $x^*(x) = \langle f^{*\prime}(\theta), g(\theta)\rangle$. We claim $z \in S \mapsto F(z) = f^{*\prime}(z)(g(z)) \in \mathbb{C}$ is holomorphic. Indeed, $\sup\{\|g(w)\|_{X_0+X_1} : w \in S\} < \infty$ by virtue of the three-line lemma (1.286) . Therefore, we have

$$\begin{aligned}
&\left|\frac{\langle f^{*\prime}(z+h), g(z+h)\rangle - \langle f^{*\prime}(z), g(z+h)\rangle - h\,\langle f^{*\prime\prime}(z), g(z+h)\rangle}{h}\right| \\
&= \left|\left\langle \frac{f^{*\prime}(z+h) - f^{*\prime}(z) - h\, f^{*\prime\prime}(z)}{h}, g(z+h)\right\rangle\right| \\
&\le \left\|\frac{f^{*\prime}(z+h) - f^{*\prime}(z) - h\, f^{*\prime\prime}(z)}{h}\right\|_{X_0{}^*+X_1{}^*} \|g(z+h)\|_{(X_0\cap X_1)^*}
\end{aligned}$$

which tends to 0 as $h \to 0$. Here we have used Theorem 4.21 for the last line. In the same way we have

$$\lim_{h\to 0} \frac{f^{*\prime}(z)(g(z+h)) - f^{*\prime}(z)(g(z)) - h\, f^{*\prime}(z)(g'(z))}{h} = 0.$$

Therefore, $z \in S \mapsto F(z) = f^*(z)(g(z)) \in \mathbb{C}$ is holomorphic.

By the three-line lemma (1.286) again we have

$$|x^*(x)| = |F(\theta)| \le \max_{z\in\partial S} |F(z)| \le (1+\varepsilon)^2 \|x^*\|_{(X_0{}^*, X_1{}^*)_\theta} \|x\|_{[X_0,X_1]_\theta}. \tag{4.84}$$

Consequently, we have $x^* \in [X_0, X_1]_\theta{}^*$.

$\boxed{[X_0, X_1]_\theta{}^* \hookrightarrow [X_0{}^*, X_1{}^*]_\theta.}$ Let $x^* \in [X_0, X_1]_\theta{}^*$. Then the mapping

$$f \in \mathscr{F}(X_0; X_1) \mapsto x^*(f(\theta)) \in \mathbb{C} \tag{4.85}$$

is continuous. Let us write

$$E \equiv \{(f_0, f_1) \in L^1(X_0) \oplus L^1(X_1) \,:\, f_j = f(j+i\star)\mu_j \text{ for some } f \in \mathscr{F}(X_0; X_1)\},$$

where each μ_j is given by (1.280). We equip E with the norm induced by the sum space $L^1(X_0) \oplus L^1(X_1)$. Then $l : (f_0, f_1) \in E \mapsto x^*(f(\theta)) \in \mathbb{C}$, where $f_j = f(j + i\tau)\mu_j$, is a continuous mapping with norm less than $\|x^*\|_{[X_0,X_1]_\theta{}^*}$. By the density argument, the Hahn–Banach theorem and a general duality formula (Lemma 1.10) $L^1(X)^* = \text{Lip}(\mathbb{R}, X^*)$, there exist $g_0 \in \text{Lip}(\mathbb{R}, X_0{}^*)$ and $g_1 \in \text{Lip}(\mathbb{R}, X_0{}^*)$ such that

$$x^*(f(\theta)) = L_{g_0}(f \cdot \mu_0) + L_{g_1}(f \cdot \mu_1)$$

for all $f \in \mathscr{F}(X_0; X_1)$ and that

$$\max(\|g_0\|_{\text{Lip}(\mathbb{R}, X_0{}^*)}, \|g_1\|_{\text{Lip}(\mathbb{R}, X_1{}^*)}) \le \|x^*\|_{[X_0,X_1]_\theta{}^*}.$$

Let $a \in X_0 \cap X_1$ be fixed. Then for any bounded continuous function h on $\bar{S}$ which is holomorphic in S,

$$h(\theta)l(a) = l(h(\theta)a) = L_{g_0}(h \cdot \mu_0 a) + L_{g_1}(h \cdot \mu_1 a).$$

Define

$$\mu(z) \equiv \frac{\exp(i\pi z) - \exp(i\pi\theta)}{\exp(i\pi z) - \exp(-i\pi\theta)} \quad (z \in \overline{S}).$$

We define $\tilde{k}_a : \partial\Delta(1) \to \mathbb{C}$ by

$$\tilde{k}_a(\mu(j + i\tau)) \equiv \frac{\mathrm{d}}{\mathrm{d}t}\langle g_j(\tau + t), a\rangle\bigg|_{t=0}.$$

Denote by $\Delta(1)$ the open unit disk in $\mathbb{C}$. Then by change of variables and the Poisson formula, we obtain

$$\int_{\partial\Delta(1)} \tilde{k}_a(z) z^k \mathrm{d}z = 0 \quad (k = 1, 2, \ldots),$$

which yields $\tilde{k}_a$ is realized as a boundary value of an analytic function k_a defined on $\Delta(1)$. Furthermore, we have

$$
\begin{aligned}
|k_a(\tau)| &\le \sup_{h\in\mathbb{R}\setminus\{0\}} \frac{\max(|\langle g_0(\tau+h)-g_0(\tau),a\rangle|, |\langle g_1(\tau+h)-g_1(\tau),a\rangle|)}{|h|} \\
&\le \max(\|g_0\|_{\mathrm{Lip}(\mathbb{R};X_0)}, \|g_1\|_{\mathrm{Lip}(\mathbb{R};X_1)})\|a\|_{X_0\cap X_1}
\end{aligned}
$$

for $\tau \in \partial\Delta(1)$. Thus, $|k_a(z)| \le \max(\|g_0\|_{\mathrm{Lip}(\mathbb{R};X_0)}, \|g_1\|_{\mathrm{Lip}(\mathbb{R};X_1)})\|a\|_{X_0\cap X_1}$ for all $z \in \Delta(1)$ by the maximum principle. We define $k(z) \in (X_0+X_1)^*$ by

$$
\langle k(z), a\rangle \equiv k_a(z) \text{ for } z \in S \text{ and } a \in X_0 \cap X_1.
$$

We define

$$
g(z) \equiv \int_{\frac{1}{2}}^{\mu^{-1}(z)} k(\zeta)\mathrm{d}\zeta
$$

for $z \in \overline{S}$. Then a passage to the nontangential limit gives us that

$$
\langle g(j+i\tau+ih)-g(j+i\tau),a\rangle = i\langle g_j(\tau+h)-g_j(\tau),a\rangle \quad (j=0,1).
$$

Let $t, h \in \mathbb{R}$. By the density assumption,

$$
g(j+it+ih)-g(j+it) = i(g_j(t+h)-g_j(t)) \in X_j{}^*.
$$

As a result, we obtain $\|g\|_{\mathscr{G}(X_0;X_1)} = \max(\|g_0\|_{L^\infty(X_0{}^*)}, \|g_1\|_{L^\infty(X_1{}^*)})$. Finally,

$$
\begin{aligned}
l(f(\theta)) &= \lim_{h\to 0}\sum_{k=0}^{1}\frac{1}{h}\int_{\mathbb{R}}\langle g_k(\tau+h)-g_k(\tau), \mu_k(\tau)f(k+i\tau)\rangle\mathrm{d}\tau \\
&= \lim_{h\to 0}\frac{1}{h}\langle ig(\theta+ih)-g(\theta), f(\theta)\rangle \\
&= -\langle g\prime(\theta), f(\theta)\rangle
\end{aligned}
$$

for all $f \in \mathscr{F}(X_0;X_1)$. Thus, $l = -g\prime(\theta)$ and belongs to $[X_0{}^*, X_1{}^*]^\theta = [X_0{}^*, X_1{}^*]_\theta$. Therefore the proof is complete.

For this second functor, we have the following ideal result:

Theorem 4.32 (Complex interpolation of function spaces) *Suppose that we have seven parameters $p_0, p_1, q_0, q_1, s_0, s_1$ satisfying*

$$
1 \le p_0, p_1, q_0, q_1 \le \infty, \quad s_0, s_1 \in \mathbb{R}.
$$

For $\theta \in (0, 1)$, define p, q, s by (4.76). Then $[A^{s_0}_{p_0 q_0}(\mathbb{R}^n), A^{s_1}_{p_1 q_1}(\mathbb{R}^n)]^\theta \approx A^s_{pq}(\mathbb{R}^n)$ with equivalence of norms.

Proof Let $G \in \mathscr{G}(A^{s_0}_{p_0 q_0}(\mathbb{R}^n), A^{s_1}_{p_1 q_1}(\mathbb{R}^n))$. Then we have $2^{-j}(G(\star + 2^j i) - G) \in \mathscr{F}(A^{s_0}_{p_0 q_0}(\mathbb{R}^n), A^{s_1}_{p_1 q_1}(\mathbb{R}^n))$. Thus, $2^{-j}(G(\theta + 2^j i) - G) \in A^s_{pq}(\mathbb{R}^n)$ and

$$\begin{aligned}
\|2^{-j}(G(\theta + 2^j i) - G)\|_{A^s_{pq}} &\lesssim \|2^{-j}(G(\theta + 2^j i) - G)\|_{[A^{s_0}_{p_0 q_0}(\mathbb{R}^n), A^{s_1}_{p_1 q_1}(\mathbb{R}^n)]_\theta} \\
&\le \|2^{-j}(G(\star + 2^j i) - G)\|_{\mathscr{F}(A^{s_0}_{p_0 q_0}, A^{s_1}_{p_1 q_1})} \\
&\le \|G\|_{\mathscr{G}(A^{s_0}_{p_0 q_0}, A^{s_1}_{p_1 q_1})}
\end{aligned}$$

from (4.81). Letting $j \to -\infty$, we obtain $\|f\|_{A^s_{pq}} \lesssim \|G\|_{\mathscr{G}(A^{s_0}_{p_0 q_0}, A^{s_1}_{p_1 q_1})}$ thanks to the Fatou property of $A^s_{pq}(\mathbb{R}^n)$. Thus, $[A^{s_0}_{p_0 q_0}(\mathbb{R}^n), A^{s_1}_{p_1 q_1}(\mathbb{R}^n)]^\theta \hookrightarrow A^s_{pq}(\mathbb{R}^n)$.

Conversely, let $f \in A^s_{pq}(\mathbb{R}^n)$. Define $\lambda_{\nu m}(z)$ by (4.78). We also define

$$\Theta_{\nu m}(z) \equiv \int_\theta^z \lambda_{\nu m}(w) \mathrm{d}w \quad (z \in \bar{S}).$$

Then similar to Theorem 4.29, using $F(j + i\star) \in A^{s_j}_{p_j q_j}(\mathbb{R}^n)$ is a bounded function, we can show that

$$G : z \in \overline{S} \mapsto \sum_{\nu=0}^\infty \sum_{m \in \mathbb{Z}^n} \Theta_{\nu m}(z) a_{\nu m} \in A^{s_0}_{p_0 q_0}(\mathbb{R}^n) + A^{s_1}_{p_1 q_1}(\mathbb{R}^n)$$

is an element in $\mathscr{G}(A^{s_0}_{p_0 q_0}(\mathbb{R}^n), A^{s_1}_{p_1 q_1}(\mathbb{R}^n))$, satisfying

$$\|G\|_{\mathscr{G}(A^{s_0}_{p_0 q_0}, A^{s_1}_{p_1 q_1})} \lesssim \|f\|_{A^s_{pq}}.$$

Thus, $A^s_{pq}(\mathbb{R}^n) \hookrightarrow [A^{s_0}_{p_0 q_0}(\mathbb{R}^n), A^{s_1}_{p_1 q_1}(\mathbb{R}^n)]^\theta$.

We are instrested in $[A^{s_0}_{p_0 q_0}(\mathbb{R}^n), A^{s_1}_{p_1 q_1}(\mathbb{R}^n)]_\theta$ where the interpolation space is given in Definition 4.15. To this end, we need to understand $[X_0, X_1]_\theta$ deeply. First we establish a general fact that $X_0 \cap X_1$ is dense in $[X_0, X_1]_\theta$. To this end, denote by $\mathscr{F}_0(X_0, X_1)$ the smallest closed subspaces of $\mathscr{F}(X_0, X_1)$ containing $\exp(a\star^2)F$ for all $a > 0$ and $F \in \mathscr{F}(X_0, X_1)$. Note that

$$[X_0, X_1]_\theta = \{F(\theta) : F \in \mathscr{F}_0(X_0, X_1)\}$$

and that for all $f \in [X_0, X_1]_\theta$

$$\|f\|_{[X_0, X_1]_\theta} = \inf\{\|F(\theta)\|_{\mathscr{F}(X_0, X_1)} : F \in \mathscr{F}_0(X_0, X_1), f = F(\theta)\}.$$

Lemma 4.11 *Let (X_0, X_1) be a compatible couple of Banach spaces. Then the set*

$$\mathscr{U} \equiv \left\{\exp(\delta z^2)\exp(\lambda z)x \,:\, x \in X_0 \cap X_1,\ \lambda \in \mathbb{R},\ \delta > 0\right\} \tag{4.86}$$

spans a dense subset of $\mathscr{F}_0(X_0, X_1)$.

Proof We have only to approximate the function $g^{2\delta}(z) \equiv \exp(2\delta z^2)f(z)$, $z \in S$, where $\delta > 0$ and $f \in \mathscr{F}(X_0, X_1)$. Let $\rho \in C_c^\infty(\mathbb{R})$ be such that $\|\rho\|_1 = 1$ and that $\rho \geq 0$. Define

$$f_t(z) \equiv t\int_{\mathbb{R}} e^{\delta(z+iu)^2} f(z+iu)\rho(t^{-1}u)\mathrm{d}u \quad (z \in \overline{S}).$$

Then we have $f_t(z) \to e^z f(z)$ uniformly over $z \in S$ and

$$\lim_{t \downarrow 0} \|g^{2\delta} - g^{2\delta,t}\|_{\mathscr{F}(X_0,X_1)} = 0,$$

where $g^{2\delta,t}(z) \equiv \exp(\delta z^2)f_t(z)$, $z \in \bar{S}$. Therefore, we can assume that $f(s+it)$ is smooth in the variable $t \in \mathbb{R}$ for all $s \in [0, 1]$ to approximate $g^{2\delta}$.

Set

$$g_j^\delta(z) \equiv \sum_{k=-\infty}^{\infty} g^\delta(z + 2\pi ikj)$$

for $j \geq 1$ and $z \in S$, so that $g_j(z) = g_j(z + 2\pi ij)$, $z \in \bar{S}$. Note that the series defining g_j converges absolutely and locally uniformly. We also observe $\lim\limits_{j\to\infty} \exp(\delta\star^2)g_j^\delta = g^{2\delta}$ in $\mathscr{F}(X_0, X_1)$. We expand g_j^δ having period $2\pi ij$ into a Fourier series. To employ knowledge of analytic functions, we expand it in a tricky manner:

$$g_j^\delta(s+it) = \sum_{k=-\infty}^{\infty} \exp\left(\frac{ks+ikt}{j}\right)x_{kj}(s),$$

where

$$x_{kj}(s) \equiv \frac{1}{2\pi}\int_{-\pi}^{\pi} \exp\left(-\frac{ks+iku}{j}\right)g_j(s+iu)\mathrm{d}u.$$

We conclude from the Cauchy integral theorem and the periodicity that $x_{kj}(s)$ is independent of $s \in [0, 1]$, which we write a_{kj}. This means that $a_{kj} = x_{kj}(0) = x_{kj}(1) \in X_0 \cap X_1$. The convergence takes place absolutely. If we write

$$g_{j,l}^{\delta}(s+it) = \sum_{k=-l}^{l} \exp\left(\frac{ks+ikt}{j}\right) x_{kj}(s),$$

then we have $\exp(\delta \star^2) g_j^{\delta} = \lim_{l \to \infty} e^{\delta \star^2} g_{j,l}^{\delta}$ in $\mathscr{F}(X_0, X_1)$.

An immediate consequence of this lemma is the following result on density.

Corollary 4.4 (Density) *Let* (X_0, X_1) *be a compatible couple of Banach spaces, and let* $0 < \theta < 1$. *Then* $X_0 \cap X_1$ *is dense in* $[X_0, X_1]_\theta$.

We want to investigate the counterpart to Corollary 4.4. To this end, we need the following lemma:

Lemma 4.12 *Let* $k \in \mathrm{Lip}(\mathbb{R})$ *and* $\theta \in [0, 1]$. *Write* $k^{\delta,\theta}(t) \equiv k(t) \exp(\delta(it-\theta)^2)$ *for* $t \in \mathbb{R}$. *Then*

$$\limsup_{\delta \downarrow 0} \|k^{\delta,\theta}\|_{\mathrm{Lip}(\mathbb{R})} \le \|k\|_{\mathrm{Lip}(\mathbb{R})}. \tag{4.87}$$

Proof By the mollification, we may also assume that $k \in C^\infty(\mathbb{R})$. Next, we consider the pointwise convergence

$$k(t) = k(0) + \lim_{j \to \infty} \int_0^t E_j k'(s) \mathrm{d}s \quad (t \in \mathbb{R}),$$

where E_j is the dyadic average operator of generation j defined in Definition 1.41. Thus, if necessary, we may instead assume that k is piecewise linear.

In this case, we can verify (4.87) by verifying $\limsup_{\delta \downarrow 0} \|k'\|_\infty \le \|k\|_{\mathrm{Lip}(\mathbb{R})}$.

Denote by

$$\mathscr{G}_0(X_0, X_1)$$

the smallest closed subspaces of $\mathscr{G}(X_0, X_1)$ containing $\exp(a\star^2)G$ for all $a > 0$ and $G \in \mathscr{G}(X_0, X_1)$.

The next lemma allows us to choose a good function G to consider $G'(\theta)$.

Lemma 4.13 *Let* (X_0, X_1) *be a compatible couple of Banach spaces. Then we have*

$$[X_0, X_1]^\theta = \{G'(\theta) : G \in \mathscr{G}_0(X_0, X_1)\}$$

and

$$\|f\|_{[X_0,X_1]^\theta} = \inf\{\|G(\theta)\|_{\mathscr{G}(X_0,X_1)} : G \in \mathscr{G}_0(X_0, X_1),\ f = G'(\theta)\}$$
$$(f \in [X_0, X_1]^\theta).$$

Proof Let $G \in \mathscr{G}(X_0, X_1)$. Then there is a decomposition $G(1) - G(0) = x_0 + x_1$, where $x_0 \in X_0$ and $x_1 \in X_1$. Define $H_\delta(z) \equiv e^{2\delta(z-\theta)^2}(G(z) - G(0) - x_0)$ for $\delta > 0$. Then by Lemma 4.12, we have $H_\delta \in \mathscr{G}(X_0, X_1)$ and

$$\|G'(\theta)\|_{[X_0,X_1]^\theta} \geq \limsup_{\delta \downarrow 0} \|H_\delta'(\theta)\|_{[X_0,X_1]^\theta}.$$

We can now approximate H_δ by using Lemma 4.11. Thus, the proof of Lemma 4.13 is complete.

If we mimic the proof of Lemma 4.11, then we have the following:

Lemma 4.14 *Let (X_0, X_1) be a compatible couple of Banach spaces. Then the set $\mathscr{U}$ defined in Lemma* 4.11 *spans a dense subset of $\mathscr{G}_0(X_0, X_1)$.*

To explain the gap between $[X_0, X_1]_\theta$ and $[X_0, X_1]^\theta$, we use the following theorem:

Theorem 4.33 *Let (X_0, X_1) be a compatible couple of Banach spaces, and let $0 < \theta < 1$. Then with coincidence of norms $[X_0, X_1]_\theta = \overline{X_0 \cap X_1}^{[X_0,X_1]^\theta}$.*

Proof Since we know that $X_0 \cap X_1$ is dense in $[X_0, X_1]_\theta$ by Corollary 4.4, we have only to prove $\|x\|_{[X_0,X_1]_\theta} = \|x\|_{[X_0,X_1]^\theta}$ for all $x \in X_0 \cap X_1$. It is trivial that $\|x\|_{[X_0,X_1]^\theta} \leq \|x\|_{[X_0,X_1]_\theta}$. Thus, we need to show the reverse inequality. To this end, setting Y_0 and Y_1 as the closure of $X_0 \cap X_1$ in X_0 and X_1, we prove $\|x\|_{[Y_0,Y_1]_\theta} \leq \|x\|_{[Y_0,Y_1]^\theta}$, keeping in mind Lemmas 4.11 and 4.14. We assume $\|x\|_{[Y_0,Y_1]^\theta} = 1$.

Choose $G \in \mathscr{G}(Y_0, Y_1)$ with norm strictly less than 1 so that $x = G'(\theta)$. By the Hahn–Banach theorem and Theorem 4.31, we have $\|x\|_{[Y_0,Y_1]_\theta} = y^*(x)$ for some $y^* \in [Y_0{}^*, Y_1{}^*]^\theta = ([Y_0, Y_1]_\theta)^*$ with norm 1. Choose $G^* \in \mathscr{G}((Y_0)^*, (Y_1)^*)$ with norm less than or equal to 1 so that $y^* = (G^*)'(\theta)$, where the derivative is understood in the sense of $Y_0^* + Y_1^*$. Thus, since $x \in X_0 \cap X_1$,

$$y^*(\theta)(x) = -i \lim_{j \to \infty} 2^j (G^*(\theta + 2^{-j}i)(x) - G^*(\theta)(x)).$$

We consider $g_j(z) \equiv -2^j i(G^*(z + 2^{-j}i)(x) - G^*(z)(x))$ for $z \in \bar{S}$. If we use the Poisson integral expression and the change of variables, then we obtain Theorem 4.33.

Theorem 4.34 (Complex interpolation of function spaces) *Suppose that we have six parameters $p_0, p_1, q_0, q_1, s_0, s_1$ satisfying $1 \leq p_0, p_1, q_0, q_1 \leq \infty$ and $s_0, s_1 \in \mathbb{R}$. For $\theta \in (0, 1)$, define p, q, s by (4.76). Then $[A^{s_0}_{p_0q_0}(\mathbb{R}^n), A^{s_1}_{p_1q_1}(\mathbb{R}^n)]_\theta$ is isomorphic to the closure of $A^{s_0}_{p_0q_0}(\mathbb{R}^n) \cap A^{s_1}_{p_1q_1}(\mathbb{R}^n)$ in $A^s_{pq}(\mathbb{R}^n)$ with equivalence of norms, where $[A^{s_0}_{p_0q_0}(\mathbb{R}^n), A^{s_1}_{p_1q_1}(\mathbb{R}^n)]_\theta$ denotes the interpolation space given by Definition 4.15 with $1 \leq p_0, p_1, q_0, q_1 \leq \infty$.*

Theorem 4.34 shows that $[A^{s_0}_{p_0q_0}(\mathbb{R}^n), A^{s_1}_{p_1q_1}(\mathbb{R}^n)]_\theta$ can be different from $A^s_{pq}(\mathbb{R}^n)$.

Example 4.6 We have $[B^{s_0}_{\infty\infty}(\mathbb{R}^n), B^{s_1}_{\infty\infty}(\mathbb{R}^n)]_\theta \subsetneq B^{(1-\theta)s_0+\theta s_1}_{\infty\infty}(\mathbb{R}^n)$ for $s_0, s_1 \in \mathbb{R}$ with $s_0 \neq s_1$ according to Theorem 4.34.

Thus, in Theorem 4.32 the case where $\max(p, q) = \infty$ must be excluded. Let $0 < p < \infty$, $0 < \theta < 1$ and $s_0, s_1 \in \mathbb{R}$. Triebel proved $[B^{s_0}_{p\infty}(\mathbb{R}^n), B^{s_1}_{p\infty}(\mathbb{R}^n)]_\theta = \overline{\mathscr{S}(\mathbb{R}^n)}^{B^{s_0+(1-\theta)s_1}_{p\infty}(\mathbb{R}^n)}$ in [99, Theorem 2.4.1(e)], a good contrast to Theorem 4.34.

It may be important to define some closed subspaces. For a quasi-Banach space $X(\mathbb{R}^n) \subset \mathscr{S}'(\mathbb{R}^n)$ of distributions, we define $\overset{\diamond}{X}(\mathbb{R}^n)$ as the closure in X of the set of all infinitely differentiable functions f such that $\partial^\alpha f \in X$ for all multi-indexes α and $\overset{\circ}{X}(\mathbb{R}^n)$ as the closure of $C^\infty_c(\mathbb{R}^n)$. These spaces describe the subtle qualitative property of the functions.

Exercises

Exercise 4.36 [790, Proposition 2.1] Let $0 < \theta < 1$, $p_j, q_j \in [1, \infty]$ and $s_j \in \mathbb{R}$ for $j = 0, 1$. Set

$$\frac{1}{p} = \frac{1-\theta}{p_0} + \frac{\theta}{p_1}, \quad \frac{1}{q} = \frac{1-\theta}{q_0} + \frac{\theta}{q_1}, \quad s = (1-\theta)s_0 + \theta s_1.$$

Let $\mathscr{M}^s_{pq}(\mathbb{R}^n)$ be the closure of the Schwartz space $\mathscr{S}(\mathbb{R}^n)$ in $M^s_{pq}(\mathbb{R}^n)$. Then show that $[\mathscr{M}^{s_0}_{p_0q_0}(\mathbb{R}^n), \mathscr{M}^{s_1}_{p_1q_1}(\mathbb{R}^n)]_\theta = \mathscr{M}^s_{pq}(\mathbb{R}^n)$ using the molecular characterization of modulation spaces. Here it will be understood that the complex interpolation $[\mathscr{M}^{s_0}_{p_0q_0}(\mathbb{R}^n), \mathscr{M}^{s_1}_{p_1q_1}(\mathbb{R}^n)]_\theta$ is given in the sense of Definition 4.15.

Exercise 4.37 In Theorem 4.29, we assume that $p, p_0, p_1, q, q_0, q_1 \in [1, \infty)$. Show that the output for the original Calderón first complex interpolation functor is the same as Theorem 4.29 via the "distorted" complex interpolation functor defined in Sect. 4.2.3.

Exercise 4.38 Let $1 \le p, p_0, p_1, q_0, q_1 \le \infty$ and $s_0, s_1 \in \mathbb{R}$.

1. Show that $M^0_{p1}(\mathbb{R}^n) \hookrightarrow L^p(\mathbb{R}^n) \hookrightarrow M^0_{p\infty}(\mathbb{R}^n)$ by mimicking the proof of Proposition 2.1.
2. Give the description of $[M^{s_0}_{p_0q_0}(\mathbb{R}^n), M^{s_1}_{p_1q_1}(\mathbb{R}^n)]^\theta$.
3. Show that $M^0_{p\,\min(p,p')}(\mathbb{R}^n) \hookrightarrow L^p(\mathbb{R}^n) \hookrightarrow M^0_{p\,\max(p,p')}(\mathbb{R}^n)$ by complex interpolation.

Exercise 4.39 Let $1 \le p_1 < p < p_0$, $1 \le q_1 \le q \le q_0$ and $s_0, s_1, s \in \mathbb{R}$ satisfy

$$\frac{1}{p} = \frac{1-\theta}{p_0} + \frac{\theta}{p_1}, \quad \frac{1}{q} = \frac{1-\theta}{q_0} + \frac{\theta}{q_1}, \quad s = (1-\theta)s_0 + \theta s_1.$$

We explain an observation made by Frazier and Jawerth in [483] and expanded in [1160, Proposition 2.6] and [529]. Let $\{\lambda_{jm}\}_{j\in\mathbb{Z},m\in\mathbb{Z}^n}$ be a sequence of complex numbers.

1. Write

$$\lambda^*_{jm} := \sum_{\nu=-\infty}^{j} \sum_{\tilde{m}\in\mathbb{Z}^n} |\lambda_{\nu\tilde{m}}|^q \chi_{Q_{\nu\tilde{m}}}(2^{-j}m) \quad ((j,m)\in\mathbb{Z}\times\mathbb{Z}^n).$$

Show that the following pointwise estimates hold for all $x\in\mathbb{R}^n$:

$$\sum_{(j,m)\in\mathbb{Z}\times\mathbb{Z}^n} |\lambda_{jm}|^q \chi_{Q_{jm}}(x)(\lambda^*_{jm})^{\frac{p}{p_0}-1} \le \frac{p_0}{p}\left(\sum_{(j,m)\in\mathbb{Z}\times\mathbb{Z}^n} |\lambda_{jm}|^q \chi_{Q_{jm}}(x)\right)^{\frac{p}{p_0}},$$

$$\sum_{(j,m)\in\mathbb{Z}\times\mathbb{Z}^n} |\lambda_{jm}|^q \chi_{Q_{jm}}(x)(\lambda^*_{jm})^{\frac{p}{p_1}-1} \le \left(\sum_{(j,m)\in\mathbb{Z}\times\mathbb{Z}^n} |\lambda_{jm}|^q \chi_{Q_{jm}}(x)\right)^{\frac{p}{p_1}}.$$

2. Write

$$S(\lambda) \equiv \left(\sum_{(\nu,m)\in\mathbb{N}_0\times\mathbb{Z}^n} |\lambda_{\nu m}|^q \chi_{Q_{\nu m}}\right)^{\frac{1}{q}}.$$

Fix $k\in\mathbb{Z}$. Let us set

$$A_k := \{x\in\mathbb{R}^n : S(\lambda)(x) > 2^k\} \subset \mathbb{R}^n \tag{4.88}$$

and

$$C_k := \{(\nu,m)\in\mathbb{Z}\times\mathbb{Z}^n : 2|Q_{\nu m}\cap A_k| > |Q_{\nu m}|, 2|Q_{\nu m}\cap A_{k+1}| \le |Q_{\nu m}|\}. \tag{4.89}$$

(a) Whenever $(\nu,m)\in\mathbb{Z}\times\mathbb{Z}^n\setminus\bigcup_{k'\in\mathbb{Z}} C_{k'}$ and $k\in\mathbb{Z}$, show that $2|Q_{\nu m}\cap A_k| \le |Q_{\nu m}|$.
(b) Whenever $(\nu,m)\in\mathbb{Z}\times\mathbb{Z}^n\setminus\bigcup_{k\in\mathbb{Z}} C_k$, show that $\lambda_{\nu m}=0$.

In particular, we have an expression:

$$S(\lambda) = \left(\sum_{k=-\infty}^{\infty} \sum_{(\nu,m)\in C_k} |\lambda_{\nu m}|^q \chi_{Q_{\nu m}}\right)^{\frac{1}{q}}. \tag{4.90}$$

3. Let $(\nu, m) \in C_k$. Then show that

$$|Q_{\nu m} \cap A_{k+1}^{c}| \geq \frac{1}{2}|Q_{\nu m}|, \tag{4.91}$$

$$\chi_{Q_{\nu m}}(x) \leq 2M(\chi_{Q_{\nu m} \cap A_{k+1}^{c}})(x) \quad (x \in \mathbb{R}^n), \tag{4.92}$$

and

$$\chi_{Q_{\nu m}}(x) \leq 2M(\chi_{Q_{\nu m} \cap A_k})(x) \quad (x \in \mathbb{R}^n). \tag{4.93}$$

4. Denote by $\dot{\mathbf{f}}^{s}_{pq}(2^{\nu(s-n/p)})$ the set of all sequences $\{\lambda_{\nu m}\}_{(\nu,m)\in\mathbb{Z}^{n+1}}$ for which $\|\lambda_{\nu m}\|_{\dot{\mathbf{f}}^{s}_{pq}(2^{\nu(s-n/p)})} = \|\{2^{\nu(s-n/p)}\lambda_{\nu m}\}_{(\nu,m)\in\mathbb{Z}^{n+1}}\|_{\dot{\mathbf{f}}^{s}_{pq}}$ is finite. Then show that

$$(\dot{\mathbf{f}}_{p_0q_0}(2^{\nu(s_0-n/p_0)}))^{1-\theta}(\dot{\mathbf{f}}^{s_1}_{p_1q_1}(2^{\nu(s_1-n/p_1)}))^{\theta} = \dot{\mathbf{f}}^{s}_{pq}(2^{\nu(s-n/p)}).$$

Textbooks in Sect. 4.2

General Facts on Interpolation

We refer to [7, 11] for interpolation theory.

Interpolation of the Boundedness of the Operators

See [31, Section 3.6] for more about interpolation of the operator norms related to Lebesgue spaces and Hardy spaces. See [46, Chapter 4] for more about interpolation of the operator norms related to Lebesgue spaces.

See [90, Chapter 5] for more about interpolation of the operator norms related to Lebesgue spaces and Lorentz spaces together with the complex interpolation.

Interpolation Inequalities

See [6, Proposition 2.22].

Real Interpolation of Quasi-Banach Spaces

See [7, Chapter 3], [10], [19, Chapter 6] and [30, Chapter 10] for fundamental results on real interpolation theory.

We refer to [7, Chapter 3] and [19, Chapter 6] for the definition of the real interpolation functors. The book [4, Chapter 7] describes the relation between Sobolev spaces and Besov spaces using real interpolation, the book [7] is an exhaustive explanation of interpolation theory and the book [19] is related to interpolation and various approximation methods such as trigonometric polynomials, spline, Haar functions and Hermite polynomials.

Complex Interpolation of Banach Spaces

See [7] and [30, Chapter 9] for complex interpolation.

Real Interpolation of $A^s_{pq}(\mathbb{R}^n)$

We refer to [71, p. 106, Theorem 6] for Theorem 4.25 in the case of $1 \le p_0, p_1 \le \infty$ and $0 < q_0, q_1 \le \infty$, which deals with Besov spaces. See [99, Section 2.4.2] for the real interpolation of $A^s_{pq}(\mathbb{R}^n)$.

Complex Interpolation of $A^s_{pq}(\mathbb{R}^n)$

See [77] for interpolation of quasi-Banach spaces.

4.3 Paraproduct and Pointwise Multipliers

The classical Hölder inequality for Lebesgue spaces asserts that

$$\|f \cdot g\|_{p_0} \le \|f\|_{p_1} \|g\|_{p_2}, \quad 0 < p_0, p_1, p_2 \le \infty, \quad \frac{1}{p_0} = \frac{1}{p_1} + \frac{1}{p_2} \tag{4.94}$$

for all $f \in L^{p_1}(\mathbb{R}^n)$ and $g \in L^{p_2}(\mathbb{R}^n)$. We are oriented to an inequality of the form:

$$\|f \cdot g\|_{A^{s_0}_{p_0q_0}} \le C\|f\|_{A^{s_1}_{p_1q_1}} \|g\|_{A^{s_2}_{p_2q_2}} \tag{4.95}$$

which is valid for all $f \in A^{s_1}_{p_1q_1}(\mathbb{R}^n)$ and $g \in A^{s_2}_{p_2q_2}(\mathbb{R}^n)$, where the parameters at least satisfy $0 < p_0, p_1, p_2, q_0, q_1, q_2 \le \infty$ and $s_0, s_1, s_2 \in \mathbb{R}$. Note that (4.94) does not contain the constant C. Since we have slight ambiguity which the system of functions causes in the definition of Besov spaces and Triebel–Lizorkin spaces, we need to add C here. We tolerate eight possible choices since we have three A's; each A can be taken as either B or F independently. Let us call an inequality of the form:

$$\|f \cdot g\|_{A^{s_0}_{p_0q_0}} \le C\|f\|_{F^{s_1}_{p_1q_1}} \|g\|_{B^{s_2}_{p_2q_2}},$$

the AFB type, the AFB case or the AFB inequality. So, using A, B, F, we can consider $3^3 = 27$ types of Hölder inequalities.

In this book we are interested in the cases:

$$A^{s_0}_{p_0q_0}(\mathbb{R}^n) = F^{s_0}_{p_0q_0}(\mathbb{R}^n), \quad A^{s_1}_{p_1q_1}(\mathbb{R}^n) = F^{s_1}_{p_1q_1}(\mathbb{R}^n), \quad A^{s_2}_{p_2q_2}(\mathbb{R}^n) = F^{s_2}_{p_2q_2}(\mathbb{R}^n),$$
$$A^{s_0}_{p_0q_0}(\mathbb{R}^n) = B^{s_0}_{p_0q_0}(\mathbb{R}^n), \quad A^{s_1}_{p_1q_1}(\mathbb{R}^n) = B^{s_1}_{p_1q_1}(\mathbb{R}^n), \quad A^{s_2}_{p_2q_2}(\mathbb{R}^n) = B^{s_2}_{p_2q_2}(\mathbb{R}^n),$$

and

$$\begin{cases} A^{s_0}_{p_0q_0}(\mathbb{R}^n) = A^{s_2}_{p_2q_2}(\mathbb{R}^n) = B^{s_0}_{p_0q_0}(\mathbb{R}^n), & A^{s_1}_{p_1q_1}(\mathbb{R}^n) = B^{s_1}_{\infty\infty}(\mathbb{R}^n), \\ A^{s_0}_{p_0q_0}(\mathbb{R}^n) = A^{s_2}_{p_2q_2}(\mathbb{R}^n) = F^{s_0}_{p_0q_0}(\mathbb{R}^n), & A^{s_1}_{p_1q_1}(\mathbb{R}^n) = B^{s_1}_{\infty\infty}(\mathbb{R}^n). \end{cases}$$

which are dealt with in Theorems 4.35, 4.36 and 4.37, respectively. Since the elements in $A^s_{pq}(\mathbb{R}^n)$ are not always regular, we may ask ourselves how $f \cdot g$ makes sense for $f \in A^{s_1}_{p_1q_1}(\mathbb{R}^n)$ and $g \in A^{s_2}_{p_2q_2}(\mathbb{R}^n)$. All we know is that $f, g \in \mathcal{S}'(\mathbb{R}^n)$ in general.

We do not consider this problem seriously, however. Let us suppose that $f, g \in \mathcal{S}(\mathbb{R}^n)$ and consider the pointwise product $f \cdot g$. The most natural idea is to approximate $f \cdot g$ with $\{\psi_j(D)f \cdot \psi_j(D)g\}_{j=1}^{\infty}$. This is the idea which we actually use in this book. The point is that we need to be careful when we control $\psi_j(D)f \cdot \psi_j(D)g$. First we deal with the question above in Lemma 4.15; for what parameters $p_0, p_1, p_2, s_0, s_1, s_2, q_0, q_1, q_2$ do we have (4.95)?

Let s_0, s_1 be integers. If $f \in C^{s_0}(\mathbb{R}^n)$ and $g \in C^{s_1}(\mathbb{R}^n)$, then $f \cdot g \in C^{\min(s_0,s_1)}(\mathbb{R}^n)$. So it is natural to assume $s_0 = s_1 = s_2 = s$. We will deal with Hölder's inequality $\|f \cdot g\|_{A^s_{p_0q_0}} \lesssim \|f\|_{A^s_{p_1q_1}} \|g\|_{A^s_{p_2q_2}}$ with this in mind in Theorems 4.35 and 4.36. We plan to organize Sect. 4.3, as follows: First we justify the definition of the product in Sect. 4.3.1. Based on Sect. 4.3.1, we deal with Besov spaces and Triebel–Lizorkin spaces in Sect. 4.3.2. One of the aims of dealing with products in the theory of function spaces is in connection with the function spaces on domains. To this end, we want to consider the pointwise product in the case of $A^s_{p_0q_0}(\mathbb{R}^n) = A^s_{p_1q_1}(\mathbb{R}^n)$. Section 4.3.3 specializes this problem. Finally, we deal with the div-curl lemma in Sect. 4.3.4. Section 4.3.4 is somewhat different from the remaining parts in this section but as an application of the pointwise product we consider the div-curl lemma and the Kato–Ponce inequality, which we formulate in Sect. 4.3.4.

4.3.1 Paraproduct

4.3.1.1 How to Consider the Product

Apart from the sharpness of the results, let us prove that the pointwise product makes sense now. As before, we fix $\psi \in \mathscr{S}(\mathbb{R}^n)$ so that $\chi_{B(1)} \le \psi \le \chi_{B(2)}$, and define $\varphi \in \mathscr{S}(\mathbb{R}^n)$ by $\varphi \equiv \psi - \psi(2\star)$. Define a bilinear function $\Pi_{j,\psi}(\star_1, \star_2)$ by $\Pi_{j,\psi}(f, g) \equiv \psi_j(D)f \cdot \psi_j(D)g$, that is

$$\Pi_{j,\psi}(f, g) \equiv \left(\psi(D)f + \sum_{k=1}^{j} \varphi_k(D)f\right)\left(\psi(D)g + \sum_{k=1}^{j} \varphi_k(D)g\right)$$

for $f, g \in \mathscr{S}'(\mathbb{R}^n)$. The elements $\psi_j(D)f$ and $\psi_j(D)g$ belong to $\mathscr{O}_M(\mathbb{R}^n)$, the set of all $C^\infty(\mathbb{R}^n)$-functions that have at most polynomial growth at infinity. Observe also $\lim\limits_{j\to\infty} \psi_j(D)f = f$ and $\lim\limits_{j\to\infty} \psi_j(D)g = g$ in the topology of $\mathscr{S}'(\mathbb{R}^n)$, as we showed in Theorem 1.24. We want to define $f \cdot g = \lim\limits_{j\to\infty} \Pi_{j,\psi}(f, g)$ but we are still faced with the two problems: First, does the limit exist? Second, does the limit depend on ψ? To tackle these problems, we consider a quasi-Banach space $E \hookrightarrow \mathscr{S}'(\mathbb{R}^n)$ satisfying

$$(\|f + g\|_E)^\mu \le (\|f\|_E)^\mu + (\|g\|_E)^\mu \quad (f, g \in E) \tag{4.96}$$

for some $\mu \in (0, 1]$ and

$$\sup_{j\in\mathbb{N}} \|\Pi_{j,\psi}(f, g)\|_E \lesssim \|f\|_{A^{s_1}_{p_1q_1}} \|g\|_{A^{s_2}_{p_2q_1}} \quad f \in A^{s_1}_{p_1q_1}(\mathbb{R}^n), \quad g \in A^{s_2}_{p_2q_1}(\mathbb{R}^n). \tag{4.97}$$

The underlying idea is that "the pointwise product" of function spaces will be contained in a Besov space or a Triebel–Lizorkin space. It remains unclear whether this quasi-Banach space E works. But anyway, we set

$$\mathscr{M}(\psi) \equiv \bigcup_E \{(p_1, q_1, s_1; p_2, q_2, s_2) : 0 < p_1, p_2, q_1, q_2 \le \infty, \ s_1, s_2 \in \mathbb{R}, \ (4.97) \text{ holds}\},$$

where E runs over all quasi-Banach spaces which are embedded into $\mathscr{S}'(\mathbb{R}^n)$ and satisfy (4.96).

Lemma 4.15 *Let $\varepsilon > 0$, $\psi \in \mathscr{S}(\mathbb{R}^n)$ satisfy $\chi_{B(1)} \le \psi \le \chi_{B(2)}$ and assume that the parameters $p_1, q_1, s_1, p_2, q_2, s_2$ satisfy $(p_1, q_1, s_1 - \varepsilon; p_2, q_2, s_2 - \varepsilon) \in \mathscr{M}(\psi)$. Let $f \in B^{s_1}_{p_1\infty}(\mathbb{R}^n)$ and $g \in B^{s_2}_{p_2\infty}(\mathbb{R}^n)$.*

1. *The limit*

$$\lim_{j\to\infty} \Pi_{j,\psi}(f,g) \tag{4.98}$$

exists in the topology of $\mathscr{S}'(\mathbb{R}^n)$.

2. *Choose* $\rho \in \mathscr{S}(\mathbb{R}^n)$ *so that* $\chi_{B(1)} \le \rho \le \chi_{B(2)}$. *Then* $\mathscr{M}(\psi) = \mathscr{M}(\rho)$ *and*

$$\lim_{j\to\infty} \Pi_{j,\rho}(f,g) = \lim_{j\to\infty} \Pi_{j,\psi}(f,g). \tag{4.99}$$

Proof We prove that the limit (4.98) exists in $\mathscr{S}'(\mathbb{R}^n)$ if

$$(p_1, q_1, s_1 - \varepsilon; p_2, q_2, s_2 - \varepsilon) \in \mathscr{M}(\psi).$$

Choose a quasi-Banach space E so that

$$\sup_{j\in\mathbb{N}} \|\Pi_{j,\psi}(\tilde{f},\tilde{g})\|_E \lesssim \|\tilde{f}\|_{A^{s_1-\varepsilon}_{p_1q_1}} \|\tilde{g}\|_{A^{s_2-\varepsilon}_{p_2q_1}}, \quad \tilde{f} \in A^{s_1-\varepsilon}_{p_1q_1}(\mathbb{R}^n), \quad \tilde{g} \in A^{s_2-\varepsilon}_{p_2q_2}(\mathbb{R}^n).$$

Observe that

$$\begin{aligned}
&\Pi_{j+1,\psi}(f,g) - \Pi_{j,\psi}(f,g)\\
&= (\psi_{j+1}(D)f - \psi_j(D)f)\cdot\psi_{j+1}(D)g + \psi_j(D)f\cdot(\psi_{j+1}(D)g - \psi_j(D)g)\\
&= \varphi_{j+1}(D)f\cdot\psi_{j+1}(D)g + \psi_j(D)f\cdot\varphi_{j+1}(D)g\\
&= \Pi_{j+2,\psi}(\varphi_{j+1}(D)f, \psi_{j+1}(D)g) + \Pi_{j+2,\psi}(\psi_j(D)f, \varphi_{j+1}(D)g).
\end{aligned}$$

From (4.97), we have

$$\begin{aligned}
&\|\Pi_{j+1,\psi}(f,g) - \Pi_{j,\psi}(f,g)\|_E\\
&\lesssim \|\Pi_{j+2,\psi}(\varphi_{j+1}(D)f, \psi_{j+1}(D)g)\|_E + \|\Pi_{j+2,\psi}(\psi_j(D)f, \varphi_{j+1}(D)g)\|_E\\
&\lesssim \|\varphi_{j+1}(D)f\|_{A^{s_1-\varepsilon}_{p_1q_1}} \|\psi_{j+1}(D)g\|_{A^{s_2-\varepsilon}_{p_2q_2}} + \|\psi_j(D)f\|_{A^{s_1-\varepsilon}_{p_1q_1}} \|\varphi_{j+1}(D)g\|_{A^{s_2-\varepsilon}_{p_2q_2}};
\end{aligned}$$

hence we have

$$\begin{aligned}
&\|\Pi_{j+1,\psi}(f,g) - \Pi_{j,\psi}(f,g)\|_E\\
&\lesssim \|2^{j(s_1-\varepsilon)}\varphi_{j+1}(D)f\|_{p_1} \|g\|_{A^{s_2-\varepsilon}_{p_2q_2}} + \|f\|_{A^{s_1-\varepsilon}_{p_1q_1}} \|2^{j(s_2-\varepsilon)}\varphi_{j+1}(D)g\|_{p_2}\\
&\lesssim 2^{-j\varepsilon}\|f\|_{B^{s_1}_{p_1\infty}} \|g\|_{B^{s_2}_{p_2\infty}}
\end{aligned}$$

by Theorem 1.53. Thus, (4.96) yields $\lim_{j\to\infty} \Pi_{j,\psi}(f,g)$ in the topology of E. Since $E \hookrightarrow \mathscr{S}'(\mathbb{R}^n)$, $\lim_{j\to\infty} \Pi_{j,\psi}(f,g)$ exists in the topology of $\mathscr{S}'(\mathbb{R}^n)$.

We prove (4.99). As we did to prove (4.98), we have

$$\begin{aligned}&\Pi_{j,\psi}(f,g) - \Pi_{j,\rho}(f,g)\\&= \Pi_{j+2,\psi}(\psi_j(D)f - \rho_j(D)f, \psi_j(D)g) + \Pi_{j+2,\psi}(\rho_j(D)f, \psi_j(D)g - \rho_j(D)g).\end{aligned}$$

By $(p_1, q_1, s_1 - \varepsilon; p_2, q_2, s_2 - \varepsilon) \in \mathscr{M}(\psi)$ and Theorem 1.53, similar to above we have

$$\begin{aligned}&\|\Pi_{j,\psi}(f,g) - \Pi_{j,\rho}(f,g)\|_E\\&\lesssim 2^{-j\varepsilon} \sum_{k=j-1}^{j+1} (\|2^{ks_1}\varphi_k(D)f\|_{p_1}\|g\|_{A^{s_2}_{p_2\infty}} + \|f\|_{A^{s_1}_{p_1\infty}}\|2^{s_2k}\varphi_k(D)g\|_{p_2}).\end{aligned}$$

This shows $\lim_{j\to\infty} \Pi_{j,\rho}(f,g) = \lim_{j\to\infty} \Pi_{j,\psi}(f,g)$.

We transform Lemma 4.15 a little. The function ψ does not affect the definition of $\mathscr{M}(\psi)$ by virtue of Lemma 4.15. So, we write $\mathscr{M}$.

Corollary 4.5 *Let $\varepsilon > 0$. Suppose that the parameters $p_0, q_0, s_0, p_1, q_1, s_1$ satisfy*

$$0 < p_0, p_1, q_0, q_1 \le \infty, s_0, s_1 \in \mathbb{R}.$$

Assume that there exist $\widetilde{p_0}, \widetilde{q_0}, \widetilde{s_0}, \widetilde{p_1}, \widetilde{q_1}, \widetilde{s_1}$ such that $(\widetilde{p_0}, \widetilde{q_0}, \widetilde{s_0} - \varepsilon; \widetilde{p_1}, \widetilde{q_1}, \widetilde{s_1} - \varepsilon) \in \mathscr{M}$ and that

$$A^{s_0}_{p_0q_0}(\mathbb{R}^n) \hookrightarrow A^{\widetilde{s_0}}_{\widetilde{p_0}\widetilde{q_0}}(\mathbb{R}^n), \quad A^{s_1}_{p_1q_1}(\mathbb{R}^n) \hookrightarrow A^{\widetilde{s_1}}_{\widetilde{p_1}\widetilde{q_1}}(\mathbb{R}^n). \tag{4.100}$$

Then $f \cdot g \equiv \lim_{j\to\infty} \psi_j(D)f \cdot \psi_j(D)g$ is an element of $\mathscr{S}'(\mathbb{R}^n)$ independent of ψ for all $f \in A^{s_1}_{p_1q_1}(\mathbb{R}^n)$ and $g \in A^{s_2}_{p_2q_2}(\mathbb{R}^n)$.

Thus, here and below, in this book we assume (4.100) so we fix (4.97). Furthermore, if we replace f and g with $\psi_j(D)$ and $\psi_j(D)g$, respectively, then

$$\|\psi_j(D)f\|_{A^{s_1}_{p_1q_1}} \lesssim \|f\|_{A^{s_1}_{p_1q_1}}, \quad \|\psi_j(D)g\|_{A^{s_2}_{p_2q_2}} \lesssim \|g\|_{A^{s_2}_{p_2q_2}}. \tag{4.101}$$

Hence generally speaking, we need only show that

$$\|f \cdot g\|_{A^{s_0}_{p_0q_0}} \lesssim \|f\|_{A^{s_1}_{p_1q_1}}\|g\|_{A^{s_2}_{p_2q_2}} \tag{4.102}$$

for all band-limited elements $f \in A^{s_1}_{p_1q_1}(\mathbb{R}^n)$ and $g \in A^{s_2}_{p_2q_2}(\mathbb{R}^n)$.

4.3.1.2 Low-Frequency Estimates

We investigate how $\psi(D)f$, $\psi(D)g$ affect the results. Here note that $\psi(D)f \cdot g \in \mathscr{S}'(\mathbb{R}^n)$ provided $f \in \mathscr{S}'(\mathbb{R}^n)$ and $g \in A^s_{pq}(\mathbb{R}^n)$. Hence it always makes sense to discuss to what Besov spaces and to what Triebel–Lizorkin spaces $\psi(D)f \cdot g$ belongs. The next proposition and Theorem 2.14 enable us to skip the argument of $\psi(D)f$, $\psi(D)g$.

Proposition 4.3 *Let* $0 < p_0, p_1, p_2, q \le \infty$, *and let* $\rho, s \in \mathbb{R}$ *satisfy*

$$\frac{1}{p_0} = \frac{1}{p_1} + \frac{1}{p_2}.$$

Then

$$\|\psi(D)f \cdot g\|_{A^s_{p_0 q}} \lesssim \|f\|_{A^\rho_{p_1 q}} \|g\|_{A^s_{p_2 q}}, \tag{4.103}$$

$$\|\psi(D)f \cdot \psi(D)g\|_{A^s_{p_0 q}} \lesssim \|f\|_{A^\rho_{p_1 q}} \|g\|_{A^s_{p_2 q}}, \tag{4.104}$$

provided that $f \in A^\rho_{p_1 q}(\mathbb{R}^n)$ *and* $g \in A^s_{p_2 q}(\mathbb{R}^n)$.

Proof By using the quarkonial decomposition, we may assume that $f \in C^\infty_c(\mathbb{R}^n)$. Recall that Proposition 1.11 yields

$$\mathrm{supp}(f * g) \subset \mathrm{supp}(f) + \mathrm{supp}(g) \tag{4.105}$$

when the convolution of compactly supported functions f and distributions g makes sense. Here $\mathrm{supp}(f) + \mathrm{supp}(g)$ denotes the Minkovski sum. Inclusion (4.105) is an elementary observation when we consider the pointwise product for Besov spaces and Triebel–Lizorkin spaces. With this, we will estimate the size of the Fourier transform of pointwise estimates.

By virtue of Theorem 1.53, we have

$$\|\psi(D)f \cdot g\|_{A^s_{p_0 q}} \lesssim \|\psi(D)f \cdot \psi(D)g\|_{p_0} + \|2^{js}\psi(D)f \cdot \varphi_j(D)g\|_{L^{p_0}(\ell^q)}.$$

Next, we use $\|\psi(D)f\|_{p_1} \lesssim \|f\|_{p_1}$ and Hölder's inequality for Lebesgue spaces to have

$$\|\psi(D)f \cdot g\|_{A^s_{p_0 q}} \lesssim \|f\|_{p_1}(\|\psi(D)g\|_{p_2} + \|2^{js}\varphi_j(D)g\|_{L^{p_2}(\ell^q)}) \lesssim \|f\|_{A^\rho_{p_1 q}} \|g\|_{A^s_{p_2 q}}.$$

Inequality (4.104) follows from (4.103) and Theorem 1.53.

Noteworthy is the fact that $\psi(D)1 = 1$. So, $\psi(D)f \cdot \varphi_j(D)g$ is a sort of product.

Exercises

Exercise 4.40 Let $1 < p < \infty$ and $m \in \mathbb{N}$. Show that $W^{m,p}(\mathbb{R}^n)$ is closed under pointwise multiplication.

Exercise 4.41

1. Show that $\mathfrak{A} \equiv \bigcup_{R>0} \mathscr{S}_{B(R)}(\mathbb{R}^n)$ is closed under pointwise multiplication.
2. Show that $\mathfrak{B} \equiv \bigcup_{R>0} \mathscr{S}'_{B(R)}(\mathbb{R}^n)$ is closed under pointwise multiplication.
3. Let $0 < p \le \infty$. Then show that $\mathfrak{C} \equiv \bigcup_{R>0} L^p_{B(R)}(\mathbb{R}^n)$ is closed under pointwise multiplication.

4.3.2 Hölder's Inequality for Besov Spaces and Triebel–Lizorkin Spaces

Proposition 4.3 shows that we do not have to consider $\psi(D)f$, $\psi(D)g$ anymore for the proof of Theorems 4.35, 4.36, and 4.37. Here and below, we consider $S(f, g) = \sum_{j,k=1}^{\infty} \varphi_j(D)f \cdot \varphi_k(D)g$. Since we can assume f, g can be replaced with $\psi_j(D)f$ and $\psi_j(D)g$, respectively, the sum of $S(f, g)$ is essentially finite. We define four linear operators $f \odot g$, $f \odot^{\pm 3} g$, $f \succeq g$, $f \preceq g$ by

$$f \odot g \equiv \sum_{j=1}^{\infty} \varphi_j(D)f \cdot \varphi_j(D)g, \quad f \odot^{\pm 3} g \equiv \sum_{j,k\in\mathbb{N}, 0<|j-k|\le 3} \varphi_j(D)f \cdot \varphi_k(D)g,$$

$$f \succeq g \equiv \sum_{j,k\in\mathbb{N}, j-k>3} \varphi_j(D)f \cdot \varphi_k(D)g = \sum_{j=5}^{\infty} \varphi_j(D)f \cdot \left(\sum_{k=1}^{j-4} \varphi_k(D)g\right),$$

$$f \preceq g \equiv \sum_{j,k\in\mathbb{N}, k-j>3} \varphi_j(D)f \cdot \varphi_k(D)g = \sum_{k=5}^{\infty} \varphi_k(D)g \cdot \left(\sum_{j=1}^{k-4} \varphi_k(D)f\right).$$

Due to symmetry, $f \succeq g$ and $f \preceq g$ can be estimated in the same manner. The same can be said for $f \odot g$ and $f \odot^{\pm 3} g$. So, we will deal with $f \odot g$ and $f \succeq g$ mainly. The operators $f \preceq g$, $f \succeq g$ and $f \odot g$ $(f \odot^{\pm 3} g)$ are referred to as the *low-high interaction*, *high-low interaction* and *high-high interaction*, respectively.

When $s \le 0$, the space contains nasty distributions; that is, distributions which are not regular. This makes matters more difficult. So let us suppose $s > 0$.

4.3.2.1 Hölder's Inequality for Triebel–Lizorkin Spaces

First we formulate Hölder's inequality for Triebel–Lizorkin spaces.

Theorem 4.35 (Hölder's inequality for Triebel–Lizorkin spaces, the FFF case) *Suppose we have parameters p_0, p_1, p_2, q, s satisfying $0 < p_0, p_1, p_2 < \infty$, $0 < q \le \infty$, $s > 0$ and*

$$0 < \frac{1}{p_0} - \frac{s}{n} < 1, \quad 0 < \frac{1}{p_1} - \frac{s}{n} < 1, \quad 0 < \frac{1}{p_2} - \frac{s}{n} < 1, \quad \frac{1}{p_0} = \frac{1}{p_1} + \frac{1}{p_2} - \frac{s}{n}.$$

Then $\|f \cdot g\|_{F^s_{p_0q}} \lesssim \|f\|_{F^s_{p_1q}} \|g\|_{F^s_{p_2q}}$ for $f \in F^s_{p_1q}(\mathbb{R}^n)$ and $g \in F^s_{p_2q}(\mathbb{R}^n)$.

In this proof we choose $\psi \in \mathcal{S}(\mathbb{R}^n)$ so that $\chi_{B(1)} \le \psi \le \chi_{B(3/2)}$ and define $\varphi \in \mathcal{S}(\mathbb{R}^n)$ by $\varphi \equiv \psi - \psi(2\star)$. Also, let $\kappa \in \mathcal{S}(\mathbb{R}^n)$ satisfy $\chi_{Q(3)} \le \kappa \le \chi_{Q(3+1/100)}$.

We need a lemma to estimate $f \odot g$.

Lemma 4.16 *Let $0 < p < \infty$, $0 < q \le \infty$.*

1. *Let $\rho \in \mathbb{R}$. Suppose that we have $f_j \in \mathcal{S}'_{B(2^{j+1})\setminus B(2^{j-1})}(\mathbb{R}^n)$ for each $j \in \mathbb{N}$. Then*

$$\left\| \sum_{j=1}^{\infty} f_j \right\|_{B^\rho_{pq}} \lesssim \|\{2^{j\rho} f_j\}_{j=1}^{\infty}\|_{\ell^q(L^p)}.$$

2. *Let $\rho \in \mathbb{R}$. Suppose that we have $f_j \in \mathcal{S}'_{B(2^{j+1})\setminus B(2^{j-1})}(\mathbb{R}^n)$ for each $j \in \mathbb{N}$. Then*

$$\left\| \sum_{j=1}^{\infty} f_j \right\|_{F^\rho_{pq}} \lesssim \|\{2^{j\rho} f_j\}_{j=1}^{\infty}\|_{L^p(\ell^q)}. \tag{4.106}$$

3. *Let $s > \sigma_p$. Suppose that we have $g_j \in \mathcal{S}'_{B(2^j)}(\mathbb{R}^n)$ for each $j \in \mathbb{N}$. Then*

$$\left\| \sum_{j=1}^{\infty} g_j \right\|_{B^s_{pq}} \lesssim \|\{2^{js} g_j\}_{j=1}^{\infty}\|_{\ell^q(L^p)}.$$

4. *Let $s > \sigma_{pq}$. Suppose that we have $g_j \in \mathcal{S}'_{B(2^j)}(\mathbb{R}^n)$ for each $j \in \mathbb{N}$. Then*

$$\left\| \sum_{j=1}^{\infty} g_j \right\|_{F^s_{pq}} \lesssim \|\{2^{js} g_j\}_{j=1}^{\infty}\|_{L^p(\ell^q)}. \tag{4.107}$$

Proof Due to similarity, we deal with Triebel–Lizorkin spaces. Assume that $\kappa \in \mathscr{S}(\mathbb{R}^n)$ satisfies $\chi_{Q(3)} \le \kappa \le \chi_{Q(3+1/100)}$. Then thanks to Thoerem 4.9

$$f_j = (2\pi)^{-\frac{n}{2}} \sum_{m \in \mathbb{Z}^n} f_j(2^{-j}m)\mathscr{F}^{-1}\kappa(2^j \star - m),$$

where the convergence takes place in $\mathscr{S}'(\mathbb{R}^n)$. Thus, by the molecular decomposition, we have

$$\left\| \sum_{j=1}^{\infty} f_j \right\|_{F^\rho_{pq}} \lesssim \left\| \left\{ 2^{j\rho} \sum_{m \in \mathbb{Z}^n} f_j(2^{-j}m)\chi_{Q_{jm}} \right\}_{j=1}^{\infty} \right\|_{L^p(\ell^q)}.$$

Let $\eta \equiv 2^{-1}\min(p, q, 1)$ as before. By Theorem 1.50, we obtain

$$\left\| \sum_{j=1}^{\infty} f_j \right\|_{F^\rho_{pq}} \lesssim \|\{2^{j\rho} M f_j\}_{j=1}^{\infty}\|_{L^p(\ell^q)}.$$

Finally, by the Fefferman–Stein vector-valued maximal inequality, we have

$$\left\| \sum_{j=1}^{\infty} f_j \right\|_{F^\rho_{pq}} \lesssim \|\{2^{j\rho} f_j\}_{j=1}^{\infty}\|_{L^p(\ell^q)}.$$

The proof of (4.107) is almost the same except in that we use

$$\left\| \sum_{j=1}^{\infty} g_j \right\|_{F^\rho_{pq}} \lesssim \|\{2^{j\rho}(Mg_1 + Mg_2 + \cdots + Mg_j)\}_{j=1}^{\infty}\|_{L^p(\ell^q)}.$$

We go back to the proof of the theorem.

Proof (The estimate of $f \odot g$) Define $r_0 > 0$ by $\dfrac{1}{r_0} = \dfrac{1}{p_0} + \dfrac{s}{n}$. Theorem 2.14 reduces the matter to proving;

$$\|f \odot g\|_{F^{2s}_{r_0\infty}} \lesssim \|f\|_{F^s_{p_1\infty}} \|g\|_{F^s_{p_2\infty}} \quad (f \in F^s_{p_1\infty}(\mathbb{R}^n), g \in F^s_{p_2\infty}(\mathbb{R}^n)). \tag{4.108}$$

This can be done by the use of Lemma 4.16 easily.

Proof (The estimate of $f \succeq g$) We can take $\mu > 1$ so that

$$\operatorname{supp}(\psi_{j-2}(D)f \cdot \varphi_j(D)g) \subset \overline{B(\mu\, 2^j) \setminus B(\mu^{-1}\, 2^j)}.$$

Thanks to Theorem 1.53, the multiplier theorem, we have

$$\|f \succeq g\|_{F^s_{pq}} \lesssim \|2^{js}\psi_{j-2}(D)f \cdot \varphi_j(D)g\|_{L^p(\ell^q)}$$
$$\lesssim \left\|2^{js} \sup_{k\in\mathbb{N}_0} |\psi_k(D)f \cdot \varphi_j(D)g|\right\|_{L^p(\ell^q)}.$$

We now define $r_1 \in (1, \infty)$ by $\frac{1}{r_1} = \frac{1}{p_1} - \frac{s}{n}$. Then observe that $\frac{1}{p_0} = \frac{1}{r_1} + \frac{1}{p_2}$. By Theorem 1.47, we have

$$\|f \succeq g\|_{F^s_{p_0q}} \lesssim \left\| \sup_{j\in\mathbb{N}_0} |\psi_j(D)f| \right\|_{r_1} \|2^{js}\varphi_j(D)g\|_{L^{p_2}(\ell^q)}. \tag{4.109}$$

Hence by the $L^{r_1}(\mathbb{R}^n)$-boundedness of the Hardy–Littlewood maximal operator M,

$$\|f \succeq g\|_{F^s_{p_0q}} \lesssim \|Mf\|_{r_1}\|g\|_{F^s_{p_2q}} \lesssim \|f\|_{r_1}\|g\|_{F^s_{p_2q}}.$$

Recall that $L^{r_1}(\mathbb{R}^n) \approx F^0_{r_12}(\mathbb{R}^n) \hookleftarrow F^s_{p_1q}(\mathbb{R}^n)$ according to Theorem 2.14. So the estimate of $f \succeq g$ is valid.

4.3.2.2 Hölder's Inequality for Besov Spaces

Following the idea of the proof of Hölder's inequality for Triebel–Lizorkin spaces, we will prove Hölder's inequality for Besov spaces.

Theorem 4.36 (Hölder's inequality for Besov spaces, the BBB case) *Suppose that the parameters $p_0, p_1, p_2, q_0, q_1, q_2, s \in (0, \infty)$ satisfy*

$$0 < \frac{1}{p_0} - \frac{s}{n} < 1, \quad 0 < \frac{1}{p_1} - \frac{s}{n} < 1, \quad 0 < \frac{1}{p_2} - \frac{s}{n} < 1,$$

$$0 < q_1 \le \frac{np_1}{n - p_1 s}, \quad 0 < q_2 \le \frac{np_2}{n - p_2 s}, \quad 0 < q_0 \le \min(q_1, q_2), \tag{4.110}$$

and $\frac{1}{p_0} = \frac{1}{p_1} + \frac{1}{p_2} - \frac{s}{n}$. Then $f \cdot g$ can be defined naturally for all $f \in B^s_{p_1q_1}(\mathbb{R}^n)$ and $g \in B^s_{p_2q_2}(\mathbb{R}^n)$ so that $\|f \cdot g\|_{B^s_{p_0q_0}} \lesssim \|f\|_{B^s_{p_1q_1}}\|g\|_{B^s_{p_2q_2}}$.

We have three parameters q_0, q_1, q_2. So, compared with the case of Triebel–Lizorkin spaces, the situation looks generalized. If the first two conditions of (4.110) are satisfied and if $0 < q_0 \le \min(q_1, q_2)$, then

$$B^s_{p_0q_0}(\mathbb{R}^n) \hookrightarrow B^s_{p_0\,\min(q_1,q_2)}(\mathbb{R}^n), \quad B^s_{p_l\,\min(q_1,q_2)}(\mathbb{R}^n) \hookrightarrow B^s_{p_lq_l}(\mathbb{R}^n), \quad l = 1, 2.$$

So, as is the case with Triebel–Lizorkin spaces, we can unify the results in terms of q.

Proof The proof is the same as Theorem 4.35. So we content ourselves with mentioning where to use (4.110), which is a strong assumption. In the last part of the estimates of $f \succeq g$, we used $F^s_{p_1 q}(\mathbb{R}^n) \hookrightarrow F^0_{u2}(\mathbb{R}^n)$. Here we need to replace this embedding with the one $B^s_{p_1 q_1}(\mathbb{R}^n) \hookrightarrow B^s_{p_1 u}(\mathbb{R}^n) \hookrightarrow F^0_{u2}(\mathbb{R}^n)$. Then we are led to the necessity of $q_1 \le \dfrac{np_1}{n - p_1 s} = u$. Furthermore, we have $B^s_{p_1 q_0}(\mathbb{R}^n) \hookrightarrow B^s_{p_1 q_1}(\mathbb{R}^n)$ by $q_0 \le q_1$. These inclusions allow us to argue as in Theorem 4.35.

4.3.2.3 Hölder's Inequality for Cut-Off Functions and Variable Coefficients

Now we look for a suitable function space X so that $\|f \cdot g\|_{A^s_{pq}} \lesssim \|f\|_X \|g\|_{A^s_{pq}}$. Such an inequality is important in differential equations with variable coefficients.

Theorem 4.37 (General multipliers, Hölder's inequality of multiplier type, the ABA case) *Suppose that four parameters p, q, s, ρ satisfy*

$$0 < p, q \le \infty, \quad s \in \mathbb{R}, \quad \rho > \max\left(s, \sigma_p - s\right) = \max\left(s, n\left(\frac{1}{p} - 1\right)_+ - s\right).$$

Then $\|f \cdot g\|_{A^s_{pq}} \lesssim \|f\|_{B^\rho_{\infty\infty}} \|g\|_{A^s_{pq}}$ for all $f \in B^\rho_{\infty\infty}(\mathbb{R}^n)$ and $g \in A^s_{pq}(\mathbb{R}^n)$.

When we considered the atomic decomposition, we considered the condition: $L \ge \max(-1, [\sigma_{pq} - s])$ and $\sigma_{pq} = \max(\sigma_p, \sigma_q)$ came into play. However, instead of σ_{pq}, σ_p appears here. The interested readers may consider why in the course of the proof.

Proof The conditions on $f \succeq g$ and $f \preceq g$ are not symmetric. We investigate $f \odot g$, $f \succeq g$ and $f \preceq g$. Furthermore, we concentrate on Triebel–Lizorkin spaces.

Let us estimate $f \odot g$. Theorem 4.9 expands $f \odot g$:

$$f \odot g \simeq_n \sum_{j=1}^{\infty} \sum_{m \in \mathbb{Z}^n} \varphi_j(D) g(2^{-j} m) \varphi_j(D) f \cdot \mathscr{F}^{-1}\kappa(2^j \star - m).$$

Let $x \in \mathbb{R}^n$ be fixed. Then Corollary 1.6 yields

$$\begin{aligned} &|\partial^\alpha [2^{-j\left(s+\rho-\frac{n}{p}\right)} \varphi_j(D) f(x) \cdot \mathscr{F}^{-1}\kappa(2^j x - m)]| \\ &\lesssim_{N,\alpha} 2^{-j\left(s+\rho-\frac{n}{p}\right)+j|\alpha|} \langle 2^j x - m\rangle^{-N} \|\varphi_j(D) f\|_\infty \end{aligned}$$

for all $x \in \mathbb{R}^n$. Since $s+\rho > \sigma_p = \sigma_{p\infty}$, $F^{s+\rho}_{p\infty}(\mathbb{R}^n)(\hookleftarrow F^s_{pq}(\mathbb{R}^n))$ admits an atomic decomposition without moment condition (Theorem 4.1) to obtain

$$\begin{aligned}
\|f \odot g\|_{F^s_{pq}}^p &\le \|f \odot g\|_{F^{s+\rho}_{p\infty}}^p \\
&\lesssim \int_{\mathbb{R}^n} \sup_{j\in\mathbb{N}} \left(2^{j\left(s+\rho-\frac{n}{p}\right)}\|\varphi_j(D)f\|_\infty |\varphi_j(D)g(2^{-j}m)|\chi_{jm}^{(p)}(x)\right)^p \mathrm{d}x \\
&\lesssim \|f\|_{B^\rho_{\infty\infty}}^p \|g\|_{F^s_{p\infty}}^p \le \|f\|_{B^\rho_{\infty\infty}}^p \|g\|_{F^s_{pq}}^p.
\end{aligned}$$

For $f \succeq g$, we use Theorem 1.53 to obtain

$$\begin{aligned}
\|f \succeq g\|_{F^s_{pq}} &= \left\| \left\{ 2^{js}\varphi_j(D)f \cdot \left(\sum_{k=1}^{j-4}\varphi_k(D)g\right)\right\}_{j\ge 5}\right\|_{L^p(\ell^q)} \\
&\lesssim \|\{2^{j\rho}\varphi_j(D)f\}_{j=1}^\infty\|_{\ell^\infty(L^\infty)} \left\| \left\{ 2^{-j(\rho-s)}\sum_{k=1}^{j-4}\varphi_k(D)g\right\}_{j\ge 5}\right\|_{L^p(\ell^q)}.
\end{aligned}$$

Since $\rho > s$, we have

$$\begin{aligned}
\sum_{j=1}^\infty \left| 2^{-j(\rho-s)}\sum_{k=1}^{j}\varphi_k(D)g\right|^q &\le \sum_{j=1}^\infty \left| 2^{-j(\rho-s)}\sum_{k=1}^{j}2^{-ks}\sup_{l\in\mathbb{Z}}|2^{ls}\varphi_l(D)g|\right|^q \\
&\le \sup_{l\in\mathbb{Z}}|2^{ls}\varphi_l(D)g|^q \left(\sum_{j=1}^\infty 2^{-j\rho q+j\max(s,0)q}\right) \\
&\simeq \sup_{l\in\mathbb{Z}}|2^{ls}\varphi_l(D)g|^q.
\end{aligned}$$

Hence $\|f \succeq g\|_{F^s_{pq}} \lesssim \|f\|_{B^\rho_{\infty\infty}}\|g\|_{F^s_{p\infty}} \simeq \|f\|_{B^\rho_{\infty\infty}}\|g\|_{F^s_{pq}}$.

We consider $f \preceq g$. We abbreviate $F_k \equiv \sum_{j=1}^{k-4}\varphi_k(D)f$ for $k \ge 5$. Then

$$\|f \preceq g\|_{F^s_{pq}} \lesssim \left\| \left\{2^{ks}\varphi_k(D)g \cdot F_k\right\}_{k\ge 5}\right\|_{L^p(\ell^q)} \lesssim \|F_k\|_{\ell^\infty(L^\infty)}\|2^{ks}\varphi_k(D)g\|_{L^p(\ell^q)}.$$

We estimate the quantity $\|F_k\|_{\ell^\infty(L^\infty)}$ carefully:

$$\|f \preceq g\|_{F^s_{pq}} \lesssim \|\varphi_k(D)f\|_{\ell^1(L^\infty)}\|2^{ks}\varphi_k(D)g\|_{L^p(\ell^q)} \lesssim \|f\|_{B^\rho_{\infty\infty}}\|g\|_{F^s_{pq}}.$$

We can handle $f \succeq g$ similarly. Thus, the proof is complete.

Remark 4.4 Apply Theorem 4.37 to $F^0_{p2}(\mathbb{R}^n) \approx h^p(\mathbb{R}^n)$, $0 < p \le 1$. Then

$$\|h \cdot f\|_{h^p} \lesssim \|h\|_{\mathscr{C}^\rho} \|f\|_{h^p}, \quad \rho > \sigma_p = n\left(\frac{1}{p} - 1\right)_+. \tag{4.111}$$

Observe that this property is characteristic to local Hardy spaces; Hardy spaces do not enjoy such a property.

4.3.2.4 Localization of Triebel–Lizorkin Spaces

Here, as an application, we consider the localization of function spaces. Localization is a technique used in the theory of function spaces and PDEs. A basic idea is to assemble local construction to obtain global information. We have the following result for Triebel–Lizorkin spaces. Let us leave it untouched for Besov spaces.

Theorem 4.38 (Localization of Triebel–Lizorkin spaces) *Let $0 < p < \infty$, $0 < q \le \infty$ and $s \in \mathbb{R}$ and $a > 0$. Choose $\rho \in \mathscr{S}(\mathbb{R}^n)$ so that*

$$\mathrm{supp}(\rho) \subset Q(2a), \quad \sum_{m \in \mathbb{Z}^n} \rho(\star - a\,m) \equiv 1.$$

Then for $f \in \mathscr{S}'(\mathbb{R}^n)$, we have

$$\|f\|_{F^s_{pq}} \sim_a \left(\sum_{m \in \mathbb{Z}^n} \|\rho(\star - a\,m) f\|_{F^s_{pq}}^p\right)^{\frac{1}{p}}.$$

Proof We scale the local means for some suitable $a > 0$; see Theorem 2.34. To describe the local means, we take a compactly supported function $K \in \mathscr{S}(\mathbb{R}^n)$ so that $\chi_{B(a)} \le K \le \chi_{B(2a)}$. According to Theorem 2.34,

$$\|f\|_{F^s_{pq}} \sim \left(\|K * f\|_p{}^p + \|2^{js}(\Delta^M K)^j * f\|_{L^p(\ell^q)}{}^p\right)^{\frac{1}{p}}$$

as long as $2M > s$.

We fix $m_0 \in \mathbb{M} \equiv \{0, 1, 2, \ldots, 9\}^n$. Define

$$R_{m_0} \equiv \sum_{m \in \mathbb{Z}^n} \rho(\star - 10a\,m - a\,m_0).$$

Then we have

$$\|f\|_{F^s_{pq}} \sim \sum_{m \in \mathbb{M}} \|R_{m_0} f\|_{F^s_{pq}} \tag{4.112}$$

by Hölder's inequality and the quasi-triangle inequality for function spaces. Furthermore, by the quasi-triangle inequality, we have

$$\|K * f\|_p \lesssim \sum_{m_0 \in \mathbb{M}} \|K * (R_{m_0} f)\|_p.$$

We decompose $K*(R_{m_0}f) = \sum_{m \in \mathbb{Z}^n} K*(\rho(\star - 10a\,m - a\,m_0)f)$ and the summands in the right-hand side are disjointly supported. This is why we use local means. Hence

$$|K * (R_{m_0} f)|^p \sim \sum_{m \in \mathbb{Z}^n} |K * (\rho(\star - 10a\,m - a\,m_0) f)|^p.$$

As a result, we have

$$\begin{aligned}
\sum_{m_0 \in \mathbb{M}} \|K * (R_{m_0} f)\|_p &\sim \sum_{m_0 \in \mathbb{M}} \left(\sum_{m \in \mathbb{Z}^n} \|K * (\rho(\star - 10a\,m - a\,m_0) f)\|_p{}^p \right)^{\frac{1}{p}} \\
&\sim \left(\sum_{m_0 \in \mathbb{M}} \sum_{m \in \mathbb{Z}^n} \|K * (\rho(\star - 10a\,m - a\,m_0) f)\|_p{}^p \right)^{\frac{1}{p}} \\
&\sim \left(\sum_{m \in \mathbb{Z}^n} \|K * (\rho(\star - a\,m) f)\|_p{}^p \right)^{\frac{1}{p}}. \qquad (4.113)
\end{aligned}$$

A similar observation to $2^{js}(\Delta K)^j * f$ together with (4.112) and (4.113) proves the theorem.

Unfortunately, the counterpart of Theorem 4.38 to Besov spaces does not hold in general; see [262, IV, p. 157] and [262, V, p. 159] for counterexamples. Nevertheless, we have a partial result. We leave the rest to interested readers. See Exercise 4.46.

4.3.2.5 Ring Structure of $M^0_{p1}(\mathbb{R}^n)$

The same idea can be used for modulation spaces. The following is an example of the application of the idea used in this section.

Theorem 4.39 (Ring structure of $M^0_{p1}(\mathbb{R}^n)$) *Let $1 \le p \le \infty$. Let $f, g \in M^0_{p1}(\mathbb{R}^n)(\hookrightarrow L^p(\mathbb{R}^n))$. Then the pointwise product $f \cdot g \in L^{\frac{p}{2}}(\mathbb{R}^n)$ belongs to $M^0_{p1}(\mathbb{R}^n)$ and it satisfies $\|f \cdot g\|_{M^0_{p1}} \lesssim \|f\|_{M^0_{p1}} \|g\|_{M^0_{p1}}$. More precisely, $\|f \cdot g\|_{M^0_{p1}} \lesssim \|f\|_{M^0_{\infty 1}} \|g\|_{M^0_{p1}}$.*

Proof Choose a function $\tau \in \mathscr{S}(\mathbb{R}^n)$ satisfying $\sum_{k=-\infty}^{\infty} \tau(\star - k) \equiv 1$. Then Lebesgue's convergence theorem and Theorem 1.53 show that

$$\lim_{N\to\infty} \sum_{k\in\mathbb{Z}\cap B(N)} \tau(D-k)f = f, \quad \lim_{N\to\infty} \sum_{k\in\mathbb{Z}\cap B(N)} \tau(D-k)g = g$$

in the topology of $\mathscr{S}'(\mathbb{R}^n)$. Thus, it can be assumed that f and g are band-limited. Then keeping in mind that the sum below is essentially finite, we have

$$f\cdot g = \sum_{k,l\in\mathbb{Z}} \tau(D-k)f \cdot \tau(D-k)g = \sum_{k=-\infty}^{\infty} \left(\sum_{l=-\infty}^{\infty} \tau(D-k+l)f \cdot \tau(D-l)g \right).$$

Again, Theorem 1.53 (or Young's inequality) yields

$$\begin{aligned} \|f\cdot g\|_{M^0_{p1}} &\lesssim \sum_{k=-\infty}^{\infty} \left\| \sum_{l=-\infty}^{\infty} \tau(D-k+l)f \cdot \tau(D-l)g \right\|_p \\ &\sim \sum_{k=-\infty}^{\infty} \|\tau(D-k)f\|_\infty \sum_{l=-\infty}^{\infty} \|\tau(D-k+l)g\|_p \\ &= \sum_{k=-\infty}^{\infty} \|\tau(D-k)f\|_\infty \sum_{l=-\infty}^{\infty} \|\tau(D-l)g\|_p = \|f\|_{M^0_{\infty 1}} \|g\|_{M^0_{p1}}, \end{aligned}$$

as was to be shown.

Exercises

Exercise 4.42 Disprove that a counterpart to (4.111) for $H^1(\mathbb{R}^n)$ holds using $H^1(\mathbb{R}^n) \subset \mathscr{P}_0(\mathbb{R}^n)^\perp$.

Exercise 4.43 Let $0 < p, q \le \infty$ and $s \in \mathbb{R}$, and let $\psi \in C^\infty_c(\mathbb{R}^n) \setminus \{0\}$ be a nonnegative function. Define $A^s_{pq,\mathrm{unif}}(\mathbb{R}^n)$ by $\|f\|_{A^s_{pq,\mathrm{unif}}} \equiv \sup_{y\in\mathbb{R}^n} \|f\cdot\psi(\star - y)\|_{A^s_{pq}}$. Then show that $A^s_{pq,\mathrm{unif}}(\mathbb{R}^n)$ is independent of the choice of admissible ψ using Theorem 4.37.

Exercise 4.44 Let $0 < p \le \infty$, and let $f \in M^0_{\infty\,\min(1,p)}(\mathbb{R}^n)$ and $g \in M^0_{p\,\min(1,p)}(\mathbb{R}^n)$. Then, by mimicking the proof of Theorem 4.39, establish that the pointwise product $f \cdot g \in L^p(\mathbb{R}^n)$ belongs to $M^0_{p\,\min(1,p)}(\mathbb{R}^n)$ and satisfies

$$\|f\cdot g\|_{M^0_{p\,\min(1,p)}} \lesssim \|f\|_{M^0_{\infty\,\min(1,p)}} \|g\|_{M^0_{p\,\min(1,p)}},$$

where the implicit constant does not depend on f and g.

Exercise 4.45 Let $1 \le p, p_1, p_2, q, q_1, q_2 \le \infty$. Assume $\dfrac{1}{p} = \dfrac{1}{p_1} + \dfrac{1}{p_2}$, and $1+\dfrac{1}{q} = \dfrac{1}{q_1}+\dfrac{1}{q_2}$. Show that $\|f\cdot g\|_{M^0_{pq}} \lesssim \|f\|_{M^0_{p_1q_1}}\|g\|_{M^0_{p_2q_2}}$ for all $f \in M^0_{p_1q_1}(\mathbb{R}^n)$ and $g \in M^0_{p_2q_2}(\mathbb{R}^n)$, mimicking the proof of Theorem 4.39.

Exercise 4.46 Let ρ and a be the same as Theorem 4.38. Mimic the proof of Theorem 4.38 to show that

$$\|f\|_{B^s_{pq}}{}^p \lesssim_a \sum_{m\in\mathbb{Z}^n} \|\rho(\star - a\,m)f\|_{B^s_{pq}}{}^p \quad (f \in B^s_{pq}(\mathbb{R}\,n)) \tag{4.114}$$

when $p \le q$ and the inequality is reversed otherwise.

Exercise 4.47 Assume that $1 \le p, q \le \infty$, $s \in \mathbb{R}$ and $M > 1$ satisfy $\|f\cdot g\|_{A^s_{pq}} \le M\|f\|_{A^s_{pq}}\|g\|_{A^s_{pq}}$ for $f, g \in A^s_{pq}(\mathbb{R}^n)$. Let $F : \mathbb{C} \to \mathbb{C}$ be an analytic function of the form $F(z) = \sum\limits_{n=1}^{\infty} a_n z^n$, $z \in \mathbb{C}$, and set $F^+(z) = \sum\limits_{n=1}^{\infty} |a_n|z^n$, $z \in \mathbb{C}$ Then show that $\|F(f)\|_{A^s_{pq}} \le F^+(M\|f\|_{A^s_{pq}})$ for all $f \in A^s_{pq}(\mathbb{R}^n)$.

Exercise 4.48 [258] Write $\mathscr{C}^s(\mathbb{R}^n) \equiv B^s_{\infty\infty}(\mathbb{R}^n)$ for $s \in \mathbb{R}$. Mimic the proof of Theorem 4.35 to prove the following:

1. Show that $\|u \preceq v\|_{\mathscr{C}^\beta} \lesssim \|u\|_\infty \|v\|_{\mathscr{C}^\beta}$ for all $u \in \mathscr{C}^\alpha(\mathbb{R}^n)$ and $v \in \mathscr{C}^\beta(\mathbb{R}^n)$ provided $\beta \in \mathbb{R}$.
2. Show that $\|u \succeq v\|_{\mathscr{C}^{\alpha+\beta}} \lesssim \|u\|_{\mathscr{C}^\alpha}\|v\|_{\mathscr{C}^\beta}$ for all $u \in \mathscr{C}^\alpha(\mathbb{R}^n)$ and $v \in \mathscr{C}^\beta(\mathbb{R}^n)$ whenever $\alpha, \beta \in \mathbb{R}$ satisfy $\beta > 0$.
3. Show that $\|u \odot v\|_{\mathscr{C}^{\alpha+\beta}} \lesssim \|u\|_{\mathscr{C}^\alpha}\|v\|_{\mathscr{C}^\beta}$ for all $u \in \mathscr{C}^\alpha(\mathbb{R}^n)$ and $v \in \mathscr{C}^\beta(\mathbb{R}^n)$ whenever $\alpha, \beta \in \mathbb{R}$ satisfy $\alpha + \beta > 0$.
4. Assume that $\alpha + \beta > 0$. Show that the mapping $(f, g) \in C^\alpha(\mathbb{R}^n) \times C^\beta(\mathbb{R}^n) \to f\cdot g \in C^{\alpha+\beta}(\mathbb{R}^n)$ is continuous.

Exercise 4.49 [524] Assume that $\alpha < 1$ and $\beta, \gamma \in \mathbb{R}$ satisfy $\beta + \gamma > 0$ and $\alpha + \beta + \gamma > 0$. Then show that $\|C(u, v, w)\|_{\mathscr{C}^{\alpha+\beta+\gamma}} \lesssim \|u\|_{\mathscr{C}^\alpha}\|v\|_{\mathscr{C}^\beta}\|w\|_{\mathscr{C}^\gamma}$ for all $u \in \mathscr{C}^\alpha(\mathbb{R}^n)$, $v \in \mathscr{C}^\beta(\mathbb{R}^n)$ and $w \in \mathscr{C}^\gamma(\mathbb{R}^n)$ by mimicking the proof of Theorem 4.35, where $C(u, v, w) = (u \preceq v) \odot w - u(v \odot w)$.

4.3.3 *Characteristic Function of the Upper Half Plane as a Pointwise Multiplier*

We consider the characteristic function of the upper half plane as a pointwise multiplier; this type of multiplier will be of importance because it allows extension by zero from the half space to the whole space.

Theorem 4.40 (Characteristic function of the upper half space $\mathbb{R}^n_+$ as a pointwise multiplier) *Let $0 < p, q \le \infty$ and $s \in \mathbb{R}$. Furthermore, assume*

$$\max\left(\frac{1}{p} - 1, \frac{n}{p} - n\right) < s < \frac{1}{p}.$$

Set $G \equiv \chi_{\mathbb{R}^n_+}$. Then $G \cdot f$ makes sense and $\|G \cdot f\|_{A^s_{pq}} \lesssim \|f\|_{A^s_{pq}}$ for $f \in A^s_{pq}(\mathbb{R}^n)$.

Proof Concentrate on Triebel–Lizorkin spaces. We use the following notation for a point $x \in \mathbb{R}^n$; $x = (x', x_n) \in \mathbb{R}^{n-1} \times \mathbb{R}$. We may assume $f \in C^\infty_{\mathrm{c}}(\mathbb{R}^n)$ using mollification and truncation. Furthermore, Theorem 4.37 reduces the matter to proving $\|G(\psi_J f)\|_{A^s_{pq}} \lesssim \|\psi_J f\|_{A^s_{pq}}$ for all $f \in A^s_{pq}(\mathbb{R}^n)$ with the constant uniform over J, where $\psi \in \mathscr{S}(\mathbb{R}^n)$ and $\psi_J = \psi(2^{-J}\star)$.

Here and below, let $f \in C^\infty_{\mathrm{c}}(\mathbb{R}^n)$. Set $g \equiv \chi_{(0,\infty)} \in L^\infty(\mathbb{R}_{x_n})$. To simplify the notation, we regard elements in $\mathscr{S}'(\mathbb{R}_{x_n})$ as those in $\mathscr{S}'(\mathbb{R}^n)$. Choose $\tau \in C^\infty(\mathbb{R}_{x_n})$ so that $\chi_{(-1,1)} \le \tau \le \chi_{(-2,2)}$ and decompose: $g(x) = \tau(x_n)g(x) + (1 - \tau(x_n))g(x)$ for $x \in \mathbb{R}^n$. Theorem 4.37 yields $\|(1-\tau)g \cdot f\|_{A^s_{pq}} \lesssim \|f\|_{A^s_{pq}}$ for $f \in A^s_{pq}(\mathbb{R}^n)$. Thus, by replacing τg with g and defining $G(x) \equiv g(x_n)$ for $x \in \mathbb{R}^n$, we need to prove

$$\|G \cdot f\|_{A^s_{pq}} \lesssim \|f\|_{A^s_{pq}}, \quad f \in A^s_{pq}(\mathbb{R}^n).$$

Since g is already replaced with τg, Example 4.1 shows that $g \in B^{\frac{1}{v}}_{v\infty}(\mathbb{R})$ for all $v \in (0, \infty)$.

Choose a function $\kappa \in C^\infty_{\mathrm{c}}(\mathbb{R})$ so that $\chi_{(-1,1)} \le \kappa \le \chi_{(-2,2)}$. Define

$$\tau \equiv \overbrace{\kappa \otimes \cdots \otimes \kappa}^{n-1 \text{ times}} \in C^\infty_{\mathrm{c}}(\mathbb{R}^{n-1}), \quad \psi \equiv \overbrace{\kappa \otimes \cdots \otimes \kappa}^{n \text{ times}} \in C^\infty_{\mathrm{c}}(\mathbb{R}^n)$$

$$\rho \equiv \kappa - \kappa_{-1} \in C^\infty_{\mathrm{c}}(\mathbb{R}), \quad \varphi \equiv \psi - \psi_{-1} \in C^\infty_{\mathrm{c}}(\mathbb{R}^n).$$

Then $\varphi_j(D)G = \rho_j(D)g$.

We estimate $f \odot g$, $f \succeq g$ and $f \preceq g$. For $f \odot g$, let $0 < \theta \ll 1$.

$$\|f \odot g\|_{F^s_{pq}}{}^{\theta}$$

$$\lesssim \left\| \sum_{j=1}^{\infty} \psi_4(D)[\varphi_j(D)f \cdot \varphi_j(D)g] \right\|_p^{\theta} + \left\| \sum_{j=1}^{\infty} 2^{ks} \varphi_{k+4}(D)[\varphi_j(D)f \cdot \varphi_j(D)g] \right\|_{L^p(\ell^q)}^{\theta}.$$

In view of the size of support, we have

$$\|f \odot g\|_{F^s_{pq}}{}^{\theta} \lesssim \left\| \sum_{j=1}^{\infty} \psi_4(D)[\varphi_j(D)f \cdot \varphi_j(D)g] \right\|_p^{\theta}$$

$$+ \left\| \sum_{j=1}^{\infty} 2^{ks} \varphi_{k+4}(D)[\varphi_{j+k}(D)f \cdot \varphi_{j+k}(D)g] \right\|_{L^p(\ell^q)}^{\theta} =: \mathrm{I} + \mathrm{II}.$$

Instead of estimating the $L^p(\mathbb{R}^n)$-norm directly, we will consider the intermediate $L^p(\mathbb{R}_{x_n})$-norm. Let $u < p$. Theorem 1.53 yields

$$\mathrm{II} = \left\| \left(\sum_{k=0}^{\infty} 2^{ksq} |\varphi_{k+4}[\varphi_{j+k}(D)f \cdot \varphi_{j+k}(D)g]|^q \right)^{\frac{1}{q}} \right\|_{L^p(\mathbb{R}_{x_n})}$$

$$\lesssim \left\| \sum_{k=0}^{\infty} \exp(2^{-k-10} \star_n i) \varphi_{k+4}[\varphi_{j+k}(D)f \cdot \varphi_{j+k}(D)g] \right\|_{F^s_{pq}(\mathbb{R}_{xn})}.$$

Set $\omega_0 \equiv s + \dfrac{1}{u} - \dfrac{1}{p}$. Then by Theorem 4.11,

$$\mathrm{II} \lesssim \left\| \sum_{k=0}^{\infty} \exp(2^{-k-10} \star_n i) \varphi_{k+4}[\varphi_{j+k}(D)f \cdot \varphi_{j+k}(D)g] \right\|_{B^{\omega_0}_{up}(\mathbb{R}_{xn})}$$

$$\lesssim \left(\sum_{k=0}^{\infty} 2^{k\omega_0 p} \|\varphi_{k+4}[\varphi_{j+k}(D)f \cdot \varphi_{j+k}(D)g]\|_{L^u(\mathbb{R}_{xn})}^p \right)^{\frac{1}{p}}.$$

Hence adding the estimate for I, we obtain:

$$\|f \odot g\|_{F^s_{pq}}{}^{\theta} \lesssim \sum_{j=1}^{\infty} \left(\sum_{k=0}^{\infty} 2^{k\omega_0 p} \|\varphi_{k+4}[\varphi_{j+k}(D)f \cdot \varphi_{j+k}(D)g]\|_{L^p_{x'} L^u_{xn}}^p \right)^{\frac{\theta}{p}}.$$

Here

$$\|F(\star', \star_n)\|_{L^p_{x'} L^u_{x_n}} = \| \, \|F(\star', \star_n)\|_{L^u_{x_n}} \, \|_{L^p_{x'}} \tag{4.115}$$

and $L^p_{x'} L^u_{x_n}(\mathbb{R}^n)$ is the mixed Lebesgue space of all the measurable functions F for which $\|F(\star', \star_n)\|_{L^p_{x'} L^u_{x_n}}$ is finite.

Now we have to consider two cases: $p > 1$ and $p \le 1$. When $p > 1$, we choose u slightly larger than 1; by the Minkovski inequality, we have

$$\begin{aligned}
\|f \odot g\|_{F^s_{pq}}{}^{\theta} &\lesssim \sum_{j=1}^{\infty} \left(\sum_{k=0}^{\infty} 2^{k\omega_0 p} \|\varphi_{j+k}(D) f \cdot \varphi_{j+k}(D) g\|^p_{L^p_{x'} L^u_{x_n}} \cdot \right)^{\frac{\theta}{p}} \\
&= \sum_{j=1}^{\infty} 2^{-k\omega_0 \theta} \left(\sum_{k=j}^{\infty} 2^{k\omega_0 p} \|\varphi_k(D) f \cdot \varphi_k(D) g\|^p_{L^p_{x'} L^u_{x_n}} \right)^{\frac{\theta}{p}} \\
&\lesssim \left(\sum_{k=0}^{\infty} 2^{k\omega_0 p} \|\varphi_k(D) f \cdot \varphi_k(D) g\|^p_{L^p_{x'} L^u_{x_n}} \right)^{\frac{\theta}{p}}.
\end{aligned}$$

Hence since $g \in B^{\frac{1}{u}}_{u\infty}(\mathbb{R})$, we have

$$\begin{aligned}
\|f \odot g\|_{F^s_{pq}} &\lesssim \left(\sum_{k=0}^{\infty} 2^{k\left(s-\frac{1}{p}\right)p} \|\varphi_k(D) f\|^p_{L^p(L^\infty(\mathbb{R}_{x_n}), \mathbb{R}^{n-1}_{x'})} \right)^{\frac{1}{p}} \\
&= \left\| \left(\sum_{k=0}^{\infty} 2^{k\left(s-\frac{1}{p}\right)p} \|\varphi_k(D) f\|^p_{L^\infty(\mathbb{R}_{x_n})} \right)^{\frac{1}{p}} \right\|_{L^p(\mathbb{R}^{n-1}_{x'})}.
\end{aligned}$$

Furthermore, Theorem 1.53 yields

$$\|f \odot g\|_{F^s_{pq}} \lesssim \left\| \left\| \sum_{k=0}^{\infty} \exp(2^{-k-10} \star_n i) \varphi_k(D) f \right\|_{B^{s-1/p}_{p\infty}(\mathbb{R}_{x_n})} \right\|_{L^p(\mathbb{R}^{n-1}_{x'})}.$$

By Theorem 4.11,

$$\begin{aligned}
\|f \odot g\|_{F^s_{pq}} &\lesssim \left\| \left\| \sum_{k=0}^{\infty} \exp(2^{-k-10} \star_n i) \varphi_k(D) f \right\|_{F^s_{pq}(\mathbb{R}_{x_n})} \right\|_{L^p(\mathbb{R}^{n-1}_{x'})} \\
&\lesssim \|\psi(D) f\|_p + \|2^{ks} \varphi_k(D) f\|_{L^p(\ell^q)} \lesssim \|f\|_{F^s_{pq}}.
\end{aligned}$$

Thus, the estimate of $f \odot g$ with $p \geq 1$ is valid.

Let $p < 1$. Then by Corollary 1.7 or more precisely (1.165),

$$\begin{aligned}
&\|\varphi_{k+4}(D)[\varphi_{j+k}(D)f \cdot \varphi_{j+k}(D)g]\|_{L^p_{x'}L^u_{x_n}} \\
&\lesssim 2^{jn\left(\frac{1}{u}-1\right)} \| \, \|\varphi_{k+4}(D) \cdot [\varphi_{j+k}(D)f \cdot \varphi_{j+k}(D)g](x - \star)\|_{L^u(\mathbb{R}^n)} \, \|_{L^p_{x'}L^u_{x_n}} \\
&\lesssim 2^{jn\left(\frac{1}{u}-1\right)} \|\varphi_{j+k}(D)f \cdot \varphi_{j+k}(D)g\|_{L^p_{x'}L^u_{x_n}}.
\end{aligned}$$

Set $\omega \equiv s + \dfrac{1}{u} - \dfrac{1}{p}$ for u slightly less than p. Since $s > \frac{n}{p} - n$, $\omega > 0$. Hence

$$\begin{aligned}
\|f \odot g\|_{F^s_{pq}}{}^{\theta} &\lesssim \sum_{j=1}^{\infty} 2^{jn\left(\frac{1}{u}-1\right)} \left(\sum_{k=0}^{\infty} 2^{k\omega p} \|\varphi_{j+k}(D)f \cdot \varphi_{j+k}(D)g\|^p_{L^p_{x'}L^u_{x_n}} \right)^{\frac{\theta}{p}} \\
&\lesssim \left(\sum_{k=0}^{\infty} 2^{k\omega p} \|\varphi_k(D)f \cdot \varphi_k(D)g\|^p_{L^p_{x'}L^u_{x_n}} \right)^{\frac{\theta}{p}}.
\end{aligned}$$

We can go through the same argument as in the case where $1 < p < \infty$. Thus, the estimate of $f \odot g$ is valid.

The estimate of $f \succeq g$ is simple; Theorem 1.53 yields

$$\|f \succeq g\|_{A^s_{pq}} \lesssim \sup_{j \in \mathbb{N}} \left\| \sum_{k=1}^{j} \varphi_k(D)g \right\|_{\infty} \|f\|_{A^s_{pq}} \lesssim \|g\|_{\infty} \|f\|_{A^s_{pq}} \leq \|f\|_{A^s_{pq}}.$$

Finally, we estimate $f \preceq g$. Concentrate on Triebel–Lizorkin spaces. If we invoke Theorem 1.53 and Fubini's theorem (Theorem 1.3),

$$\begin{aligned}
\|f \preceq g\|_{F^s_{pq}} &\lesssim \left\| \left(\sum_{k=5}^{\infty} 2^{ksq} \left| \left(\sum_{j=1}^{k-4} \varphi_j(D)f \right) \varphi_k(D)g \right|^q \right)^{\frac{1}{q}} \right\|_p \\
&= \left\| \left\| \left(\sum_{k=5}^{\infty} 2^{ksq} \left| \left(\sum_{j=1}^{k-4} \varphi_j(D)f \right) \varphi_k(D)g \right|^q \right)^{\frac{1}{q}} \right\|_{L^p(\mathbb{R}_{x_n})} \right\|_{L^p(\mathbb{R}^{n-1}_{x'})}.
\end{aligned}$$

We now freeze $x_1, x_2, \dots, x_{n-1}$. Since $f \in C_c^\infty(\mathbb{R}^n)$, we consider the Fourier transform with respect to the single variable x_n to the function $\left(\sum_{j=1}^{k-4} \varphi_j(D) f\right) \varphi_k(D) g$.

Then the Fourier transform has the support in $[-2^{k+10}, 2^{k+10}] \setminus [-2^{k-10}, 2^{k-10}]$, since g does not contain $x_1, \dots, x_{n-1}$. Hence by virtue of Theorem 1.53,

$$\|f \preceq g\|_{F^s_{pq}} \simeq \left\| \left\| \sum_{k=5}^{\infty} \left(\sum_{j=1}^{k-4} \varphi_j(D) f(\star', \star_n) \right) \varphi_k(D) g \right\|_{F^s_{pq}(\mathbb{R}_{x_n})} \right\|_{L^p(\mathbb{R}^{n-1}_{x'})}.$$

Furthermore, by letting $t \gg 1$, we set $u \equiv \dfrac{pt}{p+t}$. Theorem 4.11 yields:

$$\|f \preceq g\|_{F^s_{pq}} \lesssim \left\| \left\| \sum_{k=5}^{\infty} \left(\sum_{j=1}^{k-4} \varphi_j(D) f(\star', \star_n) \right) \varphi_k(D) g \right\|_{B^{s+\frac{1}{t}}_{up}(\mathbb{R}_{x_n})} \right\|_{L^p(\mathbb{R}^{n-1}_{x'})}.$$

By the triangle inequality

$$\|f \preceq g\|_{F^s_{pq}}$$
$$\lesssim \left(\left\| \sum_{k=5}^{\infty} 2^{k\left(s+\frac{1}{t}\right)p} \left\| \left(\sum_{j=1}^{k-4} \varphi_j(D) f(\star', \star_n) \right) \varphi_k(D) g \right\|_{L^u(\mathbb{R}_{x_n})}^p \right\|_{L^1(\mathbb{R}^{n-1}_{x'})} \right)^{\frac{1}{p}}.$$

Since $g \in B^{1/t}_{t\infty}(\mathbb{R}_{x_n})$,

$$\|f \preceq g\|_{F^s_{pq}} \lesssim \left\| \left(\sum_{k=5}^{\infty} 2^{k\left(s+\frac{1}{t}-\frac{1}{p}\right)p} \left\| \left(\sum_{j=1}^{k-4} \varphi_j(D) f(\star', \star_n) \right) \right\|_{L^t(\mathbb{R}_{x_n})}^p \right)^{\frac{1}{p}} \right\|_{L^p(\mathbb{R}^{n-1}_{x'})}$$
$$\simeq \left\| \left(\sum_{k=1}^{\infty} 2^{k\left(s+\frac{1}{t}-\frac{1}{p}\right)p} \left\| \left(\sum_{j=1}^{k} \varphi_j(D) f(\star', \star_n) \right) \right\|_{L^t(\mathbb{R}_{x_n})}^p \right)^{\frac{1}{p}} \right\|_{L^p(\mathbb{R}^{n-1}_{x'})}.$$

When $p \le 1$, we use the p-triangle inequality:

$$\|f \preceq g\|_{F^s_{pq}} \simeq \left\| \left(\sum_{k=1}^{\infty} 2^{k\left(s+\frac{1}{t}-\frac{1}{p}\right)p} \sum_{j=1}^{k} \|\varphi_j(D) f(\star', \star_n)\|_{L^t(\mathbb{R}_{x_n})}^p \right)^{\frac{1}{p}} \right\|_{L^p(\mathbb{R}^{n-1}_{x'})}$$

and Theorem 4.11 yields

$$\|f \preceq g\|_{F^s_{pq}} \lesssim \left\| \left\| \sum_{k=1}^{\infty} \exp(2^{-k-10} \star_n i) f(\star', \star_n) \right\|_{B^{s+\frac{1}{t}-\frac{1}{p}}_{tp}(\mathbb{R}_{x_n})} \right\|_{L^p(\mathbb{R}^{n-1}_{x'})}$$

$$\lesssim \left\| \left\| \sum_{k=1}^{\infty} \exp(2^{-k-10} \star_n i) f(\star', \star_n) \right\|_{F^s_{pq}(\mathbb{R}_{x_n})} \right\|_{L^p(\mathbb{R}^{n-1}_{x'})} \lesssim \|f\|_{F^s_{pq}}.$$

Meanwhile, when $1 \le p < \infty$, choose t so that $\omega \equiv s + \dfrac{1}{t} - \dfrac{1}{p} > 0$. Then

$$\|f \preceq g\|_{F^s_{pq}} \lesssim \left\| \left(\sum_{k=1}^{\infty} \left(\sum_{j=1}^{k} 2^{k\omega} \left\|\varphi_j(D) f(\star', \star_n)\right\|_{L^t(\mathbb{R}_{x_n})} \right)^p \right)^{\frac{1}{p}} \right\|_{L^p(\mathbb{R}^{n-1}_{x'})}$$

$$= \left\| \left(\sum_{k=1}^{\infty} \left(\sum_{j=1}^{k} 2^{(k-j)\omega + j\omega} \left\|\varphi_j(D) f(\star', \star_n)\right\|_{L^t(\mathbb{R}_{x_n})} \right)^p \right)^{\frac{1}{p}} \right\|_{L^p(\mathbb{R}^{n-1}_{x'})}$$

and hence

$$\|f \preceq g\|_{F^s_{pq}} = \left\| \left(\sum_{k=1}^{\infty} \sum_{j=1}^{k} 2^{\frac{k-j}{2}\omega p + j\omega p} \|\varphi_j(D) f(\star', \star_n)\|_{L^t(\mathbb{R}_{x_n})}^p \right)^{\frac{1}{p}} \right\|_{L^p(\mathbb{R}^{n-1}_{x'})}$$

$$= \left\| \left(\sum_{j=1}^{\infty} \sum_{k=j}^{\infty} 2^{\frac{k-j}{2}\omega p + j\omega p} \|\varphi_j(D) f(\star', \star_n)\|_{L^t(\mathbb{R}_{x_n})}^p \right)^{\frac{1}{p}} \right\|_{L^p(\mathbb{R}^{n-1}_{x'})}.$$

We calculate the right-hand side:

$$\|f \preceq g\|_{F^s_{pq}} \lesssim \left\| \left(\sum_{j=1}^{\infty} 2^{j\omega p} \|\varphi_j(D) f(\star', \star_n)\|_{L^t(\mathbb{R}_{x_n})}^p \right)^{\frac{1}{p}} \right\|_{L^p(\mathbb{R}^{n-1}_{x'})}.$$

If we use the 1-dimensional Besov norm, we obtain

$$\|f \preceq g\|_{F^s_{pq}} \simeq \left\| \left\| \sum_{j-1}^{\infty} \exp(2^{k+20}x_n i)\varphi_j(D)f(\star', \star_n) \right\|_{B^{\omega}_{tp}(\mathbb{R}_{x_n})} \right\|_{L^p(\mathbb{R}^{n-1}_{x'})}$$

$$\lesssim \left\| \left\| \sum_{j=1}^{\infty} \exp(2^{k+20}x_n i)\varphi_j(D)f(\star', \star_n) \right\|_{F^s_{pq}(\mathbb{R}_{x_n})} \right\|_{L^p(\mathbb{R}^{n-1}_{x'})}$$

$$\lesssim \|f\|_{F^s_{pq}}$$

by Hölder's inequality and Theorem 4.11. Thus, the estimate for $f \preceq g$ is valid.

Exercises

Exercise 4.50 Let $\psi \in C^\infty_c(\mathbb{R}^n)$ be such that

$$\sum_{l \in \mathbb{Z}^n} \psi(\star - l) \equiv 1.$$

The self-similar space $A^s_{pq:\mathrm{selfs}}(\mathbb{R}^n)$ collects all $f \in \mathscr{S}'(\mathbb{R}^n)$ for which

$$\|f\|_{A^s_{pq:\mathrm{selfs}}} \equiv \sup_{l \in \mathbb{Z}^n, k \in \mathbb{Z}} \|\psi(\star - l) f(2^{-k}\star)\|_{A^s_{pq}}$$

is finite. Show that $A^s_{pq:\mathrm{selfs}}(\mathbb{R}^n)$ does not depend on the choise of ψ as a set.

Exercise 4.51 [110, Proposition 6.5] Let $1 \le p \le \infty$.

1. Show that $\chi_{\mathbb{R}^n_+} \in B^{n/p-n}_{p\infty}(\mathbb{R}^n) \cap \dot{B}^{n/p-n}_{p\infty}(\mathbb{R}^n)$.
2. Show that $\chi_{\mathbb{R}^n_+} \in B^s_{pq}(\mathbb{R}^n) \cap \dot{B}^s_{pq}(\mathbb{R}^n)$ when $0 < q \le \infty$ and $s \in (n/p-n, 1/p)$.

Here we consider the realization when we consider homogeneous spaces.

Exercise 4.52 [1074, Theorem 6.5] Let $0 < p, q \le \infty$ and $s \in (\sigma_p, 1)$ when $A = B$, and let $0 < p < \infty$, $0 < q \le \infty$ and $s \in (\sigma_{pq}, 1)$ when $A = F$. Show that

$$\|\psi f\|_{A^s_{pq}} \lesssim (\|\psi\|_\infty + \|\psi\|_{\mathrm{Lip}})\|f\|_{A^s_{pq}}$$

for all $\psi \in \mathrm{Lip}(\mathbb{R}^n)$ and $f \in A^s_{pq}(\mathbb{R}^n)$. Hint: Refine the atomic decomposition and mimic the proof of Theorem 4.37.

Exercise 4.53 (Spaces of Kudrjavcev type) Let $\widetilde{\psi}, \widetilde{\varphi} \in \mathscr{S}(\mathbb{R}^n)$ satisfy

$$\widetilde{\psi}(\xi) = \begin{cases} 1 & (|\xi| \le 4), \\ 0 & (|\xi| \ge 8), \end{cases} \quad \widetilde{\varphi}(\xi) = \begin{cases} 1 & (2 \le |\xi| \le 4), \\ 0 & (|\xi| \le 1 \text{ or } |\xi| \ge 8) \end{cases}$$

as usual. For $j \in \mathbb{N}$, we write $\widetilde{\varphi}_j \equiv \widetilde{\varphi}(2^{-j}\star)$. Let $0 < p, q \le \infty$ and $s, \mu \in \mathbb{R}$. Define the *spaces of Kudrjavcev type* $A^s_{p,q,\mu}(\mathbb{R}^n)$ as the set of all $f \in \mathscr{S}'(\mathbb{R}^n)$ for which

$$\|f\|_{A^s_{p,q,\mu}} \equiv \|\widetilde{\psi} f\|_{A^s_{pq}} + \left(\sum_{j=1}^{\infty} 2^{j\mu p} (\|\widetilde{\varphi}_j f\|_{A^s_{pq}})^p \right)^{1/p}$$

is finite. Then using Theorem 4.37, show that the space $A^s_{p,q,\mu}(\mathbb{R}^n)$ does not depend as a set on $\widetilde{\psi}$ and $\widetilde{\varphi}$ appearing here. Note: Triebel investigated function spaces of Kudrjavcev type in [1043, 1044].

4.3.4 Applications: Div-Curl Lemma, Kato–Ponce Inequality and Riemann–Stieltjes Integral

The idea developed in this section is strong and provides us with many important estimates. Here we take up the div-curl lemma, the Kato–Ponce inequality and the class for which the Riemann–Stieltjes integral can be defined.

4.3.4.1 A Bilinear Estimate and An Application to the Div-Curl Lemma

So far, we have been considering the pointwise product. However, the same idea can be applied to more general bilinear operators of the form $(f, g) \in A^s_{p_1 q}(\mathbb{R}^n) \times A^s_{p_2 q}(\mathbb{R}^n) \mapsto f \cdot R_j g \in A^s_{pq}(\mathbb{R}^n)$, say, where R_j is the j-th Riesz transform. Since R_j is bounded on $L^p(\mathbb{R}^n)$ as long as $1 < p < \infty$, if $1 < p_2 < \infty$, we have control of $f \cdot R_j g$. This is just a mere combination of the pointwise product and the boundedness of the Riesz transform. However, the sum of the singular integral operators can be resonant to provide us with stronger estimates. As an example we take up the div-curl lemma. First in a similar way to Lemma 4.16 we can prove the following:

Corollary 4.6 *Let* $\{f_j\}_{j=-\infty}^{\infty} \in L^1(\ell^2, \mathbb{R}^n)$. *Assume that* $f_j \in \mathscr{S}'_{B(2^j)}(\mathbb{R}^n) \cap \mathscr{P}_0(\mathbb{R}^n)^\perp$ *for each* $j \in \mathbb{Z}$. *Then* $\left\| \sum_{j=-\infty}^{\infty} f_j \right\|_{H^1} \lesssim \|\{f_j\}_{j=-\infty}^{\infty}\|_{L^1(\ell^2)}$.

As we have mentioned above, we can say more than the mere pointwise product $f R_j g$. The following is a typical example:

Theorem 4.41 *Let* $1 < p < \infty$. *Then for* $k = 1, 2, \dots, n$, $f \in L^p(\mathbb{R}^n)$ *and* $g \in L^{p'}(\mathbb{R}^n)$, *we have*

$$\|g \cdot R_k f + R_k f \cdot g\|_{H^1} \lesssim \|g\|_p \|f\|_{p'}.$$

A couple of remarks may be in order.

Remark 4.5 Let $1 < p < \infty$.

1. Since R_k preserves $L^p(\mathbb{R}^n)$ and $L^{p'}(\mathbb{R}^n)$, $g \cdot R_k f, R_k f \cdot g \in L^1(\mathbb{R}^n)$. Hence, it matters that the sum $g \cdot R_k f + R_k f \cdot g$ belongs to $H^1(\mathbb{R}^n)$ although there is no guarantee that $g \cdot R_k f, R_k f \cdot g \in H^1(\mathbb{R}^n)$. One can thus say that Theorem 4.41 is a resonant estimate.
2. If $f, g \in \mathscr{S}_\infty(\mathbb{R}^n)$, then $(g \cdot R_k f + R_k f \cdot g) \perp \mathscr{P}_0(\mathbb{R}^n)$ by the Fourier transform, which is a key fact for resonance. By density and the remark above, this observation carries over to $f \in L^p(\mathbb{R}^n)$ and $g \in L^{p'}(\mathbb{R}^n)$. See Exercise 4.56.

Proof Due to symmetry, $f \preceq R_k g + R_k f \preceq g$ and $f \succeq R_k g + R_k f \succeq g$ can be estimated in the same manner; we can consider $f \preceq R_k g$, $R_k f \preceq g$, $f \succeq R_k g$ and $R_k f \succeq g$ separately and similarly to the usual operators $f \preceq g$ and $f \succeq g$. To handle $f \odot R_k g + R_k f \odot g$ and $f \odot^{\pm 3} R_k g + R_k f \odot^{\pm 3} g$, we use Corollary 4.6 instead of using the atomic decomposition, where a delicate "resonant" condition is used.

As an application of Theorem 4.41, we prove the div-curl inequality.

Theorem 4.42 *Let* $1 < p < \infty$, *and let* $F = (F_1, F_2, \dots, F_n) \in L^p(\mathbb{R}^n)^n$ *and* $G = (G_1, G_2, \dots, G_n) \in L^{p'}(\mathbb{R}^n)^n$ *satisfy*

$$\operatorname{div}(G) \equiv \sum_{k=1}^{n} \frac{\partial G_k}{\partial x_k} = 0 \in \mathscr{S}'(\mathbb{R}^n)$$

and

$$\operatorname{rot}(F) = \operatorname{curl}(F) \equiv \left(\frac{\partial F_k}{\partial x_j} - \frac{\partial F_j}{\partial x_k} \right)_{j,k=1}^{n} = 0 \in \mathscr{S}'(\mathbb{R}^n)^{n^2}.$$

Then $\|G \cdot F\|_{H^1} \lesssim \|G\|_{(L^{p'})^n} \|F\|_{(L^p)^n}$.

Proof Take $\varphi \in C^\infty_c(\mathbb{R}^n \setminus \{0\})$ so that $\sum_{j=-\infty}^{\infty} \varphi_j = \chi_{\mathbb{R}^n \setminus \{0\}}$. Then $G \cdot F = \sum_{l=1}^{n} \sum_{j=-\infty}^{\infty} G_l \cdot \varphi_j(D) F_l$ in the topology of $L^1(\mathbb{R}^n)$. Thus, by the Fatou property of $H^1(\mathbb{R}^n)$ we have only to prove

$$\left\| \sum_{l=1}^{n} \sum_{j=-J}^{J} G_l \cdot \varphi_j(D) F_l \right\|_{H^1} \lesssim \|G\|_{(L^{p'})^n} \|F\|_{(L^p)^n}$$

with the constant independent of J. Here and below we fix $J \in \mathbb{N}$. Since J is finite, we may assume $\text{supp}(\mathscr{F} F_l)$ is a compact set in $\mathbb{R}^n \setminus \{0\}$ for each $l = 1, 2, \ldots, n$.

Since F is rotation-free, there exists $f \in \mathscr{S}'(\mathbb{R}^n)$ such that $F = \text{grad}(f)$ from Example 1.21. Using the Littlewood–Paley decomposition, we may assume that $\text{supp}(f)$ is a compact set in $\mathbb{R}^n \setminus \{0\}$, so that $\|F\|_{(L^p)^n} \sim \|(-\Delta)^{\frac{1}{2}} f\|_p$ (see Exercise 4.57). Inserting $F = \text{grad}(f)$ into $G \cdot F$, we have

$$G \cdot F \simeq \sum_{j=1}^{n} G_j \partial_{x_j} f \simeq \sum_{j=1}^{n} G_j R_j [(-\Delta)^{\frac{1}{2}} f]$$

Since G is divergence free,

$$G \cdot F \simeq \sum_{j=1}^{n} \{G_j \cdot R_j [(-\Delta)^{\frac{1}{2}} f] + R_j G_j \cdot (-\Delta)^{\frac{1}{2}} f\}$$

Thus, from Lemma 4.41, we conclude

$$\|G \cdot F\|_{H^1} \leq \sum_{j=1}^{n} \|G_j\|_{p'} \|(-\Delta)^{\frac{1}{2}} f\|_p \lesssim \sum_{j=1}^{n} \|G_j\|_{p'} \|\partial_{x_i} f\|_p \lesssim \|G\|_{(L^{p'})^n} \|F\|_{(L^p)^n}.$$

We have a similar results for the endpoint case.

Theorem 4.43 *Let $F \in H^1(\mathbb{R}^n)^n$ and $G \in L^\infty(\mathbb{R}^n)^n$ fulfill* $\text{div}(G) = 0 \in \mathscr{S}'(\mathbb{R}^n)$ *and* $\text{rot}(F) = 0 \in \mathscr{S}'(\mathbb{R}^n)^{n^2}$. *Then $G \cdot F \in H^1(\mathbb{R}^n)$ and $\|G \cdot F\|_{H^1} \lesssim \|G\|_{(L^\infty)^n} \|F\|_{(H^1)^n}$.*

Proof Let $F = (F_1, F_2, \ldots, F_n)$ and $G = (G_1, G_2, \ldots, G_n)$. We may assume that $\text{supp}(\mathscr{F} F_l) \cup \text{supp}(\mathscr{F} G_l)$ is a compact subset in $\mathbb{R}^n \setminus \{0\}$ for each $l = 1, 2, \ldots, n$. Since F is rotation-free, $F = \text{grad}(f)$ for some $f \in \mathscr{S}'(\mathbb{R}^n)$ such that $\text{supp}(\mathscr{F} f)$ is a compact subset in $\mathbb{R}^n \setminus \{0\}$ (see Exercise 4.57).

We fix $\tau \in C_c^\infty(\mathbb{R}^n) \setminus \mathscr{P}(\mathbb{R}^n)^\perp$. Then we have

$$\begin{aligned}\int_{\mathbb{R}^n} G(y) \cdot F(y) t^{-n} \tau\left(\frac{x-y}{t}\right) \text{d}y &= \int_{\mathbb{R}^n} t^{-n} \tau\left(\frac{x-y}{t}\right) G(y) \cdot \text{grad}(f)(x) \text{d}y \\ &= \int_{\mathbb{R}^n} \text{div}\left(t^{-n} \tau\left(\frac{x-\star}{t}\right) G\right)(y) \cdot f(y) \text{d}y.\end{aligned}$$

Let $Y_{x,t,k} \equiv \sum_{l=1}^{n} R_k R_l \left[t^{-n} \tau \left(\frac{x-\star}{t} \right) \cdot G \right]$ for $k = 1, 2, \ldots, n$. Then we have

$$\operatorname{div} \left(t^{-n} \tau \left(\frac{x-\star}{t} \right) \cdot G \right) \simeq \operatorname{div}(Y_{x,t,1}, Y_{x,t,2}, \ldots, Y_{x,t,n}). \tag{4.116}$$

Let $r > 1$. Since G is divergence free, we have

$$\sum_{j,l=1}^{n} \|\partial_l Y_{x,t,jk}\|_r \lesssim_r \left\| \operatorname{div} \left[t^{-n} \tau \left(\frac{x-\star}{t} \right) \cdot G \right] \right\|_r = \left\| t^{-1-n} \operatorname{grad} \tau \left(\frac{x-\star}{t} \right) \cdot G \right\|_r$$

by the L^r-boundedness of the Riesz transform, so that

$$\sum_{j,l=1}^{n} \|\partial_l Y_{x,t,jk}\|_r \lesssim t^{-1-\frac{n}{r}} \|G\|_{(L^\infty)^n}. \tag{4.117}$$

Meanwhile from (4.116),

$$\begin{aligned}
&\int_{\mathbb{R}^n} G(y) \cdot F(y) t^{-n} \tau \left(\frac{x-y}{t} \right) \mathrm{d}y \\
&= \int_{\mathbb{R}^n} \operatorname{div}(Y_{x,t,1}, Y_{x,t,2}, \ldots, Y_{x,t,n})(y) f(y) \mathrm{d}y \\
&= -\int_{\mathbb{R}^n} (Y_{x,t,1}, Y_{x,t,2}, \ldots, Y_{x,t,n})(y) \cdot \operatorname{grad}(f)(y) \mathrm{d}y \\
&= -\int_{\mathbb{R}^n} (Y_{x,t,1}, Y_{x,t,2}, \ldots, Y_{x,t,n})(y) \cdot F(y) \mathrm{d}y.
\end{aligned}$$

Thus, it suffices to show that

$$\left\| \sup_{t>0} \left| \int_{\mathbb{R}^n} Y_{x,t,k}(y) F_k(y) \mathrm{d}y \right| \right\|_1 \lesssim \|G\|_{(L^\infty)^n} \|F_k\|_{H^1} \tag{4.118}$$

for any $F_k \in H^1(\mathbb{R}^n)$ after we fix $k = 1, 2, \ldots, n$. Using the atomic decomposition, we may further assume that $F_k \in L^\infty(\mathbb{R}^n) \cap \mathscr{P}_0(\mathbb{R}^n)^\perp$ satisfies $|F_k| \le |Q|^{-1} \chi_Q$. We estimate

$$Z_{x,t,k} \equiv \left| \int_{\mathbb{R}^n} Y_{x,t,k}(y) F_k(y) \mathrm{d}y \right|$$

via a couple of steps. We note that

$$Z_{x,t,k} \le \sum_{l=1}^{n} \left| \int_{\mathbb{R}^n} t^{-n} \tau \left(\frac{x-y}{t} \right) \cdot G_k(y) \cdot R_k R_l F_k(y) \mathrm{d}y \right| \lesssim \sum_{l=1}^{n} \|G_l\|_\infty M[R_k R_l F_k](x).$$

If $x \notin 5Q$ and $0 < 3t < |x - c(Q)|$, then we have

$$Z_{x,t,k} \leq \sum_{l=1}^{n} \left| \int_{\mathbb{R}^n} t^{-n} \tau\left(\frac{x-y}{t}\right) \cdot G_k(y) \cdot R_k R_l F_k(y) \mathrm{d}y \right|$$
$$\lesssim \sum_{l=1}^{n} \|G_l\|_\infty |Q|^{-1} M\chi_Q(x)^{\frac{n+1}{n}}.$$

If $x \notin 5Q$ and $3t > |x - c(Q)|$, then we have

$$Z_{x,t,k} \leq m_Q(|Y_{x,t,k} - m_Q(Y_{x,t,k})|) \leq \ell(Q) \|\nabla Y_{x,t,k}\|_{(L^1)^n}$$

by the Poincaré–Wirtinger inequality. Thus, by (4.117),

$$Z_{x,t,k} \leq m_Q(|Y_{x,t,k} - m_Q(Y_{x,t,k})|) \lesssim \frac{\ell(Q)^{\frac{1}{2}}}{t^{n+\frac{1}{2}}} \lesssim |Q|^{-1} M\chi_Q(x)^{\frac{1}{2n}+1}.$$

Thus, in total, we have

$$\sup_{t>0} Z_{x,t,k} \lesssim \sum_{l=1}^{n} \|G_l\|_\infty M[R_k R_l F_k](x)\chi_{5Q}(x) + \sum_{l=1}^{n} \|G_l\|_\infty |Q|^{-1} M\chi_Q(x)^{\frac{1}{2n}+1}.$$

If we integrate this estimate over $\mathbb{R}^n$ and use the Hölder inequality, we obtain

$$\left\| \sup_{t>0} Z_{x,t,k} \right\|_1 \lesssim \|G\|_{(L^\infty)^n} \|M[R_k R_l F_k]\|_{L^2(5Q)} |Q|^{\frac{1}{2}} + \sum_{l=1}^{n} \|G_l\|_\infty.$$

Since $\|M[R_k R_l F_k]\|_{L^2(5Q)} \leq \|M[R_k R_l F_k]\|_2 \lesssim \|R_k R_l F_k\|_2 \lesssim \|F_k\|_2 \leq |Q|^{-\frac{1}{2}}$ thanks to the boundedness of the Riesz transform, we obtain (4.118) as required.

4.3.4.2 Kato–Ponce Inequality

This section will refine the Hölder inequality for function spaces. Let $1 < p, p_1, p_2 < \infty$ satisfy $\dfrac{1}{p} = \dfrac{1}{p_1} + \dfrac{1}{p_2}$. Let $f \in W^{1,p_1}(\mathbb{R}^n)$, $g \in W^{1,p_2}(\mathbb{R}^n)$. According to the Hölder inequality we obtained in this section and $W^{1,p}(\mathbb{R}^n) \approx F^1_{p2}(\mathbb{R}^n)$, we have

$$\|f \cdot g\|_{W^{1,p}} \lesssim \|f\|_{W^{1,p_1}} \|g\|_{W^{1,p_2}}.$$

However, it is easy to see that

$$\|\nabla(f \cdot g)\|_{(L^p)^n} \le \|f\|_{p_1}\|\nabla g\|_{(L^{p_2})^n} + \|\nabla f\|_{(L^{p_1})^n}\|g\|_{p_2}.$$

Thus, we have $\|f \cdot g\|_{W^{1,p}} \lesssim \|f\|_{W^{1,p_1}}\|g\|_{p_2} + \|f\|_{p_1}\|g\|_{W^{1,p_2}}$. Here we generalize this observation.

Theorem 4.44 *Let* $0 < p, p_1, p_2, p_3, p_4, q \le \infty$ *satisfy* $\frac{1}{p} = \frac{1}{p_1} + \frac{1}{p_2} = \frac{1}{p_3} + \frac{1}{p_4}$.

1. *If* $0 < p, p_2, p_4 < \infty$ *and* $s > \sigma_p$*, then for all* $f \in B^s_{p_1 q}(\mathbb{R}^n) \cap F^0_{p_3 2}(\mathbb{R}^n)$ *and* $g \in F^0_{p_2 2}(\mathbb{R}^n) \cap B^s_{p_4 q}(\mathbb{R}^n)$, $\|f \cdot g\|_{B^s_{pq}} \lesssim \|f\|_{B^s_{p_1 q}}\|g\|_{F^0_{p_2 2}} + \|f\|_{F^0_{p_3 2}}\|g\|_{B^s_{p_4 q}}$.
2. *If* $0 < p_1, p_2, p_3, p_4 < \infty$ *and* $s > \sigma_{pq}$*, then for all* $f \in F^s_{p_1 q}(\mathbb{R}^n) \cap F^0_{p_3 2}(\mathbb{R}^n)$ *and* $g \in F^0_{p_2 2}(\mathbb{R}^n) \cap F^s_{p_4 q}(\mathbb{R}^n)$, $\|f \cdot g\|_{F^s_{pq}} \lesssim \|f\|_{F^s_{p_1 q}}\|g\|_{F^0_{p_2 2}} + \|f\|_{F^0_{p_3 2}}\|g\|_{F^s_{p_4 q}}$.

Proof We concentrate on Triebel–Lizorkin spaces; Besov spaces are somewhat easier to handle. We can handle $f \succeq g$ and $f \preceq g$ as we did before. For example, similar to (4.109) we have $\|f \preceq g\|_{F^s_{pq}} \lesssim \left\| \sup_{j \in \mathbb{N}_0} |\psi_j(D) f| \right\|_{p_3} \|2^{js}\varphi_j(D) g\|_{L^{p_4}(\ell^q)}$.

Since $\left\| \sup_{j \in \mathbb{N}_0} |\psi_j(D) f| \right\|_{p_3} \sim \|f\|_{h^{p_3}} \sim \|f\|_{F^0_{p_3 2}}$, the estimate for $f \succeq g$ is valid.

To handle $f \odot g$, we use Lemma 4.16. Using Lemma 4.16 and $\|f\|_{F^0_{p_3 \infty}} \le \|f\|_{F^0_{p_3 2}}$, we have

$$\|f \odot g\|_{F^s_{pq}} \lesssim \|\{2^{js}\varphi_j(D) f \cdot \varphi_j(D) g\}_{j=1}^\infty\|_{L^p(\ell^q)} \lesssim \|f\|_{F^0_{p_3 2}}\|g\|_{F^s_{p_4 q}}.$$

4.3.4.3 Applications to the Riemann–Stieltjes Integral

One of the fundamental facts on the integration theory is that $\int_a^b f(t) \mathrm{d}g(t)$ is understood as $\int_a^b f(t) g'(t) \mathrm{d}t$ when f is a suitable function and g is a $C^1(\mathbb{R}^n)$-function on $[a, b]$. The next theorem will allow us to relax this smoothness assumption.

Theorem 4.45 *Let* $\alpha > 0$ *and* $\beta \in \mathbb{R}$*. Assume in addition that* $\alpha + \beta > 0$*. Then there exists a bilinear mapping*

$$(f, g) \in B^\alpha_{\infty\infty}(\mathbb{R}^n) \times B^\beta_{\infty\infty}(\mathbb{R}^n) \mapsto f \cdot g \in B^{\min(\alpha,\beta)}_{\infty\infty}(\mathbb{R}^n),$$

or equivalently,

$$(f, g) \in \mathscr{C}^\alpha(\mathbb{R}^n) \times \mathscr{C}^\beta(\mathbb{R}^n) \mapsto f \cdot g \in \mathscr{C}^{\min(\alpha,\beta)}(\mathbb{R}^n)$$

such that this is a natural extension of the pointwise product if restricted to $\mathscr{S}(\mathbb{R}^n) \times \mathscr{S}(\mathbb{R}^n)$.

The proof of this theorem is similar to other results on the pointwise multipliers. See Exercise 4.58.

Here we present an example of the applications related to stochastic differential equations.

Example 4.7 One of the convenient ways to define the Riemann–Stieltjes (indefinite) integral

$$\int_a^b f(t)\,dg(t) = \int_a^b f(t)g'(t)\,du \quad (f \in B^\alpha_{\infty\infty}(\mathbb{R}),\ g \in B^\beta_{\infty\infty}(\mathbb{R}))$$

is the application of Theorem 4.45, where α, β are arbitrary real numbers satisfying $\alpha + \beta > 1$. If $\alpha > 0$ and $\beta > 1$, then there is no difficulty in defining this integral. However, when $\beta \le 1$, we need some tricks using Theorem 4.45. First, according to Theorem 4.45, we have $h = f \cdot g' \in B^{\min(\alpha,\beta-1)}_{\infty\infty}(\mathbb{R})$. We can readily define

$$\int_a^b \psi(D)h(t)\,dt,$$

since $\psi(D)h \in \mathrm{BC}(\mathbb{R})$. Meanwhile, using the Fourier transform, we can construct $k \in B^{\min(\alpha+1,\beta)}_{\infty\infty}(\mathbb{R})$ with

$$k' = (1 - \psi(D))h, \quad k(a) = 0$$

since $\mathscr{F}^{-1}[(1 - \psi(D))h]$ is supported away from a neighborhood of the origin. Thus, we can define

$$\int_a^b f(t)\,dg(t) = \int_a^b \psi(D)h(t)\,dt + k(b).$$

This is an application of the paraproduct to the Young integral [1172]. See [524, Section 3.2] for more details together with an account [524, p. 3].

Exercises

Exercise 4.54 Let $k, l = 1, 2, \dots, n$, and let $1 < p, p_1, p_2 < \infty$ satisfy $\dfrac{1}{p} = \dfrac{1}{p_1} + \dfrac{1}{p_2}$. Show that $\|\partial_{x_k} f \cdot \partial_l g\|_p \lesssim \|\Delta f\|_{p_1}\|g\|_{p_2} + \|f\|_{p_1}\|\Delta g\|_{p_2}$.

Exercise 4.55 Let $f_1, f_2, \ldots, f_n \in H^1(\mathbb{R}^n)$ and $\partial_{x_j} f_k = \partial_{x_k} f_j$ for $j, k = 1, 2, \ldots, n$, so that there exists $f \in \mathscr{S}'(\mathbb{R}^n)$ such that $\partial_{x_j} f = f_j$ for $j = 1, 2, \ldots, n$. Show that $f \in \dot{F}^1_{12}(\mathbb{R}^n)$, more precisely $[f] \in \dot{F}^1_{12}(\mathbb{R}^n)$.

Exercise 4.56 Let $f, g \in \mathscr{S}_\infty(\mathbb{R}^n)$. Show that $(g \cdot R_k f + R_k f \cdot g) \perp \mathscr{P}_0(\mathbb{R}^n)$.

Exercise 4.57 Suppose we have $\{f_j\}_{j=1}^n$ which is rotation-free. Assume in addition that the frequency support of each f_j is a compact set in $\mathbb{R}^n \setminus \{0\}$. Then show that there exists $f \in \mathscr{S}'(\mathbb{R}^n)$ such that $f_j = \partial_{x_j} f$ without recourse to Example 1.21. Hint: Use the Fourier transform.

Exercise 4.58 Prove Theorem 4.45 by mimicking the proof of Theorem 4.36 – the BBB case.

Textbooks in Sect. 4.3

Paraproduct

See [94, p. 35, p. 104] and [66, Section 16.7] for paraproducts. See the book by Edmunds and Triebel [23, Section 2.4].

Pointwise Multiplication: Theorem 4.37

See [99, Section 2.8.2] and [100, Section 42] for pointwise multiplication, as well as a new textbook [58]. Triebel took approaches via atoms [100, Section 4.2.1] and via local means [100, Section 4.2.2]. See the textbook [83] and the paper [958].

Fubini Property: Theorem 4.49

See [99, Section 2.5.13] for the Fubini property for $A^s_{pq}(\mathbb{R}^n)$ with $0 < p < \infty$, $0 < q \le \infty$ and $s > \dfrac{n}{\min(p, q)}$.

4.4 Fundamental Theorems on Function Spaces

In this section, we are interested in the properties of the function spaces, in particular diffeomorphism and boundedness of the trace operator and Fubini's property. As is easily seen from the viewpoint of partial differential equations, the first two theorems and the pointwise multiplier are essential when we develop a theory of function spaces on domains. When we consider elliptic linear differential equations on domains, we need the pointwise multiplier, the diffeomorphism (see Sect. 4.4.1), the trace operator, and the extension operator (see Sect. 4.4.2). Due the complicated definitions it is not so easy to prove them directly. We also investigate Fubini's property of Triebel–Lizorkin spaces (see Sect. 4.4.3). Our weapon, the decomposition theory, plays a key role.

We fully utilize atomic decomposition and molecular decomposition to study the boundedness of these operators.

We remark that Theorems 4.35, 4.36, and 4.37 are called the *key theorems in function spaces*.

4.4.1 *Diffeomorphism*

The technique of straightening the boundary via the diffeomorphisms is absolutely necessary when we solve the Laplace problem C^∞-domains. Here we consider isomorphisms which diffeomorphisms induce.

4.4.1.1 Regular Diffeomorphism

Recall that $\mathscr{B}^M(\mathbb{R}^n) = \mathrm{BC}^M(\mathbb{R}^n)$ is the set of all $C^M(\mathbb{R}^n)$-functions such that all the derivatives up to order M are bounded.

Definition 4.18 (Regular diffeomorphism) A $C^M(\mathbb{R}^n)$-diffeomorphism $\psi : \mathbb{R}^n \to \mathbb{R}^n$ is said to be *regular*, if ψ and its inverse belong to $\mathrm{BC}^M(\mathbb{R}^n)$.

Example 4.8

1. The diffeomorphism $f(t) \equiv e^t - \mathrm{e}^{-t}$ is not regular, while $g(t) \equiv 2t - \sin t$ is regular.
2. A diffeomorphism $h : \mathbb{R}^n \to \mathbb{R}^n$ is said to be fiber preserving if

$$h(x) = (h_1(x_1), h_2(x_2), \dots, h_n(x_n))$$

where each h_j is a strictly increasing function. If in addition $h_j \in \mathscr{B}C^M(\mathbb{R}^n)$ and $\inf h'_j > 0$ for each $j = 1, 2 \dots, n$, then h is regular.

We need the following geometric observation: Now that ψ is bi-Lipschitz; that is, both ψ and ψ^{-1} are Lipschitz continuous, there exist $I \in \mathbb{N}$ and $D > 0$ that depend on ψ such that, for each ν, $\mathbb{Z}^n$ is partitioned into $M_1^\nu, M_2^\nu, \dots, M_I^\nu$, and there exist injections

$$\iota_1^\nu : M_1^\nu \to \mathbb{Z}^n, \iota_2^\nu : M_2^\nu \to \mathbb{Z}^n, \dots, \iota_I^\nu : M_I^\nu \to \mathbb{Z}^n \tag{4.119}$$

such that, for all $i = 1, 2, \dots, I$, $\nu \in \mathbb{N}_0$ and multi-index $\beta \in \mathbb{N}_0{}^n$, we have

$$\psi^{-1}(\mathrm{supp}((\beta\mathrm{qu})_{\nu m})) \subset D\,Q_{\nu\,\iota_i^\nu(m)}.$$

Note that ι_i^ν is a bijection from M_i^ν to $\iota_i^\nu(M_i^\nu)$.

Lemma 4.17 *We have a bound independent of* ν*;* $I \lesssim 1$.

Proof Since ψ, ψ^{-1} are bounded functions,

$$(\|\nabla[\psi^{-1}]\|_\infty)^{-1}|x-y| \le |\psi(x)-\psi(y)| \le \|\nabla\psi\|_\infty|x-y|.$$

Here and below, we set

$$C_0 \equiv \max\{\|\nabla\psi\|_{(L^\infty)^n}, \|\nabla[\psi^{-1}]\|_{(L^\infty)^n}\}.$$

Then $C_0{}^{-1}|x-y| \le |\psi(x)-\psi(y)| \le C_0|x-y|$. Fix $m_0 \in \mathbb{Z}^n$. Once we show that the number of $m \in \mathbb{Z}^n$ satisfying $\psi^{-1}(\mathrm{supp}(\beta\mathrm{qu})_{\nu m_0}) \subset DQ_{\nu m}$ is bounded, we obtain the estimate of I from above.

The diameter of $\psi^{-1}(\mathrm{supp}(\beta\mathrm{qu})_{\nu m_0})$, which is given by

$$\sup\{|x-y| \,:\, x,y \in \psi^{-1}(\mathrm{supp}(\beta\mathrm{qu})_{\nu m_0})\},$$

satisfies $2^{\nu+1}r \times C_0$. Let $|\mathrm{supp}((\beta\mathrm{qu})_{\nu m})| \le (2r)^n$. Then $\{DQ_{\nu m}\}_{m\in\mathbb{Z}^n}$ overlaps at most $[D+2]^n$ times. That is,

$$\sum_{m\in\mathbb{Z}^n} \chi_{DQ_{\nu m}} \le [D+2]^n.$$

Hence the set $\psi^{-1}(\mathrm{supp}(\beta\mathrm{qu})_{\nu m_0})$ intersects at most $[\sqrt{n}\times 2^1 r\times C_0+1]^n\times[D+2]^n$ cubes belonging to $\{DQ_{\nu m}\}_{m\in\mathbb{Z}^n}$. Thus, we conclude $I \le [\sqrt{n}\times 2^1 r\times C_0+1]^n \times [D+2]^n$ and that the proof is complete.

4.4.1.2 Diffeomorphism on Function Spaces

If a function ψ is a regular $C^M(\mathbb{R}^n)$-diffeomorphism, then so is its inverse thanks to symmetry. Recall that $\mathscr{B}^M(\mathbb{R}^n)$ is given in Definition 2.3.

Theorem 4.46 (Diffeomorphism) *Suppose that we have parameters* p, q, s *satisfying* $0 < p, q \le \infty$ *and* $s \in \mathbb{R}$*. Let* $M \gg 1$ *be an integer. Also assume that* ψ *is a regular* $C^M(\mathbb{R}^n)$*-diffeomorphism. Then the composition mapping* $\varphi \in \mathscr{B}^M(\mathbb{R}^n) \mapsto \varphi\circ\psi \in \mathscr{B}^M(\mathbb{R}^n)$ *induces a continuous mapping* $f \in A^s_{pq}(\mathbb{R}^n) \mapsto f\circ\psi \in A^s_{pq}(\mathbb{R}^n)$ *and, for all* $f \in A^s_{pq}(\mathbb{R}^n)$*, we have* $\|f\circ\psi\|_{A^s_{pq}} \lesssim_\psi \|f\|_{A^s_{pq}}$.

Proof Concentrate on the Triebel–Lizorkin space $F^s_{pq}(\mathbb{R}^n)$. Suppose first that $s > \sigma_{pq}$. Then maintaining the notation of Theorem 4.8, let $\rho > r$. We will invoke the quarkonial decomposition; see Theorem 4.8. We expand

$$f = \sum_{\beta \in \mathbb{N}_0{}^n} \sum_{\nu=0}^{\infty} \sum_{m \in \mathbb{Z}^n} \lambda^{\beta}_{\nu m} (\beta \mathrm{qu})_{\nu m}.$$

Here the coefficient $\lambda = \{\lambda^{\beta}_{\nu m}\}_{\beta \in \mathbb{N}_0{}^n, (\nu,m) \in \mathbb{N}_0 \times \mathbb{Z}^n}$ satisfies

$$\|\lambda\|_{\mathbf{f}_{pq},\rho} \lesssim \|f\|_{F^s_{pq}}. \tag{4.120}$$

For $i = 1, 2, \ldots, I$ and $\nu \in \mathbb{N}_0$, we write $\theta_i^\nu \equiv (\iota_i^\nu)^{-1}$.

Set

$$\lambda^{\beta,i}_{\nu \bar{m}} \equiv \begin{cases} \lambda^{\beta}_{\nu \theta_i^\nu(\bar{m})} & \bar{m} \in \iota_i^\nu(M_i^\nu), \\ 0 & \text{otherwise,} \end{cases}$$

and

$$(\beta \mathrm{qu})^i_{\nu \bar{m}} \equiv \begin{cases} (\beta \mathrm{qu})_{\nu \theta_i^\nu(\bar{m})} \circ \psi & \bar{m} \in \iota_i^\nu(M_i^\nu), \\ 0 & \text{otherwise.} \end{cases}$$

Then we want to define

$$f \circ \psi = \sum_{i=1}^{I} \sum_{\beta \in \mathbb{N}_0{}^n} \sum_{\nu=0}^{\infty} \sum_{\overline{m} \in \mathbb{Z}^n} \lambda^{\beta,i}_{\nu \overline{m}} (\beta \mathrm{qu})^i_{\nu \overline{m}}. \tag{4.121}$$

To justify the definition, we need to verify that the infinite sum defining (4.121) makes sense as an element in $F^s_{pq}(\mathbb{R}^n)$. Set

$$f^{i,\beta} \equiv \sum_{\nu=0}^{\infty} \sum_{\overline{m} \in \mathbb{Z}^n} \lambda^{\beta,i}_{\nu \overline{m}} (\beta \mathrm{qu})^i_{\nu \overline{m}}$$

for $\beta \in \mathbb{N}_0{}^n$ and $i = 1, 2, \ldots, I$. Then the atomic decomposition theorem (Theorem 4.2) yields

$$\|f^{i,\beta}\|_{F^s_{pq}} \lesssim_\psi 2^{(r+\varepsilon)|\beta|} \|\{\lambda^{\beta,i}_{\nu \overline{m}}\}_{\nu \in \mathbb{N}_0, \overline{m} \in \mathbb{Z}^n}\|_{\mathbf{f}_{pq}} \lesssim_\psi 2^{(r+\varepsilon)|\beta|} \|\lambda^\beta\|_{\mathbf{f}_{pq}},$$

where $\lambda^\beta \equiv \{\lambda^{\beta}_{\nu m}\}_{(\nu,m) \in \mathbb{N}_0 \times \mathbb{Z}^n}$.

By the estimate of the quarkonial decompositions, $\|f^{i,\beta}\|_{F^s_{pq}} \lesssim_\psi \|f^\beta\|_{F^s_{pq}} \lesssim_\psi 2^{-\delta|\beta|}\|f\|_{F^s_{pq}}$, where $\delta = \rho - r - \varepsilon$.

Hence if we use the $\min(p, q, 1)$-triangle inequality to the sum

$$f \circ \psi = \sum_{i=1}^{I} \sum_{\beta \in \mathbb{N}_0{}^n} f^{i,\beta},$$

then $\|f \circ \psi\|_{F^s_{pq}} \lesssim \|f\|_{F^s_{pq}}$. Hence $f \circ \psi \in F^s_{pq}(\mathbb{R}^n)$.

Exercises

Exercise 4.59 Let A be an $n \times n$ matrix with nonzero determinant. Using Theorem 4.46, show that the transform $\varphi_A(x) \equiv Ax$ ($x \in \mathbb{R}^n$) induces an isomorphim $f \in A^s_{pq}(\mathbb{R}^n) \mapsto f \circ \varphi_A \in A^s_{pq}(\mathbb{R}^n)$ for any $0 < p, q \le \infty$ and $s \in \mathbb{R}$.

Exercise 4.60 Let Ω be a bounded C^∞-domain. Show that the multiplier $f \in A^s_{pq}(\mathbb{R}^n) \to \chi_\Omega f \in A^s_{pq}(\mathbb{R}^n)$ is bounded as long as $f \in A^s_{pq}(\mathbb{R}^n) \to \chi_{\mathbb{R}^n_+} f \in A^s_{pq}(\mathbb{R}^n)$ is bounded using Theorems 4.46.

4.4.2 Trace Operator

Let $n \ge 2$. We consider the trace operator (the boundary-value operator)

$$\mathrm{Tr}_{\mathbb{R}^n} f(x') \equiv f(x', 0) \quad (x' \in \mathbb{R}^{n-1}),$$

which is defined initially for $f \in \mathscr{S}(\mathbb{R}^n)$. As before, $x', m', \ldots$ stand for points in $\mathbb{R}^{n-1}$; $x \mapsto x'$ is an operation to omit the n-th coordinate, and we decompose $x \in \mathbb{R}^n$ into $x = (x', x_n)$.

4.4.2.1 The Boundedness of the Trace Operator: Subcritical Case

One of the big achievements in the theory of Besov spaces and Triebel–Lizorkin spaces is that we can completely describe the image of the trace operator as the following theorem shows:

Theorem 4.47 (Trace operator) *Let* $0 < p, q \le \infty$. *Consider the trace mapping*

$$\mathrm{Tr}_{\mathbb{R}^n} : f \in \mathscr{S}(\mathbb{R}^n) \mapsto f(\star', 0_n) \in \mathscr{S}(\mathbb{R}^{n-1}). \tag{4.122}$$

1. *Suppose that the parameters* $s \in \mathbb{R}$ *satisfy* $s > \dfrac{1}{p} + (n-1)\left(\dfrac{1}{p} - 1\right)_+$.

(a) *The trace operator* $\mathrm{Tr}_{\mathbb{R}^n}$ *extends to a continuous surjection from* $B^s_{pq}(\mathbb{R}^n)$ *to* $B^{s-1/p}_{pq}(\mathbb{R}^{n-1})$.
(b) *The trace operator* $\mathrm{Tr}_{\mathbb{R}^n}$ *extends to a continuous surjection from* $F^s_{pq}(\mathbb{R}^n)$ *to* $F^{s-1/p}_{pp}(\mathbb{R}^{n-1})$.

2. *Suppose the parameters* $s \in \mathbb{R}$ *and* $k \in \mathbb{N}$ *satisfy* $s > k + \dfrac{1}{p} + (n-1)\left(\dfrac{1}{p}-1\right)_+$.

(a) *For* $g_0 \in B^{s-1/p}_{pq}(\mathbb{R}^{n-1})$, $g_1 \in B^{s-1/p-1}_{pq}(\mathbb{R}^{n-1}), \ldots, g_k \in B^{s-1/p-k}_{pq}(\mathbb{R}^{n-1})$, *we can find* $f \in B^s_{pq}(\mathbb{R}^n)$ *such that*

$$\mathrm{Tr}_{\mathbb{R}^n}(f) = g_0, \mathrm{Tr}_{\mathbb{R}^n}(\partial_{x_n} f) = g_1, \ldots, \mathrm{Tr}_{\mathbb{R}^n}({\partial_{x_n}}^k f) = g_k.$$

(b) *For* $g_0 \in F^{s-1/p}_{pp}(\mathbb{R}^{n-1})$, $g_1 \in F^{s-1/p-1}_{pp}(\mathbb{R}^{n-1}), \ldots, g_k \in F^{s-1/p-k}_{pp}(\mathbb{R}^{n-1})$, *we can find* $f \in F^s_{pq}(\mathbb{R}^n)$ *such that*

$$\mathrm{Tr}_{\mathbb{R}^n}(f) = g_0, \mathrm{Tr}_{\mathbb{R}^n}(\partial_{x_n} f) = g_1, \ldots, \mathrm{Tr}_{\mathbb{R}^n}({\partial_{x_n}}^k f) = g_k. \tag{4.123}$$

The case where s is large enough as in Theorem 4.47 is called subcritical. The critical case will be dealt with later.

One of the great achievements in the theory of function spaces is that we can completely specify the range of the trace operator thanks to the parameters s and q. In particular, the new parameter q plays a key role.

In Triebel–Lizorkin spaces, the result is a little complicated; the value q changes into p. To understand this complicated situation, we need the following lemma:

Lemma 4.18 *Let* $0 < p < \infty$ *and* $0 < q \le \infty$. *Then for a doubly indexed complex sequence* $\lambda = \{\lambda_{\nu m}\}_{(\nu,m)\in\mathbb{N}_0\times\mathbb{Z}^n}$, *we have*

$$\|\{\lambda_{\nu\,(m',0)}\}_{(\nu,m')\in\mathbb{N}_0\times\mathbb{Z}^{n-1}}\|_{f_{pp}(\mathbb{R}^{n-1})} \sim \|\{\delta_{m_n 0}\lambda_{\nu m}\}_{(\nu,m)\in\mathbb{N}_0\times\mathbb{Z}^n}\|_{f_{pq}(\mathbb{R}^n)}.$$

Note that we wrote $\|\{\lambda_{\nu\,(m',0)}\}_{(\nu,m')\in\mathbb{N}_0\times\mathbb{Z}^{n-1}}\|_{\mathbf{f}_{pp}(\mathbb{R}^{n-1})}$ to specify the set on which the functions are defined since there are a couple of possibilities.

Proof Since each term does not contain $\lambda_{\nu m}$ as long as $m_n \neq 0$, we can suppose $\lambda_{\nu m} = 0$ unless $m_n = 0$. First take η so that $\eta \equiv 2^{-1}\min(1,p,q) < p, q$. Furthermore, define λ' so that $\lambda' \equiv \{\lambda_{\nu(m',0)}\}_{(\nu,m')\in\mathbb{N}_0\times\mathbb{Z}^{n-1}}$. We start to prove $\|\lambda'\|_{\mathbf{f}_{pp}(\mathbb{R}^{n-1})} \lesssim \|\lambda\|_{\mathbf{f}_{pq}(\mathbb{R}^n)}$.

To simplify the notation we define

$$\Lambda'_\nu(x') \equiv \sum_{m'\in\mathbb{Z}^{n-1}} \lambda_{\nu(m',0)}\chi^{(p^*)}_{\nu m'}(x'), \quad \Lambda_\nu(x) \equiv \sum_{m\in\mathbb{Z}^n} \lambda_{\nu m}\chi^{(p)}_{\nu m}(x) \quad (x'\in\mathbb{R}^{n-1}, x\in\mathbb{R}^n)$$

for $\nu \in \mathbb{N}_0$. We claim

$$\left(\int_{\mathbb{R}^{n-1}} \sum_{\nu=0}^{\infty} |\Lambda'_\nu(x')|^p \mathrm{d}x'\right)^{\frac{1}{p}} \lesssim_{p,q} \|\lambda\|_{\mathbf{f}_{pq}(\mathbb{R}^n)}.$$

To this end, for $\nu \in \mathbb{N}_0$ and $m' \in \mathbb{Z}^{n-1}$, define $\tilde{Q}_{\nu m'} \equiv Q_{\nu m'} \times [2^{-1-\nu}, 2^{-\nu}]$, the Whitney region of $Q_{\nu m'}$. Observe that

$$\begin{aligned}\int_{\mathbb{R}^{n-1}} \sum_{\nu=0}^{\infty} |\Lambda'_\nu(x')|^p \mathrm{d}x' &= 2\int_{\mathbb{R}^{n-1}} \left(\sum_{\nu=0}^{\infty} \left|\sum_{m' \in \mathbb{Z}^{n-1}} 2^{\frac{n}{p}} \lambda_{\nu m'} \chi_{\tilde{Q}_{\nu m'}}(x)\right|^p\right)^{\frac{p}{p}} \mathrm{d}x \\ &= 2\int_{\mathbb{R}^{n-1}} \left(\sum_{\nu=0}^{\infty} \left|\sum_{m' \in \mathbb{Z}^{n-1}} 2^{\frac{n}{p}} \lambda_{\nu m'} \chi_{\tilde{Q}_{\nu m'}}(x)\right|^q\right)^{\frac{p}{q}} \mathrm{d}x.\end{aligned}$$

Set $\eta_0 \equiv 2^{-1}\min(p, q, 1)$. Using the Hardy–Littlewood maximal operator, we have

$$\left|\sum_{m' \in \mathbb{Z}^{n-1}} 2^{\frac{n}{p}} \lambda_{\nu m'} \chi_{\tilde{Q}_{\nu m'}}\right| \lesssim M^{(\eta_0)}\left[\sum_{m' \in \mathbb{Z}^{n-1}} \lambda_{\nu m'} \chi^{(p)}_{\nu(m',0)}\right].$$

If we use the Fefferman–Stein vector-valued maximal inequality (see Theorem 1.49),

$$\begin{aligned}\left(\int_{\mathbb{R}^{n-1}} \sum_{\nu=0}^{\infty} |\Lambda'_\nu(x')|^p \mathrm{d}x'\right)^{\frac{1}{p}} &\lesssim \left\|M^{(\eta_0)}\left[\sum_{m' \in \mathbb{Z}^{n-1}} \lambda_{\nu m'} \chi^{(p)}_{\nu(m',0)}\right]\right\|_{L^p(\ell^q)} \\ &\lesssim \left\|\sum_{m' \in \mathbb{Z}^{n-1}} \lambda_{\nu m'} \chi^{(p)}_{\nu(m',0)}\right\|_{L^p(\ell^q)}.\end{aligned}$$

To prove the reverse inequality, we make use of the Fefferman–Stein vector-valued maximal inequality (Theorem 1.49) and

$$\left|\sum_{m' \in \mathbb{Z}^{n-1}} \lambda_{\nu m'} \chi^{(p)}_{\nu(m',0)}\right| \lesssim M^{(\eta_0)}\left[\sum_{m' \in \mathbb{Z}^{n-1}} 2^{\frac{\nu n}{p}} \lambda_{\nu m'} \chi_{\tilde{Q}_{\nu m'}}\right].$$

The proof being similar, we omit the details.

We prove Theorem 4.47. As usual, we deal with Triebel–Lizorkin spaces.

Proof (Step 1: The definition of $\mathrm{Tr}_{\mathbb{R}^n}$*)* We begin with a setup: the function ψ in the quarkonial decomposition, Theorem 4.8. We choose a smooth function $\mu : \mathbb{R} \to \mathbb{R}$ so that

$$\mathrm{supp}(\mu) \subset (-1, 1) \tag{4.124}$$

to define

$$\psi(x) \equiv \mu(x_1)\mu(x_2)\cdots\mu(x_n) \tag{4.125}$$

for $x \in \mathbb{R}^n$. We consider $f \in F^s_{pq}(\mathbb{R}^n)$ and decompose it into the sum of quarks according to Theorem 4.8: $f = \sum_{\beta\in\mathbb{N}_0{}^n} \sum_{\nu=0}^{\infty} \sum_{m\in\mathbb{Z}^n} \lambda^\beta_{\nu m}(\beta\mathrm{qu})_{\nu m}$. If we insert this quarkonial decomposition into $\mathrm{Tr}_{\mathbb{R}^n}(f)$, then we see that the natural candidate of the trace operator is

$$\mathrm{Tr}_{\mathbb{R}^n}(f) = \sum_{\beta\in\mathbb{N}_0{}^n} \sum_{\nu=0}^{\infty} \sum_{m\in\mathbb{Z}^n} \chi_{\{(0,0)\}}(\beta_n, m_n)\lambda^\beta_{\nu m}(\beta\mathrm{qu})_{\nu m}(\star', 0). \tag{4.126}$$

If $f \in \mathscr{S}(\mathbb{R}^n)$, then the convergence is uniform. Indeed, $\mathscr{S}(\mathbb{R}^n) \subset B^n_{11}(\mathbb{R}^n)$. Hence as is seen from the definitions of $\lambda^\beta_{\nu m}$, $(\beta\mathrm{qu})_{\nu m}$, we have

$$f = \sum_{\beta\in\mathbb{N}_0{}^n} \sum_{\nu=0}^{\infty} \sum_{m\in\mathbb{Z}^n} \lambda^\beta_{\nu m}(\beta\mathrm{qu})_{\nu m} \tag{4.127}$$

in $B^n_{11}(\mathbb{R}^n)$. Since $B^n_{11}(\mathbb{R}^n) \hookrightarrow B^0_{\infty 1}(\mathbb{R}^n) \hookrightarrow \mathrm{BUC}(\mathbb{R}^n)$, (4.126) converges uniformly when $f \in \mathscr{S}(\mathbb{R}^n)$.

In view of Lemma 4.18, let $\lambda^{\beta'} \equiv \{\lambda^{(\beta',0)}_{\nu(m',0)}\}_{(\nu,m')\in\mathbb{N}_0\times\mathbb{Z}^{n-1}}$. Then we have

$$\|\lambda^{\beta'}\|_{\mathbf{f}_{pp}} \lesssim \|\lambda^\beta\|_{\mathbf{f}_{pq}}.$$

See Exercise 4.62. So, we see that the definition of the trace mapping makes sense.

Proof (Step 2: Proof of the surjectivity of $\mathrm{Tr}_{\mathbb{R}^n}$ *and of (4.123))* The surjectivity of Theorem 4.47 is covered in (4.123) with $k = 0$. So, we concentrate on (4.123). Let $j = 1, 2, \ldots, k$. Choose $g_j \in F_{pp}^{s-j-\frac{1}{p}}(\mathbb{R}^{n-1})$. Decompose g_j into quarks (Theorem 4.8) to have

$$g_j = \sum_{\beta'\in\mathbb{N}_0{}^{n-1}} \sum_{\nu=0}^{\infty} \sum_{m'\in\mathbb{Z}^{n-1}} \lambda^{\beta'}_{\nu m'}(\beta'\mathrm{qu})_{\nu m'}. \tag{4.128}$$

Let $L \gg 1$. Then

$$v_j \equiv \sum_{\beta' \in \mathbb{N}_0^{n-1}} \sum_{\nu \in \mathbb{N}_0} \sum_{m' \in \mathbb{Z}^{n-1}} \frac{\lambda_{\nu m'}^{\beta'}}{(2L+j)! 2^{\nu(2L+j)}} ((\beta', 2L+j)\mathrm{qu})_{\nu(m',0)}$$

belongs to $F_{pq}^{s+2L}(\mathbb{R}^n)$ by Theorem 4.8 and Lemma 4.18. Set $h_j \equiv \partial_{x_n}{}^{2L} v_j \in F_{pq}^s(\mathbb{R}^n)$. That is, define $h_j \in F_{pq}^s(\mathbb{R}^n)$ by

$$h_j \equiv \sum_{\beta' \in \mathbb{N}_0{}^{n-1}} \sum_{\nu=0}^{\infty} \sum_{m' \in \mathbb{Z}^{n-1}} \frac{\lambda_{\nu m'}^{\beta'}}{(2L+j)!} \partial_{x_n}^{2L} [x_n{}^{2L+j}((\beta', 0)\mathrm{qu})_{\nu(m',0)}]. \tag{4.129}$$

Let $\delta \in \left(0, s-k-\frac{1}{p}-(n-1)\left(\frac{1}{p}-1\right)\right)$. Since $F_{pq}^{s-l}(\mathbb{R}^n) \hookrightarrow F_{p1}^{s-l-\delta}(\mathbb{R}^n)$, the right-hand side

$$\partial_{x_n}^l h_j = \frac{1}{(2L+j)!} \sum_{\beta' \in \mathbb{N}_0{}^{n-1}} \sum_{\nu=0}^{\infty} \sum_{m \in \mathbb{Z}^{n-1}} \lambda_{\nu m'}^{\beta'} \partial_{x_n}^{2L+l} [x_n{}^{2L+j}((\beta', 0)\mathrm{qu})_{\nu(m',0)}] \tag{4.130}$$

converges in the topology of $F_{p1}^{s-l-\delta}(\mathbb{R}^n)$. Since $\mathrm{Tr}_{\mathbb{R}^n}$ is shown to be continuous in $F_{p1}^{s-l-\delta}(\mathbb{R}^n)$, we can change the order of $\mathrm{Tr}_{\mathbb{R}^n}$ and the infinite sum. Furthermore, by using the Kronecker delta δ_{jl}, we have

$$\mathrm{Tr}_{\mathbb{R}^n}\left[\partial_{x_n}^{2L+l}[x_n{}^{2L+j}((\beta', 0)\mathrm{qu})_{\nu m'}(x')]\right] = \delta_{jl}(2L+j)!((\beta', 0)\mathrm{qu})_{\nu m'}(x'). \tag{4.131}$$

Hence it follows that

$$\begin{aligned}\mathrm{Tr}_{\mathbb{R}^n}[\partial_{x_n}^l h_j] &= \frac{\delta_{jl}}{(2L+j)!} \sum_{\beta \in \mathbb{N}_0^{n-1}} \sum_{\nu \in \mathbb{N}_0} \sum_{m' \in \mathbb{Z}^{n-1}} \lambda_{\nu m'}^{\beta'} \mathrm{Tr}_{\mathbb{R}^n}[\partial_{x_n}^{2L+l}(x_n^{2L+j}(\beta' \mathrm{qu})_{\nu m'})] \\ &= \delta_{jl}\, g_j.\end{aligned}$$

Thus, $f = \sum_{j=0}^{k} h_j$ is the desired element in Theorem 4.47 which satisfies (4.123).

Here once again we reduce what we have obtained in full generality to classical Sobolev spaces.

Example 4.9 Let $1 < p < \infty$ and $m \in \mathbb{N}$. The trace of the Sobolev space $W^{m,p}(\mathbb{R}^n) \approx F_{p2}^m(\mathbb{R}^n)$ is $B_{pp}^{m-1/p}(\mathbb{R}^{n-1})$ in the light of Theorem 4.47.

4.4.2.2 The Boundedness of the Trace Operator: Critical Case

We consider another trace theorem that does not fall under the scope of Theorem 4.47.

Lemma 4.19 *Let $0 < p < \infty$. Then any $h \in L^p(\mathbb{R}^n)$ can be decomposed:*

$$h = \sum_{\nu=0}^{\infty} \sum_{m \in \mathbb{Z}^n} \lambda_{\nu m} a_{\nu m}$$

in the topology of $L^p(\mathbb{R}^n)$.

Here the coefficients $\{\lambda_{\nu m}\}_{\nu \in \mathbb{N}_0, m \in \mathbb{Z}^n}$ and $\{a_{\nu m}\}_{\nu \in \mathbb{N}_0, m \in \mathbb{Z}^n} \subset C_c^\infty(\mathbb{R}^n)$ satisfy

$$\left(\sum_{\nu=1}^{\infty} \left(\sum_{m \in \mathbb{Z}^n} |\lambda_{\nu m}|^p \right)^{\frac{q}{p}} \right)^{\frac{1}{q}} \lesssim_q \|h\|_p \tag{4.132}$$

for all $0 < q < \infty$ and

$$|\partial^\alpha a_{\nu m}| \lesssim_\alpha 2^{-\nu\left(\frac{1-n}{p} - |\alpha|\right)} \chi_{Q_{\nu m}}. \tag{4.133}$$

Conversely, let a collection $\{a_{\nu m}\}_{(\nu,m) \in \mathbb{N}_0 \times \mathbb{Z}^n} \subset \mathscr{B}^0$ of functions and a doubly indexed sequence $\{\lambda_{\nu m}\}_{(\nu,m) \in \mathbb{N}_0 \times \mathbb{Z}^n}$. satisfy (4.133) *with $\alpha = 0$ and*

$$\sum_{\nu=1}^{\infty} \left(\sum_{m \in \mathbb{Z}^n} |\lambda_{\nu m}|^p \right)^{\frac{\min(1,p)}{p}} < \infty. \tag{4.134}$$

Then the sum $h = \sum_{\nu=0}^{\infty} \sum_{m \in \mathbb{Z}^n} \lambda_{\nu m} a_{\nu m}$ is convergent in $L^p(\mathbb{R}^n)$ and it satisfies

$$\|h\|_p \lesssim \left(\sum_{\nu=1}^{\infty} \left(\sum_{m \in \mathbb{Z}^n} |\lambda_{\nu m}|^p \right)^{\frac{\min(1,p)}{p}} \right)^{\frac{1}{\min(1,p)}}. \tag{4.135}$$

Proof Let $p_* \equiv \min(1, p)$. Approximate h with a simple function and define E_1 by

$$E_1 \equiv \sum_{m \in \mathbb{Z}^n} \lambda_{\nu_1 m} \chi_{\nu_1 m}, \quad \|E_1 - h\|_p \le 4^{-1/p_*} \|h\|_p. \tag{4.136}$$

By (4.136), $\|E_1\|_p^{p_*} \le \|E_1 - h\|_p^{p_*} + \|h\|_p^{p_*} \le 2\|h\|_p^{p_*}$. Choose a function $\kappa \in \mathscr{S}(\mathbb{R}^n)$ so that $0 \le \kappa \le \chi_{[0,1]^n}$ and that $\|\chi_{[0,1]^n} - \kappa\|_p \le 4^{-1/p_*}$. Next, set

$$\widetilde{E_1} \equiv \sum_{m \in \mathbb{Z}^n} \lambda_{\nu_1 m} \kappa(2^{\nu_1} \star - m).$$

Then since κ is supported in $[0, 1]^n$, we have

$$\begin{aligned}
\|\widetilde{E_1} - E_1\|_p &= \left(\sum_{m \in \mathbb{Z}^n} |\lambda_{\nu_1 m}|^p \int_{Q_{\nu_1 m}} |1 - \kappa(2^{\nu_1} x - m)|^p \mathrm{d}x \right)^{\frac{1}{p}} \\
&\le \left(4^{-p/p_*} \sum_{m \in \mathbb{Z}^n} |\lambda_{\nu_1 m}|^p |Q_{\nu_1 m}| \right)^{\frac{1}{p}} = 4^{-1/p_*} \|E_1\|_p.
\end{aligned}$$

Hence

$$\|h - \widetilde{E_1}\|_p^{p_*} \le \|E_1 - \widetilde{E_1}\|_p^{p_*} + \|h - E_1\|_p^{p_*} \le \frac{1}{4}\|E_1\|_p^{p_*} + \frac{1}{4}\|h\|_p^{p_*} = \frac{1}{2}\|h\|_p^{p_*}.$$

Thus, $\widetilde{E_1}$ is the element with the approximation level $1/2$. Namely, we can approximate h whose p_*-power of $L^p(\mathbb{R}^n)$-norm is halved. Likewise let $\nu_2 \ge \nu_1$ and choose a function

$$\widetilde{E_2} \equiv \sum_{m \in \mathbb{Z}^n} \lambda_{\nu_2 m} \kappa(2^{\nu_2} \star - m)$$

so that

$$\|h - \widetilde{E_1} - \widetilde{E_2}\|_p^{p_*} \le \frac{1}{2}\|h - \widetilde{E_1}\|_p^{p_*} \le \frac{1}{4}\|h\|_p^{p_*}.$$

If we repeat this procedure, we can find a sequence $\nu_1 < \nu_2 < \cdots$ of nonnegative integers, we have

$$h = \sum_{l=1}^{\infty} \sum_{m \in \mathbb{Z}^n} \lambda_{\nu_l m} \kappa(2^{\nu_l} \star - m).$$

When $\nu \in \mathbb{Z} \setminus \{\nu_l\}_{l=1}^{\infty}$, we set $\lambda_{\nu m} \equiv 0$ and $a_{\nu m} \equiv 0$ for each $m \in \mathbb{Z}^n$. Furthermore, let $a_{\nu_l m} \equiv \kappa(2^{\nu_l} \star - m)$ for $l \in \mathbb{N}$ and $m \in \mathbb{Z}^n$. Then we see that h admits the decomposition (4.132) and (4.133). For the converse assertion, we simply use the p_*-triangle inequality.

If we use this decomposition, we obtain the next theorem:

Theorem 4.48 ($B_{pq}^{1/p}(\mathbb{R}^n)$-boundedness of the trace operator) *Let $0 < p < \infty$, $0 < q \le \min(1, p)$. Then we can define the trace operator $\mathrm{Tr}_{\mathbb{R}^n} : B_{pq}^{1/p}(\mathbb{R}^n) \to L^p(\mathbb{R}^{n-1})$ so that it is continuous and surjective.*

Proof Theorem 4.47 allows us to define $\mathrm{Tr}_{\mathbb{R}^n} : B^{1/p}_{pq}(\mathbb{R}^n) \to L^p(\mathbb{R}^{n-1})$. Theorem 4.47 also shows that the trace operator is continuous. Let us prove the surjectivity. For $h \in L^p(\mathbb{R}^{n-1})$, we consider the decomposition $h = \sum_{\nu=0}^{\infty} \sum_{m' \in \mathbb{Z}^{n-1}} \lambda_{\nu m'} a_{\nu m'}$ as in Lemma 4.19.

Furthermore, for $L \gg 1$,

$$f(x) \equiv \frac{1}{L!} \sum_{\nu=0}^{\infty} \sum_{m' \in \mathbb{Z}^n} \lambda_{\nu m'} \partial_{x_n}^L (x_n^L \kappa(2^\nu x_n) a_{\nu m'}(x')) \quad (x = (x', x_n) \in \mathbb{R}^n)$$

using $\kappa \in \mathscr{S}(\mathbb{R})$ satisfying $\chi_{[-1,1]} \le \kappa \le \chi_{[-2,2]}$. Then f can be decomposed into atoms (Theorem 4.2), $f \in B^{1/p}_{pq}(\mathbb{R}^n)$. It is trivial from the definition of h that the trace of f is h. Thus, $\mathrm{Tr}_{\mathbb{R}^n} : B^{1/p}_{pq}(\mathbb{R}^n) \to L^p(\mathbb{R}^{n-1})$ is a surjection.

Theorems 4.47 and 4.48 completely determine the trace of the function spaces when s is large enough. Here we roughly describe the situation by using the Fourier transform.

Example 4.10 Let $n \ge 2$ and $s > \frac{1}{2}$. Then

$$\int_{\mathbb{R}^{n-1}} |\mathrm{Tr}_{\mathbb{R}^n} f(x', 0)|^2 \mathrm{d}x \lesssim \int_{\mathbb{R}^n} |(1 - \partial_{x_n}^2)^{\frac{s}{2}} f(x)|^2 \mathrm{d}x.$$

In fact,

$$\mathrm{Tr}_{\mathbb{R}^n} f(x) \simeq_n \int_{\mathbb{R}^{n-1}} \mathrm{e}^{ix' \cdot \xi'} \left\{ \int_{\mathbb{R}^{n-1}} \mathrm{e}^{-iy' \cdot \xi'} \left(\int_{\mathbb{R}^n} \mathrm{e}^{-i\xi_n \cdot y_n} f(y', y_n) \mathrm{d}y_n \right) \mathrm{d}y' \right\} \mathrm{d}\xi'.$$

Thus by the Plancherel theorem, we have

$$\|\mathrm{Tr}_{\mathbb{R}^n}\|^2_{L^2(\mathbb{R}^{n-1})} \simeq_n \int_{\mathbb{R}^{n-1}} \left| \int_{\mathbb{R}^n} \mathrm{e}^{-i\xi_n \cdot y_n} f(x', y_n) \mathrm{d}y_n \right|^2 \mathrm{d}x'.$$

By the Hölder inequality and $s > \frac{1}{2}$, we obtain the desired result.

Exercises

Exercise 4.61 Using Theorems 4.46 and 4.47, show that the trace operator $T_{\mathrm{diag}} : f \in A^s_{pq}(\mathbb{R}^n) \mapsto f(\star_1, \star_1, \ldots, \star_1) \in A^{s-(n-1)/p}_{pq}(\mathbb{R})$ is bounded.

Exercise 4.62 In the above proof of Lemma 4.18 and Theorem 4.47, show that, in the definition $\mathrm{Tr}_{\mathbb{R}^n} f$, we do not have to take into account the case $m_n \neq 0$ or $\beta_n \neq 0$. Hint: Use (4.124) and (4.125).

Exercise 4.63 Let $n \geq 2$ and $s > 1/2$. What is the surjective image by the trace operator $\mathrm{Tr}_{\mathbb{R}^n}$ of $H^s(\mathbb{R}^n)$? Hint: Use $H^s(\mathbb{R}^n) \approx F^s_{22}(\mathbb{R}^n) \approx B^s_{22}(\mathbb{R}^n)$ with equivalence of norms; see Theorem 2.6.

Exercise 4.64 [838, Lemma 2.4] Let $0 < p, q < \infty$. Assume that we have a disjoint collection $\{E_{j,k}\}_{j=1}^\infty$ for each $k \in \mathbb{N}$. Also assume that

$$\left| E_{j,k} \cap \bigcup_{j'=1}^{\infty} E_{j',m} \right| \lesssim 2^{i-k} |E_{j,k}|$$

if $k > i$. Then show that

$$\left\| \sum_{j=1}^{\infty} \lambda_{j,k} \chi_{E_{j,k}} \right\|_{L^p(\ell^q)} \sim \left\| \sum_{j=1}^{\infty} \lambda_{j,k} \chi_{E_{j,k}} \right\|_{L^p(\ell^p)}$$

for all complex sequences $\{\lambda_{j,k}\}_{j,k \in \mathbb{N}}$.

Exercise 4.65 Let $0 < p, q \leq \infty$, $m \in [1, n) \cap \mathbb{N}$. Consider the trace operator $T_m : f \in \mathscr{S}(\mathbb{R}^n) \mapsto f(\star_1, \star_2, \dots, \star_m, 0, 0, \dots, 0) \in \mathscr{S}(\mathbb{R}^m)$:

1. [931, Remark 2.4] Let $s > \dfrac{m-n}{p}$. Then show that T_m extends to a linear operator from $B^s_{pq}(\mathbb{R}^n)$ to $B^s_{pq}(\mathbb{R}^m)$.
2. [931, Remark 2.7] Let $s = \dfrac{m-n}{p}$. Then show that T_m extends to a linear operator from $B^s_{pq}(\mathbb{R}^n)$ to $L^p(\mathbb{R}^m)$.

4.4.3 Fubini's Property

We consider Fubini's property for Triebel–Lizorkin spaces. Fubini's theorem (Theorem 1.3) is a theorem asserting that the order of integrals can be interchanged for nonnegative functions or integrable functions. Let us rephrase this theorem.

Proposition 4.4 (Restatement of Fubini's theorem) *Let $1 < p < \infty$, let $f : \mathbb{R}^n \to \mathbb{C}$ be a Borel measurable function, and let $j = 1, 2, \dots, n$. If*

$$\int_{\mathbb{R}^{j-1}_{(x_1,\dots,x_{j-1})} \times \mathbb{R}^{n-j}_{(x_{j+1},\dots,x_n)}} \left(\int_{\mathbb{R}_{x_j}} |f(x)|^p \mathrm{d}x_j \right) \mathrm{d}x_1 \dots \mathrm{d}x_{j-1} \mathrm{d}x_{j+1} \dots \mathrm{d}x_n$$

is finite, then $f \in L^p(\mathbb{R}^n)$.

4.4.3.1 Fubini's Property for Triebel–Lizorkin Spaces

With the replacement above, we can consider Fubini's theorem for Triebel–Lizorkin spaces.

Theorem 4.49 (Fubini's property for Triebel–Lizorkin spaces) *Let* $1 < p < \infty$, $1 < q \le \infty$, $s > 0$. *Then*

$$\|f\|_{F^s_{pq}(\mathbb{R}^n)} \sim \sum_{j=1}^{n} \| \, \|f(x)\|_{F^s_{pq}(\mathbb{R}_{x_j})} \, \|_{L^p(\mathbb{R}^{j-1}_{(x_1,\dots,x_{j-1})} \times \mathbb{R}^{n-j}_{(x_{j+1},\dots,x_n)})}$$

for $f \in F^s_{pq}(\mathbb{R}^n)$.

Before the proof, a couple of remarks may be in order.

Remark 4.6

1. The result still holds for $0 < p < \infty$, $0 < q \le \infty$ and $s > \sigma_{pq}$. We omit the details; see [103].
2. In the proof of this theorem one writes $x = (x', x_n) \in \mathbb{R}^{n-1} \times \mathbb{R}$ for $x \in \mathbb{R}^n$. The same can be said for ε', m'.

Proof First of all, let us show

$$\|f\|_{F^s_{pq}(\mathbb{R}^n)} \lesssim \sum_{j=1}^{n} \| \, \|f(x)\|_{F^s_{pq}(\mathbb{R}_{x_j})} \, \|_{L^p(\mathbb{R}^{j-1}_{(x_1,\dots,x_{j-1})} \times \mathbb{R}^{n-j}_{(x_{j+1},\dots,x_n)})}. \tag{4.137}$$

Choose $\varphi \in C^\infty_{\rm c}(\mathbb{R})$ and define

$$\Phi(x) \equiv \varphi(x_1)\varphi(x_2)\dots\varphi(x_n), \quad \Psi(x) \equiv \sum_{j=1}^{n} \left(\frac{\partial}{\partial x_j}\right)^{2L} \Phi(x),$$

for $x = (x_1, x_2, \dots, x_n) \in \mathbb{R}^n$. By virtue of the local means, Theorem 2.34, we have

$$\|f\|_{F^s_{pq}} \simeq \|\Phi * f\|_p + \|\{2^{js}\Psi^j * f\}_{j=0}^{\infty}\|_{L^p(\ell^q)}$$

for $L \gg 1$. Now we write $\Theta_{(k)}(x) \equiv \prod_{l=1}^{n} \left(\frac{\partial}{\partial x_l}\right)^{2L\delta_{kl}} \varphi(x_l)$ for $x \in \mathbb{R}^n$. Then

$\|\{\Psi^j * f\}_{j=0}^{\infty}\|_{L^p(\ell^q)} \lesssim \sum_{k=1}^{n} \|\{2^{js}\Theta^j_{(k)} * f\}_{j=0}^{\infty}\|_{L^p(\ell^q)}$ by the triangle inequality.

Since $f \in L^p(\mathbb{R}^n)$ with $1 < p < \infty$, we estimate

$$|\Theta^j_{(n)} * f(x)| \le M_{x'}\left[2^j\varphi^{(2L)}(2^j x_n) * f\right](x).$$

Hence

$$\left\|\{2^{js}\Theta^j_{(n)} * f\}_{j=0}^\infty\right\|_{L^p(\ell^q)} \lesssim \left\|\|\{2^{js}M_{x'}\left[2^j\varphi^{(2L)}(2^jx_n) * f\right]\}_{j=0}^\infty\|_{\ell^q}\right\|_{L^p_{x_n}L^p_{x'}}$$

$$\lesssim \left\|\|\{2^{js}(2^j\varphi^{(2L)}(2^jx_n) * f)\}_{j=0}^\infty\|_{\ell^q}\right\|_{L^p_{x_n}L^p_{x'}}$$

$$= \left\|\|\{2^{js}(2^j\varphi^{(2L)}(2^jx_n) * f)\}_{j=0}^\infty\|_{\ell^q}\right\|_{L^p_{x'}L^p_{x_n}}.$$

Again by the local means,

$$\left\|\{2^{js}\Theta^j_{(n)} * f\}_{j=0}^\infty\right\|_{L^p(\ell^q)} \lesssim \left\|\{2^{js}(2^j\varphi^{(2L)}(2^jx_n) * f)\}_{j=0}^\infty\right\|_{L^p_{x'}L^p_{x_n}(\ell^q)}$$

$$\lesssim \|\,\|f(x)\|_{F^s_{pq}(\mathbb{R}_{x_n})}\,\|_{L^p(\mathbb{R}^{n-1}_{x'})}.$$

We can swap the roles of x_n and x_j. Hence (4.137) is proved.

Let us prove the reverse inequality. To this end, we decompose f into wavelets. We maintain the same notation as Theorem 4.7. The Fatou lemma allows us to assume

$$f = \sum_{m\in\mathbb{Z}^n\cap B(J)} \langle f, \varphi(\star - m)\rangle\varphi(\star - m)$$

$$+\sum_{j=0}^{J}\sum_{m\in\mathbb{Z}^n\cap B(J)}\sum_{\varepsilon\in E}\langle f, 2^{jn/2}\Psi^\varepsilon(2^j\star - m)\rangle 2^{jn/2}\Psi^\varepsilon(2^j\star - m)$$

and that this is a finite sum. This assumption enables us to skip the argument of the infinite sum. Furthermore, to simplify the notation, we consider

$$\sum_{j=0}^{J}\sum_{m\in\mathbb{Z}^n\cap B(J)}\sum_{\varepsilon\in E}\langle f, 2^{jn/2}\Psi^\varepsilon(2^j\star - m)\rangle 2^{jn/2}\Psi^\varepsilon(2^j\star - m).$$

The remaining term can be estimated similarly. According to the wavelet decomposition, we have

$$\|\,\|f(x)\|_{F^s_{pq}(\mathbb{R}_{x_n})}\,\|_{L^p(\mathbb{R}^{n-1}_{x'})}$$

$$\lesssim \sup_{\varepsilon\in E}\left\|2^{j(s+n)}\sum_{m\in\mathbb{Z}^n}\langle f, \Psi^\varepsilon(2^j\star - m)\rangle(\Psi^{\varepsilon'}\otimes\chi_{[0,1)})(2^j\star - m)\right\|_{L^p(\ell^q)}.$$

Hence by the triangle inequality,

$$\left| \sum_{m \in \mathbb{Z}^n} \sum_{\varepsilon \in E} \langle f, \Psi^{\varepsilon}(2^j \star - m) \rangle \Psi^{\varepsilon'}(2^j x' - m') \chi_{[0,1)}(2^j x_n - m_n) \right|$$
$$\leq \sum_{m \in \mathbb{Z}^n} \sum_{\varepsilon \in E} |\langle f, \Psi^{\varepsilon}(2^j \star - m) \rangle| \chi_{[-R,R]}(2^j x' - m') \chi_{[0,1)}(2^j x_n - m_n)$$
$$\lesssim M \left[\sum_{m \in \mathbb{Z}^n} \sum_{\varepsilon \in E} |\langle f, \Psi^{\varepsilon}(2^j \star - m) \rangle| \chi_{[0,1)^n}(2^j \star - m) \right] (x).$$

Thus, by the Fefferman–Stein vector-valued maximal inequality (Theorem 1.49), we have $\| \| f(x) \|_{F^s_{pq}(\mathbb{R}_{x_n})} \|_{L^p(\mathbb{R}^{n-1}_{x'})} \lesssim \| f \|_{F^s_{pq}(\mathbb{R}^n)}$. We can interchange x_n and x_j. Thus, the proof is complete.

Exercises

Exercise 4.66 Let $1 < p < \infty$, $1 < q \leq \infty$, $s > 0$. Then show that

$$\| f \|_{F^s_{pq}(\mathbb{R}^n)} \sim \| \| f(x) \|_{F^s_{pq}(\mathbb{R}^{n-1}_{x'})} \|_{L^p(\mathbb{R}_{x_n})} + \| \| f(x) \|_{F^s_{pq}(\mathbb{R}_{x_n})} \|_{L^p(\mathbb{R}^{n-1}_{x'})}$$

for $f \in F^s_{pq}(\mathbb{R}^n)$ by mimicking the proof of Theorem 4.49.

Exercise 4.67 In the proof of the Fubini property it is essential that Triebel–Lizorkin spaces are used. Namely, for Besov spaces we do not have such a property. Here and below we let $0 < p, q \leq \infty$ and $s \in \mathbb{R}$ with $p \neq q$.

1. Choose a function ψ which is sufficiently smooth so that

$$\| f \|_{B^s_{pq}} \sim \left(\sum_{j=1}^{\infty} |a_j|^q \right)^{\frac{1}{q}},$$

where

$$f \equiv \sum_{j=1}^{\infty} a_j 2^{-j(s-n/p)} \psi(2^j(\star - (j, j, \ldots, j))).$$

Hint: Use the wavelet decomposition.

2. Show that the Fubini property corresponding to Besov spaces does not hold in general.

Textbooks in Sect. 4.4

Diffeomorphism: Theorem 4.47

Triebel took approaches via atoms [100, Section 4.3.1] and via local means [100, Section 4.3.2]. See [99, Section 2.10].

Triebel employed a different atomic decomposition obtained in [1065], [1068, Section 2] and [100, Section 3.3.2] to prove Theorem 4.47*1* for Triebel–Lizorkin spaces in [100, p. 213 Theorem 4.4.2] and Theorem 4.47*1* for Triebel–Lizorkin spaces in [100, p. 219 Corollary 4.4.2]. The method employed in this book to prove Theorem 4.46 can be found in [482, Section 5] and [931, Theorem 2.2].

Trace Theorem: Theorems 4.47 and 4.48

Theorem 4.47 for $1 < p < \infty$, $1 \le q \le \infty$ and $s > 1/p$ can be found in [7, Theorem 6.6.1].

Triebel considered Theorem 4.48 in [100, Section 4.4.3].

See also [99, Section 3.3.3] for the trace operators. Motivated by the works [628, 629], Triebel used the characterization by means of the Peetre maximal operator in [99, Section 2.7.2].

See [4, Theorem 7.39] for a direct account without recourse to Triebel–Lizorkin spaces for the proof that $W^{m,p}(\mathbb{R}^n) \approx F^m_{p2}(\mathbb{R}^n)$ is $B^{m-1/p}_{pp}(\mathbb{R}^{n-1})$ in the light of Theorem 4.47.

Fubini Property: Theorem 4.49

We refer to [99, 103] for this section. In [99, p. 115] together with the formulation in [99, p. 114, Definition], we can find the Fubini property for Triebel–Lizorkin spaces. We can find an approach via the quarkonial decomposition in [103, pp. 34–40] together with a detailed account of the failure of the Fubini property for Besov spaces.

4.5 Notes for Chap. 4

Section 4.1

Section 4.1.1

Smooth Atomic Decomposition: Theorems 4.1 and 4.2

We refer to [482, Theorem 3.1] for the counterpart of Theorem 4.1 to $\dot{B}^s_{pq}(\mathbb{R}^n)$ and to [482, Theorem 2.6] for the counterpart of Theorem 4.2 to $\dot{B}^s_{pq}(\mathbb{R}^n)$. We refer to [482, Theorem 2.6] for the counterpart of Theorem 4.2 to $\dot{B}^s_{pq}(\mathbb{R}^n)$. We refer to the paper [1068, Section 2] for the atomic decomposition of another type. See [483, Lemma A.2] and [697, Lemma 7.1] for Lemma 4.2.

Nonsmooth Atomic Decomposition: Theorem 4.3

We refer to [203] for the idea of reducing the matter of function spaces to the vector-valued spaces. See Asami [141] for the general case.

Molecular Decomposition: Theorems 4.1 and 4.2

In [100] Triebel dealt with the atomic decomposition of $A^s_{pq}(\mathbb{R}^n)$, which we did not treat in this book. The atomic decomposition dealt with in this book appeared in [101]. In [482, 483], the atomic decomposition for homogeneous spaces is investigated very precisely. See also [292, 539].

Definition 4.2 is based upon [482, pp. 777–778]. In [482, Theorem 3.1], $\dot{B}^s_{pq}(\mathbb{R}^n)$ is dealt with. See [482, Theorem 4.1] for the case of $\mathrm{BMO}(\mathbb{R}^n)$, where Frazier and Jawerth considered the expansion theorem for $\mathrm{BMO}(\mathbb{R}^n)$.

Minimality of $\dot{B}^0_{11}(\mathbb{R}^n)$: Exercise 4.7

Meyer pointed out the minimality of $\dot{B}^0_{11}(\mathbb{R}^n)$ in [64, 775].

Section 4.1.2

General Remarks on Wavelets in Connection with This Book

The theory of wavelets dates back to as long ago as 1992 (see [386]).

Wavelet Decomposition from the Viewpoint of Decomposition Theory

We can find the wavelet decomposition of the space $\dot{A}^s_{pq}(\mathbb{R}^n)$ in [29, Chapter 7].

Triebel has dealt with the Schauder basis in $B^s_{pq}(\mathbb{R}^n)$ with $1 < p < \infty$, $0 < q \leq \infty$ and $s \in \mathbb{R}$ in [99, p. 86, 2.5.5.]. We refer to [954] for the spline wavelet decomposition. Sickel investigated the unconditional convergence of the wavelet expansion of the functions in $F^s_{pq}(\mathbb{R}^n)$ in [955, Theorem 1 and Corollary 2]. We refer to Sickel [957] and Moritoh [802] for the wavelet characterization of homogeneous Besov spaces and homogeneous Triebel–Lizorkin spaces. Sickel considered unconditionality of wavelets in [957] and Moritoh related the wave front sets and the Littlewood–Paley decomposition.

In the case of Besov spaces, the wavelet basis is not greedy except when $p = q$. In the case of Triebel–Lizorkin spaces, the wavelet basis is greedy. See [693, 694] for more about this direction of research of the wavelet expansions. See [260] for a different construction of wavelets using the bounded admissible partition of unity.

Wavelet Decomposition from the Viewpoint of Approximation

It is well known that the wavelet function cannot have compact support if it belongs to $C^\infty(\mathbb{R}^n)$. This problem creates difficulty when we consider the coupling of wavelet functions and functions in Besov spaces or Triebel–Lizorkin spaces. One of the techniques to overcome this problem is to use the Calderón reproducing formula. The work [697] provides a good approach, where Kyriazis explained how to cope with the problem in [697, Section 2] and how to develop the expansions of wavelet type in [697, Section 3]. We refer to [595, Theorem 2.3] for another quantitative approach in this direction.

Unconditionality of the Wavelet Expansion

See [522] for the criterion of unconditionality, where Gröchenig showed that the wavelet basis is unconditional when the Schwartz function space is dense in the space.

Regarding the wavelet expansion, to approximate $L^p(\mathbb{R})$-functions Jaffard, Okada and Ueno used the 1-dimensional Besov space. They used Theorem 2.39 for the Besov-norm and gave a sufficient condition for the wavelet approximation to converge in the $L^p(\mathbb{R})$-topology; see [625, Theorem 3.1].

Haar Wavelet

Among the wavelets, the Haar wavelet has a long history and attracts many mathematicians. As for the expansion of the function space $B^s_{pq}(\mathbb{R}^{n-1})$ with

$$\max\left(n\left(\frac{1}{p}-1\right),\frac{1}{p}-1\right) < s < \min\left(\frac{1}{p},1\right)$$

goes back to the work [1052]. See [1085] for more recent approach. In [108, Chapter 2] and [110, Chapter 6] we can find applications of wavelets to function spaces. Triebel proposes to use the wavelet expansion to consider the Navier–Stokes equations in [110, Chapter 6].

Section 4.1.3

φ-Transform: Theorem 4.9

Theorem 4.9 is important for the atomic decomposition as we have seen. Frazier and Jawerth developed the theory of the φ-transform in [479, 480, 482, 483]. Based on [479, 482, 483], the theory of decompositions was refined rapidly. See [482, Lemma 2.1] for more details.

Quarkonial Decomposition: Theorems 4.8 and 4.10

See [103].

Other Decomposition of Functions

Hunt passed the Carleson theorem of the almost everywhere convergence of the Fourier transform of the $L^2(\mathbb{R})$-functions to $L^p(\mathbb{R})$. Hunt used the maximal operator to this end. Based upon the technique of Lacey and Thiele, Grafakos, Terwilleger and Tao reproved the results by Hunt.

To overcome the disadvantage of wavelets, we "take the average": as an example, Lacey and Thiele reproved the almost everywhere convergence of the Fourier transform of the $L^2(\mathbb{R})$-functions, which is known as the Carleson theorem. See [135, 349, 516, 610, 701, 702] for various extensions of admissible classes of the functions. Their technique is very ingenious and we have to say that the average plays a small role in the whole proof. But even for the averaging technique, the atomic decomposition in this book or the wavelet decomposition is indispensable to understand their proof. The technique of taking the average is slightly simplified by the author; see [917].

Applications of Quarkonial Decomposition

We refer to [1073] for the application of the quarkonial decomposition, where Triebel defined the Q-operator.

Other Expansions

The wavelet expansion was obtained in many spaces. For example, Triebel obtained the wavelet characterization for Lorentz spaces and Zygmund spaces in [1079, Theorem 17]. See [142] for the unconditional basis.

Section 4.1.4

Limiting Case of the Embedding into $L^\infty(\mathbb{R}^n)$

Sato, Sawano, Morii, Wadade, and Nagayasu refined the Sobolev embedding, which claims that some classes of function spaces are "almost embedded" into $L^\infty(\mathbb{R}^n)$ in terms of Besov spaces and so on: We refer to [318] and [688, Theorem 3] for another approach using the local growth envelope. See [805–808, 819, 918].

Hardy Inequality: Theorem 4.20

We have many extensions for the Hardy inequality to the various directions. We refer to [1059] for the extension to Besov spaces.

Section 4.2

Section 4.2.1

Interpolation Inequalities

Gagliardo and Nirenberg obtained an inequality of the form

$$\sum_{|\alpha|=m} \|\partial^\alpha f\|_p \le C(\|f\|_{p_1})^\theta \left(\sum_{|\alpha|=l} \|\partial^\alpha f\|_{p_2}\right)^{1-\theta}$$

in [494, 849], where

$$m - \frac{n}{p} = (1-\theta)\left(l - \frac{n}{p_2}\right) - \theta\frac{n}{p_1}.$$

Interpolation inequalities are obtained in [242, Theorems 1 and 2]. See [188] for the case of mixed norms and [191, 194] for the anisotropic case. We refer to [352, Theorem 1] and [801, Theorem 4.4] for more about this inequality; the idea in both papers was to use Exercise 1.4.

Section 4.2.2

Real Interpolations of Function Spaces

Triebel proved Theorem 4.25 for $1 < p, q < \infty$ [1039, Theorems 8.1.3 and 8.3.3].

The general case for Besov spaces is handled in [71, p. 246, Theorem 11]. See also [1050, Theorem 2]. We remark that Besov spaces arise naturally in the context of the real interpolation if we consider the potential spaces [7, Theorem 6.2.4] and [71, p. 67, Corollary 2]. In fact, (4.57) is obtained for the special case of $1 \le p, q \le \infty$. Besov spaces can be expressed by means of the real interpolation in [1041, 1042]. Furthermore, the theory of semi-groups is developed there. In [1039, Section 8], when we consider the real interpolation of $W^{s,p}(\mathbb{R}^n)$, then we obtain Besov spaces; see [324, 521, 736, 1026]. See [338] for the real interpolation of weighted tent spaces.

Real Interpolation Functors

The definition dates back to Lions and Peetre [736].

Real Interpolation and Image Processing

See [376].

Section 4.2.3

Complex Interpolation of Lebesgue Spaces: Interpolation Inequalities by M. Riesz, F. Riesz and G. Thorin

See the works by M. Riesz [890], F. Riesz [889] and G. Thorin [1032].

Complex Interpolation Functor Defined by Calderón in 1964

Calderón's interpolation functor emerged from the three-line lemma (1.286) on complex analysis. The $L^p(\mathbb{R}^n)$-spaces are interpolated by the use of this original complex interpolation functor. See [170] for Theorem 4.33.

Complex Interpolation of Function Spaces: Theorems 4.29, 4.32 and 4.34

The complex interpolation theory of Banach spaces is in [7]. However, when $0 < p < 1$ or $0 < q < 1$, $A^s_{pq}(\mathbb{R}^n)$ is not a Banach space, we need to polish the definition of the complex interpolation. We can find some techniques in [99, 1055] to overcome this difficulty. See the papers by Paivarinta and Triebel [861, 1054, 1055, 1057]. See [99, p. 67, Definitions 1 and 2, p. 69, Theorem] for Theorem 4.29.

Corollary 4.4 comes from Stafney's work in 1970 [1006].

Generalized Three-Line Lemma: Lemma 4.10

See [325, 9.4 and 29.4] and [103, Section 2.4.6] for Lemma 4.10.

Complex Interpolation of $A^s_{pq}(\mathbb{R}^n)$: Definition 4.16 and Theorems 4.25 and 4.29

Definition 4.16 goes back to the work by Calderón-Torchinsky [329]. In [1055, p. 319], Triebel considered $M^{2\mathrm{A}}$, as well as the underlying spaces $L^p(\ell^q, \mathbb{R}^n)$ to consider the complex interpolation of $F^s_{pq}(\mathbb{R}^n)$ with $0 < p, q < \infty$ and $s \in \mathbb{R}$.

Furthermore, subharmonic function inequality (1.282) is due to [590].

Triebel proved Theorem 4.29 for $1 < p, q < \infty$ [1039, Theorem 10.1.1]. Theorems 4.25 and 4.29 concern the concrete interpolation results in [99]. Theorem 4.30 is due to Triebel [98, 182 and 185]. See also Päivärinta [861] for the complex interpolation of function spaces. Triebel obtained the theory of interpolations in the paper [1057] via [1039, 1055]. See [399, 400, 1098] for the complex interpolation of quasi-Banach spaces.

We refer to [281, Theorem 1.1] for an extrapolation result; that is, an expression of the sequence norm $\|\lambda\|_{\mathbf{f}_{p_0q_0}}$ in terms of $\|\lambda\|_{\mathbf{f}_{p_1q_1}}$ and $\|\lambda\|_{\mathbf{f}_{pq}}$ under the assumption of Theorem 4.29.

We refer to [1178, Theorem 2.28] for the results on complex interpolations including $[B^{s_0}_{p_0\infty}(\mathbb{R}^n), B^{s_1}_{p_1\infty}(\mathbb{R}^n)]_\theta \approx \overset{\circ}{B}{}^s_{p\infty}(\mathbb{R}^n)$ and $[B^{s_0}_{\infty q_0}(\mathbb{R}^n), B^{s_1}_{\infty q_1}(\mathbb{R}^n)]_\theta \approx \overset{\circ}{B}{}^s_{\infty q}(\mathbb{R}^n)$, for example.

Complex Interpolation of Modulation Spaces

We refer to [548, Section 2.2] for the complex interpolation of modulation spaces.

Section 4.3

Section 4.3.1

Paraproduct

We refer to the work by Bony for the definition of the paraproduct [258].

General Composition Operators

See [275].

Section 4.3.2

Hölder's Inequality for Besov Spaces and Triebel–Lizorkin Spaces

We refer to the paper by Sickel and Triebel [970] for the Hölder inequality for Besov spaces and Triebel–Lizorkin spaces. In [970] we can find the detailed necessary condition for Hölder's inequality.

We refer to [640, Section 3] for more details related to Lemma 4.15.

We refer to [954, Theorems 1 and 2] for the estimate of m products with $m \geq 3$, where the "$F^{m+1} = \overbrace{FFF \cdots F}^{m+1\text{times}}$" case is investigated. We refer to [810, Section 6].

Pointwise Multipliers

For multipliers, we have many results. Theorem 4.37 is one of the main results in [99, 2.8.2] and there are many other important results in connection with half spaces or domains. See [99, 2.8.8] for an example of Theorem 4.37. See also [844, 1064].

See [839] for the characterization of pointwise multipliers on $F^s_{pq}(\mathbb{R}^n)$. We refer to [437, Theorems 2.1 and 2.5] for the FBF case.

We refer to [825] for the description of the pointwise multipliers of $\mathrm{BMO}(\mathbb{R}^n)$. We refer for the characterization of the multipliers in $B^0_{\infty\infty}(\mathbb{R}^n)$ and $B^0_{\infty 1}(\mathbb{R}^n)$ to [683, Theorem 4] and [683, Theorem 5], respectively. See [1012] for $F^s_{p2}(\mathbb{R}^n)$ with $1 < p < \infty$ and $s > \dfrac{n}{p}$, [71] for $B^s_{pp}(\mathbb{R}^n) = F^s_{pp}(\mathbb{R})$ with $1 \leq p \leq \infty$ and $s > \dfrac{n}{p}$, [477] for $F^s_{pq}(\mathbb{R}^n)$ with $1 \leq p < \infty$, $1 \leq q \leq \infty$ and $s > \dfrac{n}{p}$ and [263] for $B^s_{1q}(\mathbb{R}^n)$ with $1 \leq q \leq \infty$ and $s > 0$

We refer to the survey paper [963] for more details.

The algebraic property of the function spaces is related to the embedding into $C(\mathbb{R}^n)$; see [657]. We refer to [436] for another multiplier property of $A^s_{pq}(\mathbb{R}^n)$.

We may wonder whether we really need $F^s_{p_1q}(\mathbb{R}^n)$ in the inequality $\|f\cdot g\|_{F^s_{p_0q}} \lesssim \|f\|_{F^s_{p_1q}}\|g\|_{F^s_{p_2q}}$ in Theorem 4.35 when $p_0 = p_2$. Sickel gave an answer for this problem in [962, Theorem 1].

Characteristric Function on the Half Space as a Multiplier

Extension by zero of the functions is a natural method of obtaining global functions starting from the ones defined on domains. Burenkov considered this property from the point of the size of the boundary in [298, Teorema 3]. We also refer to [211, 315] for this direction of research.

Theorem 4.40 is based on [78, 4.6.3, p. 208]. Theorem 4.40 has a history. It appeared initially in [99, 2.8.7, p. 158], which dealt with the Besov spaces. See [477, Proposition 5.1], as well as [99, 2.8.5, Remark 4, p. 154] and [78, p. 258].

As in Theorem 4.40, the characteristic function $\chi_{\mathbb{R}^n_+}$ of the upper half space $\mathbb{R}^n_+$, it is of importance to check

$$\|\chi_{\mathbb{R}^n_+} f\|_{A^s_{pq}} \lesssim \|f\|_{A^s_{pq}}. \tag{4.138}$$

Inequality (4.138) is used to define function spaces on domains; we can define the functions by setting 0 outside. For $L^p(\mathbb{R}^n) \approx F^0_{p2}(\mathbb{R}^n)$, (4.138) holds but for $W^{1,p} \approx F^1_{p2}(\mathbb{R}^n)$ with $1 < p < \infty$ (4.138) fails. We refer to [99, 1044, 1074] for the failure of (4.138). The condition on s, p in Theorem 4.40 is sharp; see [78]. The condition for the $A^s_{pq}(\mathbb{R}^n)$-boundedness of multipliers of the indicator functions of the half space carries over to those on bounded Lipschitz domains; see [1074, Proposition 5.3].

Localization of $A^s_{pq}(\mathbb{R}^n)$: Theorem 4.38 and Exercise 4.46

In the case of $p = q$, see [1012, 1013]. In the case of $1 < p, q < \infty$, we refer to [78, 656]. The final form is obtained in [1071, Theorem 2.1.2, p. 696]. See also [129].

Estimates of the Power for Besov Spaces and Triebel–Lizorkin Spaces

When we consider nonlinear terms, we include not only the pointwise multiplication but also the composition operator $f \mapsto A(f)$. One of the techniques to handle this type of operators is described in Exercise 5.33, which is originally due to Bony [259]. See also [774]. In particular, of interest is the case where $A(f) = |f|^p$ or $A(f) = |f|^{p-1}f$. For example, see [953, Theorem 4] for the operator $f \in F^s_{pq}(\mathbb{R}^n) \to f^m \in F^{s-(m-1)(n/p-s)}_{pq}(\mathbb{R}^n)$. See [267] for the cases of BMO$(\mathbb{R}^n)$, bmo$(\mathbb{R}^n)$ and cmo$(\mathbb{R}^n)$, [269] for the case of homogeneous Besov spaces, [270] for the case of homogeneous Besov/Triebel–Lizorkin spaces, [271–273, 898] for the composition operator. We refer to [959] for many counterexamples to this direction. When $A(f) = |f|^2$, as a corollary of the boundedness property of A, we have the algebraic property of $A^s_{pq}(\mathbb{R}^n)$; $f, g \in A^s_{pq}(\mathbb{R}^n)$ implies $f \cdot g \in A^s_{pq}(\mathbb{R}^n)$; see [1049]. See also [7, p. 163] for the case of the Besov space $B^s_{pq}(\mathbb{R}^n)$ such that $B^s_{pq}(\mathbb{R}^n) \hookrightarrow L^\infty(\mathbb{R}^n)$, where the author discussed the condition $B^s_{pq}(\mathbb{R}^n) \times B^s_{pq}(\mathbb{R}^n) \subset B^s_{pq}(\mathbb{R}^n)$ holds. See [960, Theorems 1 and 2] and [961, Theorem 3] for the condition of A for Triebel–Lizorkin spaces and Besov spaces, respectively.

The estimate for modulation spaces can be found in [1023, Theorem 4.1].

Truncation

Among the composition operators described above, the truncation $f \mapsto \max(f, 0)$, or equivalently the absolute value $f \mapsto |f|$, is of importance in connection with PDEs in particular with the variational problem. This line of research goes back to [12, 268, 860]. Triebel studied the truncation in [1072, Theorem 1]. Furthermore, by taking advantage of this, Triebel defined the Q-operator in [1073, Definition 2]. Triebel employed the quarkonial decomposition because the coefficients are linear. A disadvantage of taking the absolute value is that the absolute value does not preserve smoothness of functions. This implies that we cannot consider the moment condition of the function. Therefore, we need to give our attention to the regular case. As an application, Triebel investigated the semi-linear equation

$$u(x) = \int_{\mathbb{R}^n} K(y)u_+(x-y)\mathrm{d}y + h(x)$$

for $K \in L^1(\mathbb{R}^n)$ and $h \in L^p(\mathbb{R}^n)$ in [1073, Section 2.3].

Div-Curl Lemma: Theorems 4.42 and 4.43

See [154] for Theorem 4.43. See [381] for the converse, that is, the div-curl decomposition.

Kato–Ponce Inequality: Theorem 4.44

See [68, 515, 663, 680, 820] for this direction of research. See also [1088, Theorems 1.3 and 1.4] for more about the estimates of this type, where Tsutsui used the sharp maximal operator.

Let us consider the rough differential equation (RDE):

$$u'(t) = F(u(t))\xi(t), \quad u(0) = u_0, \quad t \in \mathbb{R}$$

as Prömel and Trabs did in [882, (1)]. Here F is a vector-valued function, u is the unknown function and ξ is called the signal which has a bad regularity; ξ is supposed to be a continuous function but its differentiability is low. Prömel and Trabs considered the following condition on u:

$$u = \Pi_4(u^{\dagger}, u) + u^{\sharp}.$$

Section 4.4

Section 4.4.1

Diffeomorphism: Theorem 4.46

This is traced back to [1064]. The word "regular diffeomorphism" is due to the author [919, Definition 1.6]. We refer to [274] for the diffeomorphism generated by the quasi-conformal in the 1-dimensional case.

An example of $L^p(\mathbb{R}^n)$ with $1 \le p < \infty$, presented by Bourdaud in [265, Theorem 1] tells us that some Lipschitz continuity is enough to guarantee that the homeomorphism induces isomorphism over $L^p(\mathbb{R}^n)$. When $s > 0$ is a small number, the diffeomorphism need not be C^1 as is proved in [265]. Triebel generalized the diffeomorphism in [1074, Section 4] and he introduced Lipschitz diffeomorphism, or equivalently bi-Lipschitz homeomorphism. The price to pay by allowing the domains not to be sufficiently smooth is to assume $\sigma_p < s < 1$ for Besov spaces and $\sigma_{pq} < s < 1$ for Triebel–Lizorkin spaces.

The method of the proof of Theorem 4.46 can be found in [931, Proposition 3.14].

Section 4.4.2

Trace Operators for $W^s_p(\mathbb{R}^n)$: Theorem 4.47

Let us consider the trace operator from $W^s_p(\mathbb{R}^n)$ to $F^{s-1/p}_{pp}(\mathbb{R}^{n-1})$ as a corollary of Theorem 4.47. In this case many people made great efforts; see [139, 492, 744, 1094, 1095] and the final solution was obtained by Besov [171, 172]. We refer also to [782, 1090] for more approaches in the weight setting.

Trace Operators for $B^s_{pq}(\mathbb{R}^n)$: Theorem 4.47

The homogenous Besov counterpart of Theorem 4.47 dates back to 1978 (see [629, Theorem 2.1]). Theorem 4.47 itself appeared initially in [99, Theorem p. 132].

Here let us view the results of the boundedness of the trace operator which were proved by means of the atomic decomposition. Theorem 4.47 is a typical example of such an approach. Frazier and Jawerth initially used the atomic decomposition described in this book. See [482, Section 5] for the technique used in this book for homogeneous Besov spaces. Frazier and Jawerth used the atomic decomposition to prove the counterpart of Theorem 4.47 to homogeneous Triebel–Lizorkin spaces in [483, Theorem 11.1(a)]. See [483, Theorem 11.1(c)–(e)] for negative results.

We refer to [717] for the surjectivity and the boundedness of

$$\mathrm{Tr}_{\mathbb{R}^n} : F^s_{\infty q}(\mathbb{R}^n) \to B^s_{\infty\infty}(\mathbb{R}^{n-1}).$$

See [1106] for the case of the mixed derivatives together with [1106, Section 1] explaining the difficulty of this case.

We refer to [930] for the function spaces assuming their value in a Banach space E.

Trace Operator in the Limiting Case: Theorem 4.48

Let $1 \le p < \infty$. See [306, 509, 876] for $\mathrm{Tr}_{\mathbb{R}^n}(B^{1/p}_{p1}(\mathbb{R}^n)) = L^p(\mathbb{R}^{n-1})$. Furthermore, the homogeneous counterpart of Theorem 4.48 is in [482, Theorem 5.1] and Theorem 4.48 itself appears in the last paragraph of [482, Section 7]. We refer to [931, Corollary 2.5]. We refer to [996, Theorem 5.1] for the case of noncompact hypersurfaces.

Trace Operators for Wiener Amalgam Spaces

We refer to [398] for the trace of Wiener amalgam spaces.

Trace Operator to General Sets

We refer to [931, Theorem 3.16] for the trace operator to the boundary of C^k-domains.

See [306] for another extension operator.

Section 4.4.3

Fubini Property: Theorem 4.49

See [1071, Theorem 2.1.12].

Chapter 5
Applications: PDEs, the $T1$ Theorem and Related Function Spaces

As an application of what we have been gathering, we investigate partial differential equations. In Sect. 5.1, we develop a theory of function spaces on domains so as to consider partial differential equations on domains. To consider some solution operators we take up the pseudo-differential operators in Sect. 5.2. We apply what we have obtained to various equations such as heat equations, Schrödinger equations and wave equations in Sect. 5.3. Although we will consider some estimates on the elliptic differential equations in Sect. 5.2, we take up elliptic differential equations in more depth in Sect. 5.4. Another application includes the $T1$ theorem in Sect. 5.5, which guarantees that the integral operator having the standard kernel is actually $L^2(\mathbb{R}^n)$-bounded.

5.1 Function Spaces on Domains

In this book, we have been considering function spaces on $\mathbb{R}^n$. Now we consider function spaces on domains $\Omega \subset \mathbb{R}^n$. By "a domain" we mean an open set in $\mathbb{R}^n$, which is not necessarily connected. Recall that we defined $f \mid \Omega \in \mathscr{D}'(\Omega)$, the restriction of f to Ω for $f \in \mathscr{S}'(\mathbb{R}^n)$ in Definition 1.30. We consider four different fundamental domains: $\mathbb{R}^n_+$ (see Sect. 5.1.1), bounded C^∞-domains (see Sect. 5.1.2), uniformly C^m-domains (see Sect. 5.1.3) and Lipschitz domains (see Sect. 5.1.4).

5.1.1 Function Spaces on the Half Space

5.1.1.1 Function Spaces on the Upper Half Space

Here we investigate function spaces on the upper half space on $\mathbb{R}^n_+$.

Y. Sawano, *Theory of Besov Spaces*, Developments in Mathematics 56,
https://doi.org/10.1007/978-981-13-0836-9_5

Definition 5.1 (Function spaces $B^s_{pq}(\mathbb{R}^n_+)$, $F^s_{pq}(\mathbb{R}^n_+)$ and $A^s_{pq}(\mathbb{R}^n_+)$ on the upper half space) Let $\mathbb{R}^n_+$ be the upper half space given by (5.1).

1. Let $0 < p, q \le \infty$ and $s \in \mathbb{R}$. Define the set $B^s_{pq}(\mathbb{R}^n_+)$ of distributions on $\mathbb{R}^n_+$ by

$$B^s_{pq}(\mathbb{R}^n_+) \equiv \{g \,|\, \mathbb{R}^n_+ \in \mathscr{D}'(\mathbb{R}^n_+) \,:\, g \in B^s_{pq}(\mathbb{R}^n)\}(\subset \mathscr{D}'(\mathbb{R}^n_+)).$$

   Furthermore, make $B^s_{pq}(\mathbb{R}^n_+)$ into a normed space by defining the norm by

$$\|f\|_{B^s_{pq}(\mathbb{R}^n_+)} \equiv \inf\{\|g\|_{B^s_{pq}(\mathbb{R}^n)} \,:\, g \in B^s_{pq}(\mathbb{R}^n),\ f = g|\mathbb{R}^n_+\} \quad (f \in B^s_{pq}(\mathbb{R}^n_+)).$$

2. Let $0 < p < \infty, 0 < q \le \infty$ and $s \in \mathbb{R}$. Define the set $F^s_{pq}(\mathbb{R}^n_+)$ of distributions on $\mathbb{R}^n_+$ by

$$F^s_{pq}(\mathbb{R}^n_+) \equiv \{g \,|\, \mathbb{R}^n_+ \in \mathscr{D}'(\mathbb{R}^n_+) \,:\, g \in F^s_{pq}(\mathbb{R}^n)\}(\subset \mathscr{D}'(\mathbb{R}^n_+)).$$

   Furthermore, make $F^s_{pq}(\mathbb{R}^n_+)$ into a normed space by defining the norm by

$$\|f\|_{F^s_{pq}(\mathbb{R}^n_+)} \equiv \inf\{\|g\|_{F^s_{pq}(\mathbb{R}^n)} \,:\, g \in F^s_{pq}(\mathbb{R}^n),\ f = g|\mathbb{R}^n_+\} \quad (f \in F^s_{pq}(\mathbb{R}^n_+)).$$

3. Denote by $A^s_{pq}(\mathbb{R}^n_+)$ either $B^s_{pq}(\mathbb{R}^n_+)$ or $F^s_{pq}(\mathbb{R}^n_+)$. It is tacitly understood that $p < \infty$ when $A^s_{pq}(\mathbb{R}^n_+)$ stands for the Triebel–Lizorkin space $F^s_{pq}(\mathbb{R}^n_+)$.

The next theorem is a direct consequence of the properties of the function space $A^s_{pq}(\mathbb{R}^n)$ on the whole space.

Theorem 5.1 (The elementary properties of $A^s_{pq}(\mathbb{R}^n_+)$ as a Banach space) *Let $0 < p, q \le \infty$ and $s \in \mathbb{R}$. Then $A^s_{pq}(\mathbb{R}^n_+)$ is a quasi-Banach space and for all $f, g \in A^s_{pq}(\mathbb{R}^n_+)$,*

$$\|f + g\|_{A^s_{pq}(\mathbb{R}^n_+)}{}^{\min(1,p,q)} \le \|f\|_{A^s_{pq}(\mathbb{R}^n_+)}{}^{\min(1,p,q)} + \|g\|_{A^s_{pq}(\mathbb{R}^n_+)}{}^{\min(1,p,q)}.$$

In particular, when $p, q \ge 1$, $A^s_{pq}(\mathbb{R}^n_+)$ is a Banach space.

Proof To prove that $A^s_{pq}(\mathbb{R}^n_+)$ is complete under the quasi-norm, it suffices to show that $\sum_{j=1}^{\infty} f_j$ is convergent in $A^s_{pq}(\mathbb{R}^n_+)$ as long as $\sum_{j=1}^{\infty} \|f_j\|_{A^s_{pq}(\mathbb{R}^n_+)}{}^{\min(1,p,q)} < \infty$; we content ourselves with this remark. See Exercise 5.1 for more details.

5.1.1.2 Lift Operator for the Half Space

We hope to construct a family $\{J_\sigma\}_{\sigma \in \mathbb{R}}$ of bounded linear operators which satisfies the following requirements for all $\sigma \in \mathbb{R}$:

- J_σ is an isomorphism between $A^s_{pq}(\mathbb{R}^n)$ and $A^{s-\sigma}_{pq}(\mathbb{R}^n)$.
- J_σ and $J_{-\sigma}$ are inverse to each other.
- Most importantly, if $f \in \mathscr{S}'(\mathbb{R}^n)$ is supported on $\mathbb{R}^{n-1} \times (-\infty, 0]$, then so is $J_\sigma f$.

Note that $(1-\Delta)^{\frac{\sigma}{2}}$ and J_σ enjoy the same properties other than the last property.

Let $\eta \in \mathscr{S}(\mathbb{R})$ be a nonnegative function supported on $(-2, -1)$ and having integral 2. For $0 < \varepsilon \ll 1$, we define a holomorphic function ψ_ε on $\mathbb{C}$ by

$$\psi_\varepsilon(z) \equiv \int_{-\infty}^0 \eta(t)\mathrm{e}^{-i\varepsilon t z}\mathrm{d}t - i\,z.$$

Define the *upper half plane* in $\mathbb{C}$ and its closure by $\mathbb{H} \equiv \{z \in \mathbb{C} : \Im(z) > 0\}$ and $\overline{\mathbb{H}} \equiv \{z \in \mathbb{C} : \Im(z) \geq 0\}$, respectively.

Let $\Omega \equiv \{z \in \mathbb{C} : |z| > 4, \Re(z) > 0\}$. If $z \in \mathbb{C}$ satisfies $-iz \in \Omega$, the distance from $\psi_\varepsilon(z)$ to Ω does not exceed 2. In fact, the real part of $-iz$ is $\Im(z)$ and $-iz \in \Omega$. Hence

$$\operatorname{dist}(\psi_\varepsilon(z), \Omega) \leq |\psi_\varepsilon(z) + iz| = \left|\int_{-\infty}^0 \eta(t)\mathrm{e}^{-i\varepsilon t z}\mathrm{d}t\right| < 2. \tag{5.1}$$

Conversely, if $|z| \leq 4$ and $\Im(z) \geq 0$, then $\Re(\psi_0(z)) = 2 + \Im(z)$; hence

$$\Re(\psi_\varepsilon(z)) = \int_{-\infty}^0 \eta(t)e^{\varepsilon t \Im(z)} \cos(\varepsilon\, t\, \Re(z))\mathrm{d}t + \Im(z) \geq \frac{3}{2}, \quad 0 < \varepsilon \ll 1.$$

Hence if $\varepsilon > 0$ is sufficiently small, ψ_ε maps $\overline{\mathbb{H}}$ into

$$\Omega_0 \equiv \{z \in \mathbb{C} : \Re(z) > 1\} \cup \{z \in \mathbb{C} : |\Im(z)| > 1\}.$$

Here and below, we fix such small $\varepsilon > 0$.

We write $\langle x' \rangle \equiv \sqrt{1 + {x_1}^2 + \cdots + {x_{n-1}}^2}$. Choose a branch of log on the domain $\mathbb{C} \setminus (-\infty, 0]$ so that $\log 1 = 0$. Define $a^z = \exp(a \log z)$ for $z \in \mathbb{C} \setminus (-\infty, 0]$. For $\sigma \in \mathbb{R}$, define a function $\varphi^{(\sigma)} : \mathbb{R}^{n-1} \times \overline{\mathbb{H}} \to \mathbb{C}$ by

$$\varphi^{(\sigma)}(x', z_n) \equiv \langle x' \rangle^\sigma \psi_\varepsilon\left(\frac{z_n}{\langle x' \rangle}\right)^\sigma = \left(\langle x' \rangle \psi_\varepsilon\left(\frac{z_n}{\langle x' \rangle}\right)\right)^\sigma, \ z \in \overline{\mathbb{H}}. \tag{5.2}$$

Set $\varphi \equiv \varphi^{(1)}$. If we restrict φ to $\mathbb{R}^{n-1} \times \mathbb{R} \subset \mathbb{R}^{n-1} \times \mathbb{C}$, then it behaves like $\langle \star \rangle$. This observation is quantified in the next lemma.

Lemma 5.1

1. *For all* $(x', z_n) \in \mathbb{R}^{n-1} \times \overline{\mathbb{H}}$ $\langle x' \rangle + |z_n| \sim |\varphi(x', z_n)|$.
2. *Let* $\alpha \in {\mathbb{N}_0}^n$. *Then we have* $|\partial^\alpha \varphi(x', z_n)| \lesssim_\alpha (\langle x' \rangle + |z_n|)^{1-|\alpha|}$ *for* $(x', z_n) \in \mathbb{R}^{n-1} \times \overline{\mathbb{H}}$.

Proof

1. We have to consider two cases: $|z_n| \le 4\langle x'\rangle$ and $|z_n| > 4\langle x'\rangle$.
   If $|z_n| > 4\langle x'\rangle$, then $\left|\psi_\varepsilon\left(\frac{|z_n|}{\langle x'\rangle}\right) + i\frac{z_n}{\langle x'\rangle}\right| < 2$ by (5.1). Since $\frac{|z_n|}{\langle x'\rangle} > 4$, we have $\left|\psi_\varepsilon\left(\frac{|z_n|}{\langle x'\rangle}\right) + i\frac{z_n}{\langle x'\rangle}\right| < \frac{|z_n|}{2\langle x'\rangle}$. Hence $\frac{|z_n|}{2\langle x'\rangle} < \left|\psi_\varepsilon\left(\frac{|z_n|}{\langle x'\rangle}\right)\right| < \frac{3|z_n|}{2\langle x'\rangle}$. Thus, $\langle x'\rangle + |z_n| \sim |\varphi(x', z_n)|$.
   If $|z_n| \le 4\langle x'\rangle$, then $\left|\psi_\varepsilon\left(\frac{|z_n|}{\langle x'\rangle}\right)\right| \le \int_{-\infty}^0 \eta(t)dt + \frac{|z_n|}{\langle x'\rangle} \le 2 + 4 = 6$. So we have $\langle x'\rangle + |z_n| \sim |\varphi(x', z_n)|$ trivially.
2. By definition,

$$\varphi(x', z_n) = \langle x'\rangle \int_{-\infty}^0 \eta(t)\exp\left(-it\varepsilon\frac{z_n}{\langle x'\rangle}\right)dt - iz_n.$$

Note that the partial differential of the second term can be easily controlled. Concentrate on the first term. By the Leibniz formula, we have

$$\begin{aligned}
&\left|\partial_{x'}^\alpha\left(\langle x'\rangle \int_{-\infty}^0 \eta(t)\exp\left(-it\varepsilon\frac{z_n}{\langle x'\rangle}\right)dt\right)\right| \\
&\lesssim \sum_{\gamma\le\alpha}\left|\partial_{x'}^{\alpha-\gamma}\langle x'\rangle\partial_{x'}^\gamma\left(\int_{-\infty}^0 \eta(t)\exp\left(-it\varepsilon\frac{z_n}{\langle x'\rangle}\right)dt\right)\right| \\
&\lesssim \sum_{\gamma\le\alpha}\langle x'\rangle^{1-|\alpha|+|\gamma|}\left|\partial_{x'}^\gamma\left(\int_{-\infty}^0 \eta(t)\exp\left(-it\varepsilon\frac{z_n}{\langle x'\rangle}\right)dt\right)\right|.
\end{aligned}$$

Since

$$\int_{-\infty}^0 \eta(t)\exp\left(-it\varepsilon\frac{z_n}{\langle x'\rangle}\right)dt = \sqrt{2\pi}\mathscr{F}\left[\eta\cdot\exp\left(\frac{\Im(z_n)\star}{\langle x'\rangle}\right)\right]\left(\varepsilon\frac{\Re(z_n)}{\langle x'\rangle}\right),$$

we have

$$\begin{aligned}
\left|\partial^\alpha\left(\langle x'\rangle \int_{-\infty}^0 \eta(t)\exp\left(-it\varepsilon\frac{z_n}{\langle x'\rangle}\right)dt\right)\right| &\lesssim \langle x'\rangle^{1-|\alpha|}\left(1+\frac{|z_n|}{\langle x'\rangle}\right)^{1-|\alpha|} \\
&= (\langle x'\rangle + |z_n|)^{1-|\alpha|}.
\end{aligned}$$

Thus, the proof is complete.

Restrict $\varphi^{(\sigma)}$ to $\mathbb{R}^{n-1}\times\mathbb{R}(\subset \mathbb{R}^{n-1}\times\overline{\mathbb{H}})$ and denote it again by $\varphi^{(\sigma)}$.

Proposition 5.1 *Let $0 < p, q \le \infty$, and let $s, \sigma \in \mathbb{R}$. Then $J_\sigma \equiv \varphi^{(\sigma)}(D)$ is an isomorphism from $A^s_{pq}(\mathbb{R}^n)$ to $A^{s-\sigma}_{pq}(\mathbb{R}^n)$ and $\|J_\sigma f\|_{A^{s-\sigma}_{pq}(\mathbb{R}^n)} \sim \|f\|_{A^s_{pq}(\mathbb{R}^n)}$ for all $f \in A^s_{pq}(\mathbb{R}^n)$.*

Proof Combine Theorem 1.53 and Lemma 5.1. We omit the details.

The lemma characterizes the support of band-limited distributions and it is known as the Paley-Wiener theorem.

Lemma 5.2 (Paley-Wiener) *Let $\varphi \in \mathscr{S}(\mathbb{R}^n)$ and $\varepsilon > 0$. Then $\mathrm{supp}(\varphi) \subset \mathbb{R}^{n-1} \times [\varepsilon, \infty)$ if and only if $\mathscr{F}^{-1}\varphi$ extends to a continuous function $\Psi : \mathbb{R}^{n-1} \times \overline{\mathbb{H}} \to \mathbb{C}$ with the following properties:*

1. *$\Psi(\xi', \star)$ is a holomorphic function on $\mathbb{H}$.*
2. *For all $(\xi', \xi_n + i\zeta_n) \in \mathbb{R}^{n-1} \times \overline{\mathbb{H}}$ and $N \in \mathbb{N}$,*

$$|\Psi(\xi', \xi_n + i\zeta_n)| \lesssim_N \langle \xi \rangle^{-N} (1 + \zeta_n)^{-N} \exp(-\varepsilon \zeta_n). \tag{5.3}$$

Proof It is easy to prove necessity. More precisely, define Ψ by

$$\Psi(\xi', \xi_n + i\zeta_n) \equiv (2\pi)^{-\frac{n}{2}} \int_{\mathbb{R}^n} \varphi(x) \mathrm{e}^{ix\cdot\xi - x_n\zeta_n} \mathrm{d}x$$

for $\xi' \in \mathbb{R}^{n-1}$, $\xi_n \in \mathbb{R}$ and $\zeta_n \in [0, \infty)$ and check (5.3).

Let us prove the sufficiency, which is difficult. Let $x \in \mathbb{R}^n$ be fixed. The Fourier transform of φ is the restriction of Ψ to $\mathbb{R}^n$. So the inverse Fourier transform of Ψ is φ. That is,

$$\varphi(x) = (2\pi)^{-\frac{n}{2}} \int_{\mathbb{R}^n} \Psi(\xi) \mathrm{e}^{-ix\cdot\xi} \mathrm{d}\xi.$$

Now let $x_n < \varepsilon$. Due to assumption (5.3) and the fact that $x_n < \varepsilon$, we can rewrite

$$\varphi(x) = (2\pi)^{-\frac{n}{2}} \int_{\mathbb{R}^n} \Psi(\xi', \xi_n + i\zeta_n) \mathrm{e}^{-ix\cdot\xi + x_n\zeta_n} \mathrm{d}\xi, \quad \zeta_n > 0$$

using the complex line integral of the variable ξ_n. Let $N = n + 1$. Since $x_n\zeta_n \le \varepsilon\zeta_n$ and (5.3), $|\varphi(x)| \lesssim (1 + \zeta_n)^{-n-1}$ for all $\zeta_n > 0$.

Since ζ_n is arbitrary, $\varphi(x) = 0$ for all $x_n \le \varepsilon$. That is, $\mathrm{supp}(\varphi) \subset \mathbb{R}^{n-1} \times [\varepsilon, \infty)$. Since $\varepsilon > 0$ is arbitrary, we obtain the desired result.

Finally, let us investigate the support of $J_\sigma f$.

Proposition 5.2 *Let $\sigma \in \mathbb{R}$. If $f \in \mathscr{S}'(\mathbb{R}^n)$ is supported on $\mathbb{R}^{n-1} \times (-\infty, 0]$, so is $J_\sigma f$.*

Proof Choose a test function $\psi \in \mathscr{D}(\mathbb{R}^{n-1} \times (0,\infty))$ arbitrarily. Since ψ is compactly supported in $\mathbb{R}^{n-1} \times (0,\infty)$, we can find $\varepsilon > 0$ so that $\mathrm{supp}(\psi) \subset \mathbb{R}^{n-1} \times [\varepsilon,\infty)$.

Then

$$\langle J_\sigma f, \psi\rangle = \langle f, \mathscr{F}[\varphi^{(\sigma)}\mathscr{F}^{-1}\psi]\rangle, \tag{5.4}$$

since $\mathscr{F}^{-1}\psi$ satisfies (5.3) of Lemma 5.2. By Lemma 5.1 the same can be said for $\varphi^{(\sigma)}\mathscr{F}^{-1}\psi$. Therefore, $\mathrm{supp}(\mathscr{F}[\varphi^{(\sigma)}\mathscr{F}^{-1}\psi]) \subset \mathbb{R}^{n-1} \times [\varepsilon,\infty)$.

Since f is supported on $\mathbb{R}^{n-1} \times (-\infty,0]$, we have $\langle J_\sigma f, \psi\rangle = 0$. Thus, we obtain the desired result.

Theorem 5.2 (Lift operator adapted to the upper half space) *Let $0 < p,q \le \infty$, and let $s,\sigma \in \mathbb{R}$.*

1. *Let $f \in A^s_{pq}(\mathbb{R}^n_+)$. Then $J_\sigma f \equiv (J_\sigma g)\,|\,\mathbb{R}^n_+$ does not depend on the choice of the representative $g \in A^s_{pq}(\mathbb{R}^n)$ of f satisfying $f = g|\mathbb{R}^n_+$.*
2. *The mapping J_σ is an isomorphism between $A^s_{pq}(\mathbb{R}^n_+)$ and $A^{s-\sigma}_{pq}(\mathbb{R}^n_+)$, whose inverse is $J_{-\sigma}$.*

Proof To prove the first assertion, suppose that $g_1, g_2 \in A^s_{pq}(\mathbb{R}^n)$ satisfies

$$f = g_1|\mathbb{R}^n_+ = g_2|\mathbb{R}^n_+.$$

Then since $(g_1 - g_2)|\mathbb{R}^n_+ = 0$, we have $(J^\sigma(g_1 - g_2))|\mathbb{R}^n_+ = 0$ by Proposition 5.2. Hence $(J^\sigma g_1)|\mathbb{R}^n_+ - (J^\sigma g_2)|\mathbb{R}^n_+ = (J^\sigma(g_1 - g_2))|\mathbb{R}^n_+ = 0$. This implies

$$(J^\sigma g_1)|\mathbb{R}^n_+ = (J^\sigma g_2)|\mathbb{R}^n_+. \tag{5.5}$$

Relation (5.5) shows that $J_\sigma f$ does not depend on $g \in A^s_{pq}(\mathbb{R}^n)$ satisfying $f = g|\mathbb{R}^n_+$. The second assertion follows from the property of J_σ, which acts on the function space $A^s_{pq}(\mathbb{R}^n)$ on $\mathbb{R}^n$.

5.1.1.3 Extension Operator

We depend upon the technique of Hestenes and Whitney. We define the extension operator and prove its boundedness.

Theorem 5.3 (Extension operators) *Let $N \in \mathbb{N}$. There exists a mapping, called the extension operator,*

$$\mathrm{Ext}_N : \bigcup_{(p,q,s)\in[N^{-1},\infty]\times[-N,N]} A^s_{pq}(\mathbb{R}^n_+) \to \bigcup_{(p,q,s)\in[N^{-1},\infty]\times[-N,N]} A^s_{pq}(\mathbb{R}^n) \tag{5.6}$$

with the following properties:

1. Ext_N *is continuous from* $A^s_{pq}(\mathbb{R}^n_+)$ *to* $A^s_{pq}(\mathbb{R}^n)$ *if restricted to* $A^s_{pq}(\mathbb{R}^n_+)$.
2. $(\mathrm{Ext}_N f) \mid \mathbb{R}^n_+ = f$ *for all* $f \in A^s_{pq}(\mathbb{R}^n_+)$.

We prove the theorem by steps.

Proof (Step 1: Construction of the Hestones extension operator Ext_N*)* Let $M \in \mathbb{N}$ be sufficiently large. Define $\lambda_1, \lambda_2, \dots, \lambda_M$ uniquely by a system of equations

$$\sum_{j=1}^{M} (-j)^l \lambda_j = 1, \quad l = 0, 1, \dots, M-1. \tag{5.7}$$

The determinant D is a constant multiple of the Vandermonde determinant $\det\{j^i\}_{i,j=1,\dots,M}$, which never vanishes. Hence from the system of equations (5.7), we can define $\lambda_1, \lambda_2, \dots, \lambda_M$ uniquely. For a function $f : \mathbb{R}^{n-1} \times [0, \infty) \to \mathbb{C}$, define

$$f^*(x) \equiv \begin{cases} f(x) & x_n \geq 0, \\ \displaystyle\sum_{j=1}^{M} \lambda_j f(x', -jx_n) & x_n \leq 0. \end{cases} \tag{5.8}$$

Then for a $\mathrm{BC}^M(\mathbb{R}^n)$-function f defined in a neighborhood of $\mathbb{R}^{n-1} \times [0, \infty)$, $f^* : \mathbb{R}^n \to \mathbb{C}$ given by (5.8), the differential coefficient agrees with f This f^* belongs to $\mathrm{BC}^M(\mathbb{R}^n)$.

Proof (Step 2: The definition of Ext_N *for* $s > \frac{n}{p}$*)* First of all, we assume $s > \frac{n}{p}$. As we have seen in Proposition 2.1, we have $A^s_{pq}(\mathbb{R}^n) \hookrightarrow \mathrm{BUC}(\mathbb{R}^n)$.

Consider the case where $\dfrac{1}{N} \leq p, q \leq \infty$, $\dfrac{n}{p} < s \leq N$.

Thanks to the definition of the norms of the function spaces on the half space, for any $f \in A^s_{pq}(\mathbb{R}^n_+)$ we can find $g \in A^s_{pq}(\mathbb{R}^n)$ so that $f = g \mid \mathbb{R}^n_+$ and that $\|g\|_{A^s_{pq}(\mathbb{R}^n)} \leq 2\|f\|_{A^s_{pq}(\mathbb{R}^n_+)}$. Let $\rho > \tilde{d}$. Define the coefficient λ by $\|\lambda\|_{\mathbf{a}_{pq}(\mathbb{R}^n),\rho} \lesssim \|g\|_{A^s_{pq}(\mathbb{R}^n)} \lesssim \|f\|_{A^s_{pq}(\mathbb{R}^n_+)}$. According to Theorem 4.8, we have the quarkonial decomposition:

$$g = \sum_{\beta \in \mathbb{N}_0{}^n} \sum_{\nu=0}^{\infty} \sum_{m \in \mathbb{Z}^n} \lambda^{\beta}_{\nu m} (\beta \mathrm{qu})_{\nu m} \tag{5.9}$$

for $g \in A^s_{pq}(\mathbb{R}^n)$. Since $A^s_{pq}(\mathbb{R}^n) \hookrightarrow \mathrm{BUC}(\mathbb{R}^n)$ for $s > \dfrac{n}{p}$,

$$g^* \equiv \sum_{\beta \in \mathbb{N}_0{}^n} \sum_{\nu=0}^{\infty} \sum_{m \in \mathbb{Z}^n} \lambda^{\beta}_{\nu m} ((\beta \mathrm{qu})_{\nu m})^*$$

can be defined without ambiguity of the representative g; it depends only on f. In fact, let $h \in A^s_{pq}(\mathbb{R}^n)$ be such that h is identical to g and f, if restricted to $\mathbb{R}^n_+$. Then for

$$(g, h) = \left(\sum_{\beta \in \mathbb{N}_0{}^n} \sum_{\nu=0}^{\infty} \sum_{m \in \mathbb{Z}^n} \lambda^\beta_{\nu m} (\beta\text{qu})_{\nu m}, \sum_{\beta \in \mathbb{N}_0{}^n} \sum_{\nu=0}^{\infty} \sum_{m \in \mathbb{Z}^n} \rho^\beta_{\nu m} (\beta\text{qu})_{\nu m} \right),$$

we have

$$(g^*, h^*) = \left(\sum_{\beta \in \mathbb{N}_0{}^n} \sum_{\nu=0}^{\infty} \sum_{m \in \mathbb{Z}^n} \lambda^\beta_{\nu m} ((\beta\text{qu})_{\nu m})^*, \sum_{\beta \in \mathbb{N}_0{}^n} \sum_{\nu=0}^{\infty} \sum_{m \in \mathbb{Z}^n} \rho^\beta_{\nu m} ((\beta\text{qu})_{\nu m})^* \right).$$

Let us show that $g^* = h^*$ on $\mathbb{R}^n$. In fact, g^* and h^* are uniformly continuous bounded functions. To prove that $g^* = h^*$ in $\mathcal{S}'(\mathbb{R}^n)$, it suffices to prove $g^*(x) = h^*(x)$ for all $x \in \mathbb{R}^n$. Since g and h are continuous functions, they coincide in the sense of $\mathcal{D}'(\mathbb{R}^n)$, if we restrict them to $\mathbb{R}^n_+$. For $x = (x', x_n)$ with $x_n > 0$, $g(x) = h(x)$. By the continuity, we have $g(x) = h(x)$ for $x = (x', x_n)$ with $x_n \geq 0$. Thus, for the definition of g^*, h^*, we need $g(x), h(x)$ only with $x \geq 0$. This implies $g^*(x) = h^*(x)$ for all $x \in \mathbb{R}^n$.

Consider

$$\text{Ext}_N f \equiv \sum_{\beta \in \mathbb{N}_0{}^n} \sum_{\nu=0}^{\infty} \sum_{m \in \mathbb{Z}^n} \lambda^\beta_{\nu m} ((\beta\text{qu})_{\nu m})^* \tag{5.10}$$

and its β-partial sum

$$\text{Ext}^\beta_N f \equiv \sum_{\nu=0}^{\infty} \sum_{m \in \mathbb{Z}^n} \lambda^\beta_{\nu m} ((\beta\text{qu})_{\nu m})^*. \tag{5.11}$$

Unlike (5.9), the definition of $\text{Ext}_N f$, the right-hand side of (5.11), cannot be regarded as a quarkonial decomposition. But if $\varepsilon > 0$ is sufficiently small, say, $0 < \varepsilon < \rho - \tilde{d}$, then we can use the atomic decomposition (see Theorem 4.1), $2^{-(\tilde{d}+\varepsilon)|\beta|}\text{Ext}^\beta_N f$. Therefore, by letting $\delta = \rho - \tilde{d} - \varepsilon > 0$,

$$\|\text{Ext}^\beta_N f\|_{A^s_{pq}(\mathbb{R}^n)} \lesssim 2^{(\rho+\varepsilon)|\beta|} \|\lambda^\beta\|_{\mathbf{a}_{pq}(\mathbb{R}^n)} \lesssim \frac{\|\lambda\|_{\mathbf{a}_{pq}(\mathbb{R}^n),\rho}}{2^{\delta|\beta|}} \lesssim \frac{\|f\|_{A^s_{pq}(\mathbb{R}^n_+)}}{2^{\delta|\beta|}}.$$

Hence if we sum this over $\beta \in \mathbb{N}_0{}^n$, Ext_N is a continuous mapping with the desired property.

Proof (Step 3: Ext_N *for* $s \in \mathbb{R}$*)* Let $L \gg 1$ and consider the mapping

$$\mathrm{Ext}_L : A^s_{pq}(\mathbb{R}^n_+) \to A^s_{pq}(\mathbb{R}^n)$$

in Step 2. Choose a positive integer $L \gg 1$ and $\sigma \in \mathbb{R}$ with $nN < -N + \sigma < N + \sigma \le L$. With the observation in Step 2, we see that $\mathrm{Ext}_L | A^s_{pq}(\mathbb{R}^n_+)$ is a continuous mapping from $A^s_{pq}(\mathbb{R}^n_+)$ to $A^s_{pq}(\mathbb{R}^n)$. We have $(\mathrm{Ext}_L f) \,|\, \mathbb{R}^n_+ = f$ for all $f \in A^s_{pq}(\mathbb{R}^n_+)$.

Recall that $J_{-\sigma}$ is a continuous mapping from $A^s_{pq}(\mathbb{R}^n_+)$ to $A^{s+\sigma}_{pq}(\mathbb{R}^n_+)$. Furthermore, J_σ is a continuous mapping from $A^{s+\sigma}_{pq}(\mathbb{R}^n)$ to $A^s_{pq}(\mathbb{R}^n)$. Hence the composition mapping

$$\mathrm{Ext}_N \equiv J_\sigma \circ \mathrm{Ext}_L \circ J_{-\sigma} : A^s_{pq}(\mathbb{R}^n_+) \to A^s_{pq}(\mathbb{R}^n)$$

makes sense.

Let us verify $\mathrm{Ext}_N f \,|\, \mathbb{R}^n_+ = f$. Choose a smooth test function $\varphi \in \mathscr{D}(\mathbb{R}^n_+)$. Let $g \in A^s_{pq}(\mathbb{R}^n)$ be a representative of f; $g|\mathbb{R}^n_+ = f$. Then by the property of Ext_L, we have

$$\begin{aligned}
\langle \mathrm{Ext}_N f \,|\, \mathbb{R}^n_+, \varphi \rangle &= \langle \mathrm{Ext}_N f, E\varphi \rangle \\
&= \langle \mathrm{Ext}_L J_{-\sigma} f, \mathscr{F}[\varphi^{(\sigma)} \mathscr{F}^{-1} E\varphi] \rangle \\
&= \langle J_{-\sigma} f, \mathscr{F}[\varphi^{(\sigma)} \mathscr{F}^{-1} E\varphi] \,|\, \mathbb{R}^n_+ \rangle.
\end{aligned}$$

By the property of the operator Ext_N and the definition of the restriction, we have

$$\langle \mathrm{Ext}_N f \,|\, \mathbb{R}^n_+, \varphi \rangle = \langle J_{-\sigma} g, \mathscr{F}[\varphi^{(\sigma)} \mathscr{F}^{-1} E\varphi] \rangle = \langle g, E\varphi \rangle = \langle f, \varphi \rangle.$$

Therefore, we have $\mathrm{Ext}_N f \,|\, \mathbb{R}^n_+ = f$ for all $f \in A^s_{pq}(\mathbb{R}^n_+)$.

5.1.1.4 Trace Operator on the Half Space $\mathbb{R}^n_+$

We extend Theorem 4.47 to the function spaces on $\mathbb{R}^n_+$.

Theorem 5.4 (The boundedness of the trace operator over $\mathbb{R}^n_+$) *Define $\mathrm{Tr}_{\mathbb{R}^n_+} f \equiv \mathrm{Tr}_{\mathbb{R}^n} [\mathrm{Ext}_N f\,]$ for $f \in A^s_{pq}(\mathbb{R}^n_+)$, where Ext_N is an extension operator in Theorem 5.3. Suppose that the parameters p, q satisfy $0 < p, q \le \infty$. Let also N be sufficiently large. Then:*

1. *The definition of $\mathrm{Tr}_{\mathbb{R}^n_+}$ does not actually depend on N.*
2. *Let $s > \dfrac{1}{p} + (n-1)\left(\dfrac{1}{p} - 1\right)_+$.*
   (a) *$\mathrm{Tr}_{\mathbb{R}^n_+}$ extends to a continuous mapping from $B^s_{pq}(\mathbb{R}^n_+)$ to $B^{s-1/p}_{pq}(\mathbb{R}^{n-1})$.*
   (b) *$\mathrm{Tr}_{\mathbb{R}^n_+}$ extends to a continuous mapping from $F^s_{pq}(\mathbb{R}^n_+)$ to $F^{s-1/p}_{pp}(\mathbb{R}^{n-1})$.*
3. *Let $s > k + \dfrac{1}{p} + (n-1)\left(\dfrac{1}{p} - 1\right)_+$.*

(a) *For all* $g_0 \in B_{pq}^{s-1/p}(\mathbb{R}^{n-1}), g_1 \in B_{pq}^{s-1/p-1}(\mathbb{R}^{n-1}), \dots, g_k \in B_{pq}^{s-1/p-k}(\mathbb{R}^{n-1})$ *there exists* $f \in B_{pq}^s(\mathbb{R}^n_+)$ *such that*

$$\mathrm{Tr}_{\mathbb{R}^n_+}(f) = g_0, \mathrm{Tr}_{\mathbb{R}^n_+}(\partial_{x_n} f) = g_1, \dots, \mathrm{Tr}_{\mathbb{R}^n_+}({\partial_{x_n}}^k f) = g_k. \tag{5.12}$$

(b) *For all* $g_0 \in F_{pp}^{s-1/p}(\mathbb{R}^{n-1}), g_1 \in F_{pp}^{s-1/p-1}(\mathbb{R}^{n-1}), \dots, g_k \in F_{pp}^{s-1/p-k}(\mathbb{R}^{n-1})$ *there exists* $f \in F_{pq}^s(\mathbb{R}^n_+)$ *such that*

$$\mathrm{Tr}_{\mathbb{R}^n_+}(f) = g_0, \mathrm{Tr}_{\mathbb{R}^n_+}(\partial_{x_n} f) = g_1, \dots, \mathrm{Tr}_{\mathbb{R}^n_+}({\partial_{x_n}}^k f) = g_k. \tag{5.13}$$

We prove this theorem in two auxiliary steps. Write $\mathbf{e}_n$ for the unit vector in the x_n direction.

Lemma 5.3 *We claim* $\mathrm{Tr}_{\mathbb{R}^n_+}[f \mid \mathbb{R}^n_+] = \lim_{\varepsilon \downarrow 0} \mathrm{Tr}_{\mathbb{R}^n}\left[\mathrm{Ext}_N[f|\mathbb{R}^n_+](\star + \varepsilon \boldsymbol{e}_n)\right]$ *for all* $f \in A_{pq}^s(\mathbb{R}^n)$.

Proof We decompose $\mathrm{Ext}_N[f|\mathbb{R}^n_+] \in A_{pq}^s(\mathbb{R}^n)$ into the sum of quarks using Theorem 4.8;

$$\mathrm{Ext}_N[f|\mathbb{R}^n_+] = \sum_{\beta \in {\mathbb{N}_0}^n} \sum_{\nu=0}^{\infty} \sum_{m \in \mathbb{Z}^n} \lambda^\beta_{\nu m} (\beta \mathrm{qu})_{\nu m}.$$

As we established in the proof of Theorem 4.47, we have

$$\mathrm{Tr}_{\mathbb{R}^n}[\mathrm{Ext}_N[f|\mathbb{R}^n_+]] = \sum_{\beta \in {\mathbb{N}_0}^n} \sum_{\nu=0}^{\infty} \sum_{m \in \mathbb{Z}^n} \lambda^\beta_{\nu m} \mathrm{Tr}_{\mathbb{R}^n}[(\beta \mathrm{qu})_{\nu m}].$$

Let $\delta > 0$. By virtue of Theorem 4.1, we have a rapid decrease with regard to β and ν in the norm of $A_{pq}^{s-\delta}(\mathbb{R}^n)$:

$$\mathrm{Tr}_{\mathbb{R}^n}[\mathrm{Ext}_N f] = \lim_{\varepsilon \downarrow 0} \left(\sum_{\beta \in {\mathbb{N}_0}^n} \sum_{\nu=0}^{\infty} \sum_{m \in \mathbb{Z}^n} \lambda^\beta_{\nu m} \mathrm{Tr}_{\mathbb{R}^n}[(\beta \mathrm{qu})_{\nu m}(\star + \varepsilon \mathbf{e}_n)] \right)$$

in the topology of $\mathscr{S}'(\mathbb{R}^n)$.

Lemma 5.4 *Let* $f \in A_{pq}^s(\mathbb{R}^n_+)$. *Then* $\mathrm{Tr}_{\mathbb{R}^n}[\mathrm{Ext}_N f] = \lim_{\varepsilon \downarrow 0} \mathrm{Tr}_{\mathbb{R}^n}[\mathrm{Ext}_N f(\star + \varepsilon \boldsymbol{e}_n)]$ *holds in* $\mathscr{S}'(\mathbb{R}^{n-1})$.

Proof The partial sum forms a bounded set in $A_{pq}^s(\mathbb{R}^n)$. That is, we have a uniform estimate over ν, β; for some κ

$$\left\| \sum_{m \in \mathbb{Z}^n} \lambda^{\beta}_{\nu m} \mathrm{Tr}_{\mathbb{R}^n} \left[(\beta \mathrm{qu})_{\nu m} \right] \right\|_{A^{s-1/p}_{pq}(\mathbb{R}^{n-1})} \lesssim 2^{-\kappa|\beta|}.$$

If we replace s with $s - \delta$, the definition of $(\beta \mathrm{qu})_{\nu m}$ undergoes a change. This then yields

$$\left\| \sum_{m \in \mathbb{Z}^n} \lambda^{\beta}_{\nu m} \mathrm{Tr}_{\mathbb{R}^n} \left[(\beta \mathrm{qu})_{\nu m} \right] \right\|_{A^{s-1/p-\delta}_{pq}(\mathbb{R}^{n-1})} \lesssim 2^{-\kappa|\beta|-\delta\nu}.$$

Since $A^{s-1/p-\delta}_{pq}(\mathbb{R}^{n-1}) \hookrightarrow B^{s-n/p-\delta}_{\infty\infty}(\mathbb{R}^{n-1})$, we have

$$\left\| \sum_{m \in \mathbb{Z}^n} \lambda^{\beta}_{\nu m} \mathrm{Tr}_{\mathbb{R}^n} \left[(\beta \mathrm{qu})_{\nu m} \right] \right\|_{B^{s-n/p-\delta}_{\infty\infty}(\mathbb{R}^{n-1})} \lesssim 2^{-\kappa|\beta|-\delta\nu}.$$

With the above observation in mind, let us prove that the convergence takes place in $\mathscr{S}'(\mathbb{R}^{n-1})$. Choose a test function $\varphi \in \mathscr{S}(\mathbb{R}^{n-1})$. Then

$$\left| \left\langle \sum_{m \in \mathbb{Z}^n} \lambda^{\beta}_{\nu m} \mathrm{Tr}_{\mathbb{R}^n} \left[(\beta \mathrm{qu})_{\nu m} \right], \varphi \right\rangle \right| \lesssim 2^{-\kappa|\beta|-\delta\nu}.$$

Hence for any $\varepsilon' > 0$, we can find a finite set $\mathfrak{A} \subset {\mathbb{N}_0}^n \times \mathbb{N}$ so that

$$\sum_{(\beta,\nu) \in {\mathbb{N}_0}^n \times \mathbb{N} \setminus \mathfrak{A}} \left| \left\langle \sum_{m \in \mathbb{Z}^n} \lambda^{\beta}_{\nu m} \mathrm{Tr}_{\mathbb{R}^n} \left[(\beta \mathrm{qu})_{\nu m} \right](\star + \varepsilon \mathbf{e}_n), \varphi \right\rangle \right| < \frac{\varepsilon'}{2}$$

with the constant independent of $\varepsilon \in (0, 1)$. The limit

$$\lim_{\varepsilon \downarrow 0} \left(\sum_{(\beta,\nu) \in \mathfrak{A}} \sum_{m \in \mathbb{Z}^n} \lambda^{\beta}_{\nu m} \mathrm{Tr}_{\mathbb{R}^n} \left[(\beta \mathrm{qu})_{\nu m} (\star + \varepsilon \mathbf{e}_n) \right] \right)$$

clearly exists. So, if we write

$$A_j(\beta, \nu) = \left\langle \sum_{m \in \mathbb{Z}^n} \lambda^{\beta}_{\nu m} \mathrm{Tr}_{\mathbb{R}^n} \left[(\beta \mathrm{qu})_{\nu m} \right](\star + \varepsilon_j \mathbf{e}_n), \varphi \right\rangle \quad (j = 1, 2)$$

for $0 < \varepsilon_1, \varepsilon_2 \leq 1$, we have

$$\sum_{(\beta,\nu)\in\mathfrak{A}} |A_1(\beta,\nu) - A_2(\beta,\nu)| < \frac{\varepsilon'}{2}$$

for $0 < \varepsilon_1, \varepsilon_2 \ll 1$. Thus, the limit exists.

Proof (of Theorem 5.4) Observe that $\mathrm{Tr}_{\mathbb{R}^n}\,[\,\mathrm{Ext}_N f(\star + \varepsilon \mathbf{e}_n)\,]$ is independent of N for all $\varepsilon > 0$. Consequently, the definition of $\mathrm{Tr}_{\mathbb{R}^n_+}$ does not depend on N by Lemma 5.3. Let us prove that $\mathrm{Tr}_{\mathbb{R}^n_+}$ satisfies (5.12) and (5.13). Recall that $\mathrm{Tr}_{\mathbb{R}^n}$ is surjective (see Theorem 4.47) on the whole space. So, we need only to show that

$$\mathrm{Tr}_{\mathbb{R}^n} f = \mathrm{Tr}_{\mathbb{R}^n_+}[\,f \mid \mathbb{R}^n_+] \in B^{s-1/p}_{pq}(\mathbb{R}^{n-1}) \text{ or } F^{s-1/p}_{pp}(\mathbb{R}^{n-1}) \tag{5.14}$$

as $f \in B^s_{pq}(\mathbb{R}^n)$ or $f \in F^s_{pq}(\mathbb{R}^n)$. This is achieved, as follows. Going through the same argument as Lemma 5.4, we have

$$\mathrm{Tr}_{\mathbb{R}^n} f = \lim_{\varepsilon\downarrow 0} \mathrm{Tr}_{\mathbb{R}^n} f(\star + \varepsilon \mathbf{e}_n), \quad \mathrm{Tr}_{\mathbb{R}^n_+}[f \mid \mathbb{R}^n_+] = \lim_{\varepsilon\downarrow 0} \mathrm{Tr}_{\mathbb{R}^n} \mathrm{Ext}_N[\,f(\star + \varepsilon \mathbf{e}_n) \mid \mathbb{R}^n_+] \tag{5.15}$$

in the topology of $\mathscr{S}'(\mathbb{R}^{n-1})$.

We have $f(\star', \star_n + \epsilon) = \mathrm{Ext}_N[f(\star', \star_n + \epsilon)|_{\mathbb{R}^n_+}]$ on $\{x_n \ge -\varepsilon/2\}$. As a result, we conclude that

$$\mathrm{Tr}_{\mathbb{R}^n} f = \lim_{\varepsilon\downarrow 0} \mathrm{Tr}_{\mathbb{R}^n} f(\star + \varepsilon \mathbf{e}_n) = \lim_{\varepsilon\downarrow 0} \mathrm{Tr}_{\mathbb{R}^n} \mathrm{Ext}_N[\,f(\star + \varepsilon \mathbf{e}_n) \mid \mathbb{R}^n_+] = \mathrm{Tr}_{\mathbb{R}^n_+}[f|\mathbb{R}^n_+]$$

from (5.15).

Exercises

Exercise 5.1 Show Theorem 5.1 using (1.3) and the Minkowski inequality.

Exercise 5.2 Let $0 < p, q \le \infty$ and $s, \sigma \in \mathbb{R}$. Show $\partial_{x_j} \circ J_\sigma = J_\sigma \circ \partial_{x_j} : A^s_{pq}(\mathbb{R}^n_+) \to A^{s-\sigma-1}_{pq}(\mathbb{R}^n_+)$.

5.1.2 Function Spaces on Bounded C^∞-Domains

Now we consider the function spaces on bounded C^∞-domains.

Definition 5.2 (Bounded C^∞-domain) A bounded domain Ω is said to be C^∞, if the boundary $\partial\Omega$ carries the structure of a C^∞-submanifold of $\mathbb{R}^n$.

5.1.2.1 Function Spaces on Bounded C^∞-Domains

Let Ω be an open set. For the time being, we do not assume that Ω is bounded nor that Ω is smooth. We can replace the definition above with the one in Exercise 5.3.

Definition 5.3 (Function spaces $B^s_{pq}(\Omega)$, $F^s_{pq}(\Omega)$ and $A^s_{pq}(\Omega)$ on domains) Let Ω be a domain on $\mathbb{R}^n$ which is not always bounded.

1. Let $0 < p, q \le \infty$ and $s \in \mathbb{R}$. Define the set $B^s_{pq}(\Omega)$ of functions by

$$B^s_{pq}(\Omega) \equiv \{g \mid \Omega \in \mathscr{D}'(\Omega) \,:\, g \in B^s_{pq}(\mathbb{R}^n)\}(\subset \mathscr{D}'(\Omega))$$

   and the (quasi-)norm is given by

$$\|f\|_{B^s_{pq}(\Omega)} \equiv \inf\{\|g\|_{B^s_{pq}(\mathbb{R}^n)} \,:\, g \in B^s_{pq}(\mathbb{R}^n),\, f = g|\Omega\} \quad (f \in B^s_{pq}(\Omega)).$$

2. Let $0 < p < \infty$, $0 < q \le \infty$ and $s \in \mathbb{R}$. Define the set $F^s_{pq}(\Omega)$ of functions by

$$F^s_{pq}(\Omega) \equiv \{g \mid \Omega \in \mathscr{D}'(\Omega) \,:\, g \in F^s_{pq}(\mathbb{R}^n)\}(\subset \mathscr{D}'(\Omega))$$

   and the (quasi-)norm is given by

$$\|f\|_{F^s_{pq}(\Omega)} \equiv \inf\{\|g\|_{F^s_{pq}(\mathbb{R}^n)} \,:\, g \in F^s_{pq}(\mathbb{R}^n),\, f = g|\Omega\} \quad (f \in B^s_{pq}(\Omega)).$$

3. Denote by $A^s_{pq}(\Omega)$ either $B^s_{pq}(\Omega)$ or $F^s_{pq}(\Omega)$. It is tacitly understood that $p < \infty$ when $A^s_{pq}(\Omega)$ stands for Triebel–Lizorkin space $F^s_{pq}(\Omega)$.

In terms of algebra, we can say that

$$A^s_{pq}(\Omega) = A^s_{pq}(\mathbb{R}^n)/\{f \in A^s_{pq}(\mathbb{R}^n) \,:\, f|\Omega = 0\}$$

with coincidence of norms.

The next theorem is the same as that on $\mathbb{R}^n_+$.

Theorem 5.5 (Elementary properties of $A^s_{pq}(\Omega)$ as Banach spaces) *Let Ω be a domain, and let $0 < p, q \le \infty$, $s \in \mathbb{R}$. Then:*

1. *$A^s_{pq}(\Omega)$ is a quasi-Banach space under its norm.*
2. *$A^s_{pq}(\Omega)$ is a Banach space under its norm, provided that $p, q \ge 1$.*
3. *For all $f, g \in A^s_{pq}(\Omega)$,*

$$\|f+g\|_{A^s_{pq}(\Omega)}{}^{\min(1,p,q)} \le \|f\|_{A^s_{pq}(\Omega)}{}^{\min(1,p,q)} + \|g\|_{A^s_{pq}(\Omega)}{}^{\min(1,p,q)}.$$

5.1.2.2 Extension Operator

Let us prove the extension theorem. This is where we need what we have been gathering. A formulation like Theorem 5.3 is available. But we formulate the result in a somewhat simplified manner.

Theorem 5.6 (Extension operator on bounded C^∞-domains) *Let $N \in \mathbb{N}$, and let Ω be a bounded C^∞-domain and p, q, s be parameters satisfying $N^{-1} \le p, q \le \infty$ and $|s| \le N$. Then there exists a mapping* $\mathrm{Ext}_N : A^s_{pq}(\Omega) \mapsto A^s_{pq}(\mathbb{R}^n)$ *such that* $(\mathrm{Ext}_N f)| \Omega = f$.

Proof By Exercise 5.3, we can assume that each point $x \in \partial\Omega$ has a neighborhood of U_x of the form $Q_{\hat{j}}(r)$. Note that $\Omega \cap U_x$ has expressions (5.16) and (5.17). Furthermore, for all $x \in \Omega$, we can find U_x such that $U_x \subset \Omega$. Then $\{U_x\}_{x\in\partial\Omega} \cup \{U_x\}_{x\in\Omega}$ forms an open covering of a bounded closed set $\overline{\Omega}$. We can thus find finite sets $F_1 \subset \partial\Omega$ and $F_2 \subset \Omega$ such that $\{U_x\}_{x\in F_1\cup F_2}$ is an open covering of $\overline{\Omega}$. That is, there exist $\mathscr{U} = \{U_1, U_2, \dots, U_N\} \subset \{U_x\}_{x\in\partial\Omega}$ and $\mathscr{V} = \{V_1, V_2, \dots, V_N\} \subset \{U_x\}_{x\in\Omega}$ such that $\mathscr{U} \cup \mathscr{V}$ is an open covering of $\overline{\Omega}$. Choose a partition of unity $\mathscr{A} \equiv \{\Phi_{(j)}, \Psi_{(j)}\}_{j=1,\dots,N}$ subordinate to the open covering $\mathscr{U} \cup \mathscr{V}$ of $\overline{\Omega}$. We decompose $f \in A^s_{pq}(\Omega)$; $f = \sum_{j=1}^N \Phi_{(j)} f + \sum_{j=1}^N \Psi_{(j)} f$. Then the support of $\Psi_{(j)} f$ being compact in Ω, this can be extended naturally outside the support.

Here and below, let us consider the extension of a function of the form $\Phi_{(j)} f$. Hence from the start Ω can be assumed to take the form of (5.16) or (5.17). A symmetry allows us to concentrate on the case of (5.16) without any loss of generality. Furthermore, Theorem 4.37 allows us to assume that there exists a compact set K of $Q_{\hat{j}}(r)$ such that $\mathrm{supp}(f) \subset K$.

Define a diffeomorphism from $\mathbb{R}^n$ to itself by $\varphi(x) \equiv (x', x_n - \omega(x'))$ for $x = (x', x_n)$. Then since $f \circ \varphi^{-1} \in \mathscr{D}'(\mathbb{R}^n)$, we can define $\mathrm{Ext}_N f \equiv \mathrm{Ext}_N^{\mathbb{R}^n_+}[f \circ \varphi^{-1}] \circ \varphi$.

When we consider $\varphi(x) = (x', x_n - \omega(x'))$, we use $\omega \in \mathscr{D}(\mathbb{R}^{n-1})$.

By Theorem 4.46, we obtain Ext_N satisfying the desired conditions.

Exercises

Exercise 5.3 Let $r > 0$ and $j = 1, 2, \dots, n$. Define

$$Q_{\hat{j}}(r) \equiv (x_1 - r, x_1 + r) \times \cdots \times (x_{j-1} - r, x_{j-1} + r)$$
$$\times \mathbb{R} \times (x_{j+1} - r, x_{j+1} + r) \times \cdots \times (x_n - r, x_n + r).$$

Let Ω be a bounded domain. Then show that Ω is a bounded C^∞-domain if and only if for all $x \in \partial\Omega$, there exist $j \in \{1, 2, \dots, n\}$, a $C^\infty_c(\mathbb{R}^n)$-function $\omega : \mathbb{R}^{n-1} \to \mathbb{R}$ and $r > 0$ such that

$$\Omega \cap Q_{\hat{j}}(r) = \{x \in Q_{\hat{j}}(r) \,:\, x_j > \omega(x_1, x_2, \ldots, x_{j-1}, x_{j+1}, \ldots, x_n)\} \tag{5.16}$$

or that

$$\Omega \cap Q_{\hat{j}}(r) = \{x \in Q_{\hat{j}}(r) \,:\, x_j < \omega(x_1, x_2, \ldots, x_{j-1}, x_{j+1}, \ldots, x_n)\}. \tag{5.17}$$

We refer to a book [50] for manifolds.

Exercise 5.4 Let Ω be a bounded domain. Show that $A^s_{p_0q}(\Omega) \subset A^s_{p_1q}(\Omega)$ using the wavelet decomposition whenever $0 < p_1 < p_0 < \infty$, $0 < q \le \infty$ and $s \in \mathbb{R}$.

Exercise 5.5 Let Ω be a bounded domain. Show that the embedding $A^{s+\varepsilon}_{p_0q}(\Omega) \subset A^s_{p_1q}(\Omega)$ is compact using the wavelet decomposition whenever $\varepsilon > 0$, $0 < p_1 < p_0 < \infty$, $0 < q \le \infty$ and $s \in \mathbb{R}$.

Exercise 5.6 (Poincaré's lemma for $A^s_{pq}(\mathbb{R}^n)$) Let $s \ge 1$ and $1 < p, q < \infty$, and let Ω be a bounded C^∞-domain. Define

$$A^s_{pq}(\Omega)_0 \equiv \{f \in A^s_{pq}(\mathbb{R}^n) \cap \mathscr{P}_0(\mathbb{R}^n)^\perp \,:\, \mathrm{supp}(f) \subset \Omega\}.$$

Show $\|f\|_{A^s_{pq}(\Omega)} \lesssim \sum_{j=1}^n \|\partial_{x_j} f\|_{A^{s-1}_{pq}(\Omega)}$. Hint: Assume otherwise. Then we can find a sequence $\{f_k\}_{k=1}^\infty \subset A^s_{pq}(\Omega)_0 \subset A^s_{pq}(\mathbb{R}^n)$ such that

$$\|f_k\|_{A^s_{pq}(\Omega)} = 1 \ge k \sum_{j=1}^n \|\partial_{x_j} f_k\|_{A^{s-1}_{pq}(\Omega)}.$$

Since $A^s_{pq}(\Omega) \hookrightarrow A^{s-1}_{pq}(\Omega)$ is compact, we can suppose that $\{f_k\}_{k=1}^\infty$ is convergent to f in $A^{s-1}_{pq}(\mathbb{R}^n)$. Observe also that $\partial_{x_j} f_k \to 0$ as $k \to \infty$ in $A^{s-1}_{pq}(\mathbb{R}^n)$. Thus, $f_k \to f$ in $A^s_{pq}(\mathbb{R}^n)$. Thus, by the Fatou theorem, we can conclude $\partial_{x_k} f = 0$. Keeping these in mind, what can we say?

Exercise 5.7 Let Ω_1, Ω_2 be open sets. Define a closed subspace W of $A^s_{pq}(\Omega_1) \oplus A^s_{pq}(\Omega_2)$ by

$$W \equiv \{(f, g) \in A^s_{pq}(\Omega_1) \oplus A^s_{pq}(\Omega_2) \,:\, f|\Omega_1 \cap \Omega_2 = g|\Omega_1 \cap \Omega_2\}.$$

Then show that $f \in A^s_{pq}(\Omega_1 \cup \Omega_2) \mapsto (f|\Omega_1, f|\Omega_2) \in W$ is an isomorphism using the partition of unity subordinated to the covering $\{\Omega_1, \Omega_2\}$.

Exercise 5.8 [931, Proposition 3.15] Let $U_0 = B(\sqrt{(4n)^{-1}})$ and

$$U_j^{\pm} \equiv \left\{x \in B(1) \,:\, \pm x_j > \frac{1}{2n}\right\}$$

for $j = 1, 2, \ldots, n$. Choose a partition of unity $\{\Phi_0\} \cup \{\Phi_j^{\pm}\}_{j=1}^n$ associated with $\{U_0\} \cup \{U_j^{\pm}\}_{j=1}^n$. Define

$$\psi_+^{(j)}(x) \equiv (x', \sqrt{1 - |x'|^2} - x_n), \quad \psi_-^{(j)}(x) \equiv (x', x_n + \sqrt{1 - |x'|^2}) \quad (x \in B(1))$$

for $j = 1, 2, \ldots, n$. Then show that

$$\|f\|_{A^s_{pq}(B(1))} \sim \|\Phi_0 f\|_{A^s_{pq}(B(1))} + \sum_{j=1}^n (\|(\Phi_j^+ f) \circ (\psi_+^{(j)})^{-1}\|_{A^s_{pq}(B(1))} + \|(\Phi_j^- f) \circ (\psi_-^{(j)})^{-1}\|_{A^s_{pq}(B(1))})$$

for all $f \in \mathscr{D}'(B(1))$ using Theorem 4.46.

Exercise 5.9 Let $0 < p_0, q_0, p_1, q_1 < \infty, s_0, s_1 \in \mathbb{R}$. By the wavelet decomposition and the extension operators, prove that $B^{s_0}_{p_0q_0}(B(1)) \subset B^{s_1}_{p_1q_1}(B(1))$ in the set theoretical sense and the embedding is compact if and only if $s_0 - s_1 > n(1/p_0 - 1/p_1)_+$.

Exercise 5.10 [297] Let Ω_1 and Ω_2 be domains in $\mathbb{R}^n$ with nonempty intersection.

1. By the use of the partition of unity, show that for all $(f, g) \in \mathscr{D}'(\Omega_1) \times \mathscr{D}'(\Omega_2)$ such that $f|\Omega_1 \cap \Omega_2 = g|\Omega_1 \cap \Omega_2$, there exists $h \in \mathscr{D}'(\Omega_1 \cup \Omega_2)$ such that $f = h|\Omega_1$ and that $g = h|\Omega_2$.
2. Let $f \in A^s_{pq}(\Omega_1 \cup \Omega_2)$. Using a partition of unity, show that

$$\|f\|_{A^s_{pq}(\Omega_1 \cup \Omega_2)} \lesssim \|f|\Omega_1\|_{A^s_{pq}(\Omega_1)} + \|f|\Omega_2\|_{A^s_{pq}(\Omega_2)}.$$

Exercise 5.11 Let Ω, Ω' be domains.

1. Prove Theorem 5.5 using corresponding assertions for $\mathbb{R}^n$.
2. If $\Omega' \subset \Omega$, show that $A^s_{pq}(\Omega) \hookrightarrow A^s_{pq}(\Omega')$.

5.1.3 Function Spaces on Uniformly C^m-Open Sets

As we have discussed above, we can define the function spaces on bounded C^∞-domains. But more generally, we can define function spaces on uniformly C^m-domains:

Definition 5.4 (Uniformly C^m-open sets) Let Ω be a domain.

1. A domain Ω is said to be a *uniformly* C^mt*-domain* if there exists $\{O_j\}_{j\in J}$, where $J \subset \mathbb{N}$, and a collection $\{\Phi_j = (\Phi_{j1}, \Phi_{j2}, \ldots, \Phi_{jn}) : O_j \to B(1)\}_{j\in J}$ of diffeomorphisms such that the following conditions hold:
   (a) For $j \in J$
   $$\Phi_j(O_j \cap \Omega) = \{y \in B(1) : y_n > 0\}, \quad \Phi_j(O_j \cap \partial\Omega) = \{y \in B(1) : y_n = 0\}.$$
   (b) Set $O'_j \equiv \Phi_j^{-1}(B(2^{-1}))$ for $j \in J$ and then
   $$\partial\Omega \subset \bigcup_{j\in J} O'_j, \quad \operatorname{dist}\left(\partial\Omega, \mathbb{R}^n \setminus \bigcup_{j\in J} O'_j\right) > 0.$$
   (c) $\sup_{x\in\mathbb{R}^n} \sharp\{j \in J : B(x,1) \cap O_j \neq \emptyset\} < \infty$.
   (d) Let $\Psi_j \equiv (\Psi_{j1}, \Psi_{j2}, \ldots, \Psi_{jn})$ be an inverse mapping of Φ_j. The functions $\partial^\alpha \Phi_{ji}, \partial^\alpha \Psi_{ji}$ are bounded uniformly over $j \in J$ and $i = 1, 2, \ldots, n$ and multi-indexes α with $0 < |\alpha| \le m$. Furthermore, we have
   $$|\Psi_j(y) - \Psi_j(0)| \lesssim 1, \quad |\Phi_{jn}(x)| \lesssim \operatorname{dist}(x, \partial\Omega)$$
   uniformly over $j \in J$.
2. Maintain the same setting above. Choose an auxiliary function $\kappa \in \mathscr{S}(\mathbb{R}^n)$ so that $\chi_{B(2)} \le \kappa \le \chi_{B(3)}$. For $j \in J$, define $\kappa_{(j)} \in C_c^\infty(\mathbb{R}^n)$ so that $\kappa_{(j)}(x) \equiv \kappa(4\Phi_j(x))$. Furthermore, take $\tau \in C_c^\infty(\Omega)$ so that
   $$\Psi(x) \equiv \tau(x) + \sum_{j\in J} \kappa_{(j)}(x) > 0$$
   for all x in an open set containing $\overline{\Omega}$. Define $\{\rho^{(j)}\}_{j\in J\cup\{0\}}$ by $\rho^{(0)} \equiv \dfrac{\tau}{\Psi}$, $\rho^{(j)} \equiv \dfrac{\kappa_{(j)}}{\Psi}$ for $j \in J$. Call $\{\rho^{(j)}\}_{j\in J\cup\{0\}}$ the *partition of unity subordinate to* $\{O_j\}_{j\in J}$.

5.1.3.1 Extension Operator

We can consider the boundedness of the extension operator and many other results. But here we consider the following property:

Proposition 5.3 *Let* $0 < p < \infty$, $0 < q \le \infty$ *and* $s \in \mathbb{R}$, *and let* Ω *be a uniform* C^m*-domain, where* $m \in \mathbb{N}$ *be sufficiently large according to* p, q, s. *Maintain the same notation as Definition* 5.4. *Furthermore, let* $\{\rho^{(j)}\}_{j\in J\cup\{0\}}$ *be the* *corresponding partition of unity. Then* $\|f\|_{F^s_{pq}(\Omega)} \sim \left(\sum_{j\in J\cup\{0\}} \|\rho^{(j)} f\|_{F^s_{pq}(\Omega)}{}^p\right)^{\frac{1}{p}}$ *for all* $f \in F^s_{pq}(\Omega)$ *with implicit constants independent of* f.

Proof Let E_j be the extension operator of the function space $F^s_{pq}(\Omega \cap O'_j)$. Then by multiplying a $C^\infty_c(\mathbb{R}^n)$-function, for each $j \in J$, we can find $f \in F^s_{pq}(\Omega)$ supported on $\Phi_j{}^{-1}(B(e^{-2}))$ so that

$$\mathrm{supp}(E_j f) \subset \Phi_j{}^{-1}(B(e^{-1})), \quad \mathrm{supp}(E_j f) \cap \Omega = \mathrm{supp}(f) \cap \Omega.$$

By virtue of the property of the extension operator E_j, we have

$$\|\rho^{(j)} f\|_{F^s_{pq}(\Omega)} \sim \|E_j[\rho^{(j)} f]\|_{F^s_{pq}(\Omega)}.$$

Furthermore, since E_j is an extension operator, we have

$$f = \sum_{j\in\Omega} E_j[\rho^{(j)} f]$$

on Ω. Thus, we can go through the same argument as Theorem 4.38.

5.1.3.2 Various Domains

There are many other notions of domains on which the theory of function spaces are staged. Here we content ourselves with the definitions.

Definition 5.5 (**(ε, δ)-domains**) A proper open subset Ω is an (ε, δ)-*domain* if for any $x, y \in \Omega$ satisfying $|x - y| \le \delta$, there exists a rectifiable path Γ of length $\le C_0|x - y|$, connecting x and y, such that for each $z \in \Gamma$,

$$\mathrm{dist}(z, \partial\Omega) \ge \varepsilon \min(|x - z|, |y - z|).$$

Definition 5.6 (Domains satisfying the horn condition) A domain Ω in $\mathbb{R}^n$ is said to satisfy the *flexible horn condition* if there exist $\delta \in (0, 1)$ and $T \in (0, \infty)$ with the following property: for any $x \in G$, there exists a Lipschitz continuous curve $\rho : [0, T] \to \mathbb{R}^n$ with norm less than or equal to 1 such that $\rho(0) = 0$ and that $x + \rho(t) + [-t\delta, t\delta]^n \subset G$ for all $t \in [0, T]$.

Definition 5.7 (John domain) A domain $\Omega \subset \mathbb{R}^n$ is said to be a *John domain* if there exist $c > 1$ and $x_0 \in \Omega$ such that for all $x \in \Omega$ there exists a curve $\gamma : [0, \ell] \to \Omega$, parametrized according to its length, such that $\gamma(0) = x$, $\gamma(\ell) = x_0$ and that $\mathrm{dist}(\gamma(t), \partial\Omega) \ge ct$.

Exercises

Exercise 5.12 Let $\varphi \in \mathrm{Lip}(\mathbb{R}^n) \cap C^m(\mathbb{R}^n)$. Then $\{(x, x_{n+1}) : x_{n+1} > \varphi(x)\}$ is a uniformly bounded C^m-domain.

Exercise 5.13 Let $f, g \in A^s_{pq}(\mathbb{R}^n)$ be such that $f = g$ away from a compact set $K \subset \mathbb{R}^n$, that is $f|(\mathbb{R}^n \setminus K) = g|(\mathbb{R}^n \setminus K)$. Assume that for all $x \in \mathbb{R}^n$ there exists a neighborhood U_x of x such that $f|U_x = g|U_x$. Then show that $f = g$ using Theorem 4.37.

5.1.4 Function Spaces on Lipschitz Domains

Let $\omega : \mathbb{R}^{n-1} \to \mathbb{R}$ be a Lipschitz function. Define its homogeneous Lipschitz norm by

$$\|\omega\|_{\mathrm{Lip}} \equiv \inf\{L > 0 : |\omega(x) - \omega(y)| \le L|x - y|\}.$$

Then the domain

$$\Omega = \{x = (x', x_n) \in \mathbb{R}^n : x_n > \omega(x')\} \tag{5.18}$$

is called the Lipschitz domain. We consider function spaces on such an Ω.

5.1.4.1 Universal Extension Operators

We follow the idea of Rychkov [900] here. In words of Kufner [51, p, 25, Section 4.9], the key property here is the *inner cone property*, which we describe. In a word, the inner cone property ensures that there exists a reference cone so that every point of the domain is realized as a cone which is inside the cone and is congruent to the reference cone. Define a cone $+K$ and its *reflection* $-K$ by $\pm K \equiv \{y = (y', y_n) \in \mathbb{R}^n : \pm y_n > \|\omega\|_{\mathrm{Lip}}|y'|\}$. Since ω is a Lipschitz function, we have $x + y \in \Omega$ as long as $x \in \Omega$ and $y \in K$.

Choose $\Psi \in \mathscr{S}(\mathbb{R}^n) \setminus \mathscr{P}_0(\mathbb{R}^n)^\perp$ so that $\mathrm{supp}(\Psi) \subset K$ and that

$$\Phi \equiv \Psi - \Psi^{-1} = \Psi - 2^{-n}\Psi(2^{-1}\star).$$

Let $L \gg 1$. Then there exist $\eta, \psi \in C^\infty_{\mathrm{c}}(\mathbb{R}^n)$ such that $\varphi \equiv \eta - \eta^{-1} \in \mathscr{P}_L(\mathbb{R}^n)^\perp$, and that $\psi * \Psi + \sum_{j=1}^\infty \varphi^j * \Phi^j = \delta$ thanks to Theorem 1.39. As is seen in the proof of Theorem 1.39, due to the inner cone property, we can choose ψ and φ so that they are supported on K. We define

$$\mathscr{M}^{\Omega}_{0,N} f(x) \equiv \sup_{y \in \Omega} \langle x-y \rangle^{-N} |\Psi * f(x)|, \; x \in \Omega$$

and, for $j \geq 1$, we set

$$\mathscr{M}^{\Omega}_{j,N} f(x) \equiv \sup_{y \in \Omega} \langle 2^j(x-y) \rangle^{-N} |\Phi^j * f(x)|, \; x \in \Omega.$$

Observe that $\Psi * f$ depends on the "value" of f on Ω.

Theorem 5.7 (Local means for domains-I) *Let Ω be a Lipschitz domain defined by* (5.18). *Let $f \in \mathscr{D}'(\Omega)$.*

1. *We have norm equivalence:*

$$\|f\|_{B^s_{pq}(\Omega)} \sim \left(\sum_{j=0}^{\infty} 2^{jqs} \|\mathscr{M}^{\Omega}_{j,N} f\|^q_{L^p(\Omega)} \right)^{\frac{1}{q}} \tag{5.19}$$

for $0 < p, q \leq \infty$ and $s \in \mathbb{R}$.

2. *We have norm equivalence:*

$$\|f\|_{F^s_{pq}(\Omega)} \sim \left\| \left(\sum_{j=0}^{\infty} (2^{js} \mathscr{M}^{\Omega}_{j,N} f)^q \right)^{\frac{1}{q}} \right\|_{L^p(\Omega)} \tag{5.20}$$

for $0 < p < \infty$, $0 < q \leq \infty$ and $s \in \mathbb{R}$.

More precisely, if the right-hand side of (5.19) *or* (5.20) *is finite, then $f \in A^s_{pq}(\Omega)$. Conversely, if $f \in A^s_{pq}(\Omega)$, the right-hand side of* (5.19) *or* (5.20) *is finite, Furthermore, the norm equivalence* (5.19) *or* (5.20) *holds.*

Proof By the definition of the function spaces on domains, $\gtrsim$ is clear. However, the reverse inequality is not clear. In general, for a function g defined on Ω, we write g_Ω for the zero extension of g outside Ω. Define

$$F \equiv \psi * [(\Psi * f)_\Omega] + \sum_{j=1}^{\infty} \varphi^j * [(\Phi^j * f)_\Omega]$$

for $f \in \mathscr{D}'(\Omega)$. We have $F|\Omega = f$, since

$$\psi * \Psi + \sum_{j=1}^{\infty} \varphi^j * \Phi^j = \delta \tag{5.21}$$

in $\mathscr{D}'(\Omega)$. Let us prove $\|F\|_{F^s_{pq}(\mathbb{R}^n)} \sim \|f\|_{F^s_{pq}(\Omega)}$. As in the proof of Theorem 2.34 on $\mathbb{R}^n$ we can prove

$$\|F\|_{F^s_{pq}(\mathbb{R}^n)} \lesssim \left\| \sup_{y\in\Omega} \frac{|\Psi * f(y)|}{\langle \star - y\rangle^N} \right\|_{L^p(\mathbb{R}^n)} + \left\| \left(\sum_{j=1}^{\infty} \sup_{y\in\Omega} \frac{|\Phi^j * f(y)|^q}{\langle 2^j(\star - y)\rangle^{Nq}} \right)^{\frac{1}{q}} \right\|_{L^p(\mathbb{R}^n)}.$$

Here we have

$$\begin{aligned}
&|x'-y'|^2 + (y_n + x_n - 2\omega(x'))^2 \\
&\sim |x'-y'|^2 + (y_n - \omega(y') + x_n - \omega(x'))^2 \\
&\gtrsim |x'-y'|^2 + (y_n - \omega(y') - x_n + \omega(x'))^2 + (\omega(y') - \omega(x'))^2 \\
&\gtrsim |x'-y'|^2 + |y_n - x_n|^2 = |x-y|^2
\end{aligned}$$

by the Lipschitz continuity of ω as long as $(x', x_n) \in \Omega$ and $(y', y_n) \in \Omega$.

Hence a change of variables: $(x', x_n) \in \mathbb{R}^n \setminus \Omega \mapsto (x', 2\omega(x') - x_n) \in \Omega$ yields

$$\left\| \sup_{y\in\Omega} \frac{|\Psi * f(y)|}{\langle \star - y\rangle^N} \right\|_{L^p(\mathbb{R}^n\setminus\Omega)} \lesssim \left\| \sup_{y\in\Omega} \frac{|\Psi * f(y)|}{\langle \star - y\rangle^N} \right\|_{L^p(\Omega)}$$

$$\left\| \left(\sum_{j=1}^{\infty} \sup_{y\in\Omega} \frac{|\Phi^j * f(y)|^q}{\langle 2^j(\star - y)\rangle^{Nq}} \right)^{\frac{1}{q}} \right\|_{L^p(\mathbb{R}^n\setminus\Omega)} \lesssim \left\| \left(\sum_{j=1}^{\infty} \sup_{y\in\Omega} \frac{|\Phi^j * f(y)|^q}{\langle 2^j(\star - y)\rangle^{Nq}} \right)^{\frac{1}{q}} \right\|_{L^p(\Omega)}.$$

Hence Theorem 5.7 is proved.

By Lemma 1.12, we can arrange that Φ satisfy the moment condition of order ∞. For $0 < p_1, p_2, q_1, q_2 \le \infty$ and $s_1, s_2 \in \mathbb{R}$, we abbreviate $A1 \equiv A^{s_1}_{p_1 q_1}(\mathbb{R}^n)$ and $A2 \equiv A^{s_2}_{p_2 q_2}(\mathbb{R}^n)$. It can happen that $\{A1, A2\} = \{B, F\}$.

Theorem 5.8 (Extension operator of Lipschitz domains, Universal extension operators) *Let Ω be a Lipschitz domain given by* (5.18). *Let* $0 < p, q \le \infty$, $s \in \mathbb{R}$. *We can define a bounded linear operator* $\mathrm{Ext}_{A^s_{pq}} : A^s_{pq}(\Omega) \to A^s_{pq}(\mathbb{R}^n)$ *so that we have the following compatibility:* $\mathrm{Ext}_{A1}(f) = \mathrm{Ext}_{A2}(f)$ *for* $f \in A1 \cap A2$.

Theorem 5.9 (Local means on domains-II) *Let Ω be a Lipschitz domain given by* (5.18). *Let $f \in \mathscr{D}'(\Omega)$.*

1. *We have the following norm equivalence:*

$$\|f\|_{B^s_{pq}(\Omega)} \sim \|\psi * f\|_{L^p(\Omega)} + \left(\sum_{j=1}^{\infty} 2^{jqs} \|\varphi^j * f\|^q_{L^p(\Omega)} \right)^{\frac{1}{q}} \tag{5.22}$$

for $0 < p, q \le \infty$ *and* $s \in \mathbb{R}$.

2. *We have the following norm equivalence:*

$$\|f\|_{F^s_{pq}(\Omega)} \sim \|\psi * f\|_{L^p(\Omega)} + \left\| \left(\sum_{j=1}^{\infty} 2^{jqs} |\varphi^j * f|^q \right)^{\frac{1}{q}} \right\|_{L^p(\Omega)} \tag{5.23}$$

for $0 < p < \infty$, $0 < q \le \infty$ *and* $s \in \mathbb{R}$.

More precisely, if the right-hand side (5.22) *or* (5.23) *is finite, then* $f \in A^s_{pq}(\Omega)$. *Conversely, if* $f \in A^s_{pq}(\Omega)$, *then the right-hand side of* (5.22) *or* (5.23) *is finite. Furthermore, in this case, we have* (5.22) *or* (5.23).

Proof Let $0 < r < \min(1, p, q)$. Choose smooth functions Ψ, Φ such that $0 \le \Psi \le \chi_{-K}$, that $\|\Psi\|_1 = 1$ and that $\Phi = \Psi^1 - \Psi^0$. Let $x \in \mathbb{R}^n$. Argue as we did in Theorem 1.39 to have $\mu > 0$ such that

$$2^{js} \mathcal{M}^{\Omega}_{j,N} f(x) \lesssim_r 2^{-\mu|j|} M^{(r)}[|\psi * f|_{\Omega}](x) + \sum_{k=1}^{\infty} 2^{-\mu|j-k|} M^{(r)}[|2^{ks} \varphi^k * f|_{\Omega}](x)$$

for all positive integers $j \in \mathbb{N}$ and $x \in \mathbb{R}^n$. Hence we obtain the proof of Theorem 5.9 by the Fefferman–Stein vector-valued maximal inequality (Theorem 1.49).

As an application, we obtain the following result:

Theorem 5.10 (A characterization of elements in $\mathcal{D}'(\Omega)$ obtained by the restriction of $\mathcal{S}'(\mathbb{R}^n)$ to Ω) *Let* Ω *be a Lipschitz domain. A distribution* $f \in \mathcal{D}'(\Omega)$ *can be expressed as* $f = g|\Omega$ *for some* $g \in \mathcal{S}'(\mathbb{R}^n)$ *if and only if there exists* $N \in \mathbb{N}$ *such that*

$$|\langle f, \kappa(x - \star)|\Omega\rangle| \lesssim_f p_N(\kappa)\langle x\rangle^N \tag{5.24}$$

whenever $\kappa \in C^\infty_c(\mathbb{R}^n)$ *and* $x \in \mathbb{R}^n$ *satisfies* $\operatorname{supp}(\kappa(x - \star)) \subset \Omega$.

Proof In order that there exists $g \in \mathcal{S}'(\mathbb{R}^n)$ such that $f = g|\Omega$, we need to have (5.24). Conversely, suppose that f satisfies (5.24). Then $F \equiv (1+|x|^2)^{-N} f \in \mathcal{D}'(\Omega)$ satisfies

$$|\langle f, \kappa(x - \star)|\Omega\rangle| \lesssim_f p_N(\kappa).$$

Hence it follows that

$$\mathcal{M}^{\Omega}_{j,N} f(x) \lesssim 2^{j(n+N)}. \tag{5.25}$$

Note that (5.25) means that $F \in B^{-n-N}_{\infty\infty}(\Omega) \subset \mathscr{S}'(\mathbb{R}^n)$ thanks to Theorem 5.7. In particular, there exists $G \in \mathscr{S}'(\mathbb{R}^n)$ such that $F = G|\Omega$. Set $g \equiv \langle x\rangle^{2N}G \in \mathscr{S}'(\mathbb{R}^n)$. Then $f = g|\Omega$.

We have the following variant:

Remark 5.1 Let $\mathscr{S}'(\Omega)$ be the set of all $f \in \mathscr{D}'(\Omega)$ satisfying

$$|\langle f, \varphi\rangle| \lesssim p_N(\varphi) \quad (\varphi \in \mathscr{D}(\Omega) \subset \mathscr{S}(\mathbb{R}^n))$$

for some $N \in \mathbb{N}$. Then a distribution $f \in \mathscr{D}'(\Omega)$ can be expressed as $f = g|\Omega$ for some $g \in \mathscr{S}'(\mathbb{R}^n)$ if and only if $f \in \mathscr{S}'(\Omega)$.

One of the interesting things after proving the theorems in the section is to compare the theorems in many textbooks such as [4, 51, 57].

5.1.4.2 Measuring the Strength of the Compact Embedding

There are a couple of ways to approach this.

Definition 5.8 (Entropy number, Approximation number, best m-term approximation) Let X, Y be quasi-Banach spaces such that $X \hookrightarrow Y$ in the sense of compact embedding. Denote by $\iota_{X\to Y}$ this compact embedding.

1. We define the *entropy number* $e_k = e_k(X \hookrightarrow Y)$ *with degree* $k \in \mathbb{N}$ for the compact embedding $X \hookrightarrow Y$ as the infimum of $\varepsilon > 0$ such that there exist $y_1, y_2, \ldots, y_{2^{k-1}} \in Y$ such that $\min\limits_{1\le j\le 2^{k-1}} \|x - y_j\|_Y \le \varepsilon$ for all $x \in X$ with norm less than or equal to 1.
2. We define the *approximation number* $a_k = a_k(X \hookrightarrow Y)$ *with degree* $k \in \mathbb{N}_0$ for the compact embedding $X \hookrightarrow Y$ by

$$a_k = \inf\{\|\iota_{X\to Y} - L\|_{X\to Y} \,:\, L \in B(X,Y) \text{ has rank less than or equal to } k-1\}.$$

3. Let $\{\Phi_j\}_{j=1}^\infty$ be a countable set in quasi-Banach space. Then the *best m-term approximation with respect to* $\{\Phi_j\}_{j=1}^\infty$ is given by

$$\sigma_m(f, \{\Phi_j\}_{j=1}^\infty)_X \equiv \inf\left\{\left\|f - \sum_{j\in\Lambda} c_j\Phi_j\right\|_X \,:\, \sharp\Lambda \le m, c_j \in \mathbb{C}, j \in \Lambda\right\}.$$

If Y is a quasi-Banach space which is embedded continuously into X, then we can consider

$$\sigma_m(Y, X, \{\Phi_j\}_{j=1}^\infty) \equiv \sup\{\sigma_m(f, \{\Phi_j\}_{j=1}^\infty)_X \,:\, \|f\|_Y \le 1\}.$$

Exercises

Exercise 5.14 [1074, Theorem 2.13] Suppose the parameters $p_0, p_1, q_0, q_1, s_0, s_1$ satisfy $1 \le p_0, p_1, q_0, q_1 < \infty$, $s_0, s_1 \in \mathbb{R}$. For $\theta \in (0, 1)$, define p, q, s by

$$\frac{1}{p} = \frac{1-\theta}{p_0} + \frac{\theta}{p_1}, \quad \frac{1}{q} = \frac{1-\theta}{q_0} + \frac{\theta}{q_1}, \quad s = (1-\theta)s_0 + \theta s_1.$$

Let Ω be a bounded Lipschitz domain. Then prove that $[A^{s_0}_{p_0 q_0}(\Omega), A^{s_1}_{p_1 q_1}(\Omega)]_\theta \approx A^s_{pq}(\Omega)$, invoking Theorem 4.30.

Exercise 5.15 Let $0 < p, q \le \infty$ and $s \in \mathbb{R}$, and let Ω be a bounded domain with the extension operator $E : A^s_{pq}(\Omega) \to A^s_{pq}(\mathbb{R}^n)$. Namely, $E : A^s_{pq}(\Omega) \to A^s_{pq}(\mathbb{R}^n)$ is a continuous linear mapping such that $Ef|\Omega = f$ for all $f \in A^s_{pq}(\Omega)$. Define

$$R_\Omega f \equiv f|\Omega, \quad \ker(R_\Omega) = \{f \in A^s_{pq}(\mathbb{R}^n) : R_\Omega f = 0\}.$$

Then show that $A^s_{pq}(\mathbb{R}^n) = E(A^s_{pq}(\Omega)) \oplus \ker(R_\Omega)$ in the sense of the direct sum. Hint: This is a special case of the general results on algebra.

Exercise 5.16 By using the universal extension, prove that

$$\bigcap_{s>0} B^s_{\infty\infty}(\mathbb{R}^n_+) = \bigcap_{s>0} B^s_{\infty\infty}(\overline{\mathbb{R}^n_+}),$$

where

$$B^s_{\infty\infty}(\overline{\mathbb{R}^n_+}) = \bigcup_{\Omega \supset \overline{\mathbb{R}^n_+}, \text{open set}} B^s_{\infty\infty}(\Omega).$$

Exercise 5.17 Prove (5.21) by checking the support condition carefully.

Textbooks in Sect. 5.1

Function Spaces on the Half Space

See [99, Section 2.9.1], [100, Section 4.5] and [103, Section 5].

Function Spaces on Domains

We refer to [99, Chapter 3], [100, Chapter 5], [101, Section 23], as well as [104, Section 1.11.1–1.11.3].

Function Spaces on Lipschitz Domains

See [104, Section 1.11.4–1.11.6].

Triebel considered the extension operator in [100, Section 4.5] from $\mathbb{R}^n_+$ to $\mathbb{R}^n$ via oscillation, difference and distinguished representation and in [100, Section 5.1.3] from a domain Ω to $\mathbb{R}^n$.

Others

In Sect. 5.1.4 we relied upon [900] completely.

5.2 Pseudo-differential Operators on Besov Spaces and Triebel–Lizorkin Spaces

In Sect. 5.2.1, we describe the fundamental properties of pseudo-differential operators. In Sect. 5.2.2 we investigate the boundedness of $A^s_{pq}(\mathbb{R}^n)$ on the class of $S^m_{\rho\delta}$. We then consider a wider class of pseudo-differential operators. Applications are given in Sect. 5.2.3. Finally, in Sect. 5.2.4 we collect classical results and compare them with the results up to Sect. 5.2.3.

5.2.1 Pseudo-differential Operators

The pseudo-differential operator is given by

$$a(X, D)f(x) \equiv \frac{1}{\sqrt{(2\pi)^n}} \int_{\mathbb{R}^n} a(x, \xi)\mathscr{F}f(\xi)e^{ix\cdot\xi}d\xi.$$

However, in general there is no guarantee that the integral in the right-hand side converges. So, we start with the best case: $f \in \mathscr{S}(\mathbb{R}^n)$. After that we pass to the duality to define $a(X, D)f$ for $f \in \mathscr{S}'(\mathbb{R}^n)$. Once we define $a(X, D)f$ for $f \in \mathscr{S}'(\mathbb{R}^n)$, we can consider the boundedness of $a(X, D)$ on Besov spaces and Triebel–Lizorkin spaces.

5.2.1.1 Pseudo-differential Operators on $\mathscr{S}(\mathbb{R}^n)$ and $\mathscr{S}'(\mathbb{R}^n)$

First of all, we discuss how pseudo-differential operators act on our test function space $\mathscr{S}(\mathbb{R}^n)$.

Definition 5.9 (Pseudo-differential operators of Kohn–Nirenberg type) Let $a : \mathbb{R}^{2n} \to \mathbb{C}$ be a function. Then as long as the right-hand side of (5.26) below makes sense, define $a(X, D)$, the *pseudo-differential operator* (*of Kohn–Nirenberg type*) by

$$a(X, D)f(x) \equiv \frac{1}{\sqrt{(2\pi)^n}} \int_{\mathbb{R}^n} a(x, \xi)\mathscr{F} f(\xi)e^{ix\cdot\xi}d\xi \quad (x \in \mathbb{R}^n). \tag{5.26}$$

Example 5.1 In the case of $a(x, \xi) = -|\xi|^2$, we have $a(X, D) = \Delta$.

Some prefer to denote $a(X, D)$ by $a(x, D)$. But it is a bit confusing; $a(x, D)$ makes the function $a(x, D)f$ similar to the evaluation $a(x, D)f(x)$ at x. So, we prefer to write $a(X, D)$. The function a is called the symbol.

Definition 5.10 (Symbol class $S^m_{\rho\delta}$) Let $0 \le \rho, \delta \le 1$ and $m \in \mathbb{R}$.

1. Let $\alpha, \beta \in \mathbb{N}_0{}^n$. Define the norm $\|a\|_{S^m_{\rho\delta}(\alpha,\beta)}$ by

$$\|a\|_{S^m_{\rho\delta}(\alpha,\beta)} \equiv \sup_{x,\xi\in\mathbb{R}^n} \langle\xi\rangle^{-(m+\delta|\beta|-\rho|\alpha|)}|\partial_x^\beta\partial_\xi^\alpha a(x, \xi)|$$

   for $a \in C^\infty(\mathbb{R}^{2n})$.
2. Define the set $S^m_{\rho\delta}$ of $C^\infty(\mathbb{R}^{2n})$-functions by

$$S^m_{\rho\delta} \equiv \bigcap_{\alpha,\beta\in\mathbb{N}_0{}^n} \left\{a \in C^\infty(\mathbb{R}^{2n}) \,:\, \|a\|_{S^m_{\rho\delta}(\alpha,\beta)} < \infty\right\}. \tag{5.27}$$

   The class $S^m_{\rho\delta}$ is called the symbol class.
3. Abbreviate $S^m \equiv S^m_{10}$.

Example 5.2

1. The symbol $a(x, \xi) = \xi_j$ belongs to S^1 for $j = 1, 2, \ldots, n$.
2. The class S^0_{00} is called the *Hörmander class*.

Let us check that (5.26) makes sense for $f \in \mathscr{S}(\mathbb{R}^n)$.

Proposition 5.4 *Let* $0 \le \rho, \delta \le 1$, $m \in \mathbb{R}$, *and let* $a \in S^m_{\rho\delta}$. *Then* $a(X, D)f \in \mathscr{S}(\mathbb{R}^n)$ *for all* $f \in \mathscr{S}(\mathbb{R}^n)$. *Furthermore, the mapping* $f \in \mathscr{S}(\mathbb{R}^n) \mapsto a(X, D)f \in \mathscr{S}(\mathbb{R}^n)$ *is continuous.*

Proof Since $f \in \mathscr{S}(\mathbb{R}^n)$ and $a \in S^m_{\rho\delta}$, the integral defining the right-hand side of (5.26) converges absolutely. Thus, $a(X, D)f \in \mathscr{S}(\mathbb{R}^n)$. Define a differential operator L_ξ by $L_\xi \equiv \langle x\rangle^{-2}(\mathrm{id}_{\mathscr{S}} - \Delta_\xi)$. Then a simple calculation shows that $L_\xi[e^{ix\cdot\xi}] = e^{ix\cdot\xi}$. Denote by $(L_\xi)^N$ the N-fold composition of L_ξ. Let us integrate by parts to estimate $\partial^\alpha a(X, D)f(x)$. An integration by parts yields

$$a(X, D)f(x) = \frac{1}{\sqrt{(2\pi)^n}} \int_{\mathbb{R}^n} a(x, \xi)\mathscr{F}f(\xi)(L_\xi)^N e^{ix\cdot\xi} d\xi$$
$$= \frac{1}{\sqrt{(2\pi)^n}} \int_{\mathbb{R}^n} (L_\xi)^N (a(x, \xi)\mathscr{F}f(\xi))e^{ix\cdot\xi} d\xi.$$

(This is the heart of the matter about quasi-integral operators.) By the Leibniz formula, we have

$$|\partial_x^\alpha {L_\xi}^N[(a(x, \xi)\mathscr{F}f](\xi))| \lesssim_{M,N,\alpha} \langle\xi\rangle^{-M}.$$

Hence $\sup_{x\in\mathbb{R}^n} |x^\alpha \partial^\beta a(X, D)f(x)| < \infty$. Thus, we obtain $a(X, D)f \in \mathscr{S}(\mathbb{R}^n)$ and we conclude that $a(X, D)$ is continuous.

Corollary 5.1 *Let $0 \le \rho, \delta \le 1$, $m \in \mathbb{R}$. Suppose that $\{a_\varepsilon\}_{\varepsilon\in[0,1]} \subset S^m_{\rho\delta}$ is a uniform family in the sense that*

$$|\partial_x^\beta \partial_\xi^\alpha a_\varepsilon(x, \xi)| \lesssim_{\alpha,\beta} \langle\xi\rangle^{m+\delta|\beta|-\rho|\alpha|} \tag{5.28}$$

for all $\varepsilon \in [0, 1]$ and that

$$\lim_{\varepsilon\downarrow 0} \partial_x^\beta \partial_\xi^\alpha a_\varepsilon(x, \xi) = \partial_x^\beta \partial_\xi^\alpha a_0(x, \xi), \quad (x, \xi \in \mathbb{R}^n) \tag{5.29}$$

in the sense of pointwise convergence for all $\alpha, \beta \in \mathbb{R}^n$. Then $\lim_{\varepsilon\downarrow 0} a_\varepsilon(X, D)f = a_0(X, D)f$ in the topology of $\mathscr{S}(\mathbb{R}^n)$.

Proof We use the idea of Proposition 5.4 and the Lebesgue convergence theorem.

The following is an example of the function a_ε in Corollary 5.1.

Example 5.3 Let $0 \le \rho, \delta \le 1$, $m \in \mathbb{R}$, and let $a \in S^m_{\rho\delta}$. Choose $\gamma \in C^\infty(\mathbb{R}^n)$ so that $\chi_{Q(1)} \le \gamma \le \chi_{Q(2)}$. If we set $a_\varepsilon(x, \xi) \equiv a(x, \xi)\gamma(\varepsilon x)\gamma(\varepsilon\xi)$, for $x, \xi \in \mathbb{R}^n$ and $0 \le \varepsilon \le 1$, then $\{a_\varepsilon\}_{\varepsilon\in[0,1]}$ satisfies (5.28) and (5.29) in Corollary 5.1 uniformly over $\varepsilon > 0$.

We transform Corollary 5.1 into the form which we use later.

Proposition 5.5 (Approximation formula of pseudo-differential operators) *Let $0 \le \rho \le 1$, $0 \le \delta \le 1$, $m \in \mathbb{R}$, and let $a \in S^m_{\rho\delta}$. Choose $\gamma \in C^\infty_c(\mathbb{R}^n)$ so that $\chi_{Q(1)} \le \gamma \le \chi_{Q(2)}$. Set*

$$a_\varepsilon(x, \xi) \equiv a(x, \xi)\gamma(\varepsilon x)\gamma(\varepsilon\xi) \quad (x, \xi \in \mathbb{R}^n) \tag{5.30}$$

for $0 < \varepsilon \le 1$. Then

$$a(X,D)f(x) = (2\pi)^{-n} \lim_{\varepsilon \downarrow 0} \iint_{\mathbb{R}^{2n}} a_\varepsilon(x,\xi) f(y) e^{i(x-y)\cdot\xi} d\xi dy \tag{5.31}$$

in the topology of $\mathscr{S}(\mathbb{R}^n)$ *whenever* $f \in \mathscr{S}(\mathbb{R}^n)$.

Proof Use Corollary 5.1 to approximate $a(X,D)f(x) = \lim_{\varepsilon \downarrow 0} a_\varepsilon(X,D)f(x)$. Then

$$\begin{aligned} a(X,D)f(x) &= \frac{1}{\sqrt{(2\pi)^n}} \lim_{\varepsilon \downarrow 0} \int_{\mathbb{R}^n} a_\varepsilon(x,\xi) \mathscr{F} f(\xi) e^{ix\cdot\xi} d\xi \\ &= \frac{1}{(2\pi)^n} \lim_{\varepsilon \downarrow 0} \iint_{\mathbb{R}^{2n}} a_\varepsilon(x,\xi) f(y) e^{i(x-y)\cdot\xi} d\xi dy. \end{aligned}$$

Thus, (5.31) is proved.

We want to define $a(X,D)f$ for $f \in \mathscr{S}'(\mathbb{R}^n)$. To this end, we need to consider the dual $a(X,D)^*$ of $a(X,D)$, where $a(X,D)^*$ is given by

$$a(X,D)^* g \equiv \frac{1}{(2\pi)^n} \iint_{\mathbb{R}^{2n}} a(x,\xi) e^{i\xi(\star - x)} g(x) dx d\xi \tag{5.32}$$

for $g \in \mathscr{S}(\mathbb{R}^n)$. A similar conclusion holds for $a(X,D)^*$. See Exercise 5.22 for more details. Duality entails conclusions similar to $a(X,D)$.

Lemma 5.5 (Duality) *We have*

$$\langle a(X,D)f, g \rangle = \langle f, a(X,D)^* g \rangle \tag{5.33}$$

for all $f, g \in \mathscr{S}(\mathbb{R}^n)$.

Relation (5.33) shows that the transpose of $a(X,D)$ is $a(X,D)^*$.

Proof If we write down the limit in question using the function a_ε defined by (5.30), then:

$$\begin{aligned} &\int_{\mathbb{R}^n} a(X,D)f(x) \cdot g(x) dx \\ &= \frac{1}{(2\pi)^n} \lim_{\varepsilon \downarrow 0} \int_{\mathbb{R}^n} \left(\iint_{\mathbb{R}^{2n}} a_\varepsilon(x,\xi) g(x) f(y) e^{i(x-y)\xi} d\xi dy \right) dx \\ &= \frac{1}{(2\pi)^n} \lim_{\varepsilon \downarrow 0} \int_{\mathbb{R}^n} \left(\iint_{\mathbb{R}^{2n}} a_\varepsilon(y,\xi) g(y) f(x) e^{i(y-x)\xi} d\xi dx \right) dy \\ &= \int_{\mathbb{R}^n} f(x) a(X,D)^* g(x) dx. \end{aligned}$$

Since we truncate the integral, in the second equality above, we can justify the use of Fubini's theorem (see Theorem 1.3).

Definition 5.11 **($a(X,D)f$ for $f \in \mathscr{S}'(\mathbb{R}^n)$)** Let $0 \le \rho, \delta \le 1$, $m \in \mathbb{R}$, and let $a \in S^m_{\rho\delta}$. For $f \in \mathscr{S}'(\mathbb{R}^n)$, define $a(X,D)f \in \mathscr{S}'(\mathbb{R}^n)$ by way of $a(X,D)^*$ so that (5.33) holds for all $g \in \mathscr{S}(\mathbb{R}^n)$.

Note that duality entails that $a(X,D) : \mathscr{S}'(\mathbb{R}^n) \to \mathscr{S}'(\mathbb{R}^n)$ is continuous.

5.2.1.2 Asymptotic Expansion

We can handle the operation of pseudo-differential operators in that we have the following convergence results:

Theorem 5.11 (Asymptotic expansion) *Let $0 \le \rho, \delta \le 1$. Suppose that we have a sequence $\{m_k\}_{k=0}^\infty$ divergent to $-\infty$. For each $j = 0, 1, 2, \ldots$, we suppose that we have $a_j \in S^{m_j}_{\rho\delta}$. Then there exists $a \in S^{m_0}_{\rho\delta}$ such that*

$$a - \sum_{k=0}^{j-1} a_k \in S^{m_j}_{\rho\delta} \quad (j = 1, 2, \ldots). \tag{5.34}$$

Before we come to the proof, let us explain how to read our assumption.

Remark 5.2 Since $\{m_j\}_{j=0}^\infty$ is decreasing, it follows that a decreasing class $\{S^{m_j}_{\rho\delta}\}_{j=0}^\infty$ is given. Thus, for each $k = j, j+1, \ldots$, $a_k \in S^{m_j}_{\rho\delta}$. However, it may happen that $a_0, a_1, \ldots, a_{j-1}$ fails to belong to $S^{m_j}_{\rho\delta}$. Thus, the task is to control $a_0, a_1, \ldots, a_{j-1}$ in a certain manner in order to overcome this problem.

Proof Choose an auxiliary function $\varphi \in C^\infty(\mathbb{R}^n)$ so that $\chi_{B(1)} \le \varphi \le \chi_{B(2)}$. Define

$$a(x,\xi) \equiv \sum_{k=0}^\infty (1 - \varphi(2^{-j_k}\xi)) a_k(x,\xi) \quad (x, \xi \in \mathbb{R}^n), \tag{5.35}$$

where $\{j_k\}_{k=1}^\infty$ is an increasing sequence that will be determined later. Note that the sum is locally finite over any compact set in $\mathbb{R}^{2n}$; hence we can change the order of integration and summation.

Let us prove (5.34). This amounts to proving that

$$\sup_{x \in \mathbb{R}^n, \xi \in \mathbb{R}^n} \langle \xi \rangle^{m_j + \delta|\beta| - \rho|\alpha|} \left| \partial_x^\beta \partial_\xi^\alpha \left(a(x,\xi) - \sum_{k=0}^{j-1} a_k(x,\xi) \right) \right| < \infty$$

for any multi-indexes α, β. Observe that

$$a(x,\xi) - \sum_{k=0}^{j-1} a_k(x,\xi) = \sum_{k=j}^\infty (1 - \varphi(2^{-j_k}\xi)) a_k(x,\xi) - \sum_{k=0}^{j-1} \varphi(2^{-j_k}\xi) a_k(x,\xi).$$

Hence we have

$$\begin{aligned}
&\partial_x^\beta \partial_\xi^\alpha \left[a(x,\xi) - \sum_{k=0}^{j-1} a_k(x,\xi) \right] \\
&= \sum_{k=j}^{\infty} (1 - \varphi(2^{-j_k}\xi)) \partial_x^\beta \partial_\xi^\alpha a_k(x,\xi) - \sum_{k=0}^{j-1} \varphi(2^{-j_k}\xi) \partial_x^\beta \partial_\xi^\alpha a_k(x,\xi) \\
&\quad - \sum_{k=0}^{\infty} \sum_{\gamma \in \mathbb{N}_0{}^n, 0 \neq \gamma \leq \alpha} \binom{\alpha}{\gamma} \partial_\xi^\gamma [\varphi(2^{-j_k}\xi)] \partial_x^\beta \partial_\xi^{\alpha-\gamma} [a_k(x,\xi)].
\end{aligned}$$

We estimate the first term:

$$\begin{aligned}
&\sup_{x,\xi \in \mathbb{R}^n} \langle \xi \rangle^{-(m_j + \delta|\beta| - \rho|\alpha|)} \left| \sum_{k=j}^{\infty} (1 - \varphi(2^{-j_k}\xi)) \partial_x^\beta \partial_\xi^\alpha a_k(x,\xi) \right| \\
&\leq \sup_{x,\xi \in \mathbb{R}^n} \langle \xi \rangle^{-(m_j + \delta|\beta| - \rho|\alpha|)} \sum_{k=j}^{\infty} \left| \chi_{B(2^{j_k})^c}(\xi) (1 - \varphi(2^{-j_k}\xi)) \partial_x^\beta \partial_\xi^\alpha a_k(x,\xi) \right| \\
&\leq \sup_{\xi \in \mathbb{R}^n} \sum_{k=j}^{\infty} \langle \xi \rangle^{m_k - m_j + 1} \chi_{B(2^{j_k})^c}(\xi) \leq \sum_{k=j}^{\infty} 2^{(-m_k + m_j + 1)k},
\end{aligned}$$

where we choose $\{j_k\}_{k=1}^{\infty}$ so that the second inequality holds. In total, since a decreasing sequence $\{m_k\}_{k=0}^{\infty}$ diverges to $-\infty$, we obtain

$$\sup_{x,\xi \in \mathbb{R}^n} \langle \xi \rangle^{-(m_j + \delta|\beta| - \rho|\alpha|)} \left| \sum_{k=j}^{\infty} (1 - \varphi(2^{-j_k}\xi)) \partial_x^\beta \partial_\xi^\alpha a_k(x,\xi) \right| < \infty. \tag{5.36}$$

Using $\varphi \in C_c^\infty(\mathbb{R}^n)$, we estimate the second term easily:

$$\sup_{x,\xi \in \mathbb{R}^n} \langle \xi \rangle^{-(m_j + \delta|\beta| - \rho|\alpha|)} \left| \sum_{k=0}^{j-1} \varphi(2^{-j_k}\xi) \partial_x^\beta \partial_\xi^\alpha [a_k(x,\xi)] \right| < \infty. \tag{5.37}$$

Finally, we will estimate the third term. Thanks to the fact that $0 \neq \gamma \leq \alpha$ and $a_k \in S_{\rho\delta}^{m_k}$ for $k = 0, 1, 2, \ldots$, we have

$$\begin{aligned}
&\langle \xi \rangle^{-(m_j + \delta|\beta| - \rho|\alpha|)} |\partial_\xi^\gamma [\varphi(2^{-j_k}\xi)] \partial_x^\beta \partial_\xi^{\alpha-\gamma} [a_k(x,\xi)]| \\
&\lesssim 2^{-|\gamma|k} \langle \xi \rangle^{-(m_j + \delta|\beta| - \rho|\alpha|)} \langle \xi \rangle^{m_k + \delta|\beta| - \rho|\alpha - \gamma|} \chi_{B(2) \setminus B(1)}(2^{-k}\xi) \\
&\lesssim 2^{-|\gamma|k} \langle \xi \rangle^{m_k - m_j + \rho|\gamma|} \chi_{B(2) \setminus B(1)}(2^{-k}\xi).
\end{aligned}$$

Keeping in mind that $\{m_k\}_{k=0}^{\infty}$ is a sequence divergent to $-\infty$, we define $K(j;\alpha)$ by

$$K(j;\alpha) \equiv \min\{k \in \mathbb{N}_0 \,:\, m_l - m_j + \rho|\alpha| < 0, \text{ for all } l = k, k+1, \ldots\}(\geq j).$$

When $k > K(j;\alpha)$, we use $\langle \xi \rangle^{m_k - m_j + \rho|\gamma|} \leq |\xi|^{m_k - m_j + \rho|\gamma|}$:

$$\begin{aligned}
&\langle \xi \rangle^{-(m_j+\delta|\beta|-\rho|\alpha|)} |\partial_\xi^\gamma[\varphi(2^{-j_k}\xi)]\partial_x^\beta \partial_\xi^{\alpha-\gamma}[a_k(x,\xi)]| \\
&\lesssim 2^{-|\gamma|k} |\xi|^{-m_j+m_k+\rho|\gamma|} \chi_{B(2)\setminus B(1)}(2^{-k}\xi) \\
&\lesssim 2^{(m_k-m_j+(\rho-1)|\gamma|)k} \chi_{B(2)\setminus B(1)}(2^{-k}\xi) \lesssim 2^{(m_k-m_j)k}.
\end{aligned}$$

Hence it follows that

$$\begin{aligned}
&\sup_{x,\xi\in\mathbb{R}^n} \langle \xi \rangle^{-(m_j+\delta|\beta|-\rho|\alpha|)} \left| \sum_{k=0}^{\infty} \sum_{\gamma\in\mathbb{N}_0{}^n, 0\neq\gamma\leq\alpha} \binom{\alpha}{\gamma} \partial_\xi^\gamma[\varphi(2^{-j_k}\xi)]\partial_x^\beta\partial_\xi^{\alpha-\gamma}[a_k(x,\xi)] \right| \\
&\leq \sum_{k=1}^{K(j;\alpha)} \sup_{\xi\in\mathbb{R}^n} \left(2^{-|\gamma|k}\langle\xi\rangle^{m_k-m_j+\rho|\gamma|}\chi_{B(2)\setminus B(1)}(2^{-k}\xi)\right) + \sum_{k=1}^{\infty} 2^{(m_k-m_j)k} < \infty.
\end{aligned}$$

This implies

$$\sup_{x,\xi\in\mathbb{R}^n} \frac{1}{\langle\xi\rangle^{m_j+\delta|\beta|-\rho|\alpha|}} \left| \sum_{k=0}^{\infty} \sum_{\gamma\in\mathbb{N}_0{}^n, 0\neq\gamma\leq\alpha} \binom{\alpha}{\gamma} \partial_\xi^\gamma[\varphi(2^{-j_k}\xi)]\partial_x^\beta\partial_\xi^{\alpha-\gamma}[a_k(x,\xi)] \right| < \infty. \tag{5.38}$$

Thus, the proof is complete from (5.36), (5.37) and (5.38).

In the next section we plan to use the following theorem and consider various pseudo-differential operators starting from given symbols:

Theorem 5.12 (A construction of pseudo-differential operators) *Suppose that, for each $\alpha, \beta \in \mathbb{N}_0{}^n$, we have sequences $\{m_j\}_{j=0}^{\infty}$ and $\{\mu_j(\alpha,\beta)\}_{j=0}^{\infty}$ satisfying*

$$\lim_{j\to\infty} \min(\mu_j(\alpha,\beta), -m_j) = \infty \tag{5.39}$$

and that we have a symbol $a_j \in S_{\rho\delta}^{m_j}$ for $j \in \mathbb{N}_0$. Let $a \in C^\infty(\mathbb{R}^{2n})$ satisfy

$$\sup_{x,\xi\in\mathbb{R}^n} \langle\xi\rangle^{\mu_j(\alpha,\beta)} \left| \partial_x^\alpha\partial_\xi^\beta a(x,\xi) - \sum_{k=0}^{j} \partial_x^\alpha\partial_\xi^\beta a_k(x,\xi) \right| < \infty \tag{5.40}$$

for each $\alpha, \beta \in \mathbb{N}_0{}^n$. *Then* $a - \sum_{k=0}^{j-1} a_k \in S^{m_j}_{\rho\delta}$ *for all* $j \in \mathbb{N}$.

Proof There exists a symbol $q \in S^{m_0}_{\rho\delta}$ such that $q - \sum_{k=0}^{j-1} a_k \in S^{m_j}_{\rho\delta}$ for all positive integers j in view of Theorem 5.11. Hence it suffices to show $a - q \in S^{-\infty}_{\rho\delta}$, or equivalently,

$$\sup_{x,\xi \in \mathbb{R}^n} \langle \xi \rangle^k |\partial_x^\beta \partial_\xi^\alpha a(x,\xi) - \partial_x^\beta \partial_\xi^\alpha q(x,\xi)| < \infty \tag{5.41}$$

for all positive integers $k \in \mathbb{N}$ and multi-indexes $\alpha, \beta \in \mathbb{N}_0{}^n$. With this in mind, we fix $k \in \mathbb{N}$ and multi-indexes α, β. In view of (5.39), we have

$$\min(\mu_j(\alpha, \beta), -m_j - \delta|\beta| + \rho|\alpha|) > k$$

by letting $j = j(k) \gg 1$. Below we fix such a j.

In view of (5.40) and $\mu_j(\alpha, \beta) > k$, we have

$$\begin{aligned}
&\sup_{x,\xi \in \mathbb{R}^n} \langle \xi \rangle^k \left| \partial_x^\beta \partial_\xi^\alpha a(x,\xi) - \sum_{k=0}^{j} \partial_x^\beta \partial_\xi^\alpha a_k(x,\xi) \right| \\
&\le \sup_{x,\xi \in \mathbb{R}^n} \langle \xi \rangle^{\mu_j(\alpha,\beta)} \left| \partial_x^\beta \partial_\xi^\alpha a(x,\xi) - \sum_{k=0}^{j} \partial_x^\beta \partial_\xi^\alpha a_k(x,\xi) \right| \\
&< \infty.
\end{aligned} \tag{5.42}$$

Meanwhile, since $q - \sum_{k=0}^{j-1} a_k \in S^{m_j}_{\rho\delta}$ and $-m_j - \delta|\beta| + \rho|\alpha| > k$, we have

$$\begin{aligned}
&\sup_{x,\xi \in \mathbb{R}^n} \langle \xi \rangle^k \left| \partial_x^\beta \partial_\xi^\alpha q(x,\xi) - \sum_{k=0}^{j} \partial_x^\beta \partial_\xi^\alpha a_k(x,\xi) \right| \\
&\le \sup_{x,\xi \in \mathbb{R}^n} \langle \xi \rangle^{-m_j - \delta|\beta| + \rho|\alpha|} \left| \partial_x^\beta \partial_\xi^\alpha q(x,\xi) - \sum_{k=0}^{j} \partial_x^\beta \partial_\xi^\alpha a_k(x,\xi) \right| \\
&< \infty.
\end{aligned} \tag{5.43}$$

Since $a - q = \left(a - \sum_{k=0}^{j} a_k \right) - \left(q - \sum_{k=0}^{j} a_k \right)$ for any $j \in \mathbb{N}$, (5.41) follows from (5.42) and (5.43). Hence Theorem 5.12 is proved.

5.2.1.3 Operation for Pseudo-differential Operators

First of all, let us collect some formulas on the conjugate operation.

Let $0 \le \delta \le \rho < 1$. Choose $\gamma \in C^\infty(\mathbb{R}^n)$ so that $\chi_{Q(1)} \le \gamma \le \chi_{Q(2)}$. For $a \in S^m_{\rho\delta}$, and $0 < \varepsilon_1, \varepsilon_2 \le 1$ define

$$a_{\varepsilon_1,\varepsilon_2}(x,\xi) \equiv a(x,\xi)\gamma(\varepsilon_1 x)\gamma(\varepsilon_2\xi) \quad (x,\xi \in \mathbb{R}^n).$$

Then an integration by parts shows that the limits

$$\lim_{\varepsilon_1,\varepsilon_2\downarrow 0} \iint_{\mathbb{R}^{2n}} a_{\varepsilon_1,\varepsilon_2}(y,\eta) \mathrm{e}^{i(x-y)\cdot(\xi-\eta)} \mathrm{d}y\mathrm{d}\eta,$$

$$\lim_{\varepsilon_1\downarrow 0}\left(\lim_{\varepsilon_2\downarrow 0} \iint_{\mathbb{R}^{2n}} a_{\varepsilon_1,\varepsilon_2}(y,\eta) \mathrm{e}^{i(x-y)\cdot(\xi-\eta)} \mathrm{d}y\mathrm{d}\eta\right)$$

exist and coincide; see the proof of Theorem 5.13. We denote the limit by

$$\iint_{\mathbb{R}^{2n}} a(y,\eta) \mathrm{e}^{i(x-y)\cdot(\xi-\eta)} \mathrm{d}y\mathrm{d}\eta.$$

It is understood that the integral $\iint_{\mathbb{R}^{2n}} a(x+y,\xi+\eta)\mathrm{e}^{iy\cdot\eta}\mathrm{d}y\mathrm{d}\eta$ is defined by way of a similar approximation.

We give an example of equality which can be obtained as a consequence of this rule. For multi-indexes α, β, we set $\delta_{\alpha\beta} \equiv \begin{cases} 1, \alpha = \beta, \\ 0, \alpha \ne \beta. \end{cases}$

The next lemma shows

$$\iint_{\mathbb{R}^{2n}} y^\beta \eta^\alpha \mathrm{e}^{iy\cdot\eta} \mathrm{d}y\mathrm{d}\eta = (2\pi)^n (-i)^{|\alpha|} \alpha! \delta_{\alpha\beta}$$

according to our new operation defined by means of the truncation and integration by parts.

Lemma 5.6 *If $\gamma \in C^\infty(\mathbb{R}^n)$ satisfies $\chi_{Q(1)} \le \gamma \le \chi_{Q(2)}$, then*

$$\lim_{\varepsilon_1,\varepsilon_2\downarrow 0} \iint_{\mathbb{R}^{2n}} \gamma(\varepsilon_1 y)\gamma(\varepsilon_2\eta) y^\beta \eta^\alpha \mathrm{e}^{iy\cdot\eta} \mathrm{d}y\mathrm{d}\eta = (2\pi)^n (-i)^{|\alpha|} \alpha! \delta_{\alpha\beta} \tag{5.44}$$

for all multi-indexes α, β.

Proof Denote by I the left-hand side of (5.44). According to the definition, we have

$$\begin{aligned}
\mathrm{I} &= \lim_{\varepsilon_1\downarrow 0}\lim_{\varepsilon_2\downarrow 0}\int_{\mathbb{R}^n}\gamma(\varepsilon_1 y)y^\beta\left(\int_{\mathbb{R}^n}\gamma(\varepsilon_2\eta)\eta^\alpha e^{iy\cdot\eta}d\eta\right)dy\\
&= \lim_{\varepsilon_1\downarrow 0}\lim_{\varepsilon_2\downarrow 0}\int_{\mathbb{R}^n}\gamma(\varepsilon_1 y)y^\beta\left(\frac{1}{i}\frac{d}{dy}\right)^\alpha\left(\int_{\mathbb{R}^n}\gamma(\varepsilon_2\eta)e^{iy\cdot\eta}d\eta\right)dy\\
&= \lim_{\varepsilon_1\downarrow 0}\lim_{\varepsilon_2\downarrow 0}\int_{\mathbb{R}^n}\left(-\frac{1}{i}\frac{d}{dy}\right)^\alpha[\gamma(\varepsilon_1 y)y^\beta]\left(\int_{\mathbb{R}^n}\gamma(\varepsilon_2\eta)e^{iy\cdot\eta}d\eta\right)dy.
\end{aligned}$$

The symbol δ_0 stands for the Dirac delta at the origin. Then

$$\lim_{\varepsilon_2\downarrow 0}\int_{\mathbb{R}^n}\gamma(\varepsilon_2\eta)e^{iy\cdot\eta}d\eta = (2\pi)^n\alpha!\delta_0 \tag{5.45}$$

in the topology of $\mathscr{S}'(\mathbb{R}^n)$. We leave the details to interested readers; see Exercise 5.20. Hence

$$\mathrm{I} = \lim_{\varepsilon_1\downarrow 0}\left\langle\delta_0, \left(-\frac{1}{i}\frac{d}{dy}\right)^\alpha\left[\gamma(\varepsilon_1 y)y^\beta\right]\right\rangle = (2\pi)^n(-i)^{|\alpha|}\alpha!\delta_{\alpha\beta}.$$

Thus, the proof is complete.

Theorem 5.13 (Adjoint of pseudo-differential operators) *Let $0 \le \delta < \rho \le 1$ and $m \in \mathbb{R}$. For $a \in S^m_{\rho\delta}$, we set*

$$\begin{aligned}
A(x,\xi) &\equiv \frac{1}{(2\pi)^n}\iint_{\mathbb{R}^{2n}} a(y,\eta)e^{i(x-y)\cdot(\xi-\eta)}dyd\eta\\
&= \frac{1}{(2\pi)^n}\iint_{\mathbb{R}^{2n}} a(x+y,\xi+\eta)e^{iy\cdot\eta}dyd\eta.
\end{aligned}$$

Then $A \in S^m_{\rho\delta}$. Furthermore,

$$\int_{\mathbb{R}^n} A(X,D)f(x)\cdot g(x)dx = \int_{\mathbb{R}^n} f(y)\cdot a(X,D)g(y)dy, \quad f,g \in \mathscr{S}(\mathbb{R}^n). \tag{5.46}$$

Proof First of all, approximation allows us to assume a belongs to $C^\infty_c(\mathbb{R}^n)$. It is easy to check (5.46) in this case; see Exercise 5.19. Let $L \in \mathbb{N}$ be sufficiently large. An integration by parts yields

$$\begin{aligned}
A(x,\xi) &\simeq_n \iint_{\mathbb{R}^{2n}} \frac{a(x+y,\xi+\eta)}{\langle\eta\rangle^{2L}}(1-\Delta_y)^L e^{iy\cdot\eta}dyd\eta\\
&\simeq_n \iint_{\mathbb{R}^{2n}} \frac{(1-\Delta_y)^L a(x+y,\xi+\eta)}{\langle\eta\rangle^{2L}}e^{iy\cdot\eta}dyd\eta
\end{aligned}$$

$$\simeq_n \iint_{\mathbb{R}^{2n}} \frac{(1-\Delta_y)^L a(x+y,\xi+\eta)}{\langle y\rangle^{2L}\langle \eta\rangle^{2L}}(1-\Delta_\eta)^L \mathrm{e}^{iy\cdot\eta}\mathrm{d}y\mathrm{d}\eta$$

$$\simeq_n \iint_{\mathbb{R}^{2n}} (1-\Delta_\eta)^L\left(\frac{(1-\Delta_y)^L a(x+y,\xi+\eta)}{\langle y\rangle^{2L}\langle \eta\rangle^{2L}}\right)\mathrm{e}^{iy\cdot\eta}\mathrm{d}y\mathrm{d}\eta$$

$$\simeq_n \iint_{\mathbb{R}^{2n}} \langle y\rangle^{-2L}(1-\Delta_\eta)^L\left(\frac{(1-\Delta_y)^L a(x+y,\xi+\eta)}{\langle \eta\rangle^{2L}}\right)\mathrm{e}^{iy\cdot\eta}\mathrm{d}y\mathrm{d}\eta.$$

Here the integrand is made up of the linear combination of

$$\langle y\rangle^{-2L}\partial_\xi^\alpha(1-\Delta_y)^L a(x+y,\xi+\eta)\cdot\partial^\beta\langle\eta\rangle^{-2L}, \quad |\alpha|+|\beta|\le 2L,$$

whose absolute value is bounded by the constant multiple of

$$\langle\xi+\eta\rangle^{m+2\delta L-\rho|\alpha|}\langle y\rangle^{-2L}\langle\eta\rangle^{-2L-|\beta|}.$$

Furthermore, by Peetre's formula, the integrand is bounded by the constant multiple of

$$\langle\xi\rangle^{m+2\delta L}\langle y\rangle^{-2L}\langle\eta\rangle^{-2L+|m+2\delta L|}.$$

Hence keeping in mind that $\delta<1$, we can choose $L\gg 1$ so that $2L>n$ and that $2L-|2\delta L+m|>n$. Hence we obtain

$$|A(x,\xi)|\lesssim\langle\xi\rangle^{m+2\delta L}. \tag{5.47}$$

Let us consider the higher differential of (5.47). Let $\alpha,\beta\in{\mathbb{N}_0}^n$. Then

$$\partial_x^\beta\partial_\xi^\alpha A(x,\xi)=\frac{1}{(2\pi)^n}\iint_{\mathbb{R}^{2n}}\partial_x^\beta\partial_\xi^\alpha a(x+y,\xi+\eta)\mathrm{e}^{iy\cdot\eta}\mathrm{d}y\mathrm{d}\eta.$$

Since $\partial_x^\beta\partial_\xi^\alpha a\in S_{\rho\delta}^{m+\delta|\beta|-\rho|\alpha|}$, in (5.47), we can replace $a\in S_{\rho\delta}^m$ with $\partial_x^\beta\partial_\xi^\alpha a\in S_{\rho\delta}^{m+\delta|\beta|-\rho|\alpha|}$. So, we obtain

$$|\partial_x^\beta\partial_\xi^\alpha A(x,\xi)|\lesssim_{\alpha\beta}\langle\xi\rangle^{m+\delta|\beta|-\rho|\alpha|+2\delta L}. \tag{5.48}$$

We expand a into the Taylor series at (x,ξ):

$$a(x+y,\xi+\eta)-\sum_{\substack{\alpha,\beta\in{\mathbb{N}_0}^n\\|\alpha|+|\beta|<2N}}\frac{1}{\alpha!\beta!}\partial_x^\beta\partial_\xi^\alpha a(x,\xi)y^\beta\eta^\alpha$$

$$= \frac{1}{(2N-1)!}\int_0^1 (1-t)^{2N-1}\frac{\mathrm{d}^{2N}}{\mathrm{d}t^{2N}}a(x+t\,y,\xi+t\,\eta)\mathrm{d}t.$$

This together with Lemma 5.6 yields

$$A(x,\xi) - \sum_{\substack{\alpha\in\mathbb{N}_0{}^n\\ |\alpha|<N}} \frac{(-i)^{|\alpha|}}{\alpha!}\partial_x^\alpha\partial_\xi^\alpha a(x,\xi)$$

$$\simeq_{n,N} \iiint_{[0,1]\times\mathbb{R}^{2n}} (1-t)^{2N-1}\frac{\mathrm{d}^{2N}}{\mathrm{d}t^{2N}}a(x+t\,y,\xi+t\,\eta)\mathrm{e}^{iy\cdot\eta}\mathrm{d}t\mathrm{d}y\mathrm{d}\eta.$$

Let $|\alpha|+|\beta|=2N$. Denote by $r_{\alpha\beta}(x,\xi;t)$ the remainder term:

$$r_{\alpha\beta}(x,\xi;t) \equiv \iint_{\mathbb{R}^{2n}} y^\beta\eta^\alpha\partial_x^\beta\partial_\xi^\alpha a(x+t\,y,\xi+t\,\eta)\mathrm{e}^{iy\cdot\eta}\mathrm{d}y\mathrm{d}\eta$$

for $x,\xi\in\mathbb{R}^n$ and $t\in[0,1]$. An integration by parts yields

$$\begin{aligned} r_{\alpha\beta}(x,\xi;t) &= \iint_{\mathbb{R}^{2n}} y^\beta\partial_x^\beta\partial_\xi^\alpha a(x+t\,y,\xi+t\,\eta)(-i\partial_y)^\alpha[\mathrm{e}^{iy\cdot\eta}]\mathrm{d}y\mathrm{d}\eta \\ &= \iint_{\mathbb{R}^{2n}} (i\partial_y)^\alpha[y^\beta\partial_x^\beta\partial_\xi^\alpha a(x+t\,y,\xi+t\,\eta)]\mathrm{e}^{iy\cdot\eta}\mathrm{d}y\mathrm{d}\eta. \end{aligned}$$

Observe that

$$\partial_y^\gamma y^\beta = \begin{cases} \dfrac{\beta!}{(\beta-\gamma)!}y^{\beta-\gamma} & (\beta\geq\gamma) \\ 0 & \text{otherwise.} \end{cases} \tag{5.49}$$

See Exercise 5.23. With the binomial coefficients and integration by parts, we can write

$$\begin{aligned} r_{\alpha\beta}(x,\xi;t) = &\sum_{\gamma\leq\min(\alpha,\beta)} C_{\alpha,\beta,\gamma}t^{|\alpha|-|\gamma|} \\ &\times\iint_{\mathbb{R}^{2n}} y^{\beta-\gamma}\mathrm{e}^{iy\cdot\eta}\partial_x^{\alpha+\beta-\gamma}\partial_\xi^\alpha[a(x+t\,y,\xi+t\,\eta)]\mathrm{d}y\mathrm{d}\eta \\ = &\sum_{\gamma\leq\min(\alpha,\beta)} C_{\alpha,\beta,\gamma}t^{|\alpha+\beta-\gamma|}t^{-|\gamma|} \\ &\times\iint_{\mathbb{R}^{2n}} (\partial_x^{\alpha+\beta-\gamma}\partial_\xi^{\alpha+\beta-\gamma}[a(x+t\,y,\xi+t\,\eta)])\mathrm{e}^{iy\cdot\eta}\mathrm{d}y\mathrm{d}\eta. \end{aligned}$$

Another integration by parts yields

$$\iint_{\mathbb{R}^{2n}} \partial_x^{\alpha+\beta-\gamma} \partial_\xi^{\alpha+\beta-\gamma} [a(x+t\,y, \xi+t\,\eta)] \mathrm{e}^{iy\cdot\eta} \mathrm{d}y \mathrm{d}\eta$$
$$= \iint_{\mathbb{R}^{2n}} \partial_x^{\alpha+\beta-\gamma} \partial_\xi^{\alpha+\beta-\gamma} [a(x+t\,y, \xi+t\,\eta)] \left(\frac{1-\Delta_y}{1+|\eta|^2}\right)^L \mathrm{e}^{iy\cdot\eta} \mathrm{d}y \mathrm{d}\eta.$$

If we integrate by parts once again, then we have

$$\begin{aligned}
&\iint_{\mathbb{R}^{2n}} \partial_x^{\alpha+\beta-\gamma} \partial_\xi^{\alpha+\beta-\gamma} [a(x+t\,y, \xi+t\,\eta)] \mathrm{e}^{iy\cdot\eta} \mathrm{d}y \mathrm{d}\eta \\
&= \iint_{\mathbb{R}^{2n}} \left(\frac{1-\Delta_y}{1+|\eta|^2}\right)^L \\
&\qquad \times \partial_x^{\alpha+\beta-\gamma} \partial_\xi^{\alpha+\beta-\gamma} [a(x+t\,y, \xi+t\,\eta)] \left(\frac{1-\Delta_\eta}{1+|y|^2}\right)^L \mathrm{e}^{iy\cdot\eta} \mathrm{d}y \mathrm{d}\eta \\
&= \iint_{\mathbb{R}^{2n}} \left(\frac{1-\Delta_\eta}{1+|y|^2}\right)^L \left(\frac{1-\Delta_y}{1+|\eta|^2}\right)^L \\
&\qquad \times \partial_x^{\alpha+\beta-\gamma} \partial_\xi^{\alpha+\beta-\gamma} [a(x+t\,y, \xi+t\,\eta)] \mathrm{e}^{iy\cdot\eta} \mathrm{d}y \mathrm{d}\eta.
\end{aligned}$$

Since $\partial_x^{\alpha+\beta-\gamma} \partial_\xi^{\alpha+\beta-\gamma} a \in S_{\rho\delta}^{m-(\rho-\delta)N}$ and $\rho > \delta$, when L and N are large enough, we have

$$\left|\iint_{\mathbb{R}^{2n}} \partial_x^{\alpha+\beta-\gamma} \partial_\xi^{\alpha+\beta-\gamma} [a(x+t\,y, \xi+t\,\eta)] \mathrm{e}^{iy\cdot\eta} \mathrm{d}y \mathrm{d}\eta\right| \lesssim \langle\xi\rangle^{m-(\rho-\delta)N}.$$

If we pass to the partial differentials, we obtain

$$\begin{aligned}
&\left|\partial_x^{\beta*} \partial_\xi^{\alpha*} \iint_{\mathbb{R}^{2n}} \partial_x^{\alpha+\beta-\gamma} \partial_\xi^{\alpha+\beta-\gamma} [a(x+t\,y, \xi+t\,\eta)] \mathrm{e}^{iy\cdot\eta} \mathrm{d}y \mathrm{d}\eta\right| \\
&\lesssim \langle\xi\rangle^{m+\delta|\beta*|-\rho|\alpha*|-(\rho-\delta)N}
\end{aligned}$$

for $\alpha*, \beta* \in {\mathbb{N}_0}^n$ with the constant independent of $t \in [0, 1]$. Let us set

$$A_j(x, \xi) \equiv \sum_{\substack{\alpha\in{\mathbb{N}_0}^n \\ |\alpha|=j}} \frac{(-i)^{|\alpha|}}{\alpha!} \partial_x^\alpha \partial_\xi^\alpha a(x, \xi).$$

Hence it follows that

$$\left|\partial_x^{\beta*} \partial_\xi^{\alpha*} \left(A(x, \xi) - \sum_{j=0}^{N-1} A_j(x, \xi)\right)\right| \lesssim \langle\xi\rangle^{m+\delta|\beta*|-\rho|\alpha*|-(\rho-\delta)N}.$$

Since $A_N \in S^{m-(\rho-\delta)N}$ and $\delta < \rho$, we are in the position of using Theorem 5.12 with

$$\mu_N(\alpha*, \beta*) = m - (\rho - \delta)N + \delta|\beta *| - \rho|\alpha *|, \quad m_N = m - (\rho - \delta)N$$

to have $A \in S^m_{\rho\delta}$.

Now we consider the composition of pseudo-differential operators, which we use to study the function space $A^s_{pq}(\mathbb{R}^n)$.

Theorem 5.14 (Composition of pseudo-differential operators) *Let $m_1, m_2 \in \mathbb{R}$ and $0 \le \delta < \rho \le 1$. For $a \in S^{m_1}_{\rho\delta}$ and $b \in S^{m_2}_{\rho\delta}$, we set*

$$c(x, \xi) \equiv \frac{1}{(2\pi)^n} \iint_{\mathbb{R}^{2n}} a(x, \eta) b(y, \xi) \mathrm{e}^{i(x-y)\cdot(\eta-\xi)} \mathrm{d}y \mathrm{d}\eta \quad (x, \xi \in \mathbb{R}^n).$$

Then $c \in S^{m_1+m_2}_{\rho\delta}$ and c satisfies $c(X, D) = a(X, D) \circ b(X, D)$.

Proof We prove $c \in S^{m_1+m_2}_{\rho\delta}$ using change of variables. Then

$$c(x, \xi) = \frac{1}{(2\pi)^n} \iint_{\mathbb{R}^{2n}} a(x, \eta + \xi) b(y + x, \xi) \mathrm{e}^{-iy\cdot\eta} \mathrm{d}y \mathrm{d}\eta. \tag{5.50}$$

Set $L \equiv [\max(m_1, 0) + n] + 1$. If we integrate the right-hand side of (5.50) by parts by L times, then

$$\begin{aligned} &c(x, \xi) \\ &= \frac{1}{(2\pi)^n} \iint_{\mathbb{R}^{2n}} a(x, \eta + \xi) \frac{(1 - \Delta_y)^L b(y + x, \xi)}{\langle \eta \rangle^{2L}} \mathrm{e}^{-iy\cdot\eta} \mathrm{d}y \mathrm{d}\eta \\ &= \frac{1}{(2\pi)^n} \iint_{\mathbb{R}^{2n}} \frac{1}{\langle y \rangle^{2L}} (1 - \Delta_\eta)^L \\ &\quad \left[a(x, \eta + \xi) \left(\frac{(1 - \Delta_y)^L b(y + x, \xi)}{\langle \eta \rangle^{2L}} \right) \right] \mathrm{e}^{-iy\cdot\eta} \mathrm{d}y \mathrm{d}\eta. \end{aligned}$$

Observe that the integrand is bounded from above by

$$\langle y \rangle^{-2L} \langle \eta + \xi \rangle^{m_1} \langle \xi \rangle^{m_2+2\delta L} \langle \eta \rangle^{-2L} \lesssim \langle y \rangle^{-2L} \langle \xi \rangle^{m_1+m_2+2\delta L} \langle \eta \rangle^{|m_1|-2L}. \tag{5.51}$$

We conclude

$$|c(x, \xi)| \lesssim \langle \xi \rangle^{m_1+m_2+2\delta L}. \tag{5.52}$$

As we did in (5.50), we have

$$\partial_x^\beta \partial_\xi^\alpha c(x,\xi) = \sum_{\alpha' \le \alpha, \beta' \le \beta} c_{\alpha,\beta,\alpha'\beta'} \iint_{\mathbb{R}^{2n}} \partial_x^{\alpha'} \partial_\xi^{\beta'} a(x,\eta+\xi) \partial_x^{\alpha-\alpha'} \partial_\xi^{\beta-\beta'} b(y+x,\xi) \frac{\mathrm{d}y\mathrm{d}\eta}{\mathrm{e}^{iy\cdot\eta}}.$$

Since

$$\partial_x^{\alpha'} \partial_\xi^{\beta'} a \in S_\delta^{m_1+\rho|\beta'|-\delta|\alpha'|}, \quad \partial_x^{\alpha-\alpha'} \partial_\xi^{\beta-\beta'} b \in S_{\rho\delta}^{m_2+\rho|\beta-\beta'|-\delta|\alpha-\alpha'|},$$

we have

$$|\partial_x^\beta \partial_\xi^\alpha c(x,\xi)| \lesssim \langle \xi \rangle^{m_1+m_2+(\rho-\delta)(|\alpha|+|\beta|)+2\delta L}.$$

Furthermore, expand the function $(\eta, y) \mapsto a(x,\eta+\xi)b(y+x,\xi)$ into the Taylor expansion around $(0,0)$ to have

$$a(x,\eta+\xi)b(y+x,\xi) - \sum_{|\alpha+\beta|<2N} \frac{\eta^\beta y^\alpha}{\alpha!\beta!} \partial_\xi^\beta a(x,\xi) \partial_x^\alpha b(x,\xi) = \frac{1}{(2N-1)!} \int_0^1 (1-t)^{2N-1} \frac{\mathrm{d}^{2N}}{\mathrm{d}t^{2N}} [a(x,\xi+t\,\eta)b(x+t\,y,\xi)]\mathrm{d}t.$$

Hence it follows that

$$\iint_{\mathbb{R}^{2n}} a(x,\eta+\xi)b(y+x,\xi)\mathrm{e}^{-iy\cdot\eta}\mathrm{d}y\mathrm{d}\eta - \sum_{\substack{\alpha\in\mathbb{N}_0{}^n \\ |\alpha|<N}} \frac{(2\pi)^n(-i)^{|\alpha|}}{\alpha!} \partial_\xi^\alpha a(x,\xi) \partial_x^\alpha b(x,\xi)$$
$$\simeq_N \int_0^1 \left(\iint_{\mathbb{R}^{2n}} \frac{\mathrm{d}^{2N}}{\mathrm{d}t^{2N}} a(x,\xi+t\,\eta)b(x+t\,y,\xi)\mathrm{e}^{-iy\cdot\eta}\mathrm{d}y\mathrm{d}\eta \right) (1-t)^{2N-1}\mathrm{d}t$$

For any multi-indexes $\alpha, \beta \in \mathbb{N}_0{}^n$ satisfying $|\alpha|+|\beta| = 2N$, we set

$$r_{\alpha\beta}(x,\xi) \equiv \iiint_{[0,1]\times\mathbb{R}^{2n}} y^\beta \eta^\alpha (1-t)^{2N-1} \partial_\xi^\alpha a(x,\xi+t\,\eta) \partial_x^\beta b(x+t\,y,\xi)\mathrm{e}^{-iy\cdot\eta}\mathrm{d}t\mathrm{d}y\mathrm{d}\eta.$$

By the Leibniz rule, we have

$$\iint_{\mathbb{R}^{2n}} a(x, \eta + \xi) b(y + x, \xi) e^{-iy\cdot\eta} \mathrm{d}y\mathrm{d}\eta$$

$$= \sum_{\substack{\alpha \in \mathbb{N}_0{}^n \\ |\alpha| < N}} \frac{(2\pi)^n (-i)^{|\alpha|}}{\alpha!} \partial_\xi^\alpha a(x, \xi) \partial_x^\alpha b(x, \xi) + \sum_{\alpha+\beta=2N} c_{\alpha\beta} r_{\alpha\beta}(x, \xi),$$

where $c_{\alpha\beta}$ is a suitable coefficient.

We transform the term

$$\iint_{\mathbb{R}^{2n}} y^\beta \eta^\alpha \partial_\xi^\alpha a(x, \xi + t\,\eta) \partial_x^\beta b(x + t\,y, \xi) \mathrm{e}^{-iy\cdot\eta} \mathrm{d}y\mathrm{d}\eta,$$

where $t \in [0, 1]$ is fixed.

Lemma 5.7 *There exists a collection $\{C_{\alpha,\beta,\gamma}\}_{\alpha,\beta,\gamma \in \mathbb{N}_0{}^n}$ such that*

$$\iint_{\mathbb{R}^{2n}} y^\beta \eta^\alpha \partial_\xi^\alpha a(x, \xi + t\,\eta) \partial_x^\beta b(x + t\,y, \xi) \mathrm{e}^{-iy\cdot\eta} \mathrm{d}y\mathrm{d}\eta$$

$$= \sum_{\gamma \leq \min(\alpha,\beta)} C_{\alpha,\beta,\gamma} t^{|\alpha|+|\beta|-2|\gamma|}$$

$$\times \iint_{\mathbb{R}^{2n}} \partial_\xi^{\alpha+\beta-\gamma} a(x, \xi + t\,\eta) \partial_x^{\alpha+\beta-\gamma} b(x + t\,y, \xi) \mathrm{e}^{-iy\cdot\eta} \mathrm{d}y\mathrm{d}\eta.$$

Proof In the proof we write $C_{\alpha,\beta,\gamma}$, which may be different from one occurrence to another. An integration by parts against η yields

$$\iint_{\mathbb{R}^{2n}} y^\beta \eta^\alpha \partial_\xi^\alpha a(x, \xi + t\,\eta) \partial_x^\beta b(x + t\,y, \xi) \mathrm{e}^{-iy\cdot\eta} \mathrm{d}t\mathrm{d}y\mathrm{d}\eta$$

$$= \iint_{\mathbb{R}^{2n}} \eta^\alpha \partial_\xi^\alpha a(x, \xi + t\,\eta) \partial_x^\beta b(x + t\,y, \xi) (i\partial_\eta)^\beta \mathrm{e}^{-iy\cdot\eta} \mathrm{d}y\mathrm{d}\eta$$

$$= \sum_{\gamma \leq \min(\alpha,\beta)} C_{\alpha,\beta,\gamma} t^{|\beta|-|\gamma|}$$

$$\times \iint_{\mathbb{R}^{2n}} \eta^{\alpha-\gamma} \partial_\xi^{\alpha+\beta-\gamma} a(x, \xi + t\,\eta) \partial_x^\beta b(x + t\,y, \xi) \mathrm{e}^{-iy\cdot\eta} \mathrm{d}y\mathrm{d}\eta.$$

Next, an integration by parts against y yields

$$\iint_{\mathbb{R}^{2n}} y^\beta \eta^\alpha \partial_\xi^\alpha a(x, \xi + t\,\eta) \partial_x^\beta b(x + t\,y, \xi) \mathrm{e}^{-iy\cdot\eta} \mathrm{d}t\mathrm{d}y\mathrm{d}\eta$$

$$= \sum_{\gamma \leq \min(\alpha,\beta)} C_{\alpha,\beta,\gamma} t^{|\beta|-|\gamma|}$$

$$\times \iint_{\mathbb{R}^{2n}} \partial_\xi^{\alpha+\beta-\gamma} a(x, \xi + t\eta) \partial_x^\beta b(x + ty, \xi)(i\partial_y)^{\alpha-\gamma} e^{-iy\cdot\eta} dy d\eta$$

$$= \sum_{\gamma \le \min(\alpha,\beta)} C_{\alpha,\beta,\gamma} t^{|\beta|-|\gamma|}$$

$$\times \iint_{\mathbb{R}^{2n}} (-i\partial_y)^{\alpha-\gamma} [\partial_\xi^{\alpha+\beta-\gamma} a(x, \xi + t\eta) \partial_x^\beta b(x + ty, \xi)] e^{-iy\cdot\eta} dy d\eta$$

$$= \sum_{\gamma \le \min(\alpha,\beta)} C_{\alpha,\beta,\gamma} t^{|\alpha|+|\beta|-2|\gamma|}$$

$$\times \iint_{\mathbb{R}^{2n}} \partial_\xi^{\alpha+\beta-\gamma} a(x, \xi + t\eta) \partial_x^{\alpha+\beta-\gamma} b(x + ty, \xi) e^{-iy\cdot\eta} dy d\eta.$$

Thus, Lemma 5.7 is therefore proved.

Returning to the proof, we analyze

$$\mathrm{I} \equiv \iint_{\mathbb{R}^{2n}} \partial_\xi^{\alpha+\beta-\gamma} a(x, \xi + t\eta) \partial_x^{\alpha+\beta-\gamma} b(x + ty, \xi) e^{-iy\cdot\eta} dy d\eta$$

for each fixed α, β, γ. We use a similar technique to obtain (5.51), where the symbol c is given by (5.50). Hence if $L \gg 1$, say $m_1 + m_2 + 2\delta L > 0$. Peetre's inequality yields an estimate uniform over $t \in [0, 1]$;

$$\mathrm{I} \lesssim \langle \xi \rangle^{m_1+m_2-(\rho-\delta)|\alpha+\beta-\gamma|}. \tag{5.53}$$

In particular, we are considering multi-indexes α, β, γ such that $|\alpha| + |\beta| = 2N$ and that $\gamma \le \min(\alpha, \beta)$. Thus, $|\alpha+\beta-\gamma| \ge N$ and if we integrate against $t \in [0, 1]$, Then

$$|r_{\alpha\beta}(x, \xi)| \lesssim \langle \xi \rangle^{m_1+m_2-(\rho-\delta)N}. \tag{5.54}$$

Meanwhile, we have

$$\partial_x^{\alpha*} \partial_\xi^{\beta*} \iint_{\mathbb{R}^{2n}} y^\beta \eta^\alpha \partial_\xi^\alpha a(x, \xi + t\eta) \partial_x^\beta b(x + ty, \xi) e^{-iy\cdot\eta} dy d\eta$$

$$= \sum_{\gamma \le \min(\alpha,\beta)} C_{\alpha,\beta,\gamma} t^{|\alpha|+|\beta|-2|\gamma|}$$

$$\times \iint_{\mathbb{R}^{2n}} \partial_\xi^{\alpha+\beta-\gamma+\alpha*} a(x, \xi + t\eta) \partial_x^{\alpha+\beta-\gamma+\beta*} b(x + ty, \xi) e^{-iy\cdot\eta} dy d\eta$$

$$= \iint_{\mathbb{R}^{2n}} y^\beta \eta^\alpha \partial_\xi^\alpha \partial_\xi^{\alpha*} a(x, \xi + t\eta) \partial_x^\beta \partial_\xi^{\beta*} b(x + ty, \xi) e^{-iy\cdot\eta} dy d\eta.$$

Hence it follows that

$$|\partial_x^{\beta*}\partial_\xi^{\alpha*} r_{\alpha\beta}(x,\xi)| \lesssim \langle\xi\rangle^{m_1+m_2+\delta|\beta*|-\rho|\alpha*|-(\rho-\delta)N}$$

by (5.54).

Since we assume $\delta < \rho$, we are in the position of invoking Theorem 5.12 to conclude that $c \in S^{m_1+m_2}_{\rho\delta}$.

We will prove $c(X,D) = a(X,D) \circ b(X,D)$. Let $x, z \in \mathbb{R}^n$. According to the definition of pseudo-differential operators, we have

$$a(X,D)g(x) = \frac{1}{(2\pi)^n}\iint_{\mathbb{R}^{2n}} a(x,\xi)g(z)\mathrm{e}^{i(x-z)\cdot\xi}\mathrm{d}z\mathrm{d}\xi, \quad g \in \mathscr{S}(\mathbb{R}^n),$$

$$b(X,D)f(z) = \frac{1}{(2\pi)^n}\iint_{\mathbb{R}^{2n}} b(z,\eta)f(y)\mathrm{e}^{i(z-y)\cdot\eta}\mathrm{d}y\mathrm{d}\eta, \quad f \in \mathscr{S}(\mathbb{R}^n).$$

In view of (5.31), we can assume $a, b \in C^\infty_{\mathrm{c}}(\mathbb{R}^{2n})$. This will justify that we can use Fubini's theorem freely. We calculate the composition of $a(X,D)$ and $b(X,D)$:

$$\begin{aligned} &a(X,D)[b(X,D)f](x) \\ &= \frac{1}{(2\pi)^{2n}}\iint_{\mathbb{R}^{2n}\times\mathbb{R}^{2n}} a(x,\xi)b(z,\eta)f(y)\mathrm{e}^{i(x-z)\cdot\xi+i(z-y)\cdot\eta}\mathrm{d}y\mathrm{d}z\mathrm{d}\xi\mathrm{d}\eta \\ &= \frac{1}{(2\pi)^{\frac{3n}{2}}}\iiint_{\mathbb{R}^{2n}\times\mathbb{R}^n} a(x,\xi)b(z,\eta)\mathscr{F}f(\eta)\mathrm{e}^{i(x-z)\cdot\xi+iz\cdot\eta}\mathrm{d}z\mathrm{d}\xi\mathrm{d}\eta. \end{aligned}$$

If we interchange the roles of ξ and η, then

$$\begin{aligned} &a(X,D)[b(X,D)f](x) \\ &= \frac{1}{(2\pi)^{\frac{3n}{2}}}\iiint_{\mathbb{R}^{2n}\times\mathbb{R}^n} a(x,\eta)b(z,\xi)\mathscr{F}f(\xi)\mathrm{e}^{i(x-z)\cdot\eta+iz\cdot\xi}\mathrm{d}z\mathrm{d}\xi\mathrm{d}\eta \\ &= \frac{1}{(2\pi)^{\frac{3n}{2}}}\iiint_{\mathbb{R}^{2n}\times\mathbb{R}^n} a(x,\eta)b(z,\xi)\mathscr{F}f(\xi)\mathrm{e}^{i(x-z)\cdot(\eta-\xi)}\mathrm{e}^{ix\cdot\xi}\mathrm{d}z\mathrm{d}\xi\mathrm{d}\eta. \end{aligned}$$

If we use Fubini's theorem, then $a(X,D)[b(X,D)f] = c(X,D)f$.

Exercises

Exercise 5.18 Show that the function $a(x,\xi) \equiv \langle\xi\rangle^m$ belongs to S^m. Hint: Use Lemma 2.1.

Exercise 5.19 Prove (5.46). Hint: Use Corollary 5.1 to justify that a is compactly supported.

Exercise 5.20 Prove (5.45) using the Fourier transform and its inverse.

Exercise 5.21 Using the idea of Example 5.3, show $K(x) \equiv \int_{\mathbb{R}^n} \mathrm{e}^{-ix\cdot\xi} \langle\xi\rangle^{-\sigma} \mathrm{d}\xi$ makes sense for any $\sigma \in \mathbb{R}$.

Exercise 5.22 Let $a(X, D)^*$ be the operator given by (5.32). Then show that the adjoint $a(X, D)^*$ sends $\mathscr{S}'(\mathbb{R}^n)$ continuously into itself. Hint: Integration by parts.

Exercise 5.23 Prove (5.49). Hint: Induct on the dimension.

Exercise 5.24 Let $c(X, D) = a(X, D) \circ b(X, D)$, where a, b are symbols.

1. Show that $c \in S^0_{00}$ if $a, b \in S^0_{00}$.
2. Show that $c \in M_{\infty 1}(\mathbb{R}^{2n})$ if $a, b \in M_{\infty 1}(\mathbb{R}^{2n})$. Hint: Use the molecular decomposition for $M_{\infty 1}(\mathbb{R}^{2n})$.

5.2.2 *Boundedness of Pseudo-differential Operators on Besov Spaces and Triebel–Lizorkin Spaces*

Having set down the elementary facts on pseudo-differential operators, we will investigate the boundedness on Besov spaces and Triebel–Lizorkin spaces.

In this section, we suppose that $\varphi, \psi, \kappa \in \mathscr{S}(\mathbb{R}^n)$ satisfy

$$\chi_{Q(2)} \le \kappa \le \chi_{Q(3)}, \quad \chi_{Q(1)} \le \psi \le \chi_{Q(2)}, \quad \varphi = \psi - \psi(2\star).$$

5.2.2.1 Pseudo-differential Operator with Symbol S^m_{11}

We aim here to prove the $A^s_{pq}(\mathbb{R}^n)$-boundedness of $a(X, D)$ for $a \in S^m_{11}$.

Let us recall Theorem 4.9 to this end. By the Leibniz formula,

$$\begin{aligned}
&\partial_x^\alpha \left(\int_{\mathbb{R}^n} a(x, \xi)\kappa_j(\xi) \exp(i(x - 2^{-j}k)\cdot\xi)\mathrm{d}\xi \right) \\
&= \sum_{\beta\le\alpha} c_{\alpha\beta} \int_{\mathbb{R}^n} [\partial_x^\beta a(x, \xi)]\kappa_j(\xi)(i\,\xi)^{\alpha-\beta} \exp(i(x - 2^{-j}k)\cdot\xi)\mathrm{d}\xi
\end{aligned}$$

for all $k \in \mathbb{Z}^n$. Since κ is supported away from the origin, integration by parts shows that

$$2^{-j(s+m)} \int_{\mathbb{R}^n} a(x, \xi)\kappa_j(\xi) \exp(i(x - 2^{-j}k)\cdot\xi)\mathrm{d}\xi$$

obeys the size condition of the molecules for $A^s_{pq}(\mathbb{R}^n)$.

Theorem 5.15 (Pseudo-differential operators with symbol in S^m_{11} on $A^s_{pq}(\mathbb{R}^n)$) *Let $0 < p, q \le \infty$, $s, m \in \mathbb{R}$, and let $a \in S^m_{11}$.*

1. *If $s > \sigma_p = n\left(\frac{1}{p} - 1\right)_+$, then $\|a(X, D)f\|_{B^s_{pq}} \lesssim \|f\|_{B^{s+m}_{pq}}$ for all $f \in B^{s+m}_{pq}(\mathbb{R}^n)$.*
2. *If $p < \infty$, $s > \sigma_{pq} = n\left(\frac{1}{p \wedge q} - 1\right)_+$, then $\|a(X, D)f\|_{F^s_{pq}} \lesssim \|f\|_{F^{s+m}_{pq}}$ for all $f \in F^{s+m}_{pq}(\mathbb{R}^n)$.*

5.2.2.2 Pseudo-differential Operator with Symbol $S^m_{1\delta}$ with $\delta \in [0, 1)$

Observe from Theorem 5.15 that pseudo-differential operators are bounded for any s provided that $a \in S^m_{1\delta}$ with $0 \le \delta < 1$.

Corollary 5.2 *Let $0 < p, q \le \infty$, $s, m \in \mathbb{R}$, $0 \le \delta < 1$, and let $a \in S^m_{1\delta}$. Then*

$$\|a(X, D)f\|_{A^s_{pq}} \lesssim \|f\|_{A^{s+m}_{pq}} \quad (f \in A^{s+m}_{pq}(\mathbb{R}^n)).$$

Proof Since $a \in S^m_{1\delta}$, we are in the position of using Theorem 5.14, the composition formula. Let $0 \le \delta < 1$, and let $a \in S^m_{1\delta}$. Choose N large enough, say, $N \equiv 2[|s| + 1 + \sigma_{p,q}]_+ \in \mathbb{N}_0$. Theorem 5.14 yields $b \in S^m_{1\delta}$ such that there is a decomposition:

$$a(X, D) = (1 - \Delta)^N \circ b(X, D) \circ (1 - \Delta)^{-N},$$

so that Theorem 2.12 yields $\|a(X, D)f\|_{A^s_{pq}} \sim \|b(X, D) \circ (1 - \Delta)^{-N} f\|_{A^{s+2N}_{pq}}$. Since $s + N > \sigma_{p,q}$, the function space $A^{s+N}_{pq}(\mathbb{R}^n)$ falls under the scope of the regular case. Hence combining Theorems 2.12 and 5.15, we conclude:

$$\|a(X, D)f\|_{A^s_{pq}} \lesssim \|(1 - \Delta)^{-N} f\|_{A^{s+2N+m}_{pq}} \lesssim \|f\|_{A^{s+m}_{pq}}.$$

Thus, the proof is complete.

We next consider the class $B^l_{\infty\infty}S^m_{1\delta}(\mathbb{R}^n)$ of pseudo-differential operators with $0 \le \delta \le 1$ and $m \in \mathbb{R}$. We aim here to present further applications of the molecular decomposition.

5.2.2.3 Definition of $B^l_{\infty\infty}S^m_{1\delta}(\mathbb{R}^n)$: Nonsmooth Symbols

Since the coefficients of differential equations are not always assumed C^∞, we are led to the class $B^l_{\infty\infty}S^m_{1\delta}(\mathbb{R}^n)$ instead of $S^m_{1\delta}(\mathbb{R}^n)$.

Definition 5.12 ($B^l_{\infty\infty}S^m_{1\delta}(\mathbb{R}^n)$) Let l, m, δ satisfy $l > 0$, $m \in \mathbb{R}$, $0 \le \delta \le 1$.

1. For $N \in \mathbb{N}$ and $a \in C^\infty(\mathbb{R}^n_\xi, \mathscr{S}'(\mathbb{R}^n_x))$, set

$$\|a\|_{B^l_{\infty\infty} S^m_{1\delta};N}$$
$$\equiv \sup_{\xi\in\mathbb{R}^n,\,\alpha\in\mathbb{N}_0{}^n,\,|\alpha|\le N} \left\{\langle\xi\rangle^{|\alpha|-m-\delta l}\|\partial_\xi^\alpha a(\star,\xi)\|_{B^l_{\infty\infty}} + \langle\xi\rangle^{|\alpha|-m}\|\partial_\xi^\alpha a(\star,\xi)\|_\infty\right\}.$$

2. The set $B^l_{\infty\infty} S^m_{1\delta}(\mathbb{R}^n)$ of functions is defined to be

$$B^l_{\infty\infty} S^m_{1\delta}(\mathbb{R}^n) \equiv \bigcap_{N\in\mathbb{N}} \left\{a \in C^\infty(\mathbb{R}^n_\xi, \mathscr{S}'(\mathbb{R}^n_x)) \,:\, \|a\|_{B^l_{\infty\infty} S^m_{1\delta};N} < \infty\right\}.$$

3. A symbol $a \in B^l_{\infty\infty} S^m_{1\delta}(\mathbb{R}^n)$ is said to be elementary if for $x, \xi \in \mathbb{R}^n$,

$$a(x,\xi) = \sigma_0(x)\psi(\xi) + \sum_{j=1}^\infty \sigma_j(x)\varphi_j(\xi) = \sigma_0(x)\psi(\xi) + \sum_{j=1}^\infty \sigma_j(x)\varphi(2^{-j}\xi),$$

where $\{\sigma_j\}_{j=0}^\infty \subset B^l_{\infty\infty}(\mathbb{R}^n)$ and $\{\psi,\varphi\} \subset C^\infty_c(\mathbb{R}^n)$ fulfill the conditions below:

(a) $\sup_{j\in\mathbb{N}_0} \left\{2^{-j(m+\delta l)}\|\sigma_j\|_{B^l_{\infty\infty}} + 2^{-jm}\|\sigma_j\|_\infty\right\} < \infty.$
(b) $\mathrm{supp}(\psi) \subset B(4)$ and $\mathrm{supp}(\varphi) \subset B(8) \setminus B(1)$.

For elementary symbols a, we would like to discuss how the definition

$$a(X,D)f = \sigma_0 \cdot \psi(D)f + \sum_{j=1}^\infty \sigma_j \cdot \varphi_j(D)f$$

can be justified. Once this is done, we can move on to the case of $B^l_{\infty\infty} S^m_{1\delta}(\mathbb{R}^n)$ using Theorem 5.16 below. For the class $B^l_{\infty\infty} S^m_{1\delta}(\mathbb{R}^n)$ of symbol, elementary symbols due to Coifman and Meyer are fundamental pieces, which we need to consider as the theorem shows.

Theorem 5.16 (Expansion of $B^l_{\infty\infty} S^m_{1\delta}(\mathbb{R}^n)$) *Let the parameters l, m, δ, θ satisfy*

$$l > 0, \quad 0 < \theta < 1, \quad m \in \mathbb{R}, \quad 0 \le \delta \le 1.$$

Then for each $a \in B^l_{\infty\infty} S^m_{1\delta}(\mathbb{R}^n)$, we can find a sequence $\{a_k\}_{k=1}^\infty \in B^l_{\infty\infty} S^m_{1\delta}(\mathbb{R}^n)$ of elementary symbols such that a is expanded: $a = \sum_{k=1}^\infty a_k$ in $B^l_{\infty\infty} S^m_{1\delta}(\mathbb{R}^n)$ and

$$\sum_{k=1}^\infty \|a_k\|^\theta_{B^l_{\infty\infty} S^m_{1\delta};N} \lesssim_{\theta,N} \|a\|^\theta_{B^l_{\infty\infty} S^m_{1\delta};N+\left[\frac{n+2}{2}\right]} \tag{5.55}$$

for all $N \in \mathbb{N}$.

Proof Choose $\psi \in \mathscr{S}(\mathbb{R}^n)$ and $\kappa \in \mathscr{S}(\mathbb{R}^n)$ so that $\chi_{B(1)} \le \psi \le \chi_{B(2)}$ and that $\chi_{B(2)} \le \kappa \le \chi_{B(3)}$. For $j \in \mathbb{N}$, set $\varphi_j \equiv \psi(2^{-j}\star) - \psi(2^{-j+1}\star)$, $\kappa_j \equiv \kappa(2^{-j}\star)$. Then

$$a(x,\xi) = a(x,\xi)\psi(\xi) + \sum_{j=1}^{\infty} a(x,\xi)\varphi_j(\xi) \quad (x,\xi \in \mathbb{R}^n). \tag{5.56}$$

We can incorporate $a(x,\xi)\psi(\xi)$ later and thus we do not deal with this term. Define σ_{jk} by

$$\sigma_{jk}(x) \equiv \frac{1}{(22^j\pi)^n}\int_{[-2^j\pi,2^j\pi]^n} a(x,\eta)\varphi_j(\eta)\exp(-k\cdot 2^{-j}\xi i)\mathrm{d}\eta$$

for $x \in \mathbb{R}^n$.

We freeze $x \in \mathbb{R}^n$ for the time being and expand each term of (5.56) into a Fourier series; denote by $\sigma_{jk}(x)$ the Fourier coefficient to have

$$a(x,\xi)\varphi_j(\xi) = \sum_{k\in\mathbb{Z}^n} \sigma_{jk}(x)(\kappa_j(\xi) - \kappa_{j-4}(\xi))\exp(k\cdot 2^{-j}\xi i).$$

An integration by parts yields

$$\begin{aligned}
\sigma_{jk}(x) &\simeq_n \int_{Q(2^j\pi)} a(x,\eta)\varphi_j(\eta)\exp(-k\cdot 2^{-j}\xi i)\mathrm{d}\eta \\
&= \frac{1}{\langle k\rangle^{2L}}\int_{Q(2^j\pi)} a(x,\eta)\varphi_j(\eta)(1-2^{2j}\Delta_\eta)^L\exp(-k\cdot 2^{-j}\xi i)\mathrm{d}\eta \\
&= \frac{1}{\langle k\rangle^{2L}}\int_{Q(2^j\pi)} (1-2^{2j}\Delta_\eta)^L[a(x,\eta)\varphi_j(\eta)]\exp(-k\cdot 2^{-j}\xi i)\mathrm{d}\eta.
\end{aligned}$$

Hence for all multi-indexes $\alpha, \beta \in \mathbb{N}_0{}^n$ satisfying $|\alpha| + |\beta| = 2M \le 2L$,

$$\|2^{2jM}\partial_\eta^\alpha a(x,\eta)\partial_\eta^\beta[\varphi_j(\eta)]\|_{B^l_{\infty\infty}} \lesssim 2^{(2M-|\alpha|+m+\delta l-|\beta|)j} = 2^{m+\delta l}.$$

By the triangle inequality,

$$\|\sigma_j\|_{B^l_{\infty\infty}} \lesssim \int_{Q(2^j\pi)} \frac{\|(1-2^{2j}\Delta_\eta)^L[a(x,\eta)\varphi_j(\eta)]\|_{B^l_{\infty\infty}}}{\langle k\rangle^{2L}}\mathrm{d}\eta \lesssim 2^{j(m+\delta l)}\|a\|_{B^l_{\infty\infty}S^m_{1\delta};N}.$$

Likewise, we have $2^{-jm}\|\sigma_j\|_\infty \lesssim \|a\|_{B^l_{\infty\infty}S^m_{1\delta};N}$. If we set

$$\varphi^{(k)}(\xi) \equiv (\kappa(\xi) - \kappa(16\xi)) \exp(k \cdot \xi i), \quad a_k(x, \xi) \equiv \sum_{j=1}^{\infty} \sigma_{jk}(x) \varphi_j^{(k)}(\xi),$$

then $\|a_k\|_{B^l_{\infty\infty} S^m_{1\delta}; N} \lesssim \langle k \rangle^{-2L} \|a\|_{B^l_{\infty\infty} S^m_{1\delta}; N+2L}$. If we choose $L \gg 1$, then we have (5.55) by summing this over k.

5.2.2.4 $A^s_{pq}(\mathbb{R}^n)$-Boundedness of Pseudo-differential Operators with Symbol in $B^l_{\infty\infty} S^m_{1\delta}(\mathbb{R}^n)$

In $A^s_{pq}(\mathbb{R}^n)$, we suppose s is sufficiently large; we focus upon the regular case.

Theorem 5.17 (Pseudo-differentialtors with symbol in $B^l_{\infty\infty} S^m_{1\delta}(\mathbb{R}^n)$ on $A^s_{pq}(\mathbb{R}^n)$) *Suppose that the parameters $p, q, r, s, l, \delta, m \in \mathbb{R}$ satisfy*

$$0 < s < l, \quad 1 \le p, q \le \infty, \quad 0 \le \delta \le 1.$$

Let $a \in B^l_{\infty\infty} S^m_{1\delta}(\mathbb{R}^n)$. Then $\|a(X, D) f\|_{A^s_{pq}} \lesssim \|f\|_{A^{s+m}_{pq}}$ for $f \in A^{s+m}_{pq}(\mathbb{R}^n)$.

Proof The lift operator allows us to assume that $m = 0$ and that a is an elementary symbol. Let $f \in A^s_{pq}(\mathbb{R}^n)$ and a be elementary symbol with an expression $a(x, \xi) = \sum_{j=1}^{\infty} \sigma_j(x) \varphi_j(\xi)$. We omit the part of $j = 0$ in a; we can readily incorporate it later. Let $a_{jk} \equiv \varphi_k(D) \sigma_j$. Since a is elementary,

$$\|a_{jk}\|_\infty \lesssim 2^{(j-k)l}. \tag{5.57}$$

Furthermore, we have

$$a(X, D) f(x) = \sum_{j,k=1}^{\infty} a_{jk}(x) \varphi_j(D) f(x).$$

Split $a(X, D)$ by $a(X, D) = a_1(X, D) + a_2(X, D) + a_3(X, D)$ with

$$a_1(X, D) f(x) \equiv \sum_{j=4}^{\infty} \left(\sum_{k=0}^{j-4} a_{jk}(x) \right) \varphi_j(D) f(x),$$

$$a_2(X, D) f(x) \equiv \sum_{j=0}^{\infty} \left(\sum_{k=\max(j-3,0)}^{j+3} a_{jk}(x) \right) \varphi_j(D) f(x),$$

$$a_3(X, D)f(x) \equiv \sum_{j=0}^{\infty} \left(\sum_{k=j+4}^{\infty} a_{jk}(x) \right) \varphi_j(D)f(x).$$

Since the frequency support of $\left(\sum_{k=0}^{j-4} a_{jk} \right) \varphi_j(D)f$ is concentrated on the neighborhood of $\{|\xi| = 2^j\}$, we have

$$\begin{aligned}
\|a_1(X, D)f\|_{F^s_{pq}} &\lesssim \left\| \left\{ 2^{js} \left(\sum_{k=0}^{j-4} a_{jk} \right) \varphi_j(D)f \right\}_{j=4}^{\infty} \right\|_{L^p(\ell^q)} \\
&\lesssim \sup_{j \in \mathbb{N}_0 \cap [4,\infty)} \left\| \sum_{k=0}^{j-4} a_{jk} \right\|_{\infty} \|2^{js}\varphi_j(D)f\|_{L^p(\ell^q)} \\
&\lesssim \|f\|_{F^s_{pq}}
\end{aligned}$$

by Theorem 1.53 for the case of Triebel–Lizorkin spaces; the same argument works for Besov spaces.

We can use the molecular decomposition to estimate $a_2(X, D)f$. We have

$$\left\| \partial^{\alpha} \left(\sum_{k=\max(j-3,0)}^{j+3} a_{jk} \right) \right\|_{\infty} \lesssim 2^{j|\alpha|} \left\| \sum_{k=\max(j-3,0)}^{j+3} a_{jk} \right\|_{\infty} \lesssim 2^{j|\alpha|} \quad (j \in \mathbb{N}_0)$$

by Corollary 1.6 and $L^{\infty}(\mathbb{R}^n)$-estimate (5.57) of a_{jk}. Hence it follows that

$$a_2(X, D)f = \sum_{j=0}^{\infty} \sum_{m \in \mathbb{Z}^n} \varphi_j(D)f(2^{-j}m) \left(\sum_{k=\max(j-3,0)}^{j+3} a_{jk} \right) \mathscr{F}^{-1}\kappa(2^j \star -m)$$

is the molecular decomposition of $a_2(X, D)$, which shows that the operator $a_2(X, D)$ is bounded on $A^s_{pq}(\mathbb{R}^n)$.

As for $a_3(X, D)$, we decompose

$$a_3(X, D)f = \sum_{k=4}^{\infty} \left(\sum_{j=0}^{k-4} a_{jk} \right) \varphi_j(D)f.$$

The frequency support $\sum_{j=0}^{k-4} a_{jk}(x)\varphi_j(D)f(x)$ is concentrated on a neighborhood of $\{|\xi| = 2^k\}$. This then yields

$$\|a_3(X,D)f\|_{F^s_{pq}} \lesssim \left\| \left\{ 2^{ks} \sum_{j=0}^{k-4} a_{jk}\varphi_j(D)f \right\}_{k=4}^{\infty} \right\|_{L^p(\ell^q)} \tag{5.58}$$

by Theorem 1.53 for the case of Triebel–Lizorkin spaces. We obtain

$$\|a_3(X,D)f\|_{F^s_{pq}} \lesssim \left\| \left\{ \sum_{j=0}^{k-4} 2^{(s-l)(k-j)} 2^{js} |\varphi_j(D)f| \right\}_{k=4}^{\infty} \right\|_{L^p(\ell^q)} \lesssim \|f\|_{F^s_{pq}} \tag{5.59}$$

by Hölder's inequality and the fact that $l > s$. The same argument works for Besov spaces.

Exercises

Exercise 5.25 Let $l > 0$, $0 < \delta \leq 1$, $m > 0$ and $N \gg 1$. Show that $S^N \subset B^l_{\infty\infty} S^m_{1\delta}(\mathbb{R}^n)$.

Exercise 5.26 Interpolate between the boundedness of different Triebel–Lizorkin spaces to complete the proof of Theorem 5.17.

5.2.3 Applications to Partial Differential Equations

For simplicity, we assume that $1 < p, q < \infty$ and $s \geq 0$. Recall $\mathscr{C}^s(\mathbb{R}^n) \approx B^s_{\infty\infty}(\mathbb{R}^n)$ according to Theorems 2.7 and 2.8. We consider the following second order elliptic partial differential equations of the nondivergence form:

$$L_0 u \equiv -\sum_{j,k=1}^{n} a_{jk} \cdot \partial_{x_j}\partial_{x_k} u = f \quad (f \in A^s_{pq}(\mathbb{R}^n)) \tag{5.60}$$

on $\mathbb{R}^n$. We note that an operator of the form $Lf \equiv -\sum_{i,j=1}^{n} \partial_{x_i}(a_{ij}\partial_{x_j} f)$ is called the elliptic differential operator of divergence form and that an operator of the form $Lf \equiv -\sum_{i,j=1}^{n} a_{ij}\partial_{x_i}\partial_{x_j} f$ is called the elliptic differential operator of nondivergence form.

We postulate u on the following conditions:

$$a_{jk} \in \mathscr{C}^{\infty}(\mathbb{R}^n) \equiv \bigcap_{l \in \mathbb{R}} B^l_{\infty\infty}(\mathbb{R}^n) \tag{5.61}$$

for all $j, k = 1, 2, \ldots, n$ and

$$\sum_{j,k=1}^{n} \Re(a_{jk}(x)\xi_j\xi_k) \geq \Theta|\xi|^2, \quad (\xi = (\xi_1, \xi_2, \ldots, \xi_n), x \in \mathbb{R}^n) \tag{5.62}$$

for some constant Θ independent of x. We call this Θ the uniformly elliptic constant.

5.2.3.1 Parametrix

In view of the results in Sect. 4.3.2, Theorem 4.37 shows that

$$\|L f\|_{A^s_{pq}} + \|f\|_{A^s_{pq}} \lesssim \|f\|_{A^{s+2}_{pq}}.$$

We aim here to prove the reverse inequality. First let id denote the identity operator.
To this end, we set

$$a(x, \xi) \equiv \sum_{j,k}^{n} a_{jk}(x)\xi_j\xi_k \tag{5.63}$$

and consider the corresponding symbol.

Lemma 5.8 *Let a be a symbol given by* (5.63) *via* $\{a_{ij}\}_{i,j=1,2,\ldots,n} \subset \mathscr{C}^\infty(\mathbb{R}^n)$ *satisfying* (5.61) *and* (5.62). *Then there exist $b, r \in S^{-2}_{10}$ such that*

$$b(X, D) \circ a(X, D) = \text{id} + r(X, D). \tag{5.64}$$

Equation (5.64) can be termed $b(X, D) \circ a(X, D) = \text{id} +$ "error$''$.

Proof Choose $\eta \in C^\infty_c(\mathbb{R}^n)$ so that $\chi_{B(1)} \leq \eta \leq \chi_{B(2)}$. We define $B \in C^\infty(\mathbb{R}^{2n})$ by

$$B(x, \xi) \equiv \frac{1}{a(x, \xi)} \begin{cases} 1 - \eta(\xi) & (\xi \neq 0), \\ 0 & (\text{ otherwise}). \end{cases}$$

By (5.62), $B \in S^{-2}_{10}$. We expand $B(X, D) \circ a(X, D)$ using Theorem 5.14:

$$B(X, D) \circ a(X, D) - [B \cdot a](X, D) \in S^{-1}_{10}.$$

Since $1 - B \cdot a = \eta$, we have $R(X, D) \equiv 1 - B(X, D) \circ a(X, D) \in S^{-1}_{10}$. As a result, if we set $b(X, D) \equiv \sum_{j=0}^{1} R(X, D)^j \circ B(X, D) = R(X, D) \circ B(X, D) + B(X, D)$, then

$$\begin{aligned} b(X,D)\circ a(X,D)-1&=\sum_{j=0}^{1}R(X,D)^j\circ B(X,D)\circ a(X,D)-1\\ &=\sum_{j=0}^{1}R(X,D)^j\circ(1-R(X,D))-1\\ &=-R(X,D)\circ R(X,D)\in S^{-2}_{10}. \end{aligned}$$

Thus, $r(X, D) = -R(X, D) \circ R(X, D)$ gives the desired symbol $r(x, \xi)$.

5.2.3.2 A Priori Estimate of Elliptic Differential Operators

Based on the above observation, we prove the following estimate:

Theorem 5.18 (A priori estimate of elliptic differential operators) *We suppose* $0 < p, q \leq \infty$ *and* $s \in \mathbb{R}$. *If* L_0 *is a second order elliptic differential operator defined in* (5.60), *then* $\|f\|_{A^{s+2}_{pq}(\mathbb{R}^n)} \sim \|L_0 f\|_{A^s_{pq}(\mathbb{R}^n)} + \|f\|_{A^s_{pq}(\mathbb{R}^n)}$ *for* $f \in A^s_{pq}(\mathbb{R}^n)$.

Proof We use the decomposition of Lemma 5.8; $b(X, D)\circ a(X, D) = \mathrm{id}+r(X, D)$. By Corollary 5.2, we have

$$\begin{aligned} \|f\|_{A^{s+2}_{pq}(\mathbb{R}^n)}&=\|(b(X,D)\circ a(X,D)-r(X,D))f\|_{A^{s+2}_{pq}(\mathbb{R}^n)}\\ &\lesssim\|b(X,D)\circ a(X,D)f\|_{A^{s+2}_{pq}(\mathbb{R}^n)}+\|r(X,D)f\|_{A^{s+2}_{pq}(\mathbb{R}^n)}\\ &\lesssim\|a(X,D)f\|_{A^{s}_{pq}(\mathbb{R}^n)}+\|f\|_{A^{s}_{pq}(\mathbb{R}^n)}. \end{aligned}$$

The reverse inequality is more direct and simple. Thus, the proof is complete.

This estimate can be used to obtain the interior estimate.

Corollary 5.3 *Let* Ω *be a domain. Let* Ω' *be a domain satisfying* $\Omega' \Subset \Omega \subset \mathbb{R}^n$; *that is, the closure of* Ω' *is contained in* Ω. *Let* $0 < p, q \leq \infty$ *and* $s \in \mathbb{R}$. *If* $a_{ij} \in C^\infty(\Omega)$ *satisfies*

$$\inf_{x\in\Omega'}\sum_{j,k=1}^{n}\Re(a_{jk}(x)\xi_j\xi_k)\geq c(\Omega')|\xi|^2$$

for some $c(\Omega') > 0$ *independent of* ξ, *then*

$$\|f\|_{A^{s+2}_{pq}(\Omega)}\sim_{\Omega,\Omega',\{a_{ij}\}_{i,j=1,2,\ldots,n}}\|L_0 f\|_{A^s_{pq}(\Omega)}+\|f\|_{A^s_{pq}(\Omega)}$$

for all $f \in A^s_{pq}(\Omega)$ *such that* $\mathrm{supp}(f) \subset \Omega'$.

Proof We consider f as an element in $\mathscr{S}'(\mathbb{R}^n)$ naturally; that is, $f \in A^s_{pq}(\mathbb{R}^n)$. Choose a subdomain Ω^* so that $\Omega' \Subset \Omega^* \Subset \Omega$. Choose an elliptic operator L_0^* on the whole space $\mathbb{R}^n$ of the form

$$L_0^* = \sum_{j,k=1}^n a^*_{jk}(x) \cdot \partial_{x_j} \partial_{x_k}.$$

Assume that L_0 and L_0^* coincide on Ω^*. Since the support of f is contained in Ω', we have

$$\|f\|_{A^{s+2}_{pq}(\Omega)} \sim \|f\|_{A^{s+2}_{pq}(\mathbb{R}^n)}$$

by the pointwise multiplier. Thus we have

$$\|L_0^* f\|_{A^s_{pq}(\mathbb{R}^n)} + \|f\|_{A^s_{pq}(\mathbb{R}^n)} \sim \|L_0 f\|_{A^s_{pq}(\Omega)} + \|f\|_{A^s_{pq}(\Omega)}.$$

Thus, the proof is complete.

Furthermore, we have a similar result for elliptic differential operators with lower order terms. For the sake of simplicity, we work within $\mathbb{R}^n$.

Corollary 5.4 *Let L be a differential operator*

$$L = L_0 + \sum_{j=1}^n b_j \partial_{x_j} + c, \quad b_1, b_2, \ldots, b_n, c \in \mathscr{C}^\infty(\mathbb{R}^n)$$

in $\mathbb{R}^n$. Then for $f \in A^{s+2}_{pq}(\mathbb{R}^n)$, the following a priori estimate holds:

$$\|f\|_{A^{s+2}_{pq}(\mathbb{R}^n)} \sim \|f\|_{A^s_{pq}(\mathbb{R}^n)} + \|Lf\|_{A^s_{pq}(\mathbb{R}^n)}.$$

Proof Note that for all $D > 0$ there exists $c_D > 0$ such that

$$\|(L - L_0) f\|_{A^s_{pq}(\mathbb{R}^n)} \le D^{-1} \|f\|_{A^{s+2}_{pq}(\mathbb{R}^n)} + c_D \|f\|_{A^s_{pq}(\mathbb{R}^n)} \tag{5.65}$$

and use the so-called absorbing argument. See Exercise 5.27.

Exercises

Exercise 5.27 Let $1 \le p, q \le \infty$, $s > 0$, and let $B_1, B_2, \ldots, B_n, C \in B^{s+2}_{\infty\infty}(\mathbb{R}^n)$. Define a differential operator L by $L \equiv -\Delta + \sum_{j=1}^n B_j \, \partial_{x_j} + C$. Then show that $\|f\|_{A^{s+2}_{pq}} \sim \|Lf\|_{A^s_{pq}} + \|f\|_{A^0_{pq}}$ for $f \in A^{s+2}_{pq}(\mathbb{R}^n)$ using the interpolation inequality

$$\|f\|_{A^s_{pq}} \le \varepsilon \|f\|_{A^{s+1}_{pq}} + c_\varepsilon \|f\|_{A^0_{pq}}, \quad \varepsilon > 0 \tag{5.66}$$

for $f \in A^{s+1}_{pq}(\mathbb{R}^n)$.

Exercise 5.28 Let $0 < p < \infty, 0 <, q \le \infty$ and $s > 0$. Set $a_j(x_1, x_2, \ldots, x_n) \equiv 3 + j + \sin x_j$ for $j = 1, 2, \ldots, n$. Show that $\|f\|_{A^{s+2}_{pq}} \lesssim \|f\|_{A^s_{pq}} + \sum_{j=1}^n \|a_j \partial_{x_j}{}^2 f\|_{A^s_{pq}}$ for all $f \in A^{s+2}_{pq}(\mathbb{R}^n)$ using Theorem 4.37.

Exercise 5.29 Let $n \ge 2$, and let $0 < p < \infty, 0 <, q \le \infty$ and $s > 0$. Disprove that $\|f\|_{A^{s+2}_{pq}} \lesssim \|\partial_1 \partial_2 f\|_{A^s_{pq}}$ for all $f \in A^{s+2}_{pq}(\mathbb{R}^n)$. Hint: Let $f_k(x) \equiv \prod_{j=1}^n e^{i x_j k} \mathscr{F}^{-1} \tau_0(x_j)$ for $x = (x_1, x_2, \ldots, x_n) \in \mathbb{R}^n$, where τ_0 is an appropriate bump.

5.2.4 *Examples and Classical Results*

So far, we have many results for the boundedness of various pseudo-differential operators. Now we present some examples and recall some classical results.

5.2.4.1 Pseudo-differential Operator with Symbol S^m

Using $F^0_{p2}(\mathbb{R}^n) \approx L^p(\mathbb{R}^n)$ with equivalence of norms, we review our results.

Proposition 5.6 *Any symbol $a \in S^0_{10}$ generates an $L^p(\mathbb{R}^n)$-bounded operator if $1 < p < \infty$*

Proof Since $a \in S^0 = S^0_{10}$ and $F^0_{p2}(\mathbb{R}^n) \approx L^p(\mathbb{R}^n)$, we can use Corollary 5.2.

Proposition 5.6 dates back to the work by Mihlin and Hörmander in 1950s. We note that Proposition 5.6 does not carry over to modulation spaces; see [1021, Theorem 2.1].

Next, we consider $\mathscr{C}^s(\mathbb{R}^n)$.

Proposition 5.7 *Let $s > 0$ and $a \in S^0_{11}$. Then $a(X, D)$ is bounded on $\mathscr{C}^s(\mathbb{R}^n) \approx B^s_{\infty\infty}(\mathbb{R}^n)$.*

Proof This is a special case of Theorem 5.15.

Proposition 5.8 *Let $1 < p < \infty$ and $0 \le \delta < 1$. Then $a(X, D)$ is $L^p(\mathbb{R}^n)$-bounded for any $a \in S^0_{1\delta}$.*

Proof Again resort to Corollary 5.2.

5.2.4.2 Pseudo-differential Operator with Symbol S^0_{00}: The Calderón–Vailancourt Theorem

Now we invoke a result due to Calderón and Vailancourt using modulation spaces defined in Sect. 2.3.3. In particular, recall $L^2(\mathbb{R}^n) \approx M^0_{22}(\mathbb{R}^n)$.

Theorem 5.19 (Calderón–Vailancourt) *For any $a \in S^0_{00}$, the pseudo-differential operator $a(X, D)$ is $L^2(\mathbb{R}^n)$-bounded.*

Proof Choose $\varphi \in C^\infty_c(\mathbb{R}^n)$ and $\kappa \in C^\infty_c(\mathbb{R}^n)$ so that $\sum\limits_{m\in\mathbb{Z}^n} \varphi(\star - m) \equiv 1$, and that $\chi_{B(2)} \le \kappa \le \chi_{B(3)}$. Set $\varphi_{(m)} \equiv \varphi(\star - m)$ for $m \in \mathbb{Z}^n$. First of all, we can expand f by Theorem 4.9:

$$f = \sum_{m\in\mathbb{Z}^n} \left(\sum_{l\in\mathbb{Z}^n} \mathrm{e}^{i(\star-l)\cdot m} \varphi_{(m)}(D) f(l) \mathscr{F}\kappa(\star - l) \right). \tag{5.67}$$

Furthermore, Theorems 1.49 and 1.50 yield

$$\left(\sum_{m,l\in\mathbb{Z}^n} |\varphi_{(m)}(D) f(l)|^2 \right)^{\frac{1}{2}} \lesssim \left(\sum_{m\in\mathbb{Z}^n} \|M[\varphi_{(m)}(D) f]\|_2{}^2 \right)^{\frac{1}{2}} \lesssim \|f\|_2.$$

For $m, l \in \mathbb{Z}^n$, we define $\psi_{(m;l)}(x) \equiv \mathrm{e}^{ix\cdot m} \mathscr{F}\kappa(x - l)$. Then we claim

$$|\partial^\alpha (\mathrm{e}^{-ix\cdot m} a(X, D)\psi_{(m;l)}(x))| \lesssim_{\alpha,N} \langle x - l\rangle^{-N}, \tag{5.68}$$

$$|\varphi_{(k)}(D)[a(X, D)\psi_{(m;l)}](x)| \lesssim_N \langle m - k\rangle^{-N} \langle x - l\rangle^{-N}. \tag{5.69}$$

Indeed, by inserting the definitions into $a(X, D)\Psi_{(m;l)}$, we calculate that

$$\begin{aligned} a(X, D)\psi_{(m;l)}(x) &\simeq_n \int_{\mathbb{R}^n} a(x, \xi) \mathrm{e}^{ix\cdot\xi} \mathscr{F}[\psi_{(m;l)}](\xi) \mathrm{d}\xi \\ &= \int_{\mathbb{R}^n} a(x, \xi) \mathrm{e}^{ix\cdot\xi + il\cdot(m-\xi)} \kappa(\xi - m) \mathrm{d}\xi. \end{aligned}$$

A change of variables yields

$$\begin{aligned} a(X, D)\psi_{(m;l)}(x) &= \int_{\mathbb{R}^n} a(x, \xi + m) \mathrm{e}^{ix\cdot(\xi+m) - il\cdot\xi} \kappa(\xi) \mathrm{d}\xi \\ &= \mathrm{e}^{ix\cdot m} \int_{\mathbb{R}^n} a(x, \xi + m) \mathrm{e}^{i(x-l)\cdot\xi} \kappa(\xi) \mathrm{d}\xi. \end{aligned}$$

Hence

$$e^{-ix\cdot m}a(X,D)\psi_{(m;l)}(x) \simeq_n \int_{\mathbb{R}^n} a(x,\xi+m)e^{i(x-l)\cdot\xi}\kappa(\xi)d\xi. \tag{5.70}$$

A repeated integration by parts yields

$$e^{-ix\cdot m}a(X,D)\psi_{(m;l)}(x) \simeq_n \frac{1}{\langle x-l\rangle^{2N}}\int_{\mathbb{R}^n}(1-\Delta_\xi)^N(a(x,\xi+m)\kappa(\xi))e^{i(x-l)\cdot\xi}d\xi.$$

Using this, let us prove (5.69). Setting $\eta_{(m;l)}(x) \equiv e^{-ix\cdot m}a(X,D)\psi_{(m;l)}(x)$ for $l,m \in \mathbb{Z}^n$, we write

$$\varphi_{(k)}(D)[a(X,D)\psi_{(m;l)}](x) = e^{ix\cdot k}\int_{\mathbb{R}^n}\mathscr{F}^{-1}\varphi(x-y)\eta_{(m;l)}(y)e^{iy(m-k)}dy. \tag{5.71}$$

We now recall

$$(1-\Delta_y)^N e^{iy(m-k)} = \langle m-k\rangle^{2N}e^{iy(m-k)}.$$

Then if we insert this expression into (5.71), then we have

$$\begin{aligned}&\varphi_{(k)}(D)[a(X,D)\psi_{(m;l)}](x)\\&=\frac{e^{ik\cdot x}}{\langle m-k\rangle^{2N}}\int_{\mathbb{R}^n}(1-\Delta_y)^N[\mathscr{F}^{-1}\varphi(x-y)\eta_{(m;l)}(y)]e^{iy(m-k)}dy.\end{aligned}$$

By differential inequality (5.68), we have

$$|(1-\Delta_y)^N(\mathscr{F}^{-1}\varphi(x-y)\eta_{(m;l)}(y))| \lesssim \frac{1}{\langle x-y\rangle^{2N}\langle y-l\rangle^{2N}} \lesssim \frac{1}{\langle x-l\rangle^{N}\langle y-l\rangle^{N}},$$

which yields (5.69).

Inserting (5.69) into the integral defining $\varphi_{(k)}(D)[a(X,D)f](x)$, we obtain

$$|\varphi_{(k)}(D)[a(X,D)f](x)| \lesssim_N \sum_{m,l\in\mathbb{Z}^n}\frac{|\varphi_{(m)}(D)f(l)|}{\langle m-k\rangle^{2N}\langle x-l\rangle^{2N}}$$

for all $N \in \mathbb{N}$. By the Hölder inequality,

$$\begin{aligned}&|\varphi_{(k)}(D)[a(X,D)f](x)|^2\\&\lesssim\left(\sum_{m,l\in\mathbb{Z}^n}\frac{|\varphi_{(m)}(D)f(l)|}{\langle m-k\rangle^{2N}\langle x-l\rangle^{2N}}\right)^2\end{aligned}$$

$$\lesssim \left(\sum_{m,l \in \mathbb{Z}^n} \frac{|\varphi_{(m)}(D) f(l)|^2}{\langle m-k \rangle^N \langle x-l \rangle^N} \right) \left(\sum_{m,l \in \mathbb{Z}^n} \frac{1}{\langle m-k \rangle^N \langle x-l \rangle^N} \right)$$
$$\sim \sum_{m,l \in \mathbb{Z}^n} \frac{|\varphi_{(m)}(D) f(l)|^2}{\langle m-k \rangle^N \langle x-l \rangle^N}.$$

If we sum this estimate over k and then integrate it against $x \in \mathbb{R}^n$, then we have

$$\|a(X, D) f\|_2{}^2 \lesssim \sum_{m,l \in \mathbb{Z}^n} |\varphi_{(m)}(D) f(l)|^2 \lesssim \|f\|_2{}^2.$$

Thus, the proof is complete.

We conclude this section with a negative result.

Proposition 5.9 *Symbols in S^0_{11} do not always generate $L^2(\mathbb{R}^n)$-bounded operators.*

Proof Choose $\Theta \in C^\infty_c(\mathbb{R}^n)$ so that $\chi_{B(7/3)\setminus B(5/3)} \le \Theta \le \chi_{B(4)\setminus B(1)}$ and we set

$$a(x, \xi) \equiv \sum_{j=1}^{\infty} \exp(-2^j x_1 i) \Theta(4^{-j}\xi) \quad (x = (x_1, x_2, \ldots, x_n), \xi \in \mathbb{R}^n).$$

We prove that this is the desired S^0_{11}-symbol.

First of all, we define a sequence $\{f_N\}_{N \ge 4}$ of $L^2(\mathbb{R}^n)$-bounded functions by

$$f_N(x) = \sum_{j=4}^{N} \frac{1}{j} \exp(2^j x_1 i) \mathscr{F}^{-1}\Theta\left(\frac{1}{4}x\right) \quad (x = (x_1, x_2, \ldots, x_n) \in \mathbb{R}^n). \tag{5.72}$$

Note that $\mathscr{F} f_N(\xi) \simeq_n \sum_{j=4}^{N} \frac{1}{j} \Theta(4\xi - 2^j \mathbf{e}_1)$ $(\xi \in \mathbb{R}^n)$. Thus, $\{f_N\}_{N \ge 4}$ is a bounded set in $L^2(\mathbb{R}^n)$ by the Plancherel theorem. However, if we operate $a(X, D)$, then

$$a(X, D) f_N(x) \simeq_n \left(\sum_{j=4}^{N} \frac{1}{j} \right) \mathscr{F}^{-1}\Theta\left(\frac{1}{4}x\right).$$

Thus, $\{a(X, D) f_N\}_{N \ge 4}$ is not an $L^2(\mathbb{R}^n)$-bounded set, which completes the proof.

As this example shows, S^0_{11} is difficult to handle. Stein called it the forbidden class.

Exercises

Exercise 5.30 Show that $\tilde{a}(X, D)$ is not $L^3(\mathbb{R}^n)$-bounded using Proposition 5.9.

Exercise 5.31 [682]

1. Establish the theory of molecular decomposition of $M^0_{12}(\mathbb{R}^n)$.
2. Establish that the pseudo-differential operator $a(X, D)$ is $M^0_{12}(\mathbb{R}^n)$-bounded for any $a \in S^0_{00}$.

Exercise 5.32 Let $a \in S^0_{1\rho}$ with $0 \le \rho < 1$.

1. Show that the function
$$K(x, y) = \int_{\mathbb{R}^n} e^{i(x-y)\cdot\xi} a(x, \xi)\,d\xi \quad (x, y \in \mathbb{R}^n)$$
is a Calderón–Zygmund operator.
2. Show that a generates a Calderón–Zygmund operator.

Exercise 5.33 Let $F \in C^\infty(\mathbb{R}^n)$ with $F(0) = 0$.

1. Let $f \in A^s_{pq}(\mathbb{R}^n)$ with $0 < p, q \le \infty$ and $s > \dfrac{n}{p}$. Show that the mapping from $\mathscr{S}'(\mathbb{R}^n)$ to itself defined by
$$g \mapsto F(t\psi(D)f) \cdot \psi(D)g + \sum_{j=1}^{\infty} F(\psi_{j-1}(D) + t\varphi_j(D)f) \cdot \varphi_j(D)g$$
is generated by a pseudo-differential operator with symbol in $S^0_{11}(\mathbb{R}^n)$.
2. Show that $F(f) \in A^s_{pq}(\mathbb{R}^n)$ whenever $f \in A^s_{pq}(\mathbb{R}^n)$ with $0 < p, q \le \infty$ and $s > \dfrac{n}{p}$.

Textbooks in Sect. 5.2

Elementary Properties of Pseudo-differential Operators

See [75], [35, Chapter 14], [86, Chapters 6 and 7] for more about pseudo-differential operators.

Pseudo-differential Operators on Sobolev Spaces

See the textbooks [85] and [86, Chapter 6, Section 5], for example, for the counterpart of what we obtained for the pseudo-diffrential operators to Sobolev spaces.

Pseudo-differential Operators on Besov Spaces and Triebel–Lizorkin Spaces

We can find the boundedness of the pseudo-differential operators on Besov spaces in [71, pp. 281–287]. See [100, Chapter 6] for more about pseudo-differential operators on Besov spaces and Triebel–Lizorkin spaces, which includes the exotic pseudo-differential operators. Triebel used the atoms in [100, Section 6.3]. Corollary 5.2 is in [100, 6.2.2].

The Class $B^l_{\infty\infty}S^m_{1\delta}(\mathbb{R}^n)$

See [94] for the class $B^l_{\infty\infty}S^m_{1\delta}(\mathbb{R}^n)$.

Wave Equations

See [91, Section 5.3] for more about the wave equations and the Schwartz distributions.

Schrödinger Equations

See [91, Section 5.4] for more about the Schrödinger equations and the Schwartz distributions.

5.3 Semi-groups: Applications to Heat Equations, Schrödigner Equations and Wave Equations

Our aim in Sect. 5.3 is twofold. One is to develop the theory of function spaces in the context of functional analysis. With the help of what we have obtained, we can study the fundamental property of functional analysis. Another aim in Sect. 5.3 is to prepare for some terminology needed to state and prove the Kato conjecture [660]. The structure of this big section is as follows. First we start with the fundamental facts on functional analysis in Sect. 5.3.1. Functional calculus is a fundamental tool to consider the composition $f(A)$ of the operator A. Among others, it is practical to define $\sqrt{A}$. This is taken up in Sect. 5.3.2. These two sections are mainly for the second aim mentioned above. The remaining sections are more practical and related to our function spaces. The heat semi-group is considered in Sect. 5.3.3. Section 5.3.4 considers the wave equation. We consider the relation between modulation spaces and the Schrödinger propagator in Sect. 5.3.5.

5.3.1 Bounded Holomorphic Calculus

We aim here to discuss the heat semi-group. To formulate this, we consider sectorial operators.

5.3.1.1 Sectorial Operators

Among closed operators, let us consider sectorial operators. In some sense the sectorial operators are close to positive self-adjoint operator, as we will see.

Definition 5.13 (Sectorial operator) Let X be a Banach space. A densely defined closed operator $L : X \to X$ is said to be *sectorial*, if there exist $M > 0$ and $0 < \eta < \frac{\pi}{2}$ such that $S_\eta \equiv \{z \in \mathbb{C}\setminus\{0\} : |\arg(z)| > \eta\} \subset \rho(L)$ and that the inverse mapping $(\lambda \mathrm{id}_X - L)^{-1} : X \to X$ satisfies the *resolvent estimate* $|\lambda| \cdot \|(\lambda \mathrm{id}_X - L)^{-1}\|_{B(X)} \le M$ for all $\lambda \in S_\eta$.

Example 5.4 As it turns out later in Theorem 5.32, the (positive) Laplacian $-\Delta$ is a sectorial operator in $A^s_{pq}(\mathbb{R}^n)$ for all $s \in \mathbb{R}$, $1 < p, q < \infty$.

We start with the following observation:

Lemma 5.9 *If* $L : X \to X$ *is a sectorial operator, then* $(\mathrm{id}_X + j^{-1}L)^{-1}x \to x$ *as* $j \to \infty$ *for all* $x \in X$.

Proof Note that $\|(\mathrm{id}_X + j^{-1}L)^{-1}\|_{B(X)} \lesssim 1$ with the constant independent of j from the resolvent estimate and that $\mathrm{Dom}(L)$ is dense in X. Thus, we may assume that $x \in \mathrm{Dom}(L)$. In this case, we can use $x - (\mathrm{id}_X + j^{-1}L)^{-1}x = j^{-1}(\mathrm{id}_X + j^{-1}L)^{-1}Lx = \mathrm{O}(j^{-1})$.

We can refine Lemma 5.9 by mimicking the proof.

Corollary 5.5 *Let* $L : X \to X$ *be a sectorial operator. Then* $\mathrm{Dom}(L^k)$ *is dense in* X *for all* $k \in \mathbb{N}$.

Proof Induct on k. If $k = 1$, then this is just an assumption. Once we assume that $\mathrm{Dom}(L^k)$ is dense in X, then we can show that $\mathrm{Dom}(L^{k+1})$ is dense in X by Lemma 5.9.

Unfortunately, sectorial operators are not defined everywhere on X. So, we want to approximate them in a standard manner. For our later consideration, we need the following estimate.

Lemma 5.10 *Let $L : X \to X$ be a sectorial operator on a Banach space X. Define $L_a \equiv a\mathrm{id}_X + L(\mathrm{id}_X + aL)^{-1}$ for $a > 0$. Then L_a is also a sectorial operator and $\|\lambda(\lambda \mathrm{id}_X - L_a)^{-1}\|_{B(H)} \lesssim_\mu 1$ for all $\lambda \in S_\mu$.*

Proof By the spectral mapping theorem, the condition on the spectrum is satisified; we can choose the same sector as L.

Let $\lambda \in S_\mu$ and write $Q_a \equiv L(\mathrm{id}_X + aL)^{-1}$. Then

$$(\lambda \mathrm{id}_X - a\mathrm{id}_X - Q_a)^{-1} = (\lambda \mathrm{id}_X - Q_a)(\lambda \mathrm{id}_X - a\mathrm{id}_X - Q_a)^{-1}(\lambda \mathrm{id}_X - Q_a)^{-1}.$$

Note that

$$(\lambda - \star)(\lambda - a - \star)^{-1} \in L^\infty(S_\mu).$$

Thus, we have only to consider the norm estimate of $(\lambda \mathrm{id}_X - Q_a)^{-1}$. This can be achieved as follows. First we compute

$$(\lambda \mathrm{id}_X - Q_a)^{-1} = \lambda^{-1}(\mathrm{id}_X + aL)(\mathrm{id}_X - (a + \lambda^{-1})L)^{-1}.$$

So we consider $(\mathrm{id}_X + aL)(\mathrm{id}_X - (a + \lambda^{-1})L)^{-1}$. If $|\lambda| \lesssim a$, then

$$\|(\mathrm{id}_X + aL)(\mathrm{id}_X - (a + \lambda^{-1})L)^{-1}\|_{B(H)} \lesssim 1.$$

If $|\lambda| \gtrsim a$, we can use Proposition 5.12 to follow.

5.3.1.2 Exponential of Sectorial Operators for Sectorial Operators on a Banach Space X

One of the aims of developing the theory of semi-groups is to solve evolution equations. An evolution equation is of the form:

$$\partial_t u - Lu = 0,$$

where L is a linear operator. In the case where L is multiplication by a constant, say L, the solution is simply $u = \mathrm{e}^{-tL}u_0$, where u_0 is the initial value. Based on the above observation, we expand the notion of e^{-tL} to a large extent.

Definition 5.14 (Exponential of sectorial operators) Let

$$0 < \eta < \theta < \frac{\pi}{2}, \tag{5.73}$$

and let γ_θ be a curve given by

$$\gamma_\theta(t) \equiv t\exp(i\,\mathrm{sgn}(t)\theta) \quad (t \in \mathbb{R}). \tag{5.74}$$

Maintain the same notation as Definition 5.13. The *exponential* e^{-tL} is defined to be

$$\mathrm{e}^{-tL} \equiv \frac{1}{2\pi i}\int_{\gamma_\theta}\left(\mathrm{e}^{-tz} - \frac{1}{1+tz}\right)(z\mathrm{id}_X - L)^{-1}\mathrm{d}z + (\mathrm{id}_X + tL)^{-1}. \tag{5.75}$$

Remark 5.3 The integral defining (5.75) is the integral of functions whose value is in a Banach space considered in Sect. 1.1.5. The singularity is at $t = 0$ and one has to consider improper integrals; at $t = \pm\infty$, one needs to check that the integrand decays sufficiently rapidly.

In order to check that the complex line integral converges absolutely, we add and subtract $(1 + tz)^{-1}$.

Proposition 5.10 *Let $t > 0$. In Definition 5.14, the integral defining e^{-tL} converges absolutely. Also, the definition (5.75) of e^{-tL} does not depend on θ satisfying (5.73).*

Proof Use the Cauchy integration theorem. See Exercise 5.36.

Now we investigate the additivity and the semi-group properties of the exponential e^{-tL}.

Theorem 5.20 (Semi-group property of e^{-tL}) *Let L be a sectorial operator. For $s, t > 0$, $\mathrm{e}^{-tL}\mathrm{e}^{-sL} = \mathrm{e}^{-(t+s)L}$.*

Proof We may replace L with L_a for some $a > 0$ by the Lebesgue convergence theorem. We write the left-hand side out in full. Let $\eta < \theta < \theta' < \frac{\pi}{2}$. Consider the following operator-valued integral:

$$B_1 \equiv \int_{\gamma_\theta}\left(\mathrm{e}^{-tz} - \frac{1}{1+tz}\right)(z\mathrm{id}_X - L)^{-1}\mathrm{d}z \circ \int_{\gamma_{\theta'}}\left(\mathrm{e}^{-sw} - \frac{1}{1+sw}\right)(w\mathrm{id}_X - L)^{-1}\mathrm{d}w.$$

Here $\circ$ denotes the composition. Since the integral converges absolutely, we have

$$B_1 = \iint_{\gamma_\theta\times\gamma_{\theta'}}\left(\mathrm{e}^{-tz} - \frac{1}{1+tz}\right)\left(\mathrm{e}^{-sw} - \frac{1}{1+sw}\right)(z\mathrm{id}_X - L)^{-1}(w\mathrm{id}_X - L)^{-1}\mathrm{d}w\mathrm{d}z.$$

We have

$$B_1 = \iint_{\gamma_\theta\times\gamma_{\theta'}}\frac{1}{z-w}\left(\mathrm{e}^{-tz} - \frac{1}{1+tz}\right)\left(\mathrm{e}^{-sw} - \frac{1}{1+sw}\right)(w\mathrm{id}_X - L)^{-1}\mathrm{d}w\mathrm{d}z$$

$$
\begin{aligned}
&-\iint_{\gamma_\theta \times \gamma_{\theta'}} \frac{1}{z-w}\left(\mathrm{e}^{-tz}-\frac{1}{1+tz}\right)\left(\mathrm{e}^{-sw}-\frac{1}{1+sw}\right)(z\mathrm{id}_X-L)^{-1}\mathrm{d}w\mathrm{d}z \\
&=\int_{\gamma_{\theta'}}\left(\int_{\gamma_\theta} \frac{1}{z-w}\left(\mathrm{e}^{-tz}-\frac{1}{1+tz}\right)\left(\mathrm{e}^{-sw}-\frac{1}{1+sw}\right)(w\mathrm{id}_X-L)^{-1}\mathrm{d}z\right)\mathrm{d}w \\
&\quad+\int_{\gamma_\theta}\left(\int_{\gamma_{\theta'}} \frac{1}{w-z}\left(\mathrm{e}^{-tz}-\frac{1}{1+tz}\right)\left(\mathrm{e}^{-sw}-\frac{1}{1+sw}\right)(z\mathrm{id}_X-L)^{-1}\mathrm{d}w\right)\mathrm{d}z.
\end{aligned}
$$

by the resolvent equation (1.54). In view of the position of the points z, w, we have

$$
B_1 = 2\pi i \int_{\gamma_\theta}\left(\mathrm{e}^{-tz}-\frac{1}{1+tz}\right)\left(\mathrm{e}^{-sz}-\frac{1}{1+sz}\right)(z\mathrm{id}_X-L)^{-1}\mathrm{d}z
$$

by the Cauchy integral theorem.

Next, we need to control:

$$
B_2(t,s) \equiv (\mathrm{id}_X+tL)^{-1}\frac{1}{2\pi i}\int_{\gamma_{\theta'}}\left(\mathrm{e}^{-sz}-\frac{1}{1+sz}\right)(z\mathrm{id}_X-L)^{-1}\mathrm{d}z.
$$

Again by the resolvent equation, we have

$$
B_2(t,s)=\frac{1}{2\pi i}\int_{\gamma_{\theta'}}\frac{1}{1+tz}\left(\mathrm{e}^{-sz}-\frac{1}{1+sz}\right)\left(t(\mathrm{id}_X+tL)^{-1}+(z\mathrm{id}_X-L)^{-1}\right)\mathrm{d}z.
$$

Since the decay of $(1+tz)^{-1}$ is sufficiently rapid, we have

$$
B_2(t,s)=\frac{1}{2\pi i}\int_{\gamma_{\theta'}}\frac{1}{1+tz}\left(\mathrm{e}^{-sz}-\frac{1}{1+sz}\right)(z\mathrm{id}_X-L)^{-1}\mathrm{d}z
$$

by the Cauchy integral theorem. Write

$$
\begin{aligned}
F(z,t,s) \equiv \mathrm{e}^{-(t+s)z}-\frac{1}{(1+tz)(1+sz)} &= \left(\mathrm{e}^{-tz}-\frac{1}{1+tz}\right)\left(\mathrm{e}^{-sz}-\frac{1}{1+sz}\right) \\
&\quad+\frac{1}{1+tz}\left(\mathrm{e}^{-sz}-\frac{1}{1+sz}\right)+\frac{1}{1+sz}\left(\mathrm{e}^{-tz}-\frac{1}{1+tz}\right).
\end{aligned}
$$

By adding B_1, $B_2(t,s)$, $B_2(s,t)$, we have

$$
\mathrm{e}^{-tL}\mathrm{e}^{-sL}=(\mathrm{id}_X+tL)^{-1}(\mathrm{id}_X+sL)^{-1}+\frac{1}{2\pi i}\int_{\gamma_{\theta'}}F(z,t,s)(z\mathrm{id}_X-L)^{-1}\mathrm{d}z.
$$

We have

$$(\mathrm{id}_X + tL)^{-1}(\mathrm{id}_X + sL)^{-1} - (\mathrm{id}_X + (t+s)L)^{-1} \tag{5.76}$$
$$= \frac{1}{2\pi i}\int_{\gamma_{\theta'}} \left(\frac{1}{(1+tz)(1+sz)} - \frac{1}{1+(t+s)z}\right)(z\mathrm{id}_X - L)^{-1}\mathrm{d}z$$

again by the Cauchy integral theorem. We omit the details. See Exercise 5.37. So, we conclude $\mathrm{e}^{-(t+s)L} = \mathrm{e}^{-tL}\mathrm{e}^{-sL}$.

Example 5.5

1. By the Cauchy integral formula, we have

$$\mathrm{e}^{-tL} = \frac{1}{2\pi i}\int_{\gamma_\theta}\left(\mathrm{e}^{-tz} - \frac{1}{(1+tz)^2}\right)(z\mathrm{id}_X - L)^{-1}\mathrm{d}z + (\mathrm{id}_X + tL)^{-2},$$

   which yields $\mathrm{e}^{-tL}x \in \mathrm{Dom}(L)$ for all $x \in X$. Since $\mathrm{e}^{-tL} = (\mathrm{e}^{-k^{-1}tL})^k$, $\mathrm{e}^{-tL}x \in \mathrm{Dom}(L^k)$ for all $x \in X$.
2. By differentiating (5.75) in $t > 0$, we learn that

$$\frac{\mathrm{d}}{\mathrm{d}t}\mathrm{e}^{-tL} = -\frac{1}{2\pi i}\int_{\gamma_\theta}\left(z\mathrm{e}^{-tz} + \frac{z}{(1+tz)^2}\right)(z\mathrm{id}_X - L)^{-1}\mathrm{d}z - L(\mathrm{id}_X + tL)^{-2}.$$

   By the Cauchy integral formula,

$$\frac{\mathrm{d}}{\mathrm{d}t}\mathrm{e}^{-tL} = -L\circ\frac{1}{2\pi i}\int_{\gamma_\theta}\left(\mathrm{e}^{-tz} + \frac{1}{(1+tz)^2}\right)(z\mathrm{id}_X - L)^{-1}\mathrm{d}z - L\circ(\mathrm{id}_X + tL)^{-2},$$

   since we can control the infinite integral. Once again by the Cauchy integral formula,

$$\frac{\mathrm{d}}{\mathrm{d}t}\mathrm{e}^{-tL} = -L\circ\frac{1}{2\pi i}\int_{\gamma_\theta}\left(\mathrm{e}^{-tz} - \frac{1}{1+tz}\right)(z\mathrm{id}_X - L)^{-1}\mathrm{d}z - L\circ(\mathrm{id}_X + tL)^{-1}.$$

   Thus, $\frac{\mathrm{d}}{\mathrm{d}t}\mathrm{e}^{-tL} = L\mathrm{e}^{-tL}$. In particular, for $x \in \mathrm{Dom}(L)$

$$\int_0^\infty L^2\mathrm{e}^{-tL}x\mathrm{d}t = \lim_{\varepsilon\downarrow 0, R\to\infty}\int_\varepsilon^R L^2\mathrm{e}^{-tL}x\mathrm{d}t = Lx.$$

5.3.1.3 Accretive Operators and Sectorial Operators

Here and below in this section let H be a Hilbert space. Suppose that we have a densely defined closed operator $D : H \to H$ and a bounded linear operator $A : H \to H$. Assume that A satisfies the following estimate: There exists $\lambda > 0$ such that

$$\Re(\langle Ax, x\rangle_H) \geq \lambda \|x\|_H{}^2 \tag{5.77}$$

for all $x \in H$. Such an operator A is called *accretive*. We define $L = D^* AD$. This is a generalization of the elliptic differential operators dealt with in Sect. 5.2.3. For example, let $A \equiv \{a_{ij}\}_{i,j=1,\dots,n}$ be an $n \times n$-matrix with entries of $L^\infty(\mathbb{R}^n)$-functions. Then $M_A : \{f_j\}_{j=1}^n \in L^2(\mathbb{R}^n)^n \mapsto \left\{\sum_{k=1}^n a_{ik} f_k\right\}_{k=1}^n \in L^2(\mathbb{R}^n)^n$ is an accretive operator.

Here we present an example of the sectorial operators.

Theorem 5.21 *Let H_0 and H_1 be Hilbert spaces. Suppose that we have a densely defined closed operator $D : H_0 \to H_1$ and a bounded linear operator $A : H_1 \to H_1$ such that*

$$\Re(\langle ADv, Dv\rangle_{H_1}) \geq 2\delta \|v\|_{H_0}{}^2 \tag{5.78}$$

for all $v \in \mathrm{Dom}(D)$. More precisely, assume that there exists a nonnegative constant $\theta_0 \ll 1$ such that

$$\Re(\mathrm{e}^{i\theta} \langle ADv, Dv\rangle_{H_1}) \geq \delta \|v\|_{H_0}{}^2 \tag{5.79}$$

for all $v \in \mathrm{Dom}(D)$ and $|\theta| \leq 2\theta_0$. Then for all $z \in \mathbb{C} \setminus \{0\}$ with $|\arg(z)| < \frac{\pi}{2} + \theta_0$, $z\mathrm{id}_H + D^* AD$ satisfies

$$\|(z\mathrm{id}_H + D^* AD)^{-1}\|_{H_0 \to H_0} \leq \frac{1}{|z| \sin\theta_0}. \tag{5.80}$$

In particular, $D^* AD$ is a sectorial operator.

Here we make a couple of remarks on the structure of the proof.

Remark 5.4

1. Note that (5.78) is slightly weaker than the accretivity itself. See Exercise 5.35.
2. Since the conclusion (5.80) does not depend on δ, we may assume that $|z| = 1$.
3. So, let us set $z = \mathrm{e}^{i\varphi}$ with $|\varphi| \leq \frac{\pi}{2} + \theta_0$. Furthermore, (5.78) allows us assume $z = \mathrm{e}^{i\varphi}$ with $|\varphi| \leq \frac{\pi}{2} - \theta_0$.
4. The proof is made up of the following steps.
   - Lemma 5.11 is a key step where we construct an auxiliary operator B.
   - In Lemmas 5.12 and 5.13 we show that $z\mathrm{id}_H + D^* AD : \mathrm{Dom}(D^* AD) \to H_0$ is bijective so that the notation $(z\mathrm{id}_H + D^* AD)^{-1} : H_0 \to \mathrm{Dom}(D^* AD)$ makes sense.
   - We show that $\mathrm{Dom}(D^* AD)$ is dense in H_0 in Lemma 5.14. Finally, we verify the resolvent estimate in Lemma 5.15.

Lemma 5.11 *Keep the assumption in Theorem 5.21. For any $u \in H_0$, there uniquely exists $u' = Bu \in \mathrm{Dom}(D)$ such that*

$$\langle u, v\rangle_{H_0} = \mathrm{e}^{i\varphi}\langle u', v\rangle_{H_0} + \langle ADu', Dv\rangle_{H_1} \tag{5.81}$$

for all $v \in \mathrm{Dom}(D)$ and $\varphi \in \left(\theta_0 - \frac{\pi}{2}, \frac{\pi}{2} - \theta_0\right)$.

Proof We define an inner product on $\mathrm{Dom}(D)$ by

$$\langle u, v\rangle_{\mathrm{Dom}(D)} \equiv \Re(\mathrm{e}^{i\varphi}\langle u, v\rangle_{H_0} + \langle ADu, Dv\rangle_{H_1})$$

for $u, v \in \mathrm{Dom}(D)$, so that $\mathrm{Dom}(D)$ is a "real" inner product space. Since D is closed, this inner product space is a real Hilbert space. For each $u \in H_0$, consider the linear mapping $v \in H_0 \mapsto \Re(\langle u, v\rangle_{H_0}) \in \mathbb{R}$. Then this mapping is a bounded linear operator and by the Riesz representation theorem, we have

$$\langle u', v\rangle_{\mathrm{Dom}(D)} = \Re(\mathrm{e}^{i\varphi}\langle u', v\rangle_{H_0} + \langle ADu', Dv\rangle_{H_1}) = \Re(\langle u, v\rangle_{H_0})$$

for some $u' \in \mathrm{Dom}(D)$. By considering iv as well, we have (5.81).

Lemma 5.12 *We have* $\mathrm{Ran}(B) = \mathrm{Dom}(D^*AD)$.

Proof Let $\tilde{u} \in \mathrm{Ran}(B)$. Then $\tilde{u} = Bu$ for some $u \in \mathrm{Dom}(B)$. Then

$$\langle v, u - \mathrm{e}^{-i\varphi}Bu\rangle_{H_0} = \langle A^*Dv, DBu\rangle_{H_1}.$$

Thus, $\tilde{u} = Bu \in \mathrm{Dom}(D^*AD)$ (and $D^*ADBu = u - \mathrm{e}^{-i\varphi}Bu$).

Conversely, let $\tilde{u} \in \mathrm{Dom}(D^*AD)$. Then there exists $U \in H_0$ such that for all $v \in \mathrm{Dom}(D)$ $\langle v, U\rangle_{H_0} = \langle A^*Dv, D\tilde{u}\rangle_{H_1}$. Thus, $\langle v, U + \mathrm{e}^{-i\varphi}\tilde{u}\rangle_{H_0} = \langle A^*Dv, D\tilde{u}\rangle_{H_1} + \mathrm{e}^{i\varphi}\langle v, \tilde{u}\rangle_{H_0}$, which implies $\tilde{u} = B(U + \mathrm{e}^{-i\varphi}\tilde{u}) \in \mathrm{Ran}(B)$.

Lemma 5.13 *The mapping $B : H_0 \to H_0$ is injective.*

Proof Let $u \in \ker(B)$. Then (5.81) yields $\langle u, v\rangle_{H_0} = 0$ for all $v \in \mathrm{Dom}(D)$. Since $\mathrm{Dom}(D)$ is dense in H_0, $u = 0$.

Lemma 5.14 *Let* $\mathrm{Dom}(D)$ *be the Hilbert space as above. Then* $\mathrm{Dom}(D^*AD)$ *is dense in* $\mathrm{Dom}(D)$. *In particular,* $\mathrm{Dom}(D^*AD)$ *is dense in H_0.*

Proof If $u \perp \mathrm{Dom}(D^*AD)$ in $\mathrm{Dom}(D)$, then $u \perp \mathrm{Ran}(B)$. From the definition of the inner product of $\mathrm{Dom}(D)$, we deduce $\Re(\mathrm{e}^{i\varphi}\langle u, Bv\rangle_{H_0} + \langle ADu, Dv\rangle_{H_1}) = 0$ for all $v \in \mathrm{Dom}(D)$. Consequently, $\langle u, v\rangle_{H_0} = \mathrm{e}^{i\varphi}\langle u, Bv\rangle_{H_0} + \langle ADu, Dv\rangle_{H_1} = 0$ for all $v \in \mathrm{Dom}(D)$. Since $\mathrm{Dom}(D)$ is dense in H_0, we conclude $u = 0$.

Lemma 5.15 *The mapping $B : u \in H_0 \mapsto u' \in H_0$ above is continuous. More precisely, $\|Bu\|_{H_0} \leq \dfrac{1}{\sin\theta_0}\|u\|_{H_0}$ for all $u \in H_0$ and $\varphi \in \left(\theta_0 - \frac{\pi}{2}, \frac{\pi}{2} - \theta_0\right)$.*

Proof Use the same notation as Lemma 5.11. Let $u \in H_0$. Then

$$\langle u, v\rangle_{H_0} = \mathrm{e}^{i\varphi}\langle Bu, v\rangle_{H_0} + \langle ADBu, Dv\rangle_{H_1}$$

for all $v \in \mathrm{Dom}(D)$. Since $Bu \in \mathrm{Dom}(D)$, we have

$$\langle u, Bu\rangle_{H_0} = \mathrm{e}^{i\varphi}\|Bu\|_{H_0}{}^2 + \langle ADBu, DBu\rangle_{H_1};$$

hence by the Cauchy–Schwarz inequality and accretivity (5.77) we have

$$\|Bu\|_{H_0}\|u\|_{H_0} \geq \Re(\langle u, Bu\rangle_{H_0}) \geq \Re(\mathrm{e}^{i\varphi}\langle Bu, Bu\rangle_{H_0}) \geq \|Bu\|_{H_0}{}^2 \sin\theta_0.$$

Thus we obtain Lemma 5.15.

We investigate the expression of the dual operator of D^*AD.

Proposition 5.11 *The dual of D^*AD is D^*A^*D; $(D^*AD)^* = D^*A^*D$.*

Proof We have $\langle(\mathrm{id}_H + D^*AD)^{-1}u, v\rangle_{H_0} = \langle u, (\mathrm{id}_H + D^*A^*D)^{-1}v\rangle_{H_0}$ since we know that $\mathrm{Ran}(\mathrm{id}_H + D^*AD)$ and $\mathrm{Ran}(\mathrm{id}_H + D^*A^*D)$ are dense in H_0. Thus, $((\mathrm{id}_H + D^*AD)^{-1})^* = (\mathrm{id}_H + D^*A^*D)^{-1}$. Since $((\mathrm{id}_H + D^*AD)^{-1})^* = ((\mathrm{id}_H + D^*AD)^*)^{-1}$, we conclude $\mathrm{id}_H + D^*A^*D = (\mathrm{id}_H + D^*AD)^*$; hence $D^*A^*D = (D^*AD)^*$.

5.3.1.4 Bounded Holomorphic Calculus

Denote by $i\mathbb{R}$ the imaginary axis oriented from $-i\infty$ to $i\infty$.

Proposition 5.12 *Let $L : H \to H$ be a bounded linear operator such that $\Re(z) > 0$ for all $z \in \sigma(L)$, and let f be a bounded holomorphic function defined on a neighborhood of $\Re(z) \geq 0$. Then*

$$f(L) = -\frac{1}{2\pi i}\int_{i\mathbb{R}} f(z)(z\mathrm{id}_H - L)^{-1}(L + L^*)(z\mathrm{id}_H + L^*)^{-1}\mathrm{d}z.$$

Proof Let $C_1(R)$ be a line given by $z = it$ for $-R \leq t \leq R$ and $C_2(R)$ be a curve given by $z = R\mathrm{e}^{it}$ for $-\pi \leq t \leq \pi$. Then

$$\frac{1}{2\pi i}\int_{C_2(R)} f(z)(z\mathrm{id}_H - L)^{-1}\mathrm{d}z - \frac{1}{2\pi i}\int_{C_1(R)} f(z)(z\mathrm{id}_H - L)^{-1}\mathrm{d}z = f(L) \tag{5.82}$$

as long as R is large enough. Meanwhile, by the Cauchy integral theorem, we have

$$\frac{1}{2\pi i}\int_{C_2(R)} f(z)(z\mathrm{id}_H + L^*)^{-1}\mathrm{d}z - \frac{1}{2\pi i}\int_{C_1(R)} f(z)(z\mathrm{id}_H + L^*)^{-1}\mathrm{d}z = 0. \tag{5.83}$$

Thus,

$$
\begin{aligned}
f(L) &= \frac{1}{2\pi i}\int_{C_2(R)} f(z)(z\mathrm{id}_H - L)^{-1}\mathrm{d}z - \frac{1}{2\pi i}\int_{C_2(R)} f(z)(z\mathrm{id}_H + L^*)^{-1}\mathrm{d}z \\
&\quad + \frac{1}{2\pi i}\int_{C_1(R)} f(z)(z\mathrm{id}_H + L^*)^{-1}\mathrm{d}z - \frac{1}{2\pi i}\int_{C_1(R)} f(z)(z\mathrm{id}_H - L)^{-1}\mathrm{d}z \\
&= -\frac{1}{2\pi i}\int_{C_1(R)} f(z)(z\mathrm{id}_H - L)^{-1}(L+L^*)(z\mathrm{id}_H + L^*)^{-1}\mathrm{d}z \\
&\quad + \frac{1}{2\pi i}\int_{C_2(R)} f(z)(z\mathrm{id}_H - L)^{-1}(L+L^*)(z\mathrm{id}_H + L^*)^{-1}\mathrm{d}z.
\end{aligned}
$$

Letting $R \to \infty$, we obtain

$$
\begin{aligned}
&\lim_{R\to\infty}\frac{1}{2\pi i}\int_{C_1(R)} f(z)(z\mathrm{id}_H - L)^{-1}(L+L^*)(z\mathrm{id}_H + L^*)^{-1}\mathrm{d}z \\
&= \frac{1}{2\pi i}\int_{i\mathbb{R}} f(z)(z\mathrm{id}_H - L)^{-1}(L+L^*)(z\mathrm{id}_H + L^*)^{-1}\mathrm{d}z \\
&\lim_{R\to\infty}\frac{1}{2\pi i}\int_{C_2(R)} f(z)(z\mathrm{id}_H - L)^{-1}(L+L^*)(z\mathrm{id}_H + L^*)^{-1}\mathrm{d}z \\
&= 0,
\end{aligned}
$$

which yields the desired identity.

Corollary 5.6 *Let $L : H \to H$ be a bounded linear operator such that $\Re(z) > 0$ for all $z \in \sigma(L)$. Then for all $u \in H$,*

$$
u = -\frac{1}{2\pi i}\int_{i\mathbb{R}} (z\mathrm{id}_H - L)^{-1}(L+L^*)(z\mathrm{id}_H + L^*)^{-1}u\mathrm{d}z.
$$

Proof Simply consider the case of $f \equiv 1$ in Proposition 5.12.

Since H is a Hilbert space, we can use the inner product to obtain the following conclusion:

Theorem 5.22 *Let $L : H \to H$ be a sectorial operator. Define*

$$
L_a \equiv a\mathrm{id}_H + L(\mathrm{id}_H + aL)^{-1}
$$

for $a > 0$. Let $f : U \to \mathbb{C}$ be a bounded holomorphic function defined on an open neighborhood U of $\Re(z) \geq 0$. Then $f(L_a)u$ converges to a limit for all $u \in H$ as $a \downarrow 0$.

Proof Since

$$
\|f(L_a)\|_{B(H)} \leq \|f\|_{L^\infty(\{\Re(z)\geq 0\})} \tag{5.84}
$$

and $u = \lim_{b\downarrow 0}(1+bL)^{-6}u$, we may further assume $u \in \mathrm{Ran}((\mathrm{id}_H + L)^{-6})$. We decompose $f(z) = f(0) + f'(0)z + g(z)z^2$ for z in the domain of f, where g is a bounded holomorphic function. If $u \in \mathrm{Ran}((\mathrm{id}_H + L)^{-6})$, then we easily see that $L_a u \to u$. Thus, we may assume $f(z) = g(z)z^2$, where g is a bounded holomorphic function and that $u = (\mathrm{id}_H + L)^{-3}v$ for some $v \in \mathrm{Dom}(L^3)$. We observe that

$$f(L_a)u = -\frac{1}{2\pi i}\int_{i\mathbb{R}} \frac{g(z)z^2}{(z+1)^3}(z\mathrm{id}_H - L_a)^{-1}(L_a + L_a{}^*)(z\mathrm{id}_H + L_a{}^*)^{-1}v\mathrm{d}z.$$

By the Lebesgue convergence theorem, Lemma 5.10 and (5.84), we have

$$\begin{aligned}\lim_{a\downarrow 0} f(L_a)(\mathrm{id}_H + L_a)^{-3}v &= \lim_{a\downarrow 0} g(L_a)L_a{}^2(\mathrm{id}_H + L_a)^{-3}v\\ &= -\frac{1}{2\pi i}\int_{i\mathbb{R}} \frac{g(z)z^2}{(z+1)^3}((z\mathrm{id}_H - L)^{-1}v - (z\mathrm{id}_H + L^*)^{-1}v)\mathrm{d}z,\end{aligned}$$

as required.

In view of (5.84) and Theorem 5.22, we define $f(L)$ as follows:

Definition 5.15 ***(f(L))*** Let H be a Hilbert space. For a sectorial operator L and a bounded holomorphic function f defined on a neighborhood of $\Re(z) \ge 0$, define $f(L) \in B(H)$ by

$$f(L)u \equiv -\frac{1}{2\pi i}\lim_{a\downarrow 0}\int_{i\mathbb{R}} f(z)(z\mathrm{id}_H - L_a)^{-1}(L_a + L_a{}^*)(z\mathrm{id}_H + L_a{}^*)^{-1}u\mathrm{d}z$$

for $u \in H$, where $L_a \equiv a\mathrm{id}_H + L(\mathrm{id}_H + aL)^{-1}$ for $a > 0$.

Theorem 5.23 *Let H be a Hilbert space. For a sectorial operator L and a bounded holomorphic function f defined on a neighborhood of $\Re(z) \ge 0$, $\|f(L)\|_{B(H)} \le \|f\|_{L^\infty(i\mathbb{R})}$.*

Proof Let $u, v \in H$ be arbitrary. We start with an approximation: $|\langle f(L)u, v\rangle_H| = \lim_{a\downarrow 0}|\langle f(L_a)u, v\rangle_H|$. If we use Proposition 5.12, then we have

$$|\langle f(L_a)u, v\rangle_H| \le \frac{1}{2\pi}\int_{i\mathbb{R}} |f(z)\langle (z\mathrm{id}_H - L_a)^{-1}(L_a + L_a{}^*)(z\mathrm{id}_H + L_a{}^*)^{-1}u, v\rangle_H|\mathrm{d}z.$$

If we use Corollary 5.6, then we obtain

$$\begin{aligned}&|\langle f(L_a)u, v\rangle_H|\\ &\le \frac{\|f\|_{L^\infty(i\mathbb{R})}}{2\pi}\int_{i\mathbb{R}} |\langle (z\mathrm{id}_H - L_a)^{-1}(L_a + L_a{}^*)(z\mathrm{id}_H + L_a{}^*)^{-1}u, v\rangle_H|\mathrm{d}z\end{aligned}$$

$$
\begin{aligned}
&= \frac{\|f\|_{L^\infty(i\mathbb{R})}}{2\pi} \int_{i\mathbb{R}} |\langle \sqrt{L_a + L_a{}^*}(z\mathrm{id}_H + L_a{}^*)^{-1} u, \sqrt{L_a + L_a{}^*}(z\mathrm{id}_H + L_a{}^*)^{-1} v \rangle_H| \mathrm{d}z \\
&\le \frac{\|f\|_{L^\infty(i\mathbb{R})}}{2\pi} \sqrt{\int_{i\mathbb{R}} \langle \sqrt{L_a + L_a{}^*}(z\mathrm{id}_H + L_a{}^*)^{-1} u, \sqrt{L_a + L_a{}^*}(z\mathrm{id}_H + L_a{}^*)^{-1} u \rangle_H \mathrm{d}z} \\
&\quad \times \sqrt{\int_{i\mathbb{R}} \langle \sqrt{L_a + L_a{}^*}(z\mathrm{id}_H + L_a{}^*)^{-1} v, \sqrt{L_a + L_a{}^*}(z\mathrm{id}_H + L_a{}^*)^{-1} v \rangle_H \mathrm{d}z} \\
&\le \|f\|_{L^\infty(i\mathbb{R})} \|u\|_H \|v\|_H.
\end{aligned}
$$

Thus, since $u, v \in H$ and $a > 0$ are arbitrary, we obtain the desired result.

5.3.1.5 Exponential of Sectorial Operators for Sectorial Operators on a Hilbert Space H

One of the aims of developing the theory of semi-groups is to solve evolution equations. An evolution equation is of the form: u is usually defined on $[0, \infty)$ or $(-\infty, \infty)$, assumes its value in X and satisfies

$$\partial_t u - Lu = 0, \quad u(0) = u_0,$$

where L is a sectorial operator. See Exercise 5.38 for the existence and the uniqueness. As we have mentioned before, in the case where L is multiplication by a constant, say L, the solution is simply $u = \mathrm{e}^{-tL} u_0$, where u_0 is the initial value. Based on the above observation, we expand the notion of e^{-tL} to a large extent.

Definition 5.16 (Exponential of sectorial operators revisited) Let $L : H \to H$ be a sectorial operator on a Hilbert space. Then redefine e^{-tL} following Definition 5.15.

We have two definitions of $\mathrm{e}^{t\Delta} : H \to H$. See Exercise 5.38 for uniqueness.

Now we investigate the additivity and the semi-group properties of the exponential e^{-tL}.

Theorem 5.24 (Semi-group property of e^{-tL}) *Let $L : H \to H$ be a sectorial operator on a Hilbert space H. Then for $s, t > 0$, $\mathrm{e}^{-tL}\mathrm{e}^{-sL} = \mathrm{e}^{-(t+s)L}$.*

Proof This is clear from $\mathrm{e}^{-tz}\mathrm{e}^{-sz} = \mathrm{e}^{-(t+s)z}$.

Exercises

Exercise 5.34 Let X be a Banach space $L \in B(X)$. If $\lambda \gg 1$, then show that $L + \lambda \mathrm{id}_X$ is a sectorial operator using an estimate similar to Lemma 2.1.

Exercise 5.35 Let $A = \begin{pmatrix} 0 & -1 \\ 1 & 0 \end{pmatrix}$, and let $D = \nabla : L^2(\mathbb{R}^2) \to L^2(\mathbb{R}^2)^n$ be the gradient.

1. Show that A is not accretive.
2. Show that $D^* A D = 0$.

Exercise 5.36 Prove Proposition 5.10 using the Cauchy integration theorem. Pay attention to the fact that the integrand decays sufficiently rapidly at ∞.

Exercise 5.37 Let $L : X \to X$ be a bounded accretive operator on a Hilbert space. Prove (5.76) using

$$
\begin{aligned}
&(\mathrm{id}_X + tL)^{-1}(\mathrm{id}_X + sL)^{-1} - (\mathrm{id}_X + (t+s)L)^{-1} \qquad (5.85)\\
&= \frac{1}{2\pi i}\int_{\gamma_{\theta'}} \left((\mathrm{id}_X + tL)^{-1}(\mathrm{id}_X + sL)^{-1} - (\mathrm{id}_X + (t+s)L)^{-1}\right)(z\mathrm{id}_X - L)^{-1}\mathrm{d}z.
\end{aligned}
$$

Exercise 5.38 Let H be a Hilbert space and $u_0 \in H$. Suppose that we are given an accretive operator L. We consider the integral equation:

$$
u(t) \equiv u_0 + \int_0^t Lu(s)\mathrm{d}s \quad (t \geq 0).
$$

Here $u_0 \in \mathrm{Dom}(L)$.

1. Show that the integral defining $u(t)$ makes sense.
2. Show that $u(t) \in \mathrm{Dom}(L)$ for $t > 0$.
3. Show that $\partial_t u(t) - Lu(t) = 0$ for $t > 0$.
4. Show that $\lim_{t \downarrow 0} u(t) = u_0$.
5. Show the uniqueness of the solution u of $u_t - Lu = 0$ subject to $u(0) = 0$.

5.3.2 *The Square Root of the Sectorial Operators*

5.3.2.1 Smoothing of the Sectorial Operators

Let H be a Hilbert space. We aim here to define $\sqrt{L} : H \to H$ for sectorial operators L. Choose the branch of $\sqrt{z}$ with $\Re(z) > 0$ so that $\sqrt{z} \in \mathbb{R}$ for $z > 0$ to this end. However, the function $\sqrt{z}$ does not fall under the scope of Sect. 5.3.1. For example, 0 is not in the resolvent set and $\sqrt{z}$ is unbounded So, we need to approximate L in some sense.

Lemma 5.16 *For $a, b > 0$ and $z \in \mathbb{C}$ with $\Re(z) \geq 0$ $|\sqrt{z+b} - \sqrt{z+a}| \leq \sqrt{b-a}$.*

Proof By the fundamental theorem of calculus, we have

$$\sqrt{z+b}-\sqrt{z+a}=\int_a^b \frac{1}{\sqrt{w+t}}\mathrm{d}t.$$

Thus, since $|\sqrt{w+t}|=\sqrt{|w+t|}$ if $\Re(w)>0$,

$$|\sqrt{z+b}-\sqrt{z+a}|=\int_a^b \frac{1}{2|\sqrt{w+t}|}\mathrm{d}t \le \int_a^b \frac{1}{2\sqrt{t}}\mathrm{d}t=\sqrt{b}-\sqrt{a}\le\sqrt{b-a},$$

as required.

By Theorem 5.23 and Lemma 5.16, we have the following conclusion:

Corollary 5.7 *Let L be a sectorial operator. Then*

$$\frac{\sqrt{L}}{\mathrm{id}_H+L}\equiv\lim_{\varepsilon\downarrow 0}\frac{\sqrt{L+\varepsilon}}{\mathrm{id}_H+L}$$

exists in the topology of $B(H)$ and we have

$$\left(\frac{\sqrt{L}}{\mathrm{id}_H+L}\right)^2=L(\mathrm{id}_H+L)^{-2}.$$

5.3.2.2 Square Root of the Sectorial Operators

The branch of the argument arg will be in $(-\pi,\pi)$. Our plan to define $\sqrt{L}$ is rather indirect. In fact, we start with the definition of $\dfrac{\sqrt{L}}{\mathrm{id}_H+L}$.

Definition 5.17 Let $L:H\to H$ be an accretive operator on a Hilbert space H. One defines an unbounded operator $\sqrt{L}$ by

$$\mathrm{Dom}(\sqrt{L})\equiv\left\{u\in H\,:\,\frac{\sqrt{L}}{\mathrm{id}_H+L}u\in\mathrm{Dom}(L)\right\}$$

and

$$\sqrt{L}u\equiv(\mathrm{id}_H+L)\left(\frac{\sqrt{L}}{\mathrm{id}_H+L}\right)u\quad(u\in\mathrm{Dom}(L)).$$

A direct consequence of Theorem 5.32 is:

Lemma 5.17 *We have* $\left(\dfrac{\sqrt{L}}{\mathrm{id}_H + L}\right)^2 = L(\mathrm{id}_H + L)^{-2}$.

Proof In fact, by Theorem 5.32, we have

$$\left(\frac{\sqrt{L_a}}{\mathrm{id}_H + L_a}\right)^2 = L_a(\mathrm{id}_H + L_a)^{-2}, \tag{5.86}$$

where $L_a \equiv a+L(\mathrm{id}_X+aL)^{-1}$ for $a > 0$. It remains to let $a \downarrow 0$. See Exercise 5.40.

Let us verify that $\sqrt{L} : H \to H$ is a densely defined closed operator.

Lemma 5.18 *Let $L : H \to H$ be a sectorial operator which is injective.*

1. *The operator $\sqrt{L}$ is closed.*
2. *We have* $\mathrm{Dom}(L) \subset \mathrm{Dom}(\sqrt{L})$. *In particular,* $\mathrm{Dom}(\sqrt{L})$ *is densely defined.*

Proof

1. Let $\{u_j\}_{j=1}^{\infty}$ be a sequence such that $u_j \to u$ in H and that $\sqrt{L}u_j \to v$ for some $u, v \in H$. Then

$$\frac{\sqrt{L}}{\mathrm{id}_H + L}u_j = (\mathrm{id}_H + L)^{-1}\sqrt{L}u_j \to (\mathrm{id}_H + L)^{-1}v, \tag{5.87}$$

since $(\mathrm{id}_H + L)^{-1} : H \to H$ is continuous. Meanwhile, since $\dfrac{\sqrt{L}}{\mathrm{id}_H + L}$ is also continuous,

$$\frac{\sqrt{L}}{\mathrm{id}_H + L}u_j \to \frac{\sqrt{L}}{\mathrm{id}_H + L}u. \tag{5.88}$$

Thus,

$$\frac{\sqrt{L}}{\mathrm{id}_H + L}u = (\mathrm{id}_H + L)^{-1}v, \tag{5.89}$$

meaning that $u \in \mathrm{Dom}(\sqrt{L})$. We also have

$$v = (\mathrm{id}_H + L)\frac{\sqrt{L}}{\mathrm{id}_H + L}u = \sqrt{L}u$$

from (5.89). Thus $\sqrt{L}$ is closed.
2. If $u \in \mathrm{Dom}(L)$, then

$$\frac{\sqrt{L + j^{-1}\mathrm{id}_H}}{L + \mathrm{id}_H}u \in \mathrm{Dom}(L) \tag{5.90}$$

since L is a closed operator. See Exercise 5.42. Since

$$\lim_{j\to\infty} \frac{\sqrt{L+j^{-1}\mathrm{id}_H}}{L+\mathrm{id}_H}u = \left(\frac{\sqrt{L}}{\mathrm{id}_H+L}\right)u$$

and

$$\lim_{j\to\infty} L\frac{\sqrt{L+j^{-1}\mathrm{id}_H}}{L+\mathrm{id}_H}u = \lim_{j\to\infty} \frac{\sqrt{L+j^{-1}\mathrm{id}_H}}{L+\mathrm{id}_H}Lu = \left(\frac{\sqrt{L}}{\mathrm{id}_H+L}\right)Lu, \tag{5.91}$$

it follows from the closedness of L that

$$\left(\frac{\sqrt{L}}{\mathrm{id}_H+L}\right)u \in \mathrm{Dom}(L).$$

We now show that $L \subset (\sqrt{L})^2$.

Lemma 5.19 *Let $u \in \mathrm{Dom}(L)$. Then $\sqrt{L}u \in \mathrm{Dom}(\sqrt{L})$ and $(\sqrt{L})^2u = Lu$.*

Proof To prove that $\sqrt{L}u \in \mathrm{Dom}(\sqrt{L})$, we need to show that

$$\frac{\sqrt{L}}{\mathrm{id}_H+L}\sqrt{L}u = \left(\frac{\sqrt{L}}{\mathrm{id}_H+L}\right)(\mathrm{id}_H+L)\left(\frac{\sqrt{L}}{\mathrm{id}_H+L}\right)u \in \mathrm{Dom}(L).$$

Since $u \in \mathrm{Dom}(L)$, we have

$$(\mathrm{id}_H+L)\frac{\sqrt{L+j^{-1}\mathrm{id}_H}}{\mathrm{id}_H+L} = \frac{\sqrt{L+j^{-1}\mathrm{id}_H}}{\mathrm{id}_H+L}(\mathrm{id}_H+L).$$

Thus,

$$\frac{\sqrt{L}}{\mathrm{id}_H+L}\sqrt{L}u = L(\mathrm{id}_H+L)^{-2}(\mathrm{id}_H+L)u = (\mathrm{id}_H+L)^{-1}Lu \in \mathrm{Dom}(L), \tag{5.92}$$

showing that $\sqrt{L}u \in \mathrm{Dom}(L)$. From (5.92) and Lemma 5.17 we also deduce

$$\begin{aligned}(\sqrt{L})^2u &= (\mathrm{id}_H+L)\left(\frac{\sqrt{L}}{\mathrm{id}_H+L}\right)\left(\frac{\sqrt{L}}{\mathrm{id}_H+L}\right)(\mathrm{id}_H+L)u \\ &= (\mathrm{id}_H+L)\left(\frac{L}{(\mathrm{id}_H+L)^2}\right)(\mathrm{id}_H+L)u = (\mathrm{id}_H+L)L(\mathrm{id}_H+L)^{-1}u = Lu,\end{aligned}$$

as desired.

We next show that $L \supset (\sqrt{L})^2$.

Lemma 5.20 *If $u \in \mathrm{Dom}(\sqrt{L})$ and $\sqrt{L}u \in \mathrm{Dom}(\sqrt{L})$, then $u \in \mathrm{Dom}(L)$.*

Proof By definition $\sqrt{L}u \in \mathrm{Dom}(\sqrt{L})$ if and only if

$$\left(\frac{\sqrt{L}}{\mathrm{id}_H + L}\right)\sqrt{L}u \in \mathrm{Dom}(L).$$

Note that

$$\left(\frac{\sqrt{L}}{\mathrm{id}_H + L}\right)\sqrt{L}u = \left(\frac{\sqrt{L}}{\mathrm{id}_H + L}\right)(\mathrm{id}_H + L)\left\{\left(\frac{\sqrt{L}}{\mathrm{id}_H + L}\right)u\right\}.$$

Since

$$\left(\frac{\sqrt{L}}{\mathrm{id}_H + L}\right)u \in \mathrm{Dom}(L),$$

we have

$$\begin{aligned}\mathrm{Dom}(L) \ni \left(\frac{\sqrt{L}}{\mathrm{id}_H + L}\right)\sqrt{L}u &= (\mathrm{id}_H + L)\left(\frac{\sqrt{L}}{\mathrm{id}_H + L}\right)\left\{\left(\frac{\sqrt{L}}{\mathrm{id}_H + L}\right)u\right\}\\ &= (\mathrm{id}_H + L)L(\mathrm{id}_H + L)^{-2}u\\ &= L(\mathrm{id}_H + L)^{-1}u.\end{aligned}$$

Thus, $u = (\mathrm{id}_H + L)^{-1}u + L(\mathrm{id}_H + L)^{-1}u \in \mathrm{Dom}(L)$.

Lemma 5.21 *The operator $\sqrt{L}$ is densely defined, so that $\overline{\mathrm{Dom}(\sqrt{L})} = H$. More precisely, if we set $u_j \equiv (\mathrm{id}_X + j^{-1}L)^{-1}u \in \mathrm{Dom}(L)$ for $u \in \mathrm{Dom}(\sqrt{L})$ and $j \in \mathbb{N}$, then $u_j \to u$ in H and $\sqrt{L}u_j \to \sqrt{L}u$.*

Proof Use Lemma 5.9 and $\sqrt{L}u_j = \sqrt{L}(\mathrm{id}_H + j^{-1}L)^{-1}u = (\mathrm{id}_H + j^{-1}L)^{-1}\sqrt{L}u$. See Exercise 5.42.

Combining Lemmas 5.17, 5.18, 5.19, 5.20 and 5.21, we obtain the following conclusion:

Theorem 5.25 *Let L be a sectorial operator on a Hilbert space H. Then there exists a densely defined operator $S : H \to H$ such that $S^2 = L$.*

We verify that S is an accretive operator in Corollary 5.9.

5.3.2.3 An Expression of the Square Root

So far we have investigated how to define $\sqrt{L}$. We used the line integral but we can also use the improper integral. For example, we can use

$$\int_0^\infty \frac{a\mathrm{d}x}{1+ax^2} = \sqrt{a}\pi.$$

We want to insert $a = L$ in the left-hand side to define $\pi\sqrt{L}$. Here we will solidify the idea above. For $u \in \mathrm{Dom}(L^2)$, we have the following expression:

Theorem 5.26 *For all $u \in \mathrm{Dom}(L^2)$, $\sqrt{L}u = \dfrac{16}{\pi}\displaystyle\int_0^\infty (\mathrm{id}_H + t^2L)^{-3}t^2L^2u\mathrm{d}t$.*

Proof Let $\varepsilon, t > 0$. We observe

$$\|(\mathrm{id}_H + t^2(L+2\varepsilon\mathrm{id}_H))^{-3}t^2(L+2\varepsilon\mathrm{id}_H)^2(\mathrm{id}_H+2\varepsilon\mathrm{id}_H+L)^{-2}\|_{B(H)} \lesssim \frac{1}{1+t^2}. \tag{5.93}$$

and

$$\|(\mathrm{id}_H + t^2(L_a+2\varepsilon\mathrm{id}_H))^{-3}t^2(L_a+2\varepsilon\mathrm{id}_H)^2(\mathrm{id}_H+2\varepsilon\mathrm{id}_H+L_a)^{-2}\|_{B(H)} \lesssim \frac{1}{1+t^2}. \tag{5.94}$$

Thus we are in the position of Fubini theorem to have

$$\begin{aligned}
&\int_0^\infty (\mathrm{id}_H + t^2(L_a+2\varepsilon\mathrm{id}_H))^{-3}t^2(L_a+2\varepsilon\mathrm{id}_H)^2(\mathrm{id}_H+2\varepsilon\mathrm{id}_H+L_a)^{-1}\mathrm{d}t\\
&= \int_0^\infty t^2(L_a+2\varepsilon\mathrm{id}_H)^2(\mathrm{id}_H + t^2(L_a+2\varepsilon\mathrm{id}_H))^{-3}(\mathrm{id}_H+2\varepsilon\mathrm{id}_H+L_a)^{-1}\mathrm{d}t\\
&= \frac{1}{2\pi i}\int_0^\infty\left(\int_{i\mathbb{R}} \frac{t^2(z+\varepsilon)^2}{(1+t^2(z+\varepsilon))^3(1+\varepsilon+z)}(z\mathrm{id}_H-\varepsilon\mathrm{id}_H-L_a)^{-1}\mathrm{d}z\right)\mathrm{d}t\\
&= \frac{1}{2\pi i}\int_{i\mathbb{R}}\left(\int_0^\infty \frac{t^2(z+\varepsilon)^2}{(1+t^2(z+\varepsilon))^3(1+\varepsilon+z)}(z\mathrm{id}_H-\varepsilon\mathrm{id}_H-L_a)^{-1}\mathrm{d}t\right)\mathrm{d}z.
\end{aligned}$$

Since

$$\int_0^\infty \frac{t^2(z+\varepsilon)^2}{(1+t^2(z+\varepsilon))^3}\mathrm{d}t = \sqrt{z+\varepsilon}\int_0^\infty \frac{t^2}{(1+t^2)^3}\mathrm{d}t = \frac{\pi}{16}\sqrt{z+\varepsilon} \quad (z \in i\mathbb{R}),$$

it follows that

$$\int_0^\infty (\mathrm{id}_H + t^2(L_a+2\varepsilon\mathrm{id}_H))^{-3}t^2(L_a+2\varepsilon\mathrm{id}_H)^2(\mathrm{id}_H+2\varepsilon\mathrm{id}_H+L_a)^{-1}\mathrm{d}t$$

$$= \frac{\pi}{16} \times \frac{1}{2\pi i} \int_{i\mathbb{R}} \frac{\sqrt{z+\varepsilon}}{1+\varepsilon+z} (z\mathrm{id}_H - \varepsilon\mathrm{id}_H - L_a)^{-1} \mathrm{d}z = \frac{\pi\sqrt{L_a + 2\varepsilon\mathrm{id}_H}}{(\mathrm{id}_H + 2\varepsilon\mathrm{id}_H + L_a)^2}.$$

Thus,

$$\int_0^\infty (\mathrm{id}_H + t^2(L + 2\varepsilon\mathrm{id}_H))^{-3} t^2 (L + 2\varepsilon\mathrm{id}_H)^2 (\mathrm{id}_H + 2\varepsilon\mathrm{id}_H + L)^{-2} \mathrm{d}t$$
$$= \frac{\pi\sqrt{L+2\varepsilon}}{16(1+2\varepsilon+L)^2}.$$

Thus, letting $\varepsilon \downarrow 0$, we obtain

$$\int_0^\infty (\mathrm{id}_H + t^2 L)^{-3} t^2 L^2 (\mathrm{id}_H + L)^{-2} \mathrm{d}t = \frac{\pi\sqrt{L}}{16(\mathrm{id}_H + L)^2}$$

from (5.93). Thus, from (5.94)

$$\int_0^\infty (\mathrm{id}_H + t^2 L)^{-3} t^2 L^2 u \mathrm{d}t = \int_0^\infty (\mathrm{id}_H + t^2 L)^{-3} t^2 L^2 (\mathrm{id}_H + L)^{-2} (\mathrm{id}_H + L)^2 u \mathrm{d}t$$
$$= \frac{\pi\sqrt{L}}{16(\mathrm{id}_H + L)^2} (\mathrm{id}_H + L)^2 u$$
$$= (\mathrm{id}_H + L) \frac{\pi\sqrt{L}}{16(\mathrm{id}_H + L)^2} (\mathrm{id}_H + L) u$$
$$= (\mathrm{id}_H + L) \frac{\pi\sqrt{L}}{16(\mathrm{id}_H + L)} u = \frac{\pi\sqrt{L}}{16} u,$$

as required.

Mimicking the argument above, we obtain the following formula:

Corollary 5.8 *For $u \in \mathrm{Dom}(L^2)$, $\sqrt{L}u = \frac{1}{\pi} \int_0^\infty L(\mathrm{id}_H + t^2 L)^{-1} u \mathrm{d}t$.*

Corollary 5.9 *Let L be an accretive operator. Then $\sqrt{L}$ defined in Theorem 5.25 is accretive.*

Proof Use Corollary 5.8 for $u \in \mathrm{Dom}(L)$. Since $2|\arg(\langle L(\mathrm{id}_H + t^2 L)^{-1} u, u\rangle_H)| \leq \pi$, we obtain the desired result.

5.3.2.4 Littlewood–Paley Theory for Operators

We have considered the operator $\varphi_j(D)$ for $j \in \mathbb{Z}$. We can now replace $\varphi_j(D)$ with $2^j L(\mathrm{id}_H + 2^j L)^{-2}$. Here we present a continuous version.

Theorem 5.27 *If $L : H \to H$ is a sectorial operator, then*

$$\int_0^\infty \|t^2 L(\mathrm{id}_H + t^2 L)^{-2} u\|_H{}^2 \frac{\mathrm{d}t}{t} \lesssim \|u\|_H{}^2$$

for all $u \in H$.

Proof We decompose

$$\int_0^\infty \|t^2 L(\mathrm{id}_H + t^2 L)^{-2} u\|_H{}^2 \frac{\mathrm{d}t}{t} \sim \sum_{k=-\infty}^{\infty} \|4^k L(\mathrm{id}_H + 4^k L)^{-2} u\|_H{}^2$$

using Theorem 5.23. By the Rademacher sequence, we have

$$\sum_{k=-N}^{N} \|4^k L(\mathrm{id}_H + 4^k L)^{-2} u\|_H{}^2$$

$$= \int_0^1 \left\langle \sum_{k=-N}^{N} r_k(t) 4^k L(\mathrm{id}_H + 4^k L)^{-2} u, \sum_{k=-N}^{N} r_k(t) 4^k L(\mathrm{id}_H + 4^k L)^{-2} u \right\rangle_H \mathrm{d}t.$$

Thus, it suffices to show

$$\left\| \sum_{k=-N}^{N} r_k(t) 4^k L(\mathrm{id}_H + 4^k L)^{-2} \right\|_{B(H)} \lesssim 1. \tag{5.95}$$

We set

$$f(z) \equiv \sum_{k=-N}^{N} r_k(t) 4^k z(1 + 4^k z)^{-2}$$

for $z \in \mathbb{C}$ such that the right-hand side makes sense. For $a > 0$ we also define $L_a \equiv a + L(\mathrm{id}_H + aL)^{-1}$. Since $\|f(L_a)\|_{B(H)} \le \|f\|_{L^\infty(\{\Re(z)=0\})} \lesssim 1$, we obtain

$$|\langle f(L_a)u, v\rangle_H| \le \|f\|_{L^\infty(\Re(z)=0)} \|u\|_H \|v\|_H \lesssim \|u\|_H \|v\|_H.$$

Thus, (5.95) follows.

Textbooks in Sect. 5.3.2

See the accessible source [769].

Exercises

Exercise 5.39 Identify 2×2 matrices with the linear operator in $\mathbb{C}$. Let $x = \begin{pmatrix} -2 & 0 \\ 0 & 0 \end{pmatrix}$ and $y = \begin{pmatrix} 1 & 1 \\ 1 & 1 \end{pmatrix}$. Then disprove that $\sqrt{(x+y)^*(x+y)} \le \sqrt{x^*x} + \sqrt{y^*y}$.

Exercise 5.40 Let $L : H \to H$ be an accretive operator on a complex Hilbert space H.

1. Let $0 < \mu < \pi$. Show $\sup_{0<a\le 1, z \in S_\mu} \left| \frac{\sqrt{z+a}}{z+1} \right| \lesssim 1$.
2. Prove Lemma 5.17.

Exercise 5.41 Let $L : H \to H$ be a sectorial operator. Using the resolvent equation, show that e^{-tL} and $(L - \lambda \mathrm{id}_H)^{-1}$ commute if $\lambda \in \rho(L)$.

Exercise 5.42 Let $L : H \to H$ be an accretive operator on a Hilbert space H.

1. Let $a > 0$. Show
$$\frac{\sqrt{L_a + j^{-1}\mathrm{id}_H}}{\mathrm{id}_H + L_a} u \in \mathrm{Dom}(L) \tag{5.96}$$
for all $u \in H$.
2. Prove (5.90) for $u \in \mathrm{Dom}(L)$.
3. Show that $L \frac{\sqrt{L_a + j^{-1}\mathrm{id}_H}}{\mathrm{id}_H + L_a} u = \frac{\sqrt{L_a + j^{-1}\mathrm{id}_H}}{\mathrm{id}_H + L_a} L u$ for $u \in \mathrm{Dom}(L)$.
4. Prove (5.91) for $u \in \mathrm{Dom}(L)$.
5. Prove $\sqrt{L}(\mathrm{id}_H + j^{-1}L)^{-1}u = (\mathrm{id}_H + j^{-1}L)^{-1}\sqrt{L}u$ for $u \in \mathrm{Dom}(L)$.

5.3.3 Applications of Function Spaces to the Heat Semi-group

Here we define the heat semi-group $\{\mathrm{e}^{t\Delta}\}_{t \ge 0}$ and investigate the properties. Although we depended on the theory of Hilbert spaces in the above section, we use a different method. Here we depend on the Fourier transform to define $\mathrm{e}^{t\Delta}$. The method will be rather direct. The definition of $\mathrm{e}^{t\Delta}$ coincides with the exponential of the sectorial operator, which we prove in the latter half of this section.

5.3.3.1 Heat Semi-group

Using multiplier theorem (Theorem 1.53), we study the heat semi-group. Define the heat kernel by $E(x,t) \equiv (4\pi t)^{-\frac{n}{2}} \exp\left(-\frac{|x|^2}{4t}\right)$ for $x \in \mathbb{R}^n$ and $t > 0$.

Definition 5.18 (Heat semi-group in $\mathscr{S}(\mathbb{R}^n)$) Set $e^{0\Delta} \equiv \mathrm{id}_{\mathscr{S}(\mathbb{R}^n)}$. Let $t > 0$. Then define a continuous mapping $\mathrm{e}^{t\Delta} : \mathscr{S}(\mathbb{R}^n) \to \mathscr{S}(\mathbb{R}^n)$ by

$$\mathrm{e}^{t\Delta} f \equiv E(\star, t) * f \quad (f \in \mathscr{S}(\mathbb{R}^n)).$$

The family $\{\mathrm{e}^{t\Delta}\}_{t\geq 0}$ of mappings is called the *heat semi-group*.

Lemma 5.22 *For $f \in \mathscr{S}(\mathbb{R}^n)$, $t \geq 0$ and $x \in \mathbb{R}^n$, define $u(t,x) \equiv \mathrm{e}^{t\Delta} f(x)$. Then:*

1. $u \in C^\infty((0,\infty) \times \mathbb{R}^n)$.
2. *u solves the heat equation:*

$$\partial_t u(t,x) - \Delta u(t,x) = 0 \quad (t > 0, x \in \mathbb{R}^n)$$

   subject to the boundary condition; $\lim\limits_{t\downarrow 0} u(t,\star) = \varphi$, where the convergence takes place in the topology of $\mathscr{S}(\mathbb{R}^n)$.
3. *Let $t, s \geq 0$. Then $e^{(t+s)\Delta} f = \mathrm{e}^{t\Delta}[\, e^{s\Delta} f\,]$.*

Lemma 5.23 *The conclusions in Lemma 5.22 still hold, if we replace $\mathscr{S}(\mathbb{R}^n)$ with $\mathscr{S}'(\mathbb{R}^n)$.*

Proof This is just a matter of duality, for which we leave the proof to interested readers as Exercise 5.43.

Let us show that $-\Delta$ is sectorial. We have the following fundamental resolvent estimate:

Theorem 5.28 (Spectrum of Δ) *Let $1 \leq p, q \leq \infty$ and $s \in \mathbb{R}$. The spectrum of the closed operator $\Delta : A^s_{pq}(\mathbb{R}^n) \to A^s_{pq}(\mathbb{R}^n)$ is $(-\infty, 0]$. Furthermore, when $0 < \theta < \pi$,*

$$\|(\Delta - z)^{-1}\|_{B(A^s_{pq})} \lesssim_\theta |z|^{-1} \tag{5.97}$$

for all $z \in \mathbb{C} \setminus \{0\}$ satisfying $\arg(z) > \pi - \theta$.

Proof We prove that $\mathbb{C} \setminus (-\infty, 0]$ is contained in the resolvent set to specify the spectrum. We prove $\mathbb{C} \setminus (-\infty, 0] \subset \sigma(\Delta)$. Let $z \in \mathbb{C} \setminus (-\infty, 0]$. Define $M_z(\xi) \equiv \frac{1}{|\xi|^2 + z}$. Then M_z is a $C^\infty(\mathbb{R}^n)$-function whose partial derivatives are all bounded. As a result, $M_z(D) = (-\Delta + z)^{-1}$. Thus, $\mathbb{C} \setminus (-\infty, 0]$ is contained in the resolvent set.

We specify the spectrum. Since $A^{s+2}_{pq}(\mathbb{R}^n)$ is a proper subset of $A^s_{pq}(\mathbb{R}^n)$, the spectrum cannot be bounded. Therefore, the spectrum is the unbounded set $(-\infty, 0]$. Define $I_t : f \in \mathscr{S}'(\mathbb{R}^n) \mapsto f(t\star) \in \mathscr{S}'(\mathbb{R}^n)$. Then $I_t{}^{-1}(\Delta - z)I_t = (t^2\Delta - z)$. Hence $(-\infty, 0)$ is contained in the spectrum. Since the spectrum is a closed set in $\mathbb{C}$, it follows that $\sigma(\Delta) = (-\infty, 0]$.

Finally, we can use Theorem 1.53 to prove (5.97).

The next theorem quantifies the smoothing effect of the heat kernel.

Theorem 5.29 (Time dependency of the smoothing effect of the heat kernel) *Suppose the parameters satisfy* $0 < p, q \le \infty$, $s \in \mathbb{R}$ *and* $\varepsilon \ge 0$. *Then*

$$\|e^{t\Delta} f\|_{A^{s+\varepsilon}_{pq}} \lesssim_{p,q,s,\varepsilon} t^{-\frac{\varepsilon}{2}} \|f\|_{A^s_{pq}} \quad (f \in \mathscr{S}'(\mathbb{R}^n))$$

for $0 \le t \le 1$.

Proof We will concentrate on Triebel–Lizorkin spaces as usual. Use Theorem 1.53 again. Although the proof is elementary, we have to be prudent; we want to investigate the dependency on t. Recall that $H^M(\mathbb{R}^n)$ is the potential space. To make use of Theorem 1.53, let $M \gg 1$. Also, to define the function spaces $A^s_{pq}(\mathbb{R}^n)$ and $A^{s+\varepsilon}_{pq}(\mathbb{R}^n)$, we take $\psi, \varphi \in \mathscr{S}(\mathbb{R}^n)$ so that (2.31) holds. Furthermore, to use Theorem 1.53, we choose auxiliary functions $\eta, \psi \in \mathscr{S}(\mathbb{R}^n)$ so that $\chi_{B(8)} \le \eta \le \chi_{B(16)}$ and that $\chi_{B(8)\setminus B(1/2)} \le \psi \le \chi_{B(16)\setminus B(1/4)}$. Define $\rho \in C^\infty_c(\mathbb{R}^n)$ so that $\rho(x)|x|^\varepsilon = \varphi(x)$ for $x \in \mathbb{R}^n$. We abbreviate

$$M(t) \equiv t^{-\frac{n}{2}} \left(\|\eta(\star)e^{-4^k t\,|\star|^2}\|_{H^M} + \sup_{k\in\mathbb{N}} \||2^k t^{\frac{1}{2}} \star|^\varepsilon \psi(\star)e^{-4^k t\,|\star|^2}\|_{H^M} \right).$$

Then

$$\begin{aligned} \|e^{t\Delta} f\|_{F^{s+\varepsilon}_{pq}} &= \|\psi(D)e^{t\Delta} f\|_p + \|2^{j(s+\varepsilon)}\varphi_j(D)e^{t\Delta} f\|_{L^p(\ell^q)} \\ &= \|\eta(D)\psi(D)e^{t\Delta} f\|_p + \|2^{j(s+\varepsilon)}\zeta_j(D)\varphi_j(D)e^{t\Delta} f\|_{L^p(\ell^q)} \\ &\lesssim t^{-\frac{\varepsilon}{2}} M(t)\, \|2^{js}\rho_j(D) f\|_{L^p(\ell^q)} \end{aligned}$$

by Theorem 1.53.

A change of variables shows $\sup\limits_{0\le t\le 1} M(t) < \infty$. Thus, the proof is complete.

5.3.3.2 Properties of Semi-groups

We now prove that heat semi-group$\{e^{t\Delta}\}_{t\ge 0}$ is a continuous semi-group. To simplify, we assume $1 < p, q < \infty$ and $s \in \mathbb{R}$ until the end of this section.

Theorem 5.30 (Continuity of the heat semi-group) *The family* $\{e^{t\Delta}\}_{t\ge 0}$ *is a continuous semi-group on* $A^s_{pq}(\mathbb{R}^n)$. *That is,* $\{e^{t\Delta}\}_{t\ge 0}$ *is a family of bounded operator on* $A^s_{pq}(\mathbb{R}^n)$, *which satisfies the following conditions:*

1. *Whenever* $t, s \ge 0$, $e^{t\Delta}e^{s\Delta} = e^{(t+s)\Delta}$.
2. $e^{0\Delta} = \mathrm{id}_{A^s_{pq}}$.
3. *The mapping* $t \in [0, \infty) \mapsto e^{t\Delta} \in B(A^s_{pq})$ *is continuous.*

Proof In view of Theorem 5.29, $\{e^{t\Delta}\}_{t\ge 0}$ is a family of operators which is bounded. If we go through the same argument as Theorem 5.29, we can prove that

$$\|(e^{t\Delta} - e^{t'\Delta})f\|_{A^s_{pq}} \lesssim_f |t - t'| \quad (f \in \mathscr{S}(\mathbb{R}^n))$$

when $0 \le t, t' \le 1$. Furthermore, as we proved in Lemmas 5.22 and 5.23, $e^{t\Delta}(e^{s\Delta}f) = e^{(t+s)\Delta}f$ for all $f \in \mathscr{S}'(\mathbb{R}^n)$.

Consequently, it follows that $\{e^{t\Delta}\}_{t\ge 0}$ is a continuous semi-group on $A^s_{pq}(\mathbb{R}^n)$.

Next, we specify the generator.

Theorem 5.31 (Laplacian as the generator of the heat semi-group) *The generator of the heat semi-group* $e^{t\Delta} : A^s_{pq}(\mathbb{R}^n) \to A^s_{pq}(\mathbb{R}^n)$ *is* Δ *and its domain is* $A^{s+2}_{pq}(\mathbb{R}^n)$. *That is, the domain of the operator A is defined by setting*

$$\mathrm{Dom}(A) \equiv \left\{ f \in A^s_{pq}(\mathbb{R}^n) : \lim_{t\downarrow 0} \frac{e^{t\Delta} - \mathrm{id}_{A^s_{pq}(\mathbb{R}^n)}}{t} f \text{ exists in } A^s_{pq}(\mathbb{R}^n) \right\}.$$

and define the operator A by $Af \equiv \lim_{t\downarrow 0} \dfrac{e^{t\Delta} - \mathrm{id}_{A^s_{pq}(\mathbb{R}^n)}}{t} f$. *Then* $\mathrm{Dom}(A) = A^{s+2}_{pq}(\mathbb{R}^n)$ *and* $A = \Delta : A^{s+2}_{pq}(\mathbb{R}^n) \to A^s_{pq}(\mathbb{R}^n)$.

Proof Let $f \in \mathrm{Dom}(A)$. We have $\lim_{t\downarrow 0} \dfrac{e^{t\Delta}f - f}{t} = \Delta f$; at least in the topology of $\mathscr{S}'(\mathbb{R}^n)$, the limit $\lim_{t\downarrow 0} \dfrac{e^{t\Delta}f - f}{t}$ coincides with $\Delta f \in \mathscr{S}'(\mathbb{R}^n)$. Thus, $f \in \mathrm{Dom}(A)$ if and only if $f \in A^s_{pq}(\mathbb{R}^n)$ and $\Delta f \in A^s_{pq}(\mathbb{R}^n)$. However, according to Theorem 2.12, this is equivalent to $f \in A^{s+2}_{pq}(\mathbb{R}^n)$. Thus, the proof is complete.

We summarize the above observations.

Theorem 5.32 (Elementary property of the heat semi-group) *Let* $s \in \mathbb{R}$ *and* $1 < p, q < \infty$. *Then the Laplacian* Δ *is a sectorial operator on* $A^s_{pq}(\mathbb{R}^n)$ *whose domain is* $A^{s+2}_{pq}(\mathbb{R}^n)$. *Furthermore, it generates the heat semi-group* $e^{t\Delta}$ *on* $L^p(\mathbb{R}^n)$, *which is a continuous semi-group.*

Although the next theorem is a direct consequence of the above observations and $L^p(\mathbb{R}^n) \approx F^0_{p2}(\mathbb{R}^n)$ with equivalence of norms for $1 < p < \infty$, we dare to formulate it because of its importance.

Theorem 5.33 (Elementary properties of the heat semi-group on $L^p(\mathbb{R}^n)$) *Let* $1 < p < \infty$.

1. *The Laplacian* $\Delta : L^p(\mathbb{R}^n) \to L^p(\mathbb{R}^n)$ *is a sectorial operator whose domain is* $W^{2,p}(\mathbb{R}^n)$.
2. *The Laplacian generates a continuous semi-group, which is called the heat semi-group* $\mathrm{e}^{t\Delta}$.

Exercises

Exercise 5.43 Prove Lemmas 5.22 and 5.23. See [25, Chapter 2] for Lemma 5.22. Duality entails Lemma 5.23.

Exercise 5.44 Carry out the resolvent estimate of $(\Delta - z)^{-1}$ to have (5.97) using an esimate similar to Lemma 2.1.

Exercise 5.45 Let $-\Delta$ be the Laplacian on $L^2(\mathbb{R}^n)$, which is a unbounded positive self-adjoint operator, and let $-\Delta = \int_0^\infty \lambda \mathrm{d}E_\lambda$ be its spectrum decomposition. What is the relation between $\{E_\lambda\}_{\lambda \geq 0}$ and $\varphi_j(D)$? Hint: Use the Fourier transform and the functional calculus.

Exercise 5.46 [795] Let $f \in C^2(\mathbb{R}^n)$ be a bounded harmonic function. It is well known that f is a constant; see [25] for example. The aim of this exercise is to show this property from the heat equation. Define $u(x,t) \equiv \mathrm{e}^{t\Delta} f(x)$ for $x \in \mathbb{R}^n$ and $t > 0$.

1. Show that $\Delta u(x,t) = 0$ and $(\partial_t - \Delta)u(x,t) = 0$, where Δ is the spatial Laplacian.
2. Let ∇ be the spatial gradient. Show that $\nabla u = 0$ using integration by parts and the fact f is bounded.
3. Show that u is a constant.

Exercise 5.47 Let $0 < p, q \leq \infty$ and $s > 0$. (Note that "$p = \infty$" is allowed.) Show that $\mathrm{e}^{t\Delta}u \to u$ in $B^s_{pq}(\mathbb{R}^n)$ for all $u \in B^s_{pq}(\mathbb{R}^n)$ if and only if $q < \infty$. Hint: When $q = \infty$, we can consider

$$f = \sum_{j=1}^{\infty} \frac{2^{-js}}{\|\mathscr{F}^{-1}\varphi_j\|_p} \mathscr{F}^{-1}\varphi_j \in B^s_{p\infty}(\mathbb{R}^n).$$

and we can disprove that $\mathrm{e}^{t\Delta}u \to u$ in $B^s_{p\infty}(\mathbb{R}^n)$.

5.3.4 Applications of Function Spaces to the Wave Equations

As an application to the differential equations, we consider the wave equations.

5.3.4.1 Stationary Phase Method

We use the next lemma.

Lemma 5.24 *Let $\eta \in C_c^\infty(B(1))$ be a smooth function, and let $\lambda > 0$. Then*

$$\left|\int_{\mathbb{R}^n} \mathrm{e}^{i\lambda|x|^2} x^\alpha \eta(x)\mathrm{d}x\right| \lesssim_\alpha (1+\lambda)^{-\frac{n+|\alpha|}{2}} \quad (\alpha \in {\mathbb{N}_0}^n).$$

Proof Choose $\psi \in C_c^\infty(\mathbb{R}^n)$ so that $\chi_{B(1)} \le \psi \le \chi_{B(2)}$ and we decompose

$$\int_{\mathbb{R}^n} \mathrm{e}^{i\lambda|x|^2} x^\alpha \eta(x)\mathrm{d}x$$
$$= \int_{\mathbb{R}^n} \mathrm{e}^{i\lambda|x|^2} x^\alpha \psi\left(\frac{x}{\varepsilon}\right) \eta(x)\mathrm{d}x + \int_{\mathbb{R}^n} \mathrm{e}^{i\lambda|x|^2} x^\alpha \left\{1-\psi\left(\frac{x}{\varepsilon}\right)\right\} \eta(x)\mathrm{d}x.$$

For the first term, we simply use the triangle inequality. We want to handle the second term. By partition of unity on $\mathbb{R}^n$, it can be assumed that the function η is supported on the cone $V \equiv \{x \in \mathbb{R}^n : 2nx_1 \ge |x|\}$. When $0 < \lambda \le 1$, the result is clear by the triangle inequality. We may therefore assume $\lambda > 1$. Define a differential operator D by $Df(x) \equiv \dfrac{1}{2i\lambda x_1}\dfrac{\partial f}{\partial x_1}(x)$. Then $D\exp(i\lambda|x|^2) = \exp(i\lambda|x|^2)$ and the formal adjoint D^* is given by $D^*f(x) \equiv -\dfrac{\partial}{\partial x_1}\dfrac{f(x)}{2i\lambda x_1}$. Hence if we integrate by parts N times, then

$$\int_{\mathbb{R}^n} \mathrm{e}^{i\lambda|x|^2} x^\alpha \left\{1-\psi\left(\frac{x}{\varepsilon}\right)\right\} \eta(x)\mathrm{d}x$$
$$= \int_{\mathbb{R}^n} \mathrm{e}^{i\lambda|x|^2} \left(\frac{\partial}{\partial x_1}\frac{i}{2\lambda x_1}\right)^N x^\alpha \left\{1-\psi\left(\frac{x}{\varepsilon}\right)\right\} \eta(x)\mathrm{d}x.$$

By the triangle inequality, we have

$$\left|\int_{\mathbb{R}^n} \mathrm{e}^{i\lambda|x|^2} x^\alpha \eta(x)\mathrm{d}x\right| \lesssim \varepsilon^{|\alpha|+n} + \lambda^{-N}\varepsilon^{|\alpha|-2N+n}.$$

By optimizing the right-hand side by $\varepsilon = \lambda^{-\frac{1}{2}}$, we obtain the desired result.

A point ω in S^{n-1} will be denoted by $\omega = (\omega_1, \omega_2, \ldots, \omega_n)$. Furthermore, denote by $\mathrm{d}\sigma(\omega)$ the surface area of S^{n-1}.

Lemma 5.25 *Define a function $\tau : (0,\infty) \to \mathbb{C}$ by*

$$\tau(R) \equiv \int_{S^{n-1}} \exp(iR\omega_1)\mathrm{d}\sigma(\omega), \quad (R > 0).$$

Then there is a decomposition: $\tau(R) = \tau_1^+(R) + \tau_1^-(R) + \tau_*(R)$, where

$$\left|\frac{\mathrm{d}^k}{\mathrm{d}R^k}\exp(iR)\tau_1^-(R)\right| + \left|\frac{\mathrm{d}^k}{\mathrm{d}R^k}\exp(-iR)\tau_1^+(R)\right| \lesssim_k (1+R)^{-\frac{n-1}{2}-k} \tag{5.98}$$

$$|\tau_*^{(k)}(R)| \lesssim_{k,N} (1+R)^{-N} \tag{5.99}$$

for all $k, N \in \mathbb{N}_0$. In particular $\tau(R) = \mathrm{O}(R^{-\frac{n-1}{2}})$ as $R \to \infty$.

Proof We embed S^{n-1} into $\mathbb{R}^n$. That is, we write a point in S^{n-1}, as follows: $\omega = (\omega_1, \ldots, \omega_n) =: (\omega_1, \omega')$ by the coordinate in $\mathbb{R}^n$. Let $j = 1, 2, \ldots, n$. Define open sets $U_j^\pm$ in S^{n-1} by

$$U_j^\pm \equiv \left\{\omega \in S^{n-1} : \pm 2n\omega_j > 1\right\}. \tag{5.100}$$

The pigion hole principle shows that $\mathscr{U} \equiv \{U_j^+, U_j^-\}_{j=1,2,\ldots,n}$ is an open covering in S^{n-1}. Choose a partition $\{\rho_{(j)}^+, \rho_{(j)}^-\}_{j=1,2,\ldots,n}$ of unity subordinate to the open covering $\mathscr{U}$. That is, a collection $\{\rho_{(j)}^+, \rho_{(j)}^-\}_{j=1,2,\ldots,n}$ of functions satisfies

$$\rho_{(j)}^\pm \geq 0, \quad \mathrm{supp}(\rho_{(j)}^\pm) \subset U_j^\pm, \quad j = 1, 2, \ldots, n, \quad \sum_{j=1}^n (\rho_{(j)}^+ + \rho_{(j)}^-) \equiv \chi_{S^{n-1}}. \tag{5.101}$$

With this partition of unity, we define

$$\tau_{(j)}^\pm(R) \equiv \int_{S^{n-1}} \rho_{(j)}^\pm(\omega)\exp(iR\omega_1)\mathrm{d}\sigma(\omega) \quad (j = 1, 2, \ldots, n).$$

Then $\sum_{j=1}^n \left(\tau_{(j)}^+(R) + \tau_{(j)}^-(R)\right) = \tau(R)$ from (5.101), so that we are tempted to set

$$\tau_*(R) = \sum_{j=2}^n \left(\tau_{(j)}^+(R) + \tau_{(j)}^-(R)\right). \tag{5.102}$$

Then $\tau(R) = \tau_1^+(R) + \tau_1^-(R) + \tau_*(R)$. Set $V_1 \equiv \{x' \in \mathbb{R}^{n-1} : |x'| < \sqrt{1-(2n)^{-2}}\}$. Then

$$\tau_1^+(R) = \exp(iR)\int_{V_1} \exp(-iR(1-\sqrt{1-|x'|^2}))\theta(x')\mathrm{d}x' \tag{5.103}$$

for some $\theta \in C_c^\infty(V_1)$. We expand the function $1 - \sqrt{1-|x'|^2}$ to the Taylor series:

$$1-\sqrt{1-|x'|^2}=\frac{1}{2}|x'|^2+\frac{3}{8}|x'|^4+\cdots=\sum_{k=0}^{\infty}c_k|x'|^{2k+2},$$

where $c_0, c_1, \ldots, c_k, \ldots > 0$ are constants. Then we consider the transformation:

$$x' \mapsto X'=\left(c_0+c_1|x'|^2+\cdots+c_k|x'|^{2k}+\cdots\right)x'.$$

Note that

$$|X'|^2=\sum_{k=0}^{\infty}c_k|x'|^{2k+2}=1-\sqrt{1-|x'|^2}.$$

Denote by W_1 the range of X' where $x' \in V_1$ moves. That is, $W_1 = B((2n)^{-1}) \subset \mathbb{R}^{n-1}$. The mapping $x' \in V_1 \mapsto X' \in W_1$ is a diffeomorphism. By this diffeomorphism, we have

$$\tau_1^+(R)=\exp(iR)\int_{W_1}\exp(-iR|X'|^2)\Theta(X')\mathrm{d}X'. \tag{5.104}$$

If we use Lemma 5.24, then we have (5.98).

We consider τ_1^- likewise. For $\tau_j^{\pm}$ with $j = 2, 3, \ldots, n$, we use the coordinate to express the integral defining $\tau_j^{\pm}$ and we integrate by parts to have

$$|(\tau_j^+)^{(k)}(R)| \lesssim_{k,N} (1+R)^{-N}, \quad j=2,3,\ldots,n. \tag{5.105}$$

Thus, (5.99) holds.

We record the following important corollary:

Corollary 5.10 *Let* $\mu \in C_c^\infty(\mathbb{R}^n)$. *Then* $\left\|\mathscr{F}^{-1}\left[\mu\exp(it|\star|)\right]\right\|_\infty \lesssim t^{-\frac{n-1}{2}}$ *for* $t > 0$.

Proof A scaling allows us to assume that $j = 0$. Let $\mathbf{e}_1$ be the unit vector in the x_1 direction. By the spherical coordinate, we have

$$\begin{aligned}\mathscr{F}^{-1}[\mu\exp(it|\star|)](x) &\simeq_n \int_{\mathbb{R}^n}\mu(\xi)\exp(it|\xi|+ix\cdot\xi)\mathrm{d}\xi\\ &\simeq_n \int_0^\infty\left(\int_{S^{n-1}}\mu(r\mathbf{e}_1)\exp(ir(t+x\cdot\omega))\mathrm{d}\sigma(\omega)\right)\mathrm{d}r\\ &\simeq_n \int_0^\infty\mu(r\mathbf{e}_1)\exp(irt)\left(\int_{S^{n-1}}\exp(irx\cdot\omega)\mathrm{d}\sigma(\omega)\right)\mathrm{d}r.\end{aligned}$$

According to the decomposition in Lemma 5.25, we have

$$\begin{aligned}&\mathscr{F}^{-1}\left[\mu\exp(it|\star|)\right](x)\\&\simeq_n\int_0^\infty\mu(r\mathbf{e}_1)\exp(irt)\tau_1^+(r|x|)\mathrm{d}r+\int_0^\infty\mu(r\mathbf{e}_1)\exp(irt)\tau_1^-(r|x|)\mathrm{d}r\\&\qquad+\int_0^\infty\mu(r\mathbf{e}_1)\exp(irt)\tau_*(r|x|)\mathrm{d}r.\end{aligned}$$

Here and below, we estimate the second term of the right-hand side; other terms can be estimated similarly. If $t-|x|\geq 1$ and $t-|x|\leq -1$, we can integrate by parts to obtain the desired result. Let us assume $-1\leq t-|x|\leq 1$. Then if we use Lemma 5.25 with $k=0$, $-1\leq t-|x|\leq 1$ and the triangle inequality for integrals, we obtain

$$\left|\int_0^\infty\mu(r\mathbf{e}_1)\exp(irt)\tau_1^-(r|x|)\mathrm{d}r\right|\lesssim\int_1^8(1+r|x|)^{-\frac{n-1}{2}}\mathrm{d}r\lesssim(1+t)^{-\frac{n-1}{2}}.$$

Thus, the proof is complete.

We follow the convention of the wave equation to define

$$\cos(t\sqrt{-\Delta})f\equiv\mathscr{F}^{-1}\left[\cos(t|\star|)\mathscr{F}f\right],\quad\frac{\sin(t\sqrt{-\Delta})}{\sqrt{-\Delta}}g\equiv\mathscr{F}^{-1}\left[\frac{\sin(t|\star|)}{|\star|}\mathscr{F}g\right]$$

for $f\in\mathscr{S}'(\mathbb{R}^n)$.

As the following expression shows, we have singularity at the origin. Using homogeneous function spaces, we can cancel such singularity.

Theorem 5.34 (Wave operator) *Let $f\in\dot{B}_{11}^{\frac{n+1}{2}}(\mathbb{R}^n)$, $g\in\dot{B}_{11}^{\frac{n-1}{2}}(\mathbb{R}^n)$. Then for any $t\neq 0$, the solution u to the wave equation $u_{tt}-\Delta u=0$ on $\mathbb{R}^{n+1}$, subject to $u(0)=f$ and $u'(0)=g$, which is given by*

$$u(t)\equiv\cos(t\sqrt{-\Delta})f+\frac{\sin(t\sqrt{-\Delta})}{\sqrt{-\Delta}}g,$$

satisfies $\|u(t)\|_{\dot{B}^0_{\infty 1}}\lesssim|t|^{-\frac{n-1}{2}}\|f\|_{\dot{B}_{11}^{\frac{n+1}{2}}}+|t|^{-\frac{n-1}{2}}\|g\|_{\dot{B}_{11}^{\frac{n-1}{2}}}$.

Proof Although our Fourier multipliers have singularity at the origin, since $\mathscr{S}_\infty(\mathbb{R}^n)$ is dense in $\dot{B}_{11}^{\frac{n-1}{2}}(\mathbb{R}^n)$ and $\dot{B}_{11}^{\frac{n+1}{2}}(\mathbb{R}^n)$, we may assume that $f,g\in\mathscr{S}_\infty(\mathbb{R}^n)$ and we can ignore this singularity. See Thoerem 2.30. Alternatively, by taking the Littlewood–Paley operator in the definition of the homogeneous Besov spaces $\dot{B}_{11}^{\frac{n-1}{2}}(\mathbb{R}^n)$ and $\dot{B}_{11}^{\frac{n+1}{2}}(\mathbb{R}^n)$, we can consider the Fourier multiplier has compact support in $\mathbb{R}^n\setminus\{0\}$. Choose a radial function $\varphi\in\mathscr{S}(\mathbb{R}^n)$ so that $\chi_{B(4)\setminus B(2)}\leq\varphi\leq$

$\chi_{B(8)\setminus B(1)}$. Furthermore, define $\mu \in C_c^\infty(\mathbb{R}^n)$ by $\mu(\xi) \equiv |\xi|^{-\frac{n-1}{2}}\varphi(\xi)$. Then since

$$\varphi_j(D)^2 u(t) = \mathscr{F}^{-1}[\varphi_j \cos(t|\star|)\mathscr{F}\varphi_j(D)f] + \mathscr{F}^{-1}\left[\varphi_j \frac{\sin(t|\star|)}{|\star|}\mathscr{F}\varphi_j(D)g\right],$$

we have

$$\begin{aligned}
&\|\varphi_j(D)^2 u(t)\|_\infty \\
&\le \|\mathscr{F}^{-1}[\varphi_j \cos(t|\star|)\mathscr{F}\varphi_j(D)f]\|_\infty + \left\|\mathscr{F}^{-1}\left[\varphi_j \frac{\sin(t|\star|)}{|\star|}\mathscr{F}\varphi_j(D)g\right]\right\|_\infty \\
&\lesssim 2^{-\frac{n-1}{2}j}\left\|\mathscr{F}^{-1}\left[\mu_j \cos(t|\star|)\right]\right\|_\infty \|(-\Delta)^{\frac{n-1}{2}}\varphi_j(D)f\|_1 \\
&\quad +2^{-\frac{n-1}{2}j}\left\|\mathscr{F}^{-1}\left[\mu_j \sin(t|\star|)\right]\right\|_\infty \|(-\Delta)^{\frac{n+1}{2}}\varphi_j(D)g\|_1. \qquad (5.106)
\end{aligned}$$

It remains to use Corollary 5.10.

Exercises

Exercise 5.48 Use the reduction to absurdity to show that

$$\mathbb{R}^n = \bigcup_{k=1}^n \left(\{x \in \mathbb{R}^n : 2nx_k \ge |x|\} \cup \{x \in \mathbb{R}^n : 2nx_k \le -|x|\}\right).$$

Exercise 5.49 Let $\psi \in C_c^\infty(\mathbb{R}^n)$ and let ϕ be a smooth function defined on an open set containing the support of ψ. Assume that $\nabla\phi$ never vanishes on the support of ψ. Let also $N \in \mathbb{N}$ and $\lambda > 1$. Then show that $\left|\int_{\mathbb{R}^n} e^{i\lambda\phi(x)}\psi(x)dx\right| \lesssim \lambda^{-N}$ by mimicking integration by parts in (5.105).

Exercise 5.50 (Stationary phase for nondegenerate stationary points) Let x_0 be a point in $\mathbb{R}^n$. Suppose $\phi : \mathbb{R}^n \to \mathbb{R}$ is a smooth function on a neighborhood of x_0 which has a nondegenerate stationary point at x_0. Namely, we suppose that $\nabla\phi(x_0) = 0$, but that $\det(\mathrm{Hess}\phi(x_0)) \ne 0$, where $\mathrm{Hess}\phi$ is the Hessian matrix of ϕ at x_0. Let also ψ be a bump function supported on a sufficiently small neighborhood of x_0. Then show that $\int_{\mathbb{R}^n} e^{i\lambda\phi(x)}\psi(x)dx = C\psi(x_0)e^{i\lambda\phi(x_0)}\lambda^{-\frac{n}{2}} + O(\lambda^{-\frac{n+1}{2}})$ as $\lambda \to \infty$, where C is a constant that depends on ϕ.

5.3.5 *Applications of Modulation Spaces to the Schrödinger Propagator*

Let us investigate how modulation spaces are applied to partial differential equations.

5.3.5.1 Schrödinger Propagator

Let us consider the Schrödinger equation $\partial_t u - i\Delta u = 0$ subject to initial condition $u(\star; 0) = f$. By the Fourier transform, we see that the solution is $u \simeq_n \exp(-it|D|^2)\mathscr{F}^{-1}f = e^{it\Delta}f$. If we calculate the Fourier transform, then we obtain a kernel expression:

$$e^{it\Delta}f = \frac{1}{(4\pi i t)^{\frac{n}{2}}}\int_{\mathbb{R}^n} \exp\left(-\frac{|x-y|^2}{4it}\right) f(y)dy. \tag{5.107}$$

See Exercise 5.51. Hence we obtain $\|e^{it\Delta}f\|_2 = \|f\|_2$ and $\|e^{it\Delta}f\|_\infty \lesssim t^{-\frac{n}{2}}\| f \|_1$.

The $L^2(\mathbb{R}^n)$-equivalence follows from the Plancherel theorem and this is the best; we have equality. Since the constant in the $L^1(\mathbb{R}^n)$-estimate blows up as $t \downarrow 0$, we want to improve this. This is achieved by modulation spaces:

Theorem 5.35 (Schrödinger propagator on modulation spaces) *Let $1 \le p \le 2$, $0 < q \le \infty$ and $s \in \mathbb{R}$. Then $\|e^{it\Delta}f\|_{M^s_{p'q}} \lesssim \| f \|_{M^s_{pq}}$ for $f \in M^s_{pq}(\mathbb{R}^n)$ and $|t| \le 1$.*

See Exercise 5.53 for what happens if $|t| > 1$.

Proof From the definition of the norm, it suffices to show that

$$\| \exp(-it|D|^2)\tau(D-k)^2 f \|_{p'} \lesssim \| \tau(D-k)\mathscr{F}f \|_p. \tag{5.108}$$

We write $f_k \equiv \tau(D-k)f$. Fix $x \in \mathbb{R}^n$. Then

$$\begin{aligned}
&\exp(-it|D|^2)\tau(D-k)f_k(x)\\
&\simeq_n \int_{\mathbb{R}^n} \exp(ix\cdot\xi - it|\xi|^2)\tau(\xi-k)\mathscr{F}f_k(\xi)d\xi\\
&= \int_{\mathbb{R}^n} \exp(ix\cdot(\xi+k) - it|\xi+k|^2)\tau(\xi)\mathscr{F}f_k(\xi+k)d\xi\\
&= \int_{\mathbb{R}^n} \exp(ix\cdot(\xi+k) - it|\xi+k|^2)\tau(\xi)\mathscr{F}[\exp(-ik\cdot\star)f_k](\xi)d\xi.
\end{aligned}$$

The functions $\exp(-it|D|^2)\tau(D-k)f_k$ and

$$x \mapsto c_n \int_{\mathbb{R}^n} \exp(ix \cdot \xi - it|\xi|^2 - 2it\xi \cdot k)\tau(\xi)\mathscr{F}[\exp(-ik \cdot \star) f_k](\xi) d\xi$$

have the same $L^p(\mathbb{R}^n)$-norm, as is seen from the above equality. Furthermore, a translation shows that the $L^p(\mathbb{R}^n)$-norm of $\exp(-it|D|^2)\tau(D-k)f_k$ equals that of

$$x \mapsto c_n \int_{\mathbb{R}^n} \exp(ix \cdot \xi - it|\xi|^2)\tau(\xi)\mathscr{F}[\exp(-ik \cdot \star) f_k](\xi) d\xi.$$

Set $M_t(\xi) \equiv \exp(-it|\xi|^2)\tau(\xi)$ for $\xi \in \mathbb{R}^n$. Then

$$\int_{\mathbb{R}^n} \exp(ix \cdot \xi - it|\xi|^2)\tau(\xi)\mathscr{F}[\exp(-ik \cdot \star) f_k](\xi) d\xi \simeq_n M_t(D)[\exp(-ik \cdot \star) f_k](x).$$

By Theorem 1.53, we conclude $\| \exp(-it|D|^2)\tau(D-k)f_k \|_p \lesssim \| f_k \|_p$. Thus, (5.108) is proved.

Exercises

Exercise 5.51 Verify (5.107) by the use of Example 1.15 and the analytic extension.

Exercise 5.52 [790, Lemma 2.2] Let $1 \le p, q \le \infty$, $s \in \mathbb{R}$, and let $m \in \mathscr{S}(\mathbb{R}^n)$. Then show that the Fourier multiplier $m(D)$ is bounded from $M^s_{pq}(\mathbb{R}^n)$ to $M^0_{pq}(\mathbb{R}^n)$ if and only if $\mathfrak{M} \equiv \sup\limits_{k \in \mathbb{Z}^n} \langle k \rangle^{-s} \|\psi(D-k)m(D)\|_{B(L^p)} < \infty$. Moreover, in this case show that $\|m(D)\|_{B(M^s_{pq})} \sim \mathfrak{M}$.

Exercise 5.53 Let $1 \le p \le 2$, $0 < q \le \infty$ and $s \in \mathbb{R}$.

1. Show that $\|e^{it\Delta} f\|_{p'} \lesssim |t|^{\frac{n}{p}-\frac{n}{2}} \| f \|_p$ for all $t \in \mathbb{R} \setminus \{0\}$.
2. Show that $\|e^{it\Delta} f\|_{M^s_{p'q}} \lesssim |t|^{\frac{n}{p}-\frac{n}{2}} \| f \|_{M^s_{pq}}$ for $f \in M^s_{pq}(\mathbb{R}^n)$ and $|t| > 1$.

Exercise 5.54 (Strichartz estimate, [1014]) Let $2 \le q \le \infty$ and $2 \le r < \infty$ satisfy $\dfrac{2}{q} + \dfrac{d}{r} = \dfrac{d}{2}$.

1. Show that $\|\varphi_j(D)e^{it\Delta} f \|_q \lesssim |t|^{-\frac{n}{q}+\frac{n}{2}} \| f \|_{q'}$ for all $t \in \mathbb{R} \setminus \{0\}$ and $j \in \mathbb{N}$.
2. Show that

$$\left(\int_{-T}^{T} \|\varphi_j(D)e^{it\Delta} f \|_q{}^r \, dt \right)^{\frac{1}{r}} \lesssim_{j,T} \|f\|_2 \quad (f \in L^2(\mathbb{R}^n)).$$

3. (TT^*-method) Let $T : H \to X$ be a bounded linear operator from a Hilbert space H to a Banach space X. Show that the square root of the operator norm $TT^* : X \to X$ is the operator norm of T.

4. Show that

$$\left(\int_{-\infty}^{\infty} \|\varphi_j(D)e^{it\Delta} f\|_q{}^r \, dt\right)^{\frac{1}{r}} \lesssim \|f\|_2 \quad (f \in L^2(\mathbb{R}^n)).$$

Here the implicit constant is independent of j.

5. Show that

$$\left(\int_{-\infty}^{\infty} \|e^{it\Delta} f\|_{L^q}{}^r \, dt\right)^{\frac{1}{r}} \lesssim \|f\|_2 \quad (f \in \mathscr{S}_\infty(\mathbb{R}^n)). \tag{5.109}$$

6. Show that $u(x,t) = e^{it\Delta} f(x)$ is measurable in x and t for all $f \in L^2(\mathbb{R}^n)$ and that (5.109) is valid for all $f \in L^2(\mathbb{R}^n)$.

Exercise 5.55 Let $\{r_j\}_{j=1}^{\infty}$ be the Rademacher sequence defined in Definition 3.1. For $f \in L^2(\mathbb{T})$, denote by $\hat{f}(j)$ the Fourier coefficient:

$$\hat{f}(j) \equiv \frac{1}{2\pi}\int_0^{2\pi} f(s)e^{-ijs} ds.$$

Define $Tf(x,t) \equiv \sum_{j=1}^{\infty} r_j(t)\hat{f}(j)e^{ijx}$. for $x \in \mathbb{R}^n$ and $t > 0$. Then show $\|Tf\|_{L^4(\mathbb{T}\times[0,1])} \lesssim \|f\|_{L^2(\mathbb{T})}$. Hint: Use $\left|\sum_{j=1}^{N} z_j\right|^4 = \sum_{j_1,j_2,j_3,j_4=1}^{N} z_{j_1} z_{j_2} \overline{z_{j_3} z_{j_4}}$.

Textbooks in Sect. 5.3

Semi-groups

We refer to [119, Chapter 9] for semi-groups. Tanabe [92] considered semi-groups which the elliptic differential operators generated in $L^p(\mathbb{R}^n)$ and Sobolev spaces.

Sectorial Operators

See the textbooks [55, 56] for semi-groups that sectorial operators generate.

5.4 Elliptic Differential Equations of the Second Order

In this section, we consider a priori estimates of the second order elliptic differential equations. An a priori estimate is an estimate of the form $\|f\|_{A^{s+2}_{pq}} \lesssim \|Lf\|_{A^s_{pq}} + \|f\|_{A^s_{pq}}$, where L is the second differential operator. In Sect. 5.4.1 we work on the whole space and in Sect. 5.4.2 we focus upon the half space. Furthermore, we extend our results to domains with smooth boundary in Sect. 5.4.3. We take advantage of the Schwartz distribution spaces, although there are some definitions of the notion of solutions such as the classical one and the one in Sobolev sense: The solutions are understood in the sense of $\mathscr{S}'(\mathbb{R}^n)$.

5.4.1 A Priori Estimate on the Whole Space

We consider elliptic differential operators satisfying the following conditions:

$$L = -\sum_{i,j=1}^{n} a_{ij}(x)\partial_{x_i}\partial_{x_j} + \sum_{j=1}^{n} b_j(x)\partial_{x_j} + c(x). \tag{5.110}$$

We are oriented to an a priori estimate of $A^s_{pq}(\mathbb{R}^n)$ and we assume that the coefficient are sufficiently smooth and that

$$a_{ij}, b_j, c \in \mathscr{C}^{s+2}(\mathbb{R}^n) \approx B^{s+2}_{\infty\infty}(\mathbb{R}^n) \quad (i, j = 1, 2, \ldots, n, s > 0).$$

Furthermore, a matrix of the real-valued functions $\{a_{ij}\}_{i,j=1}^n$ satisfies the following uniformly elliptic conditions: There exists a constant $\Theta \in (1, \infty)$ such that

$$\Theta^{-1}|\xi|^2 \le \sum_{i,j=1}^{n} a_{ij}(x)\xi_i\xi_j \le \Theta|\xi|^2$$

for all $\xi = (\xi_1, \xi_2, \ldots, \xi_n) \in \mathbb{R}^n$ and $x \in \mathbb{R}^n$.

We assume $1 < p, q < \infty$ and $s > 0$ for the sake of simplicity in this book. Furthermore, we concentrate on Triebel–Lizorkin spaces. Let us review Hölder's inequality for Triebel–Lizorkin spaces. When $\rho > s$, we have

$$\|f \cdot g\|_{A^s_{pq}} \lesssim \|f\|_{B^\rho_{\infty\infty}} \|g\|_{A^s_{pq}} \quad (f \in B^\rho_{\infty\infty}(\mathbb{R}^n), g \in A^s_{pq}(\mathbb{R}^n)), \tag{5.111}$$

according to Theorem 4.37.

5.4.1.1 A Priori Estimate of Elliptic Differential Operators Over the Whole Space

First we prove the next theorem, which is close to Corollary 5.4. Here we have a weak assumption on the coefficient.

Theorem 5.36 (A priori estimate of elliptic differential operators over the whole space) *Let $1 < p, q < \infty$ and $s > 0$. Assume that the boundary is smooth. Then the elliptic operator L, given by* (5.110), *satisfies*

$$\|f\|_{F^{s+2}_{pq}} \sim \|Lf\|_{F^s_{pq}} + \|f\|_{F^s_{pq}} \tag{5.112}$$

for all $f \in F^{s+2}_{pq}(\mathbb{R}^n)$.

Here we consider Triebel–Lizorkin spaces since we can localize Triebel–Lizorkin spaces, unlike Besov spaces. We remark that we can easily prove

$$\|f\|_{F^{s+2}_{pq}} \gtrsim \|Lf\|_{F^s_{pq}} + \|f\|_{F^s_{pq}}$$

for all $f \in F^{s+2}_{pq}(\mathbb{R}^n)$. Thus, the heart of the matter is to prove the reverse inequality.

5.4.1.2 Coefficients-Freezing Method

Our result will be obtained by piecing together local results on bounded domains to recover the result on the whole space. We proceed by several steps to prove Theorem 5.36.

Lemma 5.26 *Let $1 < p, q < \infty$, $s \geq 0$. Suppose that the functions $\{a_{ij}\}_{i,j=1,2,\dots,n}$ satisfy the conditions in Theorem* 5.36. *Then there exists $0 < \theta < 1$ such that*

$$\|(a_{ij} - a_{ij}(x_0))f\|_{F^s_{pq}} \lesssim r^\theta \|f\|_{F^s_{pq}} + \|f\|_{F^0_{pq}} \tag{5.113}$$

for all r and $f \in F^s_{pq}(\mathbb{R}^n)$ satisfying $\operatorname{diam}(\operatorname{supp}(f)) \leq 2r$ *and* $x_0 \in \operatorname{supp}(f)$.

Proof We prove (5.113) by induction on s. Let $0 \leq s < 1$. Choose a function $\tau \in \mathscr{S}(\mathbb{R}^n)$ so that $\chi_{B(1)} \leq \tau \leq \chi_{B(2)}$ and set $\tau_{x_0,r}(x) = \tau(r^{-1}(x - x_0))$. Let $\rho \in (s, 1)$. We have

$$\begin{aligned}
&\|(a_{ij} - a_{ij}(x_0))f\|_{F^s_{pq}} \\
&= \|\tau_{x_0,r}(a_{ij} - a_{ij}(x_0))f\|_{F^s_{pq}} \lesssim \|\tau_{x_0,r}(a_{ij} - a_{ij}(x_0))\|_{\mathscr{C}^\rho} \|f\|_{F^s_{pq}}
\end{aligned}$$

by Theorem 4.37. By the real interpolation (Theorem 4.25), we have $\mathscr{C}^\rho(\mathbb{R}^n) \approx B^\rho_{\infty\infty}(\mathbb{R}^n) \approx (B^0_{\infty\infty}(\mathbb{R}^n), B^1_{\infty\infty}(\mathbb{R}^n))_{\rho,\infty}$. Thus, if we use this interpolation relation, then we have

$$\begin{aligned}\|\tau_{x_0,r}(a_{ij}-a_{ij}(x_0))\|_{\mathscr{C}^\rho} &\lesssim \|\tau_{x_0,r}(a_{ij}-a_{ij}(x_0))\|_{\mathscr{C}^1}^{\rho}\|\tau_{x_0,r}(a_{ij}-a_{ij}(x_0))\|_{\mathscr{C}^0}^{1-\rho}\\ &\lesssim \|\tau_{x_0,r}(a_{ij}-a_{ij}(x_0))\|_{\mathscr{B}^1}^{\rho}\|\tau_{x_0,r}(a_{ij}-a_{ij}(x_0))\|_{\mathscr{B}^0}^{1-\rho}\\ &\lesssim r^{1-\rho}.\end{aligned}$$

By letting $\theta \equiv 1-\rho$, we obtain the result for $0 \le s < 1$.

Let $m \in \mathbb{N}$, and let $t \in [m-1, m)$. Our induction assumption is as follows. we have

$$\|(a-a_{ij}(x_0))g\|_{F^t_{pq}} \lesssim r^\theta \|g\|_{F^t_{pq}} + \|g\|_{F^{t-1}_{pq}} \tag{5.114}$$

for any $g \in F^t_{pq}(\mathbb{R}^n)$ and $0 < r \le 1$ satisfying $\operatorname{supp}(g) \subset B(x_0, r)$ and $x_0 \in \mathbb{R}^n$. Then let us prove (5.113) for $m \le s < m+1$.

Thanks to Theorem 2.12, we have

$$\begin{aligned}&\|(a-a_{ij}(x_0))f\|_{F^s_{pq}}\\ &\sim \|(a-a_{ij}(x_0))f\|_{F^{s-1}_{pq}} + \sum_{j=1}^n \|\partial_{x_j}[(a-a_{ij}(x_0))f]\|_{F^{s-1}_{pq}}\\ &\lesssim \|(a-a_{ij}(x_0))f\|_{F^{s-1}_{pq}} + \sum_{j=1}^n \|(a-a_{ij}(x_0))\partial_{x_j} f\|_{F^{s-1}_{pq}} + \sum_{j=1}^n \|\partial_{x_j} a \cdot f\|_{F^{s-1}_{pq}}\\ &=: \mathrm{I}+\mathrm{II}+\mathrm{III}.\end{aligned}$$

For $t = s-1$, we use induction assumption (5.114) to estimate I and II:

$$\begin{aligned}\mathrm{I}+\mathrm{II} &\lesssim r^\theta\|f\|_{F^{s-1}_{pq}} + \|f\|_{F^{s-2}_{pq}} + \sum_{j=1}^n (r^\theta \|\partial_{x_j} f\|_{F^{s-1}_{pq}} + \|\partial_{x_j} f\|_{F^{s-2}_{pq}})\\ &\lesssim r^\theta \|f\|_{F^s_{pq}} + \|f\|_{F^{s-1}_{pq}}.\end{aligned}$$

We have $\mathrm{III} \lesssim \|f\|_{F^{s-1}_{pq}}$ by the Hölder inequality (5.111) for Triebel–Lizorkin spaces.

So, we obtain the desired result for $m \le s < m+1$.

Lemma 5.27 *Under the assumption of Theorem* 5.36*, we can find* $r_0 > 0$ *with the following property: The conclusion of Theorem* 5.36 *holds for any* $f \in F^{s+2}_{pq}(\mathbb{R}^n)$ *with* $x_0 \in \mathbb{R}^n$ *and* $\operatorname{supp}(f) \subset Q(x_0, 4r_0)$.

We rely upon the coefficient-freezing method.

Proof Here and below, the constants do not depend on $r_0 > 0$ and $x_0 \in \mathbb{R}^n$. Define a constant coefficient differential operator L_0 by

$$L_0 \equiv -\sum_{i,j=1}^{n} a_{ij}(x_0)\partial_{x_i}\partial_{x_j}.$$

Then by virtue of the uniformly elliptic condition, the eigenvalue of the matrix $\{a_{ij}(x_0)\}_{i,j=1,2,\dots,n}$ lies in the closed interval $[\Theta^{-1}, \Theta]$. Thus, we obtain

$$\|f\|_{F^{s+2}_{pq}} \sim \|L_0 f\|_{F^s_{pq}} + \|f\|_{F^s_{pq}}$$

by mimicking the proof of Theorem 2.12.

Hence it follows that

$$\|f\|_{F^{s+2}_{pq}} \sim \|Lf\|_{F^s_{pq}} + \|(L-L_0)f\|_{F^s_{pq}} + \|f\|_{F^s_{pq}}. \tag{5.115}$$

Since we do not need the second term of (5.115), we consider eliminating this term.

Lemma 5.28 *Let f be a function satisfying the conditions in Lemma 5.27. Then*

$$\|(L-L_0)f\|_{F^s_{pq}} \lesssim r_0{}^{\theta}\, \|f\|_{F^{s+2}_{pq}} + \|f\|_{F^{s+1}_{pq}} \tag{5.116}$$

with the constant in $\lesssim$ independent of r_0, where θ depends on s and $0 < \theta < 1$.

We prove Lemma 5.27 by admitting Lemma 5.28 for the time being. Lemma 5.28 and (5.115) yield positive constants K_1 and $K_2 > 0$ independent of r_0 such that

$$\|f\|_{F^{s+2}_{pq}} \le K_1(\|Lf\|_{F^s_{pq}} + \|(L-L_0)f\|_{F^s_{pq}} + \|f\|_{F^s_{pq}}) \tag{5.117}$$

and that

$$\|(L-L_0)f\|_{F^s_{pq}} \le K_2(r_0{}^{\theta}\, \|f\|_{F^{s+2}_{pq}} + \|f\|_{F^{s+1}_{pq}}). \tag{5.118}$$

Hence by plugging (5.118) into (5.117), we obtain

$$\|f\|_{F^{s+2}_{pq}} \le K_2 r_0{}^{\theta}\|f\|_{F^{s+2}_{pq}} + K_1\|Lf\|_{F^s_{pq}} + K_1K_2\|f\|_{F^s_{pq}}. + K_2\|f\|_{F^{s+1}_{pq}}.$$

Note that $K_2\|f\|_{F^{s+1}_{pq}} \le \dfrac{1}{4}\|f\|_{F^{s+2}_{pq}} + K_3\|f\|_{F^s_{pq}}$. Since we are assuming $f \in F^{s+2}_{pq}(\mathbb{R}^n)$, we have $\|f\|_{F^{s+2}_{pq}} \le 2K_2\|Lf\|_{F^s_{pq}} + 2K_1K_2\|f\|_{F^s_{pq}}$ as long as $2K_2 r_0{}^{\theta} \le 1$.

So the proof is complete modulo Lemma 5.28.

Proof (of Lemma 5.28) To prove (5.116), we absorb the terms of lower order of L into $\|f\|_{F^{s+1}_{pq}}$, we can assume that L in (5.110) takes the form $\sum\limits_{i,j=1}^{n} -a_{ij}(x)\partial_{x_i}\partial_{x_j}$.

Under this assumption,

$$\|(L - L_0)f\|_{F^s_{pq}}$$

$$\lesssim \sum_{j,k=1}^{n} \left(r_0{}^{\theta} \|\partial_{x_k}\partial_{x_j} f\|_{F^s_{pq}} + \|\partial_{x_k}\partial_{x_j} f\|_{F^{s-1}_{pq}} \right) \lesssim r_0{}^{\theta} \|f\|_{F^{s+2}_{pq}} + \|f\|_{F^{s+1}_{pq}}$$

by Lemma 5.26. Hence (5.116) is proved.

We now prove Theorem 5.36.

Proof (of Theorem 5.36*)* We have

$$\|f\|_{F^{s+2}_{pq}} \gtrsim \|Lf\|_{F^s_{pq}} + \|f\|_{F^s_{pq}}$$

by Hölder's inequality. Hence let us concentrate on the reverse inequality.

By Theorem 4.38 with $a = r_0$, we have norm equivalence:

$$\|f\|^p_{F^{s+2}_{pq}} \sim \sum_{m \in \mathbb{Z}^n} \|\rho(\star - r_0\, m) f\|^p_{F^{s+2}_{pq}}.$$

Note that this does not hold for Besov spaces. We have

$$\|\rho(\star - r_0\, m) f\|_{F^{s+2}_{pq}}{}^p \sim \|L[\rho(\star - r_0\, m) f]\|_{F^s_{pq}}{}^p + \|\rho(\star - r_0\, m) f\|_{F^s_{pq}}{}^p$$

by Lemma 5.27. Observe that the differential order of the commutator $\rho(\star - r_0 m)Lf - L[\rho(\star - r_0 m)f]$ is 1. So, we obtain

$$\|\rho(\star - r_0 m)Lf - L[\rho(\star - r_0 m)f]\|_{F^s_{pq}}$$

$$\lesssim \|\rho(\star - r_0 m)L\tau(\star - r_0 m)f - L[\rho(\star - r_0 m)\tau(\star - r_0 m)f]\|_{F^s_{pq}}$$

$$\lesssim \|\tau(\star - r_0 m)f\|_{F^{s+1}_{pq}}$$

by (5.111). We now choose an function $\tau \in C^\infty_{\mathrm{c}}(\mathbb{R}^n)$ satisfying $\chi_{\mathrm{supp}(\rho)} \le \tau \le 1$ so as not to lose information on the function. Hence it follows that

$$\|\rho(\star - r_0\, m) f\|^p_{F^{s+2}_{pq}}$$

$$\lesssim \|\rho(\star - r_0\, m)Lf\|^p_{F^s_{pq}} + \|\rho(\star - r_0\, m) f\|^p_{F^s_{pq}} + \|\tau(\star - r_0\, m) f\|^p_{F^{s+1}_{pq}}.$$

The heart of the proof of $\gtrsim$ in Theorem 4.38 is that ρ is compactly supported. Thus,

$$\left(\sum_{m \in \mathbb{Z}^n} \|\tau(\star - r_0\, m) f\|^p_{F^{s+1}_{pq}} \right)^{\frac{1}{p}} \lesssim \|f\|_{F^{s+1}_{pq}}.$$

As a result, the "$\lesssim$" inequality in Theorem 4.38 yields

$$\|f\|_{F^{s+2}_{pq}} \lesssim \|Lf\|_{F^s_{pq}} + \|f\|_{F^{s+1}_{pq}} + \|f\|_{F^s_{pq}}.$$

By virtue of Theorem 1.53, we can find a constant $M > 0$ such that

$$\|f\|_{F^{s+2}_{pq}} \le \frac{1}{2}\|f\|_{F^{s+2}_{pq}} + M\,\|Lf\|_{F^s_{pq}} + M\,\|f\|_{F^s_{pq}}. \tag{5.119}$$

Since $f \in F^{s+2}_{pq}(\mathbb{R}^n)$, we can move $\frac{1}{2}\|f\|_{F^{s+2}_{pq}}$ to the right-hand side in (5.119). Hence Theorem 5.36 is proved.

5.4.1.3 Gaffney-Type Estimates

We have the following Gaffney-type estimates, which are called the off-diagonal estimates. To this end, for a closed set E, we define $L^2(E) \subset L^2(\mathbb{R}^n)$ to be the subspace of all $f \in L^2(\mathbb{R}^n)$ for which $f = \chi_E f$. Since the coefficients of L are assumed smooth, we can use the integral kernel of $(\mathrm{id}_{L^2} + t^2 L)^{-1}$. However, it matters that the following estimates result solely from the ellipticity condition:

Lemma 5.29 (Gaffney-type estimate, off-diagonal estimate) *Let $0 < \lambda' < \Lambda < \infty$, $0 < d < \infty$, and let $A \equiv \{a_{ij}\}_{i,j=1,2,\ldots,n}$ be a matrix with entries $a_{ij} : \mathbb{R}^n \to \mathbb{C}$ of measurable functions satisfying for all $(\xi_1, \xi_2, \ldots, \xi_n) \in \mathbb{C}^n$*

$$\text{Ellipticity} \quad \Re\left(\sum_{i,j=1}^n a_{ij}(\star)\xi_i\overline{\xi_j}\right) \ge \lambda'|\xi|^2, \tag{5.120}$$

$$\text{Boundedness} \quad \left\|\sum_{i=1}^n \left|\sum_{j=1}^n a_{ij}(\star)\xi_i\right|^2\right\|_\infty \le \Lambda^2 \sum_{j=1}^n |\xi_j|^2. \tag{5.121}$$

Define $L \equiv -\mathrm{div} \circ M_A \circ \nabla$. Let E, F be disjoint closed sets in $\mathbb{R}^n$ with $\mathrm{dist}(E, F) > d$. Then there exists $a > 0$ such that for all $f \in L^2(\mathbb{R}^n)$ and $\boldsymbol{F} \in \mathrm{Dom}(\mathrm{div})$ satisfying $\mathrm{supp}(f) \cup \mathrm{supp}(\boldsymbol{F}) \subset E$,

$$\|(\mathrm{id}_{L^2} + t^2 L)^{-1} f\|_{L^2(F)} \lesssim \exp\left(-a\frac{d}{t}\right)\|f\|_{L^2(E)}, \tag{5.122}$$

$$\|t\nabla(\mathrm{id}_{L^2} + t^2 L)^{-1} f\|_{L^2(F)^n} \lesssim \exp\left(-a\frac{d}{t}\right)\|f\|_{L^2(E)}, \tag{5.123}$$

$$\|t(\mathrm{id}_{L^2} + t^2 L)^{-1}\mathrm{div}(\boldsymbol{F})\|_{L^2(F)} \lesssim \exp\left(-a\frac{d}{t}\right)\|\boldsymbol{F}\|_{L^2(E)^n}. \tag{5.124}$$

Proof We define $\tilde{F} \equiv \{x \in \mathbb{R}^n : \text{dist}(x, F) \le \text{dist}(x, E)\}$. Let $t \in [d, \infty)$. In this case, the proof is simpler, since the exponential in the left-hand side does not come into play. Note that

$$\|(\text{id}_{L^2} + t^2 L)^{-1}\|_{B(L^2)} \le 1. \tag{5.125}$$

We deduce from (5.125)

$$\|(\text{id}_{L^2} + t^2 L)^{-1} f\|_{L^2(F)} \le \|(\text{id}_{L^2} + t^2 L)^{-1} f\|_2 \le \|f\|_2 = \|f\|_{L^2(E)},$$

which yields (5.122). As for (5.123) with $0 < t \le d$ using (5.125) we calculate

$$\begin{aligned}
\|t\nabla(\text{id}_{L^2} + t^2 L)^{-1} f\|_{L^2(F)} &\le \|t\nabla(\text{id}_{L^2} + t^2 L)^{-1} f\|_2 \\
&= \sqrt{\langle t\nabla(\text{id}_{L^2} + t^2 L)^{-1} f, t\nabla(\text{id}_{L^2} + t^2 L)^{-1} f\rangle_{(L^2)^n}} \\
&\lesssim \sqrt{|\langle M_A t\nabla(\text{id}_{L^2}+t^2 L)^{-1} f, t\nabla(\text{id}_{L^2}+t^2 L)^{-1} f\rangle_{(L^2)^n}|} \\
&= \sqrt{|\langle t^2 L(\text{id}_{L^2} + t^2 L)^{-1} f, (\text{id}_{L^2} + t^2 L)^{-1} f\rangle_2|} \\
&\lesssim \|f\|_{L^2(E)}.
\end{aligned}$$

We prove (5.122) for $0 < t \le d$, which is essential. We let $u^t \equiv (\text{id}_{L^2} + t^2 L)^{-1} f$. Then

$$\begin{aligned}
&\int_{\mathbb{R}^n} u^t(x)\overline{v(x)}\text{d}x + t^2 \int_{\mathbb{R}^n} A(x)\nabla u^t(x) \cdot \overline{\nabla v(x)}\text{d}x \\
&= \int_{\mathbb{R}^n} (\text{id}_{L^2} + t^2 L)u^t(x)\overline{v(x)}\text{d}x = \int_{\mathbb{R}^n} f(x) \cdot \overline{v(x)}\text{d}x.
\end{aligned}$$

Choose $\eta \in C^\infty(\mathbb{R}^n)$ so that

$$\chi_F \le \eta \le \chi_{\tilde{F}}, \quad \|\nabla\eta\|_\infty \sim d^{-1}. \tag{5.126}$$

Let $\zeta \in C^\infty(\mathbb{R}^n)$ have the same support as η. Since $\text{supp}(f) \cap \text{supp}(\zeta) = \emptyset$, we have

$$\begin{aligned}
&\int_{\mathbb{R}^n} |u^t(x)|^2 \zeta(x)^2 \text{d}x + t^2 \int_{\mathbb{R}^n} A(x)\nabla u^t(x) \cdot \overline{\nabla u^t(x)}\zeta(x)^2 \text{d}x \\
&= -2t^2 \int_{\mathbb{R}^n} A(x)\nabla u^t(x) \cdot \overline{u^t(x)}\zeta(x)\nabla\zeta(x)\text{d}x
\end{aligned}$$

for $v \equiv u^t \zeta^2$. Let $\varepsilon \equiv \dfrac{\lambda}{4\Lambda} > 0$ be a constant. By the Cauchy inequality, which asserts $2\sqrt{ab} \le a+b$ for $a, b > 0$, and the ellipticity, we have

$$\begin{aligned}
&\int_{\mathbb{R}^n} |u^t(x)|^2 \zeta(x)^2 \mathrm{d}x + \lambda t^2 \int_{\mathbb{R}^n} \nabla u^t(x) \cdot \overline{\nabla u^t(x)} \zeta(x)^2 \mathrm{d}x \\
&\le -2t^2 \int_{\mathbb{R}^n} A(x) \nabla u^t(x) \cdot \overline{u^t(x)} \zeta(x) \nabla \zeta(x) \mathrm{d}x \\
&\le \frac{2\varepsilon t^2}{\Lambda} \int_{\mathbb{R}^n} |A(x)[\zeta(x) \nabla u^t(x)]|^2 \mathrm{d}x + 2\varepsilon^{-1} \Lambda t^2 \int_{\mathbb{R}^n} |u^t(x)|^2 |\nabla \zeta(x)|^2 \mathrm{d}x \\
&\le 2\varepsilon \Lambda t^2 \int_{\mathbb{R}^n} |\zeta(x) \nabla u^t(x)|^2 \mathrm{d}x + 2\varepsilon^{-1} \Lambda t^2 \int_{\mathbb{R}^n} |u^t(x)|^2 |\nabla \zeta(x)|^2 \mathrm{d}x.
\end{aligned}$$

As a result,

$$\begin{aligned}
&\int_{\mathbb{R}^n} |u^t(x)|^2 \zeta(x)^2 \mathrm{d}x + \lambda t^2 \int_{\mathbb{R}^n} \nabla u^t(x) \cdot \overline{\nabla u^t(x)} \zeta^2(x) \mathrm{d}x \\
&\le 2\varepsilon \Lambda t^2 \int_{\mathbb{R}^n} |\zeta(x) \nabla u^t(x)|^2 + 2\varepsilon^{-1} \Lambda t^2 \int_{\mathbb{R}^n} |u^t(x)|^2 |\nabla \zeta(x)|^2 \mathrm{d}x.
\end{aligned} \tag{5.127}$$

Inserting the precise value of ε, we have

$$\int_{\mathbb{R}^n} |u^t(x)|^2 \zeta(x)^2 \mathrm{d}x \le \frac{16\Lambda^2 t^2}{\lambda} \int_{\mathbb{R}^n} |u^t(x)|^2 |\nabla \zeta(x)|^2 \mathrm{d}x. \tag{5.128}$$

Let $\alpha \equiv \dfrac{\sqrt{\lambda}}{8\Lambda t \|\nabla \eta\|_{(L^\infty)^n}}$. In (5.128) we specify ζ: $\zeta \equiv \exp(\alpha\eta) - 1$, so that

$$\int_{\mathbb{R}^n} |u^t(x)|^2 |\exp(\alpha\eta(x)) - 1|^2 \mathrm{d}x \le \frac{1}{4} \int_{\mathbb{R}^n} |u^t(x)|^2 \exp(2\alpha\eta(x)) \mathrm{d}x.$$

Assume $\alpha \ge 2$. Equating this relation and using $\eta(x) \equiv 1$ for $x \in F$ and $3e^{2k} - 8e^k + 4 \ge e^k - 1$ for $k \ge 2$, we conclude

$$e^\alpha \int_F |u^t(x)|^2 \mathrm{d}x \le \int_{\mathbb{R}^n} |u^t(x)|^2 \exp(\alpha\eta(x)) \mathrm{d}x \le \int_{\mathbb{R}^n} |u^t(x)|^2 \mathrm{d}x \le \int_E |f(x)|^2 \mathrm{d}x.$$

We can readily incorporate the case where $\alpha \le 2$; $\|u^t\|_{L^2(F)} \le \|u^t\|_2 \lesssim e^{-\alpha} \|f\|_{L^2(E)}$. So the proof of (5.122) is complete.

Likewise, letting $\varepsilon = \dfrac{\lambda}{4\Lambda}$ and $\zeta = \eta$ in (5.127), we obtain

$$\int_F |t\nabla u^t(x)|^2 \mathrm{d}x \le \int_F |t\nabla u^t(x)|^2 \eta(x)^2 \mathrm{d}x \le \frac{8\Lambda^2 t^2}{\lambda}\int_{\mathbb{R}^n} |u^t(x)|^2 |\nabla \eta(x)|^2 \mathrm{d}x.$$

From the definition of $\tilde{F}$, we learn that

$$\int_F |t\nabla u^t(x)|^2 \mathrm{d}x \le \frac{8\Lambda^2 t^2}{\lambda}\int_{\tilde{F}} |u^t(x)|^2 |\nabla \eta(x)|^2 \mathrm{d}x.$$

From (5.122), we have the desired result modulo a change of a:

$$\int_F |t\nabla u^t(x)|^2 \mathrm{d}x \lesssim t^2 d^{-2} \exp\left(-2a\frac{d}{t}\right)\int_E |f(x)|^2 \mathrm{d}x \lesssim \exp\left(-a\frac{d}{t}\right)\int_E |f(x)|^2 \mathrm{d}x.$$

Finally, we claim that (5.123) and (5.124) are dual to each other. In fact, thanks to (5.123) we have

$$\begin{aligned}
&\|t(\mathrm{id}_{L^2} + t^2 L)^{-1}\mathrm{div}(\mathbf{F})\|_{L^2(F)}\\
&= \sup_{g \in L^2(F), \|g\|_2 = 1} \left|\int_{\mathbb{R}^n} t(\mathrm{id}_{L^2} + t^2 L)^{-1}\mathrm{div}(\mathbf{F})(x)\overline{g(x)}\mathrm{d}x\right|\\
&= \sup_{g \in L^2(F), \|g\|_2 = 1} \left|\int_E \mathbf{F}(x)\overline{t\nabla[(\mathrm{id}_{L^2} + t^2 L^*)^{-1} g](x)}\mathrm{d}x\right|\\
&\le \sup_{g \in L^2(F), \|g\|_2 = 1} \|\mathbf{F}\|_{(L^2)^n} \|t\nabla(\mathrm{id}_{L^2} + t^2 L^*)^{-1} g\|_{(L^2(E))^n}\\
&\lesssim \exp\left(-a\frac{d}{t}\right)\|\mathbf{F}\|_{(L^2)^n}.
\end{aligned}$$

Note that the deduction of these estimates uses ellipticity solely.

We complexify what we obtained.

Proposition 5.13 *Let $0 < \lambda < \Lambda < \infty$, and let $A \equiv \{a_{ij}\}_{i,j=1,2,\ldots,n}$ be a matrix with entries $a_{ij} : \mathbb{R}^n \to \mathbb{C}$ of measurable functions satisfying* (5.120) *and* (5.121) *for $(\xi_1, \xi_2, \ldots, \xi_n) \in \mathbb{C}^n$. Let E, F be disjoint closed sets in $\mathbb{R}^n$ with $d > \mathrm{dist}(E, F)$. Let $0 < \theta \ll 1$. Then there exists $a > 0$ such that for all $f \in L^2(\mathbb{R}^n)$ and $\boldsymbol{F} \in \mathrm{Dom}(\mathrm{div})$ satisfying $\mathrm{supp}(f) \cup \mathrm{supp}(\boldsymbol{F}) \subset E$,*

$$\|(z\mathrm{id}_{L^2} - L)^{-1} f\|_{L^2(F)} \lesssim \exp\left(-a\frac{d}{\sqrt{|z|}}\right)\|f\|_{L^2(E)}$$

for all $f \in L^2(E)$ and $z \in \mathbb{C}$ with $\arg(z) = \frac{\pi}{2} - \theta$ or $\arg(z) = -\frac{\pi}{2} + \theta$.

Proof Simply verify that $z^{-1}L$ is also the same type.

Using Proposition 5.12, we can define e^{-tL}. For this operator, we have the following estimates:

Theorem 5.37 (Gaffney-type estimate, off-diagonal estimate) *Let* $0 < \lambda < \Lambda < \infty$, $0 < d < \infty$, *and let* $A \equiv \{a_{ij}\}_{i,j=1,2,\ldots,n}$ *be a matrix with entries* $a_{ij} : \mathbb{R}^n \to \mathbb{C}$ *of measurable functions satisfying* (5.120) *and* (5.121) *for all* $(\xi_1, \xi_2, \ldots, \xi_n) \in \mathbb{C}^n$. *Let* E, F *be disjoint closed sets in* $\mathbb{R}^n$ *with* $\mathrm{dist}(E, F) > d$. *Then there exists* $a > 0$ *such that for all* $f \in L^2(\mathbb{R}^n)$ *and* $\boldsymbol{F} \in \mathrm{Dom}(\mathrm{div})$ *satisfying* $\mathrm{supp}(f) \cup \mathrm{supp}(\boldsymbol{F}) \subset E$,

$$\|\exp(-t^2L)f\|_{L^2(F)} \lesssim \exp\left(-a\frac{d^2}{t^2}\right) \|f\|_{L^2(E)}, \tag{5.129}$$

$$\|t^2L\exp(-t^2L)f\|_{L^2(F)} \lesssim \exp\left(-a\frac{d^2}{t^2}\right) \|f\|_{L^2(E)}, \tag{5.130}$$

$$\|t\nabla\exp(-t^2L)f\|_{L^2(F)^n} \lesssim \exp\left(-a\frac{d^2}{t^2}\right) \|f\|_{L^2(E)}, \tag{5.131}$$

$$\|\exp(-t^2L)t\mathrm{div}(\boldsymbol{F})\|_{L^2(F)} \lesssim \exp\left(-a\frac{d^2}{t^2}\right) \|\boldsymbol{F}\|_{L^2(E)^n}. \tag{5.132}$$

Proof Let $\varphi_R \equiv \min(R, \mathrm{dist}(\star, E))$ for $R > 0$. Then $L_R \equiv e^{\rho\varphi_R}Le^{-\rho\varphi_R} + K\rho^2\mathrm{id}_{L^2}$ is sectorial if $K \gg 1$. In fact, since $R > 0$, the multiplication mapping $f \mapsto e^{\rho\varphi_R}f$ is a topological and linear isomorphism with its inverse $f \mapsto e^{-\rho\varphi_R}f$. So, $e^{\rho\varphi_R}Le^{-\rho\varphi_R}$ and $e^{\rho\varphi_R}Le^{-\rho\varphi_R}$ have the same spectrum. The addition of $K\rho^2\mathrm{id}_{L^2}$ pushes the spectrum to the right. Meanwhile, we can verify $\|(\lambda\mathrm{id}_{L^2} + L_R)f\|_2 \gtrsim |\lambda| \cdot \|f\|_2$ by taking $K \gg 1$. Thus,

$$\|\mathrm{e}^{-t^2L_R}f\|_2 + \|t^2L_R\exp(-t^2L_R)f\|_2 + \|t\nabla\exp(-t^2L_R)f\|_2 \lesssim e^{2K\rho^2t^2}\|f\|_{L^2}.$$

We note that $\exp(-t^2L_R)f = e^{\rho\varphi_R}\mathrm{e}^{-t^2L}[e^{-\rho\varphi_R}f]$ since $\mathrm{e}^{-t^2\varphi_R}$ is an isomorphism. Thus, we obtain $\|\exp(-t^2L)f\|_2 + \|t^2L\exp(-t^2L)f\|_2 \lesssim e^{3Kt^2\rho^2-\rho\mathrm{dist}(E,F)}\|f\|_2$ for all $f \in L^2(E)$ if we let $R \to \infty$. We optimize this estimate over $\rho > 0$ to have (5.129), (5.130) and (5.131) and then we obtain (5.132) thanks to (5.131); (5.131) and (5.132) are dual to each other.

We end this section with some estimates on composition of these types of operators.

Lemma 5.30 *Let* $\{A_t\}_{t>0}$ *and* $\{B_t\}_{t>0}$ *be two families of uniformly* $L^2(\mathbb{R}^n)$-*bounded linear operators. Assume*

$$\|A_tf\|_{L^2(F)} + \|B_tf\|_{L^2(F)} \lesssim \exp\left(-\frac{\alpha\mathrm{dist}(E, F)}{t}\right) \|f\|_{L^2(E)}$$

whenever E and F are disjoint closed sets in $\mathbb{R}^n$ with distance d and $f \in L^2(\mathbb{R}^n)$ is supported in E. Then for all $t, s > 0$,

$$\|A_t B_s f\|_{L^2(F)} \lesssim \exp\left(-\frac{\alpha \mathrm{dist}(E, F)}{2\max(s,t)}\right) \|f\|_{L^2(E)}$$

whenever E and F are disjoint closed sets in $\mathbb{R}^n$ with distance d and $f \in L^2(\mathbb{R}^n)$ is supported in E.

Proof We use the decomposition $A_t B_s f = A_t[\chi_G B_s f] + A_t[\chi_{\mathbb{R}^n \setminus G} B_s f]$, where $G = \{x \in \mathbb{R}^n : \mathrm{dist}(x, E) \geq \mathrm{dist}(x, F)\}$. Since $\mathrm{dist}(E, G), \mathrm{dist}(F, G^c) \geq \frac{1}{2}\mathrm{dist}(E, F)$, we obtain the desired result.

5.4.1.4 The Operator L for Other Lebesgue Spaces

We investigate the further mapping property of L using the *Neumann expansion* on Banach spaces X: A linear operator A with $\|\mathrm{id}_X - A\|_{B(X)} < 1$ is invertible and $A^{-1} = (\mathrm{id}_X - (\mathrm{id}_X - A))^{-1} = \sum_{j=0}^{\infty} (\mathrm{id}_X - A)^{-j}$. We also recall that any complex matrix A induces a bounded linear transformation M_A with norm $\|M_A\|_{B(\mathbb{C}^n)}$.

Let A be a matrix with entries of measurable functions satisfying (5.120) and (5.121). We know that the operator $\mathrm{id}_{L^p} + L$ maps $W^{1,p}(\mathbb{R}^n)$ to $W^{-1,p}(\mathbb{R}^n)$ boundedly, where $L = -\mathrm{div} M_A \nabla$. Since the real part of the eigenvalue of each $A(x)$ is contained in a crescent of the form $\{z \in \mathbb{C} : \Re(z) \geq \delta,\ |z| \leq \Lambda\}$ for some $\delta > 0$ and $\Lambda > 0$, there exist a large constant $\mu \gg 1$ and $\mu' \in (0, 1)$ such that

$$\|M_{A(x)} - \mu \mathrm{id}_{\mathbb{C}^n}\|_{B(\mathbb{C}^n)} \leq \mu' \mu \quad (\text{almost every } x \in \mathbb{R}^n), \tag{5.133}$$

where M_C denotes the linear operator induced by an $n \times n$ matrix C and E denotes the $n \times n$ identity matrix. For such a μ, we have the following decomposition:

$$A(x) = \mu E + \mu B(x) \quad (x \in \mathbb{R}^n),$$

where for almost every $x \in \mathbb{R}^n$

$$\|B(x)\|_{B(\mathbb{C}^n)} \leq \mu' < 1. \tag{5.134}$$

Then there is a decomposition of $\mathrm{id}_{L^p} + L$:

$$\begin{aligned}
&\mathrm{id}_{L^p} + L \\
&= \mathrm{id}_{L^p} - \mu \Delta - \mu \mathrm{div} M_B \nabla
\end{aligned}$$

$$= (\mathrm{id}_{L^p} - \mu\Delta)^{\frac{1}{2}}(\mathrm{id}_{L^p} - \mu(\mathrm{id}_{L^p} - \mu\Delta)^{-\frac{1}{2}}\mathrm{div} M_B \nabla(\mathrm{id}_{L^p} - \mu\Delta)^{-\frac{1}{2}})(\mathrm{id}_{L^p} - \mu\Delta)^{\frac{1}{2}}.$$

Since M_B is a bounded operator on $L^p(\mathbb{R}^n)$, we have

$$c(p) \equiv \|\mu(\mathrm{id}_{L^p} - \mu\Delta)^{-\frac{1}{2}}\mathrm{div} M_B \nabla(\mathrm{id}_{L^p} - \mu\Delta)^{-\frac{1}{2}}\|_{B(L^p)} < \infty.$$

From (5.134) and the Fourier transform, we deduce $c(2) < 1$. By Theorem 4.26, we see that $c(p) \le c(p_0)^{1-\theta}c(p_1)^\theta$ if $1 < p_0, p_1, p < \infty$ and $\theta \in (0,1)$ satisfy

$$\frac{1}{p} = \frac{1-\theta}{p_0} + \frac{\theta}{p_1}.$$

Thus, the function $p \in (0,1) \mapsto \log c(p^{-1})$ is a convex function. Using the Neumann expansion and (5.133), we see that

$$\mu(\mathrm{id}_{L^p} - \mu\Delta)^{-\frac{1}{2}}\mathrm{div} M_B \nabla(\mathrm{id}_{L^p} - \mu\Delta)^{-\frac{1}{2}} \in B(L^p(\mathbb{R}^n))$$

has a norm less than 1 and $\mathrm{id}_{L^p} + L$ is invertible on $L^p(\mathbb{R}^n)$ as long as p is sufficiently close to 2. Thus we have the following conclusion:

Theorem 5.38 *The operator $\iota + L$ maps $W^{1,p}(\mathbb{R}^n)$ to $W^{-1,p}(\mathbb{R}^n)$ isomorphically as long as p is sufficiently close to 2. Here ι denotes the natural inclusion.*

Theorem 5.38 together with the Gaffney estimate yields the following property of semi-groups.

Proposition 5.14 *Let $r \in (1,2]$ be such that $\iota + L$ maps $W^{1,r}(\mathbb{R}^n)$ to $W^{-1,r}(\mathbb{R}^n)$ isomorphically.*

1. *If $\dfrac{nr}{n+r} \le p \le 2$, then e^{-tL} maps $L^p(\mathbb{R}^n)$ into $L^2(\mathbb{R}^n)$ boundedly.*
2. *If $\dfrac{nr}{n+r} < p \le 2$, then e^{-tL} maps $L^p(\mathbb{R}^n)$ into $L^p(\mathbb{R}^n)$ boundedly. Furthermore, it also satisfies the off-diagonal estimate.*
3. *If $\dfrac{nr}{n+r} < p \le 2$, then e^{-tL} maps $L^p(\mathbb{R}^n)$ into $L^2(\mathbb{R}^n)$ boundedly. Furthermore, it also satisfies the off-diagonal estimate.*

Before the proof a couple of remarks may be in order.

Remark 5.5

1. According to Theorem 5.38, such an r as in Proposition 5.14 exists.
2. Let $r = 2$ in Proposition 5.14. A direct corollary of Theorem 5.38 is that e^{-tL} maps $L^p(\mathbb{R}^n)$ into $L^2(\mathbb{R}^n)$ boundedly together with the off-diagonal estimate as long as $p_0 \equiv \dfrac{2n}{n+2} \le p \le 2$.

Proof (of Proposition 5.14) We may assume that $t = 1$ using the dilation argument.

1. By the Sobolev embedding, we have $L^p(\mathbb{R}^n) \hookrightarrow W^{-1,r}(\mathbb{R}^n)$. Let $r_0 = r$ and define r_{m+1} inductively by

$$\frac{1}{r_{m+1}} + \frac{1}{n} = \frac{1}{r_m} - \frac{1}{n}$$

if $r_m < 2$, so that $W^{1,r_m}(\mathbb{R}^n)$ is embedded into $W^{-1,r_{m+1}}(\mathbb{R}^n)$. If $r_m > 2$, we stop and let $M \equiv m$. Thus, $(\iota + L)^{-M+1}$ maps $L^p(\mathbb{R}^n)$ into $W^{1,r_{M-1}}(\mathbb{R}^n)$ and $(\iota + L)^{-M}$ maps $L^p(\mathbb{R}^n)$ into $L^2(\mathbb{R}^n)$ boundedly. Thus, $e^{-L} = (\iota + L)^M e^{-L}(\iota + L)^{-M}$ maps $L^p(\mathbb{R}^n)$ into $L^2(\mathbb{R}^n)$ boundedly.
2. Interpolating between the $L^{\frac{nr}{n+r}}(\mathbb{R}^n)$-$L^2(\mathbb{R}^n)$-boundedness and the Gaffney estimates, Lemma 5.29, we have $\|\chi_{Q_{0m}} e^{-L}[\chi_{Q_{0m'}} f]\|_p \lesssim \exp(-a|m - m'|)\|f\|_2$ for $f \in L^2(\mathbb{R}^n)$ and $m, m' \in \mathbb{Z}^n$. Thus, $\|\{\|e^{-L} f\|_{L^2(Q_{0m})}\}_{m\in\mathbb{Z}^n}\|_{\ell^p} \lesssim \|f\|_p$. By the Hölder inequality we conclude that e^{-L} maps $L^p(\mathbb{R}^n)$ into itself boundedly.
3. Mimic the proof above.

Remark 5.6 Since the above estimates depend solely on λ and Λ, a dilation argument yields $\|\mathrm{e}^{-tL}\|_{B(L^q,L^2)} \lesssim t^{\frac{n}{2}-\frac{n}{q}}$. Arguing similarly, one notices that the Gaffney estimate carries over to the boundedness from $L^p(\mathbb{R}^n)$ to $L^2(\mathbb{R}^n)$, which yields the $L^p(\mathbb{R}^n)$-$L^2(\mathbb{R}^n)$ off-diagonal estimate.

Exercises

Exercise 5.56 Let $1 < p, q < \infty$, $s > 0$, and let L be an elliptic differential operator satisfying the conditions in Theorem 5.36, where we regard $L : F^s_{pq}(\mathbb{R}^n) \to F^s_{pq}(\mathbb{R}^n)$ as an unbounded linear operator. Show that L is a closed operator with $D(L) = F^{s+2}_{pq}(\mathbb{R}^n)$ using Theorem 5.36.

Exercise 5.57 Let $L = -\sum_{i,j=1}^{n} a_{ij}\partial_{x_i}\partial_{x_j}$ be an elliptic differential operator with constant coefficient. Suppose that $\{a_{ij}\}_{i,j=1}^n$ is positive definite. Define

$$\Gamma(x) \equiv \frac{1}{(n-2)\omega_n\sqrt{\det(\{a_{ij}\}_{i,j=1}^n)}}\left(\sum_{j,k=1}^{n} a_{jk}x_i x_k\right)^{\frac{2-n}{2}} \quad (x \in \mathbb{R}^n).$$

1. Let $f \in C^\infty_c(\mathbb{R}^n)$. Show that $L_0(\Gamma * f)(x) = f(x)$ for all $x \in \mathbb{R}^n$ using the Stokes theorem.
2. Let $1 \le k, l \le n$. Show that $Tf \equiv \partial_{x_k}\partial_l \Gamma * f$ is a Calderón–Zygmund operator.

Exercise 5.58 [652] This exercise motivates us to introduce the function spaces dealt with in this book. Choose $\psi \in C^\infty(\mathbb{R}^2)$ so that $\chi_{B(1)} \le \psi \le \chi_{B(2)}$. Let $f(x, y) \equiv x\psi(x, y)\log(x^2 + y^2)$ for $x, y \in \mathbb{R}^n$. Show that $f \notin C^2(\mathbb{R}^n)$ but that $\Delta f \in C(\mathbb{R}^n)$.

Exercise 5.59 Let Ω be a bounded domain.

1. Let $1 < p, q < \infty$, $s > 0$, and let $L : A^{s+2}_{pq}(\Omega) \to A^s_{pq}(\Omega)$ be an injective bounded linear operator. (Forget about the structure of L here.) Assume in addition that the estimate $\|u\|_{A^{s+2}_{pq}(\Omega)} \le C_0(\|Lu\|_{A^s_{pq}(\Omega)} + \|u\|_{A^0_{pq}(\Omega)})$ holds with some constant $C_0 > 0$ for all $u \in A^{s+2}_{pq}(\Omega)$. Let us show that there exists $C_1 > 0$ such that $\|u\|_{A^{s+2}_{pq}(\Omega)} \le C_1\|Lu\|_{A^s_{pq}(\Omega)}$ for all $u \in A^{s+2}_{pq}(\Omega)$ supported in a fixed bounded set U. To prove it by reduction to absurdity, we suppose that for any $m > 0$ there exists $u_m \in A^{s+2}_{pq}(\Omega)$ such that $m\|Lu_m\|_{A^s_{pq}(\Omega)} < \|u_m\|_{A^{s+2}_{pq}(\Omega)} = 1$.
   (a) Look for a function space whose dual is $A^{s+2}_{pq}(\Omega)$.
   (b) Use the compactness of the embedding, which we can assume $\{u_m\}_{m=1}^\infty$ convergent to some u in $A^{s+2}_{pq}(\Omega)$.
   (c) Show that $\|u\|_{A^0_{pq}(\Omega)} \ge {C_0}^{-1} > 0$ using the compactness of the embedding and show also that $Lu = 0$.
   (d) Deduce a contradiction.
2. Let X, Y, Z be Banach spaces such that X is embedded compactly into Z, and let $A \in B(X, Y)$ satisfy $\|x\|_X \le \|Ax\|_Y + \|x\|_Z$ for all $x \in X$. Then show that there exists a constant $C > 0$ such that $\|x\|_X \le C\|Ax\|_Y$ for all $x \in X$.

Exercise 5.60 Let X and Y be Banach spaces. Suppose that we have a bounded linear operator $A(t) : X \to Y$ for each $t \in [0, 1]$ such that $\|x\|_A \le \|A(t)x\|_Y$ for all $x \in X$. Assume in addition that $A : t \in [0, 1] \to A(t) \in B(X, Y)$ is continuous and that $A(0)$ is bijective.

1. Set $I \equiv \{t \in [0, 1] : A(t)$ is bijective $\}$, so that $0 \in I$. Show that I is an open set using the Neumann expansion.
2. Show that I is a closed set.
3. Show that $A(1)$ is bijective. Hint: $[0, 1]$ is connected.

Exercise 5.61 Let E, F be disjoint closed sets in $\mathbb{R}^n$ and $d > \mathrm{dist}(E, F)$. Then show that there exists a smooth function ζ satisfying (5.126). Hint: We may assume that E and F are unions of cubes having sidelength $(3n)^{-1}d$. We can use the partition of unity.

Textbooks for Sect. 5.4.1

Elliptic Differential Equations

See [37, 42] for fundamental facts. See [5] for Theorem 5.38.

5.4.2 A Priori Estimate in the Half Space

In this section, we consider a priori estimates of the elliptic differential equations on the half space $\mathbb{R}^n_+ = \{x \in \mathbb{R}^n : x_n > 0\}$.

5.4.2.1 Case of the Laplacian

First of all, we consider the fundamental solution of $1 - \Delta$ and then investigate its properties. For, in the nonhomogeneous space, $1 - \Delta$ is much easier to handle than $-\Delta$.

Proposition 5.15 *Let $1 < p < \infty$. For $f \in L^p(\mathbb{R}^n)$, the solution $u \in W^{2,p}(\mathbb{R}^n)$ to $(1 - \Delta)u = f$ can be expressed:*

$$u(x) = (4\pi)^{-\frac{n}{2}} \int_0^\infty t^{-\frac{n}{2}} \left(\int_{\mathbb{R}^n} \exp\left(-t - \frac{|x-y|^2}{4t}\right) f(y)\mathrm{d}y \right) \mathrm{d}t. \tag{5.135}$$

In (5.135), the right-hand side is given by the convolution given by the integral kernel $k(x) \equiv (4\pi)^{-\frac{n}{2}} \int_0^\infty t^{-\frac{n}{2}} \exp\left(-t - \frac{|x|^2}{4t}\right) \mathrm{d}t$.

Proof Write $v(x)$ for the right-hand side of (5.135). By virtue of the Young inequality, we have

$$\|v\|_p \le \|f\|_p; \tag{5.136}$$

hence the expression (5.135) is clearly $L^p(\mathbb{R}^n)$-convergent as a function of x. Here and below we suppose that $f \in L^1(\mathbb{R}^n) \cap L^\infty(\mathbb{R}^n)$ by density and (5.136).

Let us show that equality (5.135) makes sense as elements in $\mathscr{S}'(\mathbb{R}^n)$. Choose a test function $\tau \in \mathscr{S}(\mathbb{R}^n)$. If we use the definition of v, Then

$$\begin{aligned} &\langle v, \tau \rangle \\ &= (4\pi)^{-\frac{n}{2}} \int_{\mathbb{R}^n} f(y) \int_0^\infty t^{-\frac{n}{2}} \left\{ \left(\int_{\mathbb{R}^n} \exp\left(-t - \frac{|x-y|^2}{4t}\right) \tau(x)\mathrm{d}x \right) \mathrm{d}t \right\} \mathrm{d}y. \end{aligned} \tag{5.137}$$

See Exercise 5.62. We can check

$$\int_{\mathbb{R}^n} \exp\left(-\frac{|x|^2}{4t} - i\, x \cdot \xi\right) dx = (4\pi t)^{\frac{n}{2}} \exp(-t|\xi|^2). \tag{5.138}$$

We leave the matter of checking (5.138) using the complex line integral; see Exercise 5.63. Thus, we obtain

$$\langle v, \tau \rangle = (2\pi)^{-\frac{n}{2}} \int_{\mathbb{R}^n} f(y) \int_0^\infty \left\{ \left(\int_{\mathbb{R}^n} \exp(-t - t|\xi|^2 + i\, y\, \xi) \mathscr{F}\tau(\xi) d\xi \right) dt \right\} dy$$

by Plancherel's theorem.

We change the order of integrals and then we integrate this against $t \in (0, \infty)$. Then

$$\begin{aligned}\langle v, \tau \rangle &= (2\pi)^{-\frac{n}{2}} \int_{\mathbb{R}^n} f(y) \left\{ \int_{\mathbb{R}^n} \left(\int_0^\infty \exp(-t - t|\xi|^2 + i\, y\, \xi) \mathscr{F}\tau(\xi) dt \right) d\xi \right\} dy \\ &= (2\pi)^{-\frac{n}{2}} \int_{\mathbb{R}^n} f(y) \left(\int_{\mathbb{R}^n} \frac{1}{1 + |\xi|^2} \exp(i\, y\, \xi) \mathscr{F}\tau(\xi) d\xi \right) dy. \end{aligned} \tag{5.139}$$

See Exercise 5.62. By the inverse Fourier transform, we have

$$\langle v, \tau \rangle = \langle f, (1 - \Delta)^{-1} \tau \rangle = \langle (1 - \Delta)^{-1} f, \tau \rangle = \langle u, \tau \rangle.$$

This shows (5.135).

We now extend functions in $\mathbb{R}^{n-1}$ to the functions in $\mathbb{R}^n$. More precisely, we will solve a differential equation $(1 - \Delta) f = 0, \quad \mathrm{Tr}_{\mathbb{R}^n_+} f = f_0$ on $\mathbb{R}^n_+$. Let $E : f_0 \mapsto f$ be the solution operator, which is also called the extension operator.

We collect some properties of E.

Proposition 5.16 *Let $1 < p < \infty$. For $f_0 \in \mathscr{S}_\infty(\mathbb{R}^{n-1})$, define a Borel measurable function $E f_0$ over $\mathbb{R}^n_+$ by*

$$E f_0(x', x_n) \equiv \frac{1}{(2\pi)^n} \int_{\mathbb{R}^{n-1}} f_0(x' - y') \left(\int_{\mathbb{R}^{n-1}} \exp(-x_n \sqrt{1 + |\xi'|^2} - i\xi' \cdot y') d\xi' \right) dy'$$

for $(x', x_n) \in \mathbb{R}^n_+$. Choose $N \gg 1$ and $\lambda_1, \lambda_2, \ldots, \lambda_{N+1}$ so that

$$\sum_{j=1}^{N+1} \lambda_j (-j)^k = 1 \quad (k = 0, 1, \ldots, N). \tag{5.140}$$

Extend further $E f_0$ to $\mathbb{R}^n$ by

$$Ef_0(x) \equiv \begin{cases} Ef_0(x) & x \in \mathbb{R}^n_+, \\ f_0(x') & x \in \partial\mathbb{R}^n_+, \\ \sum_{j=1}^{N+1} \lambda_j Ef_0(x', -j\, x_n) & x \in \mathbb{R}^n_-. \end{cases} \tag{5.141}$$

Then $Ef_0 \in \mathscr{S}'(\mathbb{R}^n)$ for all $f_0 \in \mathscr{S}_\infty(\mathbb{R}^{n-1})$. More precisely $Ef_0 \in L^p(\mathbb{R}^n)$ and

$$\|Ef_0\|_{L^p(\mathbb{R}^n)} \lesssim \|f_0\|_{L^p(\mathbb{R}^{n-1})} \tag{5.142}$$

with the constant independent of f_0. In particular, if $s > 1/p$, then E extends to a continuous linear mapping from $F^{s-1/p}_{pp}(\mathbb{R}^{n-1})$ to $L^p(\mathbb{R}^n)$.

Proof From the definition (5.141) we have only to consider the integral over $\mathbb{R}^n_+$ to prove (5.142). Let $m_a(\xi') \equiv \exp(-\sqrt{a^2+|\xi'|^2})$ for $a > 0$. As is verified by induction, $\partial^\alpha m_a(\xi)$ takes the form

$$\sum_{k,l \in \mathbb{N}_0, \beta \in \mathbb{N}_0{}^n} \chi_{[k+|\beta|, k+|\beta|+|\alpha|]}(l) \frac{c_{\alpha;\beta,k,l}\, a^k \xi'^\beta}{(\sqrt{a+|\xi'|^2})^l} \exp(-\sqrt{a+|\xi'|^2}),$$

which is a finite sum. Hence

$$|\partial^{\alpha'} m_a(\xi')| \lesssim_{\alpha'} |\xi'|^{-|\alpha'|} \quad (\alpha' \in \mathbb{N}_0{}^{n-1})$$

with the constant independent of a. Thus, we have

$$|\partial^{\alpha'} \exp(-|x_n|\langle \xi' \rangle)| = |x_n|^{|\alpha'|} |\partial^{\alpha'} m_1(|x_n|\xi')| \lesssim_{\alpha'} |x_n|^{|\alpha'|} |\xi'|^{-|\alpha'|}$$

for all $\alpha' \in \mathbb{N}_0{}^{n-1}$ with the constant independent of $x_n \in (-1, 1)$.

Since $L^p(\mathbb{R}^{n-1}) \approx \dot{F}^0_{p2}(\mathbb{R}^{n-1})$, we have

$$\begin{aligned} \|Ef_0\|_{L^p(\mathbb{R}^{n-1}_{x'} \times (-1,1))} &= \| \, \|Ef_0\|_{L^p(\mathbb{R}^{n-1}_{x'})} \|_{L^p_{x_n}(-1,1)} \\ &\sim \| \, \|Ef_0\|_{\dot{F}^0_{p2}(\mathbb{R}^{n-1}_{x'})} \|_{L^p_{x_n}(-1,1)} \\ &\lesssim \|f_0\|_p \end{aligned}$$

by Theorem 1.53. When $|x_n| > 1$, the integral kernel decays fast; since

$$\|Ef_0(\star', x_n)\|_{L^p(\mathbb{R}^{n-1})} \lesssim \exp(-|x_n|/2)\|f_0\|_{L^p(\mathbb{R}^{n-1})} \quad (x_n \in \mathbb{R}^n \setminus B(1)),$$

we can integrate against x_n to conclude (5.142).

Now we solve the differential equation $(1-\Delta)f=0$ subject to the boundary condition $\mathrm{Tr}_{\mathbb{R}^n_+} f = g$ on the upper half space, keeping in mind that trace operator $\mathrm{Tr}_{\mathbb{R}^n} : W^{2+\varepsilon,p}(\mathbb{R}^n) \to B_{pp}^{2+\varepsilon-\frac{1}{p}}(\mathbb{R}^{n-1})$ is bounded for $\varepsilon > 0$.

Proposition 5.17 *Let* $1 < p < \infty$ *and* $\varepsilon > 0$, *and let* $f \in W^{2+\varepsilon,p}(\mathbb{R}^n)$. *If* $(1-\Delta)f|\mathbb{R}^n_+ = 0$, *then* $f = E[\mathrm{Tr}_{\mathbb{R}^n} f]$ *on* $\mathbb{R}^n_+$.

Proof Since the mappings

$$f \in W^{2+\varepsilon,p}(\mathbb{R}^n) \mapsto f|\mathbb{R}^n_+ \in L^{2+\varepsilon}(\mathbb{R}^n_+)$$

and

$$f \in W^{2+\varepsilon,p}(\mathbb{R}^n) \mapsto E[\mathrm{Tr}_{\mathbb{R}^n} f] \in L^{2+\varepsilon}(\mathbb{R}^n_+)$$

are bounded, we can assume that $f \in C^\infty_c(\mathbb{R}^n)$ by density. Set $g \equiv f - E[\mathrm{Tr}_{\mathbb{R}^n} f] \in L^{2+\varepsilon}(\mathbb{R}^n_+)$. Note that g is bounded since f is bounded. Observe that $(1-\Delta)g = 0$ on $\mathbb{R}^n_+$ and the boundary value $g(\star', 0_n)$ vanishes. Thus by Lemma 1.40 $g = 0$ and hence $f = E[\mathrm{Tr}_{\mathbb{R}^n} f]$ on $\mathbb{R}^n_+$.

Lemma 5.31 *Let* $1 < p, q < \infty$ *and* $s > 1/p$. *Then* $\|Ef_0\|_{F^s_{pq}(\mathbb{R}^n)} \lesssim \|f_0\|_{F^{s-1/p}_{pp}(\mathbb{R}^{n-1})}$ *for all* $f_0 \in F^{s-1/p}_{pp}(\mathbb{R}^{n-1})$.

Proof Since $\mathscr{S}_\infty(\mathbb{R}^{n-1})$ is dense in $F^{s-1/p}_{pp}(\mathbb{R}^{n-1})$ thanks to the Littlewood–Paley theory $L^p(\mathbb{R}^n) \approx F^0_{p2}(\mathbb{R}^{n-1})$, it can be assumed that $f \in \mathscr{S}_\infty(\mathbb{R}^{n-1})$. Let $k \in \mathrm{BC}(\mathbb{R})$ be a function given by

$$k(x_n) \equiv \chi_{[0,\infty)}(x_n)\exp(-x_n) + \sum_{j=1}^{N+1} \lambda_j \chi_{(-\infty,0)}(x_n)\exp(j\,x_n)$$

and let $m : \mathbb{R}^n \to \mathbb{C}$ be given by

$$\begin{aligned}
m(\xi) &\equiv \int_0^\infty \exp(-x_n\sqrt{|\xi'|^2+1} - ix_n\xi)\mathrm{d}x_n \\
&\quad + \sum_{j=1}^{N+1} \lambda_j \int_{-\infty}^0 \exp(j\,x_n\sqrt{|\xi'|^2+1} - ix_n\xi)\mathrm{d}x_n \\
&= \frac{1}{\sqrt{|\xi'|^2+1}} \int_{\mathbb{R}} k(x_n)\exp\left(-\frac{ix_n\,\xi_n}{\sqrt{|\xi|^2+1}}\right)\mathrm{d}x_n \\
&= \frac{\sqrt{2\pi}}{\sqrt{|\xi'|^2+1}} \mathscr{F}k\left(\frac{\xi_n}{\sqrt{|\xi|^2+1}}\right),
\end{aligned}$$

where $\lambda_1, \ldots, \lambda_N$ satisfies (5.140). A direct calculation shows

$$\mathscr{F} E f_0(\xi) = m(\xi)\mathscr{F}' f_0(\xi').$$

By (5.140), the partial derivative of k is compatible at $\mathbb{R}^{n-1} \times \{0\}$. Since $k \in C^N(\mathbb{R})$ satisfies the differential inequalities $|k^{(l)}(x_n)| \lesssim_N e^{-|x_n|}$ for $l = 0, 1, \ldots, N$, assuming N large enough, we have

$$|\mathscr{F} k^{\alpha_n}(\xi_n)| \lesssim_{\alpha_n} \min(1, |\xi_n|^{-N}) \quad (\alpha_n \in \mathbb{N}_0).$$

Hence

$$|\partial^\alpha m(\xi)| = |\partial^{(\alpha',0)}\partial^{(0',\alpha_n)} m(\xi)| \lesssim_\alpha \frac{\min(1, (1+|\xi'|^2)^{\frac{N}{2}}|\xi_n|^{-N})}{(1+|\xi'|^2)^{\frac{1}{2}}} \tag{5.143}$$

for all $\alpha \in {\mathbb{N}_0}^n$.

To describe the norms of Triebel–Lizorkin spaces in $\mathbb{R}^n$ and $\mathbb{R}^{n-1}$, choose $\kappa \in \mathscr{S}(\mathbb{R})$ so that $\chi_{Q(1)} \le \kappa \le \chi_{Q(2)}$. Using this κ, we define

$$\psi \equiv \overbrace{\kappa \otimes \ldots \otimes \kappa}^{n \text{ times}} \in \mathscr{S}(\mathbb{R}^n), \quad \tilde{\alpha} \equiv \overbrace{\kappa \otimes \ldots \otimes \kappa}^{n-1 \text{ times}} \in \mathscr{S}(\mathbb{R}^{n-1}).$$

Furthermore, set

$$\tau \equiv \kappa - \kappa_{-1} \in \mathscr{S}(\mathbb{R}), \quad \varphi \equiv \psi - \psi_{-1} \in \mathscr{S}(\mathbb{R}^n), \quad \tilde{\beta} \equiv \tilde{\alpha} - \tilde{\alpha}_{-1} \in \mathscr{S}(\mathbb{R}^{n-1}).$$

Let $j \ge 10$. In view of the size of supports, we have

$$\begin{aligned} \mathscr{F}[\varphi_j(D)E[\tilde{\alpha}(D)f_0]] &= \varphi_j \cdot m \cdot \tilde{\alpha}(\star')\mathscr{F}' f_0(\star') \\ &= \tau_{j-1}(\star_n) m \cdot \tilde{\alpha}(\star')\mathscr{F}' f_0(\star'). \end{aligned} \tag{5.144}$$

Let $\sigma > 0$ and $N \gg 1$ be sufficiently large. Then

$$\begin{aligned} &\left\| \{2^{js}\varphi_j(D)E[\tilde{\alpha}(D)f_0]\}_{j=10} \right\|_{L^p(\ell^q)} \\ &\lesssim \left\| \{2^{-j\sigma}\mathscr{F}^{-1}[\tau_{j-1}(\star_n)m \cdot \tilde{\alpha}(\star')\mathscr{F}' f_0(\star')]\}_{j=10}^\infty \right\|_{L^p(\ell^q)} \\ &\lesssim \left\| \{2^{-j\sigma}\mathscr{F}^{-1}[\tau_{j-1}(\xi_n)\tilde{\alpha}(\xi')\mathscr{F}' f_0(\xi')]\}_{j=10}^\infty \right\|_{L^p(\ell^q)} \\ &= \left\| \{2^{-j\sigma}\mathscr{F}^{-1}[\tau_{j-1}] \otimes \tilde{\alpha}(D) f_0\}_{j=10}^\infty \right\|_{L^p(\ell^q)} \\ &\lesssim \|\tilde{\alpha}(D)f_0\|_p \le \|f\|_{F_{pp}^{s-1/p}(\mathbb{R}^{n-1})} \end{aligned}$$

by Theorem 1.53 and (5.144). Likewise, we can prove

$$\|\psi(D)E[\alpha(D)f_0]\|_p + \sum_{j=1}^{10} \|\varphi_j(D)E[\alpha(D)f_0]\|_p \lesssim \|f_0\|_{F_{pp}^{s-1/p}(\mathbb{R}^{n-1})}.$$

We have only to deal with $E[f_0 - \tilde{\alpha}(D)f_0]$.

Furthermore, since

$$\varphi = \tilde{\alpha} \otimes \kappa - \tilde{\alpha}_{-1} \otimes \kappa_{-1} = (\tilde{\alpha} - \tilde{\alpha}_{-1}) \otimes \kappa - \tilde{\alpha}_{-1} \otimes (\kappa - \kappa_{-1}) = \tilde{\beta} \otimes \kappa - \tilde{\alpha}_{-1} \otimes \tau,$$

it suffices to treat the following sums:

$$\mathrm{I} \equiv \|2^{js}\tilde{\alpha}_{j-1} \otimes \tau_j(D)E[f_0 - \tilde{\alpha}(D)f_0]\|_{L^p(\ell^q)},$$
$$\mathrm{II} \equiv \|2^{js}\tilde{\beta}_j \otimes \kappa_j(D)E[f_0 - \tilde{\alpha}(D)f_0]\|_{L^p(\ell^q)}.$$

First of all, we deal with I. Choose $\sigma > 0$ and $N \gg 1$ sufficiently large. Then

$$\|2^{js}\tilde{\alpha}_{j-1} \otimes \tau_j(D)E[f_0 - \tilde{\alpha}(D)f_0]\|_p \le 2^{-j\sigma}\|\tilde{\alpha}_{j-1}(D)f_0\|_p$$

by Theorem 1.53.

Hence $\mathrm{I} \le \|f_0\|_{F_{pp}^{s-1/p}(\mathbb{R}^{n-1})}$.

Term II is a quantity that needs to be dealt with most seriously. If we use Theorem 1.53, then

$$\mathrm{II} \lesssim \|2^{(s-1)j}\tilde{\beta}_j(D)f \otimes \mathscr{F}_{x_n}^{-1}[\kappa_j](D)\|_{L^p(\ell^q)}.$$

By the use of the Hardy–Littlewood maximal operator $\mathrm{II} \lesssim \|2^{js}\tilde{\beta}_j(D)f \otimes M[\chi_{(2^{-j},2^{-j+1})}]\|_{L^p(\ell^q)}$. By Fubini's theorem (Theorem 1.3) and the boundedness of the Hardy–Littlewood maximal operator, we deduce

$$\begin{aligned}
\mathrm{II} &\lesssim \left\| \|2^{js}\tilde{\beta}_j(D)f \otimes M[\chi_{(2^{-j},2^{-j+1})}]\|_{L^p_{x_n}(\ell^q)} \right\|_{L^p_{x'}(\mathbb{R}^{n-1})} \\
&\lesssim \left\| \|2^{js}\tilde{\beta}_j(D)f \otimes \chi_{(2^{-j},2^{-j+1})}\|_{L^p_{x_n}(\ell^q)} \right\|_{L^p_{x'}(\mathbb{R}^{n-1})} \\
&= \left\| 2^{j(s-1/p)}\tilde{\beta}_j(D)f \right\|_{L^p_{x'}(\ell^q)}.
\end{aligned}$$

Thus, $\|Ef_0\|_{F^s_{pq}(\mathbb{R}^n)} \lesssim \|f_0\|_{F_{pp}^{s-1/p}(\mathbb{R}^{n-1})}$, as was to be shown.

With these estimates in mind, we prove the a priori estimate of the constant coefficient elliptic differential operators.

Theorem 5.39 (A priori estimate of constant coefficient elliptic differential operator on the half space) *Let* $1 < p, q < \infty$, $s > \frac{1}{p}$. *Then*

$$\|f\|_{F^{s+2}_{pq}(\mathbb{R}^n_+)} \sim \|(1-\Delta)f\|_{F^s_{pq}(\mathbb{R}^n_+)} + \|\mathrm{Tr}_{\mathbb{R}^n} f\|_{F^{s+2-\frac{1}{p}}_{pp}(\mathbb{R}^{n-1})} \tag{5.145}$$

for $f \in F^{s+2}_{pq}(\mathbb{R}^n_+)$. *In particular,*

$$\|f\|_{F^{s+2}_{pq}(\mathbb{R}^n_+)} \sim \|\Delta f\|_{F^s_{pq}(\mathbb{R}^n_+)} + \|f\|_{F^s_{pq}(\mathbb{R}^n_+)} + \|\mathrm{Tr}_{\mathbb{R}^n} f\|_{F^{s+2-\frac{1}{p}}_{pp}(\mathbb{R}^{n-1})} \tag{5.146}$$

for $f \in F^{s+2}_{pq}(\mathbb{R}^n_+)$.

Proof Since $\gtrsim$ follows trivially, we prove $\lesssim$. We choose $G \in F^s_{pq}(\mathbb{R}^n)$ so that

$$(1-\Delta)f = G|\mathbb{R}^n_+, \quad \|(1-\Delta)f\|_{F^s_{pq}(\mathbb{R}^n_+)} \le \|G\|_{F^s_{pq}(\mathbb{R}^n)} \le 2\|(1-\Delta)f\|_{F^s_{pq}(\mathbb{R}^n_+)}.$$

Set $g \equiv (1-\Delta)^{-1}G \in F^{s+2}_{pq}(\mathbb{R}^n)$. Then $\|g|\mathbb{R}^n_+\|_{F^{s+2}_{pq}(\mathbb{R}^n_+)} \le \|g\|_{F^{s+2}_{pq}(\mathbb{R}^n)} \sim \|(1-\Delta)g\|_{F^s_{pq}(\mathbb{R}^n)}$; hence

$$\|g|\mathbb{R}^n_+\|_{F^{s+2}_{pq}(\mathbb{R}^n_+)} \lesssim \|G\|_{F^s_{pq}(\mathbb{R}^n)} \sim \|(1-\Delta)f\|_{F^{s+2}_{pq}(\mathbb{R}^n_+)}. \tag{5.147}$$

Let $h \equiv f - g|\mathbb{R}^n_+ \in F^{s+2}_{pq}(\mathbb{R}^n_+)$. Then since

$$(1-\Delta)h = (1-\Delta)f - (1-\Delta)g|\mathbb{R}^n_+,$$

we have $(1-\Delta)h = G|\mathbb{R}^n_+ - (1-\Delta)g|\mathbb{R}^n_+ = (G-(1-\Delta)g)|\mathbb{R}^n_+ = 0$ on $\mathbb{R}^n_+$. By Propositions 5.16 and 5.17, $\|h\|_{F^{s+2}_{pq}(\mathbb{R}^n_+)} \lesssim \|\mathrm{Tr}_{\mathbb{R}^n_+} h\|_{F^{s+2-\frac{1}{p}}_{pp}(\mathbb{R}^{n-1})}$. Furthermore, by the relation $h = f - g|\mathbb{R}^n_+$ and by the quasi-triangle inequality,

$$\|h\|_{F^{s+2}_{pq}(\mathbb{R}^n_+)} \lesssim \|\mathrm{Tr}_{\mathbb{R}^n_+} f\|_{F^{s+2-\frac{1}{p}}_{pp}(\mathbb{R}^{n-1})} + \|\mathrm{Tr}_{\mathbb{R}^n_+}(g|\mathbb{R}^n_+)\|_{F^{s+2-\frac{1}{p}}_{pp}(\mathbb{R}^{n-1})}.$$

In view of the continuity of $\mathrm{Tr}_{\mathbb{R}^n_+} : F^{s+2}_{pq}(\mathbb{R}^n_+) \to F^{s+2-\frac{1}{p}}_{pp}(\mathbb{R}^{n-1})$ and (5.147), we have

$$\|h\|_{F^{s+2}_{pq}(\mathbb{R}^n_+)} \lesssim \|\mathrm{Tr}_{\mathbb{R}^n} f\|_{F^{s+2-\frac{1}{p}}_{pp}(\mathbb{R}^{n-1})} + \|(1-\Delta)f\|_{F^s_{pq}(\mathbb{R}^n_+)}.$$

Thus, (5.145) follows.

5.4.2.2 Elliptic Operator of the Second Order with Variable Coefficients

Let $1 < p, q < \infty$, $s > 0$. We consider the elliptic linear operator

$$Lf \equiv -\sum_{i,j=1}^{n} a_{ij}\partial_{x_i}\partial_{x_j} f + \sum_{k=1}^{n} b_k \partial_{x_k} f + c f$$

in $A^s_{pq}(\mathbb{R}^n_+)$. For simplicity, we suppose a_{ij}, b_k, c satisfy

$$a_{ij} \in \mathscr{C}^{s+2}(\mathbb{R}^n), \quad b_k \in \mathscr{C}^{s+2}(\mathbb{R}^n), \quad c \in \mathscr{C}^{s+2}(\mathbb{R}^n) \quad (i, j, k = 1, 2, \dots, n) \tag{5.148}$$

and the uniformly elliptic condition

$$\Theta^{-1}|\xi|^2 \le \sum_{i,j=1}^{n} a_{ij}(x)\xi_i\xi_j \le \Theta|\xi|^2 \quad (\xi = (\xi_1, \xi_2, \dots, \xi_n) \in \mathbb{R}^n, x \in \mathbb{R}^n) \tag{5.149}$$

with $\Theta > 1$. The constant Θ is called the uniformly bounded constant.

Proposition 5.18 *Let $1 < p, q < \infty$, $s > 0$ and suppose that the constant coefficients $\{a_{ij}\}_{i,j=1,2,\dots,n}$ satisfy (5.148) and (5.149). Let $\mathrm{Tr}_{\mathbb{R}^n_+} : F^{s+2}_{pq}(\mathbb{R}^n_+) \to F^{s+2-\frac{1}{p}}_{pp}(\mathbb{R}^{n-1})$ denote the trace operator. Then*

$$\|f\|_{F^{s+2}_{pq}(\mathbb{R}^n_+)} \sim \|Lf\|_{F^s_{pq}(\mathbb{R}^n_+)} + \|f\|_{F^s_{pq}(\mathbb{R}^n_+)} + \|\mathrm{Tr}_{\mathbb{R}^n_+} f\|_{F^{s+2-\frac{1}{p}}_{pp}(\mathbb{R}^{n-1})}$$

for all $f \in F^{s+2}_{pq}(\mathbb{R}^n_+)$.

Proof We note that

$$x \mapsto (\lambda_1 x_1, \lambda_2 x_2, \dots, \lambda_{n-1} x_{n-1}, \lambda_n x_n),$$
$$x \mapsto (x_1 + \rho_1 x_n, x_2 + \rho_2 x_n, \dots, x_{n-1} + \rho_{n-1} x_n, x_n),$$
$$x \mapsto \left(\sum_{k=1}^{n-1} p_{1k} x_k, \sum_{k=1}^{n-1} p_{2k} x_k, \dots, \sum_{k=1}^{n-1} p_{n-1\,k} x_k, x_n\right)$$

preserve $\mathbb{R}^n_+$. We now suppose

$$\lambda_1, \lambda_2, \dots, \lambda_n > 0, \quad \rho_1, \rho_2, \dots, \rho_{n-1} \in \mathbb{R}$$

and $\{p_{ij}\}_{i,j=1,2,\dots,n-1} \in \mathrm{GL}(\mathbb{R}; n-1)$, the set of all invertible of $(n-1) \times (n-1)$ matrices. Starting from the identity mapping, we can transform it into $\{a_{ij}\}_{i,j=1,2,\dots,n} = P^*P$ by a finite number of compositions. Hence we can consider f instead of $f \circ P^{-1}$ and matters are reduced to (5.146). Thus, the proof is complete.

Theorem 5.40 (A priori estimate of elliptic differential operators over half space) *Let* $1 < p, q < \infty$, $s > 0$. *Suppose that the coefficient satisfies* (5.148) *and* (5.149). *Then* $\|f\|_{F^{s+2}_{pq}(\mathbb{R}^n_+)} \sim \|Lf\|_{F^s_{pq}(\mathbb{R}^n_+)} + \|f\|_{F^{s+1}_{pq}(\mathbb{R}^n_+)} + \|\mathrm{Tr}_{\mathbb{R}^n_+} f\|_{F^{s+2-\frac{1}{p}}_{pp}(\mathbb{R}^{n-1})}$ *for all* $f \in F^{s+2}_{pq}(\mathbb{R}^n_+)$. *Here* $\mathrm{Tr}_{\mathbb{R}^n_+} : F^{s+2}_{pq}(\mathbb{R}^n_+) \to F^{s+2-\frac{1}{p}}_{pp}(\mathbb{R}^{n-1})$ *denotes the trace operator.*

As we did in $\mathbb{R}^n$, we can use the coefficient-freezing method and we omit the proof.

Exercises

Exercise 5.62

1. Prove (5.137) using $f \in L^1(\mathbb{R}^n) \cap L^\infty(\mathbb{R}^n)$.
2. Prove (5.139) using $\tau \in \mathscr{S}(\mathbb{R}^n)$.

Exercise 5.63 Prove (5.138) using the complex line integral.

5.4.3 A Priori Estimate on Domains with Smooth Boundary

With what we obtained, we consider the elliptic differential equations on smooth domains.

5.4.3.1 A Priori Estimate on Domains with Smooth Boundary

If Ω is a uniform C^m-domain, diffeomorphism can be used to prove the next theorem. For the sake of simplicity, we suppose

$$a_{ij} \in \mathscr{C}^{s+2}(\mathbb{R}^n), \quad b_k \in \mathscr{C}^{s+2}(\mathbb{R}^n), \quad c \in \mathscr{C}^{s+2}(\mathbb{R}^n) \tag{5.150}$$

on the whole space $\mathbb{R}^n$. Again we assume the uniformly elliptic condition (5.149). Since $\|f|\partial\Omega\|_{L^p(\Omega)} \lesssim \|f|\Omega\|_{F^s_{pq}(\Omega)}$ for all $f \in \mathscr{S}(\mathbb{R}^n)$, we can define $\mathrm{Tr}_{\partial\Omega} : F^{s+2}_{pq}(\Omega) \to L^p(\partial\Omega)$. Denote by $F^{s+2-1/p}_{pp}(\partial\Omega)$ its range.

Theorem 5.41 (A priori estimate of elliptic differential operator on domains) *Let* $1 < p, q < \infty$, $s \in \mathbb{R}$. *Suppose that the coefficients satisfy* (5.149) *and* (5.150). *Then*

$$\|f\|_{F^{s+2}_{pq}(\Omega)} \sim \|Lf\|_{F^s_{pq}(\Omega)} + \|f\|_{F^s_{pq}(\Omega)} + \|\mathrm{Tr}_{\partial\Omega} f\|_{F^{s+2-\frac{1}{p}}_{pp}(\partial\Omega)}$$

for all $f \in F^{s+2}_{pq}(\Omega)$.

We leave the proof of Theorem 5.41 to interested readers; see Exercise 5.64.

5.4.3.2 Hodge Decomposition

We can apply the elliptic differential operators to mathematics in general, for example, we can obtain the Hodge decomposition theorem in differential geometry and the Oka theorem in complex analysis of several variables. Let us state the Hodge decomposition theorem without proof. It will enrich mathematicians to view how elliptic differential operators or function spaces can be applied. We will not give even definitions here: See [116].

Let us formulate the Hodge decomposition theorem. Let (M, g) be a Riemannian manifold. Let $d : \Omega^k(M) \to \Omega^{k+1}(M)$ be the exterior differential operator. Then the adjoint operator $\delta : \Omega^{k+1}(M) \to \Omega^k(M)$ is defined. Define the (positive) *Laplacian* $\Delta : \Omega^k(M) \to \Omega^k(M)$ by $\Delta \equiv d \circ \delta + \delta \circ d$. Then:

Theorem 5.42 (Hodge decomposition theorem) *Let M be a compact and oriented manifold. Then* $\Omega^k(M) = d(\Omega^{k-1}(M)) \oplus \delta(\Omega^{k+1}(M)) \oplus \mathrm{Ker}(\Delta)$.

Exercises

Exercise 5.64 Prove Theorem 5.41 using Theorem 4.46.

Exercise 5.65 A vector field is a linear mapping W from $C^\infty(\mathbb{R}^n)$ to itself such that $W[f \cdot g] = Wf \cdot g + f \cdot Wg$ for all $f, g \in C^\infty(\mathbb{R}^n)$, where $\cdot$ denotes the pointwise product. For a couple of vector fields X, Y, the *bracket* $[X, Y]$ is given by $[X, Y]f = XYf - YXf$ for $f \in C^\infty(\mathbb{R}^n)$. Define vector fields X, Y, Z on $\mathbb{R}^3$ by

$$X \equiv \frac{\partial}{\partial x} + \frac{y}{2}\frac{\partial}{\partial z}, \quad Y \equiv \frac{\partial}{\partial y} - \frac{x}{2}\frac{\partial}{\partial z}, \quad Z \equiv \frac{\partial}{\partial z}.$$

Then show by a direct calculation that $[X, Y] = Z$ and that $[X, Z] = [Y, Z] = 0$.

Textbooks in Sect. 5.4

Fundamental Solutions of Elliptic Differential Operators

See [92, Section 4.1] for fundamental solutions of elliptic operators.

Elliptic Differential Operators on Function Spaces

Needless to say, there are many results on a priori estimates for elliptic differential operators. We refer to [21, 25, 93] for the elliptic differential operators in Sobolev spaces, for example. Here we list those mainly done by Besov, Lizorkin and Triebel together with those related to Besov spaces and Triebel–Lizorkin spaces.

Haroske and Triebel investigated the entropy numbers of the operator related to elliptic differential operators; see [576, 577].

A Priori Estimates of Elliptic Differential Equations on the Whole Space

We can find a priori estimates of elliptic differential equations over Besov spaces and Triebel–Lizorkin spaces in [99].

Others

I have learned a lot in this section from [92, 478].

5.5 $T1$ Theorem and Its Applications

Finally, we discuss the $T1$ theorem, which ensures the $L^2(\mathbb{R}^n)$-boundedness of operators. We also present applications to the boundedness of operators. We will see how our tools play key roles.

In Sect. 5.5.1, we formulate and prove the $T1$ theorem which asserts that the singular integral operators are $L^2(\mathbb{R}^n)$-bounded as long as $T1$ and the adjoint T^*1 are in $\mathrm{BMO}(\mathbb{R}^n)$ and T is weakly bounded. We present applications in Sect. 5.5.2.

5.5.1 $T1$-Theorem

5.5.1.1 CZ(Calderón–Zygmund)-Kernel

Here and below, given a continuous linear mapping $T : \mathscr{S}(\mathbb{R}^n) \to \mathscr{S}'(\mathbb{R}^n)$, let K be a kernel in Theorem 1.42. Define $\mathrm{diag} \equiv \{(x, x) \in \mathbb{R}^{2n} : x \in \mathbb{R}^n\}$.

Definition 5.19 (**CZ(Calderón–Zygmund)-kernel**) Let $0 < \varepsilon \le 1$. A distribution $U \in \mathscr{S}'(\mathbb{R}^n \times \mathbb{R}^n)$ is said to be equipped with the CZ-kernel (continuous degree ε) if there exists a continuous function $K : \mathbb{R}^{2n} \setminus \mathrm{diag} \to \mathbb{C}$ which satisfies the following conditions:

1. The size condition:

$$|K(x,y)| \lesssim |x-y|^{-n} \quad ((x,y) \in \mathbb{R}^{2n} \setminus \mathrm{diag}). \tag{5.151}$$

2. The Hörmander condition

$$|K(x,y) - K(x,z)| + |K(y,x) - K(z,x)| \lesssim |y-z|^{\varepsilon}|x-z|^{-n-\varepsilon} \tag{5.152}$$

whenever $|x-z| > 2|y-z|$.
3. As long as $f, g \in \mathscr{S}(\mathbb{R}^n)$ have disjoint compact support,

$$U[f \otimes g] = \iint_{\mathbb{R}^{2n}} K(x,y)f(x)g(y)\mathrm{d}x\mathrm{d}y.$$

In this case, the function K is said to be the CZ-kernel and $U \in \mathscr{S}'(\mathbb{R}^n \times \mathbb{R}^n)$ corresponds to $T : \mathscr{S}(\mathbb{R}^n) \to \mathscr{S}'(\mathbb{R}^n)$. But sometimes T is said to be equipped with the CZ-kernel K (continuous degree ε).

We understand that something bad happens due to the fact that K is not defined on the diagonal. We learn from the above definition that K is singular along the diagonal. In general, it is not guaranteed that T extends to a bounded linear operator in $L^2(\mathbb{R}^n)$. Our aim here is to consider the condition for T to extend to such an operator. When we consider the convolution operator satisfying this condition, then we can resort to the Fourier transform. However, things are not so simple. For example, if $\sigma \in S^0_{11}$, then $\sigma(X,D)$ falls under this scope; the integral kernel can be shown to be a CZ-kernel because $\sigma(X,D)$ is $L^2(\mathbb{R}^n)$. But there exists a symbol $\sigma \in S^0_{11}$ such that $\sigma(X,D)$ is not bounded as we have seen. Meanwhile, as we have seen in Theorem 5.19, any symbol $\sigma \in S^0_{00}$ generates the $L^2(\mathbb{R}^n)$-bounded linear operator. Thus, if $\sigma(X,D) \in S^0_{10}$, then $\sigma(X,D)$ has a kernel and $\sigma(X,D)$ is an $L^2(\mathbb{R}^n)$-bounded linear operator. See also Theorem 5.6.

The main question addressed in this section is what additional conditions exist for such operators to be $L^2(\mathbb{R}^n)$. The following definition is clearly necessary.

Definition 5.20 (Restricted boundedness) A linear operator $T : \mathscr{S}(\mathbb{R}^n) \to L^2(\mathbb{R}^n)$ is said to enjoy the *restricted boundedness*, if there exists $N \in \mathbb{N}$ such that

$$\|T[f(\tau \star -y)]\|_2 \lesssim \tau^{-\frac{n}{2}} p_N(f)$$

for all $\tau > 0$, $y \in \mathbb{R}^n$ and $f \in \mathscr{S}(\mathbb{R}^n)$.

If T extends to an $L^2(\mathbb{R}^n)$-bounded operator, T clearly has restricted boundedness.

Lemma 5.32 *A linear operator $T : \mathscr{S}(\mathbb{R}^n) \to L^2(\mathbb{R}^n)$ has restricted boundedness if and only if*

$$\| T[f(\tau \cdot -y)] \|_2 \lesssim \tau^{-\frac{n}{2}} p_N(f), \quad \tau > 0 \quad (y \in \mathbb{R}^n, f \in C^\infty_{\mathrm{c}}(B(1))) \tag{5.153}$$

for some $N \in \mathbb{N}$.

Proof Choose $\Psi \in C_c^\infty(B(1))$ and define $\Phi \equiv \Psi - \Psi_{-1}$. We decompose $f = \Psi \cdot f + \sum_{j=1}^\infty \Phi_j \cdot f$. Then we see that (5.153) is equivalent to restricted boundedness.

5.5.1.2 The Definition of $T1$

As the name of the theorem suggests, we need to consider the "function" $T1$, which is supposed to be an image of the constant function 1 by T.

Under the assumption of the restricted boundedness, Tf makes sense as an element in $L^2(\mathbb{R}^n)$ for any $f \in \mathscr{S}(\mathbb{R}^n)$. So, we are interested in the extra conditions for T to be $L^2(\mathbb{R}^n)$ with the restricted boundedness in mind. We benefit a lot from what we have done. As it turns out, the space $\mathrm{BMO}(\mathbb{R}^n)$ plays the key role. To connect matters with $\mathrm{BMO}(\mathbb{R}^n)$, we discuss a property related to the Hardy space $H^1(\mathbb{R}^n)$.

Lemma 5.33 *Suppose that* $\varphi, \psi \in \mathscr{S}(\mathbb{R}^n)$ *satisfy* $\chi_{B(1)} \le \varphi, \psi \le \chi_{B(2)}$. *Then*

$$\lim_{\tau_1, \tau_2 \downarrow 0} \langle T[\varphi(\tau_1 \star) - \psi(\tau_2 \star)], A \rangle = 0$$

for all $(1, \infty)$*-atoms* A.

Proof By symmetry, we can suppose that $0 < \tau_1 \le \tau_2$. Let us assume that $A \in C^\infty(\mathbb{R}^n)$ is supported on a cube Q for the time being. If $0 < \tau_1, \tau_2 \ll 1$, then we can use the property of the kernel T: for $z_A \in Q$, we have

$$\begin{aligned}
&\langle T[\varphi(\tau_1 \star) - \psi(\tau_2 \star)], A \rangle \\
&= \iint_{\mathbb{R}^{2n}} K(u, v)(\varphi(\tau_1 u) - \psi(\tau_2 u))A(v) \mathrm{d}u \mathrm{d}v \\
&= \iint_{\mathbb{R}^{2n}} (K(u, v) - K(u, z_A))(\varphi(\tau_1 u) - \psi(\tau_2 u))A(v) \mathrm{d}u \mathrm{d}v
\end{aligned}$$

by the condition of the atomic decomposition. We have

$$|\langle T[\varphi(\tau_1 \star) - \psi(\tau_2 \star)], A \rangle| \lesssim \int_{B(\tau_2{}^{-1}) \times Q} \frac{|v - z_A|^\varepsilon |A(v)| \mathrm{d}u \mathrm{d}v}{|u - z_A|^{n+\varepsilon}} \lesssim_A \tau_2{}^\varepsilon \tag{5.154}$$

by the Hörmander condition (5.152). Recall that (5.154) is proved for all $A \in C^\infty(\mathbb{R}^n)$. However, as is seen from (5.154), we can prove (5.154) for any $(1, \infty)$-atom A using the mollifier.

If we let $\tau_1, \tau_2 \downarrow 0$ in (5.154), then we obtain the desired result.

Lemma 5.34 *Suppose that $T : \mathscr{S}(\mathbb{R}^n) \to \mathscr{S}'(\mathbb{R}^n)$ is equipped with a CZ-kernel K continuous degree $\varepsilon > 0$ and that T has restricted boundedness. Furthermore, choose $\varphi \in \mathscr{S}(\mathbb{R}^n)$ so that*

$$\chi_{B(1)} \le \varphi \le \chi_{B(2)}. \tag{5.155}$$

1. *Let $\tau > 0$. Then*

$$\| T[\varphi(\tau\star)] \|_{\mathrm{BMO}} \lesssim_\varphi 1, \tag{5.156}$$

   where the implicit constant is independent of τ.
2. *The limit $B_\varphi \equiv \lim_{\tau\downarrow 0} T[\varphi(\tau\star)]$ exists in the weak-* topology of $H^1(\mathbb{R}^n)$. Furthermore, the continuous linear functional B_φ does not depend on φ satisfying* (5.155).

Proof Choose a function $\kappa \in \mathscr{S}(\mathbb{R}^n)$ so that $\chi_{B(1)} \le \kappa \le \chi_{B(2)}$. Here and below, the constant in $\lesssim$ does depend on κ but it is not immaterial. For a ball B, let us prove $m_B^{(2)}(|T[\varphi(\tau\star)](x) - m_B(T[\varphi(\tau\star)]))\lesssim_\varphi 1$ for $\tau > 0$. For κ satisfying (5.155), define $\kappa_B \equiv \kappa\left(\dfrac{\star - c(B)}{2r(B)}\right)$ for a ball B centered at $c(B)$ and of radius $r(B)$. Since K is a CZ-kernel with continuous degree $\varepsilon > 0$, note that

$$T[(1-\kappa_B)\varphi(\tau\star)](x) = \int_{\mathbb{R}^n} K(x,y)(1-\kappa_B(y))\varphi(\tau y)\mathrm{d}y$$

for $x \in B$, since $x \notin \mathrm{supp}((1-\kappa_B)\varphi(\tau\star))$. Using the Hörmander condition (5.152), we can easily prove

$$m_B^{(2)}(T[(1-\kappa_B)\varphi(\tau\star)](x) - m_B(T[(1-\kappa_B)\varphi(\tau\star)])) \lesssim_\varphi 1. \tag{5.157}$$

Meanwhile, if $0 < \tau < 1/r(B)$, observe that the function defined by setting

$$F_{\tau,B}(x) \equiv \varphi\left(\tau r(B)\left(x + \frac{c(B)}{r(B)}\right)\right)\kappa\left(\frac{1}{2}x\right) \quad (x \in \mathbb{R}^n)$$

satisfies $p_N(F_{\tau,B}) \lesssim_{N,\varphi} 1$. Hence it follows that

$$\kappa_B(x)\varphi(\tau x) = F_{\tau,B}\left(\frac{x - c(B)}{r(B)}\right) \quad (x \in \mathbb{R}^n).$$

By virtue of the restricted boundedness, Definition 5.20:

$$m_B^{(2)}(T[\kappa_B\varphi(\tau\star)](x) - m_B(T[\kappa_B\varphi(\tau\star)])) \lesssim_\varphi 1. \tag{5.158}$$

When $\tau \geq 1/r(B)$, we set

$$G_{\tau,B}(x) \equiv \varphi(x)\kappa\left(\frac{1}{2\tau}\left(\frac{x-\tau c(B)}{r(B)}\right)\right) \quad (x \in \mathbb{R}^n).$$

Then $p_N(G_{\tau,B}) \lesssim_\varphi 1$. Since $\kappa_B(x)\varphi(\tau x) = G_{\tau,B}(\tau x)$, by virtue of the restricted boundedness (see Definition 5.20),

$$m_B^{(2)}(T[\kappa_B\varphi(\tau\star)](x) - m_B(T[\kappa_B\varphi(\tau\star)])) \lesssim_\varphi \frac{1}{(\tau r(B))^n} \leq 1. \tag{5.159}$$

From (5.157) to (5.159), (5.156) is proved.

To prove 2, since the set of all finite linear combination of $(1,\infty)$-atoms is dense in $H^1(\mathbb{R}^n)$, it suffices to use Lemma 5.33 and (5.156).

Now we can explain the meaning of $T1$ in the $T1$ theorem; $T1$ in the $T1$ theorem stands for the image of 1 by the mapping T.

Definition 5.21 (**$T1 \in$ BMO($\mathbb{R}^n$) for T having restricted boundedness**) Let T have restricted boundedness and suppose that T is equipped with an integral kernel K with continuity order $\varepsilon(>0)$. Choose $\varphi \in \mathscr{S}(\mathbb{R}^n)$ so that $\chi_{B(1)} \leq \varphi \leq \chi_{B(2)}$. Then define $T1 \in \mathrm{BMO}(\mathbb{R}^n)$ by $T1 \equiv \lim\limits_{\tau\downarrow 0} T[\varphi(\tau\star)]$. Note that the convergence of the right-hand side is the weak-* limit in $H^1(\mathbb{R}^n)$. That is, $T1$ is an element $b \in \mathrm{BMO}(\mathbb{R}^n)$ satisfying $\langle b, f\rangle = \lim\limits_{\tau\downarrow 0}\langle T[\varphi(\tau\star)], f\rangle$ for $f \in H^1(\mathbb{R}^n)$.

To be able to show that the operator T is $L^2(\mathbb{R}^n)$, we need to decompose T into the cancellative part and the paraproduct part. The cancellative part is the part satisfying $T1, T^*1 = 0 \in \mathrm{BMO}(\mathbb{R}^n)$. We prove a sort of orthogonality for this part. This is where our theory of function spaces plays a key role involving the Fourier transform.

Proposition 5.19 *We let $T : \mathscr{S}(\mathbb{R}^n) \to \mathscr{S}'(\mathbb{R}^n)$ be a continuous linear mapping equipped with the CZ-kernel K of continuous degree $\varepsilon > 0$. Furthermore, let T, T^* have restricted boundedness and assume that $T1, T^*1 = 0 \in \mathrm{BMO}(\mathbb{R}^n)$. Choose an even function $\Psi \in \mathscr{S}(\mathbb{R}^n)$ so that $\chi_{B(1)} \leq \Psi \leq \chi_{B(2)}$. Furthermore, let*

$$\Phi \equiv \Psi^1 - \Psi = 2^n\Psi(2\star) - \Psi \tag{5.160}$$

and $\Theta_{j,k}(x,y) \equiv \langle T[\Phi^j(\star - x)], \Phi^k(\star - y)\rangle$. Then there exists $\delta \in (0,\varepsilon)$ such that

$$|\Theta_{j,k}(x,y)| \lesssim 2^{-|j-k|\delta + \min(j,k)n}(1 + 2^{\min(j,k)}|x-y|)^{-n-\delta}. \tag{5.161}$$

Proof Let $j \leq k$. We use $\Theta_{j,k}(x,y) = \langle \Phi^j(\star - x), T^*[\Phi^k(\star - y)]\rangle$. In view of symmetry of T^* and T, it can be assumed that $j > k$. For $|x-y| \geq 2^{-k+3}$, we have

$$|\Theta_{j,k}(x,y)| = \left|\iint_{\mathbb{R}^{2n}} K(u,v)\Phi^j(u-x)\Phi^k(v-y)\mathrm{d}u\mathrm{d}v\right|$$

$$= \left| \iint_{\mathbb{R}^{2n}} (K(u,v) - K(x,v)) \Phi^j(u-x) \Phi^k(v-y) \mathrm{d}u \mathrm{d}v \right|$$

in view of the expression of T, since $\Phi \perp \mathscr{P}_0(\mathbb{R}^n)$.

By the Hörmander condition (5.152) of K, we have

$$|\Theta_{j,k}(x,y)| \lesssim \iint_{\mathbb{R}^{2n}} \frac{|u-x|^\varepsilon}{|u-v|^{n+\varepsilon}} |\Phi^j(u-x)\Phi^k(v-y)| du dv. \tag{5.162}$$

When $u \in B(x, 2^{-j})$, $v \in B(y, 2^{-k})$, we have $|u-x| \le 2^{-j}$, $|u-v| \sim |x-y|$ since $|x-y| \ge 2^{-k+3}$. Hence we obtain

$$|\Theta_{j,k}(x,y)| \lesssim \frac{2^{-j\varepsilon}}{|x-y|^{n+\varepsilon}} \iint_{\mathbb{R}^{2n}} |\Phi^j(u-x)\Phi^k(v-y)| du dv \simeq \frac{2^{-j\varepsilon}}{|x-y|^{n+\varepsilon}}.$$

Thus, (5.161) holds.

When $|x-y| < 2^{-k+3}$ and $0 \le j-k \le 10$, we have

$$|\Theta_{j,k}(x,y)| \le \|T[\Phi^j(\star - x)]\|_2 \|\Phi^k(\star - y)\|_2 \lesssim 2^{(j+k)n/2} p_N(\Phi)^2 \lesssim 2^{jn}$$

by restricted boundedness (Definition 5.20). Hence (5.161) holds.

So, it remains to consider the case where $|x-y| < 2^{-k+3}$ and $j > k+10$. Since $T^*1 = 0$, we have $\Psi_l(\star - y)\Phi^k(\star - y) = \Phi^k(\star - y)$ for $l \gg 1$; hence

$$\Theta_{j,k}(x,y) = \lim_{l \to \infty} \langle \Phi^j(\star - x), T^*[\Psi_l(\star - y)(\Phi^k(\star - y) - \Phi^k(x-y))] \rangle.$$

Define $\theta \in (0,1)$ by $2\theta - (1-\theta)n = \min(1, 2\varepsilon)$. Let $l \ge |k| + |j| + 4$. Set

$$\mathrm{I} \equiv \langle T[\Phi^j(\star - x)], \Psi(2^{j\theta + k(1-\theta)}(\star - x))(\Phi^k(\star - y) - \Phi^k(x-y)) \rangle$$
$$\mathrm{II} \equiv \langle T[\Phi^j(\star - x)], (\Psi_l(\star - y) - \Psi(2^{j\theta + k(1-\theta)}(\star - x)))(\Phi^k(\star - y) - \Phi^k(x-y)) \rangle.$$

and we decompose

$$\langle \Phi^j(\star - x), T^*[\Psi_l(\star - y)(\Phi^k(\star - y) - \Phi^k(x-y))] \rangle = \mathrm{I} + \mathrm{II}.$$

For I, we use the mean-value theorem and the restricted boundedness (Definition 5.20) to conclude that

$$\begin{aligned} |\mathrm{I}| &\lesssim \|T[\Phi^j(\star - x)]\|_2 \|\Psi(2^{j\theta + k(1-\theta)}(\star - x))(\Phi^k(\star - y) - \Phi^k(x-y))\|_2 \\ &\lesssim 2^{jn/2} \cdot 2^{-jn/2 + kn + (k-j)((1-\theta)n/2 - \theta)} = 2^{kn - (k-j)\min(1/2, \varepsilon)}. \end{aligned}$$

Let us consider II. In view of the size of the support of the functions, we can depend upon the kernel of T. That is,

$$\begin{aligned}
\mathrm{II} &= \iint_{\mathbb{R}^{2n}} K(u,v)\Phi^j(u-x) \\
&\quad \times(\Psi_l(v-y)-\Psi(2^{j\theta+k(1-\theta)}(v-x)))(\Phi^k(v-y)-\Phi^k(x-y))\mathrm{d}v\mathrm{d}u \\
&= \iint_{\mathbb{R}^{2n}} (K(u,v)-K(x,v))\Phi^j(u-x) \\
&\quad \times(\Psi_l(v-y)-\Psi(2^{j\theta+k(1-\theta)}(v-x)))(\Phi^k(v-y)-\Phi^k(x-y))\mathrm{d}v\mathrm{d}u.
\end{aligned}$$

Observe that $|v-y| \le 2^{-k+4} \le 2^l$ if $|v-x| \le 2^{-j+(j-k)(1-\theta)}$. The integrand of the last term above becomes 0 unless $(u,v) \in B(x,2^{-j+1}) \times (\mathbb{R}^n \setminus B(x,2^{-j+(j-k)(1-\theta)}))$. Hence it follows that

$$\begin{aligned}
|\mathrm{II}| &\lesssim 2^{(j+k)n} \iint_{(u,v)\in B(x,2^{-j+1})\times(\mathbb{R}^n\setminus B(x,2^{-j+(j-k)(1-\theta)}))} \frac{|u-x|^\varepsilon \mathrm{d}u\mathrm{d}v}{|v-x|^{n+\varepsilon}} \\
&\lesssim 2^{(j+k)n-j\varepsilon} \iint_{(u,v)\in B(x,2^{-j+1})\times(\mathbb{R}^n\setminus B(x,2^{-j+(j-k)(1-\theta)}))} \frac{\mathrm{d}u\mathrm{d}v}{|v-x|^{n+\varepsilon}} \\
&\simeq 2^{kn-j\varepsilon} \times 2^{-\varepsilon(-j+(j-k)(1-\theta))} = 2^{kn-\varepsilon(j-k)(1-\theta)}.
\end{aligned}$$

Thus, (5.161) is proved for $\delta \equiv (1-\theta)\varepsilon$.

Suppose that a continuous linear mapping $T : \mathscr{S}(\mathbb{R}^n) \to \mathscr{S}'(\mathbb{R}^n)$ is equipped with a CZ-kernel K of continuous degree $\varepsilon > 0$. Then we can extend T to the space of Lipschitz continuous spaces, as follows. Recall that we have an expression of Hölder–Zygmund spaces $\dot{\mathscr{C}}^s(\mathbb{R}^n) \approx \dot{B}^s_{\infty\infty}(\mathbb{R}^n)$ with equivalence of norms, which is a homogeneous counterpart of Theorem 2.7.

Proposition 5.20 *Suppose that a continuous linear mapping $T : \mathscr{S}(\mathbb{R}^n) \to \mathscr{S}'(\mathbb{R}^n)$ is equipped with a CZ-kernel K of continuous degree $\varepsilon > 0$. Furthermore, assume that T has restricted boundedness and that $T1 = 0 \in \mathrm{BMO}(\mathbb{R}^n)$. Then*

$$\|Tf\|_{\dot{B}^s_{\infty\infty}} \lesssim_s \|f\|_{\dot{B}^s_{\infty\infty}}, \quad 0 < s < \delta, \quad f \in C^\infty_c(\mathbb{R}^n),$$

where $\delta > 0$ is from Proposition 5.19.

Proof Choose Ψ as we did in Proposition 5.19 and define Φ by (5.160). Then it suffices to show that

$$\sup\{|2^{js}\Phi^j * Tf(x)| : j \in \mathbb{Z}, x \in \mathbb{R}^n\} \lesssim \|f\|_{\dot{B}^s_{\infty\infty}}$$

by the local means (Theorem 2.34). Fix $x \in \mathbb{R}^n$. Since Φ is even,

$$\Phi^j * Tf(x) = \langle Tf, \Phi^j(\star - x)\rangle = \langle f, T^*[\Phi^j(\star - x)]\rangle. \tag{5.163}$$

Since $\left\{\sum_{j=-M}^{N} \Phi^j * f\right\}_{M,N\in\mathbb{N}}$ converges to f in the topology of $L^2(\mathbb{R}^n)$ as $M, N \to \infty$, we have $\Phi^j * Tf(x) = \sum_{k=-\infty}^{\infty} \langle \Phi^k * f, T^*[\Phi^j(\star - x)]\rangle$.

Meanwhile, Fubini's theorem (Theorem 1.3) yields

$$\int_{\mathbb{R}^n} \langle \Phi^k(\star - y), T^*[\Phi^j(\star - x)]\rangle \mathrm{d}y = \int_{\mathbb{R}^n} \left(\int_{\mathbb{R}^n} \Phi^k(z-y)\mathrm{d}y\right) T^*[\Phi^j(\star - x)](z)\mathrm{d}z$$
$$=0.$$

Since

$$\begin{aligned}\langle \Phi^k * f, T^*[\Phi^j(\star - x)]\rangle &= \int_{\mathbb{R}^n} f(y)\langle \Phi^k(\star - y), T^*[\Phi^j(\star - x)]\rangle \mathrm{d}y \\ &= \int_{\mathbb{R}^n} (f(y) - f(x))\langle \Phi^k(\star - y), T^*[\Phi^j(\star - x)]\rangle \mathrm{d}y,\end{aligned}$$

Proposition 5.19 yields

$$\|2^{js}\Phi^j * Tf\|_\infty \lesssim \|f\|_{\dot{B}^s_{\infty\infty}} \sup_{x\in\mathbb{R}^n} \sum_{k=-\infty}^{\infty} \int_{\mathbb{R}^n} \frac{2^{js-|j-k|\delta+\min(j,k)n}|x-y|^s}{(1+2^{\min(j,k)}|x-y|)^{n+\delta}} \mathrm{d}y \lesssim \|f\|_{\dot{B}^s_{\infty\infty}}.$$

Thus, Proposition 5.20 is proved.

Now we prove the $T1$ theorem under two additional assumptions.

Theorem 5.43 (**$T1$ theorem (weak form)**) *Let $T : \mathscr{S}(\mathbb{R}^n) \to \mathscr{S}'(\mathbb{R}^n)$ be a continuous linear mapping equipped with the* CZ*-kernel K. Suppose that:*

1. $T1 = 0 \in \mathrm{BMO}(\mathbb{R}^n)$.
2. $T^*1 = 0 \in \mathrm{BMO}(\mathbb{R}^n)$.
3. *T has the restricted boundedness.*

Then T extends to an $L^2(\mathbb{R}^n)$-bounded operator.

Proof Recall that $L^2(\mathbb{R}^n) \approx \dot{F}^0_{22}(\mathbb{R}^n)$ with equivalence of norms holds with equivalence of norms. We decompose $f \in L^2(\mathbb{R}^n)$ by the atomic decomposition; for each $(\nu, m) \in \mathbb{Z}^{n+1}$, it suffices to show that T maps $a_{\nu m}$ centered at $Q_{\nu m}$ as we did in Proposition 4.1, to the molecule $Ta_{\nu m} =: M_{\nu m}$. Also, as is seen from the proof of Theorem 4.2, we can assume $a_{\nu m} \in C^\infty_c(\mathbb{R}^n)$. We remark that $M_{\nu m}$ can be regarded as a function in $\dot{\mathscr{C}}^s(\mathbb{R}^n) = \dot{B}^s_{\infty\infty}(\mathbb{R}^n)$ from Proposition 5.20. First of all, we will obtain pointwise estimate of $M_{\nu m}$ on $\mathbb{R}^n \setminus 2n\,Q_{\nu m}$. We deduce

$$M_{\nu m}(x) = \int_{\mathbb{R}^n} (K(x,y) - K(x, 2^{-\nu}m))a_{\nu m}(y)\mathrm{d}y \quad (x \in \mathbb{R}^n \setminus 2n\, Q_{\nu m})$$

from the moment condition. Thus, by the Hörmander condition, we have

$$|M_{\nu m}(x)| \lesssim \int_{\mathbb{R}^n} \frac{|y - 2^{-\nu}m|^{\varepsilon}|a_{\nu m}(y)|\mathrm{d}y}{|x - 2^{-\nu}m|^{n+\varepsilon}} \lesssim \frac{2^{\frac{\nu n}{2}}}{\langle 2^{\nu}x - m\rangle^{n+\varepsilon}} \tag{5.164}$$

as long as $x \in \mathbb{R}^n \setminus 2n\, Q_{\nu m}$. Furthermore,

$$\begin{aligned} M_{\nu m}(x) - M_{\nu m}(z) = &\int_{\mathbb{R}^n} (K(x,y) - K(z,y))a_{\nu m}(y)\mathrm{d}y \\ &- \int_{\mathbb{R}^n} (K(x, 2^{-\nu}m) - K(z, 2^{-\nu}m))a_{\nu m}(y)\mathrm{d}y \end{aligned}$$

provided that $x, z \in \mathbb{R}^n \setminus 2n\, Q_{\nu m}$ satisfy $|x - z| \le 2^{-\nu}$. Hence

$$|M_{\nu m}(x) - M_{\nu m}(z)| \lesssim \int_{\mathbb{R}^n} \frac{|x - z|^{\varepsilon}|a_{\nu m}(y)|\mathrm{d}y}{|x - 2^{-\nu}m|^{n+\varepsilon}} \lesssim \frac{2^{\frac{\nu n}{2}+\nu\varepsilon}|x - z|^{\varepsilon}}{\langle 2^{\nu}x - m\rangle^{n+\varepsilon}}. \tag{5.165}$$

By the restricted boundedness, $M_{\nu m} \in L^2(\mathbb{R}^n)$. Thus, Proposition 5.20 yields $M_{\nu m} \in \dot{\mathscr{C}}^s(\mathbb{R}^n)$, $0 < s < \varepsilon$. Furthermore, by Proposition 5.20 the inequalities (5.164) and (5.165) can be extended to $\mathbb{R}^n$. That is, for $0 < s < \varepsilon$, we have

$$|M_{\nu m}(x)| \lesssim_s 2^{\frac{\nu n}{2}} \langle 2^{\nu}x - m\rangle^{-n-s} \quad (x \in \mathbb{R}^n)$$

and

$$|M_{\nu m}(x) - M_{\nu m}(z)| \lesssim_s 2^{\frac{\nu n}{2}+\nu s}|x - z|^s \langle 2^{\nu}x - m\rangle^{-n-s} \quad (x, z \in \mathbb{R}^n).$$

Finally, we need to check the moment condition in Proposition 4.1. To this end, we choose an auxiliary function Ψ so that $\chi_{B(1)} \le \Psi \le \chi_{B(2)}$. Then

$$\int_{\mathbb{R}^n} M_{\nu m}(x)\mathrm{d}x = \lim_{j\to\infty} \int_{\mathbb{R}^n} Ta_{\nu m}(x)\Psi_j(x)\mathrm{d}x = \lim_{j\to\infty} \langle Ta_{\nu m}, \Psi_j\rangle = \langle a_{\nu m}, T^*1\rangle.$$

Since we are assuming $T^*1 = 0$, it follows that $M_{\nu m} \perp \mathscr{P}_0(\mathbb{R}^n)$.

Regarding Proposition 4.1 a clarifying remark may be in order.

Remark 5.7 Note that K corresponding to T has continuous degree $\varepsilon(< 1)$. Thus, Proposition 4.1 is essential.

5.5.1.3 A Generalization of Theorem 5.43

We generalize Theorem 5.43. Theorem 5.44 generalizes Theorem 5.43 in that we do not assume $T1 = 0$ or $T^*1 = 0$ in $\mathrm{BMO}(\mathbb{R}^n)$.

Theorem 5.44 (**$T1$ theorem**) *Let $T : \mathscr{S}(\mathbb{R}^n) \to \mathscr{S}'(\mathbb{R}^n)$ be a continuous linear mapping equipped with the CZ-kernel K. If T^* and T have restricted boundedness as well, T extends to an $L^2(\mathbb{R}^n)$-bounded operator.*

In view of Theorem 5.43, it suffices to find an $L^2(\mathbb{R}^n)$-bounded CZ-operator S such that $S1 = T1$ and that $S^*1 = T^*1$. Due to the linearity, Theorem 5.44 is reduced to the next lemma involving the paraproduct, whose idea resembles that of the boundedness of the pointwise product.

Lemma 5.35 *Let $a \in \mathrm{BMO}(\mathbb{R}^n)$. Choose an even function $\psi \in \mathscr{S}(\mathbb{R}^n)$ so that $\chi_{B(1)} \le \psi \le \chi_{B(2)}$ and set $\varphi \equiv \psi - \psi_{-1} = \psi - \psi(2\star)$. Define the paraproduct operator T_a by $T_a f \equiv \sum_{j=-\infty}^{\infty} \varphi_j(D)a \cdot \psi_{j-4}(D)f$.*

1. *T_a is an $L^2(\mathbb{R}^n)$-bounded operator. More precisely, for $f \in L^2(\mathbb{R}^n)$ the infinite sum defining $T_a f$ is $L^2(\mathbb{R}^n)$-convergent and $\|T_a f\|_2 \lesssim \|a\|_{\mathrm{BMO}}\|f\|_2$. Furthermore,*

$$\|\{\varphi_j(D)a \cdot \Psi_{j-4}(D)f\}_{j=-J}^{J}\|_{L^1(\ell^2)} \lesssim \|f\|_{H^1}. \tag{5.166}$$

2. *T_a is an operator with CZ-kernel $K_a(x, y)$ of continuity degree 1. Furthermore,*

$$T_a 1 = a, \tag{5.167}$$

$$T_a^* 1 = 0. \tag{5.168}$$

Before the proof of Lemma 5.35, let us conclude the proof of Theorem 5.44. Using Lemma 5.35 twice, we can find a function $a, b \in \mathrm{BMO}(\mathbb{R}^n)$ such that $T1 = T_a 1$ and $T_a^* 1 = 0$ and that $T_b 1 = T^*1$ and $T_b^* 1 = 0$. Thus, $T - T_a - (T_b)^*$ is $L^2(\mathbb{R}^n)$-bounded. Since T_a and T_b are $L^2(\mathbb{R}^n)$-bounded from Lemma 5.35, it follows that T is $L^2(\mathbb{R}^n)$-bounded.

Proof We need to be careful because T_a is not proved to be an $L^2(\mathbb{R}^n)$-bounded operator. Let us consider the approximation of T_a;

$$T_{a,J} \equiv \sum_{j=-J}^{J} \varphi_j(D)a \cdot \psi_{j-4}(D)f.$$

In view of the size of the supports of ψ and φ, we have

$$\mathrm{supp}\left(\mathscr{F}[\varphi_j(D)a \cdot \psi_{j-4}(D)f]\right) \subset B(2^{j+2}) \setminus B(2^{j-2}). \tag{5.169}$$

We recall $\dot{F}^0_{2,2}(\mathbb{R}^n) \approx L^2(\mathbb{R}^n)$. By virtue of Theorem 3.28, we obtain

$$\|T_{a,J}\|_2 \lesssim \sqrt{\sum_{j=-J}^{J} (\|\varphi_j(D)a \cdot \psi_{j-4}(D)f\|_2)^2} \lesssim \|f\|_2. \tag{5.170}$$

Thus, $T_{a,J}$ is $L^2(\mathbb{R}^n)$-bounded. Furthermore, letting $J \to \infty$, we conclude that T_a is also $L^2(\mathbb{R}^n)$-bounded. Similar to (5.170), we can also prove (5.166).

Let us prove 2. Since $\varphi_j(D)1 = 0$ and $a \in \mathrm{BMO}(\mathbb{R}^n)$,

$$|\varphi_j(D)a(x)| = |\varphi_j(D)[a - m_{B(x,2^{-j})}(a)](x)| \lesssim \|a\|_{\mathrm{BMO}}.$$

Hence

$$\begin{aligned}\sum_{j=-\infty}^{\infty} |\varphi_j(D)a(x)(\mathscr{F}^{-1}\psi)^j(x-y)| &\lesssim \|a\|_{\mathrm{BMO}} \sum_{j=-\infty}^{\infty} \frac{2^{jn}}{(1+2^j|x-y|)^{n+1}} \\ &= \frac{\|a\|_{\mathrm{BMO}}}{|x-y|^n} \sum_{j=-\infty}^{\infty} \frac{(2^j|x-y|)^n}{(1+2^j|x-y|)^{n+1}} \\ &\sim \frac{\|a\|_{\mathrm{BMO}}}{|x-y|^n}.\end{aligned}$$

We define

$$K_a(x,y) \equiv \sum_{j=-\infty}^{\infty} \varphi_j(D)a(x)(\mathscr{F}^{-1}\psi)^j(x-y) \quad (x, y \in \mathbb{R}^n).$$

Fix $f \in C^\infty_{\mathrm{c}}(\mathbb{R}^n)$. Then $T_a f(x) = \int_{\mathbb{R}^n} K_a(x,y)f(y)\mathrm{d}y$ for almost all $x \notin \mathrm{supp}(f)$. The same argument as before works:

$$|\nabla_x K_a(x,y)| + |\nabla_y K_a(x,y)| \lesssim \|a\|_{\mathrm{BMO}}|x-y|^{-n-1} \quad (x, y \in \mathbb{R}^n). \tag{5.171}$$

Hence K_a is a CZ-kernel with continuous degree 1.

Roughly speaking, (5.167) seems trivial since $\psi_j(D)1 = 1$. Furthermore, (5.168) seems trivial from (5.169). But we need to be careful; the definition of $T_a 1$ and $T_a^* 1$ was complicated. To prove (5.167) and (5.168) we fix a $(1,\infty)$-atom A. Furthermore, we fix $\zeta \in C^\infty_{\mathrm{c}}(\mathbb{R}^n)$ so that $\chi_{B(1)} \leq \zeta \leq \chi_{B(2)}$. Then there exists a constant d_A that depends on A such that $\mathrm{supp}(A) \subset B({d_A}^{-1})$. Hence

$$\langle T_a^* 1, A\rangle = \lim_{\tau\downarrow 0}\langle T_a^*[\zeta(d_A\star)\zeta(\tau\star)], A\rangle + \lim_{\tau\downarrow 0}\int_{\mathbb{R}^n} T_a^*[(1-\zeta(d_A\star))\zeta(\tau\star)](x)A(x)\mathrm{d}x.$$

We easily handle the first term and we use the conditions of the kernel for the second term. The result is

$$\begin{aligned}\langle T_a^*1, A\rangle &= \int_{\mathbb{R}^n} T_a^*[\zeta(d_A\star)](x)A(x)\mathrm{d}x + \lim_{\tau\downarrow 0}\int_{\mathbb{R}^n} T_a^*[(1-\zeta(d_A\star))\zeta(\tau\star)](x)A(x)\mathrm{d}x \\ &= \int_{\mathbb{R}^n} T_a^*[\zeta(d_A\star)](x)A(x)\mathrm{d}x \\ &\quad + \int_{\mathbb{R}^n}\left(\int_{\mathbb{R}^n}(K_a(x,y)-K_a(0,y))(1-\zeta(d_A\cdot y))\mathrm{d}y\right)A(x)\mathrm{d}x.\end{aligned}$$

An argument used for T_a works for $T_{a,J}$ and we obtain $A = \lim_{J\to\infty}\sum_{j=-J}^{J}\varphi_j(D)A$ in the topology of $H^1(\mathbb{R}^n) \approx \dot{F}^0_{12}(\mathbb{R}^n)$. Hence we conclude

$$\lim_{J\to\infty}\sum_{j=-J}^{J}\langle\varphi_j(D)a, A\rangle = \lim_{J\to\infty}\sum_{j=-J}^{J}\langle a, \varphi_j(D)A\rangle = \int_{\mathbb{R}^n} a(x)A(x)\mathrm{d}x.$$

We go through the same argument to deal with $T_a^*, T_{a,J}^*$; by (5.171) and Lebesgue's convergence theorem

$$\langle T_a 1, A\rangle = \lim_{J\to\infty}\langle T_{a,J}1, A\rangle = \lim_{J\to\infty}\sum_{j=-J}^{J}\langle\varphi_j(D)a, A\rangle = \int_{\mathbb{R}^n} a(x)A(x)\mathrm{d}x$$

for all $(1,\infty)$-atoms A.

Let us also prove that $T_a^*1 = 0$.

Let A be a $(1,\infty)$-atom once again. From (5.166) we obtain $T_aA \in H^1(\mathbb{R}^n)$. Thus,

$$\lim_{\tau\downarrow 0}\langle T_aA, \Psi(\tau\star)\rangle = \langle T_aA, 1\rangle = 0.$$

Hence we obtain (5.167) and (5.168).

5.5.2 *Applications of T1-Theorem*

5.5.2.1 Cauchy Integral Operator

We give applications of the $T1$ theorem.

Theorem 5.45 (Example of $L^2(\mathbb{R}^n)$-bounded CZ-kernel) *Let $k \in \mathrm{Lip}(\mathbb{R}^n)$; that is, k is a continuous function satisfying $\|k\|_{\mathrm{Lip}} < \infty$. For $r > 0$ define $T_{k,r} : \mathscr{S}(\mathbb{R}^n) \to \mathscr{S}'(\mathbb{R}^n)$ by*

$$T_{k,r} f(x) \equiv \int_{\mathbb{R}^n \setminus B(x,r)} \frac{k(x) - k(y)}{|x - y|^{n+1}} f(y) \mathrm{d}y,$$

for $x \in \mathbb{R}^n$ *and* $f \in \mathscr{S}(\mathbb{R}^n)$.

1. *The limit* $T_k f \equiv \lim_{r \downarrow 0} T_{k,r} f$, *called the Cauchy integral operator, exists in* $\mathscr{S}'(\mathbb{R}^n)$ *for all* $f \in \mathscr{S}(\mathbb{R}^n)$ *and* T_k *has a* CZ-*kernel* K *of continuous degree* 1; $K(x, y) \equiv \dfrac{k(x) - k(y)}{|x - y|^{n+1}}$.
2. *The operator* T_k *extends to an* $L^2(\mathbb{R}^n)$-*bounded operator.*

Proof Let $f \in \mathscr{S}(\mathbb{R}^n)$. We will prove that $T_{k,r} f$ is an $L^2(\mathbb{R}^n)$-function and that the $L^2(\mathbb{R}^n)$-norm depends only on $f \in C_c^\infty(B(1))$ but not on r. Note that

$$\begin{aligned}
\langle T_{k,r} f, g \rangle &= \iint_{\{(x,y) \in \mathbb{R}^{2n} : |x-y| > r\}} \frac{k(x) - k(y)}{|x - y|^{n+1}} f(y) g(x) \mathrm{d}y \mathrm{d}x \\
&= \frac{1}{2} \iint_{\{(x,y) \in \mathbb{R}^{2n} : |x-y| > r\}} \frac{k(x) - k(y)}{|x - y|^{n+1}} (f(y) g(x) - f(x) g(y)) \mathrm{d}y \mathrm{d}x
\end{aligned}$$

for $g \in \mathscr{S}(\mathbb{R}^n)$. By changing variables,

$$\begin{aligned}
\langle T_{k,r} f, g \rangle &= \frac{1}{2} \iint_{\{(x,y) \in \mathbb{R}^{2n} : r < |x-y| \le 1\}} \frac{k(x) - k(y)}{|x - y|^{n+1}} (f(y) g(x) - f(x) g(y)) \mathrm{d}y \mathrm{d}x \\
&\quad + \frac{1}{2} \iint_{\{(x,y) \in \mathbb{R}^{2n} : |x-y| > 1\}} \frac{k(x) - k(y)}{|x - y|^{n+1}} (f(y) g(x) - f(x) g(y)) \mathrm{d}y \mathrm{d}x.
\end{aligned}$$

Choose $\psi \in \mathscr{S}(\mathbb{R}^n)$ so that $\chi_{B(1)} \le \psi \le \chi_{B(2)}$. Set $k_j \equiv \psi_j * k$ for each $j \in \mathbb{Z}$. Then when $r < |x - y| \le 1$,

$$\begin{aligned}
|f(y) g(x) - f(x) g(y)| &\le |f(y) g(x) - f(x) g(x)| \\
&\quad + |f(x) g(x) - f(x) g(y)| \lesssim \frac{|x - y|}{\langle x \rangle^{n+1}},
\end{aligned}$$

so from Lebesgue's convergence theorem, we deduce

$$\begin{aligned}
&\langle T_{k,r} f, g \rangle \\
&= \frac{1}{2} \lim_{j \to \infty} \iint_{\{(x,y) \in \mathbb{R}^{2n} : r < |x-y| \le 1\}} \frac{k_j(x) - k_j(y)}{|x - y|^{n+1}} (f(y) g(x) - f(x) g(y)) \mathrm{d}y \mathrm{d}x \\
&\quad + \frac{1}{2} \lim_{j \to \infty} \iint_{\{(x,y) \in \mathbb{R}^{2n} : |x-y| > 1\}} \frac{k_j(x) - k_j(y)}{|x - y|^{n+1}} (f(y) g(x) - f(x) g(y)) \mathrm{d}y \mathrm{d}x \\
&= \lim_{j \to \infty} \langle T_{k_j,r} f, g \rangle.
\end{aligned}$$

We denote by $\mathrm{d}\sigma$ the surface area of S^{n-1}. Then by the Stokes theorem, we have

$$|T_{k_j,r}f(x)| \lesssim \sum_{l=1}^{n} \left| \int_{\mathbb{R}^n} \chi_{B(x,r)^c}(y) \frac{x_l - y_l}{|x-y|^{n+1}} \partial_l k_j(y) f(y) \mathrm{d}y \right| + \|k\|_{\mathrm{Lip}} \int_{\mathbb{R}^n} \frac{|\partial_{x_i} f(y)| \mathrm{d}y}{|x-y|^{n-1}} + \frac{\|k\|_{\mathrm{Lip}}}{r^{n-1}} \int_{\partial B(r)} |f(x+y)| \mathrm{d}\sigma(y).$$

If we use the Hardy–Littlewood maximal operator,

$$|T_{k_j,r}f(x)| \lesssim \sum_{j=1}^{n} \left| \int_{\mathbb{R}^n} \frac{\partial}{\partial x_j} \left[\left\{ 1 - \psi\left(\frac{x-y}{r}\right) \right\} \frac{1}{|x-y|^{n-1}} \right] \partial_{x_i} k_j(y) f(y) \mathrm{d}y \right| + \|k\|_{\mathrm{Lip}} Mf(x) + \|k\|_{\mathrm{Lip}} \int_{\mathbb{R}^n} \frac{|\partial_{x_i} f(y)| \mathrm{d}y}{|x-y|^{n-1}} + \frac{\|k\|_{\mathrm{Lip}}}{r^{n-1}} \int_{\partial B(r)} |f(x+y)| \mathrm{d}\sigma(y).$$

Furthermore, we have $|T_{k,r}f(x)| \lesssim \langle x \rangle^{-n}$ for $|x| \geq 2$. The homogeneous version of the atomic decomposition (Theorem 4.1) yields

$$\frac{\partial}{\partial x_j} \left[\left\{ 1 - \psi\left(\frac{x-y}{r}\right) \right\} \frac{1}{|x-y|^n} \right] \in \dot{B}^0_{1\infty}(\mathbb{R}^n).$$

Thus, by virtue of Theorem 2.10, the Calderón–Zygmund theory and the Plancherel theorem,

$$\sqrt{\int_{\mathbb{R}^n} \left| \int_{\mathbb{R}^n} \frac{\partial}{\partial x_j} \left[\left\{ 1 - \psi\left(\frac{x-y}{r}\right) \right\} \frac{1}{|x-y|^{n-1}} \right] \partial_{x_i} k_j(y) f(y) \mathrm{d}y \right|^2 \mathrm{d}x} \lesssim \|k\|_{\mathrm{Lip}} \|f\|_2.$$

Hence by taking the $L^2(\mathbb{R}^n)$-norm, we have an estimate independent of j: $\|T_{k_j,r}f\|_2 \lesssim \|k\|_{\mathrm{Lip}} p_{n+1}(f)$. Letting $j \to \infty$, we have $\|T_{k,r}f\|_2 \lesssim \|k\|_{\mathrm{Lip}} p_{n+1}(f)$. By the Banach–Alaoglu theorem (Theorem 1.17), there exists a sequence $\{r_j\}_{j=1}^\infty$ decreasing to 0, which is dependent on f. such that $T_{k,r_j}f$ converges to an $L^2(\mathbb{R}^n)$-function weakly. The sequence $\{T_{k,r_j}f\}_{j=1}^\infty$ converges to $T_k f$ in $\mathscr{S}'(\mathbb{R}^n)$; we have $T_k f \in L^2(\mathbb{R}^n)$ and $\|T_k f\|_2 \lesssim \|k\|_{\mathrm{Lip}} p_{n+1}(f)$.

An affine change of variables yields

$$\begin{aligned} \|T_k[f(R \star + z)]\|_2 &= R^{-\frac{n}{2}} \|T_{R \cdot k(\frac{\star - z}{R})} f\|_2 \\ &\lesssim R^{-\frac{n}{2}} \left\| R \cdot k\left(\frac{\star - z}{R}\right) \right\|_{\mathrm{Lip}} p_{n+1}(f) \end{aligned}$$

$$\lesssim R^{-\frac{n}{2}}\|k\|_{\mathrm{Lip}}p_{n+1}(f)$$

for $R > 0, z \in \mathbb{R}^n$. Thus, T_k enjoys restricted boundedness. In view of the symmetry, so does T_k^*.

Let $f \in C_c^\infty(\mathbb{R}^n)$ and $g \in C_c^\infty(\mathbb{R}^n)$. Then whenever $r < \mathrm{dist}(\mathrm{supp}(f), \mathrm{supp}(g))$, Lebesgue's convergence theorem yields

$$\begin{aligned}
\langle T_k f, g\rangle &= \lim_{r\downarrow 0}\langle T_{k,r} f, g\rangle \\
&= \lim_{r\downarrow 0}\iint_{\{(x,y)\in\mathbb{R}^{2n}\,:\,|x-y|>r\}} K(x,y)f(x)g(y)\mathrm{d}x\mathrm{d}y \\
&= \int_{\mathbb{R}^n} K(x,y)f(x)g(y)\mathrm{d}x\mathrm{d}y.
\end{aligned}$$

Hence T_k is equipped with an integral kernel K. By Theorem 5.44, T_k extends to an $L^2(\mathbb{R}^n)$-bounded operator. Thus, the proof is complete.

Likewise, we can prove the following theorem:

Theorem 5.46 (An example of $L^2(\mathbb{R})$-bounded CZ-kernels) *Let $k \in \mathrm{Lip}(\mathbb{R})$. Let also $m \in \mathbb{N}$ and $r > 0$. Then if we set*

$$S_{k,m,r}f(x) \equiv \int_{\mathbb{R}}\left(\frac{k(x)-k(y)}{x-y}\right)^m \frac{f(y)}{x-y}\mathrm{d}y$$

for $x \in \mathbb{R}^n$ and $f \in \mathscr{S}(\mathbb{R}^n)$, then $S_{k,m}f \equiv \lim\limits_{r\downarrow 0} S_{k,m,r}f$ exists in the topology of $\mathscr{S}'(\mathbb{R})$ for all $f \in \mathscr{S}(\mathbb{R})$ and $S_{k,m}$ extends to an $L^2(\mathbb{R})$-bounded operator. More precisely, there exists $\alpha > 1$ independent of m such that

$$\|S_{k,m}\|_{B(L^2(\mathbb{R}))} \le \alpha^m \|k\|_{\mathrm{Lip}(\mathbb{R})}^m. \tag{5.172}$$

Proof We induct on m. Theorem 5.45 covers (5.172) with $m = 1$. As we did in Theorem 5.45, by considering k_j instead of k, it can be assumed that $k \in \mathscr{B}^2(\mathbb{R})$. Consider the limit as $r \downarrow 0$. By integration by parts,

$$\begin{aligned}
S_{k,m}f(x) &= \frac{1}{m}\int_{\mathbb{R}}(k(x)-k(y))^m f(y)\frac{\mathrm{d}}{\mathrm{d}y}(x-y)^{-m}\mathrm{d}y \\
&= \int_{\mathbb{R}}\frac{k'(y)(k(x)-k(y))^{m-1}}{(x-y)^m}f(y)\mathrm{d}y + \frac{1}{m}\int_{\mathbb{R}}\frac{(k(x)-k(y))^m}{(x-y)^m}f'(y)\mathrm{d}y.
\end{aligned}$$

Hence by the induction assumption (5.172) for $m-1$, we can find a constant D independent of m,

$$\begin{aligned}
\|S_{k,m,r} f\|_2 &\le D\left(\|S_{k,m-1}(k' f)\|_{L^2(\mathbb{R})} + \|k\|_{\mathrm{Lip}}^m \cdot p_3(f)\right) \\
&\le D\left(\alpha^{m-1}\|k\|_{\mathrm{Lip}}^m \cdot p_3(f) + \|k\|_{\mathrm{Lip}}^m \cdot p_3(f)\right) \\
&\le 2D\alpha^{m-1} p_3(f).
\end{aligned}$$

Then Theorem 5.44 yields a constant D' satisfying $\|S_{k,m,r}\|_{B(L^2(\mathbb{R}))} \le 2DD'\alpha^{m-1}$. Hence $\alpha \equiv 2DD'$ does the job. Thus the proof of (5.172) is complete.

To conclude this chapter, we apply the $T1$ theorem to the boundedness of Cauchy integral operators.

Corollary 5.11 (Cauchy integral operator) *Let $k : \mathbb{R} \to \mathbb{R}$ be a Lipschitz function. Assume $\|k\|_{\mathrm{Lip}(\mathbb{R})} \ll 1$ and define $\gamma(x) \equiv x + k(x)i \in \mathbb{C}$ for $x \in \mathbb{R}$. The line integral, which is called the Cauchy integral,*

$$T_\gamma f(x) = \int_{\mathbb{R}} \frac{(1 + ik'(y))f(y)\mathrm{d}y}{x - y + i(k(x) - k(y))} \quad (x \in \mathbb{R})$$

is an $L^2(\mathbb{R})$-bounded operator. The operator T_γ is called the Cauchy integral operator.

Proof By the geometric series, we have

$$\begin{aligned}
T_\gamma f(x) &= \sum_{m=0}^{\infty} \frac{1}{i^m} \int_{\mathbb{R}} \left(\frac{k(x) - k(y)}{x - y}\right)^m \frac{(1 + ik'(y))f(y)}{x - y} \mathrm{d}y \\
&= \sum_{m=0}^{\infty} \frac{1}{i^m} S_{m,k}[(1 + ik')f].
\end{aligned}$$

Now simply apply Theorem 5.46 to obtain a conclusion.

Exercises

Exercise 5.66 Let $a \in \mathrm{BMO}(\mathbb{R}^n)$, and let $b \in \dot{B}^0_{\infty\infty}(\mathbb{R}^n)$.

Choose an even function $\psi \in \mathscr{S}(\mathbb{R}^n)$ so that $\chi_{B(1)} \le \psi \le \chi_{B(2)}$ and set $\varphi \equiv \psi - \psi_{-1} = \psi - \psi(2\star)$. Define T_a by $T_a f \equiv \sum_{j=-\infty}^{\infty} \varphi_j(D)a \cdot \psi_j(D)b \cdot \psi_{j-4}(D)f$. Show that T_a is an $L^2(\mathbb{R}^n)$-bounded operator; more precisely, for $f \in L^2(\mathbb{R}^n)$ an infinite sum defining $T_a f$ is $L^2(\mathbb{R}^n)$-convergent and $\|T_a f\|_2 \lesssim \|a\|_{\mathrm{BMO}}\|b\|_{\dot{B}^0_{\infty\infty}}\|f\|_2$ using Theorems 1.57 and 3.24.

Exercise 5.67 Let $f \in H^p(\mathbb{R}^n)$ with $0 < p \le 1$, and let $g \in \mathrm{BMO}$. Then show that $\|\varphi_k(D)f \cdot \varphi_k(D)g\|_{L^p(\ell^1)} \lesssim \|f\|_{H^p}\|g\|_{\mathrm{BMO}}$ using Theorems 1.57 and 3.24.

Textbooks in Sect. 5.5

$T1$ Theorem

We refer to [22, Chapter 9], [33], [64, Chapters 7 and 8] and [86] for the $T1$ theorem.

Cauchy Integral Operators

The boundedness of the Cauchy integral operators is in [67], where Murai gave eight different proofs. See [64, Section 9.6] for more about the account of Murai's proof.

Others

Throughout this section we referred to [28, 29, 86]. We remark that Proposition 4.1 is [29, Theorem (1.14)].

5.6 Notes for Chap. 5

Section 5.1

Section 5.1.1

Lift Operator on the Half Space: Theorem 5.2

Theorem 5.2 is due to Frank and Runst. Frank and Runst considered the lift operator adapted to the half space to study elliptic differential equations in [478].

Extension Operators on the Half Line and Intervals

Besov considered the extension operator from $B^m_{q\infty}(0,a)$ to $B^m_{q\infty}(-a,a)$ using the difference operator in [175, Teorema 1]. Burenkov and Kalyabin obtained a lower estimate of the extension operator $E : W^s_p(0,\infty) \to W^s_p(\mathbb{R})$ in [309].

In [174, 176], Besov proved the following: Let $\phi \in L^p(a,b)$ with $1 \le p \le \infty$. If $(c,d) \supset [a,b]$ and $0 < r < 2$, then there exists $\psi \in L^p(a_1,b_1)$ which extends ϕ so that

$$\sup_{0<h<\frac{d-c}{4}} h^{-r}\|\phi(\star+h)-2\phi+\phi(\star-h)\|_p \lesssim \sup_{0<h<\frac{b-a}{4}} h^{-r}\|\phi(\star+h)-2\phi+\phi(\star-h)\|_p.$$

More is known about the extension of function spaces defined on intervals. See [312, Theorem 2.1] for example.

Extension Operator on the Half Space: Theorem 5.4

Theorem 5.4 is due to Triebel [1068].

Trace Operator on Manifolds

Besov proved the trace theorem in the case of Lipschitz $(n-1)$-dimensional manifolds [185, 186]. We refer to [794, Theorem 1] for the trace operator from $\mathbb{R}^2$ to the unit circle S^1.

Section 5.1.2

The Characterization of $\mathscr{S}'(\mathbb{R}^n)$ to the Lipschitz Domain Ω: Theorem 5.10

Theorem 5.10 is due to Rychkov [903, Proposition 3.1].

Density of Function Spaces on Domains

Besov discussed the density of Besov spaces in [177], where the domain Ω satisfies $x+y \in \Omega$ for all $x \in \Omega$ and $y=(y_1, y_2, \dots, y_n)$ with $y_j \geq 0$ for $j=1,2,\dots,n$. Besov discussed the density of smooth functions defined on a neighborhood of the domain in [180].

Differences and Function Spaces on Domains

One of the traditional ways to define function spaces is to use the difference operator [189].

Muramatu defined Besov spaces and potential spaces using difference and complex interpolation [817, p. 327].

Interpolation of the Function Spaces on Domains

We refer to [208–210] for interpolation of function spaces $A^s_{pq}(\Omega)$.

Homogeneous Besov Spaces on Domains

See [293, 614, 615]. In [615] the authors proposed how to define the homogeneous Besov space $\dot{B}^s_{pq}(\Omega)$ for any domain Ω.

Extension Operators on Domains: Theorem 5.6

Besov considered the extension operator on the unit cube [173, Teorema]. See also [179]. Extension operators are considered in many settings [323, 364, 846, 1004]. See [157, 243] for more. It is very important to consider what condition is necessary to obtain the extension. So we can define the notion of the A^s_{pq}-extension domain if we have a satisfactory extension operator. Jones considered the extension operator from $W^s_p(\Omega)$ to $W^s_p(\mathbb{R}^n)$ when Ω is an (ε, δ)-domain in $\mathbb{R}^n$ [650], $s \in \mathbb{N}$ and $1 \le p \le \infty$ We refer to [512, 1101, 1102] for the case where $s = 1$ and $n = 2$. See Definition 5.5 for its definition. We refer to [9, Chapter 6] for extension operators in intervals in $\mathbb{R}$. Besov considered the extension operator in [178, p. 40 Teorema]. See [786, Theorem 2].

Many Definitions of Function Spaces on Domains

It is rather difficult to define function spaces on domains. We refer, for example, to [299] for the method of defining the space $C^r(\Omega)$ naturally in connection with the boundary condition.

There are several definitions of the function spaces on a domain. For example, we have the following definition:

Definition 5.22 ($\widetilde{A}^s_{pq}(\Omega)$, $A^s_{pq\Omega}(\mathbb{R}^n)$)

1. Let Ω be an open set. Define $\widetilde{A}^s_{pq}(\Omega)$ to be the closure of $C^\infty_c(\Omega)$ in $A^s_{pq}(\mathbb{R}^n)$.
2. Let Ω be a set. Define $A^s_{pq\Omega}(\mathbb{R}^n)$ to be the set of all $f \in A^s_{pq}(\mathbb{R}^n)$ whose support is contained in $\bar{\Omega}$.

Analogously, we define $\tilde{H}^s(\Omega)$ and so on. For example in 2000 McLean showed that $\tilde{H}^s(\Omega) = H^s_{\bar{\Omega}}(\mathbb{R}^n)$ [59]. See also [353].

For $L^p(\mathbb{R}^n) \approx F^0_{p2}(\mathbb{R}^n)$, this definition coincides with the other plausible definition. But in general, we need to some assumptions in order that this definition agrees with the one in this book; see textbook [103]. The restriction to more than one open set like Exercise 5.7 is considered in [313]. See [817, 902, 903, 1086] including the recent works [237, 240] by Besov.

We refer to Besov [197, 204], Triebel [1041, 1042], Lions and Magnes [53, 733–735] and Muramatsu [816] for interpolation theory of function spaces on domains.

Wavelet Expansion of Function Spaces on Domains

We can develop a theory of wavelet expansion on domains. In connection with fractal and the atomic decomposition of function spaces, Triebel proposed the notion of exterior thick domains; see [1078, Definition 12].

Density of Function Spaces

Although the method taken up in this book is standard and traditional, it is still hard to define and investigate function spaces on domains. As we have seen, many people made much effort to consider extension operators. Due to difficulty in defining function spaces on domains, as well as extension operators to the whole space, we are interested in proving that the set of the smooth functions having compact support forms a dense subset. See [196].

Entropy Number

Edmunds and Haroske considered the entropy number of embeddings into spaces of Lipschitz type [446, 447]. We refer to [559] for more details of these Lipschitz-type embeddings. Edmunds and Triebel dealt with embeddings from Besov spaces or Triebel–Lizorkin spaces into Besov spaces or Triebel–Lizorkin spaces (unweighted case) [449]. See [557] for an attempt to use the Orlicz spaces. In [563] Haroske and Moura considered embeddings from Besov spaces or Triebel–Lizorkin spaces into Lipschitz function spaces, where the smoothness of the function spaces in the starting domain is generalized using the slowly varying functions in the sense of Karamata. A function $\varphi : (0, \infty) \to (0, \infty)$ is said to be *slowly varying* if for all $s > 0$ $\varphi(st)/\varphi(t) \to 1$ as $t \downarrow 0$. We refer to [558, Section 3] for the entropy number of the embedding of weighted Sobolev Orlicz spaces. For embeddings from Besov spaces or Triebel–Lizorkin spaces with weights having polynomial growth or logarithmic growth into unweighted Besov spaces or Triebel–Lizorkin spaces, see [567, Theorem 3.6], [569, Theorem 4.1], [570, Theorem 4.7], [578, Theorem 4.1], [699, Theorem 2], [700, Theorem 2] and [711, Theorem 6], as well as more recent works [238, 239].

Entropy Number for the Functions in Bounded Domains

There are a huge number of embeddings of function spaces. In particular, there are many results for the Sobolev embedding, even if we restrict them to the one related to function spaces in this book. For example, Edmunds and Triebel investigated and quantified compact embeddings by entropy number and approximation number in [449].

Compact Embeddings for Radial Functions

The situation is different when we restrict ourselves to the radial setting; see [698, Theorem 4].

Skrzypczak estimated the entropy number of the embedding from radial Besov space to $E_{\nu,\rho}(\mathbb{R}^n)$ [990, Theorem 2], where $E_{\nu,\rho}(\mathbb{R}^n)$ is the Orlicz space generated by $\Phi_{\nu,\rho}(t) = t^\nu \exp(t^\rho)$. Skrzypczak and Tomasz also worked in the setting of noncompact Riemannian manifolds [995, Theorem 0.1]. We refer to [1083, Theorem 16] for the case of mixed derivatives.

In principle, the proof hinges upon the decomposition methods presented in this book. Of course, the decomposition methods must be transformed in each setting. Meanwhile, Triebel proposed to use the Hardy operator in [1084, Theorem 3.4]. See also [107]. Compactness of the embedding related to A_∞ weights are investigated in [567, Proposition 2.1]. See Definition 6.6 for its definition. Leopold and Skrzypczak considered function spaces on quasi-bounded uniformly E-porous domains. They calculated the entropy number and applied their results to estimate the spectrum [712]. See [568] for more about this direction of research.

Approximation Number

Birman and Solomjak calculated the entropy number and the approximation number of compact embeddings of the function spaces defined on domains [251].

Edmunds and Triebel estimated the entropy numbers and the approximation numbers from above by dividing the proof into 11 steps in [449]. As far as the approximation numbers on the embedding of unweighted function spaces are concerned, the shape of the domains does not matter that much; see [1074, Theorem 2.14]. Haroske calculated the entropy number and the approximation number in [556]. She considered the weighted setting. Later on, Edmunds and Haroske considered the embedding from function spaces close to Lipschitz spaces in [447]. Skrzypczak and Tomasz took advantage of the symmetric setting in [994]. See [563, 567, 570, 991, 992, 997] for more about the calculation of the approximation number, as well as the entropy number in the weighted settings.

Triebel calculated the approximation on fractals in [1076] and also proposed the method using the Hardy inequality in [1084].

See [854, Theorem 26] for the relation between the approximation number and the entropy number of the embedding $\iota : A^s_{pq}(\Omega) \to L^r(\Omega)$, where Ω is a Lipschitz domain.

Envelop

Best m-Term Approximation

See [404–407] for approximation of the elliptic operators using a similar approach. See [549–551] for the calculation of this quantity when $\{\Phi_j\}_{j=1}^{\infty}$ is the wavelet system described in this book.

There are other ways of measuring the size of approximation. Keeping in mind that $B^{n/p}_{pq}(\mathbb{R}^n) \subset L^1_{\rm loc}(\mathbb{R}^n)$, we define

$$\mathcal{E}_G B^{n/p}_{pq}(t) \equiv \sup\{f^*(t) \,:\, \|f\|_{B^{n/p}_{pq}} \le 1\} \quad 0 < t < \varepsilon,$$

where $f^*(t)$ denotes the decreasing rearrangement and $\varepsilon \in (0,1)$ is a fixed positive number [103, (12.37)]. We let

$$\psi(t) \equiv \frac{1}{\mathcal{E}_G B^{n/p}_{pq}(t)}, \qquad \Psi(t) \equiv -\log \mathcal{E}_G B^{n/p}_{pq}(t)$$

for $0 < t < \varepsilon$ and let μ_Ψ be the associated Borel measure [103, p. 193]. Let $1 < q \le \infty$. Then in [103, Theorem 13.2(i)], it is shown that there exists C_v such that

$$\left(\int_0^\varepsilon \psi(t)^v f^*(t)^v \, d\mu_\Psi(t)\right)^{1/v} \le C_v \|f\|_{B^{n/p}_{pq}} \quad (f \in B^{n/p}_{pq}(\mathbb{R}^n))$$

if and only if $v \ge q$. Also, in [103, Theorem 13.2(i)], we know $\mathcal{E}_G B^{n/p}_{pq}(t)$ behaves like $|\log t|^{\frac{1}{q'}}$. So, the pair $(|\log t|^{\frac{1}{q'}}, q)$ is referred to as the envelope of $B^{n/p}_{pq}(\mathbb{R}^n)$. We refer to the papers [511, 560, 563, 564, 566] and the textbook [36] for more details of this field.

Weyl Number

See [841] for the Weyl number of the embeddings of Besov spaces of tensor type.

Section 5.1.3

Function Spaces on Uniformly C^m-Open Sets

See [55].

Section 5.1.4

Function Spaces on Lipschitz Domains

See [658, 659].

Local Means on Lipschitz Domains: Theorem 5.9

See [900, Theorem 3.2].

Calderón's Reproducing Formulas on Lipschitz Domains–(5.21)

See [900, Proposition 2.1].

Universal Extension Operators on Lipschitz Domains: Theorem 5.8

Theorem 5.8 is [900, Theorem 2.2]. On Lipschitz domains, Rychkov [900] showed that the extension operator exists. We also refer to [304, 305, 307, 308, 901] for this direction of research.

We refer to [1074] for function spaces on Lipschitz domains. In [1074], using the restriction operator and the extension operator (Theorem 5.3) skillfully, Triebel transplanted what we obtained on the whole space $\mathbb{R}^n$ to domains.

Decompositions of the Functions on Domains

Triebel [1074] described decomposition theory of function spaces such as atomic decomposition and quarkonial decomposition. See [379, 1078] for wavelet decomposition of function space on domains.

Section 5.2

Section 5.2.1

General Facts on Pseudo-differential Operators

We refer to [607].

Section 5.2.2

Boundedness of Pseudo-differential Operators: Theorem 5.15

Theorem 5.15 is in [258, 608, 896, 897, 1036]. For example, when $q < \infty$, Theorem 5.15 for Triebel–Lizorkin spaces can be found in [897]. Our technique is close to that of [1036]. See [718] for the compactness of the pseudo-differential operators.

Furthermore, in [518] Grafakos and Torres investigated the class $\dot{S}^m_{11}$ of pseudo-differential operators. This matches the molecular decomposition (Theorem 4.1); see [518].

The underlying idea of the proof of the boundedness of the pseudo-differential operators is that some fundamental properties of the atoms remain unchanged under their action. Such a notion is called "almost diagonal". This notion is presented in the fundamental paper [483] and is expanded in [120, 1110, 1175].

Section 5.2.3

Pseudo-differential Operators on Sobolev Spaces

The class $B^l_{\infty\infty}S^m_{1\delta}(\mathbb{R}^n)$ dates back to 1978 (see [94, 382]).

Section 5.2.4

Pseudo-differential Operators on Modulation Spaces

Calderón and Vailancourt [335] showed pseudo-differential operators with symbols in S^0_{00} are $L^2(\mathbb{R}^n)$-bounded (see Theorem 5.19). Sjöstrand [973] extended S^0_{00} to modulation space $M^0_{\infty 1}$. See [892, 1024, 1025] for approaches using the basis. Our proof depends on the technique of [682]. It may be interesting to note that Sjöstrand proved this result by means of the so-called T^*T-method, while the results on modulation spaces are beyond the reach of the method employed in [973]. Toft investigated the boundedness of pseudo-differential operators on modulation spaces; see [1033, 1034].

In [1021, Theorem 1.1], Sugimoto and Tomita obtained the following negative result.

Proposition 5.21 *Let $1 < p, q < \infty$, $m \in \mathbb{R}$ and $0 < \delta < 1$. If*

$$m > -\delta n \left| \frac{1}{q} - \frac{1}{2} \right|,$$

then there exists a symbol $\sigma \in S^m_{1\delta}$ such that $\sigma(X, D)$ is not bounded on $M^0_{pq}(\mathbb{R}^n)$.

The construction of such a σ is as follows. Let $\varphi, \eta \in \mathscr{S}(\mathbb{R}^n)$ be radial functions satisfying $\chi_{B(1/8)} \le \varphi \le \chi_{B(1/4)}$, $\chi_{B(\sqrt[4]{2})\setminus B(1/\sqrt[4]{2})} \le \eta \le \chi_{B(\sqrt{2})\setminus B(1/\sqrt{2})}$. Moreover, we assume that φ is radial. Define

$$\sigma(x,\xi) \equiv \sum_{j=j_0}^{\infty} 2^{jm} \left(\sum_{k \in B(2^{j\delta/2})\setminus\{0\}} \mathrm{e}^{-ik(2^{j\delta/2}x-k)} \mathscr{F}^{-1}\varphi(2^{j\delta/2}x-k) \right) \eta(2^{-j}\xi),$$

where $0 < \delta < 1$ and $j_0 \in \mathbb{N}_0$ is chosen to satisfy $\sqrt[4]{2} - 2^{j_0(\delta-1)+\frac{5}{4}} \ge 1$ and $64n \le 2^{j_0\delta}$. See [1021, Theorem 2.1] for a negative result and [1022, Theorem 1] for a characterization of the boundedness of the pseudo-differential operator with symbol $S^m_{\rho\delta}$.

Section 5.3

Section 5.3.1

Bounded Holomorphic Calculus

Yagi, M^c^Intosh, Auscher and Duong are pioneers of the sectorial operators and contributed to the theory of bounded H_∞-holomorphic calculus [128, 151, 769].

Function Spaces Associated with Operators–Hardy Spaces and BMO

When we have partial differential operators, we can consider function spaces based on the differential operators with the help of bounded H_∞-holomorphic calculus. We can apply it to partial differential equations. See Duong and Yan [441, 442]. As an example, we have the duality $H^1(\mathbb{R}^n)$-BMO$(\mathbb{R}^n)$ associated with operators; see [442].

Others

I have learned from Professor Schlag a lot about Theorem 5.34.

Section 5.3.2

Sectorial Operators and H_∞-Holomorphic Calculus and Square Root of the Sectorial Operators

In addition to the papers [128, 151, 769] see [1139].

Section 5.3.3

Heat Semi-group and Navier–Stokes Equations

The heat semi-group is applied to the Navier–Stokes equation. Recall that the Navier–Stokes equation is given by

$$\begin{cases} \partial_t u - \mu \Delta u + (u \cdot \nabla) u + \nabla p = 0 \text{ in } \Omega \times (0, T), \\ \nabla \cdot u = 0 \text{ in } \Omega \times (0, T), \\ u(\star, 0) = u_0 \text{ in } \Omega, \end{cases}$$

where Ω is an open set in $\mathbb{R}^n$ and $T > 0$. More precisely, we let

$$u(x,t) = (u^1(x,t), \ldots, u^n(x,t)), \quad u_0(x) = (u_0^1(x), \ldots, u_0^n(x))$$

and define

$$(u, \nabla) \equiv \sum_{l=1}^n u^l \frac{\partial}{\partial x_l}, \quad \nabla \cdot u \equiv \sum_{l=1}^n \frac{\partial u^l}{\partial x_l}.$$

One of the main problems for this equation is to find in terms of smoothness the border where the time local well-posedness is true or false.

1. $L^2(\mathbb{R}^n)$, Leray [707] in 1933.
2. $\dot{H}^{n/2-1}(\mathbb{R}^n)$, Kato and Ponce [663].
3. $H^{n/2-1}(\mathbb{R}^n)$, Fujita and Kato [488] in 1964.
4. $L^n(\mathbb{R}^n)$, Kato [661] in 1984 and Giga and Miyakawa [503] in 1985 independently.
5. $L^\infty(\mathbb{R}^n)$, Cannone [14] Giga, Inui and Matsui [502].
6. $L^p(\mathbb{R}^n)$ for $n \le p < \infty$, Giga [501].
7. $b_{p\infty}^{\frac{n}{p}-1}(\mathbb{R}^n) = \overline{\mathcal{S}(\mathbb{R}^n)}^{B_{p\infty}^{\frac{n}{p}-1}(\mathbb{R}^n)}$ for $n < p < \infty$, Amann in 1997 and 2000 [131, 132].
8. $b_{\infty\infty}^{-1/2}(\mathbb{R}^n) = \overline{\mathcal{S}(\mathbb{R}^n)}^{B_{\infty\infty}^{-1/2}(\mathbb{R}^n)}$, Kobayashi and Muramatsu [677].
9. $B_{p\infty}^{\alpha}(\mathbb{R}^n)$ for $0 \le \alpha \le n/p - 1$ and $n \le p < \infty$, Sawada [912].
10. $\mathrm{BMO}^{-1}(\mathbb{R}^n) = \{\nabla \cdot b : b \in \mathrm{BMO}(\mathbb{R}^n)^n\} = \dot{F}_{\infty 2}^{-1}(\mathbb{R}^n)$, Koch and Tataru [674].
11. $\dot{B}_{\infty\infty}^{-1}(\mathbb{R}^n)$, Bourgain and Pavlovic [284].
12. $\dot{B}_{p\infty}^{\frac{n}{p}-1}(\mathbb{R}^n)$ for $n < p < \infty$, Kozono and Yamazaki [687].
13. $\dot{B}_{n\infty}^{0}(\mathbb{R}^n)$, Cannone and Planchon [336].
14. $B_{\infty\infty\infty}^{(-1,1/2)}(\mathbb{R}^n)$, Yoneda [1171]; see also Example 6.14. Yoneda discussed the existence of the time local solution in the space wider than $B_{\infty 2}^{-1}(\mathbb{R}^n)$. The heart of the matter is to consider a space wider than the above space by adjusting the parameter q [1171]. See also [148].

See [351, 1028] for an application of function spaces to Euler equations.

Applications of Heat Semi-group to the Study of Function Spaces

We refer to [793, 795] for applications of the heat semi-group to the study of Besov spaces on $\mathbb{R}^n$.

Section 5.3.4

Applications of Function Spaces to the Wave Equations

See [881] for example.

Section 5.3.5

Schrödinger Propagator: Theorem 5.35

We refer to [167, Theorem 1] for Theorem 5.35. We refer to [790] for the boundedness of the Schrödinger propagator on modulation spaces. We have a counterpart of Theorem 5.35 to homogeneous Besov spaces by Mizuhara [796, p. 158, Theorem]. When $p = 2$, the result follows trivially by Plancherel's theorem. Otherwise, Mizuhara showed $e^{it\Delta}$ is unbounded. As Miyachi has shown in [783], Sobolev spaces are not sufficient to grasp the boundedness property of modulation spaces.

Applications of Modulation Spaces to PDE

Wang and Huang applied modulation spaces to partial differential equations such as the KdV equation, and nonlinear Schrödinger equations [1107].

We refer to [409] for the study of the Besov regularity of the nonlinear elliptic differential equations.

Section 5.4

Section 5.4.1

Gaffney-Type Estimates: Lemma 5.29

See [491] for the original Gaffney inequality and [149, Lemma 2.1]. We refer to [592, Lemma 2.3] for Lemma 5.30.

A Priori Estimate on the Whole Space

Exercise 5.58 nicely explains the reason why we need to introduce function spaces; C^2 is not enough to describe the smoothness property of the solution to the elliptic differential equations. For example, we refer to [190, Teorema 1] for what else is needed to control $f_{x_1x_2}$ when we have a function $f : \mathbb{R}^2 \to \mathbb{R}$ such that $f_{x_1x_1}$ and $f_{x_2x_2}$ are bounded.

Section 5.4.2

A Priori Estimate on the Half Space

We refer to [1053, 1056] for the works by Triebel. See [781] for the case of manifolds.

A Priori Estimate on Domains

We refer to [195] for the estimates for anisotropic Besov spaces. See [197, 198] for those on domains with flexible horn condition; see Definition 5.5 for the definition. We refer to [402, 403, 408] for a priori estimates for Besov spaces.

A Priori Estimate on Manifolds

Hodge's theorem is dealt with in differential geometry, but a priori estimates are subjects in analysis. So in the textbooks of differential geometry, it can frequently happen that only the results are introduced. However, the textbook [116] in differential geometry covers nicely both its geometric side and its analytic side.

A Priori Estimate and Oka's Coherent Theorem

We refer to [48] for fundamental facts on complex analysis of several variables, where we can find Oka's theorem.

Section 5.5

Section 5.5.1

$T1$ Theorem-Theorems 5.43 and 5.44

The $T1$ theorem is due to David and Journé [413]. The idea of proving Theorem 5.44 via Theorem 5.43 is also due to this paper. David and Journé used it to prove $L^2(\mathbb{R}^n)$-boundedness of the Cauchy integral operators. Later, David, Journé and Semmes got rid of Fourier analysis by proposing the Tb theorem in [414]. The theory carries over to the metric spaces; see [833].

The formula (2.107) is referred to as the *homogeneous Calderòn reproducing formula*. Han considered a counterpart on spaces of homogeneous type and applied to the $T1$ theorem there [535]. The $T1$ theorem is still in progress and it is also used to solve the Kato conjecture, the problem of specifying the domain of the square root of nonsmooth elliptic differential operators.

We refer to [417, 542, 776, 1142] for $T1$ theorems on Besov and Triebel–Lizorkin spaces on spaces of homogeneous type and their applications.

See [1037] for the systematical approach to the boundedness of the integral operators using the atomic decomposition.

Section 5.5.2

Cauchy Integral Operators

See the works [18, 365, 366, 413] for the boundedness of the Cauchy integral operators. We refer to [792, Appendix A] for a proof of the boundedness of T_a by means of the bilinear estimates of the Fourier multipliers. Theorem 5.45 is due to Calderón, while Theorem 5.46 is due to Coifman and Meyer.

Chapter 6
Various Function Spaces

In recent decades, the subject of function spaces has undergone a rapid diversification and expansion, though the decomposition of functions and operators into simpler parts, mentioned in Chap. 4, remains a central tool and theme. One of the benefits in the theory of function spaces is that we can adjust many problems of function spaces in analysis when we handle some problems after familiarizing ourselves with the fundamental theory. In this book, we have studied the huge development of the rich theory of function spaces. Here we will present how we apply this to the theory of other function spaces. Section 6.1 is a starting point, which describe the function spaces arising in many contemporary problems. Section 6.2 explains what happens to the function spaces appearing in Sect. 6.1 if we use Hardy spaces. Although the Hardy space $H^p(\mathbb{R}^n)$ is proved to be isomorphic to $\dot{F}^0_{p2}(\mathbb{R}^n)$ in Sect. 3.2.9, this approach is of independent interest. Section 6.4 answers the same questions for Besov spaces and Triebel–Lizorkin spaces. Section 6.5 views what we do in Sect. 6.4 in a different manner. In Sects. 6.3 and 6.4 we replace p with more generalized spaces. Mainly, we not only replace L^p with the weighted Lebesgue space defined in Sect. 6.1.2 but also we generalize s. In Sect. 6.6 we show that our theory can be staged over sets. Finally, we formulate and prove the Kato conjecture in Sect. 6.7.

6.1 Various Function Spaces

In recent years, it was realized that the classical function spaces are no longer appropriate spaces when we attempt to consider a number of contemporary problems arising naturally in many other branches in science such as non-linear elasticity theory, fluid mechanics, image restoration, mathematical modelling of various physical phenomena, solvability problems of non-linear partial differential equations and so on. It was evident that many such problems are naturally related to

Y. Sawano, *Theory of Besov Spaces*, Developments in Mathematics 56,
https://doi.org/10.1007/978-981-13-0836-9_6

the problems with non-standard local growth (see e.g., [73, 79, 122, 908, 1117, 1182] and references therein). It thus became necessary to introduce and study some new function spaces from various viewpoints. As examples of these spaces, we can list variable exponent spaces and grand function spaces. These spaces were intensively investigated by researchers in the last two decades.

This section is a setup for the purpose of stating the results related to Besov spaces and Triebel–Lizorkin spaces.

We can say that Besov spaces and Triebel–Lizorkin spaces are tools to investigate Lebesgue spaces more precisely. However, for band-limited distributions f, then $f \in A^s_{pq}(\mathbb{R}^n)$ if and only if $f \in L^p(\mathbb{R}^n)$, for any $0 < p, q \leq \infty$ and $s \in \mathbb{R}$. Observe also that for band-limited distributions f, $f \in F^s_{\infty q}(\mathbb{R}^n)$ if and only if $f \in L^\infty(\mathbb{R}^n)$, for any $0 < q \leq \infty$ and $s \in \mathbb{R}$. So, for such an f, Besov spaces and Triebel–Lizorkin spaces are not enough. To investigate such an f we can use many other function spaces. However, their definition is in general difficult; it involves many parameters and eventually complicates things. One of the ways is to follow the idea of Besov spaces and Triebel–Lizorkin spaces. So, we are interested in replacing $L^p(\mathbb{R}^n)$ with some other quasi-Banach function spaces, say $X(\mathbb{R}^n)$. See Definition 6.1 for the definition of quasi-Banach function spaces. We are interested in the case where $X(\mathbb{R}^n)$ is the Morrey space $\mathcal{M}^p_q(\mathbb{R}^n)$ (Sect. 6.1.5) or the weighted Lebesgue space $L^p(w)$ (Sect. 6.1.2) for example, both of which we define later. By doing this, we hope to understand the property of X more clearly. To explain this attempt more deeply, we remark that the resulting function space will have three parameters $X(\mathbb{R}^n)$, q and s. As we have seen in Propositions 2.1 and 2.6, the smaller q is, the smaller the resulting space is, or equivalently, the better the resulting space is. Therefore, we expect that we will have a new control of the linear operators by adjusting q in the new function space. Similar to the boundedness property, we expect that this attempt clarifies how strongly s is required to be large in order that the embedding is compact, for example. This section is organized as follows. First we define some terminology to classify the function spaces we consider. We consider weighted Lebesgue spaces (see Sect. 6.1.2), mixed Lebesgue spaces (see Sect. 6.1.3), variable Lebesgue spaces (see Sect. 6.1.4), Morrey spaces (see Sect. 6.1.5), Orlicz spaces (see Sect. 6.1.6) and Herz spaces (see 6.1.7).

6.1.1 Function Norms

We start from the function norms, which describe the size of functions.

To describe the quantity of functions, we will introduce the notion of quasi-Banach function spaces. Let $L^0(\mathbb{R}^n)$ be the space of all the measurable functions on $\mathbb{R}^n$.

Definition 6.1 (Quasi-Banach function space) A linear space $X(\mathbb{R}^n) \subset L^0(\mathbb{R}^n)$ is said to be a *quasi-Banach function space* if $X(\mathbb{R}^n)$ is equipped with a functional $\| \star \|_X : L^0(\mathbb{R}^n) \to [0, \infty]$ enjoying the following properties:

Let $f, g, f_j \in L^0(\mathbb{R}^n)$, $j \in \mathbb{N}$, and let $\lambda \in \mathbb{C}$.

(1) $f \in X(\mathbb{R}^n)$ holds if and only if $\|f\|_X < \infty$.
(2) (Norm property):

(A1) (Positivity): $\|f\|_X \geq 0$.
(A2) (Strict positivity) $\|f\|_X = 0$ if and only if $f = 0$ a.e.
(B) (Homogeneity): $\|\lambda f\|_X = |\lambda| \cdot \|f\|_X$.
(C) (Triangle inequality): for some $\alpha \geq 1$ $\|f + g\|_X \leq \alpha(\|f\|_X + \|g\|_X)$.

(3) (Symmetry): $\|f\|_X = \| |f| \|_X$.
(4) (Lattice property): If $0 \leq g \leq f$ a.e., then $\|g\|_X \leq \|f\|_X$.
(5) (Fatou property): If $0 \leq f_1 \leq f_2 \leq \cdots$ and $\lim_{j\to\infty} f_j = f$, then $\lim_{j\to\infty} \|f_j\|_X = \|f\|_X$.
(6) For all measurable sets E with $|E| < \infty$, we have $\|\chi_E\|_X < \infty$.

If $\alpha = 1$ and the following property:

$$\chi_E f \in L^1 \text{ for all } f \in X(\mathbb{R}^n), |E| < \infty \tag{6.1}$$

is satisfied, then $X(\mathbb{R}^n)$ is called the *Banach function space*.

Sometimes we need to extend this notion so that we include many important function spaces.

Definition 6.2 (Ball quasi-Banach space) A linear space $X(\mathbb{R}^n) \subset L^0(\mathbb{R}^n)$ is said to be a *ball quasi-Banach function space* if $X(\mathbb{R}^n)$ is equipped with a functional $\| \star \|_X : L^0(\mathbb{R}^n) \to [0, \infty]$ enjoying (1)–(5) in Definition 6.1, as well as the following properties (6)$'$ and (7):

Let $f, g, f_j \in L^0(\mathbb{R}^n)$, $j \in \mathbb{N}$, and let $\lambda \in \mathbb{C}$.

(6)$'$ For all balls B, we have $\|\chi_B\|_X < \infty$.
(7) For all balls B and $f \in X$, we have $\chi_B f \in L^1(\mathbb{R}^n)$.

If $\alpha = 1$, then X is said to be a *ball Banach function* space.

From Definitions 6.1 and 6.2 we have the following relations between the above notions:

- Banach function spaces are ball Banach function spaces and quasi-Banach function spaces.
- Ball Banach function spaces and quasi-Banach function spaces are ball quasi-Banach function spaces.

6.1.1.1 Weak Function Spaces

In general, weakening its membership condition, we can enlarge the space $X(\mathbb{R}^n)$ to obtain $\mathrm{w}X(\mathbb{R}^n)$.

Definition 6.3 (Weak function space) Let $X(\mathbb{R}^n)$ be a ball quasi-Banach space. Then the *weak X-space* $\mathrm{w}X(\mathbb{R}^n)$ is the set of all $f \in L^0(\mathbb{R}^n)$ for which $\|f\|_{\mathrm{w}X} \equiv \sup\limits_{\lambda>0} \|\lambda\chi_{(\lambda,\infty)}(|f|)\|_X < \infty$.

6.1.1.2 An Example of Herz–Morrey Spaces

Although we take up many ball Banach function spaces in this chapter, we give some examples before we go on further.

Example 6.1 Let $0 < p, q \le \infty$ and $\alpha, \lambda \in \mathbb{R}$. The *Herz–Morrey space* $K^{\alpha,\lambda}_{pq}(\mathbb{R}^n)$ is defined to be the set of all $f \in L^0(\mathbb{R}^n)$ for which the norm

$$\|f\|_{MK^{\alpha,\lambda}_{pq}} \equiv \|\chi_{Q_0} f\|_p + \sup_{L\in\mathbb{N}_0} 2^{-L\lambda}\left(\sum_{j=1}^{L}(2^{j\alpha}\|\chi_{C_j} f\|_p)^q\right)^{\frac{1}{q}}$$

is finite. It can be shown that $MK^{\alpha,\lambda}_{pq}(\mathbb{R}^n)$ is a quasi-ball Banach function space.

We learn from the results in this book that the vector-valued norms are useful. So, letting X be a ball quasi-Banach function space, we define

$$\|\{f_j\}_{j=1}^{\infty}\|_{\ell^q(X)} = \|f_j\|_{\ell^q(X)} \equiv \left(\sum_{j=1}^{\infty} \|f_j\|_X{}^q\right)^{\frac{1}{q}} \tag{6.2}$$

and

$$\|\{f_j\}_{j=1}^{\infty}\|_{X(\ell^q)} = \|f_j\|_{X(\ell^q)} \equiv \left\|\left(\sum_{j=1}^{\infty} |f_j|^q\right)^{\frac{1}{q}}\right\|_X.$$

Accordingly we can define $\ell^q(X, \mathbb{R}^n)$ and $X(\ell^q, \mathbb{R}^n)$.

6.1.1.3 Some Fundamental Properties

We can extend the duality $L^p(\mathbb{R}^n)$-$L^{p'}(\mathbb{R}^n)$ in terms of the ball Banach function spaces, as follows.

Definition 6.4 (Köthe dual) If $\|\star\|_X$ is a ball function norm, its *associate norm* $\|\star\|_{X'}$ is defined on $L^0(\mathbb{R}^n)$ by

$$\|g\|_{X'} \equiv \sup\left\{\|f\cdot g\|_1 \,:\, f \in L^0(\mathbb{R}^n), \|f\|_X \le 1\right\} \quad (g \in L^0(\mathbb{R}^n)). \tag{6.3}$$

The space $X'(\mathbb{R}^n)$ collects all measurable functions $f \in L^0(\mathbb{R}^n)$ for which the quantity $\|f\|_{X'}$ is finite. The space $X'(\mathbb{R}^n)$ is called the *Köthe dual* of $X(\mathbb{R}^n)$ or the *associated space* of $X(\mathbb{R}^n)$.

Note that $X'(\mathbb{R}^n)$ is also a ball Banach function space; see Exercise 6.3.

A trivial consequence of (6.3) is

$$\left|\int_{\mathbb{R}^n} f(x)g(x)\mathrm{d}x\right| \le \|f\|_X \|g\|_{X'} \quad (f, \in X(\mathbb{R}^n), g \in X'(\mathbb{R}^n)).$$

Among the properties of the norms, we list the notion of absolute continuity.

Definition 6.5 (Absolute continuity of norms) Let $X(\mathbb{R}^n)$ be a Banach function space. A function f in $L^0(\mathbb{R}^n)$ is said to *have an absolutely continuous norm* in $X(\mathbb{R}^n)$ if $\|f\chi_{E_k}\|_X \to 0$ for every sequence $\{E_k\}_{k=1}^\infty$ satisfying $E_k \to \emptyset$ a.e. in the sense that $\chi_{E_k}(x) \to 0$ for almost every $x \in \mathbb{R}^n$.

Exercises

Exercise 6.1 Show that any nonzero constant function does not have an absolutely continuous norm in $L^\infty(\mathbb{R}^n)$.

Exercise 6.2 Let $1 \le p \le \infty$. Show that the Köthe dual of $L^p(\mathbb{R}^n)$ is $L^{p'}(\mathbb{R}^n)$. Hint: Use the duality $L^p(\mathbb{R}^n)$-$L^{p'}(\mathbb{R}^n)$.

Exercise 6.3

1. Show that $X'(\mathbb{R}^n)$ is a Banach function space for all Banach function spaces $X(\mathbb{R}^n)$.
2. Show that $X'(\mathbb{R}^n)$ is a ball Banach function space for all ball Banach function spaces $X(\mathbb{R}^n)$.

Exercise 6.4 Let $p \ge 1$, and let X be a ball Banach function space. Define a ball Banach function space $X^p(\mathbb{R}^n)$ by $\|f\|_{X^p} \equiv (\| |f|^p \|_X)^{\frac{1}{p}}$.

1. Show that $X^p(\mathbb{R}^n)$ is a normed space by using the Köthe dual.
2. Show that $X^p(\mathbb{R}^n)$ is a ball Banach function space.

Exercise 6.5 Let X be a ball Banach function space. Show that the Köthe dual of X is also a Banach function space.

6.1.2 Weighted Lebesgue Spaces

When we consider the change of variables formula, we need to multiply the Jacobian. So, we are interested in the setting of weighted measures. By a weighted measure, we mean an absolute continuous measure with respect to the Lebesgue measure based on the Radon–Nykodim theorem.

6.1.2.1 Weighted Lebesgue Spaces and Definition of A_r with $1 \leq r \leq \infty$

The weight is a tool to control the growth of functions. By a weight we mean a measurable function which satisfies $0 < w(x) < \infty$ for almost all $x \in \mathbb{R}^n$. Let $0 < p < \infty$, and let w be a weight. One defines

$$\|f\|_{L^p(w)} \equiv \left(\int_{\mathbb{R}^n} |f(x)|^p w(x) \mathrm{d}x \right)^{\frac{1}{p}}.$$

Conventionally, we set $L^\infty(w) \equiv L^\infty(\mathbb{R}^n)$. The space $L^p(w)$ is the set of all measurable functions f for which the norm $\|f\|_{L^p(w)}$ is finite. The space $L^p(w)$ is called the *weighted Lebesgue space* or the *L^p-space with weight w*. When we consider the weak solution to

$$-\sum_{j=1}^{N} \frac{\partial}{\partial x_j} \left(\rho_j \frac{\partial u}{\partial x_j} \right) + \rho_0 \cdot u = f,$$

where ρ_j, $j = 0, 1, 2, \ldots, n$ are functions, we are faced with the weighted Sobolev space whose norm is given by $\|\nabla u\|_{L^2(\rho_i)}$.

One of the main features of this book has been that the maximal operator M plays a key role. Based on this, we consider the following problem to develop a theory of function spaces:

Question 6.1 For what weights w is it the case that

$$\int_{\mathbb{R}^n} Mf(x)^p w(x) \mathrm{d}x \lesssim \int_{\mathbb{R}^n} |f(x)|^p w(x) \mathrm{d}x$$

for $f \in L^0(\mathbb{R}^n)$ when $p \in (1, \infty)$ and that

$$\int_{\{Mf > \lambda\}} w(x) \mathrm{d}x \lesssim \frac{1}{\lambda} \int_{\mathbb{R}^n} |f(x)| w(x) \mathrm{d}x \tag{6.4}$$

for $f \in L^0(\mathbb{R}^n)$ and $\lambda > 0$?

Corollary 1.3 motivates the definition of various classes of weights for the Hardy–Littlewood maximal operator. According to Corollary 1.3 for any locally integrable functions f, g and $\lambda > 0$,

$$\int_{\{Mf > \lambda\}} |g(x)| \mathrm{d}x \lesssim \frac{1}{\lambda} \int_{\mathbb{R}^n} |f(x)| Mg(x) \mathrm{d}x.$$

We write $w(E) \equiv \int_E w(x)dx$ for a measurable set E. When $E = \{\cdots\}$, then we write $w\{\cdots\}$ instead of $w(\{\cdots\})$.

Definition 6.6 (A_p)

1. A locally integrable weight w is said be an A_1*-weight*, if w is strictly positive and finite and there exists C_0 such that

$$Mw(x) \le C_0\, w(x) \tag{6.5}$$

for a.e. $x \in \mathbb{R}^n$. The A_1*-constant*, which is denoted by $[w]_{A_1}$ below, is the minimum of C_0 satisfying (6.5).
2. Let $1 < p < \infty$. A locally integrable weight w is said to be an A_p*-weight*, if $0 < w < \infty$ almost everywhere, and $[w]_{A_p} \equiv \sup\limits_{Q \in \mathcal{Q}} m_Q(w) m_Q(w^{-\frac{1}{p-1}})^{p-1} < \infty$. The quantity $[w]_{A_p}$ is referred to as the A_p*-constant.*
3. A locally integrable weight w is said to be an A_∞*-weight*, if $0 < w < \infty$ almost everywhere, and $[w]_{A_\infty} \equiv \sup\limits_{Q \in \mathcal{Q}} m_Q(w) \exp(-m_Q(\log w)) < \infty$. The quantity $[w]_{A_\infty}$ is referred to as the A_∞*-constant.*
4. For $1 \le p \le \infty$, A_p collects all locally integrable weights w for which $[w]_{A_p}$ is finite.

Remark 6.1 It is easy to check $A_1 \subset A_p \subset A_q \subset A_\infty \subset L^1_{\rm loc}(\mathbb{R}^n)$ by the Hölder inequality whenever $1 \le p \le q \le \infty$.

6.1.2.2 The Class A_1

We now answer the question about when M satisfies (6.4).

Theorem 6.1 *A weight w belongs to A_1 if and only if M is weak $L^1(w)$-bounded; that is, $\lambda w\{Mf > \lambda\} \lesssim \|f\|_{L^1(w)}$ for $f \in L^1(w)$ and $\lambda > 0$.*

Proof Assume that w is an A_1-weight. We have $Mw(x) \lesssim w(x)$ by assumption. If we invoke Corollary 1.3, then we obtain

$$w\{Mf > \lambda\} \lesssim \frac{1}{\lambda} \int_{\mathbb{R}^n} |f(x)| Mw(x)dx \lesssim \frac{1}{\lambda} \|f\|_{L^1(w)},$$

showing the weak-$L^1(w)$ boundedness of M.

Assume instead that M is weak-$L^1(w)$ bounded. We will show that cube testing suffices to prove $w \in A_1$. We will show that, if x is a Lebesgue point of w, that is, if x is a point satisfying

$$\lim_{\substack{R \in \mathcal{Q} \\ R \downarrow x}} m_R(w) = w(x), \tag{6.6}$$

then $m_Q(w) \lesssim w(x)$ for any cube Q containing x. According to Example 1.26 we have

$$w(Q) \le w\left\{M\chi_R > \frac{|R|}{2|Q|}\right\} \lesssim \frac{w(R)|Q|}{|R|}. \tag{6.7}$$

Arranging (6.7), we obtain $m_Q(w) \lesssim m_R(w)$. Shrinking R to a point x, we obtain $m_Q(w) \lesssim w(x)$ from (6.6) which is a desired result.

Now we present some examples of A_1-weights.

Theorem 6.2 *Assume μ is a Radon measure such that $M\mu(x)$ is finite for a.e. $x \in \mathbb{R}^n$. Then for $0 < \delta < 1$, $w \equiv (M\mu)^\delta$ is an A_1-weight, where $M\mu$ is given by* (1.143). *Furthermore, $A_1(w) \lesssim_\delta 1$.*

Proof We have to establish $m_Q(w) \lesssim w(x)$ whenever Q is a cube containing x.

We decompose μ according to $10Q$. Write $(\mu|A) \equiv \mu(A \cap \star)$, the restriction of μ to a measurable subset A. Split μ by $\mu = \mu_1 + \mu_2$ with $\mu_1 \equiv \mu|10Q$ and $\mu_2 \equiv \mu|(\mathbb{R}^n \setminus 10Q)$. We will prove

$$m_Q(M\mu_1{}^\delta) \lesssim w(x), \; m_Q(M\mu_2{}^\delta) \lesssim w(x)$$

for all $x \in Q$. To deal with μ_1 we use Example 1.28, the Kolmogorov inequality to obtain

$$m_Q(M\mu_1{}^\delta) \lesssim m_{10Q}(M\mu_1{}^\delta) \lesssim \left(\frac{\mu(10Q)}{|10Q|}\right)^\delta \lesssim w(x).$$

To handle μ_2, we will make use of a pointwise estimate, which we encounter frequently. Let $y \in Q$. First we write out $M\mu_2(y)$ in full:

$$M\mu_2(y) = \sup\left\{\frac{\mu(R \setminus 10Q)}{|R|} \; : \; R \in \mathcal{Q}, y \in R\right\}.$$

A geometric observation readily shows $\ell(R) \ge 3\ell(Q)$ whenever R is a cube intersecting both Q and $\mathbb{R}^n \setminus 10Q$. Therefore such an R engulfs Q, if we triple R. In view of this fact, we deduce

$$M\mu_2(y) \le 3^n \sup\left\{\frac{\mu(S)}{|S|} \; : \; S \in \mathcal{Q}, \; Q \subset S\right\} \le 3^n \, M\mu(x).$$

This pointwise estimate readily gives $m_Q(M\mu_2{}^\delta) \lesssim w(x)$. Therefore the theorem is now completely proved.

The following proposition gives us a concrete example of A_1-weights.

Proposition 6.1 *Let $0 \le a < n$. Then $w(x) \equiv |x|^{-a}$, $x \in \mathbb{R}^n$ belongs to A_1.*

Proof When $a = 0$, this is clear. Let $a \in (0, n)$. It suffices to take $\mu = \delta_0$, the point measure massed at the origin and $\delta = \frac{a}{n}$, which is a special case of Theorem 6.2. Then w equals $M\mu$ modulo multiplicative constants.

6.1.2.3 The Class A_p

We now answer the question about when M is $L^p(w)$-bounded.

Theorem 6.3 *Let $1 < p < \infty$. Then the following conditions on a weight w are equivalent:*

1. $w \in A_p$.
2. *M is weak-$L^p(w)$ bounded; that is, for every $f \in L^p(w)$ and every $\lambda > 0$,*

$$w\{Mf > \lambda\} \lesssim \frac{1}{\lambda^p} \int_{\mathbb{R}^n} |f(x)|^p w(x) \mathrm{d}x. \tag{6.8}$$

3. *M is strong $L^p(w)$-bounded; that is, for every $f \in L^p(w)$,*

$$\|Mf\|_{L^p(w)} \lesssim \|f\|_{L^p(w)}. \tag{6.9}$$

Needless to say, the implication "$w \in A_p \Longrightarrow$ (6.9)" is significant. Indeed, (6.9) is stronger than (6.8) by virtue of the Chebychev inequality (see Theorem 1.4). Meanwhile, once we assume (6.8), then taking

$$f \equiv w^{-\frac{1}{p-1}} \chi_Q, \quad \lambda \equiv m_Q(w) \tag{6.10}$$

for a given cube Q, we can easily deduce that $w \in A_p$ including $w \in L^1_{\mathrm{loc}}(\mathbb{R}^n)$.

Proof We content ourselves with proving (6.9) when $w \in A_p$ due to the reason above. We may assume that all the cubes are dyadic. We also consider M_{dyadic} instead of M itself, where M_{dyadic} is defined in Definition 1.41. Denote by $M_{w,\mathrm{dyadic}}$ the *weighted dyadic maximal operator* given by

$$M_{w,\mathrm{dyadic}} f \equiv \sup_{Q \in \mathscr{D}} \left(\frac{1}{w(Q)} \int_Q |f(y)|\, w(y) \mathrm{d}y \right) \chi_Q.$$

Using Lemma 1.28, we easily see that $M_{w,\mathrm{dyadic}}$ is $L^p(w)$-bounded with the bound independent of w. Write $\sigma \equiv w^{-\frac{1}{p-1}}$ and define $M_{\sigma,\mathrm{dyadic}}$ analogously. By the definition of the A_p-weights we have $\dfrac{w(Q)}{|Q|} \left(\dfrac{\sigma(Q)}{|Q|} \right)^{p-1} \leq A_p(w)$. Hence

$$m_Q(|f|) \le A_p(w)^{\frac{1}{p-1}} \left(\frac{|Q|}{w(Q)} \left(\frac{1}{\sigma(3Q)} \int_Q |f(y)| \mathrm{d}y \right)^{p-1} \right)^{\frac{1}{p-1}}$$

$$\le A_p(w)^{\frac{1}{p-1}} \left(M_{w,\text{dyadic}} \left[(M_{\sigma,\text{dyadic}}[f\sigma^{-1}]^{p-1} w^{-1}) \right](x) \right)^{\frac{1}{p-1}}$$

for all $x \in Q$. Or equivalently,

$$M_{\text{dyadic}} f(x) \le A_p(w)^{\frac{1}{p-1}} \left(M_{w,\text{dyadic}} \left[w^{-1} M_{\sigma,\text{dyadic}}[|f| w^{\frac{1}{p-1}}] \right](x) \right)^{\frac{1}{p-1}} \quad (x \in \mathbb{R}^n).$$

Inserting this pointwise estimate into $\|M_{\text{dyadic}} f\|_{L^p(w)}$ and using Lemma 1.28, we obtain estimate (6.9).

The classes A_p and $A_{p'}$ are related in the following way.

Lemma 6.1 *Let $1 < p < \infty$, and let $w \in A_p$. Then for all $f \in L^{p'}(w)$,*

$$\|w^{-1} M[f \cdot w]\|_{L^{p'}(w)} \lesssim \|f\|_{L^{p'}(w)}. \tag{6.11}$$

Proof Inequality (6.11) simply rephrases $\|Mg\|_{L^{p'}(w^{1-p'})} \lesssim \|g\|_{L^{p'}(w^{1-p'})}$ for $g \in L^0(\mathbb{R}^n)$, which is certainly true since $w^{1-p'} \in A_{p'}$.

6.1.2.4 The Class A_∞

We now investigate A_∞. The following quantitative observation is the key.

Lemma 6.2 *Let $w \in A_\infty$, and let $q \equiv 1 + \dfrac{1}{2^{n+3}[w]_{A_\infty}}$. Then for all cubes Q,*

$$m_Q^{(q)}(w) \le 2 m_Q(w). \tag{6.12}$$

Inequality (6.12) is called the *reverse Hölder inequality*, since $m_Q(w) \le m_Q^{(q)}(w)$ readily follows from the Hölder inequality.

Proof We may assume that w is bounded by approximating w with a function of the form $\sum_{R \in \mathscr{T}} m_R(w) \chi_R$, where $\mathscr{T} \subset \mathscr{D}$ is a partition of Q. Let $\varepsilon \equiv q - 1$. We use Theorem 1.5 to have

$$\int_Q M_{\mathscr{D}}[\chi_Q w](x)^\varepsilon w(x) \mathrm{d}x = \varepsilon \int_0^\infty \lambda^{\varepsilon-1} w(Q \cap \{M_{\mathscr{D}} w > \lambda\}) \mathrm{d}\lambda.$$

We suppose $\lambda > m_Q(w)$. Consider the set of all maximal dyadic cubes $\{Q_j\}_{j \in J(\lambda)}$ of Q in the set $Q \cap \{M_{\mathscr{D}} w > \lambda\}$ whose average of w exceeds λ. Then thanks to the maximality of each Q_j, we have

$$\frac{w(Q \cap \{M_{\mathscr{D}}[\chi_Q w] > \lambda\})}{2^n \lambda} = \sum_{j \in J(\lambda)} \frac{w(Q_j)}{2^n \lambda} \le \sum_{j \in J(\lambda)} |Q_j| = |Q \cap \{M_{\mathscr{D}}[\chi_Q w] > \lambda\}|.$$

We also note that $Q \cap \{M_{\mathscr{D}}[\chi_Q w] > \lambda\} \subset Q$ for any $\lambda > 0$. Thus,

$$\begin{aligned}\int_Q M_{\mathscr{D}} w(x)^{\varepsilon} w(x) \mathrm{d}x &\le \varepsilon \int_0^{m_Q(w)} \lambda^{\varepsilon-1} w(Q \cap \{M_{\mathscr{D}}[\chi_Q w] > \lambda\}) \mathrm{d}\lambda \\ &\quad + 2^n \varepsilon \int_{m_Q(w)}^{\infty} \lambda^{\varepsilon} |Q \cap \{M_{\mathscr{D}}[\chi_Q w] > \lambda\}| \mathrm{d}\lambda \\ &\le |Q| \left(m_Q(w)\right)^{1+\varepsilon} + \frac{2^n \varepsilon}{1+\varepsilon} \int_Q M_{\mathscr{D}}[\chi_Q w](x)^{1+\varepsilon} \mathrm{d}x.\end{aligned}$$

Since $w \in A_\infty$, we have

$$m_Q(w) \le [w]_{A_\infty} \exp(m_Q(\log w)). \tag{6.13}$$

We define

$$M_{0,\mathscr{D}} f(x) \equiv \sup_{R \in \mathscr{D}} \chi_R(x) \exp(m_R(\log |f|)) \quad (x \in \mathbb{R}^n).$$

Here, roughly speaking, "0" stands for the maximal operator based on the $L^{0+}(\mathbb{R}^n)$-average. From (6.13), we have $M_{\mathscr{D}} w \le [w]_{A_\infty} M_{0,\mathscr{D}} w$. Thus,

$$\int_Q M_{\mathscr{D}} w(x)^{\varepsilon} w(x) \mathrm{d}x \le |Q| \left(m_Q(w)\right)^{1+\varepsilon} + \frac{2^n \varepsilon}{1+\varepsilon} ([w]_{A_\infty})^{1+\varepsilon} \int_Q M_{0,\mathscr{D}}[\chi_Q w](x)^{1+\varepsilon} \mathrm{d}x.$$

By the Lebesgue differentiation theorem and $M_{0,\mathscr{D}}[\chi_Q w^{1+\varepsilon}] = (M_{0,\mathscr{D}}[\chi_Q w])^{1+\varepsilon}$, we have

$$\int_Q w(x)^{1+\varepsilon} \mathrm{d}x \le |Q| \left(m_Q(w)\right)^{1+\varepsilon} + \frac{2^n \varepsilon}{1+\varepsilon} ([w]_{A_\infty})^{1+\varepsilon} \int_Q M_{0,\mathscr{D}}[\chi_Q w^{1+\varepsilon}](x) \mathrm{d}x.$$

We have $m^{(1+\varepsilon)}(w)^{1+\varepsilon} \le (m_Q(w))^{1+\varepsilon} + \dfrac{2^n e \varepsilon}{1+\varepsilon} ([w]_{A_\infty})^{1+\varepsilon} m^{(1+\varepsilon)}(w)^{1+\varepsilon}$ by virtue of Lemma 1.28. It counts that we have $2^n e \varepsilon$ here. Since $[w]_{A_\infty} \ge 1$, we have

$$\frac{2^n \varepsilon [w]_{A_\infty}}{1+\varepsilon} ([w]_{A_\infty})^{\varepsilon} = \frac{2^n [w]_{A_\infty}}{2^{n+3}[w]_{A_\infty} + 1} ([w]_{A_\infty})^{\frac{1}{2^{n+3}[w]_{A_\infty}}} \le \frac{1}{8} \exp\left(\frac{1}{2^{n+3} e}\right) \le \frac{1}{7}.$$

Thus, it follows that $m_Q(w^{1+\varepsilon}) \le \left(m_Q(w)\right)^{1+\varepsilon} + \frac{e}{7} m_Q(w^{1+\varepsilon})$. Since w is assumed to be bounded and $2e \le 7$, it remains to absorb the second term of the right-hand side into the left-hand side.

Corollary 6.1 *Let X be a ball Banach space. Assume that there exists $C_0 > 0$ such that $\|Mf\|_X \le C_0\|f\|_X$ for all $f \in L^0(\mathbb{R}^n)$. Then there exists $\varepsilon > 0$ depending on C_0 such that $\|M^{(1+\varepsilon)} f\|_X \le C_0\|f\|_X$ for some $C_0 > 0$.*

Proof Let $f \in X(\mathbb{R}^n)$. Note that $g = |f| + \sum_{j=1}^{\infty} \frac{M^j f}{(2C_0)^j}$ is an A_1-weight and hence an A_∞-weight. Thus, by Lemma 6.2, we obtain $M^{(1+\varepsilon)}g \le 2Mg$ as long as $2^{n+4}C_0\varepsilon \le 1$. Thus, $\|M^{(1+\varepsilon)} f\|_X \le \|M^{(1+\varepsilon)} g\|_X \le 2\|Mg\|_X \le 4C_0\|f\|_X$ as required.

Another direct consequence of (6.2) is that the class A_p with $1 < p < \infty$ enjoys the *openness property*. We will characterize the class A_p as follows:

Theorem 6.4 *If $w \in A_p$ with $1 < p \le \infty$, then $w \in A_q$ for some $1 \le q < p$. In other words $A_p = \bigcup_{q \in [1,p)} A_q$.*

Proof According to Remark 6.1, we see that $A_v \subset A_u \subset A_\infty$ whenever $1 \le v \le u \le \infty$. If $w \in A_p$ and $p < \infty$, then we are in the position of using Corollary 6.1 to have $w \in A_q$ for $q = \frac{p}{1+\varepsilon}$, where ε is the constant in Corollary 6.1. Let $w \in A_\infty$. Then according to (6.12) and the Hölder inequality, there exists $\delta > 0$ such that $\frac{w(A)}{w(Q)} \lesssim \left(\frac{|A|}{|Q|}\right)^\delta$, whenever A is a subset of a cube Q. As a consequence there exists $0 < \alpha_0, \beta_0 < 1$ such that $w(E) \le \beta_0 w(Q)$ for all measurable sets E and Q such that Q is a cube containing E and that $|E| \le \alpha_0|Q|$. If we contrapose this fact, then there exists $0 < \alpha, \beta < 1$ such that

$$E \subset Q, \quad w(E) < \alpha' w(Q) \Longrightarrow |E| < \beta'|Q| \tag{6.14}$$

for all measurable sets E and cubes Q. This implies that $w(B(x, 2r)) \lesssim w(B(x, r))$ for all $x \in \mathbb{R}^n$ and $r > 0$. To prove $w \in A_u$ for some $u \in (1, \infty)$, fix a cube Q and set $\lambda_k \equiv \left(\frac{D}{\alpha}\right)^k \lambda_0$ for each $k \in \mathbb{N}_0$, where $D \gg 1$ and

$$\lambda_0 \equiv \frac{|Q|}{w(Q)}. \tag{6.15}$$

Denote by $M_{\text{dyadic},Q}$ the dyadic maximal operator with respect to Q. We decompose $\{x \in Q : M_{\text{dyadic},Q}w(x) > \lambda_k\}$ into a disjoint union of dyadic cubes $Q_{k,j}$ such that

$$\lambda_k < \frac{|Q_{k,j}|}{w(Q_{k,j})} \le D\lambda_k = \alpha\lambda_{k+1}. \tag{6.16}$$

Fix $k \in \mathbb{N}_0$. By the Lebesgue differentiation theorem, $w(x)^{-1} \le \lambda_k$ for a.e. $x \in Q \setminus \bigcup_j Q_{k,j}$. We write $\Omega_k \equiv \bigcup_j Q_{k,j}$. Let $\Omega_{-1} \equiv Q$.

Let j_0 be fixed. Recall that two dyadic cubes are disjoint unless one is contained in the other. Therefore we have

$$w(\Omega_{k+1} \cap Q_{k,j_0}) \leq \sum_{\substack{Q_{k+1,j} \\ Q_{k+1,j} \subset Q_{k,j_0}}} w(Q_{k+1,j}) \leq \sum_{\substack{Q_{k+1,j} \\ Q_{k+1,j} \subset Q_{k,j_0}}} \frac{|Q_{k+1,j}|}{\lambda_{k+1}} \leq \frac{|Q_{k,j_0}|}{\lambda_{k+1}}.$$

If we use (6.16) once again, then $w(\Omega_{k+1} \cap Q_{k,j_0}) \leq \dfrac{|Q_{k,j_0}|}{\lambda_{k+1}} \leq \alpha w(Q_{k,j_0})$. We can pass the above inequality to the unweighted one: $|\Omega_{k+1} \cap Q_{k,j_0}| \leq \beta |Q_{k,j_0}|$. Adding the above inequality, we obtain

$$|\Omega_{k+1}| \leq \beta |\Omega_k|. \tag{6.17}$$

This implies that $\left| \bigcap_{k=0}^{\infty} \Omega_k \right| = 0$, so that $\displaystyle\int_Q \frac{dx}{w(x)^\tau} = \sum_{k=-1}^{\infty} \int_{\Omega_k \setminus \Omega_{k+1}} \frac{dx}{w(x)^\tau}$. Invoking (6.16) and (6.17), we obtain

$$\int_Q \frac{dx}{w(x)^\tau} \leq {\lambda_{k+1}}^\tau |\Omega_k| \lesssim {\lambda_0}^\tau \sum_{k=-1}^{\infty} \left(\frac{D}{\alpha}\right)^{k\tau} \beta^k |\Omega_0| \leq {\lambda_0}^\tau |Q| \sum_{k=-1}^{\infty} \left(\frac{D}{\alpha}\right)^{k\tau} \beta^k.$$

Therefore, by choosing $\tau > 0$ sufficiently small the series in the above inequality converges. Thus, we obtain $\displaystyle\frac{1}{w(Q)} \int_Q \frac{dx}{w(x)^\tau} \lesssim \frac{|Q|^\tau}{w(Q)^\tau}$ and hence $w \in A_{1+\tau^{-1}}$.

6.1.2.5 Boundedness of the Singular Integral Operators

We prove the boundedness of the singular integral operators of convolution type.

Theorem 6.5 *Let T be a singular integral operator as in Theorem 1.62, and let $1 < p < \infty$. Then $\|Tf\|_{L^p(w)} \lesssim_{[w]_{A_p}} \|f\|_{L^p(w)}$ for all $f \in L^\infty_c(\mathbb{R}^n)$.*

As a consequence of Theorem 6.5, we can extend T to $L^p(w)$.

Lemma 6.3 *Let $a, q \in [1, \infty)$. Suppose that F, G are nonnegative measurable functions and that $\{G_Q\}_{Q \in \mathscr{D}}$ and $\{H_Q\}_{Q \in \mathscr{D}}$ are collections of nonnegative measurable functions satisfying* (1.240), (1.241), *and* (1.242).

1. *Let $w \in A_\infty$. For all $\lambda > 0$, $K \gg 1$ and $\gamma < 1$, we have*

$$w\{M_{\text{dyadic}} F > K\lambda, G \leq \gamma\lambda\} \lesssim_w w\{M_{\text{dyadic}} F > \lambda\}. \tag{6.18}$$

2. *Let $1 < r < \infty$. If*

$$\min(1, M_{\text{dyadic}} F) \in L^r(w), \tag{6.19}$$

then $\|M_{\rm dyadic}F\|_{L^r(w)} \lesssim_{q,r,a} \|G\|_{L^r(w)}$.

Proof (of Lemma 6.3) Combine (1.243) with (6.14) to have (6.18). Reexamine the proof of (1.245) to have $\|M_{\rm dyadic}F\|_{L^r(w)} \lesssim_{q,r,a} \|G\|_{L^r(w)}$.

We prove Theorem 6.5. Let $\tau \in \mathscr{S}(\mathbb{R}^n)$ satisfy $\chi_{Q(1)} \le \tau \le \chi_{Q(2)}$. We let $A_r = \tau(rD)$ and η be slightly greater than 1. For a cube Q, we let

$$F \equiv |Tf|, \ G \equiv M^{(\eta)}f, \ G_Q \equiv |Tf - T \circ A_{\ell(Q)}f|, \ H_Q \equiv |T \circ A_{\ell(Q)}f|.$$

Then modulo a multiplicative constant, (1.240) and (1.242) follow from the definition and $H_Q = |A_{\ell(Q)} \circ Tf|$. Meanwhile, (1.241), modulo a multiplicative constant, is a consequence of the boundedness of T and the decomposition $f = f\chi_{3Q} + f\chi_{\mathbb{R}^n \setminus 3Q}$. It remains to check (6.19). However, since f has a compact support, say, $[-N, N]^n$, we have only to show $(1 - \chi_{[-5N,5N]^n})M_{\rm dyadic}F \in L^r(w)$, which is a consequence of $M_{\rm dyadic}[\langle\star\rangle^{-n}] \lesssim M \circ M\chi_{Q(1)} \in L^r(w)$.

6.1.2.6 Extrapolation

When we consider estimates of A_p-weights, we face the following type of estimate:

$$\|Th\|_{L^p(W)} \le N([W]_{A_p})\|h\|_{L^p(W)} \quad (h \in L^p(W)), \tag{6.20}$$

where T is a mapping from $L^p(W)$ to $L^0(\mathbb{R}^n)$ and N is a positive increasing function defined on $[1, \infty)$. Extrapolation is a technique to expand the validity of (6.20) for all $1 < p < \infty$ based on the validity of (6.20) for some p_0. Based on [397], we obtain the following extrapolation result:

Theorem 6.6 *Let* $N = N(\star) : [1, \infty) \to (0, \infty)$ *be an increasing function, and let* $1 < p_0 < \infty$. *Suppose that we have a family* $\mathscr{F}$ *of the couple of measurable functions* (f, g) *satisfying*

$$\|f\|_{L^{p_0}(w)} \le N([W]_{A_{p_0}})\|g\|_{L^{p_0}(w)} \tag{6.21}$$

for all $(f, g) \in \mathscr{F}$ *and* $w \in A_{p_0}$. *Then for all* $1 < p < \infty$ $\|f\|_{L^p(w)} \lesssim_{[w]_{A_p}} \|g\|_{L^p(w)}$, *whenever* $(f, g) \in \mathscr{F}$ *and* $w \in A_p$.

To prove Theorem 6.6, we need two lemmas.

Lemma 6.4 *Let* $1 < p < \infty$, $m \in \mathbb{N}$, *and let* $w \in A_p$. *Define*

$$\mathscr{R}k \equiv |k| + \sum_{j=1}^{\infty}(2\|M\|_{B(L^p(w))})^{-j}M^jk \quad (k \in L^0(\mathbb{R}^n)), \tag{6.22}$$

where M^j *denotes the* j*-fold composition of* M. *Then for all* $k \in L^p(w)$,

$$|k| \le \mathscr{R}k, \tag{6.23}$$

$$\|\mathscr{R}k\|_{L^p(w)} \le 2\|k\|_{L^p(w)}, \tag{6.24}$$

and

$$M\mathscr{R}k \le 2\|M\|_{B(L^p(w))}\mathscr{R}k. \tag{6.25}$$

Proof These are direct and we omit the details; see Exercise 6.8.

Lemma 6.5 *Let $1 < p < \infty$, and let $w \in A_p$. Define*

$$\mathfrak{M}_w h \equiv w^{-1} M[h \cdot w] \tag{6.26}$$

and

$$\mathscr{R}'_w h \equiv |h| + \sum_{j=1}^{\infty} (2\|\mathfrak{M}_w\|_{L^{p'}(w)\to L^{p'}(w)})^{-j} \mathfrak{M}_w^j h, \tag{6.27}$$

where $\mathfrak{M}_w^j$ denotes the j-fold composition of $\mathfrak{M}_w$. Then for all $h \in L^{p'}(w)$,

$$|h| \le \mathscr{R}'_w h, \tag{6.28}$$

$$\|\mathscr{R}'_w h\|_{L^{p'}(w)} \le 2\|h\|_{L^{p'}(w)}, \tag{6.29}$$

and

$$\mathfrak{M}_w \mathscr{R}'_w \le 2\|\mathfrak{M}_w\|_{L^{p'}(w)\to L^{p'}(w)} \mathscr{R}'_w h. \tag{6.30}$$

Proof The proof of inequalities (6.27), (6.28), (6.29), and (6.30) is straightforward and we omit the details; see Exercise 6.8 again.

We now prove Theorem 6.6. Note that the case where $\|f\|_{L^p(w)}\|g\|_{L^p(w)} = 0$ is trivial. We assume otherwise. We dualize $\|f\|_{L^p(w)}$:

$$\|f\|_{L^p(w)} = \int_{\mathbb{R}^n} f(x)h(x)w(x)\mathrm{d}x, \tag{6.31}$$

where $\|h\|_{L^{p'}(w)} = 1$. We define $k \equiv \dfrac{f}{\|f\|_{L^p(w)}} + \dfrac{g}{\|g\|_{L^p(w)}}$ keeping in mind that $\|f\|_{L^p(w)}\|g\|_{L^p(w)} > 0$. We write

$$\mathrm{I} = \left(\int_{\mathbb{R}^n} \frac{f(x)^{p_0}}{\mathscr{R}k(x)^{p_0-1}} \mathscr{R}'_w h(x) w(x)\mathrm{d}x \right)^{\frac{1}{p_0}}, \quad \mathrm{II} = \left(\int_{\mathbb{R}^n} \mathscr{R}k(x)\mathscr{R}'_w h(x) w(x)\mathrm{d}x \right)^{\frac{1}{p_0'}}.$$

Then by the Hölder inequality

$$\begin{aligned}&\int_{\mathbb{R}^n} f(x)h(x)w(x)\mathrm{d}x\\ &\le \int_{\mathbb{R}^n} f(x)\mathscr{R}'_w h(x)w(x)\mathrm{d}x\\ &= \int_{\mathbb{R}^n} f(x)\mathscr{R}k(x)^{-\frac{1}{p_0'}}\mathscr{R}'_w h(x)^{\frac{1}{p_0}}w(x)^{\frac{1}{p_0}}\mathscr{R}k(x)^{\frac{1}{p_0'}}\mathscr{R}'_w h(x)^{\frac{1}{p_0'}}w(x)^{\frac{1}{p_0'}}\mathrm{d}x\\ &\le \mathrm{I}\times\mathrm{II}.\end{aligned}$$

For I, we have

$$\mathrm{I}\le N([(\mathscr{R}k)^{1-p_0}\mathscr{R}'_w h\cdot w]_{A_{p_0}})\left(\int_{\mathbb{R}^n} g(x)^{p_0}\mathscr{R}k(x)^{1-p_0}\mathscr{R}'_w h(x)w(x)\mathrm{d}x\right)^{\frac{1}{p_0}}.$$

We note that

$$M[w\mathscr{R}'_w h] = w\mathfrak{M}_w\mathscr{R}'_w h \lesssim w\mathscr{R}'_w h \tag{6.32}$$

or equivalently, $[w\mathscr{R}'_w h]_{A_1}\lesssim 1$, which yields $[(\mathscr{R}k)^{1-p_0}\mathscr{R}'_w h\cdot w]_{A_{p_0}}\lesssim 1$. Thus,

$$\mathrm{I}\lesssim_{[w]_{A_p}}\left(\int_{\mathbb{R}^n} g(x)^{p_0}\mathscr{R}k(x)^{1-p_0}\mathscr{R}'_w h(x)w(x)\mathrm{d}x\right)^{\frac{1}{p_0}}.$$

If we insert the definition of k, then we obtain

$$\begin{aligned}\mathrm{I}&\le(\|g\|_{L^p(w)})^{\frac{1}{p_0'}}\left(\int_{\mathbb{R}^n} g(x)^{p_0}\mathscr{R}g(x)^{1-p_0}\mathscr{R}'_w h(x)w(x)\mathrm{d}x\right)^{\frac{1}{p_0}}\\ &\le(\|g\|_{L^p(w)})^{\frac{1}{p_0'}}\left(\int_{\mathbb{R}^n}\mathscr{R}g(x)\mathscr{R}'_w h(x)w(x)\mathrm{d}x\right)^{\frac{1}{p_0}}\\ &\le(\|g\|_{L^p(w)})^{\frac{1}{p_0'}}(\|\mathscr{R}g\|_{L^p(w)})^{\frac{1}{p_0}}(\|\mathscr{R}'_w h\|_{L^{p'}(w)})^{\frac{1}{p_0}}\lesssim\|g\|_{L^p(w)}.\end{aligned}$$

For II we use $g(x)^{p_0}\mathscr{R}k(x)^{1-p_0}\le\|g\|_{L^p(w)}{}^{p_0}\mathscr{R}k(x)$ for all $x\in\mathbb{R}^n$, which follows from the definition of k. Since

$$\int_{\mathbb{R}^n}\mathscr{R}k(x)\mathscr{R}'_w h(x)w(x)\mathrm{d}x\le\|\mathscr{R}k\|_{L^p(w)}\|\mathscr{R}'_w h\|_{L^{p'}(w)}\lesssim 1,$$

we conclude $\mathrm{II}\lesssim_{[w]_{A_p}}\|g\|_{L^p(w)}$. If we use (6.31) together with the observation above, we obtain the desired result.

6.1.2.7 Applications of Extrapolation

Let $0 < p < \infty$ and $0 < q \le \infty$. Given a sequence of measurable functions $\{f_j\}_{j=0}^{\infty}$, we define

$$\|\{f_j\}_{j=0}^{\infty}\|_{\ell^q(L^p(w))} \equiv \left(\sum_{j=0}^{\infty}(\|f_j\|_{L^p(w)})^q\right)^{\frac{1}{q}}$$

$$\|\{f_j\}_{j=0}^{\infty}\|_{L^p(w,\ell^q)} \equiv \left\|\left(\sum_{j=0}^{\infty}|f_j|^q\right)^{\frac{1}{q}}\right\|_{L^p(w)}.$$

Define $\ell^q(L^p(w))$ and $L^p(w,\ell^q)$, called the *weighted vector-valued Lebesgue spaces*, to be the collection $\{f_j\}_{j=0}^{\infty}$ of all the measurable functions for which the norms $\|\{f_j\}_{j=0}^{\infty}\|_{\ell^q(L^p(w))}$ and $\|\{f_j\}_{j=0}^{\infty}\|_{L^p(w,\ell^q)}$ are finite, respectively. We now apply Theorem 6.6 to obtain the vector-valued inequalities for $L^p(w,\ell^q)$.

Theorem 6.7 *Let* $1 < p < \infty$, $1 < q \le \infty$ *and let* $w \in A_p$. *Then* $\|\{Mf_j\}_{j=1}^{\infty}\|_{L^p(w,\ell^q)} \lesssim \|\{f_j\}_{j=1}^{\infty}\|_{L^p(w,\ell^q)}$ *for all* $\{f_j\}_{j=1}^{\infty} \in L^p(w,\ell^q)$.

Proof When $p = q$, the result is clear from (6.9). Thus, letting

$$\mathscr{F}_J \equiv \left\{\left(\left(\sum_{j=1}^{J} Mf_j{}^q\right)^{\frac{1}{q}}, \left(\sum_{j=1}^{J}|f_j|^q\right)^{\frac{1}{q}}\right) : f_1, f_2, \ldots, f_J \in L^p(w)\right\},$$

we define $\mathscr{F} \equiv \mathscr{F}_J$ for $J \in \mathbb{N}$. Then we are in the position of applying Theorem 6.6 and then apply the monotone convergence theorem.

We know that the fundamental operators in harmonic analysis behave well with respect to these classes of weights.

Before we conclude this section, we would like to give the orientation of the theory of weighted Lebesgue spaces. For $w \in A_\infty$, we can define the weighted function space $A^s_{pq}(w)$, which stands for either $B^s_{pq}(w)$ or $F^s_{pq}(w)$. See Sect. 6.5. The results here lead us to the weighted function spaces.

Exercises

Exercise 6.6 [829, Example 2.3] Let $\alpha > -n$, and let Q be a cube. Then show that

$$\int_Q |x|^\alpha \mathrm{d}x \sim \max(\ell(Q), |c(Q)|)^\alpha |Q|.$$

Hint: Two cases must be considered: $c(Q)$ is far from the origin and otherwise.

Exercise 6.7 Let $w \in A_p$ with $1 < p < \infty$. Then show that $L^p(w)$ is a Banach function space. Hint : To prove (6.1) use the boundedness of $M^{(\eta)}$ for some $\eta > 0$.

Exercise 6.8 Check (6.23), (6.24), (6.25) (6.28), (6.29) and (6.30) by the subadditivity of M.

6.1.3 Mixed Lebesgue Spaces

When we consider differential equations, it is natural to consider the space variables $x_1, x_2, \ldots, x_n$ equally but we want to separate the time variable t and other variables. So, it may be natural to consider the L^p-norm with respect to $x_1, x_2, \ldots, x_n$ and the L^q-norm with respect to t. In this way mixed Lebesgue spaces naturally arise.

6.1.3.1 Mixed Lebesgue Spaces

Let $0 < p_1, p_2, \ldots, p_n \le \infty$ be constants. Write $\mathbf{p} = (p_1, p_2, \ldots, p_n)$. Then define the mixed Lebesgue norm $\| \star \|_{\mathbf{p}}$ by

$$\|f\|_{\mathbf{p}} \equiv \left(\int_{\mathbb{R}} \cdots \left(\int_{\mathbb{R}} \left(\int_{\mathbb{R}} |f(x_1, x_2, \ldots, x_n)|^{p_1} \mathrm{d}x_1 \right)^{\frac{p_2}{p_1}} \mathrm{d}x_2 \right)^{\frac{p_3}{p_2}} \cdots \mathrm{d}x_n \right)^{\frac{1}{p_n}}.$$

A natural modification for x_i is made when $p_i = \infty$. We define the *mixed Lebesgue space* $L^{\mathbf{p}}(\mathbb{R}^n)$ to be the set of all $f \in L^0(\mathbb{R}^n)$ with $\|f\|_{\mathbf{p}} < \infty$.

Example 6.2 If $p_1 = p_2 = \cdots = p_{n-1}$, then we have $L^{\mathbf{p}}(\mathbb{R}^n) = L^{p_n}_{x_n} L^{p_1}_{x'}(\mathbb{R}^n)$ with norm coincidence, where the norm of the right-hand side is given by (4.115).

Note that the (Köthe) dual of $L^{\mathbf{p}}(\mathbb{R}^n)$ is $L^{(p_1', p_2', \ldots, p_n')}(\mathbb{R}^n)$. There are many works on Besov spaces based on the mixed Lebesgue norm $\| \star \|_{\mathbf{p}}$.

To analyze $L^{\mathbf{p}}(\mathbb{R}^n)$ we define the *strong maximal operator* M^S by

$$M^S f(x) \equiv \sup_{-\infty < a_j < b_j < \infty} \left(\prod_{j=1}^n \frac{\chi_{(a_j, b_j)}(x_j)}{b_j - a_j} \right) \int_{\prod_{j=1}^n (a_j, b_j)} |f(y)| \mathrm{d}y$$

for $x = (x_1, x_2, \ldots, x_n)$. Observe that $Mf \le M^S f$ for all $f \in L^0(\mathbb{R}^n)$.

To investigate the boundedness property of M^S. To this end, we need the following inequality:

Theorem 6.8 *Let* $1 < p < \infty$, $m, n \in \mathbb{N}$, $1 < q_1, q_2, \ldots, q_m \le \infty$, *and let* $w \in A_p$. *Then*

$$\left\| \left(\sum_{j_m=1}^{\infty} \left(\cdots \sum_{j_2=1}^{\infty} \left(\sum_{j_1=1}^{\infty} Mf_{j_1,\ldots,j_m}{}^{q_1} \right)^{\frac{q_2}{q_1}} \cdots \right)^{\frac{q_m}{q_{m-1}}} \right)^{\frac{1}{q_m}} \right\|_{L^p(w)}$$

$$\lesssim_w \left\| \left(\sum_{j_m=1}^{\infty} \left(\cdots \sum_{j_2=1}^{\infty} \left(\sum_{j_1=1}^{\infty} |f_{j_1,\ldots,j_m}|^{q_1} \right)^{\frac{q_2}{q_1}} \cdots \right)^{\frac{q_m}{q_{m-1}}} \right)^{\frac{1}{q_m}} \right\|_{L^p(w)} \tag{6.33}$$

for all collections $\{f_{j_1,\ldots,j_m}\}_{j_1,j_2,\ldots,j_n=1}^{\infty} \subset L^0(\mathbb{R}^n)$.

Proof We induct on m. If $m = 1$, simply apply Theorem 6.7. Assume (6.33) is true for $m - 1$. Then (6.33) for m with $p = q_m$ is true by the induction assumption. Apply Theorem 6.6 to conclude (6.33).

Denote by M^{x_k} the uncentered maximal operator for the variable x_k, $k = 1, 2, \ldots, n$. We use the following lemma:

Lemma 6.6 *Let* $\mathbf{p} = (p_1, p_2, \ldots, p_n) \in (1, \infty)^n$. *Then* $\|M^{x_n} f\|_{L^{\mathbf{p}}} \lesssim \|f\|_{L^{\mathbf{p}}}$ *for all* $f \in L^{\mathbf{p}}(\mathbb{R}^n)$.

Proof Let $E_j^{x'}$ be the averaging linear operator with respect to x', which leaves x_n untouched, given by (1.100). By (6.33) and $\liminf_{j\to\infty} M^{(x_n)}[E_j^{x'} f] \le \liminf_{j\to\infty} E_j^{x'}[M^{(x_n)} f]$, we obtain the desired result.

Theorem 6.9 *Let* $\mathbf{p} = (p_1, p_2, \ldots, p_n) \in (1, \infty]^n$. *Then* M^S *is bounded on* $L^{\mathbf{p}}(\mathbb{R}^n)$.

Proof We can induct on n. If $n = 1$, the result is clear from the usual boundedness of the Hardy–Littlewood maximal operator. Assume that the result is true for $n - 1$ dimensions. Then simply use Lemma 6.6 and $M^S f \le M^{x_1} \circ M^{x_2} \circ \cdots \circ M^{x_n}$.

Let us obtain the Fefferman–Stein inequality for $L^{\mathbf{p}}(\mathbb{R}^n)$ for later consideration. Let $0 < p_1, p_2, \ldots, p_n, r \le \infty$. We define the *vector-valued mixed Lebesgue space* $L^{\mathbf{p}}(\ell^r, \mathbb{R}^n)$ with the norm $\| \star \|_{L^{\mathbf{p}}(\ell^q)}$, whose norm is given by (6.2) with X replaced by $L^{\mathbf{p}}(\mathbb{R}^n)$. The fundamental inequality we need is the following inequality.

Theorem 6.10 *Let* $\mathbf{p} = (p_1, p_2, \ldots, p_n) \in (1, \infty)^n$ *and* $1 < q \le \infty$. *Then for all* $\{f_j\}_{j=1}^{\infty} \subset L^{\mathbf{p}}(\ell^q, \mathbb{R}^n)$, $\|\{M^S f_j\}_{j=1}^{\infty}\|_{L^{\mathbf{p}}(\ell^q)} \lesssim \|\{f_j\}_{j=1}^{\infty}\|_{L^{\mathbf{p}}(\ell^q)}$.

Proof Simply mimic the proof of Theorem 6.9.

As a corollary of this maximal inequality together with its technique, we can characterize $L^{\mathbf{p}}(\mathbb{R}^n)$ using the Littlewood–Paley decompositions.

Exercises

Exercise 6.9 Let $1 \le p_1, p_2, \dots, p_n \le \infty$.

1. Show that $L^{\mathbf{p}}(\mathbb{R}^n)$ is continuously embedded into $\mathscr{S}'(\mathbb{R}^n)$.
2. Show that $L^{\mathbf{p}}(\mathbb{R}^n)$ is a Banach function space.

Exercise 6.10 Let $1 < p_1, p_2, \dots, p_n \le \infty$. Write $\mathbf{p} = (p_1, p_2, \dots, p_n)$. Show that $\|M^{x_n} \circ M^{x_{n-1}} \circ \dots \circ M^{x_1} f\|_{\mathbf{p}} \le C\|f\|_{\mathbf{p}}$ for all $f \in L^{\mathbf{p}}(\mathbb{R}^n)$.

6.1.4 Variable Lebesgue Spaces

6.1.4.1 Variable Lebesgue Spaces

The variable Lebesgue spaces are used in order to describe the situation where the value of $p(\star)$ differs according to the position. For example, we consider the situation where $p \equiv 3$ on an open set U and $p \equiv 2$ on another open set V. In this section, by a *variable exponent* we mean any measurable function from $\mathbb{R}^n$ to a subset of $(-\infty, \infty]$.

We define $\| \star \|_{L^{p(\star)}}$ which is called the variable Lebesgue norm or the Nakano–Luxenburg norm.

Definition 6.7 (Variable Lebesgue space, Variable exponent Lebesgue space) Let Ω be a measurable set in $\mathbb{R}^n$. For a measurable function $p(\star) : \Omega \to [1, \infty]$, the variable exponent *Lebesgue space* $L^{p(\star)}(\Omega)$ *with variable exponents* is defined by

$$L^{p(\star)}(\Omega) \equiv \bigcup_{\lambda>0} \{f \in L^0(\Omega) \, : \, \rho_p(\lambda^{-1} f) < \infty\},$$

where

$$\rho_p(f) \equiv \|\chi_{p^{-1}(0,\infty)}|f|^{p(\star)}\|_1 + \|f\|_{L^\infty(p^{-1}(\infty))}.$$

Moreover, for $f \in L^{p(\star)}(\Omega)$ one defines the *variable Lebesgue norm* by

$$\|f\|_{L^{p(\star)}(\Omega)} \equiv \inf\left(\left\{\lambda \in (0, \infty) \, : \, \rho_p(\lambda^{-1} f) \le 1\right\} \cup \{\infty\}\right).$$

Let $p_+ \equiv \|p(\star)\|_\infty$ and $p_- \equiv (\|p(\star)^{-1}\|_\infty)^{-1}$ for a nonnegative variable exponent $p(\star)$. If $1 \le p_- \le p_+ \le \infty$, then $\| \star \|_{L^{p(\star)}(\Omega)}$ is a norm and thereby $L^{p(\star)}(\Omega)$ is a Banach space as the following theorem shows:

Theorem 6.11 *Let Ω be a measurable set in $\mathbb{R}^n$. Let $p(\star) : \Omega \to [1, \infty]$ be a measurable function. Then $L^{p(\star)}(\Omega)$ is a Banach space.*

Proof We content ourselves with proving that $\| \star \|_{L^{p(\star)}(\Omega)}$ is a norm. We suppose that $p(\cdot)$ is finitely valued. Let $f, g \in L^{p(\star)}(\Omega)$. We are now going to show that

$$\|f + g\|_{L^{p(\star)}(\Omega)} \le \|f\|_{L^{p(\star)}(\Omega)} + \|g\|_{L^{p(\star)}(\Omega)}. \tag{6.34}$$

Let $\varepsilon > 0$ be taken arbitrarily. Then (6.34) can be reduced to showing

$$\|f + g\|_{L^{p(\star)}(\Omega)} \le \|f\|_{L^{p(\star)}(\Omega)} + \|g\|_{L^{p(\star)}(\Omega)} + 2\varepsilon. \tag{6.35}$$

From the definition of the set defining the norm $\|f\|_{L^{p(\star)}(\Omega)}$, we deduce

$$\|f\|_{L^{p(\star)}(\Omega)} + \varepsilon \in \left\{\lambda > 0 : \int_\Omega \left(\frac{|f(x)|}{\lambda}\right)^{p(x)} \mathrm{d}x \le 1\right\},$$

that is, we have

$$\int_\Omega \left(\frac{|f(x)|}{\|f\|_{L^{p(\star)}(\Omega)} + \varepsilon}\right)^{p(x)} \mathrm{d}x \le 1.$$

The same can be said for g;

$$\int_\Omega \left(\frac{|g(x)|}{\|g\|_{L^{p(\star)}(\Omega)} + \varepsilon}\right)^{p(x)} \mathrm{d}x \le 1.$$

Let us set $\alpha = \|f\|_{L^{p(\star)}(\Omega)} + \varepsilon$, $\beta = \|g\|_{L^{p(\star)}(\Omega)} + \varepsilon$ and $\theta = \dfrac{\beta}{\alpha + \beta}$.

Since $t \in [0, \infty) \mapsto t^{p(x)} \in [0, \infty)$ is convex, we have a pointwise estimate:

$$\begin{aligned}\left(\frac{|f(x) + g(x)|}{\alpha + \beta}\right)^{p(x)} &= \left((1 - \theta)\frac{|f(x)|}{\alpha} + \theta\frac{|g(x)|}{\beta}\right)^{p(x)} \\ &\le (1 - \theta)\left(\frac{|f(x)|}{\alpha}\right)^{p(x)} + \theta\left(\frac{|g(x)|}{\beta}\right)^{p(x)}. \end{aligned} \tag{6.36}$$

Integrating (6.36) over Ω, we obtain

$$\int_\Omega \left(\frac{|f(x)+g(x)|}{\|f\|_{L^{p(\star)}(\Omega)}+\|g\|_{L^{p(\star)}(\Omega)}+2\varepsilon}\right)^{p(x)} \mathrm{d}x = \int_\Omega \left(\frac{|f(x)+g(x)|}{\alpha+\beta}\right)^{p(x)} \mathrm{d}x \le 1,$$

which is equivalent to (6.35).

Example 6.3 Let $p(\star) : \Omega \to [1, \infty]$ be a variable exponent. Then every simple function f is in $L^{p(\star)}(\Omega)$ and

$$\rho_p\left(\frac{f}{\|f\|_{L^{p(\star)}(\Omega)}}\right)=1. \tag{6.37}$$

It is worth noting that (6.37) holds even when $\tilde{p}_+=\|\chi_{p^{-1}(0,\infty)}p(\star)\|_{L^\infty(\Omega)}$ is not finite. Let us prove (6.37). Assume for the time being that f has an expression $f\equiv c\chi_E$, where $E\subset\Omega$, $0<|E|<\infty$ and $c>0$. and Then f is in $L^{p(\star)}(\Omega)$, since $\rho_p(c^{-1}f)\le|E|+1<\infty$. From the properties of modularity it follows that $\lambda\mapsto\rho_p(\lambda f)$ is continuous and strictly increasing on some interval $[0,\lambda_0)$ with $\lim_{\lambda\uparrow\lambda_0}\rho_p(\lambda f)=\infty$, where $\lambda_0\in(0,\infty]$. Since $\lim_{\lambda\downarrow 0}\rho_p(\lambda f)=0$, we have (6.37). For a general function f, we have the same conclusion, since the finite sum of continuous and strictly increasing functions is also continuous and strictly increasing.

We now investigate more in the case of $p_+<\infty$.

For a variable exponent $p(\star)\in L^0(\Omega)$ satisfying $0<p(\star)<\infty$, $L^{p(\star)}(\Omega)$ is the set of all the measurable functions f on Ω such that

$$\|f\|_{L^{p(\star)}(\Omega)}\equiv\inf\left(\left\{\lambda\in(0,\infty)\,:\,\int_\Omega\left(\frac{|f(x)|}{\lambda}\right)^{p(x)}\mathrm{d}x\le 1\right\}\cup\{\infty\}\right)<\infty.$$

Lemma 6.7 *Assume that $p(\star)$ satisfies $0<p_-\le p_+<\infty$. Then for $f\in L^0(\Omega)$,*

$$\int_\Omega|f(x)|^{p(x)}\mathrm{d}x\le 1 \tag{6.38}$$

if and only if

$$\|f\|_{L^{p(\star)}(\Omega)}\le 1. \tag{6.39}$$

Proof From the definition of $\|f\|_{L^{p(\star)}(\Omega)}$, (6.38) implies (6.39).

Conversely, assume (6.39). Then from the definition of $\|f\|_{L^{p(\star)}(\Omega)}\le 1$

$$\int_\Omega\left(\frac{|f(x)|}{1+\varepsilon}\right)^{p(x)}\mathrm{d}x\le 1,\quad\varepsilon>0.$$

Thus (6.38) follows by letting $\varepsilon\downarrow 0$. Thus the proof is complete.

It is also not so hard to prove:

Lemma 6.8 *Let $p(\star)$ be a variable exponent such that $0<p_-\le p_+<\infty$. Then*

$$\int_\Omega\left(\frac{|f(x)|}{\|f\|_{L^{p(\star)}(\Omega)}}\right)^{p(x)}\mathrm{d}x=1$$

for all $f \in L^{p(\star)}(\Omega) \setminus \{0\}$ *and* $L^{p(\star)}(\Omega)=\left\{f\in L^0(\Omega) : \int_\Omega |f(x)|^{p(x)}\mathrm{d}x<\infty\right\}$.

The proof is left as an exercise. See Exercise 6.12.

6.1.4.2 Hölder's Inequality and Köthe Dual

The aim of this section is to prove results related to duality.

Recall that $L^0(\Omega)$ is the set of all the measurable functions on Ω.

For a variable exponent $p(\star) \in L^0(\Omega)$ satisfying $p(x) \in [1, \infty]$ for almost all $x \in \Omega$, its *conjugate (variable) exponent* $p'(\star) \in L^0(\Omega)$ is defined as

$$1 = \frac{1}{p(x)} + \frac{1}{p'(x)} \quad (x \in \Omega).$$

By no means does $p'(x)$ stand for the derivative of p at x. We prove the following theorem, where we work within in the measurable set Ω:

Theorem 6.12 (Generalized Hölder's inequality) *Let* $\Omega \subset \mathbb{R}^n$ *be a measurable set, and let* $p(\star) : \Omega \to [1, \infty]$ *be a variable exponent. Then for all* $f \in L^{p(\star)}(\Omega)$ *and all* $g \in L^{p'(\star)}(\Omega)$,

$$\|f \cdot g\|_{L^1(\Omega)} \le r_p \|f\|_{L^{p(\star)}(\Omega)} \|g\|_{L^{p'(\star)}(\Omega)}, \tag{6.40}$$

where

$$r_p \equiv 1 + \frac{1}{p_-} - \frac{1}{p_+} = \frac{1}{p_-} + \frac{1}{(p')_-}. \tag{6.41}$$

Proof We define

$$\Omega_1 \equiv \{x \in \Omega : p(x) = 1\} = \{x \in \Omega : p'(x) = \infty\},$$
$$\Omega_0 \equiv \{x \in \Omega : 1 < p(x) < \infty\} = \{x \in \Omega : 1 < p'(x) < \infty\},$$
$$\Omega_\infty \equiv \{x \in \Omega : p(x) = \infty\} = \{x \in \Omega : p'(x) = 1\}.$$

We may assume that $\|f\|_{L^{p(\star)}(\Omega)} = \|g\|_{L^{p'(\star)}(\Omega)} = 1$.

If $\Omega_0 \neq \emptyset$, then

$$|f(x)g(x)| \le \frac{|f(x)|^{p(x)}}{p(x)} + \frac{|g(x)|^{p'(x)}}{p'(x)} \le \frac{|f(x)|^{p(x)}}{p_-} + \frac{|g(x)|^{p'(x)}}{p'_-} \quad \text{for a.e. } x \in \Omega_0.$$

If $\Omega_1 \neq \emptyset$, then $p_- = 1$, $p'_+ = \infty$ and

$$|f(x)g(x)| \le \|g\|_{L^\infty(\Omega_1)}|f(x)| \le |f(x)| = \frac{|f(x)|^{p(x)}}{p_-} \quad \text{for a.e. } x \in \Omega_1.$$

If $\Omega_\infty \neq \emptyset$, then $p_+ = \infty$, $p'_- = 1$ and

$$|f(x)g(x)| \le \|f\|_{L^\infty(\Omega_\infty)}|g(x)| \le |g(x)| = \frac{|g(x)|^{p'(x)}}{p'_-} \quad \text{for a.e. } x \in \Omega_\infty.$$

Therefore, we have

$$\|f \cdot g\|_{L^1(\Omega)} \le \int_{\Omega_0 \cup \Omega_1} \frac{|f(x)|^{p(x)}}{p_-} dx + \int_{\Omega_0 \cup \Omega_\infty} \frac{|g(x)|^{p'(x)}}{p'_-} dx \le \frac{1}{p_-} + \frac{1}{(p')_-} = r_p.$$

This shows (6.40).

We now describe the Köthe dual of $L^{p(\star)}(\Omega)$.

Theorem 6.13 *Let $p(\star) : \Omega \to [1, \infty]$ be a variable exponent. Then $L^{p(\star)}(\Omega)' \approx L^{p'(\star)}(\Omega)$ with equivalence of norms; more precisely,*

$$\frac{1}{3}\|f\|_{L^{p'(\star)}(\Omega)} \le \|f\|_{L^{p(\star)}(\Omega)'} \le r_p\|f\|_{L^{p'(\star)}(\Omega)}, \tag{6.42}$$

where r_p is the constant defined in (6.41).

Proof Write $q(\star) = p'(\star)$ and $Q(E) = \{x \in E : q(x) \in E\}$ for a measurable set $E \subset (0, \infty]$. Let $f \in L^{q(\star)}(\Omega)$. Then the second inequality in (6.42) holds by the definition of the norm $\|\cdot\|_{L^{p(\star)}(\Omega)'}$ and generalized Hölder's inequality, Theorem 6.12. That is, $L^{p(\star)}(\Omega)' \supset L^{q(\star)}(\Omega)$.

Conversely, let $f \in L^{p(\star)}(\Omega)'$. We may assume that $f \not\equiv 0$. Take a sequence $\{f_j\}_{j=1}^\infty$ of simple functions such that $f_j \not\equiv 0$ and that $0 \le f_1 \le f_2 \le \cdots$ and $f_j \to |f|$ a.e. as $j \to \infty$. Then each f_j is in $L^{q(\star)}(\Omega)$. We normalize f_j; set $\tilde{f}_j \equiv \dfrac{f_j}{\|f_j\|_{L^{q(\star)}(\Omega)}}$. We also use the abbreviations;

$$a_j \equiv \|\tilde{f}_j\|_{L^1(Q(\{1\}))}, \ b_j \equiv \|\,|\tilde{f}_j|^{q(\star)}\|_{L^{q(\star)}(Q(1,\infty))}, \ c_j \equiv \|\tilde{f}_j\|_{L^\infty(Q(\{\infty\}))}.$$

Then $a_j + b_j + c_j = 1$ by the definition. We consider three different cases.

1. $a_j \ge \dfrac{1}{3}$. Let $g_j \equiv \chi_{Q(1)}$. Then $\rho_p(g_j) = 1$; that is, $\|g_j\|_{L^{p(\star)}(\Omega)} = 1$, and

$$\frac{\|f_j\|_{L^{p(\star)}(\Omega)'}}{\|f_j\|_{L^{q(\star)}(\Omega)}} = \|\tilde{f}_j\|_{L^{p(\star)}(\Omega)'} \ge \|\tilde{f}_j g_j\|_{L^1(\Omega)} = a_j \ge \frac{1}{3}. \tag{6.43}$$

2. $b_j \ge \dfrac{1}{3}$. Let $g_j \equiv |\tilde{f}_j|^{q(\star)-1}\chi_{Q(1,\infty)}$. Then $\rho_p(g_j) = b_j \le 1$; that is, $\|g_j\|_{L^{p(\star)}(\Omega)} \le 1$, and

$$\frac{\|f_j\|_{L^{p(\star)}(\Omega)'}}{\|f_j\|_{L^{q(\star)}(\Omega)}} = \|\tilde{f}_j\|_{L^{p(\star)}(\Omega)'} \geq \|\tilde{f}_j g_j\|_{L^1(\Omega)} = b_j \geq \frac{1}{3}. \tag{6.44}$$

3. $c_j \geq \frac{1}{3}$. Let $g_j \equiv |U_j|^{-1}\chi_{U_j}$, where $U_j \equiv \{x \in \Omega : q(x) = \infty, \ \tilde{f}_j(x) = c_j\}$. Then $\rho_p(g_j) = 1$; that is, $\|g_j\|_{L^{p(\star)}(\Omega)} = 1$, and

$$\frac{\|f_j\|_{L^{p(\star)}(\Omega)'}}{\|f_j\|_{L^{q(\star)}(\Omega)}} = \|\tilde{f}_j\|_{L^{p(\star)}(\Omega)'} \geq \|\tilde{f}_j g_j\|_{L^1(\Omega)} = c_j \geq \frac{1}{3}. \tag{6.45}$$

Inequalities (6.43), (6.44), and (6.45) yield (6.42) with f replaced by f_j. Letting $j \to \infty$, we have the left inequality in (6.42) and $L^{p(\star)}(\Omega)' \subset L^{q(\star)}(\Omega)$.

6.1.4.3 Boundedness of the Hardy–Littlewood Maximal Operator

Unlike for the usual Lebesgue spaces $L^p(\mathbb{R}^n)$, we have to be careful with the proof of the boundedness of operators. The following lemma is a fundamental one in that this lemma can be transformed for other operators when we consider the boundedness:

Lemma 6.9 *Let $p(\star) \in L^0(\mathbb{R}^n)$ and $1 \leq p_- \leq p_+ < \infty$. Then the following are equivalent:*

1. *There exists a positive constant C such that*

$$\|Mf\|_{p(\star)} \leq C\|f\|_{p(\star)} \tag{6.46}$$

   for all $f \in L^{p(\star)}(\mathbb{R}^n)$.
2. *Whenever*

$$\|\,|f|^{p(\star)}\|_1 \leq 1, \tag{6.47}$$

$$\|\,(Mf)^{p(\star)}\|_1 < \infty. \tag{6.48}$$

Proof Assume first (6.46). If (6.47) holds, then $\|f\|_{p(\star)} \leq 1$ from Lemma 6.7. By (6.46) we have

$$\frac{1}{(1+C)^{p_+}}\int_{\mathbb{R}^n} Mf(x)^{p(x)}\mathrm{d}x \leq \int_{\mathbb{R}^n}\left(\frac{Mf(x)}{1+C}\right)^{p(x)}\mathrm{d}x \leq \int_{\mathbb{R}^n}\left(\frac{Mf(x)}{C}\right)^{p(x)}\mathrm{d}x < \infty.$$

This shows (6.48).

Assume that (6.46) fails. We disprove (6.48). Then there exists a sequence of functions $f_m \geq 0$ with $\|f_m\|_{p(\star)} \leq 1$ and $\|Mf_m\|_{p(\star)} \geq 4^m$. Set $f \equiv \sum_{m=1}^{\infty} 2^{-m} f_m$. Then $\|f\|_{p(\star)} \leq 1$ by virtue of the triangle inequality for $L^{p(\star)}(\mathbb{R}^n)$.

This implies (6.47) in view of the definition of the norm. On the other hand, we obtain $\|Mf\|_{p(\star)} \geq \|2^{-m}Mf_m\|_{p(\star)} \geq 2^m$ for each m; that is, $\|Mf\|_{p(\star)} = \infty$. Since $p_+ < \infty$, this means that $\|\,(Mf)^{p(\star)}\|_1 = \infty$.

6.1.4.4 Log-Hölder Condition

As we have seen in this book, the maximal operator M plays a key role when we develop a theory of function spaces. So we are now interested in the boundedness of the Hardy–Littlewood maximal function M. Here we give a standard sufficient condition for M to be bounded. Let $p(\star) : \mathbb{R}^n \to [0,\infty)$ be a measurable function. Here we suppose $p(x) < \infty$ for all $x \in \mathbb{R}^n$ for the sake of simplicity.

We consider the local log-Hölder continuity condition:

$$|p(x) - p(y)| \leq \frac{c_*}{\log(|x-y|^{-1})} \quad \text{for} \quad |x-y| \leq \frac{1}{2}, \ x, y \in \mathbb{R}^n, \tag{6.49}$$

and the log-Hölder-type decay condition at infinity:

$$|p(x) - p_\infty| \leq \frac{c^*}{\log(e+|x|)} \quad \text{for} \quad x \in \mathbb{R}^n, \tag{6.50}$$

where c_*, c^* and p_∞ are positive constants independent of x and y. Let

$$H_{\log,0} \equiv \{p(\star) \in L^0(\mathbb{R}^n) : p(\star) \text{ satisfies (6.49) and } p(x) \in \mathbb{R} \text{ for a.e. } x \in \mathbb{R}^n\},$$

$$H_{\log,\infty} \equiv \{p(\star) \in L^0(\mathbb{R}^n) : p(\star) \text{ satisfies (6.50) and } p(x) \in \mathbb{R} \text{ for a.e. } x \in \mathbb{R}^n\},$$

$$H_{\log} \equiv H_{\log,0} \cap H_{\log,\infty}.$$

Before we go further, a couple of helpful remarks may be in order.

Remark 6.2 The classes $H_{\log,0}$, $H_{\log,\infty}$ and $H_{\log}$ enjoy the following properties:

1. $H_{\log,0} \subset L^\infty(\mathbb{R}^n)$.
2. If $p(\star) \in H_{\log,0}$ and $p_- > 1$, then $p'(\star) \in H_{\log,0}$. If $p(\star) \in H_{\log,\infty}$ and $p_- > 1$, then $p'(\star) \in H_{\log,\infty}$.
3. From $p_+ < \infty$ and (6.49) it follows that

$$|p(x) - p(y)| \leq \frac{C}{\log(e+|x-y|^{-1})} \quad \text{for all } x, y \in \mathbb{R}^n. \tag{6.51}$$

4. From (6.50) it follows that

$$|p(x) - p(y)| \leq \frac{2c^*}{\log(e+|x|)} \quad \text{for all } x, y \in \mathbb{R}^n \text{ with } |y| \geq |x|. \tag{6.52}$$

5. Condition (6.50) reads as:

$$\frac{1}{e^{c^*}} \le \frac{(e+|x|)^{p(x)}}{(1+|x|)^{p_\infty}} \le e^{c^*} \quad \text{for all } x \in \mathbb{R}^n.$$

6. Let $B = B(x,r)$ be a ball. We write $p_B^- \equiv \inf_{x \in B} p(x)$. We notice

$$r^{\frac{1}{p(y)} - \frac{1}{p_B^-}} \lesssim 1 \quad (y \in B). \tag{6.53}$$

Indeed, if $r > 1$, then (6.53) is obvious from $p(y) \ge p_B^-$. If $r \in (0,1]$, then (6.53) results from the log-Hölder continuity.

7. For $x \in \mathbb{R}^n$, we set

$$\frac{1}{s(x)} \equiv \left| \frac{1}{p(x)} - \frac{1}{p_\infty} \right|. \tag{6.54}$$

Condition (6.50) reads that for all $m > 0$ there exists $\gamma \in (0,1)$ such that

$$\gamma^{s(x)} \lesssim \langle x \rangle^{-m}. \tag{6.55}$$

8. Assume that $x, y \in \mathbb{R}^n$ satisfy $p(x) \le p(y)$. We set

$$\frac{1}{q(x,y)} = \frac{1}{p(x)} - \frac{1}{p(y)}. \tag{6.56}$$

We fix γ so that (6.55) holds. If $|x-y| \le r < 1$ and $p(x) < p(y)$, then there exists $N > 0$ such that

$$\gamma^{q(x,y)} \lesssim r^N. \tag{6.57}$$

6.1.4.5 Log-Hölder Condition as a Sufficient Condition: Diening's Result

We now prove the following fundamental boundedness of the maximal operator.

Theorem 6.14 *If $p(\star) \in H_{\log}$ and $p_- > 1$, then M is bounded on $L^{p(\star)}(\mathbb{R}^n)$.*

This boundedness relies upon the next pointwise estimate and the boundedness of M on $L^{p_-}(\mathbb{R}^n)$ for $p_- > 1$.

Theorem 6.15 *Let $N \gg 1$. If $p(\star) \in H_{\log}$ and $p_- \ge 1$, then for all measurable functions f with $\|f\|_{p(\star)} \le 1$ and for all $x \in \mathbb{R}^n$,*

$$m_{B(x,r)}(|f|)^{p(x)} \lesssim m_{B(x,r)}(|f|^{p(\star)}) + \min(r^N, 1)\langle x \rangle^{-np_-}. \tag{6.58}$$

In particular

$$Mf(x)^{p(x)} \lesssim M[|f|^{p(\star)/p_-}](x)^{p_-} + \langle x\rangle^{-np_-}. \tag{6.59}$$

We note that (6.59) is a consequence of (6.58) by replacing $p(\star)$ with $p(\star)/p_-$. We prove Theorem 6.14 admiting Theorem 6.15. From Lemma 6.9 it is enough to prove that $\|Mf\|_{p(\star)} \lesssim 1$ for all $f \in L^{p(\star)}(\mathbb{R}^n)$ with $\|f\|_{p(\star)} \le 1$. Note that $\|f\|_{p(\star)} \le 1$ is equivalent to $\||f|^{p(\star)}\|_1 \le 1$. In this case, letting $g \equiv |f|^{p(\star)/p_-}$, we have $\|g\|_{p_-} \le 1$ according to Lemma 6.7. By Theorem 6.15 and the boundedness of M on $L^{p_-}(\mathbb{R}^n)$ for $p_- > 1$, we have

$$\int_{\mathbb{R}^n} Mf(x)^{p(x)}\mathrm{d}x \lesssim \int_{\mathbb{R}^n} Mg(x)^{p_-}\mathrm{d}x + \int_{\mathbb{R}^n} \langle x\rangle^{-np_-}\mathrm{d}x \lesssim \int_{\mathbb{R}^n} g(x)^{p_-}\mathrm{d}x + 1 \lesssim 1.$$

This shows that, for some $K > 0$,

$$\int_{\mathbb{R}^n} \left(\frac{Mf(x)}{K}\right)^{p(x)} \mathrm{d}x \le 1,$$

since $p_- > 1$. Namely, $\|Mf\|_{p(\star)} \le K$.

We give a proof of Theorem 6.15 via a couple of steps.

For a nonnegative function f and a ball $B(x,r)$, let

$$I = I(x,r) \equiv m_{B(x,r)}(f), \quad J = J(x,r) \equiv m_{B(x,r)}(f^{p(\star)}).$$

Then $Mf(x) \sim \sup_{r>0} I(x,r)$ and $M(|f(\star)|^{p(\star)})(x) \sim \sup_{r>0} J(x,r)$. Let

$$\mathscr{F}_{p(\star)} \equiv \left\{f \in L^{p(\star)}(\mathbb{R}^n) \,:\, f(x) \in \{0\} \cup [1,\infty) \text{ for a.e } x \in \mathbb{R}^n, \quad \|f\|_{p(\star)} \le 1\right\},$$

$$\mathscr{G} \equiv \{f \in L^\infty(\mathbb{R}^n) \,:\, 0 \le f(x) < 1 \text{ for a.e } x \in \mathbb{R}^n\}.$$

To prove Theorem 6.15, we state and prove two basic lemmas.

Lemma 6.10 *Let $p(\star) \in H_{\log,0}$ be such that $p_- \ge 1$. Then there exists a positive constant C, dependent only on n and $p(\star)$, such that, for all functions $f \in \mathscr{F}_{p(\star)}$ and for all balls $B(x,r)$, $I \lesssim J^{1/p(x)}$.*

Proof Let $B = B(x,r)$, and let $f \in \mathscr{F}_{p(\star)}$. By the Hölder inequality, we have $m_{B(x,r)}(f) \le m^{(p_B^-)}_{B(x,r)}(f)$. Since $f(y) \in \{0\} \cup [1,\infty)$, for almost every $y \in \mathbb{R}^n$, we have

$$m_{B(x,r)}(f) \lesssim \sqrt[p_B^-]{m_{B(x,r)}(f^{p(\star)})}. \tag{6.60}$$

Meanwhile, $\|f^{p(\star)}\chi_{B(x,r)}\|_1 \le \|f^{p(\star)}\|_1 \le 1$ implies

$$\sqrt[p_B^-]{\int_{B(x,r)} f(y)^{p(y)}\mathrm{d}y} \le \sqrt[p(x)]{\int_{B(x,r)} f(y)^{p(y)}\mathrm{d}y}. \tag{6.61}$$

Putting together (6.53), (6.60) and (6.61), we obtain Lemma 6.10.

Lemma 6.11 *Let $p(\star) \in H_{\log,\infty}$ satisfy $p_- \ge 1$, and let $N \gg 1$. Then for all functions $f \in \mathscr{G}$ and for all balls $B(x,r)$, $I \lesssim_{p(\cdot)} \sqrt[p(x)]{J} + \min(1, r^N)\langle x\rangle^{-n}$.*

Proof Fix $m \gg 1$ and then choose $\gamma \ll 1$ as in (6.55). Let $B = B(x,r)$, and let $f \in \mathscr{G}$. Choose $y \in B(x,r)$ arbitrarily.

We use the Hölder inequality to have $m_{B(x,r)}(f) \le m_{B(x,r)}^{(p(x))}(f)$. We decompose

$$m_{B(x,r)}^{(p(x))}(\gamma^4 f) \le \gamma^4 m_{B(x,r)}^{(p(x))}(\chi_{(0,p(x)]}(p(\star))f) + m_{B(x,r)}^{(p(x))}(\gamma^4 \chi_{(p(x),\infty)}(p)f).$$

For the first term, we use the fact that $f \in \mathscr{G}$:

$$m_{B(x,r)}^{(p(x))}(\chi_{(0,p(x)]}(p(\star))f) \le \sqrt[p(x)]{m_{B(x,r)}(\chi_{(0,p(x)]}(p(\star))f^{p(\star)})} \le \sqrt[p(x)]{m_{B(x,r)}(f^{p(\star)})}.$$

Let $y \in B(x,r)$. Define $q(x,y)$ and $s(x)$ by (6.54). For the second term, we suppose $p(x) < p(y)$. Since $2q(x,y) \ge \min(s(x), s(y))$, by the Young inequality

$$(\gamma^4 f(y))^{p(x)} \le f(y)^{p(y)} + \gamma^{4q(x,y)} \le f(y)^{p(y)} + \gamma^{s(x)+2q(x,y)} + \gamma^{s(y)+2q(x,y)}. \tag{6.62}$$

Since N is a prescribed number and $\gamma \le 1$, we deduce $(\gamma^4 f(y))^{p(x)} \lesssim f(y)^{p(y)} + \min(1, r^N)\gamma^{s(x)} + \min(1, r^N)\gamma^{s(y)}$ from (6.57) and the definition of q. Thus, taking average of (6.62) over the ball $B(x,r)$, we obtain

$$\begin{aligned} m_{B(x,r)}^{(p(x))}(\gamma^2 f) &\lesssim \sqrt[p(x)]{m_{B(x,r)}(f^{p(\star)})} + \min(1, r^N)\gamma^{s(x)} + \min(1, r^N)M[\gamma^{s(\star)}](x) \\ &\lesssim \sqrt[p(x)]{m_{B(x,r)}(f^{p(\star)})} + \min(1, r^N)\langle x\rangle^{-n}. \end{aligned}$$

Thus, the proof is complete.

We now prove Theorem 6.15. Let $\|f\|_{p(\star)} \le 1$. We may assume that f is nonnegative. We let $f_1 \equiv \chi_{[1,\infty)}(f)f$ and $f_2 \equiv \chi_{[0,1)}(f)f$. Let $\overline{p}(\star) \equiv p(\star)/p_-$. Then $\overline{p}(\star)$ satisfies (6.49), (6.50) and $\overline{p}_- \ge 1$. In this case $\|f_1\|_{L^{\overline{p}(\star)}} \le 1$, since $f_1(y)^{\overline{p}(y)} \le f_1(y)^{p(y)} \le f(y)^{p(y)}$; that is, $f_1 \in \mathscr{F}_{\overline{p}(\star)}$ and $f_2 \in \mathscr{G}$. Let

$$I = I(x,r) \equiv m_{B(x,r)}(f), \quad \bar{J} = \bar{J}(x,r) \equiv m_{B(x,r)}(f^{\overline{p}(\star)}),$$

$$I_i = I_i(x,r) \equiv m_{B(x,r)}(f_i), \quad \bar{J}_i = \bar{J}_i(x,r) \equiv m_{B(x,r)}(f_i{}^{\overline{p}(\star)})$$

for $i = 1, 2$. By Lemmas 6.10 and 6.11 we have

$$I = I_1 + I_2 \lesssim \sqrt[\overline{p}(x)]{\bar{J}_1} + \sqrt[\overline{p}(x)]{\bar{J}_2} + \langle x \rangle^{-n} \lesssim \sqrt[\overline{p}(x)]{\bar{J}} + \langle x \rangle^{-n}.$$

Then $I^{p(x)} \lesssim \bar{J}^{p_-} + \langle x \rangle^{-np_-}$; that is, $\left(m_{B(x,r)}(f)\right)^{p(x)} \lesssim m^{(p_-)}_{B(x,r)}(f^{p(y)}) + \langle x \rangle^{-np_-}$ for all balls $B(x, r)$. Thus we have (6.59) since $r > 0$ is arbitrary. So the proof of Theorem 6.15 is complete.

6.1.4.6 Variable Vector-Valued Lebesgue Norm and Vector-Valued Inequalities

We have considered $A^s_{pq}(\mathbb{R}^n)$ in this book. We have set forth how to fractuate the exponent p. We also want to do the same thing for q and s. It is straightforward to handle s. However, to pass to the variable case of q is difficult for Besov spaces. We present the following definition. Note that $L^{p(\star)}(\ell^{q(\star)}, \mathbb{R}^n)$ is easier to define than $\ell^{q(\star)}(\ell^{L(\star)}, \mathbb{R}^n)$ and that $q(\star)$ is independent of j.

Definition 6.8 (Vector-valued variable spaces) Let $p(\star), q(\star) \to (0, \infty)$ be measurable functions.

1. Define

$$\|\{f_j\}_{j=1}^\infty\|_{\ell^{q(\star)}(L^{p(\star)})} \equiv \inf\left(\left\{\lambda \in (0, \infty) : \int_{\mathbb{R}^n} \sum_{j=1}^\infty \frac{|f_j(x)|^{p(x)}}{\lambda^{p(x)/q(x)}} \mathrm{d}x \le 1\right\} \cup \{\infty\}\right)$$

   for a collection of measurable functions $\{f_j\}_{j=1}^\infty$. The *variable* $\ell^{q(\star)}(L^{p(\star)})$-*space* $\ell^{q(\star)}(L^{p(\star)}, \mathbb{R}^n)$ denotes the set of all sequences $\{f_j\}_{j=1}^\infty$ of measurable functions for which the quantity $\|\{f_j\}_{j=1}^\infty\|_{\ell^{q(\star)}(L^{p(\star)})}$ is finite.
2. Define $\|\{f_j\}_{j=1}^\infty\|_{L^{p(\star)}(\ell^{q(\star)})} \equiv \| \, \|\{f_j\}_{j=1}^\infty\|_{\ell^{q(\star)}} \, \|_{p(\star)}$ for a collection of measurable functions $\{f_j\}_{j=1}^\infty$. The *variable* $L^{p(\star)}(\ell^{q(\star)})$-*space* $L^{p(\star)}(\ell^{q(\star)}, \mathbb{R}^n)$ denotes the set of all sequences $\{f_j\}_{j=1}^\infty$ of measurable functions for which the quantity $\|\{f_j\}_{j=1}^\infty\|_{L^{p(\star)}(\ell^{q(\star)})}$ is finite.

Let $0 < q(\star) \equiv q \le \infty$ be a constant. We define the *vector-valued variable exponent Lebesgue space* $L^{p(\star)}(\ell^q, \mathbb{R}^n)$ with the norm $\| \star \|_{L^{p(\star)}(\ell^q)}$, whose norm is given by (6.2) with X replaced by $L^{p(\star)}(\mathbb{R}^n)$. Analogously, the space $\ell^q(L^{p(\star)}, \mathbb{R}^n)$ is defined.

We consider vector-valued inequalities for the Lebesgue space $L^{p(\star)}(\mathbb{R}^n)$ with variable exponents, where $q(\star)$ is a constant. Everything goes well as long as $q(\star)$ is a constant. Let us obtain the Fefferman–Stein inequality for $L^{p(\star)}(\mathbb{R}^n)$. We will extend the boundedness to the following vector-valued inequality:

Theorem 6.16 *Let* $1 < r \le \infty$. *Suppose that* $p(\star) \in H_\infty(\mathbb{R}^n)$ *and* $p_- > 1$. *Then* $\|\{Mf_j\}_{j=1}^\infty\|_{L^{p(\star)}(\ell^r)} \lesssim \|\{f_j\}_{j=1}^\infty\|_{L^{p(\star)}(\ell^r)}$ *for all sequences* $\{f_j\}_{j=1}^\infty$ *of measurable* *functions. When $r = \infty$, it will be understood that the conclusion reads*

$$\left\| \sup_{j \in \mathbb{N}} Mf_j \right\|_{p(\star)} \lesssim \left\| \sup_{j \in \mathbb{N}} |f_j| \right\|_{p(\star)}.$$

Proof Assuming $p_- > 1$, we have only to prove that

$$\left\| \left(\sum_{j=1}^{\infty} Mf_j{}^r \right)^{\frac{1+\varepsilon}{r}} \right\|_{p(\star)} \lesssim \left\| \left(\sum_{j=1}^{\infty} |f_j|^r \right)^{\frac{1+\varepsilon}{r}} \right\|_{p(\star)}.$$

We may assume that $f_j = 0$ with a finite number of exceptions, so that the left-hand side is finite. We dualize the left-hand side to have:

$$\left\| \left(\sum_{j=1}^{\infty} Mf_j{}^r \right)^{\frac{1+\varepsilon}{r}} \right\|_{p(\star)} \lesssim \int_{\mathbb{R}^n} \left(\sum_{j=1}^{\infty} Mf_j(x)^r \right)^{\frac{1+\varepsilon}{r}} h(x)\mathrm{d}x$$

using Theorem 6.12. Here $h \in L^{p'(\star)}(\mathbb{R}^n)$ has norm 1 and is nonnegative. Choose $\varepsilon > 0$ so small that $(p')_- > 1 + \varepsilon$. Since we know that $M^{(1+\varepsilon)}h \in A_1$, we learn that

$$\begin{aligned} \int_{\mathbb{R}^n} \left(\sum_{j=1}^{\infty} Mf_j(x)^r \right)^{\frac{1+\varepsilon}{r}} h(x)\mathrm{d}x &\le \int_{\mathbb{R}^n} \left(\sum_{j=1}^{\infty} Mf_j(x)^r \right)^{\frac{1+\varepsilon}{r}} M^{(1+\varepsilon)}h(x)\mathrm{d}x \\ &\lesssim \int_{\mathbb{R}^n} \left(\sum_{j=1}^{\infty} |f_j(x)|^r \right)^{\frac{1+\varepsilon}{r}} M^{(1+\varepsilon)}h(x)\mathrm{d}x. \end{aligned}$$

It remains to use the Hölder inequality and the $L^{p'(\star)}(\mathbb{R}^n)$-boundedness of M.

It is important that the Hardy–Littlewood maximal function does not behave well when $q(\star)$ is not a constant.

Remark 6.3 [422, p. 1746] We cannot extend the vector-valued inequality to the version where the exponent q is also a variable exponent. The simplest classical case is $p = q$ as is seen from the proof of Theorem 1.49. We are thus tempted to consider

$$\left\| \left(\sum_{j=1}^{\infty} Mf_j{}^{p(\star)} \right)^{\frac{1}{p(\star)}} \right\|_{p(\star)} \lesssim \left\| \left(\sum_{j=1}^{\infty} |f_j|^{p(\star)} \right)^{\frac{1}{p(\star)}} \right\|_{p(\star)}. \tag{6.63}$$

But this is not true.

Suppose that (6.63) holds even when $p(\star)$ is a continuous function which is not constant on $\mathbb{R}^n$. Then we can choose open sets U and V such that $m \equiv \sup_U p < M \equiv \inf_V p$. Let $x_0 \in V$. Choose $r > 0$ so that $B(x_0, r) \subset V$. Then substitute $f_j \equiv j^{-\frac{2}{m+M}} \chi_{B(x_0,r)}$ to (6.63). Then we would have

$$\sum_{j=1}^{\infty} \frac{1}{j^{\frac{2m}{m+M}}} \lesssim \left\| \left(\sum_{j=1}^{\infty} M f_j{}^{p(\star)} \right)^{\frac{1}{p(\star)}} \right\|_{p(\star)} \lesssim \left\| \left(\sum_{j=1}^{\infty} |f_j|^{p(\star)} \right)^{\frac{1}{p(\star)}} \right\|_{p(\star)} \lesssim \sum_{j=1}^{\infty} \frac{1}{j^{\frac{2M}{m+M}}}.$$

This is a contradiction, since $m < M$.

From the remark above, we have to give up using the Hardy–Littlewood maximal operator. This is a very big problem judging from what we have been doing in this book. However, if we reexamine the proofs in this book, we notice that we usually considered the case of $f_j = \varphi_j(D)f$. Also, we are led to use the Hardy–Littlewood maximal operator because we wanted to use Theorem 1.50. One of the important ideas to overcome this problem is to control the convolution, or equivalently, the Fourier multipliers directly. So we need to prepare some estimates on convolution.

As a preliminary step, we need an estimate. Using the log-Hölder continuity, we can easily prove the following estimate for the function $\eta_{\nu,m}$ defined in Example 1.31.

Lemma 6.12 *Let $\alpha(\star) \in H_{\log}$, and let $m \gg 1$ and $\nu \in \mathbb{N}$. Then*

$$2^{\nu\alpha(x)} \eta_{\nu,2m}(x-y) \lesssim 2^{\nu\alpha(y)} \eta_{\nu,m}(x-y) \quad (x, y \in \mathbb{R}^n).$$

Proof The proof is left as an exercise. See Exercise 6.13.

The following theorem substitutes for the vector-valued maximal inequality (6.63), which is false. One important idea to note regarding a vector-valued inequality is that we are considering a property of a given distribution. We do not need the general inequality. Since the distributions are decomposed into functions of certain levels, we have only to consider the inequality that is adapted to each level, or equivalently, the function f_j in Theorem 1.49 need not be arbitrary.

Theorem 6.17 *Let $m \gg 1$, and let $p(\star)$, $q(\star) \in H_{\log}$ satisfy $p_-, q_- > 1$,. Then $\|\{\eta_{j,m} * f_j\}_{j=1}^{\infty}\|_{\ell^{q(\star)}(L^{p(\star)})} \lesssim \|\{f_j\}_{j=1}^{\infty}\|_{\ell^{q(\star)}(L^{p(\star)})}$ for all $\{f_j\}_{j=1}^{\infty} \in \ell^{q(\star)}(L^{p(\star)}, \mathbb{R}^n)$.*

Proof Due to homogeneity, it suffices to show that

$$\|\{\eta_{j,m} * f_j\}_{j=1}^{\infty}\|_{\ell^{q(\star)}(L^{p(\star)})} \lesssim 1 \tag{6.64}$$

when $\|\{f_j\}_{j=1}^{\infty}\|_{\ell^{q(\star)}(L^{p(\star)})} = 1$. So, we have $\sum_{j=1}^{\infty} \| |f_j|^{q(\star)} \|_{L^{p(\star)/q(\star)}} \le 1$. Let $j \in \mathbb{N}$ be fixed. We claim

$$\| |\eta_{j,m} * f_j|^{q(\star)} \|_{L^{p(\star)/q(\star)}} \lesssim \delta_j \equiv \| |f_j|^{q(\star)} \|_{L^{p(\star)/q(\star)}} + 2^{-j}. \tag{6.65}$$

Once (6.65) is proved, (6.64) readily follows. Since $\delta_j \in [2^{-j}, 2^j]$, we can push δ into the convolution $\eta_{j,m} * f_j$: After replacing m suitably, Lemma 6.12 reduces (6.65) to the estimate

$$\|\{\eta_{j,m} * (\delta_j^{-\frac{1}{q(\star)}} |f_j|)\}^{q(\star)}\|_{L^{p(\star)/q(\star)}} \lesssim 1, \tag{6.66}$$

or equivalently,

$$\int_{\mathbb{R}^n} \left(\eta_{j,m} * (\delta_j^{-\frac{1}{q(\star)}} |f_j|)(x) \right)^{p(x)} \mathrm{d}x \lesssim 1. \tag{6.67}$$

If we add (6.67) over $j \in \mathbb{N}$, we obtain the desired result.

Estimate (6.66) is easy to prove by Theorem 6.15, since we can use the Hardy–Littlewood maximal operator M. Indeed, from Lemma 6.7 we have

$$\int_{\mathbb{R}^n} \left(\delta_j^{-\frac{1}{q(x)}} |f_j(x)| \right)^{p(x)} \mathrm{d}x = \int_{\mathbb{R}^n} \left(\frac{|f_j(x)|^{q(x)}}{\delta_j} \right)^{\frac{p(x)}{q(x)}} \mathrm{d}x \le 1.$$

From Theorem 6.14, we have

$$\int_{\mathbb{R}^n} M[\delta_j^{-\frac{1}{q(\star)}} |f_j|](x)^{p(x)} \mathrm{d}x \lesssim 1.$$

Since $\eta_{j,m} * (\delta_j^{-\frac{1}{q(\star)}} |f_j|) \lesssim M[\delta_j^{-\frac{1}{q(\star)}} f_j]$ from Theorem 1.47, we have (6.67).

We also have a substitute for the maximal inequality for $L^{p(\star)}(\ell^{q(\star)}, \mathbb{R}^n)$.

Theorem 6.18 *Let $p(\star)$, $q(\star) \in H_{\log}$ satisfy $p_-, q_- > 1$. Then for all $\{f_j\}_{j=1}^{\infty} \in L^{p(\star)}(\ell^{q(\star)}, \mathbb{R}^n)$,*

$$\|\{\eta_{j,m} * f_j\}_{j=1}^{\infty}\|_{L^{p(\star)}(\ell^{q(\star)})} \lesssim \|\{f_j\}_{j=1}^{\infty}\|_{L^{p(\star)}(\ell^{q(\star)})}.$$

We prove Theorem 6.18 via a couple of steps.

We now seek to obtain the vector-valued inequality for $L^{p(\star)}(\ell^{q(\star)}, \mathbb{R}^n)$.

Lemma 6.13 *Let $p(\star)$, $q(\star) \in H_{\log}$ satisfy $p_-, q_- > 1$, and let $m \gg 1$ and $\{f_\nu\}_{\nu=1}^{\infty} \in L^{p(\star)}(\ell^{q(\star)}, \mathbb{R}^n)$. Assume in addition that*

$$\left(\frac{p}{q}\right)_- q_- > 1. \tag{6.68}$$

Then $\|\{\eta_{\nu,m} * f_\nu\}_{\nu=1}^\infty\|_{L^{p(\star)}(\ell^{q(\star)})} \lesssim \|\{f_\nu\}_{\nu=1}^\infty\|_{L^{p(\star)}(\ell^{q(\star)})}$.

Proof We may assume that $\|\{f_\nu\}_{\nu=1}^\infty\|_{L^{p(\star)}(\ell^{q(\star)})} = 1$ and that each f_ν is nonnegative. We observe

$$\int_{\mathbb{R}^n} \left(\sum_{\nu=0}^\infty [\eta_{\nu,m} * f_\nu(x)]^{q(x)}\right)^{\frac{p(x)}{q(x)}} \mathrm{d}x$$

$$\lesssim \int_{\mathbb{R}^n} \left(\sum_{\nu=0}^\infty \left[\sum_{j=0}^\infty 2^{j(n-m)} \sum_{Q \in \mathscr{D}_{\nu-j}} m_Q(f_\nu)\chi_{3Q}(x)\right]^{q(x)}\right)^{\frac{p(x)}{q(x)}} \mathrm{d}x.$$

Since $q_- > 1$ and $\sum_{Q \in \mathscr{D}_{\nu-j}} \chi_{3Q} = 3^n$ for all $\nu, j \in \mathbb{Z}$, we have

$$\int_{\mathbb{R}^n} \left(\sum_{\nu=0}^\infty [\eta_{\nu,m} * f_\nu(x)]^{q(x)}\right)^{\frac{p(x)}{q(x)}} \mathrm{d}x$$

$$\lesssim \int_{\mathbb{R}^n} \left(\sum_{\nu=0}^\infty \sum_{j=0}^\infty 2^{j(n-m)} \left[\sum_{Q \in \mathscr{D}_{\nu-j}} m_Q(f_\nu)\chi_{3Q}(x)\right]^{q(x)}\right)^{\frac{p(x)}{q(x)}} \mathrm{d}x$$

$$\lesssim \int_{\mathbb{R}^n} \left(\sum_{\nu=0}^\infty \sum_{j=0}^\infty 2^{j(n-m)} \sum_{Q \in \mathscr{D}_{\nu-j}} m_Q(f_\nu)\chi_{3Q}(x)^{q(x)}\right)^{\frac{p(x)}{q(x)}} \mathrm{d}x$$

thanks to Example 1.32. We let

$$h \equiv \sum_{\nu=0}^\infty \sum_{j=0}^\infty 2^{j(n-m)} \sum_{Q \in \mathscr{D}_{\nu-j}} m_Q(f_\nu)^{q(\star)}\chi_{3Q},$$

$$h_1 \equiv \sum_{\nu=0}^\infty \sum_{j=0}^\infty 2^{j(n-m)} \sum_{Q \in \mathscr{D}_{\nu-j}} m_Q^{(q_-)}(|f_\nu|^{q(\star)})\chi_{3Q},$$

$$h_2 \equiv \sum_{\nu=0}^\infty \sum_{j=0}^\infty 2^{j(n-m)} \sum_{Q \in \mathscr{D}_{\nu-j}} \frac{\min(1, |Q|)}{(1+|\star|)^{nq_-}}\chi_{3Q}.$$

We insert (6.58) into the above estimate and we obtain $h \lesssim h_1 + h_2$ and hence

$$\int_{\mathbb{R}^n} \left(\sum_{\nu=0}^{\infty} [\eta_{\nu,m} * f_\nu(x)]^{q(x)} \right)^{\frac{p(x)}{q(x)}} dx$$

$$\lesssim \int_{\mathbb{R}^n} \left(\sum_{\nu=1}^{\infty} M[f_\nu{}^{\frac{q(\star)}{q_-}}](x)^{q_-} \right)^{\frac{p(x)q_-}{q(x)}} dx + \int_{\mathbb{R}^n} h_2(x)^{\frac{p(x)}{q(x)}} dx.$$

Since $h_2(x) \lesssim \langle x \rangle^{-nq_-}$, we are in the position of using (6.68) to see that the second term is a finite constant. We use the Fefferman–Stein vector-valued inequality (Theorem 6.16) to obtain the desired result.

We prove Theorem 6.18.

Proof We may assume that each f_ν is nonnegative and that

$$\|\{f_\nu\}_{\nu=1}^{\infty}\|_{L^{p(\star)}(\ell^{q(\star)})} = 1. \tag{6.69}$$

Since $p(\star)$ and $q(\star)$ are uniformly continuous and

$$p_- > 1, \quad \lim_{x\to\infty} \frac{p(x)}{q(x)} = \frac{p_\infty}{q_\infty}, \quad \lim_{x\to\infty} q(x) = q_\infty,$$

we can choose a finite open cover $\{\Omega_l\}_{l=1}^{k}$ $(k \geq 2)$ of $\mathbb{R}^n$ satisfying the following properties:

1. If $l, l' \in \{1, 2, \ldots, k\}$ and $|l - l'| > 1$, then $d(\Omega_l, \Omega_{l'}) > 0$.
2. For $1 \leq l \leq k - 1$,

$$\inf_{x\in\Omega_{l-1}\cup\Omega_l\cup\Omega_{l+1}} \frac{p(x)}{q(x)} \cdot \inf_{x\in\Omega_{l-1}\cup\Omega_l\cup\Omega_{l+1}} q(x) > 1. \tag{6.70}$$

By dilation we may assume that $\frac{1}{3} < \min_{|l-l'|>1} d(\Omega_l, \Omega_{l'}) < \frac{2}{3}$. We decompose

$$\int_{\mathbb{R}^n} \left(\sum_{\nu=0}^{\infty} [\eta_{\nu,m} * f_\nu(x)]^{q(x)} \right)^{\frac{p(x)}{q(x)}} dx \leq \sum_{i=1}^{k} \int_{\Omega_i} \left(\sum_{\nu=0}^{\infty} [\eta_{\nu,m} * f_\nu(x)]^{q(x)} \right)^{\frac{p(x)}{q(x)}} dx.$$

We use Example 1.32 to have

$$\int_{\Omega_l} \left(\sum_{\nu=0}^{\infty} [\eta_{\nu,m} * f_\nu(x)]^{q(x)} \right)^{\frac{p(x)}{q(x)}} dx$$

$$\lesssim \int_{\Omega_l} \left(\sum_{\nu=0}^{\infty} \sum_{j=0}^{\nu} 2^{j(n-m)} \sum_{Q \in \mathscr{D}_{\nu-j}} \chi_{3Q}(x) m_Q(f_\nu)^{q(x)} \right)^{\frac{p(x)}{q(x)}} \mathrm{d}x$$

$$+ \int_{\Omega_l} \left(\sum_{\nu=0}^{\infty} \sum_{j=\nu}^{\infty} 2^{j(n-m)} \sum_{Q \in \mathscr{D}_{\nu-j}} \chi_{3Q}(x) m_Q(f_\nu)^{q(x)} \right)^{\frac{p(x)}{q(x)}} \mathrm{d}x$$

$$\lesssim \int_{\Omega_l} \left(\sum_{\nu=0}^{\infty} \sum_{j=0}^{\nu} 2^{j(n-m)} \sum_{Q \in \mathscr{D}_{\nu-j}} \chi_{3Q}(x) m_Q(f_\nu)^{q(x)} \right)^{\frac{p(x)}{q(x)}} \mathrm{d}x$$

$$+ \int_{\Omega_l} \left(\sum_{\nu=0}^{\infty} \sum_{j=\nu}^{\infty} 2^{j(n-m)} M f_\nu(x)^{q(x)} \right)^{\frac{p(x)}{q(x)}} \mathrm{d}x.$$

For the first term, we use (6.70) to have

$$\int_{\Omega_l} \left(\sum_{\nu=0}^{\infty} \sum_{j=0}^{\nu} 2^{j(n-m)} \sum_{Q \in \mathscr{D}_{\nu-j}} \chi_{3Q}(x) m_Q(f_\nu)^{q(x)} \right)^{\frac{p(x)}{q(x)}} \mathrm{d}x$$

$$\lesssim \int_{\Omega_l} \left(\sum_{\nu=0}^{\infty} [\eta_{\nu,m} * f_\nu(x)]^{q(x)} \right)^{\frac{p(x)}{q(x)}} \mathrm{d}x.$$

For the second term, we observe that

$$\int_{\Omega_l} \left(\sum_{\nu=0}^{\infty} \sum_{j=\nu}^{\infty} 2^{j(n-m)} M f_\nu(x)^{q(x)} \right)^{\frac{p(x)}{q(x)}} \mathrm{d}x \sim \int_{\Omega_l} \left(\sum_{\nu=0}^{\infty} 2^{-\nu(m-n)} M f_\nu(x)^{q(x)} \right)^{\frac{p(x)}{q(x)}} \mathrm{d}x.$$

Let ε be sufficiently small. Then we have

$$\begin{aligned}
\int_{\Omega_l} \left(\sum_{\nu=0}^{\infty} \sum_{j=\nu}^{\infty} 2^{j(n-m)} M f_\nu(x)^{q(x)} \right)^{\frac{p(x)}{q(x)}} \mathrm{d}x &\sim \sum_{\nu=0}^{\infty} 2^{-\nu(m-n)\varepsilon} \int_{\Omega_l} \left(M f_\nu(x)^{q(x)} \right)^{\frac{p(x)}{q(x)}} \mathrm{d}x \\
&\lesssim \sum_{\nu=0}^{\infty} 2^{-\nu(m-n)\varepsilon} \\
&\simeq 1.
\end{aligned}$$

Here we used Theorem 6.14 for the penultimate inequality and the normalization (6.69) for the last inequality. Thus, we obtain the desired result.

Exercises

Exercise 6.11 Let $p(\star)$ be a variable exponent with $p_- \geq 1$. Show that $L^{p(\star)}(\mathbb{R}^n)$ is a Banach space.

Exercise 6.12 Prove Lemma 6.8 by mimicking the proof of Lemma 6.7.

Exercise 6.13 Let $\nu \in \mathbb{N}$.

1. Prove Lemma 6.12 when $2^\nu|x-y| \leq 1$ using the log-Hölder continuity of $\alpha(\star)$.
2. Let $x, y \in \mathbb{R}^n$ satisfy $2^\nu|x-y| > 1$. Choose the smallest integer $k \in \mathbb{N}$ such that $2^\nu|x-y| > 2^k$. Show that there exists a constant $D > 0$ such that

$$1+2^\nu|x-y| \sim 2^k, \quad \frac{\eta_{\nu,2m}(x-y)}{\eta_{\nu,m}(x-y)} \lesssim 2^{-km}, \quad \nu(\alpha(x)-\alpha(y)) \gtrsim -k-D.$$

   Complete the proof of Lemma 6.12 when $2^\nu|x-y| > 1$ using these estimates.

6.1.5 *Morrey Spaces*

6.1.5.1 Morrey Spaces

From the definition of the Morrey norm $\| \star \|_{\mathcal{M}^p_q}$, which we define below, we see that Morrey spaces can describe the local integrability and the global integrability separately.

Definition 6.9 ($\mathcal{M}^p_q(\mathbb{R}^n)$) Let $0 < q \leq p < \infty$. Define the *Morrey norm* $\| \star \|_{\mathcal{M}^p_q}$ by

$$\|f\|_{\mathcal{M}^p_q} \equiv \sup\left\{|Q|^{\frac{1}{p}-\frac{1}{q}}\|f\|_{L^q(Q)} \,:\, Q \text{ is a cube in } \mathbb{R}^n\right\}$$

for a measurable function f. The *Morrey space* $\mathcal{M}^p_q(\mathbb{R}^n)$ is the set of all the measurable functions f for which $\|f\|_{\mathcal{M}^p_q}$ is finite.

The parameter p seems to describe the global regularity of the functions, as is seen from the dilation relation

$$\|f(t\star)\|_{\mathcal{M}^p_q} = t^{-n/p}\|f\|_{\mathcal{M}^p_q} \quad (f \in \mathcal{M}^p_q(\mathbb{R}^n)). \tag{6.71}$$

Meanwhile, the parameter q seems to describe the local regularity of the functions, as is seen from the inclusions:

$$\mathcal{M}_q^p(\mathbb{R}^n) \subset L^q_{\rm loc}(\mathbb{R}^n) \tag{6.72}$$

and for any $u \in (q, \infty)$,

$$\mathcal{M}_q^p(\mathbb{R}^n) \setminus L^u_{\rm loc}(\mathbb{R}^n) \neq \emptyset. \tag{6.73}$$

We leave the proof of (6.71) to interested readers; see Exercise 6.14. Inclusion (6.73) is an easy consequence of Hölder's inequality and (6.73) requires a more delicate construction; see Exercises 6.15 and 6.16, respectively.

As the following examples show, it may be difficult to handle Morrey spaces. This is because Morrey norms contain information on the integral of balls not on any measurable sets.

Example 6.4 Let $1 < q < p < \infty$. Although $\mathcal{M}_q^p(\mathbb{R}^n)$ is a ball Banach function space, $\mathcal{M}_q^p(\mathbb{R}^n)$ is not a Banach function space. In fact, For simplicity, we let $n = 1$ and $1 < q < 2 = p$; other cases are dealt analogously. Let us consider the sequence

$$(a_1, a_2, \dots) = \left(1, \frac{1}{4}, \frac{1}{4}, \frac{1}{16}, \frac{1}{16}, \frac{1}{16}, \frac{1}{16}, \dots\right);$$

that is, a_i is a decreasing sequence and 4^{-l} appears 2^l times for $l = 0, 2, \dots$. We may write $a_j = 4^{-[\log_2 j]}$.

Let $\alpha(p, q) \gg 1$. We define $E \equiv \bigcup_{j=1}^{\infty} (\alpha(p, q)^j, \alpha(p, q)^j + a_j)$. Then $|E| = 2$.

Define

$$f(t) \equiv \sum_{j=1}^{\infty} 4^{[\log_2 j]/p} \chi_{(\alpha(p,q)^j, \alpha(p,q)^j + a_j)}(t) \quad (t \in \mathbb{R}).$$

Then f belongs to $\mathcal{M}_q^p(\mathbb{R})$. Meanwhile,

$$\int_E f(t)\mathrm{d}t = \sum_{j=1}^{\infty} 4^{j/p-j} \cdot 2^j = \infty.$$

Example 6.5 Let $1 < q < p < \infty$. Although $\mathcal{M}_q^p(\mathbb{R}^n)$ is embedded into $\mathcal{S}'(\mathbb{R}^n)$ continuously, $\mathcal{M}_q^p(\mathbb{R}^n)$ is not a subset of $L^1(\mathbb{R}^n) + L^\infty(\mathbb{R}^n)$. Let $n = 1$ for simplicity. Define

$$\delta_j \equiv \frac{1}{[\log_2 \log_2 j]}, \quad j \geq 100, \quad f = f_p \equiv \sum_{j=100}^{\infty} {\delta_j}^{1/p} \chi_{[j!, j! + \delta_j]}. \tag{6.74}$$

Then f belongs to $\mathcal{M}_q^p(\mathbb{R})$ but does not belong to $L^1(\mathbb{R}) + L^\infty(\mathbb{R})$. Let (a, b) be an interval which intersects the support of f.

1. Case 1: $b-a<2$. In this case, there exists uniquely $j \in \mathbb{N} \cap [100,\infty)$ such that $[a,b] \cap [j!, j!+\delta_j] \neq \emptyset$. Thus,

$$\|f\|_{L^q(a,b)} = \left(\int_{\max(a,j!)}^{\min(b,j!+\delta_j)} f(t)^q \mathrm{d}t\right)^{\frac{1}{q}}$$
$$\leq (\min(b,j!+\delta_j)-\max(a,j!))^{\frac{1}{p}-\frac{1}{q}} \left(\int_{\max(a,j!)}^{\min(b,j!+\delta_j)} f(t)^q \mathrm{d}t\right)^{\frac{1}{q}}$$
$$= (b-a)^{\frac{1}{q}-\frac{1}{p}} {\delta_j}^{-\frac{1}{p}} \left(\min\left(b,j!+\delta_j\right)-\max(a,j!)\right)^{\frac{1}{p}}$$
$$\leq (b-a)^{\frac{1}{q}-\frac{1}{p}}.$$

2. Case 2: $b-a>2$. Set

$$m \equiv \min([a,b] \cap \mathrm{supp}(f)), \quad M \equiv \max([a,b] \cap \mathrm{supp}(f)).$$

Choose $j_m, j_M \in \mathbb{N} \cap [100,\infty)$ so that $m \in [j_m!, j_m! + {j_m}^{-1}]$ and $M \in [j_M!, j_M! + {j_M}^{-1}]$. If $j_M - j_m \leq 2$, then we go through an argument similar to before. Assume $j_M - j_m \geq 3$. Then $b-a \geq M-m \geq j_M! - j_m! - {j_m}^{-1} \geq j_M! - j_m! - 1$. Thus,

$$(b-a)^{\frac{1}{p}-\frac{1}{q}} \left(\int_a^b f(t)^q \mathrm{d}t\right)^{\frac{1}{q}} \leq (j_M! - j_m! - 1)^{\frac{1}{p}-\frac{1}{q}} \left(\int_{j_m!}^{j_M!+1} f(t)^q \mathrm{d}t\right)^{\frac{1}{q}}$$
$$\lesssim j_M!^{\frac{1}{p}-\frac{1}{q}} \left(\sum_{j=j_m}^{j_M} {\delta_j}^{\frac{p-q}{p}}\right)^{\frac{1}{q}}$$
$$\lesssim 1.$$

Thus, $f \in \mathcal{M}^p_q(\mathbb{R})$.

Now we disprove $f \in L^1(\mathbb{R}^n) + L^\infty(\mathbb{R}^n)$. Let R be fixed. Then a geometric observation shows that

$$\|f - \min(f,R)\|_1 \leq \|f-h\|_\infty$$

for any $h \in L^\infty(\mathbb{R}^n)$ with $\|h\|_\infty \leq R$.

Let $S > 2R+2$ be an integer. Then

$$\int_{f^{-1}(\{S\})} (f(x) - \min(f(x),R))\mathrm{d}x = |f^{-1}(\{S\})|(S-R)$$

$$\geq \frac{S}{2}|f^{-1}(\{S\})|$$
$$= \frac{S}{2}\sum_{k=2^{2^S}}^{2^{2^{S+1}}}\frac{1}{k}$$
$$\gtrsim S \cdot 2^S.$$

Thus, $\|f - \min(f, R)\|_1 = \infty$. Hence $f \notin L^1(\mathbb{R}) + L^\infty(\mathbb{R})$.

6.1.5.2 Boundedness of the Maximal Operator

Now we present a typical argument about the proof of the boundedness of operators on Morrey spaces. As the proof shows, we depend heavily upon the so-called local v.s. global strategy.

Theorem 6.19 *Let* $1 < q \leq p < \infty$. *Then* $\|Mf\|_{\mathcal{M}^p_q} \lesssim_q \|f\|_{\mathcal{M}^p_q}$ *for all* $f \in \mathcal{M}^p_q(\mathbb{R}^n)$.

We use Example 1.27 here.

Proof For the proof we have only to show, from the definition, that

$$|Q|^{\frac{1}{p}-\frac{1}{q}}\|Mf\|_{L^q(Q)} \lesssim \|f\|_{\mathcal{M}^p_q}, \quad Q \in \mathcal{Q}. \tag{6.75}$$

Write $f = f_1 + f_2$, where $f_1 = f$ on $5Q$ and $f_2 = f$ outside $5Q$. The estimate of (6.75) can be decomposed into:

$$|Q|^{\frac{1}{p}-\frac{1}{q}}\|Mf_1\|_{L^q(Q)} \lesssim \|f\|_{\mathcal{M}^p_q}, \tag{6.76}$$

$$|Q|^{\frac{1}{p}-\frac{1}{q}}\|Mf_2\|_{L^q(Q)} \lesssim \|f\|_{\mathcal{M}^p_q}. \tag{6.77}$$

Since M is $L^q(\mathbb{R}^n)$-bounded, the local estimate (6.76) can be shown easily:

$$\begin{aligned}
|Q|^{\frac{1}{p}-\frac{1}{q}}\|Mf_1\|_{L^q(Q)} &\leq |Q|^{\frac{1}{p}-\frac{1}{q}}\|Mf_1\|_q \\
&\lesssim |Q|^{\frac{1}{p}-\frac{1}{q}}\|f\|_{L^q(5Q)} \\
&\simeq |5Q|^{\frac{1}{p}-\frac{1}{q}}\|f\|_{L^q(5Q)} \\
&\leq \|f\|_{\mathcal{M}^p_q}.
\end{aligned}$$

It remains to prove the global estimate (6.77). If we insert (1.138) into (6.77), then we obtain

$$|Q|^{\frac{1}{p}-\frac{1}{q}}\|Mf_2\|_{L^q(Q)} \lesssim |Q|^{\frac{1}{p}} \sup_{Q\subset R\in\mathscr{Q}} m_R(|f|).$$

Taking into account $\mathscr{M}_q^p(\mathbb{R}^n) \hookrightarrow \mathscr{M}_1^p(\mathbb{R}^n)$ with embedding constant 1, we see that

$$|Q|^{\frac{1}{p}-\frac{1}{q}}\|Mf_2\|_{L^q(Q)} \lesssim \sup_{R\in\mathscr{Q}} |R|^{\frac{1}{p}-1}\int_R |f(y)|\mathrm{d}y = \|f\|_{\mathscr{M}_1^p} \le \|f\|_{\mathscr{M}_q^p}.$$

Here for the last inequality we used (6.17). Consequently, (6.77) is proved.

We have also the vector-valued maximal inequality. This vector-valued maximal inequality will enable us to develop the theory of Triebel–Lizorkin–Morrey spaces. Let $0 < q \le p < \infty$ and $0 < r \le \infty$. We define the *vector-valued Morrey space* $\mathscr{M}_q^p(\ell^r, \mathbb{R}^n)$ with the norm $\| \star \|_{\mathscr{M}_q^p(\ell^r)}$, whose norm is given by (6.2) with X and q replaced by $\mathscr{M}_q^p(\mathbb{R}^n)$ and r, respectively.

Theorem 6.20 *Suppose that the parameters p, q, u satisfy $1 < q \le p < \infty$ and $1 < u \le \infty$. Then $\|\{Mf_j\}_{j=1}^\infty\|_{\mathscr{M}_q^p(\ell^u)} \lesssim \|\{f_j\}_{j=1}^\infty\|_{\mathscr{M}_q^p(\ell^u)}$ for every sequence $\{f_j\}_{j=0}^\infty \subset L^0(\mathbb{R}^n)$.*

Proof We modify the proof of Theorem 6.19. Instead of (6.75), we have to show that:

$$|Q|^{\frac{1}{p}-\frac{1}{q}}\left(\int_Q\left(\sum_{j=1}^\infty Mf_j(y)^u\right)^{\frac{q}{u}}\mathrm{d}y\right)^{\frac{1}{q}} \lesssim \|\{f_j\}_{j=1}^\infty\|_{\mathscr{M}_q^p(\ell^u)} \quad (Q\in\mathscr{Q}).$$

We consider a decomposition similar to that in the proof of Theorem 6.19; write $f_j = f_{1,j} + f_{2,j}$, where $f_{j,1} \equiv \chi_{5Q}f$ and $f_{j,2} \equiv f_j - f_{j,1}$. Estimates (6.76) and (6.77) correspond to

$$|Q|^{\frac{1}{p}-\frac{1}{q}}\|\{\chi_Q Mf_{j,1}\}_{j=1}^\infty\|_{L^u(\ell^q)} \lesssim \|\{f_j\}_{j=1}^\infty\|_{\mathscr{M}_q^p(\ell^u)}, \tag{6.78}$$

$$|Q|^{\frac{1}{p}-\frac{1}{q}}\|\{\chi_Q Mf_{j,2}\}_{j=1}^\infty\|_{L^u(\ell^q)} \lesssim \|\{f_j\}_{j=1}^\infty\|_{\mathscr{M}_q^p(\ell^u)}, \tag{6.79}$$

respectively.

For the local estimate (6.78), we proceed as in (6.76); first we expand the integration domain Q: $|Q|^{\frac{1}{p}-\frac{1}{q}}\|\{\chi_Q Mf_{j,1}\}_{j=1}^\infty\|_{L^u(\ell^q)} \le |Q|^{\frac{1}{p}-\frac{1}{q}}\|\{Mf_{j,1}\}_{j=1}^\infty\|_{L^u(\ell^q)}$. Next we use the Fefferman–Stein vector-valued inequality (Theorem 1.49) to have

$$\begin{aligned}|Q|^{\frac{1}{p}-\frac{1}{q}}\|\{\chi_Q Mf_{j,1}\}_{j=1}^\infty\|_{L^u(\ell^q)} &\lesssim |Q|^{\frac{1}{p}-\frac{1}{q}}\|\{f_{j,1}\}_{j=1}^\infty\|_{L^u(\ell^q)}\\ &= |Q|^{\frac{1}{p}-\frac{1}{q}}\|\{\chi_{5Q}f_j\}_{j=1}^\infty\|_{L^u(\ell^q)}\\ &\le \|\{f_j\}_{j=1}^\infty\|_{\mathscr{M}_q^p(\ell^u)}.\end{aligned}$$

For the estimate of the second term, we recall (1.138). Let $y \in Q$. By (1.138) with f replaced by f_j, we have

$$Mf_{j,2}(y) \lesssim \sup_{Q \subset R \in \mathcal{Q}} m_R(|f_j|) \tag{6.80}$$

and hence

$$Mf_{j,2}(y) \lesssim \sum_{k=1}^{\infty} m_{2^k Q}(|f_j|).$$

By the Minkowski inequality, we have

$$\left(\sum_{j=1}^{\infty} Mf_{j,2}(y)^u\right)^{\frac{1}{u}} \lesssim \sum_{k=1}^{\infty} m_{2^k Q}\left(\left(\sum_{j=1}^{\infty} |f_j|^u\right)^{\frac{1}{u}}\right) \quad (y \in Q).$$

Consequently,

$$|Q|^{\frac{1}{p}-1} \int_Q \left(\sum_{j=1}^{\infty} Mf_{j,2}(y)^u\right)^{\frac{1}{u}} \mathrm{d}y \lesssim \sum_{k=1}^{\infty} |Q|^{\frac{1}{p}} m_{2^k Q}\left(\left(\sum_{j=1}^{\infty} |f_j|^u\right)^{\frac{1}{u}}\right).$$

Going through an argument similar to (6.77) and using $p < \infty$, we obtain the global estimate (6.79).

Exercises

Exercise 6.14 Using the change of variables, show (6.71).

Exercise 6.15 Keeping in mind that the cubes are compact, show (6.72).

Exercise 6.16 [924, Proposition 4.1] Let $p > q > 0$, and let $R > 1$ solve $(R+1)^{-\frac{1}{p}} = 2^{\frac{1}{q}}(1+R)^{-\frac{1}{q}}$. For a vector $\varepsilon \in \{0,1\}^n$, we define an affine transformation T_ε by

$$T_\varepsilon(x) \equiv \frac{1}{R+1}x + \frac{R}{R+1}\varepsilon \quad (x \in \mathbb{R}^n).$$

Let $E_0 \equiv [0,1]^n$. Suppose that we have defined $E_0, E_1, E_2, \ldots, E_j$. Define

$$E_{j+1} \equiv \bigcup_{\varepsilon \in \{0,1\}^n} T_\varepsilon(E_j).$$

1. Show that $\|\chi_{E_j}\|_{\mathcal{M}^p_q} \sim (1+R)^{-jn/p}$, where the implicit constants in $\sim$ do not depend on j but can depend on p and q.
2. Using f_j, show (6.73).

Exercise 6.17

1. For $0 < q < p < \infty$, show that the *extremal function* $|x|^{-\frac{n}{p}} \in \mathcal{M}^p_q(\mathbb{R}^n)$. Note that this fails for $p = q$!
2. Let $1 \le q_1 \le q_2 \le p < \infty$. Show that

$$L^p(\mathbb{R}^n) = \mathcal{M}^p_p(\mathbb{R}^n) \hookrightarrow \mathcal{M}^p_{q_2}(\mathbb{R}^n) \hookrightarrow \mathcal{M}^p_{q_1}(\mathbb{R}^n) \hookrightarrow \mathcal{S}'(\mathbb{R}^n)$$

using the Hölder inequality.
3. Let $1 \le q \le p < \infty$, $n < p$. If $f, \nabla f \in \mathcal{M}^p_q(\mathbb{R}^n)$, show that $f \in B^{1-n/p}_{\infty\infty}(\mathbb{R}^n) = \mathcal{C}^{1-n/p}(\mathbb{R}^n)$. This result is the Morrey lemma [804], whose original statement was stated for the functions on balls. Hint: Use (1.159) for $\psi(D)f$ and $\varphi_j(D)f$ to estimate the Besov norm given by (2.3).

Exercise 6.18 Let $1 \le q < p < \infty$. Set $f_j \equiv \chi_{B(j^{-1})} \cdot |\star|^{-n/p}$ for $j \in \mathbb{N}$. Then by the scaling show that $0 < \|f_j\|_{\mathcal{M}^p_q} = \| |\star|^{-n/p} - f_j\|_{\mathcal{M}^p_q} = \| |\star|^{-n/p}\|_{\mathcal{M}^p_q} < \infty$ for all $j \in \mathbb{N}$.

6.1.6 Orlicz Spaces

6.1.6.1 Orlicz Spaces

We aim here to mix $L^p(\mathbb{R}^n)$ and $L^{\tilde{p}}(\mathbb{R}^n)$ according to the value of $|f(x)|$. One of the ideas is to use $\Phi(|f(x)|)$, where $\Phi(t) \equiv t^p + t^{\tilde{p}}$ for $t \ge 0$. To this end, we introduce the definition below:

Some literature allows the functions to take ∞. However, here for the sake of simplicity, we content ourselves with investigating the properties of functions taking finite values.

In the definition below we exclude the case where $\varphi(t) = \infty$ for some $t \in (0, \infty)$.

Definition 6.10 (Young function) A function $\Phi : [0,\infty) \to [0,\infty)$ is said to be a *Young* function, if there exists an increasing function φ which is right-continuous such that

$$\Phi(t) = \int_0^t \varphi(s)\mathrm{d}s \quad (t \ge 0). \tag{6.81}$$

Equality (6.81) is called the *canonical representation* of a Young function Φ. By convention define $\Phi(\infty) \equiv \infty$.

Observe that any Young function Φ is convex; that is,

$$\Phi((1-\theta)t_1 + \theta t_2) \le (1-\theta)\Phi(t_1) + \theta\Phi(t_2)$$

for all $t_1, t_2 \in (0, \infty)$ and $0 < \theta < 1$.

Example 6.6 The functions $e^t - 1$ and $t \log(t+1)$ are Young functions.

Lemma 6.14 *Let $\Phi : [0, \infty) \to [0, \infty)$ be a Young function.*

1. *For all $0 < \alpha < 1$ and $0 \le t < \infty$ $\Phi(\alpha t) \le \alpha\Phi(t)$.*
2. *The mapping $t \in (0, \infty) \mapsto t^{-1}\Phi(t) \in [0, \infty)$ is increasing.*

Proof

1. By the convexity we have $\Phi(\alpha t) = \Phi(\alpha t + (1-\alpha)0) \le \alpha\Phi(t) + (1-\alpha)\Phi(0) = \alpha\Phi(t)$.
2. This can be derived easily from *1*; see Exercise 6.23.

Next, we will treat the information on φ appearing in the canonical representation.

Lemma 6.15 *Let $\Phi : [0, \infty) \to [0, \infty)$ be a Young function with* (6.81), *and let $t > 0$. Then:*

$$\frac{\Phi(t)}{t} \le \varphi(t) \le \frac{\Phi(2t)}{t}, \tag{6.82}$$

$$\frac{1}{4}\varphi\left(\frac{t}{2}\right) \le \int_0^t \frac{\Phi(s)}{s} \mathrm{d}s. \tag{6.83}$$

Proof All the estimates are easy to derive:

$$\frac{\Phi(t)}{t} = \frac{1}{t}\int_0^t \varphi(s)\mathrm{d}s \le \frac{1}{t}\int_0^t \varphi(t)\mathrm{d}s = \varphi(t)$$

and

$$\varphi(t) = \frac{1}{t}\int_t^{2t} \varphi(t)\mathrm{d}t \le \frac{1}{t}\int_t^{2t} \varphi(s)\mathrm{d}s \le \frac{1}{t}\int_0^{2t} \varphi(s)\mathrm{d}s = \frac{\Phi(2t)}{t},$$

which proves (6.82).

To obtain (6.83), we use the Fubini theorem and $\log 2 = 0.69\cdots$ to have:

$$\int_0^t \frac{\Phi(s)}{s}\mathrm{d}s = \int_0^t \left(\frac{1}{s}\int_0^s \varphi(u)\mathrm{d}u\right)\mathrm{d}s = \int_0^t \varphi(u)\log\frac{t}{u}\mathrm{d}u \ge \frac{1}{4}\varphi\left(\frac{t}{2}\right).$$

Thus, the proof is therefore complete.

6.1.6.2 A Substitute for the Weak-L^1 Boundedness of the Hardy–Littlewood Maximal Operator

Orlicz spaces are used to describe the endpoint case of the boundedness of the operators. Example 1.25 disproves

$$\|Mf\|_1 \lesssim \|f\|_1 \quad (f \in L^1(\mathbb{R}^n)).$$

We want to look for a remedy for this situation. One idea is to use the weak-type inequality. This means that we look for some conditions for Mf to satisfy when we have an integrable function f. So now we are interested in Mf being locally integrable. One of the ways to have this is to assume $f \in L^p(\mathbb{R}^n)$ for $p > 1$. But this is too strong. In fact, we will have $Mf \in L^p(\mathbb{R}^n)$, which is a little far from $Mf \in L^1_{\rm loc}(\mathbb{R}^n)$. By the dilation argument, we can also disprove

$$\int_{B(1)} M[\chi_{B(1)} f](x)\mathrm{d}x \lesssim \int_{B(1)} |f(x)|\mathrm{d}x. \tag{6.84}$$

However, by refining the layer cake representation, Theorem 1.5, we can show that

$$\int_{B(1)} M[\chi_{B(1)} f](x)\mathrm{d}x \lesssim \int_{B(1)} |f(x)|\log(3 + |f(x)|)\mathrm{d}x \tag{6.85}$$

for all measurable functions f. See Exercises 6.21 and 6.22 for (6.84) and (6.84), respectively. Compared with Morrey spaces, Orlicz spaces do not require the geometric property of the underlying space. So, we let $(X, \mathcal{B}, \mu)$ be a measure space. Based on this observation we define the Orlicz space $L^\Phi(X)$, as follows.

Definition 6.11 (Orlicz space) Let $\Phi : [0, \infty) \to [0, \infty)$ be a Young function. Then define

$$\|f\|_{L^\Phi(X)} \equiv \inf\left(\left\{\lambda \in (0, \infty) : \int_X \Phi\left(\frac{|f(x)|}{\lambda}\right)\mathrm{d}\mu(x) \le 1\right\} \cup \{\infty\}\right)$$

for a μ-measurable function f. Similar to Theorem 6.11 we can show that $L^\Phi(X)$ is a Banach space. The *Orlicz space* $L^\Phi(X)$ over X is the set of μ-measurable functions f for which $\|f\|_{L^\Phi(X)}$ is finite.

Note that the definition of Orlicz norms is similar to that of variable Lebesgue norms.

6.1.6.3 Conjugate Function

A direct calculation shows:

$$ab \leq \frac{1}{p}a^p + \frac{1}{p'}b^{p'}, \tag{6.86}$$

where $1 < p < \infty$ and $a, b > 0$. Inequality (6.86) is a special case of Theorem 6.23, called the Young inequality. Changing the viewpoint, we see (6.86) can be seen as the one for the function $\Phi(t) = \frac{1}{p}t^p$.

We are led to consider the following generalization:

$$ab \leq \Phi(a) + \Psi(b),$$

where Ψ is a convex function. Our present problem is to obtain a function Ψ for a given Young function Φ.

The following definition can be regarded as a passage of the harmonic conjugate of real numbers to functions.

Definition 6.12 (Conjugate) For a Young function Φ, its (*Legendre*) *conjugate function* is defined by

$$\Phi^*(t) \equiv \int_0^t \varphi^*(s)\mathrm{d}s \quad (t \geq 0), \tag{6.87}$$

where

$$\varphi^*(v) \equiv \int_0^\infty \chi_{\varphi^{-1}([0,v])}(x)\mathrm{d}x \tag{6.88}$$

for every $v > 0$.

Concerning the definition above, we have the following observation:

Lemma 6.16 *Maintain the notation above. Let $x, v \geq 0$.*

1. *If $\varphi(x) < v$, then $x \leq \varphi^*(v)$.*
2. *If $\varphi(x) > v$, then $x \geq \varphi^*(v)$.*

Proof

1. Note that $[0, x] \subset \varphi^{-1}([0, v])$. Thus,

$$\varphi^*(v) = \int_0^\infty \chi_{\varphi^{-1}([0,v])}(t)\mathrm{d}t \geq \int_0^\infty \chi_{[0,x]}(t)\mathrm{d}t = x,$$

as required.

2. Note that $[0, x] \supset \varphi^{-1}([0, v])$. Thus,

$$\varphi^*(v) = \int_0^\infty \chi_{\varphi^{-1}([0,v])}(t)\mathrm{d}t \le \int_0^\infty \chi_{[0,x]}(t)\mathrm{d}t = x,$$

as required.

Definition 6.13 (Nice Young function) Let $\Phi : [0, \infty) \to [0, \infty)$ be a Young function with canonical representation (6.81). Then Φ is said to be a *nice Young function* if for all $t \in (0, \infty)$,

$$\varphi(0) = 0 < \varphi(t) < \lim_{t' \to \infty} \varphi(t') = \infty.$$

According to the definition above, $\Phi(t) = t$ is not a nice Young function, since $\varphi(t) \equiv 1$.

The conjugate function is important when we consider the dual space. See Exercise 6.25. First we prove that the conjugate of any nice Young function is also a nice Young function.

Proposition 6.2 *Let Φ be a Young function, and let Φ^* be its conjugate given by Definition* 6.12 *via* (6.87) *and* (6.88). *Then the following are equivalent:*

1. *Φ is a nice Young function.*
2. *Φ^* is a nice Young function.*
3. $0 < \Phi(t) < \infty$ *and* $0 < \Phi^*(t) < \infty$ *for all* $t \in (0, \infty)$.

Proof We plan to prove 1. $\Longrightarrow$ 3. $\Longrightarrow$ 2. $\Longrightarrow$ 1.

- Assume that Φ is a nice Young function. Note that Φ^* is finitely valued because $\lim_{t \to \infty} \varphi(t) = \infty$. Since $\varphi(0) = 0$ and φ is right-continuous, $\Phi^*(t) > 0$ for all $t \in (0, \infty)$. It is clear $0 < \Phi(t) < \infty$ for any $t \in (0, \infty)$, since Φ is a nice Young function.
- Assume that $0 < \Phi(t), \Phi^*(t) < \infty$ for all $t \in (0, \infty)$. From the definition of Φ it is easy to see that $\varphi^*(0) = 0$. In fact, if $\varphi^*(0) > 0$, then $\varphi^{-1}(\{0\})$ contains an interval $[0, a]$. So, $\varphi(a) = 0$. This means that $\Phi(a) = 0$, a contradiction.

  Let $t > 0$. Then from the right-continuity of φ, there exists $\delta > 0$ such that we have $\varphi(s) < t$ for all $0 \le s < \delta$. Thus, $[0, \delta) \subset \varphi^{-1}([0, t))$. Hence $\varphi^*(t) \ge \delta$. Since $\varphi(t) \uparrow \infty$ as $t \uparrow \infty$, $\varphi^*(t) < \infty$. Finally, we note that $\varphi^*(\varphi(t)) \ge t$, which yields $\lim_{t' \to \infty} \varphi^*(t') = \infty$.
- Finally, suppose that Φ^* is a nice function. If we assume $V \equiv \sup \varphi < \infty$ for a contradiction, then

$$\varphi^*(v) = \int_0^\infty \chi_{\varphi^{-1}([0,v])}(x)\mathrm{d}x = \infty$$

  for all $v \in (V, \infty)$, which is a contradiction. If we assume $V' \equiv \inf \varphi > 0$ for a contradiction, then

$$\varphi^*(v) = \int_0^\infty \chi_{\varphi^{-1}([0,v])}(x)\mathrm{d}x = \int_0^\infty \chi_{\varphi^{-1}([0,\infty)}(x)\mathrm{d}x = 0$$

for all $v \in (0, V')$, which is also a contradiction. Thus, Φ is a nice Young function.

Given a function $\Theta : (0, \infty) \to \mathbb{R}$, its *Fenchel–Legendre transform* is given by

$$t \in [0, \infty) \mapsto \sup\{st - \Theta(s) : s \in [0, \infty)\}.$$

Note that we also refer to the conclusion of Theorem 6.21 as the (Fenchel–)Legendre transform of Φ.

Theorem 6.21 *Let Φ be a Young function, and let Φ^* its conjugate. Then*

$$\Phi^*(t) = \sup\{st - \Phi(s) : s \in [0, \infty)\} \quad (t \geq 0).$$

Hence Theorem 6.21 asserts that Φ^* is the Fenchel–Legendre transform of Φ.

Proof Fix $s, t \geq 0$. We are going to show

$$\Phi(t) + \Phi^*(s) \geq st \tag{6.89}$$

for all $s, t \geq 0$ and that for each $s \geq 0$ there exist $t \geq 0$ for which equality in (6.89) holds.

If $st = 0$, then (6.89) is trivial. Therefore, we may assume that $s, t > 0$. Inserting (6.88) into the canonical representation, we have

$$\begin{aligned}
\Phi^*(s) &= \int_0^s \left(\int_0^\infty \chi_{\varphi^{-1}([0,v])}(x)\mathrm{d}x \right) \mathrm{d}v \\
&= \iint_{[0,\infty)^2} \chi_{\{(x,v)\in[0,\infty)^2 \,:\, 0\leq v\leq s,\, \varphi(x)<v\}}(x, v)\mathrm{d}v\mathrm{d}x.
\end{aligned}$$

Note that "$0 \leq v \leq s$ and $x \in \varphi^{-1}([0, v])$" is not equivalent to "$0 \leq v \leq s$ and $\varphi(x) < v$".

Meanwhile,

$$\Phi(t) = \int_0^t \varphi(x)\mathrm{d}x = \iint_{[0,\infty)^2} \chi_{\{(x,v)\in[0,\infty)^2 \,:\, 0\leq x\leq t,\, 0\leq v\leq\varphi(x)\}}(x, v)\mathrm{d}v\mathrm{d}x.$$

Observe that

$$[0, t] \times [0, s] \subset \{(x, v) : 0 \leq v \leq s, \varphi(x) \leq v \text{ or } 0 \leq x \leq t, 0 \leq v \leq \varphi(x)\}. \tag{6.90}$$

Therefore we conclude

$$\Phi^*(s) + \Phi(t) \geq \int_0^\infty \int_0^\infty \chi_{[0,t]\times[0,s]}(x, v) \mathrm{d}x\mathrm{d}v = st.$$

To show that equality holds for some t, we set $t \equiv \varphi^*(s)$. We obtain a kind of reverse inclusion in (6.90). We aim to show

$$\{(x, v) \in (0, \infty) \times (0, \infty) : 0 < v < s, \varphi(x) < v\} \subset [0, t] \times [0, s]$$

and that

$$\{(x, v) \in (0, \infty) \times (0, \infty) : 0 < x < t, 0 < v < \varphi(x)\} \subset [0, t] \times [0, s].$$

Let $0 < v < s$ satisfy $\varphi(x) < v$. Then $t = \varphi^*(s) \geq \varphi^*(v) \geq x \geq 0$ thanks to Lemma 6.16. Meanwhile, let $0 \leq x < t$ and $0 \leq v < \varphi(x)$. Since $x < t = \varphi^*(s)$, we have $\varphi(x) \leq s$ thanks to Lemma 6.16 once again. Therefore $v \leq s$.

Carrying out the same calculation as before, we conclude that $\Phi^*(s)+\Phi(t) = st$.

We now check the relation between the conjugation and the dilation.

Proposition 6.3 *Let Φ be a Young function, and let $a, b > 0$. Set $\Psi \equiv a\,\Phi(b\star)$. Then Ψ is also a Young function and the conjugate is given by $\Psi^* = a\,\Phi^*\left(\frac{\star}{ab}\right)$.*

Proof Simply resort to the definition. We leave the details; see Exercise 6.20.

The following theorem shows that the operation of conjugation is an involution: if we take the conjugate twice then we go back to the original function.

Theorem 6.22 *Let Φ be a Young function. Then $\Phi^{**} = \Phi$.*

Proof Fix $x \in [0, \infty)$. We claim that $\varphi^{**} = \varphi$ except at a countable points in $[0, \infty)$. Once this is achieved, since Φ^{**} and Φ have the same canonical representation, it follows that $\Phi^{**} = \Phi$.

Let $v > \varphi(x)$. Then $x \leq \varphi^*(v)$ thanks to Lemma 6.16. Consequently, for all $\varepsilon > 0$, $\varphi^{**}(x - \varepsilon) \leq v$ again thanks to Lemma 6.16. Since $v > \varphi(x)$ is arbitrary, $\varphi^{**}(x - \varepsilon) \leq \varphi(x)$. Since $\varepsilon > 0$ is also arbitrary, $\varphi^{**}(x) \leq \varphi(x)$ for all x at which φ^{**} is continuous.

Conversely, let $v < \varphi(x)$. Then $x \geq \varphi^*(v)$ thanks to Lemma 6.16. Thus for all $\varepsilon > 0$, $\varphi^{**}(x + \varepsilon) \geq v$ again thanks to Lemma 6.16. Since $v < \varphi(x)$ is arbitrary, $\varphi^{**}(x + \varepsilon) \geq \varphi(x)$. Since $\varepsilon > 0$ is also arbitrary and φ^{**} is right-continuous, it follows that $\varphi^{**}(x) \geq \varphi(x)$.

If we reexamine the proof, we notice that $\varphi^{**} = \varphi$ on $(0, \infty)$, since both functions are right-continuous.

Theorem 6.23 (Young's inequality) *Let Φ be a Young function and denote by Φ^* its conjugate. Then $s\,t \leq \Phi(t) + \Phi^*(s)$ for all $s, t \geq 0$.*

Proof We have only to reexamine the proof of Theorem 6.21.

Lemma 6.17 *Let Φ be a Young function and let Φ^* its conjugate. Then we have the following:*

1. *The inequality*

$$\Phi^*\left(\frac{\Phi(t)}{t}\right) \le \Phi(t) \le \Phi^*\left(\frac{2\Phi(t)}{t}\right) \tag{6.91}$$

holds for all $t > 0$.

2. *For all $t, s > 0$ such that $st \le \Phi(t)$, we have*

$$\Phi^*(s) \le st. \tag{6.92}$$

3. *For all $t, s > 0$ such that $st \ge \Phi(t)$, we have*

$$st \le \Phi^*(2s) \tag{6.93}$$

4. *Let $s, t, \lambda > 0$. If $\Phi(t) = \Phi^*(s) = \lambda$, then we have*

$$\lambda \le ts \le 2\lambda. \tag{6.94}$$

Proof Recall that $\Phi(t)/t$ is increasing in t thanks to Lemma 6.14. Thus,

$$\Phi^*\left(\frac{\Phi(t)}{t}\right) = \sup_{0<s<t} s\left(\frac{\Phi(t)}{t} - \frac{\Phi(s)}{s}\right) \le t \sup_{0<s<t}\left(\frac{\Phi(t)}{t} - \frac{\Phi(s)}{s}\right) \le \Phi(t).$$

The right inequality is easier to prove. We have only to use Theorem 6.21:

$$\Phi^*\left(\frac{2\Phi(t)}{t}\right) = \sup_{s>0} s\left(\frac{2\Phi(t)}{t} - \frac{\Phi(s)}{s}\right) \ge \Phi(t).$$

Thus, (6.91) is proved.

Let $st \le \Phi(t)$. Since $t^{-1}\Phi(t)$ is increasing in $t > 0$ again thanks to Lemma 6.14, we have

$$\frac{\Phi^*(s)}{s} = \sup_{u>0}\left(u - \frac{\Phi(u)}{s}\right) \le t \sup_{u>0}\left(\frac{u}{t} - \frac{\Phi(u)}{\Phi(t)}\right) = t \sup_{0<u<t}\left(\frac{u}{t} - \frac{\Phi(u)}{\Phi(t)}\right) \le t.$$

Thus, the proof of (6.92) is complete.

Suppose instead that $st \ge \Phi(t)$, Then

$$\frac{\Phi^*(2s)}{s} = \sup_{u>0}\left(2u - \frac{\Phi(u)}{s}\right) \ge 2t - \frac{\Phi(t)}{s} \ge t.$$

Thus, the proof of (6.93) is complete.

It follows from the Young inequality that $st \le \Phi(s) + \Phi(t) = 2\lambda$. If $\varphi(t) \le s$, then we have

$$\lambda = \Phi(t) = \int_0^t \varphi(u) du \le st.$$

If $\varphi(t) > s$, then $\varphi^*(s) \le t$ from Lemma 6.16. Thus $\lambda = \Phi^*(s) = \int_0^s \varphi^*(u) du \le st$. Therefore (6.94) is established.

6.1.6.4 Δ_2-Condition and ∇_2-Condition

To obtain the boundedness of the operators we need the following notion:

Definition 6.14 (Δ_2-condition, ∇_2-condition) Let $\Phi : [0, \infty) \to [0, \infty)$ be a Young function.

1. The function Φ is said to satisfy the Δ_2-*condition*, if there exists a constant $\mu > 1$ such that

$$\Phi(2t) \le \mu \, \Phi(t) \quad (t > 0). \tag{6.95}$$

   In this case one writes $\Phi \in \Delta_2$.
2. The function Φ is said to satisfy the ∇_2-*condition*, if

$$\Phi^* \in \Delta_2. \tag{6.96}$$

   In this case one writes $\Phi \in \nabla_2$.

The Δ_2-condition is referred to as the *doubling condition*.

Let $f_\alpha(t) \equiv \alpha^{-1} t^\alpha$ for $t \ge 0$, where $\alpha \in [1, \infty)$. It is important to note that $f_1 \in \Delta_2 \setminus \nabla_2$ but that $f_\alpha \in \Delta_2 \cap \nabla_2$ for $\alpha \in (1, \infty)$. For general case, we show that something similar to this happens.

Proposition 6.4 *Let Φ be a Young function.*

1. *The function $\Phi \in \nabla_2$ if and only if there exists a constant $A > 1$ such that*

$$\Phi(A\star) \ge 2A \, \Phi. \tag{6.97}$$

2. *If $\Phi \in \nabla_2$, then there exist $\mu > 1$ and $\varepsilon > 0$ such that*

$$\Phi(ut) \ge \mu^{-1} u^{1+\varepsilon} \Phi(t), \ \Phi(vt) \le \mu v^{1+\varepsilon} \Phi(t) \tag{6.98}$$

   whenever $0 < v \le 1 \le u$ and $t \ge 0$.

Proof As for the first assertion we note that

$$\Phi(A\star)^* = \Phi^*\left(\frac{\star}{A}\right), \quad (2A\,\Phi)^* = 2A\,\Phi\left(\frac{\star}{2A}\right)$$

from Proposition 6.3. Therefore it follows that

$$\Phi(2\star) \geq 2A\,\Phi \Longleftrightarrow \Phi^*\left(\frac{\star}{A}\right) \leq 2A\,\Phi^*\left(\frac{\star}{2A}\right) \Longleftrightarrow \Phi^*(2\star) \leq 2A\,\Phi^*$$

from Theorem 6.21.

The second assertion can be obtained by induction. Indeed, we have

$$\Phi(A^k t) \geq (2A)^k \Phi(t) \quad (t \in [0, \infty)) \tag{6.99}$$

for all $k \in \mathbb{N}$. Therefore if we set $\varepsilon = \log_A 2$, then

$$\Phi(A^k t) \geq A^{k(1+\varepsilon)} \Phi(t) \quad (t \geq 0,\ k \in \mathbb{Z}). \tag{6.100}$$

It remains to pass to the continuous valuable.

It is convenient to transform inequalities (6.95) and (6.96) into the integral form.

Proposition 6.5 *Let Φ be a Young function with canonical representation* (6.81).

1. *Assume that $\Phi \in \Delta_2$. More precisely $\Phi(2\star) \leq A\,\Phi$ for some $A \geq 2$. Set $\beta \equiv \log_2 A$. If $p > \beta + 1$, then for all $t > 0$,*

$$\int_t^\infty \frac{\varphi(s)}{s^p} \mathrm{d}s \lesssim \frac{\Phi(t)}{t^p}. \tag{6.101}$$

2. *Assume that $\Phi \in \nabla_2$. Then for all $t > 0$,*

$$\int_0^t \frac{\varphi(s)}{s} \mathrm{d}s \lesssim \frac{\Phi(t)}{t}. \tag{6.102}$$

Proof If we carry out integration by parts we obtain

$$\int_t^\infty \frac{\varphi(s)}{s^p} \mathrm{d}s = \int_t^\infty \frac{\Phi'(s)}{s^p} \mathrm{d}s = \lim_{R \to \infty} \left(\left[\frac{\Phi(s)}{s^p} \right]_t^R + p \int_t^R \frac{\Phi(s)}{s^{p+1}} \mathrm{d}s \right).$$

Now we deduce from the doubling condition $s^{-\beta}\Phi(s) \lesssim t^{-\beta}\Phi(t)$ for $s \geq t > 0$. From this estimate we obtain

$$\lim_{R \to \infty} \frac{\Phi(R)}{R^p} = 0, \quad \int_t^\infty \frac{\Phi(s)}{s^{p+1}} \mathrm{d}s \lesssim \frac{\Phi(t)}{t^p}.$$

Therefore, (6.101) follows.

Estimate (6.102) can be proved similarly. First we carry out integration by parts to have

$$\int_0^t \frac{\varphi(s)}{s}\mathrm{d}s = \left[\frac{\Phi(s)}{s}\right]_0^t + \int_0^t \frac{\Phi(s)}{s^2}\mathrm{d}s = \frac{\Phi(t)}{t} + \int_0^t \frac{\Phi(s)}{s^2}\mathrm{d}s \tag{6.103}$$

from (6.98). Now we use (6.97), which yields

$$\int_0^t \frac{\Phi(s)}{s^2}\mathrm{d}s = \sum_{j=0}^{\infty} \int_{A^{-j-1}t}^{A^{-j}t} \frac{\Phi(s)}{s^2}\mathrm{d}s \lesssim \frac{1}{t}\sum_{j=0}^{\infty} A^j \Phi(A^{-j}t) \lesssim \frac{1}{t}\sum_{j=0}^{\infty} \frac{A^j}{(2A)^j}\Phi(t) \lesssim \frac{\Phi(t)}{t}.$$

Inserting this estimate into (6.103), we obtain

$$\int_0^t \frac{\varphi(s)}{s}\mathrm{d}s \le \frac{\Phi(t)}{t} + \int_0^t \frac{\Phi(s)}{s^2}\mathrm{d}s \lesssim \frac{\Phi(t)}{t}.$$

Therefore (6.102) is established.

6.1.6.5 Maximal Inequalities

We can develop a theory of Besov spaces and Triebel–Lizorkin spaces based on Orlicz spaces. Let $0 < q \le \infty$. We define the vector-valued Orlicz space $L^\Phi(\ell^q, \mathbb{R}^n)$ with the norm $\|\star\|_{L^\Phi(\ell^q)}$, whose norm is given by (6.2) with X replaced by $L^\Phi(\mathbb{R}^n)$. To this end, the following inequality is useful:

Theorem 6.24 *If $\Phi \in \nabla_2 \cap \Delta_2$ and $1 < q \le \infty$, then we have*

$$\|\{Mf_j\}_{j=1}^{\infty}\|_{L^\Phi(\ell^q)} \lesssim_{q,\Phi} \|\{f_j\}_{j=1}^{\infty}\|_{L^\Phi(\ell^q)} \tag{6.104}$$

for all $\{f_j\}_{j=1}^{\infty} \subset L^\Phi(\mathbb{R}^n)$.

Proof The case where $q = \infty$ being simple as usual, we assume $q < \infty$. Abbreviate $\left(\sum_{j=1}^{\infty}|f_j|^q\right)^{\frac{1}{q}}$ to F and $\left(\sum_{j=1}^{\infty}Mf_j{}^q\right)^{\frac{1}{q}}$ to G. A normalization allows us to assume that the right-hand side equals 1; $\|F\|_{L^\Phi} = 1$.

Since Φ is convex, we have

$$\int_{\mathbb{R}^n} \Phi(G(x))\mathrm{d}x = \int_0^\infty \Phi'(\lambda)|\{G > \lambda\}|\mathrm{d}\lambda$$

from Theorem 1.5. For each $\lambda > 0$, we write $f_{j;\le\lambda} \equiv f_j\chi_{[0,\lambda]}(F)$ and $f_{j;>\lambda} \equiv f_j - f_{j;\le\lambda}$. Then

$$|\{G>\lambda\}| \le \left|\left\{\left(\sum_{j=1}^{\infty} Mf_{j;\le\lambda}{}^q\right)^{\frac{1}{q}} > \frac{\lambda}{2}\right\}\right| + \left|\left\{\left(\sum_{j=1}^{\infty} Mf_{j;>\lambda}{}^q\right)^{\frac{1}{q}} > \frac{\lambda}{2}\right\}\right|.$$

Since $\Phi \in \Delta_2$, we obtain

$$\Phi(at) \lesssim a^{\ell_+}\Phi(t) \quad ((a,t) \in (1,\infty)\times(0,\infty)), \tag{6.105}$$

for some ℓ_+. Likewise since $\Phi \in \nabla_2$, we obtain

$$\Phi(at) \gtrsim a^{\ell_-}\Phi(t) \quad ((a,t) \in (1,\infty)\times(0,\infty)), \tag{6.106}$$

for some $\ell_- > 1$. Let us choose ρ_1 and ρ_2 so that

$$1 < \rho_2 < \ell_- \le \ell_+ < \rho_1 < \infty. \tag{6.107}$$

If we use the Chebyshev inequality (see Theorem 1.4) and Theorem 1.49, then we have

$$\left|\left\{\left(\sum_{j=1}^{\infty} Mf_{j;\le\lambda}{}^q\right)^{\frac{1}{q}} > \frac{\lambda}{2}\right\}\right| \lesssim \lambda^{-\rho_1}\int_{\mathbb{R}^n} F(x)^{\rho_1}\chi_{[0,\lambda]}(F(x))\mathrm{d}x,$$

$$\left|\left\{\left(\sum_{j=1}^{\infty} Mf_{j;>\lambda}{}^q\right)^{\frac{1}{q}} > \frac{\lambda}{2}\right\}\right| \lesssim \lambda^{-\rho_2}\int_{\mathbb{R}^n} F(x)^{\rho_2}\chi_{(\lambda,\infty]}(F(x))\mathrm{d}x.$$

Consequently, from the doubling condition, (6.106) and (6.107), for each $x \in \mathbb{R}^n$ we conclude

$$\int_{F(x)}^{\infty} \Phi'(\lambda)\lambda^{-\rho_1}\mathrm{d}\lambda \lesssim \int_{F(x)}^{\infty} \Phi(\lambda)\lambda^{-\rho_1-1}\mathrm{d}\lambda \lesssim \Phi(F(x))F(x)^{-\rho_1}.$$

Likewise from (6.105) and (6.107) we obtain

$$\int_0^{F(x)} \Phi'(\lambda)\lambda^{-\rho_2\eta}\mathrm{d}\lambda \lesssim \int_0^{F(x)} \Phi(\lambda)\lambda^{-\rho_2\eta-1}\mathrm{d}\lambda \lesssim \Phi(F(x))F(x)^{-\rho_2\eta}.$$

Thus, we obtain

$$\int_0^{\infty} \Phi'(\lambda)\left|\left\{\left(\sum_{j=1}^{\infty} Mf_{j;\le\lambda}{}^q\right)^{\frac{1}{q}} > \frac{\lambda}{2}\right\}\right|\mathrm{d}\lambda \lesssim \int_{\mathbb{R}^n} \Phi(F(x))\mathrm{d}x \lesssim 1$$

and

$$\int_0^\infty \Phi'(\lambda) \left| \left\{ \left(\sum_{j=1}^{\infty} Mf_{j;>\lambda}{}^q \right)^{\frac{1}{q}} > \frac{\lambda}{2} \right\} \right| d\lambda \lesssim \int_{\mathbb{R}^n} \Phi(F(x)) dx \lesssim 1.$$

Since Φ satisfies the doubling condition, we obtain (6.104).

One can extend the definition of Orlicz spaces somehow. Here are some approaches. We can consider Orlicz spaces over measure spaces $(X, \mathscr{B}, \mu)$.

Definition 6.15 (Orlicz spaces) Let $(X, \mathscr{B}, \mu)$ be a measure space.

1. Let Ψ be a Young function and define $\Phi(t) \equiv \Psi(t^a)$ for $a > 0$. Then define

$$\|f\|_{L^\Phi(X)} \equiv \inf\left(\left\{ \lambda \in (0, \infty) : \int_X \Phi\left(\frac{|f(x)|}{\lambda} \right) d\mu(x) \le 1 \right\} \cup \{\infty\} \right) \tag{6.108}$$

for a μ-measurable function f. The *Orlicz space* $L^\Phi(X)$ over X is the set of μ-measurable functions f for which $\|f\|_{L^\Phi(X)}$ is finite.
2. Let $\Phi : [0, \infty) \times X \to [0, \infty)$ be a function such that $\Phi(\star, x)$ is a Young function for each $x \in X$. Then define

$$\|f\|_{L^\Phi(X)} \equiv \inf\left(\left\{ \lambda \in (0, \infty) : \int_X \Phi\left(\frac{|f(x)|}{\lambda}, x \right) d\mu(x) \le 1 \right\} \cup \{\infty\} \right) \tag{6.109}$$

for a μ-measurable function f. The *Orlicz space* $L^\Phi(X)$ over X is the set of μ-measurable functions f for which $\|f\|_{L^\Phi(X)}$ is finite.

One can mix definitions (6.108) and (6.109) whose details we omit.

Example 6.7 Let $a > 0, b \in \mathbb{R}$ be parameters, and let $X = \mathbb{R}^n$. The most important example of Orlicz spaces is generated by a function satisfying $\Phi(t) \sim t^a[\log(3 + t)]^b$ for all $t \ge 0$. Denote by $L^a[\log L]^b(\mathbb{R}^n)$ such a function space.

6.1.6.6 The Product of BMO($\mathbb{R}^n$) and $L^1(\mathbb{R}^n)$

Estimate (6.85) is a typical usage of Orlicz spaces. Another application relates to the property of the product of BMO($\mathbb{R}^n$) and $H^1(\mathbb{R}^n)$, which we describe in Sect. 6.2. The following lemma is a preparatory step to this end.

Lemma 6.18 *Let $M \ge 1$ and $s, t \ge 0$. Then $st \le (e^{t-M} + 2s)(M + \log(e + st))$.*

Proof Let $t \le M + 1$. Then

$$\frac{st}{M + \log(e + st)} \le \frac{st}{M} \le \frac{M+1}{M} s \le 2s \le e^{t-M} + 2s.$$

So the conclusion is trivial. Let us assume $t > M + 1$ here and below. Observe that

$$\frac{st}{M + \log(e + st)} = \int_0^{st} \left(\frac{u}{M + \log(e + u)} \right)' du.$$

Since $M \geq 1$, we have

$$\left(\frac{s}{M + \log(e + s)} \right)' = \frac{1}{M + \log(e + s)} - \frac{s}{(e + s)(M + \log(e + s))^2}$$
$$\leq \frac{1}{M + \log(e + s)}$$

for $u \geq 0$. Therefore, it suffices to show

$$\int_0^{st} \frac{1}{M + \log(e + u)} du \leq \mathrm{e}^{t-M} + s.$$

To this end, we define

$$\varphi(s) \equiv \int_0^{st} \frac{1}{M + \log(e + u)} du - s \quad (s \geq 0, t \geq M + 1).$$

Then

$$\varphi'(s) = \frac{t}{M + \log(e + st)} - 1,$$

showing that φ attains its maximum at $s_0 \equiv= t^{-1}(\mathrm{e}^{t-M} - \mathrm{e})$. Thus,

$$\varphi(s) \leq \varphi(s_0) \leq \int_0^{\mathrm{e}^{t-M}-\mathrm{e}} \frac{1}{M + \log(e+u)} du \leq \int_0^{\mathrm{e}^{t-M}} \frac{1}{M + \log(e + u)} du \leq \frac{\mathrm{e}^{t-M}}{M},$$

as was to be shown.

One of the advantages of considering Orlicz spaces is that we can consider the product of $L^1(\mathbb{R}^n)$-functions and $\mathrm{BMO}(\mathbb{R}^n)$ functions. Here for the sake of removing ambiguity of the additive constant in $\mathrm{BMO}(\mathbb{R}^n)$, we consider the space $\mathrm{BMO}^+(\mathbb{R}^n)$ whose norm is defined by (3.131).

Lemma 6.19 *If f and g are locally integrable functions such that*

$$\|f\|_1 + \|g\|_{\mathrm{BMO}^+} \leq 1,$$

then there exists a constant K independent of f and g such that

$$\int_{\mathbb{R}^n} \frac{|f(x)g(x)|}{\log(e + |x|) + \log(e + |f(x)g(x)|)} dx \leq K.$$

Proof According to Lemma 6.18, we have

$$\frac{|f(x)g(x)|}{(n+1)\log(e+|x|)+\log(e+|f(x)g(x)|)} \le e^{\mu|g(x)|-(n+1)\log(e+|x|)} + \frac{2}{\mu}|f(x)|.$$

Corollary 3.6, a version of the John–Nirenberg inequality, yields

$$\int_{\mathbb{R}^n} e^{\mu|g(x)|-(n+1)\log(e+|x|)}\mathrm{d}x \le B$$

as long as μ is small enough. Thus, we can take $K = B + 2\mu^{-1}$.

Exercises

Exercise 6.19 Show that t^α is a Young function, where $1 \le \alpha < \infty$.

Exercise 6.20 Prove (6.3) from the definition of the conjugate functions.

Exercise 6.21 Using scaling argument, disprove (6.84).

Exercise 6.22 Refining the layer cake representation, Theorem 1.5, show (6.85).

Exercise 6.23 Use Lemma 6.141 and prove Lemma 6.142.

Exercise 6.24 Let $f : \mathbb{R} \to [0, \infty)$ be a convex function.

1. Let $a < b$. Show that $f\left(\frac{a+b}{2}\right) \le \frac{1}{b-a}\int_a^b f(t)\mathrm{d}t \le \frac{f(a)+f(b)}{2}$.
2. Let $a \in \mathbb{R}$ and $h > 0$. Show that

$$\sum_{j=1}^k f\left(a+\left(j-\frac{1}{2}\right)h\right) \le \int_a^{a+kh} f(t)\mathrm{d}t$$

$$\le \frac{f(a)+f(a+kh)}{2} + \sum_{j=1}^{k-1} f(a+(k-1)h).$$

3. Show that $\frac{2}{3} < \log 2 = \int_1^2 \frac{\mathrm{d}x}{x} < \frac{3}{4}$.

Exercise 6.25 Let $(X, \mathscr{B}, \mu)$ be a σ-finite space, and let $\Phi \in \Delta_2$.

1. Use the doubling condition to show that $\Phi(t) \lesssim t^p$ for $t \ge 1$, where $p \ge 1$ is large enough.
2. Use (6.91) and the duality $L^p(X)$-$L^{p'}(X)$ to show that the dual space of $L^\Phi(X)$ is isomorphic to $L^{\Phi^*}(X)$, where Φ^* is the Fenchel–Legendre transform of Φ.

6.1.7 Herz Spaces

Beurling and Herz introduced some new spaces that characterize certain properties of functions [250, 588]. These new spaces are called the Herz spaces. Many studies involving these spaces can be found in the literature. One of the main reasons is that Hardy space theory based on Herz spaces is very rich.

6.1.7.1 Herz Spaces

One can expand the theory of Besov spaces and Triebel–Lizorkin spaces using Herz spaces. Such a theory was developed. in the last 1990's.

Write $Q_0 \equiv [-1,1]^n$ and $C_j \equiv [-2^j,2^j]^n \setminus [-2^{j-1},2^{j-1}]^n$ for $j \in \mathbb{Z}$.

Here we content ourselves with the definition and the maximal inequalities.

Definition 6.16 (Herz space) Let $0 < p, q \le \infty$ and $\alpha \in \mathbb{R}$.

1. The *nonhomogeneous Herz space* $K^\alpha_{pq}(\mathbb{R}^n)$ is the set of all the measurable functions f for which the norm $\|f\|_{K^\alpha_{pq}} \equiv \|\chi_{Q_0} f\|_p + \|\{2^{j\alpha}\chi_{C_j} f\}_{j=1}^\infty\|_{\ell^q(L^p)}$ is finite.
2. The *homogeneous Herz space* $\dot{K}^\alpha_{pq}(\mathbb{R}^n)$ is the set of all the measurable functions f for which the norm $\|f\|_{\dot{K}^\alpha_{pq}} \equiv \|\{2^{j\alpha}\chi_{C_j} f\}_{j=-\infty}^\infty\|_{\ell^q(L^p)}$ is finite.

6.1.7.2 Maximal Inequality

We start with the maximal inequality for Herz spaces.

Theorem 6.25 *Let* $1 < p < \infty$, $0 < q \le \infty$ *and* $\alpha \in \left(-\frac{n}{p}, n - \frac{n}{p}\right)$. *Then* M *is bounded on* $\dot{K}^\alpha_{pq}(\mathbb{R}^n)$ *and* $K^\alpha_{pq}(\mathbb{R}^n)$.

Proof Let us concentrate on $\dot{K}^\alpha_{pq}(\mathbb{R}^n)$; the proof for $K^\alpha_{pq}(\mathbb{R}^n)$ is similar. We have to prove:

$$\sum_{j=-\infty}^\infty (2^{j\alpha}\|\chi_{C_j} Mf\|_p)^q \lesssim \sum_{j=-\infty}^\infty (2^{j\alpha}\|\chi_{C_j} f\|_p)^q$$

for all $f \in K^\alpha_{pq}(\mathbb{R}^n)$. We note that

$$\sum_{j=-\infty}^\infty (2^{j\alpha}\|\chi_{C_j} M[(\chi_{C_{j-1}} + \chi_{C_j} + \chi_{C_{j+1}})f]\|_p)^q \lesssim \sum_{j=-\infty}^\infty (2^{j\alpha}\|\chi_{C_j} f\|_p)^q$$

by the boundedness of the Hardy–Littlewood maximal operator. Let $l \ge 2$. Then

$$\chi_{C_j} M[\chi_{C_{j-l}} f] \lesssim 2^{n(l-j)}\|f\|_{L^1(C_{j-l})};$$

hence

$$\begin{aligned}\sum_{j=-\infty}^{\infty}[2^{j\alpha}\|\chi_{C_j}M[\chi_{C_{j-l}}f]\|_p]^q &\lesssim \sum_{j=-\infty}^{\infty}[2^{j\left(\alpha-n+\frac{n}{p}\right)}\|f\|_{L^1(C_{j-l})}]^q\\ &\lesssim \sum_{j=-\infty}^{\infty}[2^{j\left(\alpha-n+\frac{n}{p}\right)+n(j-l)\left(1-\frac{1}{p}\right)}\|f\|_{L^p(C_{j-l})}]^q\\ &\lesssim \sum_{j=-\infty}^{\infty}[2^{j\alpha+\frac{ln}{p}-ln}\|f\|_{L^p(C_{j-l})}]^q\\ &= 2^{lq\left(\alpha+\frac{n}{p}-n\right)}\sum_{k=-\infty}^{\infty}(2^{k\alpha}\|\chi_{C_k}f\|_p)^q.\end{aligned}$$

In total, we have

$$\sum_{j=-\infty}^{\infty}(2^{j\alpha}\|\chi_{C_j}M[\chi_{C_{j-l}}f]\|_p)^q \lesssim 2^{lq\left(\alpha+\frac{n}{p}-n\right)}\|f\|_{\dot{K}^{\alpha}_{pq}}{}^q. \tag{6.110}$$

Likewise for $l \geq 2$ we have $\chi_{C_j}M[\chi_{C_{j+l}}f] \lesssim 2^{-n(l+j)}\|f\|_{L^1(C_{j+l})}$; hence

$$\sum_{j=-\infty}^{\infty}[2^{j\alpha}\|\chi_{C_j}M[\chi_{C_{j+l}}f]\|_p]^q \lesssim \sum_{j=-\infty}^{\infty}[2^{j\left(\alpha-n+\frac{n}{p}\right)-ln}\|f\|_{L^1(C_{j+l})}]^q.$$

By the Hölder inequality, we have

$$\begin{aligned}\sum_{j=-\infty}^{\infty}[2^{j\alpha}\|\chi_{C_j}M[\chi_{C_{j+l}}f]\|_p]^q &\lesssim \sum_{j=-\infty}^{\infty}[2^{j\left(\alpha-n+\frac{n}{p}\right)-ln+n(j+l)\left(1-\frac{1}{p}\right)}\|f\|_{L^p(C_{j+l})}]^q\\ &\lesssim \sum_{j=-\infty}^{\infty}(2^{j\alpha-\frac{ln}{p}}\|f\|_{L^p(C_{j+l})})^q\\ &= 2^{-lq\left(\alpha+\frac{n}{p}\right)}\sum_{k=-\infty}^{\infty}(2^{k\alpha}\|\chi_{C_k}f\|_p)^q.\end{aligned}$$

In total, we have

$$\sum_{j=-\infty}^{\infty}[2^{j\alpha}\|\chi_{C_j}M[\chi_{C_{j+l}}f]\|_p]^q \lesssim \left[2^{-l\left(\alpha+\frac{n}{p}\right)}\|f\|_{\dot{K}^{\alpha}_{pq}}\right]^q. \tag{6.111}$$

Thanks to the assumption on α, estimates (6.110) and (6.111) are summable over $l \geq 2$. Thus, we obtain the desired result.

We now state the Hölder inequality. More precisely, we have the following results on the Köthe dual:

Lemma 6.20 *Let* $1 \le p \le \infty$, $1 \le q \le \infty$ *and* $\alpha \in \mathbb{R}$. *Then*

$$\dot{K}^{\alpha}_{pq}(\mathbb{R}^n)' = \dot{K}^{-\alpha}_{p'q'}(\mathbb{R}^n), \quad K^{\alpha}_{pq}(\mathbb{R}^n)' = K^{-\alpha}_{p'q'}(\mathbb{R}^n)$$

with coincidence of norms.

Proof This is just a direct application of the Hölder inequality. We omit the details. See Exercise 6.27.

The *vector-valued homogeneous Herz space* $(\dot{K}^{\alpha}_{pq}(\ell^u, \mathbb{R}^n), \| \star \|_{\dot{K}^{\alpha}_{pq}(\ell^u)})$ is defined by (6.2) with X and u replaced by $\dot{K}^{\alpha}_{pq}(\mathbb{R}^n)$ and u, respectively. Likewise the *vector-valued nonhomogeneous Herz space* $(K^{\alpha}_{pq}(\ell^u, \mathbb{R}^n), \| \star \|_{K^{\alpha}_{pq}(\ell^u)})$ is defined by (6.2) with X and u replaced by $K^{\alpha}_{pq}(\mathbb{R}^n)$ and u, respectively. We have the following vector-valued inequality:

Theorem 6.26 *Let* $1 < p < \infty$, $0 < q \le \infty$, $1 < u \le \infty$ *and* $\alpha \in \left(-\frac{n}{p}, n - \frac{n}{p}\right)$. *Then*

$$\|\{Mf_j\}_{j=1}^{\infty}\|_{\dot{K}^{\alpha}_{pq}(\ell^u)} \lesssim \|\{f_j\}_{j=1}^{\infty}\|_{\dot{K}^{\alpha}_{pq}(\ell^u)}, \quad \|\{Mf_j\}_{j=1}^{\infty}\|_{K^{\alpha}_{pq}(\ell^u)} \lesssim \|\{f_j\}_{j=1}^{\infty}\|_{K^{\alpha}_{pq}(\ell^u)}$$

for all measurable functions $\{f_j\}_{j=1}^{\infty}$.

It is worth noting

$$\alpha \in \left(-\frac{n}{p}, n - \frac{n}{p}\right) \iff -\alpha \in \left(-\frac{n}{p'}, n - \frac{n}{p'}\right). \tag{6.112}$$

Proof Simply mimic the proof of Theorem 6.25.

Exercises

Exercise 6.26 Let $0 < p, q \le \infty$ and $\alpha \in \mathbb{R}$. Show that $\dot{K}^{\alpha}_{pq}(\mathbb{R}^n)$ and $K^{\alpha}_{pq}(\mathbb{R}^n)$ are quasi-Banach spaces using (1.4).

Exercise 6.27 Use the Hölder inequality to prove Lemma 6.20.

Textbooks in Sect. 6.1

Banach Function Spaces

See [30, Chapter 6].

Weighted Lebesgue Spaces

See [51, p. 13] for more details of weighted Lebesgue spaces together with some examples. See [31, Chapter 4] and [86, Chapter 5] for more about A_p-weights. We refer to [31, Section 2.3] for the relation between the weight class A_∞ and $\mathrm{BMO}(\mathbb{R}^n)$.

Variable Lebesgue Spaces

We refer to [16, 20, 61]. See [16, Theorem 2.26] and [20, Lemma 3.2.20] for Theorem 6.12, for example.

Morrey Spaces and Morrey–Campanato Spaces

See [1] and [100, Section 5.3] for more about Morrey spaces and Morrey–Campanato spaces.

Sobolev–Morrey Spaces

See [822, 823, 929] for Sobolev–Morrey spaces.

Mixed Lebesgue Spaces

See [4, p. 50] for the Hölder inequality for the mixed Lebsgue spaces. See [159] for the unweighted case of Theorem 6.8.

Orlicz Spaces

See [4, Chapter 8], [30, Chapter 6], [47, Chapter 2] and [74] for Orlicz spaces. We followed the idea of Cianchi [388].

Orlicz–Sobolev Spaces

See [4, Chapter 8], where the embedding of Orlicz–Sobolev spaces into some other function spaces is described: This embedding corresponds to the limiting case.

Herz Spaces

We refer to [54] for exhaustive details of Herz spaces.

6.2 Hardy Spaces Based on Ball Quasi-Banach Function Spaces

The section is a supplemental remark on general function spaces. The example of $\chi_{B(1)}$ and the Riesz transform shows that the singular integral operators are not bounded on $L^1(\mathbb{R})$. One remedy is that we instead consider $H^1(\mathbb{R})$. The same approach is useful for general Banach function spaces. It sometimes happens that the integral operators are not bounded on $X(\mathbb{R}^n)$. In this case, we seek to find a substitute. To this end, it is helpful to consider Hardy spaces based on $X(\mathbb{R}^n)$. We formulate the setting in Sect. 6.2.1 and provide some examples in Sect. 6.2.2.

6.2.1 General Definition of Hardy-Type Spaces

Starting from function spaces defined at the beginning of this chapter, we often define new function spaces by means of the grand maximal operator in harmonic analysis. Since the singular integral operators are related to fundamental solutions, we are convinced that the grand maximal operator is useful. See Exercise 6.30.

6.2.1.1 Hardy Spaces Based on Ball Quasi-Banach Function Spaces

To solidify what is stated above, we give the following definition.

Definition 6.17 ($HX(\mathbb{R}^n)$) Let $X(\mathbb{R}^n)$ be a quasi-Banach function space. Then define the Hardy space $HX(\mathbb{R}^n)$ based on $X(\mathbb{R}^n)$ to be the set of all $f \in \mathscr{S}'(\mathbb{R}^n)$ for which the quasi-norm $\|f\|_{HX} \equiv \|\mathscr{M} f\|_X$ is finite, where $\mathscr{M}$ denotes the grand (vertical) maximal operator in Definition 3.2. The space $HX(\mathbb{R}^n)$ is called the *Hardy space based on* $X(\mathbb{R}^n)$.

Based on Definition 6.17, we can investigate Hardy-type spaces.

Example 6.8

1. Let Φ be a Young function. If $X(\mathbb{R}^n) = L^\Phi(\mathbb{R}^n)$, one writes $H^\Phi(\mathbb{R}^n)$ instead of $L^\Phi(\mathbb{R}^n)$. The space $H^\Phi(\mathbb{R}^n)$ is called the *Orlicz–Hardy* space. One can also use the Orlicz space $L^\Phi(\mathbb{R}^n)$ defined in Definition 6.15.
2. Given a measurable function $p(\star) : \mathbb{R}^n \to (0, \infty)$, one can define $H^{p(\star)}(\mathbb{R}^n)$ based on $L^{p(\star)}(\mathbb{R}^n)$.

3. Let $0 < p < \infty$. Denote by $H^{p,\infty}(\mathbb{R}^n)$ the *weak Hardy space* $H^{p,\infty}(\mathbb{R}^n)$ where $X(\mathbb{R}^n) = L^{p,\infty}(\mathbb{R}^n)$. Fefferman and Soria initially introduced the weak Hardy spaces $H^{p,\infty}(\mathbb{R}^n)$ with $p \in (0, 1]$ to find out the biggest space from which the Hilbert transform is bounded to the weak Lebesgue space $L^{1,\infty}(\mathbb{R}^n)$ in [458]. As a more general class of function spaces including weak Hardy spaces, Parilov introduced and investigated the Hardy-Lorentz spaces $H^{p,q}(\mathbb{R}^n)$ which, when $p = 1$ and $q \in (1, \infty)$ [864]. In 2007, Abu-Shammala and Torchinsky studied the Hardy-Lorentz spaces $H^{p,q}(\mathbb{R}^n)$ for $p \in (0, 1]$ and $q \in (1, \infty]$ [123].
4. Let $0 < p < \infty$ and $w \in A_\infty$. Then one can define the *weighted weak Hardy space* $H^{p,\infty}(w)$ where $X(\mathbb{R}^n) = L^{p,\infty}(w)$.

Exercises

Exercise 6.28 Let Y be a ball Banach function space. Assume that Y^* is isomorphic to a ball Banach function space X such that $\|Mf\|_X \lesssim \|f\|_X$ for all $f \in X(\mathbb{R}^n)$.

1. Prove that $X(\mathbb{R}^n) \hookrightarrow \mathscr{S}'(\mathbb{R}^n)$ using $\|f\|_{L^1(B(x,1))} \lesssim \langle x\rangle^n Mf(y)$ for all $x \in \mathbb{R}^n$ and $y \in B(1)$.
2. Prove that $HX(\mathbb{R}^n) \approx X(\mathbb{R}^n)$ with equivalence of norms.

Exercise 6.29

1. Show that $\delta \notin H^p(\mathbb{R}^n)$ for $0 < p < \infty$.
2. Let $w(x) \equiv \langle x\rangle^{-n-1}$ for $x \in \mathbb{R}^n$. Show that $\delta \in H^p(w)$ for $0 < p < 1$.
3. Show that $\delta \in H^{p(\star)}(\mathbb{R}^n)$ for

$$p(x) \equiv \frac{1}{2} + \frac{|x|}{|x|+1} \quad (x \in \mathbb{R}^n).$$

4. Show that $\delta \in H^{1,\infty}(\mathbb{R}^n)$.

Exercise 6.30 Let $m \in C_c^\infty(\mathbb{R}^n)$ satisfy (3.34), and let $\Phi \in \Delta_2 \cap \nabla_2$ and $a > 0$. Define $\Psi \equiv \Phi(\star^a)$. Mimicking the proof of Theorem 3.6, show that there exists a constant $N \gg 1$ such that (3.35) holds with constants that depend only on M_α, $|\alpha| \le N$.

Exercise 6.31 Let $0 < q \le p < \infty$.

1. Define the Hardy–Morrey space $H\mathcal{M}_q^p(\mathbb{R}^n)$ [915, 1109].
2. If $q > 1$, show that $H\mathcal{M}_q^p(\mathbb{R}^n) \approx \mathcal{M}_q^p(\mathbb{R}^n)$ with equivalence of norms.

Exercise 6.32 [927] Let $0 < \theta, q < \infty$. Assume that the vector-valued inequality

$$\left\| \left(\sum_{j=1}^\infty M^{(\theta)} f_j{}^q \right)^{\frac{1}{q}} \right\|_X \lesssim \left\| \left(\sum_{j=1}^\infty |f_j|^q \right)^{\frac{1}{q}} \right\|_X$$

holds for all $\{f_j\}_{j=1}^{\infty} \subset L^0(\mathbb{R}^n)$. Then show that $HX(\mathbb{R}^n)$ can be characterized using the heat semi-group. See Example 3.1.

Exercise 6.33 Establish that the Hardy–Morrey space $H\mathcal{M}_q^p(\mathbb{R}^n)$ is continuously embedded into $\mathcal{S}'(\mathbb{R}^n)$ whenever $0 < q \le p < \infty$ using $(1-\Delta)^{-M} : H\mathcal{M}_q^p(\mathbb{R}^n) \to B^0_{\infty\infty}(\mathbb{R}^n)$ for $M > n/p$.

Exercise 6.34 Let $0 < p, q \le \infty$ and $\alpha \in \mathbb{R}$.

1. Define the *local Hardy–Herz space* $hK^{\alpha}_{pq}(\mathbb{R}^n)$.
2. Show that $\mathcal{S}(\mathbb{R}^n)$ is continuously embedded into $hK^{\alpha}_{pq}(\mathbb{R}^n)$. Hint: The local grand maximal function is not so large.

6.2.2 Hardy–Orlicz Spaces and Their Applications to Pointwise Multipliers

6.2.2.1 Hardy–Orlicz Spaces

We know that there is a natural coupling for $f \in H^1(\mathbb{R}^n)$ and $g \in \mathrm{BMO}(\mathbb{R}^n)$, which is called the duality $H^1(\mathbb{R}^n)$-$\mathrm{BMO}(\mathbb{R}^n)$. However, this is not the case that $f \cdot g \in L^1(\mathbb{R}^n)$; see Exercise 3.51.

6.2.2.2 The Product $\mathrm{BMO}^+(\mathbb{R}^n) \times H^1(\mathbb{R}^n)$

The aim of this section is to define the pointwise product $f \cdot g$ for $f \in H^1(\mathbb{R}^n)$ and $g \in \mathrm{BMO}(\mathbb{R}^n)$ in a certain sense. Although the space $\mathrm{BMO}(\mathbb{R}^n)$ is a function space modulo additive constant, using the norm defined by (3.131), we can make it into a normed space. Denote by $\mathrm{BMO}^+(\mathbb{R}^n)$ the space $\mathrm{BMO}(\mathbb{R}^n)$ endowed with the norm $\| \star \|_{\mathrm{BMO}^+}$ given by (3.131). Define

$$\Phi(x,t) \equiv \frac{t}{\log(e+|x|)+\log(e+t)}, \quad (x,t) \in (\mathbb{R}^n \times [0,\infty)).$$

Denote temporarily by $H^{\Phi}(\mathbb{R}^n)$ the set of all distributions $f \in \mathcal{S}'(\mathbb{R}^n)$ for which

$$\int_{\mathbb{R}^n} \Phi\left(x, \frac{\mathcal{M}f(x)}{\lambda}\right) dx \le 1$$

for some $\lambda > 0$ following Example 6.8.

Theorem 6.27 *Let $f \in H^1(\mathbb{R}^n)$ and $g \in \mathrm{BMO}^+(\mathbb{R}^n)$. Then there exists a decomposition $f \cdot g = h_1 + h_2$ with $h_1 \in L^1(\mathbb{R}^n)$ and $h_2 \in H^{\Phi}(\mathbb{R}^n)$ such that*

$$\|h_1\|_1 + \|h_2\|_{H^{\Phi}} \lesssim \|f\|_{H^1} \|g\|_{\mathrm{BMO}^+}.$$

Some prefer to write $H^{\log}(\mathbb{R}^n)$ for $H^{\Phi}(\mathbb{R}^n)$ in this theorem.

Proof We decompose f into the sum of atoms:

$$f = \sum_{j=1}^{\infty} \lambda_j a_j,$$

where $\{\lambda_j\}_{j=1}^{\infty} \in \ell^1(\mathbb{N})$ and a sequence $\{a_j\}_{j=1}^{\infty} \subset \mathscr{P}_0(\mathbb{R}^n)$ satisfy

$$\sum_{j=1}^{\infty} |\lambda_j| \lesssim \|f\|_{H^1}, \quad |a_j| \le \frac{\chi_{Q_j}}{|Q_j|}$$

for some cube Q_j. We let

$$h_1 \equiv \sum_{j=1}^{\infty} (g - m_{Q_j}(g)) a_j, \quad h_2 \equiv \sum_{j=1}^{\infty} m_{Q_j}(g) a_j.$$

Denote by $\mathscr{M}$ the vertical maximal operator in Definition 3.2. Then

$$\mathscr{M} a_j \lesssim |Q_j|^{-1} (M\chi_{Q_j})^{\frac{n+1}{n}} \tag{6.113}$$

according to Theorem 1.55. Hence

$$\|h_1\|_1 \lesssim \|g\|_{\mathrm{BMO}} \sum_{j=1}^{\infty} |\lambda_j| \lesssim \|f\|_{H^1} \|g\|_{\mathrm{BMO}}.$$

Meanwhile,

$$\mathscr{M} h_2 \le \sum_{j=1}^{\infty} |\lambda_j| \cdot |m_{Q_j}(g)| \mathscr{M} a_j \le \sum_{j=1}^{\infty} |\lambda_j| \cdot |g - m_{Q_j}(g)| \mathscr{M} a_j + |g| \sum_{j=1}^{\infty} |\lambda_j| \mathscr{M} a_j.$$

According to (a scaled version of) (3.141) and (6.113),

$$\sum_{j=1}^{\infty} |\lambda_j| \cdot |g - m_{Q_j}(g)| \mathscr{M} a_j \in L^1(\mathbb{R}^n) \subset L^{\Phi}(\mathbb{R}^n).$$

According to estimate (6.113), the Fefferman–Stein vector-valued inequality and Lemma 6.19, we have

$$g\sum_{j=1}^{\infty}|\lambda_j|\mathscr{M}a_j \in L^{\Phi}(\mathbb{R}^n). \tag{6.114}$$

In fact, by the Fefferman–Stein vector-valued inequality

$$\left\| g\sum_{j=1}^{\infty}|\lambda_j|\mathscr{M}a_j \right\|_{L^{\Phi}} \lesssim \|g\|_{\mathrm{BMO}} \left\| \sum_{j=1}^{\infty}|\lambda_j|\mathscr{M}a_j \right\|_1 \lesssim \|g\|_{\mathrm{BMO}} \left\| \sum_{j=1}^{\infty}|\lambda_j|\frac{\chi_{Q_j}}{|Q_j|} \right\|_1 .$$

If we use the triangle inequality, then we have

$$\left\| g\sum_{j=1}^{\infty}|\lambda_j|\mathscr{M}a_j \right\|_{L^{\Phi}} \lesssim \|g\|_{\mathrm{BMO}} \sum_{j=1}^{\infty}|\lambda_j| \lesssim \|f\|_{H^1}\|g\|_{\mathrm{BMO}}.$$

Thus, $h_1 \in L^1(\mathbb{R}^n)$ and $h_2 \in H^{\Phi}(\mathbb{R}^n)$ are the desired elements.

Exercises

Exercise 6.35 Show that $f \cdot g \in L^1(\mathbb{R}^n)$ if $f \in H^1(\mathbb{R}^n)$ and $g \in \dot{B}^0_{\infty 1}(\mathbb{R}^n)$.

Exercise 6.36 Establish (6.114) using (6.113), the Fefferman–Stein vector-valued inequality and Lemma 6.19.

Textbooks in Sect. 6.2

Weighted Hardy Spaces

See the lecture note by Strömberg and Torchinsky for weighted Hardy spaces [113].

6.3 Besov Spaces and Triebel–Lizorkin Spaces Based on Ball Quasi-Banach Function Spaces

We formulate the setting in Sect. 6.3.1 and provide some examples of $X(\mathbb{R}^n) = L^p(w)$ and $X(\mathbb{R}^n) = \mathscr{M}^p_q(w)$ in Sect. 6.3.2, which are historically long.

6.3.1 Besov Spaces and Triebel–Lizorkin Spaces Based on Ball Quasi-Banach Function Spaces

In this section we consider Besov spaces and Triebel–Lizorkin spaces based on ball quasi-Banach function spaces using two different methods. One is to replace $L^p(\mathbb{R}^n)$ with Banach function spaces and the other method is to generate Besov spaces using operators.

6.3.1.1 General Definition

Choose $\psi, \varphi \in \mathscr{S}(\mathbb{R}^n)$ so that (2.31) holds. For $j \in \mathbb{N}$, let $\varphi_j \equiv \varphi(2^{-j}\star)$ as usual. Let $0 < p, q \leq \infty$, $s \in \mathbb{R}$, and let $f \in \mathscr{S}'(\mathbb{R}^n)$. We defined the quasi-norm $\| \star \|_{B^s_{pq}}$ by

$$\|f\|_{B^s_{pq}} \equiv \|\psi(D)f\|_p + \|2^{js}\varphi_j(D)f\|_{\ell^q(L^p)},$$

and for $p < \infty$. the quasi-norm $\| \star \|_{F^s_{pq}}$ is defined by setting

$$\|f\|_{F^s_{pq}} \equiv \|\psi(D)f\|_p + \|2^{js}\varphi_j(D)f\|_{L^p(\ell^q)}.$$

Here we aim to replace $L^p(\mathbb{R}^n)$ with other function spaces listed in Sect. 6.1.1. Let X be a ball quasi-Banach function space. The quasi-norm $\| \star \|_{B^s_{X,q}}$ is defined by setting

$$\|f\|_{B^s_{X,q}} \equiv \|\psi(D)f\|_X + \|2^{js}\varphi_j(D)f\|_{\ell^q(X)} \tag{6.115}$$

and the quasi-norm $\| \star \|_{F^s_{X,q}}$ is defined by setting

$$\|f\|_{F^s_{X,q}} \equiv \|\psi(D)f\|_X + \|2^{js}\varphi_j(D)f\|_{X(\ell^q)}. \tag{6.116}$$

The spaces $B^s_{X,q}(\mathbb{R}^n)$ and $F^s_{X,q}(\mathbb{R}^n)$ denote the sets of all $f \in \mathscr{S}'(\mathbb{R}^n)$ for which $\|f\|_{B^s_{X,q}}$ and $\|f\|_{F^s_{X,q}}$ are finite, respectively. The symbol $A^s_{X,q}(\mathbb{R}^n)$ stands for either one of $B^s_{X,q}(\mathbb{R}^n)$ and $F^s_{X,q}(\mathbb{R}^n)$. Analogously, we can define the homogeneous spaces $\dot{B}^s_{X,q}(\mathbb{R}^n)$ and $\dot{F}^s_{X,q}(\mathbb{R}^n)$.

To justify that the space $B^s_{X,q}(\mathbb{R}^n)$ is independent of the choice of φ, ψ satisfying (2.31), we can use

$$\|M^{(\eta)} f_j\|_{\ell^q(X)} \lesssim \|f_j\|_{\ell^q(X)} \tag{6.117}$$

for all sequences $\{f_j\}_{j=1}^{\infty}$ of measurable functions. Likewise to justify that the space $B^s_{X,q}(\mathbb{R}^n)$ is independent of the choice of φ, ψ satisfying (2.31), we can use

$$\|M^{(\eta)} f_j\|_{X(\ell^q)} \lesssim \|f_j\|_{X(\ell^q)} \tag{6.118}$$

for all sequences $\{f_j\}_{j=1}^{\infty}$ of measurable functions. See Exercise 6.37.

We can also define the homogeneous spaces $\dot{A}^s_{X,q}(\mathbb{R}^n)$ analogously. We omit the precise definition.

We present example of the norms.

Example 6.9 We can consider the following vector-valued norms to define $\dot{B}^s_{X,q}(\mathbb{R}^n)$ and $B^s_{X,q}(\mathbb{R}^n)$. Here the parameters and the function Φ are suitably chosen.

1. $(\ell^q(L^{\mathbf{p}}, \mathbb{R}^n), \| \star \|_{\ell^q(L^{\mathbf{p}})})$. We write $\dot{A}^s_{\mathbf{p}q}(\mathbb{R}^n)$ and $A^s_{\mathbf{p}q}(\mathbb{R}^n)$ in this case.
2. $(\ell^r(\mathcal{M}^p_q, \mathbb{R}^n), \| \star \|_{\ell^r(\mathcal{M}^p_q)})$.
3. $(\ell^q(L^\Phi, \mathbb{R}^n), \| \star \|_{\ell^q(L^\Phi)})$. We write $\dot{A}^s_{\Phi q}(\mathbb{R}^n)$ and $A^s_{\Phi q}(\mathbb{R}^n)$ in this case.
4. $(\ell^u(\dot{K}^\alpha_{pq}, \mathbb{R}^n), \| \star \|_{\ell^u(\dot{K}^\alpha_{pq})})$.
5. $(\ell^u(K^\alpha_{pq}, \mathbb{R}^n), \| \star \|_{\ell^u(K^\alpha_{pq})})$.

6.3.1.2 Besov Spaces Associated with Operators

As we have seen, Besov spaces or Triebel–Lizorkin spaces can be defined by the Laplacian. Here we can mix this idea with the notion of $A^s_{pq}(\mathbb{R}^n)$ as follows:

Theorem 6.28 *Let $m \in \mathbb{N}$, $0 < \theta < 1$, and let $L : X \to X$ be a sectorial operator on the Banach space X. Then for $0 < q \le \infty$ and $N \in \mathbb{N} \cap [m, \infty)$ the following are equivalent:*

1. $x \in [X, \mathrm{Dom}(X^m)]_{\theta,q}$.
2. $x \in X$ *and* $\|x\|_X + \left(\int_0^1 \|t^{-s}(tL\mathrm{e}^{-tL})^N x\|_X{}^q \frac{\mathrm{d}t}{t}\right)^{\frac{1}{q}} < \infty.$

Note that we do not have to suppose that X consists of functions defined on $\mathbb{R}^n$.

Proof We suppose $N \ge m+1$: the case of $N = m$ will be incorporated later. Let $K \in \mathbb{N} \cap (s, \infty)$. Denote by $B^s_{X,q;K}$ the set of all $x \in X$ for which the quasi-norm $\|x\|_X + \left(\int_0^1 \|t^{-s}(tL\mathrm{e}^{-tL})^K x\|_X{}^q \frac{\mathrm{d}t}{t}\right)^{\frac{1}{q}}$ is finite. Then $B^0_{X,\infty;K} = X = B^0_{X,1;K}$ and $B^m_{X,\infty;K} \subset \mathrm{Dom}(D^m) \subset B^0_{X,1;K}$ from Example 5.5 2. We also observe that

$$\|tL\mathrm{e}^{-tL} \circ sL\mathrm{e}^{-sL}\|_{B(X)} \lesssim \min\left(\frac{t}{s}, \frac{s}{t}\right). \tag{6.119}$$

In fact, if $t > s$, we can use

$$tL\mathrm{e}^{-tL} \circ sL\mathrm{e}^{-sL} = \frac{s}{t} tL\mathrm{e}^{-tL/2} \circ tL\mathrm{e}^{-tL/2} \circ \mathrm{e}^{-sL}.$$

Estimate (6.119) will play the role of the Plancherel–Polya–Nikolski'i inequality. Thus similar to Theorem 4.25, we can prove the assertion for $N \ge m+1$.

We consider the case $N = m$. In view of the uniform boundedness of e^{-tL} on X we have only to show that

$$\left(\int_0^1 \|t^{-s}(tL\mathrm{e}^{-tL})^m x\|_X{}^q \frac{\mathrm{d}t}{t}\right)^{\frac{1}{q}} \lesssim \|x\|_X + \left(\int_0^1 \|t^{-s}(tL\mathrm{e}^{-tL})^{m+1} x\|_X{}^q \frac{\mathrm{d}t}{t}\right)^{\frac{1}{q}}$$

to complete the proof of Theorem 6.28. To this end, we may assume that $x \in \mathrm{Dom}(L^{m+2})$ by density. This is achieved by using (6.119) and

$$(tL\mathrm{e}^{-tL})^m x = \frac{1}{\Gamma(m+2)} \int_0^\infty L^{m+2}\mathrm{e}^{-sL}(tL\mathrm{e}^{-tL})^m x \mathrm{d}s,$$

which follows from Example 5.5.

6.3.1.3 Characterization of Lebesgue Spaces

Although we do not consider Triebel–Lizorkin spaces for this class of operators, we characterize $L^2(\mathbb{R}^n)$ in terms of the operators dealt with earlier. Let $0 < \lambda < \Lambda < \infty$, and let $A \equiv \{a_{ij}\}_{i,j=1,2,\ldots,n}$ be a matrix with entries $a_{ij} : \mathbb{R}^n \to \mathbb{C}$ of measurable functions satisfying (5.120) and (5.121) for $(\xi_1, \xi_2, \ldots, \xi_n) \in \mathbb{C}^n$.

6.3.1.4 Characterization of $L^2(\mathbb{R}^n)$

Let $f \in L^2(\mathbb{R}^n)$. We define

$$\mathscr{S}_L f(x) \equiv \left(\iint_{\Gamma(x)} |t^2L^2\mathrm{e}^{-tL} f(y)|^2 \frac{\mathrm{d}y\mathrm{d}t}{t^{n+1}}\right)^{\frac{1}{2}} \quad (x \in \mathbb{R}^n).$$

The following is our starting point to develop a theory, which corresponds to the Plancherel theorem:

Theorem 6.29 *For all $f \in L^2(\mathbb{R}^n)$ $\|\mathscr{S}_L f\|_2 \sim_n \|f\|_2$.*

Proof We may assume that $f \in \mathrm{Dom}(L)$. Our task amounts to proving

$$\left(\int_0^\infty (\|t^2L^2\mathrm{e}^{-tL} f\|_2)^2 \frac{\mathrm{d}t}{t}\right)^{\frac{1}{2}} = \left\|\left(\int_0^\infty |t^2L^2\mathrm{e}^{-tL} f|^2 \frac{\mathrm{d}t}{t}\right)^{\frac{1}{2}}\right\|_2 \sim \|f\|_2$$

by the Fubini theorem. First we show that

$$\left(\int_0^\infty (\|t^2L^2\mathrm{e}^{-tL} f\|_2)^2 \frac{\mathrm{d}t}{t}\right)^{\frac{1}{2}} \lesssim \|f\|_2,$$

We discretize matters:

$$\left(\int_0^\infty (\|t^2L^2\mathrm{e}^{-tL}f\|_2)^2\frac{\mathrm{d}t}{t}\right)^{\frac{1}{2}} = \left(\int_1^2 \sum_{k=-\infty}^{\infty} (\|(2^k tL)^2 e^{-2^k tL} f\|_2)^2 \frac{\mathrm{d}t}{t}\right)^{\frac{1}{2}}.$$

If we use the Rademacher sequence $\{R_k\}_{k=-\infty}^{\infty} = \{r_j\}_{j=1}^{\infty}$, where $\{r_j\}_{j=1}^{\infty}$ is the Rademacher sequence of functions in Definition 3.1, then we can reduce matters to showing

$$\left\|\sum_{k=-\infty}^{\infty} R_k(s)(2^k tL)^2 e^{-2^k tL}\right\|_{B(L^2)} \lesssim 1 \quad (s \in [0,1], t \in [1,2]),$$

which follows from Theorem 5.23. Indeed, we remark that

$$\varphi_{s,t}(z) = \sum_{k=-\infty}^{\infty} R_k(s)(2^k tz)^2 \exp(-2^k tz) \quad (s \in [0,1], t \in [1,2])$$

is a bounded continuous function on $\Re(z) \geq 0$.

Let us prove

$$\left\|\left(\int_0^\infty |t^2L^2\mathrm{e}^{-tL}f|^2\frac{\mathrm{d}t}{t}\right)^{\frac{1}{2}}\right\|_2 \gtrsim \|f\|_2$$

by duality. Since L and L^* satisfy the same elliptic condition, we have

$$\left\|\left(\int_0^\infty |(tL^*)^2\mathrm{e}^{-tL^*}f|^2\frac{\mathrm{d}t}{t}\right)^{\frac{1}{2}}\right\|_2 \lesssim \|f\|_2.$$

By duality

$$\left\|\int_0^\infty t\sqrt{t}L^2\mathrm{e}^{-tL}[F(\star,t)]\mathrm{d}t\right\|_2 \lesssim \|F\|_{L^2(\mathbb{R}^{n+1}_+)}.$$

Thus, if we let $F(x,t) \equiv t\sqrt{t}L^2\mathrm{e}^{-tL}f(x)$ for $(x,t) \in \mathbb{R}^{n+1}_+$, we obtain the desired result keeping in mind $f \in \mathrm{Dom}(L)$, since

$$\int_0^\infty t^3L^4e^{-2tL}f\mathrm{d}t \simeq f. \tag{6.120}$$

See Exercise 6.39 for the proof of (6.120).

We move on to the $L^{p_0}(\mathbb{R}^n)$ estimates.

Proposition 6.6

1. *Let* $p_0 \in [1, 2)$. *If the operator* e^{-tL} *satisfies the off-diagonal* $L^{p_0}(\mathbb{R}^n)$-$L^2(\mathbb{R}^n)$ *estimate. Then*

$$\|\mathscr{S}_L f\|_{p_0} \lesssim \|f\|_{p_0}. \tag{6.121}$$

2. *If the operator* e^{-tL} *satisfies the* $L^{p_0}(\mathbb{R}^n)$-$L^2(\mathbb{R}^n)$ *estimate, namely,*

$$\|\mathrm{e}^{-tL} f\|_{p_0} \lesssim \|f\|_2 \tag{6.122}$$

for all $f \in L^2(\mathbb{R}^n) \cap L^{p_0}(\mathbb{R}^n)$, *then*

$$\|\mathscr{S}_L f\|_{p_0} \lesssim \|f\|_{p_0}. \tag{6.123}$$

Proof Let $A_r \equiv \mathrm{id}_{L^2} - (\mathrm{id}_{L^2} - e^{-r^2 L})^m$ for $m \gg 1$ and $r > 0$. We plan to apply Theorem 1.68 by verifying (1.233), (1.234) and (1.235) to obtain (6.121) and then we plan to apply Theorem 1.69 by verifying (1.238) and (1.239) to obtain (6.122).

1. Let $f \in L^2(\mathbb{R}^n)$ and $j \in \mathbb{Z}$. Let $f_Q = f\chi_Q$. We observe

$$\begin{aligned}
&\int_{2^{j+2}Q\setminus 2^{j+1}Q} \mathscr{S}_L[(\mathrm{id}_{L^2} - A_r) f_Q](x)^2 \mathrm{d}x \\
&= \int_{2^{j+2}Q\setminus 2^{j+1}Q} \left(\iint_{\Gamma(x)} |t^2 L^2 \mathrm{e}^{-tL} (\mathrm{id}_{L^2} - e^{-r^2 L})^m f_Q(y)|^2 \frac{\mathrm{d}y\mathrm{d}t}{t^{n+1}} \right) \mathrm{d}x \\
&= \int_{2^{j+2}Q\setminus 2^{j+1}Q} \left(\iint_{\Gamma(0)} |t^2 L^2 \mathrm{e}^{-tL} (\mathrm{id}_{L^2} - e^{-r^2 L})^m f_Q(x+y)|^2 \frac{\mathrm{d}y\mathrm{d}t}{t^{n+1}} \right) \mathrm{d}x.
\end{aligned}$$

Here for the last line we changed variables: $y \to x + y$. By the Fubini theorem,

$$\begin{aligned}
&\int_{2^{j+2}Q\setminus 2^{j+1}Q} \mathscr{S}_L[(\mathrm{id}_{L^2} - A_r)[fQ]](x)^2 \mathrm{d}x \\
&= \iint_{\Gamma(0)} \left(\int_{2^{j+2}Q\setminus 2^{j+1}Q} |t^2 L^2 \mathrm{e}^{-tL} (\mathrm{id}_{L^2} - e^{-r^2 L})^m f_Q(x+y)|^2 \mathrm{d}x \right) \frac{\mathrm{d}y\mathrm{d}t}{t^{n+1}}.
\end{aligned} \tag{6.124}$$

We make use of the expression by means of the line integral

$$t^2 L^2 \mathrm{e}^{-tL} (\mathrm{id}_{L^2} - e^{-r^2 L})^m \simeq \int_{|\arg(z)| = \frac{\pi}{2} - \theta} t^2 z^2 \mathrm{e}^{-tz} (\mathrm{id}_{L^2} - e^{-r^2 z})^m (z\mathrm{id}_{L^2} - L)^{-1} \mathrm{d}z,$$

where $0 < \theta \ll 1$. Let $y \in B(t)$ be fixed. By Lemma 5.29, the off-diagonal estimate, there exists $a > 0$ such that

$$\left(\int_{2^{j+2}Q\setminus 2^{j+1}Q} |t^2L^2\mathrm{e}^{-tL}(\mathrm{id}_{L^2}-e^{-r^2L})^m f_Q(x+y)|^2\mathrm{d}x\right)^{\frac{1}{2}}$$
$$\lesssim \int_{|\arg(z)|=\frac{\pi}{2}-\theta} \frac{t^2|z|^2}{\exp(C_\theta|tz|)} \exp\left(-a2^j r\sqrt{|z|}\right)|r^2 z|^m \frac{|dz|}{|z|}\cdot \|f\|_{L^2(Q)}$$
$$= 4^{-jm}\int_{|\arg(z)|=\frac{\pi}{2}-\theta} \frac{|4^{-j}r^{-1}tz|^2}{\exp(C_\theta|4^{-j}r^{-1}tz|)} \exp\left(-a\sqrt{|z|}\right)|z|^m \frac{|dz|}{|z|}\cdot \|f\|_{L^2(Q)}$$
$$\lesssim 4^{-jm}\min(|4^{-j}r^{-1}t|^2, |4^{-j}r^{-1}t|^{-2})\|f\|_{L^2(Q)}.$$

Inserting this inequality into (6.124), we have

$$\int_{2^{j+2}Q\setminus 2^{j+1}Q} \mathscr{S}_L[(\mathrm{id}_{L^2}-A_r)f_Q](x)^2\mathrm{d}x \lesssim 4^{-2jm}\iint_{\Gamma(0)} \left(\min(|4^{-j}r^{-1}t|^2, |4^{-j}r^{-1}t|^{-2})\|f\|_{L^2(Q)}\right)^2 \frac{\mathrm{d}y\mathrm{d}t}{t^{n+1}}. \tag{6.125}$$

Thus,

$$\int_{2^{j+2}Q\setminus 2^{j+1}Q} \mathscr{S}_L[(\mathrm{id}_{L^2}-A_r)f_Q](x)^2\mathrm{d}x$$
$$\lesssim 4^{-2jm}\iint_{\Gamma(0)} \left(\min(|4^{-j}r^{-1}t|^2, |4^{-j}r^{-1}t|^{-2})\|f\|_{L^2(Q)}\right)^2 \frac{\mathrm{d}y\mathrm{d}t}{t^{n+1}}$$
$$\simeq 4^{-2jm}\int_0^\infty \left(\min(|4^{-j}r^{-1}t|^2, |4^{-j}r^{-1}t|^{-2})\|f\|_{L^2(Q)}\right)^2 \frac{\mathrm{d}t}{t}$$
$$\simeq (4^{-jm}\|f\|_{L^2(Q)})^2.$$

Thus, (1.233) is verified. Thus, assuming $m \gg n$, we have

$$m^{(2)}_{2^{j+2}Q\setminus 2^{j+1}Q}(\mathscr{S}_L[(\mathrm{id}_{L^2}-A_r)f_Q]) \lesssim 2^{-jm}m^{(2)}_Q(f).$$

Likewise, using the $L^2(\mathbb{R}^n)$-estimate, we have

$$m^{(2)}_Q(\mathscr{S}_L[(\mathrm{id}_{L^2}-A_r)f_Q]) \lesssim m^{(2)}_Q(f).$$

Estimate (1.234) follows from the $L^2(\mathbb{R}^n)$-boundedness of the operators and estimate (1.235) follows directly from Lemma 5.29, the off-diagonal estimate.

2. We need to verify the assumption in Theorem 1.69. To check (1.238) we decompose

$$
\begin{aligned}
m_Q^{(2)}(\mathscr{S}_L[(\mathrm{id}_{L^2}-A_r)f]) &\le m_Q^{(2)}(\mathscr{S}_L[(\mathrm{id}_{L^2}-A_r)(\chi_{4Q}f)]) \\
&\quad + \sum_{j=1}^{\infty} m_Q^{(2)}(\mathscr{S}_L[(\mathrm{id}_{L^2}-A_r)(\chi_{2^{j+2}Q\setminus 2^{j+1}Q}f)]) \\
&\lesssim M^{(2)}f(x) + \sum_{j=1}^{\infty} m_Q^{(2)}(\mathscr{S}_L[(\mathrm{id}_{L^2}-A_r)(\chi_{2^{j+2}Q\setminus 2^{j+1}Q}f)]).
\end{aligned}
$$

Here to obtain the last inequality, we used the $L^2(\mathbb{R}^n)$-boundedness of $\mathscr{S}_L$ and A_r. For the second term, we take a similar approach to (6.125) using Lemma 5.29, the off-diagonal estimate, to have $m_Q^{(2)}(\mathscr{S}_L[(\mathrm{id}_{L^2} - A_r)f]) \lesssim M^{(2)} f(x)$, which proves (1.239). To prove (1.240), we observe

$$
\mathscr{S}_L[e^{-kr^2L} f](x) = \left(\iint_{\Gamma(0)} |e^{-kr^2L} t^2 L^2 \mathrm{e}^{-tL} f(x+y)|^2 \frac{\mathrm{d}y\mathrm{d}t}{t^{n+1}} \right)^{\frac{1}{2}}.
$$

Thus,

$$
m_Q^{(p_0)}(\mathscr{S}_L[e^{-kr^2L} f])^2 \le \iint_{\Gamma(0)} m_{Q_0}^{(p_0)}(|e^{-kr^2L} t^2 L^2 \mathrm{e}^{-tL} f(\star + y)|)^2 \frac{\mathrm{d}y\mathrm{d}t}{t^{n+1}}.
$$

We decompose f as follows:

$$
f = \chi_{y+4Q_0} f + \sum_{j=1}^{\infty} \chi_{y+2^{j+2}Q_0\setminus y+2^{j+1}Q_0} f.
$$

Using the Fubini theorem and Lemma 5.29, the off-diagonal estimate, we have

$$
\begin{aligned}
m_Q^{(p_0)}(\mathscr{S}_L[e^{-kr^2L} f])^2 &\lesssim \sum_{j=1}^{\infty} \iint_{\Gamma(0)} \exp(-a2^j) \left(m_{y+2^jQ_0}^{(2)}(|t^2L^2\mathrm{e}^{-tL} f|) \right)^2 \frac{\mathrm{d}y\mathrm{d}t}{t^{n+1}} \\
&= \sum_{j=1}^{\infty} \exp(-a2^j) m_{2^jQ_0}^{(2)}(\mathscr{S}_L f)^2
\end{aligned}
$$

for some $a > 0$, which proves (1.239).

Exercises

Exercise 6.37 Let $s \in \mathbb{R}$ and $q \in (0, \infty)$. Assume that a ball quasi-Banach function space X satisfies (6.117). Then prove that the space $B^s_{X,q}(\mathbb{R}^n)$ is independent of the choice of φ, ψ satisfying (2.31).

Exercise 6.38 Let p, q, s, α satisfy $1 < p < \infty$, $1 < q < \infty$, $s \in \mathbb{R}$ and $\alpha > 0$. Assume that $\dfrac{1}{p} - \dfrac{\alpha}{n} = \dfrac{1}{q}$ and that $X(\mathbb{R}^n)$ satisfies (6.118). Assume in addition that $\|f(t\star)\|_X = t^n \|f\|_X$ for all $f \in X(\mathbb{R}^n)$ and $t > 0$.

1. Prove that the space $F^s_{X,q}(\mathbb{R}^n)$ is independent of the choice of φ, ψ satisfying (2.31).
2. [836, Theorem 4] Mimic the proof of Theorem 2.14 to show that $(1-\Delta)^{\alpha/2}$: $F^s_{X^p\,\infty}(\mathbb{R}^n) \to F^s_{X^s,1}(\mathbb{R}^n)$ is a bounded linear operator, where $X^p(\mathbb{R}^n)$ is a ball Banach function space defined in Exercise 6.4.
3. If X is absolutely continuous, then show that $C^\infty_c(\mathbb{R}^n)$ is dense in $A^s_{X,1}(\mathbb{R}^n)$.

Exercise 6.39 Prove (6.120) mimicking the proof of Theorem 5.26.

Exercise 6.40 Let $1 \le p \le \infty$, $0 < q \le \infty$ and $s > 0$. Define the Besov space $B^s_{pq}(\mathbb{R}^n)$ using the Laplacian Δ instead of the usual system $\{\psi\} \cup \{\varphi_j\}_{j=1}^\infty$. Here and below, let $f \in L^p(\mathbb{R}^n)$. Fix a positive integer L with $2L > s$ and define

$$\|f\|_{B^s_{pq};\Delta} \equiv \|f\|_p + \left(\int_0^\infty \lambda^{sq} \|\lambda^{2L} \Delta^L (\lambda^2 - \Delta)^{-2L} f\|_p{}^q \frac{\mathrm{d}\lambda}{\lambda}\right)^{\frac{1}{q}} \quad (f \in L^p(\mathbb{R}^n)).$$

Then use Exercise 3.9 if necessary.

1. Let $2^j \le \lambda \le 2^{j+1}$. Using Theorem 1.53 or the Young inequality, show that
$$\|\varphi_j(D) f\|_p \lesssim 2^{2jL} \|\Delta^L (\lambda^2 - \Delta)^{-2L} f\|_p.$$
2. Show that $\|f\|_{B^s_{pq}} \lesssim \|f\|_{B^s_{pq};\Delta}$ for all $f \in L^p(\mathbb{R}^n)$ with $\|f\|_{B^s_{pq};\Delta} < \infty$.
3. Let $2^j \le \lambda \le 2^{j+1}$. Using Theorem 1.53 or the Young inequality, show that
$$\|\Delta^L (\lambda^2 - \Delta)^{-2L} f\|_p \lesssim \sum_{k=-\infty}^{\infty} 2^{2kL - 4\max(j,k)L} \|\varphi_k(D) f\|.$$
4. Show that $\|f\|_{B^s_{pq};\Delta} \lesssim \|f\|_{B^s_{pq}}$.

Exercise 6.41 Show that

$$\int_{|\arg(z)| = \frac{\pi}{2} - \theta} \|(tz)^2 \mathrm{e}^{-tz} (\mathrm{id}_{L^2} - e^{-r^2 z})^m (z\mathrm{id}_{L^2} - L)^{-1}\|_{B(L^2)} \, |dz| < \infty.$$

6.3.2 Besov Spaces and Triebel–Lizorkin Spaces Based on Morrey Spaces and Herz Spaces

Based on the above framework, we provide two examples of this attempt.

6.3.2.1 Smoothness Morrey Spaces

Here we investigate $B^s_{\mathcal{M}^p_q,r}(\mathbb{R}^n)$, $F^s_{\mathcal{M}^p_q,r}(\mathbb{R}^n)$ and $A^s_{\mathcal{M}^p_q,r}(\mathbb{R}^n)$. Traditionally these spaces are written as $\mathcal{N}^s_{pqr}(\mathbb{R}^n)$, $\mathcal{E}^s_{pqr}(\mathbb{R}^n)$ and $\mathcal{A}^s_{pqr}(\mathbb{R}^n)$, respectively. Namely, for $f \in \mathcal{S}'(\mathbb{R}^n)$ based on (6.115) and (6.116), we define

$$\|f\|_{\mathcal{N}^s_{pqr}} \equiv \|\psi(D)f\|_{\mathcal{M}^p_q} + \left(\sum_{j=1}^{\infty}(2^{js}\|\phi_j(D)f\|_{\mathcal{M}^p_q})^r\right)^{\frac{1}{r}}, \tag{6.126}$$

$$\|f\|_{\mathcal{E}^s_{pqr}} \equiv \|\psi(D)f\|_{\mathcal{M}^p_q} + \left\|\left(\sum_{j=1}^{\infty}2^{jrs}|\phi_j(D)f|^r\right)^{\frac{1}{r}}\right\|_{\mathcal{M}^p_q}. \tag{6.127}$$

The spaces $\mathcal{N}^s_{pqr}(\mathbb{R}^n)$, which we call the *Besov–Morrey space* and the *Triebel–Lizorkin–Morrey space* respectively, and $\mathcal{E}^s_{pqr}(\mathbb{R}^n)$ are the sets of all $f \in \mathcal{S}'(\mathbb{R}^n)$ for which the norms $\|f\|_{\mathcal{N}^s_{pqr}}$ and $\|f\|_{\mathcal{E}^s_{pqr}}$ are finite, respectively. To unify $\mathcal{N}^s_{pqr}(\mathbb{R}^n)$ and $\mathcal{E}^s_{pqr}(\mathbb{R}^n)$ we write $\mathcal{A}^s_{pqr}(\mathbb{R}^n)$. The space $\mathcal{A}^s_{pqr}(\mathbb{R}^n)$ is called the *smoothness Morrey space*.

The trace operator seems to describe the role of the parameters q and r very well. We define the subsets $T_{\mathcal{E}}$ and $T_{\mathcal{N}}$ of $(0,\infty]^3 \times (0,\infty)$ as follows:

$$T_{\mathcal{N}} \equiv \left\{(p,q,r,s)\,:\,0<\frac{p}{n}\le q\le p\le\infty,\ 0<r\le\infty,\ \frac{1}{q}+(n-1)\left(\frac{1}{q}-1\right)_+<s\right\},$$

$$T_{\mathcal{E}} \equiv \left\{(p,q,r,s)\in T_{\mathcal{N}}\,:\,0<\frac{p}{n}<q\le p<\infty \text{ or } 0<p=q\,n<\infty,\ r=\infty\right\}.$$

Denote by $T_{\mathcal{A}}$ either $T_{\mathcal{N}}$ or $T_{\mathcal{E}}$. For $(p,q,r,s) \in T_{\mathcal{A}}$, we define $p^*, r^*, s^* \in (0,\infty]$ by

$$\frac{n-1}{p^*} = \frac{n}{p} - \frac{1}{q},\quad r^* = \begin{cases} r & \text{if } \mathcal{A} = \mathcal{N}, \\ q & \text{if } \mathcal{A} = \mathcal{E}, \end{cases} \quad s^* = s - \frac{1}{q}. \tag{6.128}$$

Before we formulate our theorem, we remark that

$$\bigcup_{(p,q,r,s)\in T_{\mathcal{E}}} \mathcal{E}^s_{pqr}(\mathbb{R}^n) \subset \bigcup_{(p,q,r,s)\in T_{\mathcal{N}}} \mathcal{N}^s_{pqr}(\mathbb{R}^n). \tag{6.129}$$

The following theorem describes the role of the parameters nicely; its proof is similar to the case of $p = q$. See Exercise 6.45.

Theorem 6.30 *The trace operator* $\mathrm{Tr}_{\mathbb{R}^n}$ *enjoys the following properties:*

1. $\mathrm{Tr}_{\mathbb{R}^n}$ *is a bounded operator from* $\mathcal{A}^s_{pqr}(\mathbb{R}^n)$ *onto* $\mathcal{A}^{s^*}_{p^*qr^*}(\mathbb{R}^{n-1})$ *for* $(p,q,r,s) \in T_{\mathcal{A}}$.

2. *Let* $k \in \mathbf{N}_0$. *Define* $\overline{\mathrm{Tr}}_{\mathbb{R}^n} f \equiv (\mathrm{Tr}_{\mathbb{R}^n} f, \mathrm{Tr}_{\mathbb{R}^n}[\partial_n f], \ldots, \mathrm{Tr}_{\mathbb{R}^n}[\partial_n{}^k f])$ *for* $f \in \mathscr{A}^s_{pqr}(\mathbb{R}^n)$ *with* $(p,q,r,s-k) \in T_{\mathscr{A}}$. *Then* $\overline{\mathrm{Tr}}_{\mathbb{R}^n}$ *is a surjective operator from* $\mathscr{A}^s_{pqr}(\mathbb{R}^n)$ *to* $\prod_{j=0}^{k} \mathscr{A}^{s^*-j}_{p^*qr^*}(\mathbb{R}^{n-1})$, *if* $(p,q,r,s-k) \in T_{\mathscr{A}}$.

6.3.2.2 Herz-Type Besov Spaces

In analogy to (6.126) and (6.127), we define Herz-type Besov spaces, where we replace $X(\mathbb{R}^n)$ with the Herz spaces in (6.115) and (6.116).

Let $0 < p, q, r \le \infty$ and $\alpha, s \in \mathbb{R}$. Let $\varphi, \psi \in C^\infty(\mathbb{R}^n)$ satisfy $\chi_{B(4)} \le \psi \le \chi_{B(8)}$ and $\chi_{B(4)\setminus B(2)} \le \varphi \le \chi_{B(8)\setminus B(1)}$. Then the *Herz-type Besov space* is given by the norm: $\|f\|_{\mathscr{N}_{K^\alpha_{pq} B^s_r}} = \|f\|_{B^s_r K^\alpha_{pq}} \equiv \|\psi(D) f\|_{K^\alpha_{pq}} + \left(\sum_{j=1}^\infty (2^{js}\|\varphi_j(D) f\|_{K^\alpha_{pq}})^r\right)^{\frac{1}{r}}$ for $f \in \mathscr{S}'(\mathbb{R}^n)$. The space $B^s_r K^\alpha_{pq}(\mathbb{R}^n)$ or $\mathscr{N}_{K^\alpha_{pq} B^s_r}(\mathbb{R}^n)$ stands for the set of all $f \in \mathscr{S}'(\mathbb{R}^n)$ for which $\|f\|_{B^s_r K^\alpha_{pq}}$ is finite. Likewise we can define the *Herz-type Triebel–Lizorkin space* $F^s_r K^\alpha_{pq}(\mathbb{R}^n) = \mathscr{E}_{K^\alpha_{pq} B^s_r}(\mathbb{R}^n)$ with $p < \infty$ and $A^s_r K^\alpha_{pq}(\mathbb{R}^n)$ or $\mathscr{A}_{K^\alpha_{pq} B^s_r}(\mathbb{R}^n)$ whose formulation we omit. It seems natural to write $A^s_{K^\alpha_{pq}, r}(\mathbb{R}^n)$ instead of $A^s_r K^\alpha_{pq}(\mathbb{R}^n)$. However, traditionally we adopt the notation $A^s_{K^\alpha_{pq}, r}(\mathbb{R}^n)$. We call $A^s_{K^\alpha_{pq}, r}(\mathbb{R}^n)$ the *smoothness Herz space*.

Exercises

Exercise 6.42 [925, 1031] Suppose that the parameters p, q, r, s satisfy $0 < q \le p < \infty$, $0 < r \le \infty$ and $s \in \mathbb{R}$. By mimicking the proof of Theorem 2.11 show that the function space $\mathscr{A}^s_{pqr}(\mathbb{R}^n)$ does not depend on the admissible choices of functions ψ and φ satisfying (2.31).

Exercise 6.43 [925] Suppose that the parameters p, q, r, s satisfy $0 < q \le p < \infty$, $0 < r \le \infty$ and $s > \dfrac{n}{p}$. Then show that $\mathscr{A}^s_{pqr}(\mathbb{R}^n)$ is embedded into $\mathrm{BC}(\mathbb{R}^n)$. Hint: Use (1.159) to obtain an inequality of the form $\|\varphi_j(D) f\|_\infty \le 2^{jn/p}\|\varphi_j(D) f\|_{\mathscr{M}^p_q}$ when $q \ge 1$. When $0 < q \le 1$, use that the powered Hardy–Littlewood maximal operator is bounded on Morrey spaces.

Exercise 6.44 Let $0 < p, q, r \le \infty$ and $\alpha, s \in \mathbb{R}$. Then show that

$$\mathscr{S}(\mathbb{R}^n) \hookrightarrow A^s_r K^\alpha_{pq}(\mathbb{R}^n) \hookrightarrow \mathscr{S}'(\mathbb{R}^n)$$

in the sense of continuous embedding.

Exercise 6.45 [919, Theorem 1.1] Prove Theorem 6.30 using the atomic decomposition.

Textbooks in Sect. 6.3

Besov–Morrey Spaces and Triebel–Lizorkin–Morrey Spaces

See [120].

Function Spaces Associated with Operators

See [102] for an exhaustive account of the Besov spaces associated with operators taken up here.

6.4 Besov-Type Spaces and Triebel–Lizorkin-Type Spaces

In this section, we develop a theory of Besov-type spaces and Triebel–Lizorkin-type spaces. One of the basic ideas is to parametrize the definition of $F^s_{\infty q}(\mathbb{R}^n)$ considered in Sect. 3.3.3. In Sect. 6.4.1 we define the underlying space; that is, we generalize $L^p(\ell^q, \mathbb{R}^n)$. Here we content ourselves mainly with the generalization of Triebel–Lizorkin spaces. In Sect. 6.4.2 we first generalize Triebel–Lizorkin spaces. Some function spaces are realized as a special case of this framework. In Sect. 6.4.3 we consider the predual.

6.4.1 The Spaces $F\dot{W}^{s,\tau}_{pq}(\mathbb{R}^{n+1}_+)$ and $F\dot{T}^{s,\tau}_{pq}(\mathbb{R}^{n+1}_+)$

We would like to expand the results of $F^s_{\infty q}(\mathbb{R}^n)$ developed in Sect. 3.3. We recall that the tent $\hat{B}$ is defined in Definition 1.3 for a ball B. For a function $f : \mathbb{R}^{n+1}_+ \to \mathbb{C}$, we sometimes write $f_k = f(\star, 2^{-k})$, so that the value of $f(x,t)$ with $x \in \mathbb{R}^n$ and $\log_2 t \in \mathbb{Z}$ matters. Recall that $\hat{\Omega}$ denotes the Carleson tent of an open set Ω.

Definition 6.18 (FW-space) Let $p, q \in (1, \infty]$, $s \in \mathbb{R}$ and $\tau \in [0, \infty)$. The FW-space $F\dot{W}^{s,\tau}_{pq}(\mathbb{R}^{n+1}_+)$ is defined to be the set of all Lebesgue measurable functions f on $\mathbb{R}^{n+1}_+$ for which $\|f\|_{F\dot{W}^{s,\tau}_{pq}(\mathbb{R}^{n+1}_+)} \equiv \sup\limits_{B \in \mathscr{B}} |B|^{-\tau} \|\{2^{js} f_j \chi_{\hat{B}}\}_{j=-\infty}^{\infty}\|_{L^p(\ell^q)}$ is finite.

6.4.1.1 Hölder-Type Inequality

We recall that the nontangential maximal function is defined in Definition 1.40. We use the following Hölder inequality and the Choquet integral:

Lemma 6.21 *Let $s \in \mathbb{R}$.*

1. *Let $p \in (1, \infty)$, and let $\tau \in (0, p^{-1})$. Then*

$$\|\{2^{ks} f_k \omega_k\}_{k=-\infty}^{\infty}\|_{L^p(\ell^p)} \lesssim \|f\|_{F\dot{W}^{s,\tau}_{pp}} \left(\int_{\mathbb{R}^n} N\omega(x)^p \mathrm{d}H^{(n\tau p)}(x) \right)^{\frac{1}{p}} \tag{6.130}$$

for all $f \in F\dot{W}^{s,\tau}_{pp}(\mathbb{R}^{n+1}_+)$ and all Borel measurable functions $\omega : \mathbb{R}^{n+1}_+ \to [0, \infty)$.

2. *Let $1 < p < q < \infty$, and let $\tau \in (0, q^{-1})$. Then*

$$\|\{2^{ks} f_k \omega_k\}_{k=-\infty}^{\infty}\|_{L^p(\ell^q)} \lesssim \|f\|_{F\dot{W}^{s,\tau}_{pq}(\mathbb{R}^{n+1}_+)} \left(\int_{\mathbb{R}^n} N\omega(x)^p \mathrm{d}H^{(n\tau p)}(x) \right)^{\frac{1}{p}} \tag{6.131}$$

for all $f \in F\dot{W}^{s,\tau}_{pq}(\mathbb{R}^{n+1}_+)$ and all Borel measurable functions $\omega : \mathbb{R}^{n+1}_+ \to [0, \infty)$.

3. *Let $1 < q < p < \infty$, and let $\tau \in (0, q^{-1})$. Then*

$$\|\{2^{ks} f_k \omega_k\}_{k=-\infty}^{\infty}\|_{L^p(\ell^q)} \lesssim \|f\|_{F\dot{W}^{s,\tau}_{pq}(\mathbb{R}^{n+1}_+)} \left(\int_{\mathbb{R}^n} N\omega(x)^q \mathrm{d}H^{(n\tau q)}(x) \right)^{\frac{1}{q}} \tag{6.132}$$

for all $f \in F\dot{W}^{s,\tau}_{pq}(\mathbb{R}^{n+1}_+)$ and all Borel measurable functions $\omega : \mathbb{R}^{n+1}_+ \to [0, \infty)$.

We remark that we have to consider (6.130), (6.131) and (6.132) according to $p = q$, $p < q$ and $p > q$. We omit the proof of (6.132) since it is similar to (6.131); see Exercise 6.46.

Proof We prove (6.131). The proof of (6.130) and (6.132) is similar.

Let $l \in \mathbb{Z}$. We set $O_l \equiv \{N\omega > 2^l\}$. Then there exists a collection of balls $\{B_{j,l}\}_{j=1}^{\infty}$ such that

$$O_l \subset \bigcup_{j=1}^{\infty} B_{j,l}, \quad \widehat{O_l} \subset \bigcup_{j=1}^{\infty} \widehat{B_{j,l}}, \tag{6.133}$$

$$\sum_{j=1}^{\infty} |B_{j,l}|^{\frac{d}{n}} \lesssim_{n,d} H^d(O_l) \tag{6.134}$$

thanks to Proposition 1.5. We set

$$A_l \equiv \{(y, 2^{-k}) \in \mathbb{R}^{n+1}_+ \, : \, k \in \mathbb{Z}, \quad 2^l < \omega(y, 2^{-k}) \le 2^{l+1}\}.$$

Then a standard argument shows

$$A_l \subset \widehat{O_l}. \tag{6.135}$$

See Exercise 6.48. Thus,

$$\begin{aligned}
&\|\{2^{ks} f_k \omega_k\}_{k=-\infty}^{\infty}\|_{L^p(\ell^q)} \\
&\lesssim \left\{ \sum_{l=-\infty}^{\infty} \int_{\mathbb{R}^n} 2^{lp} \left(\sum_{k=-\infty}^{\infty} 2^{ksq} |f_k(x)|^q \chi_{A_l}(x, 2^{-k}) \right)^{\frac{p}{q}} \mathrm{d}x \right\}^{\frac{1}{p}}
\end{aligned}$$

from the definition of A_l. From (6.133) and (6.135) we deduce

$$\begin{aligned}
&\|\{2^{ks} f_k \omega_k\}_{k=-\infty}^{\infty}\|_{L^p(\ell^q)} \\
&\lesssim \left\{ \sum_{l=-\infty}^{\infty} \int_{\mathbb{R}^n} 2^{lp} \left(\sum_{k,j=-\infty}^{\infty} 2^{ksq} |f_k(x)|^q (\chi_{\widehat{B_{j,l}}})_k(x) \right)^{\frac{p}{q}} \mathrm{d}x \right\}^{\frac{1}{p}}.
\end{aligned}$$

Since $q > p$, we can use $a^\theta + b^\theta \ge (a+b)^\theta$ for $a, b \ge 0$ and $\theta \in (0, 1)$ to obtain

$$\begin{aligned}
&\|\{2^{ks} f_k \omega_k\}_{k=-\infty}^{\infty}\|_{L^p(\ell^q)} \\
&\lesssim \left\{ \sum_{j,l=-\infty}^{\infty} \int_{\mathbb{R}^n} 2^{lp} \left(\sum_{k=-\infty}^{\infty} 2^{ksq} |f_k(x)|^q (\chi_{\widehat{B_{j,l}}})_k(x) \right)^{\frac{p}{q}} \mathrm{d}x \right\}^{\frac{1}{p}} \\
&\le \|f\|_{F\dot{W}^{s,\tau}_{pq}(\mathbb{R}^{n+1}_+)} \left\{ \sum_{j,l=-\infty}^{\infty} 2^{lp} |B_{j,l}|^{\tau p} \right\}^{\frac{1}{p}}
\end{aligned}$$

from the definition of $\|f\|_{F\dot{W}^{s,\tau}_{pq}(\mathbb{R}^{n+1}_+)}$. In total,

$$\|\{2^{ks} f_k \omega_k\}_{k=-\infty}^{\infty}\|_{L^p(\ell^q)} \lesssim \|f\|_{F\dot{W}^{s,\tau}_{pq}(\mathbb{R}^{n+1}_+)} \left\{ \sum_{j,l=-\infty}^{\infty} 2^{lp} |B_{j,l}|^{\tau p} \right\}^{\frac{1}{p}}. \quad (6.136)$$

If we use (6.134), we have

$$\sum_{j,l=-\infty}^{\infty} 2^{lp} |B_{j,l}|^{\tau p} \lesssim \sum_{l=-\infty}^{\infty} 2^{lp} H^{(n\tau p)}(O_l) \sim \int_{\mathbb{R}^n} N\omega(x)^p \mathrm{d}H^{(n\tau p)}(x). \quad (6.137)$$

If we insert (6.137) into (6.136), then we obtain (6.131).

6.4.1.2 The Space $F\dot{T}^{s,\tau}_{pq}(\mathbb{R}^{n+1}_+)$

We now aim to consider a function space whose dual is $F\dot{W}^{s,\tau}_{pq}(\mathbb{R}^{n+1}_+)$. To this end, we define the following function space:

Definition 6.19 ($F\dot{T}^{s,\tau}_{pq}(\mathbb{R}^{n+1}_+)$) Suppose we have parameters $p, q \in [1,\infty)$, $s \in \mathbb{R}$ and $\tau > 0$ satisfying $p+q > 2$ and $\tau \le (p')^{-1} \vee (q')^{-1}$. The space $F\dot{T}^{s,\tau}_{pq}(\mathbb{R}^{n+1}_+)$ is defined to be the set of all Lebesgue measurable functions f on $\mathbb{R}^{n+1}_+$ for which

$$\|f\|_{F\dot{T}^{s,\tau}_{pq}(\mathbb{R}^{n+1}_+)} \equiv \inf_{\omega} \|\{2^{ks} {\omega_k}^{-1} f_k\}_{k=-\infty}^{\infty}\}\|_{L^p(\ell^q)}$$

is finite. Here ω moves over all Borel measurable functions on $\mathbb{R}^{n+1}_+$ such that $\omega_k(x)$ is allowed to vanish only when $f_k(x) = 0$ and moves over all Borel measurable functions satisfying

$$\int_{\mathbb{R}^n} N\omega(x)^{(p' \wedge q')} dH^{n\tau(p' \wedge q')}(x) \le 1. \quad (6.138)$$

6.4.1.3 The Hölder Inequality and Duality

We prove the following Hölder-type inequality:

Theorem 6.31 *Suppose that we have parameters* $(p,q,s,\tau) \in (1,\infty]^2 \times \mathbb{R} \times (0,\infty)$ *satisfying*

$$0 < \tau \le (p')^{-1} \vee (q')^{-1}. \quad (6.139)$$

Then for all $f \in F\dot{T}^{s,\tau}_{pq}(\mathbb{R}^{n+1}_+)$ *and* $g \in F\dot{W}^{-s,\tau}_{p'q'}(\mathbb{R}^n)$,

$$\|\{f_k g_k\}_{k=-\infty}^{\infty}\|_{L^1(\ell^1)} \lesssim \|f\|_{F\dot{T}^{s,\tau}_{pq}(\mathbb{R}^{n+1}_+)}\|g\|_{F\dot{W}^{-s,\tau}_{p'q'}(\mathbb{R}^{n+1}_+)}.$$

In particular, g induces a bounded linear functional L_g on $F\dot{T}^{s,\tau}_{pq}(\mathbb{R}^{n+1}_+)$.

Proof Let $\omega : \mathbb{R}^{n+1}_+ \to [0, \infty)$ be a Borel measurable function satisfying (6.138). Then using the Hölder inequality for sequences we have

$$\begin{aligned}\|\{f_k g_k\}_{k=-\infty}^{\infty}\|_{L^1(\ell^1)} &= \int_{\mathbb{R}^n} \sum_{k=-\infty}^{\infty} |f_k(x) g_k(x)| \mathrm{d}x \\ &\le \int_{\mathbb{R}^n} \left(\sum_{k=-\infty}^{\infty} \frac{2^{ksq}|f_k(x)|^q}{\omega_k(x)^q} \right)^{\frac{1}{q}} \left(\sum_{k=-\infty}^{\infty} 2^{ksq'}|g_k(x)|^{q'}\omega_k(x)^{q'} \right)^{\frac{1}{q'}} \mathrm{d}x\end{aligned}$$

Next, using the Hölder inequality for functions we have

$$\|\{f_k g_k\}_{k=-\infty}^{\infty}\|_{L^1(\ell^1)} \le \|\{2^{kq}\omega_k{}^{-1} f_k\}_{k=-\infty}^{\infty}\|_{L^p(\ell^q)}\|\{2^{ks}\omega_k g_k\}_{k=-\infty}^{\infty}\|_{L^{p'}(\ell^{q'})}.$$

By Lemma 6.21, we have

$$\begin{aligned}\|\{f_k g_k\}_{k=-\infty}^{\infty}\|_{L^1(\ell^1)} &\lesssim \|\{2^{kq}\omega_k{}^{-1} f_k\}_{k=-\infty}^{\infty}\|_{L^p(\ell^q)} \\ &\quad \times \|g\|_{F\dot{W}^{-s,\tau}_{p'q'}(\mathbb{R}^{n+1}_+)} \left(\int_{\mathbb{R}^n} N\omega(x)^{n(p'\wedge q')} \mathrm{d}H^{n\tau(p'\wedge q')}(x) \right).\end{aligned}$$

By (6.138), we have

$$\|\{f_k g_k\}_{k=-\infty}^{\infty}\|_{L^1(\ell^1)} \lesssim \left(\int_{\mathbb{R}^n} \left(\sum_{k=-\infty}^{\infty} \frac{2^{ksq}|f_k(x)|^q}{\omega_k(x)^q} \right)^{\frac{p}{q}} \mathrm{d}x \right)^{\frac{1}{p}} \|g\|_{F\dot{W}^{-s,\tau}_{p'q'}(\mathbb{R}^{n+1}_+)}.$$

Taking the infimum over all ω, we obtain the desired result.

We next consider the case where the function f is supported in the truncated tent.

Example 6.10 Suppose that we have parameters $(p, q, s, \tau) \in (1, \infty]^2 \times \mathbb{R} \times (0, \infty)$ satisfying (6.139). Let $\varepsilon > 0$, and let B be a ball. Write

$$T^{\varepsilon}(B) \equiv \widehat{B} \cap (\mathbb{R}^n \times (\varepsilon, \infty)). \tag{6.140}$$

Then for any Borel measurable function $f : \mathbb{R}^{n+1} \to \mathbb{C}$ with $\mathrm{supp}(f) \subset T^{\varepsilon}(B)$, one can check that

$$\|f\|_{F\dot{T}^{s,\tau}_{pq}(\mathbb{R}^{n+1}_+)} \lesssim (\varepsilon^{-s_+} + r_B{}^{-s_-})|B|^{\tau}\|\{f_k(\chi_{T^{\varepsilon}(B)})_k\}_{k=-\infty}^{\infty}\}\|_{L^p(\ell^q)}$$

using a function ω by (6.141). We can estimate the norm from above. See Exercise 6.56.

We need the following density result:

Lemma 6.22 *Suppose that we have parameters* $(p, q, s, \tau) \in (1, \infty)^2 \times \mathbb{R} \times (0, \infty)$ *satisfying* (6.139). *Let* $\mathscr{X}$ *be the set of all all measurable functions* $f \in F\dot{T}^{s,\tau}_{pq}(\mathbb{R}^{n+1}_+)$ *whose support is contained in* $T^{\varepsilon}(B)$, *where* $T^{\varepsilon}(B)$ *is given by* (6.140), *for some* $\varepsilon > 0$ *and some ball* B. *Then* $\mathscr{X}$ *is dense in* $F\dot{T}^{s,\tau}_{pq}(\mathbb{R}^{n+1}_+)$.

Proof This is routine and uses the standard truncation argument. So we omit the details. See Exercise 6.47.

Based on Theorem 6.31, we establish the following duality:

Theorem 6.32 *Suppose that we have parameters* $(p, q, s, \tau) \in (1, \infty]^2 \times \mathbb{R} \times (0, \infty)$ *satisfying* (6.139). *Then any linear functional* L *on* $F\dot{T}^{s,\tau}_{pq}(\mathbb{R}^{n+1}_+)$ *is realized as* $L = L_g$ *for some* $g \in F\dot{W}^{-s,\tau}_{pq}(\mathbb{R}^{n+1}_+)$. *Here* L_g *is the linear functional in Theorem* 6.31.

Proof According to the duality $L^p(\ell^q)$-$L^{p'}(\ell^{q'})$, we can find a sequence $\{g^{(k)}\}_{k=1}^{\infty}$ of measurable functions such that $\{\chi_K \cdot g^{(k)}\}_{k=1}^{\infty} \in L^{p'}(\ell^{q'})$ for all compact sets $K \subset \mathbb{R}^{n+1}_+$ and that

$$L(f) = \int_{\mathbb{R}^n} \sum_{k=-\infty}^{\infty} f_k(x) g^{(k)}(x) \mathrm{d}x$$

for all $f \in F\dot{T}^{s,\tau}_{pq}(\mathbb{R}^{n+1}_+)$ with support in $T^{\varepsilon}(B)$ for some $\varepsilon > 0$ and some ball B. Since such functions span a dense subspace in $F\dot{T}^{s,\tau}(\mathbb{R}^{n+1}_+)$, all we have to show is that $g \in F\dot{W}^{-s,\tau}_{p'q'}(\mathbb{R}^{n+1}_+)$, where $g(x, t) \equiv g_{-[\log_2 t]}(x)$ for $(x, t) \in \mathbb{R}^{n+1}_+$.

We fix a ball $B \subset \mathbb{R}^n$. We set

$$G \equiv \sum_{k=-\infty}^{\infty} 2^{-ksq'} |g_k|^{q'} (\chi_{T^{\varepsilon}(B)})_k$$

and

$$f_{\varepsilon}(x, t) \equiv t^{sq'} |g(x, t)|^{q'-1} \chi_{T^{\varepsilon}(B)}(x, t) \overline{\mathrm{sgn}(g(x, t))} G(x)^{\frac{p'}{q'}-1}$$

for each $\varepsilon > 0$ and $(x, t) \in \mathbb{R}^{n+1}_+$. We let ω be defined by

$$\omega(x, t) \equiv \frac{c_0}{|B|^{\tau}} \min\left(1, \frac{|B|}{|x - c_B|^n + t^n}\right)^N \quad ((x, t) \in \mathbb{R}^{n+1}_+), \tag{6.141}$$

where $N \gg 1$ and $0 < c_0 \ll 1$. Then from Example 1.34 we deduce

$$N\omega(x) \lesssim \frac{c_0}{|B|^\tau} \min\left(1, \frac{|B|}{|x - c_B|^n}\right)^N \quad (x \in \mathbb{R}^n)$$

and hence (6.138). Note that

$$\int_{\mathbb{R}^n} G(x)^{\frac{p'}{q'}} \mathrm{d}x = L(f_\varepsilon) \leq \|L\|_{F\dot{T}^{s,\tau}_{pq}(\mathbb{R}^{n+1}_+)^*} \tag{6.142}$$

and that

$$\|f_\varepsilon\|_{F\dot{T}^{s,\tau}_{pq}(\mathbb{R}^{n+1}_+)} \lesssim \left\{\int_{\mathbb{R}^n}\left(\sum_{k=-\infty}^{\infty} 2^{-ksq}\omega_k(x)^{-q}|f_\varepsilon(x, 2^{-k})|^q \chi_{T^\varepsilon(B)}(x, 2^{-k})\right)^{\frac{p}{q}} \mathrm{d}x\right\}^{\frac{1}{p}}.$$

Since $\omega_k(x) \sim 1$ when $(x, 2^{-k}) \in \hat{B}$, it follows that

$$\|f_\varepsilon\|_{F\dot{T}^{s,\tau}_{pq}(\mathbb{R}^{n+1}_+)} \lesssim \|\{2^{-ks} f_\varepsilon(\star, 2^{-k})(\chi_{T^\varepsilon(B)})_k\}_{k=-\infty}^{\infty}\|_{L^p(\ell^q)}.$$

Inserting the definition of f_ε into the above expression and using (6.142), we obtain $g \in F\dot{W}^{-s,\tau}_{p'q'}(\mathbb{R}^{n+1}_+)$, as desired.

Exercises

Exercise 6.46 Prove (6.132) using the Hölder inequality to obtain a counterpart to (6.136).

Exercise 6.47 Using the Fatou lemma, prove Lemma 6.22.

Exercise 6.48 Prove (6.135) using the definition of the nontangential maximal operator.

Exercise 6.49 Prove Lemma 6.10.

6.4.2 Besov-Type Spaces and Triebel–Lizorkin-Type Spaces

6.4.2.1 Besov-Type Spaces and Triebel–Lizorkin-Type Spaces

Using the idea of function spaces $F^s_{\infty q}(\mathbb{R}^n)$, Yang and Yuan proposed a new framework containing Besov spaces, Morrey spaces, Triebel–Lizorkin spaces and the space BMO [1152, 1153], whose root is El Baraka's 2002 paper [4]. Yang and Yuan investigated Besov-type spaces and Triebel–Lizorkin-type spaces in [1152, 1153]. Choose $\psi, \varphi \in \mathscr{S}(\mathbb{R}^n)$ so that (2.31) holds. For $j \in \mathbb{N}$, let $\varphi_j \equiv \varphi(2^{-j}\star)$.

Definition 6.20 ($A^{s,\tau}_{pq}(\mathbb{R}^n)$) Let $0 < p, q \le \infty$, $s \in \mathbb{R}$, $\tau \ge 0$. For $f \in \mathscr{S}'(\mathbb{R}^n)$ one defines the *nonhomogeneous Besov-type norm* and the *nonhomogeneous Triebel–Lizorkin-type norm* with $p < \infty$ by

$$\|f\|_{B^{s,\tau}_{pq}} \equiv \sup_{Q\in\mathscr{D}} \frac{\|\psi(D)f\|_{L^p(Q)}}{|Q|^\tau} + \sup_{Q\in\mathscr{D}} \frac{\|\{2^{js}\chi_Q\varphi_j(D)f\}_{j=\max(1,-\log_2\ell(Q))}^\infty\|_{\ell^q(L^p)}}{|Q|^\tau}$$

and

$$\|f\|_{F^{s,\tau}_{pq}} \equiv \sup_{Q\in\mathscr{D}} \frac{\|\psi(D)f\|_{L^p(Q)}}{|Q|^\tau} + \sup_{Q\in\mathscr{D}} \frac{\|\{2^{js}\chi_Q\varphi_j(D)f\}_{j=\max(1,-\log_2\ell(Q))}^\infty\|_{L^p(\ell^q)}}{|Q|^\tau},$$

respectively.

The (nonhomogeneous) Besov-type space $B^{s,\tau}_{pq}(\mathbb{R}^n)$ and the (nonhomogeneous) Triebel–Lizorkin-type space $F^{s,\tau}_{pq}(\mathbb{R}^n)$ stand for the linear spaces of functions $f \in \mathscr{S}'(\mathbb{R}^n)$ for which the quantities $\|f\|_{B^{s,\tau}_{pq}}$ and $\|f\|_{F^{s,\tau}_{pq}}$ are finite, respectively. The notation $A^{s,\tau}_{pq}(\mathbb{R}^n)$ stands for either $B^{s,\tau}_{pq}(\mathbb{R}^n)$ or $F^{s,\tau}_{pq}(\mathbb{R}^n)$. Exclude the case where $0 < p < \infty$ when $A = F$.

We now aim to show that the definition of $A^{s,\tau}_{pq}(\mathbb{R}^n)$ is independent of the choice of ψ and φ satisfying (2.31).

Lemma 6.23 *Let $P \in \mathscr{D}$ be a dyadic cube, and let j be an integer such that $j \ge -\log_2\ell(P)$. Let $0 < \theta < \min(1, q)$. Then we have*

$$\|\langle 2^j(x-\star)\rangle^{-\frac{n}{\theta}} f\|_\infty \lesssim_{\theta,q} M^{(\theta)}[\chi_{10nP}f](x) + 2^{jn\left(\frac{1}{q}-\frac{1}{\theta}\right)} \sup_{k\in\mathbb{Z}^n\setminus\{0\}} \frac{\|f\|_{L^q(\ell(P)k+P)}}{\ell(P)^{\frac{n}{\theta}}\langle k\rangle^{\frac{n}{\theta}}}$$

for all $f \in \mathscr{S}'_{B(2^j)}(\mathbb{R}^n)$ and $x \in P$.

Proof Reexamine the proof of Lemma 3.16, where a modification is needed in (3.145).

Theorem 6.33 *Let $0 < p, q \le \infty$, $s \in \mathbb{R}$ and $\tau \ge 0$. Then the space $A^{s,\tau}_{pq}(\mathbb{R}^n)$ is independent of the choice of ψ and φ satisfying* (2.31).

Proof Similar to Proposition 3.13. We omit the details. See Exercise 6.51.

Naturally, we introduce the homogeneous Besov-type norm $\|\star\|_{\dot{B}^{s,\tau}_{pq}}$ defined in [1153, Definition 1.1] and the homogeneous Triebel–Lizorkin-type norm $\|\star\|_{\dot{F}^{s,\tau}_{pq}}$ defined in [1152, Definition 3.1].

Definition 6.21 ($\dot{A}^{s,\tau}_{pq}(\mathbb{R}^n)$) Let $0 < q \le \infty$, $s \in \mathbb{R}$, $\tau \ge 0$, and let φ and ψ satisfy (2.31). For $f \in \mathscr{S}'_\infty(\mathbb{R}^n)$ one defines the *homogenous Besov-type norm* and the *homogeneous Triebel–Lizorkin-type norm* by

$$\|f\|_{\dot{B}^{s,\tau}_{pq}} \equiv \sup_{Q\in\mathscr{D}} |Q|^{-\tau} \|\{2^{js}\chi_Q\varphi_j(D)f\}_{j=-\log_2 \ell(Q)}^{\infty}\|_{\ell^q(L^p)}$$

for $0 < p \le \infty$ and

$$\|f\|_{\dot{F}^{s,\tau}_{pq}} \equiv \sup_{Q\in\mathscr{D}} |Q|^{-\tau} \|\{2^{js}\chi_Q\varphi_j(D)f\}_{j=-\log_2 \ell(Q)}^{\infty}\|_{L^p(\ell^q)}$$

for $0 < p < \infty$, respectively. The spaces $\dot{B}^{s,\tau}_{pq}(\mathbb{R}^n)$ and $\dot{F}^{s,\tau}_{pq}(\mathbb{R}^n)$ stand for linear spaces of functions $f \in \mathscr{S}'_\infty(\mathbb{R}^n)$ for which the quantities $\|f\|_{\dot{B}^{s,\tau}_{pq}}$ and $\|f\|_{\dot{F}^{s,\tau}_{pq}}$ are finite, respectively.

The space $\dot{B}^{s,\tau}_{pq}(\mathbb{R}^n)$ is called the (homogeneous) *Besov-type space* and the space $\dot{F}^{s,\tau}_{pq}(\mathbb{R}^n)$ is called the (homogeneous) *Triebel–Lizorkin-type space*. The notation $\dot{A}^{s,\tau}_{pq}(\mathbb{R}^n)$ stands for either $\dot{B}^{s,\tau}_{pq}(\mathbb{R}^n)$ or $\dot{F}^{s,\tau}_{pq}(\mathbb{R}^n)$ as usual. The space $\dot{F}^{s,\tau}_{\infty q}(\mathbb{R}^n)$ is not defined.

Note that $\dot{A}^{s,0}_{pq}(\mathbb{R}^n) = \dot{A}^{s}_{pq}(\mathbb{R}^n)$ with coincidence of norms. See Exercise 6.50. We have the following fundamental fact:

Theorem 6.34 *Let* $0 < p, q \le \infty$, $s \in \mathbb{R}$ *and* $\tau \ge 0$. *Then as a set the space* $\dot{A}^{s,\tau}_{pq}(\mathbb{R}^n)$ *is independent of the choice of* ψ *and* φ *satisfying* (2.31).

Proof Similar to Proposition 3.13. We omit the details. See Exercise 6.56.

6.4.2.2 The φ-Transform for $F\dot{W}^{s,\tau}_{pq}(\mathbb{R}^{n+1}_+)$ and $\dot{F}^{s,\tau}_{pq}(\mathbb{R}^n)$

Here we investigate the φ-transform for $F\dot{W}^{s,\tau}_{pq}(\mathbb{R}^{n+1}_+)$ and $\dot{F}^{s,\tau}_{pq}(\mathbb{R}^n)$, which is useful to characterize the space $\dot{F}^{s,\tau}_{pq}(\mathbb{R}^n)$. This φ-transform will be used in order to investigate the relation between $\dot{F}^{s,\tau}_{pq}(\mathbb{R}^n)$ and Triebel–Lizorkin–Hausdorff spaces, which we defined in Sect. 6.4.3.

Proposition 6.7 *Let* $1 < p < \infty$, $1 < q \le \infty$, $s \in \mathbb{R}$ *and* $\tau \ge 0$. *We choose* $\varphi \in C^\infty_c(\mathbb{R}^n \setminus \{0\})$ *so that*

$$\sum_{j=-\infty}^{\infty} \varphi(2^{-j}\star)^2 = \chi_{\mathbb{R}^n\setminus\{0\}}. \tag{6.143}$$

1. [1152, Lemma 5.1] *The mapping*

$$\pi : F \in F\dot{W}^{s,\tau}_{pq}(\mathbb{R}^{n+1}_+) \mapsto \sum_{j=-\infty}^{\infty} \varphi_j(D)F_j \in \dot{F}^{s,\tau}_{pq}(\mathbb{R}^n) \tag{6.144}$$

is a bounded linear operator.

2. [1152, Lemma 5.4] *The mapping*

$$\Theta : f \in \dot{F}^{s,\tau}_{pq}(\mathbb{R}^n) \mapsto \Theta f \in F\dot{W}^{s,\tau}_{pq}(\mathbb{R}^{n+1}_+), \tag{6.145}$$

where

$$\Theta f(x,t) \equiv \varphi_{[-\log_2 t]}(D)f(x), \tag{6.146}$$

is a bounded linear operator.

3. *We have*

$$\pi \circ \Theta = \mathrm{id}_{\dot{F}^{s,\tau}_{pq}(\mathbb{R}^n)}. \tag{6.147}$$

Proof We concentrate on (6.144); equality (6.147) is clear and we leave the boundedness of Θ as in Exercise 6.54. Let B be a fixed ball. For $j \in \mathbb{Z} \cap [-\log_2 r(B), \infty)$, we have

$$\varphi_j(D)F_j = \varphi_j(D)[F_j \chi_{\widehat{B}}] + \sum_{l=1}^{\infty} \varphi_j(D)[F_j \chi_{\widehat{2^l B} \setminus \widehat{2^{l-1} B}}].$$

Thus, for any $N \gg 1$, we have $|\varphi_j(D)F_j| \lesssim M[F_j \chi_{\widehat{B}}] + \sum_{l=1}^{\infty} 2^{-lN} M[F_j \chi_{\widehat{2^l B} \setminus \widehat{2^{l-1} B}}]$.

We observe

$$\|f\|_{F\dot{W}^{s,\tau}_{pq}(\mathbb{R}^{n+1}_+)} \sim \sup_{B \in \mathscr{B}} \frac{1}{|aB|^\tau} \left\{ \int_{\mathbb{R}^n} \left(\sum_{k=-\infty}^{\infty} 2^{ksq} |f_k(x)|^q \chi_{\widehat{aB}}(x, 2^{-k}) \right)^{\frac{p}{q}} \mathrm{d}x \right\}^{\frac{1}{p}}$$

with the constant independent of $a > 0$. Thus, we obtain the desired result combining the Fefferman–Stein vector-valued inequality.

6.4.2.3 Triebel–Lizorkin–Morrey-Spaces and Triebel–Lizorkin-Type Spaces

It seems that the new parameter τ here complicates things in $A^{s,\tau}_{pq}(\mathbb{R}^n)$. However, we will clarify that this parameter τ can be used to recapture our function spaces which we already defined.

Theorem 6.35 *Let* $0 < p < \infty$, $0 < q \le \infty$, $0 < r < \infty$ *and* $s \in \mathbb{R}$. *Set* $\tau \equiv \dfrac{1}{q} - \dfrac{1}{p}$. *Then* $\mathscr{E}^s_{pqr}(\mathbb{R}^n) \approx F^{s,\tau}_{qr}(\mathbb{R}^n)$ *and* $\dot{\mathscr{E}}^s_{pqr}(\mathbb{R}^n) \approx \dot{F}^{s,\tau}_{qr}(\mathbb{R}^n)$ *with equivalence of norms.*

We note that the counterpart of Besov-type spaces is not available except when $r = \infty$.

Proof We concentrate on the nonhomogeneous spaces. If we compare the definition of the norm, we obtain $\mathcal{E}^s_{pqr}(\mathbb{R}^n) \hookrightarrow F^{s,\tau}_{qr}(\mathbb{R}^n)$. Thus, we need to show $\mathcal{E}^s_{pqr}(\mathbb{R}^n) \hookleftarrow F^{s,\tau}_{qr}(\mathbb{R}^n)$, which amounts to showing

$$\sup_{R\in\mathscr{D}} |R|^{-\tau}\left(\int_R\left(\sum_{j=0}^{\max(1,-\log_2 \ell(R))} 2^{jsr}|\varphi_j(D)f(x)|^r\right)^{\frac{q}{r}} dx\right)^{\frac{1}{q}}$$
$$\lesssim \sup_{Q\in\mathscr{D}} |Q|^{-\tau}\left(\int_Q\left(\sum_{j=\max(1,-\log_2 \ell(Q))}^{\infty} 2^{jsr}|\varphi_j(D)f(x)|^r\right)^{\frac{q}{r}} dx\right)^{\frac{1}{q}}.$$

In fact, we will show that

$$|R|^{-\tau}\left(\int_R\left(2^{jsr}|\varphi_j(D)f(x)|^r\right)^{\frac{q}{r}} dx\right)^{\frac{1}{q}}$$
$$\lesssim (2^{jn}|R|)^{\frac{1}{p}} \sup_{Q\in\mathscr{D}_{-\log_2 \ell(R)}} |Q|^{-\tau}\left(\int_Q\left(\frac{1}{\ell(R)^{sr}}|\varphi_{-\log_2 \ell(R)}(D)f(x)|^r\right)^{\frac{q}{r}} dx\right)^{\frac{1}{q}}$$

for any fixed cube R and $j \in \mathbb{Z}\cap[1,\max(1,-\log_2 \ell(R))]$. Once this is achieved, we can add this estimate over all $j \in \mathbb{Z}\cap[1,\max(1,-\log_2 \ell(R))]$. Choose a cube S containing R and having volume 2^{-jn}. By Corollary 1.4, we have

$$|R|^{-\tau}\left(\int_R\left(2^{jsr}|\varphi_j(D)f(x)|^r\right)^{\frac{q}{r}} dx\right)^{\frac{1}{q}}$$
$$\lesssim \left(\frac{|R|}{|S|}\right)^{\frac{1}{p}} |S|^{-\tau}\left(\int_S\left(2^{jsr}M^{(\theta)}[\varphi_j(D)f](x)^r\right)^{\frac{q}{r}} dx\right)^{\frac{1}{q}}.$$

We decompose $\varphi_j(D)f$ according to $3S$. Then

$$|S|^{-\tau}\left(\int_S\left(2^{jsr}M^{(\theta)}[\varphi_j(D)f](x)^r\right)^{\frac{q}{r}} dx\right)^{\frac{1}{q}}$$
$$\lesssim |S|^{-\tau}\left(\int_S\left(2^{jsr}M^{(\theta)}[\chi_{3S}\varphi_j(D)f](x)^r\right)^{\frac{q}{r}} dx\right)^{\frac{1}{q}}$$
$$+ |S|^{-\tau}\left(\int_S\left(2^{jsr}M^{(\theta)}[\chi_{\mathbb{R}^n\setminus 3S}\varphi_j(D)f](x)^r\right)^{\frac{q}{r}} dx\right)^{\frac{1}{q}}.$$

We use the boundedness of the Hardy–Littlewood maximal operator for the first term. For the second term, we use Lemma 1.27 to have

$$\frac{1}{|S|^{\frac{1}{q}}}\left(\int_S \left(2^{jsr}M^{(\theta)}[\chi_{\mathbb{R}^n\setminus 3S}\varphi_j(D)f](x)^r\right)^{\frac{q}{r}}\mathrm{d}x\right)^{\frac{1}{q}}$$

$$\lesssim (2^{jn}|R|)^{\frac{1}{p}}\sup_{Q\in\mathscr{D}_{-\log_2\ell(R)}}|Q|^{-\tau}\left(\int_Q\left(\frac{1}{\ell(R)^{sr}}|\varphi_{-\log_2\ell(R)}(D)f(x)|^r\right)^{\frac{q}{r}}\mathrm{d}x\right)^{\frac{1}{q}}.$$

Thus, we obtain the desired result.

6.4.2.4 *Q*-Spaces and Triebel–Lizorkin-Type Spaces

One of the advantages of introducing Triebel–Lizorkin-type spaces is that it recovers the space $Q_p^{\alpha,q}(\mathbb{R}^n)$.

Definition 6.22 **($Q_p^{\alpha,q}(\mathbb{R}^n)$)** Let $\alpha\in(0,1)$, $p\in(0,\infty]$ and $q\in[1,\infty]$. Choose $\varphi\in C_c^\infty(\mathbb{R}^n\setminus\{0\})$ such that

$$\sum_{j=-\infty}^{\infty}\varphi(2^{-j}\star)=\chi_{\mathbb{R}^n\setminus\{0\}}. \tag{6.148}$$

The Q-space $Q_p^{\alpha,q}(\mathbb{R}^n)$ is defined to be the set of all $f\in\mathscr{S}_\infty'(\mathbb{R}^n)$ such that

$$f(x)-f(y)=\lim_{L\to\infty}\left(\sum_{l=-L}^{L}(\varphi_j(D)f(x)-\varphi_j(D)f(y))\right) \tag{6.149}$$

in $L^1_{\mathrm{loc}}(\mathbb{R}^{2n})\cap\mathscr{S}_\infty'(\mathbb{R}^{2n})$ and

$$\|f\|_{Q_p^{\alpha,q}}\equiv\sup_{Q\in\mathscr{Q}}|Q|^{\frac{1}{p}-\frac{1}{q}}\left(\iint_{Q\times Q}|x-y|^{-n-q\alpha}|f(x)-f(y)|^q\mathrm{d}x\mathrm{d}y\right)^{\frac{1}{q}}$$

is finite.

Before we proceed further, a couple of helpful remarks may be in order.

Remark 6.4

1. Let $f\in\mathscr{S}_\infty'(\mathbb{R}^n)$ satisfy (6.149). A geometric observation shows that $\|f\|_{Q_p^{\alpha,q}}<\infty$ if and only if

$$\|f\|^{\dagger}_{Q_p^{\alpha,q}} \equiv \sup_{Q \in \mathcal{Q}} |Q|^{\frac{1}{p}-\frac{1}{q}} \left(\iint_{Q \times B(\ell(Q))} |y|^{-n-q\alpha} |f(x+y) - f(y)|^q \mathrm{d}x\mathrm{d}y \right)^{\frac{1}{q}}$$

is finite.

2. Assume that $\varphi \in C_c^\infty(\mathbb{R}^n \setminus \{0\})$ satisfies (6.148) and that $f \in \mathcal{S}'_\infty(\mathbb{R}^n)$ satisfies $\|f\|_{Q_p^{\alpha,q}} < \infty$, where $f(x)-f(y)$ exists in the sense of (6.149). If $\varphi^* \in C_c^\infty(\mathbb{R}^n \setminus \{0\})$ satisfies (2.116), then

$$f(x) - f(y) = \lim_{L \to \infty} \left(\sum_{l=-L}^{L} (\varphi_j^*(D)f(x) - \varphi_j^*(D)f(y)) \right)$$

in $L^1_{\mathrm{loc}}(\mathbb{R}^{2n}) \cap \mathcal{S}'_\infty(\mathbb{R}^{2n})$.

We now aim to show that $Q_p^{\alpha,q}(\mathbb{R}^n)$ is isomorphic to $\dot{F}_{qq}^{\alpha,\tau}(\mathbb{R}^n)$ when $\tau = \frac{1}{q} - \frac{1}{p} \geq 0$. To this end, we present the following definition:

Definition 6.23 **(**$Q_p^{\alpha,q,*}(\mathbb{R}^n)$**)** Let $\alpha \in (0,1)$, $p \in (0,\infty]$, $q \in [1,\infty]$, and let $\varphi \in C_c^\infty(\mathbb{R}^n)$ satisfy (6.148). Then $Q_p^{\alpha,q,*}(\mathbb{R}^n)$ collects all $f \in \mathcal{S}'_\infty(\mathbb{R}^n)$ such that

$$\|f\|_{Q_p^{\alpha,q,*}} \equiv \sup_{Q \in \mathcal{Q}} |Q|^{\frac{1}{p}-\frac{1}{q}} \|\{2^{j\alpha} \chi_Q \varphi_j(D)f\}_{j=-\log_2 \ell(Q)}^{\infty}\|_{L^q(\ell^q)}$$

is finite.

Remark 6.5 Let $\tau \equiv \dfrac{1}{q} - \dfrac{1}{p}$.

1. The case $\tau < 0$ is torelated in Definition 6.23.
2. The definition of $Q_p^{\alpha,q,*}(\mathbb{R}^n)$ does not depend on φ satisfying (2.31). See Exercise 6.55.
3. Let $0 < p < \infty$, $1 \leq q \leq \infty$ and $s \in (0,1)$. Then using $\sum_{j=0}^{\infty} \|\varphi_j(D)f\|_{Q_{0m}} \lesssim \|f\|_{Q_p^{\alpha,q,*}}$ for all $m \in \mathbb{Z}^n$ and all $f \in \dot{A}_{pq}^{s,\tau}(\mathbb{R}^n)$, we see that

$$f_+ \equiv \sum_{j=0}^{\infty} \varphi_j(D)f$$

is a locally integrable function such that $\|f_+\|_{Q_{0,m}} \lesssim \|f\|_{Q_p^{\alpha,q,*}}$. Likewise, by the mean-value theorem,

$$\int_{Q_{0m}} \sum_{j=-\infty}^{-1} |\nabla \varphi_j(D)f(x)| \mathrm{d}x \lesssim \|f\|_{Q_p^{\alpha,q,*}}.$$

Thus, $g_{k,-} \equiv \sum_{j=0}^{\infty} \partial_{x_k}[\varphi_j(D)f]$ satisfies $\|g_{k,-}\|_{L^1(Q_{0\,m})} \lesssim \|f\|_{Q_p^{\alpha,q,*}}$ for all $m \in \mathbb{Z}^n$. Thus, the limit $\lim_{L\to\infty}\left(\sum_{j=-L}^{\infty}(\varphi_j(D)f(x) - \varphi_j(D)f(y))\right)$ converges in $L^1_{\rm loc}(\mathbb{R}^{2n}) \cap \mathscr{S}'_\infty(\mathbb{R}^{2n})$.

Theorem 6.36 *Let $\alpha \in (0,1)$, $p \in (0,\infty]$, $q \in [1,\infty]$, and let $\varphi \in \mathscr{S}(\mathbb{R}^n)$ be a function satisfying* (6.148). *Then $Q_p^{\alpha,q}(\mathbb{R}^n) \approx Q_p^{\alpha,q,*}(\mathbb{R}^n)$ with equivalence of norms.*

Proof Let $f \in Q_p^{\alpha,q}(\mathbb{R}^n)$ with $1 \le q < \infty$. Choose φ satisfying (6.148). Then

$$f_{(j)} \equiv \sum_{k=-j}^{j} \varphi_k(D)f \in Q_p^{\alpha,q}(\mathbb{R}^n) \cap C^\infty(\mathbb{R}^n) \tag{6.150}$$

and by the translation invariance of the norm $\|\star\|_{Q_p^{\alpha,q}}$, we have $\|f_{(j)}\|_{Q_p^{\alpha,q}} \lesssim \|f\|_{Q_p^{\alpha,q}}$ and $\|f\|_{Q_p^{\alpha,q,*}} \le \liminf_{j\to\infty}\|f_j\|_{Q_p^{\alpha,q,*}}$. Thus, we may assume that $f \in Q_p^{\alpha,q}(\mathbb{R}^n) \cap C^\infty(\mathbb{R}^n)$ for the purpose of proving $f \in Q_p^{\alpha,q,*}(\mathbb{R}^n)$. Fix a cube $Q \in \mathscr{Q}$. We decompose $\varphi_j(D)f(x) = \varphi_j(D)[f - f(x)](x) \simeq \int_{\mathbb{R}^n} \mathscr{F}^{-1}\varphi_j(x-y)(f(y) - f(x)){\rm d}y$ into two parts: $y \in B(x,\ell(Q))$ and $y \in \mathbb{R}^n \setminus B(x,\ell(Q))$. Along this decomposition, we have

$$\begin{aligned}
&\left(\sum_{j=-\log_2\ell(Q)}^{\infty} 2^{jq\alpha}|\varphi_j(D)f(x)|^q\right)^{\frac{1}{q}}\\
&\lesssim \left(\sum_{j=-\log_2\ell(Q)}^{\infty} 2^{jq\alpha}\left|\int_{B(x,\ell(Q))}\mathscr{F}^{-1}\varphi_j(x-y)(f(y)-f(x)){\rm d}y\right|^q\right)^{\frac{1}{q}}\\
&\quad+\left(\sum_{j=-\log_2\ell(Q)}^{\infty} 2^{jq\alpha}\left|\int_{\mathbb{R}^n\setminus B(x,\ell(Q))}\mathscr{F}^{-1}\varphi_j(x-y)(f(y)-f(x)){\rm d}y\right|^q\right)^{\frac{1}{q}}.
\end{aligned}$$

We fix $x \in \mathbb{R}^n$ for a while. Let $M \gg 1$. As for the first term, we use

$$\sum_{j=-\log_2\ell(Q)}^{\infty}\frac{2^{jq\alpha-jM}}{(2^{-j}+|x-y|)^{n+M}} \le \sum_{j=-\infty}^{\infty}\frac{2^{jq\alpha-jM}}{(2^{-j}+|x-y|)^{n+M}} \sim \frac{1}{|x-y|^{n+q\alpha}}.$$

Inserting this estimate into the first term and using the Hölder inequality, we obtain

$$\left(\sum_{j=-\log_2 \ell(Q)}^{\infty} 2^{jq\alpha}\left|\int_{B(x,\ell(Q))} \mathscr{F}^{-1}\varphi_j(x-y)(f(y)-f(x))\mathrm{d}y\right|^q\right)^{\frac{1}{q}}$$

$$\lesssim \left(\sum_{j=-\log_2 \ell(Q)}^{\infty} 2^{jq\alpha}\int_{B(x,\ell(Q))} \frac{2^{-jM}|f(y)-f(x)|^q}{(2^{-j}+|x-y|)^{n+M}}\mathrm{d}y\right)^{\frac{1}{q}}$$

$$\lesssim \left(\int_{B(x,\ell(Q))} |x-y|^{-n-q\alpha}|f(x)-f(y)|^q\mathrm{d}y\right)^{\frac{1}{q}}.$$

Integrate this estimate over Q against x, we have the desired estimate. Thus, the estimate for the first term is complete.

The estimate for the second term is similar. If $q=\infty$, a similar argument works.

Assume that $f \in Q_p^{\alpha,q,*}(\mathbb{R}^n)$. Going through an argument similar to above using $f_{(j)}$ defined by (6.150), we may assume that $\mathrm{supp}(\mathscr{F}f) \subset \mathbb{R}^n \setminus \{0\}$.

Let $0<\theta<\min(1,p,q)$ be fixed as usual. Fix $y \in \mathbb{R}^n$. We first observe

$$f(\star+y)-f = \sum_{j=-\infty}^{\infty}\left(\tilde{\psi}_j(D)\varphi_j(D)f(\star+y)-\tilde{\psi}_j(D)\varphi_j(D)f\right) \tag{6.151}$$

in $\mathscr{S}'_\infty(\mathbb{R}^n)$, where $\tilde{\psi} \in C^\infty_{\mathrm{c}}(\mathbb{R}^n)$ satisfies $0 \notin \mathrm{supp}(\tilde{\psi})$ and $\sum_{j=-\infty}^{\infty} \tilde{\psi}_j\varphi_j = \chi_{\mathbb{R}^n\setminus\{0\}}$.

We write

$$V_jf(x,y,z) \equiv |\mathscr{F}^{-1}\psi_j(x+y-z)-\mathscr{F}^{-1}\psi_j(x-z)|\cdot|\varphi_j(D)f(z)| \tag{6.152}$$

for $j \in \mathbb{Z}$ and $x,y,z \in \mathbb{R}^n$. Inserting (6.151) and (6.152) into (6.149), we obtain

$$|Q|^{\frac{1}{p}-\frac{1}{q}}\left(\iint_{Q\times B(\ell(Q))} |y|^{-n-q\alpha}|f(x+y)-f(y)|^q\mathrm{d}x\mathrm{d}y\right)^{\frac{1}{q}}$$

$$\lesssim |Q|^{\frac{1}{p}-\frac{1}{q}}\left(\iint_{Q\times B(\ell(Q))}\left(\sum_{j\in\mathbb{Z}}\int_{\mathbb{R}^n} V_jf(x,y,z)\mathrm{d}z\right)^q\frac{\mathrm{d}x\mathrm{d}y}{|y|^{n+q\alpha}}\right)^{\frac{1}{q}}.$$

We now set

$$\mathrm{I} \equiv |Q|^{\frac{1}{p}-\frac{1}{q}}\left(\iint_{Q\times B(\ell(Q))}\left(\sum_{j\in\mathbb{Z}\cap(-\infty,\log_2|y|]}\int_{\mathbb{R}^n} V_jf(x,y,z)\mathrm{d}z\right)^q\frac{\mathrm{d}x\mathrm{d}y}{|y|^{n+q\alpha}}\right)^{\frac{1}{q}},$$

$$\mathrm{II} = |Q|^{\frac{1}{p}-\frac{1}{q}} \left(\iint_{Q \times B(\ell(Q))} \left(\sum_{j \in \mathbb{Z} \cap (\log_2 |y|, \infty)} \int_{\mathbb{R}^n} V_j f(x,y,z) \mathrm{d}z \right)^q \frac{\mathrm{d}x\mathrm{d}y}{|y|^{n+q\alpha}} \right)^{\frac{1}{q}}.$$

Let $y \in B(\ell(Q))$ and $x \in Q$. By the change of variables, we obtain

$$\begin{aligned} &\sum_{j \in \mathbb{Z} \cap (-\infty, \log_2 |y|]} \int_{\mathbb{R}^n} V_j(x,y,z) \mathrm{d}z \\ &= \sum_{j \in \mathbb{Z} \cap (-\infty, \log_2 |y|]} \int_{\mathbb{R}^n} |\mathscr{F}\tilde{\psi}(z + 2^j y) - \mathscr{F}\tilde{\psi}(z)| \cdot |\varphi_j(D) f(x - 2^{-j} z)| \mathrm{d}z. \end{aligned}$$

Using the Plancherel–Polya–Nikolski'i inequality, we obtain

$$\sum_{j \in \mathbb{Z} \cap (-\infty, \log_2 |y|]} \int_{\mathbb{R}^n} V_j(x,y,z) \mathrm{d}z \lesssim |y| \sum_{j \in \mathbb{Z} \cap (-\infty, \log_2 |y|]} 2^j M^{(\theta)}[\varphi_j(D) f](x)$$

due to the restriction on j. Thus, by the Stein-type dual inequality (see Proposition 1.17)

$$\mathrm{I} \lesssim \sup_{Q \in \mathscr{Q}} |Q|^{\frac{1}{p}-\frac{1}{q}} \left(\int_Q \sum_{j=-\log_2 \ell(Q)}^{\infty} 2^{jq\alpha} M^{(\theta)}[\varphi_j(D) f](x)^q \mathrm{d}x \right)^{\frac{1}{q}} \lesssim \|f\|_{Q_p^{\alpha,q,*}}. \tag{6.153}$$

Finally, we deal with II. By the mean-value theorem, we have

$$V_j f(x,y,z) \le \int_0^1 |\mathscr{F}^{-1}\psi_j(x + ty - z)| \cdot |\varphi_j(D) f(z)| \mathrm{d}t.$$

Let $t \in [0,1]$, $x \in Q$ and $y \in B(\ell(Q))$. Then

$$\begin{aligned} &\sum_{j \in \mathbb{Z} \cap (-\infty, \log_2 |y|]} \int_{\mathbb{R}^n} |\mathscr{F}^{-1}\psi_j(x + ty - z)| \cdot |\varphi_j(D) f(z)| \mathrm{d}z \\ &\lesssim \sum_{j \in \mathbb{Z} \cap (-\infty, \log_2 |y|]} M^{(\theta)}[\varphi_j(D) f](x + ty) \\ &\lesssim |y|^{-\alpha} M^{(\theta)} \left[\left(\sum_{j=-\log_2 \ell(Q)}^{\infty} (2^{j\alpha} |\varphi_j(D) f|)^q \right)^{\frac{1}{q}} \right](x + ty). \end{aligned}$$

Thus, inserting this pointwise estimate into II, we obtain

$$\mathrm{II} \lesssim |Q|^{\frac{1}{p}-\frac{1}{q}} \left\| M^{(\theta)} \left[\left(\sum_{j=-\log_2 \ell(Q)}^{\infty} (2^{j\alpha}|\varphi_j(D)f|)^q \right)^{\frac{1}{q}} \right] \right\|_{L^q(3Q)}. \tag{6.154}$$

Thus, if we use $p < \infty$ and Proposition 1.17 (the dual inequality), then we obtain the estimate for II.

Combining (6.153) and (6.154), we obtain the desired result.

We conclude this section with the remark that $A^{s,\tau}_{pq}(\mathbb{R}^n)$ reduces to a little trivial case of the Besov space $B^{s+n\tau-n/p}_{\infty\infty}(\mathbb{R}^n)$ when $\tau > 1/p$.

Theorem 6.37 *Let $0 < p, q \le \infty$ and $s \in \mathbb{R}$. Then $A^{s,\tau}_{pq}(\mathbb{R}^n) \approx B^{s+n\tau-n/p}_{\infty\infty}(\mathbb{R}^n)$ with equivalence of norms whenever $\tau > 1/p$.*

Proof Let $f \in A^{s,\tau}_{pq}(\mathbb{R}^n)$. Then $\|2^{js}\varphi_j(D)f\|_{L^p(Q)} \le \|f\|_{A^{s,\tau}_{pq}}|Q|^\tau$. By the Plancherel–Polya–Nikolski'i inequality (see Corollary 1.8), we have

$$|2^{js}\varphi_j(D)f(x)| \lesssim |Q|^{\tau-1/p}\|f\|_{A^{s,\tau}_{pq}} \quad (x \in \mathbb{R}^n).$$

Thus, $f \in B^{s+n\tau-n/p}_{\infty\infty}(\mathbb{R}^n)$.

Conversely, let $f \in B^{s+n\tau-n/p}_{\infty\infty}(\mathbb{R}^n)$. Then we have

$$2^{js}|\varphi_j(D)f(x)| \le 2^{-jn\tau+jn/p}\|f\|_{B^{s+n\tau-n/p}_{\infty\infty}}.$$

Adding this inequality over $j \in \mathbb{Z} \cap [-\log_2 \ell(Q), \infty)$, we obtain

$$\left(\sum_{j=-\log_2 \ell(Q)}^{\infty} 2^{js}|\varphi_j(D)f(x)|^q \right)^{\frac{1}{q}} \lesssim |Q|^{1/p-\tau}\|f\|_{B^{s+n\tau-n/p}_{\infty\infty}}$$

for all $Q \in \mathscr{D}$. If we integrate this estimate over Q, then we obtain $f \in F^{s,\tau}_{pq}(\mathbb{R}^n)$. Likewise we can prove $f \in B^{s,\tau}_{pq}(\mathbb{R}^n)$.

Exercises

Exercise 6.50 Let $0 < p, q \le \infty$ and $s \in \mathbb{R}$. Using the monotone convergence theorem, prove that $\dot{A}^{s,0}_{pq}(\mathbb{R}^n) \approx \dot{A}^{s}_{pq}(\mathbb{R}^n)$ with coincidence of norms.

Exercise 6.51

1. Prove Lemma 6.23 by reexamining the proof of Lemma 3.16.
2. Using Lemma 6.23, prove Theorems 6.33 and 6.34.

See [120, Corollary 2.1] for an alternate proof of Theorem 6.33.

Exercise 6.52 Let $0 < p, q \le \infty$, $s \in \mathbb{R}$ and $\tau \ge 0$.

1. Let $f \in A_{pq}^{s,\tau}(\mathbb{R}^n)$. Show that $2^{(s+\tau n)j}\|\varphi_j(D)f\|_{L^p(Q)} \lesssim \|f\|_{\dot{A}_{pq}^{s,\tau}}$ for all $Q \in \mathscr{D}_j$.
2. Use Corollary 1.4 to show that $A_{pq}^{s,\tau}(\mathbb{R}^n)$ is complete.

Exercise 6.53 Let $0 < p, q \le \infty$, $s \in \mathbb{R}$ and $\tau \ge 0$. Show that $\mathscr{S}(\mathbb{R}^n) \hookrightarrow A_{pq}^{s,\tau}(\mathbb{R}^n)$ using Theorem 1.56.

Exercise 6.54 Let $0 < p, q \le \infty$, $s \in \mathbb{R}$ and $\tau \ge 0$. By using the definition of the norm $\| \star \|_{\dot{F}_{pq}^{s,\tau}}$, prove that Θ given by (6.145) is bounded.

Exercise 6.55 Let $0 < p \le \infty$, $q \in [1, \infty]$ and $\alpha \in (0, 1)$. Using the Hardy–Littlewood maximal operator and the Stein-type dual inequality, prove that the definition of $Q_p^{\alpha,q,*}(\mathbb{R}^n)$ does not depend on φ satisfying (2.31).

Exercise 6.56 Prove Theorem 6.34 using Lemma 6.23.

Exercise 6.57 Let $0 < p, q \le \infty$, $s \in \mathbb{R}$ and $\tau \ge 0$. Show that the definition of $B_{pq}^{s,\tau}(\mathbb{R}^n)$ and $F_{pq}^{s,\tau}(\mathbb{R}^n)$ does not depend on ψ and φ satisfying (2.31).

6.4.3 Besov–Hausdorff-Type Spaces and Triebel–Lizorkin–Hausdorff Spaces

6.4.3.1 Besov–Hausdorff Spaces and Triebel–Lizorkin–Hausdorff Spaces

We define Besov–Hausdorff spaces and Triebel–Lizorkin–Hausdorff spaces, which will be the dual to Besov-type spaces and Triebel–Lizorkin-type spaces.

Definition 6.24 (Besov–Hausdorff space, Triebel–Lizorkin–Hausdorff space) Let $\varphi \in \mathscr{S}(\mathbb{R}^n)$ satisfy (2.116). Let $p \in (1, \infty)$ and $s \in \mathbb{R}$.

1. If $q \in [1, \infty)$ and $\tau \in [0, (p')^{-1} \vee (q')^{-1}]$, the *Besov–Hausdorff space* $B\dot{H}_{pq}^{s,\tau}(\mathbb{R}^n)$ is then defined to be the set of all $f \in \mathscr{S}'_\infty(\mathbb{R}^n)$ such that

$$\|f\|_{B\dot{H}_{pq}^{s,\tau}} \equiv \inf_{\omega} \left\{ \sum_{j=-\infty}^{\infty} (\|2^{js}\omega_j{}^{-1}\varphi_j(D)f\|_p)^q \right\}^{\frac{1}{q}} < \infty,$$

where ω runs over all nonnegative Borel measurable functions on $\mathbb{R}_+^{n+1}$ such that

$$\int_{\mathbb{R}^n} N\omega(x)^{(p'\wedge q')} \mathrm{d}H^{n\tau(p'\wedge q')}(x) \le 1 \tag{6.155}$$

and that ω_j is allowed to vanish only where $\varphi_j(D)f$ vanishes for any $j \in \mathbb{Z}$.

2. If $q \in (1,\infty)$ and $\tau \in [0,(p')^{-1} \vee (q')^{-1}]$, the *Triebel–Lizorkin–Hausdorff space* $F\dot{H}^{s,\tau}_{pq}(\mathbb{R}^n)$ is then defined to be the set of all $f \in \mathscr{S}'_\infty(\mathbb{R}^n)$ such that

$$\|f\|_{F\dot{H}^{s,\tau}_{pq}} \equiv \inf_{\omega} \left\| \left\{ \sum_{j=-\infty}^{\infty} \frac{|2^{js}\varphi_j(D)f|^q}{\omega(\star,2^{-j})^q} \right\}^{\frac{1}{q}} \right\|_p < \infty,$$

where ω runs over all nonnegative Borel measurable functions on $\mathbb{R}^{n+1}_+$ such that ω satisfies (6.155) and with the restriction that for any $j \in \mathbb{Z}$, $\omega(\star,2^{-j})$ is allowed to vanish only where $\varphi_j(D)f$ vanishes.

3. The symbol $A\dot{H}^{s,\tau}_{pq}(\mathbb{R}^n)$ stands for either $B\dot{H}^{s,\tau}_{pq}(\mathbb{R}^n)$ with $p \in (1,\infty)$, $q \in [1,\infty)$, $s \in \mathbb{R}$ and $\tau \in [0,(p')^{-1} \vee (q')^{-1}]$ or $F\dot{H}^{s,\tau}_{pq}(\mathbb{R}^n)$ with $p \in (1,\infty)$, $q \in (1,\infty)$, $s \in \mathbb{R}$ and $\tau \in [0,(p')^{-1} \vee (q')^{-1}]$.

Remark 6.6 Triebel–Lizorkin–Hausdorff spaces are called Hardy–Hausdorff spaces as in [1152, Definition 5.2].

The definition of $A\dot{H}^{s,\tau}_{pq}(\mathbb{R}^n)$ is independent of the choice of $\varphi \in \mathscr{S}(\mathbb{R}^n)$ satisfying (2.116).

Proposition 6.8 *Let $p \in (1,\infty)$, $q \in [1,\infty)$, $s \in \mathbb{R}$ and $\tau \in [0,(p')^{-1} \vee (q')^{-1}]$. Then the definition of $A\dot{H}^{s,\tau}_{pq}(\mathbb{R}^n)$ does not depend on $\varphi \in \mathscr{S}(\mathbb{R}^n)$ satisfying* (2.116).

Since $p \in (1,\infty)$, a crude estimate using the maximal operator suffices.

Proof Since φ satisfies (2.116), we can find ψ satisfying the same conditions as φ so that $\sum_{j=-\infty}^{\infty} \psi_j\varphi_j = \chi_{\mathbb{R}^n\setminus\{0\}}$. Let $\varphi^\dagger$ be a function satisfying the same conditions as φ. Then $\varphi^\dagger_j(D)f = \sum_{l=-1}^{1} \varphi^\dagger_j(D)\psi_{j+l}(D)\varphi_{j+l}(D)f$. Let ω be a Borel measurable function on $\mathbb{R}^{n+1}_+$ satisfying (6.155) and allowed to vanish only when $\varphi_j(D)f$ vanishes. We define

$$\tilde{\omega}(x,t) \equiv \sup_{\beta\ge 1} \beta^{-N} \left(\sup_{y\in B(x,\beta t)} \omega(y,t) \right), \tag{6.156}$$

where $N \gg 1$. Then

$$\int_{\mathbb{R}^n} N\tilde{\omega}(x)^{n(p'\wedge q')} d^{n\tau(p'\wedge q')}(x) \lesssim_N 1 \tag{6.157}$$

according to Example 1.35. Meanwhile, $\dfrac{|\varphi^\dagger(2^{-j}D)f|}{\tilde{\omega}(\star,2^{-j})} \lesssim \sum_{l=-1}^{1} M\left[\dfrac{\varphi_{j+l}(D)f}{\tilde{\omega}(\star,2^{-j})}\right]$. Using the Fefferman–Stein vector-valued inequality, we obtain the desired result.

6.4.3.2 The φ-Transform for $F\dot{T}^{s,\tau}_{pq}(\mathbb{R}^{n+1}_+)$ and $F\dot{H}^{s,\tau}_{pq}(\mathbb{R}^n)$

We aim to show that the dual of $F\dot{H}^{s,\tau}_{pq}(\mathbb{R}^n)$ is $\dot{F}^{s,\tau}_{pq}(\mathbb{R}^n)$. To this end, we need a technique from functional analysis. We depend on the φ-transform for $F\dot{T}^{s,\tau}_{pq}(\mathbb{R}^{n+1}_+)$ and $F\dot{H}^{s,\tau}_{pq}(\mathbb{R}^n)$. We concentrate on the F-scale. Of course, the B-scale can be handled similarly.

Proposition 6.9 *Let* $p \in (1,\infty)$, $q \in (1,\infty)$, $s \in \mathbb{R}$ *and* $\tau \in [0,(p')^{-1} \vee (q')^{-1}]$. *Choose* $\varphi \in C^\infty_c(\mathbb{R}^n \setminus \{0\})$ *satisfying* (6.143).

1. *The mapping*

$$\pi : F \in F\dot{T}^{s,\tau}_{pq}(\mathbb{R}^{n+1}_+) \mapsto \sum_{j=-\infty}^{\infty} \varphi_j(D)F_j \in F\dot{H}^{s,\tau}_{pq}(\mathbb{R}^n) \tag{6.158}$$

is a bounded linear operator.

2. *The mapping*

$$\Theta : f \in F\dot{H}^{s,\tau}_{pq}(\mathbb{R}^n) \mapsto \Theta f \in F\dot{T}^{s,\tau}_{pq}(\mathbb{R}^{n+1}_+), \tag{6.159}$$

where Θf *is defined by* (6.146), *is a bounded linear operator.*

3. *We have* $\pi \circ \Theta = \mathrm{id}_{F\dot{H}^{s,\tau}_{pq}(\mathbb{R}^n)}$.

Proof We concentrate on (6.158), since (6.159) follows from the definition immediately. Other assertions are easier to prove. We remark that

$$\varphi_j(D)[\pi(F)] = \sum_{l=-N_1}^{N_2} \varphi_j(D)\varphi_{j+l}(D)[F(\star, 2^{-j-l})],$$

where N_1 and N_2 are fixed nonnegative integers that depend only on φ.

We need to verify

$$\begin{aligned}\| \pi(F) \|_{F\dot{H}^{s,\tau}_{pq}} &\equiv \inf_{\omega} \|\{2^{js}{\omega_j}^{-1}\varphi_j(D)[\pi(F)]\}_{j=-\infty}^{\infty}\|_{L^p(\ell^q)} \\ &\lesssim \inf_{\omega} \|\{2^{js}{\omega_j}^{-1}F_j\}_{j=-\infty}^{\infty}\|_{L^p(\ell^q)},\end{aligned} \tag{6.160}$$

where ω moves over all Borel measurable functions on $\mathbb{R}^{n+1}_+$ satisfying (6.155) and ω is allowed to vanish only when $\varphi_j(D)[\pi(F)]$ vanishes.

Let ω be one such function. We define $\tilde{\omega}$ by (6.156). Then we have (6.157). Also, we have

$$\frac{|\varphi_j(D)[\pi(F)]|}{\tilde{\omega}(\star, 2^{-j})} \lesssim M\left[\frac{F_j}{\tilde{\omega}(\star, 2^{-j})}\right]. \tag{6.161}$$

Thus, we are in the position of applying the vector-valued maximal inequality to conclude (6.160).

6.4.3.3 Duality

To show that $\dot{F}^{-s,\tau}_{pq}(\mathbb{R}^n)$ is the dual of $F\dot{H}^{s,\tau}_{pq}(\mathbb{R}^n)$, we use Proposition 4.2.

Combining Propositions 4.2, 6.7 and 6.9, we obtain the following conclusion:

Theorem 6.38 *Let $p, q \in (1, \infty)$ and $s \in \mathbb{R}$. Suppose $0 < \tau \le (p')^{-1} \vee (q')^{-1}$. Then for any $g \in \dot{F}^{s,-\tau}_{p'q'}(\mathbb{R}^n)$, we have $|\langle g, \Phi\rangle| \lesssim \|g\|_{\dot{F}^{-s,\tau}_{p'q'}} \|\Phi\|_{F\dot{H}^{s,\tau}_{pq}}$. In particular, g induces a bounded linear functional L_g on $F\dot{H}^{s,\tau}_{pq}(\mathbb{R}^n)$. Conversely, any bounded linear functional L is realized as $L = L_g$ for some $g \in \dot{F}^{s,-\tau}_{p'q'}(\mathbb{R}^n)$.*

For Besov-type spaces, we have an analogy; see [1152, Theorem 6.1].

Exercises

Exercise 6.58 Let $p, q \in (1, \infty)$, $s \in \mathbb{R}$ and $\tau \in [0, (p')^{-1} \vee (q')^{-1}]$. Show that

$$\mathscr{S}(\mathbb{R}^n) \hookrightarrow B\dot{H}^{s,\tau}_{pq}(\mathbb{R}^n) \hookrightarrow \mathscr{S}'(\mathbb{R}^n).$$

Exercise 6.59 Let $p, q \in (1, \infty)$, $s \in \mathbb{R}$ and $\tau \in [0, (p')^{-1} \vee (q')^{-1}]$. Formulate the counterpart for $B\dot{H}^{s,\tau}_{pq}(\mathbb{R}^n)$ in Proposition 6.9. Note that we have to use (6.161) to this end.

Textbooks in Sect. 6.4

Besov-Type Spaces and Triebel–Lizorkin-Type Spaces

See the textbook [120] for the exhaustive details in the introductory facts on these spaces.

6.5 Weighted Besov Spaces and Triebel–Lizorkin Spaces

We have been showing how to generalize the parameter p. Here we consider how to generalize the parameter s in addition to the parameter p or how to generalize $\psi(D)$ and $\varphi_j(D)$. Section 6.5.1 considers how to generalize the parameter p, which can be viewed as a special case of Sect. 6.3.1. In Sect. 6.5.2 we generalize the parameters p and s. Section 6.5.3 is oriented in a somewhat different direction.

Section 6.5.3 considers what happens if the parameters p, q, s depend on the position x. Sections 6.5.4 and 6.5.5 concern how to generalize $\psi(D)$ and $\varphi_j(D)$.

6.5.1 *Besov Spaces and Triebel–Lizorkin Spaces with A_p-Weights*

If $f \in \mathscr{S}'(\mathbb{R}^n)$ satisfies $|\langle f, \tau\rangle| \lesssim \sum_{|\alpha|\le N} \|\partial^\alpha \tau\|_\infty$ for all $\tau \in \mathscr{S}(\mathbb{R}^n)$ and hence $|\langle f, \tau\rangle| \lesssim \|\varphi\|_{B^{N+1}_{\infty\infty}}$ for $\varphi \in \mathscr{S}(\mathbb{R}^n)$. Thus, if we denote by $b^{N+1}_{\infty\infty}(\mathbb{R}^n)$ the closure of $\mathscr{S}(\mathbb{R}^n)$ in $B^{N+1}_{\infty\infty}(\mathbb{R}^n)$, then $f \in b^{N+1}_{\infty\infty}(\mathbb{R}^n)^* \approx B^{-N-1}_{11}(\mathbb{R}^n)$. Generally, due to the definition of $\mathscr{S}'(\mathbb{R}^n)$, for the function $f \in \mathscr{S}'(\mathbb{R}^n)$ we can find $N \in \mathbb{N}$ such that

$$|\langle f, \tau\rangle| \lesssim \sum_{|\alpha|\le N} \sup_{x\in\mathbb{R}^n} (1+|x|)^N |\partial^\alpha \tau(x)|$$

for all $\tau \in \mathscr{S}(\mathbb{R}^n)$, so for the next step we need the weighted Besov spaces.

6.5.1.1 Weighted Besov Spaces and Weighted Triebel–Lizorkin Spaces

We use the weighted spaces as the underlying function spaces to define weighted Besov spaces or Triebel–Lizorkin spaces.

Definition 6.25 ($A^s_{pq}(w)$) Suppose that the parameters p, q, s and the weight w satisfy

$$0 < p < \infty, \quad 0 < q \le \infty, \quad s \in \mathbb{R}, \quad w \in A_\infty.$$

Let φ and ψ satisfy (2.31). Then define

$$\|f\|_{B^s_{pq}(w)} \equiv \|\psi(D)f\|_{L^p(w)} + \|\{2^{js}\varphi_j(D)f\}_{j=1}^\infty\|_{\ell^q(L^p(w))}, \tag{6.162}$$

$$\|f\|_{F^s_{pq}(w)} \equiv \|\psi(D)f\|_{L^p(w)} + \|\{2^{js}\varphi_j(D)f\}_{j=1}^\infty\|_{L^p(w,\ell^q)} \tag{6.163}$$

for $f \in B^s_{pq}(w)$ and $f \in F^s_{pq}(w)$ respectively. To simplify our formulation, denote by $A^s_{pq}(w)$ either $B^s_{pq}(w)$ or $F^s_{pq}(w)$. We exclude the possibility of $p = \infty$ when $A = F$. Analogously, the space $\dot{A}^s_{pq}(w)$ can be defined.

It counts that we do not have to suppose $w \in A_p$. One of the advantages of introducing weights is that one can handle more functions. One of the typical examples is the function $w(x) = \exp(|x|)$.

Example 6.11 Using the Fourier transform and analytic continuation, we can define $f(x + i\mathbf{e}_1)$ for $f \in L^2(\mathbb{R}^n)$ for example. In this case, we need to consider all $f \in L^2(\mathbb{R}^n)$ such that $g(\xi) \equiv e^{\xi}\mathscr{F}f(\xi), \xi \in \mathbb{R}^n$ belongs to $L^2(\mathbb{R}^n)$.

In this sense, we can define the *local-A_p* class by considering the cubes with volume less than 1 in Definition 6.6, so that we have $A_p^{\rm loc}$. This class characterizes the *local Hardy–Littlewood maximal operator* $M^{\rm loc}$,

$$M^{\rm loc}f(x) \equiv \sup_{0<R<1} m_{B(x,R)}(|f|).$$

Example 6.12 We have $e^{|\star|} \in A_1^{\rm loc}$ and $(1 + |\star|)^N \in A_1^{\rm loc}$ for $N \in \mathbb{R}$. Another example is a measurable function w satisfying

$$0 < w(x) \lesssim \exp(|x-y|^{\beta})w(y) \quad (x, y \in \mathbb{R}^n). \tag{6.164}$$

Analogously to the case of A_p, we can show that

$$A_p^{\rm loc} = \bigcup_{1\le q<p} A_q^{\rm loc} \quad (1 < p \le \infty). \tag{6.165}$$

As is the case with the function spaces with variable exponent, we are led to the idea that the Fefferman–Stein vector-valued inequality is too strong for development of our theory. If we consider such a class of weights, we have to pay attention to the class of distributions. If we consider weights that decay too rapidly in infinity, we have to go outside $\mathscr{S}'(\mathbb{R}^n)$.

Definition 6.26 ($\mathscr{S}_{\rm e}(\mathbb{R}^n)$, $\mathscr{S}_{\rm e}'(\mathbb{R}^n)$) Let $N \in \mathbb{N}$ and $\alpha \in {\mathbb{N}_0}^n$. Write temporarily $\varphi_{(N;\alpha)} \equiv e^{N|\star|}\partial^{\alpha}\varphi$ for $\varphi \in C^{\infty}(\mathbb{R}^n)$. Define $\mathscr{S}_{\rm e}(\mathbb{R}^n)$, as follows:

$$\mathscr{S}_{\rm e}(\mathbb{R}^n) \equiv \bigcap_{N\in\mathbb{N},\alpha\in{\mathbb{N}_0}^n} \left\{\varphi \in C^{\infty}(\mathbb{R}^n) \,:\, \varphi_{(N;\alpha)} \in L^{\infty}(\mathbb{R}^n)\right\}.$$

The class $\mathscr{S}_{\rm e}'(\mathbb{R}^n)$ is the dual of $\mathscr{S}_{\rm e}(\mathbb{R}^n)$.

The definition of $\mathscr{S}_{\rm e}'(\mathbb{R}^n)$ and $\mathscr{S}_{\rm e}(\mathbb{R}^n)$ is similar to those of $\mathscr{S}'(\mathbb{R}^n)$ and $\mathscr{S}(\mathbb{R}^n)$.

Definition 6.27 ($A_{pq}^s(w)$) Suppose that the parameters p, q, s and the weight w satisfy

$$0 < p < \infty, \quad 0 < q \le \infty, \quad s \in \mathbb{R}, \quad w \in A_{\infty}^{\rm loc}.$$

Let $\Psi \in C^{\infty}(\mathbb{R}^n)$ satisfy $\chi_{B(1)} \le \Psi \le \chi_{B(2)}$ and define Φ by $\Phi \equiv \Delta^L\Psi$ for $L \gg 1$. Then define

$$\|f\|_{B^s_{pq}(w)} \equiv \|\Psi * f\|_{L^p(w)} + \|\{2^{js}\Phi^j * f\}_{j=1}^\infty\|_{\ell^q(L^p(w))}, \tag{6.166}$$

$$\|f\|_{F^s_{pq}(w)} \equiv \|\Psi * f\|_{L^p(w)} + \|\{2^{js}\Phi^j * f\}_{j=1}^\infty\|_{L^p(w,\ell^q)} \tag{6.167}$$

for $f \in \mathscr{S}_{\rm e}'(\mathbb{R}^n)$. The spaces $B^s_{pq}(w)$ and $F^s_{pq}(w)$ stand for the set of all the elements $f \in \mathscr{S}_{\rm e}'(\mathbb{R}^n)$ for which the norms $\|f\|_{B^s_{pq}(w)}$ and $\|f\|_{F^s_{pq}(w)}$ are finite, respectively. The space $A^s_{pq}(w)$ stands for either $B^s_{pq}(w)$ or $F^s_{pq}(w)$. The case $p = \infty$ is excluded for $F^s_{pq}(w)$.

Note that $A^s_{pq}(w)$ is a subspace of $\mathscr{S}_{\rm e}'(\mathbb{R}^n)$. Note also that we adopted the local means. This is due to the property of the local maximal operator and the class of $A_p^{\rm loc}$. We do not go into further details. Here we content ourselves with listing the reference in the notes at the end of the chapter.

Exercises

Exercise 6.60 Let $0 < p, q \le \infty$ and $s \in \mathbb{R}$, and let $w \in A_\infty$. Show that the definition of $A^s_{pq}(w)$ is independent of the choice of φ, ψ satisfying (2.31).

Exercise 6.61 Let $w \in A_\infty$. Mimicking the idea of the local means, show that the norms (6.162) and (6.163) are equivalent to (6.166) and (6.167) respectively.

Exercise 6.62 Let $0 < p, q \le \infty$, $s \in \mathbb{R}$ and let $w \in A_\infty$. Establish the theory of atomic decomposition as in Chapter 4.

Exercise 6.63 Suppose that the parameters p, q, s satisfy $0 < p < \infty, 0 < q \le \infty$ and $s \in \mathbb{R}$. Let $w(x) \equiv \langle x \rangle^N$, $x \in \mathbb{R}^n$ with $N \in \mathbb{R}$. Then show that $A^s_{pq}(w) \hookrightarrow \mathscr{S}'(\mathbb{R}^n)$.

Exercise 6.64 Equip $\mathscr{S}_{\rm e}(\mathbb{R}^n)$ with a (natural) topology and show that $\mathscr{S}_{\rm e}(\mathbb{R}^n) \hookrightarrow \mathscr{S}(\mathbb{R}^n)$ mimicking the discussion in Sect. 1.2.2.

6.5.2 *Microlocal Besov Spaces and Triebel–Lizorkin Spaces*

The concept of 2-microlocal analysis or 2-microlocal function spaces dates back to 1974 (see Peetre's book [70]). Later Bony [259] and Jaffard [623] elaborated this concept. Like the spaces of variable exponents, it is an appropriate instrument to describe the local regularity and the oscillatory behavior of functions near singularities. We use a Fourier-analytical approach here; we will consider Littlewood–Paley analysis of distributions.

6.5.2.1 Classes of Weights

We are going toconsiderthe weighted Lebesgue spaces. But the weight is somewhat different from the ones in the class A_p. We introduce the class $\mathscr{W}^{\alpha}_{\alpha_1,\alpha_2}$.

Example 6.13 The most familiar cases, the classical Besov space $B^s_{pq}(\mathbb{R}^n)$ and the Triebel–Lizorkin space $F^s_{pq}(\mathbb{R}^n)$, are realized by letting $w_j \equiv 2^{js}$ with $j \in \mathbb{N}_0$ and $s \in \mathbb{R}$.

Another important class of weights is the class $\mathscr{W}^{\alpha_3}_{\alpha_1,\alpha_2}$, which is given as follows.

Definition 6.28 (Weight class $\mathscr{W}^{\alpha_3}_{\alpha_1,\alpha_2}$) Let $\alpha_1, \alpha_2, \alpha_3 \in [0, \infty)$. The *class* $\mathscr{W}^{\alpha_3}_{\alpha_1,\alpha_2}$ of weights is defined as the set of all the measurable functions $w : \mathbb{R}^{n+1}_+ \to (0, \infty)$ satisfying the following conditions:

1. There exists a positive constant C such that, for all $x \in \mathbb{R}^n$ and $j,\ \nu \in \mathbb{N}_0$ with $j \ge \nu$,

$$2^{-(j-\nu)\alpha_1} w(x, 2^{-\nu}) \lesssim w(x, 2^{-j}) \lesssim 2^{-(\nu-j)\alpha_2} w(x, 2^{-\nu}). \tag{6.168}$$

2. For all $x,\ y \in \mathbb{R}^n$ and $j \in \mathbb{N}_0$

$$w(x, 2^{-j}) \lesssim w(y, 2^{-j}) \langle 2^j (x - y) \rangle^{\alpha_3}. \tag{6.169}$$

Given a weight w and $j \in \mathbb{N}_0$, we write $w_j(x) \equiv w(x, 2^{-j})$ for $x \in \mathbb{R}^n$.

6.5.2.2 Microlocal Besov Spaces and Triebel–Lizorkin Spaces

The 2-microlocal spaces are defined as follows.

Definition 6.29 (Microlocal Besov spaces, Microlocal Triebel–Lizorkin spaces) Let $w \in \mathscr{W}^{\alpha_3}_{\alpha_1,\alpha_2}$ with $\alpha_2 \ge \alpha_1 \ge 0$ and $\alpha_3 \ge 0$, and let ψ and φ satisfy (2.31).

1. Let $0 < p, q \le \infty$. Then for $f \in \mathscr{S}'(\mathbb{R}^n)$ define

$$\|f\|_{B^w_{pq}} \equiv \|w_0 \cdot \psi(D) f\|_p + \|\{w_j \cdot \varphi_j(D) f\}_{j=1}^{\infty}\|_{l_q(L^p)}.$$

   The 2-*microlocal Besov space* $B^w_{pq}(\mathbb{R}^n)$ is the set of all $f \in \mathscr{S}'(\mathbb{R}^n)$ for which $\|f\|_{B^w_{pq}}$ is finite.
2. Let $0 < p < \infty$ and $0 < q \le \infty$. Then for $f \in \mathscr{S}'(\mathbb{R}^n)$ define

$$\|f\|_{F^s_{pq}(w)} \equiv \|w_0 \cdot \psi(D) f\|_p + \|\{w_j \cdot \varphi_j(D) f\}_{j=1}^{\infty}\|_{L^p(l_q)}.$$

   The 2-*microlocal Triebel–Lizorkin space* $F^w_{pq}(\mathbb{R}^n)$ is the set of all $f \in \mathscr{S}'(\mathbb{R}^n)$ for which $\|f\|_{F^w_{pq}}$ is finite.

3. The space $A^w_{pq}(\mathbb{R}^n)$ stands for either $B^w_{pq}(\mathbb{R}^n)$ or $F^w_{pq}(\mathbb{R}^n)$, where the case of $p = \infty$ is not allowed when $A^w_{pq}(\mathbb{R}^n) = F^w_{pq}(\mathbb{R}^n)$.

Analogously, we can define $\dot{B}^w_{pq}(\mathbb{R}^n)$ with $0 < p \le \infty$ and $\dot{F}^w_{pq}(\mathbb{R}^n)$ with $0 < p < \infty$. We omit further details.

Theorem 6.39 *Let $0 < p, q \le \infty$, and let $w \in \mathscr{W}^{\alpha_3}_{\alpha_1,\alpha_2}$ with $\alpha_2 \ge \alpha_1 \ge 0$ and $\alpha_3 \ge 0$. Then the definition of $A^w_{pq}(\mathbb{R}^n)$ does not depend on the choice of ψ and φ satisfying* (2.31).

The proof of Theorem 6.39 hinges upon the modification of Corollary 1.4, which we formulate now.

Corollary 6.2 *For $j \in \mathbb{N}$, $\eta > 0$ and a function $\varphi \in \mathscr{S}_{B(2^j)}(\mathbb{R}^n)$, we have*

$$2^{-j} \sup_{z \in \mathbb{R}^n} \langle 2^j z \rangle^{-\frac{n}{\eta}} |w_j(x-z) \nabla \varphi(x-z)| \tag{6.170}$$

$$\lesssim \sup_{z \in \mathbb{R}^n} \langle 2^j z \rangle^{-\frac{n}{\eta}} |w_j(x-z)\varphi(x-z)| \lesssim_\eta M^{(\eta)}[w_j \varphi](x).$$

The proof of Corollary 6.2 is just the examination of the proof of Corollary 1.4 together with a related estimate, Theorem 1.50. See Exercise 6.65.

Using this type of estimate, we can develop the theory of the atomic decomposition and the wavelet decomposition and so on; see the notes for this chapter.

Example 6.14 Let $0 < p, q \le \infty$, $s, \alpha \in \mathbb{R}$. Using ψ, φ for the space $B^s_{pq}(\mathbb{R}^n)$, we define $\|f\|_{B^{s,\alpha}_{pq}} \equiv \|\psi(D) f\|_p + \left\{ \sum_{j=1}^{\infty} (2^{js} j^{-\alpha} \|\varphi_j(D) f\|_p)^q \right\}^{\frac{1}{q}}$ for $f \in \mathscr{S}'(\mathbb{R}^n)$.
We can show that the function spaces $B^{s,\alpha}_{pq}(\mathbb{R}^n)$ will be defined despite the ambiguity of the choice of admissible functions ψ and φ by reexamining the proof of Theorem 2.1. Analogously we can define $F^{s,\alpha}_{pq}(\mathbb{R}^n)$ and $A^{s,\alpha}_{pq}(\mathbb{R}^n)$. These function spaces can be used to measure the "zero" smoothness.

Exercises

Exercise 6.65 [99, Section 6.3.1] Prove (6.170) by reexamining Corollary 1.4 and Theorem 1.50.

Exercise 6.66

1. Let $w_j(x) \equiv 2^{js}$ for some $s \in \mathbb{R}$ and all $x \in \mathbb{R}^n$. Then show that $w \in \mathscr{W}^0_{ss}$.
2. Let $w \in \mathscr{W}^{\alpha_3}_{\alpha_1 \alpha_2}$ with $\alpha_3 \in \mathbb{R}$ and $\alpha_1 < \alpha_2$, and let $s \in \mathbb{R}$. Let $\tilde{w}_j(x) \equiv 2^{js} w_j(x)$ for $x \in \mathbb{R}^n$ and $j \in \mathbb{N}_0$. Then show that $\tilde{w}$ belongs to the class $\mathscr{W}^{\alpha_3}_{(\alpha_1 - s)_+, (\alpha_2 + s)_+}$.

Exercise 6.67 Let $f \in \mathscr{S}'(\mathbb{R}^n)$. Then using Theorem 4.37 show that the following are equivalent:

1. For all $\psi \in C_c^\infty(\mathbb{R}^n)$, $\psi \cdot f \in A^s_{pq}(\mathbb{R}^n)$.
2. For each open set U, there exists $g_U \in A^s_{pq}(\mathbb{R}^n)$ such that $f|U = g_U|U \in \mathscr{D}'(\mathbb{R}^n)$.

6.5.3 *Function Spaces with Variable Exponents*

Here we allow the smoothness of the functions to vary according to the point. For example, we consider functions such that f is C^2 at one point but f is C^3 at another point.

6.5.3.1 Function Spaces with Variable Exponents

We are now interested in the case where the exponents p, q, s are dependent on the position $x \in \mathbb{R}^n$. Based on the vector-valued norms defined in Sect. 6.1.4, we define Besov spaces and Triebel–Lizorkin spaces with variable exponents.

Definition 6.30 (Variable exponent Besov spaces, Variable exponent Triebel–Lizorkin spaces) Let $p(\star), q(\star) \in C^{\log}(\mathbb{R}^n) \cap \mathscr{P}_0(\mathbb{R}^n)$ and $s(\star) \in C^{\log}(\mathbb{R}^n)$. Let φ and ψ satisfy (2.31).

1. The *variable exponent Besov space* $B^{s(\star)}_{p(\star)q(\star)}(\mathbb{R}^n)$ is the collection of $f \in \mathscr{S}'(\mathbb{R}^n)$ such that

$$\|f\|_{B^{s(\star)}_{p(\star)q(\star)}} \equiv \|\psi(D)f\|_{p(\star)} + \left\| \left\{ 2^{js(\star)}\varphi_j(D)f \right\}_{j=1}^{\infty} \right\|_{\ell^{q(\star)}(L^{p(\star)})} < \infty.$$

2. The *variable exponent Triebel–Lizorkin space* $F^{s(\star)}_{p(\star)q(\star)}(\mathbb{R}^n)$ is the collection of all $f \in \mathscr{S}'(\mathbb{R}^n)$ such that

$$\|f\|_{F^{s(\star)}_{p(\star)q(\star)}} \equiv \|\psi(D)f\|_{p(\star)} + \left\| \left\{ 2^{js(\star)}\varphi_j(D)f \right\}_{j=1}^{\infty} \right\|_{L^{p(\star)}(\ell^{q(\star)})} < \infty.$$

3. As before we write $A^{s(\star)}_{p(\star)q(\star)}(\mathbb{R}^n)$ to denote either $B^{s(\star)}_{p(\star)q(\star)}(\mathbb{R}^n)$ or $F^{s(\star)}_{p(\star)q(\star)}(\mathbb{R}^n)$.

We do not define the homogeneous counterparts.

By combining Exercise 1.100 and Theorems 6.17 and 6.18, we obtain the following conclusion:

Theorem 6.40 *Let $p(\star), q(\star) \in C^{\log}(\mathbb{R}^n) \cap \mathscr{P}_0(\mathbb{R}^n)$ and $s(\star) \in C^{\log}(\mathbb{R}^n)$. Then in the sense of sets the definition of $A^{s(\star)}_{p(\star),q(\star)}(\mathbb{R}^n)$ does not depend on the starting functions ψ and φ satisfying* (2.31).

In Theorem 6.40, if in addition $p(\star)$ and $q(\star)$ are constant, then $A^{s(\star)}_{p(\star)q(\star)}(\mathbb{R}^n)$ is a special case of the microlocal spaces.

Exercises

Exercise 6.68 By reexamining the proof of similar assertions showing that the definition of function spaces is independent of the choice of Φ, prove Theorem 6.40.

Exercise 6.69 Let $p(\star), q(\star) \in C^{\log}(\mathbb{R}^n) \cap \mathscr{P}_0(\mathbb{R}^n)$ and $s(\star) \in C^{\log}(\mathbb{R}^n)$.

1. Show that $A^{s(\star)}_{p(\star)q(\star)}(\mathbb{R}^n) \subset A^{s_-}_{p(\star)\infty}(\mathbb{R}^n) \subset B^{s_-}_{p(\star)\infty}(\mathbb{R}^n) \subset B^{s_- - \frac{n}{p_-}}_{\infty\infty}(\mathbb{R}^n)$.
2. Show that $A^{s(\star)}_{p(\star)q(\star)}(\mathbb{R}^n)$ is complete.

6.5.4 Function Spaces with Mixed Smoothness

We allow the smoothness of the functions to vary according to the variables $x_1, x_2, \ldots, x_n$. For example, we consider functions such that f is differentiable with respect to x_1 but f is twice differentiable with respect to x_2. When we consider function spaces with several variables, we are faced with the situation where we do not need to consider all derivative whose information we have. Here is an example: Let $f = f(x, y) : \mathbb{R}^2 \to \mathbb{R}$ be a $C^2(\mathbb{R}^2)$-function. To approximate $\partial_{xy} f$, we may use

$$\frac{f(\star_1 + h, \star_2 + k) - f(\star_1 + h, \star_2) - f(\star_1, \star_2 + k) + f}{hk}.$$

In this case, we do not have to consider f_{xx}, f_{yy}; $f_{xy}(= f_{yx})$ suffices. So sometimes we need to control some restricted collections of functions. Based on this example, we introduce the function spaces with mixed smoothness.

6.5.4.1 Function Spaces with Mixed Smoothness

We content ourselves with the definition.

Definition 6.31 (Function Spaces with Mixed Smoothness) Let $0 < p, q \le \infty$, $s_1, s_2 \in \mathbb{R}$, and let ψ and φ satisfy (2.31). For $f \in \mathscr{S}'(\mathbb{R}^{2n})$, one defines

$$\|f\|_{B^{s_1,s_2}_{pq}} \equiv \|\psi \otimes \psi(D) f\|_p + \|\{2^{js}\varphi_j \otimes \psi(D) f\}_{j=1}^{\infty}\|_{\ell^q(L^p)}$$

$$+\|\{2^{js}\psi \otimes \varphi_j(D)f\}_{j=1}^{\infty}\|_{\ell^q(L^p)}$$
$$+\|\{2^{(j+k)s}\varphi_j \otimes \varphi_k(D)f\}_{j,k=1}^{\infty}\|_{\ell^q(L^p)}$$

and

$$\|f\|_{F_{pq}^{s_1,s_2}} \equiv \|\psi \otimes \psi(D)f\|_p + \|\{2^{js}\varphi_j \otimes \psi(D)f\}_{j=1}^{\infty}\|_{L^p(\ell^q)}$$
$$+ \|\{2^{js}\psi \otimes \varphi_j(D)f\}_{j=1}^{\infty}\|_{L^p(\ell^q)}$$
$$+ \|\{2^{(j+k)s}\varphi_j \otimes \varphi_k(D)f\}_{j,k=1}^{\infty}\|_{L^p(\ell^q)}.$$

Denote by $B_{pq}^{s_1,s_2}(\mathbb{R}^{2n})$ and $F_{pq}^{s_1,s_2}(\mathbb{R}^{2n})$ the subspaces of $\mathscr{S}'(\mathbb{R}^{2n})$ consisting of all f satisfying $\|f\|_{B_{pq}^{s_1,s_2}} < \infty$ and $\|f\|_{F_{pq}^{s_1,s_2}} < \infty$, respectively. To unify $B_{pq}^{s_1,s_2}(\mathbb{R}^{2n})$ and $F_{pq}^{s_1,s_2}(\mathbb{R}^{2n})$, one defines $A_{pq}^{s_1,s_2}(\mathbb{R}^{2n})$ as usual.

Remark 6.7 One can also define $B_{pq}^{s_1,s_2}(\mathbb{R}^m \times \mathbb{R}^n)$, $F_{pq}^{s_1,s_2}(\mathbb{R}^m \times \mathbb{R}^n)$ and $A_{pq}^{s_1,s_2}(\mathbb{R}^m \times \mathbb{R}^n)$ but the details are omitted here.

Remark 6.8 Although we considered a decomposition $\mathbb{R}^{2n} = \mathbb{R}^n \times \mathbb{R}^n$ in this section, we can generalize the decomposition: If $N = n_1 + n_2 + \cdots + n_k$, then we can use $\mathbb{R}^N = \mathbb{R}^{n_1} \times \mathbb{R}^{n_2} \times \cdots \times \mathbb{R}^{n_k}$. The details are omitted here.

Exercises

Exercise 6.70 Let $s_1, s_2 \in \mathbb{R}$. Then show that $f \otimes g \in B_{\infty,\infty}^{s_1,s_2}(\mathbb{R}^{2n})$ for all $f \in B_{\infty\infty}^{s_1}(\mathbb{R}^n)$ and $g \in B_{\infty\infty}^{s_2}(\mathbb{R}^n)$. Hint: $\psi \otimes \psi(D)[f \otimes g](x, y) = \psi(D)f(x)\psi(D)g(y)$ for $x, y \in \mathbb{R}^n$.

Exercise 6.71 Let $s_1, s_2 > 0$, $k_1, k_2 \in \mathbb{N}_0$ and $0 < p, q \le \infty$. Show that

$$\|f\|_{A_{pq}^{s_1+k_1,s_2+k_2}} \sim \sum_{|\alpha|\le k_1} \sum_{|\beta|\le k_2} \|\partial^{(\alpha,\beta)} f\|_{A_{pq}^{s_1,s_2}}$$

for all $f \in \mathscr{S}'(\mathbb{R}^n)$ mimicking the proof of the boundedness property of the lifting operator.

Exercise 6.72 Let $s_1, s_2 > 0$ and $0 < p, q \le \infty$. Show that

$$A_{pq}^{s_1+s_2}(\mathbb{R}^n) \hookrightarrow A_{pq}^{s_1,s_2}(\mathbb{R}^{2n}) \hookrightarrow A_{pq}^{\min(s_1,s_2)}(\mathbb{R}^n).$$

Exercise 6.73 Let $0 < p < \infty$, $0 < q \le \infty$ and $s > 0$. Show that $\mathrm{Tr}_{\mathbb{R}^n} F_{pq}^{s,1/p}(\mathbb{R}^n) \mapsto B_{pp}^{s}(\mathbb{R}^n)$ is a surjection.

Exercise 6.74 Let $f \in \mathscr{S}'_{Q(r_1)\times Q(r_2)}(\mathbb{R}^n \times \mathbb{R}^n)$ for $r_1, r_2 > 0$. Then show that

$$\sup_{(y_1,y_2)\in\mathbb{R}^n\times\mathbb{R}^n}(1+r_1{}^{-1}|y_1|+r_2{}^{-1}|y_2|)^{-\frac{n}{\eta}}|f(y_1,y_2)| \lesssim M^{(\eta)}f(x_1,x_2)$$

for $(x_1, x_2) \in \mathbb{R}^n \times \mathbb{R}^n$.

Exercise 6.75 Let $0 < p, q \leq \infty$, $s_1, s_2, \rho_1, \rho_2 \in \mathbb{R}$. Then show that $(1 - \Delta^{(1)})^{\rho_1}(1 - \Delta^{(2)})^{\rho_2}$ is an isomorphism from $A_{pq}^{s_1+\rho_1,s_2+\rho_2}(\mathbb{R}^n)$ to $A_{pq}^{s_1,s_2}(\mathbb{R}^n)$.

Exercise 6.76

1. Let $0 < p, q \leq \infty$, $s_1, s_2 \in \mathbb{R}$. If $f_1 \in B_{pq}^{s_1}(\mathbb{R}^n)$ and $f_2 \in B_{pq}^{s_2}(\mathbb{R}^n)$, then show that $f_1 \otimes f_2 \in B_{pq}^{s_1,s_2}(\mathbb{R}^n)$.
2. Let $0 < p, q < \infty$, $s_1, s_2 \in \mathbb{R}$. If $f_1 \in F_{pq}^{s_1}(\mathbb{R}^n)$ and $f_2 \in F_{pq}^{s_2}(\mathbb{R}^n)$, then show that $f_1 \otimes f_2 \in F_{pq}^{s_1,s_2}(\mathbb{R}^n)$.

6.5.5 Anisotropic Function Spaces

6.5.5.1 Dilation Matrix

We are oriented to (homogeneous) anisotropic Besov spaces and (homogeneous) anisotropic Triebel–Lizorkin spaces. Non-homogeneous anisotropic Besov spaces and nonhomogeneous anisotropic Triebel–Lizorkin spaces are defined analogously. The thrust of defining such function spaces is that we want to consider the heat equation $u_t(x,t) - \Delta u(x,t) = 0$. In this heat equation, the time variable t and the space variable x need to be considered separately. So we may ask ourselves whether we can make these variables play equal roles. The anisotropic spaces enable this.

Definition 6.32 (Expansive matrices)

1. A real $n \times n$-matrix A is said to be an *expansive matrix*, sometimes called a *dilation*, if the absolute value of the all eigenvalues exceeds 1.
2. Let A be an expansive matrix. The *quasi-norm adapted to* A is a measurable mapping $\rho_A : \mathbb{R}^n \to [0, \infty)$ satisfying $\rho_A{}^{-1}(0) = \{0\}$, $\rho_A(Ax) = |\det(A)|\rho_A(x)$ and $\rho_A(x+y) \leq H(\rho_A(x)+\rho_A(y))$ for all $x, y \in \mathbb{R}^n$. Here $H \geq 1$ is a constant independent of x and y. If $H = 1$, then ρ_A is said to be a *norm adapted to* A.

Example 6.15 Let $A \equiv \text{diag}(4, 2, 2, \ldots, 2)$. Then

$$\rho_A(x) \equiv \sqrt[2n]{(t+x_1{}^2+x_2{}^2+\cdots+x_n{}^2)^{n+1}} \quad (x = (x_1, x_2, \ldots, x_n) \in \mathbb{R}^n)$$

is a quasi-norm associated with A.

The notion of expansive matrices is natural. In fact, it generalizes the condition $\psi(2^{-k}D)f \to f$ as $k \to \infty$ to the matrix setting.

Example 6.16 Let $A \equiv \begin{pmatrix} 1.2 & 0 \\ -1 & 1.2 \end{pmatrix} = \frac{1}{5}\begin{pmatrix} 6 & 0 \\ -5 & 6 \end{pmatrix}$. Then the eigenvalue of A is 1.2 and 1.2 counted according to multiplicity. But A maps (1, 1) to (1.2, 0.2). So it is not always the case that expansive matrices increases the length of the vector.

6.5.5.2 Anisotropic Besov Spaces and Triebel–Lizorkin Spaces

Keeping in mind that A^j corresponds to 2^j, we define anisotropic function spaces.

Definition 6.33 (Anisotropic function spaces)

1. Choose a function $\varphi \in C_c^\infty(\mathbb{R}^n)$ so that

$$\sum_{j=-\infty}^{\infty} \varphi(A^{-j}\star)^2 \equiv \chi_{\mathbb{R}^n\setminus\{0\}}. \tag{6.171}$$

   Define $\varphi_j^A \equiv \varphi(A^{-j}\star)$.
2. Let $0 < p, q \le \infty$ and $s \in \mathbb{R}$.
   (a) Define $\|f\|_{\dot{B}^s_{pq}(A)} \equiv \|\{2^{js}\varphi_j^A(D)f\}_{j=-\infty}^{\infty}\|_{\ell^q(L^p)}$ for $f \in \mathcal{S}'_\infty(\mathbb{R}^n)$.
   (b) Let $0 < p < \infty$. Define $\|f\|_{\dot{F}^s_{pq}(A)} \equiv \|\{2^{js}\varphi_j^A(D)f\}_{j=-\infty}^{\infty}\|_{L^p(\ell^q)}$ for $f \in \mathcal{S}'_\infty(\mathbb{R}^n)$.
   (c) The notation $\|f\|_{\dot{A}^s_{pq}(A)}$ stands for either one of them. As usual, the case of $p = \infty$ is excluded when $\dot{A}^s_{pq}(A)$ denotes $\dot{F}^s_{pq}(A)$.
3. Let $0 < p, q \le \infty$ and $s \in \mathbb{R}$. The homogeneous anisotropic Besov space $\dot{B}^s_{pq}(A)$ and the homogeneous anisotropic Triebel–Lizorkin space $\dot{F}^s_{pq}(A)$ are defined to be the set of all $f \in \mathcal{S}'_\infty(\mathbb{R}^n)$ for which the quasi-norm $\|f\|_{\dot{B}^s_{pq}(A)}$ and $\|f\|_{\dot{F}^s_{pq}(A)}$ are finite, respectively. But exclude the case where $p = \infty$ for the space $\dot{F}^s_{pq}(\mathbb{R}^n, A)$. The space $\dot{A}^s_{pq}(\mathbb{R}^n, A)$ stands for $\dot{B}^s_{pq}(\mathbb{R}^n, A)$ or $\dot{F}^s_{pq}(\mathbb{R}^n, A)$.

We can define $A^s_{pq}(\mathbb{R}^n, A)$ analogously, which unifies $B^s_{pq}(\mathbb{R}^n, A)$ or $F^s_{pq}(\mathbb{R}^n, A)$. We omit the details.

Exercises

Exercise 6.77 Let $r_1, r_2, \ldots, r_n > 0$. Define

$$E_{(r_1,r_2,\ldots,r_n)} \equiv \left\{ x = (x_1, x_2, \ldots, x_n) \in \mathbb{R}^n : \sum_{j=1}^{n} \frac{{x_j}^2}{{r_j}^2} \le 1 \right\}.$$

Then show that for all $f \in \mathscr{S}'_{E(r_1,r_2,\dots,r_n)}(\mathbb{R}^n)$,

$$\sup_{y\in\mathbb{R}^n} \frac{|f(x-y)|}{(1+r_1{}^{-1}|y_1|+r_2{}^{-1}|y_2|+\cdots+r_n{}^{-1}|y_n|)^{n/\theta}} \lesssim_\theta M^{(\theta)}f(x) \quad (x \in \mathbb{R}^n) \tag{6.172}$$

and that

$$\|f\|_q \lesssim_{pq} (r_1r_2\cdots r_n)^{\frac{1}{p}-\frac{1}{q}}\|f\|_p \tag{6.173}$$

whenever $0 < p \le q \le \infty$. Hint: Rescale Theorem 1.50 to prove (6.172) and reexamine the proof of Corollary 1.6 to prove (6.173).

Exercise 6.78 Let A be a dilation matrix.

1. Let $\Theta \in \mathscr{S}(\mathbb{R}^n)$ chosen so that $\chi_{B(1)} \le \Theta \le \chi_{B(2)}$. Define $\theta \equiv \Theta(A^{-1}\star) - \Theta$. Show that $\sum_{j=-\infty}^{\infty} \theta(A^{-j}\star) = \chi_{\mathbb{R}^n\setminus\{0\}}$.
2. Construct φ satisfying (6.171).
3. Let $f \in \mathscr{S}'_\infty(\mathbb{R}^n)$. Show that $\sum_{j=-\infty}^{\infty} \theta(A^{-j}D)f = f$ in $\mathscr{S}'_\infty(\mathbb{R}^n)$.

Exercise 6.79 Define the *nonhomogeneous anisotropic Besov space* $B^s_{pq}(\mathbb{R}^n, A)$ for $0 < p, q \le \infty$ and $s \in \mathbb{R}$ and the *nonhomogeneous anisotropic Triebel–Lizorkin space* $F^s_{pq}(\mathbb{R}^n, A)$ for $0 < p < \infty$, $0 < q \le \infty$ and $s \in \mathbb{R}$.

Textbooks in Sect. 6.5

Weighted Sobolev Spaces

We refer to [4, Chapter 4] and [51, 52, 72] for weighted Sobolev spaces.

Weighted Besov Spaces and Triebel–Lizorkin Spaces

See [80, Section 5.1] and [99, Chapters 6 and 7] for weighted spaces, where Triebel considered a class of weights similar to $\mathscr{W}^{\alpha}_{\alpha_1,\alpha_2}$. More precisely, [80, Section 5.1] and [99, Chapter 7] for weighted Besov spaces and Triebel–Lizorkin spaces with the weight satisfying (6.164).

Function Spaces of Mixed Smoothness

We refer to [108, Section 1.2] for a more detailed survey of function spaces of mixed smoothness. See the servey for the explanation of the dominating mixed smoothness. The spaces of the dominating mixed smoothness go back to the paper by Nikolski'i [847]. Later Schmeisser considered using Besov spaces and Triebel–Lizorkin spaces; see [932–934]. Later Vibíral obtained the atomic and wavelet decompositions. See [843] for more about the pointwise multiplier on this space.

The space $B^{s,\alpha}_{pq}(\mathbb{R}^n)$ See the papers [36, 319, 370, 710] for the emtropy number and interpolation [249, 371, 373, 378, 453] as well as the paper for [1171] the application to the Navier–Stokes equations.

Anisotropic Function Spaces

See [99, Section 10. 1] and [84].

6.6 Function Spaces on Various Sets

In this book, we have been considering function spaces in Euclidean spaces and on open sets in Euclidean spaces. We can develop a theory of function spaces even on sets. We may get out of the Euclidean spaces to metric measure spaces. Metric (measure) spaces play a prominent role in many fields of mathematics such as probability theory and differential geometry. In particular, they constitute natural generalizations of manifolds admitting all kinds of singularities and still providing rich geometric structure. Here as some typical examples of application of our function spaces, we list function spaces on the torus in Sect. 6.6.1 and on the fractals in Sect. 6.6.2.

6.6.1 Function Spaces on the Torus

Recall that $\mathscr{D}(\mathbb{T}^n)$ stands for the set of all $2\pi\mathbb{Z}^n$-periodic functions in $C^\infty(\mathbb{R}^n)$; see Definition 1.23. The topology of $\mathscr{D}(\mathbb{T}^n)$ is given by the family of semi-norms in Definition 1.23. We remark that

$$\left\| \left(\sum_{j=1}^{\infty} Mf_j{}^q \right)^{\frac{1}{q}} \right\|_{L^p(\mathbb{T}^n)} \lesssim \left\| \left(\sum_{j=1}^{\infty} |f_j|^q \right)^{\frac{1}{q}} \right\|_{L^p(\mathbb{T}^n)} \tag{6.174}$$

for all measurable functions $\{f_j\}_{j=1}^{\infty} \subset L^p(\mathbb{T}^n)$ when $1 < p < \infty$ and $1 < q \leq \infty$. The proof is similar to Theorem 1.49. See Exercise 6.81.

Let $0 < p < \infty$ and $0 < q \le \infty$. Here and below we define

$$\|\{f_j\}_{j=1}^{\infty}\|_{L^p(\ell^q,\mathbb{T}^n)} = \|f_j\|_{L^p(\ell^q,\mathbb{T}^n)} \equiv \left\| \left(\sum_{j=1}^{\infty} |f_j|^q \right)^{\frac{1}{q}} \right\|_{L^p(\mathbb{T}^n)}$$

and

$$\|\{f_j\}_{j=1}^{\infty}\|_{\ell^q(L^p,\mathbb{T}^n)} = \|f_j\|_{\ell^q(L^p,\mathbb{T}^n)} \equiv \left(\sum_{j=1}^{\infty} \|f_j\|_{L^p(\mathbb{T}^n)}{}^q \right)^{\frac{1}{q}}$$

for $\{f_j\}_{j=1}^{\infty} \subset L^p(\mathbb{T}^n)$.

6.6.1.1 Function Spaces on the Torus

Here we will define the nonhomogeneous Besov space $B^s_{pq}(\mathbb{T}^n)$ and the nonhomogeneous Triebel–Lizorkin space $F^s_{pq}(\mathbb{T}^n)$. We do not define the homogeneous Besov space $\dot{B}^s_{pq}(\mathbb{T}^n)$ and the homogeneous Triebel–Lizorkin space $\dot{F}^s_{pq}(\mathbb{T}^n)$. We recall that $\mathscr{D}'(\mathbb{T}^n)$ is the subset of $\mathscr{S}'(\mathbb{R}^n)$ as we have pointed out in Remark 1.3.

Definition 6.34 ($A^s_{pq}(\mathbb{T}^n)$) Let $0 < p, q \le \infty$ and $s \in \mathbb{R}$.

1. Define $\|f\|_{B^s_{pq}(\mathbb{T}^n)} \equiv \|\psi(D)f\|_{L^p(\mathbb{T}^n)} + \|\{2^{js}\varphi_j(D)f\}_{j=1}^{\infty}\|_{\ell^q(L^p,\mathbb{T}^n)}$ for $f \in \mathscr{D}'(\mathbb{T}^n)$. The *nonhomogeneous Besov space* $B^s_{pq}(\mathbb{T}^n)$ is the set of all $f \in \mathscr{D}'(\mathbb{T}^n)$ for which the quasi-norm $\|f\|_{B^s_{pq}(\mathbb{T}^n)}$ is finite.
2. Let $0 < p < \infty$. For $f \in \mathscr{D}'(\mathbb{T}^n)$, define

$$\|f\|_{F^s_{pq}(\mathbb{T}^n)} \equiv \|\psi(D)f\|_{L^p(\mathbb{T}^n)} + \|\{2^{js}\varphi_j(D)f\}_{j=1}^{\infty}\|_{L^p(\ell^q,\mathbb{T}^n)}.$$

   The *nonhomogeneous Triebel–Lizorkin space* $F^s_{pq}(\mathbb{T}^n)$ is the set of all $f \in \mathscr{D}'(\mathbb{T}^n)$ for which the quasi-norm $\|f\|_{F^s_{pq}(\mathbb{T}^n)}$ is finite.
3. The symbol $A^s_{pq}(\mathbb{T}^n)$ is used to denote $B^s_{pq}(\mathbb{T}^n)$ or $F^s_{pq}(\mathbb{T}^n)$. The case $p = \infty$ is excluded when $A^s_{pq}(\mathbb{T}^n) = F^s_{pq}(\mathbb{T}^n)$.

We recall that sinc is a continuous function defined on $\mathbb{R}$ so that

$$\operatorname{sinc}(t) = \frac{\sin t}{t} \quad (t \in \mathbb{R} \setminus \{0\});$$

see Example 1.18. The next theorem will connect the function spaces $A^s_{pq}(\mathbb{R}^n)$ and $A^s_{pq}(\mathbb{T}^n)$.

Theorem 6.41 *Let* $0 < p, q \le \infty$, $s \in \mathbb{R}$, *and let*

$$\tau_0(x) \equiv \prod_{j=1}^{n} \operatorname{sinc}^{2N}(N^{-1}x_j) \quad (x = (x_1, x_2, \dots, x_n) \in \mathbb{R}^n),$$

where $N \gg 1$. *Let* $f \in \mathscr{D}'(\mathbb{T}^n)$. *Then the following are equivalent:*

1. *We have*

$$\tau_0 f \in A^s_{pq}(\mathbb{R}^n). \tag{6.175}$$

2. *For any function* $\tau \in \mathscr{S}(\mathbb{R}^n)$ *such that* $\tau \ge 0$ *and that* $\sum_{k\in\mathbb{Z}^n} \tau(\star - 2\pi k) > 0$,

$$\tau \cdot f \in A^s_{pq}(\mathbb{R}^n). \tag{6.176}$$

3. *We have*

$$f \in A^s_{pq}(\mathbb{T}^n). \tag{6.177}$$

Proof In view of Theorem 4.37, (6.175) and (6.176) are clearly equivalent. We concentrate on Triebel–Lizorkin spaces since the case of Besov spaces is similar. To describe the norms $\| \star \|_{F^s_{pq}(\mathbb{T}^n)}$ and $\| \star \|_{F^s_{pq}(\mathbb{R}^n)}$, let us choose ψ and φ so that

$$\chi_{B(1)} \le \psi \le \chi_{B(2)}, \quad \varphi = \psi - \psi_{-1}.$$

We adopt the following definition of the norms: we define the Triebel–Lizorkin norm $\| \star \|_{F^s_{pq}}$ on $\mathbb{R}^n$ by $\|f\|_{F^s_{pq}} \equiv \|\psi(D)f\|_p + \|\{2^{js}\varphi_j(D)f\}_{j=1}^{\infty}\|_{L^p(\ell^q)}$ and the one on $\mathbb{T}^n$ by $\|f\|_{F^s_{pq}(\mathbb{T}^n)} \equiv \|\psi(D)f\|_{L^p(\mathbb{T}^n)} + \|\{2^{js}\varphi_j(D)f\}_{j=1}^{\infty}\|_{L^p(\ell^q,\mathbb{T}^n)}$. Assume first that (6.177) holds. We decompose $\tau_0 f = \tau_0 \cdot \psi(D)f + \sum_{j=1}^{\infty} \tau_0\varphi_j(D)f$. Thus, if $k \ge 3$, then we deduce from (4.105) $\varphi_k(D)[\tau_0 f] = \sum_{j=k-2}^{k+2} \varphi_k(D)[\tau_0\varphi_j(D)f]$. If we use Corollary 1.4, then we obtain $|\varphi_k(D)[\tau_0 f]| \lesssim \sum_{j=k-2}^{k+2} M^{(\eta)}[\tau_0\varphi_j(D)f]$. Thus, for $k \ge 3$,

$$\left(\sum_{k=3}^{\infty}(2^{ks}|\varphi_k(D)[\tau_0 f]|)^q\right)^{\frac{1}{q}} \lesssim \left(\sum_{j=1}^{\infty}(2^{ks}M^{(\eta)}[\tau_0\varphi_j(D)f])^q\right)^{\frac{1}{q}}.$$

The terms for $k = 1, 2$ and $\psi(D)[\tau_0 f]$ are readily incorporated and we have

$$|\psi(D)[\tau_0 f]| + \left(\sum_{k=1}^{\infty}(2^{ks}|\varphi_k(D)[\tau_0 f]|)^q\right)^{\frac{1}{q}}$$

$$\lesssim M^{(\eta)}[\tau_0 \cdot \psi(D)f] + \left(\sum_{j=1}^{\infty}(2^{ks}M^{(\eta)}[\tau_0\varphi_j(D)f])^q\right)^{\frac{1}{q}}.$$

Thus, if we use the Fefferman–Stein inequality (Theorem 1.49), we have

$$\begin{aligned}\|\tau_0 f\|_{F^s_{pq}} &\lesssim \|\tau_0\psi(D)f\|_p + \|\{2^{js}\tau_0\varphi_j(D)f\}_{j=1}^{\infty}\|_{L^p(\ell^q)} \\ &\lesssim \|\psi(D)f\|_{L^p(\mathbb{T}^n)} + \|\{2^{js}\varphi_j(D)f\}_{j=1}^{\infty}\|_{L^p(\ell^q,\mathbb{T}^n)} \\ &= \|f\|_{F^s_{pq}(\mathbb{T}^n)}.\end{aligned}$$

Assume (6.175). Then we have $\varphi_j(D)f = \sum_{l=j-4}^{j+4}\sum_{m\in\mathbb{Z}^n}\varphi_j(D)[\tau_0(\star - m)\varphi_l(D)f]$ for $j \geq 5$. If we use the Plancherel–Polya–Nikolski'i inequality, then we obtain the desired result.

Exercises

Exercise 6.80 Let $0 < q \leq \infty$ and $s \in \mathbb{R}$. Using the fact that $\|f\|_\infty = \|f\|_{L^\infty(\mathbb{T}^n)}$ for all $f \in C(\mathbb{T}^n)$, show that $B^s_{\infty q}(\mathbb{T}^n) = B^s_{\infty q}(\mathbb{R}^n) \cap \mathscr{D}'(\mathbb{T}^n)$.

Exercise 6.81 By mimicking the proof of Theorem 1.49 prove (6.174).

Exercise 6.82 Let $f \in \mathscr{D}'(\mathbb{T}^n)$.

1. Via the embedding $\mathscr{D}'(\mathbb{T}^n) \hookrightarrow \mathscr{S}'(\mathbb{R}^n)$, prove that $|\langle f, \varphi\rangle| \leq Cp_N(\varphi)$ for all $\varphi \in \mathscr{S}'(\mathbb{R}^n)$.
2. Prove that $f \in B^s_{\infty\infty}(\mathbb{T}^n)$ for some $s \in \mathbb{R}$.

Exercise 6.83 Let M be a compact manifold. Then define the space $A^s_{pq}(M)$ properly for $0 < p, q \leq \infty$ and $s \in \mathbb{R}$. If necessary, justify the definition using Theorems 4.37 and 4.46.

6.6.2 Function Spaces on Fractals

6.6.2.1 Function Spaces on D-Sets

We can consider the function spaces on sets. Here we work within a subset E in $\mathbb{R}^n$. But we can generalize it and we can work on the metric measure spaces. The following geometric structures play important roles.

Definition 6.35 **(D-set)** A *D-set* is a subset Γ of $\mathbb{R}^n$ equipped with a Radon measure μ such that $\mu(B(x,r)\cap\Gamma)\sim r^D$ for all $x\in\Gamma$ and $r\in(0,1)$.

The first problem that arises in this generalized setting is how we give the meaning of the distributions or functions. Here we do not consider this problem in depth. We content ourselves with considering $L^p(\Gamma,\mu)$-functions on Γ with $1\le p\le\infty$.

Theorem 6.42 *Let Γ be a D-set with $0<D\le n$, and let $1\le p\le\infty$. Choose a function $\psi\in\mathscr{S}(\mathbb{R}^n)$ so that $\chi_{B(1)}\le\psi\le\chi_{B(2)}$. Define $\varphi\equiv\psi-\psi_{-1}$. Then the limit*

$$\mathrm{Tr}_{\mathbb{R}^n\to\Gamma}f=f|\Gamma\equiv\psi(D)f|\Gamma+\lim_{J\to\infty}\sum_{j=1}^{J}\varphi_j(D)f|\Gamma \tag{6.178}$$

exists in $L^p(\Gamma,\mu)$ and satisfies

$$\|f|\Gamma\|_{L^p(\Gamma,\mu)}\lesssim\|f\|_{B_{p1}^{\frac{n-D}{p}}} \tag{6.179}$$

for all $f\in B_{p1}^{\frac{n-D}{p}}(\mathbb{R}^n)$.

Proof The proof is simple. Since $1\le p\le\infty$, we have only prove that

$$\|\varphi_j(D)f|\Gamma\|_{L^p(\Gamma,\mu)}\lesssim 2^{\frac{j(n-D)}{p}}\|\varphi_j(D)f\|_p \tag{6.180}$$

with the constant independent of j and f, which also includes the convergence of (6.178). First we partition Γ to have

$$\|\varphi_j(D)f|\Gamma\|_{L^p(\Gamma,\mu)}=\left(\sum_{m\in\mathbb{Z}^n}\int_{\Gamma\cap Q_{jm}}|\varphi_j(D)f(x)|^p\mathrm{d}\mu(x)\right)^{\frac{1}{p}}.$$

Since Γ is a D-set, we have

$$\|\varphi_j(D)f|\Gamma\|_{L^p(\Gamma,\mu)} \le \left(\sum_{m\in\mathbb{Z}^n} \sup_{x\in Q_{jm}} |\varphi_j(D)f(x)|^p \mu(\Gamma\cap Q_{jm})\right)^{\frac{1}{p}}$$

$$\lesssim 2^{-\frac{jD}{p}} \left(\sum_{m\in\mathbb{Z}^n} \sup_{x\in Q_{jm}} |\varphi_j(D)f(x)|^p\right)^{\frac{1}{p}}.$$

By the Plancherel–Polya–Nikolski'i inequality, we obtain

$$\|\varphi_j(D)f|\Gamma\|_{L^p(\Gamma,\mu)} \lesssim 2^{\frac{j(n-D)}{p}} \|M^{(1/2)}[\varphi_j(D)f]\|_p.$$

Finally, using the $L^{2p}(\mathbb{R}^n)$-boundedness we obtain (6.180). If we add (6.180) over $j \in \mathbb{N}$ and further add a similar estimate $\|\psi(D)f|\Gamma\|_{L^p(\Gamma,\mu)} \lesssim \|\psi(D)f\|_p$, then we obtain (6.179).

Let $0 < p, q \le \infty$ and $s > p^{-1}(n-D)$. Based on the embedding $A^s_{pq}(\mathbb{R}^n) \hookrightarrow B^{\frac{n-D}{p}}_{p1}(\mathbb{R}^n)$, we present the following definition:

Definition 6.36 ($A^{s-\frac{n-D}{p}}_{pq}(\Gamma,\mu)$) Let Γ be a D-set together with the associated measure μ. For $1 \le p, q \le \infty$ and $s > p^{-1}(n-D)$, define

$$A^{s-\frac{n-D}{p}}_{pq}(\Gamma,\mu) \equiv \{f \in L^p(\Gamma,\mu) \,:\, f = \mathrm{Tr}_{\mathbb{R}^n\to\Gamma}F \text{ for some } F \in A^s_{pq}(\mathbb{R}^n)\},$$

where $\mathrm{Tr}_{\mathbb{R}^n\to\Gamma} : B^{\frac{n-D}{p}}_{p1}(\mathbb{R}^n) \to L^p(\Gamma)$ is the trace operator in (6.178).

We do not go into the detail of the analysis of this function space. See the reference cited in the notes for this chapter.

6.6.2.2 Hausdorff Measure and Hausdorff Dimension

Here we present examples of the D-sets. The underlying measure will be always the Hausdorff measure.

Definition 6.37 (Hausdorff measure) Let $A \subset \mathbb{R}^n$.

1. A δ-covering of A is the collection of balls having radius less than δ whose union covers A.
2. Let $s \ge 0$. Then define

$$\mathscr{H}^s_\delta(A) \equiv \inf\left\{\sum_{j=1}^{\infty} \omega_s {r_j}^s \,:\, \{B(x_j,r_j)\}_{j=1}^{\infty} \text{ is a } \delta\text{-covering of } A\right\}.$$

Here ω_s is a constant given by $\omega_s \equiv \pi^{\frac{s}{2}} \Gamma\left(\frac{s+2}{2}\right)^{-1}$, where Γ denotes the Gamma function.

3. Let $s \geq 0$. Then define $\mathscr{H}^s(A) \equiv \lim\limits_{\delta \downarrow 0} \mathscr{H}^s_\delta(A)$.
4. The *Hausdorff dimension* of A is given by $\dim_{\mathscr{H}}(A) \equiv \inf\{s \geq 0 : \mathscr{H}^s(A) = 0\}$.

The next theorem is useful when we want to know the Hausdorff dimension of the sets.

Theorem 6.43 *Let $\Gamma \subset \mathbb{R}^n$ be a compact set. Assume that μ is a measure such that $\mu(B(x,r)) \sim r^D$ holds uniformly over $0 < r < 1$ and $x \in \Gamma$. Then $\dim_{\mathscr{H}}(\Gamma) = D$ and the D-dimensional Hausdorff measure of Γ is equivalent to μ in the sense that $\mathscr{H}_D(E \cap \Gamma) \sim \mu(E)$ for all Borel sets E.*

Proof Let $A \subset \Gamma$ be a Borel set. Firstly let us cover A with balls. Suppose that we are given an r-cover $\{B(x_j, r_j)\}_{j=1}^\infty$ of A. Then we have

$$\mu(A) \leq \sum_{j=1}^\infty \mu(A \cap B(x_j, r_j)) \lesssim \sum_{j=1}^\infty {r_j}^D. \tag{6.181}$$

Since the r-cover is arbitrary, this implies that $\mu(A) \lesssim \mathscr{H}_r{}^D(A)$ hence a passage to the limit yields $\mu(A) \lesssim \mathscr{H}^D(A)$.

We prove the reverse inequality. Suppose again that we are given an r-cover $\{B(x_j, r_j)\}_{j=1}^\infty$ of A. By the $5r$-covering lemma we have $x_1, x_2, \ldots$ such that

$$A \subset \bigcup_{j=1}^\infty B(x_j, 5r_j) \text{ and that } \{B(x_j, r_j)\}_{j=1}^\infty \text{ is disjoint.}$$

Conversely, we also have

$$\mathscr{H}_r{}^D(A) \lesssim \sum_{j=1}^\infty {r_j}^D \lesssim \sum_{j=1}^\infty \mu(A \cap B(x_j, r_j)) \leq \mu(A).$$

Thus we obtain the desired assertion.

A measure μ on $\mathbb{R}^n$ is said to be a *Frostman measure* if there exists $D > 0$ such that

$$\mu(B(x,r)) \leq r^D \tag{6.182}$$

for all $r > 0$ and x. If we go through the same argument using (6.181), then we obtain the following partial but useful conclusion.

Corollary 6.3 (Frostman's lemma) *If there exists a Frostman measure* μ *satisfying* (6.182) *on* E *such that* $E \supset \mathrm{supp}(\mu)$, *then* $\dim_{\mathscr{H}}(E) \geq D$.

6.6.2.3 Hausdorff Distance

We want to construct D-sets. To this end, we use the following completeness to guarantee the existence of the underlying set itself, not the measure.

The Hausdorff distance (Hausdorff–Pompeiu distance) serves as a measure of vicinity; we can ask ourselves what the distance between two sets is.

Definition 6.38 (Hausdorff distance, Hausdorff–Pompeiu distance) Denote by $\mathscr{K}(X)$ the set of all compact sets in X.

1. Let $K \in \mathscr{K}(X)$ and $\delta > 0$. The δ-*body* of K, which is denoted by K_δ, is the set of all points in X whose distance from K is less than δ.
2. Define a metric function $d_{\mathscr{K}}$ on $\mathscr{K}(X)$ by

$$d_{\mathscr{K}}(K, L) \equiv \begin{cases} \inf\{\delta > 0 \,:\, L \subset K_\delta,\ K \subset L_\delta\}, & K, L \neq \emptyset, \\ 1, & L \neq K = \emptyset \text{ or } K \neq L = \emptyset, \\ 0, & K = L = \emptyset \end{cases}$$

for $K, L \in \mathscr{K}(X)$. The number $d_{\mathscr{K}}(K, L)$ is called the *Hausdorff distance* between K and L.

We note that $d_{\mathscr{K}}$ is a distance function over $\mathscr{K}(X)$.

Introducing the Hausdorff distance results in two surprising consequences. First we can discuss the sequence of sets $\{A_k\}_{k=1}^\infty$ having the limit A_∞ in a very precise sense. Second, in Theorem 6.45 to follow, some important fractal sets can be shown to exist as the limit of iterated function systems.

Theorem 6.44 *The metric space* $(X, d_{\mathscr{K}})$ *is complete.*

Proof The axiom of the complete metric space is clear other than completeness. We concentrate on completeness of $(X, d_{\mathscr{K}})$.

Let $\{K_j\}_{j=1}^\infty$ be a Cauchy sequence: We need to construct a compact set K to which $\{K_j\}_{j=1}^\infty$ converges. We define

$$K \equiv \left\{ x \in X \,:\, \lim_{j\to\infty} x_j = x, \text{ where } \{x_j\}_{j=1}^\infty \text{ satisfies (6.183) below} \right\}.$$

Here condition (6.183) is:

$$x_1 \in K_{n_1},\ x_2 \in K_{n_2},\ \ldots \text{ for some increasing sequence } n_1 < n_2 < \cdots. \tag{6.183}$$

We claim that K, defined above, is actually compact and that $\{K_j\}_{j=1}^\infty$ tends to K.

We are assuming X is complete. So, to prove that K is compact, it suffices to prove that K is totally bounded and that K is closed.

To prove that K is totally bounded, we take $\varepsilon > 0$. Then there exists $J_0 > 0$ such that $d_{\mathscr{K}}(K_j, K_l) < 3\varepsilon$ for all $j, l \geq J_0$ from the definition of the Cauchy sequence. Consequently, we have

$$K_l \subset (K_{J_0})_{3\varepsilon} \quad (l \geq J_0).$$

In view of the fact that K_{J_0} is compact, we can cover the set K_{J_0} with a finite number of open balls $B_1, B_2, \ldots, B_N$ of radius 4ε. Then the collection $\{4B_1, 4B_2, \ldots, 4B_N\}$ is an open cover of $\overline{\bigcup_{j=J_0}^{\infty} K_j}$. Therefore, we can cover K with N open balls of radius 7ε; hence K is totally bounded.

To prove that K is closed, we take a sequence $\{x_j\}_{j=1}^{\infty}$ in K convergent to $x \in X$. Let $\{x_{j,k}\}_{k=1}^{\infty}$ be a sequence corresponding to x_j such that $x_{j,k} \in K_{l(j,k)}$, where $l(j,1) < l(j,2) < \cdots$. First we take j_1 so that $d(x, x_{j_1}) < 1$ and we choose k_1 and l_1 so large that $d(x_{j_1}, x_{j_1,k_1}) < 1$ with $x_{j_1,k_1} \in K_{l(j_1,k_1)}$. Next, we take $j_2 > j_1$ so that $2d(x, x_2) < 1$ and we choose $k_1 > k_1$ and $l_1 > l_2$ so large that $2d(x_{j_2}, x_{j_2,k_2}) < 1$ with $x_{j_2,k_2} \in K_{l(j_2,k_2)}$ and $l(j_2,k_2) > l(j_1,k_1)$. Repeat this procedure and then we will obtain three increasing sequences $\{j_m\}_{m=1}^{\infty}$, $\{k_m\}_{m=1}^{\infty}$, $\{l_m\}_{m=1}^{\infty}$ of positive integers so that they satisfy $x_{j_m,k_m} \in K_{l_m}$ and $md(x_{j_m,k_m}, x) \leq 2$. Therefore, it follows that $x \in K$.

Consequently, K is compact. It remains to show that K is a limit of the sequence $\{K_j\}_{j=1}^{\infty}$.

Let $\varepsilon > 0$ be fixed. Then there exists $J_0 \in \mathbb{N}$ such that $d_{\mathscr{K}}(K_j, K_l) < \varepsilon$ for all $j, l \geq J_0$. Let $j \geq J_0$ and prove that $d_{\mathscr{K}}(K, K_j) \leq 4\varepsilon$.

Let $x \in K$. Then there exists a sequence $\{x_j\}_{j=1}^{\infty}$ convergent to x with (6.183). If we take l large enough, then $d(x, x_l) < \varepsilon$. Since $d_{\mathscr{K}}(K_j, K_l) < \varepsilon$, we can choose $y_j \in K_j$ so that $d(y_j, x_l) < \varepsilon$. Therefore, we conclude $d(x, y_j) < 2\varepsilon$; hence $x \in (K_j)_{2\varepsilon}$. Conversely, we let $x \in K_j$. Then from the fact that $d_{\mathscr{K}}(K_j, K_l) < \varepsilon$ for all $j, l \geq J_0$ and $\lim_{j,k\to\infty} d_{\mathscr{K}}(K_j, K_l) = 0$, there exists a sequence $\{x_j\}_{j=1}^{\infty}$ convergent to $y \in \overline{B(x, 2\varepsilon)}$ with (6.183) as before. Since $y = \lim_{l\to\infty} x_l \in \overline{\bigcup_{l\geq j} K_l} \subset K_{2\varepsilon}$, we conclude that $K_j \subset K_{4\varepsilon}$. Therefore, it follows that $d_{\mathscr{K}}(K, K_j) \leq 4\varepsilon$ for all $j \geq J_0$.

Hence $\{K_j\}_{j=1}^{\infty}$ converges to K.

Definition 6.39 (Iterating function system, IFS) A system of Lipschitz functions $\{S_j\}_{j=1}^{m}$ is said to be an iterating function system (IFS for short) if each S_j satisfy

$$|S_j(x) - S_j(y)| \leq r|x - y|, \tag{6.184}$$

where $r \in (0, 1)$ is independent of $j = 1, 2, \ldots, n$.

The contraction mapping principle can be traced back to Stefan Banach and we apply this important principle to prove the following theorem:

Theorem 6.45 *Suppose that* $\{S_j\}_{j=1}^m$ *is an IFS as in* (6.184). *Then there is a unique compact set* $E \in \mathcal{K}$ *such that* $E = \bigcup_{j=1}^m S_j(E)$.

Proof For the proof we reformulate the problem: Define a mapping $\Phi : \mathcal{K} \to \mathcal{K}$ by the formula $\Phi(F) \equiv \bigcup_{j=1}^m S_j(F)$. Our task is to show the unique existence of a fixed point in $\mathcal{K}$. Let $x \in K$, $y \in L$. Then from (6.184) we deduce that $d_{\mathcal{K}}(\Phi(K), \Phi(L)) \le r\, d_{\mathcal{K}}(K, L)$. Hence Φ is a contraction and the desired result follows from the Banach fixed point theorem of contractions.

Definition 6.40 (Attractor) The unique set Γ, whose existence is guaranteed in Theorem 6.45, is called an attractor $\{S_j\}_{j=1}^m$ of IFS.

Following the original work of Hutchinson, we introduce notation. Suppose that $\{S_j\}_{j=1}^m$ is an IFS. Let $I = (i_1, i_2, \ldots, i_k)$ with $1 \le i_j \le m$. We write

$$S_I(A) \equiv S_{i_k} \circ S_{i_{k-1}} \circ \cdots \circ S_{i_1}(A)$$

for all $A \in \mathcal{K}$.

As a special case of IFS we will consider similitude functions.

Definition 6.41 (Similitude) A *similitude* is a continuous function S satisfying

$$|S(x) - S(y)| = r|x - y| \quad (x, y \in \mathbb{R}^n)$$

for some $0 < r < 1$.

The following theorem is proved by a geometric observation.

Theorem 6.46 *Suppose that* $S : \mathbb{R}^n \to \mathbb{R}^n$ *is a similitude. That is,*

$$|S(x) - S(y)| = r|x - y| \quad (x, y \in \mathbb{R}^n).$$

Then we can write $S(x) = rAx + b$ *for all* $x \in \mathbb{R}^n$*, where* $r \in (0, \infty)$*,* $b \in \mathbb{R}^n$ *and* $A \in \mathrm{O}(n)$*, where* $\mathrm{O}(n)$ *denotes the set of all orthogonal matrices.*

Proof Simply observe that $S - b$ is a linear map, since

$$S\left(\frac{x+y}{2}\right) = \frac{1}{2}(S(x) + S(y)).$$

Let us apply Theorem 6.44 to obtain compact sets of interest such as the Cantor set, Sierpinski gasket and so on. Below is the general procedure to obtain such an interesting set.

Definition 6.42 (The open set condition) An IFS $\{S_j\}_{j=1}^m$ is said to satisfy the *open set condition* if there exists an open set U such that $\sum_{j=1}^m S_j(U) \subset U$, where the left-hand side indicates the disjoint sum. One calls $\{S_j\}_{j=1}^m$ a similitude IFS.

Here and below let $M(\mathbb{R}^n)$ be the set of all positive finite measures.

Definition 6.43 (V_I) Suppose that a similitude IFS $\{S_j\}_{j=1}^m$ satisfies the open set condition. Define $\mu_k \in M(\mathbb{R}^n)$ for $k \in \mathbb{N}$ in the following way: Let U be an open set associated with the open set condition. Take a nonempty compact set V contained in U. (It may be arbitrary as long as it is contained in U.) Firstly, we temporarily define index sets. Let

$$J_k \equiv \{I = (i_1, i_2, \dots, i_k) \,:\, 1 \le i_1, i_2, \dots, i_k \le m\}$$

and $J \equiv \bigcup_{k=1}^\infty J_k$. Write $V_I \equiv S_I(V)$, where $I = (i_1, i_2, \dots, i_k)$. Let μ_k be a probability Borel measure on $\bigcup_{I\in J_k} V_I$ whose restriction to V_I is $c_n {r_{i_1}}^{s-n} {r_{i_2}}^{s-n} \dots {r_{i_k}}^{s-n} \mathrm{d}x$, where $\mathrm{d}x$ denotes the Lebesgue measure, $I \in J_k$ and c_n is a normalization constant.

Lemma 6.24 *Let $a, b, r > 0$. Suppose that $\{V_i\}_{i\in I}$ is a family of disjoint open sets such that any V_i with $i \in I$ contains a ball with its radius $a \cdot r$ and is contained in a ball with its radius $b \cdot r$. Then any ball $B(x, r)$, $x \in X$, can intersect at most $a^{-n}(1 + b)^n$ open sets of V_i $(i \in I)$.*

Proof Define $I_0 \equiv \{i \in I \,:\, B(x, r) \cap V_i \neq \emptyset\}$ For each $j \in I_0$ there exists a ball $B(x_j, ar) \subset V_j$. Using this ball, we have

$$\sharp I_0 \cdot |B(1)|(a\, r)^n = \sum_{j\in I_0} |B(x_j, ar)| \le \left|\bigcup_{j\in I_0} V_j\right| \le |B(x, br+r)| = |B(1)|(b+1)^n r^n.$$

This implies $\sharp I_0 \le a^{-n}(1 + b)^n$.

Theorem 6.47 *Let $S_j(x) = r_j A_j x + b_j$, $x \in \mathbb{R}^n$, $j = 1, 2, \dots, k$, be the IFS on $\mathbb{R}^n$ as in Theorem 6.46, and let $D > 0$ be a solution to $\sum_{j=1}^k {r_j}^D = 1$. Under the same notation as Definition 6.43, the sequence $\{\mu_m\}_{m=1}^\infty$ converges to a measure*

$\mu \in M(\mathbb{R}^n)$. *Furthermore,* $\mu(E \cap B(x,r)) \sim r^D$ *uniformly over* $x \in E$ *and* $r \in (0,1)$, *where* E *is given in Theorem* 6.45.

Proof For the proof we fix a continuous function f defined on $\mathbb{R}^n$ with compact support and fix $\varepsilon > 0$. Then by uniform continuity of f there exists $\delta > 0$ such that $|f(x) - f(y)| \le \varepsilon$ for all x, y with $|x - y| \le \delta$.

If m is sufficiently large, say $m \ge M$, then we have $\mathrm{diam}(V_I) \le \delta$ for all $I \in J_m$. Using this observation, we have

$$\left|\int_{\mathbb{R}^n} f(x)\mathrm{d}\mu_m(x) - \int_{\mathbb{R}^n} f(x)\mathrm{d}\mu_k(x)\right|$$

$$\le \sum_{I \in J_k} \left| \sum_{(a_1,\dots,a_{m-k}) \in \{1,2,\dots,k\}^{m-k}} \int_{V_{(I,a_1,\dots,a_{m-k})}} f(x)\mathrm{d}\mu_m(x) - \int_{V_I} f(x)\mathrm{d}\mu_k(x)\right|$$

$$< 2\varepsilon$$

for all $k < m$. Thus, the limit $\lim\limits_{m\to\infty} \mu_m$ exists in the weak-* topology, and the first assertion follows.

For the proof of the second assertion we introduce some notation. We fix $x \in E$ and $r \in (0,1)$. We set $r_{\max} \equiv \max(r_1, r_2, \dots, r_m)$ and $r_{\min} \equiv \min(r_1, r_2, \dots, r_m)$.

We select an integer p so that ${r_{\max}}^p < r \le {r_{\max}}^{p-1}$. Define the set J_r by

$$J_r \equiv \{J \equiv (j_1, j_2, \dots, j_q) \,:\, q \le p,\ {r_{\max}}^p < \mathrm{diam}(F_J(O)) \le {r_{\max}}^{p-1}\}.$$

Notice that $\{F_J(O)\}_{J \in J_r}$ satisfies the hypothesis of Lemma 6.24. The number of $J \in J_r$ satisfying $B(x,r) \cap F_J(O) \ne \emptyset$ is majorized by a constant c depending not on r, x but on O. For each $m \gg 1$, we consider J with $\sharp J = m$ and $B(x,r) \cap F_J(O) \ne \emptyset$. We observe

$$\begin{aligned}\mu_m(B(x,r)) &= \sum_{J:\sharp J = m,\, B(x,r)\cap F_J(O) \ne \emptyset} \mu_m(F_J(O) \cap B(x,r)) \\ &\le \sum_{J' \in J_r : F_{J'}(O) \cap B(x,r) \ne \emptyset} \left(\sum_{J:\sharp J = m,\, F_J(O) \subset F_{J'}(O),\, B(x,r) \cap F_J(O) \ne \emptyset} \mu_m(F_J(O)) \right) \\ &\lesssim \sum_{J' \in J_r : F_{J'}(O) \cap B(x,r) \ne \emptyset} r^D \sim r^D.\end{aligned}$$

Thus, we obtain $\mu(B(x,r)) = \liminf\limits_{m\to\infty} \mu_m(B(x,r)) \le \mu_m(F_J(O)) \lesssim r^D$.

It remains to show the estimate below: $\mu(B(x,r)) \gtrsim r^D$ for all $r > 0$ and x. But it is easy. We can find I such that $F_I(E) \subset B(x,r)$. We take such an I minimally in the sense that $\sharp I$ is minimal. Then we claim $2\mathrm{diam}(F_I(E)) \ge r_{\min} r$. Assume otherwise. Decomposing $I = (I', i_n)$, we would have $x \in F_{I'}(E)$ and

$\mathrm{diam}(F_{I'}(E)) \leq \frac{r}{2}$, yielding $F_{I'}(E) \subset B(x, r)$. This is a contradiction to the minimality of I.

Thus, we conclude that $\mu(B(x,r)) \geq \mu(F_I(E)) \gtrsim r^D$.

A Borel measure $\mu \in \mathcal{M}(\mathbb{R}^n)$ is self-similar if there exist an IFS $\{S_1, S_2, \ldots, S_N\}$ and numbers $r_1, r_2, \ldots, r_N \in (0, 1)$ such that $N \geq 2$, $\sum_{j=1}^{N} r_j = 1$ and

$$\mu(E) = \sum_{j=1}^{N} r_j \mu(S_j(E)).$$

It thus follows that the measure obtained in Theorem 6.47 is self-similar.

6.6.2.4 Examples of Attractors

In the examples below it is convenient to identify $\mathbb{R}^2$ with $\mathbb{C}$. We will present examples of attractors having special names. We remark that all of these sets satisfy the open set condition.

Example 6.17 (Cantor set) Set

$$F_1(z) \equiv \frac{1}{3}z, \ F_2(z) \equiv \frac{1}{3}z + \frac{2}{3}. \tag{6.185}$$

The attractor is called the *Cantor set*, which appeared in 1883 [337].

Example 6.18 (Koch curve) Set

$$F_1(z) \equiv \frac{z}{3}, \ F_2(z) \equiv \frac{e^{\frac{\pi}{3}i}}{3}z + \frac{1}{3}, \ F_3(z) \equiv \frac{e^{-\frac{\pi}{3}i}}{3}z + \frac{1}{2} + \frac{i}{2\sqrt{3}}, \ F_4(z) \equiv \frac{z+2}{3}.$$

The attractor is called the *Koch curve*.

Example 6.19 (Sierpinski gasket, Sierpinski triangle) Define $F_1, F_2, F_3 : \mathbb{C} \to \mathbb{C}$ by

$$F_1(z) \equiv \frac{z+p_1}{2}, \ F_2(z) \equiv \frac{z+p_2}{2}, \ F_3(z) \equiv \frac{z+p_3}{2}$$

for distinct points p_1, p_2, p_3. The attractor is defined to be the *Sierpinski gasket*, or the *Sierpinski triangle*.

Let $V_0 = \{p_1, p_2, p_3\}$, and let $V_{j+1} \equiv F_1(V_j) \cup F_2(V_j) \cup F_3(V_j)$ for $j \in \mathbb{N}_0$. Sometimes it is of use to consider each V_j.

Example 6.20 (Cantor dust) Define four functions $F_1, F_2, F_3, F_4 : \mathbb{C} \to \mathbb{C}$ by

$$F_1(z) \equiv \frac{1}{4}z,\ F_2(z) \equiv \frac{1}{4}z + \frac{3}{4},\ F_3(z) \equiv \frac{1}{4}z + \frac{3}{4}i,\ F_4(z) \equiv \frac{1}{4}z + \frac{3}{4}(1+i).$$

The attractor is referred to as the *Cantor dust*.

Example 6.21 (Hata's tree) Define two functions $F_1, F_2 : \mathbb{C} \to \mathbb{C}$ by

$$F_1(z) \equiv \frac{1}{4}e^{\frac{\pi}{3}i}z,\ F_2(z) \equiv \frac{1}{4}e^{-\frac{\pi}{3}i}z + \frac{1}{4}. \tag{6.186}$$

The set $K = \overline{\bigcup_{j=1}^{\infty} K_j}$ is called *Hata's tree*. Here we start from $K_0 = [0, 1]$ and we define $K_{j+1} \equiv F_1(K_j) \cup F_2(K_j)$ inductively.

6.6.2.5 Various Measure Spaces

Here we define a typical measure space on which the theory of function spaces is staged. We content ourselves with the definition of this space and we do not define the function spaces.

A metric measure space (X, d, μ) is a metric space (X, d) equipped with a measure μ.

Definition 6.44 (Space of homogeneous type) A metric measure space (X, d, μ) is said to be a *space of homogeneous type* if there exists a constant $C > 0$ such that $\mu(B(x, 2r)) \le C\mu(B(x, r))$. In this case the measure μ is called a *doubling measure*.

Exercises

Exercise 6.84 Let $I \equiv [0, 1] \subset \mathbb{C}$.

1. Display $\Phi(I)$ and $\Phi(\Phi(I))$, where $\Phi : \mathscr{K}(\mathbb{C}) \to \mathscr{K}(\mathbb{C})$ is given by $\Phi(K) \equiv F_1(K) \cup F_2(K)$ via (6.185).
2. Display $\Phi(I)$ and $\Phi(\Phi(I))$, where $\Phi : \mathscr{K}(\mathbb{C}) \to \mathscr{K}(\mathbb{C})$ is given by $\Phi(K) \equiv F_1(K) \cup F_2(K)$ via (6.186).

Exercise 6.85 In this exercise, we let $X \equiv \mathbb{R}^2$.

1. Let $K \equiv \{x = (x_1, x_2) \in \mathbb{R}^2 \ : \ |x_1| + |x_2| \le 2\}$. Display $K_1 = \bigcup_{x \in K} B(x, 1)$ in the (x_1, x_2)-plane.
2. Let $L \equiv Q(2)$. Then calculate $d_{\mathscr{K}}(K, L)$.

Exercise 6.86 Show that the notion of the n-dimensional Hausdorff measure coincides with the Lebesgue measure.

Exercise 6.87 Consider the Cantor set, the Koch curve, the Sierpinski gasket, the Cantor dust and the Hata tree.

1. Let $I^2 \equiv \{z \in \mathbb{C} : 0 \le \Re(z) \le 1,\ 0 \le \Im(z) \le 1\}$. Then display $\Phi(I^2)$ and $\Phi(\Phi(I^2))$, where $\Phi : \mathscr{K}(\mathbb{C}) \to \mathscr{K}(\mathbb{C})$ is given by $\Phi(K) \equiv F_1(K) \cup F_2(K) \cup F_3(K) \cup F_4(K)$. Here F_3 in the case of Hata's tree and F_4 in the case of the Sierpinski gasket and Hata's tree will be understood as F_1.
2. Calculate the Hausdorff dimension of the Cantor set, the Koch curve, the Sierpinski gasket and the Cantor dust after verifying that the system of contractions satisfies the open set condition.

Exercise 6.88 [62] Let $\mathscr{H}^{n-1}_{S^{n-1}} \in \mathscr{S}'(\mathbb{R}^n)$ be the Hausdorff measure of S^{n-1}; that is,

$$\langle \mathscr{H}^{n-1}_{S^{n-1}}, \varphi \rangle = \int_{S^n} \varphi(\xi) \mathrm{d}\mathscr{H}^{n-1}_{S^{n-1}}(\xi) \quad (\varphi \in \mathscr{S}(\mathbb{R}^n)).$$

1. Show that $\mathscr{H}^{n-1}_{S^{n-1}}$ is nothing but the surface measure of S^{n-1}
2. Show that $\left| \mathscr{F}(\mathscr{H}^{n-1}_{S^{n-1}})(\xi) - C_1 \mathrm{e}^{i|\xi|}|\xi|^{-\frac{n-1}{2}} - C_2 \mathrm{e}^{-i|\xi|}|\xi|^{-\frac{n-1}{2}} \right| \lesssim |\xi|^{-\frac{n}{2}}$ for $\xi \in \mathbb{R}^n \setminus B(1)$ using Lemma 5.25.
3. Suppose that there exists an estimate: $\|\mathscr{F} f\|_{L^p(\mathscr{H}^{n-1}_{S^{n-1}})} \lesssim \|f\|_p$ for $f \in \mathscr{S}(\mathbb{R}^n)$. Then show that $\|\mathscr{F}(\mathscr{H}^{n-1}_{S^{n-1}})\|_{p'} \lesssim 1$ by the duality argument and hence $(n+1)p < 2n$.

Textbooks in Sect. 6.6

Analysis on Metric Measure Spaces

See [26, 38] for elementary facts on analysis in metric spaces. See [49, 1089] for more.

Analysis on Group

See [115].

Hausdorff Measure

See [88, Chapter 7] for the Hausdorff measure together with the example of iterated function systems.

Function Spaces on Sets

The textbook [101] develops what we have considered in Sect. 6.6. In particular, [101, Chapter 7] considers function spaces on manifolds and Lie groups. See also [103, Section 8] See [104, Sections 1.12–1.17] and [103, Section 9] for more about function spaces on sets. The book [38] is a textbook in this field.

Hardy Spaces on Homogeneous Groups

See [27] for Hardy spaces on homogeneous groups.

Hardy Spaces on $\mathbb{R}^n$ with a Measure Satisfying $\mu(B(x,r)) \lesssim r^D$ for Some $0 < D \le n$

See [118] for various function spaces, in particular Hardy spaces.

Upper Gradient

See [39] for the analysis on metric measure spaces using the upper gradient.

6.7 Applications of Function Spaces to the Kato Theorem

One of the aims in analysis is to handle equations that describe some phenomenon. The Laplace operator is one of the fundamental operators. However, we are interested in other operators such as $L = -\Delta + V$, which acts on $L^2(\mathbb{R}^n)$, although we cannot handle operators of this type. As we have seen in this book, the operator Δ can define function spaces. Likewise other operators such as $-\Delta + V$ also define function spaces. Such function spaces match the operator in question. In this section, as an application of the theory of function spaces, we consider the Kato conjecture proved by Auscher, Hofmann, Lacey, McIntosh and Tchamitchian in 2002 and some related facts.

First we formulate the Kato conjecture in Sect. 6.7.1. Section 6.7.2 simplfies and solves the problem. Finally, we investigate some similar assertions for other Lebesgue spaces in Sect. 6.7.3.

6.7.1 Kato Conjecture

We now formulate the Kato conjecture.

6.7.1.1 Formulation of the Kato Conjecture

Let $A(x) = \{a_{ij}(x)\}_{i,j=1,2,\dots,n}$ be a matrix with entries $a_{ij} : \mathbb{R}^n \to \mathbb{C}$ of measurable functions satisfying

$$\text{Ellipticity} \quad \Re\left(\sum_{i,j=1}^{n} a_{ij}(\star)\xi_i\overline{\xi_j}\right) \geq \lambda|\xi|^2, \quad (\xi_1, \xi_2, \dots, \xi_n) \in \mathbb{C}^n, \tag{6.187}$$

$$\text{Boundedness} \quad \left\|\sum_{i=1}^{n}\left|\sum_{j=1}^{n} a_{ij}(\star)\xi_i\right|^2\right\|_\infty \leq \Lambda^2|\xi|^2, \quad (\xi_1, \xi_2, \dots, \xi_n) \in \mathbb{C}^n \tag{6.188}$$

for some $0 < \lambda < \Lambda < \infty$. For a matrix A satisfying (6.187) and (6.188) define the pointwise multiplication $M_A : L^2(\mathbb{R}^n)^n \to L^2(\mathbb{R}^n)^n$ by

$$M_A(\{f_j\}_{j=1,2,\dots,n}) \equiv \left\{\sum_{k=1}^{n} a_{jk} f_k\right\}_{j=1,2,\dots,n}.$$

We also set $\operatorname{div} \equiv -\nabla^*$, the adjoint of ∇. Define the elliptic differential operator L of divergence form by $L \equiv -\operatorname{div} \circ M_A \circ \nabla$. Recall that $\sqrt{L}$ is an accretive operator according to Corollary 5.9. The Kato conjecture asserts the following:

Theorem 6.48 (Kato conjecture, Kato theorem) *Let* λ, Λ *satisfy* (6.187) *and* (6.188). *Then* $\operatorname{Dom}(\sqrt{L}) = H^1(\mathbb{R}^n)$ *and for all* $f \in H^1(\mathbb{R}^n)$ $\|\sqrt{L}f\|_2 \simeq \|\nabla f\|_{(L^2)^n}$.

This theorem was proposed by T. Kato in 1961 and was proved by Auscher, Hofmann, Lacey, McIntosh and Tchamitchian in 2002.

We remark that the following properties hold for any $f \in \operatorname{Dom}(\sqrt{L})$ and $g \in \operatorname{Dom}(L)$:

- $(\mathrm{id}_{L^2} + j^{-1}L)^{-2} f \in \operatorname{Dom}(L^2)$ for all $j \in \mathbb{N}$,
- $(\mathrm{id}_{L^2} + j^{-1}L)^{-2} f \to f$ in $L^2(\mathbb{R}^n)$ as $j \to \infty$,
- in $L^2(\mathbb{R}^n)^n$ as $j \to \infty$

$$\nabla(\mathrm{id}_{L^2} + j^{-1}L)^{-2} g \to \nabla g, \tag{6.189}$$

- $\sqrt{L}(\mathrm{id}_{L^2} + j^{-1}L)^{-2} f \to \sqrt{L} f$ in $L^2(\mathbb{R}^n)$ as $j \to \infty$.

We also have an expression:

$$\sqrt{L}f = \frac{16}{\pi}\int_0^\infty (\mathrm{id}_{L^2}+t^2L)^{-3}t^3L^2 f\frac{\mathrm{d}t}{t} \quad (f \in \mathrm{Dom}(L^2))$$

according to Theorem 5.26. See Exercise 6.89 for (6.189).

For such an L we want to show that

$$\mathrm{Dom}(\sqrt{L}) \supset H^1(\mathbb{R}^n), \quad \|\sqrt{L}f\|_2 \lesssim \|\nabla f\|_{(L^2)^n} \quad (f \in H^1(\mathbb{R}^n)). \tag{6.190}$$

Once (6.190) is proved, we have the same for the adjoint L^*

$$\mathrm{Dom}(\sqrt{L^*}) \supset H^1(\mathbb{R}^n), \quad \|\sqrt{L^*}f\|_2 \lesssim \|\nabla f\|_{(L^2)^n} \quad (f \in H^1(\mathbb{R}^n)) \tag{6.191}$$

according to Proposition 5.11. By duality, we have

$$\mathrm{Dom}(\sqrt{L}) \subset H^1(\mathbb{R}^n), \quad \|\sqrt{L}f\|_2 \gtrsim \|\nabla f\|_{(L^2)^n} \quad (f \in \mathrm{Dom}(\sqrt{L})). \tag{6.192}$$

See Exercise 6.90. Putting (6.190) and (6.192) together, we conclude that the Kato conjecture is true. Thus, it suffices to prove (6.190) to solve the Kato conjecture.

6.7.1.2 A Reformulation Without the Square Root

Let $L = -\mathrm{div} \circ M_A \circ \nabla \in B(L^2(\mathbb{R}^n))$ be an accretive operator as above. According to Theorem 5.27, we have

$$\int_0^\infty \|t^2L^*(\mathrm{id}_{L^2}+t^2L^*)^{-2}g\|_2{}^2\frac{\mathrm{d}t}{t} \lesssim \|g\|_2{}^2 \quad (g \in L^2(\mathbb{R}^n)). \tag{6.193}$$

Thus, once we can show that

$$\int_0^\infty \|tL(\mathrm{id}_{L^2}+t^2L)^{-1}f\|_2{}^2\frac{\mathrm{d}t}{t} \left(= \int_0^\infty \|(\mathrm{id}_{L^2}+t^2L)^{-1}tLf\|_2{}^2\frac{\mathrm{d}t}{t}\right) \lesssim \|\nabla f\|_{(L^2)^n}{}^2 \tag{6.194}$$

for $f \in \mathrm{Dom}(L^2)$, then the Kato conjecture will have been solved. In fact, we note

$$\|\sqrt{L}f\|_2 \lesssim \|\nabla f\|_{(L^2)^n}, \quad (f \in \mathrm{Dom}(L^2)), \tag{6.195}$$

which results from (6.193), (6.194) and

$$\langle\sqrt{L}f, g\rangle_2 = \frac{16}{\pi}\int_0^\infty \langle(\mathrm{id}_{L^2}+t^2L)^{-3}t^3L^2f, g\rangle_2\frac{\mathrm{d}t}{t}.$$

By using the approximation sequence $\{(\mathrm{id}_{L^2}+j^{-1}L)^{-1}f\}_{j=1}^{\infty}$, we can prove

$$\|\sqrt{L}f\|_2 \lesssim \|\nabla f\|_{(L^2)^n}, \quad (f \in \mathrm{Dom}(L^1)) \tag{6.196}$$

from (6.195). For $f \in H^1(\mathbb{R}^n)$ we can approximate it with the elements in $\mathrm{Dom}(L^1)$. Using (6.196) we want to show that $f \in \mathrm{Dom}(\sqrt{L})$ and that $\|\sqrt{L}f\|_2 \lesssim \|\nabla f\|_{(L^2)^n}$ when $f \in H^1(\mathbb{R}^n)$. Namely, let $\{f_j\}_{j=1}^{\infty}$ be a sequence in $\mathrm{Dom}(L^1)$ such that $\lim_{j\to\infty} f_j = f$ in the topology of $H^1(\mathbb{R}^n)$. We can find such a sequence with the help of Lemma 5.14. From (6.196), we deduce

$$\|\sqrt{L}f_j - \sqrt{L}f_k\|_2 \lesssim \|\nabla f_j - \nabla f_k\|_{(L^2)^n},$$

which shows that the limit $g \equiv \lim\limits_{j\to\infty} \sqrt{L}f_j$ exists in $L^2(\mathbb{R}^n)$. Since $\sqrt{L}$ is a closed operator, $f \in \mathrm{Dom}(\sqrt{L})$ and $g = \sqrt{L}f$. Thus,

$$H^1(\mathbb{R}^n) \subset \mathrm{Dom}(\sqrt{L}), \quad \|\sqrt{L}f\|_2 \lesssim \|\nabla f\|_{(L^2)^n} \quad (f \in H^1(\mathbb{R}^n)) \tag{6.197}$$

and (6.190) follows.

6.7.1.3 A Reduction to the Smooth Coefficients

At first glance, the nonsmooth coefficients complicate matters. So, as a preparatory step, we would like to mollify the coefficients. Let $\rho \in C^{\infty}_{\mathrm{c}}(\mathbb{R}^n)$ satisfy $\rho \geq 0$ and $\|\rho\|_1 = 1$. Define $\rho_\varepsilon \equiv \varepsilon^{-n}\rho(\varepsilon^{-1}\star)$ and $a_{ij,\varepsilon} \equiv \rho_\varepsilon * a_{ij}$. Note the $a_{ij,\varepsilon}$ satisfy (6.187) and (6.188). We define $A_\varepsilon \equiv \{a_{ij,\varepsilon}\}_{i,j=1,\ldots,n}$ and $L_\varepsilon \equiv -\mathrm{div} \circ M_{A_\varepsilon} \circ \nabla$. We have

$$\mathrm{Dom}(L_\varepsilon) = H^2(\mathbb{R}^n),$$

as is seen from Exercise 6.91. Suppose that we have shown that

$$\mathrm{Dom}(\sqrt{L_\varepsilon}) = H^1(\mathbb{R}^n), \quad \|\sqrt{L_\varepsilon}f\|_2 \sim_{\lambda,\Lambda} \|\nabla f\|_{(L^2)^n} \quad (f \in H^1(\mathbb{R}^n)) \tag{6.198}$$

for all $\varepsilon > 0$. Letting $\varepsilon \downarrow 0$, we claim

$$\mathrm{Dom}(\sqrt{L}) = H^1(\mathbb{R}^n), \quad \|\sqrt{L}f\|_2 \sim_{\lambda,\Lambda} \|\nabla f\|_{(L^2)^n} \quad (f \in H^1(\mathbb{R}^n)). \tag{6.199}$$

We start by showing that the approximation and the resolvent commute.

Lemma 6.25 *In the strong topology,*

$$\lim_{\varepsilon\downarrow 0}(\mathrm{id}_{L^2}+t^2L_\varepsilon)^{-1} = (\mathrm{id}_{L^2}+t^2L)^{-1};$$

namely, for all $u \in L^2(\mathbb{R}^n)$, $\lim_{\varepsilon \downarrow 0} (\mathrm{id}_{L^2} + t^2 L_\varepsilon)^{-1} u = (\mathrm{id}_{L^2} + t^2 L)^{-1} u$ *in* $L^2(\mathbb{R}^n)$.

Proof Let $u \in L^2(\mathbb{R}^n)$. Then

$$\begin{aligned} &\|(\mathrm{id}_{L^2} + t^2 L_\varepsilon)^{-1} u - (\mathrm{id}_{L^2} + t^2 L)^{-1} u\|_2 \\ &= \sup_{\substack{v \in L^2(\mathbb{R}^n) \\ \|v\|_2 = 1}} t^2 |\langle (M_{A_\varepsilon} - M_A) \nabla (\mathrm{id}_{L^2} + t^2 L)^{-1} u, \nabla (\mathrm{id}_{L^2} + t^2 L_\varepsilon^*)^{-1} v \rangle_{(L^2)^n}| \\ &\le \sup_{\substack{v \in L^2(\mathbb{R}^n) \\ \|v\|_2 = 1}} t^2 \|(M_{A_\varepsilon} - M_A) \nabla (\mathrm{id}_{L^2} + t^2 L)^{-1} u\|_{(L^2)^n} \|\nabla (\mathrm{id}_{L^2} + t^2 L_\varepsilon^*)^{-1} v\|_{(L^2)^n} \end{aligned}$$

by the Cauchy–Schwartz inequality. Since

$$\begin{aligned} \|\nabla (\mathrm{id}_{L^2} + t^2 L_\varepsilon^*)^{-1} v\|_{(L^2)^n} &\lesssim |\langle M_{A_\varepsilon^*} \nabla (\mathrm{id}_{L^2} + t^2 L_\varepsilon^*)^{-1}, \nabla (\mathrm{id}_{L^2} + t^2 L_\varepsilon^*)^{-1} v \rangle_{(L^2)^n}| \\ &\lesssim \|v\|_2 \end{aligned}$$

and $\nabla(\mathrm{id}_{L^2} + t^2 L)^{-1} u \in L^2(\mathbb{R}^n)^n$, we have, by the dominated convergence theorem,

$$\lim_{\varepsilon \downarrow 0} (\mathrm{id}_{L^2} + t^2 L_\varepsilon)^{-1} u = (\mathrm{id}_{L^2} + t^2 L)^{-1} u$$

in $L^2(\mathbb{R}^n)$ for all $u \in L^2(\mathbb{R}^n)$. Thus, Lemma 6.25 is proved.

Using (6.198) and Lemma 6.25 we conclude (6.199), as follows. As we have explained, estimate (6.198) results from

$$\int_0^\infty \|(\mathrm{id}_{L^2} + t^2 L_\varepsilon)^{-1} t L_\varepsilon f\|_2{}^2 \frac{dt}{t} \lesssim \|\nabla f\|_{(L^2)^n}{}^2 \tag{6.200}$$

for all $f \in \mathrm{Dom}(L_\varepsilon) = H^2(\mathbb{R}^n)$. Thanks to Lemma 6.25 and

$$t^2 L_\varepsilon (\mathrm{id}_{L^2} + t^2 L_\varepsilon)^{-1} f = f - (\mathrm{id}_{L^2} + t^2 L_\varepsilon)^{-1} f,$$

$$\lim_{\varepsilon \downarrow 0} \|t L_\varepsilon (\mathrm{id}_{L^2} + t^2 L_\varepsilon)^{-1} f\|_2 = \|t L (\mathrm{id}_{L^2} + t^2 L)^{-1} f\|_2 \quad (f \in H^2(\mathbb{R}^n)). \tag{6.201}$$

Thus, by the Fatou lemma and (6.201), we learn that (6.197) follows.

Consequently, we justified that the coefficients are infinitely differentiable and that they are bounded.

Exercises

Exercise 6.89 Prove (6.189) using (6.187).

Exercise 6.90 Prove (6.192) using (6.191).

Exercise 6.91 Let $A(x) = \{a_{ij}(x)\}_{i,j=1,2,\dots,n}$ be a matrix with entries $a_{ij} : \mathbb{R}^n \to \mathbb{C}$ of smooth functions satisfying (6.187) and (6.188) such that $\partial^\alpha a_{ij} \in L^\infty(\mathbb{R}^n)$ for all $\alpha \in \mathbb{N}_0{}^n$.

1. For $f \in L^2(\mathbb{R}^n)$, define Lf to be an element in $\mathscr{S}'(\mathbb{R}^n)$.
2. Use Theorem 5.18 and an estimate similar to (5.65) to have $\|f\|_{H^2} \sim \|f\|_2 + \|Lf\|_2$ for all $f \in H^2(\mathbb{R}^n)$.
3. Let $f \in L^2(\mathbb{R}^n)$ satisfy $Lf \in L^2(\mathbb{R}^n)$ in the sense that $\langle Lf, \varphi\rangle_2 = \langle g, \varphi\rangle_2$ for some $g \in L^2(\mathbb{R}^n)$. Choose a sequence $\{f_j\}_{j=1}^\infty \subset C^\infty_c(\mathbb{R}^n)$ that converges to f in $H^2(\mathbb{R}^n)$.
   (a) Use the Banach–Alaoglu theorem to show that a subsequence $\{Lf_{j_k}\}_{k=1}^\infty$ of $\{Lf_j\}_{j=1}^\infty$ converges to g in the weak topology of $L^2(\mathbb{R}^n)$; that is,
   $$\lim_{k\to\infty} \langle Lf_{j_k}, \varphi\rangle_2 = \langle g, \varphi\rangle_2$$
   for all $\varphi \in H^2(\mathbb{R}^n)$.
   (b) Prove that $f \in H^2(\mathbb{R}^n)$.

6.7.2 *Kato Conjecture (Kato Theorem): Some Reductions*

6.7.2.1 A Geometric Observation in Complex Hilbert Spaces

Here we prove a geometric estimate.

Lemma 6.26 *Let H be a Hilbert space and $W, \mathring{W} \in H$ be unit vectors. Then for all $U, V \in H$ and $\varepsilon \in (0, 1]$ satisfying*

$$\|U - \langle U, \mathring{W}\rangle_H W\|_H \le \varepsilon|\langle U, \mathring{W}\rangle_H|, \tag{6.202}$$

$$\Re(\langle V, W\rangle_H) \ge \frac{3}{4}, \tag{6.203}$$

$$\|V\|_H \le \frac{1}{4\varepsilon}, \tag{6.204}$$

we have $\|U\|_H \le 4|\langle U, V\rangle_H|$.

We observe that (6.202) implies that U is parallel to W in some sense.

Proof We observe that

$$\|U\|_H \le (1+\varepsilon)|\langle U, \mathring{W}\rangle_H| \le 2|\langle U, \mathring{W}\rangle_H| \tag{6.205}$$

from (6.202) and

$$\frac{3}{4}|\langle U, \overset{\circ}{W}\rangle_H| \le |\langle U, \overset{\circ}{W}\rangle_H \langle V, W\rangle_H| \tag{6.206}$$

from (6.203). Meanwhile, from (6.202) and (6.204),

$$|\langle U, V\rangle_H - \langle U, \overset{\circ}{W}\rangle_H \langle W, V\rangle_H| \le \|V\|_H \|U - \langle U, \overset{\circ}{W}\rangle_H W\|_H \le \frac{1}{4}|\langle U, \overset{\circ}{W}\rangle_H|. \tag{6.207}$$

Hence we conclude that

$$\begin{aligned} |\langle U, V\rangle_H| &\ge |\langle U, \overset{\circ}{W}\rangle_H \langle V, W\rangle_H| - |\langle U, V\rangle_H - \langle U, \overset{\circ}{W}\rangle_H \langle W, V\rangle_H| \\ &\ge \frac{|\langle U, \overset{\circ}{W}\rangle_H|}{2} \\ &\ge \frac{\|U\|_H}{4} \end{aligned}$$

from (6.205), (6.206), and (6.207), as required.

6.7.2.2 Integral Operator with Kernel

One of the main ideas in solving the Kato conjecture is to go beyond the Littlewood–Paley theory. More precisely, the Littlewood–Paley theory handled in this book is not enough since in our current situation we cannot deduce any information from the ellipticity condition. Here as a preparatory step, we discuss some properties of integral operators.

Fix a nonnegative function $p \in C_c^\infty(B(1))$ so that $\|p\|_1 = 1$. Define P_t by

$$P_t f \equiv p_t * f, \quad P_t F \equiv p_t * F, \quad p_t \equiv \frac{1}{t^n} p\left(\frac{\star}{t}\right) \tag{6.208}$$

for $t > 0$, a function f and a vector function F.

Set $\psi \equiv \Delta\psi^0$, where $\psi^0 \in C_c^\infty(B(1)) \setminus \{0\}$. Define

$$Q_s g \equiv \frac{1}{s^n}\psi\left(\frac{\star}{s}\right) * g, \quad Q_s G \equiv \frac{1}{s^n}\psi\left(\frac{\star}{s}\right) * G \tag{6.209}$$

for $g \in L^2(\mathbb{R}^n)$ and $G \in L^2(\mathbb{R}^n)^n$, where the convolution is taken componentwise in the second formula. Write $Q_s^2 = Q_s \circ Q_s$. Then for all $g \in L^2(\mathbb{R}^n)$,

$$\int_0^\infty Q_s^2 g \frac{ds}{s} = \lim_{R\to\infty, \varepsilon\downarrow 0} \int_\varepsilon^R Q_s^2 g \frac{ds}{s} = g$$

in $L^2(\mathbb{R}^n)$.

Let us prove the following lemma:

Lemma 6.27 *Let $m > n$. We assume that the operator $U_t : L^2(\mathbb{R}^n) + L^\infty(\mathbb{R}^n) \to L^2_{\rm loc}(\mathbb{R}^n)$ satisfies the following conditions:*

1. *For each $t > 0$ there exists a kernel $U_t : \mathbb{R}^n \times \mathbb{R}^n \to \mathbb{C}$ such that the uniformly local estimate*

$$\int_{\mathbb{R}^n} \langle t^{-1}(x-y)\rangle^{2m} |U_t(x,y)|^2 {\rm d}x \le t^{-n} \tag{6.210}$$

holds for almost all $y \in \mathbb{R}^n$, that $U_t(x,\star) \in L^2(\mathbb{R}^n)$ for each $x \in \mathbb{R}^n$, and that for all $f \in L^2(\mathbb{R}^n)$

$$U_t f(x) = \int_{\mathbb{R}^n} U_t(x,y) f(y) {\rm d}y \tag{6.211}$$

for almost all $x \in \mathbb{R}^n$ together with the estimate

$$\|U_t f\|_2 \le \|f\|_2 \quad (f \in L^2(\mathbb{R}^n)). \tag{6.212}$$

2. *Let $t > 0$. If we restrict U_t to $L^\infty(\mathbb{R}^n)$, then*

$$m^{(2)}_{B(y,t)}(U_t f) \le \|f\|_\infty \quad (f \in L^2(\mathbb{R}^n),\ y \in \mathbb{R}^n). \tag{6.213}$$

3. *In $L^2_{\rm loc}(\mathbb{R}^n)$*

$$\lim_{R\to\infty} U_t(\chi_{B(R)}) = 0. \tag{6.214}$$

In other words, for all compact sets K,

$$\lim_{R\to\infty} \int_K |U_t(\chi_{B(R)})(x)|^2 \, dx = 0.$$

Then

$$\|U_t \circ P_t \circ Q_s\|_{B(L^2)} \lesssim \sqrt{\min(st^{-1}, s^{-1}t)}. \tag{6.215}$$

One can understand that (6.212) is the global $L^2(\mathbb{R}^n)$-estimate and that (6.210) is the uniform local $L^2(\mathbb{R}^n)$-estimate.

Proof From (6.212) we deduce $\|U_t \circ P_t \circ Q_s\|_{B(L^2)} \le \|P_t \circ Q_s\|_{B(L^2)}$. Thus,

$$\|U_t \circ P_t \circ Q_s\|_{B(L^2)} \lesssim t^{-1}s \tag{6.216}$$

is clear from the structure of P_t and Q_s; see Theorem 1.56 for example or take the Fourier transform of $P_t \circ Q_s f$.

Let us show another estimate. To this end we fix $t > 0$. First of all, we set

$$K_t(x, y) \equiv \int_{\mathbb{R}^n} \overline{U_t(z, x)} U_t(z, y) \mathrm{d}z \quad (x, y \in \mathbb{R}^n).$$

Then

$$\begin{aligned} \langle t^{-1}(x-y)\rangle^m |K_t(x, y)| &\lesssim \int_{\mathbb{R}^n} \langle t^{-1}(x-z)\rangle^m |U_t(z, x)| \cdot \langle t^{-1}(z-y)\rangle^{2m} |U_t(z, y)| \mathrm{d}z \\ &\leq t^{-n}. \end{aligned} \tag{6.217}$$

by the Cauchy–Schwarz inequality and (6.210). We claim that $U_t^* \circ U_t$ has a kernel K_t. Indeed, for all $f, g \in L^2_{\mathrm{c}}(\mathbb{R}^n)$,

$$\langle U_t f, U_t g\rangle_2 = \int_{\mathbb{R}^n} \left(\int_{\mathbb{R}^n} U_t(x, y) f(y) \mathrm{d}y \right) \overline{\left(\int_{\mathbb{R}^n} U_t(x, z) g(z) \mathrm{d}z \right)} \mathrm{d}x$$

and

$$\int_{\mathbb{R}^n} \|U_t(x, \star) \cdot f\|_1 \|U_t(x, \star) \cdot g\|_1 \mathrm{d}x \leq t^{-n} \iint_{\mathbb{R}^n \times \mathbb{R}^n} \frac{|f(y) g(z)|}{\langle t^{-1}(y-z)\rangle^m} \mathrm{d}y \mathrm{d}z < \infty.$$

Thus, we are in the position of using the Fubini theorem to see that $U_t^* \circ U_t$ has a kernel K_t. We set $W_t \equiv U_t^* \circ U_t \circ P_t$. Then if we set

$$W_t(x, y) \equiv \int_{\mathbb{R}^n} K_t(x, z) p_t(z-y) \mathrm{d}z \quad (x, y \in \mathbb{R}^n),$$

then $|W_t(x, y)| \lesssim t^{-n} \langle t^{-1}(y-z)\rangle^{-m}$ for $x, y \in \mathbb{R}^n$ and $t > 0$ from (6.217). Let us verify the following relation:

$$W_t(1) = U_t^* \circ U_t(1) = 0 \tag{6.218}$$

in the sense that $\lim\limits_{R \to \infty} \langle \varphi, W_t(\chi_{B(R)})\rangle = 0$ for all $\varphi \in L^2_{\mathrm{c}}(\mathbb{R}^n)$. Admiting (6.218), let us consider the absolute value of $W_t \circ Q_s$. We notice that the kernel is bounded by the constant times

$$\frac{t}{s} \cdot t^{-n} \langle t^{-1}(x-y)\rangle^{-m}, \quad 0 < t \leq s.$$

Thus

$$\|W_t \circ Q_s\|_{B(L^2)} \lesssim t s^{-1}. \tag{6.219}$$

Putting together (6.216) and (6.219), we obtain (6.215).

It remains to check (6.218). For $\varphi \in L^2_c(\mathbb{R}^n)$, we have

$$\langle W_t(1), \varphi\rangle_2 = \lim_{R\to\infty} \langle [U_t^* \circ U_t](\chi_{B(R)}), \varphi\rangle_2 = \lim_{R\to\infty} \langle U_t(\chi_{B(R)}), U_t\varphi\rangle_2,$$

using $\langle W_t(1), \varphi\rangle_2 = \langle [U_t^* \circ U_t](1), \varphi\rangle_2$.

By the triangle inequality, we have

$$\begin{aligned} |\langle U_t(\chi_{B(R)}), U_t\varphi\rangle_2| &= \left| \iint_{\mathbb{R}^n\times\mathbb{R}^n} U_t(\chi_{B(R)})(x)\overline{U_t(x,y)\varphi(y)}\mathrm{d}y\mathrm{d}x \right| \\ &\le \iint_{\mathbb{R}^n\times\mathbb{R}^n} |U_t(\chi_{B(R)})(x)U_t(x,y)\varphi(y)|\mathrm{d}y\mathrm{d}x. \end{aligned}$$

Here we have used the Fubini theorem in the above. This will be justified later.

From (6.210),

$$\begin{aligned} |\langle U_t(\chi_{B(R)}), U_t\varphi\rangle_2|^2 &\le \iint_{\mathbb{R}^n\times\mathbb{R}^n} \frac{|U_t(\chi_{B(R)})(x)|^2|\varphi(y)|}{\langle t^{-1}(x-y)\rangle^{2m}}\mathrm{d}y\mathrm{d}x \\ &\quad\times \iint_{\mathbb{R}^n\times\mathbb{R}^n} \langle t^{-1}(x-y)\rangle^{2m}|U_t(x,y)|^2|\varphi(y)|\mathrm{d}y\mathrm{d}x \\ &\le \|\varphi\|_1 \iint_{\mathbb{R}^n\times\mathbb{R}^n} \frac{|U_t(\chi_{B(R)})(x)|^2|\varphi(y)|}{\langle t^{-1}(x-y)\rangle^{2m}}\mathrm{d}y\mathrm{d}x. \end{aligned}$$

We want to decompose the last integral. By decomposing $\mathbb{R}^n$ dyadically, we have

$$\begin{aligned} &\iint_{\mathbb{R}^n\times\mathbb{R}^n} \frac{|U_t(\chi_{B(R)})(x)|^2|\varphi(y)|t^{2m}}{(t+|x-y|)^{2m}}\mathrm{d}y\mathrm{d}x \\ &\lesssim \sum_{k=0}^{\infty} \frac{1}{2^{2km}} \iint_{|x-y|<2^k t} |U_t(\chi_{B(R)})(x)|^2|\varphi(y)|\mathrm{d}y\mathrm{d}x. \end{aligned}$$

Note that there exists $M_\varphi > 1$ depending on φ such that $\mathrm{supp}(\varphi) \subset B(M_\varphi)$. Thus,

$$\begin{aligned} &\frac{1}{2^{2km}} \iint_{|x-y|<2^k t} |U_t(\chi_{B(R)})(x)|^2|\varphi(y)|\mathrm{d}y\mathrm{d}x \\ &\lesssim \frac{1}{2^{2km}} \iint_{B(2^k t+M_\varphi)\times\mathbb{R}^n} |U_t(\chi_{B(R)})(x)|^2|\varphi(y)|\mathrm{d}y\mathrm{d}x \\ &= \frac{1}{2^{2km}}\|\varphi\|_1 \int_{B(2^k t+M_\varphi)} |U_t(\chi_{B(R)})(x)|^2\mathrm{d}x. \end{aligned}$$

Using the covering argument and the uniformly local estimate (6.210), we can show that

$$\int_{B(2^k t+M_\varphi)} |U_t(\chi_{B(R)})(x)|^2 \mathrm{d}x \lesssim (2^k t + M_\varphi)^n.$$

Thus, since φ has compact support, we can use the Lebesgue convergence theorem (with respect to the variable k) to have

$$\lim_{R\to\infty} \iint_{\mathbb{R}^n\times\mathbb{R}^n} \frac{|U_t(\chi_{B(R)})(x)|^2|\varphi(y)|}{\langle t^{-1}(x-y)\rangle^{2m}} \mathrm{d}y\mathrm{d}x = 0.$$

Thus, (6.218) follows.

6.7.2.3 Applications of Gaffney-Type Estimates to Commutators

Having simplified matters to a large extent, we now consider the Kato problem. We now investigate the properties of the elliptic differential operators.

Denote by M_h the pointwise multiplication by a function h.

Lemma 6.28 *Let $t>0$ and L be an operator in the Kato conjecture. Let $h:\mathbb{R}^n\to\mathbb{C}$ be a Lipschitz function which is also C^∞. Then*

$$\|[(\mathrm{id}_{L^2}+t^2L)^{-1}, M_h]\|_{B(L^2)} \lesssim t\|\nabla h\|_{(L^\infty)^n}, \tag{6.220}$$

$$\|\nabla[(\mathrm{id}_{L^2}+t^2L)^{-1}, M_h]\|_{B(L^2,(L^2)^n)} \lesssim \|\nabla h\|_\infty. \tag{6.221}$$

Proof We decompose the commutator:

$$[(\mathrm{id}_{L^2}+t^2L)^{-1}, M_h] = t^2(\mathrm{id}_{L^2}+t^2L)^{-1}\circ(M_hL-LM_h)\circ(\mathrm{id}_{L^2}+t^2L)^{-1}.$$

Since

$$LM_hf = \mathrm{div}\left(\left(\sum_{k=1}^n a_{j,k}\frac{\partial h}{\partial x_k}\right)_{j=1,\ldots,n}\right)\cdot f + 2\sum_{j,k=1}^n a_{j,k}\frac{\partial h}{\partial x_j}\frac{\partial f}{\partial x_k} + M_hLf,$$

we have

$$\begin{aligned}&(\mathrm{id}_{L^2}+t^2L)[(\mathrm{id}_{L^2}+t^2L)^{-1}, M_h]\\ &= -t\left\{t\mathrm{div}[(\mathrm{id}_{L^2}+t^2L)^{-1}[\star]\cdot M_A\nabla h]\right\} - t\left\{(M_A\nabla h)\cdot t\mathrm{div}(\mathrm{id}_{L^2}+t^2L)^{-1}\right\}\\ &\quad - 2tM_{{}^tA\nabla h}\circ[t\nabla(\mathrm{id}_{L^2}+t^2L)^{-1}].\end{aligned}$$

Thus, $[(\mathrm{id}_{L^2}+t^2L)^{-1}, M_h]$ is $L^2(\mathbb{R}^n)$-bounded and its operator norm is less than or equal to $Ct\|\nabla h\|_{(L^\infty)^n}$, proving (6.220). From this calculation $\nabla[(\mathrm{id}_{L^2}+t^2L)^{-1}, M_h]$ is also $L^2(\mathbb{R}^n)$-bounded and its operator norm is less than or equal to $C\|\nabla h\|_{(L^\infty)^n}$, proving (6.221).

6.7.2.4 $(\mathrm{id}_{L^2}+t^2L)^{-1}$ Acting on $C^\infty(\mathbb{R}^n)\cap\mathrm{Lip}(\mathbb{R}^n)$

As another application of the Gaffney-type estimate, we obtain the following approximation property:

Lemma 6.29 *Let Q be a cube, let $0<t\le\ell(Q)$ and let $f\in C^\infty(\mathbb{R}^n)\cap\mathrm{Lip}(\mathbb{R}^n)$. Let $\psi\in C^\infty_{\mathrm{c}}([-2,2]^n)$ satisfy $\sum_{k\in\mathbb{Z}^n}\psi_k\equiv 1$, where $\psi_k\equiv\psi(\star-k)$. Then*

$$(\mathrm{id}_{L^2}+t^2L)^{-1}f\equiv\sum_{k\in\mathbb{Z}^n}(\mathrm{id}_{L^2}+t^2L)^{-1}[\psi_k f]$$

converges locally in $L^2(\mathbb{R}^n)$ and the sum does not depend on the choice of ψ. Furthermore,

$$m_Q^{(2)}(f-(\mathrm{id}_{L^2}+t^2L)^{-1}f)\lesssim_{n,\lambda,\Lambda} t\|\nabla f\|_{(L^\infty)^n}, \tag{6.222}$$

$$m_Q^{(2)}(|\nabla f-\nabla(\mathrm{id}_{L^2}+t^2L)^{-1}f|)\lesssim_{n,\lambda,\Lambda}\|\nabla f\|_{(L^\infty)^n}. \tag{6.223}$$

Proof We omit the proof of the convergence of the sum defining $(\mathrm{id}_{L^2}+t^2L)^{-1}f$ and the independence of the choice of ψ: This is straightforward from the Gaffney estimate. See Exercise 6.93.

A dilation and translation allows us to suppose that $Q=Q(2)$. Let $k\in\mathbb{Z}^n\setminus\{0\}$. The case of $k=0$ is readily incorporated later. We note that

$$|k|\cdot\|(\mathrm{id}_{L^2}+t^2L)^{-1}[\psi_k(f-f(k))]\|_{L^2(Q)}\lesssim\|\nabla f\|_{(L^\infty)^n} \tag{6.224}$$

thanks to the Gaffney estimate. Then

$$\begin{aligned}&\|(\mathrm{id}_{L^2}+t^2L)^{-1}[\psi_k(f-f(k))]\|_{L^2(Q)}\\&\lesssim\|[(\mathrm{id}_{L^2}+t^2L)^{-1},M_{f-f(k)}]\psi_k\|_2+|k|\cdot\|\nabla f\|_{(L^\infty)^n}\|(\mathrm{id}_{L^2}+t^2L)^{-1}\psi_k\|_{L^2(Q)}.\end{aligned}$$

If we use Lemma 5.29, then we have a, which depends on λ,Λ,n, such that

$$\begin{aligned}&\|(\mathrm{id}_{L^2}+t^2L)^{-1}[\psi_k(f-f(k))]\|_{L^2(Q)}\\&\lesssim\|[(\mathrm{id}_{L^2}+t^2L)^{-1},M_{f-f(k)}]\psi_k\|_2+|k|\cdot\|\nabla f\|_{(L^\infty)^n}\exp\left(-\frac{8a|k|}{t}\right)\|\psi_k\|_{L^2(9Q_{k_0})}\end{aligned}$$

$$\lesssim |k| \cdot \|\nabla f\|_{(L^\infty)^n} \exp\left(-\frac{8a|k|}{t}\right) \|\psi_k\|_{L^2(9Q_{k_0})}.$$

Thus, since $t \le \ell(Q) = 4$,

$$\begin{aligned}\|(\mathrm{id}_{L^2} + t^2 L)^{-1}[\psi_k(f - f(k))]\|_{L^2(Q)} &\lesssim |k| \cdot \|\nabla f\|_{(L^\infty)^n} \exp\left(-\frac{8a|k|}{t}\right) \\ &\lesssim t\|\nabla f\|_{(L^\infty)^n} \exp\left(-\frac{4a|k|}{t}\right) \\ &\lesssim t\|\nabla f\|_{(L^\infty)^n} \exp(-a|k|).\end{aligned}$$

If we add this estimate over $k \in \mathbb{Z}^n$, we obtain (6.222). We can prove (6.223) similarly; see Exercise 6.92.

We apply the result as follows. For a cube Q, $\varepsilon \in (0, 1)$ and a vector $w \in \mathbb{C}^n$ we define

$$f^\varepsilon_{Q,w} \equiv (\mathrm{id}_{L^2} + \varepsilon^2 \ell(Q)^2 L)^{-1}(\langle \Phi_Q(\star), w\rangle),$$

where $\Phi_Q(x) \equiv x - c(Q)$ for $x \in \mathbb{R}^n$ and $\langle \star_1, \star_2\rangle$ denotes the Hermite inner product in $\mathbb{C}^n$. From Lemma 6.28 we have

$$m^{(2)}_{5Q}(f^\varepsilon_{Q,w} - \langle \Phi_Q(\star), w\rangle) \lesssim \varepsilon \ell(Q), \tag{6.225}$$

$$m^{(2)}_{5Q}(\nabla f^\varepsilon_{Q,w} - \nabla\langle \Phi_Q(\star), w\rangle) \lesssim 1. \tag{6.226}$$

Lemma 6.30 *Let $w \in \mathbb{C}^n$ be the unit vector. We have*

$$|m_Q(1 - w \cdot \nabla f^\varepsilon_{Q,w})| \lesssim \sqrt{\varepsilon}. \tag{6.227}$$

Proof Keeping in mind that Φ_Q is a vector function, define

$$g \equiv \langle \Phi_Q(\star), w\rangle - f^\varepsilon_{Q,w} = \langle \Phi_Q(\star), w\rangle - (\mathrm{id}_{L^2} + \varepsilon^2 \ell(Q)^2 L)^{-1}(\langle \Phi_Q(\star), w\rangle).$$

Then Lemma 6.28 yields $\|g\|_{L^2(Q)} \lesssim \varepsilon \ell(Q)\sqrt{|Q|}$ and $\|\nabla g\|_{L^2(Q)^n} \lesssim \sqrt{|Q|}$. Thus, since $|w| = 1$, $|m_Q(1 - w \cdot \nabla f^\varepsilon_{Q,w})| = |m_Q(\nabla g \cdot w)| \lesssim \sqrt{\varepsilon}$, which proves (6.227).

6.7.2.5 Applications of Gaffney-Type Estimates to Orthogonality

Although we are supposing that the coefficients are smooth here, we still have to consider the coefficient functions as somewhat singular. Indeed, all we have is the information on the ellipticity. One idea is that the Kato problem is solved once we can replace a_{ij} with δ_{ij}, the constant function. Based on the universal estimates above, we take such an approach.

We let $\psi^0 \in C_c^\infty(B(1)) \setminus \{0\}$ be a function which is radially symmetric; namely $\psi^0(A\star) = \psi^0$ for all $A \in O(n)$. Let $Q_s, s > 0$ be the operator as in (6.209).

Remark 6.9 Define the vector-valued function $\mathbf{R}_s$ by

$$\mathbf{R}_s \equiv \frac{1}{s^n}\mathrm{grad}\psi^0\left(\frac{\star}{s}\right) = \left(\frac{1}{s^n}\partial_{x_j}\psi^0\left(\frac{\star}{s}\right)\right)_{j=1,\ldots,n}. \tag{6.228}$$

Then using convolution, we have

$$\mathbf{R}_s * f = \left(\frac{1}{s^n}\partial_{x_j}\psi^0\left(\frac{\star}{s}\right) * f\right)_{j=1,\ldots,n}, \quad Q_s f = s\,\mathrm{div}[\mathbf{R}_s * f]$$

for $f \in L^2(\mathbb{R}^n)$.

In a word, Q_s is the Fourier frequency at level s and V_t is the frequency of "something" at level t. We claim that these two things are close to each other.

Define an operator $V_t, t > 0$ by $V_t h \equiv t^2\mathrm{div}(M_{A^*} \circ \nabla[(\mathrm{id}_{L^2} + t^2L^*)^{-2}h])$ for $h \in L^2(\mathbb{R}^n)$. In principle, V_t and Q_s are types of projection at levels t and s, respectively. Despite the nonsmoothness of the matrix A, we can say that V_t and Q_s are orthogonal to a large extent as the following lemma shows:

Lemma 6.31 *Let $s, t > 0$. Then $\|V_t \circ Q_s\|_{B(L^2)} \lesssim \min(st^{-1}, s^{-1}t)$.*

Proof We express $V_t \circ Q_s f$ in two different ways. First we use $\mathbf{R}_s$ defined by (6.228). Since

$$\begin{aligned} V_t \circ Q_s f &= t^2\mathrm{div}(M_{A^*} \circ \nabla[(\mathrm{id}_{L^2} + t^2L^*)^{-2}Q_s f]) \\ &= \left[st^2L^*(\mathrm{id}_{L^2} + t^2L^*)^{-2}\mathrm{div}\right](\mathbf{R}_s * f), \end{aligned}$$

we have

$$\begin{aligned} \|V_t \circ Q_s\|_{B(L^2)} &\lesssim st^{-1}\,\|t^2L^*(\mathrm{id}_{L^2} + t^2L^*)^{-2}t\mathrm{div}\|_{B((L^2)^n,L^2)}\|R_s\|_{B(L^2,(L^2)^n)} \\ &\lesssim st^{-1}\|t^2L^*(\mathrm{id}_{L^2}+t^2L^*)^{-1}\|_{B(L^2)}\|(\mathrm{id}_{L^2}+t^2L^*)^{-1}t\mathrm{div}\|_{B((L^2)^n,L^2)} \\ &\lesssim st^{-1} \end{aligned}$$

thanks to the fact that L^* is a sectorial operator and the Gaffney estimate. Meanwhile, if we let $f \in \mathrm{Dom}(L^*)$, then $V_t \circ Q_s f = t(\mathrm{id}_{L^2} + t^2L^*)^{-2}t\mathrm{div}(M_{A^*}\nabla(Q_s f))$. Thus,

$$\|V_t\circ Q_s\|_{B(L^2)} \lesssim t\,\|(\mathrm{id}_{L^2}+t^2L^*)^{-2}t\mathrm{div}\circ M_{A^*}\|_{B((L^2)^n,L^2)}\|\nabla\circ Q_s\|_{B(L^2,(L^2)^n)} \lesssim t\,s^{-1},$$

which proves Lemma 6.31.

6.7.2.6 The Operator θ_t and the Function γ_t

We set

$$\theta_t \equiv -[t(\mathrm{id}_{L^2}+t^2L)^{-1}\mathrm{div}] \circ M_A \in B(L^2(\mathbb{R}^n)^n, L^2(\mathbb{R}^n)).$$

Note that θ_t makes sense since $a_{ij} \in W^{1,\infty}(\mathbb{R}^n)$. Let $\mathbf{1} \equiv {}^t(1,1,\dots,1)$. We set

$$\gamma_t \equiv \left(-\sum_{i=1}^n t(\mathrm{id}_{L^2}+t^2L)^{-1}[\partial_{x_i}a_{ij}]\right)_{j=1}^n = \theta_t\mathbf{1} \in L^2(\mathbb{R}^n).$$

Lemma 6.32 *For $y \in \mathbb{R}^n$ and $t>0$, $m^{(2)}_{B(y,t)}(\gamma_t) \lesssim_{\lambda,\Lambda,n} 1$.*

Proof We define $\chi_k \equiv \chi_{B(y,2^kt)\setminus B(2^{k-1}t)}$ for $k \in \mathbb{N}$ and $\chi_0 \equiv \chi_{B(y,t)}$. Then using Lemma 6.29 and the Gaffney estimate, we obtain

$$\begin{aligned} m^{(2)}_{B(y,t)}(\gamma_t) &= m^{(2)}_{B(y,t)}((\mathrm{id}_{L^2}+t^2L)^{-1}t\mathrm{div}[M_A(\mathbf{1})]) \\ &\le \sum_{k=0}^\infty m^{(2)}_{B(y,t)}((\mathrm{id}_{L^2}+t^2L)^{-1}t\mathrm{div}[M_A(\chi_k\mathbf{1})]) \\ &\lesssim \sum_{k=0}^\infty \exp(-2^ka)\sqrt{\frac{1}{t^n}\int_{B(y,2^{k+3}t)}\mathrm{d}x} \simeq 1, \end{aligned}$$

where $a>0$ is a positive constant, proving Lemma 6.32.

Let $p \in C^\infty_{\mathrm{c}}(B(1))$ be a compactly supported smooth function. We set

$$U_t\mathbf{F}(x) \equiv \left(\int_{\mathbb{R}^n} U_t(x,y)f_j(y)\mathrm{d}y\right)_{j=1}^n \quad (x \in \mathbb{R}^n)$$

for $t>0$ and $\mathbf{F}=(f_1,f_2,\dots,f_n) \in L^2(\mathbb{R}^n)^n$, where

$$U_t(x,y) \equiv p_t(x-y)\gamma_t(x) - \theta_t[p_t(\star-y)\mathbf{1}](x) \quad (x,y \in \mathbb{R}^n).$$

Note that U_t is a vector-valued operator but it satisfies (6.211).

We first control $U_t \circ P_t$, as follows.

Lemma 6.33 *Let $\boldsymbol{F} \in L^2(\mathbb{R}^n)^n$. Then* $\displaystyle\int_0^\infty (\|U_t \circ P_t\boldsymbol{F}\|_{(L^2)^n})^2\frac{\mathrm{d}t}{t} \lesssim_{\lambda,\Lambda} \|\boldsymbol{F}\|^2_{(L^2)^n}.$

Proof Let $S>0$ and $R>2t+2S$. Then

$$
\begin{aligned}
U_t(\chi_{B(R)}\mathbf{1})(x) &= \theta_t[\mathbf{1}](x) \cdot P_t * \chi_{B(R)}(x)\mathbf{1} - \theta_t[P_t * \chi_{B(R)}\mathbf{1}](x) \\
&= -\theta_t[\mathbf{1}](x) \cdot P_t * \chi_{B(R)^{\rm c}}(x)\mathbf{1} + \theta_t[P_t * \chi_{B(R)^{\rm c}}\mathbf{1}](x) \\
&= \theta_t[P_t * \chi_{B(R)^{\rm c}}\mathbf{1}](x).
\end{aligned}
$$

Thus, by the Gaffney estimate there exists $a > 0$ such that

$$
\|U_t(\chi_{B(R)}\mathbf{1})\|_{L^2(B(S))} \leq \sum_{k=1}^{\infty} \|\theta_t[P_t * \chi_{B(2^k R)\setminus B(2^{k-1}R)}\mathbf{1}]\|_{L^2(B(S))} \lesssim \exp(-aR\,2^k)2^{kn},
$$

which implies $\lim_{R\to\infty} U_t(\chi_{B(R)}\mathbf{1}) = 0$ in $L^2_{\rm loc}(\mathbb{R}^n)^n$. Thus, (6.214) is true. Furthermore,

$$
t^{-2n}\int_{\mathbb{R}^n}\left(1+\frac{|x-y|}{t}\right)^{2m}\left|\gamma_t(x)p\left(\frac{x-y}{t}\right)\right|^2 {\rm d}x \lesssim t^{-2n}\int_{B(y,t)}|\gamma_t(x)|^2{\rm d}x \lesssim t^{-n}.
$$

Likewise,

$$
\begin{aligned}
&\int_{\mathbb{R}^n}\left(1+\frac{|x-y|}{t}\right)^{2m}\left|\theta_t\left[p\left(\frac{\star-y}{t}\right)\right](x)\right|^2{\rm d}x \\
&\lesssim \int_{\mathbb{R}^n}\left|\theta_t\left[p\left(\frac{\star-y}{t}\right)\right](x)\right|^2{\rm d}x + \sum_{k=1}^{\infty}2^{2km}\int_{B(y,2^k t)}\left|\theta_t\left[p\left(\frac{\star-y}{t}\right)\right](x)\right|^2{\rm d}x \\
&\lesssim \int_{\mathbb{R}^n}p\left(\frac{x-y}{t}\right)^2{\rm d}x + \sum_{k=1}^{\infty}2^{2km}\exp(-a\cdot 2^k)\int_{\mathbb{R}^n}p\left(\frac{x-y}{t}\right)^2{\rm d}x \\
&\lesssim t^n.
\end{aligned}
$$

Combining the above estimates, we obtain (6.210). Finally, (6.213) follows from Lemma 6.32.

Thus, we are in the position of using Lemma 6.27 to have

$$
\|U_t \circ P_t \circ Q_s\|_{B(L^2)} \lesssim \sqrt{\min\left(\frac{t}{s},\frac{s}{t}\right)},
$$

where Q_s is given by (6.209). Thus we conclude Lemma 6.33 from Proposition 2.12.

Finally, we reduce matters to the estimate of γ_t, as follows. By the Plancherel theorem and $\|t^2L({\rm id}_{L^2}+t^2L)^{-1}\|_{B(L^2)} \lesssim 1$, we have

$$\sqrt{\int_0^\infty (\|tL(\mathrm{id}_{L^2}+t^2L)^{-1}(\mathrm{id}_{L^2}-P_t{}^2)g\|_2)^2\frac{\mathrm{d}t}{t}} \lesssim \sqrt{\int_0^\infty (\|(\mathrm{id}_{L^2}-P_t{}^2)g\|_2)^2\frac{\mathrm{d}t}{t^3}}$$
$$\lesssim \|\nabla g\|_{(L^2)^n}.$$

Let us write E for the identity matrix of size n. We define $P_t^n \in B(L^2(\mathbb{R}^n)^n)$ naturally from $P_t \in B(L^2(\mathbb{R}^n))$, using the identity

$$\begin{aligned}\theta_t\nabla g &= \theta_t\nabla g - \gamma_t\cdot(P_t{}^2\nabla g) + \gamma_t\cdot(P_t{}^2\nabla g)\\ &= \theta_t(\mathrm{id}_{(L^2)^n} - (P_t^n)^2)\nabla g + U_t\circ P_t g + \gamma_t\cdot(P_t{}^2\nabla g)\\ &= \theta_t\nabla(\mathrm{id}_{L^2} - P_t{}^2)g + U_t\circ P_t g + \gamma_t\cdot(P_t{}^2\nabla g)\\ &= tL(\mathrm{id}_{L^2}+t^2L)^{-1}(\mathrm{id}_{L^2}-P_t{}^2)g - U_t\circ P_t g + \gamma_t\cdot(P_t{}^2\nabla g).\end{aligned}$$

Recall that we considered $U_t \circ P_t$ in Lemma 6.33. Thus, we have only to handle $\gamma_t\cdot(P_t{}^2\nabla g)$. Namely, we need to show that

$$\int_0^\infty \|\gamma_t\cdot P_t{}^2\nabla g\|_2{}^2\frac{\mathrm{d}t}{t} \lesssim \|\nabla g\|_{(L^2)^n}{}^2. \tag{6.229}$$

6.7.2.7 The Conclusion of the Proof of the Kato Conjecture

So far, we have reduced matters to (6.229). We next reduce matters to the Carleson measure, which is the crucial step of the solution to the Kato problem.

Proposition 6.10 *The solution of the Kato conjecture is reduced to the Carleson norm estimate of $\gamma_t(x)$:*

$$\sup_{Q\in\mathcal{D}}\frac{1}{|Q|}\iint_{Q\times(0,\ell(Q)]}|\gamma_t(x)|^2\frac{\mathrm{d}x\mathrm{d}t}{t} < \infty. \tag{6.230}$$

Proof We recall the Carleson embedding:

$$\int_{\mathbb{R}^{n+1}_+} F(x,t)\mathrm{d}\mu(x,t) \le 3^n\|\mu\|_{\text{Carleson}}\int_{\mathbb{R}^n}\left(\sup_{(y,t)\in\Gamma(x)} F(y,t)\right)\mathrm{d}x$$

for all nonnegative Borel functions $F : \mathbb{R}^{n+1}_+ \to [0,\infty)$. See Theorem 1.57. If we let

$$F(x,t) \equiv |P_t{}^2\nabla g(x)|^2 \quad ((x,t)\in\mathbb{R}^{n+1}_+)$$

and

$$\mu(x,t) = |\gamma_t(x)|^2 \frac{\mathrm{d}x\mathrm{d}t}{t},$$

then assuming (6.230), we obtain

$$\int_{\mathbb{R}^{n+1}_+} |P_t{}^2\nabla g(x)|^2 |\gamma_t(x)|^2 \frac{\mathrm{d}x\mathrm{d}t}{t} \lesssim \int_{\mathbb{R}^n} \left(\sup_{(y,t)\in\Gamma(x)} |P_t{}^2\nabla g(y)|^2 \right) \mathrm{d}x.$$

Meanwhile, Theorem 1.47 yields $\sup\limits_{(y,t)\in\Gamma(x)} |P_t{}^2\nabla g(y)| \lesssim M[\nabla g](x)$. It remains to use the $L^2(\mathbb{R}^n)$-boundedness of M, which proves (6.229).

The proof of (6.230) is postponed and is decomposed into Lemmas 6.35 and 6.36.

Recall that $\mathscr{D}(t)$ and S_t are defined in Example 1.33. We define $S_t^n \in B(L^2(\mathbb{R}^n)^n)$ naturally.

We show that θ_t and $\gamma_t \cdot S_t$ is close as the following lemma shows:

Lemma 6.34 *For all vector functions $\mathbf{G} \in H^1(\mathbb{R}^n)^n$,*

$$\|\theta_t \mathbf{G} - \gamma_t \cdot S_t^n \mathbf{G}\|_2 \lesssim t^2 \|\nabla \mathbf{G}\|_{(L^2)^n}. \tag{6.231}$$

Proof We observe

$$\begin{aligned}
\|\theta_t \mathbf{G} - \gamma_t \cdot S_t^n \mathbf{G}\|_2{}^2 &= \sum_{Q\in\mathscr{D}(t)} \int_Q |\theta_t \mathbf{G}(x) - \gamma_t(x) \cdot m_Q(\mathbf{G})|^2 \mathrm{d}x \\
&= \sum_{Q\in\mathscr{D}(t)} \int_Q |\theta_t(\mathbf{G} - m_Q(\mathbf{G}))(x)|^2 \mathrm{d}x
\end{aligned}$$

from the definition of the operators. Thus,

$$\begin{aligned}
\|\theta_t \mathbf{G} - \gamma_t \cdot S_t^n \mathbf{G}\|_2{}^2 &\lesssim \sum_{Q,R\in\mathscr{D}(t)} \exp\left(-a\frac{|c(Q)-c(R)|}{t}\right) \int_R |\mathbf{G}(x) - m_Q(\mathbf{G})|^2 \mathrm{d}x \\
&\lesssim \sum_{Q,R\in\mathscr{D}(t)} \exp\left(-a\frac{|c(Q)-c(R)|}{2t}\right) \int_R |\mathbf{G}(x) - m_R(\mathbf{G})|^2 \mathrm{d}x \\
&\lesssim \sum_{R\in\mathscr{D}(t)} \int_R |\mathbf{G}(x) - m_R(\mathbf{G})|^2 \mathrm{d}x \\
&\lesssim t^2 \|\nabla \mathbf{G}\|_2{}^2
\end{aligned}$$

by the Poincaré inequality and Lemma 5.29, the off-diagonal estimate, proving (6.231).

The following observation paves the way to the Carleson estimate.

Lemma 6.35 *Let $\varepsilon \in (0,1)$, $w \in \mathbb{C}^n$ with $|w| = 1$ and Q be a cube. Then*

$$\iint_{Q\times(0,\ell(Q)]} |\gamma_t(x)\cdot S_t\nabla f^{\varepsilon}_{Q,w}(x)|^2\frac{\mathrm{d}x\mathrm{d}t}{t} \lesssim |Q|. \tag{6.232}$$

Proof Choose $\psi \in C^\infty_{\rm c}(\mathbb{R}^n)$ so that $\|\psi\|_\infty + \ell(Q)\|\nabla\psi\|_\infty \lesssim 1$ and that $\chi_{2Q} \le \psi_Q \le \chi_{4Q}$. Then

$$\begin{aligned}&\iint_{Q\times(0,\ell(Q)]} |\gamma_t(x)\cdot S_t\nabla f^{\varepsilon}_{Q,w}(x)|^2\frac{\mathrm{d}x\mathrm{d}t}{t}\\ &\lesssim \int_{4Q}|\nabla[\psi_Q\cdot f^{\varepsilon}_{Q,w}](x)|^2\mathrm{d}x + \iint_{Q\times(0,\ell(Q)]}|\theta_t\nabla[\psi_Q\cdot f^{\varepsilon}_{Q,w}](x)|^2\frac{\mathrm{d}x\mathrm{d}t}{t}\end{aligned}$$

thanks to (6.231). According to Lemma 6.28, we have

$$\begin{aligned}&\iint_{Q\times(0,\ell(Q)]} |\gamma_t(x)\cdot S_t\nabla f^{\varepsilon}_{Q,w}(x)|^2\frac{\mathrm{d}x\mathrm{d}t}{t}\\ &\lesssim |Q| + \iint_{Q\times(0,\ell(Q)]}|\theta_t\nabla[\psi_Q\cdot f^{\varepsilon}_{Q,w}](x)|^2\frac{\mathrm{d}x\mathrm{d}t}{t}.\end{aligned} \tag{6.233}$$

We need to handle the right-hand side. We decompose

$$\begin{aligned}&\theta_t(\nabla\psi_Q\cdot f^{\varepsilon}_{Q,w})\\ &= t(\mathrm{id}_{L^2}+t^2L)^{-1}\left[M_{\psi_Q}Lf^{\varepsilon}_{Q,w} - \mathrm{div}(M_{f^{\varepsilon}_{Q,w}}A\nabla\psi_Q) - (M_A\nabla f^{\varepsilon}_{Q,w})\cdot\nabla\psi_Q\right].\end{aligned}$$

As for $(\mathrm{id}_{L^2}+t^2L)^{-1}[M_{\psi_Q}Lf^{\varepsilon}_{Q,w}]$, since $Lf^{\varepsilon}_{Q,w} = \varepsilon^{-2}\ell(Q)^{-2}(f^{\varepsilon}_{Q,w} - \langle\Phi_Q(\star), w\rangle)$, we can use the boundedness of $(\mathrm{id}_{L^2}+t^2L)^{-1}$ and (6.225) to have

$$\int_{\mathbb{R}^n}|t(\mathrm{id}_{L^2}+t^2L)^{-1}[M_{\psi_Q}Lf^{\varepsilon}_{Q,w}](x)|^2\mathrm{d}x \lesssim t^2\int_{\mathbb{R}^n}|Lf^{\varepsilon}_{Q,w}(x)|^2\mathrm{d}x \lesssim t^2\ell(Q)^{n-2}.$$

Thus, the estimate of the first term is valid:

$$\iint_{Q\times(0,\ell(Q)]}|t(\mathrm{id}_{L^2}+t^2L)^{-1}[M_{\psi_Q}Lf^{\varepsilon}_{Q,w}](x)|^2\frac{\mathrm{d}x\mathrm{d}t}{t} \lesssim \ell(Q)^{n-2}\int_0^{\ell(Q)}t\mathrm{d}t \simeq |Q|. \tag{6.234}$$

As for the second term, since $\mathrm{supp}(\nabla\psi_Q)$ and Q are torn apart, we can use Lemmas 5.29 and 6.28 to have

$$
\begin{aligned}
&\|(\mathrm{id}_{L^2}+t^2L)^{-1}t\mathrm{div}[M_{f^\varepsilon_{Q,w}A}\nabla\psi_Q]\|_{L^2(Q)}\\
&\lesssim \exp\left(-a\frac{\ell(Q)}{2t}\right)\|M_{f^\varepsilon_{Q,w}A}\cdot\nabla\psi_Q\|_{L^2(4Q)}\\
&\lesssim \frac{1}{\ell(Q)}\exp\left(-a\frac{\ell(Q)}{2t}\right)\|f^\varepsilon_{Q,w}\|_{L^2(4Q)}\lesssim \exp\left(-a\frac{\ell(Q)}{2t}\right)\sqrt{|Q|},
\end{aligned}
$$

which implies

$$
\begin{aligned}
&\iint_{Q\times(0,\ell(Q)]}|(\mathrm{id}_{L^2}+t^2L)^{-1}t\mathrm{div}(M_{f^\varepsilon_{Q,w}A}\nabla\psi_Q)(x)|^2\frac{\mathrm{d}x\mathrm{d}t}{t} \qquad (6.235)\\
&\lesssim \int_0^{\ell(Q)}\exp\left(-a\frac{\ell(Q)}{t}\right)|Q|\frac{\mathrm{d}t}{t}\simeq|Q|.
\end{aligned}
$$

For the third term $t(\mathrm{id}_{L^2}+t^2L)^{-1}[M_A\nabla f^\varepsilon_{Q,w}\cdot\nabla\psi_Q]$, we will make use of the $L^2(\mathbb{R}^n)$-boundedness of $(\mathrm{id}_{L^2}+t^2L)^{-1}$. That is, we move from Q to $\mathbb{R}^n$ and use the $L^2(\mathbb{R}^n)$-boundedness of $(\mathrm{id}_{L^2}+t^2L)^{-1}$ to have

$$
\begin{aligned}
\|t(\mathrm{id}_{L^2}+t^2L)^{-1}[M_A\nabla f^\varepsilon_{Q,w}\cdot\nabla\psi_Q]\|_{L^2(Q)}&\le t\|(\mathrm{id}_{L^2}+t^2L)^{-1}[M_A\nabla f^\varepsilon_{Q,w}\cdot\nabla\psi_Q]\|_2\\
&\lesssim t\|M_A\nabla f^\varepsilon_{Q,w}\cdot\nabla\psi_Q\|_2\\
&\lesssim t\ell(Q)^{-1}\|M_A\nabla f^\varepsilon_{Q,w}\|_2.
\end{aligned}
$$

We use Lemma 6.28 again to have

$$
\|t(\mathrm{id}_{L^2}+t^2L)^{-1}[M_A\nabla f^\varepsilon_{Q,w}\cdot\nabla\psi_Q]\|_{L^2(Q)}\lesssim t\ell(Q)^{n/2-1};
$$

hence

$$
\iint_{Q\times(0,\ell(Q)]}|t(\mathrm{id}_{L^2}+t^2L)^{-1}[M_A\nabla f^\varepsilon_{Q,w}\cdot\nabla\psi_Q](x)|^2\frac{\mathrm{d}x\mathrm{d}t}{t}\lesssim|Q|. \qquad (6.236)
$$

If we combine (6.233), (6.234), (6.235), and (6.236), we obtain (6.232).

Proposition 6.11 *Let $\varepsilon \ll_{n,\lambda,\Lambda} 1$ and $w\in\mathbb{C}^n$. Then there exists a packing $\mathscr{D}_w(Q)$ satisfying the following conditions:*

1. *There exists $\mu>0$ such that*

$$
\sum_{R\in\mathscr{D}_w(Q)}|R|=\left|\bigcup_{R\in\mathscr{D}_w(Q)}R\right|\le(1-\mu\sqrt{\varepsilon})|Q|. \qquad (6.237)
$$

2. *If $Q\in\mathscr{D}(Q)$ is not included in any cube in $\mathscr{D}_w(Q)$, then*

$$m_R(\Re(w \cdot \nabla f^{\varepsilon}_{Q,w})) \geq \frac{3}{4} \tag{6.238}$$

and

$$m^{(2)}_R(\nabla f^{\varepsilon}_{Q,w}) \leq (4\varepsilon)^{-1}. \tag{6.239}$$

Proof Let $\varepsilon \ll_{n,\lambda,\Lambda} 1$. Then $m_Q(\Re(w \cdot \nabla f^{\varepsilon}_{Q,w})) \geq \frac{7}{8}$ from (6.227). Likewise since $m^{(2)}_Q(\nabla f^{\varepsilon}_{Q,w}) \leq C$ for some $C > 0$, we may assume that $m^{(2)}_Q(\nabla f^{\varepsilon}_{Q,w}) \leq (4\varepsilon)^{-2}$. Here and below fix such an ε. Define $\mathscr{D}_w(Q)$ to be the maximal collection of cubes R satisfying either

$$m_R(\Re(w \cdot \nabla f^{\varepsilon}_{Q,w})) \leq \frac{3}{4} \tag{6.240}$$

or

$$m^{(2)}_R(\nabla f^{\varepsilon}_{Q,w}) \geq (4\varepsilon)^{-1}. \tag{6.241}$$

Then we have clearly (6.238) and (6.239). It remains to show (6.237). Denote by B_1 and B_2 the union of the cubes satisfying (6.240) and (6.241), respectively.

Define $b(x) \equiv 1 - \Re\left(\nabla f^{\varepsilon}_{Q,w}(x) \cdot w\right)$ for $x \in \mathbb{R}^n$. Then

$$|B_1| \leq -4\int_{B_1} b(x)\mathrm{d}x = -4\int_Q b(x)\mathrm{d}x + 4\int_{Q\setminus B_1} b(x)\mathrm{d}x$$

from (6.240). We have $\|b\|_{L^1(Q)} \lesssim \sqrt{\varepsilon}|Q|$ thanks to the Cauchy–Schwarz inequality. Meanwhile,

$$\left|\int_{Q\setminus B_1}(b(x)-1)\mathrm{d}x\right| \leq \sqrt{|Q\setminus B_1|\int_Q |b(x)-1|^2\mathrm{d}x} \lesssim \sqrt{|Q\setminus B_1|\cdot|Q|}$$

from (6.240). Thus, there exists $M > 1$ such that

$$\begin{aligned} |B_1| &\leq M\varepsilon^{\frac{1}{2}}|Q| + 4|Q| - 4|B_1| + M(|Q|-|B_1|)^{\frac{1}{2}}|Q|^{\frac{1}{2}} \\ &\leq 2M^2\varepsilon^{\frac{1}{2}}|Q| + (4+\sqrt{\varepsilon^{-1}})|Q| - (4+\sqrt{\varepsilon^{-1}})|B_1| \end{aligned}$$

thanks to the Cauchy inequality $a^2 + b^2 \geq 2ab$ for $a, b > 0$. As a consequence,

$$|B_1| \leq \frac{2M^2\varepsilon^{\frac{1}{2}} + 4 + \sqrt{\varepsilon^{-1}}}{5+\sqrt{\varepsilon^{-1}}}|Q| = \frac{1+4\sqrt{\varepsilon}+2M^2\varepsilon}{1+5\sqrt{\varepsilon}}|Q|. \tag{6.242}$$

Meanwhile, there exists $M' > 1$ such that

$$|B_2| \leq (4\varepsilon\|\nabla f^{\varepsilon}_{Q,w}\|_{L^2(Q)})^2 \leq M'\varepsilon^2|Q| \tag{6.243}$$

from (6.226) and (6.241). Thus, combining (6.242) and (6.243), we obtain $|B_1| + |B_2| \leq (1 - \mu\sqrt{\varepsilon})|Q|$ for some $\mu > 0$, proving (6.237) for $0 < \varepsilon \ll 1$.

Denote by S^{2n-1} the unit sphere of $\mathbb{C}^n$. Finally, we prove the following lemma to finish the proof of the Kato conjecture. We equip $H = \mathbb{C}^n$ with the standard Hermite inner product to apply Lemma 6.26. Choose $\varepsilon > 0$ small enough and consider the cone

$$C_w \equiv \left\{u \in \mathbb{C}^n \, : \, |u - \langle u, w\rangle_{\mathbb{C}^n} w| \leq \varepsilon|\langle u, w\rangle_{\mathbb{C}^n}|\right\}.$$

Since S^{2n-1} is compact, we can find a finite set W such that

$$S^{2n-1} \subset \bigcup_{w \in W} C_w, \quad \sharp W \lesssim_{\varepsilon} 1.$$

Lemma 6.36 *There exists $\varepsilon \ll 1$ such that*

$$\sup_{Q \in \mathscr{D}} \frac{1}{|Q|} \iint_{Q \times (0, \ell(Q)]} |\gamma_t(x)|^2 \frac{dx dt}{t}$$
$$\lesssim \sum_{w \in W} \sup_{R \in \mathscr{D}} \frac{1}{|R|} \iint_{R \times (0, \ell(R)]} |\gamma_t(x) \cdot S_t(\nabla f^{\varepsilon}_{R,w})(x)|^2 \frac{dx dt}{t}.$$

If we combine Lemmas 6.35 and 6.36, then we immediately conclude (6.230).

Proof (of Lemma 6.36) We may replace $\gamma_t(x)$ with $\gamma_{t,w}(x) \equiv \chi_{C_w}(\gamma_t(x))\gamma_t(x)$ in the left-hand side. Note that

$$|\gamma_{t,w}(x) - \langle \gamma_{t,w}(x), w\rangle_{\mathbb{C}^n} w| \leq \varepsilon|\langle \gamma_{t,w}(x), w\rangle_{\mathbb{C}^n}|. \tag{6.244}$$

We set

$$A_{w,\delta} \equiv \sup_{Q} \frac{1}{|Q|} \iint_{Q \times (\delta, \max(\delta, \delta^{-1} \wedge \ell(Q))]} |\gamma_t(x)|^2 \frac{dx dt}{t} \quad (\delta \in (0, 1)).$$

Fix a cube Q and consider $\mathscr{D}_w(Q)$ defined in Proposition 6.11. Let R be a cube which is not contained in any cube in $\mathscr{D}_w(Q)$. We set $v \equiv \overline{m_R(\nabla f^{\varepsilon}_{Q,w})} \in \mathbb{C}^n$. Then from (6.238), we have

$$\Re(\langle v, w\rangle_{\mathbb{C}^n}) \geq \frac{3}{4} \tag{6.245}$$

and from (6.239), we have

$$|v| \le \frac{1}{4\varepsilon}. \tag{6.246}$$

Estimates (6.244), (6.245), and (6.246) guarantee that we can use Lemma 6.26 with

$$U = \gamma_{t,w}(x), \quad V = v, \quad W = \overset{\circ}{W} = w$$

to conclude that

$$|\gamma_{t,w}(x)| \le 4|\gamma_{t,w}(x) \cdot S_t(\nabla f^{\varepsilon}_{Q,w})(x)| \tag{6.247}$$

for any $x \in \mathbb{R}^n$ provided R is a unique dyadic cube satisfying $\frac{1}{2}\ell(R) < t \le \ell(R)$ and $x \in R$.

Here and below we fix $w \in W$. We decompose

$$Q \times (0, \ell(Q)] = \bigcup_{R \in \mathscr{D}_w(Q)} R \times (0, \ell(R)] \cup \bigcup_{\substack{R \in \mathscr{D}(Q),\ R \text{ is not} \\ \text{included in any } S \in \mathscr{D}_w(Q)}} R \times (\ell(R)/2, \ell(R)]$$

using Lemma 1.4. Note that the second term in the right-hand side is a disjoint union. Using (6.232), (6.237) and (6.247), we have

$$\begin{aligned}
&\frac{1}{|Q|}\iint_{Q \times (\delta, \max(\delta, \delta^{-1} \wedge \ell(Q)))]} |\gamma_{t,w}(x)|^2 \frac{\mathrm{d}x\mathrm{d}t}{t} \\
&\le \sum_{R \in \mathscr{D}_w(Q)} \frac{1}{|Q|}\iint_{R \times (\delta, \max(\delta, \delta^{-1} \wedge \ell(R)))]} |\gamma_{t,w}(x)|^2 \frac{\mathrm{d}x\mathrm{d}t}{t} \\
&\quad + \sum_{\substack{R \in \mathscr{D}(Q),\ R \text{ is not} \\ \text{included in any } S \in \mathscr{D}_w(Q)}} \frac{1}{|Q|}\iint_{R \times (\ell(R)/2, \ell(R))]} |\gamma_{t,w}(x)|^2 \frac{\mathrm{d}x\mathrm{d}t}{t} \\
&\le (1 - \mu\sqrt{\varepsilon})A_{Q,\delta} + \frac{16}{|Q|}\iint_{Q \times (0, \ell(Q))]} |\gamma_t(x) \cdot S_t(\nabla f^{\varepsilon}_{Q,w})(x)|^2 \frac{\mathrm{d}x\mathrm{d}t}{t}.
\end{aligned}$$

Summing this estimate over in W, taking the supremum over all cubes Q and then tending $\delta \downarrow 0$, we obtain the desired result.

Thus, the Kato conjecture is solved.

Exercises

Exercise 6.92 Prove (6.223) by reexamining the proof of (6.222).

Exercise 6.93 Let $f \in \mathrm{Lip}(\mathbb{R}^n)$. Prove the sum defining $(\mathrm{id}_{L^2}+t^2L)^{-1}f$ converges and that the limit is independent of the choice of ψ. Hint: Use the Gaffney estimate.

6.7.3 *Kato Conjecture for Other Lebesgue Spaces*

We have proved that the quantities $\|\sqrt{L}f\|_2$ and $\|\nabla f\|_2$ are equivalent for $f \in H^1(\mathbb{R}^n) = \mathrm{Dom}(\sqrt{L})$. As the next step, we wonder what is the next possible generalization. To this end, we are oriented to other Lebesgue spaces, say, $L^p(\mathbb{R}^n)$ with $p \neq 2$.

Let

$$p_0 = \frac{2n}{n+2} \tag{6.248}$$

in Sect. 6.7.3, so that $\dot{W}^{1,p_0}(\mathbb{R}^n)$ is a subspace of $L^2(\mathbb{R}^n)$ and $\{e^{-tL}\}_{t>0}$ satisfies the $L^\rho(\mathbb{R}^n)$-$L^2(\mathbb{R}^n)$ off-diagonal estimate at least for $p_0 < \rho \leq 2$ according to Proposition 5.14.

Theorem 6.49 *Let $1 \leq \rho < p \leq 2$. Assume that $\{e^{-tL}\}_{t>0}$ satisfies the $L^\rho(\mathbb{R}^n)$-$L^2(\mathbb{R}^n)$ off-diagonal estimate. Set $\rho_* \equiv \frac{n\rho}{n+\rho}(< p_0)$. Then we have*

$$\|\sqrt{L}f\|_{L^{p,\infty}} \lesssim \|\nabla f\|_{(L^p)^n} \quad (f \in H^1(\mathbb{R}^n) \cap \dot{W}^{1,p}(\mathbb{R}^n))$$

if $1 \leq \rho_*$ and

$$\|\sqrt{L}f\|_{L^{1,\infty}} \lesssim \|\nabla f\|_{(L^1)^n} \quad (f \in H^1(\mathbb{R}^n) \cap \dot{W}^{1,p}(\mathbb{R}^n))$$

if $\rho_* < 1$. In particular $\|\sqrt{L}f\|_p \lesssim \|\nabla f\|_{(L^p)^n}$ for all $f \in H^1(\mathbb{R}^n) \cap \dot{W}^{1,p}(\mathbb{R}^n)$ and for $\rho < p \leq 2$.

When $p = 2$, this is a direct consequence of the Kato conjecture.

To prove Theorem 6.49, we need a lemma, where the Littlewood–Paley theory comes into play.

Lemma 6.37 *Let $1 \leq \rho \leq 2$. Assume that $\{e^{-tL}\}_{t>0}$ satisfies the $L^\rho(\mathbb{R}^n)$-$L^2(\mathbb{R}^n)$ off-diagonal estimates. Let*

$$\psi(z) \equiv z \int_1^\infty e^{-t^2 z} dt \quad (z \in \mathbb{C}, \Re(z) > 0). \tag{6.249}$$

Then for $\rho < q \leq 2$,

$$\left\| \sum_{k=-\infty}^{\infty} \psi(4^k L) f_k \right\|_q \lesssim \|\{f_k\}_{k=-\infty}^{\infty}\|_{L^q(\ell^2)} \quad (\{f_k\}_{k=-\infty}^{\infty} \in L^q(\ell^2, \mathbb{R}^n)). \tag{6.250}$$

Proof Duality entails the $L^2(\mathbb{R}^n)$-$L^{\rho'}(\mathbb{R}^n)$ off-diagonal estimates of $\{e^{-tL^*}\}_{t>0}$. Hence an argument similar to (6.123) yields $\|\{\psi(4^k L^*) f\}_{k=-\infty}^{\infty}\|_{L^{q'}(\ell^2)} \lesssim \|f\|_{q'}$ for all $f \in L^{q'}(\mathbb{R}^n) \cap L^2(\mathbb{R}^n)$, since $\rho < q \leq 2$. By duality we obtain (6.250).

Proof (of Theorem 6.49) Let $f \in H^1(\mathbb{R}^n) \cap \dot{W}^{1,p}(\mathbb{R}^n)$ and $\lambda > 0$. We have to prove

$$|\{|\sqrt{L} f| > \lambda\}| \lesssim (\lambda^{-1} \|\nabla f\|_{(L^p)^n})^p.$$

Form the Calderón–Zygmund decomposition $f = g + \sum_{j \in J} b_j$ of f at height λ^p; see Theorem 1.61. We know that the sum converges in $W^{1,2}(\mathbb{R}^n) = H^1(\mathbb{R}^n)$. See Remark 1.8. Thus, we may assume that J is finite as we did in (1.237). For the good part, we can go through the same argument as the singular integral operators, since the Kato conjecture was solved affirmatively. Let us concentrate on the bad part. For $j \in J$, we let

$$T_j \equiv \int_0^{\ell(Q_j)} L \exp(-t^2 L) \mathrm{d}t, \quad U_j \equiv \int_{\ell(Q_j)}^{\infty} L \exp(-t^2 L) \mathrm{d}t,$$

so that $T_j + U_j \simeq \sqrt{L}$ on $\mathrm{Dom}(\sqrt{L})$ according to Theorem 5.26. Along this decomposition, we need to show

$$\left| \left\{ \left| \sum_{j \in J} T_j b_j \right| > \lambda \right\} \setminus \bigcup_{j^\dagger \in J} 4Q_{j^\dagger} \right| \lesssim (\lambda^{-1} \|\nabla f\|_{(L^p)^n})^p, \tag{6.251}$$

$$\left| \left\{ \left| \sum_{j \in J} U_j b_j \right| > \lambda \right\} \setminus \bigcup_{j^\dagger \in J} 4Q_{j^\dagger} \right| \lesssim (\lambda^{-1} \|\nabla f\|_{(L^p)^n})^p. \tag{6.252}$$

We start with T_j, where we need to use the off-diagonal estimates. For (6.251), we use the Chebyshev inequality (see Theorem 1.4) to have

$$\left| \left\{ \left| \sum_{j \in J} T_j b_j \right| > \lambda \right\} \setminus \bigcup_{j^\dagger \in J} 4Q_{j^\dagger} \right| \leq \frac{1}{\lambda^2} \int_{\mathbb{R}^n \setminus \bigcup_{j^\dagger \in J} 4Q_{j^\dagger}} \left| \sum_{j \in J} T_j b_j(x) \right|^2 \mathrm{d}x.$$

Choose $u \in L^2(\mathbb{R}^n)$ having the unit norm so that

$$\int_{\mathbb{R}^n\setminus\bigcup_{j^\dagger\in J}4Q_{j^\dagger}}\left|\sum_{j\in J}T_jb_j(x)\right|^2\mathrm{d}x \le 2\left(\left\|\sum_{j\in J}uT_jb_j\right\|_{L^1(\mathbb{R}^n\setminus\bigcup_{j^\dagger\in J}4Q_{j^\dagger})}\right)^2$$
$$\le 2\left(\sum_{j\in J}\|uT_jb_j\|_{L^1(\mathbb{R}^n\setminus\bigcup_{j^\dagger\in J}4Q_{j^\dagger})}\right)^2.$$

We take out the summation and estimate:

$$\sum_{j\in J}\|uT_jb_j\|_{L^1(\mathbb{R}^n\setminus\bigcup_{j^\dagger\in J}4Q_{j^\dagger})} \le \sum_{j\in J}\int_{\mathbb{R}^n\setminus 4Q_j}|u(x)T_jb_j(x)|\mathrm{d}x$$
$$=\sum_{l=1}^{\infty}\sum_{j\in J}\int_{2^{l+2}Q_j\setminus 2^{l+1}Q_j}|u(x)T_jb_j(x)|\mathrm{d}x.$$

By the Cauchy–Schwarz inequality, we have:

$$\sum_{j\in J}\|uT_jb_j\|_{L^1(\mathbb{R}^n\setminus\bigcup_{j^\dagger\in J}4Q_{j^\dagger})}\le\sum_{l=1}^{\infty}\sum_{j\in J}\|T_jb_j\|_{L^2(2^{l+2}Q_j\setminus 2^{l+1}Q_j)}\|u\|_{L^2(2^{l+2}Q_j\setminus 2^{l+1}Q_j)}.$$

By our $L^2(\mathbb{R}^n)$-$L^2(\mathbb{R}^n)$ off-diagonal estimate assumption of $\mathrm{e}^{-t^2L/2}$ and the $L^{p_0}(\mathbb{R}^n)$-$L^2(\mathbb{R}^n)$ off-diagonal estimate (5.131) of $t^2L\mathrm{e}^{-t^2L/2}$

$$\|t^2L\exp(-t^2L)b_i\|_{L^2(2^{l+2}Q_j\setminus 2^{l+1}Q_j)} \lesssim t^{\frac{n}{2}-\frac{n}{p_0}}\exp\left(-2a\frac{2^l\ell(Q_j)}{t}\right)\|b_j\|_{p_0} \tag{6.253}$$

for some $a>0$. We now recall the $\dot{W}^{1,\infty}(\mathbb{R}^n)$-*condition* (1.200). By the Poincaré inequality, or the Sobolev embedding, and (6.248),

$$\|b_j\|_{p_0}\lesssim \ell(Q_j)^{1-\frac{n}{p}+\frac{n}{p_0}}\|\nabla b_j\|_{(L^p)^n}\lesssim \lambda\ell(Q_j)^{1+\frac{n}{p_0}}. \tag{6.254}$$

As a consequence, combining (6.253) and (6.254), we obtain

$$\|T_jb_j\|_{L^2(2^{l+2}Q_j\setminus 2^{l+1}Q_j)} \le \int_0^{\ell(Q_j)}\|L\exp(-t^2L)b_i\|_{L^2(2^{l+2}Q_j\setminus 2^{l+1}Q_j)}\mathrm{d}t$$
$$\lesssim \lambda\ell(Q_j)^{1+\frac{n}{p_0}}\int_0^{\ell(Q_j)}t^{-\frac{n}{p_0}+\frac{n}{2}-2}\exp\left(-2a\frac{2^l\ell(Q_j)}{t}\right)\mathrm{d}t$$
$$\le \lambda\ell(Q_j)^{1+\frac{n}{p_0}}\int_0^{\infty}t^{-\frac{n}{p_0}+\frac{n}{2}-2}\exp\left(-a\frac{2^l\ell(Q_j)}{t}-a2^l\right)\mathrm{d}t$$

$$\leq \lambda \ell(Q_j)^{1+\frac{n}{p_0}} \int_0^\infty t^{-\frac{n}{p_0}+\frac{n}{2}-2} \exp\left(-a\frac{\ell(Q_j)}{t} - a2^l\right) \mathrm{d}t$$

$$\lesssim \lambda |Q_j|^{\frac{1}{2}} \cdot \exp(-a2^l).$$

Thus,

$$\sum_{j \in J} \|uT_j b_j\|_{L^1(\mathbb{R}^n \setminus \bigcup_{j^\dagger \in J} 4Q_{j^\dagger})} \lesssim \lambda \int_{\bigcup_{j \in J} Q_j} M^{(2)} u(y) \mathrm{d}y.$$

If we use the Kolmogorov inequality, Example 1.29, then we have (6.251).

To deal with U_j we need the Littlewood–Paley theory. To prove (6.252), we note $U_j = \dfrac{1}{\ell(Q_j)} \displaystyle\int_1^\infty \ell(Q_j)^2 L e^{-t^2 \ell(Q_j)^2 L} \mathrm{d}t$ by a change of variables. With this in mind, we define $\psi(z)$ by (6.249). Then $\psi(4^k L) = 2^k \displaystyle\int_{2^k}^\infty L \exp(-t^2 L) \mathrm{d}t$. To apply Lemma 6.37, observe that the definitions of r_i and U_i yield

$$\sum_{j \in J} U_j b_j = \sum_{k=-\infty}^{\infty} \psi(4^k L) f_k, \tag{6.255}$$

where $f_k \equiv 2^{-k} \displaystyle\sum_{j \in J,\, \ell(Q_j) = 2^k} b_j$. Using the bounded overlapping property of $\{Q_j\}_{j \in J}$, (6.255) and Lemma 6.37, we have

$$\left\| \sum_{j \in J} U_j b_j \right\|_{L^p} \lesssim \|\{f_k\}_{k=-\infty}^{\infty}\|_{L^p(\ell^2)} \lesssim \|\ell(Q_j)^{-1} b_j\|_{L^p(\ell^p)}. \tag{6.256}$$

We now use the $\dot{W}^{1,\infty}(\mathbb{R}^n)$*-condition* (1.200). We deduce from Exercise 1.83 and the Sobolev embedding together with the fact that b_j is supported on Q_j that

$$\int_{\mathbb{R}^n} \sum_{j \in J} \ell(Q_j)^{-p} |b_j(x)|^p \mathrm{d}x \lesssim \int_{\mathbb{R}^n} \sum_{j \in J} |\nabla b_j(x)|^p \mathrm{d}x \lesssim \lambda^p \sum_{j \in J} |Q_j|. \tag{6.257}$$

Hence from (1.196), (6.256) and (6.257) we obtain

$$\left| \left\{ \left| \sum_{j \in J} U_j b_j \right| > \lambda \right\} \right| \leq \left(\lambda^{-1} \left\| \sum_{j \in J} U_j b_j \right\|_p \right)^p \lesssim \sum_{j \in J} |Q_j| \lesssim (\lambda^{-1} \|\nabla f\|_{(L^p)^n})^p,$$

proving (6.252).

6.7.3.1 The Reverse Estimate

The affirmative solution of the Kato conjecture yields $\|\nabla f\|_{(L^2)^n} \sim \|\sqrt{L}f\|_2$ for all $f \in H^1(\mathbb{R}^n) = \mathrm{Dom}(\sqrt{L})$. We also note that $\sqrt{L}$ is invertible since L is invertible. Observe that $\mathrm{Ran}(L)$ and hence $\mathrm{Ran}(\sqrt{L})$ is dense in $L^2(\mathbb{R}^n)$. Thus, the operator $\nabla L^{-1/2} : \sqrt{L}f \in L^2(\mathbb{R}^n) \mapsto \nabla f \in (L^2(\mathbb{R}^n))^n$, which is defined initially on $\mathrm{Ran}(\sqrt{L})$, extends to an $L^2(\mathbb{R}^n)$ bounded operator. By the solution of the Kato conjecture it is bounded. We prove the following estimate:

Theorem 6.50 *Let* $p_0 \equiv \dfrac{2n}{n+1}$. *Then*

$$|\{|\nabla L^{-1/2}f| > \lambda\}| \lesssim (\lambda^{-1}\|f\|_{p_0})^{p_0} \tag{6.258}$$

for all $\lambda > 0$ *and* $f \in L^{p_0}(\mathbb{R}^n) \cap L^2(\mathbb{R}^n)$. *In particular, if* $p_0 < p < 2$, *then*

$$\|\nabla f\|_{(L^p)^n} \lesssim \|\sqrt{L}f\|_p \tag{6.259}$$

for all $f \in \mathrm{Dom}(\sqrt{L}) = H^1(\mathbb{R}^n)$ *such that* $\sqrt{L}f \in L^p(\mathbb{R}^n)$.

We note that (6.259) follows from interpolation between the solution of the Kato conjecture and (6.258). In fact, we have (6.259) for $f \in L^p(\mathbb{R}^n) \cap L^2(\mathbb{R}^n)$. This means that

$$\|\mathbf{H} \cdot \nabla L^{-1/2}f\|_1 \lesssim \|\mathbf{H}\|_{(L^{p'})^n}\|f\|_{L^2} \tag{6.260}$$

for $f \in L^p(\mathbb{R}^n) \cap L^2(\mathbb{R}^n)$ and $\mathbf{H} \in (L^\infty_{\mathrm{c}}(\mathbb{R}^n))^n$. If we pass to the limit, (6.260) is valid for $f \in L^2(\mathbb{R}^n)$ and $\mathbf{H} \in (L^\infty_{\mathrm{c}}(\mathbb{R}^n))^n$. Thus, if we replace f by $\sqrt{L}f$ with $f \in \mathrm{Dom}(\sqrt{L}) = H^1(\mathbb{R}^n)$, then we have

$$\|\mathbf{H} \cdot \nabla f\|_1 \lesssim \|\mathbf{H}\|_{(L^{p'})^n}\|\sqrt{L}f\|_{L^2} \tag{6.261}$$

for $f \in H^1(\mathbb{R}^n)$ and $\mathbf{H} \in (L^\infty_{\mathrm{c}}(\mathbb{R}^n))^n$. Since $\mathbf{H}$ is arbitrary, (6.259) follows for all $f \in H^1(\mathbb{R}^n)$. Let us concentrate on (6.258). We need the following estimate:

Theorem 6.51 *Let* $m \geq 1$ *be an integer. Let* E *and* F *be disjoint closed sets in* $\mathbb{R}^n$ *with distance* d. *Then for all* $\boldsymbol{F} \in \mathrm{Dom}(\mathrm{div})$,

$$\|t\nabla L^{-1/2}(1-\exp(-t^2L))^m\mathrm{div}(\boldsymbol{F})\|_{L^2(F)} \lesssim t^{m+1/2}d^{-2m-1}\|\boldsymbol{F}\|_{L^2(E)^n}. \tag{6.262}$$

Proof The key idea is to decompose $L^{-1/2}$: We use $\nabla L^{-1/2}h \simeq \int_0^\infty \nabla e^{-(m+2)sL}h \frac{\mathrm{d}s}{\sqrt{s}}$ for $h \in \mathrm{Dom}(L^{-1/2}) = \mathrm{Ran}(\sqrt{L})$. Thus, since $\mathrm{Ran}((\mathrm{id}_{L^2} - \mathrm{e}^{-t^2L})^m) \subset \mathrm{Ran}(\sqrt{L})$,

$$\nabla L^{-1/2}(\mathrm{id}_{L^2} - e^{-^2 L})^m \mathrm{div}(\mathbf{F}) \simeq \int_0^\infty \nabla e^{-(m+2)sL}(\mathrm{id}_{L^2} - \mathrm{e}^{-t^2 L})^m \mathrm{div}(\mathbf{F}) \frac{\mathrm{d}s}{\sqrt{s}}.$$

We set

$$\mathrm{I} \equiv \int_0^{t^2} \nabla e^{-msL} \circ (\mathrm{id}_{L^2} - \mathrm{e}^{-t^2 L})^m e^{-2sL} \mathrm{div}(\mathbf{F}) \frac{\mathrm{d}s}{\sqrt{s}},$$

$$\mathrm{II} \equiv \int_{t^2}^\infty \nabla \mathrm{e}^{-sL} \circ (\mathrm{e}^{-sL} - \mathrm{e}^{-(t^2+s)L})^m \circ \mathrm{e}^{-sL} \mathrm{div}(\mathbf{F}) \frac{\mathrm{d}s}{\sqrt{s}},$$

so that $\nabla L^{-1/2}(\mathrm{id}_{L^2} - \mathrm{e}^{-t^2 L})^m \mathrm{div}(\mathbf{F}) \simeq \mathrm{I} + \mathrm{II}$. As for I, we simply use the off-diagonal estimate after we expand $(\mathrm{id}_{L^2} - \mathrm{e}^{-t^2 L})^m$. If we use Lemma 5.30, then we obtain

$$\begin{aligned}
\|\mathrm{I}\|_{L^2(F)} &\le \int_0^{t^2} \|\nabla e^{-msL} \circ (\mathrm{id}_{L^2} - \mathrm{e}^{-t^2 L})^m e^{-2sL} \mathrm{div}(\mathbf{F})\|_{L^2(F)} \frac{\mathrm{d}s}{\sqrt{s}} \\
&\lesssim t^{-1} \int_0^{t^2} \exp\left(-a\frac{d^2}{t^2}\right) \|f\|_{L^2(E)} \frac{\mathrm{d}s}{\sqrt{s}} \\
&\simeq \exp\left(-a\frac{d^2}{t^2}\right) \|f\|_{L^2(E)}
\end{aligned}$$

for some $a > 0$. As for II, from (5.130) we have

$$\|(\mathrm{e}^{-sL} - \mathrm{e}^{-(s+t^2)L})g\|_{L^2(F)} \le \int_0^{t^2} \|L\mathrm{e}^{-(s+r)L} g\|_{L^2(F)} \mathrm{d}r \lesssim \frac{t^2}{s} \exp\left(-a\frac{d^2}{s}\right) \|g\|_{L^2(E)}$$

for all $g \in L^2(E)$ for some $a > 0$ again. Thus

$$\begin{aligned}
\|\mathrm{II}\|_{L^2(F)} &\le \int_{t^2}^\infty \|\nabla e^{-msL} \circ (\mathrm{id}_{L^2} - \mathrm{e}^{-t^2 L})^m e^{-2sL} \mathrm{div}(\mathbf{F})\|_{L^2(F)} \frac{\mathrm{d}s}{\sqrt{s}} \\
&\lesssim t^{2m+1} \int_0^\infty s^{-m} \exp\left(-a\frac{d^2}{s}\right) \|f\|_{L^2(E)} \frac{\mathrm{d}s}{\sqrt{s}} \\
&\simeq t^{2m+1} d^{-2m-1} \|f\|_{L^2(E)}.
\end{aligned}$$

Using these estimates, we obtain (6.262).

We dualize what we have obtained.

Corollary 6.4 *Let E and F be disjoint closed sets. Then for all $t > 0$ and $m \in \mathbb{N}$*

$$\|M_{\chi_F} t\nabla((\nabla L^{-1/2}) \circ (\mathrm{id}_{L^2} - \mathrm{e}^{-t^2 L})^m)^*\|_{B(L^2(E)^n, L^2(F)^n)} \lesssim \frac{t^{2m+1}}{\mathrm{dist}(E, F)^{2m+1}}.$$

Proof Let $F' \equiv \{x \in \mathbb{R}^n : \mathrm{dist}(x, F) \le \mathrm{dist}(x, E)\}$. Note that

$$\begin{aligned}
&\|M_{\chi_F} t\nabla(\nabla L^{-1/2}(\mathrm{id}_{L^2} - \mathrm{e}^{-t^2 L})^m)^*\|_{B(L^2(E)^n, L^2(F)^n)} \\
&\le \sup\{|\langle t\nabla(\nabla L^{-1/2}(\mathrm{id}_{L^2} - \mathrm{e}^{-t^2 L})^m)^* \mathbf{F}, \mathbf{G}\rangle_{(L^2)^n}| \\
&\qquad : \mathbf{F} \in L^2(E)^n, \mathbf{G} \in H^1(\mathbb{R}^n)^n, \|\mathbf{F}\|_{(L^2)^n} = \|\mathbf{G}\|_{(L^2)^n} = 1, \mathrm{supp}(\mathbf{G}) \subset F'\}.
\end{aligned}$$

Let $\mathbf{F} \in L^2(E)^n$ supported in E, and let $\mathbf{G} \in H^1(\mathbb{R}^n)^n \cap L^2(F')$. Then we have

$$\begin{aligned}
&|\langle t\nabla(\nabla L^{-1/2}(\mathrm{id}_{L^2} - \mathrm{e}^{-t^2 L})^m)^* \mathbf{F}, \mathbf{G}\rangle_{(L^2)^n}| \\
&= |\langle (\nabla L^{-1/2}(\mathrm{id}_{L^2} - \mathrm{e}^{-t^2 L})^m)^* \mathbf{F}, t\mathrm{div}(\mathbf{G})\rangle_2| \\
&= |\langle \mathbf{F}, (\nabla L^{-1/2}(\mathrm{id}_{L^2} - \mathrm{e}^{-t^2 L})^m) t\mathrm{div}(\mathbf{G})\rangle_{(L^2)^n}| \\
&\le \|\mathbf{F}\|_{(L^2)^n} \|(\nabla L^{-1/2}(\mathrm{id}_{L^2} - \mathrm{e}^{-t^2 L})^m) t\mathrm{div}(\mathbf{G})\|_{(L^2(E))^n}.
\end{aligned}$$

Thus, it remains to use the new off-diagonal estimate Theorem 6.51.

Proof (of (6.258)) We decompose $f \in L^{p_0}(\mathbb{R}^n) \cap L^2(\mathbb{R}^n)$ as in Lemma 1.27. We need to show

$$|\{|\nabla L^{-1/2} g| > \lambda\}| + |\{|\nabla L^{-1/2} b| > \lambda\}| \lesssim (\lambda^{-1}\|f\|_{p_0})^{p_0}.$$

For the good part g, we go through the same argument as we did for the singular integral operators in (1.219) keeping in mind that $p_0 < 2$ and that $\nabla L^{-1/2}$ is shown to be $L^2(\mathbb{R}^n)$-$L^2(\mathbb{R}^n)^n$ bounded.

Let us deal with the bad part b. We set $A_j \equiv \mathrm{id}_{L^2} - (\mathrm{id}_{L^2} - e^{-\ell(Q_j)^2 L^*})^m$ for $m \in \mathbb{N}$ and define $b^\dagger \equiv \sum_{j \in J} A_j^* b_j$ and $b^\dagger \equiv b - b^\flat$. Matters are reduced to the estimates:

$$|\{|\nabla L^{-1/2} b^\dagger| > \lambda\}| \lesssim (\lambda^{-1}\|f\|_{p_0})^{p_0}, \tag{6.263}$$

$$|\{|\nabla L^{-1/2} b^\flat| > \lambda\}| \lesssim (\lambda^{-1}\|f\|_{p_0})^{p_0}. \tag{6.264}$$

We start with $b^\dagger$. We can say that $b^\dagger$ is not so bad. By the Chebyshev inequality (see Theorem 1.4), and the boundedness of $\nabla L^{-1/2}$ from $L^2(\mathbb{R}^n)$ to $L^2(\mathbb{R}^n)^n$, we have

$$|\{|\nabla L^{-1/2} b^\dagger| > \lambda\}| \lesssim (\lambda^{-1}\|\nabla L^{-1/2} b^\dagger\|_2)^2 \lesssim (\lambda^{-1}\|b^\dagger\|_2)^2. \tag{6.265}$$

We dualize the most right-hand side. Choose $h \in L^2(\mathbb{R}^n)$ with norm 1 such that $\|b^\dagger\|_2 = \langle b^\dagger, h\rangle_2$. We insert the definition of $b^\dagger$ to have $\|b^\dagger\|_2 = \sum_{j\in J}\langle b_j, A_h\rangle_2$. Since each $b_j \in L^1(Q_j)\cap \mathscr{P}_0^\perp$, we have $\|b^\dagger\|_2 = \sum_{j\in J}\int_{\mathbb{R}^n} b_j(x)\left(\overline{A_j h(x)} - m_{Q_j}(\overline{A_j h})\right)\mathrm{d}x$. Thus we have

$$\|b^\dagger\|_2 \le \sum_{j\in J}\|b_j\|_{p_0}\|A_j h - m_{Q_j}(A_j h)\|_{L^{p_0'}(Q_j)}$$
$$\lesssim \sum_{j\in J}\|b_j\|_{p_0}|Q_j|^{\frac{1}{p_0'}-\frac{1}{2}+\frac{1}{n}}\|\nabla A_j h\|_{L^2(Q_j)^n}$$

by the Hölder inequality and the Poincaré inequality. Using (1.194), we obtain

$$\|b^\dagger\|_2 \lesssim \lambda\sum_{j\in J}|Q_j|^{\frac{1}{2}+\frac{1}{n}}$$
$$\times\left(\|\nabla A_j[\chi_{Q_j}h]\|_{L^2(Q_j)^n} + \sum_{l=1}^\infty \|\nabla A_j[\chi_{2^l Q_j\setminus 2^{l-1}Q_j}h\|_{L^2(Q_j)^n}\right).$$

Again by Lemma 5.29, the off-diagonal estimate, we have

$$\|b^\dagger\|_2 \lesssim \lambda\sum_{j\in J}|Q_j|\inf_{x\in Q_j} M^{(2)}h(x) \le \lambda\int_{\bigcup_{j\in J}Q_j} M^{(2)}h(x)\mathrm{d}x.$$

By the Kolmogorov inequality, Example 1.29, we have

$$\|b^\dagger\|_2 \lesssim \lambda\sqrt{\left|\bigcup_{j\in J}Q_j\right|}. \tag{6.266}$$

Putting together (1.195), (6.265) and (6.266), we obtain (6.263).

We move on to the proof of (6.264). We have only to prove

$$\left|\left\{|\nabla L^{-1/2}b^\flat| > \lambda\right\}\setminus\bigcup_{j^\dagger\in J}4Q_{j^\dagger}\right| \lesssim (\lambda^{-1}\|f\|_{p_0})^{p_0}. \tag{6.267}$$

By the Chebyshev inequality (see Theorem 1.4), we have

$$\left|\left\{|\nabla L^{-1/2}b^\flat| > \lambda\right\}\setminus\bigcup_{j^\dagger\in J}4Q_{j^\dagger}\right| \le (\lambda^{-1}\|\nabla L^{-1/2}b^\flat\|_{L^2(\mathbb{R}^n\setminus\bigcup_{j^\dagger\in J}4Q_{j^\dagger})^n})^2.$$

We dualize the right-hand side. To this end, we choose again $\mathbf{H} \in L^2(\mathbb{R}^n)^n$ with norm 1 so that

$$\|\nabla L^{-1/2} b^\flat\|_{L^2(\mathbb{R}^n \setminus \bigcup_{j^\dagger \in J} 4Q_{j^\dagger})} = \int_{\mathbb{R}^n \setminus \bigcup_{j^\dagger \in J} 4Q_{j^\dagger}} \nabla L^{-1/2} b^\flat(x) \cdot \overline{\mathbf{H}(x)} \mathrm{d}x. \tag{6.268}$$

Inserting the definition of $b^\flat$ into (6.268), we obtain

$$\begin{aligned}
&\int_{\mathbb{R}^n \setminus \bigcup_{j^\dagger \in J} 4Q_{j^\dagger}} \nabla L^{-1/2} b^\flat(x) \cdot \overline{\mathbf{H}(x)} \mathrm{d}x \\
&= \sum_{j \in J} \int_{\mathbb{R}^n \setminus \bigcup_{j^\dagger \in J} 4Q_{j^\dagger}} \nabla L^{-1/2} [(\mathrm{id}_{L^2} - e^{-\ell(Q_j)^2 L})^m b_j](x) \cdot \overline{\mathbf{H}(x)} \mathrm{d}x \\
&= \sum_{j \in J} \int_{\mathbb{R}^n} b_j(x) \overline{[\nabla L^{-1/2} (\mathrm{id}_{L^2} - e^{-\ell(Q_j)^2 L})^m]^* [\chi_{\mathbb{R}^n \setminus \bigcup_{j^\dagger \in J} 4Q_{j^\dagger}} \mathbf{H}](x)} \mathrm{d}x.
\end{aligned}$$

Let $l = 1, 2, \ldots$ and $j \in J$. We write

$$k_{j,l} \equiv [\nabla L^{-1/2} (\mathrm{id}_{L^2} - e^{-\ell(Q_j)^2 L})^m]^* [\chi_{2^{l+2} Q_j \setminus \bigcup_{j^\dagger \in J} (4Q_{j^\dagger} \cup 2^{l+1} Q_j)} \mathbf{H}].$$

Then since $b_j \perp \mathscr{P}_0$,

$$\|\nabla L^{-1/2} b^\flat\|_{L^2(\mathbb{R}^n \setminus \bigcup_{j^\dagger \in J} 4Q_{j^\dagger})} = \sum_{j \in J} \sum_{l=1}^{\infty} \int_{\mathbb{R}^n} b_j(x) (k_{j,l}(x) - m_{Q_j}(k_{j,l})) \mathrm{d}x.$$

By the Poincaré inequality and (6.268), we have

$$\|\nabla L^{-1/2} b^\flat\|_{L^2(\mathbb{R}^n \setminus \bigcup_{j^\dagger \in J} 4Q_{j^\dagger})} \lesssim \sum_{j \in J} \sum_{l=1}^{\infty} |Q_j|^{\frac{1}{p_0'} - \frac{1}{2} + \frac{1}{n}} \|b_j\|_{p_0} \|\nabla k_{j,l}\|_{L^2(Q_j)^n}.$$

By Lemma 5.29, the off-diagonal estimate, and (1.194), we obtain

$$\begin{aligned}
\|\nabla L^{-1/2} b^\flat\|_{L^2(\mathbb{R}^n \setminus \bigcup_{j^\dagger \in J} 4Q_{j^\dagger})} &\lesssim \sum_{j \in J} |Q_j|^{\frac{1}{p_0'}} \|b_j\|_{p_0} \inf_{x \in Q_j} M^{(2)}[|\mathbf{H}|](x) \\
&\leq \sum_{j \in J} |Q_j| \inf_{x \in Q_j} M^{(2)}[|\mathbf{H}|](x).
\end{aligned}$$

Using the Kolmogorov inequality, Example 1.29 as before, we obtain (6.267).

Textbooks in Sect. 6.7

Kato Conjecture (Kato Theorem)

See the textbook [33, Section 4.7] for Theorem 6.48 and the comprehensive textbook [5] for more about Kato's conjecture, where the special cases of $n = 1, 2$ and the case of having the kernel satisfying the Gaussian estimates can be found.

Exercises

Exercise 6.94 Show that the function ψ defined by (6.249) is bounded and holomorphic on the right-half plane of $\mathbb{C}$.

Exercise 6.95 Let $A \equiv \{a_{ij}\}_{i,j=1}^n \in L^\infty(\mathbb{R}^n)^{n^2}$ and set $L = -\mathrm{div} \circ M_A \circ \nabla$.

1. Show that $\|\sqrt{\mathrm{id}_{L^2} + L} f\|_2 \sim \|(1 - \Delta)^{\frac{1}{2}} f\|_2$ for all $f \in H^1(\mathbb{R}^n)$.
2. Using Theorem 4.27, show that there exists an open interval I containing 2 such that $\|\sqrt{\mathrm{id}_{L^p} + L} f\|_p \sim \|(1 - \Delta)^{\frac{1}{2}} f\|_p$ for all $f \in H^1_p(\mathbb{R}^n)$ if $p \in I$.

6.8 Notes for Chap. 6

Section 6.1

Section 6.1.1

Banach Function Spaces

See the dissertation by Luxemberg [757] of 1955.

Section 6.1.2

Weighted Lebesgue Spaces

Theorem 6.7, the Fefferman–Stein vector-valued inequality for weighted Lebesgue spaces, dates back to 1980 (see [136]). The Littlewood–Paley theory for weighted Lebesgue spaces also dates back to 1980 (see [696]).

The proof of Theorem 6.3 presented in this book is by Lerner [708].

A_p Weights

See [811–813].

Section 6.1.3

Mixed Lebesgue Spaces

See [496]. See [1038] for wavelets on mixed Lebesgue spaces.

Section 6.1.4

Variable Lebesgue Spaces: The Work of Hidegoro Nakano

The variable Lebesgue space dates back to 1931; see the papers by Birnbaum and Orlicz [252] and [859, p. 211]. In [831, 832] Nakano considered variable Lebesgue norms. See [620] for a detailed explanation of the old books [831, 832].

The Boundedness of the Hardy–Littlewood Maximal Operator: Theorem 6.15

Theorem 6.15 proved by Diening [421] and Cruz-Uribe, Fiorenza and Neugebauer [389, 390]. But the proof was improved a couple of times. The method of the proof is due to Adamowicz, Harjulehto, Hästö, Mizuta and Shimomura (see [125, 799, 800]). We remark that a similar technique is used in [797, 798]. In [797, Lemma 3.5] and [489], an estimate was obtained with the help of the Hardy operator.

A similar technique to Theorem 6.15, the proof of the boundedness of the Hardy–Littlewood maximal operator in variable Lebesgue spaces, is used to prove the boundedness of the one-sided maximal operator, see [448].

See [423, Lemma 3.3 and Corollary 3.4] for (6.58).

Section 6.1.5

Morrey Spaces

Morrey spaces are function spaces which date back to 1938 (see the paper by Morrey [804]). Peetre established the position of the Banach spaces of Morrey spaces in [871]. Example 4.1 shows that $B^s_{pq}(\mathbb{R}^n)$ describes the singularity of the function $|x|^{-\alpha}$ at $x = 0$. We can use Morrey spaces as well for this purpose. We refer to [362] for the boundedness of the various fundamental operators. See [926] for Example 6.4.

Interpolation of Morrey Spaces

Let us explain why the interpolation of Morrey spaces is complicated, unlike Lebesgue spaces. Noteworthy is the fact that the first complex interpolation functor behaves differently from Lebesgue spaces. This problem comes basically from the fact that the Morrey norm $\mathcal{M}^p_q(\mathbb{R}^n)$ involves the supremum over all balls $B(a,r)$. Due to this fact, we have many difficulties when $1 < q < p < \infty$, namely:

1. The Morrey space $\mathcal{M}^p_q(\mathbb{R}^n)$ is not reflexive; see [926, Example 5.2] and [1155, Theorem 1.3].
2. The Morrey space $\mathcal{M}^p_q(\mathbb{R}^n)$ does not have $C^\infty_{\rm c}(\mathbb{R}^n)$ as a dense subspace, that is $\overline{C^\infty_{\rm c}(\mathbb{R}^n)}^{\mathcal{M}^p_q} \neq \mathcal{M}^p_q(\mathbb{R}^n)$; see [110, Proposition 2.16].
3. The Morrey space $\mathcal{M}^p_q(\mathbb{R}^n)$ is not separable; see [110, Proposition 2.16].
4. The Morrey space $\mathcal{M}^p_q(\mathbb{R}^n)$ is not included in $L^1(\mathbb{R}^n) + L^\infty(\mathbb{R}^n)$; see [532, Section 6].

The earlier result about the interpolation of Morrey spaces can be traced back to [1007]. In [377, p. 35] Cobos, Peetre, and Persson pointed out that

$$[\mathcal{M}^{p_0}_{q_0}(\mathbb{R}^n), \mathcal{M}^{p_1}_{q_1}(\mathbb{R}^n)]_\theta \subset \mathcal{M}^p_q(\mathbb{R}^n)$$

whenever $1 \le q_0 \le p_0 < \infty$, $1 \le q_1 \le p_1 < \infty$, and $1 \le q \le p < \infty$ satisfy

$$\frac{1}{p} = \frac{1-\theta}{p_0} + \frac{\theta}{p_1}, \quad \frac{1}{q} = \frac{1-\theta}{q_0} + \frac{\theta}{q_1}. \tag{6.269}$$

A counterexample by Blasco, Ruiz, and Vega [286, 895], shows that if one assumes (6.269) only, then there exists a bounded linear operator T from $\mathcal{M}^{p_k}_{q_k}(\mathbb{R}^n)$ $(k = 0, 1)$ to $L^1(\mathbb{R}^n)$, but T is unbounded from $\mathcal{M}^p_q(\mathbb{R}^n)$ to $L^1(\mathbb{R}^n)$. By using the counterexample by Ruiz and Vega in [895], Lemarié-Rieusset [705, Theorem 3(ii)] showed that if an interpolation functor F satisfies

$$F[\mathcal{M}^{p_0}_{q_0}(\mathbb{R}^n), \mathcal{M}^{p_1}_{q_1}(\mathbb{R}^n)] = \mathcal{M}^p_q(\mathbb{R}^n)$$

under the condition (6.269), then

$$\frac{q_0}{p_0} = \frac{q_1}{p_1} \tag{6.270}$$

holds. In [705, 706] under the condition (6.269) Lemarié-Rieusset also showed that the Morrey scale is closed under the second complex interpolation method, introduced by Calderón [325] in 1964. Meanwhile, as for the interpolation result under (6.269) and (6.270), by using the first complex interpolation functor by Calderón [325], Lu, Yang, and Yuan obtained the following description:

$$[\mathcal{M}_{q_0}^{p_0}(\mathbb{R}^n), \mathcal{M}_{q_1}^{p_1}(\mathbb{R}^n)]_\theta = \overline{\mathcal{M}_{q_0}^{p_0}(\mathbb{R}^n) \cap \mathcal{M}_{q_1}^{p_1}(\mathbb{R}^n)}^{\mathcal{M}_q^p(\mathbb{R}^n)}$$

in [756, Theorem 1.2]. Their result is in the setting of a metric measure space. The generalization of the result of Lu et al. and Lemarié-Rieusset in the setting of generalized Morrey spaces and generalized Orlicz–Morrey spaces can be seen in [531]. See also [529, 530] for the case of diamond spaces. See [771] for the case of variable exponent Morrey spaces. As for the real interpolation results, Burenkov and Nursultanov obtained an interpolation result in local Morrey spaces [311] and their results are generalized by Nakai and Sobukawa to $B^u_w(\mathbb{R}^n)$ setting [828].

Section 6.1.6

Orlicz Spaces

See [252, 859] for the original definition.

Section 6.1.7

Herz Spaces

Herz spaces initially appeared in the paper of Herz [588] to study the absolute convergence of the Fourier series. We remark that the Fefferman–Stein-type maximal inequality for homogeneous Herz spaces was proved in [1030, Theorems 2.1 and 2.4 and Remark 2.4]. Herz spaces can be used to characterize the multipliers in Hardy spaces [164]. They can be also used for the study of the regularity theory for elliptic equations [884, 885]. See [684] for the boundedness of the singular integral operators on Herz spaces. See [1030] for Theorem 6.26.

Section 6.2

Section 6.2.1

Weak Hardy Spaces

Weak Hardy spaces form an interpolation scale for real methods [456] and play an important role in the study of the boundedness of operators in Hardy spaces in the endpoint case. In [457], Fefferman and Soria introduced the weak Hardy space $WH^1(\mathbb{R}^n)$ via the radial maximal function and established its various characterizations in terms of the Lusin-area function, the nontangential maximal function, the grand maximal function, Riesz transforms and atoms. Fefferman and

Soria also established the boundedness of some singular integrals on $WH^1(\mathbb{R}^n)$. See the works by Liu [743, 883] for extensions of these results. The work [1141] considers weak Hardy spaces $WH^{p(\cdot)}(\mathbb{R}^n)$ with variable exponent. See [123, 749] for Hardy spaces generated by Lorentz spaces.

Hardy Spaces Associated with Beurling Algebra

The idea of introducing Hardy spaces associated with quasi-Banach function spaces dates back to the paper [361]. Chen and Lau investigated Hardy spaces associated with Beurling algebra in [361]. They specified the dual space in [361, Theorem A] and obtained the maximal function characterization in [361, Theorem B].

Hardy–Sobolev Spaces

Miyachi studied Hardy–Sobolev spaces in [785]. See [360] for Hardy and Hardy–Sobolev spaces on strongly Lipschitz domains.

Weighted Hardy Spaces

García-Cuerva initially considered weighted Hardy spaces in [497], where the weights belong to A_∞.

Strömberg and Torchinsky [1016] generalized the class of weights, they considered the doubling weights, that is, weights w satisfying $w(B(x, 2r)) \lesssim w(B(x, r))$. See [787, 789] for weighted Hardy spaces on domains.

Herz-Type Hardy Spaces

García-Cuerva and Herrero defined Herz-type Hardy spaces and obtained, in particular, the atomic decomposition [498].

Lu and Yang proved the boundedness of the Fourier multiplier in [755, Theorem 1] and [788, Theorem 6.2]. We refer to [1087, Theorem 3.1] for the boundedness of the pseudo-differential operators on Herz-type Hardy spaces. The key tool for these results is the atomic decomposition obtained in [753, Theorem 2.1] and [788, Theorem 4.1]. In particular, the paper [788] extends the existing results to the case $0 < p, q \leq \infty$. See also [752, 1088] for more about Herz-type Hardy spaces.

Hardy–Morrey Spaces

Sawano obtained the Littlewood–Paley characterization of Hardy–Morrey spaces [915]. Jia and Wang obtained the atomic decomposition of the space $H\mathcal{M}^p_q(\mathbb{R}^n)$ with $0 < q \le 1$ [631]. Jia and Wang investigated Hardy–Morrey spaces to apply them to the Navier–Stokes equations [1109].

Orlicz–Hardy Spaces

The Orlicz–Hardy space $H^{\Phi}(\mathbb{R}^n)$ was introduced by Strömberg [1015] and Janson [626] for a certain class of Φ. Serra [946] further characterized these Orlicz–Hardy spaces, respectively, via molecules on $\mathbb{R}^n$. Viviani [1099] considered the spaces of homogeneous type in the sense of Coifman and Weiss [385]. Later the relation between the Jacobian and the Orlicz–Hardy spaces was pointed out by Iwaniec and Onninen [613]. See [827] for Orlicz–Hardy spaces, where Φ does not depend on x but we can remove the convexity assumption of Φ.

Musielak–Orlicz–Hardy Spaces

The Musielak–Orlicz–Hardy space is a function space which unifies the Hardy space, the weighted Hardy space, the Orlicz–Hardy space and the weighted Orlicz–Hardy space, in which the spatial and the time variables may not be separable. We refer to [713] for the molecular characterization of anisotropic Musielak–Orlicz–Hardy spaces. Dachun Yang and Sibei Yang characterized the Musielak–Orlicz–Hardy spaces using the maximal function in [1149], while Yang and Liang characterized the Musielak–Orlicz–Hardy spaces using the intrinsic square function characterizations [724]. Musielak–Orlicz–Hardy spaces can be characterized by means of the Riesz transform; see [339]. See [725] for weak Musielak–Orlicz–Hardy spaces.

Musielak–Orlicz–Hardy spaces naturally appear in many applications; see [253, 254, 256, 422, 691], for example.

Product of $H^1(\mathbb{R}^n)$ and $\mathrm{BMO}(\mathbb{R}^n)$

Ky introduced the Orlicz–Hardy space $H^{\log}(\mathbb{R}^n)$ defined in Theorem 6.27 [691].

Hardy Spaces with Variable Exponents

Cruz-Uribe, Wang, Nakai and Sawano defined Hardy spaces with variable exponents in [394, 826]. The atomic decomposition is expanded in [920]. See [1189, Proposition 2.1] for the decomposition a dapted to the real interpolation. See [1186] for the intrinsic square function characterizations. See [638] for the setting of the martingales.

Section 6.2.2

Pointwise Multiplication of $H^1(\mathbb{R}^n)$ and $\mathrm{BMO}(\mathbb{R}^n)$

See the paper [692], where Ky used the Hardy–Orlicz spaces defined in [691].

Pointwise Multiplication of $H^1(X)$ and $\mathrm{BMO}(X)$ in Metric Measure Spaces X

The pointwise multiplication of $H^1(X)$ and $\mathrm{BMO}(X)$ in metric measure spaces (X, d) goes back to the paper [255]. See [255, p. 1416] for why we need to consider the pointwise product of $H^1(X)$ and $\mathrm{BMO}(X)$ from the viewpoint of PDEs and so on. See [470], where Feuto worked in the setting of spaces of homogeneous type. Unfortunately his construction was not bilinear. Bonami, Grellier and Ky overcame this problem using wavelets. Fu, Yang and Liang used wavelets on the spaces of homogeneous type. Auscher and Hytönen built an orthonormal basis of Hölder continuous wavelets with exponential decay via developing randomized dyadic structures and properties of spline functions over general spaces of homogeneous type; see [150]. Based on these wavelets, Fu, Yang and Liang constructed a linear mapping $H^1(\mathbb{R}^n) \times \mathrm{BMO}(\mathbb{R}^n)$ to $H^{\log}(\mathbb{R}^n)$, where $H^{\log}(\mathbb{R}^n)$ stands for the Hardy space defined in Theorem 6.27; see [486, 538, 1146].

Section 6.3.1

Axiomatic Approach of Generalized Besov/Triebel–Lizorkin Spaces

Theorems 3.2 and 3.3 are analogously transformed into those for many other function spaces such as Morrey spaces, Orlicz spaces, Herz spaces and so on. Triebel replaced $L^p(\mathbb{R}^n)$ with a general Banach space X to define the space $B^s_q(X)$ with $s > 0$ and $1 \leq q < \infty$; when $X = L^p(\mathbb{R}^n)$ the space will become $B^s_{pq}(\mathbb{R}^n)$. Triebel defined $B^s_q(X)$ in [1045, p. 169, Definition] and investigated the interpolation in [1045, Theorem 3]. More and more attention has been paid to these function spaces. We refer also to [71, p. 205, Definition 3] for this direction of extension. See [836] for some embedding results.

An increasingly axiomatic approach was taken in [584, 598, 600, 602].

Peetre Maximal Functions and Generalized Besov/Triebel–Lizorkin Spaces

Exercise 2.26 is widely known as the Peetre characterization. This type of characterization is useful because we can handle functions without recourse to the Hardy–Littlewood maximal operator to a large extent. A series of studies stems from the generalized coorbit theory. The coorbit theory was related to modulation spaces and the group structure of the underlying spaces [387, 465–467, 523, 886]. However, Fornasier and Rauhut noticed that the group structure is not necessary in order to develop a theory, as is pointed out by introducing the notion of continuous frames in locally compact spaces in [472, Section 2]. Rauhut and Ullrich pointed out that the Peetre maximal operator makes function spaces more transparent [668, 887]. The problems presented in [887, Section 1] were settled in [727, 728]; Besov-type spaces and Triebel–Lizorkin-type spaces are handled in this framework in [727] and quasi-Banach function spaces in general are shown to fall under this scope in [727].

Other Approaches

In [598], Triebel–Lizorkin spaces can be generalized by replacing the Lebesgue space $L^p(\mathbb{R}^n)$ and the sequence space $\ell^q(\mathbb{N})$ by Banach function spaces invariant under rearrangement and sequence spaces with the UMD property respectively; see [598, Definition 4.5]. In [603, Theorem 4.2], Ho considered the condition under which to have $\mathscr{X}$ possess the Littlewood–Paley characterization. Ho also obtained the atomic decomposition for these spaces.

Besov Spaces Associated with Operators: Theorem 6.28

Theorem 6.28 is due to Peetre [866] and Triebel [1041, 1042]. See [521, 729–731, 736, 869] for more.

Let N be an integer and $a(x,\xi)$ be a symbol satisfying the following properties

$$|\partial^{(\alpha,\beta)}\varphi(x,\xi)| \lesssim_{\alpha,\beta} \langle\xi\rangle^{-|\alpha|+|\beta|\delta}, \quad \langle\xi\rangle^{m'} \lesssim a \lesssim \langle\xi\rangle^{m}.$$

In [709] Leopold considered the following partition $\{\varphi_j(x,\xi)\}_{j=0}^{\infty} \subset C^\infty(\mathbb{R}^n)$ of unity to define Besov spaces associated with the pseudo-differential operator $a(X,D)$.

1. $\varphi_j \geq 0$.
2. Let $j = 0, 1, 2, \ldots, N$. If $|a(x,\xi)| \geq 2^{J+N+j}$, then $\varphi_j(x,\xi) = 0$.
3. Let $j = N+1, N+2, \ldots$. If $|a(x,\xi)| \leq 2^{J-N+j}$ or $|a(x,\xi)| \geq 2^{J+N+j}$, then $\varphi_j(x,\xi) = 0$.
4. $|\partial^{(\alpha,\beta)}\varphi(x,\xi)| \lesssim_{\alpha,\beta} \langle\xi\rangle^{-|\alpha|+|\beta|\delta}$.

5. $\sum_{j=0}^{\infty} \varphi_j(x, \xi) = 1.$

Here J is a constant which is fixed so that $|a(x, \xi)| \leq 2^{J+1}$ whenever $(x, \xi) \in \mathbb{R}^n \times B(R_a)$.

Elliptic Operators and Fundamental Estimates

We refer to [169, 293, 294, 296, 420, 669] and the references therein for function spaces generated by the elliptic operators on manifolds, or Hermite operators.

Section 6.3.2

Besov–Orlicz Spaces

See [878].

Besov–Morrey Spaces

In 1984, Netrusov defined Besov–Morrey spaces [835]. Netrusov obtained some embedding results. Later. Kozono and Yamazaki shed light on Besov–Morrey spaces from the context of differential equations [687]. It is Kozono and Yamazaki that considered the Morrey space $\mathcal{M}_q^p(\mathbb{R}^n)$ as a candidate for $X(\mathbb{R}^n)$ in 1994 [687] to investigate the Cauchy problem for the Navier–Stokes equation. Najafov considered Besov–Morrey spaces of the mixed derivative in [821]. Motivated by this, Tang and Xu [1031] defined nonhomogeneous Triebel–Lizorkin–Morrey spaces, or equivalently, nonhomogeneous Morrey-type Triebel–Lizorkin spaces in terms of the original paper [687]. After this, Sawano and Tanaka named homogeneous Triebel–Lizorkin–Morrey spaces [925]. Wang considered atomic decomposition for these function spaces. As an application of two different atomic decompositions Sawano and Wang obtained the trace theorem independently in [919, Theorem 1.1] and [1108, Proposition 1.10], respectively. The wavelet characterization of smoothness Morrey spaces can be found in [894, 914]. Triebel–Lizorkin–Morrey spaces cover Hardy–Morrey spaces; see [915, Theorem 4.2]. We refer to [571–573, 916, 925, 1031] for embedding relations of these function spaces. See [573] for the Frank–Jawerth Sobolev embedding for smoothness Morrey spaces. See also [126, 830] for more about generalized Besov–Morrey spaces.

Equivalent Norms of $\mathscr{A}^s_{pqr}(\mathbb{R}^n)$

We refer to [1183] for the characterizations of Triebel–Lizorkin–Morrey spaces via ball averages.

Applications of $\mathscr{A}^s_{pqr}(\mathbb{R}^n)$

See [768] and [915, 1031] for applications of $B^l_{\infty\infty}S^m_{1\delta}(\mathbb{R}^n)$ to partial differential equations with initial value in Besov–Morrey spaces.

Herz-type Besov/Triebel–Lizorkin Spaces

Although Kozono and Yamazaki introduced Besov–Morrey spaces in 1994, much more was investigated for Herz spaces; Xu defined Herz-type Besov spaces. These spaces cover Herz-type Sobolev spaces due to Lu and Yang [754]. Herz-type Triebel–Lizorkin spaces date back to 2005 [1137, Definition 3.1]. Xu considered the boundedness property of the Fourier multipliers in [1119] for Herz-type Triebel–Lizorkin spaces and proved the boundedness property of the lift operator, as well as the embedding property of the Schwatz class in [1120]. The boundedness property of the pointwise multiplier is obtained in [1121, 1125, 1136]. Xu proved the boundedness property of the pseudo-differential operators in [1122, 1125, 1136]. Related to Herz-type Triebel–Lizorkin spaces, Yang and Xu defined Herz-type Hardy spaces [1135, Definition 4.1]. We say that a quasi-normed space X is called admissible, if for every compact subset $E \subset X$ and for every $\varepsilon > 0$, there exists a continuous map $T : E \to X$ such that $T(E)$ is contained in a finite-dimensional subset of X and $\|Tx - x\|_X \le \varepsilon$ for all $x \in E$. Xu characterized the Herz-type Besov spaces by means of the Peetre maximal operator in [1123, 1124] and used this characterization to prove the admissibility in [1129, 1131]. Xu obtained the atomic decomposition, the molecular decomposition, and the wavelet decomposition in [1128, 1133]. We can find applications of Herz-type Triebel–Lizorkin spaces to partial differential equations, more precisely, to the Beal–Kato–Mazya-type and the Moser-type inequalities in [1132]. See [1126, Definition 2] for their definition. Likewise we can consider Herz–Morrey spaces. Drihem investigated the embedding properties of these spaces in [433, Theorems 5.9, 5.14 and 5.19]. Drihem also obtained the characterization by means of the local means in [434, Theorem 2.8] and the atomic decomposition in [434, Theorems 3.12 and 3.17].

Weak Besov Spaces and Weak Triebel–Lizorkin Spaces

The space $L^p(\mathbb{R}^n)$ is replaced by the weak $L^p(\mathbb{R}^n)$ space in [556, 567].

Lorentz Besov Spaces and Lorentz–Triebel–Lizorkin Spaces

See [1105] for an extension of Theorems 4.17 and 4.19, where Vybíral used Lorentz spaces, which cover $L^p(\mathbb{R}^n)$ spaces and used the atomic decomposition to show that the results are sharp.

Mixed Lebesgue–Besov Spaces and Mixed Lebesgue–Triebel–Lizorkin Spaces

Besov investigated some embeddings and extension operators in the mixed norm setting using the characterization by means of the difference in [186]. See [425, Theorem 2.4] for the case of Lipschitz domains.

Applications

Matsumoto and Ogawa replaced $L^p(\mathbb{R}^n)$ with some other function space $X(\mathbb{R}^n)$ in the definition of Besov spaces and applied it to semilinear evolution equations in [765].

Section 6.4

Section 6.4.1

Besov-Type Spaces and Triebel–Lizorkin-Type Spaces

Theorem 6.34 is due to Yang and Yuan; see [1153, Corollary 3.1] for Besov-type spaces and Triebel–Lizorkin-type spaces and [1152, Proposition 3.1] solely for Triebel–Lizorkin-type spaces. Theorem 6.37 can be found in [1156]. The atomic decomposition of these new spaces is investigated in [928] and Theorem 6.35 is proved in [928, Theorem 1.1]. See [1185] for the characterization by means of balls.

Application of Besov-Type Spaces to Partial Differential Equations

We refer to [1170].

The Spaces $F\dot{W}^{s,\tau}_{pq}(\mathbb{R}^{n+1}_+)$ and $F\dot{T}^{s,\tau}_{pq}(\mathbb{R}^{n+1}_+)$

See [1152] for an account. See [1153] for Besov-type spaces.

Section 6.4.3

Besov-Type Spaces and Triebel–Lizorkin-Type Spaces

Decomposition results for Besov/Triebel–Lizorkin-type spaces and for Hausdorff Besov/Triebel–Lizorkin-type spaces can be found in [1153, Theorem 4.3] and [1175, Theorem 3.2], respectively. Definition 6.18 goes back to [1151, Definition 4.2]. These spaces are also characterized in terms of the local means; see [1154, Theorem 1.1]. Saka proposed other scale containing all of these function spaces in [907]. Proposition 6.8 is due to Yang and Yuan; see [1153, Lemma 6.1(ii)] for Besov–Hausdorff spaces and [1152, Lemma 5.1] for Triebel–Lizorkin–Hausdorff spaces.

More was investigated on the spaces. The scale $A^{s,\rho}_{pq}(\mathbb{R}^n)$ arises naturally when we consider the localized space proposed by Triebel in his book [109]; see [1176, Theorem 1.4]. The boundedness of the Fourier multipliers under a weaker smoothness assumption is investigated in [1159, Theorem 1.5]. The compact embedding results are investigated in [1177, Theorem 1.4], where the radial functions play the key role. As we have seen, when ρ is large enough, more precisely, $\rho > 1/p$, then $A^{s,\rho}_{pq}(\mathbb{R}^n)$ is the Hölder–Zygmund space of order $s + n(\rho - 1/p)$; see [1156, Theorem 1], as well as the paper [435] which handles the case of variable exponents in [435, Theorem 3.8]. We refer to [163, 685] for the Littlewood–Paley characterization of Campanato spaces and $\dot{B}_\sigma$–Morrey spaces. We refer to [1161] for a passage to Musielak–Orlicz spaces, which are related to Orlicz–Morrey spaces. See [1169] for the case of variable exponents.

See the surveys [964, 965, 1157] for some further progress.

Yuan and Yang showed that Hausdorff–Triebel–Lizorkin spaces admit the predual spaces [1155, Theorem 1.3]. Drihem obtained the characterization by means of differences [431, Theorems 4.1 and 4.2]. Zhuo, Yang and Yuan clarified that Triebel–Lizorkin–Hausdorff spaces cover the predual space in [1188, Theorem 1.11].

Lemma 6.21 is [1151, Lemma 4.3], while Definition 6.19 is [1151, Definition 4.2].

Besov–Hausdorff-Type Spaces and Triebel–Lizorkin–Hausdorff Spaces

Yuan, Sawano and Yang obtained the boundedness of the trace operator by means of the atomic decomposition of Besov–Hausdorff-type spaces and Triebel–Lizorkin–Hausdorff spaces in [1175, Theorem 4.2]. See [1153, Definition 6.1] for the definition of $B\dot{H}^{s,\tau}_{pq}(\mathbb{R}^n)$ and [1153, Definition 5.1] for the definition of $F\dot{H}^{s,\tau}_{pq}(\mathbb{R}^n)$. We refer to [1152, Theorem 5.1] for Theorem 6.38. See [1152, Lemma 5.4] for the mapping $\pi : F \in F\dot{T}^{s,\tau}_{pq}(\mathbb{R}^{n+1}_+) \mapsto \sum_{j=-\infty}^{\infty} \varphi_j(D)F_j \in F\dot{H}^{s,\tau}_{pq}(\mathbb{R}^n)$ in (6.158). Lemma 6.22 is obtained in the proof of [1151, Theorem 4.1]; see [1152, p. 2795]. We refer to [1151, Theorem 4.1] for Theorem 6.32. We have an analogy of Theorems 3.2 and 3.3 to Morrey spaces; see [767, 768, 925, 1108]. Besov–Morrey spaces are applied to to 2D dissipative quasi-geostrophic equations [1134].

The inclusion from $A^{s,\tau}_{pq}(\mathbb{R}^n)$ into $\mathrm{bmo}(\mathbb{R}^n)$ or $C(\mathbb{R}^n)$ is discussed in [1173, 1179]. We refer to [1168] for a passage to the variable exponent.

The Spaces $Q_p(\mathbb{R}^n)$ and $Q^{\alpha,q}_p(\mathbb{R}^n)$

See [450, 451, 627, 1116] for Q_p-spaces and [722, 723] for their applications to Navier–Stokes equations. In connection with the Triebel–Lizorkin-type spaces, see [1152, Theorem 3.1] for Theorem 6.36. We refer to [1152, Definition 3.3] for Definition 6.22.

Complex Interpolation of $\dot{A}^{s,\tau}_{pq}(\mathbb{R}^n)$

In [1160, Theorems 1.3, 1.6 and 1.8], the complex interpolation spaces are obtained for Besov-type and Triebel–Lizorkin-type spaces. As a corollary, we have the complex interpolation of the space $\mathcal{M}^{p_0}_{q_0}(\mathbb{R}^n)$ and $\mathcal{M}^{p_1}_{q_1}(\mathbb{R}^n)$ with $p_0q_1 = p_1q_0$; see [1160, Corollary 1.4]. For related spaces and related approaches, we refer to [495].

Section 6.5

Section 6.5.1

$\mathcal{S}_{\mathrm{e}}(\mathbb{R}^n)$ and $\mathcal{S}'_{\mathrm{e}}(\mathbb{R}^n)$: Definition 6.26 and Exercise 6.64

The space $\mathcal{S}_{\mathrm{e}}(\mathbb{R}^n)$ is a type of Gelfand–Silov spaces investigated by M. Hasumi. The class $\mathcal{S}'_{\mathrm{e}}(\mathbb{R}^n)$, the dual of $\mathcal{S}_{\mathrm{e}}(\mathbb{R}^n)$, is such an example. See also [940].

Weighted Besov Spaces and Triebel–Lizorkin Spaces

When we consider weighted function spaces, we are interested in the conditions of the weights. Triebel considered weighted function spaces in [1046], where the weight w satisfies (6.164) with $0 < \beta < 1$. See [940–942] for weighted Besov spaces and Triebel–Lizorkin spaces, where the weight satisfies (6.164) with $\beta = 1$. Triebel made such an attempt in [1046, Section 5].

Meanwhile, Bui investigated weighted function spaces [289–291], where the weight belongs to A_∞. See also [294, 295].

Rychkov considered the class A^{loc}_p and defined $A^{s,w}_{pq}(\mathbb{R}^n)$. See (6.165) for [904, Lemmas 1.1 and 1.3], the openness property of A^{loc}_p. Note that Tang [1029] considered $h^1_w(\mathbb{R}^n)$ and we note that $h^1_w(\mathbb{R}^n) \approx F^{0,w}_{12}(\mathbb{R}^n)$.

Haroske and Schmeisser investigated the trace operator in the weighted setting; [565, Theorem 3.5] is a weighted counterpart of Theorem 4.47, while [565, Proposition 3.1] is a weighted counterpart of Theorem 4.48.

See [618, Theorem 16] for the weighted cases of the discussion of greediness, where Izuki and Sawano also studied greediness of the wavelet basis. See also [760] in a similar setting, where the authors obtained various characterizations including the Haar wavelet.

Atomic Decomposition of Weighted Besov Spaces and Triebel–Lizorkin Spaces

The atomic decomposition and the related decompositions are obtained in many papers in various weighted settings including the 2-microlocal setting; see [618, 618, 621, 664, 665, 728]. We refer to [618, 1115] for decompositions of the spaces with $A_\infty^{\rm loc}$-weights. We refer to [969] for the complex interpolation of the function spaces with $A_\infty^{\rm loc}$-weights, where we can find the technique of decomposing sequences. See [529] for this technique of decomposition. See [1093] for the approximation of functions with dominating mixed smoothness by means of the quasi-Monte Carlo methods. See [612] for the decomposition using the Jacobi polynomials.

Weighted Smoothness Morrey Spaces

We refer to [343, 621] for weighted smoothness Morrey spaces.

2-Microlocal Besov Spaces and Triebel–Lizorkin Spaces

Bony proposed the concept of 2-microlocal analysis or 2-microlocal function spaces [259]. It is an appropriate tool to describe the local regularity and the oscillatory behavior of functions near singularities. The class $\mathscr{W}_{\alpha_1\alpha_2}^{\alpha_3}$ is related to 2-microlocal Besov spaces and Triebel–Lizorkin spaces. The theory has been elaborated and widely used in fractal analysis and signal processing by several authors. We refer to [65, 623, 719]. See also [624] for the approach using wavelets to measure the pointwise smoothness. Many authors generalized these works in different directions; see [133, 777, 803].

Rotation Invariant Spaces

One of the ways to obtain the compactness of the embedding is to consider rotation invariant spaces; that is, radial symmetry. We refer to [988, Theorem 2], where the action of $\mathrm{O}(n)$ is replaced by the one of the closed subgroups. We refer to [968] for the surjective boundedness property of the trace operator from the radial function spaces to weighted function spaces, where the idea of the Strauss lemma is used. The Strauss lemma itself can be improved via the radial Besov space $RB^s_{p\infty}(\mathbb{R}^n)(\subset B^s_{p\infty}(\mathbb{R}^n))$; see [967, p. 549, Main Theorem]. In view of [966, Theorem 2], we learn that the critical value of s is $1/p$.

Section 6.5.2

2-Microlocal Setting (Including the Variable Exponent Case)

Kempka developed the theory of boundedness of the 2-microlocal Besov spaces and Triebel–Lizorkin spaces in [664, 665]. We refer to [852] for the boundedness of the trace operators.

Generalized Plancherel–Polya–Nikolski'i Inequality

See [99, Section 6.3.1] for Corollary 6.2.

2-Microlocal Besov Spaces and Triebel–Lizorkin Spaces with Variable Exponents

Kempka defined $B^{w,\mathrm{mloc}}_{p(\star)q(\star)}(\mathbb{R}^n) \equiv B^w_{p(\star)q(\star)}(\mathbb{R}^n)$ in [664, Definition 3] and analogously $A^w_{p(\star)q(\star)}(\mathbb{R}^n)$ in [665, Definition 2.2] for $w \in \mathscr{W}^{\alpha_1,\alpha_2}_{\alpha_3}$. Kempka obtained the atomic decomposition of $A^w_{p(\star)q(\star)}(\mathbb{R}^n)$ in [665, Theorem 3.12]. We refer to [809, 852] for the trace theorem.

Section 6.5.3

Besov/Triebel–Lizorkin Spaces with Variable Exponents, Where $s(\star)$ Is a Variable

The case where only s is a variable dates back to Beauzamy [165]. Triebel defined $A^{g(x)}_{pq}(\mathbb{R}^n)$ in [1047, Definition 2.3/3], where g satisfies a certain condition; see [1047, Definition 2.1/1]. Triebel investigated the boundedness of the multiliers, duality and embeddings [1048]. In [207] Besov considered embedding results on domains with a flexible-cone condition. Gurka, Harjulehto and Nekvinda considered the variable exponent Bessel potential spaces [527]. We also refer to [206, 213, 214, 226, 227].

It has not been easy to replace p and q with variable exponents because it is hard to analyze the behavior of the Hardy–Littlewood maximal operator when p and/or q is replaced by variable exponents. In fact, the boundedness fails.

Besov/Triebel–Lizorkin Spaces with Variable Exponents, Where $p(\star)$ Is a Variable

See [1126, Definition 2] and [1130, Theorem 2] for the definition of the spaces with some elementary properties and their atomic decomposition of $A^s_{p(\star)q}(\mathbb{R}^n)$, respectively. Xu also obtained the Littlewood–Paley characterization of the space $F^s_{p(\star)2}(\mathbb{R}^n)$ in [1127, Theorem 1]. Kopaliani obtained a characterization of variable Lebesgue space $L^{p(\star)}(\mathbb{R})$ in [679, Theorem 3]. In this case, we can use the Fefferman–Stein vector-valued inequality, which is obtained in [391, Corollary 2.1].

Besov/Triebel–Lizorkin Spaces with Variable Exponents, Where $p(\star), s(\star)$ Are Variables

Dong and Xu considered $A^{s(\star)}_{p(\star)q}(\mathbb{R}^n)$. in [427].

Besov/Triebel–Lizorkin Spaces with Variable Exponents, Where $p(\star), q(\star), s(\star)$ Are Variables

Finally, let us consider the most complicated case. Theorem 1.49, the Fefferman–Stein vector-valued inequality, has been an important tool to analyze Triebel–Lizorkin space. However, this inequality cannot be used for these function spaces, even for Besov spaces. Instead, we prove the vector-valued inequality for a sequence of convolution operators [422, Theorem 3.2] and [130, Lemma 4.7]. In this setting, the Fefferman–Stein vector-valued inequality is not available in general. Nevertheless, Almeida and Hästo defined the variable exponent Besov space $B^{s(\star)}_{p(\star)q(\star)}(\mathbb{R}^n)$ [130] and Diening, Hästo and Roudenko defined the variable exponent Triebel–Lizorkin space $F^{s(\star)}_{p(\star)q(\star)}(\mathbb{R}^n)$ [422]. For example, as for Theorem 6.40 we refer to [130, Theorem 5.5] for Besov spaces with variable exponents and [422, Theorem 3.10] for Triebel–Lizorkin spaces with variable exponents. See [667] for the case where the variable exponent Besov space $B^{s(\star)}_{p(\star)q(\star)}(\mathbb{R}^n)$ is a Banach space. See [432, Theorem 3] for the atomic decomposition of $B^{s(\star)}_{p(\star)q(\star)}(\mathbb{R}^n)$. See [853, Theorems 1.4 and 1.11] for the complex interpolation of $A^{s(\star)}_{p(\star)q(\star)}(\mathbb{R}^n)$. Izuki and Noi characterized the dual space of $A^{s(\star)}_{p(\star)q(\star)}(\mathbb{R}^n)$ in [617, Theorems 1 and 2]. See [666, Section 3] and [666, Section 4] for the characterization by the local means and the ball difference, respectively. We refer to [850] and [851] for the duality and the boundedness of the trace operator, respectively. The definition of variable exponent Besov spaces [130, Definition 5.2], while the definition of variable exponent Triebel–Lizorkin spaces goes back to [422, Definition 3.3]. Lemma 6.12, Theorem 6.18 and Theorem 6.13 [422, Lemma 6.11], [422, Theorem 3.2] and [422, Lemma 5.4], respectively, while Theorem 6.17 is [130, Lemma 4.7].

Variants of Besov/Triebel–Lizorkin Spaces with Variable Exponents

Fu and Xu considered Morrey-type Besov and Triebel–Lizorkin spaces when two exponents of Morrey spaces are both variable exponents [485]. Based on Izuki's paper on variable exponent Herz spaces [616] Shi and Xu considered Herz-type Triebel–Lizorkin spaces with s and p variable in [950, 951]. Dong and Xu considered Herz–Morrey-type Besov/Triebel–Lizorkin spaces when s and p are variable exponents in [428]. In [603] the notions of $F^{\alpha(\star)}_{p(\star),q(\star)}(\mathbb{R}^n)$ and weighted Morrey spaces are extended by the notion of families of variable Banach sequence spaces.

Section 6.5.4

Function Spaces with Zero Smoothness

The function spaces dealt with in this book can be used to consider functions with "zero" smoothness. Besov considered various expressions of spaces having "zero" smoothness in [236]. Let us use $\mathscr{C}^s(\mathbb{R}^n)$ or $B^s_{pq}(\mathbb{R}^n)$ to realize this idea. If we start with $\mathscr{C}^s(\mathbb{R}^n)$, we replace $|x-y|^s$ with $[\log(1+|x-y|^{-1})]^a$ with $a > 0$. As for $B^s_{pq}(\mathbb{R}^n)$, replace 2^{js} with j^a. This can be further generalized. We refer to the paper [826] for details on how to generalize $|x-y|^s$ and 2^{js}. The space $\mathrm{Lip}^{1,-\alpha}_{p,q}(\mathbb{R}^n)$ is a space used for the purpose with smoothness near 1; see [560, 1171]. See [321, 372–375, 921] for more recent approaches, as well as the papers [369, 452] on the real interpolation functor.

Function Spaces of Mixed Smoothness

See [182] for the embedding results. We refer to [200] for the embedding of Sobolev type on domains with the flexible horn condition. We refer to [200, p. 9 Teorema] for the interpolation inequality of Besov spaces defined on domains. Besov also investigated anisotropic Besov spaces defined on domains [203].

A recent exhaustive explanation in this field is [1103]. Johnsen and Sickel extended Corollary 1.7 to the setting of mixed smoothness in [648, Theorem 5] and extended (2.64); see [648, Theorem 6]. These results are extended also to the anisotropic setting [648, Section 3]. See also [551, 552, 716, 842] for the function spaces with mixed smoothness.

See the servey [935] for the explanation of the dominating mixed smoothness. The spaces of the dominating mixed smoothness go back to the paper by Nikolski'i [847]. Later Schmeisser considered using Besov spaces and Triebel–Lizorkin spaces; see [932–934]. Later Vibíral obtained the atomic and wavelet decompositions. See [842, 843] for more about the pointwise multiplier on this space.

The space $B^{s,\alpha}_{pq}(\mathbb{R}^n)$

See the papers [36, 319, 370, 375, 563, 710] for the emtropy number and interpolation [249, 371, 373, 374, 378, 453] as well as the paper [1171] for the application to the Navier–Stokes equations.

Function Spaces of Mixed Integrability

We refer to [199, p. 32 Teorema] for the Littlewood–Paley characterization of the space.

Besov investigated the anisotropic case in [202, Teorema A and Teorema B] where the underlying domain satisfies the flexible horn condition. A priori estimates of elliptic equations in function spaces of mixed integrability can be found in [183, Teorema 1].

We refer to [316] for the embedding of Sobolev type for function spaces on $\mathbb{R}^n$. Johnsen and Sickel applied the technique used in the proof of (2.64) to the setting of mixed integrability in [648, Section 2]. We refer to [840, Theorem 2.4] for Bernstein numbers of embeddings of the function spaces of dominating mixed smoothness.

Function Spaces of Mixed Smoothness and Integrability

Both integrability and smoothness are mixed. Besov considered some embedding estimates in [184, Teorema 1]. We also refer to [205].

Section 6.5.5

Anisotropic Function Spaces

Seeger dealt with characterizations of anisotropic Triebel–Lizorkin spaces on $\mathbb{R}^n$ and on domains, via local oscillations and differences of functions [943].

The notion of the dilation matrix dates back to [276, Definition 2.1]. Besov considered the mixed normed space, especially in inclusion relationships in [171, Teorema 3]. The motivation of introducing anisotropic function spaces is the desire to solve linear equations like: $u_{x_1x_1x_1x_1} + u_{x_2x_2} = f$ as is described in [1059]. See [261] for such an approach. Triebel considered Besov spaces and Hardy spaces on $\mathbb{R}^2$: the Hardy inequality [1059, Theorem 1], duality [1059, Theorem 3] and the localization property of Besov spaces in the diagonal case [1059, Section 4]. We refer to [212, Theorem 1] for some estimates of integral operators adapted to an anisotropic setting. Calderón and Torchinsky initiated the study of Hardy spaces on $\mathbb{R}^n$ with anisotropic dilations in [327–329].

Johnsen and Sickel discusses the trace of anisotropic Triebel–Lizorkin spaces of the mixed norm makes sense as an element in $\mathscr{D}'(\mathbb{R}^{n-1})$; see [649, Theorem 2.1].

I learned Example 6.16 from Baode Li.

Properties of Anisotropic Function Spaces

We refer to [276, 278, 279, 282] for anisotropic function spaces. The duality can be found in [280, Theorem 1.2].

Wavelet Decomposition of Anisotropic Function Spaces

We refer to [575, Theorem 2.1] for the wavelet decomposition.

Characterization of Weighted Anisotropic Spaces

See [721, Theorem 1.5] for a characterization of weighted anisotropic spaces.

We refer to [939] for comparison of function spaces of dominating mixed smoothness and spaces of best approximation with respect to hyperbolic crosses. In the anisotropic and mixed settings, Johnsen, Munch and Sickel considered the local means in [646] and diffeomorphic mappings in Triebel–Lizorkin spaces in [647].

We refer to [424] for the boundedness of the Fourier multipliers between weighted anisotropic function spaces.

See [454] for the traces of anisotropic Besov-Lizorkin-Triebel spaces in the borderline cases.

Weighted Anisotropic Besov and Triebel–Lizorkin Spaces

Li, Bownik, Yang and Yuan considered weighted anisotropic Besov and Triebel–Lizorkin spaces in [720, 721].

We remark that [597] can be regarded as a sort of study of this direction, where Ho considered the group action $t \cdot (x, y, z) \mapsto (tx, ty, t^2 z)$.

We refer to [234, 235, 244] for the Sobolev embedding of anisotropic Sobolev spaces.

Section 6.6

Section 6.6.1

Periodic Smoothness Morrey Spaces

Schmeisser and Sickel developed the theory of function spaces on an n-dimensional torus [936–938]. We refer to [911, Definition 4] for periodic smoothness Morrey spaces. Using periodic smoothness Morrey spaces, we can investigate the convergence of the Fourier series. See also [161] for a more recent approach.

Paraproducts on the Function Spaces on the torus

Using the embedding caused by the compactness of the n-dimensional torus, Funaki and Hoshino considered a coupled KPZ equation [487].

Section 6.6.2

Function Spaces on Fractals

See the long paper by Han, Müller and Yang [541] for the analysis on spaces of homogeneous type. Recall that a metric measure space (X, d, μ) is a space of homogeneous type if $\mu(B(x, 2r)) \le C\mu(B(x, r))$ for all $x \in X$ and $r > 0$. If the space of homogeneous type (X, d, μ) satisfies $C\lambda^{\kappa}\mu(B(x, r)) \le \mu(B(x, \lambda r)) \le C\lambda^{n}\mu(B(x, r))$ for all $\lambda > 1$, $r > 0$ and $x \in X$ with $\lambda r \le \operatorname{diam}(X)$, then (X, d, μ) is a space of homogeneous type. We refer to [537, 540, 541, 632, 815, 1165] for Besov spaces and Triebel–Lizorkin spaces on RD spaces. In particular, Han develops the Littlewood–Paley theory of the functions defined on sets; see [536]. The equality (2.15) is referred to as the inhomogeneous Calderòn–Zygmund reproducing formula. A space of homogeneous type is the metric measure space (X, d, μ) such that μ satisfies the doubling condition $\mu(B(x, 2r)) \le C\mu(B(x, r))$. See [536, Theorem 2] for the case of the space of homogeneous type, which paved the way for analysis on spaces of homogeneous type. We refer to [441] for BMO on spaces of homogeneous type. We refer to [948] for Newtonian Besov spaces and Newtonian Triebel–Lizorkin spaces.

Function Spaces on Gaussian Measure Spaces

Define the Gaussian measure γ by $\gamma(E) \equiv \int_E \exp(-\pi|x|^2)\mathrm{d}x$ for a measurable set E. We refer to [766, Section 5] and [766, p. 281] for the definition of $H^1(\gamma)$ and $\mathrm{BMO}(\gamma)$, respectively. We refer to [879] for the Besov space $B^s_{pq}(\gamma)$ with

$1 \le p, q \le \infty$ and $s \ge 0$ and Triebel–Lizorkin spaces $F^s_{pq}(\gamma)$ with $1 \le p, q < \infty$ and $s \ge 0$ for the Gaussian measure, where the Poisson–Hermite subgroup $\{P_t\}_{t \ge 0}$ is used in the definition; see [879, Definition 2.1] and [879, Definition 2.2], respectively. See [499] for the lifting operators on Gaussian Triebel–Lizorkin spaces.

Function Spaces on Spaces of Homogeneous Type

Coifman and Weiss introduced atomic Hardy spaces $H^p_{\rm at}(X)$ for $p \in (0, 1]$ when X is a general space of homogeneous type in the sense of Coifman and Weiss [15]. Moreover, under the assumption that the measure of any ball in X is equivalent to its radius, Coifman and Weiss [385] further established a molecular characterization of $H^1_{\rm at}(X)$ and $p \in (1/2, 1]$. Macías and Segovia gave a maximal function characterization of $H^p_{\rm at}(X)$. For p in this range, Han characterized $H^p_{\rm at}(X)$ using the Lusin-area functions [535].

Han and Yang defined $A^s_{pq}(X)$ in [547]. Applications can be found in [546, 1144, 1145]. In [1145] Yang considered the $T1$-theorem and the real interpolation. Deng, Han and Yang considered the counterpart of the Plancherel–Polya–Nikolski'i inequality [418]. See [545] for the reproducing formula. Christ defined the notion of the dyadic cubes in the spaces of homogeneous type [364]. See [758] for Lipschitz spaces on spaces of homogeneous type.

In [440] Duong and Yan characterized these atomic Hardy spaces in terms of Lusin-area functions associated with certain Poisson semi-groups.

We refer to [1184] for Hardy spaces with variable exponents on RD-spaces.

Function Spaces on Sets Having More Singular Boundaries

More and more people study function spaces on open sets. Open sets can have bad boundaries. Besov has been considering function spaces on irregular domains. In [221, 222, 224] Besov considered the function spaces of generalized smoothness. More recently, Besov defined and investigated the spaces of fractional smoothness on irregular domains in [215, 217, 218, 223, 229, 231, 232, 241]. Based on the integral representation, Besov obtained an extension of the Sobolev embedding for the functions defined on domains with the flexible σ cone property [230, 233]. For more works by Besov see [216, 219, 220]. See also [13, 642, 643]. As one such notion, we can consider (ε, δ)-domains. We refer to [420, Theorem 6.1] for the extension operator and [931, Theorem 3.10] for the decomposition results.

It may be convenient that we consider weighted function spaces on domains. See [245, 246] for this aspect.

Function Spaces on Lie Groups

The function spaces can be defined on many types of sets such as Lie groups, manifolds and fractals. The definition of function spaces on Lie groups dates back to Folland, Stein Krantz and Saka [27, 473, 474, 689, 905, 1062]. Skrzypczak obtained the characterization by the heat extension and the Poisson extension in [986, Theorem 2] and [986, Theorem 3], respectively. We refer to [484, 978, 985, 987, 989] for more about the function spaces on Lie groups.

Function Spaces on Riemannian Manifolds

The definition of function spaces on manifolds was given by Triebel [1060, 1061, 1066]. Triebel defined function spaces on Riemannian manifold having bounded geometry in [1061] and those on Lie groups in [1062]. A Riemannian manifold (M, g) is said to have bounded geometry if the injectivity radius is bounded from below. Let (M, g) be a Riemannian manifold having bounded geometry. Then Skrzypczak defined the Triebel–Lizorkin space $F^s_{pq}(M)$ using the partition to the unity and the exponential mapping on M. By means of the real interpolation, the Besov space $B^s_{pq}(M)$ is defined.

Skrzypczak considered the trace operator in [975, Theorem 1]. These function spaces are characterized by means of the heat kernel [989, Theorem 3] and wavelets [993, Theorem 1].

One of the reasons for these complicated definitions of Besov spaces on manifolds is that the localization property for Besov spaces is not available, although it is a fundamental idea to consider the local coordinates when the analysis on manifolds is staged. See [1118] for function spaces on quantum tori.

The function space of Sobolev-type was defined in [976] for Riemannian symmetric manifolds. Skrzypczak investigated pointwise multipliers in [977].

Function spaces on symmetric Riemannian manifolds are investigated in [980–982] based on the Fourier multiplier estimates [979]. The atomic decomposition of Besov spaces on symmetric manifolds is obtained in [983, Theorem 1] and [984].

See [116] for the fundamental facts on Riemannian manifolds.

Function Spaces on Lipschitz Manifolds

Function spaces on Lipschitz manifolds can be found in [1074], where a delicate approach is required. We can consider manifolds having less smoothness, say Lipschitz manifolds [1074, Section 6].

Fractals

Koch curve can be traced back to two papers [672, 673]. For the original papers on the Sierpinski gasket, we refer to [971, 972].

Function Spaces on Fractals

See [611] for the exhaustive account of the Hausdorff measures and self-similar sets. We can consider function spaces even on fractals. One of the advantages of function spaces considered in this book is that we can measure the smoothness of functions in a very delicate manner and this suits the fact that many fractal sets have noninteger dimension. Here we envisage the fractal sets such as the Koch curve, the Cantor dust and so on. But let us consider the case where the (fractal) set K are equipped with a nontrivial Radon measure μ. Motivated by Theorem 4.40, the pointwise multiplier generated by the indicator function of domains is studied furthermore. We say that a nonempty Borel set Γ is said to satisfy the *ball condition* if there is a number $0 < \eta < 1$ with the following property: For any ball $B(x, t)$ there is a ball $B(y, \eta)$ such that $B(y, \eta t) \subset B(x, \eta) \setminus \overline{\Gamma}$. We refer to [101, 18.10, p. 142], [103, 9.16, p. 138] and [1074, Definition 5.5].

Haroske and Piotrowska studied the connection between fractals and atomic decomposition; see [574].

We can use function spaces to measure complexity of sets. For example, Netrusov defined the capacity in Besov spaces [837].

Function Spaces on d-Sets

One of the generalizations of the fractals is the notion of D-sets. See the papers [644, 645]. Yang considered function spaces on d-sets using the trace operator and the quarkonial decomposition [1144]. Triebel also proposed to use the "Euclidean charts" for D-sets [1077, Theorem 3.6]. For applications of the atomic decomposition to the function spaces defined on fractal sets, see [574, Corollary 3.13], [651] for the trace, from Besov spaces to some small sets [1076, Theorem 1], and from the function spaces to D-sets [1080]. Triebel proposed to use quarks in [103] and [1077, Section 2], while in an older textbook [101], Triebel defined the function spaces on d-sets by means of the trace. See [106] for more details.

Function Spaces on Metric Measure Spaces

Analysis on metric measure spaces has been studied quite intensively; see, for example, Semmes's survey [945] for a more detailed discussion and references. See also [1017, 1018]. In [533] Hajłasz introduced Sobolev spaces on any metric measure space. In fact, he introduced the notion of Hajłasz gradients, which serves as a powerful tool to develop the first order Sobolev spaces on metric measure spaces. This idea is expanded in [476]. Later, Shanmugalingam introduced another type of the first order Sobolev space by means of upper gradients in [947], where Shanmugalingam defined Sobolev spaces on metric measure spaces. [534, 543] are important monographs. [544] is a continuation of [543]. A series of papers has been devoted to the construction and investigation of Sobolev spaces of various types on metric measure spaces. See [585, 858] for Sobolev spaces on metric measure spaces. Via the fractional version of Hajłasz gradients, Hu [609] and Yang [1143] introduced Sobolev spaces with smoothness order $\alpha \in (0, 1)$ on fractals and metric measure spaces, respectively. See [504, 581–583, 671, 1174] for more about this direction.

Section 6.7

Section 6.7.1

Kato Conjecture

Tosio Kato conjectured Theorem 6.48 in [660].

Section 6.7.2

Kato Conjecture for $L^2(\mathbb{R}^n)$

It was proved by Kato [660] and solved by Auscher, Hofmann, Lacey, M^c^Intosh and Tchamitchian; see [149, 605]. See [155] for a more general approach.

Kato Conjecture for the Degenerate Elliptic Operators

See [392, 393, 1163] for the degenerate case, where the weighted Davies–Gaffney estimate was obtained.

Section 6.7.3

Kato Conjecture for Other Lebesgue Spaces

As we have seen, the operator $\nabla L^{-1/2}$, called the Riesz transform associated with the operator L, plays a key role to the extension of the range of p for which the analogy of the Kato theorem is true. In [591], the boundedness of $\nabla L^{-1/2}$ from $H^1_L(\mathbb{R}^n)$ to $H^1(\mathbb{R}^n)$ is proved. Hofmann, Mayboroda and McIntosh proved that

$\nabla L^{-1/2}$ is bounded from $H^p_L(\mathbb{R}^n)$ to $H^p(\mathbb{R}^n)$ if $\frac{n}{n+1} < p \le 1$. See [949] for the case of $L = -\Delta + V$. Jiang and Yang extended this result in [633] to Orlicz–Hardy spaces, while Song and Yan considered the extension to weighted Hardy spaces [1005]. This result is extended to the Orlicz–Hardy space $H^\Phi_L(\mathbb{R}^n)$ and its dual space in [635], where Φ belongs to a suitable class. Our argument of the boundedness of the resolvent essentially hinges on the paper [475] in that we used the amalgam spaces. We refer to [592, Lemma 2.2] for Theorem 6.51 and Corollary 6.4.

Riesz Transform Associated with Operators

See [145] in the case of manifolds.

Function Spaces Associated with Operators with Integral Kernel

Grigor'yan and Liu used the heat kernel to define Besov spaces and Lipschitz spaces in [520]. Zhuo and Yang considered the Hardy spaces with variable exponents for the operators satisfying the Gaussian estimates [1166, 1187]. Liu, Yang and Yuan considered Besov-type and Triebel–Lizorkin-type spaces associated with heat kernels [741]. See [726] for applications to the boundedness of the operators on Orlicz–Hardy spaces including Lebesgue spaces. See [354–356] and the survey article [359] for more.

Function Spaces Associated with Operators Without Integral Kernel

The study of Hardy spaces and their generalizations associated with various different differential operators inspires great interests. In particular, in [147] Auscher, Duong and McIntosh first introduced the Hardy space $H^1_L(\mathbb{R}^n)$ associated with the operator L, where the heat kernel generated by L satisfies a pointwise Poisson-type upper bound. See [153] for the case of domains. Later, in [441, 442] Duong and Yan introduced the space $\mathrm{BMO}_L(\mathbb{R}^n)$ and proved that the dual space of $H^1_L(\mathbb{R}^n)$ is $\mathrm{BMO}_{L^*}(\mathbb{R}^n)$. Yan further introduced the Hardy space $H^p_L(\mathbb{R}^n)$ in [1140], where p is slightly less than 1. In [593, 594] Hofmann, Mayboroda and McIntosh introduced the Hardy and Sobolev spaces associated with a second order divergence form elliptic operator L on $\mathbb{R}^n$ with complex bounded measurable coefficients. See [438] for more extensions. A theory of the Orlicz–Hardy $H^\Phi_L(\mathbb{R}^n)$ space and its dual space associated with L was also developed in [636, 637]. See [439] for the application to Morrey spaces. See [416, 634] for the vanishing mean oscillation, whose definition is based on the observation by Sarason [909].

One of the key ideas of the solution of the Kato problem is the Gaffney estimate. We can generalize the notion of the Gaffney estimate. Under the assumption of the k-Davies–Gaffney estimates, Cao, Jiang and Yang considered the Hardy spaces associated with operators satisfying [346] and Cao, Chang, Wu and Yang considered their weak variant. Weak Hardy spaces are defined in [340].

See [341, 342] for the case of higher order elliptic differential operators. See [344] for the case of more than one commuting operator.

Function Spaces Associated with Operators with Integral Kernel: Schrödinger Operators

Dziubański and Zienkiewicz initially introduced the Hardy spaces $H^p_{-\Delta+V}(\mathbb{R}^n)$ for the Schrödinger operator $-\Delta + V$ with the nonnegative potential V belonging to the reverse Hölder class in [444, 445]. See [1147, 1148] for the case of the magnetic Schrödinger operators. See [357, 1150] for the case of domains. See [345] for the case of $L = \Delta^2 + V^2$.

Function Spaces Associated with Operators Without Integral Kernel: Degenerate Case

We refer to [1162] for the boundedness of the Riesz transform for the degenerate elliptic operators. See [1181] for the characterization of Hardy spaces associated with this class of operators.

More generally, for nonnegative self-adjoint operators L satisfying the Davies–Gaffney estimates, the Hardy space $H^1_L(\mathbb{R}^n)$ is introduced in [591]. Gaussian function spaces are typical examples of the function spaces associated with operators. For example, see [296, 741, 1148] for more about the function spaces associated with the operators. Cao, Mayboroda and Yang considered the local Hardy spaces associated with inhomogeneous higher order elliptic operators in [342]. See [741] for Besov-type spaces and Triebel–Lizorkin-type spaces associated with the operators having the Gaussian estimates.

Function Spaces Associated with Operators on Manifolds

Auscher, McIntosh and Russ worked in the setting of manifolds [152]. Later, Badr and Ben Ali followed their idea to develop a theory of the Schrödinger operators on manifolds [158].

References

Textbooks

1. Adams, D.R.: Morrey Spaces. Lecture Notes in Applied and Numerical Harmonic Analysis. Birkhäuser/Springer, Cham (2015)
2. Adams, D.R., Hedberg, L.I.: Function Spaces and Potential Theory.Grundlehren der Mathematischen Wissenschaften, vol. 314 Springer, Berlin (1996)
3. Adams, D.R.: Sobolev Spaces. Pure and Applied Mathematics, A Series of Monographs and Textbooks, vol 65. Academic Press, New York/San Francisco/London (1975)
4. Adams, R.A., Fournier, J.J.F.: Sobolev Spaces. Pure and Applied Mathematics, V, vol 140, 2nd edn. Elsevier/Academic Press, New York (2003)
5. Auscher, P., Tchamitchian, P.: Square root problem for divergence operators and related topics. Astérisque **249**, viii+172 (1998)
6. Bahouri, H., Chemin, J.-Y., Danchin, R.: Fourier Analysis and Nonlinear Partial Differential Equations. Springer, Berlin/Heidelberg (2011)
7. Bergh, J., Löfström, J.: Interpolation Spaces. An Introduction. Grundlehren der Mathematischen Wissenschaften, vol. **223**. Springer, Berlin/New York (1976)
8. Besov, O.V., Il'in, V.P., Nikolski'i, S.M.: Integral Representations of Functions and Imbedding Theorems, vol. I+II. V. H. Winston and Sons, Washington, DC (1978, 1979). Transalated from the Russian. Scripta Series in Mathematics, Edited by Mitchell H. Taibleson, Halsted Press [John Wiley and Sons], New York/Toronto/London (1978). viii+345pp
9. Burenkov, V.I.: Sobolev Spaces on Domains. Teubner-Texte zur Mathematik [Teubner Texts in Mathematics], vol. 137, 312pp. B. G. Teubner Verlagsgesellschaft mbH, Stuttgart (1998)
10. Bennett, C., Sharpley, R.: Interpolation of Operators. Academic Press, Boston (1988)
11. Brudnyĭ, Y.A., Krugljak, N.: Interpolation Functors and Interpolation Spaces, vol. I. Translated from the Russian by Natalie Wadhwa. With a preface by Jaak Peetre. North-Holland Mathematical Library, vol. 47, xvi+718pp. North-Holland Publishing Co., Amsterdam (1991)
12. Bourdaud, G.: The functional calculus in Sobolev spaces. In: Function Spaces, Differential Operators and Nonlinear Analysis. Teubner-Texte Math., vol. 133, pp. 127–142. Teubner, Stuttgart/Leipzig (1993)
13. Caetano, A., Hewett, D.P., Moiola, A.: Density Results for Sobolev, Besov and Triebel-Lizorkin Spaces on Rough Sets. In preparation

Y. Sawano, *Theory of Besov Spaces*, Developments in Mathematics 56,
https://doi.org/10.1007/978-981-13-0836-9

14. Cannone, M.: Ondelettes, Paraproduits et Navier–Stokes, Diderot Editeur. Arts et Sciences, Paris/New York/Amsterdam (1995)
15. Coifman, R.R., Weiss, G.: Analyse Harmonique Non-Commutative sur Certains Espaces Homogènes. Lecture Notes in Mathematics, vol. 242. Springer, Berlin (1971)
16. Cruz-Uribe D.V., Fiorenza, A.: Variable Lebesgue Spaces: Foundations and Harmonic Analysis. Applied and Numerical Harmonic Analysis. Birkhäuser/Springer, Heidelberg (2013)
17. Daubechies, I.: Ten Lectures on Wavelets. CBMS-NSF Regional Conference Series in Applied Mathematics, vol. 61. Society for Industrial and Applied Mathematics (SIAM), Philadelphia (1992)
18. David, G.: Wavelets and Singular Integrals on Curves and Surfaces. Lecture Notes in Mathematics, vol. 1465. Springer, Berlin/New York (1991)
19. DeVore, R., Lorentz, G.G.: Constructive Approximation. Grundlehren der Mathematischen Wissenschaften, vol. 303. Springer, Berlin (1993)
20. Diening, L., Harjulehto, P., Hästö, P., Růžička, M.: Lebesgue and Sobolev Spaces with Variable Exponents. Lecture Notes in Mathematics, vol. 2017. Springer, Berlin (2011)
21. Dunford, N., Schwartz, J.: Linear Operators. Part I. General Theory. With the assistance of William G. Bade and Robert G. Bartle, Reprint of the 1958 original. Wiley Classics Library. A Wiley-Interscience Publication. Wiley, New York (1988)
22. Duoandikoetxea, J.: Fourier Analysis. Translated and revised from the 1995 Spanish original by D. Cruz-Uribe. Graduate Studies in Mathematics, vol. 29. American Mathematical Society, Providence (2001)
23. Edmunds, D.E., Triebel, H.: Function Spaces. Entropy Numbers and Differential Operators. Cambridge University Press, Cambridge (1996)
24. Edwards, R.E., Gaudry, G.I.: Littlewood–Paley and Multiplier Theory. Springer, Berlin (1977)
25. Evans, C.: Partial Differential Equations. Graduate Studies in Mathematics, vol. 19, American Mathematical Society, Providence (1998)
26. Federer, H.: Geometric Measure Theory. Die Grundlehren der mathematischen Wissenschaften, vol. 153. Springer, New York (1969)
27. Folland, G.B., Stein, E.M.: Hardy Spaces on Homogeneous Groups. Mathematical Notes, vol. 28. Princeton University Press/University of Tokyo Press, Princeton/Tokyo (1982)
28. Frazier, M.: The $T1$ Theorem for Triebel–Lizorkin Spaces. Harmonic Analysis and Partial Differential Equations. Lecture Notes in Mathematics, vol. 1384, pp. 168–191. Springer, Berlin (1987)
29. Frazier, M., Jawerth, B., Weiss, G.: Littlewood–Paley Theory and the Study of Function Spaces. CBMS Regional Conference Series in Mathematics, vol. 79. Published for the Conference Board of the Mathematical Sciences, Washington, DC. American Mathematical Society, Providence (1991)
30. Garling, D.J.H.: Inequalities: A Journey into Linear Analysis. Cambridge University Press, Cambridge (2007)
31. García-Cuerva, J., Rubio de Francia, J.L.: Weighted Norm Inequalities and Related Topics. North-Holland Mathematics Studies, vol. 116. North-Holland, Amsterdam/New York (1985)
32. Grafakos, L.: Classical Fourier Analysis. Graduate Texts in Mathematics, vol. 249. Springer, New York (2008)
33. Grafakos, L.: Modern Fourier Analysis. Graduate Texts in Mathematics, vol. 250. Springer, New York (2009)
34. Gelbaum, B.R., Olmsted, J.M.H.: Counterexamples in Analysis, Corrected reprint of the second (1965) edition, xxiv+195pp. Dover Publications, Mineola (2003)
35. Gröchenig, K.: Foundations of Time-Frequency Analysis. Applied and Numerical Harmonic Analysis. Birkhäuser, Boston (2001)
36. Haroske, D.D.: Envelopes and Sharp Embeddings of Function Spaces. Chapman & Hall/CRC Research Notes in Mathematics, vol. 437, x+227pp. Chapman & Hall/CRC, Boca Raton (2007)

37. Haroske, D.D., Triebel, H.: Distributions, Sobolev Spaces, Elliptic Equations. EMS Textbooks in Mathematics, x+294pp. European Mathematical Society (EMS), Zurich (2008)
38. Heinonen, J.: Lectures on Analysis on Metric Spaces. Universitext. Springer, New York (2001)
39. Heinonen, J., Koskela, P., Shanmugalingam, N., Tyson, J.: Sobolev Spaces on Metric Measure Spaces: An Approach Based on Upper Gradients. New Mathematical Monographs. Cambridge University Press, Cambridge (2015)
40. Hernández, E., Weiss, G.: A First Course on Wavelets. CRC Press, Boca Raton (1996)
41. Holschneider, M.: Wavelets. An Analytic Tool. Clarendon Press, Oxford (1995)
42. Hörmander, L.: Linear Partial Differential Operators. Springer, Berlin/Göttingen/Heidelberg (1962)
43. Igari, S.: Real Analysis: With an Introduction to Wavelet Theory. Translated by S. Igari, Translations of Mathematical Monographs, vol. 177. American Mathematical Society, Providence (1998)
44. Lebesgue, H.: Lecons sur l'Integration et la Recherche des Fonctions Primitives. Gauthier-Villars, Paris. Available online at http://www.archive.org/details/LeconsSurLintegration (1904)
45. Kato, T.: Ryoshirikigakuno Suugaku Riron edited by Kuroda, S. (in Japanese). Kindai Kagakusha
46. Katznelson, Y.: An Introduction to Harmonic Analysis. Cambridge University Press, Cambridge/New York (2004)
47. Krasnosel'skii, M., Rutickii, Y.: Convex Functions and Orlicz Spaces. P. Noordhoff, Groningen (1961)
48. Krantz, S.G.: Function Theory of Several Complex Variables. The Wadsworth & Brooks/Cole Mathematics Series, 2nd edn. Pacific Grove: Wadsworth & Brooks/Cole Advanced Books & Software (1992)
49. Jonsson, A., Wallin, H.: Function Spaces on Subsets of $\mathbb{R}^n$, xiv + 221pp. Mathematical Report Series, vol. 2(1). Harwood Academic, Chur (1984)
50. Jost, J.: Riemannian Geometry and Geometric Analysis, 5th edn. Springer, Berlin/New York. ISBN:978-3-540-77340-5
51. Kufner, A.: Weighted Sobolev Spaces. Teubner-Texte zur Mathematik, vol. 31. Teubner, Leipzig (1980)
52. Kufner, A., John, O., Fučík, S.: Function Spaces. Academia, Publishing House of the Czechoslovak Academy of Sciences, Prague (1977)
53. Lions, J.L., Magnes, E.: Problems aux limites non homogenes et applications I and II. Dunod, Paris (1968)
54. Lu, S., Yang, D., Hu, G.: Herz Type Spaces and Their Applications. Science Press, Beijing (2008)
55. Lunardi, A.: Analytic Semi-groups and Optimal Regularity in Parabolic Problems. Progress in Nolninear Differential Equations and Their Applications, vol. 16. Birkhäuser, Basel (1995)
56. Lunardi, A.: Interpolation Theory, 2nd edn. Appunti. Scuola Normale Superiore di Pisa, Edizioni della Normale, Pisa (2009)
57. Maz'ya, V.G.: Sobolev Spaces. Springer Series in Soviet Mathematics. Springer, Berlin (1985). Translated from the Russian
58. Maz'ya, V.G., Shaposhnikova, T.O.: Theory of Sobolev Multipliers with Applications to Differential and Integral Operators for a Description of the Set of All Pointwise Multipliers $M(B^s_{p,p})$. Springer, Berlin (2009)
59. McLean, W.: Strongly Elliptic Systems and Boundary Integral Equations, xiv+357pp. Cambridge University Press, Cambridge (2000)
60. Megginson, R.: An Introduction to Banach Space Theory. Graduate Texts in Mathematics, vol. 183. Springer, New York (1998)
61. Kokilashvili, V., Meskhi, A., Rafeiro, H., Samko, S.: Integral Operators in Non-standard Function Spaces. Operator Theory: Advances and Applications (2017)

62. Mattila, P.: Geometry of Sets and Measures in Euclidean Spaces Fractals and Rectifiability. Cambridge University Press, Cambridge (2016)
63. Meyer, Y.: Régularité des solutions des équations aux dérivées partielles non linéaires (d'après J.-M. Bony). (French) [Regularity of the solutions of nolninear partial differential equations (according to J. M. Bony)] Bourbaki Seminar, vol. 1979/80, pp. 293–302. Lecture Notes in Mathematics, vol. 842. Springer, Berlin/New York (1981)
64. Meyer, Y.: Wavelets and Operators. Translated by D.H. Salinger, Cambridge Studies in Advanced Mathematics, vol. 37. Cambridge University Press, Cambridge (1992)
65. Meyer, Y.: Wavelets, Vibrations and Scalings. CRM Monograph Series, vol. 9. AMS, Providence (1997)
66. Meyer, Y., Coifman, R.R.: Calderón–Zygmund and Multilinear Operators. Translated by D.H. Salinger, Cambridge Studies in Advanced Mathematics, vol. 48. Cambridge University Press, Cambridge (1997)
67. Murai, T.: A Real Variable Method for the Cauchy Transform, and Analytic Capacity. Lecture Notes in Mathematics, vol. 1307. Springer, Berlin (1988)
68. Muscalu, C., Schlag, W.: Classical and Multilinear Harmonic Analysis, vol. II. Cambridge University Press, Cambridge (2013)
69. Nikolski'i, S.M.: Approximation of Functions of Several Variables and Imbedding Theorems, 2nd edn. (Russian) Nauka, Moskva (1977). (English translation of the first edition: Springer, Berlin/Heidelberg/New York, 1975)
70. Peetre, J.: H^p-Spaces. Lecture Notes. University of Lund, Lund (1974)
71. Peetre, J.: New Thoughts on Besov Spaces. Duke University Mathematics Series, vol. I. Mathematics Department, Duke University, Durham (1976)
72. Pick, L., Kufner, A., John, O., Fučík, S.: Function Spaces. De Gruyter Series in Nonlinear Analysis and Applications, vol. 14, extended edn. Walter de Gruyter Co., Berlin (2013)
73. Rădulescu, V.D., Repovš, D.D.: Partial Differential Equations with Variable Exponents. Taylor and Francis, Boca Raton/London/New York (2015)
74. Rao M.M., Ren Z.D.: Theory of Orlicz Spaces. Monographs and Textbooks in Pure and Applied Mathematics, vol. 146. Marcel Dekker Inc., New York (1991)
75. Raymond, X.S.: Elementary Introduction to the Theory of Pseudodifferential Operators. Studies in Advanced Mathematics, viii+108pp. CRC Press, Boca Raton (1991)
76. Reiter, H.: Classical Harmonic Analysis and Locally Compact Groups. Oxford University Press, Oxford (1968)
77. Rochberg, R., Tabacco, V.A., Vignati, M., Weiss, G.: Interpolation of quasinormed spaces by the complex method. In: "Function Spaces and Applications". Proceedings of the Conference, Lund, 1986. Lecture Notes in Mathematics, vol. 1302, pp. 91–98. Springer, Berlin (1988)
78. Runst, T., Sickel, W.: Sobolev Spaces of Fractional Order. Nemytskii Operators, and Nolninear Partial Differential Equations. de Gruyter Series in Nolninear Analysis and Applications, vol. 3. Walter de Gruyter Co., Berlin (1996)
79. Ružička, M.: Electrorheological Fluids: Modeling and Mathematical Theory. Lecture Notes in Mathematics, vol. 1748. Springer, Berlin (2000)
80. Schmeisser, H.J., Triebel, H.: Topics in Fourier Analysis and Function Spaces. Wiley, Chichester (1987)
81. Schwartz, L.: Théoriè des Distributions. Hermann, Paris (1950)
82. Schwartz, L.: Théoriè des Distributions. Hermann, Paris (1951)
83. Sickel, W.: On pointwise multipliers in Besov–Triebel–Lizorkin spaces. In: "Seminar Analysis Karl-Weierstrass Institute 1985/86," Teubner-Texte Math., vol. 96, pp. 45–103. Teubner, Leipzig (1987)
84. Skrzypczak, L.: Anisotropic Sobolev Spaces on Riemannian Symmetric Manifolds. Function Spaces (Poznań, 1989). Teubner-Texte Math., vol. 120, pp. 252–264. Teubner, Stuttgart (1991)
85. Stein, E.M.: Singular Integral and Differential Property of Functions. Princeton University Press, Princeton (1970)

86. Stein, E.M.: Harmonic Analysis: Real-Variable Methods, Orthogonality, and Oscillatory Integrals. Princeton University Press, Princeton (1993)
87. Stein, E.M., Shakarchi, R.: Fourier Analysis. An Introduction. Princeton Lectures in Analysis, vol. 1. Princeton University Press, Princeton (2003)
88. Stein, E.M., Shakarchi, R.: Real Analysis, Measure Theory, Integration, and Hilbert Spaces. Princeton Lectures in Analysis, vol. 3. Princeton University Press, Princeton (2005)
89. Stein, E.M., Shakarchi, R.: Functional Analysis. Introduction to Further Topics in Analysis. Princeton Lectures in Analysis, vol. 4. Princeton University Press, Princeton (2011)
90. Stein, E.M., Weiss, G.: Introduction to Fourier Analysis on Euclidean Spaces. Princeton Mathematical Series, vol. 32. Princeton University Press, Princeton (1971)
91. Strichartz, R.: A Guide to Distribution Theory and Fourier Transforms. Studies in Advanced Mathematics, x+213pp. CRC Press, Boca Raton (1994)
92. Tanabe, H.: Functional Analytic Methods for Partial Differential Equations (English Summary). Monographs and Textbooks in Pure and Applied Mathematics, vol. 204, pp. x+414. Marcel Dekker, New York (1997)
93. Taylor, M.: Pseudodifferential Operators. Princeton Mathematical Series, vol. 34. Princeton University Press, Princeton (1981)
94. Taylor, M.: Tools for PDE. Pseudodifferential Operators, Paradifferential Operators, and Layer Potentials. Mathematical Surveys and Monographs, vol. 81. American Mathematical Society, Providence (2000)
95. Treves, F.: Topological Vector Spaces, Distributions and Kernels. Academic Press, New York (1967)
96. Triebel, H.: Fourier Analysis and Function Spaces. Teubner-Texte Math., vol. 7. Teubner, Leipzig (1977)
97. Triebel, H.: Spaces of Besov-Hardy-Sobolev type. Teubner-Texte zur Mathematik. With German, French and Russian summaries. BSB B. G. Teubner Verlagsgesellschaft, Leipzig (1978)
98. Triebel, H.: Interpolation Theory, Function Spaces, Differential Operators. North-Holland, Amsterdam (1978)
99. Triebel, H.: Theory of Function Spaces. Birkhäuser, Basel (1983)
100. Triebel, H.: Theory of Function Spaces II. Birkhäuser, Basel (1992)
101. Triebel, H.: Fractal and Spectra. Birkhäuser, Basel (1997)
102. Triebel, H.: Interpolation Theory Function Spaces Differential Operators. 2nd Revised and Enlarged edn. Birkhäuser, Basel (1998)
103. Triebel, H.: The Structure of Functions. Birkhäuser, Basel (2000)
104. Triebel, H.: Theory of Function Spaces III. Birkhäuser, Basel (2006)
105. Triebel, H.: Function Spaces and Wavelets on Domains. EMS Tracts in Mathematics (ETM), vol. 7. European Mathematical Society (EMS), Zürich (2008)
106. Triebel, H.: Fractals and Spectra. Related to Fourier Analysis and Function Spaces. Modern Birkhauser Classics, viii+271pp. Birkhauser Verlag, Basel (2011)
107. Triebel, H.: Entropy Numbers of Quadratic Forms and Their Applications to Spectral Theory. In: Brown, B.M., Lang, J., Wood, I.G. (eds.) Spectral Theory, Function Spaces and Inequalities. Operator Theory, Advances and Applications, vol. 219, pp. 243–262. Birkhauser/Springer, Basel (2012)
108. Triebel, H.: Faber Systems and Their Use in Sampling, Discrepancy, Numerical Integration. EMS Series of Lectures in Mathematics, viii+107pp. European Mathematical Society (EMS), Zurich (2012)
109. Triebel, H.: Local Function Spaces, Heat and Navier–Stokes Equations. EMS Tracts in Mathematics, vol. 20, x+232pp. European Mathematical Society (EMS), Zurich (2013)
110. Triebel, H.: Hybrid Function Spaces, Heat and Navier–Stokes Equations. EMS Tracts in Mathematics, vol. 24, x+185pp. European Mathematical Society (EMS), Zürich (2014)
111. Sagher, Y., Zhou, K.C.: A local version of a theorem of Khinchin. In: Analysis and Partial Differential Equations. Lecture Notes in Pure and Applied Mathematics, vol. 122, pp. 327–330. Dekker, New York (1990)

112. Simon, B.: A Comprehensive Course in Analysis, Part 3, xviii+759pp. American Mathematical Society, Providence (2015)
113. Strömberg, J.O., Torchinsky, A.: Weighted Hardy spaces. Lecture Notes in Mathematics, vol. 1381. Springer, Berlin (1989)
114. Uchiyama, A.: Hardy Spaces on the Euclidean Space, With a foreword by Nobuhiko Fujii, Akihiko Miyachi and Kozo Yabuta and a personal recollection of Uchiyama by Peter W. Jones. Springer Monographs in Mathematics. Springer, Tokyo (2001)
115. Varopoulos, N.T., Saloff-Coste, L., Coulhon, T.: Analysis and Geometry on Groups. Cambridge Tracts in Mathematics, vol. 100. Cambridge University Press, Cambridge (1992)
116. Warner, F.W.: Foundations of Differentiable Manifolds and Lie Groups. Corrected reprint of the 1971 edition, Graduate Texts in Mathematics, vol. 94. Springer, New York/Berlin (1983)
117. Wojtaszczyk, P.: A Mathematical Introduction to Wavelets. Cambridge University Press, Cambridge (1997)
118. Yang, D., Yang, D., Hu, G.: The Hardy Space H^1 with Non-doubling Measures and Their Applications. Lecture Notes in Mathematics, vol. 2084. Springer, Berlin (2013)
119. Yoshida, K.: Functional Analysis. Springer, Berlin (1995)
120. Yuan, W., Sickel, W., Yang, D.: Morrey and Campanato Meet Besov, Lizorkin and Triebel. Lecture Notes in Mathematics, vol. 2005, xi+281pp. Springer, Berlin (2010)
121. Zygmund, A.: Trigonometric Series. Cambridge University Press, Cambridge (1959)

Research Papers

122. Aboulaich, R., Meskine, D., Souissi, A.: New disscusion models in image processing. Comput. Math. Appl. **56**(4), 874–882 (2008)
123. Abu-Shammala, W., Torchinsky, A.: The Hardy-Lorentz spaces $H^{p,q}(\mathbb{R}^n)$. Studia Math. **182**(3), 283–294 (2007)
124. Agmon, S., Doulis, A., Nirenberg, L.: Estimates near the boundary for solutions of elliptic partial differential equations satisfying general boundary conditions, I. Commun. Pure Appl. Math. **12**, 623–727 (1959)
125. Adamowicz, T., Harjulehto P., Hästö, P.: Maximal operator in variable exponent Lebesgue spaces on unbounded quasimetric measure spaces. Math. Scand. **116**(1), 5–22 (2015)
126. Akbulut, A., Guliyev, V.S., Noi, T., Sawano, Y.: Generalized Morrey spaces–revisited. Zeit. Anal. Anwend. **32**, 301–321 (2017)
127. Alabern, R., Mateu, J., Verdera, J.: A new characterization of Sobolev spaces on $\mathbb{R}^n$. Math. Ann. **354**(2), 589–626 (2012)
128. Albrecht, D., Duong, D., M[c]Intosh, A.: Operator theory and harmonic analysis. Workshop in Analysis and Geometry (1995); Proceedings of the Centre for Mathematics and Its Applications. ANU **34**, 77–136 (1996)
129. Allaoui, S.E., Bourdaud, G.: Localisation uniforme des espaces de Besov et de Lizorkin–Triebel. (French) [Uniform localisation of Besov and Lizorkin-Triebel spaces] Arch. Math. (Basel) **109**(6), 551–562 (2017)
130. Almeida, A., Hästö, P.: Besov spaces with variable exponent and integrability. J. Funct. Anal. **258**, 1628–1655 (2010)
131. Amann, H.: Operator-valued Fourier multipliers, vector-valued Besov spaces, and applications. Math. Nachr. **186**, 5–56 (1997)
132. Amann, H.: On the strong solvability of the Navier–Stokes equations. J. Math. Fluid. Mech. **2**, 16–98 (2000)
133. Andersson, P.: Two-microlocal spaces, local norms and weighted spaces, Paper 2 in PhD Thesis. University of Göteborg, pp. 35–58 (1997)
134. Annoni, M., Grafakos, L., Honík, P.: On an inequality of Sagher and Zhou concerning Stein's lemma. Collect. Math. **60**(3), 297–306 (2009)

135. Antonov, N.Y.: Convergence of Fourier series. Proceedings of the XX Workshop on Function Theory, Moscow (1995). East J. Approx. **2**(2), 187–196 (1996)
136. Andersen K.F., John, R.T.: Weighted inequalities for vector-valued maximal functions and singular integrals. Studia Math. **69**, 19–31 (1980)
137. Aoki, T.: Locally bounded linear topological spaces. Proc. Imp. Acad. Tokyo **18**, 588–594 (1942)
138. Aron, R.M., Lacroux, M., Ryan, R., Tonge, A.M.: The generalized Rademacher functions. Note Mat. **12**, 15–25 (1992)
139. Aronszajn, N.: Boundary values of functions with finite Dirichlet integral. Technical Report, vol. 14, University of Kansas, pp. 77–94 (1955)
140. Aronszajn, N., Smith, K.T.: Theory of Bessel potentials, I. Ann. Inst. Fourier **11**, 385–476 (1961)
141. Asami, K.: Non-smooth decomposition of homogeneous Triebel-Lizorkin spaces with applications to the Marcinkiewicz integral. Int. J. Appl. Math. **30**(6), 547–568 (2017)
142. Ashino, R., Mandai, T.: Wavelet bases for microlocal filtering and the sampling theorem in $L^p(\mathbb{R}^n)$. Appl. Anal. **82**(1), 1–24 (2003)
143. Astashkin, S.V.: Rademacher functions in symmetric spaces. (Russian) Sovrem. Mat. Fundam. Napravl. **32**, 3–161 (2009); translation in J. Math. Sci. (N. Y.) **169**(6), 725–886 (2010)
144. Astashkin, S.V., Leibov, M., Maligranda, L.: Rademacher functions in BMO. Studia Math. **205**(1), 83–100 (2011)
145. Assaad, J., Ouhabaz, E.M.: Riesz transforms of Schrödinger operators on manifolds. J. Geom. Anal. **22**, 1108–1136 (2012)
146. Auscher, P.: On necessary and sufficient conditions for L^p-estimates of Riesz transforms associated to elliptic operators on $\mathbb{R}^n$ and related estimates. Mem. Am. Math. Soc. **186**(871), xviii+75 (2007)
147. Auscher, P., Duong, X.T., McIntosh, A.: Boundedness of Banach space valued singular integral operators and Hardy spaces. (2005, Unpublished preprint)
148. Auscher, P., Frey, D.: On the well-posedness of parabolic equations of Navier-Stokes type with BMO^{-1} data. J. Inst. Math. Jussieu **16**(5), 947–985 (2017)
149. Auscher, P., Hofmann, S., Lacey, M., McIntosh, A., Tchamitchian, P.: The solution of the Kato square root problem for second order elliptic operators on $\mathbb{R}^n$. Ann. Math. (2) **156**(2), 633–654 (2002)
150. Auscher, P., Hytonen, T.: Orthonormal bases of regular wavelets in spaces of homogeneous type. Appl. Comput. Harmon. Anal. **34**(2), 266–296 (2013) and Addendum to Orthonormal bases of regular wavelets in spaces of homogeneous type [Appl. Comput. Harmon. Anal. **34**(2), 266–296 (2013)]. Appl. Comput. Harmon. Anal. **39**(3), 568–569 (2015)
151. Auscher, P., McIntosh, A., Nahmod, A.: Holomorphic functional calculi of operators, quadratic estimates and interpolation. Indiana Univ. Math. J. **46**, 375–403 (1997)
152. Auscher, P., McIntosh, A., Russ, E.: Hardy spaces of differential forms on Riemannian manifolds. J. Geom. Anal. **18**, 192–248 (2008)
153. Auscher, P., Russ, E.: Hardy spaces and divergence operators on strongly Lipschitz domains of $\mathbb{R}^n$. J. Funct. Anal. **201**, 148–184 (2003)
154. Auscher, P., Russ, E., Tchamitchian, T.P.: Hardy Sobolev spaces on strongly Lipschitz domains of $\mathbb{R}^n$. J. Funct. Anal. **218**(1), 54–109 (2005)
155. Axelsson, A., Keith, S., McIntosh, A.: Quadratic estimates and functional calculi of perturbed Dirac operators. Invent. Math. **163**(3), 455–497 (2006)
156. Babenko, K.I.: On conjugate functions. Doklady Akad. Nauk SSSR (N.S.) **62**, 157–160 (1948)
157. Babich, V.M.: On the extension of functions (Russian). Uspehi Matem. Nauk (N.S.) **8**(2(54)), 111–113 (1953)
158. Badr, N., Ben Ali, B.: L^p boundedness of the Riesz transform related to Schrödinger operators on a manifold. Ann. Sc. Norm. Super. Pisa Cl. Sci. (5) **8**, 725–765 (2009)

159. Bagby, R.J.: An extended inequality for the maximal function. Proc. Am. Math. Soc. **48**, 419–422 (1975)
160. Banach, S.: Sur la convergence presque partout de fonctionnelles linéaires. Bull. Soc. Math. France **50**(2), 27–32, 36–43 (1926)
161. Baituyakova, Z., Sickel, W.: Strong summability of Fourier series and generalized Morrey spaces. Anal. Math. **43**(3), 371–414 (2017)
162. El Baraka, A.: Function spaces of BMO and Campanato type. Function spaces of BMO and Campanato type. Electron. J. Diff. Equ. Conf. **9**, 109–115 (2002)
163. El Baraka, A.: Littlewood–Paley characterization for Campanato spaces. J. Funct. Spaces Appl. **4**(2), 193–220 (2006)
164. Baernstein, A., Sawyer, E.: Embedding and multiplier theorems for $H^p(\mathbb{R}^n)$. Mem. Am. Math. Soc. **53**(318), iv+82 (1985)
165. Beauzamy, B.: Espaces de Sobolev et de Besov dódre variable définis sur L^p. C. R. Acad. Sci. Paris (Ser. A) **274**, 1935–1938 (1972)
166. Benedek, A., Calderón, A., Panzone, R.: Convolution operators on Banach space valued functions. Proc. Nat. Acad. Sci. USA **48**, 356–365 (1962)
167. Bényi, A., Gröchenig, K., Okoudjou, K., Rogers, L.G.: Unimodular Fourier multipliers for modulation spaces. J. Funct. Anal. **246**, 366–384 (2007)
168. Bényi, A., Oh, T.: Modulation spaces, Wiener amalgam spaces, and Brownian motions. Adv. Math. **228**(5), 2943–2981 (2011)
169. Benedetto, J., Zheng, S.: Besov spaces for the Schrödinger operator with barrier potential. Complex Anal. Oper. Theory **4**(4), 777–811 (2010)
170. Bergh, J.: Relation between the 2 complex methods of interpolation. Indiana Univ. Math. J. **28**(5), 775–778 (1979)
171. Besov, O.V.: On a family of function spaces. Embedding theorems and extensions. Dokl. Acad. Nauk SSSR **126**, 1163–1165 (1959)
172. Besov, O.V.: Investigation of a class of function spaces in connection with imbedding and extension theorems. Trudy Mat. Inst. Steklov **60**, 42–81 (1961)
173. Besov, O.V.: An example in the theory of imbedding theorems (Russian). Dokl. Akad. Nauk SSSR **143**, 1014–1016 (1962)
174. Besov, O.V.: On the continuation of functions with preservation of the second-order integral modulus of smoothness (Russian). Mat. Sb. (N.S.) **58**(100), 673–684 (1962)
175. Besov, O.V.: Extension of functions with preservation of differential-difference properties in L^p (Russian). Dokl. Akad. Nauk SSSR **150**, 963–966 (1963)
176. Besov, O.V.: Extension of functions to the frontier, with preservation of differential-difference properties in L^p (Russian). Mat. Sb. (N.S.) **66**(108), 80–96 (1965)
177. Besov, O.V.: On the density of finitary functions and the extension of classes of differentiable functions (Russian). Dokl. Akad. Nauk SSSR **165**, 738–741 (1965)
178. Besov, O.V.: Continuation of certain classes of differentiable functions beyond the boundary of a region (Russian). Trudy Mat. Inst. Steklov. **77**, 35–44 (1965)
179. Besov, O.V.: Continuation of functions from L^{pl} and W^{pl} (Russian). Trudy Mat. Inst. Steklov. **89**, 5–17 (1967)
180. Besov, O.V.: The density of finite functions in $L^{p,\theta}$ and the extension of functions. (Russian), Trudy Mat. Inst. Steklov. **89**, 18–30 (1967)
181. Besov, O.V.: Behavior of differentiable functions at infinity, and density of the functions with compact support (Russian). Trudy Mat. Inst. Steklov. **105**, 3–14 (1969)
182. Besov, O.V.: Classes of functions with a generalized mixed Hölder condition (Russian). Trudy Mat. Inst. Steklov. **105**, 21–29 (1969)
183. Besov, O.V.: Estimates of derivatives in the mixed L_p norm on a region, and the extension of functions (Russian). Mat. Zametki **7**, 147–154 (1970)
184. Besov, O.V.: Inequalities for moduli of continuity of functions given on a domain, and imbedding theorems (Russian). Dokl. Akad. Nauk SSSR **202**, 507–510 (1972)
185. Besov, O.V.: The behavior of differentiable functions on a nonsmooth surface. (Russian), Studies in the theory of differentiable functions of several variables and its applications, IV. Trudy Mat. Inst. Steklov. **117**, 3–10, 343 (1972)

186. Besov, O.V.: Estimates of moduli of smoothness of functions on domains, and imbedding theorems (Russian). Studies in the theory of differentiable functions of several variables and its applications, IV. Trudy Mat. Inst. Steklov. **117**, 22–46, 343 (1972)
187. Besov, O.V.: On traces on a nonsmooth surface of classes on differentiable functions. Proc. Steklov Inst. Math. **117**, 11–23 (1972)
188. Besov, O.V.: Multiplicative estimates for integral norms of differentiable functions of several variables (Russian). Studies in the theory of differentiable functions of several variables and its applications, V. Trudy Mat. Inst. Steklov. **131**, 3–15, 244 (1974)
189. Besov, O.V.: Estimates for the moduli of continuity of abstract functions defined in a domain (Russian). Studies in the theory of differentiable functions of several variables and its applications, V. Trudy Mat. Inst. Steklov. **131**, 16–24, 244 (1974)
190. Besov, O.V.: The growth of the mixed derivative of a function in $C^{(l_1,l_2)}$ (Russian). Mat. Zametki **15**, 355–362 (1974)
191. Besov, O.V.: Multiplicative estimates of integral moduli of smoothness (Russian). Studies in the theory of differentiable functions of several variables and its applications, VI. Trudy Mat. Inst. Steklov. **140**, 21–26, 286 (1976)
192. Besov, O.V.: Estimations of the errors of cubature formulas by the smoothness of functions (Russian). C. R. Acad. Bulgare Sci. **31**(8), 949–952 (1978)
193. Besov, O.V.: Estimates of the error of cubature formulas with respect to smoothness of functions (Russian). Studies in the theory of differentiable functions of several variables and its applications, VII. Trudy Mat. Inst. Steklov. **150**, 11–23, 321 (1979)
194. Besov, O.V.: Multiplicative estimates of integral moduli of smoothness. Proc. Steklov Inst. Math. **1** (1979)
195. Besov, O.V.: Weight estimates of mixed derivatives in a domain (Russian). Studies in the theory of differentiable functions of several variables and its applications, VIII. Trudy Mat. Inst. Steklov. **156**, 16–21, 262 (1980)
196. Besov, O.V.: Density of compactly supported functions in a weighted Sobolev space (Russian). Studies in the theory of differentiable functions of several variables and its applications, IX. Trudy Mat. Inst. Steklov. **161**, 29–47 (1983)
197. Besov, O.V.: Integral representations of functions in a domain with the flexible horn condition, and imbedding theorems (Russian). Dokl. Akad. Nauk SSSR **273**(6), 1294–1297 (1983)
198. Besov, O.V.: Integral representations of functions and embedding theorems for a domain with a flexible horn condition (Russian). Studies in the theory of differentiable functions of several variables and its applications, X. Trudy Mat. Inst. Steklov. **170**, 12–30, 274 (1984)
199. Besov, O.V.: The Littlewood–Paley theorem for a mixed norm (Russian). Studies in the theory of differentiable functions of several variables and its applications, X. Trudy Mat. Inst. Steklov. **170**, 31–36, 274 (1984)
200. Besov, O.V.: Estimates of integral-moduli of continuity and imbedding theorems for a domain with the flexible horn condition (Russian). Trudy Mat. Inst. Steklov. **172**, 4–15 (1985); Proc. Steklov Inst. Math. **172**, 1–13 (1987)
201. Besov, O.V.: Hörmander's theorem on Fourier multipliers (Russian). Studies in the theory of differentiable functions of several variables and its applications **11** (Russian). Trudy Mat. Inst. Steklov. **173**, 3–13, 270 (1986)
202. Besov, O.V.: Embeddings of an anisotropic Sobolev space for a domain with a flexible horn condition (Russian). Translated in Proc. Steklov Inst. Math. **4**, 1–13 (1989); Studies in the theory of differentiable functions of several variables and its applications, XII (Russian). Trudy Mat. Inst. Steklov. **181**, 3–14, 269 (1988)
203. Besov, O.V.: Application of integral representations of functions to interpolation of spaces of differentiable functions and Fourier multipliers. Trudy Mat. Inst. Steklov. **192**, 20–34 (1990); English translation in Proc. Steklov Inst. Math. **3**, 192 (1993)
204. Besov, O.V.: Interpolation of spaces of differentiable functions defined in a domain (Russian). Trudy Mat. Inst. Steklov. **201**, Issled. po Teor. Differ. Funktsii Mnogikh Peremen. i ee Prilozh. **15**, 26–42 (1992); Translation in Proc. Steklov Inst. Math. **2**(201), 21–34 (1994)

205. Besov, O.V.: Embeddings of Sobolev-Liouville and Lizorkin-Triebel spaces in a domain (Russian). Dokl. Akad. Nauk **331**(5), 538–540 (1993); Translation in Russian Acad. Sci. Dokl. Math. **48**(1), 130–133 (1994)
206. Besov, O.V.: Embeddings of spaces of functions of variable smoothness (Russian). Dokl. Akad. Nauk **347**(1), 7–10 (1996)
207. Besov, O.V.: Embeddings of spaces of differentiable functions of variable smoothness (Russian). Tr. Mat. Inst. Steklova **214**, Issled. po Teor. Differ. Funkts. Mnogikh Perem. i ee Prilozh. **17**, 25–58 (1997); Translation in Proc. Steklov Inst. Math. **3**(214), 19–53 (1996)
208. Besov, O.V.: Interpolation of spaces of differentiable functions on a domain (Russian). Tr. Mat. Inst. Steklova **214**, Issled. po Teor. Differ. Funkts. Mnogikh Perem. i ee Prilozh. **17**, 59–82 (1997); Translation in Proc. Steklov Inst. Math. **3**(214), 54–76 (1996)
209. Besov, O.V.: Interpolation and embeddings of the spaces of generalized functions $B^s_{p,q}$ and $F^s_{p,q}$ on a domain (Russian). Tr. Mat. Inst. Steklova **219**. Teor. Priblizh. Garmon. Anal. 80–102 (1997); Translation in Proc. Steklov Inst. Math. **4**(219), 73–95 (1997)
210. Besov, O.V.: Interpolation and embeddings of function spaces B^s_{pq} and F^s_{pq} on a domain (Russian). Dokl. Akad. Nauk **357**(6), 727–730 (1997)
211. Besov, O.V.: On the continuation by zero of functions of several variables (Russian). Mat. Zametki **64**(3), 351–365 (1998); Translation in Math. Notes **64**(3–4), 303–315 (1998)
212. Besov, O.V.: Estimates for some integral operators (Russian). Tr. Mat. Inst. Steklova **227**. Issled. po Teor. Differ. Funkts. Mnogikh Perem. i ee Prilozh. **18**, 75–77 (1999); Translation in Proc. Steklov Inst. Math. **4**(227), 70–72 (1999)
213. Besov, O.V.: On spaces of functions of variable smoothness defined by pseudodifferential operators (Russian). Tr. Mat. Inst. Steklova **227**, Issled. po Teor. Differ. Funkts. Mnogikh Perem. i ee Prilozh. **18**, 56–74 (1999); Translation in Proc. Steklov Inst. Math. **4**(227), 50–69 (1999)
214. Besov, O.V.: On function spaces defined by pseudodifferential operators (Russian). Dokl. Akad. Nauk **367**(6), 730–733 (1999)
215. Besov, O.V.: The Sobolev embedding theorem for a domain with irregular boundary (Russian). Dokl. Akad. Nauk **373**(2), 151–154 (2000)
216. Besov, O.V.: On the works of S.M. Nikolski'i–in the theory of function spaces and its applications (Russian). Tr. Mat. Inst. Steklova **232**. Funkts. Prostran., Garmon. Anal. Differ. Uravn. 25–30 (2001); Translation in Proc. Steklov Inst. Math. **1**(232), 19–24 (2001)
217. Besov, O.V.: On the compactness of embeddings of weighted Sobolev spaces on a domain with an irregular boundary (Russian). Tr. Mat. Inst. Steklova **232**, Funkts. Prostran., Garmon. Anal. Differ. Uravn., 72–93 (2001); Translation in Proc. Steklov Inst. Math. **1**(232), 66–87 (2001)
218. Besov, O.V.: On the compactness of embeddings of weighted Sobolev spaces on a domain with an irregular boundary (Russian). Dokl. Akad. Nauk **376**(6), 727–732 (2001)
219. Besov, O.V.: Sobolev's embedding theorem for a domain with an irregular boundary (Russian). Mat. Sb. **192**(3), 3–26 (2001); Translation in Sb. Math. **192**(3–4), 323–346 (2001)
220. Besov, O.V.: Spaces of functions of fractional smoothness on an irregular domain (Russian). Dokl. Akad. Nauk **383**(5), 586–591 (2002)
221. Besov, O.V.: Equivalent normings of spaces of functions of variable smoothness (Russian). Dokl. Akad. Nauk **391**(5), 583–586 (2003)
222. Besov, O.V.: Equivalent normings of spaces of functions of variable smoothness (Russian). Tr. Mat. Inst. Steklova **243**. Funkts. Prostran., Priblizh., Differ. Uravn., 87–95 (2003); Translation in Proc. Steklov Inst. Math. **243**(4), 80–88 (2003)
223. Besov, O.V.: Spaces of functions of fractional smoothness on an irregular domain (Russian). Mat. Zametki **74**(2), 163–183 (2003); Translation in Math. Notes **74**(1–2), 157–176 (2003)
224. Besov, O.V.: Equivalent norms in spaces of functions of fractional smoothness on an arbitrary domain (Russian). Mat. Zametki **74**(3), 340–349 (2003); Translation in Math. Notes **74**(3–4), 326–334 (2003)
225. Besov, O.V.: A test for the uniform convergence of a trigonometric Fourier series (Russian). Dokl. Akad. Nauk **395**(6), 727–732 (2004)

226. Besov, O.V.: On the interpolation, embedding, and extension of spaces of functions of variable smoothness (Russian). Dokl. Akad. Nauk **401**(1), 7–11 (2005)
227. Besov, O.V.: Interpolation, embedding, and extension of spaces of functions of variable smoothness (Russian). Tr. Mat. Inst. Steklova **248**, Issled. po Teor. Funkts. i Differ. Uravn., 52–63 (2005); Translation in Proc. Steklov Inst. Math. **1**(248), 47–58 (2005)
228. Besov, O.V.: An estimate for the approximation of periodic functions by Fourier sums (Russian). Mat. Zametki **79**(5), 784–787 (2006); Translation in Math. Notes **79**(5–6), 726–728 (2006)
229. Besov, O.V.: Lizorkin-Triebel-type function spaces on an irregular domain (Russian). Tr. Mat. Inst. Steklova **260**, Teor. Funkts. i Nelinein. Uravn. v Chastn. Proizvodn., 32–43 (2008); Translation in Proc. Steklov Inst. Math. **260**(1), 25–36 (2008)
230. Besov, O.V.: Estimates for L^p-moduli of continuity on domains with an irregular boundary, and embedding theorems (Russian). Sovrem. Mat. Fundam. Napravl. **25**, 21–33 (2007); Translation in J. Math. Sci. (N. Y.) **155**(1), 18–30 (2008)
231. Besov, O.V.: Function spaces of Lizorkin-Triebel type on an irregular domain. Nonlinear Anal. **70**(8), 2842–2845 (2009)
232. Besov, O.V.: Spaces of functions of fractional smoothness on an irregular domain (Russian). Dokl. Akad. Nauk **425**(4), 439–442 (2009); Translation in Dokl. Math. **79**(2), 223–226 (2009)
233. Besov, O.V.: Integral estimates for differentiable functions on irregular domains (Russian). Dokl. Akad. Nauk **430**(5), 583–585 (2010); Translation in Dokl. Math. **81**(1), 87–90 (2010)
234. Besov, O.V.: Sobolev embedding theorem for anisotropically irregular domains (Russian). Dokl. Akad. Nauk **438**(5), 586–589 (2011); Translation in Dokl. Math. **83**(3), 367–370 (2011)
235. Besov, O.V.: Sobolev's embedding theorem for anisotropically irregular domains. Eurasian Math. J. **2**(1), 32–51 (2011)
236. Besov, O.V.: On spaces of functions of smoothness zero (Russian). Mat. Sb. **203**(8), 3–16 (2012); Translation in Sb. Math. **203**(7–8), 1077–1090 (2012)
237. Besov, O.V.: Embedding of a weighted Sobolev space and properties of the domain. ISSN:1064–5624. Doklady Math. **90**(3), 754–757 (2014); Doklady Akademii Nauk **459**(6), 663–666 (2014)
238. Besov, O.V.: Embedding of Sobolev space in the case of the limit exponent (Russian). Mat. Zametki **98**(4), 498–510 (2015)
239. Besov, O.V.: Embedding of a Sobolev space in the case of a limiting exponent. Dokl. Akad. Nauk **462**(2), 131–134 (2015); Translation in Dokl. Math. **91**(3), 277–280 (2015)
240. Besov, O.V.: Spaces of functions of positive smoothness on irregular domains. ISSN:1064–5624, Doklady Math. **93**(1), 13–15 (2016); Doklady Akademii Nauk **466**(2), 133–136 (2016)
241. Besov, O.V.: Embeddings of spaces of functions of positive smoothness on irregular domains in Lebesgue spaces. (Russian) Mat. Zametki **103**(3), 336–345 (2018)
242. Besov, O.V., ẑabrailov, A.D.D., Interpolation theorems for certain spaces of differentiable functions (Russian). Trudy Mat. Inst. Steklov. **105**, 15–20 (1969)
243. Besov, O.V., Il'in, V.P.: A natural extension of the class of domains in imbedding theorems (Russian). Mat. Sb. (N.S.) **75**(117), 483–495 (1968)
244. Besov, O.V., Il'in, V.P.: An imbedding theorem for the limit exponent (Russian). Mat. Zametki **6**, 129–138 (1969)
245. Besov, O.V., Kadlec, J., Kufner, A.: Certain properties of weight classes (Russian). Dokl. Akad. Nauk SSSR **171**, 514–516 (1966)
246. Besov, O.V., Kufner, A.: The density of smooth functions in weight spaces (Russian). Czechoslov. Math. J. **18**(93), 178–188 (1968)
247. Besov, O.V., Lizorkin, V.P.: The L_p-estimates of a certain class of non-isotropically singular integrals (Russian). Dokl. Akad. Nauk SSSR **169**, 1250–1253 (1966)
248. Besov, O.V., Lizorkin, V.P.: Singular integral operators and sequences of convolutions in L_p-spaces (Russian). Mat. Sb. (N.S.) **73**(115), 65–88 (1967)

249. Besoy, B.F., Cobos, F.: Duality for logarithmic interpolation spaces when $0 < q < 1$ and applications. J. Math. Anal. Appl. **466**(1), 373–399 (2018)
250. Beurling, A.: Construction and analysis of some convolution algebra. Ann. Inst. Fourier **14**, 1–32 (1964)
251. Birman, M.Š., Solomjak, M.Z.: Piecewise polynomial approximations of functions of classes W_p^α. Mat. Sb. (N.S.) **73**(115), 331–355 (1967)
252. Birnbaum, Z., Orlicz, W.: Über die Verallgemeinerung des Begriffes der zueinander konjugierten Potenzen. Studia Math. **3**, 1–67 (1931)
253. Bonami, A., Grellier, S.: Hankel operators and weak factorization for Hardy-Orlicz spaces. Colloq. Math. **118**, 107–132 (2010)
254. Bonami, A., Grellier, S., Ky, L.D.: Paraproducts and products of functions in BMO($\mathbb{R}^n$) and $H^1(\mathbb{R}^n)$ through wavelets. J. Math. Pures Appl. (9) **97**, 230–241 (2012)
255. Bonami, A., Iwaniec, T., Jones, P., Zinsmeister, M.: On the product of functions in BMO and H^1. Ann. Inst. Fourier (Grenoble) **57**, 1405–1439 (2007)
256. Bonami, A., Feuto, J., Grellier, S.: Endpoint for the DIV-CURL lemma in Hardy spaces. Publ. Mat. **54**, 341–358 (2010)
257. Bonk, M., Saksman, E., Soto, T.: Triebel–Lizorkin spaces on metric spaces via hyperbolic fillings. arXiv:1411.5906, Indiana Univ. Math. J. (to appear)
258. Bony, J.M.: Calcul symbolique et propagation des singularités pour les équations aux dérivées partielles non linéaires (French). [Symbolic calculus and propagation of singularities Quantitative analysis in Sobolev imbedding theorems for and applications to spectral theory, nonlinear partial differential equations] Ann. Sci. École Norm. Sup. (4) **14**(2), 209–246 (1981)
259. Bony, J.M.: Second microlocalization and propagation of singularities for semi-linear hyperbolic equations. In: Taniguchi Symposium HERT, Katata, pp. 11–49 (1984)
260. Borup, L., Nielsen, M.: Frame decomposition of decomposition spaces. J. Fourier Anal. Appl. **13**(1), 39–70 (2007)
261. Borup, L., Nielsen, M.: On anisotropic Triebel–Lizorkin-type spaces, with applications to the study of pseudo-dierential operators. J. Funct. Spaces Appl. **6**(2), 107–154 (2008)
262. Bourdaud, G.: Une algèbre maximale d'opérateurs pseudo-différentiels. Commun. PPDE **13**(9), 1059–1083 (1988)
263. Bourdaud, G.: Localizations des espaces de Besov. Studia Math. **90**, 153–163 (1988)
264. Bourdaud, G.: Réalisations des espaces de Besov, homogènes. Ark. Mat. **26**, 41–54 (1988)
265. Bourdaud, G.: Changes of variable in Besov spaces. II. Forum Math. **12**(5), 545–563 (2000)
266. Bourdaud, G.: Realizations of homogeneous Sobolev spaces. Complex Var. Elliptic Equ. **56**(10–11), 857–874 (2011)
267. Bourdaud, G., Cristoforis, M.L., Sickel, W.: Functional calculus on BMO and related spaces. J. Funct. Anal. **189**(2), 515–538 (2002)
268. Bourdaud, G., Meyer, Y.: Fonctions qui opèrent sur les espaces de Sobolev. J. Funct. Anal. **97**, 351–360 (1991)
269. Bourdaud, G., Meyer, Y.: Le calcul fonctionnel sous-linèaire dans les espaces de Besov homogèenes. [Sublinear functional calculus in homogeneous Besov spaces] Rev. Mat. Iberoam. **22**(2), 725–746, loose erratum (2006)
270. Bourdaud, G., Moussai, M., Sickel, W.: An optimal symbolic calculus on Besov algebras. Ann. Inst. H. Poincaré Anal. Non Lineaire **23**(6), 949–956 (2006)
271. Bourdaud, G., Moussai, M., Sickel, W.: Towards sharp superposition theorems in Besov and Lizorkin-Triebel spaces. Nonlinear Anal. **68**(10), 2889–2912 (2008)
272. Bourdaud, G., Moussai, M., Sickel, W.: Composition operators on Lizorkin-Triebel spaces. J. Funct. Anal. **259**(5), 1098–1128 (2010)
273. Bourdaud, G., Moussai, M., Sickel, W.: Composition operators acting on Besov spaces on the real line. Ann. Mat. Pura Appl. (4) **193**(5), 1519–1554 (2014)
274. Bourdaud, G., Sickel, W.: Changes of variable in Besov spaces. Math. Nachr. **198**, 19–39 (1999)

275. Bourdaud, G., Sickel, W.: Composition operators on function spaces with fractional order of smoothness. In: Harmonic Analysis and Nonlinear Partial Differential Equations. RIMS Kokyuroku Bessatsu, vol. B26, pp. 93–132. Research Institute for Mathematical Sciences (RIMS), Kyoto (2011)
276. Bownik, M.: Anisotropic Hardy spaces and wavelets. Mem. Am. Math. Soc. **164**(781), vi+122 (2003)
277. Bownik, M.: Boundedness of operators on Hardy spaces via atomic decompositions. Proc. Am. Math. Soc. **133**(12), 3535–3542 (2005)
278. Bownik, M.: Atomic and molecular decompositions of anisotropic Besov spaces. Math. Z. **250**, 539–571 (2005)
279. Bownik, M.: Anisotropic Triebel–Lizorkin spaces with doubling measures. J. Geom. Anal. **17**, 387–424 (2007)
280. Bownik, M.: Duality and interpolation of anisotropic Triebel–Lizorkin spaces. Math. Z. **259**(1), 131–169 (2008)
281. Bownik, M.: Extrapolation of discrete Triebel–Lizorkin spaces. Math. Nachr. **286**(5–6), 492–502 (2013)
282. Bownik, M., Ho, K.P.: Atomic and molecular decompositions of anisotropic Triebel–Lizorkin spaces. Trans. Am. Math. Soc. **358**(4), 1469–1510 (2006)
283. Bourgain, J., Li, D.: Strong ill-posedness of the incompressible Euler equation in borderline Sobolev spaces. Invent. Math. **201**(1), 97–157 (2015)
284. Bourgain, J., Pavlovic, N.: Ill-posedness of the Navier–Stokes equations in a critical space in 3D. J. Funct. Anal. **255**, 2233–2247 (2008)
285. Brundnyj, J.A., Krejn, S.G., Semenov, E.M.: Interpolation of linear operators (Russian). In: Itogi nauki i techniki, Se. mat. analiz. vol. 24, pp. 3–163. Moskva, Akademija nauk SSSR (1986)
286. Blasco, O., Ruiz, A., Vega, L.: Non-interpolation in Morrey-Campanato and block spaces. Ann. Scuola Norm. Sup. Pisa Cl. Sci. **28**, 31–40 (1999)
287. Brezis, H., Mironescu, P.: Gagliardo-Nirenberg, composition and products in fractional Sobolev spaces. J. Evol. Equ. **1**, 387–404 (2001)
288. Butzer, P.L., Ferreia, P.J.S.G., Higgins, J.R., Saitoh, S., Schmeisser, G., Steins, R.L.: Interpolations and sampling: E.T. Whittaker, K. Ogura and their followers. J. Fourier Anal. Appl. **17**, 320–354 (2011)
289. Bui, H.Q.: Some aspects of weighted and non-weighted Hardy spaces. Kokyuroku Res. Inst. Math. Sci. **383**, 38–56 (1980)
290. Bui, H.Q.: Weighted Hardy spaces. Math. Nachr. **103**, 45–62 (1981)
291. Bui, H.Q.: Weighted Besov and Triebel spaces: interpolation by the real method. Hiroshima Math. J. **3**, 581–605 (1982)
292. Bui, H.Q.: Representation theorems and atomic decomposition of Besov spaces. Math. Nachr. **132**, 301–311 (1987)
293. Bui, H.Q., Duong, X.T., Yan, L.: Calderón reproducing formulas and new Besov spaces associated with operators. Adv. Math. **229**(4), 2449–2502 (2012)
294. Bui, H.Q., Paluszyński, M., Taibleson, M.H.: A maximal function characterization of weighted Besov-Lipschitz and Triebel–Lizorkin spaces. Studia Math. **119**, 219–246 (1996)
295. Bui, H.Q., Paluszyński, M., Taibleson, M.H.: Characterization of the Besov-Lipschitz and Triebel–Lizorkin spaces. The case $q < 1$. J. Fourier Anal. Appl. **3**, 837–846 (1997). Special issue
296. Bui, T.A., Duong, X.T.: Besov and Triebel–Lizorkin spaces associated to Hermite operators. J. Fourier Anal. Appl. **21**(2), 405–448 (2015)
297. Burenkov, V.I.: Additivity of the spaces W_p^r and B_p^r, and embedding theorems for domains of general form (Russian). Trudy Mat. Inst. Steklov. **105**, 30–45 (1969)
298. Burenkov, V.I.: The approximations of functions from Sobolev spaces by functions with compact support in the case of an arbitrary open set (Russian). Dokl. Akad. Nauk SSSR **202**, 259–262 (1972)

299. Burenkov, V.I.: The approximation of functions in the space $C^r(\Omega)$ by functions of compact support, for an arbitrary open set Ω (Russian). Studies in the theory of differentiable functions of several variables and its applications, IV. Trudy Mat. Inst. Steklov. **117**, 62–74, 343 (1972)
300. Burenkov, V.I.: Sobolev's integral representation and Taylor's formula (Russian). Studies in the theory of differentiable functions of several variables and its applications, V. Trudy Mat. Inst. Steklov. **131**, 33–38, 244 (1974)
301. Burenkov, V.I.: The density of infinitely differentiable functions in Sobolev spaces for an arbitrary open set (Russian). Studies in the theory of differentiable functions of several variables and its applications, VI. Trudy Mat. Inst. Steklov. **131**, 39–50, 244–245 (1974)
302. Burenkov, V.I.: On partition of unity. Proc. Steklov Inst. Math. **4**, 25–31 (1981)
303. Burenkov, V.I.: Estimates for Fourier transforms and convolutions in Nikolski'i–Besov spaces (Russian). Translated in Proc. Steklov Inst. Math. **3**, 35–44 (1990); Studies in the theory of differentiable functions of several variables and its applications **13** (Russian). Trudy Mat. Inst. Steklov. **187**, 31–38 (1989)
304. Burenkov, V.I.: Extension theory for Sobolev spaces on open sets with Lipschitz boundaries. In: Nonlinear Analysis, Function Spaces and Applications, Prague, vol. 6, pp. 1–49 (1998). Academy of Sciences of the Czech Republic, Prague (1999)
305. Burenkov, V.I.: Extension theorems for Sobolev and more general spaces for degenerate open sets. In: Proceedings of the Second ISAAC Congress, Fukuoka, vol. 2, pp. 1135–1141 (1999); International Society of Analysis and Applied Computations, vol. 8. Kluwer Academic Publishers, Dordrecht (2000)
306. Burenkov, V.I., Gol'dman, M.L.: On the extensions of functions of L^p. Trudy Math. Inst. Steklov **150**, 31–51 (1979) (English transl. **4**, 33–53 (1981))
307. Burenkov, V.I., Gorbunov, A.L.: A two-sided estimate for the minimal norm of the extension operator for Sobolev spaces (Russian). Dokl. Akad. Nauk **330**(6), 680–682 (1993); Translation in Russian Acad. Sci. Dokl. Math. **47**(3), 589–592 (1993)
308. Burenkov, V.I., Gorbunov, A.L.: Sharp estimates for the minimal norm of extension operators for Sobolev spaces (Russian). Izv. Ross. Akad. Nauk Ser. Mat. **61**(1), 3–44 (1997); Translation in Izv. Math. **61**(1), 1–43 (1997)
309. Burenkov, V.I., Kalyabin, G.A.: Lower estimates of the norms of extension operators for Sobolev spaces on the halfline. Math. Nachr. **218**, 19–23 (2000)
310. Burenkov, V.I., Kudryavtsev, L.D., Neverov, I.V.: On an identity for differences of arbitrary order. Mat. Zametki **64**(2), 302–307 (1998); Translation in Math. Notes **64** (1998); (1–2), 256–261 (1999)
311. Burenkov, V.I., Nursultanov, E.D.: Description of interpolation spaces for local Morrey-type spaces (Russian). Tr. Mat. Inst. Steklova **269**, Teoriya Funktsii i Differentsialnye Uravneniya, 52–62 (2010); Translation in Proc. Steklov Inst. Math. **269**, 46–56 (2010)
312. Burenkov, V.I., Schulze, B.W., Tarkhanov, N.N.: Extension operators for Sobolev spaces commuting with a given transform. Glasgow Math. J. **40**(2), 291–296 (1998)
313. Burenkov, V.I., Senusi, A.: Estimates for constants in additivity inequalities for function spaces. (Russian) Sibirsk. Mat. Zh. **35**(1), 24–40 (1994); Translation in Siberian Math. J. **35**(1), 21–36 (1994)
314. Burenkov, V.I., Tuyakbaev, M.S.: Multipliers of the Fourier integral in weighted L^p-spaces with an exponential weight (Russian). Dokl. Akad. Nauk SSSR **320**(1), 11–14 (1991); Translation in Soviet Math. Dokl. **44**(2), 365–369 (1992)
315. Burenkov, V.I., Verdiev, T.V.: Extension by zero of functions from spaces with generalized smoothness for degenerate domains (Russian). Tr. Mat. Inst. Steklova **227**, Issled. po Teor. Differ. Funkts. Mnogikh Perem. i ee Prilozh. **18**, 78–91 (1999); Translation in Proc. Steklov Inst. Math. **4**(227), 73–86 (1999)
316. Burenkov, V.I., Viktorova, N.B.: On an embedding theorem for Sobolev spaces with a mixed norm for limit exponents (Russian). Mat. Zametki **59**(1), 62–72, 158 (1986); Translation in Math. Notes **59**(1–2), 45–51 (1986)

317. Caetano, A.M.: On the type of convergence in atomic representations. Complex Var. Elliptic Equ. **56**(10–11), 875–883 (2011)
318. Caetano, A.M., Farkas, W.: Local growth envelopes of Besov spaces of generalized smoothness. Z. Anal. Anwend. **25**, 265–298 (2006)
319. Caetano, A., Gogatishvili, A., Opic, B.: Sharp embeddings of Besov spaces involving only logarithmic smoothness. J. Approx. Theory **152**(2), 188–214 (2008)
320. Caetano, A.M., Lopes, S., Triebel, H.: A homogeneity property for Besov spaces. J. Funct. Spaces Appl. **5**(2), 123–132 (2007)
321. Caetano, A.M., Leopold, H.G.: On generalized Besov and Triebel-Lizorkin spaces of regular distributions. J. Funct. Anal. **264**, 2676–2703 (2013)
322. Caetano, A.M., Haroske, D.D.: Embeddings of Besov spaces on fractal h-sets. Banach J. Math. Anal. **9**(4), 259–295 (2015)
323. Calderón, A.P.: Lebesgue spaces of differentiable functions and distributions. In: Partial Differential Equations. Proceedings of Symposia in Pure Mathematics, vol. 4, pp. 33–49. American Mathematical Society, Providence (1961)
324. Calderón, A.P.: Intermediate spaces and interpolation. Studia Math. Seria specjalna **1**, 31–34 (1963)
325. Calderón, A.P.: Intermediate spaces and interpolation, the complex method. Studia Math. **14**(1), 113–190, 46–56 (1964)
326. Calderón, A.P.: Spaces between L^1 and L^∞ and the theorem of Marcinkiewicz. Studia Math. **26**, 273–299 (1966)
327. Calderón, A.P.: An atomic decomposition of distributions in parabolic H^p spaces. Adv. Math. **25**, 216–225 (1977)
328. Calderón, A.P., Torchinsky, A.: Parabolic maximal functions associated with a distribution. I. Adv. Math. **16**, 1–64 (1975)
329. Calderón, A.P., Torchinsky, A.: Parabolic maximal functions associated with a distribution. II. Adv. Math. **24**, 101–171 (1977)
330. Calderón, A.P., Zygmund, A.: On the existence of certain singular integrals. Acta Math. **88**, 85–139 (1952)
331. Calderón, A.P., Zygmund, A.: A note on the interpolation of sublinear operations. Am. J. Math. **78**, 282–288 (1956)
332. Calderón, A.P., Zygmund, A.: On singular integrals. Am. J. Math. **78**, 289–309 (1956)
333. Calderón, A.P., Zygmund, A.: On higher gradients of harmonic functions. Studia Math. **24**, 211–226 (1964)
334. Calderón, A.P., Zygmund, A.: Singular integral operators and differential equations. Am. J. Math. **79**, 901–921 (1957)
335. Calderón, A.P., Vaillancourt, R.: On the boundedness of pseudo-differential operators. J. Math. Soc. Japan **23**, 374–378 (1971)
336. Cannone, M., Planchon, F.: Self-similar solutions for Navier–Stokes equations in $\mathbb{R}^3$. Commun. PDE **21**, 179–193 (1996)
337. Cantor, G.: Über unendliche, lineare Punktmannigfaltigkeiten V. Math. Annal. **21**, 545–591 (1883)
338. Cao, J., Chang, D.C., Fu, Z., Yang, D.: Real interpolation of weighted tent spaces. Appl. Anal. **95**(11), 2415–2443 (2016)
339. Cao, J., Chang, D.C., Fu, Z., Yang, D., Yang, S.: Riesz transform characterizations of Musielak–Orlicz–Hardy spaces. Trans. Am. Math. Soc. **368**(10), 6979–7018 (2016)
340. Cao, J., Chang, D.C., Wu, H., Yang, D.: Weak Hardy spaces $WH^p_L(\mathbb{R}^n)$ associated to operators satisfying k-Davies-Gaffney estimates. J. Nonlinear Convex Anal. **16**(7), 1205–1255 (2015)
341. Cao, J., Mayboroda, S., Yang, D.: Maximal function characterizations of Hardy spaces associated to homogeneous higher order elliptic operators. Forum Math. **28**(5), 823–856 (2016)
342. Cao, J., Mayboroda, S., Yang, D.: Local Hardy spaces associated with inhomogeneous higher order elliptic operators. Anal. Appl. (Singap.) **15**(2), 137–224 (2017)

343. Cao, Y., Jiang, Y.: Weighted Morrey type Besov and Triebel–Lizorkin spaces and pseudo-differential operators with non-regular symbols. Adv. Math. (China) **38**(5), 629–640 (2009)
344. Cao, J., Fu, Z., Jiang, R., Yang, D.: Hardy spaces associated with a pair of commuting operators. Forum Math. **27**(5), 2775–2824 (2015)
345. Cao, J., Liu, Y., Yang, D.: Hardy spaces $H_L^1(\mathbb{R}^n)$ associated to Schrödinger type operators $(-\Delta)^2 + V^2$. Houston J. Math. **36**, 1067–1095 (2010)
346. Cao, J., Yang, D.: $H_L^p(\mathbb{R}^n)$ associated to operators satisfying k-Davies-Gaffney estimates. Sci. China Math. **55**, 1403–1440 (2012)
347. Carleson, L.: An interpolation problem for bounded analytic functions. Am. J. Math. **80**, 921–930 (1958)
348. Carleson, L.: Interpolation by bounded analytic functions and the corona problem. Ann. Math. (2nd Ser.) **76**(3), 547–559 (1962)
349. Carro, M.J., Mastyło, M., Rodríguez-Piazza, L.: Almost everywhere convergent Fourier series. J. Fourier Anal. Appl. **18**, 266–286 (2012)
350. Cascante, C., Ortega, J.M., Verbitsky, I.E.: Nonlinear potentials and two weight trace inequalities for general dyadic and radial kernels. Indiana Univ. Math. J. **53**(3), 845–882 (2004)
351. Chae, D.: On the well-posedness of the Euler equations in the Triebel-Lizorkin spaces. Commun. Pure Appl. Math. **55**, 654–678 (2002)
352. Chamorro, D., Lemarié-Rieusset, P.G.: Real interpolation method, Lorentz spaces and refined Sobolev inequalities. J. Funct. Anal. **265**(12), 3219–3232 (2013)
353. Chandler-Wilde, S.N., Hewett, D.P., Moiola, A.: Sobolev spaces on non-Lipschitz subsets of $\mathbb{R}^n$ with application to boundary integral equations on fractal screens. Integr. Equ. Oper. Theory **87**(2), 179–224 (2017)
354. Chang, D.C., Dafni, G., Stein, E.M.: Hardy spaces, BMO and boundary value problems for the Laplacian on a smooth domain in $\mathbb{R}^n$. Trans. Am. Math. Soc. **351**, 1605–1661 (1999)
355. Chang, D.C., Krantz, S.G., Stein, E.M.: Hardy spaces and elliptic boundary value problems. In: The Madison Symposium on Complex Analysis, Madison, pp. 119–131 (1991). Contemporary Mathematical, vol. 137. American Mathematical Society, Providence (1992)
356. Chang, D.C., Krantz, S.G., Stein, E.M.: H^p theory on a smooth domain in $\mathbb{R}^N$ and elliptic boundary value problems. J. Funct. Anal. **114**, 286–347 (1993)
357. Chang, D.C., Fu, Z., Yang, D., Yang, S.: Real-variable characterizations of Musielak–Orlicz-Hardy spaces associated with Schrödinger operators on domains. Math. Methods Appl. Sci. **39**(3), 533–569 (2016)
358. Chang, D.C., Liu, J., Yang, D., Yuan, W.: Littlewood–Paley characterizations of Hajłasz–Sobolev and Triebel–Lizorkin spaces via averages on balls. Potential Anal. **46**(2), 227–259 (2017)
359. Chang, D.C., Yang, D., Yagn, S.: Real-variable theory of Orlicz-type function spaces associated with operators–a survey. In: Some Topics in Harmonic Analysis and Applications. Advanced Lectures in Mathematics (ALM), vol. 34, pp. 27–70. International Press, Somerville (2016)
360. Chen, X., Jiang, R., Yang, D.: Hardy and Hardy–Sobolev spaces on strongly Lipschitz domains and some applications. Anal. Geom. Metr. Spaces **4**, 336–362 (2016)
361. Chen, Y.Z., Lau, K.S.: Some new classes of Hardy spaces. J. Funct. Anal. **84**, 255–278 (1989)
362. Chiarenza, F., Frasca, M., Morrey spaces and Hardy–Littlewood maximal function. Rend. Mat. **7**, 273–279 (1987)
363. Choquet, G.: Theory of capacities. Ann. Inst. Fourier Grenoble **5**, 1953–1954, 131–295 (1955)
364. Christ, M.: The extension problem for certain function spaces involving fractional order s of differentiability. Ark. Mat. **22**, 63–81 (1984)
365. Christ, M.: A $T(b)$ theorem with remarks on analytic capacity and the Cauchy integral. Colloq. Math. **39–40**, 601–628 (1990)

366. Christ, M.: Lectures on Singular Integral Operators. CBMS Regional Conference Series in Mathematics, vol. 77. American Mathematical Society, Providence (1990)
367. Christ, M., Seeger, A.: Necessary conditions for vector-valued operator inequalities in harmonic analysis. Proc. London Math. Soc. **93**, 447–473 (2006)
368. Cho, Y.K.: Continuous characterization of the Triebel–Lizorkin spaces and Fourier multipliers. Bull. Korean Math. Soc. **47**, 839–857 (2010)
369. Cobos, F., Fernandez-Cabrera, L.M., Kühn, T., Ullrich, T.: On an extreme class of real interpolation spaces. J. Funct. Anal. **256**(7), 2321–2366 (2009)
370. Cobos, F., Kühn, T.: Approximation and entropy numbers in Besov spaces of generalized smoothness. J. Approx. Theory **160**(1–2), 56–70 (2009)
371. Cobos, F., Kühn, T.: Equivalence of K- and J-methods for limiting real interpolation spaces. J. Funct. Anal. **261**(12), 3696–3722 (2011)
372. Cobos, F., Domínguez, O.: Approximation spaces, limiting interpolation and Besov spaces. J. Approx. Theory **189**, 43–66 (2015)
373. Cobos, F., Domínguez, O.: On Besov spaces of logarithmic smoothness and Lipschitz spaces. J. Math. Anal. Appl. **425**(1), 71–84 (2015)
374. Cobos, F., Domínguez, O.: On the relationship between two kinds of Besov spaces with smoothness near zero and some other applications of limiting interpolation. J. Fourier Anal. Appl. **22**, 1174–1191 (2016)
375. Cobos, F., Domínguez, O., Triebel, H.: Characterizations of logarithmic Besov spaces in terms of differences, Fourier-analytical decompositions, wavelets and semi-groups. J. Funct. Anal. **270**(12), 4386–4425 (2016)
376. Cobos, F., Kruglyak, N.: Exact minimizer for the couple (L^∞, BV) and the one-dimensional analogue of the Rudin-Osher-Fatemi model. J. Approx. Theory **163**, 481–490 (2011)
377. Cobos, F., Peetre, J., Persson, L.E.: On the connection between real and complex interpolation of quasi-Banach spaces. Bull. Sci. Math. **122**, 17–37 (1998)
378. Cobos, F., Segurado, A.: Description of logarithmic interpolation spaces by means of the J-functional and applications. J. Funct. Anal. **268**(10), 2906–2945 (2015)
379. Cohen, A., Dahmen, W., DeVore, R.A.: Multiscale decompositions on bounded domains. Trans. Am. Math. Soc. **352**(8), 3651–3685 (2000)
380. Coifman, R.: A real variable characterization of H^p. Studia Math. **51**, 269–274 (1974)
381. Coifman, R.R., Lions, P.L., Meyer, Y., Semmes, P.: Compensated compactness and Hardy spaces. J. Math. Pures Appl. (9) **72**, 247–286 (1993)
382. Coifman, R.R., Meyer, Y.: Au delà des opérateurs pseudo-différentiels. Astérisque, vol. 57. Société Mathématique de France, Paris (1978)
383. Coifman, R.R., Meyer, Y., Stein, E.M.: Some new function spaces and their applications to harmonic analysis. J. Funct. Anal. **62**, 304–335 (1985)
384. Coifman, R.R., Rochberg, R.: Representation theorems for holomorphic and harmonic functions in L^p. Astérisque **77**, 11–66 (1980)
385. Coifman, R.R., Weiss, G.: Extensions of Hardy spaces and their use in analysis. Bull. Am. Math. Soc. **83**, 569–645 (1977)
386. Cohen, A., Daubechies, I., Feauveau, J. C.: Biorthogonal bases of compactly supported wavelets. Commun. Pure Appl. Math. **45**, 485–560 (1992)
387. Christensen, J.G., Mayeli, A., Ólafsson, G.: Coorbit description and atomic decomposition of Besov spaces. Numer. Funct. Anal. Optim. **33**(7–9), 847–871 (2012)
388. Cianchi, A.: An optimal interpolation theorem of Marcinkiewicz type in Orlicz spaces. J. Funct. Anal. **153**, 357–381 (1998)
389. Cruz-Uribe, D., Fiorenza, A., Neugebauer, C.J.: The maximal function on variable L^p spaces. Ann. Acad. Sci. Fenn. Math. **28**(1), 223–238 (2003)
390. Cruz-Uribe, D., Fiorenza, A., Neugebauer, C.J.: Corrections to "The maximal function on variable L^p spaces". Ann. Acad. Sci. Fenn. Math. **29**(1), 247–249 (2004)
391. Cruz-Uribe, D., Fiorenza, A., Martell, J., Pérez, C.: The boundedness of classical operators on variable L^p spaces. Ann. Acad. Sci. Fenn. Math. **31**, 239–264 (2006)

392. Cruz-Uribe, D., Rios, C.: The solution of the Kato problem for degenerate elliptic operators with Gaussian bounds. Trans. Am. Math. Soc. **364**(7), 3449–3478 (2012)
393. Cruz-Uribe, D., Rios, C.: The Kato problem for operators with weighted degenerate ellipticity. Trans. Am. Math. Soc. **367**(7), 4727–4756 (2015)
394. Cruz-Uribe, D., Wang, D.L.: Variable Hardy spaces. Indiana Univ. Math. J. **63**(2), 447–493 (2014)
395. Cunanan, J.: On L^p-boundedness of pseudo-differential operators of Sjöstrand's class. J. Fourier Anal. Appl. **23**(4), 810–816 (2017)
396. Cunanan, J., Kobayashi, M., Sugimoto, M.: Inclusion relations between L^p-Sobolev and Wiener amalgam spaces. J. Funct. Anal. **268**(1), 239–254 (2015)
397. Curbera, G.P., García-Cuerva, J., Martell, J.M., Pérez, C.: Extrapolation with weights, rearrangement-invariant function spaces, modular inequalities and applications to singular integrals. Adv. Math. 20, **203**(1), 256–318 (2006)
398. Cunanan, J., Tsutsui, Y.: Trace operators on Wiener amalgam spaces. J. Funct. Spaces **2016**, Article ID 1710260, 1–6 (2006)
399. Cwikel, M., Milman, M., Sagher, Y.: Complex interpolation of some quasi-Banach spaces. J. Funct. Anal. **65**, 339–347 (1986)
400. Cwikel, M., Sagher, Y.: Analytic families of operators on some quasi-Banach spaces. Proc. Am. Math. Soc. **102**, 979–984 (1988)
401. Dafni, G., Xiao, J.: Some new tent spaces and duality theorems for fractional Carleson measures and $Q_\alpha(\mathbb{R}^n)$. J. Funct. Anal. **208**, 377–422 (2004)
402. Dahlke, S.: Besov regularity for elliptic boundary value problems in polygonal domains. Appl. Math. Lett. **12**(6), 31–36 (1999)
403. Dahlke, S., DeVore, R.A.: Besov regularity for elliptic boundary value problems. Commun. PDE **22**(1–2), 1–16 (1997)
404. Dahlke, S., Novak, E., Sickel, W.: Optimal approximation of elliptic problems by linear and nonlinear mappings. I. J. Complexity **22**(1), 29–49 (2006)
405. Dahlke, S., Novak, E., Sickel, W.: Optimal approximation of elliptic problems by linear and nonlinear mappings. II. J. Complexity **22**(4), 549–603 (2006)
406. Dahlke, S., Novak, E., Sickel, W.: Optimal approximation of elliptic problems by linear and nonlinear mappings. III. Frames. J. Complexity **23**(4–6), 614–648 (2007)
407. Dahlke, S., Novak, E., Sickel, W.: Optimal approximation of elliptic problems by linear and nonlinear mappings. IV. Errors in L^2 and other norms. J. Complexity **26**(1), 102–124 (2010)
408. Dahlke, S., Sickel, W.: Besov regularity for the Poisson equation in smooth and polyhedral cones. In: Vladimir Maz'ya (ed.) Sobolev Spaces in Mathematics. II. International Mathematical Series (New York), vol. 9, pp. 123–145. Springer, New York (2009)
409. Dahlke, S., Sickel, W.: On Besov regularity of solutions to nonlinear elliptic partial differential equations. Rev. Mat. Complut. **26**(1), 115–145 (2013)
410. Dai, F., Gogatishvili, A., Yang, D., Yuan, W.: Characterizations of Sobolev spaces via averages on balls. Nonlinear Anal. **128**, 86–99 (2015)
411. Dai, F., Gogatishvili, A., Yang, D., Yuan, W.: Characterizations of Besov and Triebel–Lizorkin spaces via averages on balls. J. Math. Anal. Appl. **433**(2), 1350–1368 (2016)
412. Dai, F., Liu, J., Yang, D., Yuan, W.: Littlewood–Paley characterizations of fractional Sobolev spaces via averages on balls. Proc. Roy. Soc. Edinburgh Sect. A. (To appear)
413. David, G., Journé, J.L.: A boundedness criterion for generalized Calderón–Zygmund operators. Ann. Math. **120**, 371–397 (1984)
414. David, G., Journé, J.L., Semmes, S.: Opérateurs de Calderón–Zygmund, fonctions para-accrótives et interpolation (French). [Calderón–Zygmund operators, para-accretive functions and interpolation] Rev. Mat. Iberoamericana **1**(4), 1–56 (1985)
415. Day, M.M.: The spaces L^p with $0 < p < 1$. Bull. Am. Math. Soc. **46**, 816–823 (1940)
416. Deng, D., Duong, X.T., Song, L., Tan, C., Yan, L.: Functions of vanishing mean oscillation associated with operators and applications. Michigan Math. J. **56**, 529–550 (2008)
417. Deng, D., Han, Y.: $T1$ theorems for Besov and Triebel–Lizorkin spaces. Sci. China Ser. A **48**(5), 657–665 (2005)

418. Deng, D., Han, Y., Yang, D.: Inhomogeneous Plancherel-Pôlya inequalities on spaces of homogeneous type and their applications. Commun. Contemp. Math. **6**(2), 221–243 (2004)
419. DeVore, R.A., Sharpley, R.C.: Maximal functions measuring smoothness. Mem. Am. Math. Soc. **47**(293), 1–115 (1984)
420. DeVore, R.A., Sharpley, R.C.: Besov spaces on domains in $\mathbb{R}^d$. Trans. Am. Math. Soc. **335**(2), 843–864 (1993)
421. Diening, L.: Maximal functions on generalized $L^{p(\star)}$ spaces. Math. Inequal. Appl. **7**, 245–253 (2004)
422. Diening, L., Hästö, P., Roudenko, S., Spaces of variable integrability and differentiability. J. Funct. Anal. **256**, 1731–1768 (2009)
423. Diening, L., Harjulehto, P., Hästö, P., Mizuta, Y., Shimomura, T.: Maximal functions in variable exponent spaces: limiting cases of the exponent. Ann. Acad. Sci. Fenn. Math. **34**(2), 503–522 (2009)
424. Dintelmann, P.: Fourier multipliers between weighted anisotropic function spaces. Part II. Besov-Triebel spaces. Z. Anal. Anwend. **15**(4), 799–818 (1996)
425. Dispa, S.: Intrinsic characterizations of Besov spaces on Lipschitz domains. Math. Nachr. **260**, 21–33 (2003)
426. Doetsch, G.: Über die obere Grenze des absoluten Betrages einer analytischen Funktion auf Geraden. Math. Z. **8**, 237–240 (1920)
427. Dong, B.H., Xu, J.S.: New Herz type Besov and Triebel–Lizorkin spaces with variable exponents. J. Funct. Spaces Appl., Art. ID 384593, 1–27 (2012)
428. Dong, D., Xu, J.S.: Herz–Morrey type Besov and Triebel–Lizorkin spaces with variable exponents. Banach J. Math. Anal. **9**(1), 75–101 (2015)
429. Dorronsoro, J.R.: A characterization of potential spaces. Proc. Am. Math. Soc. **95**, 21–31 (1985)
430. Dorronsoro, J.R.: Poisson integrals of regular functions. Trans. Am. Math. Soc. **297**, 669–685 (1986)
431. Drihem, D.: Characterizations of Besov-type and Triebel–Lizorkin-type spaces by differences. J. Funct. Spaces Appl., Art. ID 328908, 1–24 (2012)
432. Drihem, D.: Atomic decomposition of Besov spaces with variable smoothness and integrability. J. Math. Anal. Appl. **389**(1), 15–31 (2012)
433. Drihem, D.: Embeddings properties on Herz-type Besov and Triebel–Lizorkin spaces. Math. Inequal. Appl. **16**(2), 439–460 (2013)
434. Drihem, D.: Atomic decomposition of Besov-type and Triebel–Lizorkin-type spaces. Sci. China Math. **56**(5), 1073–1086 (2013)
435. Drihem, D.: Some properties of variable Besov-type spaces. Funct. Approx. Comment. Math. **52**(2), 193–221 (2015)
436. Drihem, D., Moussai, M.: Some embeddings into the multiplier spaces associated to Besov and Lizorkin-Triebel spaces. Z. Anal. Anwend. **21**(1), 179–184 (2002)
437. Drihem, D., Moussai, M.: On the pointwise multiplication in Besov and Lizorkin–Triebel spaces. Int. J. Math. Math. Sci., Art. ID 76182, 1–18 (2006)
438. Duong, X.T., Li, J.: Hardy spaces associated to operators satisfying bounded holomorphic functional calculus and Davies-Gaffney estimates. J. Funct. Anal. **264**, 1409–1437 (2013)
439. Duong, X.T., Xiao, J., Yan, L.: Old and new Morrey spaces with heat kernel bounds. J. Fourier Anal. Appl. **13**, 87–111 (2007)
440. Duong, X.T., Yan, L.: Hardy spaces of spaces of homogeneous type. Proc. Am. Math. Soc. **131**(10), 3181–3189 (2003)
441. Duong, X.T., Yan, L.: New function spaces of BMO type, the John–Nirenberg inequality, interpolation, and applications. Commun. Pure Appl. Math. **58**, 1375–1420 (2005)
442. Duong, X.T., Yan, L.: Duality of Hardy and BMO spaces associated with operators with heat kernel bounds. J. Am. Math. Soc. **18**(4), 943–973 (2005)
443. Dvoretzky, A., Erdös, P., Kakutani, S.: Nonincrease everywhere of the Brownian motion process. Proc. 4th Berkley Symp. Math. Stat. Probab. **2**, 103–106 (1961)

444. Dziubański, J., Zienkiewicz, J.: Hardy space H^1 associated to Schrödinger operator with potential satisfying reverse Hölder inequality. Rev. Mat. Ibero. **15**, 279–296 (1999)
445. Dziubański, J., Zienkiewicz, J.: H^p spaces for Schrödinger operators, In: Fourier Analysis and Related Topics (Bedlewo, 2000), vol. 56, pp. 45–53. Banach Center Publication. Institute of Mathematics of the Polish Academy of Sciences, Warsaw (2002)
446. Edmunds, D.E., Haroske, D.D.: Spaces of Lipschitz type, embeddings and entropy numbers. Diss. Math. (Rozprawy Mat.) **380**, 1–43 (1999)
447. Edmunds, D.E., Haroske, D.D.: Embeddings in spaces of Lipschitz type, entropy and approximation numbers, and applications. J. Approx. Theory **104**(2), 226–271 (2000)
448. Edmunds, D.E., Kokilashvili, V., Meskhi, A.: One-sided operators in $L^{p(x)}$ spaces. Math. Nachr. **281**(11), 1525–1548 (2008)
449. Edmunds, D.E., Triebel, H.: Entropy numbers and approximation numbers in function spaces. Proc. London Math. Soc. **58**(3), 137–152 (1989)
450. Essén, M., Xiao, J.: Some results on Q_p spaces, $0 < p < 1$. J. Reine Angew. Math. **485**, 173–195 (1997)
451. Essén, M., Janson, S., Peng, L., Xiao, J.: Q spaces of several real variables. Indiana Univ. Math. J. **49**(2), 575–615 (2000)
452. Evans, W.D., Opic, B.: Real interpolation with logarithmic functors and reiteration. Canad. J. Math. **52**(5), 920–960 (2000)
453. Evans, W.D., Opic, B., Pick, L.: Real interpolation with logarithmic functors. J. Inequal. Appl. **7**(2), 187–269 (2002)
454. Farkas, W., Johnsen, J., Sickel, W.: Traces of anisotropic Besov-Lizorkin-Triebel spaces–a complete treatment of the borderline cases. Math. Bohemica **125**, 1–37 (2000)
455. Fefferman, C.: Characterizations of bounded mean oscillation. Bull. Am. Math. Soc. **77**, 587–588 (1971)
456. Fefferman, C., Riviére, N.M., Sagher, Y.: Interpolation between H^p spaces: the real method. Trans. Am. Math. Soc. **191**, 75–81 (1974)
457. Fefferman, C., Soria, F.: The space weak H^1. Studia Math. **85**, 1–16 (1986)
458. Fefferman, R., Soria, F.: The space weak H^1. Studia Math. **85**(1), 1–16 (1987)
459. Fefferman, C., Stein, E.: Some maximal inequalities. Am. J. Math. **93**, 107–115 (1971)
460. Fefferman, C., Stein, E.: H^p spaces of several variables. Acta Math. **129**, 137–193 (1971)
461. Feichtinger, H.G.: Banach convolution algebras of Wiener's type. In: Proceedings of the Conference Function, Series, Operators, Colloquia Mathematica Societatis János Bolyai, Rumania, pp. 509–524 (1980)
462. Feichtinger, H.G.: Banach spaces of distributions of Wiener's type and interpolation. In: Functional Analysis and Approximation, vol. 60, pp. 153–165. Birkhäuser, Basel (1981)
463. Feichtinger, H.G.: Modulation spaces on locally compact Abelian groups. Technical report, University of Vienna (1983)
464. Feichtinger, H.G.: Atomic characterization of modulation spaces through Gabor-type representation. In: Proceedings of Conference on Constructive Function Theory, Edmonton. Rocky Mountain J. Math. **19**(1), 113–125 (1989)
465. Feichtinger, H.G., Gröchenig, K.H.: A unified approach to atomic decompositions via integrable group representations. In: Function Spaces and Applications, Lund, pp. 52–73, 1986. Lecture Notes in Mathematics, vol. 1302. Springer, Berlin (1988)
466. Feichtinger, H.G., Gröchenig, K.H.: Banach spaces related to integrable group representations and their atomic decompositions. I. J. Funct. Anal. **86**(2), 307–340 (1989)
467. Feichtinger, H.G., Gröchenig, K.H.: Banach spaces related to integrable group representations and their atomic decompositions. II. Monatsh. Math. **108**(2–3), 129–148 (1989)
468. Feichtinger, H.G., Gröchenig, K.H.: Gabor wavelets and the Heisenberg group: Gabor expansions and short tiem fourier transform from the group theorical point of view. In: Chui, C.K. (ed.) Wavelets: A Tutorial in Theory and Applications, pp. 359–398. Academic Press, Boston (1992)
469. Feichtinger, H.G., Gröchenig, K.H.: Gabor frames and time-frequency analysis of distributions. J. Funct. Anal. **146**, 464–495 (1997)

470. Feuto, J.: Products of functions in BMO and H^1 spaces on spaces of homogeneous type. J. Math. Anal. Appl. **359**(2), 610–620 (2009)
471. Flett, T.M.: Lipschitz spaces of functions on the circle and the disc. J. Math. Anal. Appl. **39**, 125–158 (1972)
472. Fornasier, M., Rauhut, H.: Continuous frames, function spaces, and the discretization problem. J. Fourier Anal. Appl. **11**, 245–287 (2005)
473. Folland, G.B.: Subelliptic estimates and function spaces on nilpotent Lie groups. Ark. Mat. **13**(2), 161–207 (1975)
474. Folland, G.B.: Lipschitz classes and Poisson integrals on stratified groups. Studia Math. **66**(1), 37–55 (1979)
475. Fournier, J.J.F., Stewart, J.: Amalgams of L^p and ℓ^q. Bull. Am. Math. Soc. **13**(1), 1–21 (1985)
476. Franchi, B., Hajłasz, P., Koskela, P.: Definition of Sobolev classes on metric spaces. Ann. Inst. Fourier (Grenoble) **49**, 1903–1924 (1999)
477. Franke, J.: On the spaces F^s_{pq} of Triebel–Lizorkin-type: pointwise multipliers and spaces on domains. Math. Nachr. **125**, 29–68 (1986)
478. Franke, J., Runst, T.: Regular elliptic boundary value problems in Besov-Triebel-Lizorkin space. Math. Nachr. **174**, 113–149 (1995)
479. Frazier, M., Jawerth, B.: φ-transform and applications to distribution spaces. In: Function Spaces and Applications, Lund, 1986. Lecture Notes in Mathematics, vol. 1302, pp. 223–246. Springer, Berlin (1988)
480. Frazier, M., Jawerth, B., Weiss, G.: Littlewood–Paley theory and the study of function spaces. CBMS-AMS Reg. Conf. Ser. **79**, 129–132 (1991)
481. Frazier, M., Torres, R., Weiss, G.: The boundedness of Calderón–Zygmund operators on the spaces $\dot{F}_p^{\alpha,q}$. Rev. Mat. Iberoam. **4**(1), 41–72 (1988)
482. Frazier, M., Jawerth, B.: Decomposition of Besov spaces. Indiana Univ. Math. J. **34**(4), 777–799 (1985)
483. Frazier, M., Jawerth, B.: A discrete transform and decompositions of distribution spaces. J. Funct. Anal. **93**(1), 34–170 (1990)
484. Furioli, G., Melzi, C., Veneruso, A.: Littlewood–Paley decompositions and Besov spaces on Lie groups of polynomial growth. Math. Nachr. **279**(9–10), 1028–1040 (2006)
485. Fu, J.J., Xu, J.S.: Characterizations of Morrey type Besov and Triebel–Lizorkin spaces with variable exponents. J. Math. Anal. Appl. **381**(1), 280–298 (2011)
486. Fu, X., Yang, D., Liang, Y.: Products of functions in BMO($\mathscr{X}$) and $H^1_{\rm at}(\mathscr{X})$ via wavelets over spaces of homogeneous type. J. Fourier Anal. Appl. **23**(4), 919–990 (2017)
487. Funaki, T., Hoshino, M.: A coupled KPZ equation, its two types of approximations and existence of global solutions. J. Funct. Anal. **273**(3), 1165–1204 (2017)
488. Fujita, H., Kato, T.: On the Navier–Stokes initial value problem I. Arch. Ration. Mech. Anal. **16**, 269–315 (1964)
489. Futamura, T., Mizuta, Y., Shimomura, T.: Integrability of maximal functions and Riesz potentials in Orlicz spaces of variable exponent. J. Math. Anal. Appl. **366**, 391–417 (2010)
490. Gabisoniya, O.D.: On the absolute convergence of double Fourier series and Fourier integrals. Soobshch. Akad. Nauk. Gruzin. SSR **42**, 3–9 (1966)
491. Gaffney, M.P.: The conservation property of the heat equation on Riemannian manifolds. Commun. Pure Appl. Math. **12**, 1–11 (1959)
492. Gagliardo, E.: Caratterizzazione delle tracce sulla frontiera relative ad alcune classi di funzioni in n variabili. Rend. Sem. Mat. Univ. Padova **27**, 284–305 (1957)
493. Gagliardo, E.: Proprietà di alcune classi di funzioni in più variabli. Ricerche Mat. **7**, 102–137 (1958)
494. Gagliardo, E.: Ulteriori proprietà di alcune classi di funzioni in più variabili, Ricerche. Mat. **8**, 24–51 (1959)
495. Gala, S., Sawano, Y.: Wavelet characterization of the pointwise multiplier space $\dot{X}_r$, Functiones et Approximatio **43**, 109–116 (2010)

496. Galmarino, A.R., Panzone, R.L.: L^p-spaces with mixed norm, for P a sequence. J. Math. Anal. Appl. **10**, 494–518 (1965)
497. García-Cuerva, J.: Weighted H^p spaces. Diss. Math. **12**, 1–63 (1979)
498. García-Cuerva, J., Herrero, M.J.L.: A theory of Hardy spaces associated to Herz spaces. Proc. London Math. Soc. **69**(3), 605–628 (1994)
499. Gatto, E.A., Pineda, E., Urbina, W.O.: Riesz potentials, Bessel potentials and fractional derivatives on Triebel–Lizorkin spaces for the Gaussian measure. J. Math. Anal. Appl. **422**(2), 798–818 (2015)
500. Georgiadis, A.G., Johnsen, J., Nielsen, M.: Wavelet transforms for homogeneous mixed-norm Triebel–Lizorkin spaces. Monatosh. Math. **183**(4), 587–624 (2017)
501. Giga, Y.: Solutions for semilinear parabolic equations in L^p and regularity of weak solutions of the Navier–Stokes system. J. Differ. Equ. **61**, 186–212 (1986)
502. Giga, Y., Inui, K., Matsui, S.: On the Cauchy problem for the Navier–Stokes equations with nondecaying initial data. Quaderni di Matematica **4**, 28–68 (1999)
503. Giga, Y., Miyakawa, T.: Solutions in L^r of the Navier–Stokes initial value problem. Arch. Ration. Mech. Anal. **89**, 267–281 (1985)
504. Gogatishvili, A., Koskela, P., Shanmugalingam, N.: Interpolation properties of Besov spaces defined on metric spaces. Math. Nachr. **283**, 215–231 (2010)
505. Gogatishvili, A., Koskela, P., Zhou, Y.: Characterizations of Besov and Triebel–Lizorkin spaces on metric measure spaces. Forum Math. **25**, 787–819 (2013)
506. Goncalves, H.F., Kempka, H., Non-smooth atomic decomposition of 2-microlocal spaces and applications to pointwise multipliers. J. Math. Anal. Appl. **434**, 1875–1890 (2016)
507. Goldberg, D.: A local version of real Hardy spaces. Duke Math. J. **46**(1), 27–42 (1979)
508. Goldberg, D.: Local Hardy spaces. Harmonic analysis in Euclidean spaces (Proceedings of Symposium on Pure Mathematics, Williams College, Williamstown, 1978), Part 1, pp. 245–248, Proceedings of Symposium on Pure Mathematics, XXXV, Part. American Mathematical Society, Providence (1979)
509. Gol'dman, M.L.: On the extension of functions of $L^p(\mathbb{R}^n)$ in spaces with a large number of dimensions (Russian). Mat. Zametki **25**, 513–520 (1979)
510. Gol'dman, M.L.: A covering theorem for describing general spaces of Besov type. Trudy Mat. Inst. Steklov. **156**, 51–87 (1980)
511. Gol'dman, M.L., Haroske, D.D.: Estimates for continuity envelopes and approximation numbers of Bessel potentials. J. Approx. Theory **172**, 58–85 (2013)
512. Goldŝteîn, V.M.: Extension of functions with first generalized derivatives from plane domains (Russian). Dokl. Akad. Nauk SSSR **257**(2), 268–271 (1981)
513. Golovkinm, K.K., Solonnikov, V.A.: Estimates of convolution operators (Russian). Zap. Nauĉn. Sem. Leningrad. Otdel. Mat. Inst. Steklov. (LOMI) **7**, 6–86 (1968)
514. Gordon, M., Loura, L.: Exponential generalized distributions, Math. J. Okayama Univ. **52**, 159–177 (2010)
515. Grafakos, L., Oh, T.: The Kato-Ponce inequality. Commun. PDE **39**, 1128–1157 (2014)
516. Grafakos, L., Tao, T., Terwilleger, E.: L^p bounds for a maximal dyadic sum operator. Math. Z. **246**(12), 321–337 (2004)
517. Grafakos, L., Kalton, N.: Multilinear Calderón–Zygmund operators on Hardy spaces. Collect. Math. **52**(2), 169–179 (2001)
518. Grafakos, L., Torres, R.H.: Pseudodifferential operators with homogeneous symbols. Michigan Math. J. **46**, 261–269 (1999)
519. Grevholm, B.: On the structure of the spaces $\mathscr{L}_k^{p,\lambda}$, Math. Scand. **26**, 241–254 (1970)
520. Grigor'yan, A., Liu, L.: Heat kernel and Lipschitz-Besov spaces. Forum Math. **27**(6), 3567–3613 (2015)
521. Grisvard, P.: Commutativité de deux foncteurs d'interpolation et applications. J. Math. Pures. Appl. **45**, 143–290 (1966)
522. Gröchenig, K.: Unconditional bases in translation and dilation invariant function spaces on $\mathbb{R}^n$. In: Constructive Theory of Functions, Varna, 1987, pp. 174–183. Publ. House Bulgar. Acad. Sci., Sofia (1988)

523. Gröchenig, K.: Describing functions: atomic decompositions versus frames. Monatsh. Math. **112**, 1–42 (1991)
524. Gubinelli, M., Imkeller, P., Perkowski, N.: A Fourier approach to pathwise stochastic integration. Electron. J. Probab. **21**(2016), paper no. 2, 1–37 (2016)
525. Guliyev, V., Omarova, M., Sawano, Y.: Boundedness of intrinsic square functions and their commutators on generalized weighted Orlicz–Morrey spaces. Banach J. Math. Anal. **9**(2), 44–62 (2015)
526. Gustavsson, J., Peetre, J.: Interpolation of Orlicz spaces. Studia Math. **60**(1), 33–59 (1977)
527. Gurka, P., Harjulehto, P., Nekvinda, A.: Bessel potential spaces with variable exponent. Math. Inequal. Appl. **10**, 661–676 (2007)
528. Hadamard, J.: Sur les fonctions entières. Bull. Soc. Math. France **24**, 186–187 (1986)
529. Hakim, D.I., Nakamura, S. Sawano, Y.: Interpolation of generalized Morrey spaces. Constr. Approx. **46**(3), 489–563 (2017)
530. Hakim, D.I., Nogayama, T., Sawano, Y.: Complex interpolation of smoothness Triebel-Lizorkin-Morrey spaces. Math. J. Okayama Univ (To appear)
531. Hakim, D.I., Sawano, Y.: Interpolation of generalized Morrey spaces. Rev. Mat. Complut. **29**(2), 295–340 (2016)
532. Hakim, D.I., Sawano, Y.: Calderón's first and second complex interpolations of closed subspaces of Morrey spaces. J. Four. Anal. Appl. **23**(5), 1195–1226 (2017)
533. Hajłasz, P.: Sobolev spaces on an arbitrary metric space. Potential Anal. **5**(4), 403–415 (1996)
534. Hajłasz, P., Koskela, P.: Sobolev met Poincaré. Mem. Am. Math. Soc. **145**(688), 1–101 (2000)
535. Han, Y.S.: Calderón-type reproducing formula and the Tb theorem. Rev. Mat. Ibero. **10**, 51–91 (1994)
536. Han, Y.S.: Inhomogeneous Calderón reproducing formula on spaces of homogeneous type. J. Geom. Anal. **7**, 259–284 (1997)
537. Han, Y.S.: Embedding theorem for inhomogeneous Besov and Triebel–Lizorkin spaces on RD-spaces. Canad. Math. Bull. **58**(4), 757–773 (2015)
538. Han, Y.S., Li, J., Ward, L.A.: Hardy space theory on spaces of homogeneous type via orthonormal wavelet bases. Appl. Comput. Harmon. Anal. **45**(1), 120–169 (2018)
539. Han, Y.S., Paluszyński, M., Weiss, G.: A new atomic decomposition for the Triebel–Lizorkin spaces. Contemp. Math. **189**, 235–249 (1995)
540. Han, Y.S., Müller, D., Yang, D.: Littlewood–Paley characterizations for Hardy spaces on spaces of homogeneous type. Math. Nachr. **279**(13–14), 1505–1537 (2006)
541. Han, Y.S., Müller, D., Yang, D.: A theory of Besov and Triebel–Lizorkin spaces on metric measure spaces modeled on Carnot-Carathéodory spaces. Abstr. Appl. Anal., Art. ID 893409, 1–250 (2008)
542. Han, Y.S., Hofmann, S.: $T1$ Theorem for Besov and Triebel–Lizorkin spaces. Trans. Am. Math. Soc. **337**, 839–853 (1993)
543. Han, Y.S., Sawyer, E.T.: Littlewood–Paley theory on spaces of homogeneous type and the classical function spaces. Mem. Am. Math. Soc. **110**(530), 1–126 (1994)
544. Han, Y.S., Lu, S., Yang, D.: Inhomogeneous Besov and Triebel-Lizorkin spaces on spaces of homogeneous type. Approx. Theory Appl. (N.S.) **15**(3), 37–65 (1999)
545. Han, Y.S., Lu, S., Yang, D.: Inhomogeneous discrete Calderón reproducing formulas for spaces of homogeneous type. J. Fourier Anal. Appl. **7**, 571–600 (2001)
546. Han, Y.S., Yang, D.: New characterizations and applications of inhomogeneous Besov and Triebel–Lizorkin spaces on homogeneous type spaces and fractals. Diss. Math. **403**, 1–102 (2002)
547. Han, Y.S., Yang, D.: Some new spaces of Besov and Triebel–Lizorkin-type on homogeneous spaces. Studia Math. **156**(1), 67–97 (2003)
548. Han, J., Wang, B.: α-modulation spaces (I) scaling, embedding and algebraic properties. J. Math. Soc. Japan **66**(4), 1315–1373 (2014)

549. Hansen, M., Sickel, W.: Best m-term approximation and tensor product of Sobolev and Besov spaces–the case of noncompact embeddings. East J. Approx. **16**(4), 345–388 (2010)
550. Hansen, M., Sickel, W.: Best m-term approximation and Lizorkin-Triebel spaces. J. Approx. Theory **163**(8), 923–954 (2011)
551. Hansen, M., Sickel, W.: Best m-term approximation and Sobolev-Besov spaces of dominating mixed smoothness–the case of compact embeddings. Constr. Approx. **36**(1), 1–51 (2012)
552. Hansen, M., Vybíral, J.: The Jawerth-Franke embedding of spaces with dominating mixed smoothness. Georgian Math. J. **16**(4), 667–682 (2009)
553. Hardy, G.H.: The mean value of the modulus of an analytic function. Proc. London Math. Soc. **14**, 269–277 (1914)
554. Hardy, G.H., Littlewood, J.: A maximal theorem with function-theoretic applications. Acta Math. **54**(1), 81–116 (1930)
555. Hardy, G.H., Littlewood, J.: Generalizations of a theorem of Paley. Q. J. **8**, 161–171 (1937)
556. Haroske, D.D.: Approximation numbers in some weighted function spaces. J. Approx. Theory **83**(1), 104–136 (1995)
557. Haroske, D.D.: Some logarithmic function spaces, entropy numbers, applications to spectral theory. Diss. Math. (Rozprawy Mat.) **373**, 1–59 (1998)
558. Haroske, D.D.: Logarithmic Sobolev spaces on $\mathbb{R}^n$; entropy numbers, and some applications. Forum Math. **12**(3), 257–313 (2000)
559. Haroske, D.D.: On more general Lipschitz spaces. Z. Anal. Anwend. **19**(3), 781–799 (2000)
560. Haroske, D.D.: Growth envelope functions in Besov and Sobolev spaces, local versus global results. Math. Nachr. **280**(9–10), 1094–1107 (2007)
561. Haroske, D.D., Triebel, H.: Embeddings of function spaces: a criterion in terms of differences. Complex Var. Elliptic Equ. **56**, 931–944 (2011)
562. Haroske, D.D., Triebel, H.: Some recent developments in the theory of function spaces involving differences. J. Fixed Point Theory Appl. **13**, 341–358 (2013)
563. Haroske, D.D., Moura, S.D., Continuity envelopes of spaces of generalised smoothness, entropy and approximation numbers. J. Approx. Theory **128**(2), 151–174 (2004)
564. Haroske, D.D., Moura, S.D., Continuity envelopes and sharp embeddings in spaces of generalized smoothness. J. Funct. Anal. **254**(6), 1487–1521 (2008)
565. Haroske, D.D., Schmeisser, H.-J., On trace spaces of function spaces with a radial weight: the atomic approach. Complex Var. Elliptic Equ. **55**(8–10), 875–896 (2010)
566. Haroske, D.D., Schneider, C.: Besov spaces with positive smoothness on $\mathbb{R}^n$, embeddings and growth envelopes. J. Approx. Theory **161**(2), 723–747 (2009)
567. Haroske, D.D., Skrzypczak, L.: Entropy and approximation numbers of embeddings of function spaces with Muckenhoupt weights. I. Rev. Mat. Complut. **21**(1), 135–177 (2008)
568. Haroske, D.D., Skrzypczak, L.: Spectral theory of some degenerate elliptic operators with local singularities. J. Math. Anal. Appl. **371**(1), 282–299 (2010)
569. Haroske, D.D., Skrzypczak, L.: Entropy numbers of embeddings of function spaces with Muckenhoupt weights, III. Some limiting cases. J. Funct. Spaces Appl. **9**(2), 129–178 (2011)
570. Haroske, D.D., Skrzypczak, L.: Entropy and approximation numbers of embeddings of function spaces with Muckenhoupt weights, II. General weights. Ann. Acad. Sci. Fenn. Math. **36**(1), 111–138 (2011)
571. Haroske, D.D., Skrzypczak, L.: Continuous embeddings of Besov–Morrey function spaces. Acta Math. Sin. **28**(7), 1307–1328 (2012)
572. Haroske, D.D., Skrzypczak, L., Embeddings of Besov–Morrey spaces on bounded domains. Studia Math. **218**(2), 119–144 (2013)
573. Haroske, D.D., Skrzypczak, L.: On Sobolev and Franke-Jawerth embeddings of smoothness Morrey spaces. Rev. Mat. Complut. **27**(2), 541–573 (2014)
574. Haroske, D.D., Piotrowska, I.: Atomic decompositions of function spaces with Muckenhoupt weights, and some relation to fractal analysis. Math. Nachr. **281**(10), 1476–1494 (2008)

575. Haroske, D.D., Tamasi, E.: Wavelet frames for distributions in anisotropic Besov spaces. Georgian Math. J. **12**(4), 637–658 (2005)
576. Haroske, D.D., Triebel, H.: Entropy numbers in weighted function spaces and eigenvalue distributions of some degenerate pseudodifferential operators. I. Math. Nachr. **167**, 131–156 (1994)
577. Haroske, D.D., Triebel, H.: Entropy numbers in weighted function spaces and eigenvalue distributions of some degenerate pseudodifferential operators. II. Math. Nachr. **168**, 109–137 (1994)
578. Haroske, D.D., Triebel, H.: Wavelet bases and entropy numbers in weighted function spaces. Math. Nachr. **278**(1–2), 108–132 (2005)
579. He, Z., Yang, D., Yuan, W.: Littlewood–Paley characterizations of second-order Sobolev spaces via averages on balls. Canad. Math. Bull. **59**(1), 104–118 (2016)
580. Hedberg, L.I.: On certain convolution inequalities. Proc. Am. Math. Soc. **36**(2), 505–510 (1972)
581. Heikkinen, T., Ihnatsyeva, L., Tuominen, H.: Measure density and extension of Besov and Triebel-Lizorkin functions. J. Fourier Anal Appl. **22**, 334–382 (2016)
582. Heikkinen, T., Koskela, P., Tuominen, H.: Approximation and quasicontinuity of Besov and Triebel-Lizorkin functions. Trans. Am. Math. Soc. **369**, 3547–3573 (2017)
583. Heikkinen, T., Tuominen, H.: Approximation by Hölder functions in Besov and Triebel-Lizorkin spaces. Constr. Approx. **44**, 455–482 (2016)
584. Hedberg, L., Netrusov, Y.: An axiomatic approach to function spaces, spectral synthesis, and Luzin approximation. Mem. Am. Math. Soc. **188**(882), vi+97 (2007)
585. Heinonen, J., Koskela, P., Shanmugalingam, N., Tyson, J.T.: Sobolev classes of Banach space-valued functions and quasiconformal mappings. J. Anal. Math. **85**, 87–139 (2001)
586. Helly, E.: Über Mengen konvexer Körper mit gemeinschaftlichen Punkten, Jahresbericht der Deutschen Mathematiker-Vereinigung. **32**, 175–176 (2013)
587. Hencl, S., Koskela, P.: Composition of quasiconformal mappings and functions in Triebel–Lizorkin spaces. Math. Nachr. **286**, 669–678 (2013)
588. Herz, C.: Lipschitz spaces and Bernstein's theorem on absolutely convergent Fourier transforms. J. Math. Mech. **18**, 283–324 (1968)
589. Hewett, D.P., Moiola, A.: On the maximal Sobolev regularity of distributions supported by subsets of Euclidean space. Anal. Appl. (Singap.) **15**(5), 731–770 (2017)
590. Hirschman, I.: A convexity theorem for certain groups of transformations. J. Analyse Math. **2**, 209–218 (1953)
591. Hofmann, S., Lu, G., Mitrea, D., Mitrea, M., Yan, L.: Hardy spaces associated to non-negative self-adjoint operators satisfying Davies-Gaffney estimates. Mem. Am. Math. Soc. **214**(1007), vi+78 (2011)
592. Hofmann, S., Martell, J.: L^p bounds for Riesz transforms and square roots associated to second order elliptic operators. Publ. Mat. **47**, 497–515 (2003)
593. Hofmann, S., Mayboroda, S.: Hardy and BMO spaces associated to divergence form elliptic operators. Math. Ann. **344**, 37–116 (2009) and Hofmann, S., Mayboroda, S.: Correction to Hardy and BMO spaces associated to divergence form elliptic operators, arXiv:0907.0129
594. Hofmann, S., Mayboroda, S., McIntosh, A.: Second order elliptic operators with complex bounded measurable coefficients in L^p, Sobolev and Hardy spaces. Ann. Sci. École Norm. Sup. (4) **44**, 723–800 (2011)
595. Ho, K.P.: Remarks on Littlewood–Paley analysis. Canad. J. Math. **60**(6), 1283–1305 (2008)
596. Ho, K.P.: Characterization of BMO in terms of rearrangement-invariant Banach function spaces. Expo. Math. **27**(4), 363–372 (2009)
597. Ho, K.P.: Littlewood–Paley theory for the differential operator $\frac{\partial^2}{\partial x_1{}^2}\frac{\partial^2}{\partial x_2{}^2} - \frac{\partial^2}{\partial x_3{}^2}$. Z. Anal. Anwend. **29**(2), 183–217 (2010)
598. Ho, K.P.: Littlewood–Paley spaces. Math. Scand. **108**(1), 77–102 (2011)
599. Ho, K.P.: Characterizations of BMO by A_p weights and p-convexity. Hiroshima Math. J. **41**(2), 153–165 (2011)

600. Ho, K.P.: Wavelet bases in Littlewood–Paley spaces (English summary). East J. Approx. **17**(4), 333–345 (2012)
601. Ho, K.P.: Atomic decomposition of Hardy spaces and characterization of BMO via Banach function spaces. Anal. Math. **38**(3), 173–185 (2012)
602. Ho, K.P.: Generalized Boyd's indices and applications. Analysis (Munich) **32**(2), 97–106 (2012)
603. Ho, K.P.: Vector-valued singular integral operators on Morrey type spaces and variable Triebel–Lizorkin–Morrey spaces. Ann. Acad. Sci. Fenn. Math. **37**(2), 375–406 (2012)
604. Ho, K.P.: Atomic decompositions of weighted Hardy–Morrey spaces. Hokkaido Math. J. **42**(1), 131–157 (2013)
605. Hofmann, S., Lacey, M., M^c^Intosh, A.: The solution of the Kato problem for divergence form elliptic operators with Gaussian heat kernel bounds. Ann. Math. (2) **156**(2), 623–631 (2002)
606. Hörmander, L.: Estimates for translation invariant operators in L^p spaces. Acta Math. **104**, 93–140 (1960)
607. Hörmander, L.: Pseudo-differential operators and hypoelliptic equations. In: Singular Integrals. Proceedings of Symposia in Pure Mathematics. vol. 10. American Mathematical Society, Providence (1967)
608. Hörmander, L.: Pseudo-differential operators of type 1, 1. Commun. PDE **13**(9), 1085–1111 (1988)
609. Hu, J.: A note on Hajłasz-Sobolev spaces on fractals. J. Math. Anal. Appl. **280**, 91–101 (2003)
610. Hunt, R.A.: On the convergence of Fourier series, in 1967. In: Orthogonal Expansions and Their Continuous Analogues, pp. 235–255. (Proceedings of Conference, Edwardsville). Southern Illinois University Press, Carbondale (1968)
611. Hutchinson, J.E.: Fractals and self similarity. Indiana Math. J. **30**, 713–747 (1981)
612. Ivanov, K., Petrushev, P., Xu, Y.: Decomposition of spaces of distributions induced by tensor product bases. J. Funct. Anal. **263**(5), 1147–1197 (2012)
613. Iwaniec, T., Onninen, J.: H^1-estimates of Jacobians by subdeterminants. Math. Ann. **324**, 341–358 (2002)
614. Iwabuchi, T., Matsuyama, T., Taniguchi, K.: Boundedness of spectral multipliers for Schrödinger operators on open sets. Rev. Mat. Iberoam. **34**(3), 1277–1322 (2018)
615. Iwabuchi, T., Matsuyama, T., Taniguchi, K.: Besov spaces on open sets. arxiv
616. Izuki, M.: Vector-valued inequalities on Herz spaces and characterizations of Herz-Sobolev spaces with variable exponent. Glas. Mat. Ser. III **45**(2(65)), 475–503 (2010)
617. Izuki, M., Noi, T.: Duality of Besov, Triebel–Lizorkin and Herz spaces with variable exponents. Rend. Circ. Mat. Palermo **63**, 221–245 (2014)
618. Izuki, M., Sawano, Y.: Wavelet bases in the weighted Besov and Triebel–Lizorkin spaces with $A_p^{\rm loc}$-weights. J. Approx. Theory **161**, 656–673 (2009)
619. Izuki, M., Sawano, Y.: Variable Lebesgue norm estimates for BMO functions. Czechoslov. Math. J. **62**(3(137)), 717–727 (2012)
620. Izuki, M., Nakai, E., Sawano, Y.: Function spaces with variable exponents–an introduction. Sci. Math. Jpn. **77**(2), 187–315 (2014)
621. Izuki, M., Sawano, Y., Tanaka, H.: Weighted Besov–Morrey spaces and Triebel–Lizorkin spaces. In: Harmonic Analysis and Nonlinear Partial Differential Equations.RIMS Kôkyûroku Bessatsu, vol. B22, pp. 21–60. Research Institute for Mathematical Sciences (RIMS), Kyoto (2010)
622. Izuki, M., Sawano, Y., Tsutsui, Y.: Variable Lebesgue norm estimates for BMO functions. II. Anal. Math. **40**(3), 215–230 (2014)
623. Jaffard, S.: Pointwise smoothness, two-microlocalisation and wavelet coefficients. Publ. Mat. **35**, 155–168 (1991)
624. Jaffard, S., Meyer, Y.: Wavelet methods for pointwise regularity and local oscillations of functions. Mem. Am. Math. Soc. **123**(587), x+110 (1996)

625. Jaffard, S., Okada, M., Ueno, T.: Approximate sampling theorem and the order of smoothness of the Besov space. In: Harmonic Analysis and Nonlinear Partial Differential Equations. RIMS Kôkyûroku Bessatsu, vol. B18, pp. 45–56. Research Institute for Mathematical Sciences (RIMS), Kyoto (2010)
626. Janson, S.: Generalizations of Lipschitz spaces and an application to Hardy spaces and bounded mean oscillation. Duke Math. J. **47**, 959–982 (1980)
627. Janson, S.: On the space Q_p and its dyadic counterpart. In: Proceedings of Symposium Complex Analysis and Differential Equations, June 1997, Uppsala, vol. 158; Acta Universitatis Upsaliensis C, vol. 64, (C. Kiselman, ed.) pp. 194–205. Uppsala University, Uppsala (1999)
628. Jawerth, B.: Some observations on Besov and Lizorkin-Triebel spaces. Math. Scand. **40**, 94–104 (1977)
629. Jawerth, B.: The trace of Sobolev and Besov spaces if $0 < p < 1$. Studia Math. **62**(1), 65–71 (1978)
630. Jensen, J.L.W.V.: Sur les fonctions convexes et les inégalités entre les valeurs moyennes. Acta Math **30**, 175–193 (1906)
631. Jia, H., Wang, H.: Decomposition of Hardy–Morrey spaces. J. Math. Anal. Appl. **354**, 99–110 (2009)
632. Jiang, X.J., Yang, D., Yuan, W.: Real interpolation for grand Besov and Triebel–Lizorkin spaces on RD-spaces. Ann. Acad. Sci. Fenn. Math. **36**(2), 509–529 (2011)
633. Jiang, R., Yang, D.: New Orlicz-Hardy spaces associated with divergence form elliptic operators. J. Funct. Anal. **258**, 1167–1224 (2010)
634. Jiang, R., Yang, D.: Generalized vanishing mean oscillation spaces associated with divergence form elliptic operators. Integral Eq. Oper. Theory **67**, 123–149 (2010)
635. Jiang, R., Yang, D.: Orlicz-Hardy spaces associated with operators satisfying Davies-Gaffney estimates. Commun. Contemp. Math. **13**, 331–373 (2011)
636. Jiang, R., Yang, D.: Predual spaces of Banach completions of Orlicz-Hardy spaces associated with operators. J. Fourier Anal. Appl. **17**, 1–35 (2011)
637. Jiang, R., Yang, D., Zhou, Y.: Orlicz-Hardy spaces associated with operators. Sci. China Ser. A **52**, 1042–1080 (2009)
638. Jiao, Y., Zhou, D., Hao, Z.W., Chen, W.: Martingale Hardy spaces with variable exponents. Banach J. Math. Anal. **10**(4), 750–770 (2016)
639. John, F., Nirenberg, L.: On function of bounded mean oscillation. Commun. Pure Appl. Math. **14**, 415–426 (1961)
640. Johnsen, J.: Pointwise multiplication of Besov and Triebel–Lizorkin spaces. Math. Nachr. **175**, 85–133 (1995)
641. Johnsen, J.: Simple proofs of nowhere-differentiability for Weierstrass's function and cases of slow growth. J. Fourier Anal. Appl. **16**(1), 17–33 (2010)
642. Jonsson, A.: Besov spaces on closed sets by means of atomic decompositions, Research Reports 7. Department of Mathematics, University of Umeå, Umeå (1993)
643. Jonsson, A.: Besov spaces on closed subsets of $\mathbb{R}^n$. Trans. Am. Math. Soc. **341**(1), 355–370 (1994)
644. Jonsson, A., Wallin, H.: Function spaces on subsets of $\mathbb{R}^n$. Math. Rep. **2**(1), 1–221 (1984)
645. Jonsson, A., Wallin, H.: Boundary value problems and Brownian motion on fractals. Chaos Solitons Fractals **8**, 191–205 (1997)
646. Johnsen, J., Hansen, H.M., Sickel, W.: Characterisation by local means of anisotropic Lizorkin-Triebel spaces with mixed norms. Z. Anal. Anwend. **32**(3), 257–277 (2013)
647. Johnsen, J., Hansen, H.M., Sickel, W.: Anisotropic, mixed-norm Lizorkin-Triebel spaces and diffeomorphic maps. J. Funct. Spaces, Art. ID 964794, 1–15 (2014)
648. Johnsen, J., Sickel, W.: A direct proof of Sobolev embeddings for quasi-homogeneous Lizorkin-Triebel spaces with mixed norms. J. Funct. Spaces Appl. **5**(2), 183–198 (2007)
649. Johnsen, J., Sickel, W.: On the trace problem for Lizorkin-Triebel spaces with mixed norms. Math. Nachr. **281**(5), 669–696 (2008)
650. Jones, P.W.: Quasiconformal mappings and extendability of functions in Sobolev spaces. Acta Math. **147**(1–2), 71–88 (1981)

651. Jonsson, A.: Besov spaces on closed sets by means of atomic decomposition. Complex Var. Elliptic Equ. **54**(6), 585–611 (2009)
652. Judoviĉ, V.I.: Some estimates connected with integral operators and with solutions of elliptic equations (Russian). Dokl. Akad. Nauk SSSR **138**, 805–808 (1961)
653. Kalton, N., Mitrea, M.: Stability results on interpolation scales of quasi-Banach spaces and applications. Trans. Am. Math. Soc. **350**(10), 3903–3922 (1998)
654. Kalyabin, G.A.: Characterizations of function spaces of Besov-Lizorkin-Triebel type. Dokl. Acad. Nauk. SSSR **236**, 1056–1059 (1977)
655. Kalyabin, G.A.: Multiplier conditions of function spaces of Besov and Lizorkin-Triebel type. Dokl. Acad. Nauk SSSR **251**, 25–26 (1980)
656. Kalyabin, G.A.: The description of functions of classes of Besov-Lizorkin-Triebel type (Russian). Trudy Mat. Inst. Steklov **156**, 82–109 (1980)
657. Kalyabin, G.A.: Criteria of the multiplication property and the embeddings in C of spaces of Besov-Lizorkin-Triebel type. Mat. Zametki **30**, 517–526 (1981)
658. Kalyabin, G.A.: Functional classes of Lizorkin-Triebel type in domains with Lipschitz boundary (Russian). Dokl. Akad. Nauk SSSR **271**, 795–798 (1983)
659. Kalyabin, G.A.: Theorems on extensions, multipliers and diffeomorphisms for generalized Sobolev–Liouville classes in domains with Lipschitz boundary. Trudy Mat. Inst. Steklov **172**, 173–186 (1985)
660. Kato, T.: Fractional powers of dissipative operators. J. Math. Soc. Japan **13**, 246–274 (1961)
661. Kato, T.: Strong L^p-solutions of Navier–Stokes equations in $\mathbb{R}^n$ with applications to weak solutions. Math. Z. **187**, 471–480 (1984)
662. Kato, T.: The inclusion relations between α-modulation spaces and L^p-Sobolev spaces or local Hardy spaces. J. Funct. Anal. **272**(4), 1340–1405 (2017)
663. Kato, T., Ponce, G.: Commutator estimates and the Euler and Navier–Stokes equations. Commun. Pure Appl. Math. **41**, 891–907 (1988)
664. Kempka, H.: Atomic, molecular and wavelet decomposition of generalized 2-microlocal Besov spaces. J. Funct. Spaces Appl. **8**, 129–165 (2010)
665. Kempka, H.: Atomic, molecular and wavelet decomposition of 2-microlocal Besov and Triebel–Lizorkin spaces with variable integrability. Funct. Approx. Comment. Math. **43**, 171–208 (2010)
666. Kempka, H., Vybíral, J.: Spaces of variable smoothness and integrability: characterizations by local means and ball means of differences. J. Fourier Anal. Appl. **18**(4), 852–891 (2012)
667. Kempka, H., Vybíral, J.: A note on the spaces of variable integrability and summability of Almeida and Hästö. Proc. Am. Math. Soc. **141**(9), 3207–3212 (2013)
668. Kempka, H., Schäfer, M., Ullrich, T.: General coorbit space theory for quasi–Banach spaces and inhomogeneous Function spaces with variable smoothness and integrability. **23**(6), 1348–1407 (2017)
669. Kerkyacharian, G., Petrushev, P.: Heat kernel based decomposition of spaces of distributions in the framework of Dirichlet spaces. Trans. Am. Math. Soc. **367**(1), 121–189 (2015)
670. Khintchine, A.: Über dyadische Brüche (German). Math. Z. **18**(1), 109–116 (1923)
671. Klainerman, S., Rodnianski, I.: A geometric approach to the Littlewood–Paley theory. Geom. Funct. Anal. **16**(1), 126–163 (2006)
672. von Koch, H.: Sur une courbe continus sans tangente, obtenue par une construction géometrique élémentaire. Arkiv för Matematik **1**, 681–704 (1904)
673. von Koch, H.: Une méthode géometrique élémentaire pour l'étude de certaines questions de la théorie des courbes planes. Acta Math. **30**, 145–174 (1906)
674. von Koch, H., Tataru, D.: Well-posedness for the Navier–Stokes equations. Adv. Math. **157**, 22–35 (2001)
675. Koch, H., Koskela, P., Saksman, E., Soto, T.: Bounded compositions on scaling invariant Besov spaces. J. Funct. Anal. **266**, 2765–2788 (2014)
676. Kobayashi, M.: Modulation spaces $M^{p,q}$ for $0 < p, q \leq \infty$. J. Funct. Spaces Appl. **4**(3), 329–341 (2006)

677. Kobayashi, T., Muramatu, T.: Abstract Besov space approach to the nonstationary Navier–Stokes equations. Math. Methods Appl. Sci. **15**(9), 599–620 (1992)
678. Kolmogoroff, A.: Sur les fonctions harmoniques conjuguées et les séries de Fourier. Fund. Math. **7**, 24–29 (1925)
679. Kopaliani, T.S.: Littlewood–Paley characterization on spaces $L^{p(t)}(\mathbb{R}^n)$. Ukraín. Mat. Zh. **60**(12), 1709–1715 (2008); Translation in Ukrainian Math. J. **60**(12), 2006–2014 (2008)
680. Koezuka, K., Tomita, N.: Bilinear pseudo-differential operators with symbols in $BS^m_{1,1}$ on Triebel–Lizorkin spaces. J. Fourier Anal. Appl. **24**, 309–319 (2018)
681. Kobayashi, M., Miyachi, A., Tomita, N.: Embedding relations between local Hardy and modulation spaces. Studia Math. **192**(1), 79–96 (2009)
682. Kobayashi, M., Sawano, Y.: Molecular decomposition of the modulation spaces $M^{p,q}$ and its application to the pseudo-differential operators. Osaka J. Math. **47**(4), 1029–1053 (2010)
683. Koch, H., Sickel, W.: Pointwise multipliers of Besov spaces of smoothness zero and spaces of continuous functions. Rev. Mat. Iberoam. **18**, 587–626 (2002)
684. Komori, Y.: Notes on commutators on Herz-type spaces. Arch. Math. (Basel) **81**(3), 318–326 (2003)
685. Komori-Furuya, Y., Matsuoka, K., Nakai, E., Sawano, Y.: Applications of Littlewood–Paley theory for $\dot{B}_\sigma$–Morrey spaces to the boundedness of integral operators. J. Funct. Spaces Appl., Art. ID 859402, 1–21 (2013)
686. Koskela, P., Yang, D., Zhou, Y.: Pointwise characterizations of Besov and Triebel–Lizorkin spaces and quasiconformal mappings. Adv. Math. **226**, 3579–3621 (2011)
687. Kozono, H., Yamazaki, M.: Semilinear heat equations and the Navier–Stokes equation with distributions in new function spaces as initial data. Commun. PDE **19**, 959–1014 (1994)
688. Krbec, M., Schmeisser, H.J.: Refined limiting imbeddings for Sobolev spaces of vector-valued functions. J. Funct. Anal. **227**(2), 372–388 (2005)
689. Krantz, S.G.: Lipschitz spaces on stratified groups. Trans. Am. Math. Soc. **269**(1), 39–66 (1982)
690. Kree, P.: Sur les multiplicateurs dans $\mathscr{F} L^p$ (French). Ann. Inst. Fourier (Grenoble) **16**, 31–89 (1966)
691. Ky, L.D.: New Hardy spaces of Musielak–Orlicz type and boundedness of sublinear operators. Integral Equ. Oper. Theory **78**(1), 115–150 (2014)
692. Ky, L.D.: On the product of functions in BMO and H^1 over spaces of homogeneous type. J. Math. Anal. Appl. **425**(2), 807–817 (2015)
693. Kyriazis, G., Petrushev, P.: New bases for Triebel–Lizorkin and Besov spaces. Trans. Am. Math. Soc. **354**(2), 749–776 (2002)
694. Kyriazis, G., Petrushev, P.: On the construction of frames for Triebel–Lizorkin and Besov spaces. Proc. Am. Math. Soc. **134**(6), 1759–177 (2006)
695. Krotov, V.G., Prokhorovich, M.: A. Functions from Sobolev and Besov spaces with maximal Hausdorff dimension of the exceptional Lebesgue set (Russian). Fundam. Prikl. Mat. **18**(5), 145–153 (2013); Translation in J. Math. Sci. (N.Y.) **209**(1), 108–114 (2015)
696. Kurtz, D.: Littlewood–Paley and multipliers theorems on weighted L^p spaces. Trans. Am. Math. Soc. **259**, 235–254 (1980)
697. Kyriazis, G.: Decomposition systems for function spaces. Studia Math. **157**, 133–169 (2003)
698. Kuhn, T., Leopold, H.G., Sickel, W., Skrzypczak, L.: Entropy numbers of Sobolev embeddings of radial Besov spaces. J. Approx. Theory **121**(2), 244–268 (2003)
699. Kuhn, T., Leopold, H.G., Sickel, W., Skrzypczak, L.: Entropy numbers of embeddings of weighted Besov spaces. Constr. Approx. **23**(1), 61–77 (2006)
700. Kuhn, T., Leopold, H.G., Sickel, W., Skrzypczak, L.: Entropy numbers of embeddings of weighted Besov spaces. III. Weights of logarithmic type. Math. Z. **255**(1), 1–15 (2007)
701. Lacey, M.: Carleson's theorem: proof, complements, variations. Publ. Mat. **48**, 251–307 (2004)
702. Lacey M., Thiele, C.: A proof of boundedness of the Carleson operator. Math. Res. Lett. **7**, 361–370 (2000)
703. Latter, R.: A characterization of $H^p(\mathbb{R}^n)$ in terms of atoms. Studia Math. **62**, 93–101 (1978)

704. Lebesgue, H.: Sur l'integration des fonctions discontinues. Ann. Sci. Ecole Norm. Sup. **27**(3), 361–450 (1910)
705. Lemarie-Rieusset, P.G.: Multipliers and Morrey spaces. Potential Anal. **38**(3), 741–752 (2013)
706. Lemarie-Rieusset, P.G.: Erratum to: multipliers and Morrey spaces. Potential Anal. **41**(4), 1359–1362 (2014)
707. Leray, J.: Étude de diverses équations intégrales non linéaires et de quelques problèmes que pose l'hydrodynamique. J. Math. Pures Appl. **12**, 1–82 (1933)
708. Lerner, A.K.: An elementary approach to several results on the Hardy–Littlewood maximal operator. Proc. Am. Math. Soc. **136**(8), 2829–2833 (2008)
709. Leopold, H.G.: On function spaces of variable order and differentiation. Forum Math. **3**(1), 1–21 (1991)
710. Leopold, H.G.: Embeddings and entropy numbers in Besov spaces of generalized smoothness. In: Function Spaces (Poznań, 1998). Lecture Notes in Pure and Applied Mathematics, vol. 213, pp. 323–336. Dekker, New York (2000)
711. Leopold, H.G., Skrzypczak, L.: Entropy numbers of embeddings of some 2-microlocal Besov spaces. J. Approx. Theory **163**(4), 505–523 (2011)
712. Leopold, H.G., Skrzypczak, L.: Compactness of embeddings of function spaces on quasi-bounded domains and the distribution of eigenvalues of related elliptic operators. Proc. Edinb. Math. Soc. (2) **56**(3), 829–851 (2013)
713. Li, B.D., Fan, X.Y., Fu, Z.W., Yang, D.: Molecular characterization of anisotropic Musielak–Orlicz Hardy spaces and their applications. Acta Math. Sin. (Engl. Ser.) **32**(11), 1391–1414 (2016)
714. Maeda, F.-Y., Sawano, Y., Shimomura, T.: Some norm inequalities in Musielak–Orlicz spaces. Ann. Acad. Sci. Fenn. Math. **41**(2), 721–744 (2016)
715. Machihara, S., Ozawa, T.: Interpolation inequalities in Besov spaces. Proc. Am. Math. Soc. **131**(5), 1553–1556 (2003)
716. Markhasin, L.: Discrepancy of generalized Hammersley type point sets in Besov spaces with dominating mixed smoothness. Unif. Distrib. Theory **8**(1), 135–164 (2013)
717. Marschall, J.: Some remarks on Triebel spaces. Studia Math. **87**, 79–92 (1987)
718. Marschall, J.: On the boundedness and compactness of nonregular pseudo-differential operators. Math. Nachr. **175**, 231–262 (1995)
719. Lévy, J.V., Seuret, S.: The 2-microlocal formalism. In: Fractal Geometry and Applications: A Jubilee of Benoit Mandelbrot. Proceedings of Symposium on Pure Mathematics, vol. 72(2). AMS, Providence (2004)
720. Li, B., Bownik, M., Yang, D., Yuan, W.: Duality of weighted anisotropic Besov and Triebel–Lizorkin spaces. Positivity **16**(2), 213–244 (2012)
721. Li, B., Bownik, M., Yang, D., Yuan, W.: A mean characterization of weighted anisotropic Besov and Triebel–Lizorkin spaces. Z. Anal. Anwend. **33**(2), 125–147 (2014)
722. Li, P., Zhai, Z.: Generalized Navier–Stokes equations with initial data in local Q-type spaces. J. Math. Anal. Appl. **369**, 595–609 (2010)
723. Li, P., Zhai, Z.: Well-posedness and regularity of generalized Navier–Stokes equations in some critical Q-spaces. J. Funct. Anal. **259**, 2457–2519 (2010)
724. Liang, Y., Yang, D.: Intrinsic square function characterizations of Musielak–Orlicz Hardy spaces. Trans. Am. Math. Soc. **367**(5), 3225–3256 (2015)
725. Liang, Y., Yang, D., Jiang, R.: Weak Musielak–Orlicz Hardy spaces and applications. Math. Nachr. **289**(5–6), 634–677 (2016)
726. Liang, Y., Yang, D., Yang, S.: Applications of Orlicz–Hardy spaces associated with operators satisfying Poisson estimates. Sci. China Math. **54**, 2395–2426 (2011)
727. Liang, Y., Sawano, Y., Ullrich, T., Yang, D., Yuan, W.: New characterizations of Besov-Triebel–Lizorkin–Hausdorff spaces including coorbits and wavelets. J. Fourier Anal. Appl. **18**(5), 1067–1111 (2012)
728. Liang, Y., Sawano, Y., Ullrich, T., Yang, D., Yuan, W.: A new framework for generalized Besov-type and Triebel–Lizorkin-type spaces. Diss. Math. (Rozprawy Mat.) **489**, 1–114 (2013)

729. Lions, J.L.: Un théorème de traces. Compt. Rend. Acad. Sci. Paris **249**, 2259–2261 (1959)
730. Lions, J.L.: Sur certains théorèmes d'interpolation. Compt. Rend. Acad. Sci. Paris **250**, 2104–2106 (1960)
731. Lions, J.L.: Sur les espaces d'interplation; dualité. Math. Scand. **9**, 147–177 (1961)
732. Lions, J.L.: Symétrie et compacité dans les espaces de Sobolev (French). J. Funct. Anal. **49**(3), 315–334 (1982)
733. Lions, J.L., Magnes, E.: Problemi ai limiti non omogenei III. Ann. Scuola Norm. Sup. Pisa **15**, 41–103 (1961)
734. Lions, J.L., Magnes, E.: Problems aux limites non homogenes, IV. Ann. Scuola Norm. Sup. Pisa **15**, 311–326 (1961)
735. Lions, J.L., Magnes, E.: Problemi ai limiti non omogenei V. Ann. Scuola Norm. Sup. Pisa **16**, 1–44 (1962)
736. Lions, J.L., Peetre, J.: Sur une classe d'epaces d'interpolation. Inst. Hautes Études Sci. Publ. Math. **19**, 5–68 (1964)
737. Littlewood, J.E.: On bounded bilinear forms in an infinite number of variables. Quart. J. Math. Oxford **1**, 164–174 (1930)
738. Littlewood, J.E., Paley, R.E.A.C.: Theorems on Fourier series and power series. Part I. J. London Math. Soc. **6**, 230–233 (1931)
739. Littlewood, J.E., Paley, R.E.A.C.: Theorems on Fourier series and power series. Part II. J. London Math. Soc. **42**, 52–89 (1936)
740. Littlewood, J.E., Paley, R.E.A.C.: Theorems on Fourier series and power series. Part III. J. London Math. Soc. **43**, 105–126 (1937)
741. Liu, L., Yang, D., Yuan, W.: Besov-type and Triebel-Lizorkin-type spaces associated with heat kernels. Collect. Math. **67**(2), 247–310 (2016)
742. Littman, W.: Multipliers in L^p and interpolation. Bull. Am. Math. Soc. **71**, 764–766 (1965)
743. Liu, H.: The weak H^p spaces on homogenous groups. In: Cheng, M.-T., Zhou, X.-W., Deng, D.-G. (eds.) Harmonic Analysis, Tianjin, 1988. Lecture Notes in Mathematics, vol. 1494, pp. 113–118. Springer, Berlin (1991)
744. Lizorkin, P.I.: Boundary properties of functions in weighted classes (Russian). Dokl. Akad. Nauk SSSR **132**, 514–517 (1960); English transl. in Soviet Math. Dokl. **1** (1960)
745. Lizorkin, P.I.: On Fourier multipliers in the space $L_{p,\theta}$ (Russian). Trudy Mat. Inst. Steklov **89**, 231–248 (1967)
746. Lizorkin, P.I.: Operators connected with fractional derivatives and classes of differentiable functions. Trudy Mat. Inst. Steklov **117**, 212–243 (1972)
747. Lizorkin, P.I.: Generalized Hölder classes of functions in connection with fractional derivatives (Russian). Trudy Mat. Inst. Steklov **128**, 172–177 (1972)
748. Lizorkin, P.I.: Properties of functions of the spaces $\Gamma^r_{p\theta}$. Trudy Mat. Inst. Steklov **131**, 158–181 (1974)
749. Liu, J., Yang, D., Yuan, W.: Anisotropic Hardy-Lorentz spaces and their applications. Sci. China Math. **59**(9), 1669–1720 (2016)
750. Lorentz, G.: Some new functional spaces. Ann. Math. (2) **51**, 37–55 (1950)
751. Lorentz, G.: On the theory of spaces Λ. Pac. J. Math. **1**, 411–429 (1951)
752. Lu, S.Z., Yang, D.: Some Hardy spaces associated with the Herz spaces and their wavelet characterizations (in Chinese). Beijing Shifan Daxue Xuebao (= J. Beijing Normal Univ. (Natur. Sci.)) **29**, 10–19 (1993)
753. Lu, S.Z., Yang, D.: The local versions of $H^p(\mathbb{R}^n)$ spaces at the origin. Studia Math. **116**, 103–131 (1995)
754. Lu, S.Z., Yang, D.: Herz-type Sobolev and Bessel potential spaces and their applications. Sci. China Ser. A **40**, 113–129 (1997)
755. Lu, S.Z., Yang, D.: Multiplier theorems for Herz type Hardy spaces. Proc. Am. Math. Soc. **126**, 3337–3346 (1998)
756. Lu, Y.F., Yang, D., Yuan, W.: Interpolation of Morrey spaces on metric measure spaces. Canad. Math. Bull. **57**, 598–608 (2014)
757. Luxenberg, W.A.J.: Banach function spaces. Thesis, Delft (1955)

758. Macías, R.A., Segovia, C.: Lipschitz functions on spaces of homogeneous type. Adv. Math. **33**(3), 257–270 (1979)
759. Madych, W.R.: Absolute continuity of Fourier transforms on $\mathbb{R}^n$. Indiana Univ. Math. J. **25**, 467–479 (1976)
760. Malecka, A.: Haar functions in weighted Besov and Triebel–Lizorkin spaces. J. Approx. Theory **200**, 1–27 (2015)
761. Maligranda, L.: Marcinkiewicz interpolation theorem and Marcinkiewicz spaces. Wlad. Math. **48**(2), 157–171 (2012)
762. Maligranda, L., Sabourova, N.: Real and complex operator norms between quasi L^p-L^q spaces. Math. Ineq. Appl. **14**(2), 247–270 (2011)
763. Marcinkiewicz, J.: Sur l'interpolation d'operations. C. R. Acad. Sci. Paris **208**, 1272–1273 (1939)
764. Masaki, S.: Local existence and WKB approximation of solutions to Schrödinger-Poisson system in the two-dimensional whole space. Commun. Partial Differ. Equ. **35**(12), 2253–2278 (2010)
765. Matsumoto, T., Ogawa, T.: Interpolation inequality of logarithmic type in abstract Besov spaces and an application to semilinear evolution equations. Math. Nachr. **283**, 1810–1828 (2010)
766. Mauceri, G., Meda, S.: BMO and H^1 for the Ornstein-Uhlenbeck operator. J. Funct. Anal. **252**(1), 278–313 (2007)
767. Mazzucato, A.L.: Decomposition of Besov–Morrey spaces. In: Harmonic Analysis at Mount Holyoke, South Hadley, 2001. Contemporary Mathematical, vol. 320, pp. 279–294. American Mathematical Society, Providence (2003)
768. Mazzucato, A.L.: Besov–Morrey spaces: function space theory and applications to non-linear PDE. Trans. Am. Math. Soc. **355**(4), 1297–1364 (2003)
769. M[c]Intosh, A.: Square roots of operators and applications to hyperbolic PDEs. In: Proceedings of Centre for Mathematical Analysis, vol. 5, pp. 124–136. Australian National University, Canberra (1984)
770. Meda, S., Sjögren, P., Vallarino, M.: On the H^1-L^1 boundedness of operators. Proc. Am. Math. Soc. **136**, 2921–2931 (2008)
771. Meskhi, A., Rafeiro, H., Muhammad, A.: Interpolation on variable Morrey spaces defined on quasi-metric measure spaces. J. Funct. Anal. **270**(10), 3946–3961 (2016)
772. Meyers, G.N.: Mean oscillation over cubes and Hölder continuity. Proc. Am. Math. Soc. **15**, 717–721 (1964)
773. Meyers, G.N., James, S.: $H = W$. Proc. Nat. Acad. Sci. USA **51**, 1055–1056 (1964)
774. Meyer, Y.: Remarques sur un téorèm de J.-M. Bony. In: Proceedings of the Seminar on Harmonic Analysis, Pisa, 1980. Rend. Circ. Mat. Palermo (2) **suppl. 1**, 1–20 (1981)
775. Meyer, Y.: La Minimalité de l'Espace de Besov $B_1^{0,1}$ et la Continuité des Opérateurs Definis par des IntÂt'egrales Singulières. Monografias de Matematicas, vol. 4. Universidad Autónoma de Madrid (1986)
776. Meyer, Y., Yang, Q.X.: Continuity of Calderón–Zygmund operators on Besov and Triebel–Lizorkin spaces. Anal. Appl. **6**(1), 51–81 (2008)
777. Meyer, Y., Xu, H.: Wavelet analysis and chirps. Appl. Comput. Harmonic Anal. **4**, 366–379 (1997)
778. Michlin, S.G.: Singular Integral Equations. American Mathematical Society Translation, vol. 24, pp. 1–116. American Mathematical Society, New York (1950)
779. Michlin, S.G.: On multipliers of Fourier integrals (Russian). Dokl. Akad. Nauk. SSSR **109**, 701–703 (1956)
780. Michlin, S.G.: Fourier integrals and multiple singular integrals (Russian). Vestnik Leningrad, Univ. Mat. Meh. Astronom. **7**, 143–155 (1957)
781. Mitrea M., Taylor, M.: Sobolev and Besov space estimates for solutions to second order PDE on Lipschitz domains in manifolds with Dini or Hölder continuous metric tensors. Commun. PDE **30**(1–3), 1–37 (2005)

782. Mironescu, P., Russ, E.: Traces of weighted Sobolev spaces. Old and new. Nonlinear Anal. **119**, 354–381 (2015)
783. Miyachi, A.: On some estimates for the wave equation in L^p and H^p. J. Fac. Sci. Univ. Tokyo Sect. IA Math. **27**(2), 331–354 (1980)
784. Miyachi, A.: Weak factorization of distributions in H^p spaces. Pac. J. Math. **115**(1), 165–175 (1984)
785. Miyachi, A.: Hardy–Sobolev spaces and maximal functions. J. Math. Soc. Japan **42**(1), 73–90 (1990)
786. Miyachi, A.: On the extension properties of Triebel–Lizorkin spaces. Hokkaido Math. J. **27**, 273–301 (1998)
787. Miyachi, A.: Weighted Hardy spaces on a domain. In: Proceedings of the Second ISAAC Congress, Fukuoka, vol. 1, pp. 59–64, 1999. International Society for Analysis, Applications and Computation, vol. 7. Kluwer Academic Publishers, Dordrecht (2000)
788. Miyachi, A.: Remarks on Herz-type Hardy spaces. Acta Math. Sinica. English Series. **17**, 339–360 (2001)
789. Miyachi, A.: Change of variables for weighted Hardy spaces on a domain. Hokkaido Math. J. **38**, 519–555 (2009)
790. Miyachi, A., Nicola, F., Rivetti, S., Tabacco, A., Tomita, N.: Estimates for unimodular Fourier multipliers on modulation spaces. Proc. Am. Math. Soc. **137**(11), 3869–3883 (2009)
791. Miyachi, A., Tomita, N.: Calderón-Vaillancourt-type theorem for bilinear operators. Indiana Univ. Math. J. **62**(4), 1165–1201 (2013)
792. Miyachi, A., Tomita, N.: Boundedness criterion for bilinear Fourier multiplier operators. Tohoku Math. J. **66**, 55–76 (2014)
793. Miyazaki, Y.: New proofs of the trace theorem of Sobolev spaces. Proc. Japan Acad. Ser. A Math. Sci. **84**(7), 112–116 (2008)
794. Miyazaki, Y.: Sobolev trace theorem and the Dirichlet problem in the unit disk. Milan J. Math. **82**(2), 297–312 (2014)
795. Miyazaki, Y.: Liouville's theorem and heat kernels. Expo. Math. **33**, 101–104 (2015)
796. Mizuhara, T.: On Fourier multipliers of homogeneous Besov spaces. Math. Nachr. **133**, 155–161 (1987)
797. Mizuta, Y., Shimomura, T.: Sobolev's inequality for Riesz potentials with variable exponent satisfying a log-Hölder condition at infinity. J. Math. Anal. Appl. **311**, 268–288 (2005)
798. Mizuta, Y., Shimomura, T.: Maximal functions, Riesz potentials and Sobolev's inequality in generalized Lebesgue spaces. In: Potential Theory in Matsue. Advanced Studies in Pure Mathematics, vol. 44, pp. 255–281. Mathematical Society of Japan, Tokyo (2006)
799. Mizuta, Y., Ohno, T., Shimomura, T.: Sobolev's inequalities and vanishing integrability for Riesz potentials of functions in the generalized Lebesgue space $L^{p(\cdot)}(\log L)^{q(\cdot)}$. J. Math. Anal. Appl. **345**(1), 70–85 (2008)
800. Mizuta, Y., Nakai, E., Ohno, T., Shimomura, T.: Maximal functions, Riesz potentials and Sobolev embeddings on Musielak–Orlicz–Morrey spaces of variable exponent in $\mathbb{R}^n$. Rev. Mat. Complut. **25**(2), 413–434 (2012)
801. Mizuta, Y., Nakai, E., Sawano, Y., Shimomura, T.: Littlewood–Paley theory for variable exponent Lebesgue spaces and Gagliardo-Nirenberg inequality for Riesz potentials. J. Math. Soc. Japan **65**(2), 633–670 (2013)
802. Moritoh, S.: Wavelet transforms in Euclidean spaces – their relation with wave frontsets and Besov, Triebel–Lizorkin spaces. Tohoku Math. J. **47**, 555–565 (1995)
803. Moritoh, S., Yamada, T.: Two-microlocal Besov spaces and wavelets. Rev. Mat. Iberoam. **20**, 277–283 (2004)
804. Morrey, C.B.: On the solutions of quasi linear elliptic partial differential equations. Trans. Am. Math. Soc. **43**, 126–166 (1938)
805. Morii, K., Sato, T., Sawano, Y.: Certain equalities concerning derivatives of radial homogeneous functions and a logarithmic function. Commun. Math. Anal. **9**(2), 51–66 (2010)
806. Morii, K., Sato, T., Sawano, Y., Wadade, H.: Sharp constants of Brézis-Gallouët-Wainger type inequalities with a double logarithmic term on bounded domains in Besov and Triebel–Lizorkin spaces. Bound. Value Probl., Art. ID 584521, 1–38 (2010)

807. Morii, K., Sato, T., Wadade, H.: Brézis-Gallouët-Wainger type inequality with a double logarithmic term in the Hölder space: its sharp constants and extremal functions. Nonlinear Anal. **73**(6), 1747–1766 (2010)
808. Morii, K., Sato, T., Wadade, H.: Brézis-Gallouet-Wainger inequality with a double logarithmic term on a bounded domain and its sharp constants. Math. Inequal. Appl. **14**(2), 295–312 (2011)
809. Moura, S.D., Neves, J.S., Schneider, C.: On trace spaces of 2-microlocal Besov spaces with variable integrability. Math. Nachr. **286**(11–12), 1240–1254 (2013)
810. Moussai, M.: Realizations of homogeneous Besov and Triebel–Lizorkin spaces and an application to pointwise multipliers. Anal. Appl. (Singap.) **13**(2), 149–183 (2015)
811. Muckenhoupt, B.: Weighted norm inequalities for the Hardy maximal function. Trans. Amer. Math. Soc. **165**, 207–226 (1972)
812. Muckenhoupt, B.: The equivalence of two conditions for weight functions. Studia Math. **49**, 101–106 (1974)
813. Muckenhoupt, B., Wheeden, R.: Weighted bounded mean oscillation and the Hilbert transform. Studia Math. **54**, 221–237 (1976)
814. Müller, D.: Hardy space methods for nolinear partial defferential equations. Tatra Mt. Math. Publ. **4**, 159–168 (1994)
815. Müller, D., Yang, D.: A difference characterization of Besov and Triebel–Lizorkin spaces on RD-spaces. Forum Math. **21**(2), 259–298 (2009)
816. Muramatu, T.: On Besov spaces of functions defined in general regions. Publ. Res. Inst. Math. Sci. Kyoto Univ. **6**, 515–543 (1970/1971)
817. Muramatu, T.: On Besov spaces and Sobolev spaces of generalized functions definded on a general region. Publ. Res. Inst. Math. Sci. **9**, 325–396 (1973/1974)
818. Muramatu, T.: On the dual of Besov Spaces. Publ. Res. Inst. Math. Sci. **10**, 123–140 (1976)
819. Nagayasu, S., Wadade, H.: Characterization of the critical Sobolev space on the optimal singularity at the origin. J. Funct. Anal. **258**, 3725–3757 (2010)
820. Naibo, V.: On the bilinear Hormander classes in the scales of Triebel–Lizorkin and Besov spaces. J. Fourier Anal. Appl. **21**(5), 1077–1104 (2015)
821. Najafov, A.M.: Some properties of functions from the intersection of Besov–Morrey type spaces with dominant mixed derivatives. Proc. A. Razmadze Math. Inst. **139**, 71–82 (2005)
822. Najafov, A.M.: On some properties of the functions from Sobolev–Morrey type spaces. Cent. Eur. J. Math. **3**(3), 496–507 (2005)
823. Najafov, A.M.: Embedding theorems in the Sobolev–Morrey type spaces $S^l_{p,a,\kappa,r}W(G)$ with dominant mixed derivatives. Sib. Math. J. **47**(3), 613–625 (2006)
824. Nakai, E.: Construction of an atomic decomposition for functions with compact support. J. Math. Anal. Appl. **313**, 730–737 (2006)
825. Nakai, E., Yabuta, K.: Pointwise multipliers for functions of bounded mean oscillation. J. Math. Soc. Japan **37**, 207–218 (1985)
826. Nakai, E., Sawano, Y.: Hardy spaces with variable exponents and generalized Campanato spaces. J. Funct. Anal. **262**, 3665–3748 (2012)
827. Nakai, E., Sawano, Y.: Orlicz-Hardy spaces and their duals. Sci. China Math. **57**(5), 903–962 (2014)
828. Nakai, E., Sobukawa, T.: B^u_w-function spaces and their interpolation. Tokyo J. Math. **39**(2), 483–517 (2016)
829. Nakamura, S.: Generalized weighted Morrey spaces and classical operators. Math. Nachr. **289**(17–18), 2235–2262 (2016)
830. Nakamura, S., Noi, T., Sawano, Y.: Generalized Morrey spaces and trace operator. Sci. China Math. **59**(2), 281–336 (2015)
831. Nakano, H.: Modulared Semi-ordered Linear Spaces. Maruzen Co. Ltd., Tokyo, i+288pp (1950)
832. Nakano, H.: Topology of Linear Topological Spaces. Maruzen Co. Ltd., Tokyo, viii+281pp (1951)

833. Nazarov, F., Treil, S., Volberg, A.: The Tb-theorem on non-homogeneous spaces. Acta Math. **190**(2), 151–239 (2003)
834. Neĉas, J.: Sur une méthode pour résoudre les équations aux dérivées partielles du type elliptique, voisine de la variationnelle (French). Ann. Scuola Norm. Sup. Pisa (3) **16**, 305–326 (1962)
835. Netrusov, Y.V.: Some imbedding theorems for spaces of Besov–Morrey type (Russian). In: Numerical Methods and Questions in the Organization of Calculations, vol. 7. Zap. Nauchn. Sem. Leningrad. Otdel. Mat. Inst. Steklov. (LOMI) **139**, 139–147 (1984)
836. Netrusov, Y.V.: Embedding theorems for Lizorkin–Triebel spaces (Russian). Zap. Naučn. Sem. Leningrad. Otdel. Mat. Inst. Steklov (LOMI) **159**, 103–112 (1987)
837. Netrusov, Y.V.: Metric estimates for the capacities of sets in Besov spaces (Russian). Translated in Proc. Steklov Inst. Math. **1**, 167–192 (1992). Theory of functions, Amberd, 1987. Trudy Mat. Inst. Steklov. **190**, 159–185 (1989)
838. Netrusov, Y.V.: Sets of singularities of functions in spaces of Besov and Lizorkin-Triebel type (Russian). Translated in Proc. Steklov Inst. Math. **3**, 185–203 (1990). Studies in the theory of differentiable functions of several variables and its applications, vol. 13. Trudy Mat. Inst. Steklov. **187**, 162–177 (1989)
839. Netrusov, Y.V.: Theorems on traces and multipliers for functions in Lizorkin-Triebel spaces (Russian). Zap. Nauchn. Sem. S.-Peterburg. Otdel. Mat. Inst. Steklov. (POMI) 200, Kraev. Zadachi Mat. Fiz. Smezh. Voprosy Teor. Funktsii. **24**, 132–138, 189–190 (1992); Translation in J. Math. Sci. **77**(3), 3221–3224 (1995)
840. Nguyen, V.K.: Bernstein numbers of embeddings of isotropic and dominating mixed Besov spaces. Math. Nachr. **288**(14–15), 1694–1717 (2015)
841. Nguyen, V.K., Sickel, W.: Weyl numbers of embeddings of tensor product Besov spaces. J. Approx. Theory **200**, 170–220 (2015)
842. Nguyen, V.K., Sickel, W.: Pointwise multipliers for Sobolev and Besov spaces of dominating mixed smoothness. J. Math. Anal. Appl. **452**(1), 62–90 (2017)
843. Nguyen, V.K., Sickel, W.: Pointwise multipliers for Besov spaces of dominating mixed smoothness II. Sci. China Math. **60**(11), 2241–2262 (2017)
844. Nguyen, V.K., Sickel, W.: On a problem of Jaak Peetre concerning pointwise multipliers of Besov spaces. Studia Math. **243**(2), 207–231 (2018)
845. Nikolski'i, S.M.: Inequalities for entire function of finite order and their application in the theory of differentiable functions of several variables (Russian). Trudy Mat. Inst. Steklov **38**, 244–278 (1951)
846. Nikolski'i, S.M.: On the solution of the polyharmonic equation by a variational method (Russian). Dokl. Akad. Nauk SSSR **88**, 409–411 (1953)
847. Nikolski'i, S.M.: Functions with dominant mixed derivative, satisfying a multiple Hölder condition, (Russian) Sibirsk. Mat. Z. **4**, 1342–1364 (1963)
848. Nikolski'i, S.M., Lions, J.L., Lizorkin, P.I.: Integral representation and isomorphism properties of some classes of functions. Ann. Scuola Norm. Sup. Pisa **19**, 127–178 (1965)
849. Nirenberg, L.: On elliptic partial differential equations. Ann. Scuola Norm. Sup. Pisa (3) **13**, 115–162 (1959)
850. Noi, T.: Duality of variable exponent Triebel-Lizorkin and Besov spaces. J. Funct. Spaces Appl. Article ID 361807, 1–19 (2012). https://doi.org/10.1155/2012/361807
851. Noi, T.: Trace and extension operators for Besov spaces and Triebel–Lizorkin spaces with variable exponents. Rev. Mat. Complut. **29**(2), 341–404 (2016)
852. Noi, T.: Trace operator for 2-microlocal Besov spaces with variable exponents. Tokyo J. Math. **39**(1), 293–327 (2016)
853. Noi, T.: Sawano, Y.: Complex interpolation of Besov spaces and Triebel–Lizorkin spaces with variable exponents. J. Math. Anal. Appl. **387**, 676–690 (2012)
854. Novak, E., Triebel, H.: Function spaces in Lipschitz domains and optimal rates of convergence for sampling. Constr. Approx. **23**(3), 325–350 (2006)
855. Nursultanov, E.D., Ruzhansky, M., Tikhonov, S.Y.: Nikolskii inequality and functional classes on compact Lie groups. Funct. Anal. Appl. **49**(3), 226–229 (2015). Translated from Funktsionalńyi Analiz i Ego Prilozheniya **49**(3), 83–87 (2015)

856. Nualtaranee, S.: On least harmonic majorants in half-spaces. Proc. London Math. Soc. (3) **27**, 243–260 (1973)
857. Ogura, K.: On a certain transcendental integral function in the theory of interpolation. Tohoku Math. J. **17**, 64–72 (1920)
858. Ohno, T., Shimomura, T.: Musielak-Orlicz-Sobolev spaces on metric measure spaces. Czechoslovak Math. J. **65**(140), 435–474 (2015)
859. Orlicz, W.: Über konjugierte Exponentenfolgen. Studia Math. **3**, 200–212 (1931)
860. Oswald, P.: On the boundedness of the mapping $f \longrightarrow |f|$ in Besov spaces. Comment. Univ. Carolinae **33**, 57–66 (1992)
861. Päivärinta, L.: On the spaces $L_p^A(l_q)$: maximal inequalities and complex interpolation. Ann. Acad. Sci. Fenn. Ser. AI Math. Dissertationes 25, Helsinki (1980)
862. Päivärinta, L.: Equivalent quasi-norms and Fourier multipliers in the Triebel spaces $F_{p,q}^s$. Math. Nachr. **106**, 101–108 (1982)
863. Paley, R.E.A.C., Wiener, N., Zygmund, A.: Notes on random functions. Math Z. **37**, 647–668 (1933)
864. Parilov, D.V.: Two theorems on the Hardy-Lorentz classes $H^{1,q}$ (in Russian). J. Math. Sci. (N.Y.). **139**(2), 6447–6456 (2006)
865. Peetre, J.: Théorèmes de regularite pour quelques espaces d'operateurs differentiels. Thèse, Lund (1959)
866. Peetre, J.: A theory of interpolation of normed spaces. Notes Universidade de Brasilia (1963)
867. Peetre, J.: Espaces d'interpolation et th'eorème de Soboleff (French). Ann. Inst. Fourier (Grenoble) **16**, 279–317 (1961)
868. Peetre, J.: Applications de la théorie des espaces d'interpolation dans l'analyse harmonique (French). Ricerche Mat. **15**, 3–36 (1966)
869. Peetre, J.: Interpolation i abstracta rum. Lecture Notes. Lund (1966)
870. Peetre, J.: Sur le espaces de Besov. C. R. Acad. Sci. Paris. Sér. A–B **264**, 281–283
871. Peetre, J.: On the theory of $\mathscr{L}_{p,\lambda}$. J. Funct. Anal. **4**, 71–87 (1969)
872. Peetre, J.: Remarques sur les espaces de Besov. Le cas $0 < p < 1$. C. R. Acad. Sci. Paris Sér. A-B **277**, 947–950 (1973)
873. Peetre, J.: Remark on the dual of an interpolation space. Math. Scand. **34**, 124–128 (1974)
874. Peetre, J.: On spaces of Triebel–Lizorkin-type. Ark. Mat. **13**, 123–130 (1975)
875. Peetre, J.: Correction to the paper: "On spaces of Triebel–Lizorkin-type" (Ark. Mat. **13**, 123–130 (1975)). Ark. Mat. **14**(2), 299 (1976)
876. Peetre, J.: The trace of Besov space–a limiting case. Technical report, Lund (1975)
877. Pérez, C.: Sharp L^p-weighted Sobolev inequalities. Ann. Inst. Fourier (Grenoble) **45**, 809–824 (1995)
878. Pick, L., Sickel, W.: Several types of intermediate Besov-Orlicz spaces. Math. Nachr. **164**, 141–165 (1993)
879. Pineda, E., Urbina, W.: Some results on Gaussian Besov-Lipschitz spaces and Gaussian Triebel–Lizorkin spaces. J. Approx. Theory **161**(2), 529–564 (2009)
880. Plancherel, M., Pólya, M.G.: Fonctions entières et intégrales de fourier multiples (French). Comment. Math. Helv. **10**(1), 110–163 (1937)
881. Planchon, F.: Self-similar solutions and semi-linear wave equations in Besov spaces. J. Math Pures Appl. **79**(8), 809–820 (2000)
882. Prömel, D.J., Trabs, M.: Rough differential equations driven by signals in Besov spaces. J. Diff. Equations, **260**(6), 5202–5249 (2016)
883. Quek, T., Yang, D.: Calderón–Zygmund-type operators on weighted weak Hardy spaces over $\mathbb{R}^n$. Acta Math. Sin. (Engl. Ser.) **16**, 141–160 (2000)
884. Ragusa, M.A.: Homogeneous Herz spaces and regularity results. Nonlinear Anal. **71**, 1–6 (2009)
885. Ragusa, M.A.: Parabolic Herz spaces and their applications. Appl. Math. Lett. **25**(10), 1270–1273 (2012)
886. Rauhut, H.: Banach frames in coorbit spaces consisting of elements which are invariant under symmetry groups. Appl. Comput. Harmon. Anal. **18**(1), 94–122 (2005)

887. Rauhut, H., Ullrich, T.: Generalized coorbit space theory and inhomogeneous function spaces of Besov-Lizorkin-Triebel type. J. Funct. Anal. **260**, 3299–3362 (2011)
888. Ricci, F., Taibleson, M.: Boundary values of harmonic functions in mixed norm spaces and their atomic structure. Ann. Scuola Norm. Sup. Pisa Cl. Sci. (4) **10**, 1–54 (1983)
889. Riesz, F.: Sur les valeurs moyennes des fonctions. J. Lond. Math. Soc. **5**, 120–121 (1930)
890. Riesz, M.: Sur les maxima des formes bilinéaires et sur les fonctionelles linéaires. Acta Math. **49**, 465–497 (1926)
891. Rivière, N., Sagher, Y.: On two theorems of Paley. Proc. Am. Math. Soc. **42**, 238–242 (1974)
892. Rochberg, R.R., Tachizawa, K.: Pseudodifferential operators, Gabor frames, and local trigonometric bases. In: Gabor Analysis and Algorithms. Applied and Numerical Harmonic Analysis, pp. 171–192. Birkhäuser, Boston (1998)
893. Rolewicz, S.: On a certain class of linear metric spaces. Bull. Acad. Polon. Sci. **5**, 471–473 (1957)
894. Rosenthal, M.: Local means, wavelet bases and wavelet isomorphisms in Besov–Morrey and Triebel–Lizorkin–Morrey spaces. Math. Nachr. **286**(1), 59–87 (2013)
895. Ruiz, A., Vega, L.: Corrigenda to unique continuation for Schrödinger operators with potential in Morrey spaces and a remark on interpolation of Morrey spaces. Publ. Mat. **39**, 405–411 (1995)
896. Runst, T.: Paradifferential operators in spaces of Triebel–Lizorkin and Besov type. Z. Anal. Anwend. **4**(6), 557–573 (1985)
897. Runst, T.: Pseudodifferential operators of the "exotic" class $L^0_{1,1}$ in spaces of Besov and Triebel–Lizorkin-type. Ann. Glob. Anal. Geom. **3**, 13–28 (1985)
898. Runst, T.: Mapping properties of non-linear operators in spaces of Triebel–Lizorkin and Besov type. Anal. Math. **12**, 313–346 (1986)
899. Runst, T., Sickel, W.: On strong summability of Jacobi-Fourier-expansions and smoothness properties of functions. Math. Nachr. **99**, 77–85 (1980)
900. Rychkov, V.S.: On restrictions and extension of the Besov and Triebel–Lizorkin spaces with respect to Lipschitz domains. J. London Math. Soc. (2) **60**, 237–257 (1999)
901. Rychkov, V.S.: Intrinsic characterizations of distribution spaces on domains. Studia Math. **127**, 277–298 (1998)
902. Rychkov, V.S.: On a theorem of Bui, Paluszyński, and Tailbeson. Proc. Steklov Inst. Math. **227**, 280–292 (1999)
903. Rychkov, V.S.: Intrinsic characterizations of distribution spaces on domains. Studia Math. **127**(3), 277–298 (1998)
904. Rychkov, V.S.: Littlewood–Paley theory and function spaces with $A_p^{\rm loc}$ weights. Math. Nachr. **224**, 145–180 (2001)
905. Saka, K.: Besov spaces and Sobolev spaces on a nilpotent Lie group. Tohoku Math. J. (2) **31**(4), 383–437 (1979)
906. Saka, K.: Scaling exponents of self-similar functions and wavelet analysis. Proc. Am. Math. Soc. **133**(4), 1035–1045 (2005)
907. Saka, K.: A new generalization of Besov-type and Triebel–Lizorkin-type spaces and wavelets. Hokkaido Math. J. **40**(1), 111–147 (2011)
908. Samko, S.: On a progress in the theory of Lebesgue spaces with variable exponent: maximal and singular operators. Transf. Spec. Funct. **16**(5–6), 461–482 (2005)
909. Sarason, D.: Functions of vanishing mean oscillation. Trans. Am. Math. Soc. **207**, 391–405 (1975)
910. Sato, S.: Littlewood–Paley operators and Sobolev spaces. Illinois J. Math. **58**(4), 1025–1039 (2014)
911. Sautbekova, M., Sickel, W.: Strong summability of Fourier series and Morrey spaces. Anal. Math. **40**(1), 31–62 (2014)
912. Sawada, O.: On time-local solvability of the Navier–Stokes equations in Besov spaces. Adv. Diff. Eq. **8**(4), 385–412 (2003)
913. Sawano, Y.: Sharp estimates of the modified Hardy–Littlewood maximal operator on the nonhomogeneous space via covering lemmas. Hokkaido Math. J. **34**, 435–458 (2005)

914. Sawano, Y.: Wavelet characterization of Besov–Morrey andTriebel–Lizorkin–Morrey spaces. Funct. Approx. Comment. Math. **38**, Part 1, 93–107 (2008)
915. Sawano, Y.: A Note on Besov–Morrey Spaces and Triebel–Lizorkin–Morrey Spaces. Acta Math. Sinica, **25**(8), 1223–1242 (2009)
916. Sawano, Y.: Identification of the image of Morrey spaces by the fractional integral operators. Proc. A. Razmadze Math. Inst. **149**, 87–93 (2009)
917. Sawano, Y.: Maximal operator for pseudodifferential operators with homogeneous symbols. Michigan Math. J. **59**(1), 119–142 (2010)
918. Sawano, Y.: Brézis-Gallouët-Wainger type inequality for Besov–Morrey spaces. Studia Math. **196**(1), 91–101 (2010)
919. Sawano, Y.: Besov–Morrey spaces and Triebel–Lizorkin–Morrey spaces on domains. Math. Nachr. **283**(10), 1456–1487 (2010)
920. Sawano, Y.: Atomic decompositions of Hardy spaces with variable exponents and its application to bounded linear operators. Integr. Equ. Oper. Theory **77**, 123–148 (2013)
921. Sawano, Y.: A new Brézis-Gallouët-Wainger inequality from the viewpoint of the real interpolation functors. Math. Nachr. **287**(2–3), 352–358 (2014)
922. Sawano, Y.: Survey homogeneous Besov spaces. Kyoto J. Math. (To appear)
923. Sawano, Y., Hakim, D.I., Gunawan, H.: Non-smooth atomic decomposition for generalized Orlicz–Morrey spaces. Math. Nachr. **288**(14–15), 1741–1775 (2015)
924. Sawano, Y., Sugano, S., Tanaka, H.: Generalized fractional integral operators and fractional maximal operators in the framework of Morrey spaces. Trans. Am. Math. Soc. **363**(12), 6481–6503 (2011)
925. Sawano, Y., Tanaka, H.: Decompositions of Besov–Morrey spaces and Triebel–Lizorkin–Morrey spaces. Math. Z. **257**(4), 871–905 (2007)
926. Sawano, Y., Tanaka, H.: The Fatou property of block spaces. J. Math. Sci. Univ. Tokyo. **22**, 663–683 (2015)
927. Sawano, Y., Ho, K.P., Yang, D., Yang, S.: Hardy spaces for ball Quasi-Banach function spaces. Diss. Math. **525**, 1–102 (2017)
928. Sawano, Y., Yang, D., Yuan, W.: New applications of Besov-type and Triebel–Lizorkin-type spaces. J. Math. Anal. Appl. **363**, 73–85 (2010)
929. Sawano, Y., Wadade, H.: On the Gagliardo-Nirenberg type inequality in the critical Sobolev–Morrey space. J. Fourier Anal. Appl. **19**, 20–47 (2013)
930. Scharf, B., Schmeisser, H.J., Sickel, W.: Traces of vector-valued Sobolev spaces. Math. Nachr. **285**(8–9), 1082–1106 (2012)
931. Schneider, C.: Trace operators in Besov and Triebel–Lizorkin spaces. Z. Anal. Anwend. **29**(3), 275–302 (2010)
932. Schmeisser, H.J.: On spaces of functions and distributions with mixed smoothness properties of Besov-Triebel-Lizorkin type. I. Basic properties. Math. Nachr. **98**, 233–250 (1980)
933. Schmeisser, H.J.: On spaces of functions and distributions with mixed smoothness properties of Besov-Triebel-Lizorkin type. II. Fourier multipliers and approximation representations. Math. Nachr. **106**, 187–200 (1982)
934. Schmeisser, H.J.: An unconditional basis in periodic spaces with dominating mixed smoothness properties. Anal. Math. **13**(2), 153–168 (1987)
935. Schmeisser, H.J.: Recent developments in the theory of function spaces with dominating mixed smoothness. In: NAFSA Nonlinear Analysis, Function Spaces and Applications, vol. 8, p. 144–204. Czech Academy of Sciences, Prague (2007)
936. Schmeisser, H.J., Sickel, W.: On strong summability of multiple Fourier series and smoothness properties of functions. Anal. Math. **8**(1), 57–70 (1982)
937. Schmeisser, H.J., Sickel, W.: On strong summability of multiple Fourier series and approximation of periodic functions. Math. Nachr. **133**, 211–236 (1987)
938. Schmeisser, H.J., Sickel, W.: Characterization of periodic function spaces via means of Abel-Poisson and Bessel-potential type. J. Approx. Theory **61**(2), 239–262 (1990)
939. Schmeisser, H.J., Sickel, W.: Spaces of functions of mixed smoothness and approximation from hyperbolic crosses. J. Approx. Theory **128**(2), 115–150 (2004)

940. Schott, T.: Function spaces with exponential weights I. Math. Nachr. **189**, 221–242 (1998)
941. Schott, T.: Function spaces with exponential weights II. Math. Nachr. **196**, 231–250 (1998)
942. Schott, T.: Pseudodifferential operators in function spaces with exponential weights. Math. Nachr. **200**, 119–149 (1999)
943. Seeger, A.: A note on Triebel–Lizorkin spaces. In: Approximation and Function Spaces. Banach Center Publications, vol. 22, pp. 391–400. PWN Polish Scientific Publishers, Warsaw (1989)
944. Semmes, S.: A primer on Hardy spaces, and some remarks on a theorem of Evans and Müller. Commun. PDE **19**, 277–319 (1994)
945. Semmes, S.: An introduction to analysis on metric spaces. Notices Am. Math. Soc. **50**(4), 438–443 (2003)
946. Serra, C.F.: Molecular characterization of Hardy-Orlicz spaces. Rev. Un. Mat. Argentina **40**, 203–217 (1996)
947. Shanmugalingam, N.: Newtonian spaces: an extension of Sobolev spaces to metric measure spaces. Revista Matemática Iberoamericana **16**(2), 243–279 (2000)
948. Shanmugalingam, N., Yang, D., Yuan, W.: Newton-Besov spaces and Newton-Triebel–Lizorkin spaces on metric measure spaces. Positivity **19**(2), 177–220 (2015)
949. Shen, Z.: L^p estimates for Schrödinger operators with certain potential. Ann. Inst. Fourier (Grenoble) **45**, 513–546 (1995)
950. Shi, C., Xu, J.S.: A characterization of Herz-Besov-Triebel spaces with variable exponent. Acta Math. Sinica (Chin. Ser.) **55**(4), 653–664 (2012)
951. Shi, C., Xu, J.S.: Herz type Besov and Triebel–Lizorkin spaces with variable exponent. Front. Math. China **8**(4), 907–921 (2013)
952. Sickel, W.: Periodic spaces and relations to strong summability of multiple Fourier series. Math. Nachr. **124**, 15–44 (1985)
953. Sickel, W.: On boundedness of superposition operators in spaces of Triebel–Lizorkin-type. Czechoslovak Math. J. **39**(2(114)), 323–347 (1989)
954. Sickel, W.: Spline representations of functions in Besov-Triebel–Lizorkin spaces on $\mathbb{R}^n$. Forum Math. **2**(5), 451–475 (1990)
955. Sickel, W.: A remark on orthonormal bases of compactly supported wavelets in Triebel–Lizorkin spaces. The case $0 < p, q < \infty$. Arch. Math. (Basel) **57**(3), 281–289 (1991)
956. Sickel, W.: Some remarks on trigonometric interpolation on the n-torus. Z. Anal. Anwend. **10**(4), 551–562 (1991)
957. Sickel, W.: Characterization of Besov-Triebel–Lizorkin spaces via approximation by Whittaker's cardinal series and related unconditional Schauder bases. Constr. Approx. **8**(3), 257–274 (1992)
958. Sickel, W.: Pointwise multiplication in Triebel–Lizorkin spaces. Forum Math. **5**(1), 73–91 (1993)
959. Sickel, W.: Necessary conditions on composition operators acting on Sobolev spaces of fractional order. The critical case $1 < s < n/p$. Forum Math. **9**(3), 267–302 (1997)
960. Sickel, W.: Conditions on composition operators which map a space of Triebel–Lizorkin-type into a Sobolev space. The case $1 < s < n/p$. II. Forum Math. **10**(2), 199–231 (1998)
961. Sickel, W.: Necessary conditions on composition operators acting between Besov spaces. The case $1 < s < n/p$. III. Forum Math. **10**(3), 303–327 (1998)
962. Sickel, W.: On pointwise multipliers for $F^s_{p,q}(\mathbb{R}^n)$ in case $\sigma_{p,q} < s < n/p$. Ann. Mat. Pura Appl. **176**, 209–250 (1999)
963. Sickel, W.: Pointwise multipliers for Lizorkin-Triebel spaces. Oper. Theory Adv. Appl. **110**, 295–321 (1999)
964. Sickel, W.: Smoothness spaces related to Morrey spaces–a survey I. Eurasian Math. J. **3**(3), 110–149 (2012)
965. Sickel, W.: Smoothness spaces related to Morrey spaces–a survey II. Eurasian Math. J. **4**(1), 82–124 (2013)
966. Sickel, W., Skrzypczak, L.: Radial subspaces of Besov and Lizorkin-Triebel classes: extended Strauss lemma and compactness of embeddings. J. Fourier Anal. Appl. **6**(6), 639–662 (2000)

967. Sickel, W., Skrzypczak, L.: On the interplay of regularity and decay in case of radial functions II. Homogeneous spaces. J. Fourier Anal. Appl. **18**(3), 548–582 (2012)
968. Sickel, W., Skrzypczak, L., Vybíral, J.: On the interplay of regularity and decay in case of radial functions I. Inhomogeneous spaces. Commun. Contemp. Math. **14**(1), Art ID. 1250005, 1–60 (2012)
969. Sickel, W., Skrzypczak, L., Vybíral, J.: Complex interpolation of weighted Besov and Lizorkin-Triebel spaces. Acta Math. Sin. (Engl. Ser.) **30**(8), 1297–1323 (2014)
970. Sickel, W., Triebel, H.: Hölder inequalities and sharp embeddings in function spaces of B^s_{pq} and F^s_{pq} type. Z. Anal. Anwend. **14**(1), 105–140 (1995)
971. Sierpinski, W.: Sur une courbe dont tout point est un point de ramification. C. R. Acad. Paris **160**, 302 (1915)
972. Sierpinski, W.: Sur une courbe cantorienne qui contient une image biunivoquet et continue detoute courbe donnée. C. R. Acad. Paris **162**, 629–632 (1916)
973. Sjöstrand, J.: An algebra of pseudodifferential operators. Math. Res. Lett. **1**(2), 185–192 (1994)
974. Sjöstrand, J.: Wiener type algebras of pseudodifferential operators. In Séminaire sur les Équations aux Dérivées Partielles, 1994–1995, Exp. No. IV. École Polytech., Palaiseau (1995)
975. Skrzypczak, L.: Traces of function spaces of $F^s_{p,q}$–$B^s_{p,q}$ type on submanifolds. Math. Nachr. **146**, 137–147 (1990)
976. Skrzypczak, L.: Function spaces of Sobolev type on Riemannian symmetric manifolds. Forum Math. **3**(4), 339–353 (1991)
977. Skrzypczak, L.: Remark on pointwise multipliers for Triebel scales on Riemannian manifolds. Funct. Approx. Comment. Math. **21**, 3–6 (1992)
978. Skrzypczak, L.: Besov spaces and function series on Lie groups. Comment. Math. Univ. Carolin. **34**(1), 139–147 (1993)
979. Skrzypczak, L.: Vector-valued Fourier multipliers on symmetric spaces of the noncompact type. Monatsh. Math. **119**(1–2), 99–123 (1995)
980. Skrzypczak, L.: Some equivalent norms in Sobolev-Besov spaces on symmetric Riemannian manifolds. J. Lond. Math. Soc. (2) **53**(3), 569–581 (1996)
981. Skrzypczak, L.: Heat semi-group and function spaces on symmetric spaces on non-compact type. Z. Anal. Anwend. **15**(4), 881–899 (1996)
982. Skrzypczak, L.: Besov spaces on symmetry manifolds. Hokkaido Math. J. **25**(2), 231–247 (1996)
983. Skrzypczak, L.: Besov spaces on symmetric manifolds–the atomic decomposition. Studia Math. **124**(3), 215–238 (1997)
984. Skrzypczak, L.: Atomic decompositions on manifolds with bounded geometry. Forum Math. **10**(1), 19–38 (1998)
985. Skrzypczak, L.: On Besov spaces and absolute convergence of the Fourier transform on Heisenberg groups. Comment. Math. Univ. Carolin. **39**(4), 755–763 (1998)
986. Skrzypczak, L.: Heat and harmonic extensions for function spaces of Hardy–Sobolev–Besov type on symmetric spaces and Lie groups. J. Approx. Theory **96**(1), 149–170 (1999)
987. Skrzypczak, L.: Besov spaces and Hausdorff dimension for some Carnot-Caratheodory metric spaces. Canad. J. Math. **54**(6), 1280–1304 (2002)
988. Skrzypczak, L.: Rotation invariant subspaces of Besov and Triebel–Lizorkin space: compactness of embeddings, smoothness and decay of functions. Rev. Mat. Iberoamericana **18**(2), 267–299 (2002)
989. Skrzypczak, L.: Heat extensions, optimal atomic decompositions and Sobolev embeddings in presence of symmetries on manifolds. Math. Z. **243**(4), 745–773 (2003)
990. Skrzypczak, L.: Entropy numbers of Trudinger-Strichartz embeddings of radial Besov spaces and applications. J. Lond. Math. Soc. (2) **69**(2), 465–488 (2004)
991. Skrzypczak, L.: On approximation numbers of Sobolev embeddings of weighted function spaces. J. Approx. Theory **136**(1), 91–107 (2005)

992. Skrzypczak, L.: Approximation and entropy numbers of compact Sobolev embeddings. In: Approximation and Probability. Banach Center Publications, vol. 72, pp. 309–326. Polish Academy of Sciences, Warsaw (2006)
993. Skrzypczak, L.: Wavelet frames, Sobolev embeddings and negative spectrum of Schrodinger operators on manifolds with bounded geometry. J. Fourier Anal. Appl. **14**(3), 415–442 (2008)
994. Skrzypczak, L., Tomasz, B.: Approximation numbers of Sobolev embeddings of radial functions on isotropic manifolds. J. Funct. Spaces Appl. **5**(1), 27–48 (2007)
995. Skrzypczak, L., Tomasz, B.: Entropy of Sobolev embeddings of radial functions and radial eigenvalues of Schrodinger operators on isotropic manifolds. Math. Nachr. **280**(5–6), 654–675 (2007)
996. Skrzypczak, L., Tomasz, B.: Remark on borderline traces of Besov and Triebel–Lizorkin spaces on noncompact hypersurfaces. Comment. Math. **53**(2), 293–309 (2013)
997. Skrzypczak, L., Vybíral, J.: Corrigendum to the paper: "On approximation numbers of Sobolev embeddings of weighted function spaces" [J. Approx. Theory **136** 91–107 (2005)][165121]. J. Approx. Theory **156**(1), 116–119 (2009)
998. Slobodeckij, L.N.: Generalized Sobolev spaces and their applications to boundary value problems of partial differential equations (Russian). Leningrad. Gos. Ped. Inst. Uĉep. Zap. **197**, 54–112 (1958)
999. Smith, K.T.: Inequalities for formally positive integro-differential forms. Bull. Am. Math. Soc. **67**, 368–370 (1961)
1000. Sobolev, S.L.: The Cauchy problem in a function space (Russian). Dokl. Akad. Nauk. SSSr **3**, 291–294 (1935)
1001. Sobolev, S.L.: Méthode nouvelle à resoudre le problème de Cauchy pour les équations linéaires hyperboliques normales. Math. Sb. **1**, 39–72 (1936)
1002. Sobolev, S.L.: On a theorem of functional analysis (Russian). Math. Sb. **4**, 471–497 (1938)
1003. Sobolev, S.L.: The density of compactly supported functions in the space $L_p^{(m)}(E^n)$ (Russian). Sibirsk. Mat. Z. **4**, 673–682 (1963)
1004. Solncev, J.K.: On the estimation of a mixed derivative in $L_p(G)$. Trudy Mat. Inst. Steklov **64**, 211–238 (1961). English translation in Am. Math. Soc. Transl. (2) **79** (1969)
1005. Song, L., Yan, L.: Riesz transforms associated to Schrödinger operators on weighted Hardy spaces. J. Funct. Anal. **259**, 1466–1490 (2010)
1006. Stafney, J.D.: Analytic interpolation of certain multiplier spaces. Pac. J. Math. **32**, 241–248 (1970)
1007. Stampacchia, G.: The spaces $\mathscr{L}^{(p,\lambda)}$, $N^{(p,\lambda)}$ and interpolation. Ann. Sc. Norm. Super. Pisa Cl. Sci. **19**(3), 443–462 (1965)
1008. Stein, E.M.: Note on singular integrals. Proc. Am. Math. Soc. **8**, 250–254 (1957)
1009. Stein, E.M.: On the functions of Littlewood–Paley, Lusin, and Marcinkiewicz. Trans. Am. Math. Soc. **88**, 430–466 (1958)
1010. Stein, E.M., Weiss, G.: On the theory of harmonic functions of several variables: I. The theory of H^p-spaces. Acta Math. **103**, 25–62 (1960)
1011. Strauss, W.A.: Existence of solitary waves in higher dimensions. Commun. Math. Phys. **55**, 149–162 (1977)
1012. Strichartz, R.S.: Multipliers on fractional Sobolev spaces. J. Math. Mech. **16**, 1031–1060 (1967)
1013. Strichartz, R.S.: Fubini-type theorems. Annali Scuola Norm. Sup. Pisa **22**, 399–408 (1968)
1014. Strichartz, R.S.: Restriction of Fourier transform to quadratic surfaces and decay of solutions of wave equations. Duke Math. J. **44**, 705–713 (1977)
1015. Strömberg, J.O.: Bounded mean oscillation with Orlicz norms and duality of Hardy spaces. Indiana Univ. Math. J. **28**, 511–544 (1979)
1016. Strömberg, J.O., Torchinsky, A.: Weighted Hardy Spaces. Lecture Notes in Mathematics, vol. 1381. Springer, Berlin/New York (1989)
1017. Sturm, K.T.: On the geometry of metric measure spaces. I. Acta Math. **196**(1), 65–131 (2006)

1018. Sturm, K.T.: On the geometry of metric measure spaces. II. Acta Math. **196**(1), 133–177 (2006)
1019. Sugimoto, M.: Pseudo-differential operators on Besov spaces. Tsukuba J. Math. **12**, 43–63 (1988)
1020. Sugimoto, M., Tomita, N.: The dilation property of modulation spaces and their inclusion with Besov spaces. J. Funct. Anal. **248**(1), 79–106 (2007)
1021. Sugimoto, M., Tomita, N.: A counterexample for boundedness of pseudo-differential operators on modulation spaces. Proc. Am. Math. Soc. **136**(5), 1681–1690 (2008)
1022. Sugimoto, M., Tomita, N.: Boundedness properties of pseudo-differential operators and Calderón–Zygmund operators on modulation spaces. J. Fourier Anal. Appl. **14**(1), 124–143 (2008)
1023. Sugimoto, M., Tomita, N., Wang, B.: Remarks on nonlinear operations on modulation spaces. Integral Transf Spec. Funct. **22**(4–5), 351–358 (2011)
1024. Tachizawa, K.: The boundedness of pseudodifferential operators on modulation spaces. Math. Nachr. **168**, 263–277 (1994)
1025. Tachizawa, K.: The pseudodifferential operators and Wilson bases. J. Math. Pures Appl. (9) **75**(6), 509–529 (1996)
1026. Taibleson, M.H.: On the theory of Lipschitz spaces of distributions on Euclidean n-space. I. Principal properties. J. Math. Mech. **13**, 407–479 (1964)
1027. Taibleson, M.H., Weiss, G.: The molecular characterization of certain Hardy spaces. Astérisque **77**, 67–149 (1980)
1028. Takada, R.: Counterexamples of commutator estimates in the Besov and the Triebel-Lizorkin spaces related to the Euler equations. SIAM J. Math. Anal. **42**, 2473–2483 (2010)
1029. Tang, L.: Weighted local Hardy spaces and their applications. Illinois J. Math. **56**(2), 453–495 (2012)
1030. Tang, L., Yang, D.: Boundedness of vector-valued operators on weighted Herz spaces. Approx. Theory Appl. (N.S.) **16**(2), 58–70 (2000)
1031. Tang, L., Xu, J.S.: Some properties of Morrey type Besov-Triebel spaces. Math. Nachr. **278**, 904–917 (2005)
1032. Thorin, G.O.: An extension of a convexity theorem due to M. Riesz, Kungl. Fysiografiska Sällskapets i Lund Förhandlingar **8**(14), 166–170 (1938); Medd. Lunds Univ. Mat. Sem. **4**, 1–5 (1939)
1033. Toft, J.: Continuity properties for modulation spaces, with applications to pseudo-differential calculus–I. J. Funct. Anal. **207**, 399–429 (2004)
1034. Toft, J.: Continuity and Schatten properties for pseudo-differential operators on modulation spaces. Oper. Theory Adv. Appl. **172**, 173–206 (2006)
1035. Tomita, N.: On the Hörmander multiplier theorem and modulation spaces. Appl. Comput. Harmon. Anal. **26**(3), 408–415 (2009)
1036. Torres, R.H.: Continuity properties of pseudodifferential operators of type 1, 1. Commun. PDE **15**(9), 1313–1328 (1990)
1037. Torres, R.H.: Boundedness results for operators with singular kernels on distribution spaces. Mem. Am. Math. Soc. **90**(442), viii+172 (1991)
1038. Torres, R.H., Ward, E.L.: Leibniz's rule, sampling and wavelets on mixed Lebesgue spaces. J. Fourier Anal. Appl. **21**(5), 1053–1076 (2015)
1039. Triebel, H.: Spaces of distributions of Besov type on Euclidean n-space. Duality, interpolation. Ark. Mat. **11**, 13–64 (1973)
1040. Triebel, H.: A remark on embedding theorems for Banach spaces of distributions. Ark. Mat. **11**, 65–74 (1973)
1041. Triebel, H.: Interpolation theory for function spaces of Besov type defined in domains. I. Math. Nachr. **57**, 51–85 (1973)
1042. Triebel, H.: Interpolation theory for function spaces of Besov type defined in domains. II. Math. Nachr. **58**, 63–86 (1973)
1043. Triebel, H.: Spaces of Kudrjavcev type I. J. Math. Anal. Appl. **56**, 253–277 (1976)
1044. Triebel, H.: Spaces of Kudrjavcev type II. J. Math. Anal. Appl. **56**, 278–287 (1976)

1045. Triebel, H.: General function spaces. I. Decomposition methods. Math. Nachr. **79**, 167–179 (1977)
1046. Triebel, H.: General function spaces. II. Inequalities of Plancherel-Pólya-Nikolski'j-type, L_n-spaces of analytic functions, $0 < p \leq \infty$. J. Approx. Theory **19**(2), 154–175 (1977)
1047. Triebel, H.: General function spaces. III. Spaces $B_{p,q}^{g(x)}$ and $F_{p,q}^{g(x)}$, $1 < p < \infty$: basic properties. Anal. Math. **3**(3), 221–249 (1977)
1048. Triebel, H.: General function spaces. IV. Spaces $B_{p,q}^{g(x)}$ and $F_{p,q}^{g(x)}$, $1 < p < \infty$: special properties. Anal. Math. **3**(4), 299–315 (1977)
1049. Triebel, H.: Multiplication properties of the spaces $B_{p,q}^s$ and $F_{p,q}^s$. Quasi-Banach Algebras of functions. Ann. Mat. Pura Appl. (4) **113**, 33–42 (1977)
1050. Triebel, H.: On spaces of $B_{\infty,q}^s$ type and $\mathscr{C}^s$ type. Math. Nachr. **85**, 75–90 (1978)
1051. Triebel, H.: A note on quasi-normed convolution algebras of entire analytic functions of exponential type. J. Approx. Theory **22**, 368–373 (1978)
1052. Triebel, H.: On Haar basis in Besov spaces. Serdica **4**(4), 330–343 (1978)
1053. Triebel, H.: On Besov-Hardy-Sobolev spaces in domains and regular elliptic boundary value problems. The case $0 < p \leq \infty$. Commun. PDE **3**, 1083–1164 (1978)
1054. Triebel, H.: Theorems of Littlewood–Paley type for BMO and for anisotropic Hardy spaces. In: Constructive Function Theory, vol. 77, pp. 525–532. Publishing House of the Bulgarian Academy of Sciences, Sofia (1980)
1055. Triebel, H.: On $L_p(l_q)$-spaces of entire analytic functions of exponential type: complex interpolation and Fourier multipliers. The case $0 < p < \infty$, $0 < q < \infty$. J. Approx. Theory **28**(4), 317–328 (1980)
1056. Triebel, H.: On the spaces $F_{p,q}^s$ of Hardy-Sobolev type: Equivalent quasinorms, multipliers, spaces on domains, regular elliptic boundary value problems. Commun. PDE **5**, 245–291 (1980)
1057. Triebel, H.: Complex interpolation and Fourier multipliers for the spaces $B_{p,q}^s$ and $F_{p,q}^s$ of Besov-Hardy-Sobolev type: the case $0 < p \leq \infty$, $0 < q \leq \infty$. Math. Z. **176**(4), 495–510 (1981)
1058. Triebel, H.: Characterizations of Besov-Hardy-Sobolev spaces via harmonic functions, temperatures and related means. J. Approx. Theory **35**, 275–297 (1982)
1059. Triebel, H.: Anisotropic function spaces. I: Hardy's inequality, decompositions. Anal. Math. **10**, 53–77 (1984)
1060. Triebel, H.: Spaces of Besov-Hardy-Sobolev type on complete Riemannian manifolds. Ark. Mat. **24**, 299–337 (1986)
1061. Triebel, H.: Characterizations of function spaces on a complete Riemannian manifold with bounded geometry. Math. Nachr. **130**, 321–346 (1987)
1062. Triebel, H.: Function spaces on Lie groups, the Riemannian approach. J. London Math. Soc. (2) **35**(2), 327–338 (1987)
1063. Triebel, H.: Characterization of Besov-Hardy-Sobolev spaces: a unified approach. J. Approx. Theory **52**, 162–203 (1988)
1064. Triebel, H.: Diffeomorphism properties and pointwise multipliers for function spaces. In: Function Spaces. Proceedings of Conference, Poznań, 1986. Teubner-Texte Mathematics, vol. 103, pp. 75–84. Teubner, Leipzig (1988)
1065. Triebel, H.: Atomic representations of F_{pq}^s spaces and Fourier integral operators. In: Seminar Analysis Karl–Weierstrass Institute 1986/87. Teubner-Texte Math., vol. 106, pp. 297–305. Teubner, Leipzig (1988)
1066. Triebel, H.: How to measure smoothness of distributions on Riemannian symmetric manifolds and Lie groups. Z. Anal. Anwend. **7**, 471–480 (1988)
1067. Triebel, H.: Local approximation spaces. Z. Anal. Anwend. **8**, 261–288 (1989)
1068. Triebel, H.: Atomic decompositions of F_{pq}^s spaces. Applications to exotic pseudodifferential and Fourier integral operators. Math. Nachr. **144**, 189–222 (1989)
1069. Triebel, H.: Inequalities in the theory of function spaces: a tribute to Hardy, Littlewood and Polya. In: Inequalities. Proceedings of Conference Birmingham (UK). Lecture Notes Applications in Mathematics, vol. 119, pp. 231–248. Marcel Dekker, New York (1991)

1070. Triebel, H.: A localization property for B_{pq}^{s} and F_{pq}^{s} spaces. Studia Math. **109**(2), 183–195 (1994)
1071. Triebel, H.: Decompositions of function spaces. Progress Nonlinear Diff. Equ. Appl. **35**, 691–730 (1999)
1072. Triebel, H.: Truncations of functions. Forum Math. **12**(6), 731–756 (2000)
1073. Triebel, H.: Regularity theory for some semi-linear equations: the Q-method. Forum Math. **13**(1), 1–19 (2001)
1074. Triebel, H.: Function spaces in Lipschitz domains and on Lipschitz manifolds. Characteristic functions as pointwise multipliers. Rev. Mat. Complut. **15**(2), 475–524 (2002)
1075. Triebel, H.: Lacunary measures and self-similar probability measures in function spaces. Acta Math. Sin. (Engl. Ser.) **20**(4), 577–588 (2004)
1076. Triebel, H.: Approximation numbers in function spaces and the distribution of eigenvalues of some fractal elliptic operators. J. Approx. Theory **129**(1), 1–27 (2004)
1077. Triebel, H.: A new approach to function spaces on quasi-metric spaces. Rev. Mat. Complut. **18**(1), 7–48 (2005)
1078. Triebel, H.: Wavelet para–bases and sampling numbers in function spaces on domains. J. Complexity **23**, 468–497 (2007)
1079. Triebel, H.: Wavelet bases in Lorentz and Zygmund spaces. Georgian Math. J. **15**(2), 389–402 (2008)
1080. Triebel, H.: The dichotomy between traces on d-sets Γ in $\mathbb{R}^n$ and the density of $\mathscr{D}(\mathbb{R}^n, \Gamma)$ in function spaces. Acta Math. Sin. (Engl. Ser.) **24**(4), 539–554 (2008)
1081. Triebel, H.: Sobolev-Besov spaces of measurable functions. Studia Math. **201**, 69–86 (2010)
1082. Triebel, H.: Limits of Besov norms. Arch. Math. (Basel) **96**(2), 169–175 (2011)
1083. Triebel, H.: Entropy numbers in function spaces with mixed integrability. Rev. Mat. Complut. **24**(1), 169–188 (2011)
1084. Triebel, H.: Entropy and approximation numbers of limiting embeddings; an approach via Hardy inequalities and quadratic forms. J. Approx. Theory **164**(1), 31–46 (2012)
1085. Triebel, H.: Characterizations of some function spaces in terms of Haar wavelets. Comment. Math. **53**(2), 135–153 (2013)
1086. Triebel, H., Winkelvoss, H.: Intrinsic atomic characterizations of function spaces on domains. Math. Z. **221**(4), 647–673 (1996)
1087. Tsutsui, Y.: Pseudo-differential operators of class $S_{0,0}^{m}$ on the Herz-type spaces. Hokkaido Math. J. **38**(2), 283–302 (2009)
1088. Tsutsui, Y.: Sharp maximal inequalities and its application to some bilinear estimates. J. Fourier Anal. Appl. **17**(2), 265–289 (2011)
1089. Tuominen, H.: Orlicz-Sobolev spaces on metric measure spaces. Dissertation, University of Jyväskylä, Jyväskylä, 2004. Ann. Acad. Sci. Fenn. Math. Diss. No. **135**, 1–86 (2004)
1090. Tyulenev, A.I.: Traces of weighted Sobolev spaces with Muckenhoupt weight. The case $p = 1$. Nonlinear Anal. **128**, 248–272 (2015)
1091. Uchiyama, A.: A constructive proof of the Fefferman–Stein decomposition of BMO($\mathbb{R}^n$). Acta. Math. **148**, 215–241 (1982)
1092. Ullrich, T.: Continuous characterizations of Besov–Lizorkin–Triebel spaces and new interpretations as coorbits. J. Funct. Space Appl., Article ID 163213, 1–47 (2010)
1093. Ullrich, T.: Optimal cubature in Besov spaces with dominating mixed smoothness on the unit square. J. Complexity **30**(2), 72–94 (2014)
1094. Uspenskii, S.V.: An embedding theorem for the fractional order classes of S.L. Sobolev. Dokl. Akad. Nauk SSSR **130**, 992–993 (1960) (Russian); English transl. in Soviet Math. Dokl. **1** (1960)
1095. Uspenskii, S.V.: Properties of the classes $W_{p}^{(r)}$ with fractional derivatives on differentiable manifolds (Russian). Dokl. Akad. Nauk SSSR **132**, 60–62 (1960); English transl. in Soviet Math. Dokl. **1** (1960)
1096. Uspenskii, S.V.: Imbedding theorems for classes with weights (Russian). Trudy Mat. Inst. Steklov. **60**, 282–303 (1961)
1097. Vitali, G.: Sui gruppi di punti e sull funzioni di variabili reali. Torino Att **43**, 229–246 (1908)

1098. Vignati, T.A.: Complex interpolation for families of quasi-Banach spaces. Indiana Univ. Math. J. **37**, 1–21 (1988)
1099. Viviani, B.E.: An atomic decomposition of the predual of BMO(ρ). Rev. Mat. Ibero. **3**, 401–425 (1987)
1100. Volevic, L.R., Panejah, B.P.: Some spaces of generalized functions and embedding theorem. Usp. Mat. Nauk. **20**(1), 3–74 (1965)
1101. Vodop'yanov, S.K., Gol'dstein, V.M., Latfulllin, T.G.: A criterion for the extension of functions of class L_2^1, from unbounded plane domains. Sibirsk. Mat. Zh. **20**, 416–419 (1979). English transl. in Siberian Math. J. **20** (1979)
1102. Vodop'yanov, S.K., Gol'dstein, V.M., Reshetnyak, Y.G.: Geometric properties of functions with generalized first derivatives. Uspekhi Mat. Nauk **34**(1(205)), 17–65 (1979). English transl. in Russian Math. Surveys **34** (1979)
1103. Vybíral, J.: Function spaces with dominating mixed smoothness. Diss. Math. (Rozprawy Mat.) **436**, 1–73 (2006)
1104. Vybíral, J.: A new proof of Jawerth–Franke embedding. Rev. Mat. Complut. **21**(1), 75–82 (2008)
1105. Vybíral, J.: On sharp embeddings of Besov and Triebel–Lizorkin spaces in the subcritical case. Proc. Am. Math. Soc. **138**(1), 141–146 (2010)
1106. Vybíral, J.: Sickel, W.: Traces of functions with a dominating mixed derivative in $\mathbb{R}^3$. Czechoslovak Math. J. **57**(4(132)), 1239–1273 (2007)
1107. Wang, B., Huang, C.: Frequency-uniform decomposition method for the generalized BO, KdV and NLS equations. J. Differ. Equ. **239**(1), 213–250 (2007)
1108. Wang, H.: Decomposition for Morrey type Besov–Triebel spaces. Math. Nachr. **282**(5), 774–787 (2009)
1109. Wang, H., Jia, H.: Singular integral operator, Hardy–Morrey space estimates for multilinear operators and Navier Stokes equations. Math. Methods Appl. Sci. **33**(14), 1661–1684 (2010)
1110. Weimar, M.: Almost diagonal matrices and Besov-type spaces based on wavelet expansions. J. Fourier Anal. Appl. **22**(2), 251–284 (2016)
1111. Whitney, H.: Analytic extensions of differentiable functions defined in closed sets. Trans. Am. Math. Soc. **36**(1), 63–89 (1934)
1112. Wiener, N.: The ergodic theorem. Duke Math. J. **5**, 1–18 (1939)
1113. Wilson, M.: A simple proof of the atomic decomposition for $H^p(\mathbb{R}^n)$, $0 < p \leq 1$. Studia Math. **74**(1), 25–33 (1982)
1114. Wilson, M.: The intrinsic square function. Rev. Mat. Ibero. **23**(3), 771–791 (2007)
1115. Wojciechowska, A.: Local means and wavelets in function spaces with local Muckenhoupt weights. In: Function Spaces IX. Banach Center Publications, vol. 92, pp. 399–412. Institute of Mathematics of the Polish Academy of Sciences, Warsaw (2011)
1116. Wu, Z., Xie, C.: Decomposition theorems for Q_p spaces. Ark. Mat. **40**(2), 383–401 (2002)
1117. Wunderli, T.: On time flows of minimizers of general convex functionals of linear growth with variable exponent in BV space and stability of pseudosolutions. J. Math. Anal. Appl. **364**(2), 591–598 (2010)
1118. Xiong, X., Xu, Q., Yin, Z.: Sobolev, Besov and Triebel–Lizorkin spaces on quantum tori. Mem. Amer. Soc. **252**(1203), 1–118 (2018)
1119. Xu, J.S.: Some equivalent quasi-norms in the Herz-type Triebel–Lizorkin spaces. Beijing Shifan Daxue Xuebao **37**(6), 715–719 (2001)
1120. Xu, J.S.: Some properties on the Herz-type Besov spaces. Hunan Daxue Xuebao **30**(5), 75–78 (2003)
1121. Xu, J.S.: Pointwise multipliers of Herz-type Besov spaces and their applications. Math. Appl. (Wuhan) **17**(1), 115–121 (2004)
1122. Xu, J.S.: A discrete characterization of Herz-type Triebel–Lizorkin spaces and its applications. Acta Math. Sci. Ser. B Engl. Ed. **24**(3), 412–420 (2004)
1123. Xu, J.S.: Equivalent norms of Herz-type Besov and Triebel–Lizorkin spaces. J. Funct. Spaces Appl. **3**(1), 17–31 (2005)

1124. Xu, J.S.: A characterization of Morrey type Besov and Triebel–Lizorkin spaces. Vietnam J. Math. **33**(4), 369–379 (2005)
1125. Xu, J.S.: Point-wise multipliers of Herz-type Besov spaces and their applications. Front. Math. China **1**(1), 110–119 (2006)
1126. Xu, J.S.: Variable Besov and Triebel–Lizorkin spaces. Annales Academiae Scientiarum Fennicae Mathematica **33**, 511–522 (2008)
1127. Xu, J.S.: The relation between variable Bessel potential spaces and Triebel–Lizorkin spaces. Integral Transforms Spec. Funct. **19**(7–8), 599–605 (2008)
1128. Xu, J.S.: Atomic decomposition of Herz-type Besov and Triebel–Lizorkin space. Acta Math. Sci. Ser. A Chin. Ed. **29**(6), 1500–1507 (2009)
1129. Xu, J.S.: An admissibility for topological degree of Herz-type Besov and Triebel–Lizorkin spaces. Topol. Methods Nonlinear Anal. **33**(2), 327–334 (2009)
1130. Xu, J.S.: An atomic decomposition of variable Besov and Triebel–Lizorkin spaces. Armen. J. Math. **2**(1), 1–12 (2009)
1131. Xu, J.S.: An admissibility for topological degree of variable Besov and Triebel–Lizorkin spaces. Georgian Math. J. **18**(2), 365–375 (2011)
1132. Xu, J.S.: The Beal-Kato-Majda type and the Moser type inequalities for Morrey type Besov spaces with variable exponents. Math. Appl. (Wuhan) **27**(2), 346–354 (2014)
1133. Xu, J.S.: Decompositions of non-homogeneous Herz-type Besov and Triebel–Lizorkin spaces. Sci. China Math. **57**(2), 315–331 (2014)
1134. Xu, J.S., Fu, J.: Well-posedness for the 2D dissipative quasi-geostrophic equations in the Morrey type Besov space. Math. Appl. (Wuhan) **25**(3), 624–630 (2012)
1135. Xu, J.S., Yang, D.: Vector-valued Herz spaces and Herz-type Hardy spaces. Southeast Asian Bull. Math. **26**(6), 1053–1073 (2003)
1136. Xu, J.S., Yang, D.: Applications of Herz-type Triebel–Lizorkin spaces. Acta Math. Sci. Ser. B **23**, 328–338 (2003)
1137. Xu, J.S., Yang, D.: Herz-type Triebel–Lizorkin spaces, I. Acta Math. Scinica **21**, 643–654 (2005)
1138. Yabuta, K.: A remark on the (H^1, L^1) boundedness. Bull. Fac. Sci. Ibaraki Univ. Ser. A **25**, 19–21 (1993)
1139. Yagi, A.: Coïncidence entre des espaces d'interpolation et des domaines de puissances fractionaires d'opérateurs. C. R. Acad. Sci. Paris (Sér. I) **299**, 173–176 (1984)
1140. Yan, L.: Classes of Hardy spaces associated with operators, duality theorem and applications. Trans. Am. Math. Soc. **360**, 4383–4408 (2008)
1141. Yan, X., Yang, D., Yuan, W., Zhuo, C.: Variable weak Hardy spaces and their applications. J. Funct. Anal. **271**(10), 2822–2887 (2016)
1142. Yang, D.: $T1$ theorems on Besov and Triebel–Lizorkin spaces on spaces of homogeneous type and their applications. Z. Anal. Anwend. **22**(1), 53–72 (2003)
1143. Yang, D.: New characterizations of Hajłasz-Sobolev spaces on metric spaces. Sci. China Ser. A **46**, 675–689 (2003)
1144. Yang, D.: Besov spaces and applications on homogeneous type spaces and fractals. Studia Math. **156**(1), 15–30 (2003)
1145. Yang, D.: Real interpolations for Besov and Triebel-Lizorkin spaces on spaces of homogeneous type. Math. Nachr. **273**, 96–113 (2004)
1146. Yang, D., Liang, Y.: Products of functions in BMO(X) and $H^1_{\mathrm{at}}(\mathscr{X})$ via wavelets over spaces of homogeneous type. J. Fourier Anal. Appl. **23**(4), 919–990 (2016)
1147. Yang, D., Yang, D.: Maximal function characterizations of Musielak–Orlicz–Hardy spaces associated with magnetic Schrödinger operators. Front. Math. China **10**(5), 1203–1232 (2015)
1148. Yang, D., Yang, S.: Second-order Riesz transforms and maximal inequalities associated with magnetic Schrödinger operators. Canad. Math. Bull. **58**(2), 432–448 (2015)
1149. Yang, D., Yang, S.: Maximal function characterizations of Musielak–Orlicz-Hardy spaces associated to non-negative self-adjoint operators satisfying Gaussian estimates. Commun. Pure Appl. Anal. **15**(6), 2135–2160 (2016)

1150. Yang, D., Yang, S.: Regularity for inhomogeneous Dirichlet problems of some Schrödinger equations on domains. J. Geom. Anal. **26**(3), 2097–2129 (2016)
1151. Yang, D., Yuan, W.: A note on dyadic Hausdorff capacities. Bull. Sci. Math. **132**(6), 500–509 (2008)
1152. Yang, D., Yuan, W.: A new class of function spaces connecting Triebel–Lizorkin spaces and Q spaces. J. Funct. Anal. **255**, 2760–2809 (2008)
1153. Yang, D., Yuan, W.: New Besov-type spaces and Triebel-Lizorkin-type spaces including Q spaces. Math. Z. **265**, 451–480 (2010)
1154. Yang, D., Yuan, W.: Characterizations of Besov-type and Triebel–Lizorkin-type spaces via maximal functions and local means. Nonlinear Anal. **73**, 3805–3820 (2010)
1155. Yang, D., Yuan, W.: Dual properties of Triebel-Lizorkin-type spaces and their applications. Z. Anal. Anwend. **30**, 29–58 (2011)
1156. Yang, D., Yuan, W.: Relations among Besov-type spaces, Triebel–Lizorkin-type spaces and generalized Carleson measure spaces. Appl. Anal. **92**(3), 549–561 (2013)
1157. Yang, D., Yuan, W.: Function spaces of Besov-type and Triebel–Lizorkin-type–a survey. Appl. Math. J. Chinese Univ. Ser. B **28**(4), 405–426 (2013)
1158. Yang, D., Yuan, W., Zhou, Y.: A new characterization of Triebel–Lizorkin spaces on $\mathbb{R}^n$. Publ. Mat. **57**(1), 57–82 (2013)
1159. Yang, D., Yuan, W., Zhuo, C.: Fourier multipliers on Triebel–Lizorkin-type spaces. J. Funct. Spaces Appl., Art. ID 431016, 37pp (2012)
1160. Yang, D., Yuan, W., Zhuo, C.: Complex interpolation on Besov-type and Triebel–Lizorkin-type spaces. Anal. Appl. (Singap.) **11**(5), 1350021, 1–45 (2013)
1161. Yang, D., Yuan, W., Zhuo, C.: Musielak–Orlicz Besov-type and Triebel–Lizorkin-type spaces. Rev. Mat. Complut. **27**(1), 93–157 (2014)
1162. Yang, D., Zhang, J.: Riesz transform characterizations of Hardy spaces associated to degenerate elliptic operators. Integral Equ. Oper. Theory **84**(2), 183–216 (2016)
1163. Yang, D., Zhang, J.: Weighted L^p estimates of Kato square roots associated to degenerate elliptic operators. Publ. Mat. **61**(2), 395–444 (2017)
1164. Yang, D., Zhou, Y.: A boundedness criterion via atoms for linear operators in Hardy spaces. Constr. Approx. **29**, 207–218 (2009)
1165. Yang, D., Zhuo, Y.: New properties of Besov and Triebel–Lizorkin spaces on RD-spaces. Manuscripta Math. **134**(1–2), 59–90 (2011)
1166. Yang, D., Zhuo, C.: Molecular characterizations and dualities of variable exponent Hardy spaces associated with operators. Ann. Acad. Sci. Fenn. Math. **41**(1), 357–398 (2016)
1167. Yang, D., Zhuo, C., Nakai, E.: Characterization of variable exponent Hardy spaces via Riesz transforms. Rev. Mat. Complut. **29**(2), 245–270 (2016)
1168. Yang, D., Zhuo, C., Yuan, W.: Besov-type spaces with variable smoothness and integrability. J. Funct. Anal. **269**(6), 1840–1898 (2015)
1169. Yang, D., Zhuo, C., Yuan, W.: Besov-type spaces with variable exponents. Banach J. Math. Anal. **9**(4), 146–202 (2015)
1170. Yang, M.: On analyticity rate estimates to the magneto-hydrodynamic equations in Besov-Morrey spaces. Bound. Value Probl. 2015:155, 1–19 (2015)
1171. Yoneda, T.: Ill-posedness of the 3D-Navier–Stokes equations in a generalized Besov space near BMO^{-1}. J. Funct. Anal. **258**(10), 3376–3387 (2010)
1172. Young, L.C.: An inequality of Hölder type, connected with Stieltjes integration. Acta Math. **67**(1), 251–282 (1936)
1173. Yuan, W., Haroske, D.D., Skrzypczak, L., Yang, D.: Embedding properties of weighted Besov-type spaces. Anal. Appl. (Singap.) **13**(5), 507–553 (2015)
1174. Yuan, W., Lu, Y.F., Yang, D.: Several equivalent characterizations of fractional Hajłasz–Morrey–Sobolev spaces. Appl. Math. J. Chinese Univ. Ser. B **31**(3), 343–354 (2016)
1175. Yuan, W., Sawano, Y., Yang, D.: Decompositions of Hausdorff–Besov and Triebel–Lizorkin–Hausdorff spaces and their applications. J. Math. Anal. Appl. **369**(2), 736–757 (2010)

1176. Yuan, W., Sickel, W., Yang, D.: On the coincidence of certain approaches to smoothness spaces related to Morrey spaces. Math. Nachr. **286**(14–15), 1571–1584 (2013)
1177. Yuan, W., Sickel, W., Yang, D.: Compact embeddings of radial and subradial subspaces of some Besov-type spaces related to Morrey spaces. J. Approx. Theory **174**, 121–139 (2013)
1178. Yuan, W., Sickel, W., Yang, D.: Interpolation of Morrey-Campanato and related smoothness spaces. Sci. China Math. **58**(9), 1835–1908 (2015)
1179. Yuan, W., Haroske, D.D., Skrzypczak, L., Yang, D.: Embedding properties of Besov-type spaces. Appl. Anal. **94**(2), 319–341 (2015)
1180. Zhang, J., Chang, D.C., Yang, D.: Characterizations of Sobolev spaces associated to operators satisfying off-diagonal estimates on balls. Math. Methods Appl. Sci. **40**, 2907–2929 (2017)
1181. Zhang, J., Cao, J., Jiang, R., Yang, D.: Non-tangential maximal function characterizations of Hardy spaces associated with degenerate elliptic operators. Canad. J. Math. **67**(5), 1161–1200 (2015)
1182. Zhikov, V.V.: Averaging of functionals of the calculus of variations and elasticity theory. Izvestiya Akademii Nauk SSSR Seriya Matematicheskaya **50**(4), 675–710 (1986)
1183. Zhang, J., Zhuo, C., Yang, D., He, Z.: Littlewood–Paley characterizations of Triebel–Lizorkin–Morrey spaces via ball averages. Nonlinear Anal. **150**, 76–103 (2017)
1184. Zhuo, C., Sawano, Y., Yang, D.: Hardy spaces with variable exponents on RD-spaces and applications. Diss. Math. (Rozprawy Mat.) **520**, 1–74 (2016)
1185. Zhuo, C., Sickel, W., Yang, D., Yuan, W.: Characterizations of Besov-type and Triebel-Lizorkin-type spaces via averages on balls. Canad. Math. Bull. **60**, 655–672 (2017)
1186. Zhuo, C., Yang, D., Liang, Y.: Intrinsic square function characterizations of Hardy spaces with variable exponents. Bull. Malays. Math. Sci. Soc. **39**(4), 1541–1577 (2016)
1187. Zhuo, C., Yang, D.: Maximal function characterizations of variable Hardy spaces associated with non-negative self-adjoint operators satisfying Gaussian estimates. Nonlinear Anal. **141**, 16–42 (2016)
1188. Zhuo, C., Yang, D., Yuan, W.: Hausdorff Besov-type and Triebel–Lizorkin-type spaces and their applications. J. Math. Anal. Appl. **412**(2), 998–1018 (2014)
1189. Zhuo, C., Yang, D., Yuan, W.: Interpolation between $H^{p(\cdot)}(\mathbb{R}^n)$ and $L^\infty(\mathbb{R}^n)$: real method. J. Geom. Anal. online
1190. Zygmund, A.: Smooth functions. Duke Math. J. **12**, 47–76 (1945)
1191. Zygmund, A.: On a theorem of Marcinkiewicz concerning interpolation of operators. J. Math. Pures Appl. (9) **35**, 223–248 (1956)

Index

Symbols

$(-\varDelta)^{\alpha}$, 280
$(\beta\mathrm{qu})_{\nu m}$, 454
$(\beta\mathrm{qu})_{\nu m}^{(L)}$, 462
$A_r^s K_{pq}^{\alpha}(\mathbb{R}^n)$, 784
$A_{X,q}^s(\mathbb{R}^n)$, 775
$A_{pq:\mathrm{selfs}}^s(\mathbb{R}^n)$, 529
$A_{pq\varOmega}^s(\mathbb{R}^n)$, 697
$A_{pq}^s(w)$, 806
$A_{pq}^s(\mathbb{R}^n)$, 241
$A_{pq}^w(\mathbb{R}^n)$, 810
$A_{pq}^{s,\alpha}(\mathbb{R}^n)$, 810
$A_{pq}^{s-\frac{n-D}{p}}(\varGamma, \mu)$, 822
$A_{pq}^{s_1,s_2}(\mathbb{R}^m \times \mathbb{R}^n)$, 813
$A_{pq}^{s_1,s_2}(\mathbb{R}^{2n})$, 813
A_1, 715
A_p, 715
A_p^{loc}, 807
A_∞, 715
$A_{pq}^s(\varOmega)$, 577
$A_{pq}^s(\mathbb{R}_+^n)$, 566
$A_{pq}^{s,\tau}(\mathbb{R}^n)$, 792
$A_{p(\star)q(\star)}^{s(\star)}$, 811
$B(X)$, 38
$B(X, Y)$, 38
$B(r)$, xiii
$B(x, r)$, xiii
$B\dot{H}_{pq}^{s,\tau}(\mathbb{R}^n)$, 802
$B_{\infty\infty}^l S_{1\delta}^m(\mathbb{R}^n)$, 609
$B_r^s K_{pq}^{\alpha}(\mathbb{R}^n)$, 784
$B_{X,q}^s(\mathbb{R}^n)$, 775
$B_{pq}^s(w)$, 806
$B_{pq}^s(\mathbb{R}^n)$, 210, 241
$B_{pq}^w(\mathbb{R}^n)$, 810
$B_{pq}^{s,\alpha}(\mathbb{R}^n)$, 810
$B_{pq}^{s_1,s_2}(\mathbb{R}^m \times \mathbb{R}^n)$, 813
$B_{pq}^{s_1,s_2}(\mathbb{R}^{2n})$, 813
$B_{pq}^s(\varOmega)$, 577
$B_{pq}^s(\mathbb{R}_+^n)$, 566
$B_{pq}^{s,\tau}(\mathbb{R}^n)$, 792
$B_{p(\star)q(\star)}^{s(\star)}$, 811
$C(\mathbb{T}^n)$, 76
$C^k(\mathbb{T}^n)$, 76
$C_{\mathrm{c}}^{\infty}(\varOmega)$, 16, 77
$C_{\mathrm{c}}^{\infty}(\varOmega; K)$, 77
$C_{\mathrm{c}}^{\infty}(\mathbb{R}^n)$, 9
C_k^{ν}, 175
$C_{\mathrm{c}}(\varOmega)$, 16
$E(x, t)$, 643
E_j, 147
$F\dot{H}_{pq}^{s,\tau}(\mathbb{R}^n)$, 803
$F\dot{T}_{pq}^{s,\tau}(\mathbb{R}^n)$, 788
$F\dot{W}_{pq}^{s,\tau}(\mathbb{R}^n)$, 785
$F_r^s K_{pq}^{\alpha}(\mathbb{R}^n)$, 784
$F_{X,q}^s(\mathbb{R}^n)$, 775
$F_{\infty q}^s(\mathbb{R}^n)$, 411
$F_{pq}^s(w)$, 806
$F_{pq}^s(\mathbb{R}^n)$, 241
$F_{pq}^w(\mathbb{R}^n)$, 810
$F_{pq}^{s,\alpha}(\mathbb{R}^n)$, 810
$F_{pq}^{s_1,s_2}(\mathbb{R}^m \times \mathbb{R}^n)$, 813
$F_{pq}^{s_1,s_2}(\mathbb{R}^{2n})$, 813
$F_{pq}^s(\varOmega)$, 577
$F_{pq}^s(\mathbb{R}_+^n)$, 566
$F_{pq}^{s,\tau}(\mathbb{R}^n)$, 792
$F_{p(\star)q(\star)}^{s(\star)}$, 811

Y. Sawano, *Theory of Besov Spaces*, Developments in Mathematics 56,
https://doi.org/10.1007/978-981-13-0836-9

$F_{pp}^{s+2-1/p}(\partial\Omega)$, 677
$HX(\mathbb{R}^n)$, 770
$H^1(Q)$, 55
$H^1(\mathbb{R}^n)$, 55
$H^p(\mathbb{R}^n)$, 337
$H^s(\mathbb{R}^n)$, 132
$H_p^s(\mathbb{R}^n)$, 132
$H^{\Phi}(\mathbb{R}^n)$, 770, 772
$H^{\log}(\mathbb{R}^n)$, 773
$H^{p(\star)}(\mathbb{R}^n)$, 770
$H^{p,\infty}(w)$, 771
$H^{p,\infty}(\mathbb{R}^n))$, 771
$H_{\log,0}(\mathbb{R}^n)$, 734
$H_{\log,\infty}(\mathbb{R}^n)$, 734
$H_{\log}(\mathbb{R}^n)$, 734
I_α, 107
$K_{pq}^{\alpha,\lambda}(\mathbb{R}^n)$, 712
$K_{pq}^{\alpha}(\ell^u, \mathbb{R}^n)$, 768
$K_{pq}^{\alpha}(\mathbb{R}^n)$, 766
$L^0(\mathbb{R}^n)$, 710
$L^1_{\mathrm{loc}}(\mathbb{R}^n)$, 50
$L^a[\log L]^b(\mathbb{R}^n)$, 763
$L^p(\ell^q, \mathbb{R}^n)$, 120
$L^p(w, \ell^q)$, 725
$L^p(\mathbb{T}^n)$, 76
L^p_Ω, 124
$L^p_{x'}L^u_{x_n}(\mathbb{R}^n)$, 525
$L^{\Phi}(X)$, 753, 763
$L^{\Phi}(\ell^q, \mathbb{R}^n)$, 761
$L^\infty(\mathbb{R}^n)$, 4
$L^\infty_c(\mathbb{R}^n)$, 10
$L^{p(\star)}(\ell^{q(\star)})$, 738
$L^{p(\star)}(\ell^{q(\star)}, \mathbb{R}^n)$, 738
$L^{p(\star)}(\ell^q, \mathbb{R}^n)$, 738
$L^{p,\infty}(\mathbb{R}^n)$, 110
$L^{\mathbf{p}}(\ell^r, \mathbb{R}^n)$, 727
$M(\mathbb{R}^n)$, 827
$M^{(\eta)}$, 125
M^{loc}, 807
M_0, 155
M_B, 333
$M_{\mathrm{dyadic}} f$, 147
$M_{\mathscr{D}(Q)}$, 155
N, 143
N_β, 143, 144
$Q(x, r)$, xiii
$Q_p^{\alpha,q,*}$, 797
S, 484
$S^m_{\rho\delta}$, 590
S^{2n-1}, 853
S^{n-1}, 88
S_n, 172
S_η, 623
$T1$, 683
$T^*(T \in \mathscr{S}'(\mathbb{R}^n \times \mathbb{R}^n))$, 95
$\Delta(1)$, 49
Δ_2, 759
$\Gamma(x)$, 142
$\Gamma_\beta(x)$, 381
Φ_Q, 844
$\approx$, 214
χ, xiii
$\chi_{\nu m}^{(p)}$, 431
$\dfrac{\sqrt{L}}{\mathrm{id}_H + L}$, 635
$\dim_{\mathscr{H}}(A)$, 823
$\dot{A}_{pq}^{s}(\mathbb{R}^n, A)$, 815
$\dot{A}_{pq}^{s*}(\mathbb{R}^n)$, 286
$\dot{A}_{pq}^{s,\tau}(\mathbb{R}^n)$, 793
$\dot{B}_{pq}^{s}(\mathbb{R}^n, A)$, 815
$\dot{B}_{pq}^{s,\tau}(\mathbb{R}^n)$, 793
$\dot{F}_{\infty q}^{s}(\mathbb{R}^n)$, 411
$\dot{F}_{pq}^{s}(\mathbb{R}^n, A)$, 815
$\dot{F}_{pq}^{s,\tau}(\mathbb{R}^n)$, 793
$\dot{K}_{pq}^{\alpha}(\ell^u, \mathbb{R}^n)$, 768
$\dot{K}_{pq}^{\alpha}(\mathbb{R}^n)$, 766
$\dot{L}^{1,p}(\mathbb{R}^n)$, 287
$\dot{W}^{1,\infty}(\mathbb{R}^n)$-condition, 151
$\dot{W}^{1,p}(\mathbb{R}^n)$-condition, 151
$\ell^q(L^p(w))$, 725
$\ell^q(L^p, \mathbb{R}^n)$, 120
$\ell^q(L^{\Phi}, \mathbb{R}^n)$, 776
$\ell^q(L^{\mathbf{p}}, \mathbb{R}^n)$, 776
ℓ^q_s, 478
$\ell^r(\mathscr{M}^p_q, \mathbb{R}^n)$, 776
$\ell^u(K^{\alpha}_{pq}, \mathbb{R}^n)$, 776
$\ell^u(\dot{K}^{\alpha}_{pq}, \mathbb{R}^n)$, 776
$\ell^{q(\star)}(L^{p(\star)})$, 738
$\ell^{q(\star)}(L^{p(\star)}, \mathbb{R}^n)$, 738
$\ell^q(L^{p(\star)}, \mathbb{R}^n)$, 738
$\eta_{j,m}$, 114
$\hat{O}$, 17
$\langle a \rangle$, 43
div, 55
∇, 55
∇_2, 759
$\underline{\omega_s}$, 823
$\overline{S}$, 484
$\overset{\circ}{X}(\mathbb{R}^n)$, 504
$\overset{\diamond}{X}(\mathbb{R}^n)$, 504
∂^α, 16
$\partial^\beta f$, 41

$\perp$, 83
ψ^{β}, 454
$\rho(L)$, 38
$\sigma(L)$, 38
$\sqrt{L}$, 635
$\tilde{C}_0(\mathbb{R}^n)$, 63
φ^*, 754
φ_j, 207
$\underset{\sim}{\varphi_j}$ (for anisotropic function spaces), 815
$\underset{\sim}{A}^s_{pq}(\Omega)$, 697
$\mathscr{S}(\mathbb{R}^n)_L$, 90
$a(X, D)$, 590
$a(X, D)^*$, 592
$a(X, D)f$ for $f \in \mathscr{S}'(\mathbb{R}^n)$, 593
a_k, 587
$a_k(X \hookrightarrow Y)$, 587
e_k, 587
$e_k(X \hookrightarrow Y)$, 587
$f^{\varepsilon}_{Q,w}$, 844
$hK^{\alpha}_{pq}(\mathbb{R}^n)$, 772
$h^p(\mathbb{R}^n)$, 388
$i\mathbb{R}$, 630
m_Q, 399
p_+, 728
p_-, 728
p_N, 42
$p_{\alpha,\beta}$, 43
p_{α}(for $\mathscr{D}(\mathbb{T}^n)$), 76
r_j, 322
x^{β}, 41
$\mathbb{T}^n$, 75
$\mathscr{A}^s_{pqr}(\mathbb{R}^n)$, 783
$\mathscr{B}^m$, 217
$\mathscr{C}^s(\mathbb{R}^n)$, 522
$\mathscr{D}(Q)$, 13
$\mathscr{D}(\mathbb{T}^n)$, 76
$\mathscr{D}, \mathscr{D}$, 13
$\mathscr{D}'(\Omega)$, 81
$\mathscr{D}'(\mathbb{T}^n)$, 76
$\mathscr{E}^s_{pqr}(\mathbb{R}^n)$, 783
$\mathscr{F}_N$, 331
$\mathscr{H}^s(A)$, 823
$\mathscr{K}(\Omega)$, 77
$\mathscr{M}^p_q$, 745
$\mathscr{M}^p_q(\ell^r, \mathbb{R}^n)$, 749
$\mathscr{M}_{A,j}f$, 290
$\mathscr{M}f$, 336
$\mathscr{N}^s_{pqr}(\mathbb{R}^n)$, 783
$\mathscr{O}_M(\mathbb{R}^n)$, 52, 509
$\mathscr{P}(\mathbb{R}^n)$, 50, 173
$\mathscr{P}_d(\mathbb{R}^n)$, 50
$\mathscr{P}_{-1}(\mathbb{R}^n)$, 101
$\mathscr{S}'(\Omega)$, 587
$\mathscr{S}'_{\Omega}(\mathbb{R}^n)$, 124
$\mathscr{S}_L(\mathbb{R}^n)$, 89
$\mathscr{S}_{\Omega}(\mathbb{R}^n)$, 124
$\mathscr{S}_{\infty}(\mathbb{R}^n)$, 269
$\mathscr{S}'_{\infty}(\mathbb{R}^n)$, 269
$\mathscr{W}^{\alpha_3}_{\alpha_1,\alpha_2}$, 809
BMO($\mathbb{R}^n$) (space), 399
BMO$^+(\mathbb{R}^n)$, 399
Dom(A), 645
Dom(L), 38
Hom$_{\mathbb{C}}(V, W)$, 49
Lip(θ), 235
Lip($\mathbb{R}$), 37
Lip($\mathbb{R}^n$), 225, 690
O(n), 826
bmo($\mathbb{R}^n$), 407
diam(K), 131
$e^{it\Delta}$, 652
osc$^M_1$, 302
$\mathbf{a}_{pq}(\mathbb{R}^n)$, 431
$\mathbf{b}_{pq}(\mathbb{R}^n)$, 431
$\mathbf{f}_{pq}(\mathbb{R}^n)$, 431
2-microlocal Besov space, 810
2-microlocal Triebel–Lizorkin space, 810
5r-covering lemma, 14

A

abbreviation of the elements in $\mathscr{S}'(\mathbb{R}^n)/\mathscr{P}(\mathbb{R}^n)$, 275
absolutely continuous norm, 713
accretive, 628
additive inequality, 102
adjoint of pseudo-differential operators, 598
A_{∞}-constant, 715
A_{∞}-weight, 715
approximation number with degree k, 587
associated norm, 713
A_1-constant, 715
A_1-weight, 715
A_p-constant, 715
A_p-weight, 715
atom, 432, 482
atoms for Besov spaces and Triebel–Lizorkin spaces, 432
atoms in Hardy spaces, 349

B

ball Banach function space, 711
ball means of differences, 301
ball quasi-Banach function space, 711
Banach–Alaoglu theorem, 37
Banach algebra, 263

Banach function space, 711
band-limited distribution, 71
Bernstein's lemma, 128
Besov–Hausdorff space, 802
Besov-type space, 793
Besov norm, 207
Besov norm (on the whole space), 239
Besov norm of the ball means of differences, 301
Besov space, 210
bounded set in $\mathscr{D}(\Omega)$, 80

C

Calderón's first complex interpolation functor, 484
Calderón's first complex interpolation space, 484
Calderon's second complex interpolation space, 497
Calderón–Zygmund decomposition, 149
Calderón–Zygmund decomposition for $\mathscr{S}'(\mathbb{R}^n)$, 363
Calderón's reproducing formula, 74
canonical ball cover with respect to Hausdorff capacity, 20
canonical representation, 752
Cantor dust, 830
Cantor function, 225
Cantor set, 829
Carleson box, 17
Carleson measure, 142
Carleson tent, 17
Cauchy integral, 694
Cauchy integral, 691, 694
Choquet integral, 30
$C^\infty(\mathbb{R}^n)$-function that has at most polynomial growth at infinity, 52
C^∞-domain, 576
classical Besov norm, 320
closed strip domain, 484
coefficient mapping, 439
compatible couple, 473
complex interpolation functor, 484, 490, 497
complex interpolation space, 484, 497
composition of pseudo-differential operators, 602
conjugate (variable) exponent, 731
conjugate function, 754
containing space, 473
continuous function over $\overline{S}$, 484
continuous semi-group, 644
convolution, 6, 9
convolution of $f \in \mathscr{S}(\mathbb{R}^n)$ and $g \in \mathscr{S}'(\mathbb{R}^n)$, 62
CZ(Calderón–Zygmund)-kernel, 680

D

Δ_2-condition, 759
δ-body, 824
densely defined closed operator, 38
density argument, 36
difference operator, 96
differential index, 219, 248
differentiation in $\mathscr{S}'(\mathbb{R}^n)$, 54
dilation matrix, 814
distributional Fourier transform, 68
diversity of function spaces, 254
Doetsch's three-line lemma, 194
domain, 38
domains satisfying the horn condition, 582
doubling condition, 759
doubing measure, 830
D-set, 821
$d\sigma(\omega)$, 647
dual inequality of Stein-type, 111, 112
dyadic average operator, 147
dyadic child, 13
dyadic cube, 13
dyadic cubes of j-th generation, 13
dyadic maximal operator, 147
dyadic parent, 13

E

elementary symbol, 609
entropy number with degree k, 587
(ε, δ) domains, 582
η-triangle inequality, 36
η-function, 114
expansive matrix, 814
exponential of sectorial operators, 625, 633

F

Fatou lemma, 2
Fatou property, 250
Fefferman–Stein vector-valued maximal inequality, 120
Fenchel–Legendre transform, 756
$5r$-covering lemma, 14
Fourier multiplier, 131
Fourier space, 66
Fourier transform, 65

Fourier transform for Schwartz distributions, 68
fractional integral operator, 107
Frazier–Jawerth φ-transform, 456
frequency support, 71, 124
Frostman measure, 823
Fubini's property, 537
Fubini's theorem, 3

G
Gaffney-type estimate, 660
Gaussian, 42
generalized Legendre polynomial, 175
generalized Weyl's lemma, 91
generator, 645
gradient condition, 156
grid, 108
g_λ^*-function, 381

H
Hölder–Zygmund space, 224, 227
Hölmander–Michlin multiplier theorem, 161
Hadamard gap, 236
Hardy–Littlewood maximal inequality, 109
Hardy–Littlewood maximal operator, 107
Hardy inequality, 469
Hardy space, 337
Hardy spaces based on $X(\mathbb{R}^n)$, 770
Hata's tree, 830
Hausdorff–Pompeiu distance, 824
Hausdorff capacity, 20
Hausdorff distance, 824
heat kernel, 643
heat semi-group, 643
heat semi-group in $\mathscr{S}(\mathbb{R}^n)$, 643
Herz–Morrey space, 712
high frequency part, 281
high pass filter, 281
Hilbert transform, 157
holomorphic function over S, 484
homogeneous Besov-type space, 793
homogeneous Besov space, 279
homogeneous Herz space, 766
homogeneous Triebel–Lizorkin-type norm, 793
homogeneous Triebel–Lizorkin-type space, 793
homogeneous Triebel–Lizorkin space, 279
homogenous Besov-type norm, 793
Hörmander class, 590
Hörmander condition, 680

I
inner regularity, 9
integral kernel, 157
intersection subspace, 474
inverse Fourier transform, 65

J
John–Nirenberg inequality, 400, 402
John domain, 582

K
Kato conjecture, 833
Kato theorem, 833
kernel, 38
key theorems in function spaces, 538
K-functional, 476
Khintchine's inequality, 323
Koch curve, 829
Kolmogorov inequality, 111
Köthe dual, 713

L
$L^1(\mathbb{R}^n)$-condition, 148
lacunary, 310
Laplacian, 678
Lebesgue's convergence theorem, 2
Lebesgue differentiation theorem, 114
Legendre conjugate function, 754
lift operator, 214, 246
linear spline, 225
$L^\infty(\mathbb{R}^n)$-condition, 149
Littlewood–Paley g-function, 381
Littlewood–Paley operators, 207
Lizorkin distribution, 269
Lizorkin functions, 83
local bmo($\mathbb{R}^n$) space, 407
local Hardy–Littlewood maximal operator, 807
local Hardy–Herz space, 772
local Hardy space, 388
localization of function spaces, 519
local means, 289
local reproducing formula, 84
logarithmic convexity formula, 478, 485
lower half space, xiv
low frequency part, 281
low pass filter, 281
$L^p(\mathbb{R}^n)$-inequality of the Hardy–Littlewood maximal operator, 113
L^p-space with weight, 714
Luzin function, 381

M

Marcinkiewicz interpolation theorem, 160
maximal inequality for local means, 291, 294
measurable rectangular, 3
metrizable topological space, 45
minimality of $\dot{B}^0_{11}(\mathbb{R}^n)$, 446
Minkovski sum, xvii
mixed Lebesgue space, 726
modified dyadic Hausdorff capacity, 22
modulation space, 263
molecule, 432
molecules for Besov spaces and Triebel–Lizorkin spaces, 433
moment condition, 83
monotone convergence theorem, 2
Morrey norm, 745
Morrey space, 745
multiplicative inequality, 103

N

∇_2-condition, 759
Nakano–Luxenburg norm, 728
Neumann expansion, 665
nice Young function, 754
node, 225
nonhomogeneous Besov space (over $\mathbb{T}^n$), 818
nonhomogeneous Herz space, 766
nonhomogeneous Triebel–Lizorkin-type norm, 792
nonhomogeneous Triebel–Lizorkin space (over $\mathbb{T}^n$), 818
nonhomogenous Besov-type norm, 792
nontangential maximal function, 143
nontrivial, 218

O

off-diagonal estimate, 660, 667
of trace zero, 187
openness property, 720
open set condition, 827
open strip domain, 484
optimal polynomials, 101
order of difference operator, 96
oscillation, 101, 302
overlap, 13

P

(p, q)-block centered at Q, 389
packing, 15
paraproduct operator, 688
Peetre's inequality, 43
Peetre's maximal operator, 125, 128, 291
φ-transform, 456
physical space, 66
Plancherel's theorem (for $\mathscr{S}(\mathbb{R}^n)$), 69
Plancherel's theorem (for $L^2(\mathbb{R}^n)$), 69
Plancherel–Polya–Nikolski'i inequality, 125
Poincaré–Wirtinger inequality, 105
Poincaré inequality, 105
pointwise product of C^∞-function that has at most polynomial growth at infinity and distribution, 52
potential space, 132
powered Hardy–Littlewood maximal operator, xvi
pseudo-differential operators of Kohn–Nirenberg type, 590
ψ for the quarkonial decomposition, 454

Q

Q-space, 796
quark, 454
quark for general case, 462
quarkonial decomposition for the regular case, 455
quasi-Banach function space, 711
quasi-norm associated with an expansive matrix, 814
quasi-triangle inequality, 36
quotient topology, 277

R

Rademacher sequence, 322
range, 38
real interpolation functor, 477
realization of $\dot{A}^{s*}_{pq}(\mathbb{R}^n)$, 287
realization of $\dot{A}^{s}_{pq}(\mathbb{R}^n)$, 284
reference point, 17
reflection, 583
reflection principle, 184
regular diffeomorphism, 538
regular elements in $\mathscr{S}'(\mathbb{R}^n)$, 50
regular quark, 454
resolvent estimate, 623
resolvent set, 38
restricted boundedness, 680
restriction of $\mathscr{S}'(\mathbb{R}^n)$ to open sets, 81
reverse Hölder inequality, 718
Riesz transform, 146, 157

S

sampling theorem, 129
Schwartz's kernel theorem, 93

Schwartz distribution (space), 49
Schwartz function space $\mathscr{S}(\mathbb{R}^n)$, 42
second complex interpolation functor, 497
second complex interpolation space, 497
sectorial operator, 623
Segal algebra, 263
sequence spaces for quarkonial decomposition, 454
Sierpinski gasket, 829
Sierpinski triangle, 829
singular integral operator, 157
size condition, 156, 679
Sjöstrand class, 263
slowly varying, 698
smoothness Herz space, 784
smoothness Morrey space, 783
Sobolev norm, 151, 205
Sobolev embedding of Frank–Jawerth-type, 464
Sobolev index, 248
space of homogeneous type, 830
spaces of Kudrjavcev type, 529
spectrum set, 38
stationary phase for nondegenerate stationary points, 651
step, 96
strip domain S, 484
strong maximal operator, 726
strong subadditivity, 24
sum space, 474
support, 16
support condition, 149
support of distributions, 58
symbol class, 590
symmetric, 187

T

$T1$ theorem, 686, 688
tensor-valued function, 186
tensor product, 93, 186
test functions, 42
topological vector space, 268, 473
topology of $\mathscr{S}'(\mathbb{R}^n)$, 52
topology of $\mathscr{S}'(\mathbb{R}^n)/\mathscr{P}(\mathbb{R}^n)$, 274
trace free, 187
Triebel–Lizorkin–Hausdorff space, 803
Triebel–Lizorkin-type space, 793
Triebel–Lizorkin norm (on the whole space), 239
Triebel–Lizorkin norm of ball means of differences, 301
TT^*-method, 653

U

Uchiyama class, 230
uniformly C^m-open set, 581
uniformly elliptic condition, 655
unitary, 69
universal estimate, 155
upper half space, xiv

V

validity of Besov spaces and Triebel–Lizorkin spaces, 239
variable exponent, 728
variable exponent Besov space, 811
variable exponent Lebesgue space, 728
variable exponent Triebel–Lizorkin space, 811
variable Lebesgue norm, 728
vector-valued norm, 120
vertical maximal operator, 332, 336

W

weak Hardy space, 771
weak L^p space, 110
Weierstrass function, 231
weighted dyadic maximal operator, 717
weighted Lebesgue space, 714
weighted weak Hardy space, 771
Weyl's lemma, 91
Whitney covering, 15
Whitney decomposition, 15
Whitney region, 141, 543
Wirtinger inequality, 105

Y

Young function, 752

Printed by Printforce, the Netherlands